W0080477

A. Heuberger · H. Ryssel · P. Lange (Eds.)

ESSDERC '89

19th European Solid State
Device Research Conference, Berlin

With 869 Figures

Springer-Verlag Berlin Heidelberg NewYork
London Paris Tokyo Hong Kong

Professor Dr. Anton Heuberger

Fraunhofer-Institut für Mikrostrukurtechnik
Dillenburger Straße 53
1000 Berlin 33

Professor Dr. Heiner Ryssel

Lehrstuhl für Elektronische Bauelemente
Universität Erlangen-Nürnberg
Artilleriestraße 12
8520 Erlangen

Dr. Peter Lange

Fraunhofer-Institut für Mikrostrukurtechnik
Dillenburger Straße 53
1000 Berlin 33

ISBN 978-3-642-52316-8   ISBN 978-3-642-52314-4 (eBook)
DOI 10.1007/978-3-642-52314-4

© Springer-Verlag Berlin Heidelberg 1989

Softcover reprint of the hardcover 1st edition 1989

Offsetprinting: Mercedes-Druck, Berlin; Bookbinding: Lüderitz & Bauer, Berlin
2068/3020 54321 – Printed on acid-free paper

Preface

This volume contains the extended abstracts of all papers presented at the ESSDERC 89, held in Berlin in September, 11 to 14, 1989. Twelve invited papers were selected among the proposals by the Program Committee. On the basis of a two page abstract the Program Committee also selected 188 contributions from 327 submitted papers for oral presentation. These proceedings include, however, four-page extended abstracts which were reproduced as submitted by the respective authors and have not been subjected to an additional reviewing procedure. Therefore, the responsibility for these extended abstracts rests exclusively with the authors.

Over 300 authors from sixteen European and four overseas countries including the United States and Japan submitted contributions. The growing interest of European authors in mutual cooperation, as demonstrated by participation in ESSDERC, together with the high quality of the papers, represents an important step towards securing Europe a place in the world of advanced research for microelectronic devices. We hope that this conference will further stimulate the cooperative efforts between researchers and professionals from European industries, research institutions and universities.

The sponsors, who have supported the Conference are gratefully acknowledged. The editors would like to thank especially the Program Committee for selecting the invited and contributed papers. Finally, we would like to express our sincerest gratitude to the Conference Secretary, Hans-Christian Petzold, and to the local Organizing Committee, for doing the hard work of organizing the conference.

A. Heuberger, H. Ryssel, P. Lange

Supporting Organizations

AIXTRON GmbH

Applied Materials GmbH

Balzers Hochvakuum GmbH

Bio-Rad Laboratories GmbH

Cambridge Instruments GmbH

Carl Zeiss, Oberkochen

Deutsche Bank Berlin AG

H. v. Frankenberg (Manager VDE/VDI, GME)

Dr. E. Hofmeister (Chairman VDE/VDI, GME)

IBM Deutschland GmbH

Karl Süss KG

LAM Research GmbH

Leybold AG

Micrion GmbH

Motorola GmbH, Geschäftsbereich Tegal

Nordiko GmbH

Philips GmbH

Plasma Technology GmbH

Robert Bosch GmbH

Siemens AG

Varian GmbH

VG Instruments GmbH

Wacker Chemitronic GmbH

Organizing Committee

A. Heuberger	Chairman (IMT, Berlin)
O. Manck	Vice Chairman (TU Berlin)
H. Ryssel	Program Chair (Univ. Erlangen)
H.C. Petzold	Secretary (IMT, Berlin)
P. Jespers	Chairman Steering Committee
J.P. Nougier	Chairman ESSDERC 88
P. Selway	Chairman ESSDERC 90

International Program Committee

W. Arden	Siemens	W.Germany
P. Balk	TU Delft	Netherlands
D. Bois	CNET	France
K. Brack	IBM	W.Germany
U. Bürker	Siemens	W.Germany
G. Declerck	IMEC	Belgium
W. Fallmann	TU Wien	Austria
H.C. de Graaff	Philips	Netherlands
M. Ilegems	Lausanne IT	Switzerland
J.P. Lazzari	LETI	France
H. Luginbühl	CSEM	Switzerland
M. Montier	SGS-Thomson	France
F.A. Myers	Plessey	U.K.
J.P. Nougier	Univ. Montpellier	France
A. Paoletti	IESS-CNR	Italy
M. Plihal	Siemens	W.Germany
I.S. Radelaar	Delft Univ.	Netherlands
M. Razeghi	Thomson-CSF	France
J.M. Robertson	Edingburgh Univ.	U.K.
M. Roche	SGS-Thomson	France
S. Selberherr	TU Wien	Austria
P.R. Selway	STC Technology	U.K.
G. Soncini	IRST Trento	Italy
P. Weissglas	Swedish IME	Sweden
A. Wieder	Siemens	W.Germany
F. van de Wiele	Univ. Louvain	Belgium
B. Wilson	Plessey	U.K.
K. Wörner	Telefunken	W.Germany
G. Zimmer	IMS	W.Germany

Representatives

USA:	U. Gösele, Duke University
Japan:	M. Takai, Osaka University
China:	P.-H. Tsien, Tsinghua University

Table of Contents

Device simulation

SESSION 2

Invited paper

Process and device modelling I

SESSION 3

S/D technology

Compound semiconductor devices

Process and device modelling II

SESSION 4

Invited papers

Thin dielectrics and interfaces

Optoelectronic devices

SESSION 5

Invited paper

Bipolar/BICMOS technology

Quantum well devices and optoelectonic devices

Analogue and high voltage devices

SESSION 6

Technology (miscellaneous)

New devices

Devices for special applications

SESSION 7

Invited paper

Interconnect technology

VLSI MOS

Hot carriers

SESSION 8

Invited papers

SOI 1

VLSI bipolar

Reliability

SESSION 9

Invited papers

SOI 2

Memories

Characterization

Session 1

Invited Papers

GaAs Electronic Devices

W. Kellner

Siemens Research Laboratories
Otto-Hahn-Ring 6, D-8000 Munich 83

Abstract

Unipolar, bipolar and resonant tunnelling transistors based on III-V materials are competing for various applications in high speed electronic systems. The principles of operation are described briefly. The examination of equivalent circuits shows that parasitic elements are limiting the performance of the most advanced devices. The state of the art is discussed by presenting data for low-noise-amplifiers, high power amplifiers, digital circuits and analog to digital converters.

System requirements

The need for semiconductor devices with improved high frequency performance is stimulated by systems /1,2/:

Consumer electronics is already using low-noise GaAs-MESFETs and HEMTs in TV receivers for Direct Broadcast Satellites (DBS), operating at 12 GHz. The need to reduce the size and therefore the cost of TV set-antennas is a strong driving force to develop devices with extremely low noise figure: with lower noise figure the antenna gain can be reduced and the antenna will become smaller and cheaper. Other applications include cellular mobile telephones, high definition TV (HDTV), and auto anti-collision radar. Consumer electronics will need mainly analog circuits with a possibility of integrating some digital circuits whenever this offers advantages in performance or cost of the system.

Computers: Cray 3 - a 16 Giga-Flop Supercomputer - will use about 48000 logic GaAs IC's and 40000 memory Si IC's /3/. As 32-bit RISC microprocessors on GaAs approach clock rates of 200 MHz /4,5/ suppliers of workstations will get interested in using these devices. In addition to high speed GaAs also offers the possibility of integrating optoelectronic components. A complete optoelectronic receiver operating at a speed of 1 Gb/s has been demonstrated /6/. It will be used in computer networks.

In communications fibre optic systems with bit rates up to 10 Gb/s are being demonstrated in the laboratories. A variety of analog and digital functions are needed. GaAs transimpedance amplifiers, decision circuits, laser drivers as well as multiplexer (4:1) and demultiplexers (1:4) are used in 1.7 Gb/s systems in the USA and in Japan /2/. Future applications for GaAs analog circuits in communciations are seen in Multiple Distribution Systems (MDS) as an alternative for remote cable installation, Global Positioning Satellite (GPS) and Microwave Landing System (MLS) for advanced commercial aircraft.

In modern test instruments GaAs IC's are employed as amplifiers, pulse modulators, switches, prescalers, frequency dividers and sample and hold circuits /7/. Examples for instruments using GaAs circuits are a pattern generator for 5 Gb/s (Anritsu) and a digitizing oscilloscope capable of sampling nonrepetitive data at 1 Gsample/s with 6-bit accuracy (Hewlett Packard).

Military systems are also pushing for higher speed. The frequencies are determined by the absorption windows (35 and 94 GHz) and absorption peak (60 GHz) of the atmosphere. Low-noise and high-power devices and analog ICs are needed as well as digital functions (e. g. for radar signal processors or digital RF-memories).

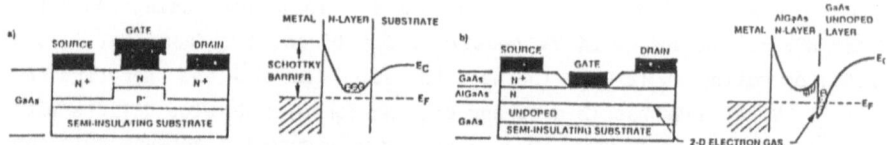

Fig. 1: Schematic cross section and band diagram of
a) MESFET   b) HEMT

Devices

High electron velocities in III-V devices are achieved by high electron mobility and by nonstationary electron transport /8/. In GaAs the transfer of an electron from the Γ to the L valley takes place with a time constant of about 1 ps. When e. g. the gate length of a FET or the drift region in a bipolar transistor is short enough, the electrons will pass the region without suffering from electron transfer. They will remain "light" electrons and their velocities will be much higher than their static saturated value. The average velocity $\bar{v}$ is given by the length of the critical region L, by the low field mobility μ_o, and by the energy gap between Γ and L-valley $\Delta E_{\Gamma L}$ as follows /9/: $\bar{v} = L/\tau = \mu_o \Delta E_{\Gamma L}/qL$. This equation is based on simple physical considerations and can be used to compare different materials: GaInAs e. g. shows an average velocity three times larger than GaAs. However, one should be careful when interpreting the above equation: First, the high field electron velocity in devices is not constant. It increases from the source to the drain side under the gate /9/. Second, doping and other material defects can reduce μ_o. Third, the electron velocity as extracted from device parameters depends also on parasitic elements.

In **U n i p o l a r   D e v i c e s** such as MESFET (Metal Semiconductor Field Effect Transistor), JFET (Junction FET), HEMT (High Electron Mobility Transistor) or MODFET (Modulation Doped FET) only one type of carrier (in most cases electrons) is involved in the charge transport. Fig. 1 shows schematic cross sections and band diagrams for a MESFET and a HEMT. In MESFETs and JFETs the drain current is controlled by the width of the space charge region extending under the gate. In HEMTs the donors are concentrated in the N-AlGaAs layer. As the electron affinity of GaAs is higher than that of the AlGaAs, the electrons diffuse towards the GaAs and remain in the potential well at the interface. The separation of ionized donors from the electrons reduces electron scattering and enhances mobility. In practical devices, sheet concentration n_s and mobility μ_o must be optimized in order to get high current and transconductance. Varying the gate voltage modulates n_s and drain current in a way similar to MOSFETs. In /10/ an overview of variations of HEMTs is given which is summarized in Table 1. Permeable Base Transistors (PBT) /11,12/ are FETs with current flow vertical to the surface. The gate consists of a tungsten grating embedded within the semiconductor. This enables the fabrication of very small gate lenths (30 nm). Due to its vertical current flow, the PBT features large gate widths within small areas which is ideal for high power devices provided the thermal resistance can be made small enough. (A similar argument also favours bipolar transistors over FETs).

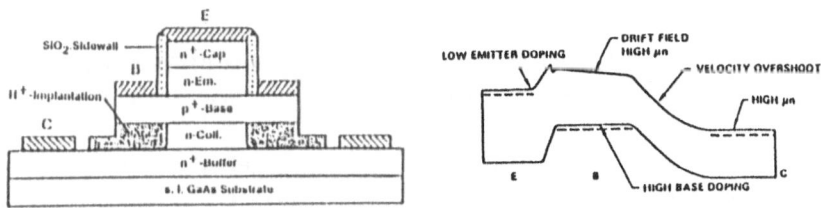

Fig. 2: Schematic cross section and band diagram of HBT

Table 1: Types of HEMTs or Heterostructure FETs. Layer sequence from gate to substrate, without n^+ contact layers.

Type	Layer Sequence	Advantage
Planar doped HEMT	Doping spike in n-AlGaAs (otherwise conventional)	Avoids I-V collapse at low temperatures due to DX-center
Inverted HEMT	Gate n-GaAs (undoped) n^+-AlGaAs n-GaAs-buffer s. i. GaAs	high g_m, low g_o due to confinement of carriers
SISFET (Semic.-Ins.-Semic.) MISFET (Metal-Ins.-Semic.) HIGFET (Heterostr.-Ins.-Gate)	Gate (n^+ GaAs or Ge or metal) u-AlGaAs u-GaAs s. i. GaAs	Control of threshold voltage self aligned contact implants n- and p-channel FET for complementary logic
DMT (doped channel-MIS-like FET)	n-GaAs instead of u-GaAs in MISFET	Increased drain current and power
Multichannel HEMT	several HEMT-layer-sequences stacked	Increased drain current and power
Pseudomorphic HEMT	Gate AlGaAs | InAlAs with planar doping u-AlGaAs | u-InAlAs spacer u-InGaAs | u-InGaAs channel u-GaAs | u-InAlAs buffer GaAs | InP substrate	Best low noise and power devices for mm-waves. Superlattice buffer improves material quality

H e t e r o j u n c t i o n   B i p o l a r   T r a n s i s t o r s (HBT) are distinguished from conventional Si bipolar transistors by several features (see Fig. 2) /13/:

- The wide gap emitter prevents hole injection from the base and allows the use of a highly doped base, without degrading the emitter efficiency.

- The emitter doping can be lowered, reducing base emitter capacitance.

- High base doping reduces output conductance and high injection effects.

- High electron mobility, built-in fields and velocity overshoot combine to short electron transit times.

- Semi-insulating substrates reduce parasitic capacitances and enable convenient integration of devices.

The layer structure of HBTs is more complex than indicated by Fig. 2 and consists typically of 9 to 10 layers /14-21/. The highly doped $(2 \cdot 10^{19}$ cm$^{-3})$ n$^+$-cap is composed of four layers: A InGaAs top layer for lowest contact resistance, a graded n$^+$-layer (In:0.75→0), a n$^+$-GaAs layer, and a graded AlGaAs layer (Al 0→0.3). The AlGaAs-emitter (Al:0.3) is doped at $3...5 \cdot 10^{17}$ cm^{-3} and can include a thin GaAs-spacer at the E-B-interface /16/. The base has a high doping level of up to $1 \cdot 10^{20}$ cm^{-3} and consists of graded AlGaAs (Al:0.12→0) or GaAs or InGaAs. Base grading produces a built in field which accelerates the electrons. The collector consists of a drift region which is necessary to obtain a sufficiently high breakdown voltage. Instead of a n-type GaAs layer (doping ≈ $5 \cdot 10^{16}$ cm^{-3}) layer combinations for optimizing the velocity overshoot have been used such as undoped-p$^+$-n$^+$ /14,15/ or p-n$^+$ /18,19/. The highly doped (≈$5 \cdot 10^{18}$ cm^{-3}) n$^+$-buffer is necessary to minimize the collector resistance. Recently, new HBT structures showed excellent performance: InGaAs/InP HBTs achieved f_{max} of 110 GHz with rather large emitters (3.2x3.2 µm^2) /20/, and collector-up HBT's with 2.6 µm wide collectors showed an f_{max} of 102 GHz. Both structures promise f_{max} > 300 GHz for HBT's scaled down to 1 µm.

R e s o n a n t   T u n n e l i n g   D e v i c e s   have at least one double barrier quantum well (e. g. AlGaAs/GaAs/AlGaAs) within their layer structure. The resonance manifests itself as peak in the current and occurs when the Fermi level crosses one of the electron states in the quantum well /22/. Incorporation of such a double barrier in the emitter of a HBT leads to the Resonant tunneling Bipolar Transistor (RBT) /23,24/. Fig. 3 shows a schematic band diagram. The devices can be operated at room temperature and showed a f_t of 12.4 GHz /23/ and 24 GHz /24/. As the double barrier causes an additional delay as compared to a HBT, the main interest in RBT's is stimulated by using them as multi-functional devices e. g. as frequency multiplier /24/ or as parity generator /25/.

Two figures of merit are used to characterize high frequency devices: The cut-off-frequency f_t, obtained from extrapolation of current gain h_{21} (f) to $h_{21} = 1$ and the maximum frequency of oscillation obtained from maximum unilateral gain. Both f_t and f_{max} are determined from measurements of S-Parameters. While f_t is more closely related to the intrinsic device, f_{max} is strongly dependent on parasitic resistance and capacitances. Fig. 4 presents f_{max}-and f_t-data for HBT's /15-21/, MESFETs /26-28/ and HEMTs /29-34/. With the exception one data point with f_{max}/f_t = 175/50 /17/ HBT's show rather low f_{max} but high f_t. On the other hand FETs show high f_{max} (up to 380 GHz) and rather low f_t.

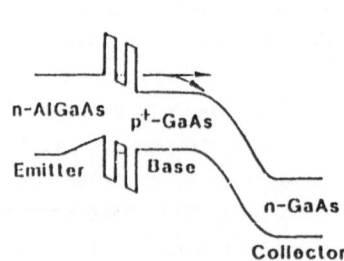

Fig. 3: Band diagram of Resonant
tunneling Bipolar Transistor

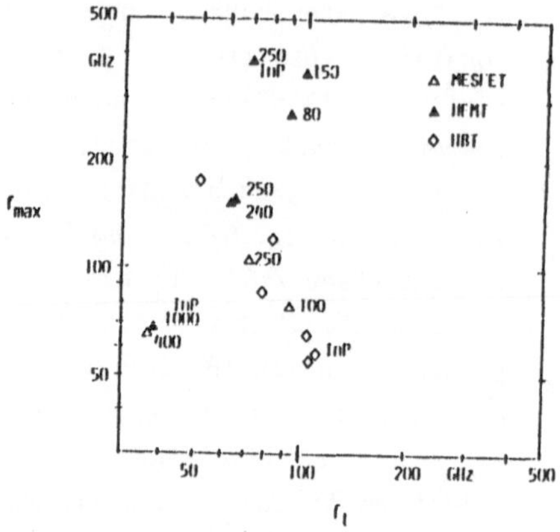

Fig. 4 f_{max} versus f_t for
HBT, MESFET and HEMTs

Applications

The GaAs MESFET technology is now well
established. Products ranging from
FETs and sensors to analog and digital IC's are being produced for a
highly competitive marketplace. Recently, HEMT's for low-noise amplifiers
have become commercially available. All other heterostructure devices and
integrated circuits are in the state of research and development.

L o w n o i s e F E T s have shown remarkable improvements with the
reduction of gate length from 1.0 to 0.1 µm. Fig. 5 includes older data
for noise figure and associated gain /35/ as well as recent data /29-31,
33, 36, 37/. 0.55 dB / 15.2 dB / 18 GHz and 1.8 dB / 6.4 dB / 60 GHz for
$NF/G_{ass}/f$ for a double-heterostructure pseudomorphic HEMT on GaAs with a
gatelength of 150 nm have been reported /36/. Even better performance was
achieved with InP-based HEMTs with a gatelength of 250 nm /37/: 0.5 dB /

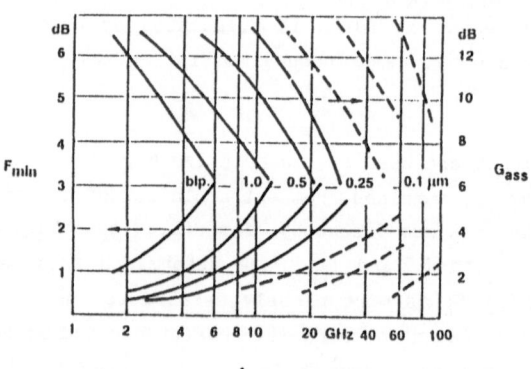

Fig. 5 Minimum noise figure F_{min} and associated gain G_{ass} for Si bipolar, MESFETs (L = 1.0/0.5/0.25 µm) and HEMTs (L =
0.25/0.1/0.1 µm). The outer right curve is an extrapolation.

15.2 dB / 18 GHz and 1.2 dB / 8.5 dB / 60 GHz and 2.1 dB / 6.4 dB / 94 GHz. These data are already approaching the curve at the outer right of Fig. 5 based on projections from /33/. InP-HEMTs will therefore be the first choice for mm-wave low-noise amplifiers if their reliability problem can be solved /37/. For p o w e r a p p l i c a t i o n s the GaAs MESFET showed a rapid development up to 1980 with slower rate of improvement after 1980. This is illustrated in Fig. 6 /35/ where recent data of output power versus frequency have been included /17, 36, 38-42/. Table 2 presents output power P_o, gain (at P_o), power added efficiency η and frequency. It appears that HEMTs are the leading devices for frequencies above 60 GHz. The interest in HBTs is stimulated by its high power density (up to 4 W per mm emitter length) and high efficiency, but this device will have to show improved performance to compete with the MESFET.

Table 2: Performance of power devices and MMICs.

Device	P_o (W)	G (dB)	η (%)	f (GHz)	Ref
MESFET	13.2	5.8	25	11.2	/38/
	12.6	5.0	28	14.2	/38/
- MMIC*	1.0	4.2	11	29.5	/39/
- MMIC	1.0	5.0	5.8	34	/40/
- MMIC	0.095	5.0	11	55	/41
HEMT	0.137	7.6	40	35	/36/
	0.125	4.5	32	60	/36/
	0.057	2.0	16	94	/36/
HBT	0.4	7.0	48	10	/17/
	0.045	2.5	25	59	/17/
	1.0	6.0	49	5	/42/

* MMIC = Monolithic Microwave Integrated Circuit

In d i g i t a l a p p l i c a t i o n s GaAs is used mainly because of its high speed. At room temperature the minimum gate delays (associated power per gate) are 6.9 ps (31 mW) for MESFETs /26/, 6 ps (25 mW) for HEMT's /43/ and 1.9 ps (44 mW) for HBTs /15/. GaAs SRAM's achieved top performances such as 4 K / 0.5 ns / 5.7 W for complexity / access time / power consumption with HEMTs /44/ and 16 K / 4 ns / 1.5 W with MESFETs /45/. 16 K SRAMs and 32-bit RISC microprocessors /5/ are VLSI-circuits with more than 10^5 transistors per chip. Available products range from small scale high-speed ($\approx$ 6 GHz) flip-flops (Fujitsu) to 4 K SRAMs (Gigabit) and 10 K Gate Arrays (Vitesse).

10

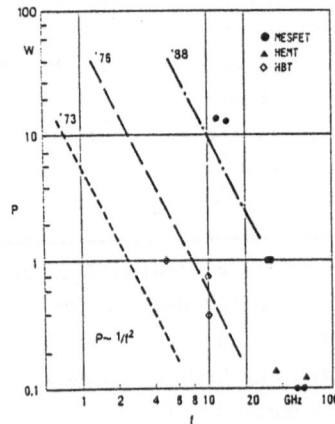

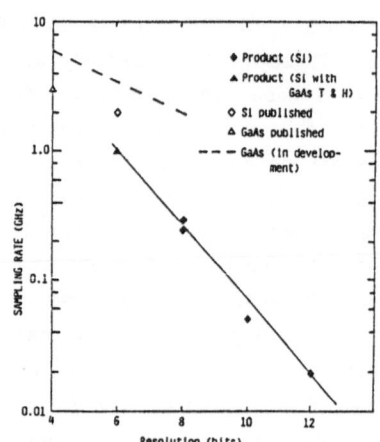

Fig. 6   Output Power P versus
         frequency for MESFETs:
         State of the art 1973,
         1976 and 1988 including
         some recent data.

Fig. 7   Sampling rate versus
         resolution for A/D con-
         verters. (Data courtesy
         R. Hagelauer, Fraunhofer
         Ges./AIS, Erlangen

A n a l o g   t o   d i g i t a l   c o n v e r t e r s   are very challenging circuits to develop.   They require a technology with tight parameter control, high speed devices and integration level of LSI. Fig. 7
shows sampling rate versus resolution for the best ADCs /46-49/. Bipolar
transistors are used in high speed ADC's on Si, and HBT's will be used on
GaAs. The high power consumption of the bipolar transistor limits the
level of integration. On the Si side the answer to this problem is
BICMOS. A BICFET heterostructure device has been reported recently /50/.
The implementation of a BICFET LSI technology will pose a very serious
challenge to the technology.

I n   c o n c l u s i o n,   Si will continue to be the dominating material for digital VLSI circuits. For special applications, however, system
designers are already using GaAs circuits with digital and analog functions. For analog microwave circuits, GaAs is first choice. Today the
GaAs MESFET fulfils most needs. In addition, Heterostructure FETs and
HBTs will be used in mm-wave-systems.

A c k n o w l e d g e m e n t s :   The author is grateful to many colleagues in the USA and Japan for sending their latest reprints. He is
also indebted to R. D. Schnell, H. J. Siweris, U. Schaper, E. Kohn for
helpful discussions and to H. Kniepkamp for critical reading of the
manuscript.

References

/ 1/ Liechti, C. A.: Heterostructure microwave transistors for ultra-high-speed systems. Europ. Microwave Conf. Stockholm (Sept. 1988) 92-100.

/ 2/ Gladstone, J.: Commercial applications of GaAs ICs. IEEE MTT-S (1988) 93-97.

/ 3/ Welch, B.M., et al: GaAs IC technology, circuits, and systems. IEEE Internat. Conf. on Circuits and Systems, Helsinki (June 1988).

/ 4/ Whitmire, D. A., et al: A 32 b GaAs RISC microprocessor. IEEE Int. Solid-St. Circ. Conf. (1988) 34, 35, 289.

/ 5/ Zuleeg, R., et al: VLSI potential of GaAs E-JFET technology. CMC 89 Colorado Springs, Colorado (March 1989).

/ 6/ Crow, J. D., et al: A GaAs MESFET IC for optical multiprocessor networks. IEEE Trans. ED 36 (Feb. 1989) 263-268.

/ 7/ Peterson, V.: Applications of GaAs ICs in instruments. IEEE GaAs IC Symp. (1988) 191-194.

/ 8/ Ruch, J. B.: Electron dynamics in short channel field effect transistors. IEEE Trans ED-19 (1972) 652-654.

/ 9/ Cappy, A., et al: Comparative potential performance of Si, GaAs, GaInAs, InAs submicrometer FET's. IEEE Trans. ED-27 (1980) 2158-2160.

/10/ Salmer, G.: Heterostructure field effect transistor, physical analysis and new structures. ESSDERC 1988, Journ. de Physique 49 (Sept. 1988) C4 125 - C4 134.

/11/ Bozler, C. O., et al: Fabrication and numerical simulation of the permeable base transistor. IEEE Trans. ED-27 (1980) 1128-1141.

/12/ Kushner, L. J., et al: 22 GHz performance of the permeable base transistor. IEEE MTT-S (1988) 525-528.

/13/ Asbeck, P. M., et al: Heterojunction bipolar transistors for microwave and millimeter-wave integrated circuits. IEEE Trans. ED-34 (Dec. 1987) 2571-2578.

/14/ Ishibashi, T., et al: A possible near-ballistic collection in an AlGaAs/GaAs HBT with a modified collector structure. IEEE Trans. ED-35 (Apr. 1988) 401-404.

/15/ Ishibashi, T., et al: Ultra-high speed AlGaAs/GaAs HBTs. IEEE Int. Electr. Dev. Meeting (1988) 826-829.

/16/ Ota, Y., et al: AlGaAs/GaAs HBT with GaInAs cap layer fabricated by multiple-self-alignment process using one mask. El. Lett. 25 (1989) 610-612.

/17/ Higgins, J. A.: Heterojunction bipolar transistors for high efficiency power amplifiers. IEEE GaAs IC Symp. (1988) 33-36.

/18/ Morizuka, K., et al: Transit time reduction in AlGaAs/GaAs HBT's using velocity overshoot in the p-type collector region. IEEE El. Dev. Lett. 9 (Nov. 1988) 585-587.

/19/ Morizuka, K., et al: AlGaAs/GaAs HBT's fabricated by a self-alignment technology using polyimide for electrode separation. IEEE El. Dev. Lett. 9 (Nov. 1988) 598-600.

/20/ Nottenburg, R. N.: Hot-electron InGaAs/InP heterostructure bipolar transistors with f_t of 110 GHz. IEEE El. Dev. Lett. (Jan 1989) 30-32.

/21/ Chang, M. F.: Self aligned AlGaAs/InGaAs/GaAs collector-up HBT's for microwave application. Device Res. Conf. Boston (June 1989).

/22/ Chang, L. L., et al: Resonant tunneling in semiconductor double barriers. Appl. Phys. Lett. 24 (1975) 593-595.

/23/ Futatsugi, T., et al: Resonant tunneling bipolar transistors using InAlAs/InGaAs heterostructures. J. Appl. Phys. 65 (Feb. 1989) 1771-1775.

12

/24/ Lunardi, L. M., et al: Microwave multiple-state resonant-tunneling bipolar transistors. IEEE El. Dev. Lett. 10 (May 1989) 219-221.

/25/ Sen, S., et al: Parity generator circuit using a multistate resonant tunneling bipolar transistor. El. Lett. 24 (Nov. 1988) 1506-1567.

/26/ Yamane, Y., et al: 5,9 ps/gate operation with 0.1 µm gate length GaAs MESFETs. Int. Electr. Dev. Meeting (1988) 894-896.

/27/ Shih, H.-D., et al: High-performance $In_{0.08}Ga_{0.92}As$ MESFET's on GaAs (100) substrates. IEEE El. Dev. Lett. 9 (Nov. 1988) 604-605.

/28/ Wang, G. W., et al: 0.25 µm gate millimeter-wave ion-implanted GaAs MESFET's. IEEE El. Dev. Lett. 10 (May 1989) 186-188.

/29/ Hanyu, I., et al: Super low-noise HEMTs with a T-shaped WSi_x Gate. Electron. Lett. 24 (1988) 1327-1328.

/30/ Ng, G. I., et al: Improved strained HEMT characteristics using double heterojunction $In_{0.52}Ga_{0.48}As$ design. IEEE El. Dev. Lett. 10 (March 1989) 114-116.

/31/ Chao, P. C., et al: DC and microwave characteristics of sub-0.1-µm gate-length planar-doped pseudomorphic HEMTs, IEEE Trans. ED 36 (March 1989) 461-473.

/32/ Kim, B., et al: AlGaAs/InGaAs/GaAs quantum-well power MESFET at mm-wave frequencies. IEEE El. Dev. Lett. 9 (Nov. 1988) 610-612.

/33/ Ho, P., et al: Extremely high gain, low noise InAlAs/InGaAs HEMTs grown by MBE. Int. El. Dev. Meeting (1988) 184-186.

/34/ Lester, L. F., et al: 0.15 µm gate-length double recess pseudomorphic HEMT with f_{max} of 350 GHz. Int. El. Dev. Meeting (1988) 172-175.

/35/ Kellner, W., et al: GaAs-Feldeffekttransistoren 2nd ed. (1989) Springer, Berlin, Heidelberg, New York.

/36/ Smith, P. M., et al: A 0.15 µm gate-length pseudomorphic HEMT. IEEE MTT-S Long Beach (June 1969).

/37/ Duh, K. H. G., et al: High performance InP-based HEMT millimeter-wave low noise amplifiers. MTT-S Long Beach (June 1989).

/38/ Yamada, Y., et al: X and Ku band high power GaAs FETs. IEEE MTT-S (1988) 847-850.

/39/ Oda, Y., et al: Ka Band 1 Watt power Gas MMICs IEEE MTT-S (1988) 413-416.

/40/ Camilleri, N., et al: Ka band monolithic GaAs FET power amplifier modules. IEEE MTT-S (1988) 179-182.

/41/ Hegazi, G., et al: V-band monolithic power MESFET amplifiers. IEEE MTT-S (1988) 499-412.

/42/ Tsuda, K., et al: AlGaAs/GaAs HBTs for microwave power application. Solid State Dev. Mat. Conf. Tokyo (1987) 271-274.

/43/ Hida, H., et al: Novel self aligned gate process technology for i-AlGaAs/n-GaAs doped-channel hetero MISFET (DMT) LSIs based on E/D logic gates. Int. El. Dev. Meeting (1988) 688-691.

/44/ Notomi, S., et al: A high speed 1K x 4 bit static RAM using 0.5 µm-gate HEMT. IEEE GaAs IC Symp. (1987) 177-180.

/45/ Hirayama, M., et al: A 16-kbit static RAM using dislocation free crystal. IEEE Trans. ED 33 (Jan. 1986) 104-110.

/46/ Poulton, K., et al: A 1-GHz 6-bit ADC system. IEEE J. Solid St. Circ. 22 (Dec. 1987) 962-970.

/47/ Data sheets from Honeywell, Tektronix, Analog Devices.

/48/ Wakimoto, T., et al: Si bipolar 2 Gs/s 6 b flash AD conversion LSI. Int. Solid St. Circ. Conf. (1988) 232-233, 283.

/49/ Ducourant, T., et al: 3 GHz, 150 mW 4 bit GaAs analogue to digital converter. IEEE GaAs IC Symp. (1986) 209-212.

/50/ Taylor, G. W., et al: An n-channel BICFET in the GaAs/AlGaAs material system. IEEE El. Dev. Lett. 10 (Feb. 1989) 88-90.

Deep Submicron Dry Etching

Ph. LAPORTE

LETI, a division of Commisariat à l'Energie Atomique
CENG 85X-38041 GRENOBLE CEDEX FRANCE

1. INTRODUCTION

In electronic device manufacturing the trend is to reduce the size of individual devices for electrical performance and cost reduction.

Etching is one of the basic techniques used for pattern definition. Since 1970, various plasma etching systems (table 1) have been developed by several manufacturers to improve anisotropy. Using the PRIME process [1] today, 0.2µm feature etching has been demonstrated at the laboratory level, as shown in photo 1. These results have been obtained with conventional reactors and processes in the three basic materials of microelectronics, polysilicon, silicon oxide and aluminium. Nevertheless, before introducing deep submicron etching into production, many improvements have to be made on :
- Selectivity between the etched layer and the underlying layer.
- Perfect line width control of Critical Dimension (CD).
- Uniformity, especially for planarisation processes.
- Reduction of particle and chemical contamination
- Reduction of induced electrical damage.

These characteristics are already studied extensively in today's technologies [2], but major improvements have to be obtained to produce up to 10E8 transistors on a single chip by 1996.

Name	Introduction year	Interest	Limitation
Tubular Diode	1972	Low cost	No thermal control
plasma	1976	Selectivity	Anisotropy, Uniformity
RIE	1976	Anisotropy	Selectivity
low gap	1981	SiO2	Reproducibility
triode	1983	Flexibility	
magnetron	1984	Etch rate	Uniformity
Microwave			
DECR, ECR, RIPE ...	72-87	Low energy	

Table 1. Plasma systems overview

2. PATTERN TRANSFER CONSTRAINTS

2.1. Selectivity requirements

For deep submicron technologies, the selectivity value is no longer sufficient to describe the consumption of the underlying layer. The latter must be defined according to the overetch time, to take in account local loading effects when reaching the interface.

Gate oxide thickness is 15 nm for today's 0.8µm technologies, and selectivity, between poly-Si and oxide less than 10:1 is acceptable. For 0.3µm technologies gate oxide thickness will be 8nm and selectivity must be more than 15:1, even with a good end point detection, to overcome uniformity problems.

Furthermore overetch time must be increased according to variations in the thickness of the layer. This is especially important in the case of contact etching (Figure 2.1.) where 100 percent overetch time in the gate contact is necessary to etch the source and drain contacts.

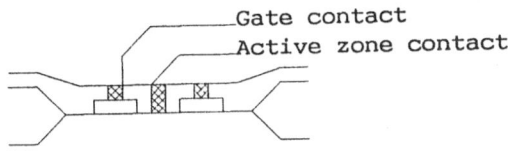

Figure 2.1. Influence of planarisation on thickness variation

When etching polysilicon with fluorinated forms, selectivity to the oxide is due to the natural chemical reactivity difference of each material. In contrast, to selectively etch the oxide over silicon, the natural reactivity difference has to be overcome by a passivation layer. In that case CxFy forms are used to etch oxide selectively with respect to silicon. Carbon forms are gettered by oxygen ($C+O2 \rightarrow CO2$) during the oxide etch. When the oxide etch is finished, carbon forms produce a teflon film (C2F2)n on silicon surfaces. The teflon film acts as a mask for the silicon etching and provides a good selectivity.

The passivation layer is necessary to insure no underlaying layer consumption, but a further cleaning step must be added to remove the passivation layer.

2.2. Critical dimension control

Line shape and CD's are controlled by directional plasma bombardment and sidewall passivation [3].Sidewall passivation can be obtained by adsorption of a gas form, such Cl2 during poly etching. In other cases a polymer passivation layer is formed in the plasma bulk [4] and sidewall passivation is governed by the equilibrium between deposition and ionic sputtering on exposed surfaces. Vertical surfaces are not sputtered and polymers can remain. Polymer formation also depends on the etched material. For example, etching TaSi2 with a SF6/Cl2 gas mixture creates a non volatile TaCl component which condenses on the sidewall.

Polymer formation is complex to handle. If the polymer becomes too thick during etching, the profile may be enlarged (Photo 2.1.). Often the photoresist consumption participates in the polymer formation. In that case some local degradation may appear at the interface between the etched layer and the underlying layer. So called "mouse holes" and inverse slopes are created when the polymer is partially destroyed during the etching (Photo 2.2).

The various shape configurations are observed using SEM. This is a good technique for precise observation but the sample is destroyed. With low voltage SEM it is possible to observe a full wafer without metallisation. CD measurement is possible by a top view of a periodic structure. Today the precision of this measurement is only 50nm (14% of a 0.35µm nominal line) and the line profiles are difficult to estimate. Electrical measurements can provide a more accurate estimate of various feature dimensions [5]. But electrical measurements involve several technological steps and recovering the etching step influence is not evident. Nevertheless, by careful statistical experiment planning and analysis, electrical line width measurement (PROMETRIX) can give valuable information on CD variation of conducting layers from the etching step only. Considering the lithography technique, it is possible to isolate the influence of the etching step only. This technique has to be calibrated with a SEM observation and assumes a vertical profile of the feature.

3. UNPATTERNED ETCHING

Planarisation and cleaning are two unpatterned etching techniques of major importance for deep submicron devices.

3.1. Uniformity requirements

The key parameter for planarisation processes is uniformity. The intermetal oxide planarisation is the most common application. Typically a 2µm oxide deposit is used to fill 1.5µm wide geometries (Figure 3.1). Since final oxide thickness above metal is 600nm, 1.4µm of oxide have to be etched back. For a deposition uniformity of ±5%, initial oxide thickness variation is 200nm. In the same way, a ±5% etch uniformity will produce a variation of 140nm of etched oxide. The overall variation of final thickness will be 340nm, more than half of the remaining thickness. Etching *and* deposition uniformity need to be improved significantly. For this issue both processes and reactor design must be optimised to obtain the optimum uniformity.

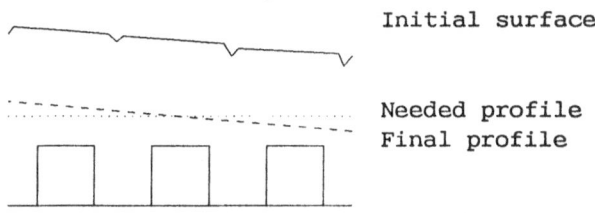

Figure 3.1. Absolute uniformity influence

Sometimes, etchback processes may be complicated by the vertical structure of the device. A typical and very important process where this difficulty appears is the tungsten plug etch back process. When the etchback is finished on the gate contacts, there is remaining tungsten on the lower region (photo 3.1). Overetching the remaining tungsten will remove the tungsten in the upper plugs. A possible solution could be to clear these residues after interconnection etching.

3.2. Cleaning and defect considerations

3.2.1. Contamination

Particle generation from gas flow and wafer handling will be reduced by improved equipment design and manufacturing [6]. Avoiding particle formation from the plasma chemistry on the other hand, requires making difficult trade offs, since polymerisation is intrinsic needed for anisotropy and selectivity in many processes. These thin polymers may also disturb further processing if not removed.

For example, after etching the LOCOS mask, stacking faults are generated in the silicon during the oxidation step, if the teflon film producing the selectivity to silicon (see 2.1.) is not removed. In the same way, after the oxide spacer etching during the oxidation step, the oxidation rate will be reduced by such a film. In both cases an oxygen plasma can be used to remove the fluorocarbon film.

3.2.2. Cleaning

Downstream plasma provides a plasma with reactive gas species with low kinetic energy. These plasmas can be used to strip both organic materials and thin inorganic layers, without chemical pollution or physical stresses.

This latter feature can be used to replace some wet chemical etching. During the oxidation mask stripping, local stress increases the etch rate of oxide in a HF solution. Microwave chlorine discharges provide a stress independant selective etching of the nitride over oxide.

Wet chemical etching will remain necessary for elimination of sidewall passivation. Metal atoms of the sidewall polymer are oxidized in classical oxygen resist stripping. Metal oxides are very often non volatile and cannot be removed after oxidation. Wet stripping can provide a solution.

4. CONCLUSION

New plasma generators ECR, DECR, RIPE and downstream, appear very promising in the way they can create a strongly dissociative plasma [7][8]. They are damage free and very good uniformity of etching rates can be achieved. In the near future the actual performances of these new reactors should be demonstrated in process integration.

At the same time an extensive study needs to be performed to understand basic physical and chemical mechanisms involved in plasma etching and polymer formation [9]. Understanding of these mechanisms may be used to create etching models [10] and to develop clean reliable processes.

References

1. C. Pierrat, F. Vinet, J.C. Guibert. PRIME process for deep UV lithography. Submitted to M.E. 1989 Cambridge U.K.

2. L.Peccoud, Ph.Laporte. New trends and limits in plasma etching. J.Appl.Physics 20(87),851-857.

3. A.Manenschyn, G.Janssen, E.Van der Drift, S.Radelaar. The etching mechanism of titanium polycide in a mixture of SF6+O2. J.Appl.Phys. 65(8) 15 april 1989.

4. G.S.Selwyn. Laser-based particulate measurements in etching plasmas. TEGAL seminar 1989.

5. G.Freeman, W.Lukaszet. Gate dimension characteristation using the inversion layer. Proc IEEE 1989 Vol 2 n°1 March

6. V.Ramakrishna, J.Harrigan. Defect learning requirements. Solid State Tech. Jan 89 p.103.

7. B.Bouyer, G.Ravel, P.Baussand, F.Mondon, L.Peccoud. Gravure selective et sans domages du silicium polycristallin, application aux circuits CCD. CIPG/SFV antibe 1989.

8. J.Pelletier, M.J.Cooke. Anisotropy control in CF4 microwave plasma etching. J.Appl.Phys 65(2) 15 jan 89.

9. P.Schoenborn, R.Patrick. Numerical simulation of a CF4/O2 plasma and corelation with spectroscopic and etch rate data. J. Elec. Soc. Vol 136 n°1 jan 1989.

10. J.Ignacio Ulacia Fresnedo. Theorical and experimental consideration necessary to build a dry etching process simulator. Technical report n° G833-1 ICL Standford june 1988.

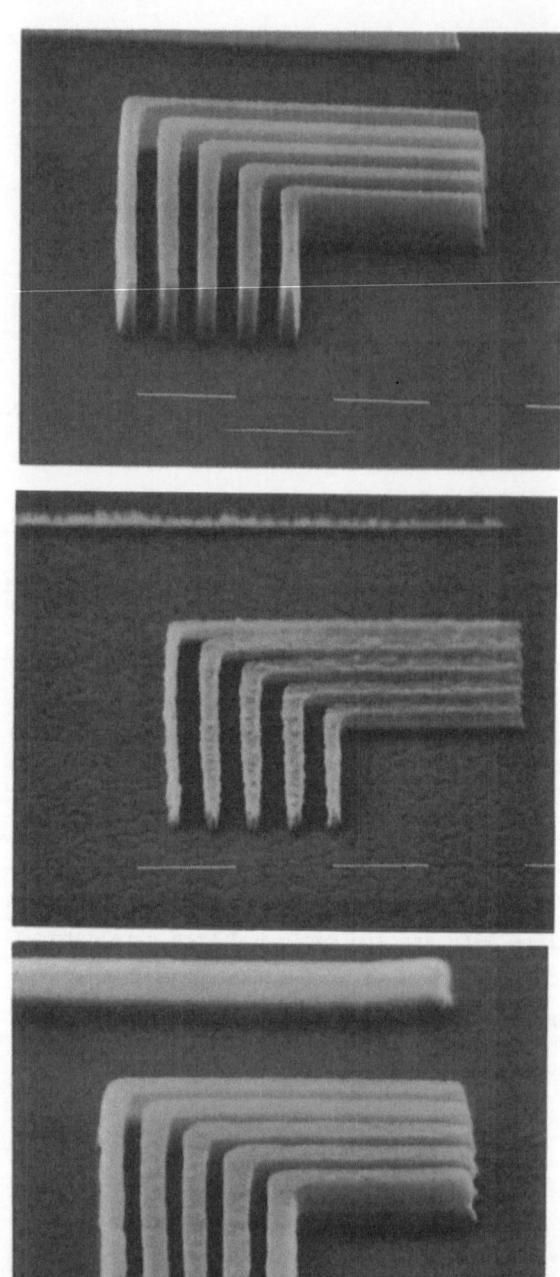

Photo 1. 0.2μm lines and spaces etching in
a:600nm oxide, b:300nm polysilicon, c:500nm aluminium

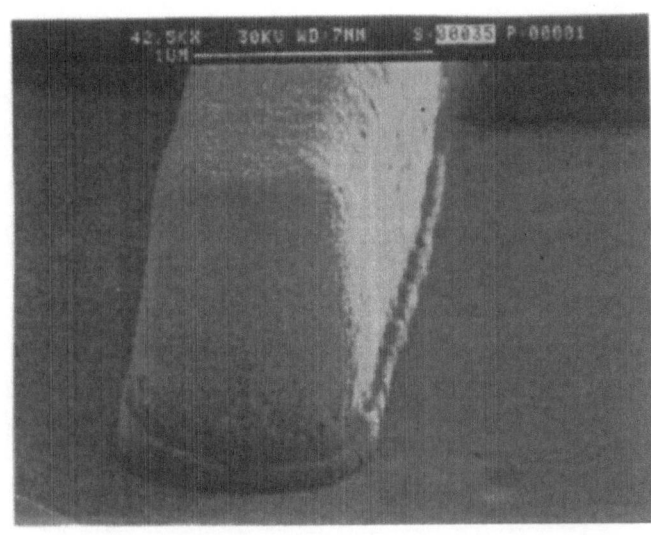

Photo 2.1. Profile deformation due to thick sidewall polymer formation.

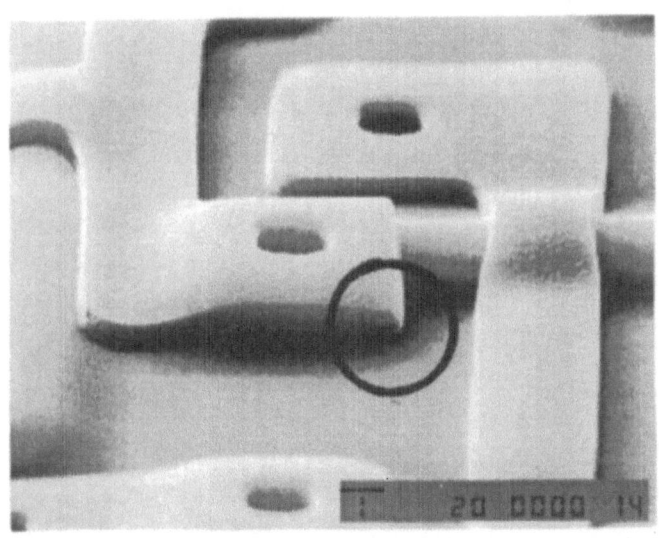

Photo 2.2. "Mouse hole" in aluminium lines

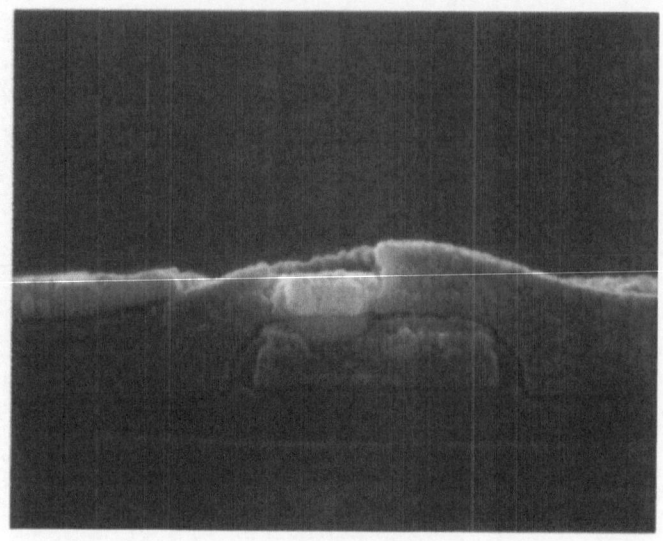

X 20.000

Photo 3.1. Extra planarisation during tungsten etch back

MOSFET Technology

Characterization of Hot Carrier Trapping in the Gate Oxide of MOSFETs

Mahnkopf, R.; Przyrembel, G.; Seifert, W; Wagemann, H.G.

Institut für Werkstoffe der Elektrotechnik
der Technischen Universität Berlin
Jebensstraße 1, D-1000 Berlin 12, West-Germany

Abstract - The degradation of MOSFET's due to hot carrier stress is described quantitatively by trapping of the injected carriers at a density of traps N_T with a capture cross section σ. The temperature- and field-dependence of σ strongly affects the stress behavior for electron and hole injection. The time dependence of the generation of fixed oxide charge and interface states suggests different mechanisms for both processes.

1 - INTRODUCTION

MOSFET degradation due to hot carrier stress strongly depends on the quality of the gate oxide. The damage process can be described by the density of trap centers within the gate oxide and their capture cross sections. The temperature dependence of the stress effect is correlated with the one of the capture cross sections. The degradation is also strongly affected by the oxide field dependence of the capture cross sections. The buildup of fixed oxide charges N_{ox} and interface states N_{it} shows a different dependence from the injected charge N_{inj} for both processes. In this study the results of our experiments on p- and n-channel MOSFET's will be shown.

2 - EXPERIMENTS

The devices used were polysilicon gated MOSFET's with a gate oxide of $d_{ox}=38nm$ and a channel width of W= 500µm. The effective channel length was L_{eff} = 2.7µm for the pMOS devices and L_{eff}=3.2 µm for the nMOS devices.

The temperature dependence was investigated on devices stressed at various temperatures ranging from 100K to 300K. The oxide field dependence was investigated applying different gate and drain stress voltages at room temperature. At pMOSFET's the stress bias causes the injection of hot electrons into the oxide, at nMOSFET's hot holes are injected from the depletion region of the reverse biased drain-substrate junction.

For the characterization of the damage the high-frequency capacitance characteristic as well as the charge pumping characteristic were taken before and after stress at stress temperature. From special HF-CV-measurements the density of stress induced fixed oxide charges ΔN_{ox} and its spatial distribution can be determined /1/. From the charge pumping measurements the density of generated interface states ΔN_{it} can be evaluated.

3 - RESULTS AND DISCUSSIONS

A pre-existing neutral density of traps N_T is assumed within the oxide layer responsible for the generation of stress induced oxide charges /2/. This trap density N_T and its capture cross section σ will be determined by our experiments. In Fig.1a the amount of stress induced fixed oxide charge ΔN_{ox} compared to the density of traps N_T is shown for pMOSFET's as a function of stress time for various stress temperatures. At higher temperatures the buildup of charge is more slowly increasing with time. According to the initial assumption the buildup of trapped charge is saturating by the equation

$$\Delta N_{ox}(t) = N_T \cdot (1 - \exp(-\sigma \cdot N_{inj}(t))) \qquad (1)$$

where $N_{inj}(t)$ is the density of injected carriers determined by an integration of the gate current over the extension of the damaged region /1/.

At each temperature the trap density N_T and its capture cross section σ has been deduced from the measured values of $N_{inj}(t)$. This was done for the injection of electrons at pMOSFET's and for the injection of holes at nMOSFET's. The trap densities N_{Tp} and N_{Tn} in the oxide of the n- and p-channel devices are comparable ($N_T \sim 0.5 \cdot 10^{12} cm^{-2}$). They agree well with values in literature /i.e.3/ and show no significant dependence on temperature. For both devices Fig.1b shows the determined values for the capture cross sections σ_p and σ_n for holes and electrons respectively vs. temperature.

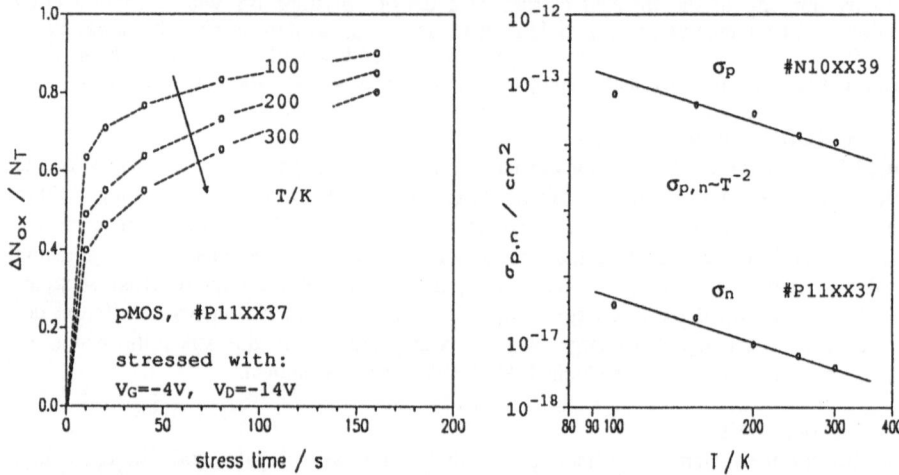

Fig.1 (a) Normalized density of $\Delta N_{ox}/N_T$ as a function of stress time for different stress temperatures (pMOS) (b) Capture cross sections $\sigma_{p,n}$ as a function of T.

As a result the capture cross sections σ_p and σ_n differ by more than 3 decades. The values $\sigma_p = 1 \cdot 10^{-14} cm^2$ for holes and $\sigma_n = 4 \cdot 10^{-18} cm^2$ for electrons at room temperature suggest to be characteristic for water related traps /4/. Both capture cross sections are strongly dependent on temperature. Assuming the eqn. $\sigma \propto T^{-n}$ an exponent $n=2$ can be determined from Fig.1b for the trapping of holes and electrons.

Furthermore the degradation of MOSFET's strongly depends on the electric field E_{ox} within the oxide. In Fig.2a the time dependence of the buildup of fixed oxide charge $\Delta N_{ox}/N_T$ is shown for pMOSFET's at various oxide fields E_{ox}. The different values of E_{ox} were realized by different gate and drain stress voltages. For higher values of E_{ox} a slower increase of the trapped charge can be seen. In Fig.2b the oxide field dependence of the capture cross sections is shown for n- and p-channel devices. For the capture of electrons now an oxide field dependence $\sigma \propto E_{ox}^{-n}$ with an exponent $n=3$ can be found in accordance with results in literature /5/. In contrast for the capture of holes an exponent $n=20$ has been evaluated. For this behavior the different properties of holes and electrons in the oxide may be responsible.

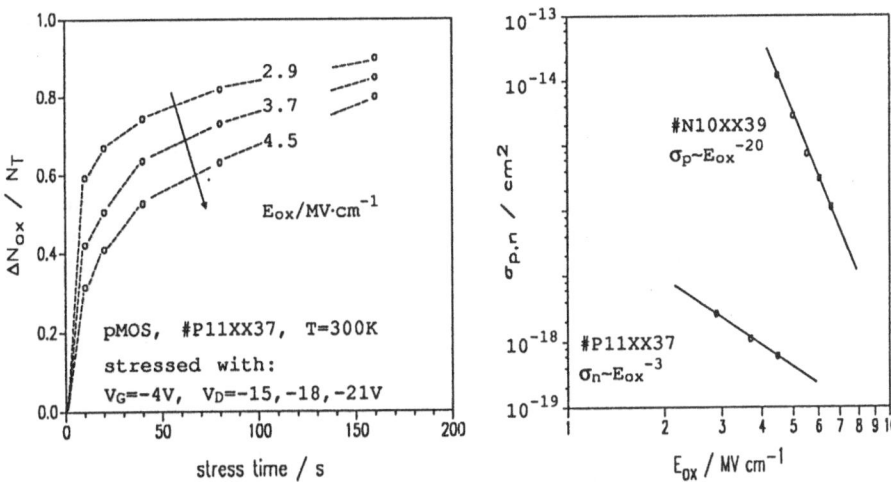

Fig.2 (a) Normalized density of $\Delta N_{ox}/N_T$ as a function of stress time for different oxide fields (p-MOS), (b) Capture cross section $\sigma_{p,n}$ as a function of E_{ox}.

In addition to the buildup of fixed charge ΔN_{ox} the time dependent generation of interface states ΔN_{it} was investigated. In Fig.3 the buildup of both, ΔN_{ox} and ΔN_{it}, is shown for the injection of electrons (pMOS) and holes (nMOS) as a function of the density of injected carriers N_{inj}.

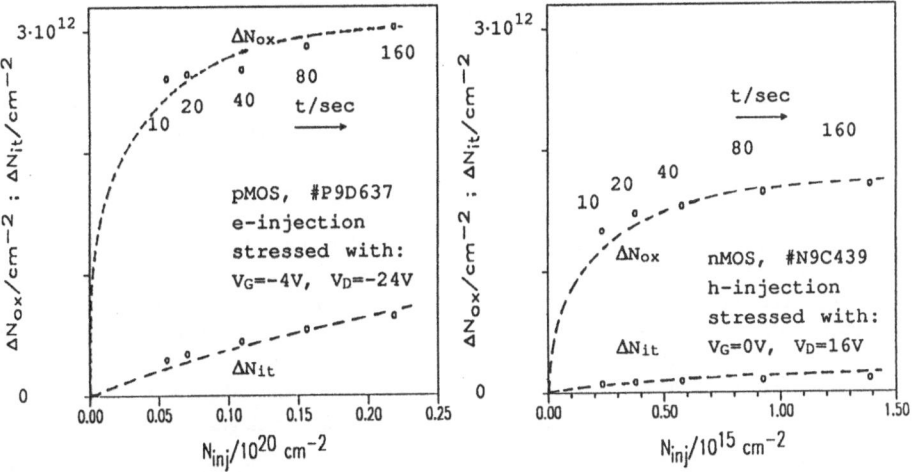

Fig.3 Density of fixed oxide charges ΔN_{ox} and interface states ΔN_{it} as a function of the density of injected carriers N_{inj}.

From Fig.3 the following statements can be pointed out:
- For a comparable stress effect the density N_{inj} differs by more than 4 decades for the two types of carriers at n- and p-channel MOSFET's.
- The course of ΔN_{it} is nearly proportional to N_{inj} in contrast to the saturation behavior of ΔN_{ox}. This may indicate different processes for the buildup of N_{it} and N_{ox}.

28

- The proportional increase of ΔN_{it} and the saturation behavior of ΔN_{ox} with N_{inj} is the same for electrons and holes. On the contrary the relation $\Delta N_{it} / \Delta N_{ox}$ is higher in the case of electron injection.

In Fig.4 the density of generated interface states ΔN_{it} is shown as a function of the density of induced fixed oxide charge ΔN_{ox} for different stress temperatures and drain stress voltages. The relation between ΔN_{it} and ΔN_{ox} can be described by the equation $\Delta N_{it} = a \cdot (\Delta N_{ox})^n$, where n is approximately 13 and independent of stress temperature. In contrast to other stress methods the exponent n is unequal 1 due to the local distribution of hot carrier damage. The constant a is strongly dependent on temperature, describing a more pronounced increase of N_{it} at elevated temperatures.

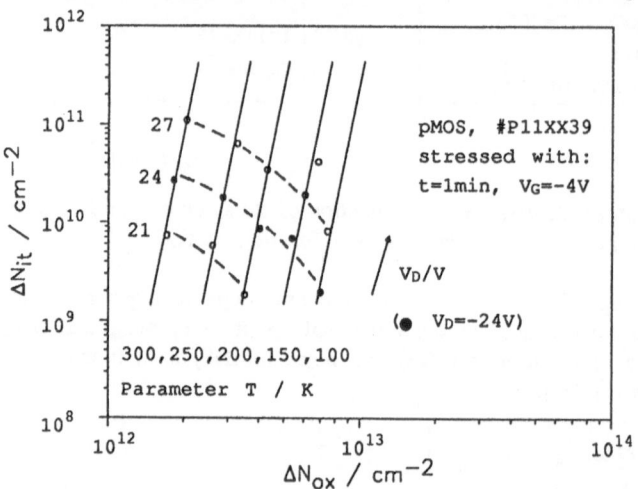

Fig. 4 The density ΔN_{it} of hot carrier stress generated interface states as a function of the density ΔN_{ox} of generated fixed oxide charges at different stress temperatures.

4 - CONCLUSION

In conclusion, the strength of damage due to hot carrier stress depends on the density N_T and the capture cross section σ of the traps within the gate oxide. Both values can be determined experimentally from stress characteristics as a function of time. The capture cross sections σ_p and σ_n show a strong temperature dependence, which allows to extract the same exponent n=2 for the power law relation. The same relation holds for the oxide field dependence with an exponent n=3 for electron- and n=20 for hole-trapping. The different buildup of the stress induced densities ΔN_{ox} and ΔN_{it} with N_{inj} can be explained by a different origin of the two types of degradation.

ACKNOWLEDGEMENTS - We wish to thank Dr. D. Haack, Fa. R. Bosch, Reutlingen, for technological support and the DFG/Deutsche Forschungsgemeinschaft Bonn as well as the FhG-Institute IMT Berlin for financial aid.

REFERENCES
/1/ Mahnkopf, R.; Przyrembel, G.; Wagemann, H.G.: ESSDERC '88, Montpellier
/2/ Nicollian, E.H.; Brews J.R.: MOS Physics and technology, Wiley,p. 535, 1982
/3/ Satoh, Y; et al.: Japanese Journal of Appl. Phys., vol. 22, No.4, pp. L221, 1983
/4/ Ning, T.H.; Yu, H.N.: Journal of Applied Physics, vol. 45, No.12, p. 5373, 1974
/5/ Ning, T.H.: Journal of Appl. Phys., vol. 47, No.7, pp. 3203, 1976

The Use of Boron Doped Polysilicon in PMOS Devices

A.J. Walker and P.H. Woerlee

Philips Research Laboratories
5600 JA Eindhoven
The Netherlands.

Abstract

Problems, such as short channel effects, arise when the buried channel PMOS device is scaled down to submicron dimensions. These can be alleviated by using boron doped polysilicon as gate material which results in a surface channel device. The effects of p-type polysilicon on capacitor and transistor parameters are presented. It is shown that, despite capacitor gate oxide instabilities, the boron doped polysilicon transistors are more hot-carrier resistant.

INTRODUCTION

In the dominant phosphorus doped polysilicon gate technology, PMOS devices require a compensating boron threshold implant in the n-well or n-type substrate in which these devices are located. This leads to devices which are prone to short-channel effects such as threshold voltage roll-off and drain-induced barrier lowering [1] as dimensions are further reduced. To alleviate these difficulties, p-type polysilicon can be used which results in a surface channel device. This paper reports work done on the use of boron doped polysilicon as gate material for the p-channel transistor.

EXPERIMENTAL

Two FZ $< 100 >$ n-type 10-15 Ω.cm batches of wafers were processed for this study. The first involved the fabrication of large capacitor structures to study boron penetration of the gate oxide and gate oxide stability. In this, polysilicon gate electrodes were formed by deposition and etching upon a 12.5nm gate oxide. Doping of the polysilicon was done by implantation as follows: (I) some wafers were implanted with phosphorus and annealed at 900°C in dry nitrogen; (II) others were similarly implanted with 16keV boron $(5 \times 10^{15} cm^{-2})$ and annealed; (III) some of the wafers from (II) were further implanted with phosphorus $(10^{16} cm^{-2})$ and annealed. Therefore, there were three types of gate electrodes: (I) n-type with only phosphorus present, (II) p-type with only boron present and, finally, (III) n-type with both phosphorus and boron. In (III), the boron is distributed in the polysilicon and gate oxide in the same way as in (II) where there is only boron present. The only difference is that in (III), the polysilicon is overdoped with phosphorus to make the gate n-type. For some wafers, where the polysilicon had been doped only with phosphorus or boron, different anneals were carried out after polysilicon doping to test boron penetration. These were 850°C, 900°C, 925°C, 950°C all in dry nitrogen.

The second batch involved the fabrication of PMOS devices with minimum polysilicon gatelengths of 0.7μm and minimum effective channel lengths of 0.4μm. The surface channel device had surface and well concentrations of $2 \times 10^{17} cm^{-3}$ and $5 \times 10^{16} cm^{-3}$ respectively while the buried channel device had an arsenic anti-punchthrough concentration of $2 \times 10^{17} cm^{-3}$. A 12.5nm gate oxide was grown followed by a threshold voltage adjust implant. Polysilicon was then deposited and doped in the case of the n-type polysilicon wafers with a phosphorus implant ($10^{16} cm^{-2}$). After patterning and etching, the source/drain implant (boron $5 \times 10^{15} cm^{-2}$ 16keV) was carried out doping the polysilicon in the case of the p-type polysilicon wafers. Annealing, dielectric deposition and metalisation were followed by a forming gas anneal at 450°C.

The transistors were stressed at maximum gate current (-1.5V source/gate voltage) with various source/drain voltages. with the maximum linear transconductance, $\beta_\square$, being monitored since this parameter changed most rapidly. The device was defined to be out of specification if $\beta_\square$ had changed (increased) by ten percent. Extrapolation of the lifetime v inverse source/drain voltage to 30 years gives the maximum V_{ds} for 30 years operation within specification.

RESULTS AND DISCUSSION

Fig.1 shows the difference in flatband voltage for the capacitor samples (I) and (II), measured using the high frequency capacitance - voltage method, as a function of thermal budget after implantation doping of the polysilicon. 900°C 60 minutes in dry nitrogen seems to be the maximum budget to limit boron diffusion through the 12.5nm gate oxide. The effect of a positive bias temperature stress (+5V, 250°C, 5min) can be seen in Fig.2 in which quasi-static C-V curves are shown from capacitors with n- and p-type gates. Only the type (II) capacitors with p-type gates were affected and only with a positive bias. This was true even if boron was present in the gate oxide and the polysilicon as in (III). Therefore, this instability seems to be independent of whether boron is present in the polysilicon or gate oxide. Rather, it appears that the polysilicon workfunction is the important parameter determining the gate oxide stability. This suggests that hole injection from the gate may be the degradation mechanism. Fig.3 shows the increase of interface states for a p-poly capacitor as a function of inverse absolute stress temperature. The activation energy of 0.9eV agrees quite well with [2].

The transistor properties will be treated next. Fig.4 shows the linear and saturation currents in the transistors with either n- or p-type polysilicon gates. The 15% lower current drive capability of the p-poly pMOS can be seen and is due to the lower hole mobility caused by the higher effective field in the inversion layer [3]. The transistors were stressed as described above. Fig.5 shows the transistor lifetime (time required for a 10% shift in the linear transconductance). Extrapolation to 30 years operation gives the maximum source/drain voltage for 30 years operation within specification. Fig.6 shows this voltage plotted as a function of effective channel length for both types of pMOS transistor. The p-poly devices are more hot-carrier resistant. This could be due to the higher lateral field near the drain due to the heavy anti-punchthrough implant in the case of the n-type poly device. Also, the difference in workfunction between the n- and p-type gates leads to a larger oxide field which in turn eases electron tunneling into the gate oxide in the case of the n-type gated device.

CONCLUSIONS

Disadvantages of using p-type polysilicon in pMOS devices include smaller current drive capability and a positive bias temperature stress instability. The former will become less important as devices are shrunk because of saturation of the hole velocity. Despite the apparent gate oxide degradation as indicated in Fig.2, Figs.5 and 6 show that the p-type polysilicon devices are more hot-carrier resistant.

References
[1] G.J. Hu and R.H. Bruce, IEEE Trans. Elect. Dev., ED-32, pp.584-588, March 1985.
[2] Y. Hiruta, F. Matsuoka. K. Hama, H. Iwai, K. Maeguchi and K. Kanzaki, IEDM, pp.578-581, 1987.
[3] A.J. Walker and P.H. Woerlee, J. de Physique, Colloque C4, supplément au n^o9, Tome 49, Sept. 1988.

32

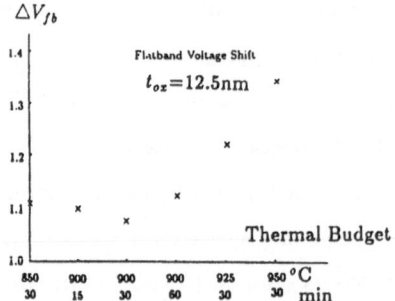

Fig.1 Difference in flatband voltage between p-type polysilicon capacitors and equivalent n-type capacitors as a function of temperature budget after polysilicon boron doping.

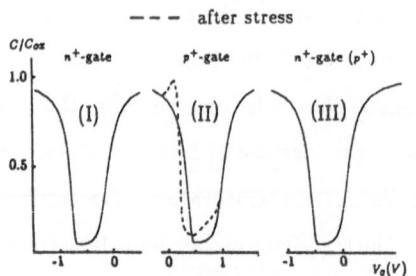

Fig.2 Changes in the quasi-static C-V curves due to a BT stress (+5V at 250°C, 5min) for three types of gate electrode: (I)purely phosphorus doped; (II)purely boron doped; (III)boron present but overdoped with phosphorus.

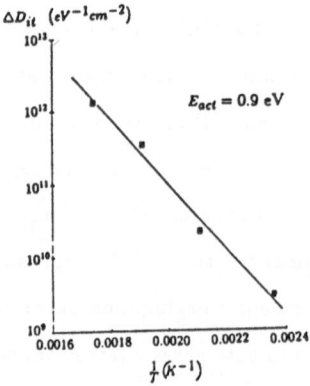

Fig.3 Increase of interface state density for p-type gated capacitor as a function of inverse absolute temperature at which BT stress was carried out. The activation energy is also given.

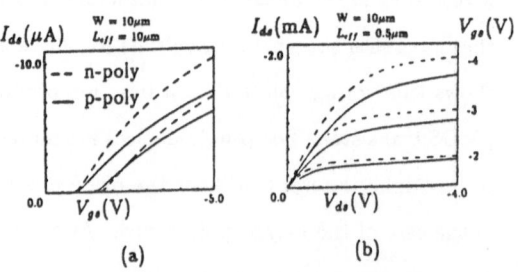

Fig.4 (a) The linear characteristics of 10μm × 10μm transistors with either n- or p-type polysilicon. The substrate bias was 0 and -3 volts. (b) The saturation characteristics for similar sub-micron devices.

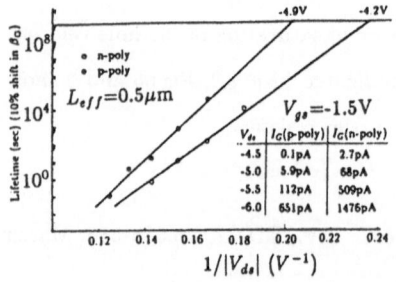

Fig.5 Plot of transistor lifetime for a 10% shift in linear transconductance as a function of the inverse source/drain voltage at which each stress occurred. Inset is the gate current as a function of the source/drain voltage for each type of device $V_{gs} = -1.5V$.

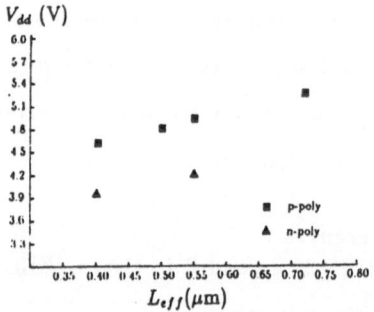

Fig.6 Maximum extrapolated source/drain voltage for 30 years operation as a function of effective channel length.

Semiconductor Device Fabrication with High Energy Ion Implantation

T.Harms[*],K.Goser[*],U.Hilleringmann[*],W.Fahrner[**],K.Oppermann[***]

*   University of Dortmund, Lehrstuhl Bauelemente der Elektro-
    technik, Postfach 500500, D-4600 Dortmund
** Fernuniversität Hagen, Lehrstuhl Bauelemente der Elektro-
    technik, Postfach 940, D-5800 Hagen
*** Siemens AG, Forschungslabor (ZFE EL PT 11), D-8000 München,
    Otto-Hahn-Ring 6

Abstract
A high energy ion implantation realized in a BiCMOS technology
reduces the npn-transistor collector series resistance R_C and
improves the Latch-up behavior.

Up to now there are two approaches to realize CMOS-transistors
and bipolar transistors on the same chip, namely

(a) using p-substrate, the n-well for PMOS forms the collector
    for vertical integrated npn-bipolar transistors [1],
(b) using p-substrate, a buried layer is formed by ion implan-
    tation followed by an epitaxial deposition of a monocrys-
    talline silicon layer [2].

Qualitative comparison of the two mentioned alternatives:

property	version (a)	version (b)
additional required masks	1	2
Latch-up immunity	low	high
collector series resistance	high	low

In the next paragraph a third version will be described which
combines the advantages of the two former ones: our approach

consists in integrating an npn-transistor into a retrograde n-well CMOS structure. The additional effort for the vertical npn-bipolar transistor is one boron implantation for the active base. Bulk material is ⟨100⟩-orientated silicon with a specific resistance of 10 - 90 Ω cm. The bipolar transistors use the n-well of the CMOS-process as collector as mentioned in version (a). In that version the n-well was realized by the following steps:

i) low energy P^{31} ion implant (150 keV, dose $5 \cdot 10^{12}$ cm^{-2}) and

ii) 17 hours annealing and drive-in at 1170°C in a N_2 ambient.

Due to diffusion the doping concentration of the n-well steadily decreases from the surface to the n-well - to - substrate pn-junction, approximately 6 μm below the silicon surface.

In the present version the well is formed by a 6.5MeV/10MeV - triply charged phosphorus ion implantation with doses in the range from $1 \cdot 10^{13}$ cm^{-2} to $2 \cdot 10^{14}$ cm^{-2}. The particle current is about 400 nA. The average ion penetration depth calculated by the method of Ziegler and Littmark (program TRIM 85 [3]) is 3.1 μm and 4.1 μm, respectively (fig. 1). For this reason the time-consuming drive-in is unnecessary. Furthermore the concentration maximum is far below the surface and forms a buried layer.

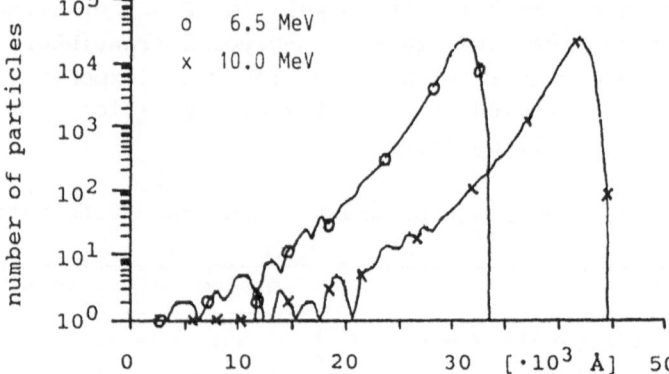

fig. 1: Monte-Carlo-simulation of particle profile vs. penetration depth (100000 particles P^{31} in Si)

Because of these ion ranges conventional ion implantation masks as photoresist, silicon dioxide or silicon nitride are unfavorable.

The mask should fulfill several requirements [4]:
- it should have a sufficient stopping power for the incoming ions;
- it should not contaminate the surface of the wafer;
- it should be easily removable;
- it must exhibit a good pattern transfer.

Common masking materials are: gold, aluminium, silicon nitride, silicon dioxide, photo resist. It is known that metals provide the best stopping power and, additionally, they transfer charge collected on the mask off the wafer, thus preventing distortion of the incoming ion beam.

Thus for implantation in the MeV range a 5.9 μm thick aluminium layer was deposited for the following reasons: The maximum range of phosphorus ions (10 MeV implant) in Al is about 4.4 μm [3] and there is an absolute selectivity of the Al etchant consisting of a composition of phosphoric acid and nitric acid to the underlying silicon and silicon dioxide layers - thus an etch of these layers is impossible.

After phosphorus implantation at the Hahn-Meitner-Institute in Berlin the masking Al layer is removed and the silicon is annealed at 1024°C for 120 minutes. This temperature treatment recovers the irradiated silicon from implantation damages. During further processing steps the strongest heat treatment did not exceed a temperature of 1024°C.

The concentration of charge carrier density was measured with the spreading-resistance-method before further processing to obtain values of the doping concentration from the surface into the substrate. The implanted n-well exhibits low surface dopant concentration, namely $1.5 \cdot 10^{14}$ cm^{-3} (6.5 MeV P-implant, dose $2 \cdot 10^{14}$ cm^{-2}) and a dopant concentration of $1 \cdot 10^{18}$ cm^{-3} in a depth of about 4.0 μm below the MOS channel region (fig. 2). A comparison of fig. 1 and fig. 2 shows that the predicted and measured penetration depths are different by roughly 30% [5].

The current gain of the bipolar transistors is unchanged with respect to those transistors fabricated with version's (a) technology mentioned above [6]. R_C is reduced to 200 Ω for the lowest dose and highest energy ($2 \cdot 10^{13}$ cm^{-2}, 10 MeV); the

36

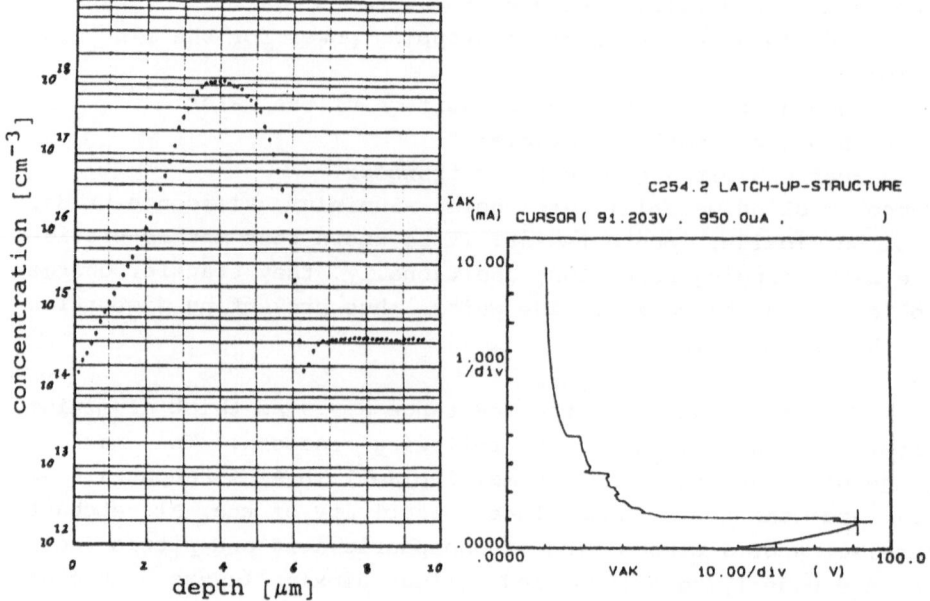

fig. 2: n-well in Si bulk
(Spreading-Resi-
stance-Measurement)

fig. 3: Latch-up behavior of a
p^+npn^+ - structure

values for higher doses and the lower energy are even better.
In addition, the implanted n-well also improves the Latch-up
immunity of CMOS circuitries due to the lowered sheet resistance in the well. The critical voltage of the parasitic thyristor, related to punch-through, is larger than 90 V (fig. 3).
So the overall performance of an integrated BiCMOS-circuit can
be improved by a single process step, namely the high energy
p^{31} implantation of the n-well.

References
[1] D.Widmann, H.Mader, H.Friedrich: Technologie hochintegrier-
ter Schaltungen ; Springer-Verlag 1988, p.301 ff.
[2] J.Arndt: BICMOS 1 - Der erste Schritt zur optimalen Sy-
stemintegration; ITG Fachbericht 1987, p.153 ff.
[3] J.F.Ziegler, J.P.Biersack, U.Littmark: The Stopping and
Range of Ions in Solids; Pergamon Press 1985
[4] J.L.Stone, J.C.Plunkett: Ion Implantation Processes in
Silicon, in: Materials Processing - Theory and Practice
Vol. 2, North Holland 1981, p.78 ff.
[5] W.R.Fahrner, K.G.Oppermann, T.Harms: 5-100 MeV Ion Implantation and its Simulation by the MARLOWE program (submitted)
[6] T.Harms, K.Goser: Influence of Implantation Energies and
Doses on the Electrical Behavior of BICMOS Devices; DPG-
Frühjahrstagung Münster 1989

A Self-aligned Gate Definition Process with Submicron Gaps

L.F.P. Warmerdam, A.A.I. Aarnink, J. Holleman, H. Wallinga

University of Twente, IC Technology and Electronics Department,
P.O. Box 217, 7500 AE Enschede, Netherlands

Abstract

A self-aligned gate definition process is proposed. Spacings between adjacent gates of 0.5 μm and smaller are fabricated. The spacing is realized by an edge-etch technique, combined with anisotropic plasma etching of the single poly-silicon layer. Straight gaps with minor width variation are fabricated. Minority carrier life-time and breakdown voltage are not affected.

Introduction

In order to completely integrate complex CCD-based functions, a combined BCCD-CMOS process has been developed. The main research area for circuits fabricated in this process is video frequency filter applications. The low-voltage n-channel BCCD-CMOS process is fully ion-implanted and uses a self-aligned gate definition process. In contrast to an overlapping gate technology, this reduces the inter-electrode capacitances considerably and avoids electrical isolation problems with the dielectric between first and second poly-silicon layer. Furthermore all four phases are identical using this approach. Definition of the sub micron gaps is obtained by technological means, rather than by advanced optical lithography or electron-beam lithography [1]. In fact, the demands on the lithographic process are not determined by the dimensions of the gap but by the minimum feature size used.

A gap between adjacent gates leads to a local maximum in the potential of a depleted BCCD channel when both gate voltages are equal. This maximum functions as a well in which a fraction of the transported charge packets is trapped, causing an increase in the charge-transfer inefficiency. In figure 1 the simulated values of the well potential as a function of gap width is shown. The potential maximum is somewhat bias dependent and vanishes rapidly with increasing potential difference between adjacent gates. Due to this effect are small well potentials tolerable. Simulations have shown that well potentials up to 0.5 V originating from gap widths up to 0.5 μm do not seriously degrade charge transfer performance.

Processing

For the fabrication <100> p-type epitaxial wafers were used. Standard LOCOS isolation is combined with high energy ion implantation of both p- and n-well. The gate definition process is based on the edge-etch technique [2,3]. Figure 2 shows the process flow of the gate definition. Gate oxide thickness is 25 nm, poly-silicon thickness is 500 nm. Key step is the isotropical back-etch of the silicon dioxide layer, particularly regarding the lateral etching of the silicon dioxide at the edges (step 4, fig. 2a). The amount of etching at this step determines the width of the etch-mask of the gap. Therefore the gap width can be monitored easily by adjusting the etch time; spacings in the deep sub micron regime are feasible. The etch of the silicon dioxide and silicon nitride stack (step 3, GD mask) is performed anisotropically in a CHF_3-O_2 plasma. The lateral back-etch of the silicon dioxide layer (step 4) is performed in a thermostated high-purity HF-NH_4F mixture and is completely isotropic. The fact that lateral etching is performed at a nearly perpendicular oxide side-wall results in homogeneous and reproducible etch behavior. This implementation yields straight gaps with minor width variation.

All features defined in step 3 using the GD mask automatically result in poly-silicon patterns. Single poly-silicon features are defined in step 7 by etching a pattern in the masking oxide. An overlap of PS and GD masks results in self-aligned poly-silicon gates. Separation of the gates is warranted by the gaps, which are present at the edges of every GD feature. The gates for BCCD devices and CMOS transistors are etched in a single step. The poly-silicon is etched anisotropically in a Cl_2-$SiCl_4$ RIE plasma with high selectivity over silicon dioxide. The integrity of the gate oxide in the gap is restored by a thermal oxidation (fig. 2d). Simultaneously the side walls of the poly-silicon gates are oxidized. The gaps are filled with a planarizing oxide. Finally a standard single-layer metal back-end process is used to complete wafer fabrication.

Results

The figures 3 and 4 show SEM pictures of 0.5 μm gap devices on a wafer, extracted from processing directly after the gate definition process. Figure 3 presents a detail of a test device with gaps running meander-like over the active area. Clearly the gaps run in straight lines and show minor width variation. Corners and topographic steps at the bird's beak do not influence gap dimensions. Gaps always end at the edge of poly-silicon features. At those points a direct transition exists from sub-micron spacing to open area.

In figure 4 a cross-section of the gap is shown. A poly-silicon capping layer has been deposited for contrast enhancement only. Due to the thermal oxidation treatment after the gap fabrication a firm passivating oxide layer exists on side walls and exposed substrate. The silicon nitride between the silicon dioxide layer and poly-silicon gate (situated at the top right side of the gap, the silicon nitride layer is emphasized by means of a drawn line) prevents the complete encapsulation in silicon dioxide of the gate. This poses no problems concerning electrical isolation. In normal processing the gap is filled with an insulator.

Several wafers with test devices have been realized. Devices had gap widths of 0.2-0.5 μm. C-t measurements have been performed on capacitances with and without gaps meandering over the surface. For both type of devices minority carrier lifetimes of 20 μs have been found. Also no difference in breakdown voltage has been observed. Breakdown voltage is 23 V. With a gate-oxide of 25 nm a breakdown field strength of 9.10^6 Vcm^{-1} is found. It is concluded that the damage caused by plasma etching is removed sufficiently.

Acknowledgements

This work has been supported by the Dutch Foundation for Fundamental Research on Matter, FOM. The authors would like to thank G.Boom and B.Otten for their assistance with SEM photography and F.W.Ragay and R.C.M.Wijburg for numerous encouraging discussions.

References

1- J.W. Slotboom et. al.; "Sub micron CCD Memory structures fabricated by electron-beam lithography"; IEDM Tech.Dig. 1984; 308-311
2- H.H. Hosack and R.H. Dyck; "Sub micron Patterning of Surfaces"; IEEE Journal of Solid-State Circuits; SC-12(4) aug. 1977; 363-367
3- V.J. Kapoor; "Charge-Coupled Devices with Sub micron Gaps"; IEEE Electron Device Letters; EDL-2(4) april 1981; 92-94

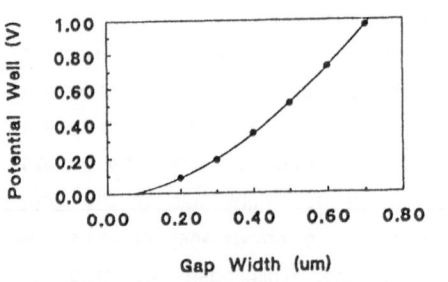

Figure 1: Simulated values of the potential well below the gap as a function of the gap width. Gate voltage on both gates was 0V.

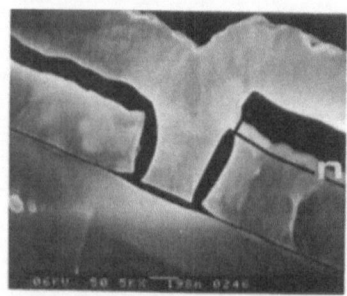

Figure 3: SEM photograph of a detail of a test device.

Figure 4: SEM photograph of a cross section of the gap after the gate definition process. A poly-silicon capping layer has been deposited. On top of the right hand gate, the presence of a silicon nitride layer is emphasized by the drawn line.

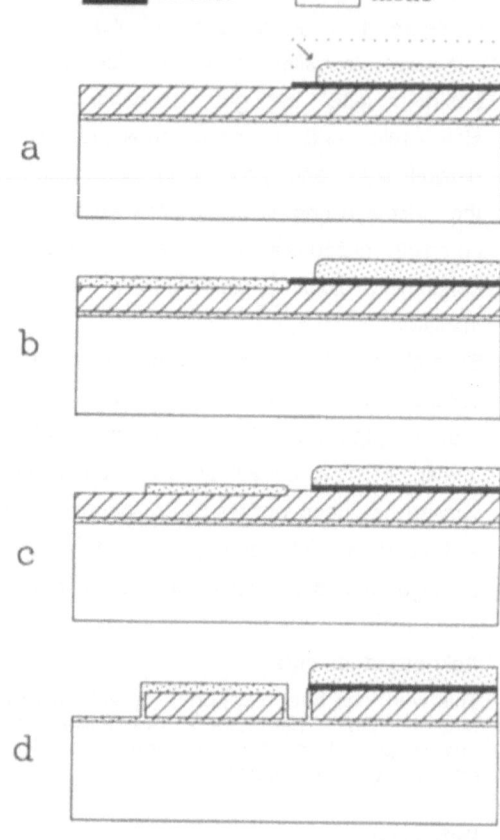

Figure 2: Process flow of the gate definition.
1. Starting point: LOCOS, implanted wells, 25 nm gate oxide
2. Deposition of poly-silicon, doping, deposition of silicon nitride and silicon dioxide
3. Definition of the silicon dioxide and silicon nitride stack, Gap Definition (GD) mask, resist strip
4. Back etch of silicon dioxide (fig. a)
5. Growth of a thermal oxide on the exposed poly-silicon. (fig. b)
6. Etch of the exposed silicon nitride rims
7. Definition of the masking oxide, Poly-Silicon (PS) mask, resist strip (fig. c)
8. Etch of poly silicon
9. Growth of passivating oxide (fig. d)

Novel Submicron Processes for Shallow p$^+$n-Junctions

L. Mader, M. Orlowski, and I. Weitzel

Siemens AG, Research and Development Laboratories, Otto-Hahn-Ring 6, 8000 München 83, FRG

Abstract - Shallow p$^+$-n junctions have been achieved employing substrate preamorphization with gallium implantation prior to high dose boron or BF$_2$ implantation. The resulting p$^+$-n junctions are investigated and compared with junctions obtained with conventional techniques. The impact of different amorphization procedures on boron diffusivity is explained by analytical models.

INTRODUCTION

Shallow source/drain junctions are one of the key requirements for the scaled CMOS technologies. In case of p$^+$-n junctions boron poses problems due to its deep penetration and its high channeling probability at low implantation energies in crystalline silicon. Boron channeling can be suppressed by use of preamorphization techniques (Si, Ge, In) [1,2,3] and to some degree by using BF$_2$ implants. As an alternative to boron gallium implants for furnace and rapid thermal anneal (RTA) have been considered in the literature [4,5]. From those experiments it has been concluded that the diffusivity of Ga which is large in Si and SiO$_2$ and secondary defects remaining after furnace anneal make Ga unattractive as a substituting element for B. On the other hand RTA processing of Ga-implant has been shown to be an appealing technique to produce very shallow defect free Ga-doped layers.

In this paper we propose new techniques [6] using dual implantation being compatible with furnace anneal employed in most CMOS technologies. First Ga-implantation amorphizes the substrate and the subsequent high dose BF$_2$ or B implantation followed by an anneal at 900°C for 40 min establish the final properties of the p$^+$-n junctions. This technique makes use of the following features: a) the preamorphization suppresses B channeling, b) the presence of Ga retards B diffusivity [7,8], c) preamorphization and presence of Ga enhance the electric activity of B [7].

EXPERIMENTAL

All samples have been implanted with the same boron dose of 5E15 cm^{-2} at 25 keV for B and at 50 keV for BF$_2$ through 15 nm of a protective oxide thickness. Five samples corresponding to the following configurations have been processed:

S1: B into crystalline Si (B)
S2: B into Si-preamorphized substrate with Si-dose = 5E14 cm^{-2} at 150 keV [1,9] (B(Si))
S3: B into Ga-preamorphized substrate with Ga-dose = 1E15 cm^{-2} at 200 keV (B(Ga))
S4: BF$_2$ into crystalline Si (BF$_2$)
S5: BF$_2$ into Ga-preamorphized substrate with Ga-dose = 1E15 cm^{-2} at 200 keV (BF$_2$(Ga))

For further investigation the sample S1 is used as a reference. The resulting boron profiles have been evaluated by the SIMS method and the sheet resistance by the four point probe measurements. Unlike B, a Ga implantation for doses higher than 5E14 cm^{-2} at room temperature results in an amorphous region. As a consequence, the annealing of Ga-implanted Si is accompanied by a solid epitaxial regrowth of the amorphized Si. At higher temperatures this regrowth occurs within seconds or less.

RESULTS AND DISCUSSION

The as-implanted profiles for the above five samples are shown in Fig.1a. It can be seen that the suppression of the boron channeling is the same for both preamorphization procedures with Si and Ga (almost identical boron profiles). It can be also seen that in case of BF$_2$ implant preamorphization suppresses channeling considerably. Apparently the gradual amorphization during BF$_2$ implantation is insufficient to suppress the channeling completely. In Table I the junction depths are compared for different background n-type concentrations. The shallowest junction is obtained for BF$_2$ implant into Ga-preamorphized substrate. The maximum gain in junction depth x$_j$ (compared with the reference S1) ranges between 85 and 165 nm depending on the values of the background concentration. The as-implanted profiles are particularly relevant for RTA treatment.

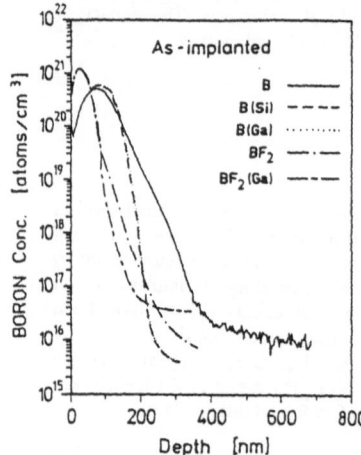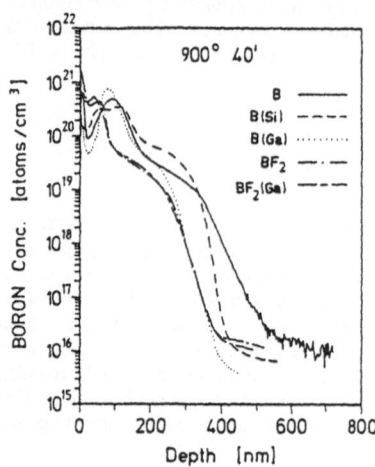

Fig.1. Boron profiles: as-implanted (a) and after 900°C anneal for 40 min (b) for different configurations S1 - S5.

In Fig.1b. The corresponding SIMS profiles of annealed samples at 900°C in N$_2$ ambient for 40 min are shown, and in Table II the pertinent junction depths are displayed. One observes that only for very high background concentration configuration S1 yields shallower junctions than other variants. However, for all background concentrations the Ga-preamorphization yields significantly shallower junctions than Si-preamorphization. This is due to the mutual retardation of B and Ga diffusion [7,8,10]. The corresponding junction depths are smaller by 50 - 75 nm. However, it is interesting to note that in case of BF$_2$ no retardation of B due to Ga can be observed. In order to study the diffusion effects versus implantation effects the junction depths differences $\Delta x_j = x_j$ (900°C, 40 min) - x$_j$ (as implanted) are displayed in Table III. Table IV exhibits the corresponding sheet resistance values. The following observations can be made:

i) Although pure boron implantation (S1) leads to the deepest junctions, its Δx_j are smaller than Δx_j for all other variants (except for S3 at 3E18 cm^{-2}), implying that the boron diffusivity on the average is smaller than the boron diffusivity for all other variants. The enhanced boron diffusivity is due to the initial generation of point defects due to implantation and preamorphization procedures. Thus the diffusivity enhancement counteracts the advantages of shallower as-implanted profiles. The enhancement of the diffusivity of S4 over that of S1 can be modeled by $D_{BF2} = D_B \{1 + \alpha \exp(-t/t^*)\}$. We find from additional experiments $\alpha = 3.2$ and $t^* = 320$ min.

ii) In case of S2 and S3 the retardation effect of Ga on B-diffusion is clearly seen. However, in case of BF$_2$, variants S4 and S5, the presence of Ga seems to enhance B diffusivity. This conflicting findings might be resolved by the coexistence of the retardation and drag effects discussed in Ref.[10]. Both effects depend strongly on the initial conditions, i.e. on relative position of the as-implanted B and Ga profiles.

Variant	B	B(Si)	B(Ga)	BF$_2$	BF$_2$(Ga)
$x_j(1)$	325	200	200	200	160
$x_j(2)$	260	190	190	150	115
$x_j(3)$	175	150	150	90	90

Table I: Junction depths x_j in nm for as-implanted B profiles at different n-type background concentrations: $x_j(1)$ at 2E17, $x_j(2)$ at 3E18, $x_j(3)$ at 3E19 cm^{-2} Variant B B(Si)

Variant	B	B(Si)	B(Ga)	BF$_2$	BF$_2$(Ga)
$x_j(1)$	450	390	340	340	340
$x_j(2)$	380	360	285	280	275
$x_j(3)$	210	285	210	160	160

Table II: Junction depths x_j for B profiles after 900°C anneal for 40 min. The same notation as in Table I is used.

Variant	B	B(Si)	B(Ga)	BF$_2$	BF$_2$(Ga)
$x_j(1)$	125	190	140	140	180
$x_j(2)$	120	170	95	130	160
$x_j(3)$	45	135	60	70	70

Table III: Junction depth differences $x_j = x_j$ (900°C) - x_j (as-impl) corresponding to Table I and Table II.

Variant	B	B(Si)	B(Ga)	BF$_2$	BF$_2$(Ga)
$R_S(\Omega/\square)$	74	34	58	52	46

Table IV: Sheet resistance for variants S1-S5

In Ref.[10] it has been assumed that in a two-component system each of the two species, A and B, can diffuse partly by itself and partly in pairs AB. The system is then characterized by four quantities: a pairing probability α_{AB}, and the diffusivities D_A, D_B and D_{AB}, for the unpaired and paired diffusion, respectively. In case of sufficiently large α_{BGa} between B and Ga and assuming that the diffusivity of B-Ga pairs D_{BGa} is smaller than the unpaired diffusivity D_B, i.e. $D_{BGa} < D_B$, the overall B-diffusivity will be retarded as shown in Fig.2. The corresponding equations and comprehensive discussion of the dynamics involved can be found in Ref.[10].

iii) The largest enhancement of B-diffusivity is found for Si-preamorphization (S2). This finding establishes a serious limitation of this technique especially for longer furnace anneals.

iv) However as seen from Table IV, the Si-preamorphization S2 yields the lowest sheet resistance of 34 $\Omega/\square$. This indicates that Si-preamorphization is more effective than Ga-preamorphization in incorporating B-atoms into the regrown silicon lattice and thus in dissolving B-precipitates or clusters. Incidentally, the higher electric activity for S2 might correlate with higher enhancement of B-diffusivity.

v) The sheet resistance data show that amorphization either due to preamorphization or to amorphization during BF$_2$-implantation is conducive to enhance boron electric activity.

vi) A surprising result, contradicting the findings by Aronowitz [7] is that Si-preamorphization is more effective in incorporating B-atoms into Si-lattice then the Ga-preamorphization.

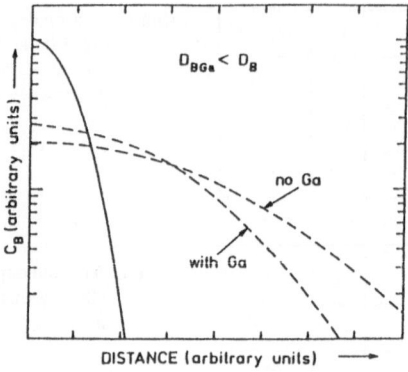

Fig.2. Simulated boron diffusion in presence
and absence of Ga (uniform distribution
according to the model in Ref.[6]).

Finally we wish to comment on Ga-profiles which might increase the effective junction depths, since Ga is also acceptor and by itself possesses larger diffusivity than B. In order to make sure that x_j is determined by the B profile the energy and dose $> 6E14$ cm^{-2} of Ga has to be carefully optimized. For example, for B-implantation with $5E15$ cm^{-2} at 50 keV and Ga-implantation with $1E15$ cm^{-2} at 200 keV and after anneal at 900°C for 40 min the Ga and B profiles intersect each other at $3E17$ cm^{-3}. Changing Ga-implantation data to $6E14$ cm^{-3} and 140 keV yields a corresponding intersection at $1E16$ cm^{-3}.

CONCLUSION

In an exploratory study towards alternative shallow p$^+$-n junctions for furnace anneals compatible with standard CMOS technologies new techniques using Ga-preamorphization prior to high dose B/BF$_2$ implantation have been investigated and compared with conventional ones. The most promising results have been obtained with BF$_2$-implants into preamorphized substrates. However, depending on the kind of application each of the techniques has its merits and drawbacks. Moreover this study quantifies the dependence of B diffusion on the kind of preamorphization and in presence of Ga.

REFERENCES

[1] Mazuré C. et al: ESSDERC'87 Proceedings (1987) 585.
[2] Ganin E. et al: Materials Research Symposium A 74 (1987) 717.
[3] Cohen S.S. et al: J.Appl.Phys. 57 (1985) 1200.
[4] Eriksson L. et al: J.Appl.Phys. 40 (1969) 842.
[5] Maex K. et al: Proc. ULSI'87 87-11 (1987) 330.
[6] Orlowski M.: patent pending
[7] Aronowitz, S.: J.Appl.Phys. 61 (1987) 2495.
[8] Orlowski, M.; Mader, L.: in preparation
[9] Orlowski, M. et al: J.Phys. (Paris) Colloq. CH. Suppl. 9 (1988) 557.
[10] Orlowski, M.: Phys.Lett.A 137 (1989) 115.

Enhanced Process Window for BPSG Flow in a Salicide Process Using a LPCVD Nitride Cap Layer

R.G.M. Penning de Vries and K. Osinski

Advanced MOS Process Development Group

Philips Research Laboratories, P.O. Box 80.000

5600 JA Eindhoven, The Netherlands

Abstract

Optimisation of BPSG flow and $TiSi_2$ degradation in a 1.0 μm CMOS process is studied. It is shown that a nitride cap layer allows to flow the BPSG in a steam ambient without excessive degradation of the silicide. A simple model for the degradation of the $TiSi_2$ sheet resistance is derived. The observed degradation of the $TiSi_2$ film is independent of the substrate, but is a function of the film thickness.

1 Introduction

The compatibility of $TiSi_2$ and poly planarisation by means of BPSG flowglass in advanced IC processes has been questioned many times in the literature [1,2,3]. It is known that the relatively high temperature budget that is needed to flow the BPSG leads to agglomeration in the $TiSi_2$ film, thereby increasing the $TiSi_2$ R_{sh} [1,2] and enhancing the risk for junction leakage [2] and degradation of the p-channel transistor performance [3].

Apart from the temperature budget, also the ambient of the BPSG flow is a relevant factor. Given a temperature budget, a steam ambient is known to enhance the flow of the BPSG [4], while at the same time degrading the $TiSi_2$ even more [1]. In this paper we report on the interplay between BPSG flow and $TiSi_2$ degradation in a salicide 1.0 μm CMOS process.

Experimental details are given in section 2, section 3 deals with the BPSG flow and impact on process yield and metal stepcoverage and in section 4 the degradation of the $TiSi_2$ film is treated. Brief remarks on device performance are made in section 5.

2 Experimental

The planarisation properties of the BPSG were investigated by SEM X-section analysis of test and product wafers fabricated in our 150 mm class 1 pilot line. Thickness of the polycrystalline silicon is $500nm$, spacer is $300nm$ and $100nm$ TEOS was deposited followed by $600nm$ LPCVD BPSG. In some cases a $40nm$ LPCVD Si_3N_4 layer, sandwiched between TEOS and BPSG, is also used. After annealing of the BPSG film, wafers were covered with resist, and sent to SEM X-section analysis to determine the flow angle, as defined in the insert of fig. 1. All flow angle data are averages of 4 readings, obtained from two samples.

The stability of the $TiSi_2$ film was investigated on testmodules, which were implemented on 64k SRAM wafers. Measurements of the $TiSi_2$ R_{sh} were done immediately after salicidation (R_{sh_i}) and after full processing (R_{sh_f}) in both cases using van der Pauw structures.

Obviously, the sheet resistance after salicidation depends (among other factors) on thickness of the as deposited Ti layer and the anneal conditions. As we are only concerned here with the degradation properties of the as-formed $TiSi_2$ film, we just state that salicidation is done by means of rapid thermal processing of $40nm$ Ti. Pertinent details of the salicidation are given elsewhere [5,6]. A description of the complete 1.0 μm CMOS process can also be found in [6].

3 BPSG Planarisation Properties

Fig. 1 gives an overview of BPSG flow related experiments. Furnace anneals in dry and wet ambients, RTA anneals in nitrogen and higher concentration BPSG (3.0 wt% B, 7.0 wt% P) have been examined. Because of compatibility with other processes in the fab the standard BPSG composition was chosen to be 2.5 wt% B, 5.5 wt% P. Temperature budgets were selected on the basis of expected minimal effect on $TiSi_2$ R_{sh_f}.

Contrary to our expectations, the higher concentration BPSG turned out to be not attractive because the degree of planarisation depends significantly on the spacing of the polysilicon. This 'proximity' effect is most pronounced for the higher doped BPSG film but is also observed using RTA BPSG processing. Based upon these flow angle data and associated acceptable $TiSi_2$ degradation (see section 4) it was decided that ramping the temperature to $890°C$, allowing stabilisation at that temperature for 10', followed by 5' at $900°C$ in N_2 was the optimal BPSG flow process in a dry ambient.

In spite of this optimisation it became clear that the achieved BPSG flow (see fig. 2) is not without risk, because of the following two reasons. First of all, the step coverage of the metal between tightly spaced polysilicon is marginal and hence a reliability hazard. Secondly, the steep slope of the BPSG results in (spacerlike) TiW residues after the highly anisotropic etch of the metal system. These residues are difficult to remove wet chemically, due to oxidation of the TiW surface during resist strip [7] and are a potential yield detractor.

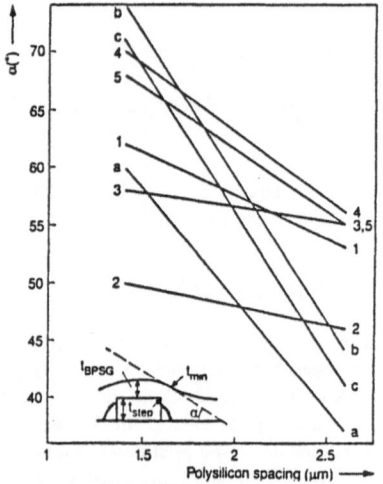

Fig. 1. Overview of flow angles. Curves '1' to '5' is standard BPSG, 'a' to 'c' higher concentration BPSG.
'1' = 30', $850°C$ in steam ambient
'2' = 60', $850°C$ in steam ambient
'3', 'a' = 10', $890°C$, 5' $900°C$, N_2
'4', 'b' = RTA, 60", $1000°C$, N_2
'5', 'c' = RTA, 20", $1050°C$, N_2

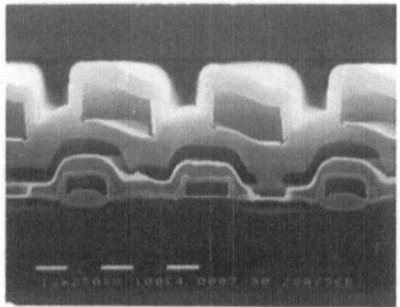
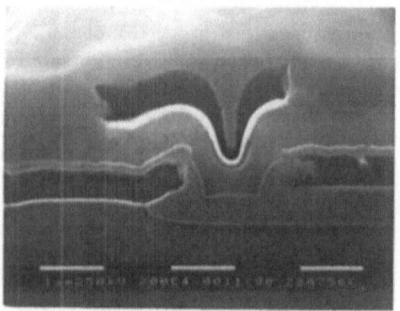

Fig. 2. SEM cross section for the 10', $890°C$, 5' $900°C$ N_2 flow process. Bad metal step-coverage and steep BPSG slopes, which lead to problems in etching of the TiW/AlSiCu metal stack.

In an attempt to alleviate those problems a BPSG flow step in steam ambient was examined. Because of the expected enhanced degradation of the $TiSi_2$, a nitride layer underneath the BPSG was introduced to shield the salicide from the ambient. SEM X-sections in fig. 3 indicate that the topography related problems are solved indeed. $TiSi_2$ degradation data are presented in the next section.

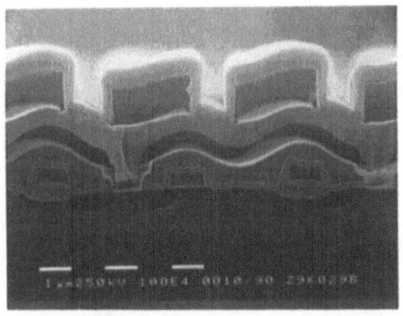 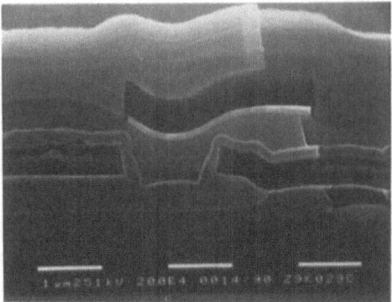

Fig. 3. SEM cross section at same location as in fig. 2. Note the improved planarisation and the almost perfect metal stepcoverage between the tightly spaced polysilicon tracks.

4 Thermal Stability of $TiSi_2$

$TiSi_2$ sheet resistance data after full processing (R_{sh_f}) as a function of the as-formed sheet resistance (R_{sh_i}) in fig. 4^a show that the nitride layer acts as an effective shield. The remaining degradation in the nitride approach, which is due to the temperature budget, is comparable to the degradation that was observed using the default dry process (fig. 4^b).

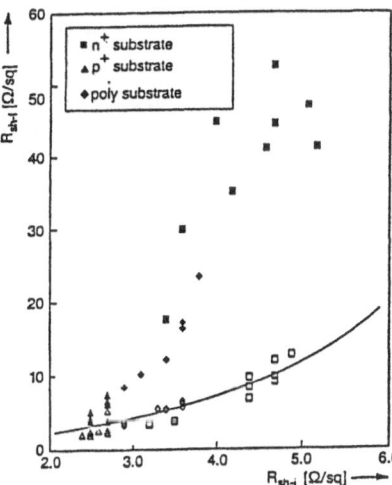 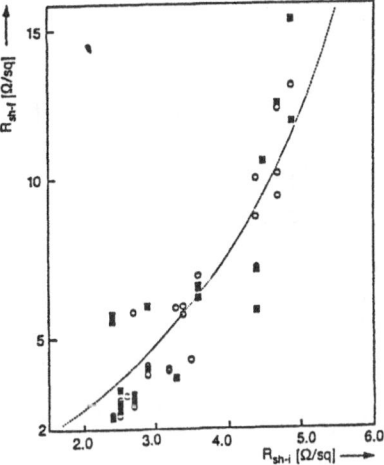

Fig. 4^a. R_{sh_f} as a function of R_{sh_i}, with (open symbols) and without 40 nm Si_3N_4 layer, on n^+, p^+ and polysilicon substrates. Solid curve according to eq. (1) with $\Delta = 21$ nm.

Fig. 4^b. R_{sh_f} as a function of R_{sh_i} for the steam (open symbols) and dry BPSG flow process.

The data in fig. 4^a also indicate that the degree of degradation does not depend on the substrate of the film. Datapoints obtained from $TiSi_2$ on n^+, p^+ or polysilicon substrate are on the same curve, and are only dependent on the *value* of the sheet resistance. This observation can be understood mathematically if we assume that a $TiSi_2$ layer of *effective* thickness Δ is no longer conducting after high temperature backend processing. It is easy to calculate the R_{sh_f} with that assumption:

$$R_{sh_f} = \frac{R_{sh_i}}{1 - \frac{\Delta}{\rho_{TiSi_2}} R_{sh_i}} \tag{1}$$

with ρ_{TiSi_2} the specific resistivity of the $TiSi_2$, which is assumed to be identical for silicides on different substrates.

The data presented in fig. 4 are obtained while starting up the salicidation process. Recent observations show even less degradation of the $TiSi_2$ film, thereby again improving the compatibility of BPSG planarisation and salicides.

5 Device Performance

In order to prevent any impact of the nitride layer on devices, the thickness and deposition temperature of the nitride need carefull attention [8]. Using 40 nm Si_3N_4 deposited at 820 $^{\circ}C$ no detrimental effects on device performance have been observed and fully functional 64k SRAM devices have been fabricated.

6 Acknowledgement

The authors wish to thank Alex Jonkers, Nitin Parekh and John Tibbe for the contributions related to the salicide process, Harrie van Houtum for valuable suggestions, Rex Stone for furnace anneals, the FAB1 pilot line for wafer processing and Jan Veldhuis for excellent SEM X-sections.

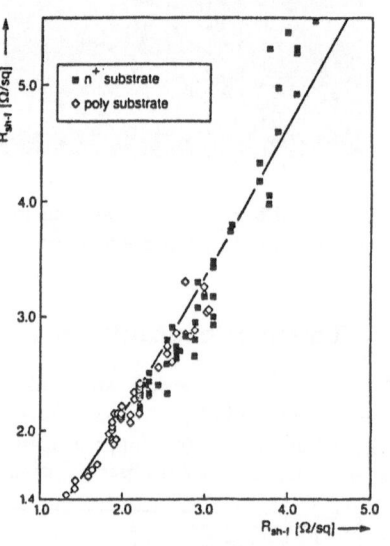

Fig. 5. R_{sh_f} as a function of R_{sh_i} on n^+ and polysilicon substrate. Solid curve according to eq. (1) with $\Delta = 7$ nm.

7 References

[1] R. Shukla and J. Multani, Proc. V-MIC 1987, pp. 470-479

[2] C. Ting et. al., J. Electro. Soc., **133**, 1986, pp. 2621-2625

[3] R. Haken, J. Vac. Sci. Techn., 1985, pp. 1657-1663

[4] J. Tong et. al., Solid State Techn. 1984, pp. 161

[5] A. Jonkers et. al., Les Couches Minces, 42, 1987, pp. 103

[6] K. Osinski et. al., to be published in Philips Journal of Research, **44**, 1989

[7] J. Hackenberg and J. Linn, J.Vac. Sci. Technol. A6(3), 1988, pp. 1388-1391

[8] K. Kuesters et. al., Inter. Symp. on ULSI Integr. Science and Techn., Philadelphia, Pennsylvania, 1987

Silicon Oxidation Rate Dependence on Dopant Pile-up

E. Biermann

Institut für Mikroelektronik
Technische Universität Berlin
Jebensstr. 1, D-1000 Berlin 12, West Germany

ABSTRACT

Oxide growth on highly As-doped silicon has been monitored during steam oxidation at temperatures of 700°C and 800°C. Basically, it was found that enhanced oxide growth occurred, which is consistent with previous observations. The generally accepted model attributes the enhancement entirely to the electrically active surface doping prior to oxidation. In contrast hereto, detailed analysis of the growth data showed that the enhancement is not only if at all dependent on this parameter. Instead, a very strong dependence on the dopant pile-up has been observed.

INTRODUCTION

Silicon oxidation has been successfully modelled by Deal and Grove [1] using a linear-parabolic relationship (Fig.1):

$$\frac{z_{ox}}{L} + \frac{z_{ox}^2}{P} = t \tag{1}$$

Where L and P represent the linear and parabolic growth rates that dominate the growth process for very thin and very thick oxides, respectively. According to the model, the oxidation starts with a linear growth ($z_{ox} \sim t$) where the linear rate is proportional to the reaction rate of oxygen and silicon at the SiO_2/Si-interface. However, the oxidant must diffuse through the already grown oxide layer to the SiO_2/Si-interface, where the oxidizing reaction occurs. This process successively limits the oxide growth velocity leading to a parabolic growth ($z_{ox}^2 \sim t$). The transition from linear to parabolic growth can be characterized by the time $t_{char} = P/4L^2$ or the thickness $z_{ox.char} = (\sqrt{2}-1)P/2L$ (Fig.2).

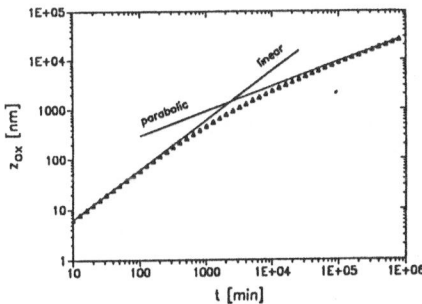

Fig.1 Steam oxidation at 800°C with linear and parabolic regions. Rate data were taken from [6].

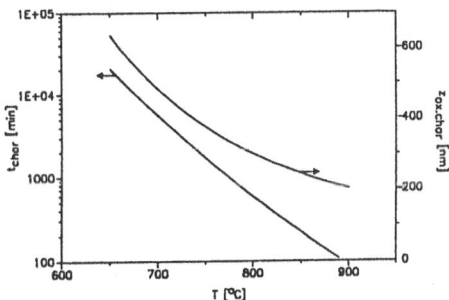

Fig.2 Characteristic time and -oxide thickness for steam oxidation. Rate data from [6].

The enhancement of the oxidation rate of highly doped Silicon (Concentration Dependent Oxidation, CDO) has been widely observed. The CDO model most frequently implemented in process simulation programs (Ho and Plummer) [2] is based on the shift of the Fermi-level produced by heavy doping. For N-type dopants, this shift increases the concentration of vacancies. This in turn is assumed to enhance the surface reaction rate of the oxidation reflected by an increase of the linear rate L in the Deal-Grove oxidation model (1). It is of major importance in this context that the model is based on an electrical effect and thus requires the dopant to be electrically active.

The impact of this effect on oxide thickness is most pronounced where the linear rate L dominates the oxidation process, i.e. thin oxides at low temperatures ($t \ll t_{char}$). An often overlooked fact is, however, that an increase in L leads to a drastic reduction in $t_{char} = P/4L^2$: an oxidation process with enhanced linear rate may be dominated by the parabolic rate far earlier than the normal process.

Lacking more detailed data of the experiments used for calibration of the CDO model [2], the electrically active surface doping prior to the oxidation was taken as the input parameter. In the model, this figure is assumed to be constant during the oxidation process. The model is thus limited by definition to cases where the doping level is constant in the depth range consumed by the oxidation. This is clearly not the case in structures with shallow doping profiles. Consequently, vast disparities have been found between simulated and measured oxide thickness with shallow-doped samples [3,4]. As a test for the basic assumptions of the CDO model, however, the following experiment is intendedly restricted to the case of constant doping, which should be covered by the model.

EXPERIMENT

<100> 100mm CZ Si-wafers with a bulk doping of ≈50Ωcm (Boron) were implanted with $5 \times 10^{14}...2 \times 10^{16}/cm^2$ As at 75 keV. Together with an undoped control wafer per set, they were subjected to a drive-in process at 1100°C that insured a flat doping profile (within 5%) to a depth of 0.5 micron. The resulting doping concentrations near the surface were ≈$5 \times 10^{18}...1.6 \times 10^{20}/cm^3$. After etching of the oxide formed during the drive-in process, every set of wafers covering the above doping range was oxidized at temperatures of 700°C or 800°C in 1atm steam for the times listed in Tab. 1.

From comparison to Fig.2 it can be seen that all oxidation times are within the linear region, at least for undoped samples.

RESULTS AND DISCUSSION

All results presented are from the implant dose of $1 \times 10^{16}/cm^2$ resulting in a doping concentration of ≈$1 \times 10^{20}/cm^3$. Wafers implanted with other doses show similar results. Oxide thickness measurements were taken with an ellipsometer (< 40nm) and/or spectrometer (> 20nm). At 800°C, the ramping period of the furnace caused some oxidation in excess of the times listed in Table 1. This is accounted for by adding 4min to the oxidation times for the graphical representations.

700°C			800°C		
Time [min]	Thickness [nm] undoped	doped	Time [min]	Thickness [nm] undoped	doped
60	6.15	10.1	30	21.1	42.1
120	12.5	25.0	60	39.8	88.8
300	31.3	81.9	120	75.1	196
600	59.7	224	240	145	448
780	76.5	302			
960	92.7	399			

Table 1: Oxide thickness determined by optical measurements

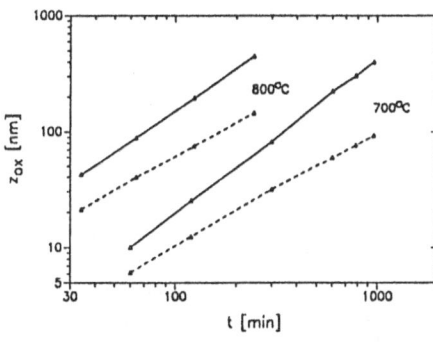

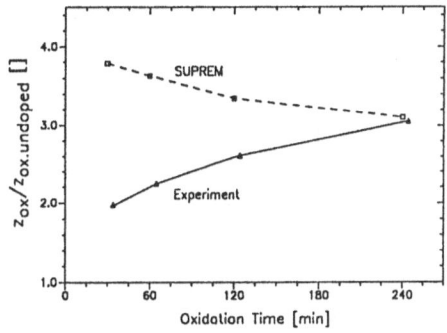

Fig.3 Oxidation results. Broken lines represent undoped samples, solid lines $N_d \approx 1\times10^{20}/cm^3$.

Fig.4 Comparison of measured and simulated enhancement ratio ($N_d \approx 1\times10^{20}/cm^3$, 800°C).

From Fig.3 it can be seen that the undoped samples exhibit a nearly ideal linear slope (slope of 1 in the double-log scale). The doped samples, however, do not only show greater oxide thickness but also a slope of more than unity. This observation can only be explained in terms of the Deal-Grove model (where the double-log slope is always 1.0 ... 0.5) if the oxidation rate is a function of time (or thickness).

For comparison with the existing model [2], Fig.4 shows the enhancement factor (doped/undoped) thickness) of the 800°C data together with data from the process simulator SUPREM III fed with a constant substrate doping of $1\times10^{20}/cm^3$.

The measured data exhibit a rising slope. This supports the assumption that the oxidation rate is not constant but rather increasing with time during this oxidation. The falling tendency of the simulated curve represents the situation of a timely (nearly) constant enhancement of the linear rate L shifting the inset of parabolic oxide growth to smaller times: while the oxide on the undoped Si continues to grow almost linearly, the enhanced process reaches the region of sublinear growth which detracts from the enhancement factor.

For a more detailed analysis, Fig.5 shows the enhancement factor of the doped samples. Using the oxide thickness as the abscissa has proven to be more demonstrative of the effect considered in the following.

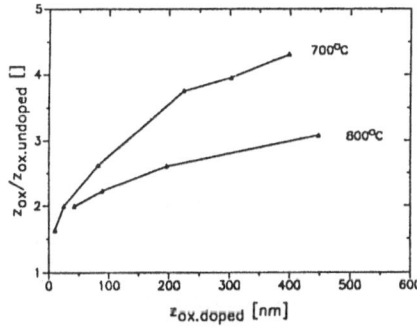

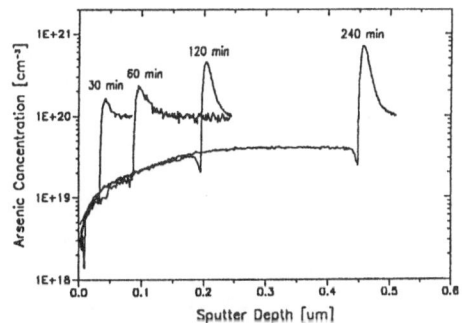

Fig.5 Enhancement ratio vs. oxide thickness obtained
on doped ($N_d \approx 1\times10^{20}/cm^3$) samples.

Fig.6 SIMS profile ($N_d \approx 1\times10^{20}/cm^3$, 800°C) of the
oxide and Si-interface region (5keV O_2+).

From comparison of Fig.5 and Fig.6 (which shows the doping profile at the 4 points of the 800°C-curve), it can be seen that the increase of the enhancement ratio (caused by an increase in the linear oxidation rate) is most likely coupled with the pile-up. This situation is similar for the two temperatures used. Since the pile-up is most probably not electrically active, the applicability of the CDO model for these processing conditions is in question. Furthermore, the measured enhancement factor tends to unity just where it should be greatest (for $z_{ox} \rightarrow 0$). Other authors [5] have also doubted the electrical nature of this effect and have proposed lattice deformation as the main source for the enhancement.

In summary, precise simulation of the evolution of the pile-up seems necessary in order to model the oxidation kinetics under these processing conditions. This is synonymous with the proper simulation of the doping concentration in the oxide. Most simulators integrate the effect of the moving boundary on diffusion and redistribution into the solution of the diffusion differential equations. The results for the oxide doping concentration (and the pile-up) obtained are not very satisfactory.

In a first-order approach to a different scheme, any diffusion of dopant has been neglected. The dopant is simply redistributed across the moving interface in accordance with mass-conservation and segregation (m = 25). Fig.7 shows the rightmost profile of Fig.6 in comparison with the simulation result. Very good agreement is obtained, which might not have been expected with such a simple model. Note that despite the different peak concentrations, the integral of dopant in the pile-up is almost identical (as can also be concluded from the good agreement in the oxide). However, there is still some uncertainty of the oxide concentration measured by SIMS (RBS showed a slightly lower level, but is near the resolution limit). Nevertheless, the identical shape of the profile supports the basic approach.

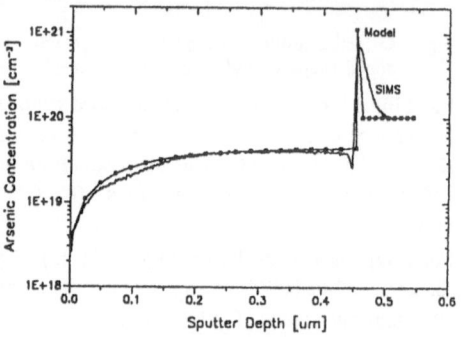

Fig.7 Comparison of measured and simulated As profile ($N_d \approx 1 \times 10^{20}/cm^3$, 800°C, 240min).

The author is indebted to Prof. H.H. Berger for his constant encouragement and to B. Müller for his experimental support. This work has been supported by the Deutsche Forschungsgemeinschaft (DFG).

[1] B.E. Deal and A.S. Grove, "General Relationship for the Thermal Oxidation of Silicon", J.Appl.Phys. 36, No.12, pp.3770 (Dec 65).

[2] C.P. Ho and J.D. Plummer, "Si/SiO2 Interface Oxidation Kinetics: A Physical Model for the Influence of High Substrate Doping Levels", J.Electrochem.Soc. 126, No.9, pp.1516 and pp.1523 (Sep 79).

[3] D.B. Slater and J.J. Paulos, "Fabrication of High Quality Capacitors by Impurity Enhanced Oxidation", J.Electrochem.Soc. 135, No.12, pp.3098 (Dec 88).

[4] E. Biermann, 5th Interim Technical Report for ESPRIT No.243, Vol.4, pp.8 (Sep 87).

[5] J.F. Götzlich, K. Haberger, H. Ryssel, et al., "Dopant Dependence of the Oxidation Rate of Ion Implanted Silicon", Rad. Eff. 47, pp.203 (1980).

[6] R.R. Razouk, L.N. Lie, and B.E. Deal, "Kinetics of High Pressure Oxidation of Silicon in Pyrogenic Steam", J.Electrochem.Soc. 128, No.10, pp.2214-2220 (Oct 81).

Compound Semiconductor Technology and Devices

Beryllium and Manganese Diffusion in Ga$_{0.47}$In$_{0.53}$As During MBE-growth

H. KÜNZEL, R. BOCHNIA, R. GIBIS, P. HARDE, AND W. PASSENBERG

Heinrich-Hertz-Institut für Nachrichtentechnik GmbH, Einsteinufer 37, D-1000 Berlin 10, FRG

Introduction

Beryllium (Be) is the p-type dopant best suited for GaAs/AlAs /1, 2/. For GaInAs Be doping is claimed to yield highly compensated material at low and medium doping levels /3/. A potential alternative is manganese (Mn), since due to the relatively narrow band gap of GaInAs, as compared to GaAs, the acceptor binding energy is low and the activation at room temperature is practicably high /4/.

Mn and Be doping during MBE is determined by the vapour pressure of the elements and a unity sticking coefficient. Smooth surfaces of MBE grown GaInAs up to acceptor levels of $5 \cdot 10^{18}$ cm^{-3} and $5 \cdot 10^{19}$ cm^{-3} are obtained for Mn and Be, respectively. Higher dopant fluxes lead to acceptor incorporation in agreement with the corresponding vapour pressure but to an increase in deterioration of surface morphology. Low vapour pressure elements like Be and Mn are expected to exhibit low diffusivity. Even for Be, diffusion during MBE growth of GaAs/AlAs is not negligible /5, 6/.

We have studied diffusion of Mn and Be during MBE growth in the GaInAs homomaterial and across the layer/substrate heterointerface using SIMS- and CV-measurements.

Diffusion in GaInAs:Mn

Fig. 1 shows a SIMS profile of two homogeneously Mn doped 2 μm thick GaInAs layers with two different doping levels applied, as indicated. The drop of the countrate by approximately two orders of magnitude is related to the layer/substrate interface, indicated by an arrow. Besides, slight accumulation of Mn at the interface and, for the highly doped layer, strong outdiffusion of Mn is observed, whereas it is below the detection limit for the low doped sample. To verify that the results presented are not an artefact of the SIMS measurement, CV-profile measurements on the same wafer have been performed. Outdiffusion in this case leads to a free hole concentration in the substrate about a factor of 40 lower than in the GaInAs layer. The difference of roughly a factor of 2, as compared to the SIMS measurement, indicates the fact that the free hole concentration does not quantitavely reflect the Mn outdiffusion in the part of the substrate converted to p-type conductivity. The amount of outdiffusion is strongly dependent on acceptor concentration. The diffusion depth is larger than the depth examined during our measurements (approximately 10 μm) for a growth temperature of 500° C.

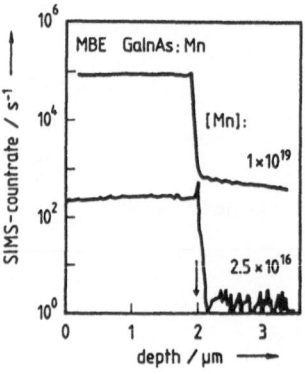

Fig 1: SIMS profile of homogeneously doped MBE GaInAs:Mn on InP:Fe substrate.

To separate the effect of the built-in electric field at the layer/substrate interface from diffusion in the homomaterial, we have grown a sequence of undoped GaInAs layers at different growth temperatures with Mn doping spikes interspersed. Fig. 2 shows a typical SIMS spectrum of a 0.3 µm Mn doped layer embedded between two 0.6 µm thick undoped GaInAs layers. The rectangular profile corresponds to the intended acceptor distribution. The measured Mn profile shows a symmetric broadening due to diffusion in the homomaterial. In addition, an asymmetric contribution towards the surface is evident. This exponential decay of the Mn concentration is attributed to Mn being accumulated and carried foreward on the growing surface, as already detected by Miller et al. /5/ for MBE GaAs:Be. This effect is more pronounced at higher growth temperatures and increases by an order of magnitude varying growth temperature from 450 to 550°C.

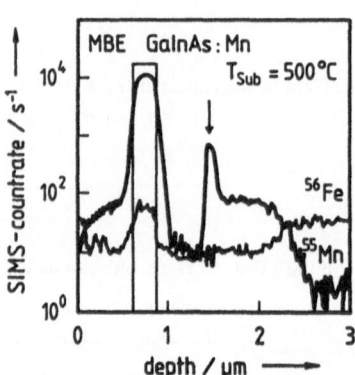

Fig. 2: Mn and Fe SIMS profile of $5 \cdot 10^{18}$ cm^{-3} spike doped MBE GaInAs on InP:Fe substrate. The rectangular profile corresponds to the intended profile.

With respect to the InP:Fe substrate, three major effects can clearly be seen. First, as already mentioned for homogeneously doped material, accumulation at the interface, indicated by an arrow, takes place. Second, the amount of Mn in the substrate is about two orders of magnitude below the level in the intentionally doped region. This outdiffusion is spatially limited and falls to the detection limit at a depth of about 2.5 µm. The depth of the diffusion front varies from below the spatial resolution of the measurement at 450°C and is in excess of 10 μm for 550° C. Third, in the manganese diffused part of the substrate outdiffusion of Fe dopants into the GaInAs layer is clearly detected. For increased growth temperature the Fe level even falls to below the detection limit. In the undiffused part of the substrate the Fe level corresponds to a countrate characteristic for the semi-insulating substrate. The outdiffused Fe species accumulate in the Mn doped part of the layer and for higher growth temperatures also at the layer/substrate interface.

Diffusion in MBE GaInAs:Be

Fig. 3 demonstrates, that the amount of diffusion for Be dopants is drastically enhanced at doping levels in excess of $5 \cdot 10^{19}$ cm^{-3} in conjunction with surface roughening. Clearly non-uniform doping redistribution within the uniformly doped 1.5 μm thick layer and accumulation at the interface, marked by an arrow, is demonstrated. In addition, strong outdiffusion across the interface, with the diffusion front (second arrow) extending several micron into the substrate, is detected. As in the case of Mn doping, Fe redistribution from the substrate into the layer takes place. Fe accumulation at the layer/substrate interface and depletion in the diffused part of the layer are obvious.

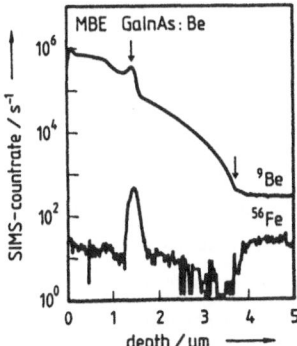

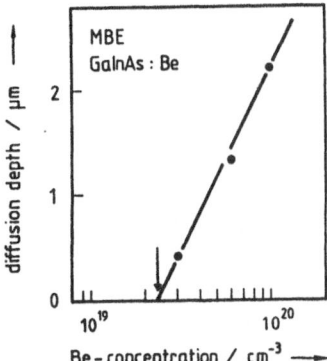

Fig. 3: Be and Fe SIMS profile of homogeneously $1 \cdot 10^{20}$ cm^{-3} doped MBE GaInAs:Be on InP:Fe substrate.

Fig. 4: Dependence of diffusion depth on Be doping level in homogeneously doped MBE GaInAs:Be on InP:Fe substrate.

Fig. 4 demonstrates that the extension of the diffusion front logarithmically depends on the Be concentration. From this figure the onset of outdiffusion can be determined by extrapolating the measured values to zero diffusion depth. A value of $2.5 \cdot 10^{19}$ cm^{-3} is determined, which roughly corresponds to the onset of surface roughening.

To study diffusion in the homomaterial we have analyzed measurements of spike doped MBE GaInAs:Be layers. Fig. 5 shows the measured Be profiles for two representative doping levels. Increasing Be concentration by roughly an order of magnitude, no influence on diffusion in GaInAs towards the substrate and on accumulation at the layer/substrate interface is detected. The marked effect observed is the dramatic

58

difference in redistribution of Be towards the surface. For the higher doping level the maximum Be concentration is far below the level intended and the width of the doping spike is markedly increased. Because this part of the layer is exposed shorter to the growth temperature than the part faced towards the substrate, this effect clearly is related to Be being carried forward during growth, as already published for GaAs:Be /5/ and also found for GaInAs:Mn. The dramatic influence of doping level on this effect is evident from Fig. 5 since for the $1 \cdot 10^{19}$ cm^{-3} Be-doped spike the effect is below the detection limit. Growth temperature in the range of 450 to 550° C, praticable for MBE growth of GaInAs, has no influence on the effects observed, in contrast to Mn doping.

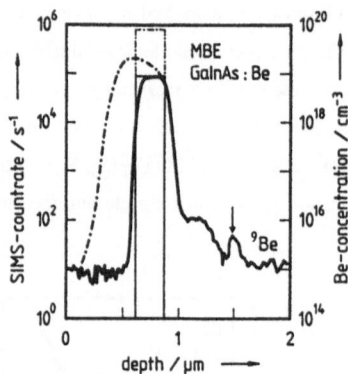

Fig. 5: Doping level dependence of Be SIMS profile of spike doped MBE GaInAs:Be on InP:Fe substrate (——$1 \cdot 10^{19}$ cm^{-3}, —·—$8 \cdot 10^{19}$ cm^{-3}). The rectangular profile corresponds to the intended Be distribution.

Conclusion

Redistribution of Mn and Be doping profiles during growth of MBE GaInAs have been analyzed by SIMS and CV-measurements. Contribution of diffusion in the homomaterial, diffusion across the heterointerface, and surface accumulation are clearly detected. The doping concentration related to the onset of the effects observed is in good agreement with the onset of surface roughening. Diffusion and accumulation is more pronounced in the case of Mn doping.

References

/1/   M. Ilegems: J. Appl. Phys. 48, 17 (1977)

/2/   K. Ploog, A. Fischer, and H. Künzel: J. Electrochem. Soc. 128, 400 (1981)

/3/   A. LeCorre, J. Caulet, M. Gauneau, S. Loualiche, H. L'Hardion, and D. Lecrosnier, A. Roizes, and J.P. David: Appl. Phys. Lett. 57, 1597 (1987)

/4/   H. Künzel, R. Bochnia, R. Gibis, P. Harde, and W. Passenberg: presented at the '5th European Workshop on Molecular Beam Epitaxy', Grainau, 1989

/5/   D.L. Miller and P.M. Asbeck: J. Appl. Phys. 57, 1816 (1985)

/6/   R.L.S. Devine, C.T. Foxon, B.A. Joyce, J.B. Clegg, and J.P. Gowers: Appl. Phys. A44, 195 (1987)

Quality and Applications of In(Ga)AlAs-Layers

C. SCHRAMM, H. KÜNZEL, C. BORNHOLDT, L.M. SU, H.H. WEHMANN*

Heinrich-Hertz-Institut für Nachrichtentechnik Berlin GmbH,
Einsteinufer 37, D-1000 Berlin 10, FRG

*Technische Universität Braunschweig, Institut für Halbleiter-
technik, Hans-Sommer-Str. 66, D-3300 Braunschweig, FRG

Summary

In the last few years the InGaAlAs material system received increasing interest for application in integrated optoelectronic
devices. We present an analysis of layer and surface quality of
MBE-grown InAlAs material, a systematic improvement of Schottky
contacts characterized by electrical measurements and finally,
the fabrication of MeSFET (Metal Semiconductor Field Effect
Transistors), DHBT (Double Heterostructure Bipolar Transistors)
and rib waveguide devices.

Advantages of In(Ga)AlAs

The advantages of the InGaAlAs material system compared to
InGaAsP are: The use of only one group V element (As) and the
fact that no phosphorus is contained ease the epitaxial growth
of quaternary material by advanced growth techniques like MBE
(Molecular Beam Epitaxy) or MOVPE (Metal Organic Vapour Phase
Epitaxy). Moreover it offers improved compositional control,
simplifies growth of graded layers, exhibits an extended adjustable bandgap from 0.76 eV to 1.46 eV and provides the possibility of high quality Schottky contacts with barrier heights in excess of 0.7 eV for InAlAs.

Surface and layer quality

Despite its favourable properties two essential problems exist
for application of as grown InAlAs layers. One is the typically
microscopic surface roughness of this material grown directly on
InP substrates. We investigated two possible solutions for this
problem according to the device requirements:

1. Replacement of InAlAs layers by the quaternary InGaAlAs material; narrow half-widths of x-ray diffraction peaks and narrow

photoluminescence lines as well as SEM (Scanning Electron Microscope) photographs indicate that the layer and surface quality
improve even at low gallium content. Deep Level Transient
Spectroscopy (DLTS) shows a deep midgap level in Si-doped InAlAs
(E_A=760meV), but there are anomalously broad emission spectra
(Fig.1). The deep level to free carrier concentration ratio is
reduced from 5% to 2.5% in Si-doped InGaAlAs. In applications of
high barriers to adjacent materials the InGaAlAs is disadvanta-
geous with increasing gallium concentration, where in bipolar
applications the ohmic contacts are improved.

2. Use of InGaAs buffer layers; a strong improvement of the
crystalline quality of the ternary InAlAs occurs when a thin
InGaAs layer with a minimum thickness in the order of 50 nm is
grown between InAlAs and the InP substrate. SEM pictures reveal
an absolutely smooth surface.

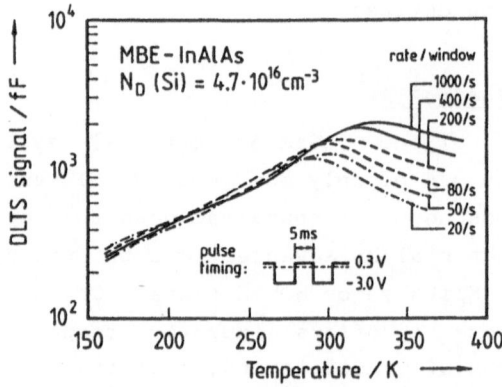

Fig.1. DLTS spectra of Si-doped MBE-grown InAlAs

The second problem is the strong susceptibility of aluminium to
oxygen, especially for uncontrolled surface oxidation processes,
which prevent formation of high quality Schottky contacts. As a
solution we generally grow a GaAs cap layer of 2 nm thickness on
top of the InAlAs. Furthermore, we investigated several surface
treatments before evaporation of the metals. The differences in
the resulting electrical parameters, e.g. Schottky barrier
height or current blocking capability of the diodes, are listed
in Table 1. The choice of the metal turned out to be uncritical
in this case. The breakdown voltage of unintentionally doped

InAlAs is considerably higher than that of Si-doped InAlAs. Doping profiles obtained by C-V measurements demonstrate the homogeneity of the MBE growth.

Treatment	Uniformity across the sample	Schottky-barrier eV	Leakage current density at −25V $\mu A \cdot mm^{-2}$
$H_3PO_4/H_2O/H_2O_2$	+	0.73	0.22
$H_2SO_4/H_2O/H_2O_2$ *	+	0.74	0.19
$C_6H_8O_7/H_2O/H_2O_2$	+/−	0.50	0.15
without	+/−	0.69	2.04

* gold-spots after lift-off process

Table 1. Influence of surface treatment before metal evaporation on the electrical behaviour of InAlAs Schottky diodes

Applications

The suitability of InAlAs layers for electrical and optical applications is demonstrated by three examples of devices not yet optimized.

The wave guiding capability of InAlAs on InP is shown by a rib waveguide (insert Fig.2). The rib was etched by Reactive Ion Etching. The losses (Fig.2) of these waveguides are typically in the range 1.5 - 2 dB/cm at 1.3 μm and 1.55 μm (best value is

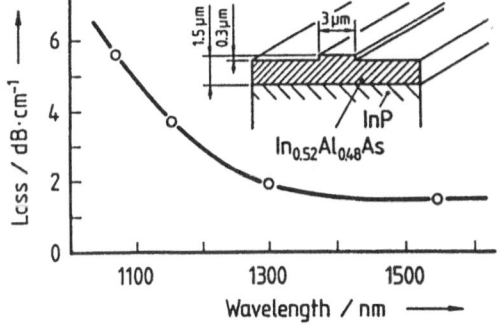

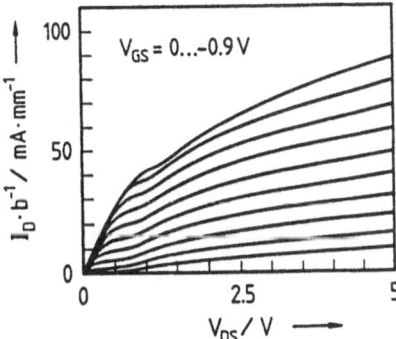

Fig.2. Attenuation and cross section of InAlAs/InP rib waveguide

Fig.3. Output characteristics of an InAlAs/InGaAs MeSFET

1.3 dB/cm at 1.55 μm). This is comparable to InGaAsP waveguides (E_g=1.28eV) with similar structure.

We fabricated an air bridge connected InAlAs/InGaAs MeSFET with low leakage current (<1μA at -5V), high modulation (Fig.3) and a transconductance of 90 mS/mm at 2 μm gate length and 270 μm gate width. The structure is designed especially for optoelectronic integration. Only one type of non-alloyed metallization (Au-Ti) is used for all contacts.

Fig.4 shows the schematic cross section of an InAlAs/InGaAs-DHBT, which was grown by MBE at 500°C. Instead of a single p^+-base two adjacent spacer layers have been inserted to avoid the electron repelling effect [1] caused by the conduction band spike at the base-collector interface and in order to minimize accceptor outdiffusion. The DHBT exhibits a common emitter current gain of 200 (I_C=1mA, V_{CE}=2V).

We conclude that the InGaAlAs material system is an excellent candidate for the fabrication of OEIC devices with high performance.

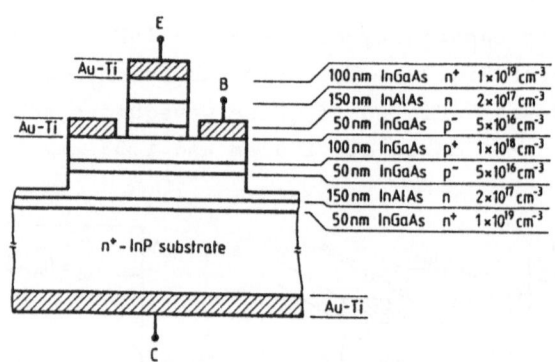

Fig.4. Schematic cross section of an InAlAs/InGaAs-DHBT

Reference

1. Su, L.M.; Grote,N.; Kaumanns,R.; Schroeter-Janßen,H.: NpnN double-heterojunction bipolar transistor on InGaAsP/InP. Appl. Phys. Lett. 47(1) (1985) 28-30.

On the High-frequency Behaviour of Ohmic Contacts

H. Zirath

Department of Applied Electronphysics
Chalmers University of Technology, Göteborg, Sweden

Introduction

In the modelling of microwave and mm-wave components such as MESFETs, MODFETs, HBTs, and quantum well diodes etc , it is generally assumed that ohmic contacts to the semiconductor are frequency independent and purely resistive in their nature. Furthermore, the performance of these components is critically dependent on keeping the parasitics small. In some heterojunction FET-devices one can suspect that the source and drain resistance may be frequency dependent due to capacitive coupling between conducting epitaxial layers / 1 /, / 2 /. The purpose of this work is to show that ohmic contacts on GaAs are not always purely resistive, how to model the contact, and to present a suitable structure for the determination of the contact impedance up to millimeter wavelengths together with experimental data for frequencies up to 25 GHz.

Theory

It is generally assumed that the resistance of an ohmic contact on highly n-doped ($> 10^{18}$ cm^{-3}) GaAs is determined by field emission, i. e. electrons tunnel through a thin Schottky barrier, see Fig 1 below.

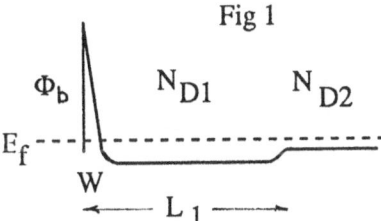

Fig 1

Φ_b is the barrier height, E_f is the Fermilevel, W is the depletion layer thickness, N_{D1} and L_1 are the impurity concentration and the depth of the alloyed, highly doped region, and N_{D2} is

64

the impurity concentration below the contact region. If $N_{D2} > N_c$, the resistance of the contact is mainly determined by the resistance of the Schottky barrier and not by the 'high-low' junction barrier between region 1 and 2, /3/.

Since there is a *capacitance* associated with the depletion region, $C_b = \varepsilon\, A/W$ (ε is the dielectric constant and A the contact area), the contact impedance should be *frequency dependent*. Neglecting the spreading resistance, the barrier impedance is

$$Z_b = r_b/(1+j\omega\tau_o) \tag{1}$$

if the current flow is vertical, since the barrier capacitance and resistance are in parallel. ω is the angular frequency, and τ_o is $\rho_c \cdot C_u$ where ρ_c is the specific contact resistance (in $\Omega \cdot cm^2$) and C_u the capacitance per unit area (in F/cm^2). Neither N_{D1} nor Φ_b are easy to determine which makes it hard to calculate the roll-off frequency $f_o = 1/2\pi\tau_o$. In Fig 2a,b , ρ_c and f_o are plotted versus N_{D1} with Φ_b as a parameter. ρ_c is calculated from Padovani-Stratton's Field-Emission theory /4/, /5/ and the associated barrier capacitance is calculated by using the Poisson equation. For simplicity it is assumed that the barrier height and impurity concentration both are laterally constant, an assumption which however may not always be valid /6/.

For a contact having a lateral current flow such as the source contact of a MESFET or the base contact of an HBT, one has to consider the current crowding effect in the contact. H.H. Berger /7/ has investigated this case and calculated the contact-impedance by considering the contact as a transmission line. The contact impedance is for this case

$$Z_{TLM} = \frac{1}{W}\sqrt{R_{sh}\,\rho_c}\; \frac{1}{\sqrt{1+j\omega C_u \rho_c}} \tag{2}$$

W is the width of the contact, and R_{sh} is the sheet resistance of the resistive layer under the contact.

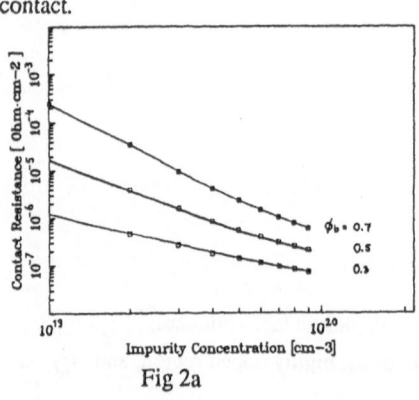

Fig 2a

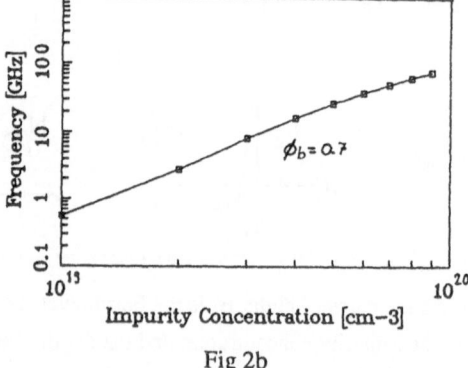

Fig 2b

Experiments

A test pattern consisting of two planar contacts located on top of a mesa was fabricated, see fig 3. The width of the mesa is 60 µm. Fig 4 shows the structure of the epitaxial GaAs-material. The sheet resistance of this material was measured to be 65 Ω per square. The ohmic contact metalization is 900 Å AuGe/ 200 Å Ni / 2000 Å Au. The contacts were annealed under different conditions in order to obtain different values of the impurity concentration beneath the metalization, N_{D1}. Structures with three different gap sizes (2, 5, 10µm) were fabricated and the capacitive coupling between the two contacts was determined separately by fabricating the structure also on semiinsulating GaAs. The two contacts were probed by coplanar wafer-probes (Cascade) and the scattering parameters were measured with an HP 8510 network analyzer in the frequency range 100 MHz to 25 GHz. The network in fig 5 was used to model the contact

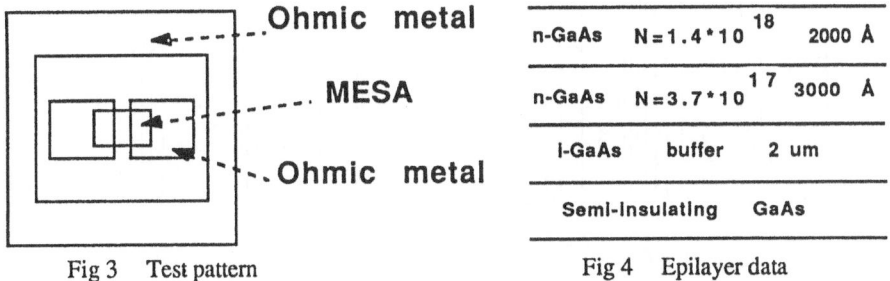

Fig 3 Test pattern	Fig 4 Epilayer data

Parameter extraction was done with a commercial microwave CAD-program (Touchstone). The equivalent circuit consists of coplanar transmission lines (CPWG) at the input and output of the circuit, corresponding to the actual pattern layout, they are necessary in order to get a good phase correspondence between the measured and modelled S-parameters. LTL is the lossy transmission line used in Bergers theory, R_s is the series resistance between the contacts, and C_s is the capacitive coupling between the electrodes In fig 6 a,b, the magnitude and phase of the measured and modelled S-parameters are plotted for two different contacts having a dc-contact resistance as measured by the tlm-method [7] of 1.6 Ωmm, and 0.1 Ωmm.

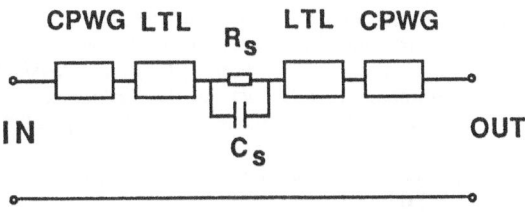

Fig 5   Equivalent network of the test structure

66

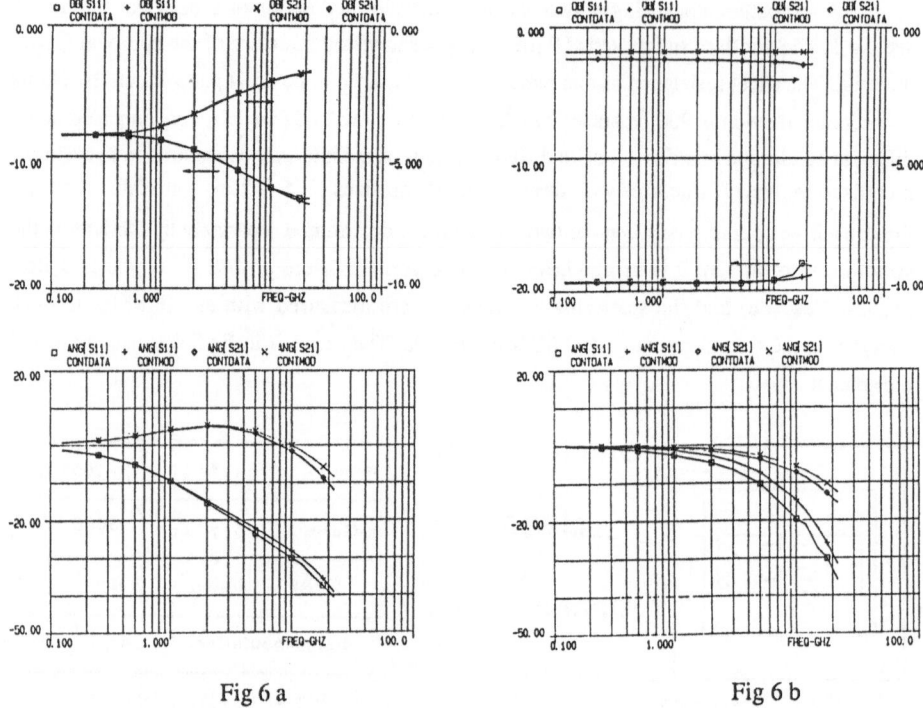

Fig 6 a   Fig 6 b

Evidently, excellent correspondence to Bergers model is obtaned in case a) with an approximate roll-off frequency of 1 GHz. In 6 b) no frequency dependence of this kind was observed indicating that the roll-off frequency is far beyond the upper frequency of the measuring-equipment as the theory predicts. The latter test structure was modelled without the LTL.

Acknowledgement

The author wants to acknowledge Prof E. Kollberg for his encouragement, S. Nilsen, H. Grönqvist, and N. Rorsman for help with the device fabrication, and The Swedish National Board for Technical Development (STU) for financial support.

References

/1/   C. Versnaeyen et .al.: Frequency dependence of source access resistance of heterojunction field-effect transistor. Electronic Letters. 21 (1985) 539-540.
/2/   Kim, B.; Matyi; R.J.; Wurtele, M.; Tserng, H.Q.: AlGaAs/InGaAs/GaAs quantum-well power MISFET at millimeter-wave frequences. IEEE Electr. Dev. Lett. 9 (1988) 610-612
/3/   Dingfen, W.; Heime, K.: New explanation of N_D dependence of specific contact resistance for n-GaAs. Electronics Letters 18 (1982) 940-941.
/4/   Stratton, R.; Padovani, F.A.: Field and thermionic-field emission in Schottky barriers. Solid-St. Electron. 9 (1966) 695-707.
/5/   Yu, A.Y.: Electron tunneling and contact resistance of metal-silicon contact barriers. Solid-St. Electron. 13 (1970) 239-247.
/6/   Braslau, N.:Alloyed ohmic contacts to GaAs. J. Vac. Sci. Technol. 19 (1981) 803-807
/7/   Berger, H.H.: Models for contacts to planar devices. Solid-St. Electron 15 (1972) 145-158.

Assessment of Semi-insulating InP: Fe Layers for Substrate Applications

N. GROTE, H.G. BACH, TH. FEIFEL[*], D. FRANKE, P. HARDE,
B. SARTORIUS, P. WOLFRAM

Heinrich-Hertz-Institut für Nachrichtentechnik Berlin GmbH, Einsteinufer 37, D-1000 Berlin (FRG)
[*]Permanent address: ANT Nachrichtentechnik GmbH, Gerberstraße 33, D-7150 Backnang (FRG)

Summary

Semi-insulating InP epi-substrates formed by epitaxial InP:Fe layers on suitable base substrates are proposed and investigated. Compared to standard commercial s.i. bulk substrates, such epi-substrate offer lower defect densities, well-controllable Fe-levels and considerably higher specific resistivities.

Introduction

Standard commercially available semi-insulating InP:Fe substrates generally do not satisfactorily meet the stringent requirements of an advanced device technology. Firstly, such substrates exhibit relatively high densities of defects which may be characterized by etch-pit densities of the order of 10^4 cm^{-2}, and above, and which may have a deleterious effect on the reproducibility of the various fabrication processes and on the performance of the devices. Secondly, the Fe concentration is known to vary from wafer to wafer and across a slice as a consequence of the small segregation coefficient of Fe. Unless the wafers are selected, such unstable material conditions again impose reproducibility problems upon device technology.

Here we propose and investigate an alternative substrate scheme relying on the use of thick epitaxial InP:Fe layers grown on suitable base substrates. Such epi-substrates will be shown to yield substantially improved s.i. substrate material. Based on a comprehensive study on InP:Fe layers, some results and conclusions relevant to this application will be given.

Experimental

The s.i. InP-layers studied were grown on low-defect density InP:S bulk substrates by atmospheric pressure MOVPE at 650°C using TMIn and PH_3 (15%). A V/III ratio of 100 and a growth rate of 1.4 μm/h were adjusted. Residual doping levels of undoped layers were guaranteed to be $<10^{15}$ cm^{-3}. As source for the Fe-doping ferrocene kept at -15°C was employed. Fe concentrations in the layers were determined by SIMS using O$^+$ as primary ions. For calibration Fe-implants were utilized. Under optimized conditions the detection limit of Fe in InP was at around $2 \cdot 10^{15}$ cm^{-3}, and the accuracy in the 10^{16} cm^{-3} range was estimated to be +/-50%.

Results

For the use as epi-substrates knowledge on the material properties of InP:Fe layers is of primary importance. Fig. 1 shows a typical I/V-characteristic and the correlated resistivity (ϱ) curve for a 7.5 μm thick layer. Three current regimes can be noticed within the useful electrical isolation range: At bias voltages <1 V the curves appear to be ohmic whereas at high bias space-charge-limited-currents ($I \approx V^2$) are prevailing. In the intermediate range a subohmic region can be observed which tends to be more and more pronounced with increasing layer thicknesses. In this regime the current/voltage dependence obeys an $I \approx V^m$ dependency with an exponential factor between 1 and 0.5. This current regime is reflected in the specific resistivity curve by a broad maximum.

Fig. 2 gives the dependence of ϱ on the measured Fe-concentrations (bias: 1 V). The resistivity can be seen to gradually decrease from a saturation level of around 10^9 Ωcm to about $2-5 \cdot 10^7$ Ωcm at [Fe] = $1 \cdot 10^{16}$ cm^{-3} which was the minimum adjustable concentration. The Fe-density at which ϱ starts to saturate almost coincides with the solubility limit of Fe in InP assumed to be $7 \cdot 10^{16}$ cm^{-3} [1]. It can be stated that, at given Fe-levels, the measured ϱ-values are by one to two orders of magnitude higher than those of bulk substrates.

Fig. 3 shows the I/V-characteristics as a function of layer thickness. The curves were measured on the same sample ([Fe] = $4 \cdot 10^{16}$ cm^{-3}) after consecutive thinning. According to theory the space-charge-limited current should be proportional to V^2/d^3 (d = thickness). The measured curves confirm this correlation as illustrated in fig. 4, which represents the bias voltage at a given current density (10^{-4} A/cm^2) as a function of thickness. Good linearity between $V(J_{scl}$ = const.) and d with a slope of 3/2 in this log-log representation is found which satisfies the theory. Hence, by extrapolation the I/V curves can be roughly modeled for larger thicknesses (see dashed curve in fig. 3) as required for epi-substrates.

To demonstrate the potential of epi-substrates to make available low-defect density substrates, photoluminescence measurements were performed to assess the crystalline quality. As InP:Fe layers do not yield PL radiation these layers were indirectly investigated by examining overgrown n-InP layers. Fig. 5 displays PL microscopy images of such a layer which was concurrently deposited on an InP:Fe bulk substrate and on an InP:Fe/InP:S epi-substrate for comparison. A drastically reduced density of defects can be noticed in the latter case (fig. 5) from the amount of dark spots as a consequence of the low defect level in the InP:S base substrate used.

Discussion

The present study shows that epi-substrates offer a viable way to overcome the major problems encountered with today's ordinary s.i. substrates outlined above: It has been demonstrated that substrates with (i) very low defect densities and with (ii) well-controlled Fe-concentrations can be easily produced. A Fe-level of $5 \cdot 10^{16}$ cm^{-3} being below the solubility limit would be advisable to avoid the formation of precipitates and clusters. At this concentration a specific resistivity of as high as $5 \cdot 10^8$ Ωcm is attainable exceeding that of bulk substrates by at least one order of magnitude. Moreover, at such levels Fe-diffusion was found to be negligible [2] thus giving good thermal stability. From the electrical point of view (resistance, capacitance) a thickness of the InP:Fe layer of a few tens of μm will be sufficient for most applications. In order to make feasible low-cost production of epi-substrates low-pressure VPE has to be preferred over MOVPE because of its very rapid growth capability ($\approx$50 μm/h) [3].

References

[1]   Nakai, K.; Ueda, O.; Odagawa, T.; Takanoshi, T.; Yamakoshi, S.: Inst. Phys. Conf. Ser. 91, pp. 199-202

[2]   Feifel, Th.; Franke, D.; Harde, P.; Wolfram, P.; Grote, N.: Abstracts "3rd Europ. Workshop on Metalorganic Vapour Phase Epitaxy", poster presentation B78, Montpellier, June 1989

[3]   Deschler, M.; Grüter, K.; Schlegel, A.; Beccard, R.; Jürgensen, H.; Balk, P.: Journal de Physique, Colloquie C4, Supplement au No. 9, Tome 49 (1988), pp. 689-92

This work was conducted under the ESPRIT programme (project 263).

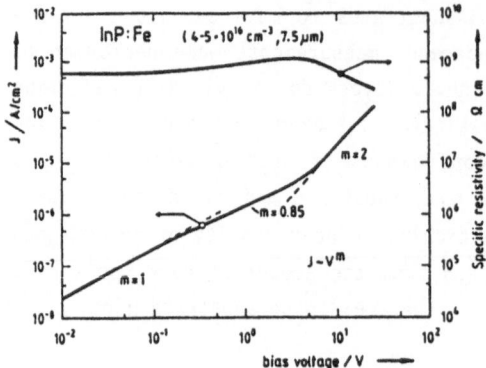

Fig. 1　Current density and correlated specific resistivity of MOVPE InP:Fe layers vs. bias voltage

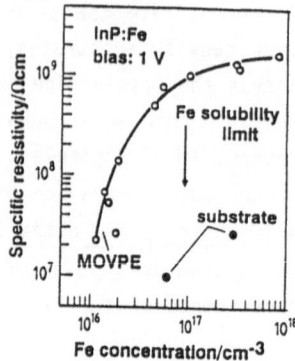

Fig. 2　Dependence of specific resistivity on Fe-concentration (SIMS)

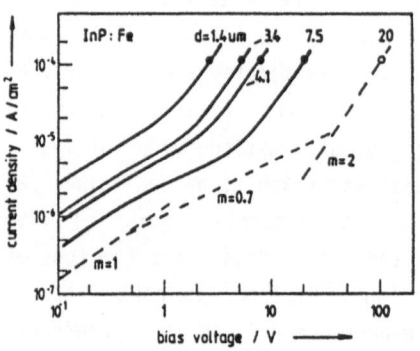

Fig. 3　J/V-characteristics at different layer thicknesses; dashed curve extrapolated

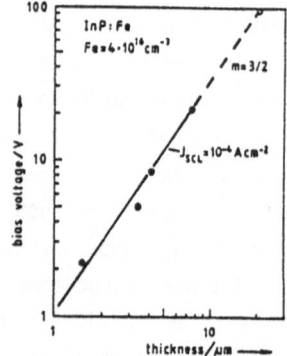

Fig. 4　Bias voltages at $J = 10^{-4}$ Acm^{-2} in the space-charge-limited current regime as a function of thickness

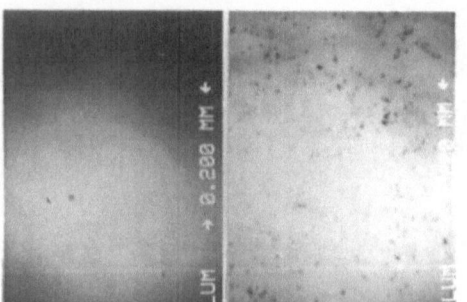

Fig. 5　Photoluminescence images of n-InP concurrently grown on InP:Fe/InP:S epi-substrate (left) and InP:Fe bulk substrate (right)

Investigation on P-buried GaAs MESFETs

Chang-Xin Shi, Xiao-Ming Li, and Shang-Heng Xin

Institute of Microelectronic Technique

Shanghai Jiao Tong University, Shanghai 200030, CHINA

Summary

A p-buried GaAs MESFET(PB-GaAsMESFET) has been fabricated by means of implanting Be into semi-insulating GaAs substrates to form a p-type buried layer under the channel-active layer. The experimental results show that PB-GaAs
MESFET not only improve the uniformity of the pinch-off voltage butincrease
the transconductances and reduce the effects of back gate as well. The theoretical analysis and process design of the PB-GaAs MESFET are also presented.

INTRODUCTION

The fabrication of GaAs MESFET on semi-insulating GaAs substrates by using the
ion implantation technique is the foundation of super-high-speed GaAsIC.[1][2]
However, in the active layer of implanted Si on semi-insulating substrates
there is generally a distribution tail for the carrier concentration. This is
the so-called "tail-fin" phenomenon, which would lead to an increase in the
pinch-off voltage.[3] This was commonly believed to be caused by the redistribution of Cr upon annealing.[4] The extent of this tailing effect is different
on the substrates, causing thepinch-off voltage of GaAs MESFET to vary in its
distribution on the same substrate.[5] Concerning this nonuniformity of pinch-
off voltage, we implanted Be onto GaAs substrates, introduced a p-type buried
layer in the channel-active layer, and successfully fabricated GaAs MESFET
with a p-type buried structure(PB-GaAs MESFET). We also investigated the characteristics, the processing design and fabrication of the PB-GaAs MESFET device.
The experimental result showed that the uniformity of the pinch-off voltage
was greatly improved, the transconductance was increased and the back-gate
effect reduced.

MODELING AND PROCESSING DESIGN

Device characteristics

PB-GaAs MESFET structure and the transverse distribution of the impurities are shown in Fig.1.

By using a two-region model,[6] a computer model for the device with a buried layer was developed. The basic equations used were:

$$qz \int_{p}^{t} n(x)\mu E_y dx = qU_s z \int_{p}^{t} n(x) dx = I, \tag{1}$$

$$V_{gs} = V_g + V_B + IR_s, \tag{2}$$

$$V_{ds} = V_{L_1} + V_{L_2} + I(R_s + R_D), \tag{3}$$

and

$$V_{L_2} = \frac{2aE_s}{\pi} \sinh\left(\frac{\pi L_2}{2a}\right), \tag{4}$$

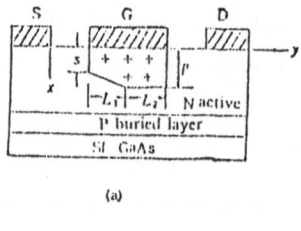

(a)

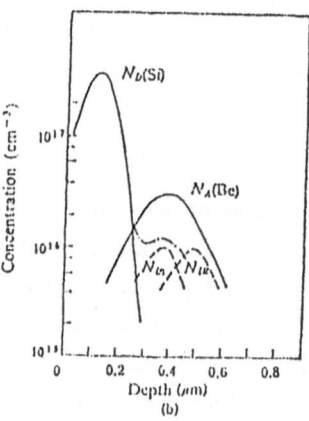

(b)

FIG.1 PB-GaAs MESFET structure(a) and impurity distribution(b). N_D is the impurity concentration in the active layer, N_A the impurity concentration in the buried layer, and N_{D1} and N_{D2} are modeled distributions of different small "tail-fins." The dash-and-dot line shows $N_{D1} + N_{D2}$.

where $n(x)$ is the net carrier concentration in the active layer (cm^{-3}); E_y the electric field along the channel, taken as $E_y = 2900$ V·cm^{-1}; U_s the saturation speed, taken as $1.3 \times 10^7 cm \cdot s^{-1}$; V_{gs} the gate partial voltage (V); V_{L1} and V_{L2} the voltage drop (V) in zone L_1 and zone L_2, respectively; V_B the potential drop between the gate and the GaAs contact, taken as 0.8 V; I the channel current (mA); μ the transfer rate of the lower field (cm^2·V^{-1} s^{-1}); R_S and R_D the source resistance and the leakage resistance in mho, respectively, (including the contact resistances). a is the active layer thickness. When there is a p buried layer, the active layer termination is taken at the PN junction. When there is no buried layer, the active layer termination is taken at the point where the impurity concentration is 10^{16} cm^{-3}. z is the gate width (cm).

Fig.2 shows the results of modeling. The solid line is for the case with tail-finning and buried layer, whereas the dotted line represents the case of no buried layer. It is seen that the introduction of the buried layer leads to a slight reduction in current, but no degradation of the characteristics. In fact, at large negative gate voltage, the transconductance increases.

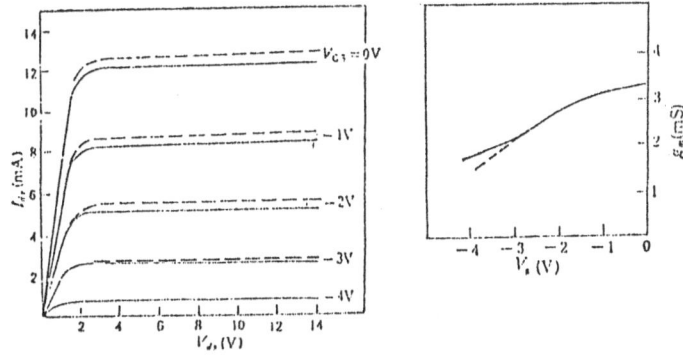

FIG.2 The output (a) and the transconductance characteristic (b) of PB-GaAs MESFET.

Processing design

In order to select the best implantation conditions, the supressing effect on tail-finning by the introduction of a p buried layer was also studied by using computer modeling . As shown in Fig.1 (b), the nonuniformity of the tail-finning on different points of the substrate was modeled by Gaussian distribution(N_{D1} ,N_{D2},....) with different peak heights and different peak positions. Figures 3(a) and (b) show the pinch-off voltage (V_p) with tail-finning but no buried layer, and the relationship between tailing dosage and peak positions, respectively. It can be seen that when the dosage varies from $5 \times 10^{10} cm^{-2}$ to $4 \times 10^{11} cm^{-2}$, the pinch-off voltage changed by 1.4 V, a deviation of 23%. The variation of peak position has less effect on the pinch-off voltage, only 5%.

When the p-buried layer was introduced, this deviation was greatly improved.

Figure 4(a) shows the effect of p layers with different dosages at an input energy of 160 keV on the uniformity of the pinch-off voltage. It is seen that as the dosage varies from 2.5 to 5 x 10^{12} cm^{-2}, the improving effect is not large. Figure 4(b) shows the effect of the input energy of the p layer. A better set of input parameters are: Si: 150keV, 5x$10^{12}$$cm^{-2}$; Be: 175keV, 2x10^{12} cm^{-2} . In this case, the voltage fluctuation was only 0.17V, otherwise 0.2 V.

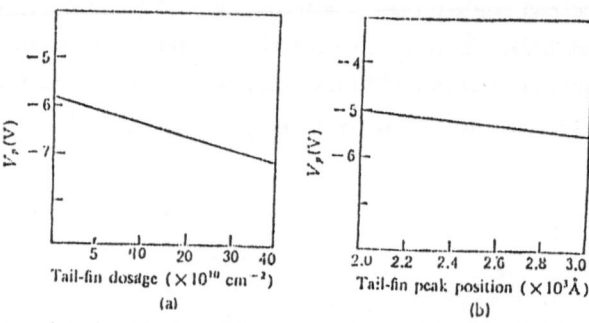

FIG. 3. The effect of tail-fin height and peak position variation on the pinch-off voltage

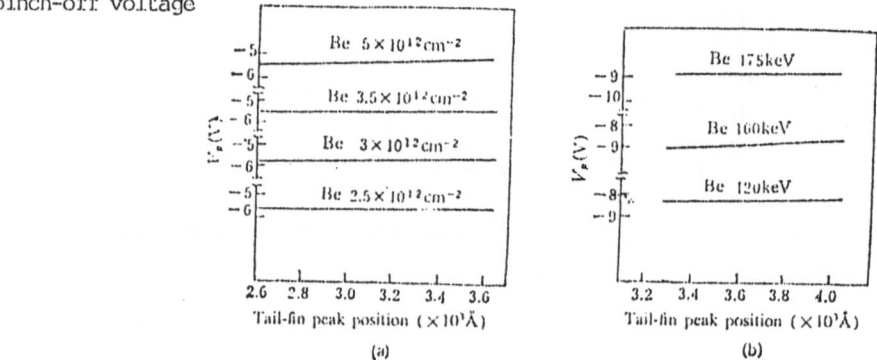

FIG. 4 The pinch-off voltage under different conditions of p buried layer implantation.

EXPERIMENTAL RESULTS

Device fabrication was accomplished by using the planar selective implantation and platform technique. The important processing steps included the following: Si was implanted at about 120 keV with a dosage of 4x$10^{12}$$cm^{-2}$ Annealing was then carried out at 800 C for30 min under a nitrogen atmosphere. Be was subsequently implanted at about 160 keV with a dosage of 2x$10^{12}$$cm^{-2}$, then annealed at 700 C for 20 min Si implantation could also be annealed by using infrared for better results. AuGeNi/Au drain source contact and Al gate contact were then fabricated by using a the lift-off technique. The annealing of.Be was carried out after the Si annealing,

because it avoided the serious redistribution of Be under theannealing conditions of 800 C and 30 min for Si. But, using rapid thermal annealing (RTA)
technique anneals Be can not obtain good results. Be experiences an obvious enhancement diffusion at temperature of $960\overset{\circ}{C}$ and time of 6 seconds.

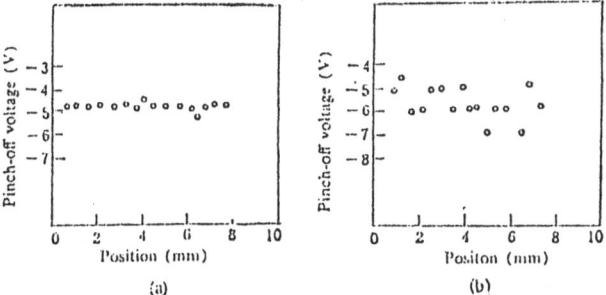

(a)   (b)

FIG.5. Comparison of the pinch-off voltages of PB-GaAs MESFET(a) , and ordinary GaAs MESFET (b).

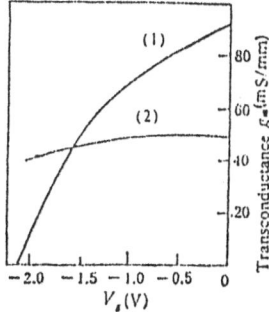

FIG. 6. The variation of the transconductance of PB-GaAs MESFET(1), and ordinary GaAs MESFET.

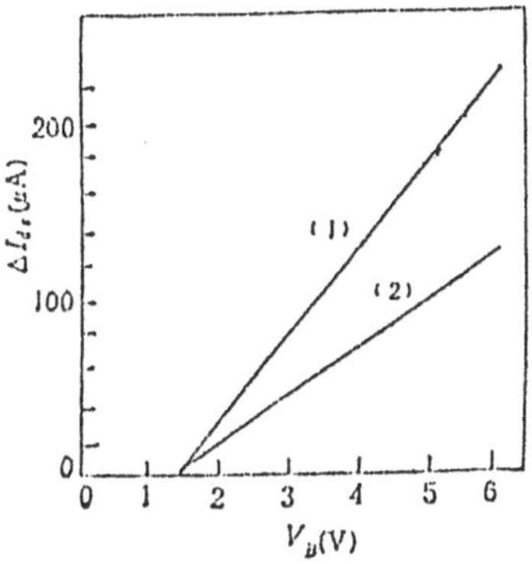

FIG.7. The improved back-gate effect; (1) for PB-GaAs MESFET and
(2) for an ordinary one.

Figures 5(a) and (b) show the pinch-off voltage distribution of two devices
fabricated on the same substrate with and without the buried layer.The device without the buried layer had and average deviation of 15%, whereas
the buried layer device had only 5%. Fig. 6. demonstrates the variation of
the transconductance of with and without p-buried layer.
The measured back gate effect[7] is shown in Fig. 7. The abscissa was the
back gate voltage, and the ordinate was the leakage current reduction. It
shows that the slope for the PB-GaAs MESFET is less than that without the
buried layer, indicating an improved back-gate effect.

76

References

1. Osamu Ryuzan and Takahiko Misugi Physica B 117-118, 50 (1983).

2. Gerard Nuzillat, Ernesto H. Perea, Georges Bert, Fatma Damay-Kavala, Maurice Gloance, Michel Peltier, Tung Pham Ngu, and Christian Arnodo, IEEE Journal of Solid-State Circuits SC-17, 569 (1982).

3. Wang Wei-Yuan et al, Proceeding on GaAs and Related Compounds, 299 (1983), Publ. Chinese Metal Society Semi-conductor Materials Committee.

4. P.N.Favennec and H.L'Haridon, Appl. Phys. Lett. 35, 699 (1979).

5. NIKKEL Electronics, 129 (1983).

6. John A Higgius, IEEE Transaction in Electron Devices ED-27, 1066 (1980).

7. Christopher Kocot and Charles A Stolts, IEEE Transactions on Electron Devices ED-29, 1059 (1982).

An InGaAs/GaAs Strained Superlattice MSM Photodiode for Fast Light Detection at 1.3 µm

M. Zirngibl, J.C. Bischoff, R. Sachot and M. Ilegems
Institute for Micro- and Optoelectronics, Swiss Federal Institute of
Technology, 1015 Lausanne, Switzerland
P. Beaud and W. Hodel
Institute of Applied Physics, University of Bern, 3012 Bern, Switzerland

ABSTRACT

A metal-semiconductor-metal (MSM) detector on GaAs sensitive up to 1.4 µm is presented. The active layer consists of a MBE grown InGaAs/GaAs superlattice. The large size detector (2500 µm²) is defined by a simple one-level lift-off process. Despite the lattice mismatch, the detector shows a response time below 30 ps full width at half maximum at 1.3 µm and a relatively low dark current of 1 µA at 10 V bias.

INTRODUCTION

In modern optical fiber telecommunications, there is a demand for fast and low noise photodetectors with response in the 1.3 to 1.6 µm wavelength range. $In_xGa_{1-x}As$ is the appropriate material to detect these wavelengths because of its small and direct bandgap and its good transport properties. $In_{0.53}Ga_{0.47}As$, grown lattice matched on InP, shows excellent optical and electrical properties and is therefore a prime candidate for photodetector applications. As an alternative, $In_xGa_{1-x}As$, grown lattice mismatched, on GaAs has the advantage to allow the integration of detectors with AlGaAs/GaAs based devices leading to the realization of high performance integrated photoreceivers. The difficulty consists in the control over dislocation density due to the lattice mismatch. Superlattice (SL) $In_xGa_{1-x}As/GaAs$ structures have been shown to be effective in the accommodation of the lattice mismatch by homogeneous strain so that the in-plane lattice parameter remains constant throughout the structure as long as the single layers are thinner than a critical layer thickness.[1]

The purpose of this work is to study the device characteristics of a planar photodetector on a $In_xGa_{1-x}As/GaAs$ SL.

GROWTH AND FABRICATION

The structure (Fig. 1) is MBE grown at 450 °C with growth rates of 1.5 µm/h and 0.5 µm/h for the $In_xGa_{1-x}As$ and GaAs respectively. Without taking into account the lattice mismatch, the InAs content is estimated at x=0.66. The SL consists of 60 periods of 8 nm $In_xGa_{1-x}As$ and 5.5 nm GaAs grown on a GaAs buffer on a semiinsulating GaAs substrate; the Schottky barrier height is enhanced by a 100 nm undoped GaAs cap layer. The surface appears mirror-like smooth under low magnification (40x); small defects appear at 400x (Fig.2). The morphology depends critically on the substrate temperature, the individual layer thicknesses, the InAs content and the As-pressure. A cross sectional TEM analysis on a SL, grown under comparable conditions to those used for the structure reported here, has shown that the lattice mismatch between the substrate and the SL is mainly accommodated in the buffer layer. The TEM electron diffraction patterns indicate that the in-plane lattice parameter in the SL is constant so that the single layers are homogeneously strained.

The metal-semiconductor-metal (MSM) detector consists of two interdigitated electrodes which are defined by a one-level lift-off process (Fig. 1). The advantages of the MSM geometry compared to vertical structures such as p-i-n diodes are the very simple technology, the low capacitance even for large active areas (30 fF and 50x50 μm^2 in our case) and the ease of integration with field effect transistors. The devices are individually mounted on broadband Wiltron-K connectors.

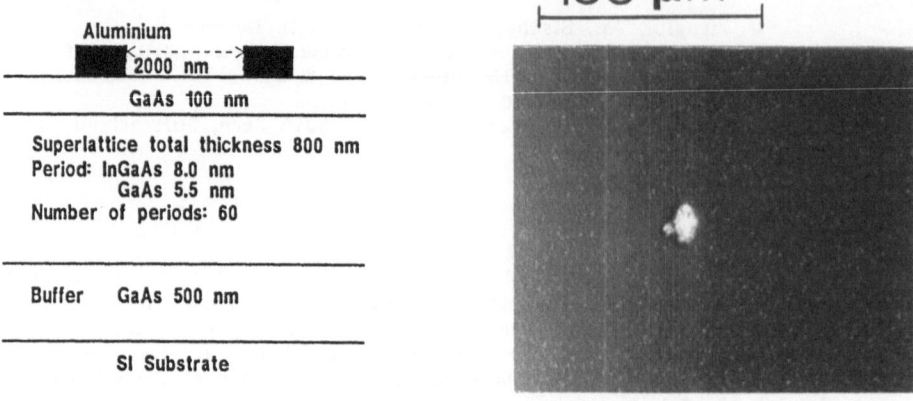

Fig. 1: Cross section of the detector.

Fig. 2: Nomarski micrograph (400x) of the active layer surface.

DEVICE RESULTS

The dark current increases exponentially with the bias voltage. At 10 V and 20 V, we measure 1 μA and 40 μA. These values are much lower than for a bulk $In_{0.4}Ga_{0.6}As$ on GaAs MSM[2] and are comparable to MSM's on lattice matched $In_{0.53}Ga_{0.47}As$[3,4,5] with InAlAs or GaAs cap layers. We have observed an increase of several orders of magnitude of the dark current with detectors where the SL was replaced by bulk $In_{0.66}Ga_{0.34}As$. The MSM shows a bias dependent dc sensitivity; above 13 V bias the internal dc quantum efficiency exceeds 100 %. The responsivity (Fig. 3) extends up to 1.4 μm.

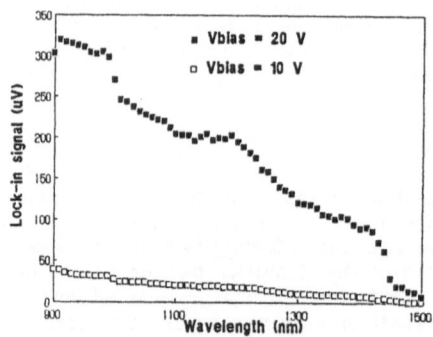

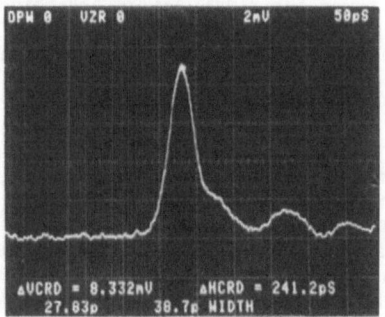

Fig. 3: Spectral responsivity measured by a lock-in amplifier and corrected for the emission spectrum of the white light source.

Fig. 4: Response at 15 V bias to a 0.82 μm wavelength pulse.

A gain switched AlGaAs/GaAs laser was used to measure the response speed at 0.82 μm. The detector is polarized with a broadband bias T and the signal is analyzed with Tektronix S4 sampling head. The averaged signal (Fig. 4) indicates a full width at half maximum (FWHM) of 38 ps. When using 300 fs pulses at 1.3 μm from a dye laser[6], the measured response time is decreased to around 30 ps FWHM (Fig. 5). In both cases, the measured responses are dominated by the 25 ps risetime of the sampling head and, for the measurement at 0.82 μm, by the non negligible optical pulse duration. The variation of the detector characteristics versus bias voltage at 0.82 μm excitation is studied in Fig. 6. Between 9 and 20 V bias, the signal amplitude increases linearly without a significant change in the pulse shape. At higher voltages, a long tail appears in the detector fall time.

In an ideal case, the dark current is given by the reverse saturation current through the Al Schottky contacts. The measured dark current is much higher than that of similar Al-GaAs MSM's[7]. We attribute the increased dark current to a lower barrier height on the strained SL material. Comparing structures with different geometries, we find that a non negligible contribution to the dark current arises at the contact patterns edges. Therefore, it should be possible to improve the dark characteristics by insulating the contact area from the active layer surface.

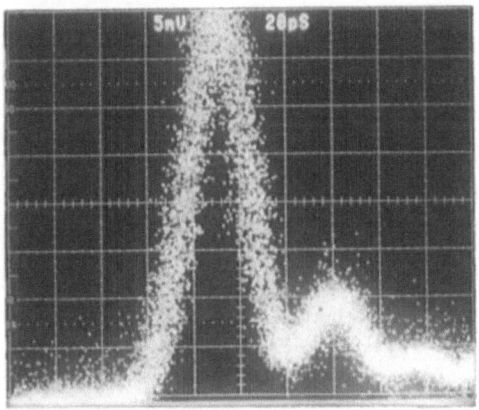

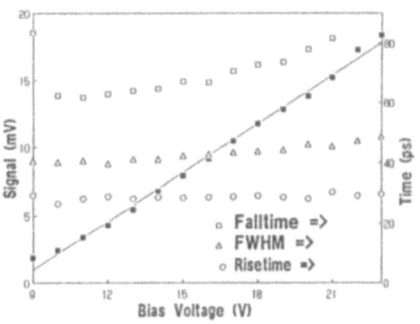

Fig. 5: Response at 15 V bias to a 1.3 μm wavelength pulse.

Fig. 6: Signal amplitude and pulse shape characteristics versus bias voltage.

The spectral responsivity shows the effect of the SL. In bulk $In_{0.66}Ga_{0.34}As$, light is absorbed up to a wavelength of 2 μm; this value is lowered by quantum size and strain effects. The relatively soft cut-off in the spectral response indicates that several transitions contribute to the overall absorption. Comparing the MSM signal to that of a p-i-n diode with 60 % external quantum efficiency[8], an internal quantum efficiency of 25% is estimated at 20 V bias under pulse excitation conditions. The longer falltimes at high voltages and the dc gain are attributed to trapped carriers which modify the electrical field under the Schottky contacts and so increase the reverse saturation current.

CONCLUSION

We have presented the characteristics of the first MSM detector on a $In_xGa_{1-x}As/GaAs$ strained superlattice exhibiting a response time below 30 ps at 1.3 µm. Thereby, we demonstrate the feasibility of developping optoelectronic components on GaAs substrates for the optical telecommunications wavelength range.

ACKNOWLEDGEMENTS

We thank F. Bobard for her excellent technical assistance and Ph. Buffat and D. Laub for the TEM measurements. This work is supported by the Swiss Post Telephone and Telegraph company PTT.

REFERENCES

[1] P.L. Gourley, I.J. Fritz, L.R. Dawson: Controversy of critical layer thickness for InGaAs/GaAs strained-layer epitaxy. Appl. Phys. Lett 52(1988) 377-379.

[2] D.L. Rogers, J.M. Woodall, G.D. Pettit, D. McInturff: High speed 1.3-µm GaInAs detectors fabricated on GaAs substrates. IEEE Electron Device Lett. 9(1988) 515-517.

[3] W.P. Hong, G.K. Chang, R. Bhat: High performance $Al_{0.15}Ga_{0.85}As/In_{0.53}Ga_{0.47}As$ MSM photodetectors grown by OMCVD. IEEE Trans. Electron Devices 36(1989), 659-661.

[4] H. Schumacher, H.P. Leblanc, J. Soole, R. Bhat: An investigation of the optoelectronics response of GaAs/InGaAs MSM photodetectors. IEEE Electron Device Lett. 9(1988) 607-609.

[5] O. Wada, H. Nobuhara, H. Hamaguchi, T. Mikawa, A. Tackeuchi, T. Fujii: Very high speed GaInAs metal-semiconductor-metal photodiode incorporating an AlInAs/GaInAs graded superlattice. Appl. Phys. Lett. 54(1989) 16-18.

[6] P. Beaud, B. Zysset, H.P. Weber: Stable 1.3 µm subpicosecond pulses from a dye laser excited by compressed Nd:YAG laser pulses. SPIE Vol. 701 ECOOSA 86 p446-450.

[7] M. Zirngibl, R. Sachot, M. Ilegems: Characterization of an AlGaAs/GaAs metal-semiconductor-metal photodetector. Journal de Physique 49(1988) C4-325-328.

[8] M. Zirngibl, Y. Hu, R. Sachot, M. Ilegems: Characterization of an top-illuminated p-i-n diode with a non alloyed ITO contact. Appl. Phys. Lett. 54 22 may (1989).

Scaling-down of Submicrometer GaAs MESFETs

N. Bannov, K. Valiev, V. Ryzhii, G. Khrenov

Institute of Physics and Technology
USSR Academy of Sciences, Moscow

Summary
Monte Carlo particle method is used to investigate the submicrometer GaAs MESFET's electrical characteristics transformation when
all dimensions of transistor have been decreased proportionate.
Essential influence of semiinsulated substrate parameters on characteristics transformation is demonstrated.

GaAs MESFET on semiinsulated substrate is one of the key
devices of high-speed digital and analog integrated circuits. The
essential peculiarities of submicrometer MESFET behaviour are
extremly nonequilibrium electron transport (for n-channel MESFETs)
and two-dimensional potential distribution. It leads to different
short-channel effects, such as velocity overshoot, nonlocal dependance semiconductor plasma parameters on the electric field and
others. Under such conditions, optimization of design, calculation
and prediction of submicrometer MESFET characteristics are
possible only by using numerical simulation. In this work scaling-
down n-channel GaAs MESFET on semiinsulated substrate with submicrometer gate length is investigated.

Kinetic approach is used to model MESFET including the
essential nonequilibrium of its electron plasma. This approach is
based on simultaneous solving the Boltzman's equations for the one
particle distribution functions of different conducting band's
valleys and Poisson's equation for the selfconsistent electric
field potential. The electron spectrum in GaAs is considered to
have three isotropic, nonparabolic and nonequivalent valleys.
The electron scattering on ionized impurities, acoustic polar
optical and intervalley phonons are taken into consideration. To
solve this kinetic equations the particle method is used where

the scattering processes are simulated by the Monte Carlo technique. The special procedure is used to minimize the number of self-scattering events. In this procedure the total scattering possibility (including self-scattering) is represented by self-adjusting step-like function. From ten to twenty thousands particles is employed in simulation. A Poisson's equation is solved with the aid of marshing algoritms and capacitance matrix method.

The structure of MESFET and simulation region are shown in Fig.1. One can see on the Fig.1 that the simulation region contains not only n-channel, but semiinsulated substrate also. Earlier, in MESFET modelling the effect of semiinsulated substrate was either neglected [1,2] or taken into account as an ideal compensated semiconductor [3]. At the same time the semiinsulated substrate constitutes as a rule an over-compensated semiconductor with

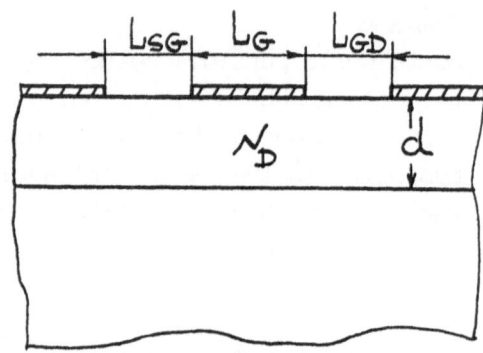

Fig.1

deep acceptor's levels (traps). In this work new numerical model taking into account the influence of deep traps recharge effects in substrate on MESFET characteristics is used (for more detail see [4]). The analysis of simulation results show that the electron channel in MESFET is squeezed by potential barriers not only from the gate side, but also from the substrate side. The latter causes the experimentally observed effects of "backgating".

Two sets of submicrometer MESFET with proportionally decreased dimensions are simulated. The parameters of the modelling devices are represented in the Table. One can see on the Table that the k time diminishing of the device sizes (gate length L_G, space between electrodes $L_{SG} = L_{GD}$, channel thickness d) leads to k^2 time increase of the donor's concentration in the channel N_D. The common feature of the devices under investigation is a conservation of the threshold voltage V_{th}^0, calculated by gradual channel approximation. The main difference between two sets of MESFET is the following. In the first set the noncompensated deep acceptors concentration in semiinsulated substrate N_t is remained constant, while in the second set N_t is changed as well as N_D.

no MESFET	L_G μm	$L_{SG}-L_{GD}$ μm	d μm	N_t cm^{-3}	N_D cm^{-3}
1-1	1.0	0,4	0,3	$2,5 \cdot 10^{16}$	$2,5 \cdot 10^{15}$
1-2	0,75	0,3	0,21	$5,0 \cdot 10^{16}$	$2,5 \cdot 10^{15}$
1-3	0,50	0.2	0,15	$1,0 \cdot 10^{17}$	$2,5 \cdot 10^{15}$
1-4	0,25	0,1	0,075	$4,0 \cdot 10^{17}$	$2,5 \cdot 10^{15}$
1-5	0,10	0,04	0,030	$2,5 \cdot 10^{18}$	$2,5 \cdot 10^{15}$
2-1	1,00	0,50	0,3	$2,5 \cdot 10^{16}$	$1,25 \cdot 10^{15}$
2-2	0,50	0,25	0,15	$1,0 \cdot 10^{17}$	$5,0 \cdot 10^{15}$
2-3	0,25	0,125	0,075	$4,0 \cdot 10^{17}$	$2,0 \cdot 10^{16}$
2-4	0,125	0,0625	0,0375	$1,6 \cdot 10^{18}$	$8,0 \cdot 10^{16}$

The transconductance g_m, cutoff frequency f_T and maximum frequency of oscillation f_{max} via gate length for both sets of devices are shown on the Fig. 2-3, respectively. Diminishing of the size for

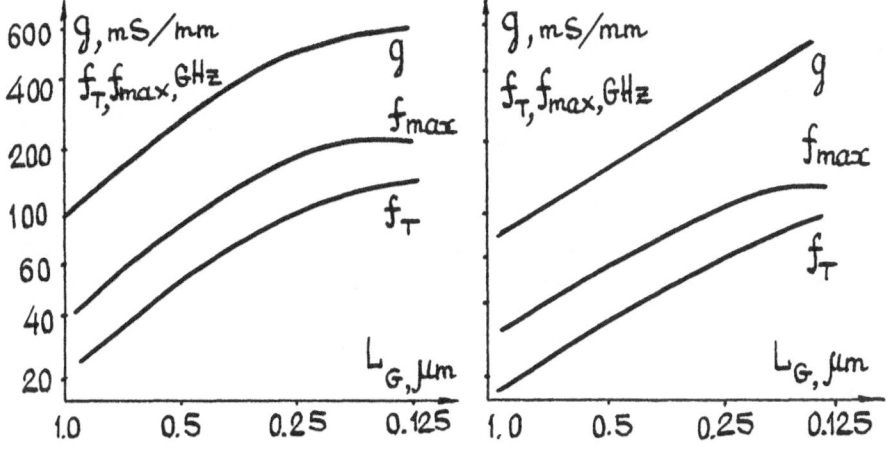

Fig. 2   Fig. 3

the first device set results in increase of the two-dimensional potential distribution influence on device operation. The greater part of channel current runs through the substrate. This effect causes essential slowing-down g_m and f_T, and saturation f_{max} when $L < 0,25$ μm (see Fig. 2). In the second transistor set potential distributions are almost similar and differences are explained only by the gain of the transport nonequilibrium while sizes are diminished. The greater heating of the electron gas in channel leads to deviation g_m, f_T and f_{max} from the linear dependence (see Fig. 3).

Saturation of the maximum frequency of oscillation for the

first device set is explained mainly by the essential gain of the channel conductance g_c at the drain current saturation region (see Fig. 4). This gain is mainly due to the penetration of the channel electron into the substrate. Two-dimensional potential distribution and heating of the electron gas are one of the main reasons of

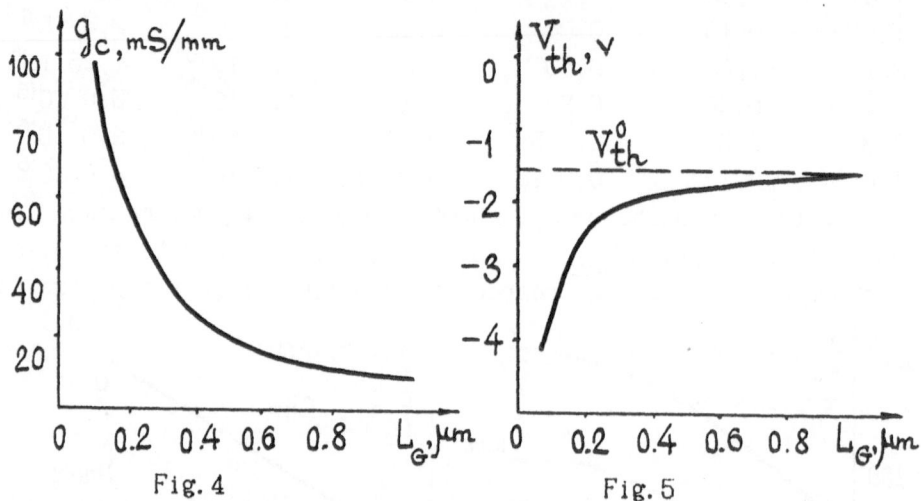

Fig. 4   Fig. 5

this penetration. Considerable shift of the threshold voltage V_{th} to more negative values (see Fig. 5) is also due to the strong electron penetration into the substrate.

References

1. Awano, Y.; Tomizawa, K.; Hashizume, N.: Principles of operation short-channel gallium arsenide field-effect transistor determinated by Monte Carlo method. IEEE Trans. Electron Dev. ED-31 (1984) 448-450.
2. Williams, C. K.; Glisson, T. H.; Hauser, J. R.; Littlejohn, M. A.; Abusaid, M. F.: Two dimensional Monte Carlo simulation of a submicron GaAs MESFET with a nonuniformly doped channel. Solid State Electron. 28 (1985) 1105-1109.
3. Awano, Y.; Tomizawa, K.; Hashizume, N.; Kawashima, M.: Monte Carlo simulation of GaAs short channel MESFET's. Electron. Lett. 19 (1983) 20-21.
4. Bannov, N. A.; Valiev, K. A.; Ryzhii, V. I.; Khrenov, G. Yu.: Influence of the charge effects in the semiinsulated substrate on the GaAs MESFET operating. Microelectronica 17 (1988) 395-398 (in Russian).

Device Simulation

The Mobility Model in MINIMOS

J.W. Slotboom and G. Streutker

Philips Research Laboratories,
P.O.B. 80000, 5600 JA Eindhoven, The Netherlands

Abstract
Several authors have shown that the dependence of the mobility on the electric field perpendicular to the gate oxide is described by a "universal curve" if an effective normal electric field is used. In this paper we show that the MINIMOS-4 local mobility model agrees very well with these empirical data over the whole range of electric fields. The numerical extraction method for determining the effective mobility and effective electric field are given.

Introduction
Modelling of the carrier mobility in the inversion layer of MOSFET transistors is of crucial importance for accurate device characterization. Several authors [1,2,3,4] have shown that the normal electric field dependence of the mobility is described by a "universal curve" if the measurements are analysed in terms of an effective normal electric field. It appeared that the substrate impurity concentration and the back-bias voltage are of little influence on this curve and it can be considered as a reproducible property of the Si/SiO_2 interface.

The effective normal electric field is the averaged electric field over the electron distribution in the inversion layer (see fig. 1):

$$E_{eff} = \frac{\int n \, E \, dy}{\int n \, dy} \, . \tag{1}$$

Because the mobilities are measured as a function of this effective electric field, it seems straightforward to implement these data by means of a "global" mobility model [5] into a device simulator and assuming that the electrons in the inversion layer feel the same effective electric field. However, such a nonlocal approach introduces numerical problems because in other regions of the device local quantities are used. It is also numerically inefficient compared to local models.
The question arises: "If we use a local mobility model, how well does it agree with the empirical "universal mobility curve" if the calculations are interpreted, just as the measurements, in terms of an effective electric field?"

Local mobility model
The local mobility model [6] for the normal electric field dependence in MINIMOS-4 is given by:

$$\mu_{SLI} = \frac{\mu_o + (\mu_{LI} - \mu_o)(1 - F(y))}{1 + F(y)(E_y/E_{ref})^\gamma}, \tag{2}$$

and is embedded in the total mobility model where also the lateral electric field dependence is taken into account. The transition from surface mobility towards bulk mobility is taken into account by the function F(y). This function is shown in fig. 2; at the surface $F(y) = 1$ and in the bulk $F(y) = 0$; μ_{Li} is the bulk mobility which is modelled in the usual way as a function of lattice and impurity scattering; μ_o, E_{ref} and γ are three parameters describing the surface mobility.

Numerical extraction method

In a post-processor the calculated local quantities are analysed in terms of the effective normal electric field and effective mobility. In this way we simulate the experimental procedure. The total channel current is given by:

$$I_n = q\,\mu_{eff}\,Q_{inv}\,\frac{d\,\varphi_n}{dx}, \tag{3}$$

where Q_{inv} is the inversion layer charge ($\int n dy$) and φ_n is the electron quasi-fermi potential. The effective mobility is easily derived from eq. (3) because I_n, Q_{inv} and $d\varphi_n/dx$ are known. The calculations are done for $V_{DS} = 0.1$ V.

Assuming that in the inversion channel $\varphi_n(y) = $ constant eq. (1) can be integrated and gives

$$E_{eff} = \frac{kT}{q}\,n(0)\,/\,Q_{inv}, \tag{4}$$

where n(0) is the electron concentration at the interface. Fig. 3 is an example of the calculations where both the variation of the local quantities over the channel as well as the effective mobility and effective electric field are shown.

Results

A number of transistors, varying in substrate dope concentration between $4.10^{15}\mathrm{cm}^{-3}$ to $2.10^{18}\mathrm{cm}^{-3}$ and in gate oxide thickness between 300 Å and 100 Å, have been simulated with MINIMOS-4. As described above, we have extracted the effective mobility curves for n-channel and p-channel transistors up to very high electric fields, and compared with experimental data [1,2,3,4].

The optimized mobility parameters are given in table 1:

	$\mu_o[\mathrm{cm}^2/\mathrm{Vs}]$	$E_{ref}[\mathrm{V/cm}]$	γ
Table 1 Electrons	638	$7 \cdot 10^5$	1.69
Holes	240	$2.7 \cdot 10^5$	1

In fig. 4 and fig. 5 the simulated effective mobilities are compared with these experimental data. It is clear that not only the phonon scattering region for electric fields below 0.5 MV/cm but also the roll-off at higher fields due to surface roughness scattering [3,4] are described

well for the electron mobility. The hole mobilities measured by Takagi e.a. are smaller than the data published by Watt/Plummer and Walker/Woerlee.
In fig. 6 and fig. 7 simulations of $I_D V_{DS}$-characteristics with MINIMOS-4 are compared to measurements [8].

Conclusions
The local mobility model in MINIMOS-4 agrees well with the empirical effective mobility vs. effective field "universal mobility curve". Calibrated model parameters are given. Although the physical meaning of the local quantities in the inversion layer itself can be disputed, the calculated effective mobilities and effective electric fields do agree with the measurements. Having shown the large validity range of this calibrated local mobility model, and because of the numerical efficiency, it also is recommended for other device simulators.

Acknowledgement: The authors highly appreciated the support and discussions with W. Hänsch on the MINIMOS-program and with P.H. Woerlee and A.J. Walker on the mobility measurements.

References

[1] A.G. Sabnis, J.T. Clemens, IEDM Dec. 1979, p 18.
[2] J.T. Watt, J.D. Plummer, VLSI Symp. May 1987, p 81.
[3] A.J. Walker, P.H. Woerlee, ESSDERC Sept. 1987, p 667.
[4] S. Takagi e.a. IEDM Dec. 1988, p 378.
[5] A.J. Walker, P.H. Woerlee, ESSDERC Sept. 1988, p 265.
[6] S. Selberherr, "MOS Device modeling at 77 K", to be published in IEEE Trans. Electron. Dev., Aug. 1989 special issue on low temperature electronics.
[7] S. Selberherr: MINIMOS-4 Manual 1988.
[8] P.T.J. Biermans, M.J.B. Bolt and P.H. Woerlee: Private communications.

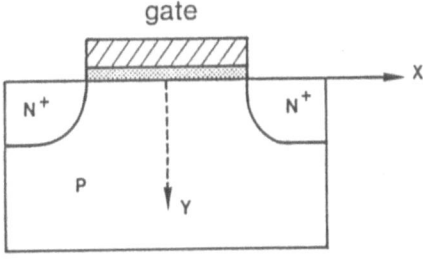

Figure 1.

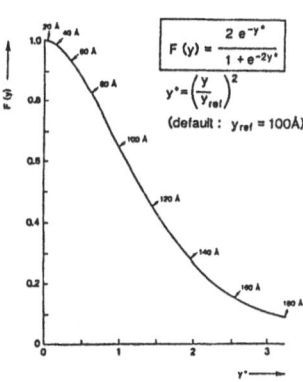

Figure 2.

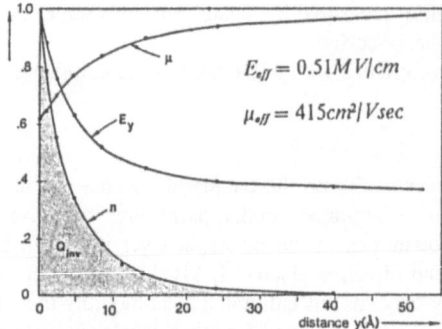

fig.3 Calculated electron density, normal electric field
and mobility in the inversion layer; the norm factors
are resp. $5.8 \ 10^{19} cm^{-3}$, $0.72 MV/cm$ and $506 cm^2/Vsec$.
 $d_{ox} = 175 \text{Å}$, $N_{SUB} = 1.8 \ 10^{17} cm^{-3}$, $V_g = 5V$, $V_{DS} = 0.1V$

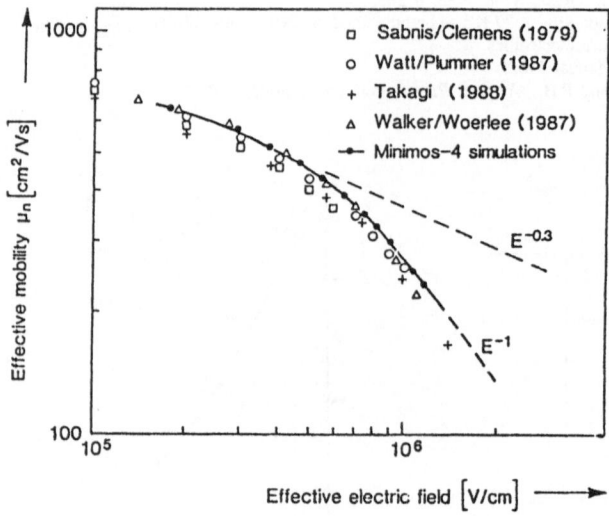

Figure 4. Effective electron mobility.

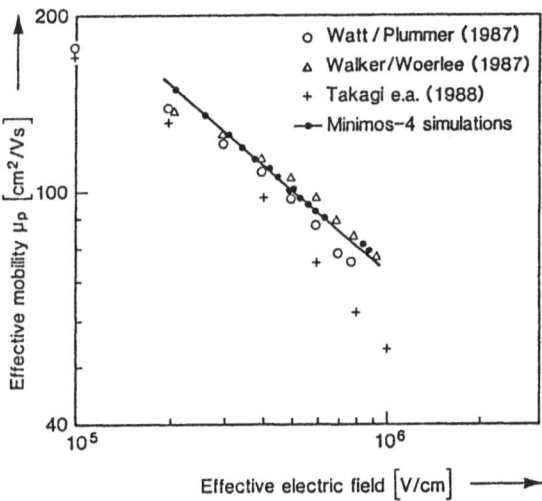

Figure 5. Effective hole mobility.

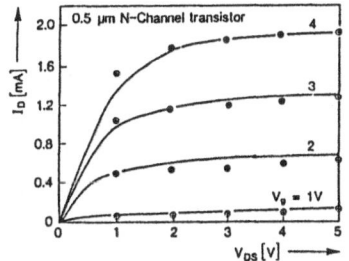

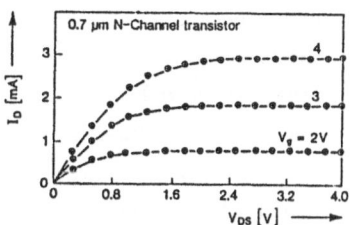

Figure 6. Measured $I_D V_{DS}$-characteristic for a 0.5 μm n-channel transistor compared with MINIMOS-4 simulations ⊕.

Figure 7. Measured $I_D V_{DS}$ characteristic for a 0.7 μm n-channel transistor compared with MINIMOS-4 simulations ⊕.

MESFET Analysis with MINIMOS

Ph. Lindorfer, S. Selberherr

Institut für Mikroelektronik
Technische Universität Wien
Gußhausstraße 27–29
A-1040 WIEN, AUSTRIA

Summary

This paper presents the implementation of models allowing the simulation of silicon as well as GaAs MESFETs with MINIMOS 5 – an integrated 2D and 3D device simulator for silicon MOSFETs. Models for the Schottky contact, device geometries and first results are shown.

Introduction

During the last years MESFETs have become more and more a serious alternative to MOSFETs. The main advantages MESFETs offer, compared to MOSFETs, are the simpler fabrication processes because of low temperature process steps, the much smaller radiation sensitivity because of the missing gate oxide, the higher carrier mobilities because of current transport deeper in the bulk and no minority carrier storage effects which result in faster speeds for high frequency applications. Especially GaAs MESFETs have become an increasing factor for high speed analog and digital cicuits since their introduction in 1970. GaAs FET amplifiers, oscillators, mixers, etc. are widely used in microwave applications whereas very fast digital circuits have been developed based on GaAs MESFET logic.

With the increasing importance of MESFETs the need for efficient simulation of these devices has become apparent. Based on MINIMOS 5, which is our integrated two– and three–dimensional device simulator for silicon MOSFETs with transient and small signal analysis capabilities we implemented models allowing the simulation of MESFETs and models for III–V compound semiconductors.

Basic equations

MINIMOS 5 solves the basic semiconductor equations in two or three space dimensions. The set of equations consists of the Poisson equation (1) and the continuity equations for electrons (2) and holes (3):

$$div \ grad \ \psi = \frac{q}{\varepsilon} \cdot (n - p - C) \tag{1}$$

$$div \ J_n = q \cdot R \tag{2}$$

$$div \ J_p = -q \cdot R \tag{3}$$

The current relations for electrons and holes differ slightly from the classical formulations, which can be found in [1]:

$$J_n = q \cdot \mu_n \cdot n \cdot (-grad \ \psi + \frac{1}{n} \cdot grad \ (U_{t_n} \cdot n)) \tag{4}$$

$$J_p = q \cdot \mu_p \cdot p \cdot (-grad \ \psi - \frac{1}{p} \cdot grad \ (U_{t_p} \cdot p)) \tag{5}$$

In equation (4) and (5) the second term accounts for carrier heating effects by field dependent modelling of the carrier temperatures. Detailed information about the derivation of these formulations can be found in [2]. These equations have proven to work well for silicon MOSFETs and MESFETs with gate length down to 0.1 microns. For GaAs devices the classical drift–diffusion approach has proven to be suitable only for relatively long devices whereas for very small feature sizes nonstationary transport effects can become apparent which are usually claimed to be properly modelled by either Monte Carlo methods or hydrodynamic equations based on higher moments of the Boltzmann equation [3]. We have evidence that our current relations (4) and (5) in connection with appropriate models for the carrier voltages U_{tn} and U_{tp} push the limit of applicability of these enhanced drift-diffusion equations significantly towards smaller feature sizes. Final investigations which give a sound proof will have to be carried out.

Boundary Conditions

To allow the simulation of MESFETs with MINIMOS a boundary condition for the Schottky gate contact had to be implemented.

A Dirichlet boundary condition is used for the potential ψ

$$\psi = \psi_{app} - \psi_s \tag{6}$$

where ψ_{app} is the applied voltage and ψ_s is the surface potential at the interface. ψ_s depends on the barrier height ϕ_B in the following way

$$\psi_s = \phi_B - \frac{E_g}{2} - \frac{kT}{2q} \cdot ln \left(\frac{N_v}{N_c} \right) \tag{7}$$

where the barrier height is allowed to change in case of large reverse bias

$$\Delta\phi_B = \sqrt{\frac{qE}{4\pi\varepsilon_s}} \tag{8}$$

to account for the image force lowering [4]. ε_s is the permittivity of the semiconductor, E the electric field at the interface, E_g the bang gap of the semiconductor and N_c and N_v the density of states of the conduction and the valence band.

Implicit boundary conditions are implemented for the carrier densities n and p by using boundary conditions for the current densities J_n and J_p perpendicular to the interface.

$$J_n = -q \cdot v_n \cdot (n - n_0) \tag{8}$$

$$J_p = q \cdot v_p \cdot (p - p_0) \tag{9}$$

n_0 and p_0 are the equilibrium carrier concentrations at the surface defined by

$$n_0 = n_i \cdot exp\left(-\frac{\psi_s}{U_T}\right) \tag{10}$$

$$p_0 = n_i \cdot exp\left(\frac{\psi_s}{U_T}\right) \tag{11}$$

The surface recombination velocities v_n and v_p are modelled current dependent.

$$v_{n,p} = v_d + \sqrt{\frac{2\,k\,T}{m_{n,p}^* \pi\, \eta_{n,p}}} \cdot \frac{exp\left(-v_d^2\left(\frac{m_{n,p}^*\,\eta_{n,p}}{2\,k\,T}\right)\right)}{1 + erf\left(v_d\sqrt{\frac{m_{n,p}^*\,\eta_{n,p}}{2\,k\,T}}\right)} \tag{12}$$

Here m^* is the effective mass for either electrons or holes, $\eta_{n,p}$ is an 'non-ideality' factor, which accounts for changes in the band structure at the interface and v_d is the drift velocity defined by

$$v_d = \frac{J_{n,p}}{q \cdot (n,p)} \tag{13}$$

This formulation avoids an unrealistic accumulation of carriers at large forward biased Schottky contacts and so leads to better results in those cases. A detailed derivation of this model can be found in [5].

Process Models

To provide user-friendly use of MINIMOS for GaAs simulation various process parameters like implantation statistics and diffusion coefficients for various dopants have been implemented based on Monte Carlo calculations with PROMIS [6].

Device Geometries

For an efficient simulation of GaAs devices it is necessary to simulate different nonplanar geometries like recessed gate or T-gate MESFETs. Various enhancements had to be made to allow the simulation of these nonplanarities with MINIMOS 5 which was able to simulate nonplanar MOS structures [7]. Fig. 1 shows shape of the source, the gate and the drain contact of a T-gate MESFET whereas one can see a recessed gate device with the recessed semiconductor surface in Fig. 2. The figures are not on the same scale in all three space dimensions.

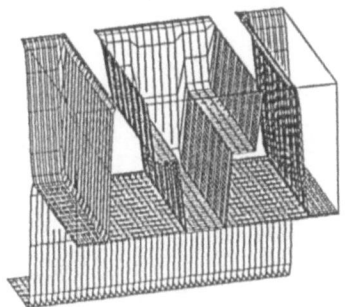

Fig. 1

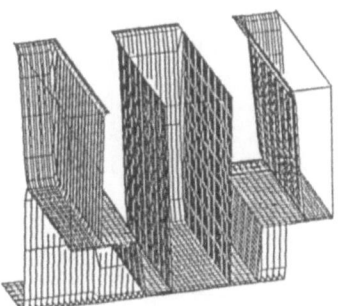

Fig. 2

Results

As an example an n-channel MESFET with 2.5 micron long recessed gate was chosen. The depth of the recess is 0.07 micron and the thickness of the semiconductor layer is 0.35 micron. The device was biased with 0.6 V on drain and −1.0 V on gate. The Schottky barrier height is 0.85 V. Fig. 3 shows the doping profile of the device. This MESFET was simulated with silicon as substrate material on the one hand and with GaAs on the other hand.

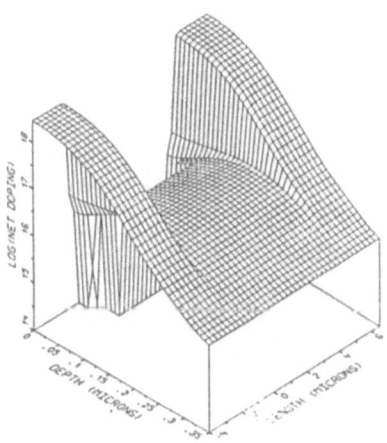

Fig. 3

Fig. 4 shows the electron distribution in the silicon MESFET with the depletion region under the gate. One can see that this device has a well conducting channel with an electron concentration in the order of 10^{16}. The GaAs MESFET works in the subthreshold regime at this bias. Fig 5. shows that the channel is completely depleted down to the interface.

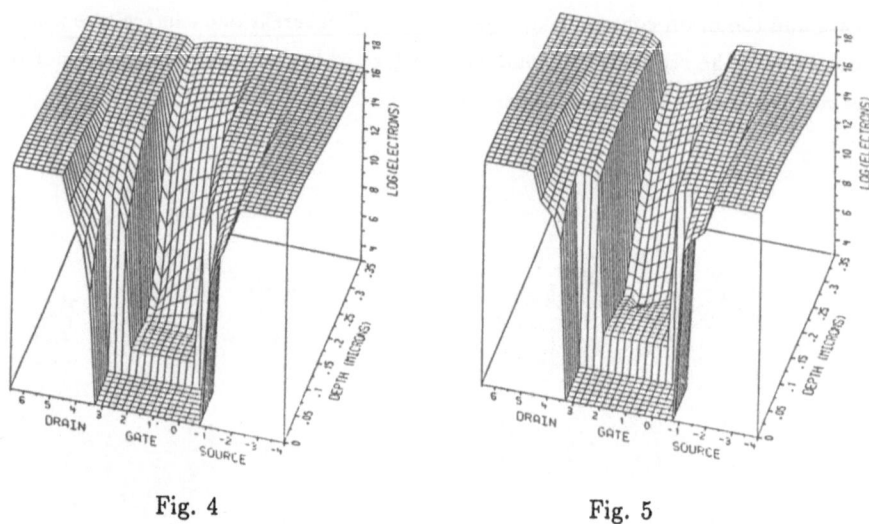

Fig. 4   Fig. 5

Acknowledgement

This work was supported by the research laboratories of Siemens AG in Munich, FRG and by DIGITAL EQUIPMENT CORPORATION in Hudson, USA.

References

[1] SELBERHERR S., *Analysis and Simulation of Semiconductor Devices*, ISBN 3-211-81800-6, Springer, Wien–New York, 1984.

[2] HÄNSCH W., SELBERHERR S., "MINIMOS 3: A MOSFET Simulator that Includes Energy Balance", *IEEE ED-34, No. 5*,1987.

[3] FENG Y., HINTZ A.,"Simulation of Submicrometer GaAs MESFET's Using a Full Dynamic Transport Model",*IEEE ED-35, No. 9*,1988.

[4] SZE S.M., *Physics of Semiconductor Devices*,ISBN 0-07-062686-3, McGraw-Hill, 1983.

[5] NYLANDER J.O., MASSZI F., SELBERHERR S., BERG S. "Computer Simulations of Schottky Contacts with a Non-Constant Recombination Velocity", *Solid-State Electronics, Vol. 32, No. 5*, 1989.

[6] HOBLER G.,"Multiple Use of Trajectories in Monte Carlo Ion Implantation Simulation", *Proc. NASECODE VI Conf.*, 1989.

[7] THURNER M., "Dreidimensionale Modellierung von MOS Transistoren", *Dissertation*, Technical University Vienna, 1988.

Numerical Analysis of Breakdown in Silicon Diodes

W. Quade[1] and *M. Rudan*

Dipartimento di Elettronica, Informatica e Sistemistica
Università di Bologna, viale Risorgimento 2, 40136 Bologna, Italy

Using the hydrodynamic model [1] for both electrons and holes together with a mean-energy dependent impact-ionization formulation [2] we simulate the behavior of one-dimensional silicon diodes near breakdown.[2] It is known that the hydrodynamic model provides a basis for describing hot-electron phenomena such as velocity overshoot [3], substrate current and oxide injection [4]. To investigate the possibilities and limitations of the model, we apply it here in the very high field regime near breakdown.

Our analysis is based on the following set of equations

$$\mathrm{div}(\mathbf{D}) = q(p - n + N) \tag{1}$$

$$\mathrm{div}(\mathbf{J}_n) = q\,U \qquad \mathrm{div}(\mathbf{J}_p) = -q\,U \tag{2}$$

$$\mathrm{div}(\mathbf{Q}_n) = \mathbf{J}_n \bullet \left[\frac{3}{2}\,\mathrm{grad}(r) - r\,\frac{\mathrm{grad}(n)}{n} \right] + \frac{5}{2}\,r\,q\,U + q\,\frac{n}{\tau_{qn}} + C(U_{ii}^{el}, U_{ii}^{ho})_n \tag{3}$$

$$\mathrm{div}(\mathbf{Q}_p) = -\mathbf{J}_p \bullet \left[\frac{3}{2}\,\mathrm{grad}(s) - s\,\frac{\mathrm{grad}(p)}{p} \right] + \frac{5}{2}\,s\,q\,U + q\,\frac{p}{\tau_{qp}} + C(U_{ii}^{el}, U_{ii}^{ho})_p \tag{4}$$

and

$$\mathbf{D} = -\frac{k_B T_0}{q}\,\epsilon_s\,\mathrm{grad}(u) \tag{5}$$

$$\mathbf{J}_n = q\,D_n\,[r\,\mathrm{grad}(n) + n\,\mathrm{grad}(r - u)] \qquad \mathbf{J}_p = -q\,D_p\,[s\,\mathrm{grad}(p) + p\,\mathrm{grad}(s + u)] \tag{6}$$

$$\mathbf{Q}_n = -\frac{q\kappa_n}{k_B}\,\mathrm{grad}(r) \qquad \mathbf{Q}_p = -\frac{q\kappa_p}{k_B}\,\mathrm{grad}(s)\,, \tag{7}$$

where n and p are the concentrations, $\mathbf{J}_n$ and $\mathbf{J}_p$ the current densities, and $\mathbf{Q}_n$ and $\mathbf{Q}_p$ the normalized heat flows for electrons and holes, respectively. The normalized electric potential is $u = q\varphi/(k_B T_0)$, where k_B is the Boltzmann constant, T_0 the lattice temperature and q the electric charge. The symbols r and s represent the normalized electron and hole temperatures T_e/T_0 and T_h/T_0, while the diffusion coefficients and the thermal conductivities are expressed by D_n, D_p and κ_n, κ_p. The derivation of the model has been shown in [1]. The net recombination rate U is made up by impact ionization and by the SRH and Auger generation-recombination mechanisms. Here, the usual form [5] for the SRH process is used, while a standard expression [5] for the Auger process is combined with a model similar to that proposed in [2], to give the total recombination rate

$$U = U_{SRH} + U_{ii}^{el} + U_{ii}^{ho}\,,$$

---

[1] W. Quade avails himself of a fellowship provided by the European Community. His present address is: Dipartimento di Fisica, Università di Modena, via G. Campi 213/A, 41100 Modena, Italy

[2] This work was supported in part by the Italian National Council of Research under the "Progetto Finalizzato MADESS", by the Italian Ministry of Foreign Affairs, by SGS-Thomson Microelectronics and by the German Acad. Exchange Service (DAAD)

where the terms U_{ii} embody the Auger and the impact ionization mechanisms for electrons and holes, respectively. Each term U_{ii} vanishes at equilibrium as required by the principle of detailed balance.

In contrast to the treatment in [1] and [3], we assume for the SRII process that the electrons and holes are created or destroyed at their band extrema, which eliminates the contribution of this process from the collision terms in Eqs. (3), (4), and (6). As for Auger recombination and impact ionization, we assume that they influence the scattering term of the energy-balance equations via the two functions C_n and C_p [2]. For a deeper discussion of carrier heating/cooling induced by generation and recombination processes we refer to [6].

By means of an improved version of program Q6 [7] we have solved the whole set of equations (1–7) at different applied biases for one-sided abrupt junctions. At the end of each bias step, the electron and hole current densities have been used to evaluate (a *posteriori*) the impact-ionization integrals [8] which, in turn, show how the breakdown condition is approached. In Fig. 1, the impact-ionization integrals for electron and holes are drawn as a function of the inverse voltage, at $N_B = 6 \times 10^{15}$ cm^{-3} and $N_B = 2.5 \times 10^{16}$ cm^{-3} background doping, respectively (these values were chosen to make a comparison with the data shown in [8, Ch. 3.5] and [9] possible). In the heavily-doped region, a Gaussian profile with the same variance for both diodes has been used.

Since the simulations were voltage controlled, it was not possible to increase the bias up to breakdown. The breakdown voltage, however, has been estimated from a theoretical reasoning [10], stating that the impact-ionization integral approaches 1 linearly as the bias increases (dotted lines in Fig. 1). This estimate was then used to tune some parameters of the impact-ionization model, such as the overlap integrals and the energy threshold [2].

In Fig. 2, the current density for the same devices and in the same voltage range as in Fig. 1 is shown together with the curve corresponding to the case where impact ionization is switched off (dotted lines). From the comparison of the two curves it is seen that the current is mostly due to SRII generation at lower voltages, and that the "onset" of impact ionization corresponds to the change in curvature taking place at higher voltages.

To further discuss the behavior of the carrier temperature, we take the diode with the higher background doping. The carrier concentration and temperature are shown in Fig. 3, while the electric field and the impact-ionization rates α_n and α_p are shown in Fig. 4. Both figures refer to a 28 V reverse bias. Electrons move from right to left and experience a continuously increasing electric field while crossing the depleted region. Holes move from left to right and experience a step-like and then decreasing electric field. The different histories of the two species of carriers are reflected by their temperature behaviour. In fact, it is seen in Fig. 4 that the impact-ionization rates are shifted with respect to each other, and that the electric field is shifted with respect to both. See the corresponding figure in [11], where a field-dependent model for the impact-ionization rates is used. Clearly, in a field-dependent model the doping distribution is expected to have a stronger effect on the impact-ionization rates because it influences the shape of the electric field. In our simulations, instead, we saw that the temperature-dependent impact-ionization model is less sensitive to the doping profile, because thermal diffusion tends to smear out strong local dependencies.

To determine the influence of impact ionization on the temperature distribution, we made a simulation without impact ionization. It was found that cooling by impact ionization gives a significant contribution long before breakdown occurs, as was already observed in the study of chaos in semiconductors [12]. This has also an interesting effect from the numerical point of view, which becomes apparent in the decoupled solution scheme used here. It is easily seen, in fact, that the energy-continuity equation embodies a negative feedback in the collision term C. This results in a good convergence rate, notwithstanding the complexity of the model.

Finally, we carried out a simulation where impact ionization by holes was prevented by letting $U_{ii}^{ho} = 0$. The result confirms what can also be derived from the theoretical breakdown condition on the impact ionization integral $Y_n = 1 - 1/M_n$ [8]. In fact, if $\alpha_p \to 0$ it is

$$Y_n = \int_0^l \alpha_n \exp\left[-\int_0^x (\alpha_n - \alpha_p)dx'\right] dx \to \int_0^l \alpha_n \exp\left[-\int_0^x \alpha_n dx'\right] dx = 1 - \exp\left[-\int_0^l \alpha_n dx\right]$$

i.e., the breakdown can only be achieved in an infinitely long device ($l \to \infty$). This makes it more obvious that breakdown is a chain reaction running over two different types of particles and does not occur if only one type of carrier is able to impact ionize.

To conclude, it is also worth adding that the form of the impact-ionization model used here lends itself to a straightforward numerical implementation in the frame of the hydrodynamic model.

Acknowledgement. We wish to thank Prof. G. Baccarani and Prof. E. Schöll for many helpful discussions.

[1] Rudan, M., and Odeh, F., *Multi-Dimensional Discretization Scheme for the Hydrodynamic Model of Semiconductor Devices*, COMPEL, vol. 5, no. 3, 1986, pp. 149–183

[2] Schöll, E., and Quade, W., *Effect of impact ionization on hot-carrier energy and momentum relaxation in semiconductors*, J. Phys. C: 20 (1987) L861–L867

[3] Baccarani, G., and Wordeman, M. R., *An Investigation of Steady-State Velocity Overshoot in Silicon*, Solid-State Electronics, 1985, Vol. 28, No. 4, pp. 407-416

[4] Meinerzhagen, B., *Consistent Gate and Substrate Current Modelling based on Energy Transport and the Lucky Electron Concept*, IEDM'88

[5] Baccarani, G., Rudan, M., Guerrieri, R., and Ciampolini, P., *Physical Models for Numerical Device Simulation*, in "Process and Device Modelling" edited by Engl, W.L., North-Holland, Amsterdam, 1986

[6] Bimberg, D. and Mycielski, J., *The Recombination-induced Temperature Change of Non-equilibrium Charge Carriers*, J. Phys. C: Solid State Phys. 10 (1986) 2363-2373

[7] Rudan, M., Odeh, F., and White, J., *Numerical Solution of the Hydrodynamic Model for a One-dimensional Semiconductor Device*, COMPEL, Vol. 6, 1987, pp. 151–170

[8] Sze, S. M., *Physics of Semiconductor Devices*, Wiley and Sons, New York, 1969

[9] Grant, W.N., *Electron and Hole Ionization Rates in Epitaxial Silicon at High Electric Fields*, Solid-State Electronics, Vol. 16, 1973, pp. 1189–1203

[10] Shockley, W., *Problems related to p-n Junctions in Silicon*, Solid-State Electronics, Vol. 2, No. 1, 1961, pp. 35–67

[11] Van Overstraeten, R., and De Man, H., *Measurements of the Ionization Rates in Diffused Silicon p-n Junctions*, Solid-State Electronics, Vol. 13, 1970, pp. 583–608

[12] Schöll, E., *Nonlinear Energy Relaxation Oscillations and Chaotic Dynamics of Hot Carriers*, Solid-State Electronics Vol. 31, No. 3/4, 1988, pp. 539–542

100

Fig. 1:
Impact ionization integrals
versus reverse voltage. The
arrows indicate the breakdown
voltages for similar devices
reported in [8]

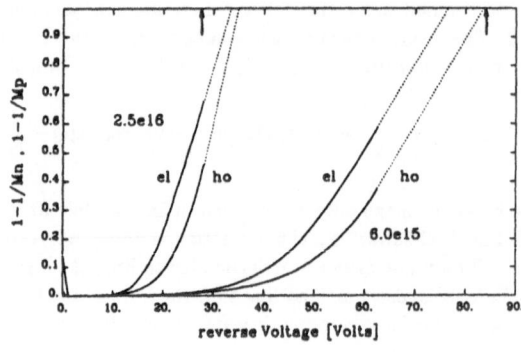

Fig. 2:
Current densities versus
reverse voltage. The dotted
lines refer to simulations
in which impact ionization was
switched off

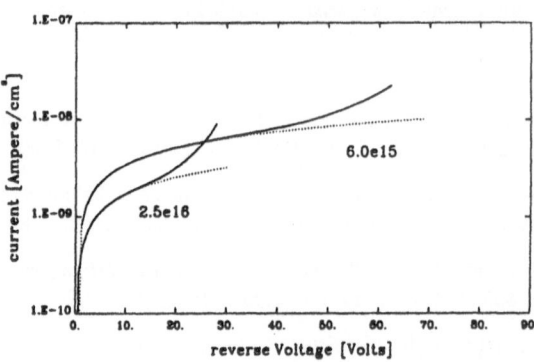

Fig. 3:
Carrier concentrations (left
scale) and norm. temperatures
(right scale) versus position.
The dotted line indicates the
metallurgical junction

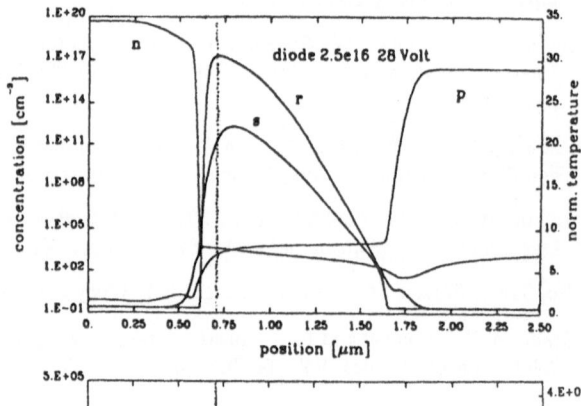

Fig. 4:
Electric field (left scale) and
ionization rates (right scale)
versus position. The dotted
line indicates the
metallurgical junction

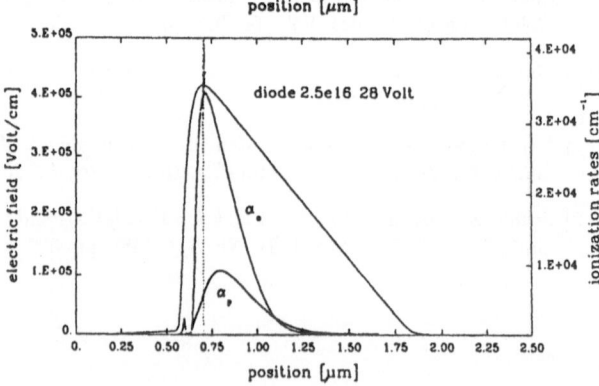

A New Algorithm to Accelerate Convergence in the Simulation of Semiconductor Devices

H. BRAND[1] and R. KIRCHER[2]

[1] SIEMENS AG Österreich, ETG215, Gudrunstr. 11, A-1101 Vienna, Austria

[2] SIEMENS AG, ZFE ELPT33, Otto-Hahn-Ring 6, D-8000 Munich, FRG

Summary

A new algorithm for an initial guess of the solution of the semiconductor equations has been developed. It is based on a 'quasi-equilibrium' approximation of the electrostatic potential and the electron and hole densities. The algorithm has been implemented into the two dimensional general purpose device simulator BAMBI [1,2] and applied to various MOS and bipolar structures. It works in a large range of bias conditions, limited only by high injection, and exhibits excellent convergence properties in combination with the hybrid solution method.

The Initial Solution in the Quasi-Equilibrium Condition

The basic semiconductor equations consist of Poisson's equation for the electrostatic potential ψ and two continuity equations for the carrier densities n and p [3]. For a numerical solution a system of nonlinear algebraic equations has to be solved. Because the solution of the linearized problem is a correction to the nonlinear one, an initial guess of the solution is necessary.

The device simulator BAMBI uses a hybrid solver. Thus, Newton's solution method is accelerated by providing an initial solution by Gummel's method which on it's turn is enhanced by the quasi-equilibrium initial solution for ψ, n and p. It is the scope of this work to show, that this solution strategy proofs to be very efficient [4], especially when a first solution (usually the solution in the equilibrium case calculated with reduced numerical effort by solving only Poisson's equation) is provided as an input.

With Boltzmann statistics the current relations in the drift-diffusion approximation can be expressed in terms of quasi-Fermi potentials φ_n, φ_p for the carrier densities:

$$\vec{J}_n = -qn\mu_n\vec{E}_n + qD_n\mathrm{grad}\,n = -qn\mu_n\mathrm{grad}\,\varphi_n \tag{1}$$

$$\vec{J}_p = -qp\mu_p\vec{E}_p - qD_p\mathrm{grad}\,p = -qp\mu_p\mathrm{grad}\,\varphi_p \tag{2}$$

In equilibrium the drift- and the diffusion currents compensate each other. Therefore the resulting current is zero. The quasi-equilibrium approximation of a non-zero bias condition treats the non-equilibrium case similar to the equilibrium [5]. The consequence of the zero-current assumption is that the quasi-Fermi levels must be constant all over the device. As quasi-Fermi levels of the majority carriers on one side of a junction appear as quasi-Fermi levels of the minority carriers on the other side of a junction, there are inconsistencies if the simulation area is contacted more than twice. To avoid contradictions for the quasi-Fermi levels the minority carriers are neglected. Then Poisson's equation for any given device structure with i contacted n- and j contacted p-regions can be written in the form:

$$\operatorname{div}\operatorname{grad}\psi = \frac{q}{\epsilon}(n - p - C) = \frac{q}{\epsilon}(n_{ie}\exp(\frac{\psi - \varphi_{n_i}}{U_{th}}) - C) \tag{3}$$

$$\operatorname{div}\operatorname{grad}\psi = \frac{q}{\epsilon}(n - p - C) = \frac{q}{\epsilon}(-n_{ie}\exp(\frac{\varphi_{p_j} - \psi}{U_{th}}) - C) \tag{4}$$

As no current flow is assumed in quasi-equilibrium, there is no potential drop in the neutral regions. The voltages applied at the contacts appear at the junctions themselves, and $\varphi_{n_i} = V_{n_i}$, $\varphi_{p_j} = V_{p_j}$ holds, where the V's denote the applied voltages. From Boltzmann statistics one obtains for the electrostatic potential in the subregions:

$$\psi_{n_i} = \varphi_{n_i} + U_{th}\ln\frac{n}{n_{ie}} = V_{n_i} + U_{th}\ln\frac{n}{n_{ie}} \tag{5}$$

$$\psi_{p_j} = \varphi_{p_j} - U_{th}\ln\frac{p}{n_{ie}} = V_{p_j} - U_{th}\ln\frac{p}{n_{ie}} \tag{6}$$

Note, that because of the zero-current assumption the new distribution of the potential according to equations (5) and (6) is fully consistent with the old carrier distributions. The potential is determined by the applied voltages and the majority carrier density given by the doping concentration. The extraction and injection of the minority carriers can be neglected in the quasi equilibrium. These effects are counted for during the solution procedure. The quasi-equilibrium initial solution approximates the solution very well in neutral regions and is qualitativly correct even in the space charge regions and at their boundaries. The assumptions made for the quasi-equilibrium solution are fulfilled under low-injection conditions and at reversed biased junctions.

The algorithm finds all gridpoints which belong to a contact. It takes advantage of the finite boxes discretisation [2] in BAMBI. In contrast to a classical finite differences grid the graph representation of the finite boxes grid provides information about the connections of the grid points on a logical level without refering to the values of coordinates. The electrostatic potential ψ is updated if the voltage at any contact has been changed. If a region is contacted twice consistency checks are made. The assignment of gridpoints to contacts can be done without restrictions to profiles and geometry.

Applications and Results

The first application is a npn-transistor. The solution for the equilibrium is provided as an input for the first bias point with Vbe = 0.5V and Vce = 0.3V. Fig. 1 shows the old initial solution, where only the boundary values of the potential at the contacts are updated. Fig. 2 shows the quasi-equilibrium solution for the same bias condition and Fig. 3 the numerically exact solution. The agreement of the potential is excellent.

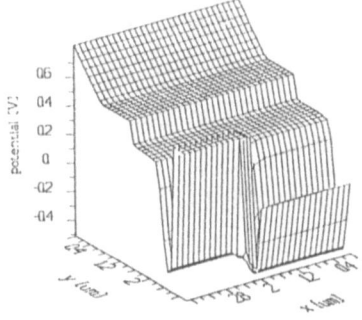

Fig. 1: Conventional initial solution of npn-transistor, Vbe=0.5V,Vce=0.3V

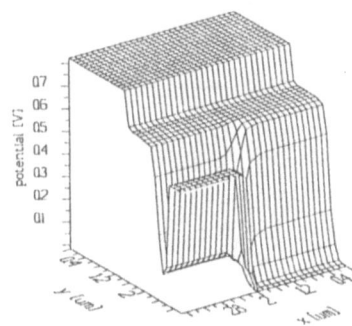

Fig. 2: Quasi-equilibrium approximation of npn-transistor, Vbe=0.5V,Vce=0.3V

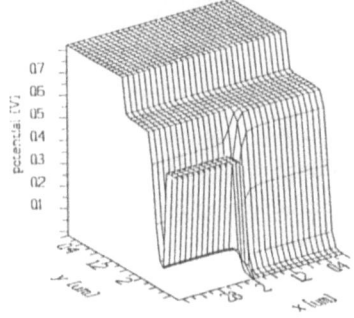

Fig. 3: Numerically exact solution of npn-transistor, Vbe=0.5V,Vce=0.3V

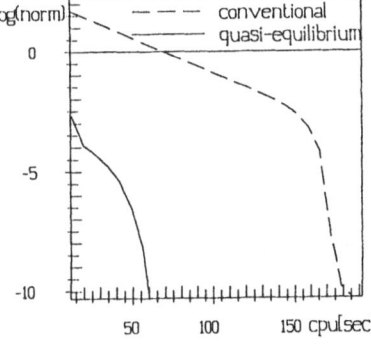

Fig. 4: Comparison of iteration history for npn-transistor, Vbe=0.5V,Vce=0.3V

It is characteristic for the quasi-equilibrium solution that the potential is significantly wrong only in the space charge regions. The improvement of convergence and the corresponding CPU-time reduction is demonstrated in Fig. 4. It shows the evolution of the Euclidean norm with the CPU-time before the first grid update. Note, that the Euclidean norm in Fig. 4 and in the following plots is the norm of the system of all three semiconductor equations. After a grid update, there is no significant difference in the iteration history due to different initial solutions. With the default parameters for the adaptive grid two grid updates are performed. Therefore the final enhancement factor is 2.

The enhancement factor is increased for a reverse biased npn-transistor, e.g. with Veb=2.5V. The quasi-equilibrium solution is again based on the equilibrium. Fig. 5 reveales striking differences of the iteration process due to the initial solutions. Because reverse currents are small the assumptions of the quasi-equilibrium are fulfilled very well. The grid is updated 3 times. Nevertheless, the final enhancement factor is 4.

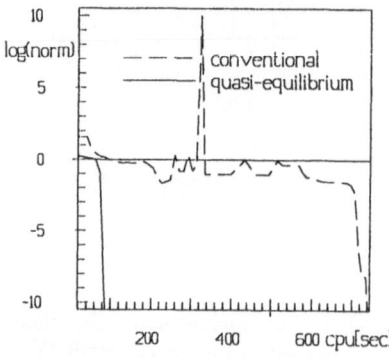 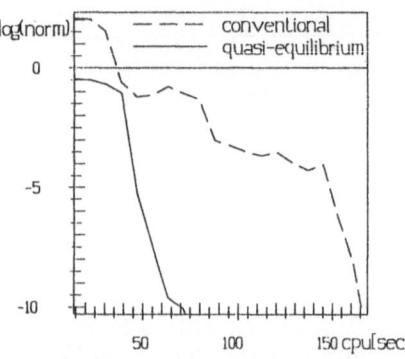

Fig. 5: Comparison of iteration history for npn-transistor, Veb=2.5V

Fig. 6: Comparison of iteration history for n-channel MOSFET, Vgs=2V, Vds=1V

Another interesting example is a n-channel MOSFET. After computing the equilibrium solution the gate voltage Vgs = 2V and the drain voltage Vds = 1V is applied. Fig. 6 shows the history of convergence for the new initial solution compared with the original one. Although 2 grid updates are necessary the reduction of CPU-time is 40 percent. An increment of the electrostatic potential at the gate contact does not affect the iteration process. The convergence behaviour, however, is very sensitive, whether Gauß's law for the electrostatic potential at the interface is fulfilled or not.

References

[1] Franz A.F., Franz G.A.: BAMBI — A design model for power MOSFET's; IEEE Trans. on Computer-Aided Design, Vol. CAD-4, No. 3, pp. 177–189, (1985).

[2] Franz A.F., Franz G.A., Selberherr S., Markowich P.: Finite Boxes — A Generalisation of the Finite-Difference Method Suitable for Semiconductor Device Simulation; IEEE Trans. on Electron Devices, Vol. ED-30, No. 9, pp. 1070–1083, (1983).

[3] Selberherr S.: Analysis and Simulation of Semiconductor Devices; Wien, New York: Springer 1984

[4] Dirks H.K.: Numerical Model Hierarchies; in Engl.W.L.(Editor): Process and Device Modelling; Elsevier Science Publishers B.V (North-Holland), pp. 329–363, (1986)

[5] Polak S.J., Wachters A., Vaes H. M. J., de Beer A., den Heyer C.: A Continuation Method for the Calculation of electrostatic potentials in Semiconductors; Browne B. T., Miller J. J. H.(Editors): Numerical Analysis of Semiconductor Devices; Boole Press Dublin, pp. 65-93, (1979)

Narrow-width Effects in Submicron MOS ICs

F.M. Klaassen, P.A. van der Plas, R.J.W. Debets*, N.A.H. Wils and M.G. Pitt

Philips Research Laboratories, P.O.Box 80000, 5600 JA Eindhoven, the Netherlands

Abstract

The effects of an (almost) birds beak free LOCOS isolation configuration on the characteristics of narrow-width MOSFETs has been investigated by 2-D device simulation and measurements of realized structures. Owing to the occurrence of excess inversion charge peaks at the LOCOS corners, an anomalous behaviour of the subthreshold characteristics and the threshold voltage is observed for device widths smaller than 1 μm. In addition a change of the birds beak length with the active area width results in a nonlinear dependence of the gain factor.

Introduction

Since the birds beak in conventional LOCOS isolation makes it impossible to realize devices with a precise width of submicron dimensions, it has been attempted to reduce its size. Generally, this has been achieved by suppressing the birds beak formation during LOCOS growth and applying a back etch [1]. Fig. 1 shows that the initial birds beak is small. However, owing to increased resistance to bending of the oxidation stack, its thickness γ decreases as the active area width W is reduced (fig.2, [2]). Consequently after etch back three different shapes of field isolation can result: one showing a small birds beak, one with zero-birdsbeak and an overetched field oxide (fig.3a, b and c). Since the latter two configurations are expected to lead to a deterioration of the subthreshold characteristics, their turn-on behaviour has been studied by 2-D simulation and experiment.

Results of 2-D simulation

The 2-D simulator used allows us to take into account the exact LOCOS shape together with the process doping profiles [3]. Fig. 4 gives the calculated mobile charge distribution at threshold condition for the field isolation shapes of fig. 3. For reasons of presentation, however, in fig. 4 the width W has been kept constant. Since in the latter two cases the gate of an active device and the LOCOS edge intersect with a sharp corner, the gate charge induces an excess inversion charge at the corners. Furthermore the charge peaks depend on the length of the gate overlap and the angle of the above corner. In contrast, the presence of a birds beak causes a decrease of charge at the same position (case a). When the gate bias is increased (strong inversion regime) the above peaks become less pronounced. Despite the high value of the corner charges, their effect on the total subthreshold current is relatively small, even for narrow-width devices. Fig. 5 gives the calculated characteristics (dots) for several values of the MOSFET width W. In this case a zero birds beak occurs at $W_0 = 0.6$ μm with an overetched situation

---

*Technical University Eindhoven, 5612 AZ Eindhoven, the Netherlands

at $W = 0.4$ μm. While at high gate bias the current is proportional to the width W, the opposite is true in the subthreshold regime. The latter result has been called the inverse narrow-width effect [4]. In the same figure it is further shown that the calculated results are confirmed by measurements (fully-drawn lines).

Experimental results

Measured results for devices of several batches with different back etch conditions are presented and interpreted on the basis of the previous simulations. Three cases a, b and c are considered, which can be distinguished by a characteristic width W_o, at which the zero-birds beak occurs. The values of W as determined from TEM micrographs are 0.4, 0.6 and 1 μm, respectively.

In comparison to fig. 5 (case b), fig. 6 shows that for case c the inverse narrow-width effect becomes more pronounced. Next, fig. 7 gives the threshold voltage as a function of transistor width for all cases considered. Due to the presence of a birds beak in case a the conventional V_T-behaviour is observed. Since in case c the birds beak is absent for $W < 1$ μm, V_T decreases in the whole range. In fact case b is a transition between the two extremes. Applying a back bias causes the inverse narrow-width effect to disappear. This is caused by a suppression of the excess corner charges due to the effect of the channel stop implant (at larger depth) on the shape of the depletion layer [5]. This generally causes the body coefficient to increase, if W is decreased.

Finally, the birds beak spread also causes some anomaly in the dependence of another major MOSFET device parameter on the active device width. This is the gain constant $\beta = \mu_s C_{ox} W_{eff}/L_{eff}$, where W_{eff} and L_{eff} are the effective electrical transistor width and length, respectively. While in conventional LOCOS processes β is a linear function of the drawn width W with some constant offset ΔW, in the present batches a nonlinear dependence is observed. This is shown in fig. 8, where in all cases considered the β-slope changes for submicron size devices. Generally for $W > 1.5$ μm, a substantial birds beak is present to cause that the drawn transistor width is 0.25 μm larger than the effective electrical width. In this case β is a linear function of W with a constant offset. However for $W < 1$ μm, the birds beak is considerably reduced. Consequently the electrical width gradually approaches the drawn width. However, even in case c the peak corner charges never become so large that W_{eff} becomes larger than W_{drawn}.

Conclusion

Although the above effects are minor from a practical point of view and therefore can be controlled during processing, their existence complicates the determination of the effective transistor width and the formulation of an accurate compact transistor model.

References

1. P.A. van der Plas, Symposium VLSI Technology, p.19 (1987).
2. P.A. van der Plas et al. ESSDERC 89.
3. SEMMY, proprietary non-planar 2-D device simulator.
4. L.A. Akers, IEEE Trans. **ED-35**, p.325 (1988).
5. F.M. Klaassen, Ch.12 in Process and Device Modelling, Elsevier (1986).

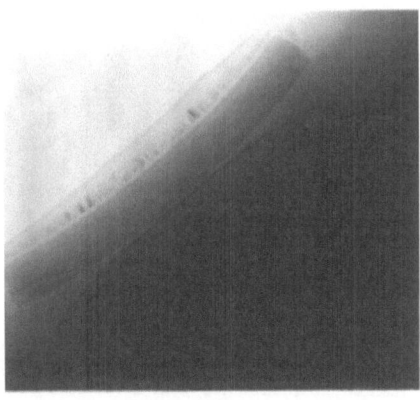

Fig. 1. Birds beak for 600 nm field oxide with an oxide/oxynitride/nitride oxidation stack (magnification 90.000x).

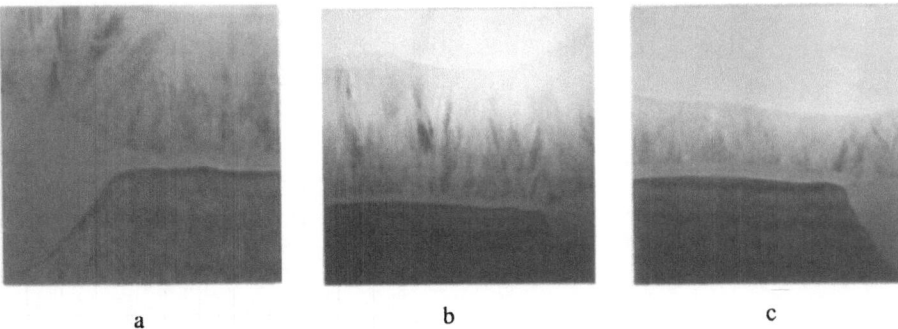

a   b   c

Fig. 3. TEM micrographs of final field oxide-gate region for several widths of the oxidation stack (a: 2 μm, b: 1 μm, c: 0.5 μm; 250.000x).

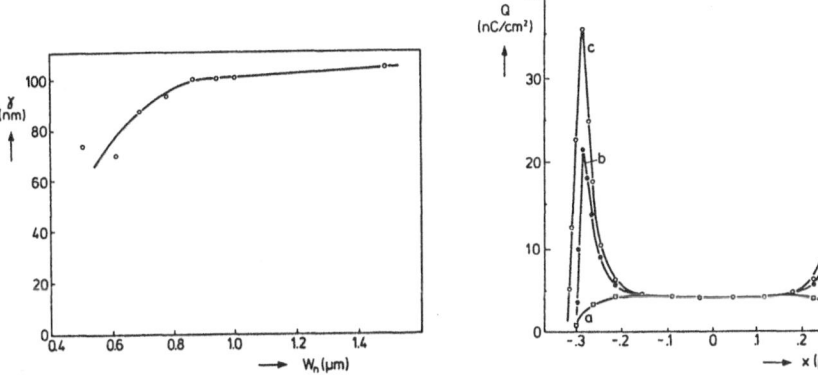

Fig. 2. Thickness of the birds beak as a function of oxidation stack width.

Fig. 4. Charge distribution at threshold condition for isolation configurations shown in fig. 3 with W set to 0.6 μm.

108

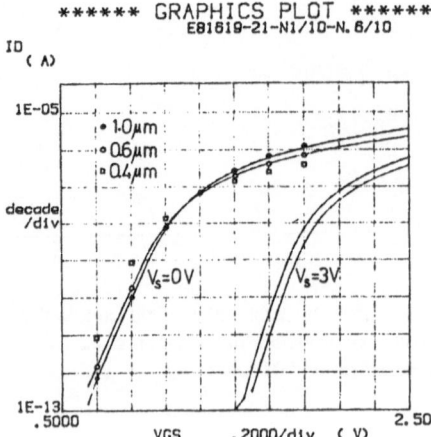

Fig. 5. Simulated (dots) and measured sub-threshold characteristics with zero birds beak occuring at 0.6 μm.

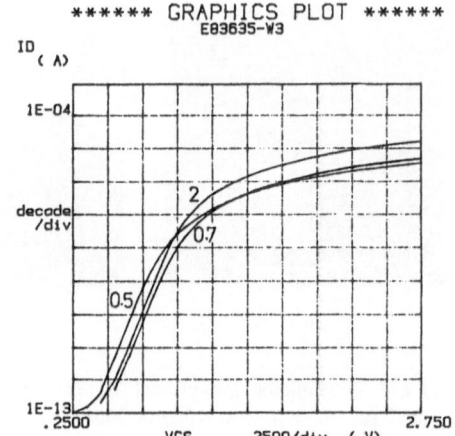

Fig. 6. Measured subthreshold characteristics with zero birds beak occuring at 1 μm.

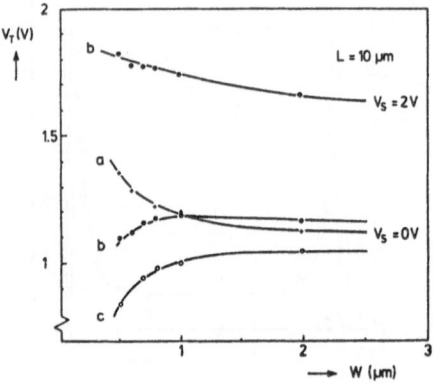

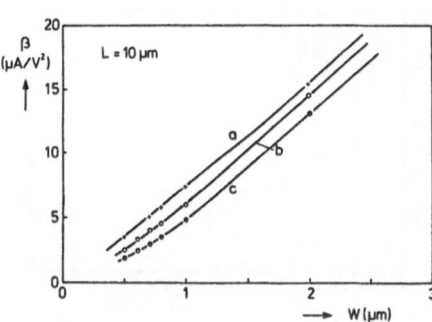

Fig. 7. Threshold voltage vs active area width for different back etch conditions.

Fig. 8. Gain constant vs active area width for different values of back etch.

Simulation of Parasitic Currents and Other Effects in Narrow Channel Devices

M. Thurner

Campusbased Engineering Center
Digital Equipment Corporation
Favoritenstraße 7, A–1040 Wien, AUSTRIA

Abstract – An investigation was carried out into how device isolation will influence the device characteristics. Either a field implant and/or field oxidation is applied to isolate the devices from each other and to define the extent of the channel. The results of that investigation are presented here. The investigations have been carried out by MINIMOS 5, a fully three–dimensional simulation program. The three–dimensional effect of the threshold shift due to parasitic currents for small channel devices has been successfully modeled.

1 Introduction

In today's IC processes, the field oxide growth and field implantation for the channel-stop drastically influence the electrical behaviour of narrow channel MOS devices. For characterizing MOSFET devices, the threshold voltage and current characteristics are most important. We have investigated the dependency of these quantities, including the possible generation of parasitic currents, on the finite channel width as defined by the field implant and field oxide for the channel stop.

A number of authors have reported on the influence of the narrow channel effects on the MOSFET device characteristics [1], [2], [4], [5], [11]. This paper will demonstrate the shift of the threshold voltage due to field implant and field oxide. With different technologies the resulting current densities may increase or decrease.

Some physical and mathematical aspects are presented in chapter 2. The results of our investigations are given in chapter 3 and a short discussion in chapter 4.

2 Physical and Mathematical Aspects

The basic equations which are implemented in the simulation program MINIMOS 5 are essentially 'the' established basic equations which are in common use ([9],[10]). Some modifications due to hot electron transport are reported in, e.g. [7]. The well known boundary conditions and some additions for the nonrectangular interfaces (field oxide) can be found in, e.g. [9], [11]. A detailed discussion on the nonrectangular interfaces which are implemented in MINIMOS 5 can be found in, e.g. [6], [12]. The coupled partial differential equations are essentially solved by Gummel's iterative scheme. The linearized Poisson's equation is solved by a modified CJG solver [3] and the continuity equation by Gaussian elimination.

The doping profile of MINIMOS 5 is described in [9] for two dimensional simulation.

A natural extension in the third direction has been applied for three–dimensional profiles. We have now included in MINIMOS 5 the ability to model field implantation by an additional analytical function in order to investigate the influence on the device characteristics of the doping in the field region, as demonstrated in the next section.

3 Results

For demonstrating the effects at the channel edge we have simulated four devices: with and without field implantation and with and without field oxide. A 3D view of the MOSFET examples is given in Fig. 1: an n–channel MOSFET with $1.2\mu m$ x $1\mu m$ channel, gate oxide $15nm$, n^+- doping of source and drain $1.5 \cdot 10^{20} cm^{-3}$, bulk doping $5.2 \cdot 10^{16} cm^{-3}$. Figs. 1 a and 1 b respectively are without and with field oxide. The profile has been approximated by analytical functions (Fig. 2 a, b). The field implant doping is about $1.5 \cdot 10^{20} cm^{-3}$. The field oxide thickness is $0.15\mu m$. The threshold voltage for the devices without field oxide is $U_{th} = 0.84V$ and for the device with field oxide $U_{th} = 1.1V$ without any channel width effects. The bias conditions for these devices are: $U_S = U_B = 0.0V$, $U_D = 2.0V$, $U_G = 0.5 - 2.0V$. The device characteristics (drain current vs. gate voltage) of the different devices are shown in Fig. 3a ,b and Fig. 4a, b. The solid lines are the results from the 2D simulations and the dashed lines are from the 3D simulations. The horizontal dashed line in the pictures denotes the threshold current.

The characteristic of device 1 which has no field implant and no field oxide is plotted in Fig. 3 a. Due to parasitic currents arising from the changing profile at the channel edge we get a higher 3D current compared to the 2D results. We can see that the threshold voltage of that device is lower than predicted by 2D simulations (intersection of the curves with the horizontal dashed line). The characteristic of device 2 which has a field implant but no field oxide is plotted in Fig. 3 b. The parasitic currents at the channel edge are suppressed by the increased doping so we get a lower 3D current compared to the 2D results. The threshold voltage of that device is higher than predicted by 2D simulations (intersection of the curves with the horizontal dashed line). The characteristic of device 3 which has no field implant but field oxide is plotted in Fig. 4 a. The expected parasitic currents at the channel edge are suppressed by the field oxide. The field oxidation has a very low gradient in shape. The 3D current is only slightly different compared to the 2D results. The threshold voltage of that device is nearly the same as that predicted by 2D simulations, and depends strongly on the position and the shape of the field oxide. Some investigations on the influence of the oxide shape can be found in e.g. [8], [11]. The characteristic of device 4 which has both field implant and field oxide is plotted in Fig. 4 b. The parasitic currents at the channel edge are supressed by the increased doping and the field oxide thus the 3D current is lower compared to the 2D results. The threshold voltage of that device is higher than predicted by 2D simulations (intersection of the curves with the horizontal dashed line). The shift of the threshold voltage to higher values even at 2D simulations without changing the 2D input is due to the reduced channel region by the field oxide.

4 Conclusion

We have demonstrated how the MOSFET device characteristic will be influenced by field implantation and field oxide for narrow channel devices. The parasitic currents at the channel edge can be either increased or decreased by both implant and oxide due to different profiles of field implant and shapes of field oxide. It has been shown that three–dimensional simulation is very useful for the prediction of the device behaviour.

References:

[1] L.A.Akers, **Characterization of the Inverse Narrow Width Effect**, IEEE, Vol.ED-34,No. 12,pp 2476-2484, 1987

[2] L.A.Akers, **Inverse Narrow Width Effects**, IEEE, Vol.EDL-7,No. 7,pp 419-421, 1986

[3] O.Axelsson, **Solution of linear systems of equations;Iterative methods**, Lecture notes in mathematics 574, SMT, 1976

[4] P.Ciampolini,A.Gnudi,R.Guerrieri,M.Rudan,G.Baccarani, **3D simulation of a narrow-width MOSFET.**, ESSDERC Proc., pp 413-416, Bologna, 1987

[5] J.L.Coppee,E.Figueras,B.Goffin,D.Gloesener,F.Van De Wiele, **Narrow Channel Effect on n– and p–Channel Devices Fabricated with the SILO and BOX Isolation Techniques.**, ESSDERC Proc., pp 749-752, Montpellier, 1988

[6] G.E. Forsythe,W.R. Wasaw, **Finite Difference Methods for Partial Differential Equations.**, Wiley, NEW-YORK, 1960

[7] W. Hänsch, M. Miura–Mattausch, **A new current relation for hot electron transport.**, Proc. NASECODE IV Conf.,pp.311-314, Boole Press, Dublin, 1985.

[8] K.K.Husueh,J.J.Sanchez,T.A.Demassa,L.A.Akers, **Inverse Narrow Width Effects and Small Geometry MOSFET Threshold Voltage Model**, IEEE, Vol.ED-35,No. 3,pp 325-338, 1988

[9] S. Selberherr, **Analysis and simulation of semiconductor devices.**, ISBN 3-211-81800-6, Springer, WIEN NEW-YORK, 1984

[10] S. Selberherr, **The status of MINIMOS.**, Proc. Simulation of semiconductor devices and processes, pp 2-15, Swansea, 1986

[11] M.Thurner,S.Selberherr, **3D MOSFET Device Effects due to Field Oxide**, ESSDERC Proc., pp 245-248, Montpellier, 1988

[12] M.Thurner,P.Lindorfer,S.Selberherr, **Numerical Treatment of Nonrectangular Field–Oxide for 3D MOSFET Simulation**, SISDEP Proc., pp 375-381, Bologna, 1988

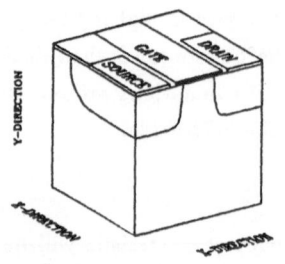 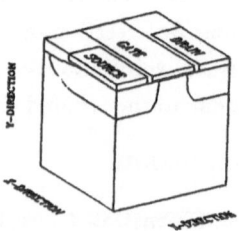

Fig.1: Perspective view of the three–dimensional MOSFET structure without (a) and with (b) field oxide.

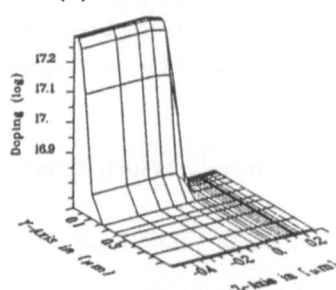 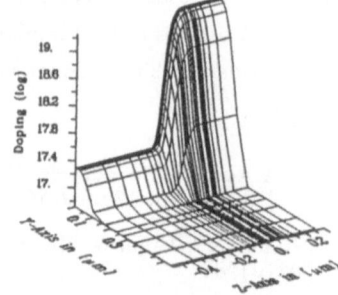

Fig.2: 3D–plot showing the doping in channel–width and depth direction, without (a) and with (b) field implant.

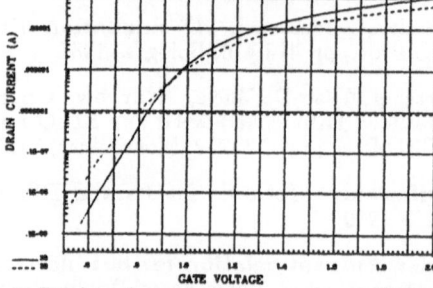 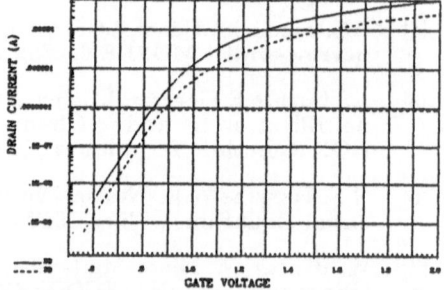

Fig.3: Device characteristics at threshold region (with $U_{DS} = 4.0$ V, $I_{TH} = 8.85 \cdot 10^{-8}$) without field oxide, without (a) and with (b) field implant.

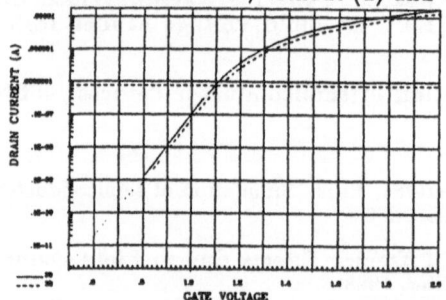 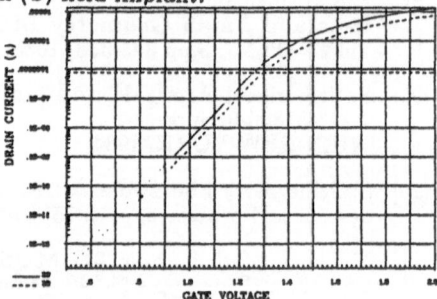

Fig.4: Device characteristics at threshold region (with $U_{DS} = 4.0$ V, $I_{TH} = 7.70 \cdot 10^{-8}$) with field oxide, without (a) and with (b) field implant.

3D Simulation of Parasitic Effects in EPROM Cells

P. Ciampolini, A. Pierantoni, M. Rudan and G. Baccarani

Dipartimento di Elettronica, Informatica e Sistemistica
Università di Bologna, viale Risorgimento 2, I-40136, Bologna

1. Introduction

General-purpose two-dimensional device simulators have become useful design tools, allowing a wide variety of practical structures to be thoroughly investigated. However, semiconductor devices are inherently three dimensional, and a 2D approach may sometimes involve unacceptable simplifications or even suppress the main effect under investigation. This situation typically occurs under the following circumstances:

- Due to the difficulty of properly scaling all device dimensions by the same scaling factor, edge effects are playing an increasingly-important role in miniaturized devices. As an example, the narrow-channel effect in MOSFETs is a typical three-dimensional effect;

- The increased packing density may lead to an unwanted coupling between neighbouring devices and to insufficient device isolation; again this requires a three-dimensional approach to the problem;

- Inherently three-dimensional devices are being devised, such as the buried-electrode dynamic RAM cell, which cannot be represented in two dimensions.

In this paper we present some simulation results obtained using HFIELDS-3D, a 3D device simulator derived from a previous 2D version [1]. In the next section, we discuss the adopted discretization scheme and the efficiency problems associated with the typical problem size. In section 3, some simulation results are presented: a fairly complex and realistic structure is investigated to highlight the occurence of an intrinsically three-dimensional behaviour. Finally, conclusions are drawn in section 4.

2. Physical and numerical aspects

In the present version of the program, the classical drift-diffusion equations are solved using the box discretization scheme [1]. In this sense, the extension of the existing 2D code has been fairly streightforward; nevertheless, some relevant points are worth addressing.

As already stressed, an accurate description of the device geometry is often crucial for a realistic device simulation. Hence, a proper choice of the basic elements of the mesh is important, as it greatly influences the resulting program flexibility. We concede that a fully general approach would require the use of tetrahedral meshes. Our program, instead, adopts prismatic elements, which allow for a drastic simplification of mesh generation and geometry management, without inducing a too severe loss of generality. Moreover, the typical 3D problem size (in the range of some tens of thousands of unknowns) makes the solution of the linear system, which is repeatedly invoked by the program, much more critical than in 2D. Direct solvers can no longer be employed for this purpose, since their performance collapses dramatically as the number of unknowns exceeds a few thousands. Therefore, the direct solver used in the 2D version of the code has been replaced by two different iterative solvers, more suitable for large systems.

At present, the Gummel decoupled procedure is used to solve the Poisson and the electron continuity equations: the linearized form of the former is solved by means of the ICCG [2] method, while the PCGS [3] method is adopted for the latter. Due to the use of iterative solvers, the overall performance improves by an order of magnitude, this making the simulation of realistic devices possible in a reasonable amount of time: the simulation to be presented in the next section, performed on a DEC Vax Station 3200 (which is rated about 3 MIPS), has required roughly half a CPU hour per bias point.

3. Simulation of the dual-EPROM structure

As a test vehicle for our program, we have chosen the complex structure shown in Fig. 1, made of two EPROM cells placed side by side, as a part of a larger array (due to symmetry, only half of each cell is accounted for). The discretization mesh shown in Fig. 1 has been obtained by replicating a 666-node triangular grid over 20 successive planes, this resulting in 13,320 nodes and 23,788 prisms. By properly choosing the orientation of the prisms, this scheme accomplishes a careful description of the transitions between gate- and field-oxide and of the multi-layer gates structure as well: nevertheless, generating such a mesh is fairly easy, simply requiring the use of the standard 2D mesh generator.

To check for the insulation among different cells of the array, we have investigated the behaviour of the parasitic MOSFET located under the field oxide (which has been, on purpose, slightly undersized). With reference to Fig. 2, we have positively biased D_1 ($V_{D1} = 1$ V), keeping all the remaining ohmic contacts grounded, and then we have computed the terminal currents as functions of the 2nd-level gate voltage.

It is worth noting that the behaviour of such a structure can hardly be simulated adopting a conventional two-dimensional scheme. In fact,

- The main flows of the carriers inside the cell and the parasitic MOSFET are roughly orthogonal. Moreover, the structure of the parasitic MOSFET is inherently three-dimensional.

- Being the channel of each cell about 1 μm wide, the accurate evaluation of the current flow must take the actual channel cross-section into account. As shown in [4], the effective channel width is modulated by the applied gate voltage, moving the channel edges well below the bird's beak and giving raise to the so-called narrow channel effect.

Fig. 3 shows the computed turn-on characteristics for both the EPROM cell and the parasitic MOSFET: the solid lines refer the left EPROM cell current, while the dashed lines sum up the currents at both the terminals of the right one (the parasitic current flow is split into two nearly symmetrical parts). As shown, by applying sufficiently high voltages at the common gate, the parasitic MOSFET can be turned on, inducing a significant leakage current: such a current may not be negligible, especially if it drives a sensing device.

In order to account for the electrons injected during the write cycle, three sets of curves are shown, related to different values of the charge stored in the floating gate of the left cell. This results, as expected, in a shift of the threshold voltage of both devices; however, the influence of the stored charge is quite different in the two cases; i.e., threshold shifts are much higher for the EPROM cell than for the parasitic device. Since charge is stored only in the left floating electrode, the right cell remains unperturbed. In other terms the stored charge does not affect the injection of carriers into the parasitic channel (unlike it does for the EPROM cell), but simply pinches off the channel under the left floating gate, as shown in Figs. 4 and 5, which compare the evolution of the inversion layer at the Si-SiO$_2$ interface. As the left floating gate is charged, the right part of the two figures does not change appreciably, this showing that the electron concentrations are similar in the two cases. Because of this even if the charge in the left floating gate increases the carriers can spread from right to left along the parasitic channel thanks to diffusion.

Figs. 6 through 8 illustrate the distribution of the electric potential inside the device, computed at the non-planar Si-SiO$_2$ interface. Fig. 6 refers to the $V_G = 3$ V, $Q_{FG1} = Q_{FG2} = 0$ case and shows inversion layers arising below the gate oxide in the two cells, while the small ridge below the field oxide stands for some carrier accumulation, still insufficient to provide a significant current flow. Raising V_G up to 8 V, the potential distribution at the same interface turns out as in Fig. 7, where the rise of the parasitic current path, connecting the channel region of the two cells , is fairly evident. Finally, Fig. 8 refers to the $V_G = 3$ V, $Q_{FG1} < 0$, $Q_{FG2} = 0$ case, again showing the influence, on both devices, of the charge stored in the floating gate.

4. Conclusions

HFIELDS-3D has been tested on a realistic example: the simulation of the dual-EPROM structure has proved both its geometrical flexibility and its algorithmic efficiency. The obtained results allowed for a deeper insight within the structure behaviour, highlighting intrinsically three-dimensional effects and improving our understanding of the coupling between the cells. The latter may impose severe limits to the integration of similar devices.

As a final remark, one can observe that the tipycal device size is now approaching the validity limits of the physical models. The proposed scheme is suitable in itself for accomodating more refined physical models: nevertheless, we believe that the most critical point for 3D tools is now to achieve a better computational performance and an easier interaction with the user. The adoption of the prism-based discretization scheme allows for a simplified management of geometries and meshes, without limiting the variety of suitable devices too severely , while a remarkable reduction of the solution time has been obtained by means of iterative linear solvers.

References

[1] G. Baccarani, R. Guerrieri, P. Ciampolini, M. Rudan, HFIELDS: a *Highly Flexible* 2-D *Semiconductor-Device Analysis Program*, Proc. NASECODE IV Conf., Dublin, 1985

[2] D.S. Kershaw, *The Incomplete Cholesky Conjugate Gradient Method for the Iterative Solution of Systems of Linear Equations*, J. of Computational Physics, n. 26, 1978

[3] G. Markham, *The CG-Squared method for solving asymmetric systems of linear equations*, TPRD/L/ APM/012 report, 1983

[4] P. Ciampolini, A. Gnudi, R. Guerrieri, M. Rudan, G. Baccarani, *Three-Dimensional Simulation of a Narrow-Width* MOSFET Proc. of ESSDERC 87, pp. 413–416, Bologna, 1987

Acknowledgement This work was supported in part by SGS-Thomson Microelectronics and by the National Council of Research under the "Progetto Finalizzato MADESS". Dr. Peter Mole and coworkers from STC are also gratefully acknowledged for making the ICCG and CGS solvers available in the frame of the EEC ESPRIT-962 (EVEREST) Project.

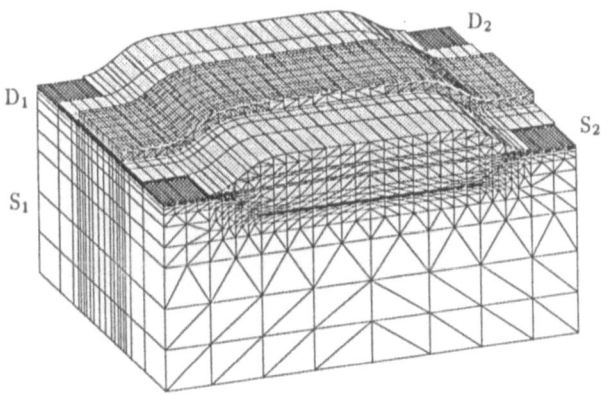

Fig.1

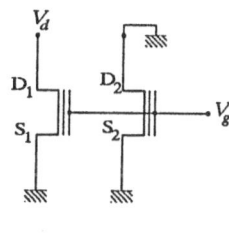

Fig.2

116

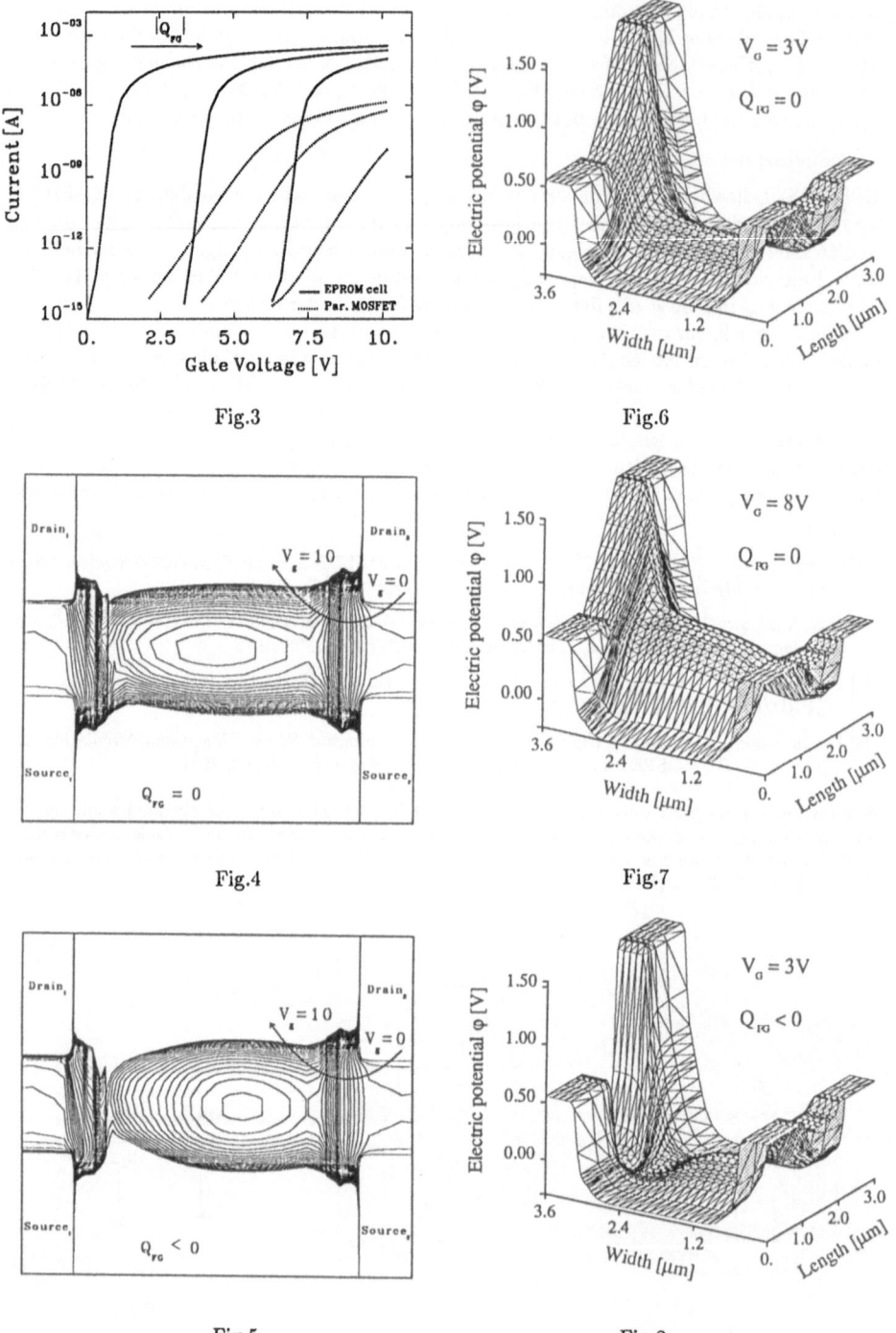

Fig.3

Fig.4

Fig.5

Fig.6

Fig.7

Fig.8

Session 2

Invited Paper

Requirements for Device and Process Modeling of Submicron Devices

S.J. Polak, R.J.G. Goossens *, and N.E.B. Cowern

*Nederlandse Philips Bedrijven B.V., CFT Automation-CPS,
Building SAQ-2, P.O. Box 218, 5600 MD Eindhoven, The Netherlands*

Abstract

In this paper the requirements for device and process modelling for submicron devices are discussed. Some examples are presented to illustrate the origins of these requirements. A framework for continuum-physics simulations is used to further outline the necessary facilities. Throughout the paper problems are considered from a numerical as well as a device-engineer's point of view.

1 Introduction

Any industrial activity is ultimately profit-oriented. In the IC business in particular, it is hard to achieve this goal owing to the very competitive nature of the market. Over the last decades this has led to a spiral of innovation which is unprecedented in industrial history, and which has pushed the minimum dimension of device features into the submicron realm. In such a quickly changing world, the industrial methods and tools also have to change. The miniaturisation of semiconductor devices makes it more and more difficult to understand both what happens during the production process and how the devices function. Also the costs involved in development and production are becoming so astronomical that even large industries have great difficulties bearing them. A third aspect is development time. It is well known that, in this field, being very early in the market is a decisive factor for profits. This paper outlines the role that process and device modelling can play in these problem areas. Of course, modelling forms just one step on the way to industrial success.

The complexity of the physics and chemistry of the processes involved in the production and functioning of semiconductor devices is such that modelling is primarily used to provide insight. This is contrary, for example, to the situation of IC design and circuit simulation, where exact predictions are expected. However, with appropriate background information, there are a number of situations where predictive modelling is successful. The relative abstinence from predictive use of process and device modelling tools seems characteristic of our European culture.

In this paper we shall focus on the requirements stemming from the decrease of the channel length of CMOS transistors into the submicron range. We shall adopt a numerical-software point of view and format the paper accordingly. In every applicable section we shall also explicitly discuss the process or device engineer's point of view. Of course we can only hope to cover some ground and do not make any pretence of treating the subject exhaustively. To set the stage first, in sect. 2, two real-life problems are discussed which are then used as a "connecting thread" throughout the paper. Next we shall present a framework for continuum physics modelling in general, in sect. 3. In the subsequent sect. 4 through 9 the

Philips Research Laboratories, P.O. Box 80000, 5600 JA Eindhoven, The Netherlands; presently with Philips Research Laboratories Sunnyvale and as a Research Associate with Stanford Electronics Laboratory, Dept. of Electrical Engineering, Stanford University.

various elements of this framework are discussed. Section 10 deals with specific software aspects. Finally in sect. 11 we formulate conclusions and present an outlook into the future.

2 Examples

In order to develop a viable submicron CMOS process not only the transistors themselves need to be accurately optimized, but also adequate control of parasitic effects is strictly necessary. For both instances it becomes increasingly important to outline potential difficulties as early as possible during the process development. One would like to use process and device simulation to achieve this. A distinction between two classes of applications is made. In the first example the focus will be on a relatively simple problem within a geometrically complex situation. The second example describes a complex phenomenon under very simplified circumstances.

With decreasing device dimensions the size of the isolation regions is also scaled down. A lot of technological effort has been invested to make this possible. In the case of locos isolation the limiting factor was the size of the bird's beak. A change in the growth and etching conditions minimized the lateral extent under the mask edge. At corners it is much more difficult to control the lateral oxidation, giving rise to a layout dependence. On implanting the source and drain regions, the field oxide and the bird's beak act as a mask for the ions, leading in these corners to an integral amount of dopant ions that is lower than expected. In a subsequent high-temperature step the diffusion in combination with the three-dimensional geometry results in a further very strong decrease of the dopant level at such a locos corner. Consequently depletion layers are much wider and large leakage currents may occur. Ultimately, even a short circuit can occur between the source/drain metallisation and the substrate. Reliable predictions on the nature and the extent of this problem require an accurate, three-dimensional description of the implantation and diffusion as well as a non-planar three-dimensional device simulation.

A different situation holds for the hot-electron-related reliability problem of submicron MOST's. The accumulation of electrons in the gate oxide due to injection from the silicon disturbs the internal charge distribution and hence the electrical characteristics of a MOST. No quick cure may be expected from an analysis of the injection mechanism, because at present this phenomenon is barely understood. Therefore, at present world-wide the effort is aimed at preventing the generation of hot electrons. This is most effectively done by minimizing the electrical field. In particular the peak value is of importance, as the injection probability is likely to depend exponentially on the energy of the hot electrons and hence on the field. Minimizing the peak electrical field requires detailed engineering of the dopant profiles.

In the past the final dopant profile was determined by diffusion over relatively large distances. With decreasing feature size, the amount of diffusion was reduced and hence the initial dopant distribution after implantation and the anomalies occurring during the initial stages of diffusion became progressively important. These anomalies consist of a surprisingly high transient rate of diffusion caused by interaction with the very high density of point defects originating from the earlier implantation. In this case the microscopic processes are still incompletely understood, making it difficult to write down the equations governing the process. This is already true for a simple one-dimensional geometry. For deep submicron processes tight control over the lateral diffusion becomes even more important. One tries to achieve this by rapid thermal anneal, but there the same complications hold.

3 Continuum-physics simulation

Continuum-physics modelling such as device and process modelling is characterised by four topics, namely (a) physical modelling, (b) numerical analysis, (c) software, and (d) hardware. The *physical model* contains our knowledge of the problem apart from our ability to analyse the situation by computer simulation. In mathematical terms, it can be broken into (a) a description of the geometry of the domain of the simulation, (b) the partial differential equations, boundary and interface conditions, which govern the problem, and finally (c) parameters and parametric functions.

The differential equations cannot be solved directly. But by segmenting the simulation domain into subregions (mesh elements) and by making assumptions on the behaviour of the solution, one can replace the small set of differential equations with an infinite number of unknowns by a large set of non-linear algebraic equations with a finite number of unknowns. Generally, one uses an iterative scheme to find the solution of the latter set. Each iteration consists of linearising the equations around the estimated solution and of finding an update of the estimated solution by solving the linearized system of equations. Summarizing, the *numerical analysis* for this class of problems always contains in some form or another (a) discretisation, (b) non linear solving, (c) linear solving, and (d) in the future, but rarely at present, optimisation.

The *software* involved may be subdivided into (a) a preprocessor, such as a geometric modeller with the facility to easily parameterise shapes, (b) the numerical software containing the implementation of the algorithms involved, and (c) visualiser to display the results either in numerical or in graphical form. We shall not discuss the hardware development because for continuum-physics simulations the development is not a key issue. This point of view holds even though the advent of cheap powerful workstations is changing the field of computer simulations dramatically.

4 Geometries

Any modelling activity starts with building a physical model of the reality. This model is then converted into a mathematical model. Based on this model a software package will be chosen or made. The first true modelling action then is the definition of the geometry of the simulation domain.

In the past it has been customary for process and device simulation software such as Pisces [1], Curry [2], Suprem IV [3], and Composite [4], to have a preprocessing module specially developed within the package. Also in other fields, like stress analysis with Ansys [5] or Nastran [6], or computational magnetics with packages such as Tosca [7], Carmen [8], and Paddy [9], it is customary to have a package-specific preprocessor for the definition of geometries and parameters.

However, the definition of geometric entities is of a very general nature. Consequently, geometric modelling is an important part of today's computer-aided design. There is a very large effort in money and manpower in the field of geometric modelling. As a result software such as Geomod [10] is far more powerful than the geometric modelling facilities in the preprocessors of the continuum-physics modelling software. On the other hand, those preprocessors, as well as handling the geometry, also facilitate the specification of problem-specific parameters, functions and meshes. Basically the specification of parameters is a question of labelling geometric entities, possibly by pointing at the screen, and connecting the parameters to the labels. The missing facilities require little programming compared to the large effort involved in producing geometric modelling software. Interfacing data structures of totally different software environments, which is needed to hook up the numerical software to the geometric-modelling software, requires a considerable effort. Yet we expect that in the future geometric-modelling software will replace the application-specific preprocessors. Already we see more and more interfacing of CAD systems such as Intergraph [11] or Unigraphics [12] with simulation packages like Ansys or Nastran. For submicron devices such modelling facilities would be very useful to represent and visualise the shapes of field oxides, trenches, interconnect etc.

Our first example of section 2 illustrates this point precisely. From SEM pictures the shape of oxide regions can easily be extracted and described, for instance, as a polygon. To describe a 3-D locos corner, one would first like to translate this 2-D polygon along one edge, subsequently rotate it with a specified curvature through 90° to produce the corner, and next translate it again to produce the other edge. During the 90° rotation some parameters describing the polygon may have to be changed to account for excessive bird's beak formation in the corners. These actions belong to the standard suite of options offered by good geometric modellers.

Summary of requirements interfacing (3-D) simulation software with non-specific geometric modellers, including a solution to the parameter/labelling problem.

5 Equations, boundary and interface conditions

Here, device and process modelling have to be discussed separately. For device modelling, the critical step is the inclusion of energy-related physical effects such as avalanche multiplication, hot-electron injection, velocity saturation, ballistic effects, and tunneling. Until recently, device-simulation software only solved Poisson's equation and the two continuity equations for the carriers. It is generally felt that the inclusion or modification of just a few equations should cope with the above-mentioned effects. In the process modelling field, the physical situation is generally too complex to allow a detailed description of all the microscopic processes involved in a given problem. In general, the most common limitation is the finite available knowledge of the basic physics and chemistry involved. This is particularly true in the context of sub-micron technology, where we see the emergence of new physical effects whose origins are not always understood. This situation calls for the formulation of new equations, within a rather modest approach which reflects the basic limitations of current knowledge.

5.1 Device Modelling

For a large number of years the mathematical model for device simulation has been basically constant, consisting of Poisson's equation and the continuity equations for the carriers. Progress was made by adding features like transient simulation, ac analysis, external circuitry, adaptive meshing, and non-planar structures. Recently, one has (successfully?) tried to add avalanche generation to the continuity equations.

The last two years the emphasis is on the the effects of local heating of the carriers and ultimately on non-Maxwellian energy distributions. Various approaches exist to tackle this problem. One can explicitly write down the extra equations describing the "flow of the solution" in energy space in the form of a Boltzmann equation. In general the resultant set of equations is too complex to be manageable. Assuming simplifying conditions, one can derive the old set of three equations plus two new equations for the effective temperature of the carriers [13]. Alternatively, the whole carrier-transport problem can be solved by Monte-Carlo techniques. These calculations involve massive amounts of CPU-time on the fastest computers and, therefore, are not practicle for engineering purposes. Summarizing, concensus seems to exist on the simplest and the most complex formulation of the equations, but differences exist in the approach to fill the gap between these two extremes.

5.2 Process Modelling

In the first investigative phase of model development, quite complicated sets of interacting model equations may have to be solved numerically. The solutions may provide a deeper insight during research, or may simply represent a means of reproducing experimental observations. In either case, it is likely that a number of alternative formulations will be attempted in the initial stage. Once sufficient insight has been obtained from this investigation, a simplified model having the required general behaviour is derived for use in practical technology simulation.

This procedure is not, of course, limited to silicon technology modelling. It is used in many situations where predictions are required even where one's understanding is incomplete - e.g. weather forecasting, enhanced oil recovery, nuclear-reactor modelling. However, it plays a significant, but often unrecognized role in the simulation of submicron silicon processes. On the one hand the physical effects at this length scale are very complex, yet on the other hand process engineers have a need to extensively use the simplified model within its validity range.

In the brief sections below, we describe the requirements for simulation of three types of silicon processing step: ion implantation, diffusion and oxidation. We arbitrarily treat the processes of layer deposition, etching and lithography as outside the scope of the present discussion.

5.3   Ion implantation

Full modelling of the process of ion implantation requires a detailed description of all scattering processes. This would be used to generate the spatial distribution of ions implanted into an arbitrary structure, together with the spatial distribution of atoms displaced by the implantation process. These latter are significant, for example, for modelling later processing steps such as diffusion. The implantation solution can be carried out either by numerical solution of the Boltzmann transport equation for the ions and recoils, or by the equivalent Monte Carlo method. The increased role of channeling effects at low implant energies will require the inclusion of crystal models within these transport calculations.

Approximate solutions applicable to many situations can be obtained by solving for the spatial distributions formed by a point source, and using a modified convolution method to simulate implantation into real structures. This approach provides a substantial reduction in computational requirements which becomes essential for routine, 3-D simulations.

From a physical point of view, ion implantation is one of the best understood processes within silicon technology. Computational limitations such as in handling implantation into complex geometrical structures, are the main constraints on accuracy in this case. Further work on removing these limitations is important, for example, for simulating high-energy implantation into device structures.

5.4   Diffusion

The requirements for modelling dopant diffusion in silicon are becoming much clearer as a result of recent research activities. Considerable success is being achieved in understanding diffusion processes in terms of the detailed interactions which take place between dopant atoms and point defects in silicon, and a major review of this topic has recently been presented by Fahey and coworkers [14].

The essential requirement is the simultaneous solution of diffusion equations for the transport of dopant atoms, point defects, and of complexes between the two. Such equations can be simplified by the use of quasi-steady-state assumptions for the point-defect distribution. This is motivated by the observation that the diffusivity of point defects is many (8-10) orders of magnitude larger than the one for the dopant atoms. This approach can be used to eliminate explicit use of the point-defect diffusion equations. However in the submicron regime, it is not generally satisfactory to assume steady-state conditions. This arises because the low thermal budget, used to restrict the lateral dopant redistribution, can also prevent diffusion processes from relaxing from their highly non-equilibrium post-implantation conditions. Adequate simulation of transient diffusion, enhanced diffusion due to process steps such as oxidation, nitridation, silicidation and doping from polysilicon layers, are important requirements for submicron technology. All these cases demand an improved approach to the modelling of dopant diffusion under non-equilibrium conditions. This point is covered by our second example and will be elaborated on in section 5.6.

5.5   Oxidation

The local oxidation of silicon to form isolation structures continues to be a standard approach for submicron technology. However, sophisticated schemes are increasingly used to reduce bird's beak formation, and structures such as oxidised trenches have a number of important applications in the submicron regime.

Temperature and partial water pressure are important parameters in tailoring the shape of grown isolation structures. This illustrates that mechanical stresses play an important role in determining the oxide growth. They may remain after processing and can then also play a significant role in later processing steps such as diffusion. Furthermore, stresses can affect device reliability.

Requirements for oxide-growth simulation are the solution of the oxide-flow problem, the solution of the oxidant-diffusion problem in the oxide, and a treatment of the actual oxidation reaction at the interface. All these problems may be coupled to the mechanical-stress field, which exists in the total, multilayer device structure. Considerable work remains to be done to clarify the required physical models. Furthermore, the medium itself may be chemically inhomogeneous, due for example to the use of a pre-nitrided pad oxide prior to locos growth.

5.6 Transient boron diffusion: an example

Assuming that one is going to solve the equations for both dopant ions and point defects, it is necessary to recognise the "stiff" nature of the problem: as said, diffusion coefficients for the mobile point defects involved in the problem are many orders of magnitude (about 8-10) larger than the effective diffusion coefficients used in traditional diffusion simulations.

A set of diffusion equations has to be solved for the dopant atoms, point defects and any immobile complexes present in the problem. Dopant atoms exist in substitutional form, in mobile interstitial (pair) form, and as part of immobile clusters. In general these three components have to be treated separately, as do the free self-interstitials and vacancies which drive the diffusion process. This gives us five coupled diffusion equations, a possible formulation of which is

$$\frac{\partial C_s}{\partial t} = \nabla \cdot (D_s \nabla C_s + Z_s C_s \nabla \log n) - G_t \tag{1}$$

$$\frac{\partial C_t}{\partial t} = \nabla \cdot (D_t \nabla C_t) + G_t - G_c \tag{2}$$

$$\frac{\partial C_c}{\partial t} = G_c \tag{3}$$

$$\frac{\partial C_i}{\partial t} = \nabla \cdot (D_i \nabla C_i + Z_i C_i \nabla \log n) + G_i - R_{iv} \tag{4}$$

$$\frac{\partial C_v}{\partial t} = \nabla \cdot (D_v \nabla C_v + Z_v C_v \nabla \log n) + G_v - R_{iv} \tag{5}$$

where the C_s, C_t and C_c are the concentrations of dopant in substitutional, interstitial and cluster form respectively and they occur once for every dopant species. The quantities C_i and C_v are the concentrations of silicon self-interstitials of ionic charge Z_i and vacancies of charge Z_v. The D_s, D_t, D_c, D_i, D_v are the corresponding diffusivities, which depend only on the diffusion temperature. The n is the concentration of electrons, given by

$$n = \frac{1}{2}\left(C_{net} + \sqrt{C_{net}^2 + 4n_i^2}\right) \tag{6}$$

$$C_{net} = \sum_p (Z_p C_p) \tag{7}$$

where n_i is the intrinsic carrier concentration, and C_p and Z_p are respectively the concentration and charge state of the p'th impurity or point defect ($Z_p = -1$ for acceptors, $p = +1$ for donor).

Additional equations are required to describe clustering (G_c), the formation of mobile interstitial dopant atoms (G_t), the recombination rate of interstitial dopant atoms and vacancies (R_{iv}), and boundary conditions for the problem. As an example, the quantity G_t is the transition rate for conversion of dopant atoms from substitutional to interstitial form. This is important because interstitial dopant atoms diffuse many orders of magnitude faster than substitutional atoms. The rate G_t depends on two possible reactions, namely "kick-out" and dissociation and is therefore related to C_i and C_v.

The problem can be solved in two ways: (1) in full generality allowing one to investigate the mathematical behaviour of different formulations of the problem, and (2) by making approximations to allow a simpler formulation, like solving in only one dimension and not explicitly allowing the point defects to diffuse.

In our initial (traditional) approach we have used the second method. Good agreement can be obtained with 1-D experimental results on boron transient diffusion, for the first time with such a model. However, technology requirements (e.g. drain-engineering) call for 2-D predictions. This implies a full solution for which the general approach is required.

Summary of requirements For device modelling the various approaches to include energy-related effects have to be evaluated. Validity ranges of assumptions have to be established. Expressions for the coefficients involved have to be found. For process-modelling research the full set of equations should be explored and verified by experiment in order to outline simplified models for engineering purposes. Totally new models have to be developed for new processing steps.

6 Discretisation

Discretising a partial differential equation means two things, namely constructing a grid and replacing the derivatives by differences, e.g. with box methods, finite-element methods (FEM), mixed FEM, boundary-element methods etc.

Both in device and process modelling two kinds of grids are predominantly used, namely those based on triangles and those based on quadrilaterals. Three types of grid construction exist, by hand, or a priori generated, or adaptively generated. The first method implies that with the geometry specification also a mesh has to be given in the user input. This has been customary in process and device modelling software.

"A priori" generated meshes are constructed by the software on density information given by the software user. However, the user often does not have sufficient knowledge about the solution to give adequate information for the generation of a mesh. During the solution process increasingly relevant information is acquired about the solution. With this information it is possible to improve the mesh. This procedure is termed "adaptive meshing". It implies that problem definition only means specifying regions and parameters. Again this explains the possibilities of using general geometric-modelling software. The example in sect. 2 clearly shows how difficult it is to predict the behaviour of the solution for submicron devices. It is therefore obvious that such adaptive meshing is an absolute necessity in this field. One simply cannot give the meshes by hand for such problems. At present adaptive meshing is developing into the standard procedure. However, as with all software dependent developments in this field, progress is slow. In practice the engineer will, as long as good adaptive meshing is not available, be served by an *a priori* mesh generation for 3-D problems. This at least would be helpful.

On a mesh we have to use differences in place of derivatives. We may distinguish three methods for this, namely finite differences (based on Taylor series), box methods (based on a Green's Theorem), and finite-element methods. It is impossible to discuss these methods for process and device modelling in depth in this paper. We shall however make some fundamental observations.

For the device modelling, i.e., Poisson's and two continuity equations, an extensive discussion can be found in [15]. At the time there seemed to be only one usable method (most used as well) that satisfies some form of discrete-current-conservation principle. In the meantime it is shown in [16] that a mixed FEM in one dimension, with zero recombination, is exactly the Gummel scheme. In [17] the problems with non-zero recombination in 1-D is treated and in [18] it is reported how the same problems in the 2-D case are solved.

The mixed FEM also satisfies a "strong discrete-current-conservation" principle. This method has two advantages over the box method. Firstly, it solves for current densities as a primary unknown and, secondly, it defines current densities and carrier concentrations consistently over the whole device. The first fact implies that fewer nodes are necessary for the same accuracy in the current densities. However, the situation is not so simple as to make this a conclusive observation. The same number of nodes gives more unknowns (about twice in the case of triangles) than in the box method. Balancing this is the fact that the coupling in the matrix is much smaller, namely five points in 2-D for triangles and seven points in 3-D for tetrahedra.

The second advantage is essential for adaptive meshing. Of course, one may always interpolate, smooth or use some other method to find intermediate values. We think, however, that a consistent way of discretising and interpolating is essential to understand the behaviour of the method. In box methods it can easily be seen that an extension of the 1-D reasoning will lead to a non-unique interpolated charge density.

More generally speaking, the mixed finite-element method gives a system of possibilities for problem solving. One may, for instance, choose different basis or test functions or vary the quadratures used.

Another example is the simulation of coupled As/P diffusion in a MOS source-drain structure. In a highly doped n-type source-drain structure, it is usual for a low-concentration of phosphorus (P) to diffuse in the presence of a very steep As concentration front. The interaction of the P with the heavy As doping leads to a strong drift flux of P atoms. Assuming simple equilibrium diffusion, the P concentration C can be written in one space dimension in the form of a parabolic equation

$$\frac{\partial C}{\partial t} = \nabla \left\{ \lambda \left(\nabla C + ZC \nabla \log n \right) \right\} \tag{8}$$

where n is the electron concentration. This equation can be re-written as

$$\frac{\partial C}{\partial t} = \epsilon \nabla^2 C + \nabla C + \theta C \tag{9}$$

$$\epsilon^{-1} = \nabla \log \lambda + Z \nabla \log n \tag{10}$$

$$\theta = \nabla \cdot (\lambda \nabla \log n) \cdot \frac{\epsilon}{\lambda} \tag{11}$$

$$\lambda = \sum_{j=0}^{2} \lambda_j \left(\frac{n}{n_i} \right)^j \tag{12}$$

and n_i is the intrinsic carrier concentration at the anneal temperature. For a typical source-drain diffusion one may have $\epsilon \sim 0.0025$ on the length scale of the simulation ($\sim 1\mu m$).

The essence of the space-discretisation problem is already present in the simplified equation $\epsilon u_{xx} + u_x = 0$, with this $\epsilon = 0.0025$. If we discretise this with a central difference method for a mesh width h, one row of the matrix, apart from a factor h^2, will be of the form (..0.. $\epsilon - h$ -2ϵ $\epsilon + h$ ..0..). To ensure that accuracy is reasonable and that no spurious oscillations are found, it is necessary that the off-diagonal elements $\epsilon - h$ and $\epsilon + h$ are positive for all situations. Also the matrix should be row-diagonally dominant. Together this implies the requirement $h < 0.0025$. In one dimension with a homogeneous mesh this results in 400 mesh points for the length of $1\mu m$. This is not necessary of course because inhomogeneous meshes can be used. Nevertheless a large number of points will be necessary. For one dimension this presents no problem. But for submicron channel lengths, 3-D process modelling will be required in the near future. Even if merely the equivalent of fifty points in the x, y, z directions were needed this would only permit simulations to be made in extremely rare instances.

In device modelling a similar situation exists and is handled using the so-called Scharfetter-Gummel scheme. This scheme is very well understood in 1-D from the point of view of numerical mathematics. In more than one dimension we come back to the mixed FEM (in one dimension exactly Scharfetter-Gummel). In any case special discretisation schemes have to be used in more then one dimension and to our knowledge little attention has been paid to this so far.

Summary of requirements: Good mesh generators in 3-D and good error control for adaptive grid generation. Discretisations giving better-conditioned matrices for the linearised problems, giving good accuracy with a smaller number of nodes. As this is the basis for all simulations a continuing effort is needed from the numerical-analysis community.

7 Non-linear solving

If a sufficiently good estimate of the solution of the problem is known, the Newton algorithm for non-linear solving converges quadratically. Then a very accurate solution can be found within a reasonable computing time. However, if such a "good estimate" is not at hand, numerical analysis has surprisingly little to offer that can be used to construct fast algorithms. It is more common sense than numerical mathematics that is used in such algorithms. Often, the Newton method is extended by damping strategies such as presented in [19]. However, if physical knowledge can be used to construct good estimates, it is superioir to any other method to get close enough to the solution for Newton's method to converge quadratically.

Many examples from the field of continuum-physics simulations can be given to demonstrate this point. One example is provided by Poisson's equation as used in device modelling. In its simplified form the problem reads $10^{-3} \cdot u_{xx} = \exp(40u) - 10^9$. The behaviour of the nonlinear equation is already characterised by $0 = \exp(40u) - 10^9$, where now u is a single number. Suppose we were to use $u_0 = 0$ as an initial estimate for the solution. The Newton corrections are given by $(1/40)\{1 - \exp(-40u) \cdot 10^9\}$, where u is the last approximation. One therefore finds $u_1 = (1/40) \cdot 10^9$. Next, the updates will practically be $1/40$ for a large number (10^9) of iterations. Of course a correction of 10^9 should not be accepted. This is where common sense comes in. In practice, charge neutrality is assumed to construct the initial estimate. In the simplified problem this means taking $u_0 = (1/40) \cdot \log 10^9$ instead of zero, in this case immediately yielding the correct solution. However, also for the actual problem this has proven to be a good strategy. For completeness' sake: via asymptotic-expansion theory a mathematician might have come to the same initial estimate.

Another central issue is the choice between solving the equations involved simultaneously or by successive substitution (e.g. Gummel, [20]). The decisive factor in this choice is the "coupling" between the different partial differential equations or (in physics terminology) the sensitivity of the different phenomena with respect to small changes in other phenomena. If the sensitivity is high, a successive substitution method may take infinitely many iterations to converge. If the sensitivity is low a few iterations may suffice, thus making successive substitution very advantageous. As usual in numerical analysis "high" and "low" are black magic and it is difficult to make them more precise.

Summary of requirements: Better initial estimates for starting the nonlinear solution processes. Such initial estimates are best found from knowledge about the physics! It is also important to establish criteria for switching from decoupled to coupled solving in order to achieve much faster software. Here also, knowledge about the physics has to play an important role.

8 Linear solving

In the heart of most (non-multigrid) numerical software for continuum physics simulation is a linear solver. This code contains an algorithm for the solution of a set of linear equations, $Ax = b$, with A a matrix that describes the linearised dependencies, x the unknown update vector, and b the right hand side containing the residuals. Until the mid 1970's some variant of Gaussian elimination was the most robust method available to solve this equation. Not much development seemed likely in that situation. Then, initiated in [21], a new class of methods was discovered, the so called preconditioned iterative methods. The basic idea was to approximate A^{-1} in a cheap way by a matrix that we denote $\tilde{A}^{-1}$. Then for $\tilde{A}^{-1}Ax = \tilde{A}^{-1}b$ an iterative method is used. Iterative methods that were slow on $Ax = b$ or only fast on a small class of such problems, were found to become very fast and reliable for wide classes of problems. However, mathematically rigorous foundations are far more difficult to find than good or reasonable preconditioned iterative methods. At the moment we are confronted with a new, ten-year-old form of black magic, i.e. finding good combinations of preconditioners and iterative methods. As the speed of this algorithm is one of the decisive factors (the others are nonlinear solving and setting up the matrix and right hand side) it is crucial to pay considerable attention to this problem. For the solution of the discrete equations arising from Poisson's equation in device modelling we employ the usual ICCG algorithm [21]. This in our opinion is optimal within the class of methods under consideration. For the coupled Poisson's and continuity equations or for the continuity equations separately, we use a preconditioned CGS [22]. However, for all other problems involved fast new methods are needed. This linear algebra problem, probably because it is quite separate from the physics involved, does not receive sufficient attention.

It should be stressed that the process of finding such methods is slow; it may take a year or more before one or two expert researchers in the field of linear algebra find a really good combination (if at all) for a certain problem class. However, without a fast linear algebra kernel a simulation package as a whole can not be fast.

Summary of requirements: in the coming years one may hope for the availability of preconditioned iterative solvers in the form of library software. General theory can only be expected to evolve slowly. Much effort should be invested in the optimization of specific preconditioned solvers for particular problem classes.

9   Visualisation

This also is a subject that is not specific for device or process modelling. In continuum physics in general visualisation is the displaying of scalar fields in the form of contours or mountain landscapes, of vector fields with arrows, the 3-D displaying of cross-sections such as cube surfaces in perspective, and 1-D graphs such as I-V characteristics.

Although there are no special requirements with respect to visualisation for device and process modelling, the present facilities known to us are not sufficient for a fast and easy investigation of results. The field of visualisation is also evolving as a subject in its own right. The coming years will certainly bring package- and application-independent visualisation software.

10   Software

Contrary to the revolutionary hardware developments over the past decade, software development is slow. This is where we find the largest bottleneck in satisfying the needs of submicron modelling. As mentioned in sect. 3, we may subdivide the software in three functional parts: preprocessing/geometric modelling, the numerical kernel, and visualisation.

Preprocessing and visualisation are not specific for process or for device modelling. Independent software is developed outside the device- and process-modelling community. Therefore we should concentrate on the numerical software forming the kernel of the application packages and obtain the non-specific parts elsewhere. The problems here are that in developing software one seems always too slow, too expensive, too late, and too inflexible.

We have developed a fourth-generation-package development environment that alleviates these problems to a high degree. The environment called MAMMY [23] consists of libraries, a Package Designer Language (PDL), and a compiler for the PDL. In the library we have an interactive preprocessor, a postprocessor, data management routines, and a FORTRAN core manager.

The PDL facilitates the definition of data structures such that, with the help of the data management routines, it is easy to define the interface with other pre- and postprocessing software. The PDL further makes it possible to program algorithms without the usual overhead of data handling. The compiler is in C and translates PDL into FORTRAN. Furthermore it makes the changing of models involved very simple. At present we are testing this environment.

A separate issue is the integration of process modelling, device modelling and circuit simulations. The interfacing and ultimately automatic combination of the different simulation software is not specific for submicron modelling. However it should be stressed that for this problem class it is needed on a relatively large scale.

Summary of requirements: Making software requires huge amounts of mantime. This has to be recognized. Much more effort is needed to make the software that satisfies the requirements that are the subject of this paper. Process modelling, device modelling and circuit simulation form a suite that should be available in an integrated way.

11 Conclusions

So far we have considered a number of technical requirements. However there is a less technical aspect that in our opinion is crucial for the optimal further development of process and device modelling. In the past it has been customary in our culture to develop a thorough understanding of the physics first. Then a complete physical model is defined. With this model numerical methods are sought and software is developed. However, acquiring a complete understanding of the basic physics usually takes a long time. Often it is possible to develop algorithms and software long before this knowledge has been fully gathered. It then has to be recognised that certain parameters are not yet known. It is possible to a certain extent to fit these parameters to give results coinciding with known, preferably measured data.

The procedure has two effects. Already useful predictions can be made with software that in this way contains combinations of physical knowledge and fitting. Also a synergetic interaction between the growing knowledge about the physics and the results of the computer simulations is possible. Some very successful electronics companies are – perhaps unconciously – using such procedures in a very effective way. It is necessary to learn to handle the combination of physical knowledge, data fitting and software development in a conscious manner.

References

[1] M.R. Pinto, C.S. Rafferty and R.W. Dutton, 'PISCES II, Poisson and continuity-equation solver', Sept. 1984, Stanford Electronics Laboratory, Dept. of Electrical Engineering, Stanford University.

[2] 'CURRY user reference manual, version 8.1', Dept. for Continuum-Physics Simulation, CFT, Philips, Eindhoven, The Netherlands.

[3] M.E. Law, R.W. Dutton, 'Verification of Analytic Point Defect Models using Suprem IV', IEEE Trans. Computer-Aided Design, **CAD-7(2)**, 181(1988).

[4] J. Lorenz, J. Pelka, H. Ryssel, A. Sachs, A. Seidl and M. Svoboda, IEEE Trans. Electron Devices **ED-32**, 1977 (1985).

[5] G.J. De Salvo and R.W. Gorman, 'ANSYS, Engineering Analysis System User Manual', May 1, 1989, Swanson Analysis Systems, P.O. Box 65, Houston, PA 15342, USA.

[6] 'The NASTRAN Theoretical Manual: Level 15.5', ed. R.H. MacNeal, The MacNeal-Schendler Corporation, Los Angeles, 1972.

[7] J. Simkin, C.W. Trowbridge, 'Three-dimensional nonlinear electromagnetic-field computations using scalar potentials', IEE Proc.(B), **127(6)**, 368-374(1980).

[8] 'The CARMEN Reference Manual', Vector Fields Ltd., 24 Bankside, Kidlington, Oxford OX51JE, England.

[9] 'PADDY Reference Manual', Dept. for Continuum-Physics Simulation, CFT, Philips, Eindhoven, The Netherlands.

[10] 'GEOMOD Reference Manual', IDEAS Level-3, SDRC, SDRC Headquarters, Milford, OH.45150-2789.

[11] 'Intergraph, Mechanical Engineering Design System User Guide', Intergraph Corporation, Huntsville, Alabama 35807.

[12] 'Unigraphics, Design Module, Version 6.0', MDC, Cyprus, CA 90630.

[13] M. Rudan, F. Odeh, and J. White, 'Numerical solution of the hydrodynamic model for a one-dimensional semiconductor device', COMPEL, **6(3)**, 151-170(1987).

[14] P. Fahey, P.B. Griffin and J.D. Plummer, Rev. Mod. Phys. **61**, 289 (1989).

[15] S.J. Polak, C.J. den Heijer, W.H.A. Schilders, and P. Markowich, IJNME, **24**, 763-838(1987).

[16] F. Brezzi, L.D. Marini, and P. Pietra (1987), 'Two-dimensional exponential fitting and applications to semiconductor device equations', Publ. No. 957, Consiglio Della Ricerche, Pavia, Italy.

[17] S.J. Polak, W.H.A. Schilders, and H.D. Couperus, 'A finite element method with current conservation', Proc. SISDEP-88 Conf. Bologna, G. Baccarani and M. Rudan (eds), pp.453-462.

[18] S.J. Polak, 'Mixed FEM for $\Delta u = \alpha u$', Proc. Conf. Oberwolfach, Mathematische Modellierung und Simulation elektrischer Schaltungen, 1988.

[19] R.E. Bank and D.J. Rose, 'Global approximate Newton methods', Numer. Math.,**37**,279-295(1981).

[20] H.K. Gummel, 'A self-consistent iterative scheme for one-dimensional steady-state transistor calculations', IEEE Trans. Electron Devices, **ED-11**, 455-465(1964).

[21] J.A. Meijerink and H.A. van der Vorst, 'An iterative method for linear systems of which the coefficient matrix is a symmetric M-matrix', Math. Comp., **31**, 148-162(1977).

[22] C. den Heijer, 'Preconditioned iterative methods for nonsymmetric linear systems', Proc. Int. Conf. Simulation of Semiconductor Devices and Processes, Pineridge Press, Swansea, 1984, pp.276-285.

[23] W.P.M. van der Linden, J.W.J.M. van der Heijden, S.J. Polak, and A.J.H. Wachters, 'The PDL Package Generator MAMMY', IEEE Trans. on Magnetics, **24(1)**, 1988.

Isolation Technology

Geometry Dependent Bird's Beak Formation for Submicron LOCOS Isolation

P.A. van der Plas, N.A.H. Wils, R. de Werdt

Philips Research Laboratories, P.O. BOX 80000, 5600 JA Eindhoven, The Netherlands

Abstract
Bird's beak formation is one of the major problems for LOCOS field isolation. In this paper we demonstrate that the bird's beak length is not a constant, but depends strongly on geometry of the oxidation mask for submicron mask dimensions. The bird's beak length can vary up to a factor of 4, dependent on mask geometry. Four independent geometry effects are distinguished and their impact on an IC-process is discussed.

1 - Introduction

LOCOS technology is widely used for its outstanding isolation properties and process simplicity. Much effort is invested to reduce the bird's beak length by modified processes [1,2], but the bird's beak has been studied as a two dimensional oxidation phenomenon. In practice the bird's beak formed in concave and convex corners of the oxidation mask has to be considered as a (complex) three dimensional oxidation phenomenon.

2 - Experimental

In the experiments S(uppressed)-LOCOS technology is used [2]. The basic feature of the SLO-COS process is the replacement of the padoxide layer by oxynitride to reduce lateral encroachment of field oxide. The field oxidations were performed at 1000 °C, using an oxidation mask of 50 nm oxynitride and 100 nm nitride. Bird's beak lengths, defined as the total encroachment of field oxide under the mask, were analysed by top-view and cross-section SEM and TEM micrographs.

3 - Results and discussion

Fig. 1 shows a top-view SEM of a part of a 1M-SRAM after the oxidation mask has been removed. The bird's beak length (white arrows) and thickness depends strongly on the geometry of the mask. Four different cases are distinguished that cause variation of the bird's beak dimensions:

1 - Convex corners

As a result of the lateral growth underneath the oxidation mask, the mask bends at the edges. Convex corners of the oxidation mask are less resistant to bending than the long edges. Due to this, enlarged bird's beaks (EBB) are formed in convex corners, while even larger bird's beaks are found at the end of narrow mask stripes (Fig. 2). The length EBB can be 2 to 3 times the normal bird's beak length BB. Fig. 3 shows that the length EBB depends strongly on the width of the mask W_{om} and on field oxide thickness D_{oz}.

Enlarged bird's beak formation imposes serious limitations on the packing density, e.g. 2 to 3 times more overlap area is needed for contacts to the end of small active areas. Furthermore, the EBB-region is susceptible to junction leakage due to the long, thin wedge-shaped beak. An etchback of this beak after junction formation can easily expose the junction; this, in combination with a silicide process, can result in increased junction leakage in the EBB-region. To investigate this, special test structures were used with 400 bars of different width, ranging from 0.6 to 3.0 μm. The junction leakage I_l strongly depends on the width of the active area (Fig.4 , open markers). With deeper junctions and a longer etch of the enlarged bird's beak (black markers), this leakage problem can be avoided.

2 - Concave corners

Shortened bird's beaks (SBB) are formed in the concave corners of the oxidation mask. Because the mask is hard to bend for this geometry, bird's beak formation is suppressed in the concave corners. The smallest and thinnest beaks are formed at the end of narrow oxide bars (Fig. 5). The bird's beak at the tip of the narrow field oxide varies from 0.2 to 0.4 μm if the opening in the oxidation mask is varied from 1.0 to 3.0 μm respectively.

Gate oxide thinning will occur in the SBB-region, because the thinner beaks are overetched during etchback of the normal bird's beak. No extra junction leakage was measured in the SBB-region.

3 - Narrow mask features

In Fig. 6 the bird's beak length BB is plotted versus the mask width W_{om} for several thicknesses of field oxide. The data were obtained from test structures with long parallel mask-stripes of various width with 1 μm wide field oxide in between. Since a narrow mask is more resistant to bending, a reduced length and thickness of the bird's beak is observed for decreasing mask width.

The spread in bird's beak dimensions results in a spread in gate oxide thickness and effective transistor width [3] after etching of the bird's beak, as is illustrated in Fig. 7. The TEM micrographs show gate oxide grown on active areas of various width. For the wide areas a small remaining bird's beak is seen, but in the 0.4 μm wide active area the beak is clearly overetched. The gate oxide thinning at the edges causes a reduced gate oxide quality.

In practice, the etchback of the beak has been adjusted in such a way that a small bird's beak will remain in the wide transistors, to avoid overetching in the narrow transistors.

4 - Narrow mask openings

The field oxide thickness decreases for narrow mask openings. Fig. 8 shows the thinning effect for a range of field oxide thicknesses. Oxide thinning occurs at mask openings below 1.5 μm and about 30% thinning is observed for 0.5 μm wide field oxide ($D_{oz} = 600$ nm). Considerably less thinning is observed for thinner field oxides. The thinning effect has been studied more extensively recently [4].

The thinning effect has consequences for the control of channelstopper outdiffusion, while in case of retrograde channelstoppers a reduction in oxide thickness will result in a reduced concentration at the Si-SiO$_2$ interface.

4 - Conclusions

We have demonstrated that bird's beak formation depends strongly on the geometry of the oxidation mask. Four geometry dependent effects have been shown: enlarged bird's beak formation in convex corners, shortened bird's beak in concave corners, thinned bird's beaks for narrow mask features and field oxide thinning for narrow mask openings.

The geometry dependent bird's beak imposes limits on the packing density of LOCOS technologies, on control of gate oxide quality and on control of dopant levels in the parasitic channels.

It is shown that reduction of field oxide thickness reduces all the geometry dependent effects described.

Acknowledgements

The authors wish to acknowledge M. Geyselaars for TEM work, M. Pitt and F.M. Klaassen for helpful discussions.

Literature

1　N. Hoshi, et al., IEDM Techn. Dig., p.300, 1986

2　P.A van der Plas, W.C.E. Snels, A. Stolmeijer, H.J. den Blanken, R. de Werdt
　　Dig. 1987 Symp. VLSI techn., p.19, 1987.

3　F.M. Klaassen, P.A. van der Plas, N.A.H. wils, R. Velghe, M. Pitt
　　to be published ESSDERC-89.

4　Betty Coulman, C.N.A. Aussems, P.A. van der Plas, N.A.H. Wils
　　Proc. Symp. on VLSI, Electrochem. Soc., May, 1989

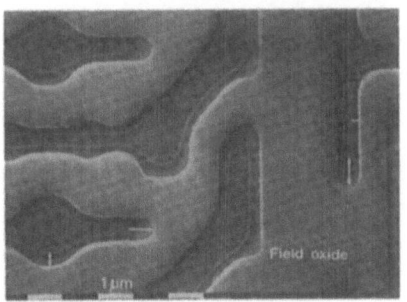

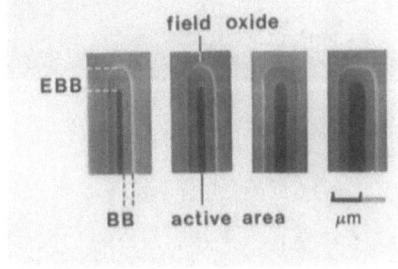

Fig.1 - Top-view SEM micrograph of a part of a 1M-SRAM cel. The bird's beak length (white arrows) is different in concave and convex corners.

Fig.2 - Top-view SEM micrographs of the bird's beak length EBB at the end of narrow active areas. BB refers to the normal bird's beak length (D_{oz}=700 nm).

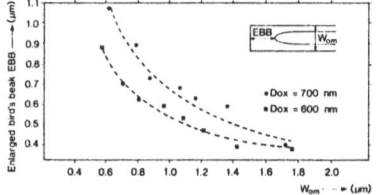

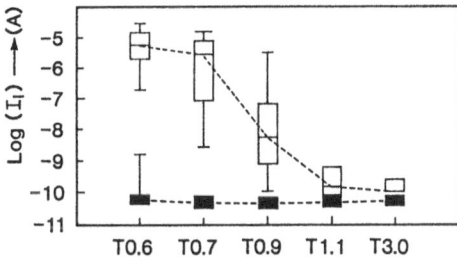

Fig.3 - Enlarged bird's beak length EBB as a function of the width of the oxidation mask W_{om} for two values of field oxide thickness. The inset shows the measured parameters. (compare Fig. 2).

Fig.4 - Junction leakage I_l of N+-Pwell junctions at reverse bias of 10 V. The symbol T refers to the width of the active area in the teststructure. (□ = before process optimisation, ■ = after process optimisation)

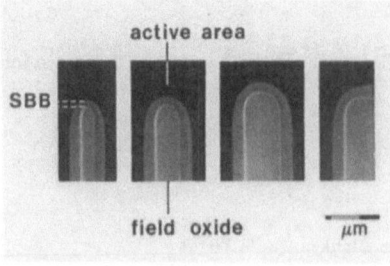

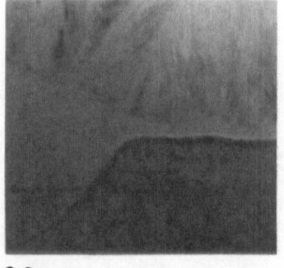

Fig.5 - Top-view SEM micrographs of shortened bird's beak lengths SBB at the tip of narrow mask openings (D_{oz}=700 nm).

Fig.6 - Length of the bird's beak BB versus the width of the mask W_{om} for various field oxide thicknesses.

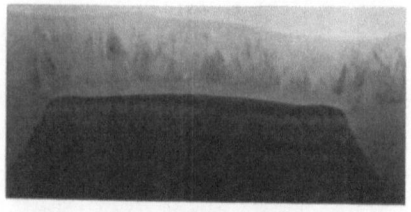

0.4 μm

0.56 μm

3.0 μm

Fig.7 - Cross-section TEM micrographs of gate oxide grown on active area of various width. The width of the active area is indicated near the figures.

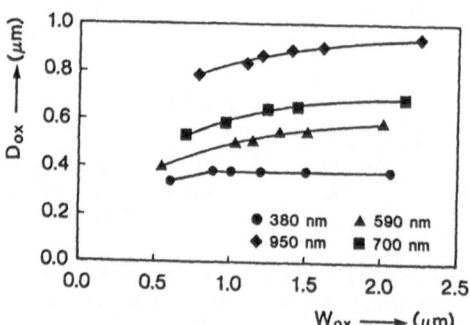

Fig.8 - Field oxide thickness D_{oz} versus the width of the opening in the oxidation mask W_{oz}. The nominal thickness has been varied.

Combination of LOCOS and BOX Isolation
for Submicron CMOS Technology

Ch. Zeller, F.X. Stelz

Siemens AG
Otto-Hahn-Ring 6, 8000 München 83, West-Germany

Abstract:

A new isolation method for submicron CMOS technology is presented, which combines the advantages of LOCOS and BOX isolation using one photolithographic step only. The technique is independent of circuit layout and the minimum isolation width is limited only by photolithographic resolution.

Introduction:

In submicron CMOS technologies both large and very narrow isolation regions have to be realized on the same chip. LOCOS isolation is easily applicable for defining large isolation regions and thick field oxides, but is limited by bird's beak growth and sausage effect for narrow isolation widths. The BOX technique is ideal for submicron isolation widths, but requires an additional photomasking step to planarize large isolation regions /1/, or is sensitive to circuit layout /2/ and shows sidewall parasitic transistor action. We present an isolation technique suitable for ULSI technologies, which combines the advantages of LOCOS and BOX without additional photomask and avoids the problems of both techniques.

Process flow:
The process flow is explained in fig. 1:
1. A nitride mask is patterned and boxes are etched into the silicon.
   After growth of a thin oxide polysilicon, nitride and CVD-oxide are deposited successively.

2. The CVD-oxide is etched anisotropically leaving the narrow boxes filled and forming a spacer at the sidewalls of large boxes. Nitride II is etched away, wheresoever it is not covered by CVD-oxide.

3. The spacer is removed and a field oxide is grown in large isolation areas.

4. The remaining nitride II and polysilicon is etched away. A second CVD-oxide and a resist layer are deposited and etched back anisotropically filling the narrow boxes and the edges of large isolation regions. The nitride I acts as an etch stop.

Discussion:

The thickness of the isolation oxide is width-dependent. It is larger in wide isolation areas than in narrow boxes. The thick oxide spacer moves the bird's beak of the field oxide away from the sidewall and prevents possible defect generation and junction leakage associated with SWAMI /3/ and other recessed oxidations. The sidewall spacer technique is insensitive to layout, which assures constant box sidewall to field oxide spacing (see Fig. 2 /4/).

Fig. 3 shows examples of isolation oxides with different widths at the end of the process. Case a is a box like isolation oxide. Narrow boxes can easily be filled with oxide independent of box depth and a flat surface can be achieved. In the case of wide isolation areas the necessary field oxide thickness depends on the box depth. Fig. 3b shows an example with a box depth of 300 nm and a field oxide thickness of 850 nm. For thinner field oxides or a deeper box a kink in the silicon surface at the box edge is observed (Fig. 4), which leads to a thinning of the gate oxide and may reduce device reliability. The thick field oxide reduces the parasitic metal to substrate capacitances in large isolation regions. In the narrow boxes no field oxide is grown. Thus the doping concentration is not reduced by segregation effects and isolation is improved in the critical regions. In a range of intermediate widths (about 1-2 μm) the field oxide does not reach the full thickness, because the window in the nitride layer is too small (sausage effect). In Fig. 5a one can see the situation directly after field oxidation. These regions are filled completely with oxide, however, during the planarization step (Fig. 5b).

Electrical results:

First electrical results show that isolation lengths smaller than one micron can be realized with our technique. Fig. 6 shows the threshold voltage of thick - oxide - transistors with metal gate in comparison with the standard LOCOS technique. No reduction of the threshold voltage is found down to 0.8 μm isolation length. For thin - oxide - transistor threshold voltage reduction is found for transistor widths below one micron. This effect can be avoided for example by an optimized sidewall implantation.

Acknowledgement:

We want to thank Mrs. Bunge and Mr. Voith for SEM preparation.

References:

[1] T.Shibata et al., IEDM Tech. Dig., 27 (1983)
[2] G.Fuse et al., IEEE Trans. Electron Devices, ED-34, 356 (1987)
[3] C.Teng et al., IEEE J.. Solid State Circuits, SC-20, 40 (1985)
[4] Note: before SEM preparation a thin polysilicon layer has been deposited (valid for all pictures).

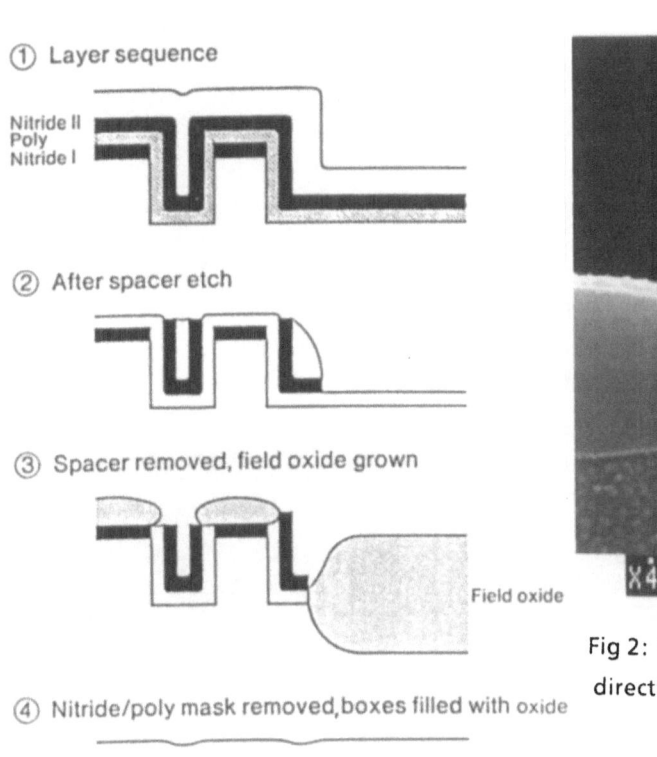

① Layer sequence

Nitride II
Poly
Nitride I

② After spacer etch

③ Spacer removed, field oxide grown

Field oxide

④ Nitride/poly mask removed, boxes filled with oxide

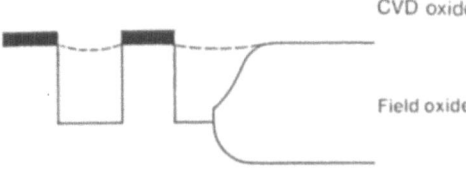

CVD oxide

Field oxide

Fig.1: Process flow of combined
LOCOS - and BOX - isolation

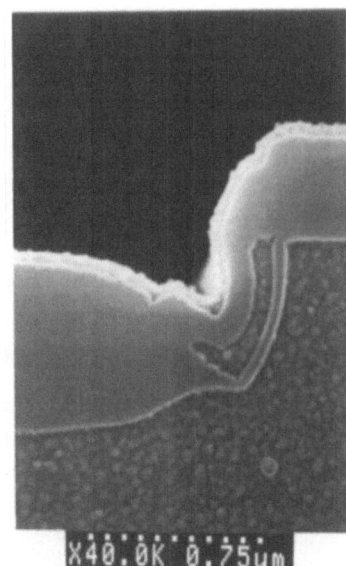

Fig 2: Wide isolation region
directly after field oxidation

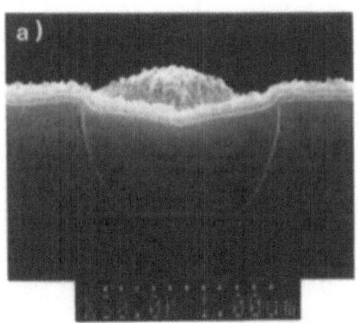
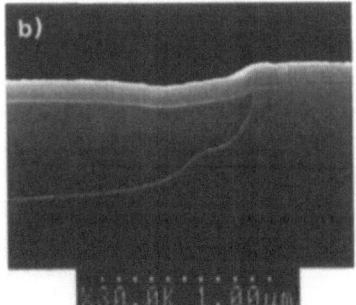

Fig.3: SEM of BOX-like (a) and LOCOS-like (b) isolation (Note: in these examples
box-depths were different for BOX- and LOCOS-like isolation).

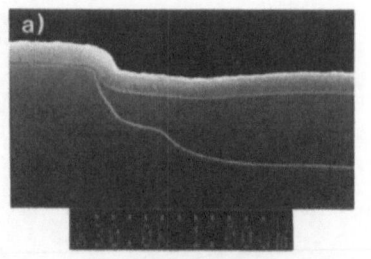 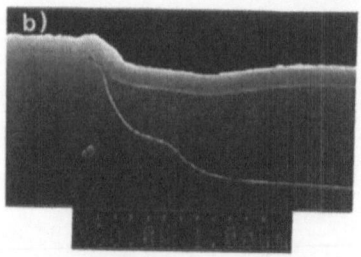

Fig.4: (a) box depth 300 nm, field oxide 700 nm
(b) box depth 400 nm, field oxide 850 nm

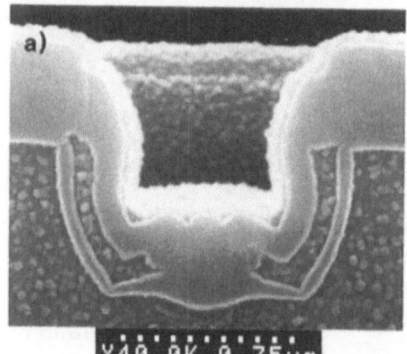 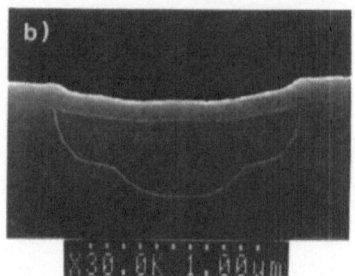

Fig.5: Intermediate case between BOX - and LOCOS-isolation
(a) after field oxidation
(b) at process end

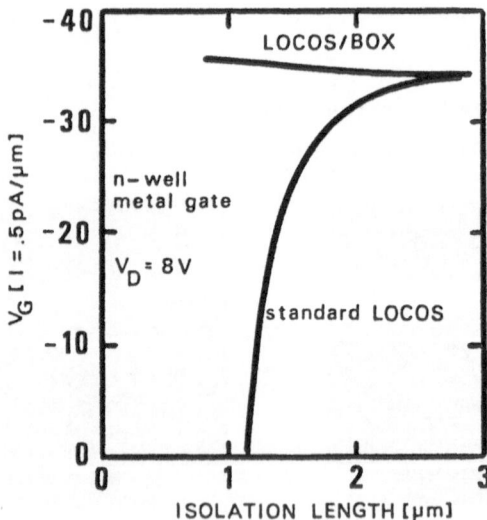

Fig.6: Comparison of isolation characteristics of standard LOCOS and combined
LOCOS-BOX-isolation

Impact of Oxide Thinning at the LOCOS Edge of MOS Capacitors on Constant Current Stress

M.Kerber and Ch.Zeller

SIEMENS AG., Corporate Research and Development,

Otto-Hahn-Ring 6, 8000 Munich 83, F.R.Germany

Summary

The oxide thinning at the field oxide edge significantly affects the oxide reliability. Statistical analysis of charge to breakdown shows that dielectric failure occurs in the corners rather than at the edges of MOS capacitors. The dependence of Q_{BD} on J_{inj} is sensitive to oxide thinning at the LOCOS edge. This limits the current density during accelerated tests in order to achieve reliable results.

1. INTRODUCTION

Reducing the feature size in MOS technology requires aggravated isolation processes like advanced LOCOS, BOX or trench isolation. Depending on the process flow a more or less thinned gate oxide appeares at the isolation oxide edge [1]. This oxide thinning consequently affects the gate oxide reliability. However, the impact on accelerated tests for dielectric breakdown has not yet been investigated in detail.

In this work we focuse on consequences of the gate oxide thinning at the LOCOS edge for the charge to breakdown (Q_{BD}) of MOS capacitors. Simply by statistical arguments the primary location of dielectric failure can be identified. The influence of pronounced oxide thinning and injection current density on Q_{BD} is discussed.

2. EXPERIMENTAL

The experiments were performed with MOS capacitors fabricated on 8 Ohm cm (100) p-type substrates. The active gate areas ranging from 10^{-4} to 10^{-2} cm^2 were defined by an advanced LOCOS process described previously [2]. A 500 nm n$^+$ poly was used as the gate electrode. Birds beak lengths ranging from 0.1 to 0.35 μm were achieved by different wet etch times to remove the pad oxide at the end of the LOCOS processing block. A shorter birds beak is accompanied by a pronounced gate oxide thinning at the LOCOS edge. TEM cross sections of the birds beak region for the samples used in this study are shown in fig 1. The process parameters are summarized in table 1.

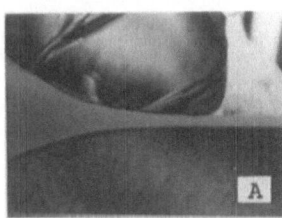 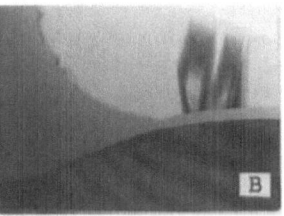 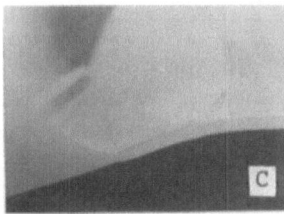

Fig 1 TEM cross sections of the region between gate oxide (20nm) and field oxide with different stages of thinning

140

The gate oxide integrity was evaluated by means of the constant current injection method with negatively
biased gate and 125°C substrate temperature. Three test structures with variable gate areas, LOCOS edge
lengths and number of LOCOS corners were investigated in order to separate the different contributions to
the breakdown behavior. The capacitor parameters are listed in table 2.

Table 1: Process parameters

process	wet etch time	oxide thinning (from TEM)
A	75sec	20nm
B	105sec	16nm
C	145sec	14.5nm

Table 2: Test structure parameters

test structure	gate area (cm²)	edge length (cm)	corners
#1	0.01	33.5	27440
#2	0.01	0.435	4
#3	0.0001	0.04	4

3. BREAKDOWN STATISTICS

Randomly distributed breakdown events lead to a cumulative failure probability P after injection of the charge
Q_{BD} according to Poisson statistics [3]

$$P(Q_{BD}) = 1-\exp(-A*D)$$

where A is the sample area and D the density of events leading to breakdown. In order to compare the
breakdown behavior of ensembles with different gate areas it is convenient to express explicitly:
$D = -\ln(1-P(Q_{BD}))/A$ and plot the area independent quantity D versus Q_{BD}. If the assumption of random
defect distribution over the test area is valid no dependence on the sample area should be found.

The breakdown characteristics of the sample without thinning (process A) is shown in fig 2a where the three
test devices are normalized to the gate area. A very good coincidence is obtained for the intrinsic failure mode.
It shows that the gate area indeed is the primary location where the intrinsic breakdown occurs. In terms of the
extrinsic branch the test structure with the long LOCOS edge (#1) exhibits a somewhat higher defect density.
Here the overetch time was kept very low, thus residues possibly were left near the LOCOS edge.

To study the impact of the gate oxide thinning, the breakdown distribution of process B is shown in fig 2b. In
this case the intrinsic defect branches for the investigated test structures are clearly different, indicating that
breakdown does not simply occur in the planar gate area. The same analysis with the total edge length and the

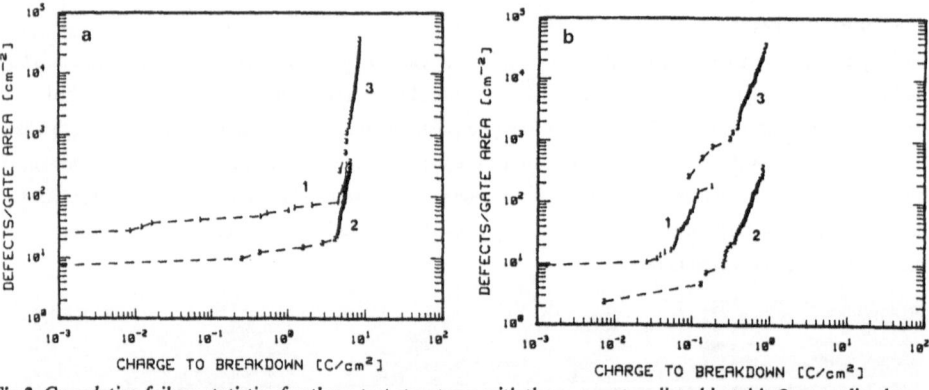

Fig 2 Cumulative failure statistics for three test structures with the parameters listed in table 2 normalized to
the gate area: a) Process A without thinning ($J_{inj} = 0.01$ A/cm²) b) Process B ($J_{inj} = 0.0003$ A/cm²)

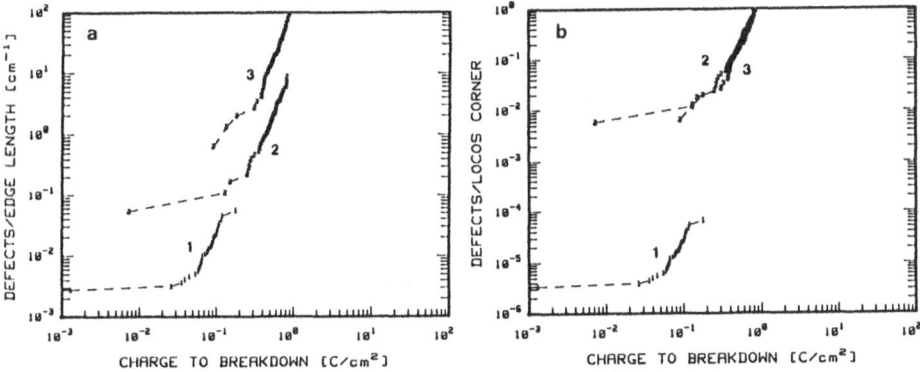

Fig 3 Cumulative failure statistics for process B as in fig 2b but with the defect density normalized to the LOCOS edge length (a) and number of LOCOS corners (b)

number of LOCOS corners as the scaling quantities is depicted in figs 3a and 3b. Again with the edge length only the statistics can not be explained sufficiently. The number of corners finally leads to the desired result of an intrinsic branch common to all test samples. Therefore, based on purely statistical analysis we conclude that for technologies with significant oxide thinning the intrinsic breakdown primarily occurs at the corners.

4. CURRENT DENSITY DEPENDENCE

If the injection current density J_{inj} in constant current stress is increased beyond a critical value J_{crit}, Q_{BD} becomes very sensitive to J_{inj} [4,5]. The LOCOS induced oxide thinning affects both the saturation level Q_{BDsat} as well as J_{crit}. This is illustrated in fig 4. Q_{BDsat} decreases with enhanced thinning and J_{crit} is shifted from 10^{-1} A/cm^2 for process A to 10^{-3} and 10^{-4} A/cm^2 for processes B and C respectively. Test devices with the same oxide thinning but different LOCOS edge lengths, however, show almost similar J_{inj} dependence as illustrated in fig 5. This indicates that the shift of J_{crit} is caused by the thinning.

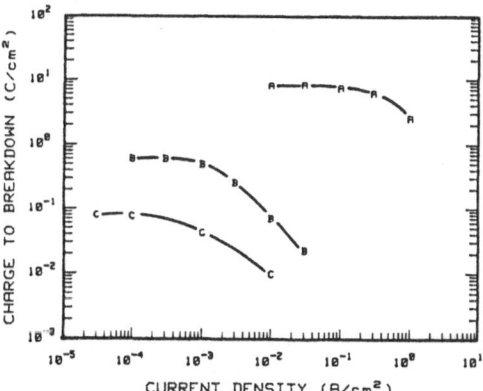

Fig 4 Dependence of Q_{BD} on J_{inj} for different oxide thinnings. The symbols refer to the items in fig 1 and table 1. Q_{BD} is for 63% failure of test device #3 (D = 10^{-4}/cm^2 or 0.25/corner)

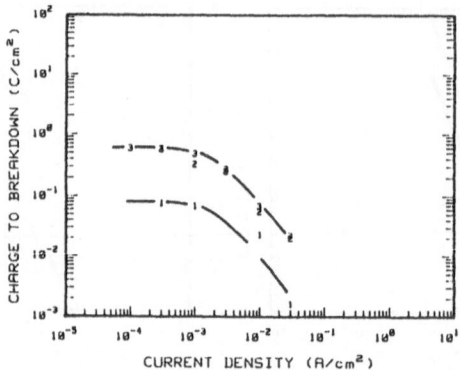

Fig 5 Dependence of Q_{BD} on J_{inj} for different test structures of process B. Q_{BD} is for 63% failure corresponding to a defect density of 0.25 for device #2,#3 and $3.6*10^{-5}$ for #1 respectively.

In order to explain the reduced Q_{BDsat} (fig 4) we consider the effective thickness of the thinned regions. Supposing equal Q_{BD}, the desired minimal oxide thickness for processes B and C can be estimated to 18.0 and 16.8 nm respectively. However, the oxide thickness observed in the TEM cross sections is smaller. To explain this deviation the area dependence of Q_{BD} has to be considered, too. The effective area of the LOCOS corners (of processes B and C), which are relevant for breakdown, is much smaller than the total gate area of the devices without thinning. If Q_{BD} is extrapolated to equal active areas, the Q_{BDsat}-differences become even larger. This results in a smaller estimated oxide thickness and therefore qualitative agreement with the minimal thicknesses found in the TEM cross sections of fig 1.

A considerable thinning appears at thermally grown oxides in DRAM trench capacitors leading to a very low Q_{BD} compared to planar capacitors [6]. This result is consistent with the arguments presented here.

5. CONCLUSION

Statistical analysis of charge to breakdown shows that in capacitors without thinning at the edges the dielectric failure occurs in the planar area. In MOS capacitors with significant thinning, however, the breakdown appears primarily in the corners. Furthermore, the range where Q_{BD} is independent of J_{inj} is strongly influenced by the degree of the thinning. This limits the maximum current density of accelerated tests in order to achieve reliable results. In modern CMOS technologies oxide thinning could affect the results for $J_{inj} > 10^{-3}$ A/cm^2. The intrinsic Q_{BD} is significantly lowered by the oxide thinning due to current crowding at the edges.

References

[1] Ch.Zeller and F.X.Stelz, this conference.

[2] R.Burmester et. al., ESSDERC88, Conf. Proc. in J. de Physique, vol.49, Suppl.9, p.545 (1988)

[3] D.R.Wolters and J.J.van der Schott, Philips J.Res.40 p115 (1985)

[4] D.R.Wolters and J.J.van der Schott, Philips J.Res.40 p137 (1985)

[5] M.Kerber and U.Schwalke, Proc. 27th Int.Rel.Phys.Symp., p.17 (1989)

[6] K.v.Sichart et. al., in "Proceedings of the symposium on reliability of semiconductor devices", ed. H.Rathore, G.Schwartz and R.A.Susku, (ECS fall meeting, Chicago 1988), p.227 (1989)

A Self-aligned Isolation-oxide Process Before Gate Formation for Enhanced Packing Density of Megabit DRAMs

P.Kuepper, B.Hasler, S.Roehl, T.Bolze, C.Diekmann, H.G.Mohr, W.Starflinger;

Siemens AG, Dept. HL T, Otto-Hahn-Ring 6, 8000 Mnchen 83, West Germany

Summary: A new completely selfaligned isolation oxide etch process prior to gate oxidation was developped with resulting gate oxide quality, isolation properties and topology comparable to a process with mask.

One of the most limiting points in increasing the packing density of DRAMs are the photolithographic processes. This is not only because of the minimum featuresize approaching the physical limits of the equipment in use, but also because linewidth variations and misregistration lead to overlay uncertainties which impose severe restrictions on designrule tightening. Selfaligning techniques do do reduce tese demands vastly.

Here we report the complete elimination of a masking step by the implementation of a new self-aligned etch process for the isolation oxide (IOX) between Poly1 (capacitor plate) and Poly2 (gate) in a 4M trench capacitor DRAM process. Table 1 compares the relevant part of the process where the change took place, while the complete process has been discussed in /1,2/.

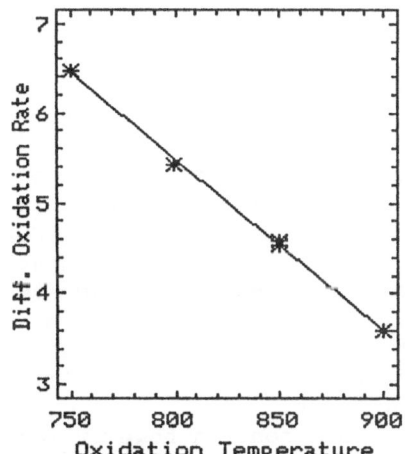

Fig.1 Differential oxidation rate of high doped Poly-Si in relation to bulk Si

TABLE 1
Comparison of the IOX part
of the process flow

with mask	selfaligned
-Poly1 patterning	Poly1 patterning
-cleaning	cleaning
-IOX growth	IOX growth
-photolithography for defining the IOX window	
-wet IOX etch	IOX etch
-resist strip	
-cleaning	cleaning
-GOX growth	GOX growth

The IOX growth process was optimized first with respect to the
differential oxidation rate between the high doped Poly1 and
the substrate and second with respect to the gate oxide quality
following the IOX etch. Results in fig. 1 show the differential
oxidation rate as function of oxidation temperature for the
applied wet oxidation. While at 900 C the differential oxidation rate is too low for the process requirements, the increase with reduced oxidation temperature enables selfaligned
etching. Fig. 2a shows a REM picture of the relevant topology
right after a 850 C IOX growth without backetch and with Poly2
serving for preparational purposes. The oxide thickness on top
of Poly1 will ensure sufficient isolation thickness after a
planar backetch, but the critical point is at the Poly1 edge.
This is further confirmed in fig. 2b and c where the results of
a mere wet etch and of a dry+wet etch combination are shown.
Finally in fig. 2d the result of a completely dry backetch can
be seen. The comparison of these REM pictures shows that wet
etch alone and also a combination of wet and dry etch give very
poor, insufficient isolation at the Poly1 edge. Besides, an
undercut is formed at this edge causing troublesome Poly2-
stringers after Poly2-etch. Only a complete dry etch results in
an acceptable isolation thickness and edge shape.

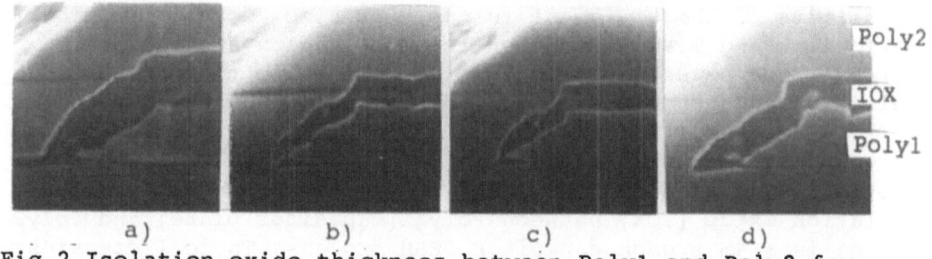

Fig.2 Isolation oxide thickness between Poly1 and Poly2 for
a) no etch; b) wet etch; c) dry+wet etch d) dry etch

The second important point investigated for introducing this
process simplification was the gate oxide (GOX) quality after
the IOX backetch. Fig. 3 shows the GOX quality for an IOX
growth temperature of 800 and 850 C respectively. The gate
oxide was
tested as the dielectric of a polisilicon-substrate capacitor
stressed with a voltage staircase of 0.5 MV/cm stepheight and
50 msec. steplength. The GOX quality is determined by the
(relative) numbers of capacitors surviving the different indicated breakdown field strengths. The results of fig. 3 show
that lowering the IOX growth temperature from 850 to 800 C results in a quality reduction of about 30 percent points (with
respect to the 8MV/cm limit) which will certainly lead to yield
problems.
Of main concern however was the influence of the IOX etch process on the GOX quality because in this process step bare substrate is exposed to plasma just before gate oxidation. For

this reason several combinations of etch processes and post-treatments, described in table 2 and 3, were investigated.

TABLE 2
dry etch processes investigated for selfaligned IOX
etching

code	flow of CHF3/O2	SiO2 etchrate
E1	high/high	medium
E2	high/low	high
E3	medium/low	low
E4	low/high	low

TABLE 3
posttreatments after
IOX etching

code	process
PTR1	NF3/Ar sputter
PTR2	Ar sputter
PTR3	medium O2 plasma in barrel
PTR4	short O2 plasma in barrel

The GOX qualities resulting from the corresponding process combinations are shown in fig. 4. Several findings are interesting from these results. First a NF3/Ar posttreatment gives the worst results. Second a high etch rate i.e. a short overall plasma time gives good results, but leads to problems with controlled overertch times. Finally a proper choice of etch process and posttreatment results in nearly the same GOX quality as a mere wet etch which was chosen as reference in fig. 4.

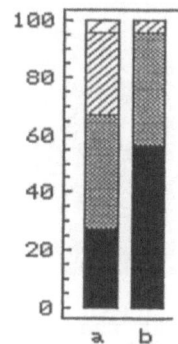

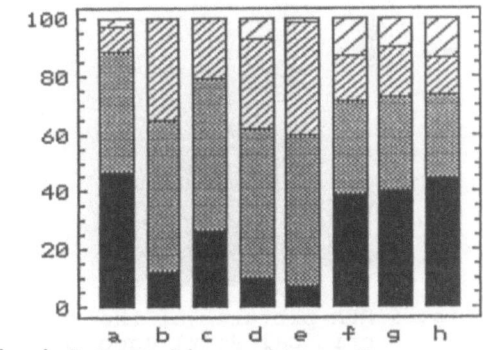

Fig.3 Gate oxide quality
      for IOX growth tem-
      peratures of 800 C
      (a) and 850 C (b)

Fig.4 Gate oxide quality after
      IOX etching, etch process :
      (a):wet; (b):E1+PTR1;
      (c):E2+PTR1; (d):E3+PTR1;
      (e):E4+PTR1; (f):E3+PTR2;
      (g):E3+PTR3; (h):E3+PTR4;

Legend for fig. 3 and 4
▨ % failing below 2 MV/cm,   ▨ % failing between 2 and 8 MV/cm
▨ % failing between 8 and 12 MV/cm,   ■ % surviving 12 MV/cm

This new process part was implemented in the complete 4M DRAM process described in /1,2/. Fig. 5 shows the leakage currents between Poly1 and Poly2 for fully processed wafers on a test-structure representing the cell array of the DRAM. The leakage current levels arte identical, only the breakdown voltage for the conventional process is about 5 V higher (however with some early breakdowns). But as the breakdown voltage for the new

146

process flow is still about 22 V and the Poly1-Poly2 voltage difference never exceeds 10 V this is thought to be a minor problem.

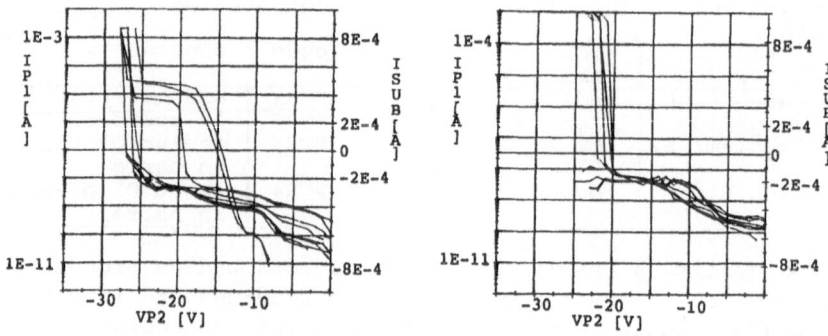

a)   b)

Fig. 5 Poly1-Poly2 Leakage current for wafers with a complete 4M process with the conventional (a) and new (b) IOX process

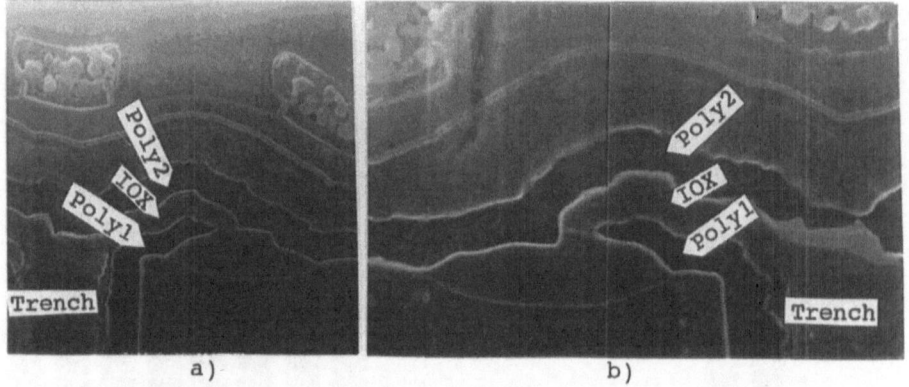

a)   b)

Fig. 6 REM-pictures of the Poly1-Poly2 isolation at the critical Poly1 edge for wafers with the complete 4M process with the conventional (a) and new (b) IOX process (different scales, 0.4 μm a) ⊢——┤ b) ⊢———┤)

Finally in fig. 6 the critical Poly1-Poly2-isolation at the Poly1 edge is compared for wafers with the new selfaligned etch process against the standard process. For the new process the isolation thickness on top of Poly1 is comparable to the old one, the isolation oxide edge shape is smoother and the vertical Poly2 thickness at this step (important for Poly2-overetch) is increased by only 14%.

This work has been supported by the federal Department of Research and Technology of the Federal Republic of Germany (sign NT 2696)

/1/ K.H.Kuesters et al.: 1987 Symposium on VLSI Technology, Karuizawa 1987, Dig. of Techical Papers, p. 93-94
/2/ L.Risch et al, ntz 3/1987, p.192-197

Field Isolation Using Shallow Trenches for Submicron CMOS Technology

G. Zwicker, P. Lange, P. Staudt-Fischbach and W. Windbracke

Fraunhofer-Institut für Mikrostrukturtechnik
Dillenburger Str. 53
D-1000 Berlin 33, Federal Republic of Germany

Abstract

This paper presents a shallow trench isolation technique using plasma etching, LPCVD oxide fill, and planarization. This novel planarization technique applying X-ray lithography and isotropic O_2-plasma etch needs no additional mask for block resist patterning over large isolation areas. MOSFETs have been fabricated showing nearly zero channel width loss and no threshold voltage shift down to 0.8 μm channel width.

Introduction

Devices with sub-μm design rules require innovative isolation techniques, since conventional LOCOS isolation is no longer applicable. This is due to the formation of the Bird's Beak, which limits the spacing between adjacent active areas as well as the minimal gatewidth of MOS devices. Aside from new techniques such as a selective epitaxial growth and approaches for Bird's Beak reduction, trench etching and subsequent refilling with a dielectric is a promising alternative isolation technique. However, not only very small trenches, but also large areas have to be filled. Several attempts have been made to overcome these problems, such as various trench filling techniques /1,2/, buried insulator isolation /3/, or global planarization /4,5/. All these solutions are promising, but the ultimate isolation scheme has yet to be developed.

Fabrication Process

In order to characterize the shallow trench isolation technique, NMOS test devices were fabricated using a mix and match technique where the significant trench layer was defined by X-ray lithography whereas the poly-Si, contact and metal layers were defined by conventional optical lithography. The process flow of the trench isolation which replaces LOCOS is depicted in Fig. 1. Starting with p-type (100) Si wafers (1-3 Ωcm), a 50 nm thick buffer oxide was grown, followed by the deposition of a Si_3N_4-layer with 140 nm thickness. The etch mask was defined by X-ray lithography using RAY-PF X-ray sensitive resist. Exposure was done at the BESSY storage ring with a dose of 12 Asec at 805 MeV, using the MAX I X-ray stepper /6/. After reactive ion etching of Si_3N_4 and SiO_2 in an AME 8111 hexode reactor, trench etching was performed on a LAM 680 plasma etcher employing chlorine chemistry.

After resist stripping and cleaning, a thermal oxide of 30 nm thickness was grown to obtain a well defined interface, followed by a boron field implant of $3 \cdot 10^{13}/cm^2$ at 20 keV. The SiO_2 trench isolation layer was fabricated in a low pressure chemical vapor deposition (LPCVD) system, using a TEOS source. This leads to an excellent conformal step coverage, thus preventing the formation of voids in small size features (Fig. 2a).oFor the following planarization sequence it was necessary to protect the large oxide areas from being etched back. This was done by patterning 2.1 μm of negative tone X-ray resist (RAY-PN) with the first (trench) mask and isotropic etch-back in an O_2-plasma. The isotropic behaviour of this plasma etch step shrinks the vertical and lateral dimensions of the resist patterns, thus removing the small patterns completely. Final planarization was done by spinning on a second 1.75 μm thick RAY-PN layer and an RIE etchback with 1:1 (resist:oxide) selectivity. Endpoint detection was carried out by means of laser interferometry (Figs. 2b,c).

The following NMOS sequence consists of a two step boron channel implant, a 10 nm gate oxide and a 250 nm thick Poly-Si gate. Source/Drain and Poly-Si doping was accomplished by implantation of $1 \cdot 10^{16}/cm^2$ As at 80 keV. As an intermediate dielectric, a layer of 650 nm CVD-SiO$_2$ was deposited.   After definition and RIE of contact-holes, 1 μm aluminium was sputter deposited and finally structured by plasma etching /6/.

Electrical Characteristics and Discussion

Electrical measurements were done using an HP 4145 A. Maximum transconductance g_m was derived from the incremental slope of the transition curve while V_t was determined by extrapolation of the g_m curve, with V_d fixed at 50 mV and zero bulk bias. Channel width and length data were acquired by SEM measurement of the X-ray mask. The channel length of all test devices was 1.2 μm.

The transconductance as a function of the channel width shows a channel width loss (Δw) of 1.15 μm for LOCOS devices due to the formation of the Bird's Beaks and the subdiffusion of the field implantation (Fig. 3). For shallow-trench isolated (STI) devices, Δw is less than 0.1 μm, thus making this process applicable for sub-μm isolation. Additionally the threshold voltage shift (narrow-channel effect) does not appear for STI devices down to 0.8 μm mask channel width Fig. 4). The subthreshold leakage current of STI devices is in the range of 10^{-13}A and is comparable to that of LOCOS devices.

In summary, the electrical results of the fabricated MOS devices show that STI leads to isolation properties corresponding to those of LOCOS. Although the number of process steps is increased, this technique allows sub-μm spacing and, therefore, higher packing densities than in conventional isolation schemes. Thus, STI is a promising candidate as a substitute for LOCOS.

Acknowledgements

We would like to thank S. Klinkenberg and K. Tomkowiak for their support in wafer processing, S. Seedorf for X-ray exposures, H. Bernt for valuable discussions and B. Bielenberg for her help in preparing the figures. This research was financially supported by the Bundesministerium für Forschung und Technologie (Ministry of Research and Technology) of the Federal Republic of Germany.

References

/1/ B. Davari, C. Koburger, T. Furukawa, Y. Taur, W. Noble, A. Megdanis, J. Warnock, J. Mauer, IEDM Tech. Dig., (1988), 92.

/2/ G. Fuse, M. Fukumoto, A. Shinohara, S. Odanaka, M. Sasago, T. Ohzone, IEEE Trans. Electron Devices, Vol. ED-34, No. 2, Part 1 (1987), 356.

/3/ M. Shimizu, M. Inuishi, T. Ogawa, H. Miyatake, K. Tsukamoto, and Y. Akasaka, IEDM Tech. Dig., (1988), 96.

/4/ G. Fuse, H. Ogawa, K. Tateiwa, I. Nakao, S. Odanaka, M. Fukumoto, H. Iwasaki, T. Ohzone, IEDM Tech. Dig., (1987), 732.

/5/ H. Mikoshiba, T. Homma, K. Hamano, IEDM Tech. Dig., (1984), 578.

/6/ W. Windbracke, H.-L. Huber, P. Staudt, and G. Zwicker, Microelectron. Eng., 9 (1989), 109

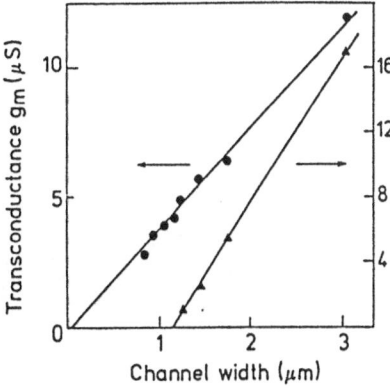

Fig. 3:
Determination of channel width loss (Δw) for LOCOS (Δw = 1.15 μm) and trench isolated devices (Δw<0.1 μm)

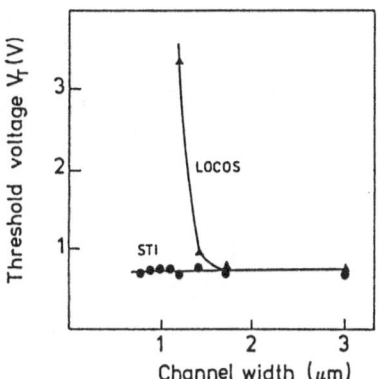

Fig. 4:
Threshold voltage shift due to narrow-channel effect for LOCOS and trench isolated devices

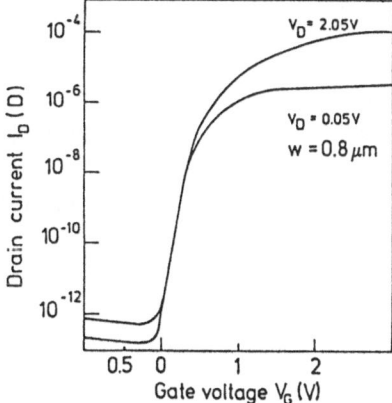

Fig. 5:
Subthreshold characteristic of a trench-isolated n-MOSFET

150

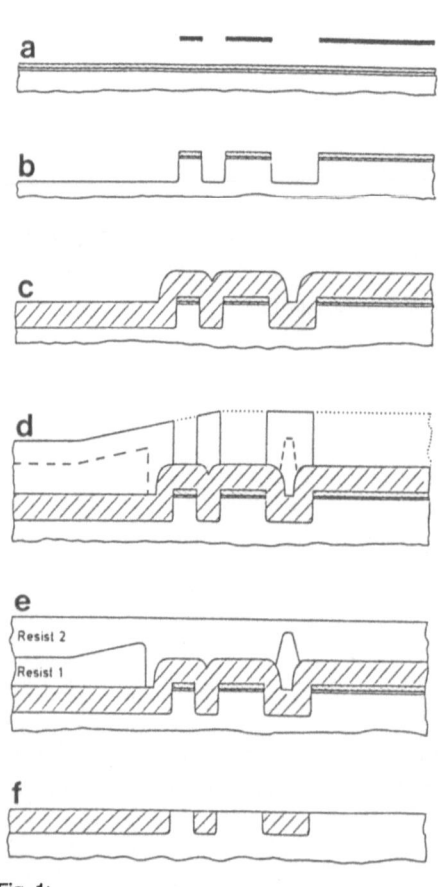

Fig. 1:
Process flow of the field isolation using shallow trenches

a) 50 nm SiO_2, 150 nm Si_3N_4,
   positive tone resist, x-ray lithography LOCOS mask

b) Si trench etch

c) CVD-SiO_2

d) negative tone resist, x-ray lithography LOCOS mask, isotropic O_2-etch, resist hardbake

e) second resist (planarization) resist hardbake

f) backetch (selectivity 1:1) Si_3N_4 removal

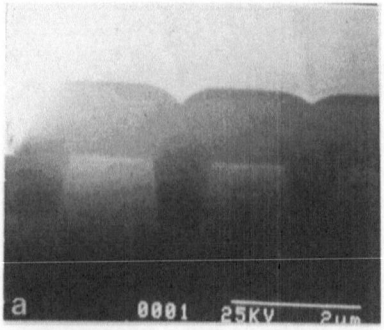

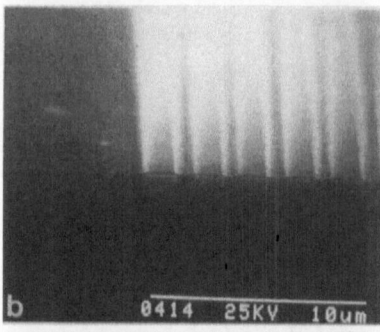

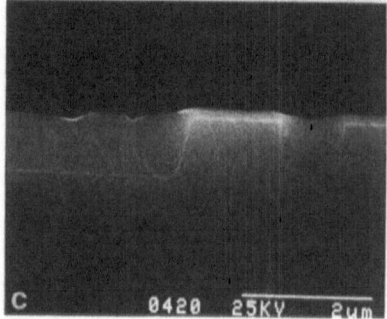

Fig. 2:
SEM micrographs of etched trenches:

a) after SiO_2 deposition

b, c) after planarization, before Si_3N_4 removal

High Performance Submicron SILO Process for High Density EPROM Memories Using Rapid Thermal Nitridation of Silicon

S.Deleonibus,P.Molle,J.Lajzerowicz,B.Guillaumot(*),
P.Laporte,A.Bergemont(*)
   LETI   CENG Avenue des Martyrs
   38041 Grenoble CEDEX France
   (*)S.G.S.-THOMSON Avenue des Martyrs
   38031 Grenoble France

INTRODUCTION   We have developed an isolation process for high performance submicron CMOS (0.6 micron design rules) compatible with 13V supply voltage.The former electrical needs as well as a large bird's beak angle are necessary for achieving high performance,high density EPROM devices: our SILO process using rapid thermal nitridation can satisfy those requirements.

PROCESS   Because classical LOCOS technology limits have been reached for submicron design rules ,new approaches have to be taken for a good control of lateral isolation.In this new isolation technique , Rapid Thermal Nitridation of silicon (R.T.N.) is used to perform a sealed Nitride-Silicon interface.Following the R.T.N. step, a subsequent nitride /oxide/nitride multilayer deposition is used to perform a sealed nitride /silicon interface .The masking structure is obtained by a LPCVD nitride/oxyde /nitride trilayer.The first nitride which has an average thickness of 10nm is directly deposited onto the R.T.N. nitride.Subsequent deposited films oxide and nitride have respectively thicknesses in the range of 20nm to 50nm and 40nm to 100nm.This process so called SILO/RTN (Sealed Interface Local Oxidation by Rapid Thermal Nitridation) can provide a structure with a bird's beak length of less than 0.15 micron for a field oxidation of 0.7 micron as grown at 950°C(fig. 1a).

        For such a technique ,specific RIE plasma etching processes were developed :
        1) Mask etch before oxidation.
        -SF6 nitride etch ,stop in oxide in a classical low pressure RIE reactor(ALCATEL GIR220)

-Selective to Si etch of remaining oxide and thin nitride in a high pressure low gap RIE reactor(ALCATEL RGV 220 using a CHF3 ,C2F6 and Ar mixture)

-Pure O2 RIE plasma cleaning process (Nextral 550 batch system) followed by a FH dip.

2) Mask removal after oxidation

-Full dry etch process with nitride etch rate 1.5 higher than the oxide rate and 2.5 times higher than the bulk silicon etch rate in a high pressure low gap RIE reactor (ALCATEL RGV 220 using a CHF3,C2F6,Ar and O2 mixture).This process leads to a field oxide consumption of 500 to 1000Å. This low selectivity process results in smooth steps at the field oxide edge.Figure 1b shows the field oxide after mask has been removed ,gate oxide performed and poly deposited. A smooth step coverage of poly is obtained.

-Pure O2 RIE plasma cleaning followed by a FH dip

RESULTS,DISCUSSION The difference between mask and finished electrical width of N channel and P channel narrow transistors is 0.3 micron which confirms that the bird's beak is 0.15 micron for 0.7 micron of as grown field oxide.

No narrow channel effects are observed for both n and p type transistors (fig.2).Our field doping conditions lead to larger than 13 V VGOFF values for 1.2 micron finished N and P type field transistors(VGOFF is defined as the gate voltage for sustaining leakage current less than 10pA/micron of transistor width).Figure 3 shows the Ids(Vgs) characteristics for N and P 50/1.2 micron W/L poly 2 gate field transistors.No short channel effects are observed on VGOFF or punch through voltage (figures 4a&b).Poly 2 gate DMOS transistors of both types n+/N- and p+/P- are in the same specifications.In worst case n+/N- devices,no short channel effect is observed down to 4 microns of n+ to N- spacing where VGOFF and BVDSS drops are started to be observed (see figure 5).

The above processing conditions are compatible with larger

than 14 V avalanche breakdown n+/p and p+/n diodes.
Perimeter leakage is less than 50pA/cm at 10V for both type
of diodes .Oxide integrity is controlled and larger than
10MV/cm breakdown field values are obtained.

The same type of process has been used for EPROM
type process flows (13 V sustaining is needed) as well as
5V 0.8 to 0.6 micron gate length devices.

ACKNOWLEDGEMENTS Authors are indebted to F.Martin from LETI
for his main contribution to RTN studies and fruitful
discussions.Special thanks to the people of silicon process
facility (SAME) as well as microelectronics research
laboratory of LETI (SMSC) for the support in this work.

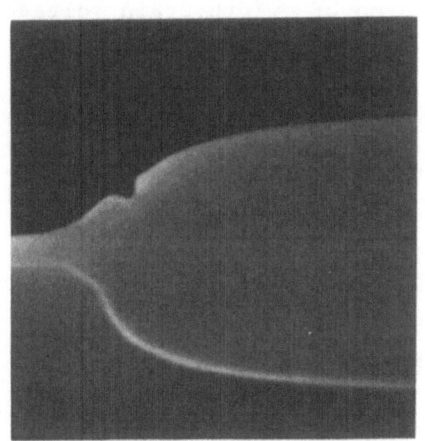
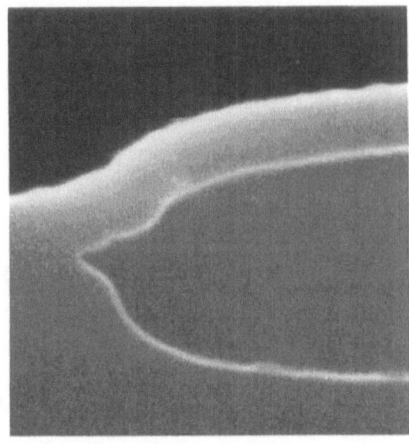

a   b

Figure 1 Field oxide SEM cross sections.0.15 micron
bird's beak is obtained for 0.7 micron as grown field
oxide.a) after oxide growth b) after poly 2 deposition

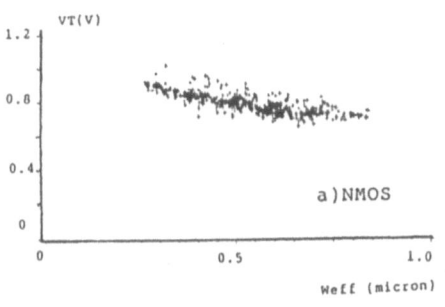
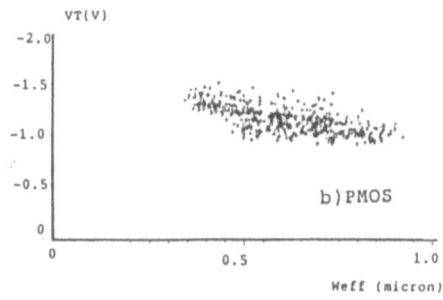

Figure 2 Threshold voltage VT of a)NMOS b)PMOS transistors
as a function of effective narrow channel width Weff.

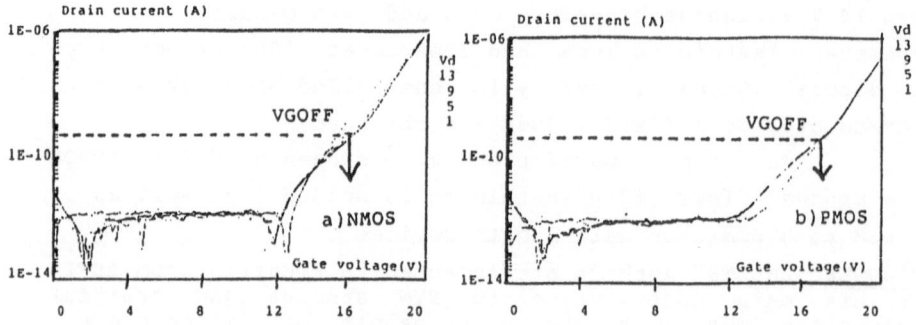

Figure 3 Ids(Vgs) characteristics of 50/1.2 micron
a) NMOS and b) PMOS poly 2 field transistors.VGOFF is the
sustaining voltage for 10pA/micron of width.Drain voltage
values are 1,5,9 and 13 V.

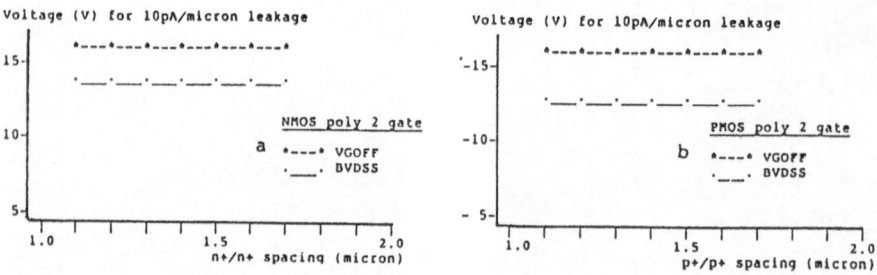

Figure 4 VGOFF and punch-through (BVDSS) voltage
for poly 2 gate parasitic transistor as a function of
spacing between active areas : a) NMOS b) PMOS

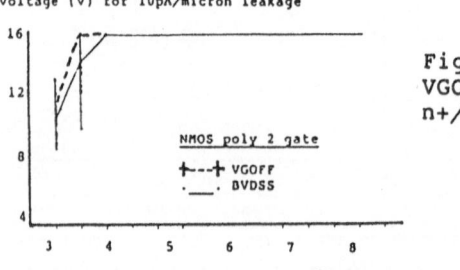

Figure 5   DMOS transistors (n+/N wel.
VGOFF and BVDSS as a function of
n+/N well spacing.

High Speed BICMOS Technology with Emitter-base Self-aligned Structure

Tunenori Yamauchi, Shin-ichi Yamada
Yoshiki Shimauchi, Katsuyuki Inayoshi

Bipolar IC Division FUJITSU LIMITED
1015 Kamikodamaka,Nakahara-ku,Kawasaki 211,Japan

S u m m a r y—Recently, we have developed a high performance BiCMOS technology,which is a composition of double
polysilicon emitter-base self-aligned bipolar and 0.8 μm
gate CMOS. With this process,15 GHz cut-off frequency(f_T)
of bipolar transistor and 60 pS of ECL basic delay time
(tpd) at Ics =1 mA have been achieved. For the BiCMOS gate,
220 pS of basic delay time and 325 pS of delay time with
0.5 pF load capacitance have been obtained.

1 - INTRODUCTION

BiCMOS technology is widely regarded as providing a possible
means for realizing high performance LSIs with multiple functions (1-3). Recently,BiCMOS LSIs are required to achieve ECL
interface. In order to achieve high speed operation for those
requirement. High performance ECL circuit must be implement in
BiCMOS LSIs. Therefore,it is profitable to combine double polysilicon self-aligned structure (4) with submicrom CMOS. We have
developed ESBiC(Emitter-base self-aligned BiCMOS) suitable for
high performance BiCMOS LSIs.

2 - DEVICE STRUCTURE AND PROCESS

A cross sectional view of the device is shown in Fig.1. An innovative structure is used in the emitter and base of NPN transistor. To reduce the parasitic capacitance of collector-base junction,and to increse its f_T value, its polysilicon emitter and
polysilicon base electrode are self-aligned. And to be simplify
process steps, the base electrodes, gate electrodes and resistors are made by polysilicon in the same mask process. Moreover,
the side walls of LDD structure and those between emitter and
base are also made in the same RIE process.   The emitter size
is 0.3 x 10.5 μm^2 . The base-collector junction area is
1.7 x 11.4 μm^2 . Table 1 indecates the parameters of ESBiC.

The process steps for fabrication are shown in Fig.2. Conventional bipolar and CMOS techniques are used in the process steps
from P-substrate to gate oxidation. After gate oxidation, the
gate oxide is removed at the bsae area of bipolar transistor.

Next, a polysilicon layer is grown. Phosphorous ions are implanted to form the MOS transistor's gate electrodes. Boron ions are
implanted to form the bipolar transistor's base electrodes and
resistors. After annealing,these electrodes are formed by the sa
me photo-etching process. And the side walls of LDD structure
and those between emitter and base are also made in the same RIE
process. The emitter area, collector and source/drain of NMOS
contact area are opened by the same photo-etching process.

Arsenic DOPOS(Doped Polysilicon) (5) layer are formed around
these contact areas. After the PSG is grown,reflow and emitter
diffusion by rapid thermal annealing is employed.

The metallization steps follow that complete device fabrication.

3 - DEVICE CHARACTERISTICS

Cut-off frequency as a function of collector current is shown
in Fig.3 . At a VCE of 3.0 V,the maximum cut-off frequency is
15 GHz. Fig.4 shows the delay time of CMOS and BiCMOS as a function of the load capacitance. The basic delay time of a BiCMOS
and CMOS gates are 220 pS and 235 pS respectivety. With conventional BiCMOS process the basic delay time of a BiCMOS gate is
longer than that of CMOS gate. This progress is achieved by
decreasing the parastic capacitance,mainly the base-collector
capacitance of bipolar transistor. The ECL basic delay time of
60 pS at Ics = 1 mA has been achieved.

Fig.5 is circuit diagram of 2 input NAND BiCMOS gate,and Fig.6
shows the photograph of the basic gate.

4 - CONCLUSION

We have developed a high performance BiCMOS technology named
ESBiC,which is a composition of double polysilicon emitter-base
self-aligned bipolar and 0.8μ m gate CMOS. This technology is
very useful for high speed LSIs,especially for those including
ECL circuits.

We are planning high speed gate arrays with ECL interface,
standard cells including ECL cells,and high speed ECL RAMs.

Reference

(1) K.Ogiue,et.al., 1986 ISSCC Digest of Technical Papers,
     pp.212-213 (Feb.1986)

(2) I.Fukushi,et.al.,1988 ISSCC Digest of Technical Papers,
     pp.134-135 (Feb.1988)

(3) Tzu-Yin Chiu,et.al., 1988 IEDM Technical Digest Papers,
     pp.752-755 (Dec.1988)

(4) K.Ueno,et.al., 1987 IEDM Technical Digest Papers,
     pp.371-374 (Dec.1987)

(5) M.Takagi,et.al., Journal of the Jpan Society of Applied
     Phsics,Vol.42,pp101-109, (1973)

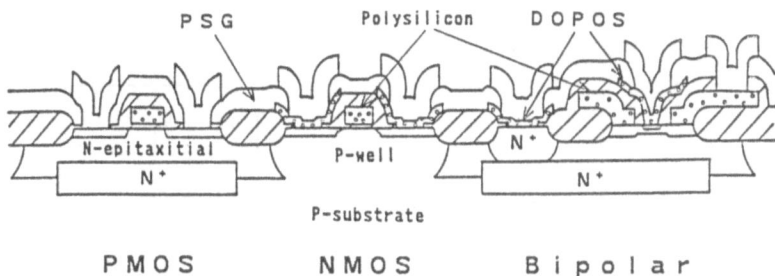

Fig. 1 Cross-section of ESBiC

Table 1 Process parameters

(1) Epitaxitial Thickness	1.4	μm
(2) LOCOS Thickness	500	nm
(3) NMOS Gate Length	0.8	μm
(4) PMOS Gate Length	1.0	μm
(5) Drain Structure	LDD	
(6) Emitter Size	0.3 × 1 0.5	μm²

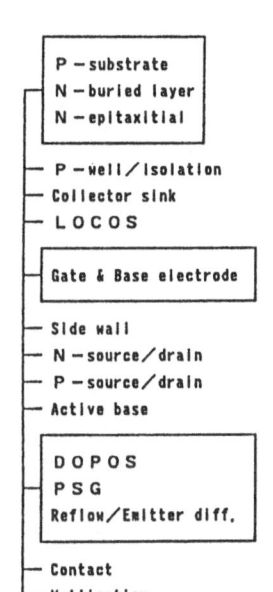

Fig. 2 Process steps of ESBiC

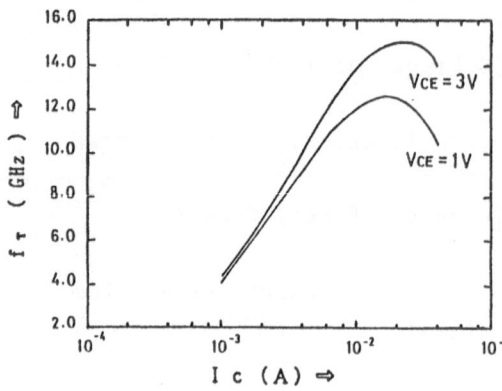

Fig. 3 Cut-off frequency (f_T) as a function
of collector current(Ic)

Table 2 Device parameters of ESBiC

Device	parameter	Value	Unit
Bipolar	h_{FE}	1 0 0	
	BV_{CEO}	7	V
	f_T	1 5	GHz
NMOS $W=20 \mu m$ $L=0.8 \mu m$	V_{th}	0.6	V
	BV_{SD}	1 3	V
	I_{ds}	1 0	mA
PMOS $W=20 \mu m$ $L=1.0 \mu m$	V_{th}	− 0.6	V
	BV_{SD}	− 1 4	V
	I_{ds}	− 5.1	mA

(Ids at Vds = 5 V, Vgs = 5 V)

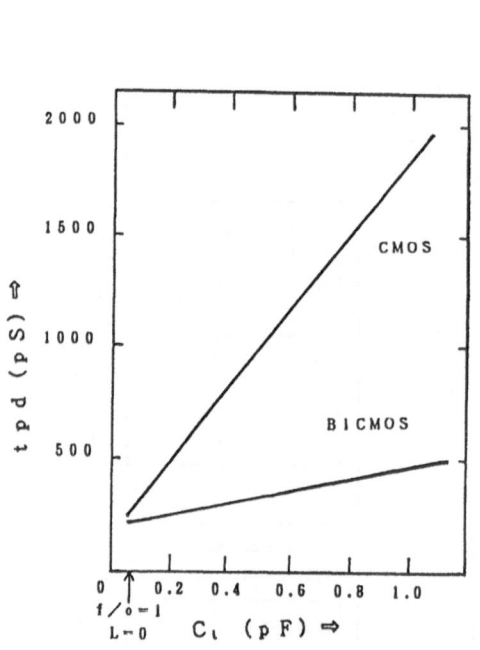

Fig. 4 Gate delay time(tpd) as
a funtion of load capacitance

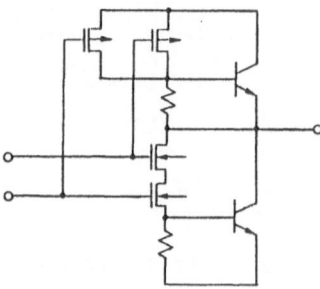

Fig. 5 Circuit of the basic gate

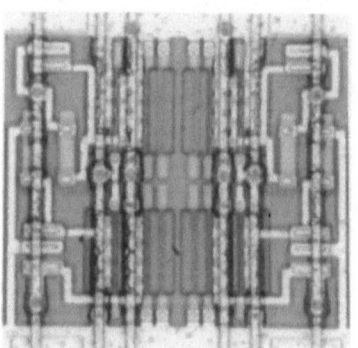

Fig. 6 Photograph of the basic gate

Compound
Semiconductor Technology

Ohmic Contacts to n-type GaAs Using GeMoW Metallization for Self-aligned Processing

C.Dubon-Chevallier,   P.Henoc,   F.Glas,   Y.Gao,   J.F.Bresse,   P.Blanconnier, C.Besombes.

Centre National d'Etudes des Télécommunications,
Laboratoire de Bagneux,
196 avenue Henri Ravera, 92220 Bagneux, France

Summary

We have formed GeMoW and Ge(As)MoW ohmic contacts to n-type GaAs using different types of annealing techniques, under either a forming gas or an As overpressure. A comprehensive study of both contacts is presented using electrical testing, Auger electron spectroscopy, secondary ion-mass spectrometry, transmission electron microscopy and scanning transmission electron microscopy. Very low specific contact resistivities, in the range of a few 10^{-7} Ωcm^2, have been obtained, when the contact included an As doped Ge layer with a doping level of 10^{20} cm^{-3} and was annealed using the semi-closed box technique under an As overpressure.

I-Introduction

High-reliability refractory ohmic contacts are required to process devices such as self-aligned Heterojunction Bipolar Transistors (HBTs) [1]. Indeed, a stable contact is necessary to allow the annealing step (800°C for a few seconds), used to activate the p-type implant after the contact metal deposition. The classical AuGeNi alloyed ohmic contact, which has been extensively studied for n-type GaAs, cannot withstand such temperature cycles. Molybdenum germanide has already been investigated, because it provides a low specific contact resistivity and a good thermal stability on GaAs [2,3]. Moreover, it can be etched selectively on GaAs, thus allowing a good dimensional control. We have investigated the formation of the GeMoW contact and correlated the electrical results and the metallurgical analysis. This study has led to the optimization of a new refractory ohmic contact, using an As doped Ge layer.

II-Experimental procedure

Ohmic contact specimens were formed on Si-doped n-type GaAs layers (doping level 4×10^{18} cm^{-3}) grown by molecular beam epitaxy on a semi-insulating substrate. The metallic layers were deposited in a chamber where electron beam evaporation and sputtering can be performed sequentially, without breaking the vacuum. The first layer, which thickness was 150 Å, was deposited by electron beam evaporation; it was either pure Ge or As-doped Ge, with an As doping level in the range 10^{19} - 10^{20} cm^{-3}. Mo (150 Å) and W (3000 Å) were RF sputtered. We have investigated two different annealing techniques: rapid thermal annealing (RTA) under a forming gas and the semi-open box technique under an As overpressure, obtained with an InAs powder. The specific contact resistivity was determined by the transmission line method. The metallurgical analysis was performed with Auger Electron Spectroscopy (AES), Secondary Ion-Mass Spectrometry (SIMS), Transmission Electron Microscopy (TEM) and Scanning Transmission Electron Microscopy (STEM).

III-Results

III.1-Contact optimization

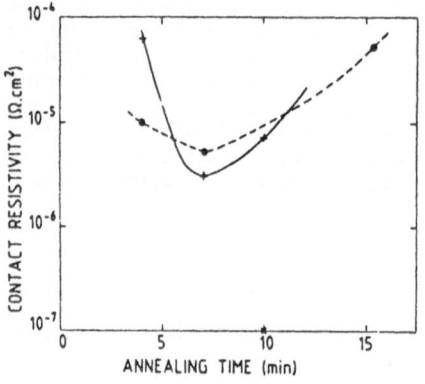

Fig.1 The specific contact resistivity of Ge(As)MoW is plotted as a function of the annealing time for different As doping level in the Ge film. The annealing temperature is 800°C.

——— pure Ge

- - - 10^{19} cm^{-3} As doped Ge

x 10^{20} cm^{-3} As doped Ge.

The influence of the annealing atmosphere on the formation of the GeMoW contact has first been investigated. With pure Ge-Mo-W, the contact was found to present a non ohmic behaviour unless it was annealed under an arsenic overpressure. The influence of the annealing atmosphere has also been investigated with the As doped Ge-Mo-W contact. In this case, it was possible to obtain an ohmic contact without the arsenic overpressure. However, the minimum resistivity achieved was 10^{-5} Ωcm^{2}. Better results were obtained in experiments combining the As overpressure and the As doped Ge layer. In figure 1, the resistivity is plotted for different As doping level. There is a decrease of the resistivity when the sample is annealed under an As overpressure; there is also a decrease of the resistivity when the As doping level in the Ge target is increased, the minimum resistivity is 5 10⁻⁶ Ωcm² for pure Ge, 3 10⁻⁶ Ωcm² when the As doping level is 10^{19} cm⁻³ and as low as 10^{-7} Ωcm² when the As doping level is 10^{20} cm⁻³.

III.2-Metallurgical evaluation

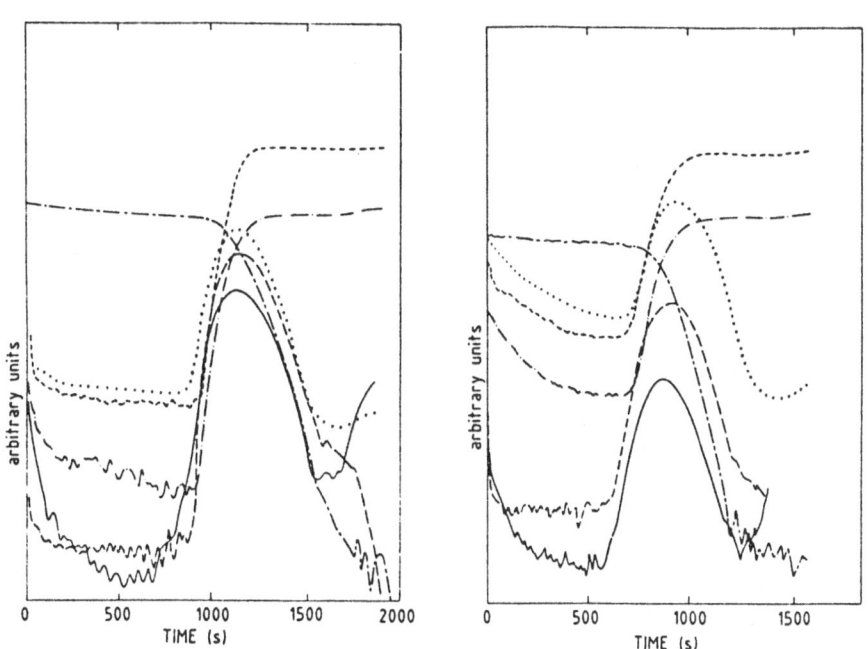

Fig.2 SIMS profiles after annealing under an As overpressure (a) pure Ge-Mo-W, (b) As doped Ge-Mo-W
- - - - Ga, —·—·—As, —·—·— W, - - - - - Mo, ———— Ge, ·········· In.

Pure Ge-Mo contacts annealed under an arsenic overpressure were studied by AES. Mo and Ge intermixed completely during the annealing treatment suggesting the formation of a stable phase. SIMS has also been carried out on pure Ge-Mo-W (fig 2.a) and As doped Ge-Mo-W (fig.2.b) samples before and after annealing under an As overpressure. The presence of the W layer degrades dramatically the depth resolution. The In profile, which presence is related to the InAs powder used for the As overpressure, has also been investigated. After annealing, the intermixing between Ge and Mo is not detected due to the poor depth resolution. However three points can be made: in both case, there is a notable diffusion of Ge which is more important in the case of As doped Ge-Mo-W, there is also an accumulation of In at the metal/GaAs interface. But, the main difference between the spectra is the important diffusion of Ga and As in the W film in the case of the As doped Ge-Mo-W contacts.

TEM has also been carried out on samples annealed either under a forming gas or under an As overpressure. With pure Ge-Mo-W contact, there is formation of a stable phase at the interface, which has been identified as $Mo_{13}Ge_{23}$ when the sample was annealed under a forming gas, and Mo_5Ge_3 when the sample was annealed under the As overpressure. In the case of As doped Ge-Mo-W, there is formation of two phases which have been identified as $MoGe_2$ and Mo_5As_4. In all cases, there was no phase including In.

IV-Conclusion

We have presented a study on the formation of the refractory ohmic contact GeMoW for n-type GaAs. The critical influence of the As overpressure on the contact formation has been investigated. In the case of pure GeMoW contact, the contacts were found to be ohmic only when the annealing was achieved under an As overpressure. The influence of As was further proved by the realization of ohmic contacts without the As overpressure, when using As doped Ge. Very low contact resistivities have been obtained, when the As doped GeMoW contact was annealed under an As overpressure. Moreover, the contact resistivity decreased when the As doping level in the Ge layer was increased; the lowest resistivity (10^{-7} Ωcm^2) was obtained with an As doping level of 10^{20} cm^{-3} in the Ge layer. An accumulation of In at the metal/GaAs interface has been detected, but it has not been proved that In played any active role in the formation of the contact, since no phase including In has been detected by TEM. Moreover, ohmic contacts have been obtained without In presence in the case of As doped GeMoW contact annealed under a forming gas. Very low specific contact resistivities, in the 10^{-7} Ωcm^2 range , have been obtained with an As doped Ge-Mo-W contact annealed with the semi-open box technique under an As overpressure.

References

1. K.Daoud-Ketata; J.F.Bresse; C.Dubon-Chevallier, Inst.Phys.Conf.Ser., 83 (1986) 301-306.

2. M.Murakami; W.H.Price; Y.C.Shih, J.Appl.Phys. 62 (1987) 3288-3294.

3. K.Daoud-Ketata; C.Dubon-Chevallier; C.Besombes; J.F.Bresse; P.Henoc, Electron.Lett. 23 (1987) 17-18.

Reliable, High Temperature Stable Schottky Contacts to GaAs Based on LaB$_6$ Diffusion Barriers

J. Würfl, J.K. Singh*, H.L. Hartnagel

TH Darmstadt, Institut für Hochfrequenztechnik, Merckstr. 25, 6100 Darmstadt, FR Germany

* on leave from:
Central Electronic Engineering Research Institute, Pilani,333031,India

Abstract

The thermal stability of Au-LaB$_6$ Schottky-diodes on n-GaAs has been investigated using I/V-measurements and XPS-analysis. The e-beam evaporated contact systems exhibit good electrical properties. It could be shown that the ideality factor and the barrier height improves after a certain annealing step and remains almost stable even after prolonged operation at 400°C. This demonstrates the suitability of lanthanum hexaboride to be used as a high temperature stable diffusion barrier in Au-LaB$_6$ Schottky contacts to GaAs.

1.   Introduction

Schottky contacts employing various types of diffusion barriers are most commonly used in GaAs microelectronics to provide stable operation conditions. Since GaAs has a relatively large bandgap in comparison to silicon, it is straight forward to exceed the save operation conditions of GaAs devices to much higher temperatures such as 400°C for example. The thermal stability of the metal contacts employed in these devices constitutes the prerequisit for a successful development of high temperature GaAs microelectronics. Temperature stable Schottky contacts on GaAs require effective diffusion barriers to prevent interdiffusion between the GaAs and the top metallization layer (mostly Au). These interdiffusion effects generally result in a rapid contact degradation at elevated temperatures. There are many approaches to realize high temperature stable Schottky contacts on GaAs using various types of diffusion barriers, but mostly sophisticated deposition and structurization processes are required. In this paper we consider the realization of a stable Schottky-metallization to GaAs based on the closely packed, chemically inert rare earth-boron compound Lanthanum hexaboride (LaB$_6$) as a diffusion barrier. LaB$_6$ can be easily electron beam evaporated and structurized using lift-off techniques [1, 2].

2. Experimental details

The Schottky diode structures were fabricated on bulk n-GaAs (n ≈ 1·10^{16}/cm^3). Since conventional AuGe based ohmic contacts cannot withstand temperatures up to 400°C without significant degradation [3] a high temperature stable ohmic contact system based on a tungsten silicide diffusion barrier [4] was employed for the common backside contact of the structures. The free GaAs surface between different diode structures (diameters 150μm) on the chip was passivated with PECVD Si$_3$N$_4$ to avoid surface oxidation. During lifetime testing the structures were mounted on a special ceramic test socket and electrically connected using thermocompression bonding. All high temperature tests have been performed in vacuum ambient (≈ 10^{-2} mbar).

3. Results and discussions

Figures 1(a) and (b) show the forward and the reverse I/V characteristics respectively in dependence on the different subsequent sample treatments.

- The diode characteristics change very significantly after annealing an as deposited contact at 200°C for several hours. Detailed calculations reveal that the barrier height increases from 0.75 eV to about 0.9 eV whereas the ideality factor drops from 1.2 to about 1.15.

- After the initial annealing process at 200°C the structures were temperature stressed at 400°C. The barrier height increases slightly but stabilises after 80 hours stressing at 400°C. The ideality factor remains nearly constant.

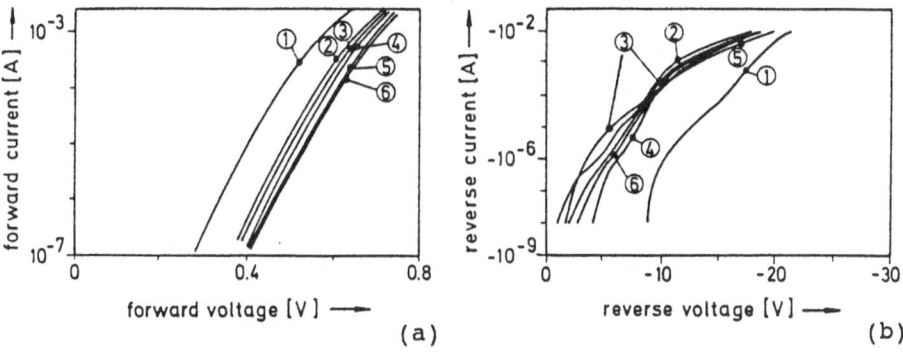

Fig.1: Forward (a) and reverse (b) I/V characteristics of Au-LaB$_6$ Schottky contacts after different types of sample treatments. (1) as deposited; (2) 200°C 17h; (3) 400°C 2h; (4) 400°C 21h; (5) 400°C 88h; (6) 400°C 151h

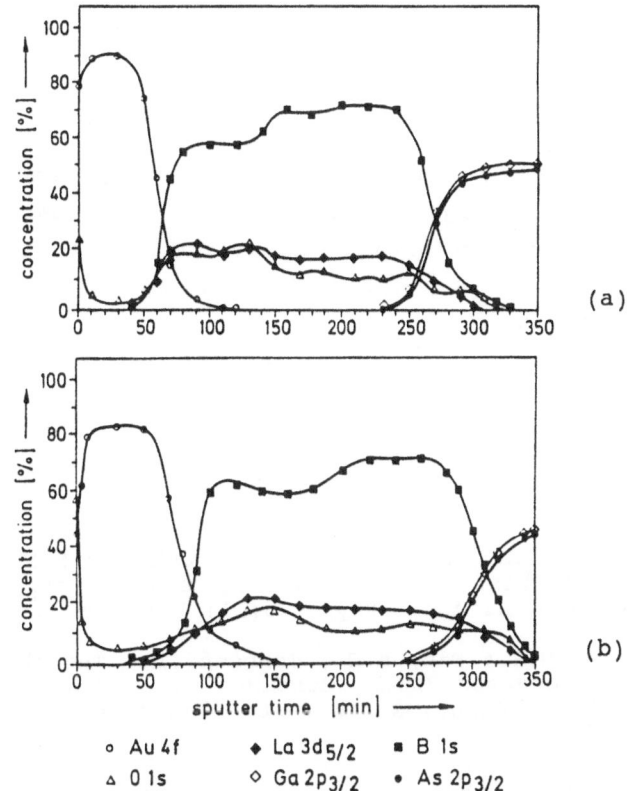

Fig.2: XPS concentration profile of a Au-LaB$_6$ Schottky contact. (a) as deposited; (b) thermal stressing at 400°C for 150 hours

The significant changes of the diode properties after the first annealing step (200°C) can be attributed to recrystallisation, to relaxation of mechanical strain in the metal layers and to the reduction of native Ga and As oxides at the LaB_6-GaAs interface [1]. The observed stable electrical properties are supported by corresponding XPS-concentration profiles obtained by successive Ar^+ ion sputtering and XPS-analysis. Figure 2 shows the concentration profiles of an as deposited (not thermally stressed) (a) and a thermally stressed (400°C, 150 hours) sample (b). No apparent interdiffusion effects can be detected even after prolonged thermal stressing at 400°C. The observed "broadening" of the LaB_6 profile of the thermally stressed sample (figure 2b) is probably due to thermally induced relaxation and recrystallisation processes in the films. In this case the Ar^+-ion sputter rate during XPS-profiling will be different for thermally stressed and as deposited samples.

Conclusions

The paper demonstrates the suitability of Lanthanum hexaboride to be used as a high temperature stable diffusion barrier in $Au-LaB_6$-Schottky contacts to GaAs. These contact systems show good electrical properties and can be fabricated using conventional techniques.

References

/1/ Uchida, Y.; Yokotsuka, T.; Nakashima, H.; Takatani, S.: Electrical properties of thermally stable LaB_6/GaAs Schottky diodes, Appl, Phys. Lett. 50 (11), 1984, pp. 670-672.

/2/ Takatani, S.; Uchida, Y.; Yokotsuka, T.; Nakashima, H.: GaAs MESFETs with a thermally stable LaB_6 self-aligned gate, Jap. J. Appl. Phys. 26(11) 1987, pp. L1770-L1773.

/3/ Marlow, G.S.; Das, M.B.; Tongson, L.: The characteristics of Au-Ge-Based Ohmic Contacts to n-GaAs Including the Effects of Aging. Solid State Electronics 26 (1983) 4, S. 259-266

/4/ Nassibian, A.G.; Würfl, J.; Hartnagel, H.L.: RTA Non Alloyed Ge-W/Si-Au and Ge-NiW/Si-Au High Temperature Contacts on GaAs. 12th Workshop on Compound Semiconductor Circuits, Cadro-Lugano, Schweiz (1988)

GaAs/InAs Heterostructures Grown by Atomic Layer Epitaxy

J. Ahopelto, H. P. Kattelus, J. Saarilahti, and I. Suni

Technical Research Centre of Finland
Otakaari 7 B, SF–02150 Espoo, Finland

Summary

Gallium arsenide layers and gallium arsenide/indium arsenide single quantum wells have been grown by Atomic Layer Epitaxy (ALE) using gallium and indium chlorides and arsine as source materials. The RBS/channeling technique gives minimum yields of 4 % for GaAs layers and 15 % for strained InAs layers compared to corresponding random spectra indicating reasonable crystalline quality. This is further confirmed by observation of a PL peak with 12 meV FWHM at 12 K originating from InAs single quantum well.

Introduction

InAs/GaAs is a promising material combination for heterostructure devices such as high electron–mobility transistors, heterojunction bipolar transistors and quantum well lasers. Lattice mismatched epitaxy has been succesfully used for thin layer thicknesses to yield pseudomorphic InAs quantum wells in GaAs host material. Although many growth techniques have been reported to produce such layers most of them suffer from poor thickness control or layer nonuniformity. The ultimate technique to achieve a monolayer control is to deposit one atomic layer at the time allowing the surface to limit the growth. This is feasible by exposing the substrate surface to each constituent of the compound in an alternating sequence of source materials. This approach of atomic layer epitaxy, originally demonstrated for II–VI compounds [1], has been applied to many well established techniques such as MOVPE and MBE or MOMBE. However, the selflimiting mechanism of ALE is not expected to be perfect for metalorganic source materials which tend to decompose and react in gas phase. This has been shown to lead to a significant temperature dependence in the thickness per growth cycle [2]. A more stable source material configuration based on group III halides and group V hydrides is therefore of interest for achieving well controlled epitaxy. Steric problems and the equilibrium of adsorbed species with the gas phase limit the growth rate even in this case. However the thickness per growth cycle is now always restricted to a maximum of one monolayer. This can be used to increase the process latitude by widening the temperature window in which the selflimiting growth occurs.

Experimental

The process used in this work is a modification of the conventional hydride vapor phase epitaxy method. The source materials are arsine and gallium and indium chlorides formed by the reaction of the elemental metal with HCl. The reactants are pulsed sequentially into the carrier gas, which is argon. The reactor walls are maintained hot to avoid the condensation of the chlorides to the walls in the growth zone. The tube has a rectangular cross–section to optimize the flow pattern and the source material consumption. Two types of substrates are used: GaAs (100) and GaAs (100) covered with 500 nm thick patterned SiO_2 deposited by plasma–CVD. The substrates are degreased in ultrasonic baths in TCE, acetone and isopropyl alcohol for 10 minutes in each and dipped in 1:10 mixture of HF:ethanol before loading into the growth chamber. The reactor pressure is maintained at 500 Pa (5 mbar) with a rotary vane pump. The total flow rate of argon is adjusted with needle valves and is about 3.2 cm^3/s. The gas flow rate and the gas velocity are limited by the throttling action of the arsine cracker used to eliminate the toxic waste from the exhaust lines. The doses of reactants, 10^{17}–10^{18} molecules per pulse, are controlled by the pulse durations and the line pressures of HCl and AsH_3. The pulse durations are typically 30–50 ms for HCl and 1 s for AsH_3 with 3–5 s purge between the pulses. The total cycle time to deposit one monolayer is 8–12 s due to the relatively low gas velocity. To keep the HCl and AsH_3 pulses from overlapping the process is monitored and adjusted with a mass spectrometer. The substrate temperature, measured by a thermocouple, is varied in our experiments between 350 and 550 ºC. The grown layers are analyzed by Rutherford backscattering spectrometry (RBS) and low temperature photoluminescence (12 K). The thicknesses are obtained by a Dektak profilometer from selectively grown layers on the patterned substrates after removing the silicon dioxide layer.

Results and discussion

The grown gallium arsenide layers typically display a smooth surface morphology providing that the reactant pulses do not overlap. If overlapping occurs the morphology degrades. Under optimized conditions, *i.e.* substrate temperature around 450 ºC and 3–5 s purge between the reactant pulses, the growth rate per cycle is 0.8–1 monolayers per cycle. RBS/channeling of 1.33 MeV He⁺–ions has been used to investigate the layer quality. For ALE grown homoepitaxial GaAs layers the minimum yield obtained from channeling measurements is χ_{min}= 4 %, which favorably compares with the minimum yield in a perfect GaAs crystal.

The usefulness of ALE for GaAs/InAs heteroepitaxy was studied in structures conforming to a layer sequence of GaAs/InAs/GaAs/InAs with a corresponding

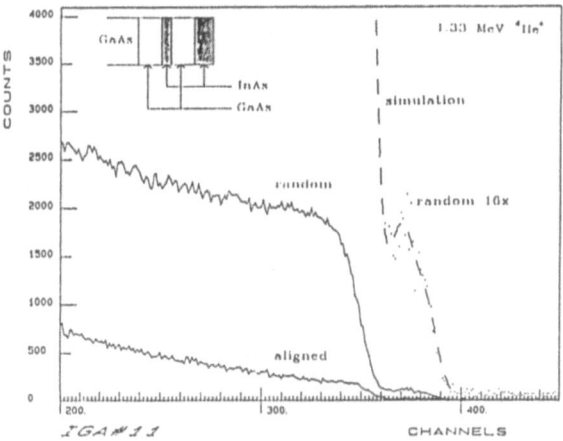

Fig. 1. Random and aligned RBS spectra from ALE grown GaAs/InAs heterostructure

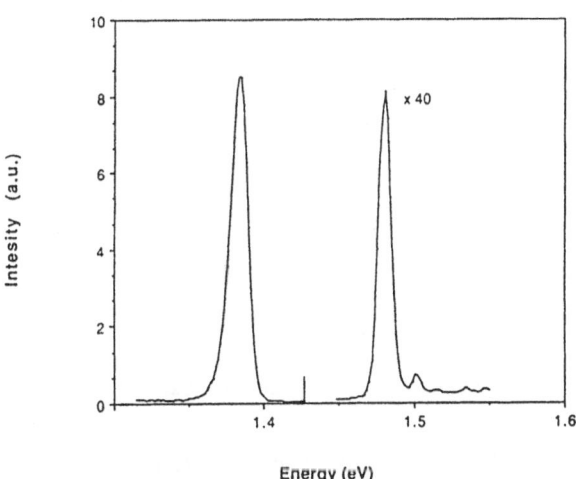

Fig. 2. 12 K photoluminescence spectrum for an InAs quantum well 5...6 monolayers thick. In the figure the GaAs related peaks are shown in 40× magnification.

sequence of 200/10/200/20 deposition cycles. A RBS analysis of such a multilayer (spectra shown in fig. 1) indicates that the approximate amount of In in the deeper InAs layer is 3.5 ($\pm$ 0.5)$\times 10^{15}$ atoms/cm^2. This is equivalent of 5 to 6 monolayers suggesting that full monolayer coverage has not been achieved. Minimum channeling yields from this strained multilayer in <100> direction are χ_{min}=11 % for GaAs and χ_{min}=15 % for InAs providing evidence of a reasonable crystal quality. The PL spectrum obtained from this sample at 12 K using the 488 nm line of argon laser is shown in figure 2. The exitation power is 5 mW focused to a 0.2 mm spot. The peaks observed at 1.501 and 1.480 eV are GaAs related and the intense peak at 1.383 eV with a 12 meV FWHM originates from the quantum well. The result is in agreement with the single quantum well energies reported by other authors for the same material system [3].

Conclusion

Epitaxial gallium arsenide and indium arsenide films have been grown on GaAs substrates by atomic layer epitaxy using gallium or indium chlorides and arsine as source materials. RBS/channeling and photoluminescence provide evidence of a reasonable crystal quality.

Acknowledgement

The work was partially supported by the Technology Development Centre (TEKES).

References

[1] T. Suntola and M. J. Antson, US Patent No. 4–058–430 (1977).

[2] J. Nishizawa, T. Kurabayashi, H. Abe, and N. Sakurai, J. Electrochem. Soc., 134 (1987) 945.

[3] J. Y. Marzin and J. M. Gerard, Superlattices and Microstructures, 5 (1989) 51.

High Temperature LPCVD of Dielectrics on III-V Substrates for Device Applications

Y. I. NISSIM, J.M. MOISON, C. LICOPPE, G. POST

C.N.E.T. Laboratoire de Bagneux
196 Avenue Henri Ravera
92220 Bagneux (France)

Summary

Silicon dioxide as well as silicon nitride films deposited on a III-V substrate (InP, GaAs, InGaAs) are obtained in a reduced pressure, air and water cooled CVD reactor, with a rapid thermal heating. Deposition rates as high as 100 Å/sec are obtained for deposition of SiO_2 at 700°C and 30 Å/sec for Si_3N_4 at 800°C. High temperature deposition is thus obtained on a III-V substrate without any surface damage. The layers display excellent structural and electrical properties suited for passivation and MISFET applications. The properties of III-V semiconductor/insulator interfaces can be improved under an in-situ treatment of the substrate with the reactive gases. Interface studies show that both silane and ammonia can be utilized prior to deposition to reduce the native oxide of InP.

1. Introduction

Silicon dioxide and silicon nitride films are essential in the fabrication of III-V devices. Beside their masking properties they are utilized as active layers for passivation or gate dielectrics in MISFET structures. Since III-V materials are unstable at elevated temperatures, there is a need to develop appropriate deposited dielectrics with deposition temperature below 300°C. The techniques proposed to overcome this difficulty are direct or indirect plasma [1, 2], low temperature pyrolitic CVD [3] and photolytic CVD [4, 5]. However when the thermal exposure of a III-V substrate is reduced to few seconds, the maximum temperature allowed before surface degradation can be considerably increased. Deposition temperature as high as 700°C for InP substrates and 800°C for GaAs substrates can be used in a system combining LPCVD and rapid thermal heating. This flash CVD technique gives precise control of thermally driven surface reactions. In this case the substrate temperature rather than the flux of reactive gas, is used as switch to turn the CVD reaction on and off. The thermal exposure of the semiconductor substrate is thus minimized. This technique is a unique opportunity for III-V material to have access to high temperature LPCVD processes, suited for device applications.

2. Experimental

The flash LPCVD reactor consists of an horizontal quartz chamber mounted above a row of halogen lamps [6]. The reactor walls are constantly air and water cooled to insure fast temperature cycling on the substrate, and avoid unwanted contamination. The reactant gases utilized are silane and oxygen for SiO_2 deposition and silane and ammonia for Si_3N_4 deposition. All the gases are diluted in nitrogen. The total pressure of the system is regulated with a throttle valve to a value of few Torr. The overall temperature cycles last for few tens of seconds with a rise time of about 50°C/sec and a natural cooling time of few seconds. Ellipsometry and IR absorption spectra are utilized to characterize the deposited layers. Capacitance and conductance versus voltage characteristic at 1 MHz are systematically measured on MIS diodes fabricated on the III-V semiconductor/deposited insulators structure to qualify the structural and interface electronic properties of the insulator.

3. In situ surface preparation

One of the major advantages of the flash LPCVD technique is that in-situ processing can be easily accomplished. The substrate is hot only during deposition and not during purging or gas flow stabilization. Different processes can thus be made in-situ without cross contamination of one to the other. The reactant gases have been utilized in order to clean the surface of InP from contamination and native oxides. The results obtained when the cleaning procedure of the surface of InP is made under a silane flow has been reported elsewhere [7, 8]. Silane reacts at 300°C with the oxidized InP surface restores a covalent InP-Si bonding at the interface.

When ammonia is used to prepare the oxidized InP surface the scheme is different [9]. When the substrate is held at 300°C the cleaning procedure occurs in two stages. The first stage is obtained under exposure of few Torr of unexcited ammonia. In this case ammonia removes the weakly bonded oxygen atoms down to a monolayer coverage. Removal of the remaining strongly bonded oxygen atoms can be obtained with ammonia excited by a hot filament. The resulting surface is free of all contaminants and of nitrogren, and its crystal structure is fairly good. It is mostly InP covered by adsorbed hydrogen with about 5 % of the surface occupied by a small quantity of indium atoms liberated by the reduction of the oxyde and gathered as metallic clusters.

4. Dielectric deposition

The silicon dioxide deposition is optimized on InP for a deposition temperature of 700°C. In a mixture of silane and oxygen diluted in nitrogen under 5 Torr of total pressure, the deposition rate is 100 Å/sec. The detailed characterization of MOS capacitor fabricated on this structure and reported before [8] indicates that this technology is perfectly suited for MISFET applications. In situ surface preparation prior to SiO_2 deposition results in a reduction by a factor 2 of the hysteresis cycle. This is observed either with the silane or the unexited ammonia cleaning procedure described above. Two microns gate length MISFET were fabricated on this structure. These transistors display a transconductance of 70 ms/mm, a small drift behaviour of the I(V) characteristics (2.7 % in 3 hours for I_{DS} at $V_G = 3$ V). Since the deposition process is made at high temperature, the transistors are stable under aging experiments.

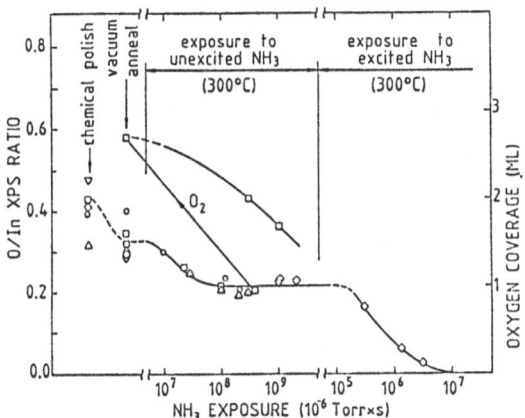

Fig. 1: Evolution of XPS-derive ratio i.e. oxygen coverage with successive treatments : chemical polish, vacuum anneal and exposures to unexcited and excited ammonia.

The case of Si_3N_4 deposition using a flash LPCVD technique is reported here for the first time. Using a nixture of silane and ammonia diluted in nitrogen, Si_3N_4 films could be deposited on GaAs substrates at 800°C with a deposition rate of 30 Å/sec. Structural and optical characterization of the layers indicate that the gas decomposition process is complete. This is well illustrated in the I.R. absorbance spectrum of the SiO_2 and Si_3N_4 layer deposited with this technique. As can be seen in these spectra no trace of residual bond such as Si-H, or Si-O-H are observed. The measured index of refraction (2.2), and the low etching rate (7 A/min) confirm the excellent structural quality of the Si_3N_4 film. Passivation tests are underway and results will be reported later.

5. Conclusions

It has been shown that LPCVD can be controlled by radiant heat. This technique permits high temperature processing of III-V materials. Furthermore multiple processing steps can be easily carried out in-situ within such a reactor.

176

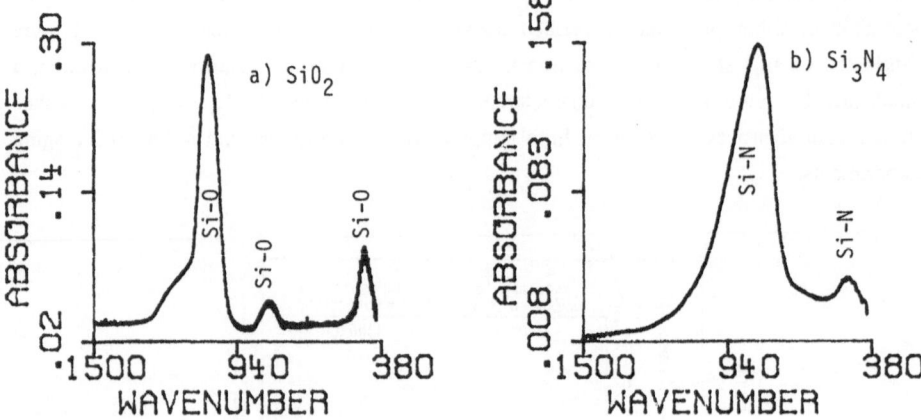

Fig. 2: Infrared absorption spectra of a deposited layer of a) SiO$_2$ (700°C) and b) Si$_3$N$_4$ (800°C) using the rapid thermal LPCVD technique.

References

1. Meiners, L.G. : Indirect plasma deposition of silicon dioxide, J. Vac. Sci. Tech., 21 (1982) 665

2. Pande K.P. and Nair, V.K.R. : High mobility n-channel MOSFETs based on SiO$_2$-InP interface, J. Appl. Phys., 55 (1984) 3109

3. Bennett, B.R., Lorenzo, J.P. and Vaccaro, K. : Electrical properties of low-temperature pyrolytic SiO$_2$, Electron. Lett. 24 (1978) 172

4. Nissim Y.I., Regolini, J.L. Bensahel, D. and Licoppe C. : Low pressure photo-CVD of SiO$_2$ on InP substrates, Electron Lett. 24 (1988) 488

5. Dimitriou P., Post G. and A. Scavennec : UV photon assisted CVD of SiO$_2$ for low drift InP MISFET's. Journal de Physique, colloque C5, Tome 50 (1989) 675

6. The reactor is designed after french patent n° 8614896; an industrial version is being designed under licence

7. Licoppe C., Moison J.M., Nissim Y.I., Regolini J.L., and Bensahel D.: Surface reactions of silane with oxidized InP and their application to the improvement of CVD grown InP based MIS devices. Appl. Phys. Lett 53(1988) 1291

8. Nissim Y.I., Licoppe C., Moison J.M. and C. Meriadec : in-situ processing of InP by flash LPCVD for surface preparation and oxide deposition, J. Phys. Coll. C 4, Tome 49 (1988) 213

9. Moison J.M., Nissim Y.I., Licoppe C. : Stabilization and removal of the native oxides of InP by low pressure exposure to NH$_3$. Submitted to J. Appl. Phys.

10.Nissim Y.I., Bensoussan M., Post G., Bensahel D., Regolini J.L.: Use of UV or IR lamps E for photothermal deposition of thin film materials, E-MRS proceedings vol. XV (1987) 213

Short-time Diffusion of Zinc into InP and InGaAsP from Spin-on Films

U. SCHADE[a], P. ENDERS[a], G. KÜHN[b], B. UNGER[c], K. VOGEL[a]

[a]Central Institute of Optics and Spectroscopy, Berlin, GDR
[b]Karl-Marx-University, Leipzig, GDR
[c]Central Institute of Inorganic Chemistry, Berlin, GDR

Summary

Zn-containing sols are available for short-time diffusion of Zn into InP and InGaAsP. It is shown that the net acceptor concentration depends on the temperature cycle due to outdiffusion of interstitial Zn. In SI InP interstitial Zn acts as single-charged donor. Short-time diffusion is able to produce p^+-regions in InGaAsP emitting at 1.3 µm.

Introduction

Zinc is the most commonly used p-type dopant in the $A_{III}B_V$-technology to form p-n junctions or p^+-regions for ohmic contacts. Much investigation is spent on the diffusion from solid sources based on emulsions by spinning on, e.g. /1/. To form shallow diffusion profiles short time diffusion is successfully applied and high surface concentrations of Zn has been achieved /2/. In InP and InGaAsP the net acceptor concentration is much smaller than that of Zn due to the donor behaviour of interstitial Zn. Additional heat treatments decrease the concentration of interstitial Zn and increase the net acceptor concentration. /3/

Experimental

The substrates are S-doped n-type and Fe-doped SI (001) InP and Zn-doped p-type $In_{0.73}Ga_{0.27}As_{0.46}P_{0.54}$ LPE-layers lattice matched on (001) InP. The coating procedure, especially, the preparation of the Zn-containing sols, the cleaning treatment of the semiconductor surfaces and the analytical methods are

described earlier /4/. The short-time diffusion was carried out
in a stripe heater system with various cooling rates under
purified nitrogen atmosphere (Fig. 1). After diffusion the spun
on films having a typical thickness of 200 nm are etched in
buffered HF.

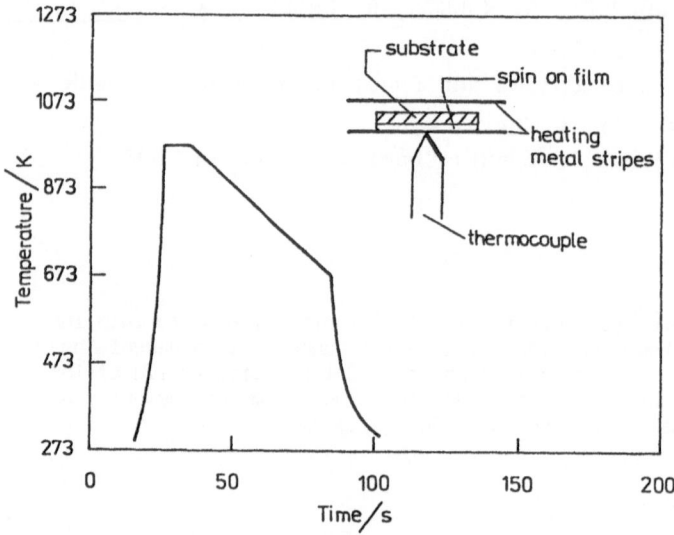

Fig. 1. A typical temperature cycle with a cooling-rate of
6 K/s for a diffusion at 700°C, 10 s

Results

A micrograph of the top and cleaved surface of n-type InP which
was diffused at 700°C, 5 s is shown in Fig. 2. The homogeneous
front of the p-type layer and the surface being smooth up to a
temperature of 700°C are seen. The spin-on source is not infinite
at temperatures of 700°C and due to out-diffusion the maximum Zn-
concentration decreases from 6×10^{19} cm^{-3} (5 s) to 1×10^{19} cm^{-3}
(30 s) in n-type InP (7×10^{17} cm^{-3}), both at a cooling rate of
55 K/s.
Fig. 3 shows the Zn and net acceptor profiles in dependence of
the cooling rate for SI InP. At lower cooling rates the net
acceptor concentration is roughly equal to that of Zn and lies
in the middle of these concentrations at higher rates.

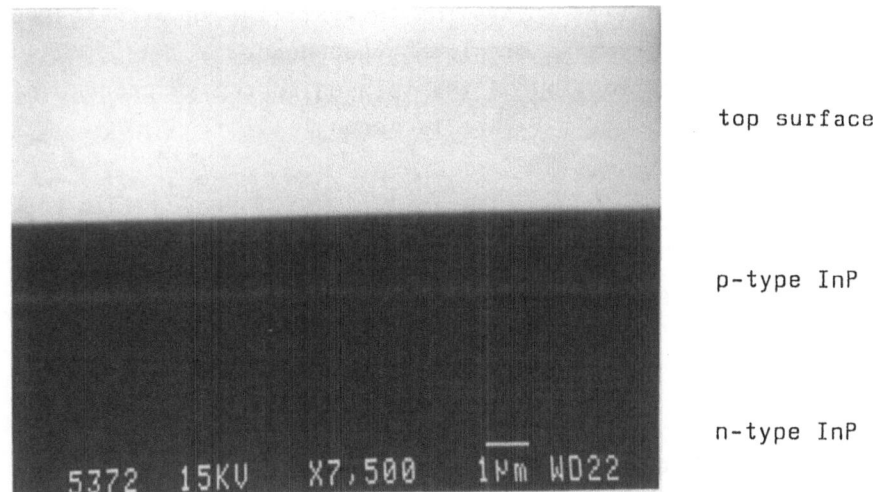

Fig. 2. SEM micrograph of diffused n-type InP (7×10^{17} cm^{-3})

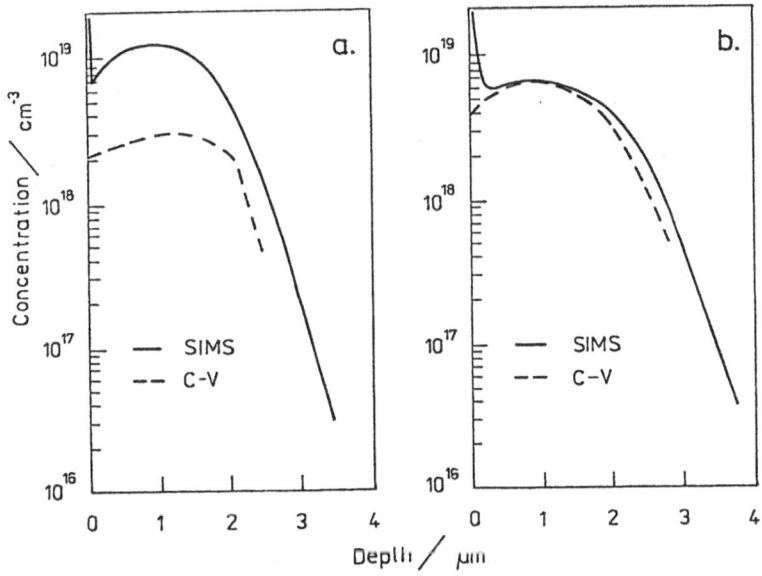

Fig. 3. Zn and net acceptor concentration for a diffusion in SI
InP at 700°C, 10 s with a cooling rate of a) 55 K/s and b) 6 K/s

180

This confirms model 2 in /3/. The interstitial Zn diffuses out
at lower rates acting as single-charged donor.
In Fig. 4 the application of the spin-on source to produce highly
p-doped InGaAsP contact layers is shown.

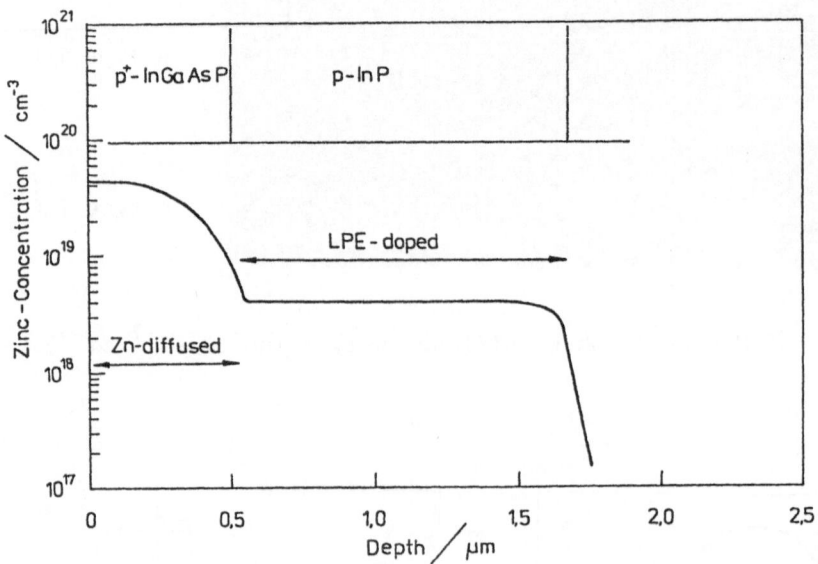

Fig. 4. Zn profile after diffusion at 675°C, 5 s and a cooling
rate of 6 K/s in InGaAsP

References

1. Arnold, N.; Schmitt, R.; Heime, K.: Diffusion in III-V semiconductors from spin-on sources.
   J. Phys. D: Appl. Phys. 21 (1983) 443-474.

2. Amann, M.-Ch.; Stegmüller, B.: Narrow-stripe metal-clad
   ridge-wave guide laser for 1.3 µm wave length.
   Appl. Phys. Lett. 48 (1986) 1027-1029.

3. van Gurp, G.J.; van Dongen, T.; Fontijn, G.M.; Jacobs, J.M.;
   Tjaden, D.L.A.: Interstitial and substitutional Zn in InP
   and InGaAsP.
   J. Appl. Phys. 65 (1989) 553-560.

4. Unger, B.; Schade, U.; Hähnert, M.; Vogel, K.:
   Glassy dopant deposits on semiconductor surfaces prepared by
   sol gel technique.
   Int. Congr. Opt. sci. Eng., Topical Conf. "Glasses for
   Optoelectronics", Paris, 1989 (in ress).

Schottky Barrier Enhancement on InP Using Pseudomorphic GaInP MBE Layers

S. LOUALICHE, A. GINUDI, A. LE CORRE, D. LECROSNIER,

C. VAUDRY, L. HENRY,

Centre National d'Etudes des Télécommunications (CNET) Lannion B Route de Trégastel BP 40 22301 LANNION FRANCE CEDEX

ABSTRACT

The GaP and GaInP materials are used as high gap semiconductors on InP to fabricate Schottky diodes. The devices exhibit excellent electrical properties when the ternary strained layer is below the critical thickness. The best device (0.8 eV barrier height ; 1.14 ideality factor ; 0.1 nA reverse current at -1V and over 250V breakdown voltage) is obtained on a sample where 11 Å (4 monolayers) of GaP are used as a high gap material to increase the Schottky barrier height on InP.

Indium Phosphide has excellent transport properties that makes it suitable for high speed electronic applications. Due to its low Schottky barrier height, (0.43 eV for Au/InP), it cannot be used as a gate in Field Effect transistor (FET) fabrication. InP has its Fermi level pinned by surface traps [1]. The Fermi level position at the surface (ψ_S) stays almost constant and the barrier height Φ_O of the Schottky diode on InP is almost independent of the metal work function and can be expressed as $\Phi_O = Eg - \psi_S$ (Eg : Bandgap). To increase the Schottky barrier height since there is presently no solution to avoid surface Fermi level pinning, we have to find a high bandgap semiconductor. The use of a heterostructure made of a high gap on a low gap (InP) material increases the Schottky barrier height by, $\Delta\Phi$ which is equal to the conduction band offset ΔE_C [2] between the two materials.

There is not a great number of III-V compounds with bandgap higher than that of InP and with the same lattice

constant. Semiconductors with Aluminium as a group III element
are to be avoided because of the oxidation of these compounds
in the ambient. Gallium Phosphide has a bandgap of 2.27 eV at
room temperature and has two interesting properties : 1) GaP
on InP has a band offset ΔEg with ΔE_C = 0.8 ΔEg [3] which is
favorable for Schottky barrier height increase and 2) this
material does not have surface Fermi level pinning [4]. The main
problem is the high value of the lattice mismatch between InP
and GaP (7.1 %). Experimental results show that, the condition
to obtain a good Schottky diode (with high Schottky barrier
height : Φ and ideality factor : $n \simeq 1$) is to avoid the
deposition of a material with a thickness exceeding the
critical thickness. The assumption that $\Phi = \Phi_0 + \Delta E_C$ in the
case of a heterostructure[2] GaP/InP is only valid if there is
no tunneling through the high gap material. To avoid
tunneling, the thickness of the high bandgap material must be
greater than about 50 Å. This condition cannot be fulfilled
for GaP because the critical thickness for a material of 7.1%
lattice mismatch is about 15 Å [5]. To increase the critical
thickness, ternary semiconductors like $Ga_x In_{1-x} P$ material are
to be used. The critical thickness of 100 Å is obtained for
GaInP with 25 % Gallium content, this material has a bandgap
of 1.57 eV at room temperature [6].

The samples used in the present experiments are grown
by gas source molecular beam epitaxy (GS MBE, where only the
group V element is provided by a gas source). The growth
temperature is 500°C and two different Indium sources are used
for InP and GaInP. The growth on a semi-insulating InP
substrate, begins with a 2 μm thick InP layer Silicon doped to
2×10^{16} cm^{-3} or n type undoped ($\sim 10^{16}$ cm^{-3}), followed by the
GaInP strained layer. The Schottky device is fabricated by
evaporating gold in vacuum (10^{-6} Torr) through a metallic mask
directly on the GaInP surface. An excellent device is obtained
for a GaInP layer with 18 % Gallium content, 150 Å thickness
and 1.5eV bandgap. The Schottky diode of 0.2 mm diameter has a
barrier height of 0.64 eV, an ideality factor of 1.08 and a
reverse current of 7 nA at -1V (Fig. 1, Table 1). These
measures are performed by current-voltage (I(V))
characteristics at 300 K. A similar diode on a material with

higher Gallium composition gives better results, the best diode is obtained with 100 % Ga (Table 1).

Table 1

X Ga	0	18 %	25 %	100 %	100 %	GaAs
e($\AA$)		150	100	11	100	100
Eg_{eV}	1.35	1.50	1.57	2.27	2.27	1.42
I(-1V)	10μA	7nA	0.3nA	0.1nA	6μA	10μA
Φ_{meV}	434	640	726	800	510	430
n	1.1	1.08	1.1	1.1	1.9	2
		Pseudomorphic			relaxed	

Au/Ga$_x$In$_{1-x}$P/InP Schottky diode.

X is the Ga composition of the high gap material, Eg its gap and e its thickness.

The barrier height is Φ, the ideality factor is n and I(-1V) is the reverse current at -1V for a diode diameter of 0.2 mm.

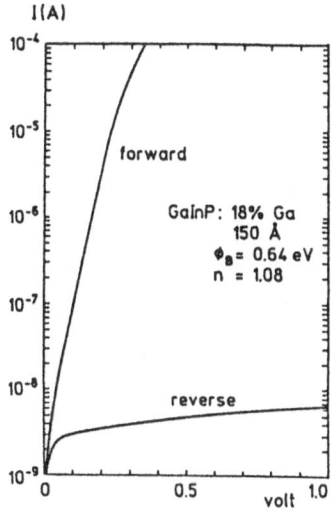

Figure 1 : I(V) room temperature characteristics of a 0.2 mm diameter Schottky device Au/GaInP/InP with 18 % Ga.

These Schottky diodes exhibit better I(V)

characteristics when they are compared to similar experiments where high gap mismatched materials above critical thickness are used to increase the surface Schottky barrier height. To fabricate FET'S on InP Chen et al [7] have used GaAs deposition (lattice mismatch 3.8 %) with thicknesses of 600 Å to 1300 Å well above the critical thickness of $\sim$ 40 Å, and obtain reverse current in the range of 0.5 µA to 200 µA at -1V. In this study we have performed the same experiment with 100 Å GaAs on InP and we have obtained a barrier height of 0.42 eV and a reverse current of $\sim$ 10 µA (Table 1). A more recent work [8] with a GaInP layer of 45 % Ga content and of 1000 Å thickness deposited on InP leads to a leakage current of 10 µA at -1V. This result is consistent with our experiments where the GaInP thicknesses (e_c) are chosen greater than the critical thickness. For a 18 % Ga content ($e_c \simeq$ 150 Å) a material with 400 Å thickness gives a Schottky diode with Φ = 0.51 eV, an ideality factor of 2 and a reverse current of 6 µA at -1V compared to the values of : 0.64 eV, 1.08 and 7nA respectively obtained on the same material with the thickness kept below the critical value (Table 1). For the reverse current the degradation is about 3 order of magnitude. These results indicate that dislocations are to be avoided to obtain Schottky contacts exhibiting good I(V) characteristics (high Φ, n $\sim$ 1 and low reverse current).

In conclusion, Schottky devices with a high barrier height (0.8 eV) and a good ideality factor (1.1) are obtained on a pseudomorphic GaInP/InP heterostructure. It was been found that the thickness of the high gap strained layer must be kept well below the critical value to avoid the destruction of the device during experimental measurements or other thermal treatments.

REFERENCES

1. T. Kendelewicz, N. Newman, R.S. List, I. Lindau, and W.E. Spicer,
J. Vac. Sci. Technol. B3, 1206, 1985.

2. M. Eizenberg, M. Heiblum, M.I. Nathan, N. Braslau and P.M. Mooney,
J. Appl. Phys. <u>61</u>, 1516, 1987.

3. Jerzy M. Langer, Helmut Heinrich, Physical Review Letters, <u>55</u>, 1414, 1985.

4. S.M. Sze, in "Physics of Semiconductors devices", John Wiley and Sons, New York, p. 275 and 541, 1981.

5. J.W. Matthews and A. Blakeslee, J. Cryst. Growth <u>27</u>, 118, 1974.

6. Landolt, Bornstein, Group 3, Volume 17, Semiconductors Physics of Group IV Elements and III-V Compounds, p. 613, Springer Verlag Berlin, 1982.

7. C.Y. Chen, S.N.G. Chu, A.Y. Cho,
Appl. Phys. Lett. <u>46</u>, 1145, 1985.

8. A. Hosseini Therani, D. Decoster, J.P. Vilcot, M. Razeghi,
J.A.P <u>64</u>, 2215, 1988.

Control of the Si_3N_4-InGaAs Interfaces by Constant Capacitance Transients

J.M.LOPEZ-VILLEGAS, J.SAMITIER, J.R.MORANTE and J.ANTON
P.BOHER* and M.RENAUD*

  Càtedra d'Electrònica. Fac.Física. Universitat de Barcelona
  Diagonal 645-647. 08028 Barcelona
* Laboratoire d'Electronique et de Physique Appliquée (LEP),
  3 Avenue Descartes, F-94451 Limeil-Brévannes Cedex,France.
  LEP is a member of the Philips research organization.

Summary

  An experimental technique to control the defects distribution in Si_3N_4-$In_{0.53}Ga_{0.47}As$ strained interfaces is presented on the basis of the analysis of the constant capacitance transients. This method allows to study the time kinetics in a wide range from $10^{-2}s$ up to 10^3s. The results indicate that the behaviour of the slow interface states involves other mechanisms than a simple direct tunnel, like a charge transfer by hopping.

Introduction

  The control of the insulator-semiconductor interface is a crucial factor in device processing. Moreover, as the ratio of the device surface to the volume increases due to the size reduction, the electrical quality of the interface becomes more important to obtain a high performance, uniformity, reliability, etc. Unfortunately, bad interfaces are often obtained in III-V based devices, especially, when native oxides can appear between the insulator and the semiconductor with an unsuitable electrical quality. This situation happens in InGaAs. In this case, the removal of native oxide and posterior passivation becomes an interesting alternative, /1/. However, this procedure can create defects located at even several monolayers underneath the semiconductor surface, which can determine the posterior growth of the insulating layer. This work is devoted to analyze the slow interface states related to these defects.

  While the fast interface states have already been analyzed /2/, the slow states are hardly known in spite of their importance. The relaxation mechanism trough them give rise to large drifts and hysteresis effects in the electrical characteristics of field effect based devices, like MISFET's.

Experimental details

The measurements have been performed on InGaAs samples passivated as described in a previous paper /1/ considering different deposition rates of the Si_3N_4 layer. The measurements were made by using the constant capacitance technique, which allows to keep the electric field and the majority carrier concentration constant at the interface. The kinetics were analyzed in the temperature range between 77 and 175 K as a function of the low and high capacitance values, respectively C_1 and C_h, as well as a function of the capture time, t_c, figure 1.

In these experiments the slow interface states contribution has been found to be significative for capture times higher than 10ms. Moreover, it should be remarked that the complete emission of the charge trapped in the slow interface states requires emission times much longer than the capture times. Figure 2 shows the dependence of the emission transient amplitude due to the slow states on the capture time for different values of C_1. This result indicates us that the slow interface states contribution to the transients is not related to the Fermi level position at the interface, which is determined by the C_1 value.

In order to obtain a detailed time resolved kinetics we have used the Isothermal Transient Spectroscopy, ITS, /3/. This technique is more suitable than the DLTS technique to analyze slow relaxation phenomena, like the present case. Each ITS spectrum is directly related to the relaxation time constants distribution. Figure 3 corresponds to the ITS spectra of the transients shown in figure 1. For the initial values of the relaxation time (lower than $10^{-1}s$) the amplitude of ITS spectra reaches a saturation value, when the capture time is increased. Moreover, when the contribution of the fast interface states is subtracted, the saturation values of the ITS spectra are proportional to the majority surface concentration during the capture process, figure 4. On the contrary, this saturation is not reached for times higher than $10^{-1}s$ and the ITS amplitude shows a linear decrease with the relaxation time logarithm.

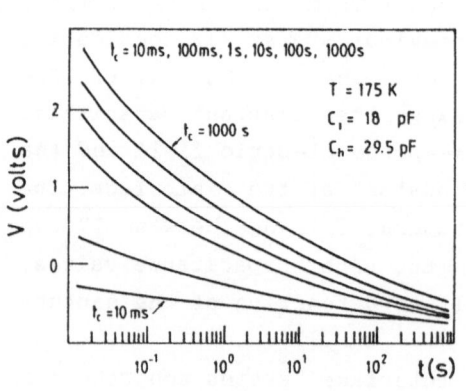

Figure 1. Voltage transients at constant capacitance as a function of capture time

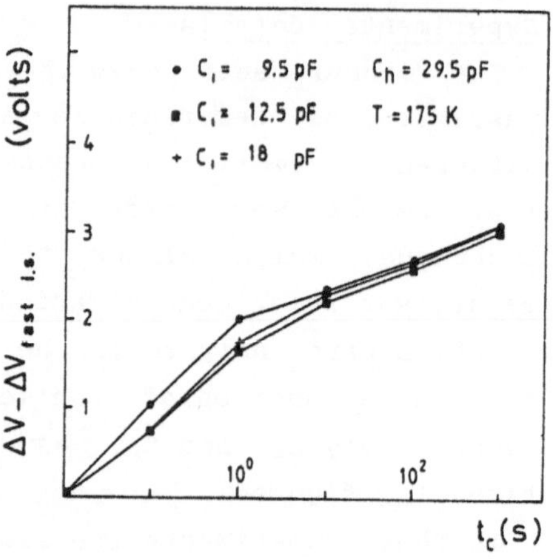

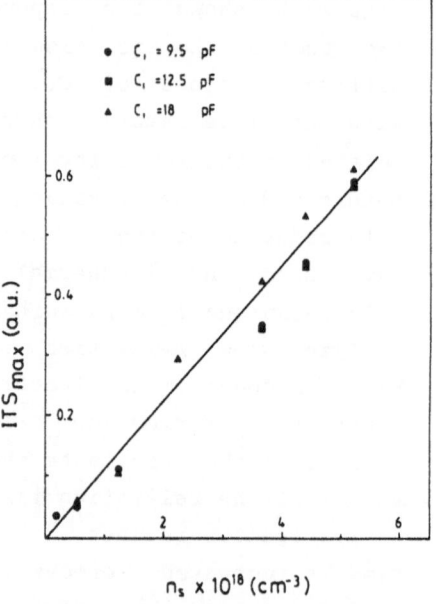

Figure 2. Variation of transients amplitude due to slow interface states as a function of capture time.

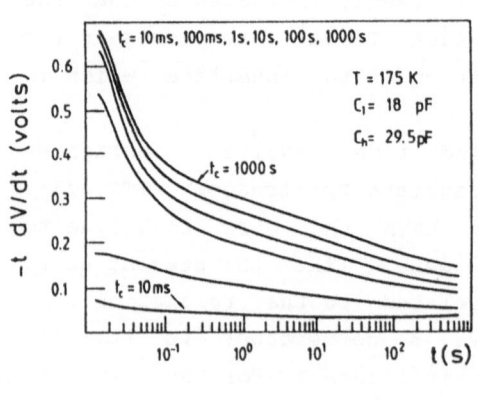

Figure 3. ITS spectra of the transients shown in figure 1.

Figure 4. Saturation value of ITS spectra as a function of surface majority carrier concentration.

Discussion

The above data show a time dependence which do not correspond with the time logarithm dependence expected for a direct tunnel mechanism. Unlike the previous models /4,5,6,7/, our measurements, figure 2, indicate that the relaxation process occurs for states which are not localized bellow the semiconductor conduction band. In agreement with this feature, we have not found any thermal activation of the relaxation process. Moreover, as the carrier concentration at the interface is maintained constant, the injected charge during the capture process can become higher than the necessary to screen the semiconductor charge during the emission. This aspect is corroborated when both, high and low, capacitance correspond to accumulation region. Consequently, we consider the ITS spectra as the addition of two mechanisms which determine the relaxation phenomena. The charge trapped at the zone near the interface (about 30 Å) is emitted by tunnel while the emission process of further zones rather seems to indicate a charge transference by hopping, which changes the trapped charge distribution and, hence, the potential necessary to keep the capacitance constant. Such a behaviour complete the results found by using techniques based on constant voltage. Nevertheless, the present analysis allows a more detailed view of the slow relaxation process which is directly related to the C-V curves hysteresis. Both phenomena, tunnel and hopping, have been found in all the samples even though the injected charge is related to the insulator deposition rate.

References

1.- P.BOHER,M.RENAUD,J.M.LOPEZ-VILLEGAS,J.SCHNEIDER,J.P.CHANE
    Appl.Surf. Scien., 30, 100-107 (1987).
2.- E.YAMAGUCHI
    Japn. J. Appl. Phys.,21, 1628 (1982).
3.- J.SAMITIER,J.R.MORANTE,L.GIRAUDET,S.GOURIER
    Appl. Phys. Lett.,48, 1138 (1.986).
4.- P.VAN STAA,H.ROMBACH,R.KASSING
    J. Appl. Phys.,54, 4014 (1.983).
5.- J.ESTEVE,J.SAMITIER,H.ALTELARREA,A.HERMS,J.R.MORANTE
    Appl.Surf. Scien.,30, 120 (1987).
6.- L.HE,H.HASEGAWA,T.SAWADA,H.OHNO.
    J.Appl.Phys.,63, 2120 (1988)
7.- F.P.HEIMAN,G.WARFIELD
    IEEE Trans.Electron.Devices,ED-12, 167 (1965)

Process and Device Modelling I

Simulation of Ion Implantation into Multilayer Structures

R.J. Wierzbicki, J. Lorenz and A. Barthel

Fraunhofer Arbeitsgruppe für Integrierte Schaltungen
Artilleriestrasse 12, 8520 Erlangen, West-Germany

Summary
For the analytical description of doping profiles after ion implantation, good multilayer models are required to describe the implantation through thin layers, e.g. a scattering oxide or non-vertical mask edges. In this paper, benefits and drawbacks of some published models are discussed, and an improvement is suggested. Comparisons with results from Monte Carlo simulations and SIMS measurements are shown.

Multilayer Models

Various models have been published [1-4] which describe the implantation into multilayer structures by analytical expressions. These models, however, fail in either the limiting case of a very thin or very thick masking layer, do not approximate the different stopping powers adequately [1] as it has been dicussed in [2], do not use the proper range distributions for the different layers [3,4], or do not take into consideration the change of profile width in the substrate due to the different stopping powers of layers on top of it [2]. Proper range distributions are only used in the Numerical Range Scaling model [2,5], which is the starting point of this work. This model reads, e.g., for a two-layer structure [2]:

$$C(x) = C_1(x) , \qquad\qquad 0 \leq x \leq t$$
$$C(x) = \alpha\, C_2[\, x - t(\, 1 - \frac{R_{p2}}{R_{p1}}\,)] , \qquad x > t. \tag{1}$$

Here, C_1 und C_2 are the dopant distributions in a semiinfinite target 1 and 2, respectively, R_{p1} and R_{p2} are the projected ranges in these targets, and α is a factor to assure conservation of the dose $N_\square$.

An important point is to use proper range distributions for C_1 und C_2. For crystalline targets, Pearson IV distributions have been shown to be appropriate [6]; for amorphous targets, however, the combination of skewness and kurtosis which may be obtained, e.g., from Boltzmann Transport calculations [7] falls outside the range of applicability of Pearson IV, so that Pearson I or Pearson VI distributions should be used [7].

Improved Model

The basic idea of the work carried out was to combine the approach of a range straggling depending on the thickness of a masking layer as discussed in [3,4] with the Numerical Range Scaling Model described above. In [4] in case of a two-layer structure, in the substrate a Joint Half Gaussian distributions is used with range straggling:

$$
\Delta R_{p2}' = \left\{
\begin{array}{ll}
\Delta R_{pa} = (R_m - t) - (R_{p1} - \Delta R_{p1} - t)\, \dfrac{R_{p2} - \Delta R_{p2}}{R_{p1} - \Delta R_{p1}}, & t < x < R_m \\[4mm]
\Delta R_{pb} = (t - R_m) + (R_{p1} + \Delta R_{p1} - t)\, \dfrac{R_{p2} + \Delta R_{p2}}{R_{p1} + \Delta R_{p1}}, & R_m < x.
\end{array}
\right. \tag{2}
$$

Here, ΔR_{p1} and ΔR_{p2} are the range stragglings in layer 1 and 2, respectively. R_m is chosen to be equal to the projected range R_{p2}' used for layer 2 in the Numerical Range Scaling Model:

$$
R_m = R_{p2}' = t + (R_{p1} - t)\, \frac{R_{p2}}{R_{p1}}. \tag{3}
$$

If one tries to use the mean value of ΔR_{pa} and ΔR_{pb} in the Numerical Range Scaling model, the modification of the width of a profile in silicon due to a masking layer, e.g. titanium, would be overestimated, see Fig.1. The reason is that the range straggling results from the nuclear stopping power which has its maximum at low energies, e.g. $\approx$ 5 keV for boron, whereas the projected range results from the sum of the electronic and the nuclear straggling which is increasing with energy. Therefore, the bigger part of the spread of the implanted ions is happening when the ions have slowed down to low energies and the effect of the masking layer on the range straggling is decreased compared to the linear approach of eq.2.

In our approach, we consider the range straggling in layer 2 to be a function of the mask thickness and the stopping powers of mask and substrate material. To deal with this dependence we use the ratio of the range straggling to the range $\Delta R_{p1,2}/R_{p1,2}$. As we need a correct limit for very thin masking layers, we assume:

$$\Delta R_{p2}' = \Delta R_{p2} + t \left(\frac{\Delta R_{p1}}{R_{p1}} - \frac{\Delta R_{p2}}{R_{p2}} \right) \tag{4}$$

In fact, eq.4 may yield negative results for a thick masking layer if $\Delta R_{p2}/R_{p2} > \Delta R_{p1}/R_{p1}$ holds; this, however, does not lead to any problems as in all cases investigated up to now, t must be much larger than R_{p1} to result in $\Delta R_{p2}' < 0$, which means that the profile does not enter layer 2 at all.

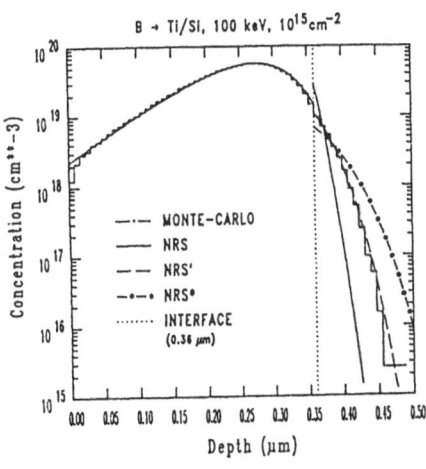

B → Ti/SI, 100 keV, $10^{15} cm^{-2}$

In Fig.1, a comparison of the Numerical Range Scaling model using a modified range straggling $\Delta R_{p2}' = (\Delta R_{pa} + \Delta R_{pb})/2$ as extracted from eq.2 (NRS*), the improved model discussed above (NRS') and results from a Monte Carlo simulation using TRIM [8] is shown. Only NRS' agrees well to the profile in layer 2.

Fig.1: Comparison of the conventional Numerical Range Scaling Model (NRS), modified model (NRS*) and improved model (NRS') with Monte Carlo results using TRIM for implantation of 100 keV boron into 0.36 μm Ti on Si.

Examples

In Fig.2 and 3, comparisons of the improved Range Scaling Model NRS' with Monte Carlo simulations and SIMS measurements are shown. Boron was implanted with an energy of 100 keV and a dose of $10^{15} cm^{-2}$ through nitride layers of various thicknesses into amorphous (Fig.2) and crystalline (Fig.3) silicon. For C_1, C_2 and $R_{p1,2}$ and $\Delta R_{p1,2}$ in eq. 4, fixed parameters for semiinfinite targets were used. In both examples a very good agreement between the analytical profiles and the Monte Carlo simulation in Fig.2 or the SIMS measurements in Fig.3 is obtained. In Fig.2, the change of the range straggling due to the masking nitride thickness is also depicted showing that $\Delta R_{p2}'$ remains positive as long as t is not very much larger than R_{p1}.

Conclusion

An modified multilayer model has been presented which improves the Numerical Range Scaling Model by a suitable description of the influence of a masking layer on the dopant profile width in silicon. Comparisons with Monte Carlo simulations and SIMS measurements show good agreement for various thicknesses of the masking layer.

196

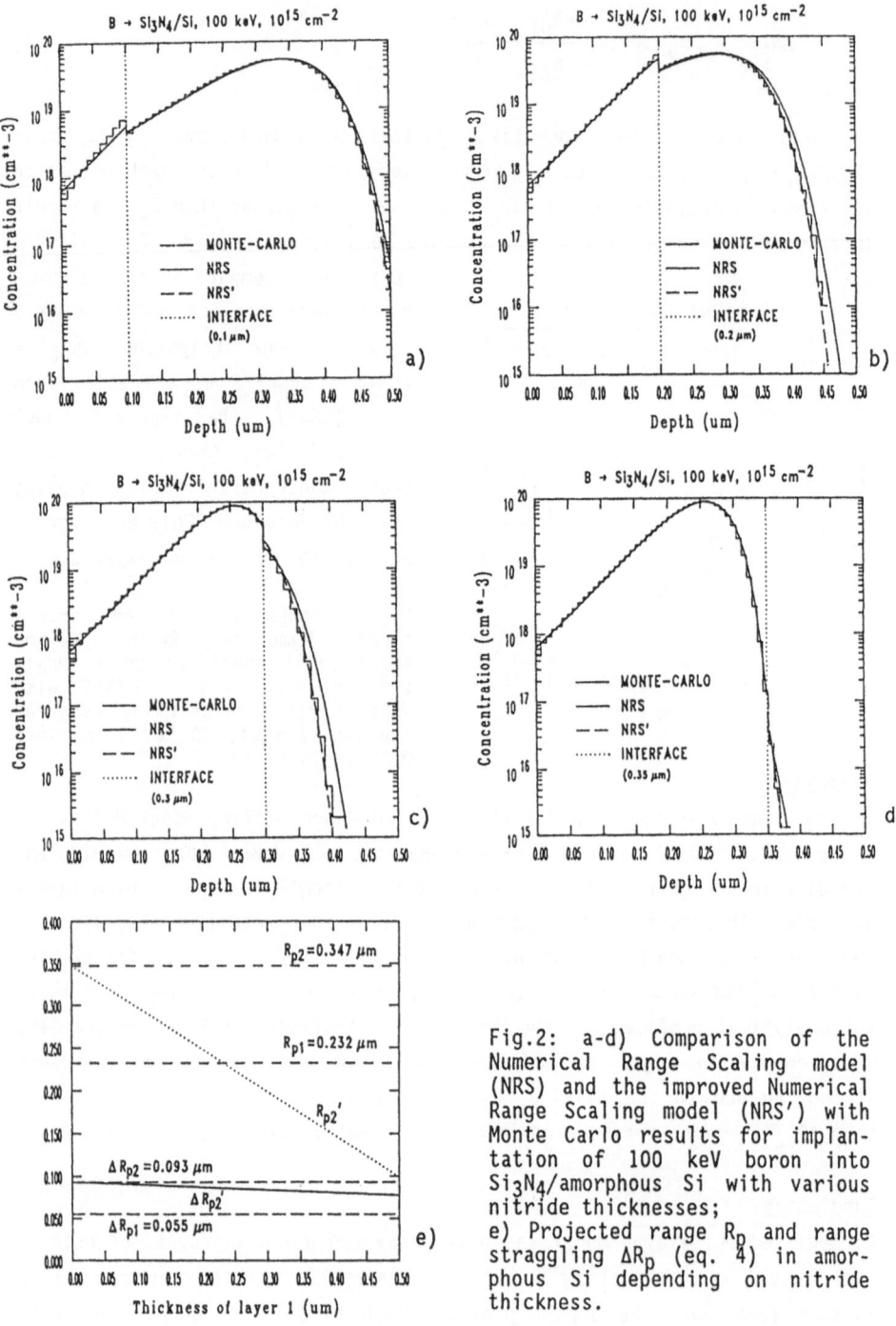

Fig.2: a-d) Comparison of the Numerical Range Scaling model (NRS) and the improved Numerical Range Scaling model (NRS′) with Monte Carlo results for implantation of 100 keV boron into Si_3N_4/amorphous Si with various nitride thicknesses;
e) Projected range R_p and range straggling ΔR_p (eq. 4) in amorphous Si depending on nitride thickness.

Acknowledgement
This work has been supported by the Bundesministerium für Forschung und Technologie under research contract No. NT 2755 3. The authors wish to thank H.Ryssel and P.Pichler for helpful comments.

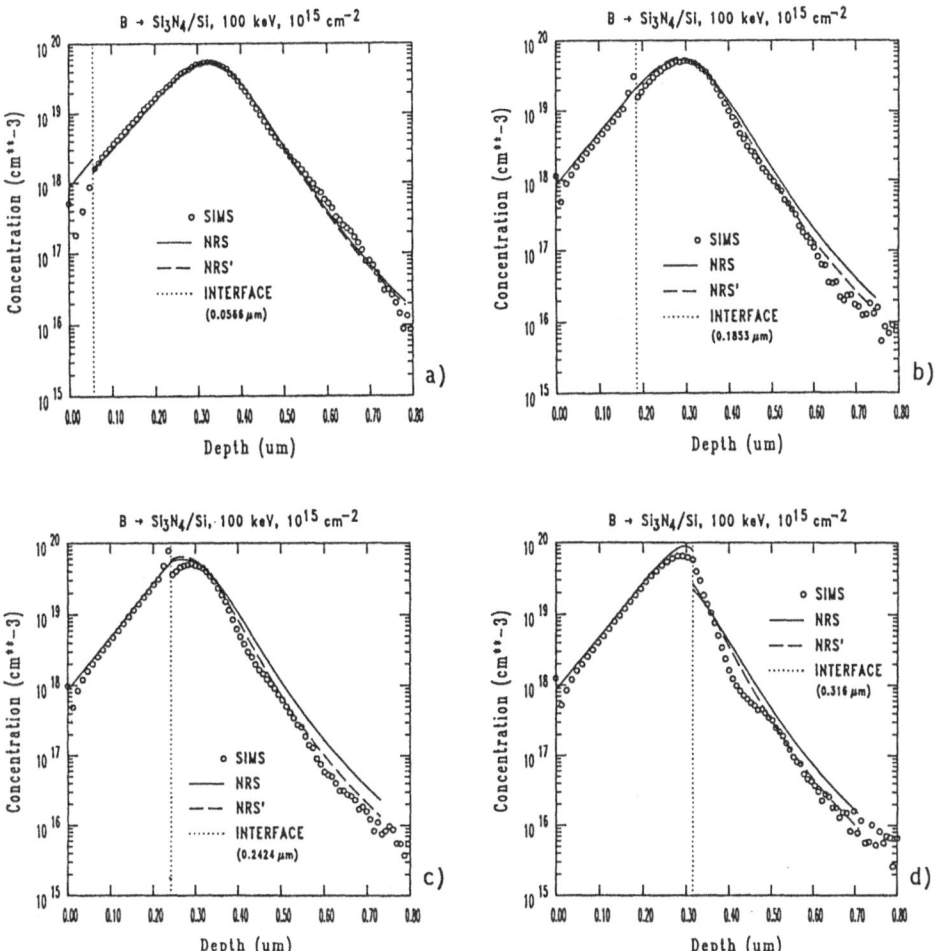

Fig.3: a-d) Comparison of the Numerical Range Scaling model (NRS) and the improved Numerical Range Scaling model (NRS') with SIMS measurements implantation of 100 keV boron into Si₃N₄/crystalline Si with various nitride thicknesses.

References

[1]   S. Furukawa, H. Matsumara, H. Ishiwara, Jap. J. Appl. Phys. 11 (1972) 134
[2]   H. Ryssel, J. Lorenz, K. Hoffmann, Appl. Phys. A 41 (1986) 201
[3]   D.K. Brice, Nucl. Instr. Meth. B17 (1986) 289
[4]   R.P. Webb, E. Maydell, Nucl. Instr. Meth. B33 (1988) 117
[5]   H. Ryssel, L. Gong, J. Lorenz, Proc. 1989 International Symposium on VLSI Technology, Systems and Applications, Taipeh, R.O.C., May 17-19, 1989
[6]   F. Jahnel, H. Ryssel, G. Prinke, K. Hoffmann, K. Müller, J.P. Biersack, R. Henkelmann, Nucl. Instr. Meth. 182/183 (1981) 223
[7]   J. Lorenz, W. Krüger, A. Barthel, Proc. NASECODE VI, Dublin, Ireland, July 11-14, 1989
[8]   J.P. Biersack, L.G. Haggmark, Nucl. Instr. Meth. 174 (1980) 257

Simulation of the Lateral Spread of Implanted Ions: Experiments

L. Gong, A. Barthel, J. Lorenz, and H. Ryssel[1]

Fraunhofer Arbeitsgruppe für Integrierte Schaltungen
Artilleriestrasse 12, 8520 Erlangen, West-Germany

[1] also: Lehrstuhl für Elektronische Bauelemente
Artilleriestrasse 12, 8520 Erlangen, West-Germany

Summary

For the measurement of dopant concentrations in two dimensions (2d), a delineation technique has been optimized. The method yields up to three dopant equiconcentration lines in one sample. The concentrations of these lines can be changed by modification of the etching conditions. This technique is applied to the investigation of 2d ion implantation profiles.

Introduction

As device dimensions have shrinked to below 1μm, a good knowledge of lateral dopant profiles is very important in order to yield realistic values for channel length of MOS devices. Physical methods such as 2d SIMS and electrical methods [1,2] for the measurement of dopant profiles in two dimensions (2d) have only recently become available. Anodic oxidation of suitable test structures, e.g. yields a resolution of about 60 nm [1]. Decoration and delineation techniques have been known since several years. They usually indicate the position of the pn-junctions in 2d. In this paper, an optimized delineation technique is presented which yields up to three different equiconcentration lines in one sample.

Experimental Technique

The method consists of short time etching of cross sections. On <100>-oriented, 1-2 Ωcm n-type silicon, a pad oxide of about 400 Å was grown and about 1.7μm of polysilicon was deposited. By optical lithography and RIE etching, the polysilicon layer was structured with a periodical test patter. Then by wet etching the photoresist and the pad oxide in the polysilicon mask window were removed. Boron was implanted with an energy of 200 keV and a dose of $5 \cdot 10^{15}$ cm^{-2}. Rapid thermal annealing (RTA) at 1000°C for 10 sec was used for electrical activation. After this, about 300 nm of polysilicon was deposited to avoid edge effects in the SEM investigation. The test patter was then used to properly align the wafer for sawing from the backside to a residual thickness of about 50 μm. After this, the wafer was broken and the vertical surface was then etched under strong UV-light for about four to seven sec to make several equiconcentration lines visible in the SEM.

Results

Various etching conditions were investigated. The best results were obtained with a mixture of HF(50%), HNO_3(70%) and CH_3COOH(100%). Because of the very short etching time, up to three equiconcentration lines could be delineated in one sample. By changing the concentration of CH_3COOH, the etching time and the etching temperature, the levels of these lines could be changed. Calibration of the concentration levels was performed using SIMS. In Fig.1, the SEM micrograph shows three lines of about $1.7 \cdot 10^{20}$, $1.5 \cdot 10^{19}$ and $6 \cdot 10^{16}$ cm^{-3}, respectively, obtained by using a ratio of the etchants of 1:3:8. In Fig 2, a ratio of 1:3:10 yields equiconcentration lines of $1.7 \cdot 10^{20}$, $2 \cdot 10^{18}$ and $3 \cdot 10^{17}$ cm^{-3}, respectively. The asymmetry of the both figures results from the 7° wafer tilt used in the implantation.

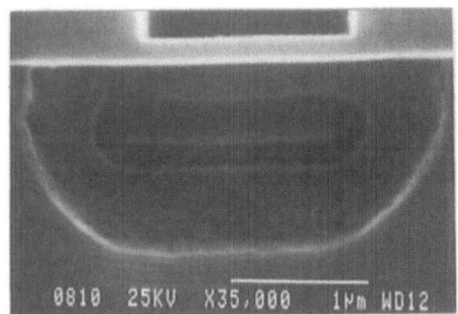

Fig.1 SEM-micrograph of a boron implantation(200 keV, $5 \cdot 10^{15}$ cm^{-2}) with 7° tilt. The ratio of the etching solution was 1:3:8. The equiconcentration lines are about $1.7 \cdot 10^{20}$ cm^{-3}, $1.5 \cdot 10^{19}$ cm^{-3} and $6 \cdot 10^{16}$ cm^{-3}, respectively.

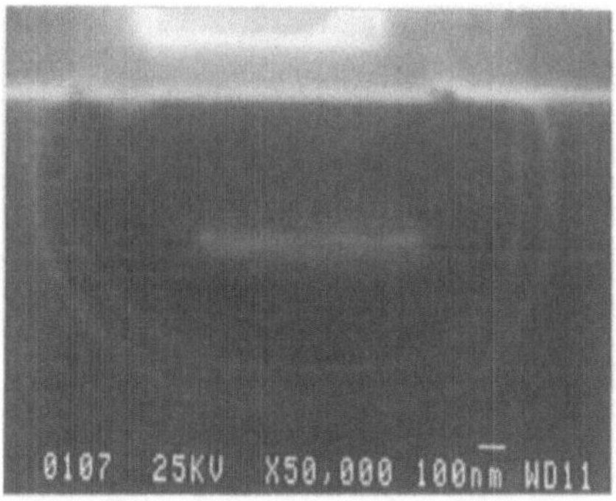

Fig.2 SEM-micrograph of a boron implantation (200 keV, $5 \cdot 10^{15}$ cm^{-2}) with 7° tilt. The ratio of the etching solution was 1:3:10. The equiconcentration lines are about $2 \cdot 10^{20}$ cm^{-3}, $2 \cdot 10^{18}$ cm^{-3} and $3 \cdot 10^{17}$ cm^{-3}, respectively.

By modifying the etching conditions as discussed above, seven equiconcentration lines were delineated from four samples made from the same wafer after implantation of boron at 200 keV and a dose of $5 \cdot 10^{15}$ cm^{-2} with wafer tilt parallel to the windows. Lateral profiles were extracted from these equiconcentration lines, and least square fits with error functions [3] were performed. This approach assumes that the lateral distributions for implantation through an infinitesimal narrow mask window are Gaussian. Fig.3 shows as an example the fit of the lateral distribution in a depth of 0.66 μm. From these fits, lateral range stragglings were extracted, and the parameters for the parabolic approach of the depth dependent range straggling discussed elsewhere [4] were calculated again by least square

200

fitting:

$$\sigma_y^2(x) = \bar{\sigma}_y^2 \left[1 + a + b(\frac{x - R_p}{\sigma_x}) - a(\frac{x - R_p}{\sigma_x})^2 \right] \qquad (1).$$

Here, R_p is the projected range, σ_x the range straggling, $\sigma_y(x)$ the depth dependent lateral range straggling, and $\bar{\sigma}_y$ its mean value. Fig.4 shows the depth dependent lateral spread extracted from the experiments and the fit using Eq.1. In Fig.5, a comparison between the analytical model using the 2d Numerical Range Scaling Model [4] and Eq.1 and a SEM micrograph is depicted, showing good agreement between model and measurement.

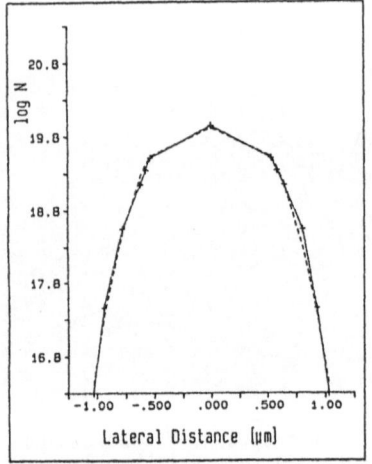

Fig.3: Lateral dopant distribution at a depth of 0.66 μm extracted from SEM micrographs and corresponding error function fit. The dashed line is the fitted curve.

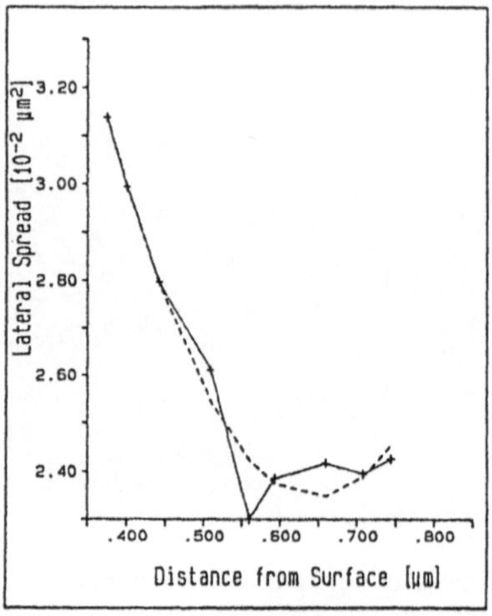

Fig.4 Depth dependent lateral spread for 200 keV boron implanted into silicon extracted from experiment and approximation using Eq.1. The dashed line is the fitted curve.

The reason for the visibility of the different equiconcentration lines is the rapid change of the etching rate of silicon in the etching solutions used. In the early years of semiconductor technology, many authors have found that the etching rate of silicon in the mixture with ratio 1:3:8 changes rapidly within the doping level of 10^{18} and 10^{19} cm^{-3} [5] without a pn-junction. We have found that the etching rate of silicon changes rapidly three times in the p-type semiconductor with a pn-junction.

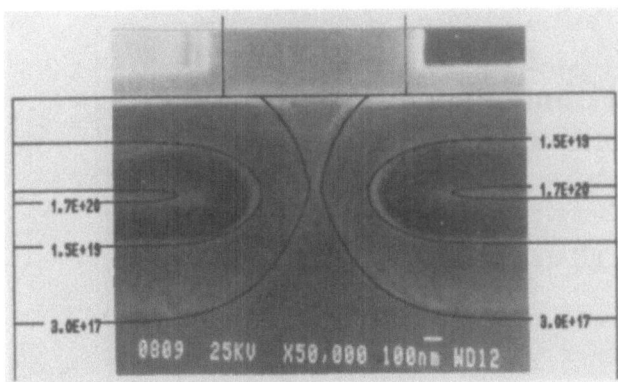

Fig:5: Comparison between the analytical model consisting of the 2d
Numerical Range Scaling Model [4] and Eq.1 with SEM-measurements for
implantation of boron at an energy of 200 keV and a dose of $5 \cdot 10^{15}$ cm^{-2}.

Conclusion

An optimized etching technique has been developed which is suitable to
delineate up to three equiconcentration lines in one sample, and seven
equiconcentration lines extracted from one wafer. First results of the
application of this technique to investigate 2d dopant profiles were
discussed.

Acknowledgement

This work was supported by the Bundesministerium für Forschung und
Technologie of Germany under research grant no. NT 2755 3. The authors
thank Dipl.Ing. W. Stark, W. Ergele and K. Hartlieb for their condiderable
help throughout the experiments.

References

1. C. Hill and P.J. Pearson: 2D Boron Distributions After Ion Implant and
   Transient Anneal, Proc. ESSDERC 87, pp. 923-926, TECNOPRINT, Bologna,
   Italy

2. Scott H. Goodwin-Johansson, Ravi Subrahmanyan, Carey E. Floyd, and
   Hisham Z. Massoud: Two Dimensional Impurity Profiling with Emission
   Computed Tomography Techniques, IEEE Transactions on Computer-Aided
   Design, Vol.8 No.4,1989

3. S. Furukawa, II. Matsumura, H. Ishiwara: Theoretical Considerations on
   Lateral Spread of Implanted Ions, Jap. J. Appl. Phys., Vol. 11, No.2,
   pp.134-142,1972

4. J. Lorenz, W.Krüger and A. Barthel: Simulation of the Lateral Spread of
   Implanted Ions: Theory, proc. NASECODE 89, Ireland

5. H. Muraoka, T. Ohhashi, Y. Sumitomo: Controlled Preferential Etching
   Technology, Semiconductor Silicon 1973, Electrochem. Soc., Princeton,
   N.J., pp. 327-338

The Modeling and Simulation of Reactive Ion Etching Rate Using Statistical Method

Xiangming FENG and Gang RUAN

Microelectronics Institute
Fudan University
Shanghai, P.R.China

ABSTRACT

A method to obtain experiential formula of reactive ion etching (RIE) rate based on statistical mathematics combined with few experiments is presented. The precision of the model is 5% for CF_4/O_2 etching SiO_2 and SF_6/O_2 etching P-doped poly-Si.The model is analytical and especially suitable for process simulation.

I. Introduction

Since 70's, various dry etching technologies have been widely used due to their particular excellent characters. However,there are many complicated physical and chemical processes simultaneously involved in etching process so that there have been no analytical dry etching models based on the fundamental mechanisms to obtain the etching parameters, for example, the etching rate and etched profile among the most important.[1] This has limited the development of dry etching simulation. Even in the famous etching simulator SAMPLE Version 1.6, no dry etching models are included.

A kind of method to acquire the experiential formula of reactive ion etching rate based on statistical mathematics combined with few experiments is presented. It shuns the basic reaction properties unrealized by human being and starts from the practical point of view.Experiment results for CF_4/O_2 etching SiO_2 and SF_6/O_2 etching P-doped poly-Si indicate that the model has high creditability and the difference between simulated and tested rate is approximately 5% in practical operation range. It's suitable especially for process simulation and will evidently reduce the time and cost in process optimization.

II. Establishment of the Model

1. Factors Screening

More than 10 factors influence the etching rate of RIE.It is unnecessary and unrealistic to establish a complete model which includes every factors. So it becomes the first problem to be solved in the build-up of a practical model to screen out the unsignificant factors without any compromises to the precision of the simulation.

According to statistical mathematics on screening design, two experiments set at the higher and lower part of every factors are enough to obtain the sufficient information of their relative contribution to final result,if the influence of the high order terms of every factors and their interactions is neglected.[2]Our experiments are conducted on RIE 80 system by Plasma Technology Company,U.K. The controllable factors are listed in the left column of Table 1. The test number will be $2^6=64$ for the 6 controllable factors for full 2-level design. Resolution-IV design can be used if the interaction between the linear term of every factor is omitted.[3] Thus,only a quarter of full 2-level design is sufficient, resulting in test number of 16.

Table 1.Screening Design Result

Factors	Results
power	274
gas flow	27
O_2 fraction	-218
pressure	166
chip position	17
etching time	73

The relative contribution of every factor to the test results are calculated with the following formula,

$$a_k = \frac{1}{N} * \prod_{i=1}^{N}(D_{ki} * R_i), k = 1...8. \tag{1}$$

a_k is the relative contribution of kth factor to the experiment results and D_{ki} is the value of ith test for factor k; R_i $(i=1...N)$ is the tested RIE rate of ith experiment; N is the total experiment number. The 3 most influential factors (those with maximum $|a_i|$) can be screened out for next step.

2. Experiment Design

Approximated to the second order,the dependence of etching rate R on 3 screened-out factors z_1, z_2, z_3 can be written as

$$R = b_0 + \sum_{i=1}^{3}b_i x_i + \sum_{i=1,3,j=2,3 \atop i \neq j} b_{ij}x_i x_j + \sum_{i=1}^{3}b_{ii}x_i^2. \tag{2}$$

here, 10 coefficients are to be determined.The number of the test value of every factor should be at last 3 when the second order term is considered. The 3-level full experiment for 3 factors needs $3^3=27$ tests. Using Box-Bohnken design method [4], only a part of these tests can fully represent the characters in this area. Detailedly, the test conditions are chosen at the center of the representing cube and the middle points of every edge of the cube; the central test point are repeated 5 times to obtain the necessary information required for later use, resulting in a total test number of 17.(Fig. 1)

3. Result Fitting

Least square regression is used to obtain the coefficients in (2) which gives out the RIE rate on whole operation region.

4. Statistical Check

Statistical check is carried out to give out the creditability of the model and analyses the source of deviation between the prediction of the model and the real value.

The creditability of the model is checked first.If the model is valid, the f-ratio

$$f = \left(\frac{S_{reg}}{f_{reg}}\right) / \left(\frac{S_{res}}{f_{res}}\right). \tag{3}$$

should serve $F(f_{reg}, f_{res})$ distribution. Here, S_{reg} and S_{res} in (3) denote the total regression error and residual error, respectively; f_{reg} and f_{res} is the corresponding freedom.[5]

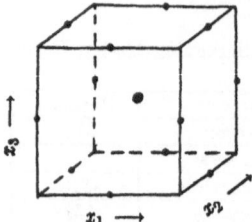

Fig. 1 Geometric representation of Box-Bohnken design with center point replication

To denote the extent of the fitness of the model to the real case, relative residual error α is defined as

$$\alpha = 1 - \left(\frac{S_{reg}}{f_{reg}}\right) / \left(\frac{S_{tot}}{f_{tot}}\right). \tag{4}$$

here S_{tot} is the total error and f_{tot} is its freedom.In case of complete fitness, $\alpha = 1$; the more the model and the reality fits, the more nearer of α to 1.

III. Experiment Results and Analysis

Three kinds of chip samples have been prepared.(Table 2) Results of screening experiment for CF_4/O_2 etching SiO_2 are listed in Table 1. Evidently,three most influential factors to the etching rate are power, O_2 fraction in gas and gas pressure.

Table 2. Preparation of Chip Samples

Lot No.	Material	Preparation	Etching Gas
1	SiO_2	145 min. 1180°C	CF_4/O_2
2	P-poly Si	LPCVD 45 min., 800°C	SF_6/O_2
		P doping 4 min.,1100°C	
3	P-poly Si	LPCVD 45 min.,800°C	SF_6/O_2
		P doping 12 min.,1100°C	

Box-Bohnken tests are made for 3 lots of sample chips. The regression calculation and statistical analyses of the results are listed in Table 3. The significance of the experiential formula are all above 95%; for poly-silicon sample, it is even greater than 99%. The nearness of α value to 1 indicates a quadratic polynomial formula to be a good representation of real condition.Residual error is divided into lack-of-fit error and duplication error.The F-check of the residual error results in a very large f-value which indicates the good steadiness of the test system and the creditability of the measurement of the test results.

Table 3. Statistical Analysis of the Model

	Lot 1	Lot 2	Lot 3
total error	82.47	71.88	58.36
regression error	74.78	70.16	56.36
residual error	7.69	1.69	1.95
F-ratio	6.0	23.0	16.0
level of significance	98%	99%	99%
lack-of-fit	6.43	1.59	1.78
replication error	1.27	0.10	0.17
F-ratio	7.7	23.0	13.0

The comparison of the test result [6] to the model is shown in Fig. 2. The average relative error is around 5% in practical operation region.

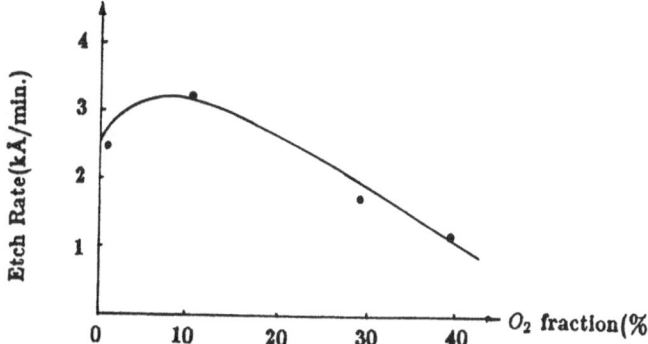

Fig. 2 Comparison of the model with the real etch rate for poly-Si
(100 W, 0.045 torr, 26 °C)

IV. Conclusion

It can be learned from the previous introduction that the using of various statistical method enables us to handle every factors changing simultaneously in RIE system and significantly reduce the time and cost in establishing a practical RIE model with high precision and reliability.This model fills the vacancy of RIE model in etching process simulation and will be sure to enlarge the usage of etching simulation tool with this model.

References
[1] S.J.Fonash,"Advances in Dry Etching Process",Solid State Technology,Jan.1985.
[2] E.I. du Pont de Nemours & Co.,"Strategy of Experimentation Course", F&FP,Applied Technology,Wilmington,DE.
[3] J.Kleijnen,"Statistical Techniques in Simulation", Part 2, Marcel Dekker,Inc.,1975.
[4] G.Box, et al.,"Statistics for Experimenters",John Wiley & Sons,Inc.,1978.
[5] S. Mou, "Regression and Experiment Design",East China Normal University Publication,1981.
[6] S. Zou, Private Communication

Two-dimensional Aspects of Ion Enhanced Reactive Etching of Silicon with SF$_6$

A. Gerodolle

CNET/CNS, chemin du Vieux-Chene , BP98
38243 Meylan CEDEX

Summary

A model for plasma silicon etching by SF6 is presented; based on the work of Petit- Pelletier, it has been implemented in a simulator, so as to explain two-dimensional effects. It is shown how this model, with its very simple assumptions on fluxes, allow a variety of observed profiles to be simulated.

Introduction

R.I.E. modelling implies various aspects - such as chemical reactions in plasma, transport equations, thermodynamics, surface reactions -, which are more or less coupled. The surface reactions are perhaps the least well known phenomena, insofar as it is difficult to measure anything else than total fluxes, pressure, etching kinetics, and reaction products. The existing surface reaction models can be easily applied to simple geometries, where the etching depth is small with regard to the width of the device and thus two-dimensional effects can be neglected. In this case, active species fluxes can be reduced to macroscopic values by integration, and models can be expressed analytically. On the other hand, when the width of a trench becomes small with respect to the depth, two-dimensional effects appear, which may have several explanations.The purpose of the present work is to show that a unique model, which implies a few parameters, allows the simulation of a variety of two-dimensional effects in the case of silicon etching by SF6.

Model description

The model is mainly based on Petit and Pelletier's work (1). The principle is to calculate at any moment the equilibrium of adsorption and desorption. The variable is θ, the adatom coverage. The adsorption rate is supposed to be proportional to the neutral flux (that is the pressure in the case of a planar structure). An adatom threshold coverage (θ_C) exists, below which the spontaneous desorption of reaction products is negligible. Ion bombardment allows this threshold to be decreased, so that the desorption rate is the sum of the spontaneous (proportional to θ-θ$_C$) and ion induced desorption rates (the latter being proportional to ionic current density). Finally, adsorbed particles may move at the surface. This possibility is summarized by the surface diffusivity. Working in two dimensions, the surface being thus one-dimensional, the equation to be solved is:

$$K.p(F).\frac{s-\theta}{s} = \frac{1}{\tau}\max(\theta-\theta_c,0) + \eta.j.\theta - D\frac{\partial^2\theta}{\partial x^2} \tag{1}$$

where x is the curvilinear abscissa, s the maximum coverage, τ the average lifetime of an adsorbed adatom, D the surface diffusivity; the random neutral flux ,p(F), and ionic current density , j, depend on the position x.

In the case of a planar structure, the adatom coverage is constant over the whole surface and can be easily derivated from equation 1, insofar as the diffusion term vanishes. Assuming the diffusion length is long with respect to the etching depth, the computed adatom coverage applies under the mask, so that the lateral etching rate can be computed. It can thus be verified that lateral etching vanishes with increasing ionic current density. But these assumptions lead to simple profiles as shown in Figure 1, with a perfectly horizontal shape in the light zone. Obviously, this does not correspond to real profiles, where various effects like "dove-tail" or round shapes (Figure 3) can be observed. To explain these shapes, the diffusion equation has to be solved numerically in the program.

Implementation

The model has been implemented in the TITAN process simulator (2). As a first approach, in order to limit the number of parameters, several assumptions have been made : the *ionic flux* is unidirectional so that ηj maintains a constant value, except in the dark regions; simulations can also be performed with equal distribution of the ionic fluxes in a given solid angle. *Neutral fluxes* are supposed to come from all directions, but only species coming directly from the plasma are taken into account: no attention is paid to particles after rebound. *Only silicon* is etched, and reflecting conditions are applied to the fluxes at the silicon/dioxide interface.

Results and discussion

Typical S.E.M. measurements of profiles have been taken from a set of experiments carried out at CNET. They do not show ideal trench profiles, and thus give a good idea of the problems of anisotropic transition.

Figure 2 shows the simulation of a "dove-tail" effect. This effect is often said to be due to rebounds of ions on the sidewalls forcing la ocal accumulation at the bottom of the sidewalls (3). Using the Petit-Pelletier model, the shape is due to the diffusion of fluorine adatoms towards the bottom: the high ionic current induces a high desorption in the bottom of the trench, so that the particles adsorbed on the sidewalls move to the bottom; they desorb before diffusing to the center. This effect can be decreased by increasing the diffusion length

Figure 3. shows a "field effect". Here, the ionic current is almost neglegible. The trench depth depends on its width because the neutral flux decreases when the depth to width ratio increases. Figure 4 shows the corresponding simulation.

Figures 5 and 6 show a difference between two trenches in the same reaction: in the small trench, the etching seems to become more anisotropic while there is no visible transition in the large trench. Experimental conditions are same as for Figure 3, but the pressure is divided by 2. In this case, the anisotropic part plays a more important role; at the beginning of the etching , the neutral flux is sufficient to equilibrate the ion induced desorption; then, the flux decreases and is no longer sufficient in the small trench, while remaining sufficient in the large one. Thus, in the small trench, species move from

208

the hidden part to the bottom, which explains the transition towards anisotropic regime, not observed in the large trench.

Conclusion

A model for plasma etching of silicon by SF6 has been presented. Making very simple assumptions on neutral and ionic fluxes, it allows a variety of two-dimensional profiles to be simulated. Results show that observed effects can be explained by surface diffusion, without involving any mechanical effects or rebounds.Foreseen improvements are first multilayer etching, with the problem of modelling interfaces; the influence of several active particles in the plasma has to be studied, in order to simulate passivation. Finally, the influence of the ionic flux distribution, in terms of energy and angles of incidence, has to be investigated.

Acknowledgements
The author would like to thank Michel Pons for theSEM photographs.

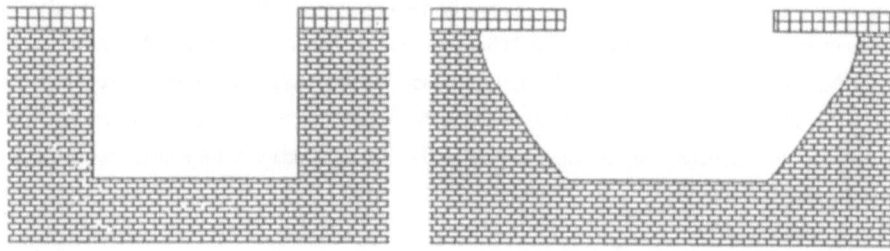

Fig.1 . profiles obtained with infinite diffusion length

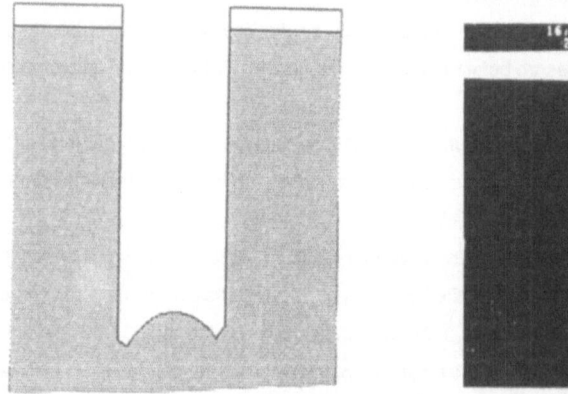
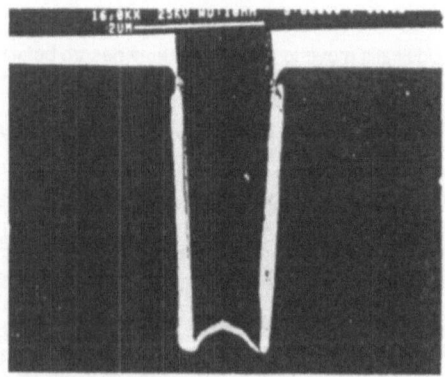

Fig.2. simulation and photo of "dove-tail" effect

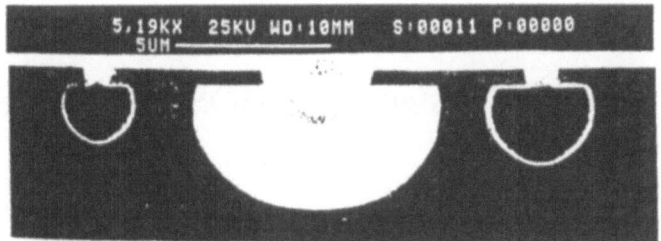

Fig. 3 SEM photograph of a profile obtained with a slightly ionic enhanced etching, pressure 10 Atm.

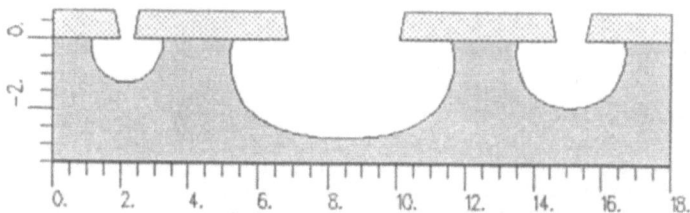

Fig. 4 simulation corresponding to Fig. 3

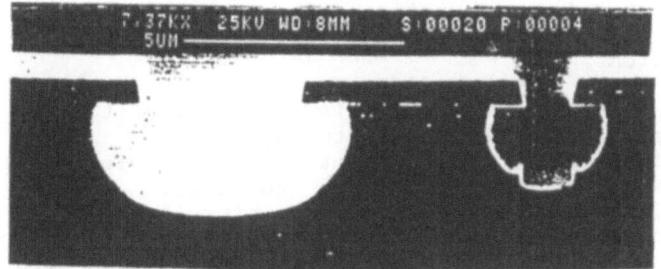

Fig. 5. Profile obtained using the same reactor as Fig. 3; pressure 5 Atm; all other parameters
unchanged.

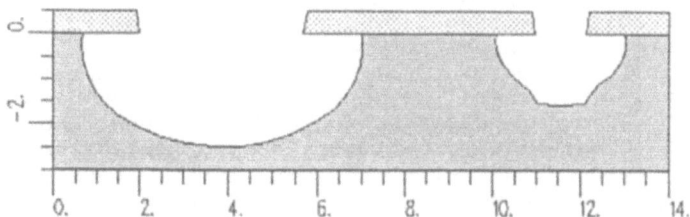

Fig. 6 simulation corresponding to Fig. 5.

References

1. B. Petit and J. Pelletier, Jpn.J.Appl.Phys. 26,825 (1987)

2. A. Gérodolle and al. , NASECODE VI (1989)

3 T.J. Cotler and al., J. Vac. Sci. Technol.B,Vol6,N°2,Mars/Apr 1988

A Yield Modelling System for NMOS and CMOS IC

He Yie   Shen Yanhui

Microelectronics Center, Southeast University
Nanjing 210018, Nanjing, China

Abstract

In this paper,a multi-level modelling system, which takes into
account the statistical fluctuations inherent in the IC
manufacturing process, is described. In the modelling system,
a reverse statistical simulation method and a yield model
expressed in quadratic form are used, which effectively
reduces the simulation time in statistical design of integrated
circuit and makes it possible to perform an optimum yield
design with a low computation cost.

Introduction

From the earliest days of integrated circuit, it has been
recognized that the manufacturing process has unavoidable
fluctuations which cause variations in the electrical behavior
of integrated circuits. Due to these variations, the
manufacturing process, device and circuit should be simulated
in a stochastic manner. The main objective of this work is to
present a multi-level modelling system, which adopts a new
yield prediction method based on advanced "design of
experiment" and a reverse statistical simulation method. In
previous works, the statistical simulation of IC manufacturing
process has been adopting forward approach, i.e. the simulation
procedure is from process modelling to device modelling and
then to circuit modelling [1]. In this case, the identification
and turning of input parameters required by statistical
simulator might consume a large amount of computation time.
Also, the assumption of the process disturbances having either
normal or log-normal distribution made in these simulators is
not always appropriate, especially for the case of small sample
size. In comparision with previous works, there are some
advantages of this modelling system: It can be used for timely
statistical analysis of process, such as for analyzing the
trial manufacturing process of a limited number of circuits or
determining the variations of process parameters in a short
fabrication period. Also, the number of process, device and
circuit analyses required in the Monto Carlo statistical design
and optimization is limited to a small number by assuming a
quadratic expression between yield and design parameters.

Multi-Level Modelling System

Our multiple-level modeling system is composed of three
parts,a fabrication process simulator,a device simulator and a
circuit simulator. The model implemented in the modelling

system allow for the accurate statistical analysis of NMOS or CMOS fabracation process. The process simulator can simulate many important process step, such as oxidation, diffusion and ion implantation etc. The combined device and circuit simulator can analyze the direct-current and alternating-current parameters of device and circuit, such as output high-level, output low-level, noise margin, threshold voltage and gate delay time etc. If the analyzed circuit is composed of resistances, capacitors and other components, the modified CAP—Circuit Analysis Program, which is developed in Berkaley University, is used as statistical circuit simulator. With the modelling system, the yield model can be obtained by extracting some useful results and circuit performance sensitivitis upon process disturbances can be simulated.

Reverse Simulation Approach

The conception of reverse simulation approach are based on the mapping from device characteristics space X into device structure parameter space D. In production line, the device and circuit characteristics are measured. By sampling the measured distributions and listing the linear regression formulas of device performance versus device structure parameters in each sub-domain over entire device structure parameter space, many sets of linear equations can be derived, where device structure parameters are taken as unknown variables. The distributions of device structure parameter can be uniquely determined from solving these equation systems. The procedure described here is named reverse simulation. The IC performance variations can be estimated in this way with an error of less than 10 percent.

Once the variations in device structure parameter are obtained, the device and circuit Monto Carlo simulation can be performed to determine statistical characteristics of device and circuit [2]. This is to say, if only the device characte-ristics are measured, the process step which variation is the main cause of fluctuations in IC performance can be pointed out. Also, the manufacturing yield and performance variabilities to each process step can be calculated.

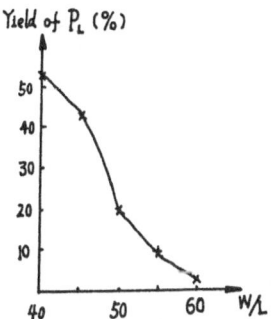

Fig.1   Yield of P_L varing with W/L

Applying the modelling system to analyze "NO" Gate fabricated in 5μm NMOS process, Fig.1 describes the calculated behavior of power dissipation yield varing with aspect ratio of W/L.

Yield Prediction Method

A yield prediction method based on quadratic orthogonal regression algorithm is used in the modelling system. The regression accuracy of yield model can be estimated. The form of regression model is written as below:

$$Y_i = b_0 + \sum_{j=1}^{P} b_j x_j + \sum_{h<j} b_{hj} x_h x_j + \sum_{j=1}^{P} b_{jj} x_j^2 + \varepsilon_i \quad (i=1,2,\cdots, N) \tag{1}$$

where P is the factor number, ε_i is the error, N is the "experiment point" number. As an example, considering 5 μm CMOS process, a table of factor and level for NMOS device threshold yield is shown (table 1). From the coding and rotary formulas:

$$X_i = \frac{2(Z-Z_2)}{Z_2 - Z_1} + 1 \tag{2}$$

$$X_i' = X_i - 0.594 \tag{3}$$

where Z denotes the real variable, the orthogonal rotary combination design of three factors is obtained (table 2).

Table 1   Factor and Level Table

Factor	t_{ox} (oxide thickness)	W/L (aspect ratio)	Dose (implantation dose)
Code	X_1	X_2	X_3
Zero Level	100nm	25	10^{11}
Range	10nm	5	10^{10}
Superior Level	110nm	30	1.1×10^{11}
Inferior Level	90nm	20	9×10^{10}

Table 2   The Orthogonal Rotary Combination Design of Three Factors

	X0	X1	X2	X3	X1X2	X1X3	X2X3	X1'	X2'	X3'	YN(%)
1	1	1	1	1	1	1	1	0.406	0.406	0.406	45
2	1	1	1	-1	1	-1	-1	0.406	0.406	0.406	27.5
3	1	1	-1	1	-1	1	-1	0.406	0.406	0.406	45
4	1	1	-1	-1	-1	-1	1	0.406	0.406	0.406	22.5
5	1	-1	1	1	-1	-1	1	0.406	0.406	0.406	20
6	1	-1	1	-1	-1	1	-1	0.406	0.406	0.406	50
7	1	-1	-1	1	1	-1	-1	0.406	0.406	0.406	20
8	1	-1	-1	-1	1	1	1	0.406	0.406	0.406	50
9	1	1.682	0	0	0	0	0	2.234	-0.594	-0.594	37.5
10	1	-1.682	0	0	0	0	0	2.234	-0.594	-0.594	17.5
11	1	0	1.682	0	0	0	0	-0.594	2.234	-0.594	30
12	1	0	-1.682	0	0	0	0	-0.594	2.234	-0.594	37.5
13	1	0	0	1.682	0	0	0	-0.594	-0.594	2.234	25
14	1	0	0	-1.682	0	0	0	-0.594	-0.594	2.234	30
15	1	0	0	0	0	0	0	-0.594	-0.594	-0.594	25
16	1	0	0	0	0	0	0	-0.594	-0.594	-0.594	30
17	1	0	0	0	0	0	0	-0.594	-0.594	-0.594	25
18	1	0	0	0	0	0	0	-0.594	-0.594	-0.594	30
19	1	0	0	0	0	0	0	-0.594	-0.594	-0.594	25
20	1	0	0	0	0	0	0	-0.594	-0.594	-0.594	30
21	1	0	0	0	0	0	0	-0.594	-0.594	-0.594	25
22	1	0	0	0	0	0	0	-0.594	-0.594	-0.594	30
23	1	0	0	0	0	0	0	-0.594	-0.594	-0.594	22.5

As a result, a regression formula is derived as follows:

$$YN = 0.3 + 0.9 \times 10^{-10} X_1 + 0.98 \times 10^{-2} X_2 + 0.94 \times 10^{-1} X_3 + 0.63 \times 10^{-2} X_1 X_2 + 0.15 \times 10^{-1} X_1 X_3$$

$$- 0.63 \times 10^{-2} X_2 X_3 + 0.31 \times 10^{-2} X_1^2 - 0.3 \times 10^{-1} X_2^2 - 0.4 \times 10^{-2} X_3^2 \qquad (4)$$

When one of the parameters is variable and other parameters are at zero levels, the regression yield model only depends on the variable parameter. Fig.2 and Fig.3 give the relationships of yield versus process design parameter X_1 and X_2, respectively.

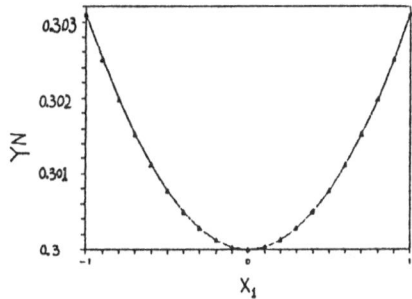

Fig.2   The relationship of YN with X_1

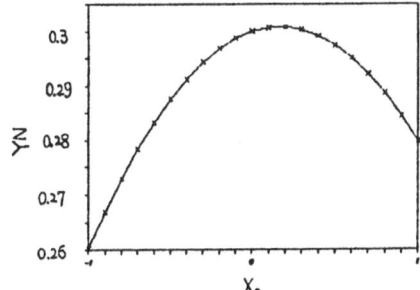

Fig.3 The relationship of YN with X_2

The experiment results show that simulation accuracy is below 15%, which satisfies the need of engineering.

Conclusions

Investigation demonstrates that the modelling system developed by us is characterized by high simulation accuracy, reasonable computation cost. With the modelling system, we can quantitatively estimate process disturbance and its impact on circuit performances, statistically analyze process on time, calculate circuit yield and optimize process, device and circuit design.

References

[1] Yukio, Aoki et al: IEEE Trans. Electron Devices. Vol.ED-31 (1984) 1462-1467.
[2] Yie, He: Simulation technique of Semiconductor Device. Beijing, China: Science Press.

Calculation of Bipolar Transistor Base Resistance Using Finite Element Method

Reinhold Vahrmann, University Hannover
Erich Barke, Siemens Semiconductor Division Munich

1. Summary

To model NPN-transistor behaviour at high currents or high frequencies, an exact description of base resistance is needed. In this paper a new method for base resistance calculation is presented, that takes into account three-dimensional physical effects at high currents. A distributed equivalent circuit is generated for each NPN-transistor layout using the finite element method. Calculating the power dissipation in the base region, the corresponding base resistance is obtained from equivalent power consumption.

2. Base resistance at high currents

With increasing collector current I_C, base resistance R_{Bb} decreases, because of the following physical mechanisms:

- Base Width Modulation,   - Base Conductivity Modulation,
- Base Pushout,   - Emitter Current Crowding.

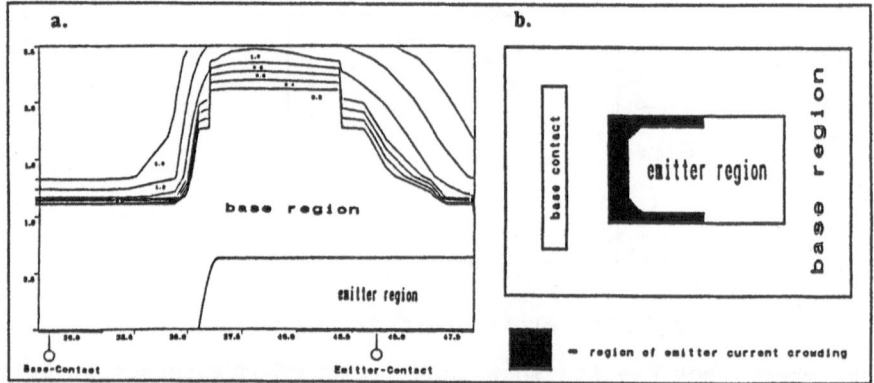

Figure 1: a.NPN-transistor electric potential (U_{be} = 850 mV, U_{ce} = 2 V). Base pushout occurs under the emitter edge facing the base contact (BAMBI-plot).
b.Current carrying region at high injection level within the emitter area.

At high injection levels all these mechanisms interact. Base conductivity modulation and base pushout reduce the intrinsic base resistance. Therefore, the base voltage drop, which causes emitter current crowding, is attenuated. Emitter current crowding, which occurs at the peripheral emitter edges (Figure 1b), determines the active regions for base width modulation, base conductivity modulation and base pushout. Because of these aspects the transistor layout, in particular the position of the base contact, facing the emitter, limits the current carrying region.

3. Modeling

To analyze NPN-transistor behaviour a three dimensional electrical circuit is built up, which models the physical effects at high injection levels. The placement of each network element and the choice of element models have been verified by using the device simulator BAMBI /1/. Figure 2 shows the equivalent circuit, which consists of the following parts:

- Emitter area
Emitter resistance is described by resistors R_e'. Contact resistance is modeled by additional vertical contact resistors R_{ec}'.
- Interior transistor and intrinsic base
A good approximation for this region is obtained by transistors T_i' (Gummel-Poon-Model) and intervening resistors R_{bi}', with resistance values depending on the Gummel-number.

- Sidewall transistor and external base
To model the current injection from emitter sides into the base region, additional transistor elements T_s' are used. Resistors R_{ba}' and contact resistors R_{bc}' describe the external base resistance.

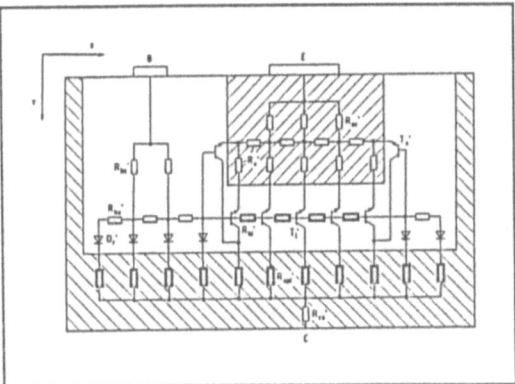

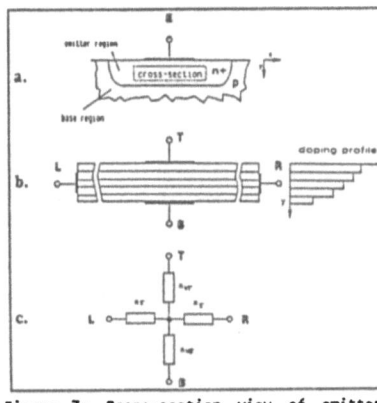

Figure 2: Three dimensional equivalent transistor circuit (cut C-C' see Figure 5b).

Figure 3: Cross-section view of emitter region. The resistivity varies with depth y. An approximation of the cross-section could made by using adjacent regions of different resistivities. Figure 3c shows a simple equivalent representation.

- PN-junction and external base
This region is represented by diodes D_r'.

- Collector area
To consider the effects in the epitaxial layer at quasi-saturation, nonlinear resistors R_{epi}' are used. The path from buried layer to collector contact is represented by a single lumped resistor R_{Ca}'.

The third dimension y is not modeled in detail, but in an average sense. For explanation Figure 3a shows a part of the resistive layer within the emitter region. The cross-section may be simulated using a approximating representation. Each horizontal rectangle represents a region with constant resistivity, corresponding to the doping profile (Figure 3b). From device simulation (Figure 4) it is found, that within the emitter and base region the current flow is mostly either vertical or horizontal. Due to this perception the cross-section is simply modeled (Figure 3c) by horizontal resistors R_S' (sheet-resistance), vertical resistors R_{VT}' (including contact resistance) and R_{VB}'.

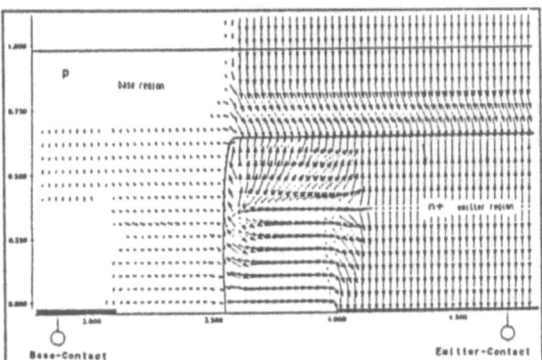

Figure 4: Current distribution in the emitter and base region at moderate conditions (U_{be} = 600 mV, U_{ce} = 2V) (BAMBI-plot).

4.   Generation of the equivalent circuit

The complete equivalent circuit consists of several layers to model the three dimensional current flow of the real transistor. For each layer the following 3D-2D transformation is executed:

$$\text{Div}\,(j_{xy}) = \pm |j_z|$$

For the vertical connections (bars) between the layers, well known /2/ electrical models can be employed:

Ohmic contact $\qquad\qquad j_z = U_{co} / r_{co}^*$,

PN-junction $\qquad\qquad j_z = j_s * (\exp(U_{pn}/v_t) - 1)$

Epitaxial resistance $\qquad j_z = \begin{matrix} U_{Epi}*Y_{Epio'} \\ U_{Epi}*Y_{Epio'}*(|j_z|/J_o) \end{matrix}$ for $\begin{matrix} j_z<j_o \\ j_z>j_o \end{matrix}$ with j_o : collector current, where base pushout begins.

Base and emitter region are automatically split into triangles (Figure 5). Each triangle vertex is copied to the upper (emitter) and the lower (collector) layer and is then connected by bars, to build the three dimensional structure. All elements in this structure are substituted by the described network element models. In doing this, within each layer the equation

$$j_{xy} = 1/r_{sg} * grad(v(x,y)),$$

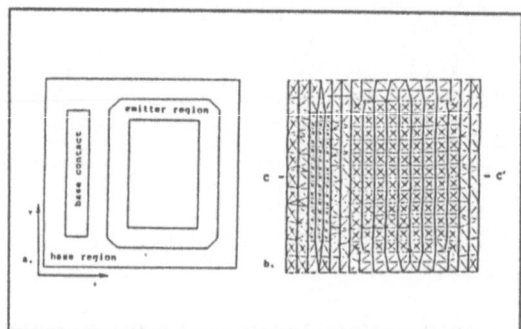

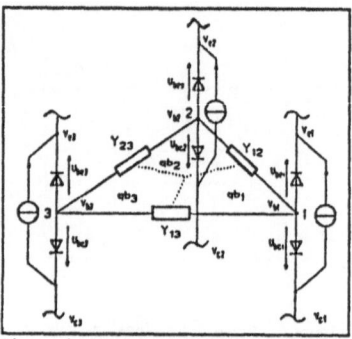

Figure 5: a. Transistor layout (emitter and base region)
b. Triangulation of emitter and base region.

Figure 6: Representation of a triangle in the active base with modified Gummel-Poon-models.

where r_{sg} is the sheet resistance of the specified layer, is solved using the finite element method. Below the Emitter area, r_{sg} is calculated to

$$r_{sg} = r_{sgo}/q_b \quad with \quad q_b = (q_{b1}+q_{b2}+q_{b3}),$$

where q_b is the normalized Gummel-number for each triangle referring to Figure 6.

A suitable approximation function, discribing the potential distribution for the finite triangle (Figure 7a), is

$$v(x,y) = a_0 + a_1*x + a_2*y.$$

The coefficients a_i can be determined by using the potentials of the element nodes V_p, V_q and V_r.

$$\begin{pmatrix} V_p \\ V_q \\ V_r \end{pmatrix} = \begin{pmatrix} 1 & X_p & Y_p \\ 1 & X_q & Y_q \\ 1 & X_r & Y_r \end{pmatrix} \cdot \begin{pmatrix} a_0 \\ a_1 \\ a_2 \end{pmatrix}$$

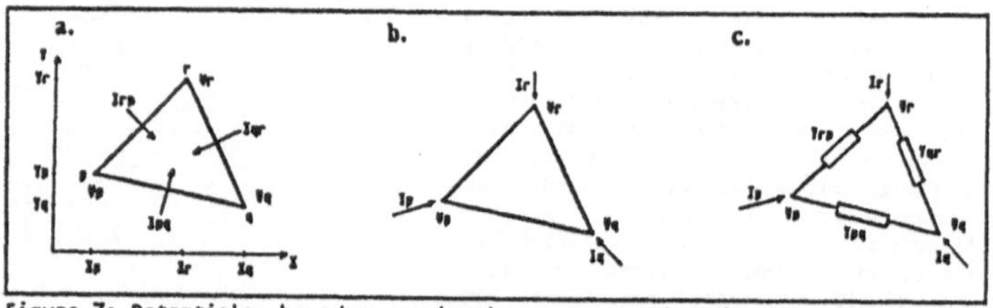

Figure 7: Potentials, boundary- and node-currents and equivalent network representation for a triangle.

Figure 7a shows the potentials V_i and the boundary-currents I_{pq}, I_{qr} and I_{rp}. These currents are used to define the equivalent node-currents (Figure 7b) I_p, I_q and I_r. They are defined as the half sum of the boundary currents at the adjacent edges of the triangle. This yields the following equations

$$\begin{pmatrix} I_p \\ I_q \\ I_r \end{pmatrix} = Y_e \begin{pmatrix} V_p \\ V_q \\ V_r \end{pmatrix}$$

$$I_p = \frac{1}{2} I_{pq} + \frac{1}{2} I_{rp}$$

$$I_q = \frac{1}{2} I_{pq} + \frac{1}{2} I_{qr}$$

$$I_r = \frac{1}{2} I_{rp} + \frac{1}{2} I_{qr}$$

$$A_0 = \frac{1}{2} \begin{vmatrix} 1 & X_p & Y_p \\ 1 & X_q & Y_q \\ 1 & X_r & Y_r \end{vmatrix}$$

$$I_{rp} = -\frac{1}{2 R_0 |A_0|} \left[\begin{array}{l} [(Y_r - Y_p)(Y_q - Y_r) - (X_r - X_p)(X_r - X_q)] V_p + \\ [(Y_r - Y_p)^2 + (X_r - X_p)^2] V_q + \\ [(Y_r - Y_p)(Y_p - Y_q) - (X_r - X_p)(X_q - X_p)] V_r \end{array} \right]$$

$$I_{pq} = -\frac{1}{2 R_0 |A_0|} \left[\begin{array}{l} [(Y_p - Y_q)(Y_q - Y_r) - (X_p - X_q)(X_r - X_q)] V_p + \\ [(Y_p - Y_q)(Y_r - Y_p) - (X_p - X_q)(X_r - X_p)] V_q + \\ [(Y_p - Y_q)^2 + (X_p - X_q)^2] V_r \end{array} \right]$$

$$I_{qr} = -\frac{1}{2 R_0 |A_0|} \left[\begin{array}{l} [(Y_q - Y_r)^2 + (X_q - X_r)^2] V_p + \\ [(Y_q - Y_r)(Y_r - Y_p) - (X_q - X_r)(X_r - X_p)] V_q + \\ [(Y_q - Y_r)(Y_p - Y_q) - (X_q - X_r)(X_q - X_p)] V_r \end{array} \right]$$

The symmetrical matrix Y_e can be interpreted as the element matrix of a three-pole element, shown in Figure 7c.

5. Calculation of potential and current distribution

The resulting equivalent transistor circuit, can be calculated using a common network analyzer /3/, provided that the analyzer contains suitable models for nonlinear base and epitaxial resistors. Direct solving of the network equations using a Newton iteration is possible as well. In order to save CPU-time a LU-factorization and a sparce-matrix technique is used. The computed potential and current distribution is shown in Figure 8.

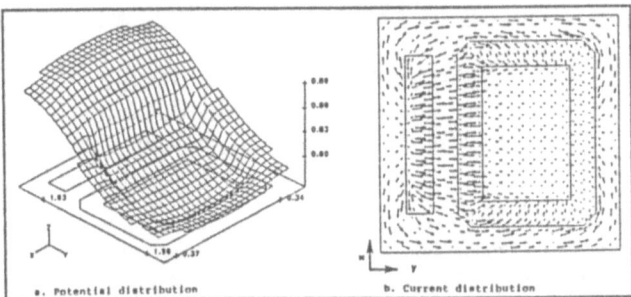

Figure 8: Calculated potential and current distribution in the base region (Ib = 1 mA)

6. Extraction of base resistance model parameters

To calculate the base resistance R_{Bb}, the power dissipation method is used. This results in

$$R_{Bb} = P_B / I_B^2 ,$$

where P_B is the total power dissipation within the base region. Seperating the active and passive base area, the active and passive part of the base resistance can be determined, so that

$$R_{Bb} = R_{Bba} + R_{Bbp} \text{ holds.}$$

SPICE parameters RB,RBM,IRB are obtained by fitting them to a calculated R_{Bb} vs. I_B curve (Figure 9).

7. Results

The program CABARET (CAlculation of BAse REsisTance) has been developed at the University of Hannover. It handles any kind of NPN-transistor layout, including multi emitter structures. CABARET is written in PASCAL and runs under VAX-VMS.

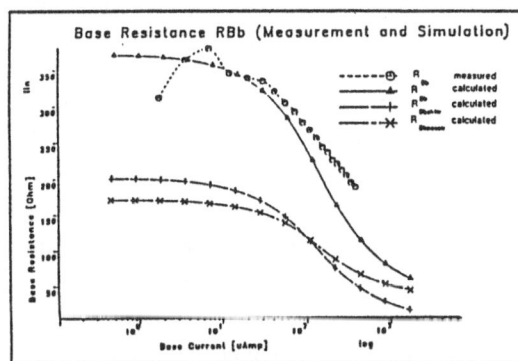

Figure 9: Measured and calculated base resistance R_{Bb}, internal and external fraction of base resistance.

To verify the CABARET results, specific test transistors were fabricated and base resistances were measured at Siemens Semiconductor Division, Munich. The calculated base resistance data show good correspondence with these measurements. Slight deviations were found (Figure 9) due to the noise measurement method used, which tends to overestimate resistance values.

8. Literature

/1/   Franz, A.F.; Franz, G.A.; Kausel, W.;
        Dickinger, P.; Nanz, G.
        BAMBI 2.0
        Basic Analyzer of MOS and Bipolar
        Devices
        Institut für Allgemeine Elektrotechnik
        und Elektronik
        Technische Universität Wien

/2/   Sze, S.M.
        Physics of Semiconductor Devices
        John Wiley & Sons, New York, 1981

/3/   Sibbert, H.; Herbst, D.
        DEMONA - Analyse von diskreten und integrierten Schalt.
        Ing.-Büro Dr.-Ing. D. Herbst, Unna, 1985

Efficient Integration of Device and Circuit Simulation

P. O'Sullivan, C.G. Cahill, C. Lyden
National Microelectronics Research Centre,
University College, Cork, Ireland.

Abstract : Integration of process, device and circuit simulation tools is desirable in modern process development and circuit design, as it allows rapid assessment of the impact of a new process or process change on circuit performance, [1]. This paper presents an efficient scheme for the extraction of MOSFET linear region SPICE level 3 parameters from numerical device simulations. Carrier concentrations, potential and mobility are used to derive a total of seven linear region parameters from three off-state bias points. This is in contrast to standard curve fitting techniques where twenty or more on-state bias points would typically be used to obtain the required parameters.

1 Introduction

Process and device simulations play a major role in process development. The accuracy of these simulators in predicting device performance prior to fabrication is continually improving. To minimise time to market it is essential that circuit design commence as early in the process development cycle as possible. It is also important to be able to assess the impact of process changes or process spreads on existing circuit designs. Hence, there is a requirement for an efficient link between device and circuit simulation.

Strategies and software for the extraction of circuit simulation parameters from measured device data are widely available, [2,3]. These techniques generally require curve fitting to device terminal characteristics which involves the use of many data points. In the case where a new process or a process modification is being evaluated, the terminal characteristics of the device must be numerically simulated. However, there is a large computational overhead associated with generating large numbers of data points with a numerical simulator. Hence, in developing techniques to link device and circuit simulators it is imperative that the number and cost of the simulations required to obtain the circuit simulation parameters are minimised. This paper presents techniques for the calculation of the SPICE level 3 linear region parameters [4], from three data points simulated using MINIMOS4, [5].

2 Numerical Simulation

The devices simulated in this paper are N-channel devices from a 3um CMOS process. The gate oxide thickness is 450 Angstroms. The source/drain regions are doubly implanted with a shallow arsenic and deeper phosphorous implant. The junction depth is 0.3 microns.

Table 1 lists the data points used for the extraction of the linear region parameters. The points are chosen such that the device is in strong inversion. The first two points are used for the extraction of the threshold voltage (VT_0), low field mobility (U_0), mobility rolloff (θ) and the channel length reduction (LD). The third point gives the body effect parameter (γ) and the parameter which describes the effect of back bias on mobility (ζ). The ζ parameter is an extension to the SPICE level 3 model, [6]. Substrate doping $NSUB$ is taken from the doping profile.

3 Extraction Technique

Threshold Voltage: The technique used to find the threshold voltage is based on that proposed by Wan et al, [1]. In strong inversion, at zero drain voltage the inversion charge, Qinv, is given by:

$$Qinv = -Cox(Vgs - VT_0) \tag{1}$$

where, Cox is the oxide capacitance. Threshold voltage may be calculated from the intercept of a line fitted to the inversion charge as a function of gate voltage. If Cox is known this requires the simulation of just one point. However, in practice, it is best to fit a line to two points and to calculate oxide thickness from the slope and threshold voltage from the intercept. Inversion charge may be calculated by integration of the carrier concentration over the the channel region as follows:

$$Qinv = -qW \sum_i \sum_j n(x_i, y_j).\Delta x_i.\Delta y_j \tag{2}$$

Mobility Parameters: In the SPICE level 3 model, mobility in the linear region is given by:

$$U_s = \frac{U_0}{1 + \theta(Vgs - Vth)} \tag{3}$$

where U_s is effective mobility, U_0 is low field mobility, θ is mobility rolloff. This expression may be rearranged as follows:

$$\frac{1}{U_s} = \frac{1}{U_0} + \frac{\theta}{U_0}(Vgs - Vth) \tag{4}$$

Hence, low field mobility and mobility rolloff can be calculated from a line fit to the average mobility in the channel at two gate biases. Average mobility at each bias point is calculated from the following integration over the channel region:

$$U_s = \frac{\sum_i \sum_j n(x_i, y_j)\mu(x_i, y_j).\Delta x_i \Delta y_j}{\sum_i \sum_j n(x_i, y_j).\Delta x_i \Delta y_j} \tag{5}$$

The modified Spice level 3 expression taking account of the mobility reduction with increase in back bias is given by, [6]:

$$U_s = \frac{U_0}{1 + \theta(Vgs - Vth) + 2\zeta\gamma(\sqrt{\phi - Vbs} - \sqrt{\phi})} \tag{6}$$

As U_0 and θ have been calculated at zero Vbs, ζ can be calculated from one extra point at a given back bias.

Channel Length Reduction: In the paper by Wan et al, [1], the channel length reduction is taken from the doping profile. Fig. 1 shows the fit to a device of drawn length 3um when LD is found in this manner. The threshold voltage and mobility parameters were extracted as described above and clearly, the resulting fit is quite poor. A novel scheme for the extraction of the channel length reduction has been developed. This is illustrated in Fig. 2, which shows a cross section of potential distribution in the channel parallel to the gate at zero volts Vds. Electrical channel length is considered to be the length of the "flat" region of the potential. This is found by fitting a line at the maximum slope of the potential distribution at the source end. The intercept of this line with the "flat" region defines the start of the channel. The distance between the start of the channel and the drawn gate edge is the channel length reduction, LD. Fig. 3 shows the fit to the 3um device using the new scheme. Fig. 4 shows the fit to this device using standard curve fitting techniques. Clearly, very good fitting is achieved using the new scheme.

Body Effect: The body effect parameter γ is given by the shift in threshold voltage between Vbs=0V and Vbs=-5V. Threshold voltage is calculated by summing the inversion charge as before. Fig. 5 shows the fit to a 3 micron long device simulated at -5V Vbs with the parameters extracted using this scheme.

Physical Parameters: The substrate doping, NSUB, is taken directly from the doping profile. The built in potential, ϕ is calculated from NSUB.

4 Conclusion

A scheme has been presented to extract all of the linear region SPICE parameters from just three bias points with a numerical device simulator. This contrasts with standard curve fitting techniques which require many data points. The scheme produces physically significant parameters, which are often not found with traditional curve fitting techniques. This is achieved by fitting directly to internal device quantities such as charge, mobility and potential rather than to terminal current characteristics. Furthermore, because terminal currents are not required for the parameter extraction, the points chosen have zero drain voltage and hence, may be simulated with an offstate model, leading to a further reduction in computational cost.

Acknowledgements

The authors would like to thank the Commission of the European Community for partial funding of this work under ESPRIT Project No. 962 entitled "Three Dimensional Algorithms for a Robust and Efficient Semiconductor Simulator with Parameter Extraction". The authors acknowledge the contribution of Dr. W.A. Lane, Analog Devices, Limerick, Ireland to this work.

References

[1] C.P. Wan et al, "Device and Circuit Simulation Interface for an Integrated VLSI Design Environment", IEEE Trans. on CAD, Vol.7, p. 998 (1988)
[2] C.G. Cahill et al, "MOS Model Parameter Extraction Techniques : A Comparison", Simulation of Semiconductor Devices and Processes Vol. 2, p.16, Pineridge Press, 1986.
[3] W. Maes et al, "SIMPAR: A Versatile Technology Independent Parameter Extraction Program using a New Optimised Fit Strategy", IEEE Trans. on CAD, Vol. 5 p.320, (1986)
[4] A. Vladmirescu and S. Liu, "The Simulation of MOS Integrated Circuits using SPICE2", Memorandum No. UCB/ERL M80/7, Feb. 1980
[5] S. Selberherr et al, "MINIMOS - A Two Dimensional MOS Transistor Analyzer", IEEE Trans. on Electron Devices, Vol. 27, p. 1770 (1980)
[6] W. Maes et al, "DC Characterisation of MOS Transistors, SPICE Model Level 3", Internal Report, Department Elektrotechniek, Katholieke Universiteit Leuven, Dec. 1984

221

	Vgs	Vds	Vbs
1	2.0	0.0	0.0
2	5.0	0.0	0.0
3	5.0	0.0	-5.0

Table 1 Simulated Bias Points for Parameter Extraction.

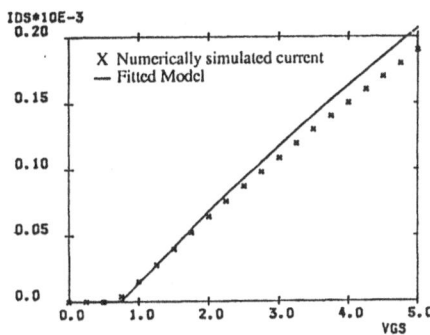

Fig. 1 Fit to linear region of a 3um long device at Vbs=0.0V, Vds=0.1V when LD is taken from the doping profile. The points are the terminal currents simulated using MINIMOS4 and the smooth curve is the SPICE level 3 model.

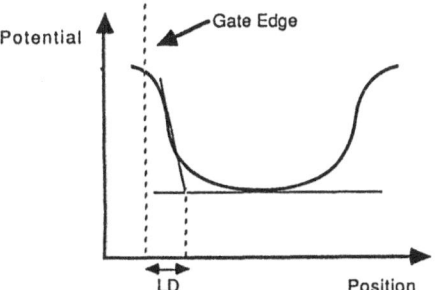

Fig. 2 Calculation of channel length reduction (LD) from potential. The figure shows a cross section of the potential in the channel parallel to the gate. LD is calculated from the intercept of the line of maximum slope with the minimum potential.

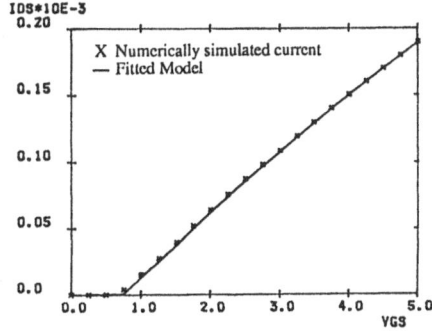

Fig. 3 Fit to the linear region of a 3um long device at Vbs=0.0V, Vds=0.1V with the new scheme for extraction of LD.

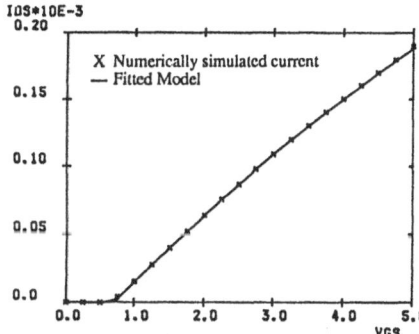

Fig. 4 Fit to the linear region of a 3um long device at Vbs=0.0V, Vds=0.1V using standard curve fitting techniques.

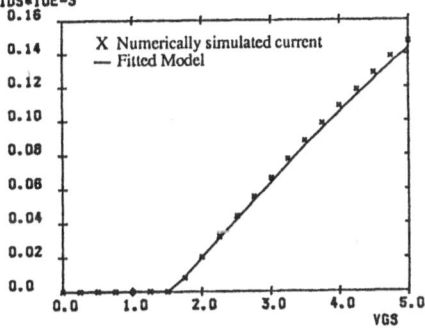

Fig. 5 Fit to the linear region of a 3um long device at Vbs=-5.0V, Vds=0.1V using the new extraction scheme.

Session 3

S/D Technology

Dopant Activation and Defect Annihilation of Heavily Doped Arsenic Implanted Silicon Layers

J. SAID*, H. JAOUEN*, G. GHIBAUDO*, I. STOEMENOS**,P. ZAUMSEIL***

*Laboratoire de Physique des Composants à Semiconducteurs, ENSERG, 23 Rue des Martyrs, B.P. 257, 38016 Grenoble Cedex, France
**Department of Physics, University of Thessaloniki, Thessaloniki, Greece
***Akademie der Wissenchaften der DDR, Institut fur Halbeiterphysik, Frankfurt, Krosingstr 2, 1200-Frankfurt, GDR

Introduction

A study of the effect of thermal annealing on defect annihilation in heavily arsenic implanted silicon has been carried out.

We investigate p-type (ρ=3.9Ω.cm) silicon wafers with the <100> orientation. Prior to the implantation, the wafers were provided with a 40 nm thick SiO_2 thermally grown passivating layer. The wafers were implanted at room temperature with As^+ ions of E=200keV at doses ϕ=10^{13}–$10^{15}cm^{-2}$. After implantation the silicon samples were subsequently thermally annealed under a dry N_2 flow at temperature ranging from 200°C to 1100°C for various durations(15-900mn).

The resistivity profiles of the implanted layers were deduced from the gross Spreading Resistance curves by applying the Local Slope Approximation of the Dickey model [1]. In addition, the samples were analyzed by TEM and TCD in order to determine the spatial extension of the defected region and the amorphous layer thickness at the surface of the silicon sample both before and after thermal annealing.

Results and Discussion

The resistivity profiles and TEM micrographs for a N^+/P sample implanted at $2 \times 10^{14}cm^{-2}$ and isochronally annealed (60mn) at various temperatures are shown on Fig.1. After annealing, the surface resistivity of a low temperature annealed sample (400°C) drops by more than 7 decades, though the surface layer is still amorphous.

The above results, correlated with the sheet resistance data of [2], lead us to conclude that this decrease of the surface resistivity occurs at an annealing temperature around 420–430°C mainly because of the electrical activation of the majority of the impurity atoms within a highly damaged surface region. By contrast, the recrystallization of the amorphous layer, that becomes increasingly efficient at higher annealing temperatures (T_a=550-600°C), does not induce a significant reduction of the resistivity even around the amorphous-crystal interface where no resistivity discontinuity appears (see Fig.1).

This last point suggests that a local impurity activation undergoes both in the crystalline and the amorphous regions via a local reconstruction process of which the relaxation time constante $\tau(x)$ can be represented by an expression of the form [3]:

$$\tau(x) = \frac{\epsilon k T_a}{4 \pi q^2} \frac{1}{D(x) N(x)} \qquad (1)$$

with

$$D\ (x)\ =\ D_o(x)\ \exp\ (-\ \frac{E_a(x)}{k\ T_a}\ )\quad, \tag{2}$$

where $D(x)$ and $N(x)$ are respectively the diffusivity and concentration of the point defects ensuring the electrical activation of the impurities, ε is the dielectric constant, q is the absolute electron charge and k is the Boltzmann constant.

The activation energy E_a has been determined experimentally by a proper analysis of the Arrhenius plot of the normalized resistivity[4]. As a result, E_a is found to be weakly depth dependent and is always smaller than 0.5 eV (Fig.2).

Likewise, the variation of the relaxation time with depth, $\tau(x)$, has been also deduced (Fig.2). $\tau(x)$ essentially increases when going from the surface towards the substrate. This feature is consistent with the adopted local relaxation process in which it is shown that the relaxation time is inversely proportional to the diffusivity and the concentration of the point defect species involved in the doping activation.

Besides, Triple crystal diffraction experiments have been conducted on the same samples. As is well known [5], X ray method are sensitive to a relative change of the lattice constant ε and to the static lattice disorder through the Debye-Waller factor $\exp(-L_H)$ [6]. Fig.3 shows the results of TCD analysis for a low temperature annealed sample and a well annealed one. It must be pointed out that, since the tetrahedral radius of arsenic is nearly the same as that of silicon, substitutional arsenic does not change the lattice constant of the doped layer. Therefore, the TCD data are actually indicative of the lattice distorsion due to implantation induced defects. In particular, the negative surface strain is likely related to an excess of vacancies while the positive strain is associated with an excess of interstitials.

Moreover, the comparison of the values of the activation energy of Fig.3 with those of usual point defect migrations (0.25eV for a vacancy—interstitial pair, 0.51eV for silicon interstitials, 0.8eV for a vacancy—As$^+$ complex and 0.33eV for vacancy [7,8]) suggests that the migrating species involved in the annealing process should be associated with a vacancy and/or an interstitial diffusion. More precisely, in view of the evolution of $E_a(x)$, it can be concluded, in agreement with our TCD data, that vacancy-like defects are presumably the dominant species in the first 100nm, while interstitials seem to prevail above 100nm.

Besides, Fig.4 displays the values of the activation energy both of the relaxation time constant associated with the electrical local activation process and that of the regrowth velocity corresponding to the recrystallization mechanism as obtained from our TEM analysis and from the litterature [9]. It is clear from Fig.4 that the activation energies deduced from our resistivity data, $E_a{\simeq}0.1$-0.5eV, are in any case much smaller than those for a regrowth process, $E_a{\simeq}2.5$-2.7eV. In the former case, the electrical activation of the impurity undergoes via a low energy consumming mechanism whereas, in the latter one, the Solid Phase Epitaxy involves a higher energy consumming process in which the re-ordering at the interface requires the breaking of silicon bonds [10].

Conclusion

It has been shown that the dopant electrical activation in highly doped arsenic implanted silicon is independent of the reconstruction mechanism occuring by Solid Phase Epitaxy. In contrast, the impurity activation process has been shown to be satisfactorily described by a local relaxation model in which the relaxation time is inversely proportional to the diffusivity and the concentration of the point defects leading to the dopant incorporation. Regarding the values of the activation energies found for this process, it is suggested that the migration of vacancies and/or silicon interstitials presumably plays a prevailing role in the impurity activation process.

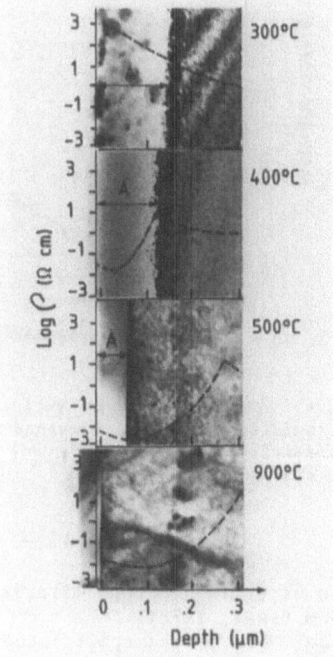

Fig.1. Resistivity profiles and TEM micrographs as obtained after various isochronal (1h) anneals for N^+/P type arsenic implanted samples ($2 \times 10^{14} As/cm^2$).

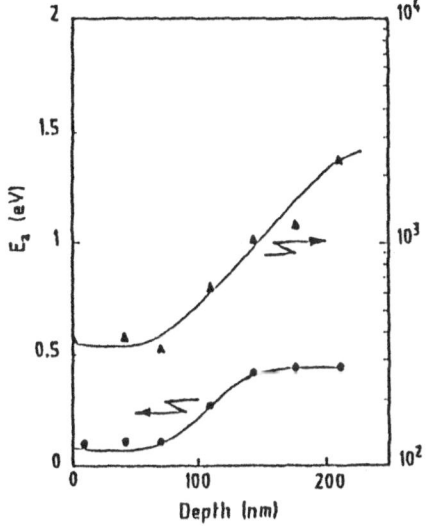

Fig.2. Variations of activation energy and time constant with depth for a N^+/P sample implanted at $2 \times 10^{14} cm^{-2}$.

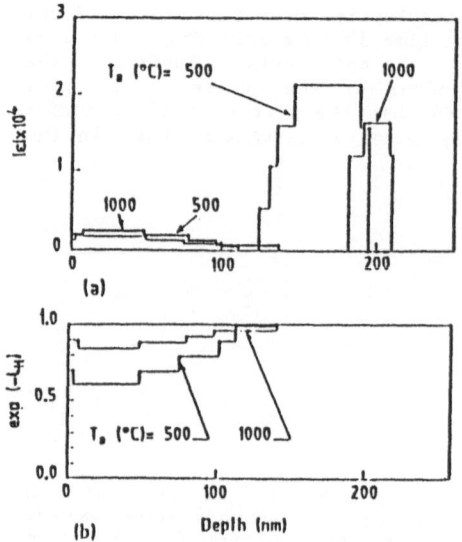

(a)

(b)   Depth (nm)

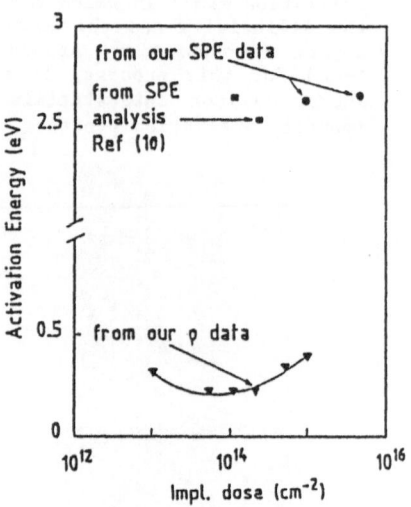

Fig.3. Mechanical relative strain (a) and Debye-Waller factor (b) variations with depth as obtained by TCD measurements (10^{15}As/cm^2, 200KeV).

Fig.4. Activation energy variations with implantation dose as obtained from our resistivity data (depth= 80nm) and from SPE analysis.

References

1. Ehrstein, J.R.: Nondestructive evaluation of semiconductors materials and Devices. Edited by J. N. Zemel (Plenum Press, 1978).
2. Christofidès, C.; Ghibaudo, G.; Jaouen, H.: A transport study of arsenic implanted silicon, influence of thermal annealing. Rev; Phys. Appl., 22, p. 407, 1987.
3. Fuller,C.S.: Defects Interactions in Semiconductors. American Chemi cal Society Monograph Series, No 140, edited by N. B. Hannay (Reinhold Publishing Corporation, New York, 1959) p. 209;
4. Said, J.; Ghibaudo, G.; Jaouen, H.; Stoemenos, I.; Zaumseil, P.: Electrical activation of heavily doped arsenic implnated silicon. MRS, Full meeting, Boston, (1988).
5. Zaumseil,P.: On the increased sensivity of x-ray rocking curves measurements by triple crystal diffraction.Phys. Stat. Sol. (a), 91, K31 (1985).
6. Zaumseil,P.; Winter,U.; Cembali,F.; Servidori,M.; Sourek,Z.: Determi nation of dislocation loop size and density in ion implanted and annealed silicon by simulation of triple crystal x-ray rocking curves. Phys. Stat. Sol. (a), 100, 95 (1987).
7. Boltaks,B.: Diffusion et Défauts Ponctuels dans les Semiconducteurs. (Mir, Moscou, 1977).
8. Watkins,G.D.; Corbett,J.W.: Phys. Rev. A, 134, 1359 (1964).
9. Kokorowski,S.A.; Olson, G.L.; Hoss,L.D.: Kinetics of laser induced solid phase epitaxy in amorphous silicon films. J. Appl. Phys., 53, 921 (1982).
10. Picraux,S.T.: Defects in semiconductors. Edited by J. Narayan and T. Y. Tan (North Holland, New York, 1981) p. 135.

Self-aligned CoSi$_2$ in a Submicron CMOS Process

R.D.J.Verhaar, A.A.Bos, H.Kraaij, R.A.M.Wolters K.Maex*, L. Van den hove*

Philips Research Laboratories, P.O. Box 80000, 5600JA Eindhoven, The Netherlands
* Interuniversity Microelectronics Center (IMEC v.z.w.) Kapeldreef 75 3030 Leuven, Belgium

Abstract

The integration aspects of a self-aligned CoSi$_2$ technology in a submicron CMOS process are described. The effect of substrate type and doping on the final sheet resistance of CoSi$_2$ was investigated. No significant influence on the sheet resistance of the finally formed CoSi$_2$ was measured, appart from an effect on the formation of the intermediate CoSi phase. The stability of CoSi$_2$ at high temperature was found to be significantly better on mono-Si than on poly-Si. Using amorphous Si as gate material a considerable improvement of the silicide stability was achieved.

Overgrowth of CoSi$_2$ (bridging) was detected by an electrical testing method and located by Voltage Contrast SEM analysis. The use of the CoSi$_2$ salicide process did not provoke any serious degradation of transistor performance or gate oxide integrity. The results are comparable with those of TiSi$_2$.

Introduction

Self-aligned silicidation technologies are receiving increasing attention. TiSi$_2$ is by far the most widely accepted material for this technology [1]. However, the interest for CoSi$_2$ has increased rapidly over the last few years due to its potential advantages over TiSi$_2$ [2].

In this paper the implementation of self-aligned CoSi$_2$ in an advanced submicron CMOS process is described.

Cobalt silicidation reaction

The Co silicide process does not suffer from competing reactions of Co with the N$_2$ ambient, in contrast to the nitridation of Ti during Ti silicidation. As a result the final sheet resistance of the CoSi$_2$ is hardly affected by dopant concentrations or substrate type.

To study the Co - Si reaction, p-type Si substrates were implanted with either As$^+$ 5 x 10^{15}cm^{-2} at 100 keV or B$^+$ 5 x 10^{15}cm^{-2} at 25 keV and subsequently annealed at 900 ^{0}C for 30 min. After an HF-dip a 38 nm Co layer was deposited in a rf-diode sputter system.

The sheet resistance was measured as a function of the annealing temperature for non-implanted as well as implanted substrates. A graphic representation of these results will be published elsewhere [6]. A maximum in the sheet resistance was observed at 500 ^{0}C, corresponding to the formation of CoSi. The maximum shifts slightly for the various doping conditions. At low temperatures (e.g. 500 ^{0}C) high dopant concentrations will retard the reaction of Co with Si. A 30 sec anneal time, however, has proved to be long enough for complete reaction of Co in all practical cases. At higher temperatures, CoSi is converted into CoSi$_2$. At 700 - 850 ^{0}C a stable sheet resistance is obtained, and no effect of dopant was observed.

In contrast to Ti, Co does not reduce silicon dioxide. Therefore the reaction of Co with silicon is more sensitive to the condition of the Si-Co interface. In the spacer formation process the Si surface is contaminated and prevents the reaction between Co and Si. Descumming or cleaning does not solve this problem. However, a sacrificial oxidation (typically 10 nm) prior to Co deposition has proven to be a good technique to "clean" the Si surface.

Thermal stability

High temperature stability of the silicide is important to permit glass reflow and to enable diffusion of dopants into the Si using the implanted silicide as the diffusion source. The stability of silicides on poly-Si is known to be limited by solid phase epitaxial regrowth (SPE-regrowth) between the silicide grains, which results in complete ball-up of the silicide film [3]. The stability of $CoSi_2$ on poly, amorphous and mono-Si is investigated when subjected to high temperatures, particularly using rapid thermal processing (RTP).

For this purpose substrates were covered with thermal oxide, CVD-Si (α-Si grown at 540 ^{0}C or poly-Si grown at 620 ^{0}C) and 38 nm Co. Some of the α-Si layers were doped by solid source doping (POCl$_3$) or P$^+$ implantation (1.2 x 10^{16}cm^{-2} at 50 keV + 1100 ^{0}C 10 sec). After $CoSi_2$ formation at 700 ^{0}C the layers were capped with 200 nm TEOS oxide and annealed for 10 sec. at 900, 1000 or 1100 ^0C. The samples were analysed by 4-point probe resistance measurements and RBS.

In fig. 1 the sheet resistance values are given as a function of RTP temperature. From this graph it can be concluded that the stability of $CoSi_2$ on MOS-gates can be significantly improved by using amorphous Si. Doping this layer has a negative effect on the stability (contrary to the effect of dopants on the stability of silicides on poly-Si [3]); most likely because the activation of the dopants causes recrystallisation of the α-Si.

Sheet resistance measurements did not provide a good indication of stability as the resistivity only increased where the silicide was nearly discontinuous. Therefore RBS analyses were carried out. In fig 2. RBS spectra of $CoSi_2$ (130 nm) on top of P$^+$ implanted α-Si and poly-Si are shown before and after a 10 sec. 1000 ^{0}C anneal. The spectrum of the α-Si sample shows some roughening of the $CoSi_2$ - Si interface where the $CoSi_2$ on the poly-Si has completely mixed with the Si.

On mono-Si a 72 nm $CoSi_2$ layer shows no increase in sheet resistance after a 1000 ^{0}C anneal for 30 seconds, similar to the measurements on α-Si. However, TEM cross-sections showed a variation in thickness of $\pm$ 25 nm.

Self-Alignment

One of the most important features of the salicidation process is its self-alignment. In many papers $CoSi_2$ is considered to be self-aligned because Co is assumed to be the main moving species and hence no overgrowth is expected to occcur [4]. However, $CoSi_2$ is formed via Co_2Si and CoSi formation. Since Si is the main moving species in the CoSi phase formation [5], overgrowth is a concern. The tremendous length of poly silicon gate edges on active area in the 1 and 4 Megabit SRAM range (up to 30 meters) requires dedicated measurements to determine shorts due to possible overgrowth (bridging) of silicide in the self-aligned silicidation process.

An electrical method to check overgrowth has been developed [6]. In Fig.3 a schematic diagram is given of such a bridging monitor. The structure consists of a n$^+$ poly-Si layer on active area with 6400 square holes, in which isolated p$^+$/n$^-$ diodes has been formed. The total length of poly edge on active area is 30 cm per structure. Bridging is easily detected by measuring the isolation between the poly-Si layer and the substrate.

In a set of experiments a 20 nm Co layer was deposited on bridging monitors followed by a first RTP step varying between 500 - 900 ^{0}C for 30 sec. After the first RTP step the wafers were subjected to a 2 min H_2SO_4/H_2O_2 etch to remove the unreacted Co. Finally all samples were submitted to a second 30 sec. RTP step at 700 ^{0}C.

In Fig.4 the yield (number of non-bridging sites/total number of sites) on the bridging monitor is given as a function of the first annealing temperature. Each point in the graph represents the result of 151 measurements on one wafer. The total tested poly Si gate edge length was 45 m per wafer.

Obviously in these experiments the silicidation is not 100 % self-aligned and the yield is low at high temperatures. As a consequence one short on every 10 cm poly-Si edge can be expected statistically, applying a one-step RTP process at 650-750 ^{0}C. No significant differences were observed when HCl/H_2O_2 mixtures were used to remove the unreacted Co instead of H_2SO_4/H_2O_2. Since HCl/H_2O_2 is known to etch CoSi, this etch procedure has been used only where the first RTP temperature was above 600 ^{0}C.

The nature of bridging has been studied using Voltage Contrast SEM. The poly-Si electrode of the bridging-monitor is biased with a small negative potential with respect to the substrate, inside a SEM. A clear black and white contrast is observed between the poly-Si gate electrode and the square diodes. In case of bridging the diode will adapt the poly-gate potential and its brightness will change. With this technique the location of shorts can easily be detected. In Fig.5 a high resolution SEM-picture is shown, which has been made from an identified location. From this picture it is clear that in this case overgrowth of $CoSi_2$ occurs from the poly-Si. The dimple in the poly-Si layer results from the silicon consumption during the $CoSi_2$ overgrowth.

Electrical data

In our process $CoSi_2$ shows a spread in resistance of $\pm$ 5 % within a batch. Based upon the available data a batch to batch reproducibility in the order of $\pm$ 7 to 10 % is expected for volume production, since the reproducibility of the sheet resistance is mainly determined by the Co deposition process. The actual transistor characteristics for $CoSi_2$ were similar to the $TiSi_2$ case. In Fig. 6 an example of an Id-Vds characteristic of a n-channel MOS transistor with a geometrical gate length of 0.7 μm is shown. The influence of salicide processing on gate oxide integrity has been studied. Simultaneous processing of wafers with $TiSi_2$, $CoSi_2$ and without salicide, showed an increase of the gate oxide defect density by a factor of three for both $CoSi_2$ and $TiSi_2$.

Conclusions

The silicidation rate of Co on Si was found to be dependent on substrate type. However an annealing process of minimum 30 sec. at 500 ^{0}C provides complete reaction of 20 nm Co. No dopant effect was observed on the final sheet resistance of $CoSi_2$ formed at 700 ^{0}C. The Co - Si reaction is very sensitive to the Si - Co interface condition. A sacrificial oxidation has proven to be a suitable method to ensure a proper reaction. The stability of $CoSi_2$ at high temperature was found to be significantly better on α-Si than on poly-Si. The stability on MOS-gates can be significantly increased by the application of α-Si.

Electrical testing, by means of a bridging monitor, was a very sensitive method to detect overgrowth in the salicide process. Voltage Contrast SEM analysis was used to locate overgrowth. Using the $CoSi_2$ salicide process good transistor and diode behaviour was obtained and the influence of $CoSi_2$ on the gate oxide quality was comparable with that of $TiSi_2$.

Acknowledgement

The authors wish to thank J.Vrehen for facilitating the voltage contrast work, W.Vandervorst(IMEC) for the RBS work, and H.Pomp and M.Pitt for helping with the electrical characterization of the CoSi$_2$ process. K.Maex and L.Van den hove are indebted to the Belgian National Fund for Scientific Research (NFWO) for their fellowship as Senior Research Assistants.

References

[1] M.E.Alperin et.al., IEEE Trans. Electron Dev. ED-32 141, 1985

[2] L.Van den hove et.al., Le Vide, Les Couches Minces, Vol. 42, N 236, 1987, p. 111.

[3] P.Lippens et.al. , Proc. of the 18th ESSDERC, Montpellier, sept. 13-16, 1988, p C4-191

[4] M.Tabasky et.al. , IEEE Trans. on Electr.Dev. Vol ED-34 No 3 p.548

[5] S.S.Lau et.al J.Appl.Phys. 49(7), July 1978 p.4005

[6] R.D.J.Verhaar et.al., Proc. of RMS workshop Houthalen 1989, to be published in Appl.Surf.Sci.

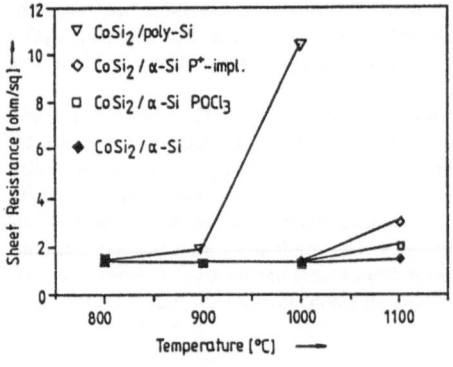

Figure 1: Sheet resistance of $CoSi_2$ on top of poly and α-Si versus anneal temperature.

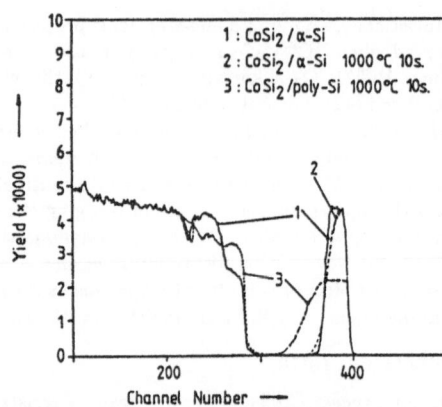

Figure 2: RBS spectra of $CoSi_2$ on top of poly and α-Si before and after 10 sec. 1000 $^{\circ}$C anneal.

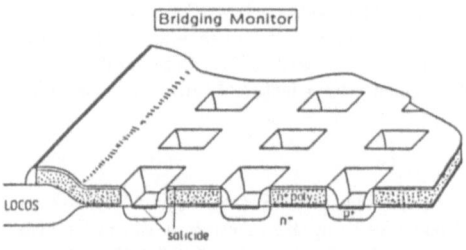

Figure 3: Schematic representation of the bridging monitor.

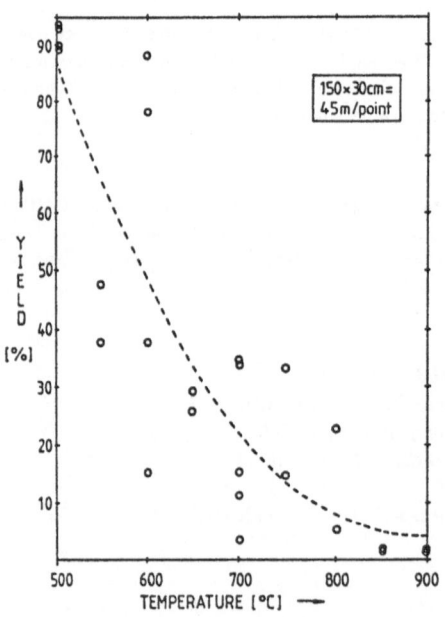

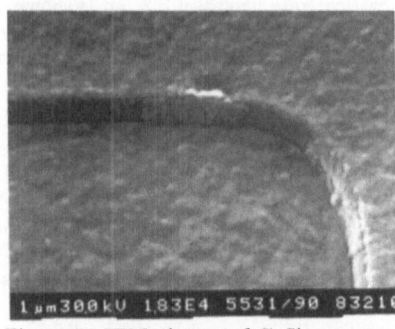

Figure 5: SEM picture of $CoSi_2$ overgrowth.

n-channel MOSFET 10/0.7

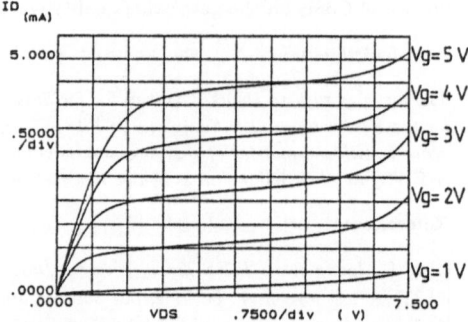

Figure 6: Id-Vds characteristic of a n-channel MOSFET with a geometrical gatelength of 0.7 μm.

Figure 4: Yield of bridging monitor as function of first RTP temperature. Annealing time 30 sec.

Reduction of Titanium Silicide Degradation During Borophosphosilicate Glass Reflow

R. Burmester, H. Joswig, A. Mitwalsky

Siemens AG, Corporate Research and Development,
Otto-Hahn-Ring 6, D-8000 München 83, F.R.G

Summary

The application of a self-aligned silicide ("salicide") process with Ti requires reduction of following excessive heat-treatment [1]. E. g. a commonly used borophosphosilicate glass (BPSG) reflow process (900 $^{\circ}$C, 40 min, N_2) has been found to cause severe morphological degradation of the $TiSi_2$ layer [2] resulting in higher sheet resistance (R_S). On the other hand the use of BPSG is advantageous because of its planarization capabilities and its gettering function of mobile alkali ions [3]. In this contribution the degradation mechanism and the influence of process parameters are investigated. As a result a Ti salicide process was established which was able to withstand following isolation oxide processes with temperatures of 900° C.

1-Experimental

In our experiments we performed a $TiSi_2$ salicide process [4] on differently doped <100> oriented silicon substrates as follows: After cleaning the substrate with a standard HF dip 40-60 nm thick Ti layers were deposited by sputtering in commercial DC-magnetron sputter systems: a batch system capable of processing 8 wafers and a single wafer system. Some of the wafers received an additional in-situ RF backsputter etch before Ti deposition. Backsputter conditions have to be chosen in order to avoid large crystal damage or argon incorporation near substrate surface because both effects slow down the silicide reaction of Ti with the underlying substrate. Therefore we have put some emphasis on the examination of the damage induced by the RF sputter etch.

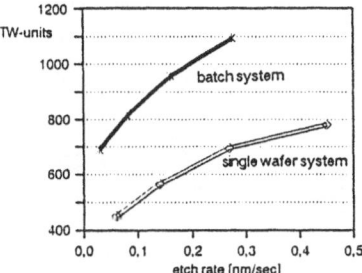

Fig. 1: Etch damage measured in thermawave units vs. etch rate. Wafers were processed at the lowest possible etchrates to keep surface damage low.

For this purpose bare silicon wafers were treated with the sputter etch cleaning step and the induced damage was measured via the method of modulated optical reflectance ("thermawave") [5, 6]. Fig. 1 illustrates that for salicide application backsputtering with low etch rate i. e. low RF power is preferable. For silicidation a two

234

step rapid thermal processing (RTP) was carried out. After the first RTP (650 °C, 20 s, N$_2$) the nonreacted Ti and TiN was removed from underlying oxide or partly silicided regions by selective wet etching (NH$_4$OH/ H$_2$O$_2$/H$_2$O). A second RTP (850 °C, 10 s, N$_2$) finished the salicide process by transforming the silicide into C54 TiSi$_2$ with its high electrical conductivity. The excessive temperature treatment during BPSG reflow was simulated by depositing a 50 nm LPCVD SiO$_2$ layer and by annealing in nitrogen ambient at different temperatures and for various periods (5 min - 60 min). Finally the capping oxide was removed by a selective dry etch step (CHF$_3$/O$_2$). A first characterization was done by four-point probe measurments of R$_S$ after significant process steps. In addition scanning electron microscope (SEM) and transmission electron microscope (TEM) analyses were performed for selected samples.

2-Results and Discussion

With increasing heat treatment generally rising R$_S$ is observed. Fig. 2 shows the dependence of R$_S$ on annealing temperature and annealing time for As doped as well as BF$_2$ doped substrates. This increase of R$_S$ is due to a morphological degradation of the TiSi$_2$ layer as illustrated in Fig. 3. During the heat treatment the surface is roughened by TiSi$_2$ grains tending to form a globular shape in order to reduce surface energy. After prolonged annealing the TiSi$_2$ layer is cracked at the grain boundaries forming epitaxial silicon islands between the TiSi$_2$ grains. As depicted in Fig. 2 TiSi$_2$ degradation is more critical on As doped substrates. Therefore, the following investigations were performed with this kind of substrate.

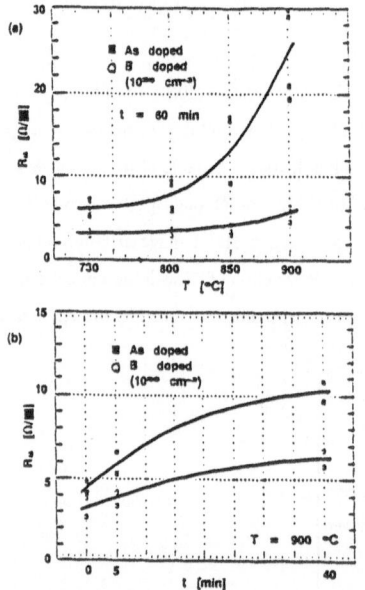

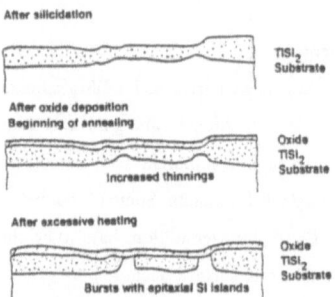

Fig. 3: Model for TiSi$_2$ degradation during high temperature treatment.

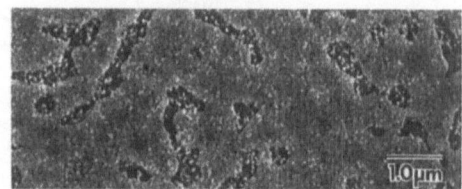

Fig. 2: TiSi$_2$ sheet resistance vs. post-silicidation annealing temperature (a) and time (b) (Ti thickness: 40 nm).

Fig. 4: SEM of TiSi$_2$ surface after annealing (900 °C, 5 min) for Ti (40 nm) deposited without backsputtering.

The surface SEM micrograph of the annealed (900 °C, 5 min, N$_2$) sample shown in Fig. 4 demonstrates the amount of silicon islands formed in the TiSi$_2$ layer. A TEM cross section of the same sample is shown in Fig.

5a. Aside from complete interruptions in the TiSi$_2$ layer, there are also regions with reduced thickness of about 25 nm indicating possible sites of breaking up after further heat treatment. The analogous sample processed without additional annealing after silicidation shows already locally very thin TiSi$_2$ regions with a thickness of only 15 nm compared to an avarage value of about 60 nm (Fig. 5b). Those regions are very susceptible to breaking up during further heat treatment because initially present thickness variations are enhanced up to the above described local bursts and epitaxial Si regrowths.

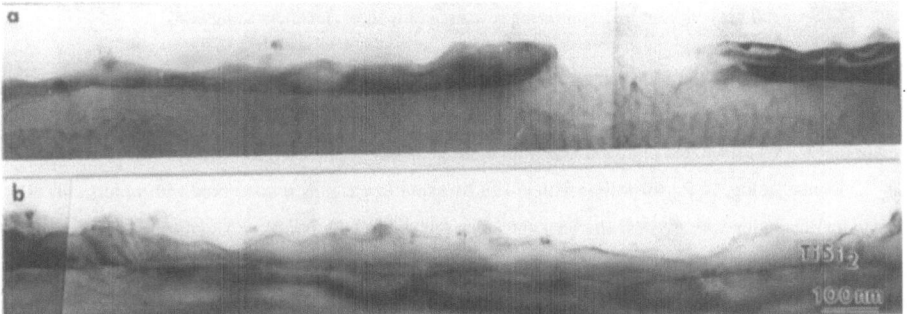

Fig. 5: TEM cross section of TiSi$_2$ layers with "standard" salicide process (40 nm Ti, no backsputtering) and (a) following heat treatment (900 °C, 5 min) forming a discontinuous silicide layer with epitaxial Si islands and (b) like (a) without post-silicidation annealing showing already local thickness reduction of the silicide.

In order to achieve a uniform TiSi$_2$ layer after silicide reaction a clean silicon surface before Ti deposition is required. This can be obtained by an in-situ backsputter etching additional to the HF dip before Ti deposition. Fig. 6 shows a TEM cross section of a sample which received this additional in-situ backsputter process. The silicided layer reveals a nearly uniform thickness of about 50 nm. The sample was not annealed after silicidation and, therefore, can be directly compared with the specimen cleaned only by HF dip (Fig. 5b). In fact after high temperature treatment (900 °C, 5 min, N$_2$) a SEM image of the TiSi$_2$ surface (Fig. 7) exhibits a continuous layer without Si islands in contrast to the sample shown in Fig. 4.

Fig. 6: TEM micrograph of an as-silicided sample (compare Fig. 5b) with backsputtering before Ti deposition showing nearly constant silicide thickness.

Fig. 7: SEM of TiSi$_2$ surface after annealing (900 °C, 5 min) for Ti (40 nm) deposited with backsputtering (compare Fig. 4).

Additionally, interruptions of the TiSi$_2$ layer can be reduced by deposition of thicker Ti layers (e. g. 60 nm) resulting in increased TiSi$_2$ thicknesses. Fig. 8 shows a TEM cross section of such a TiSi$_2$ layer after annealing (900 °C, 40 min, N$_2$). The average TiSi$_2$ thickness is about 60 nm with local reductions to 25 nm. However, no Si islands can be detected.

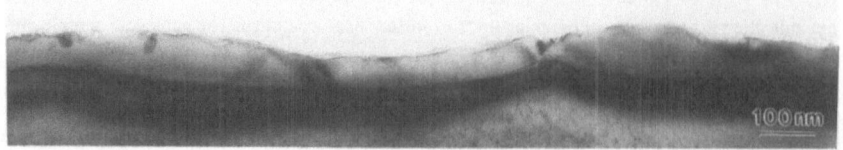

Fig. 8: Cross sectional TEM image of an annealed specimen (900 °C, 40 min) with thicker initially deposited Ti layer (60 nm) without backsputtering revealing only local thickness reductions.

Corresponding to TEM and SEM analyses R_S measurements reveal the efficiency of both described means. In Fig. 9 the effect on R_S of two annealing processes after silicidation is shown for different Ti thicknesses. It is evident that for a Ti thickness of 60 nm nearly no increase of R_S occurs even for an annealing process of 900 °C, 40 min. In Fig. 10 R_S for wafers with in-situ backsputter cleaning is compared with wafers only HF cleaned before Ti deposition. After the final annealing (900 °C, 5 min, N_2) R_S is significantly lower for the sputter etched wafers although it was higher before. Both diagrams describe the sheet resistance measured with four point probe on unpatterned silicided wafers. In the case of patterned lines the formation of epitaxial Si islands can lead to a more drastic increase of R_S during annealing due to interruptions of the silicide connection (see arrow in Fig. 10).

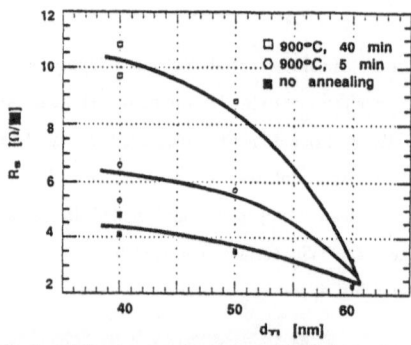

Fig. 9: TiSi$_2$ sheet resistance vs. Ti thickness for different post-silicidation annealings.

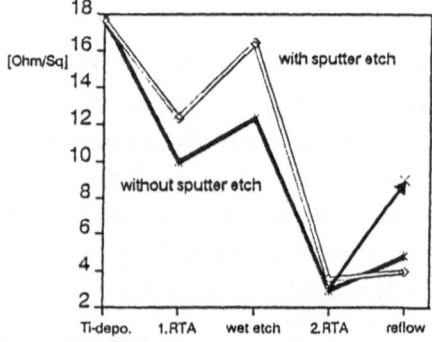

Fig.10: TiSi$_2$ sheet resistance during silicide process flow with additional heat treatment. The arrow marks the sheet resistance of a patterned line (1 um) without sputter etch preclean.

3-Conclusion

A Ti salicide process with increased Ti thickness and in-situ backsputter cleaning was established. In several experiments it proved to withstand the temperature budget of a BPSG reflow step (900 °C, 40 min, N_2) without severe morphological degradation.

References

[1] C. Y. Ting et al., J. Electrochem. Soc., Vol. 133, No. 12, 1985, p. 2621
[2] P. Lippens et al., Proc. 18 th ESSDERC, 1988, p. 191
[3] P. Balk, J. M. Eldridge, Proc. IEEE, 57, 1969, p. 1558
[4] M. E. Alperin et al., IEEE Trans. ED, Vol. 32, No. 2, 1985, p. 141
[5] R. S. Iverson, R. Reif, J. Appl. Phys., Vol. 62, 1987, p. 1675
[6] G. L. Patton et al., IEEE Trans. ED, Vol. 33, 1986, p. 1754

Shallow Junctions Fabrication by Using Molibdenum Silicide and Rapid Thermal Annealing

R.Angelucci, M.Merli, L.Dori, G.Pizzochero and S.Solmi

CNR - Istituto LAMEL, Via Castagnoli 1, 40126 Bologna, Italy

R.Canteri

Istituto per la Ricerca Scientifica e Tecnologica, Povo - Trento, Italy

Summary
Good quality p+/n and n+/p shallow junctions (~ 0.15 µm) suitable for VLSI technology have been fabricated by using Mo silicide and Rapid Thermal Annealing (RTA). The processes of implantation through Mo films and into MoSi$_2$ layers have been comparatively analysed on the basis of SIMS and carrier concentration profiles, contact resistivity measurements and electrical characterization of the junctions. Better electrical results are exhibited by the second technology which allows one to fabricate devices with low contact resistivities and low leakage currents.

Gate electrodes, interconnections, and direct ohmic contacts to source and drain regions have been so far the most widespread applications of the refractory metal silicides in VLSI technology owing to their low resistivity and superior chemical and thermal properties [1]. Recently, dopant by ion implantation through refractory metal (ITM) films and implantation in silicide layers have been explored as very promising technologies for the formation of shallow junctions for VLSI applications. Encouraging results have been obtained by using titanium [2], cobalt [3] and tungsten [4]. This work is devoted to an investigation on molybdenum.

n+/p and p+/n circular diodes with 10^{-2} cm^2 area and test patterns for contact resistivity measurements have been fabricated by implanting As and B through a film of 40 nm of Mo deposited by e-gun evaporation. A second set of devices was prepared by implanting the dopants into a 100 nm thick MoSi$_2$ layer obtained by RTA (tungsten halogen lamp) at 700°C for 60 s in a nitrogen atmosphere. As was implanted with dose of 7x10^{15} cm^{-2} at an energy of 160 keV, while B was implanted at the same dose with an energy of 30 keV. Post implantation treatments have been carried out at 1000°C, 10 s and 1100°C, 2s in nitrogen atmosphere.

Dopant distribution both in silicide and silicon was measured by SIMS, while carrier concentration profiles were determined by incremental sheet resistance and Hall effect measurements. The sectioning technique of MoSi$_2$ by anodic oxidation was calibrated. A current density of 8 mA/cm^2 and a net forming voltage of 19 V remove 16 nm of MoSi$_2$.The silicide resistivity values obtained for the different sample preparation conditions are reported in Table I. The resistivity depends on the silicide formation technology (ITM process gives always a resistivity 15% lower than implantation into silicide), and on the annealing temperature (30% smaller at 1100°C than at 1000°C). Since MoSi$_2$ is a multiband conductor, no reliable value for carrier concentration can be derived from Hall effect alone. A more exhaustive inquiry by using the magnetoresistance technique should be performed in order to separate the different contributions to the conductivity [5,6]. Nevertheless a change in sign of the Hall coefficient from positive to negative was detected: a predominantly p-type conduction was measured for intrinsic and B implanted silicide, whereas As implanted silicide exhibited a n-type conduction.

TABLE I

Technology	Junction type	Annealing cycle	ρ (MoSi$_2$) ($\mu\Omega$ cm)	ρ_s (Ω/□)	ρ_c (Ω cm^2)	J_R (nA/cm^2)
	n$^+$/p	1000°C,10 s	220	26.1	7x10^{-7}	20
		1100°C, 2 s	151	20.1	4x10^{-6}	40
I T Mo						
	p$^+$/n	1000°C,10 s	255	31.1	--	1
		1100°C, 2 s	176	23.2	--	1
	n$^+$/p	1000°C,10 s	245	25.4	8x10^{-7}	2
Implant into		1100°C, 2 s	182	19.0	3x10^{-7}	5
MoSi$_2$	p$^+$/n	1000°C,10 s	290	30.0	--	0.5
		1100°C, 2 s	208	23.7	--	1

Carrier and dopants profiles after annealing for As and B doped samples are shown in Figs. 1a and 1b, respectively, for the case of dopants implanted into silicide layer. Similar results were obtained on the specimens prepared with ITM technology (not reported). The data show junction depth in the range between 130 and 200 nm. The higher values are obtained for B, due to the ion channelling occurring during implantation. As does not show any significant diffusion in silicon during the annealing (the diffusion length associated to the

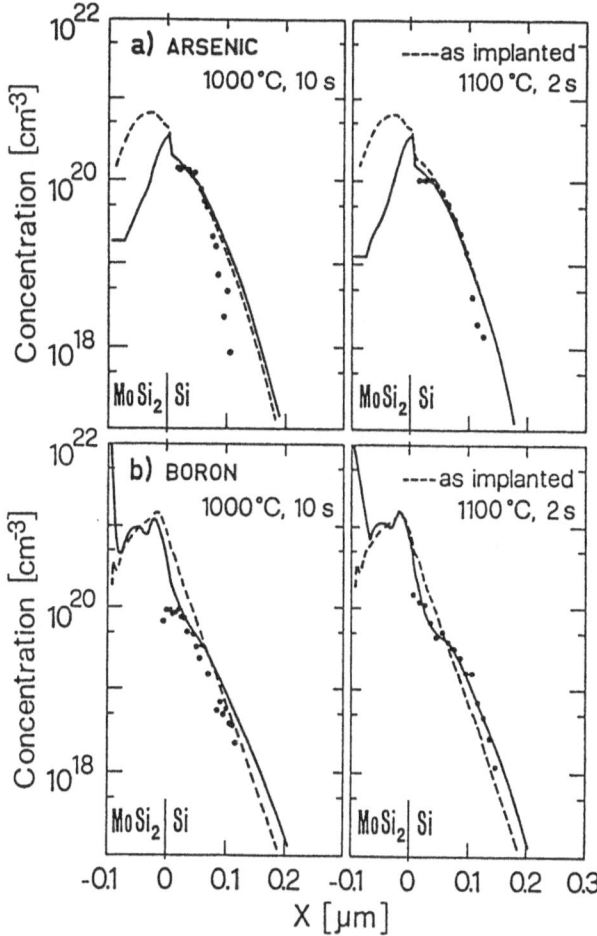

Fig. 1. SIMS (continuous and dashed lines) and carrier concentration profiles
(•) on samples obtained by dopant implantation into 100 nm thick MoSi₂ layer.
Dashed curves represent the as-implanted dopant distribution before RTA.

two annealing cycles being 7 and 16 nm for 1000°C, 10 s, and 1100°C, 2s respectively). A strong out diffusion is, however, observed due to the high As diffusivity along the grain boundary of the silicide. In spite of this, the heavy
implanted dose and the shortness of the annealing allow us to preserve a high
surface concentration; this is essential in order to keep low the contact
resistance. Surface concentrations higher than 1×10^{20} cm^{-3} and the practically
complete electrical activation of the dopant are confirmed by carrier profiles.
On the contrary, B profiles do not show significant out diffusion indicating a
low diffusivity of this dopant in the silicide. A lower B mobility compared with
As was observed also in Ti silicides [7]. However, due to the higher B diffusivity
in Si compared with As, a moderate redistribution of this dopant occurs during

the annealing, giving rise to a small increase of the junction depth and to a migration into the silicides. Also in this case the carrier profiles confirm the high surface concentration and the complete electrical activation.

The electrical characteristic of the junctions are summarized in Table I, where silicides resistivity ρ (MoSi$_2$), sheet resistivity ρ_s, contact resistivity ρ_c and reverse current density J_R measured at -1 V, are reported for the different fabrication procedures. Better reverse currents and contact resistivities are obtained for devices fabricated by implantation into Mo silicide. The worst performance for ITM procedure should not be related to Mo recoil [8], but rather to different generation and evolution of the implantation damage. Electron microscopy observations and double crystal X-ray analysis are under way in order to correlate residual damage and leakage current. Measurements carried out on diodes with different perimeter/area ratio pointed out the relevant contribution of the perimeter current on the total leakage current. For the best devices we obtained J_P = 15 pA/cm and J_A = 0.5 nA/cm^2 for n$^+$/p junctions and J_P = 16 pA/cm and J_A = 0.1 pA/cm^2 for p$^+$/n junctions.

In conclusion, the feasibility of Mo silicide for shallow junction fabrication has been demonstrated. Implantation of the dopant into self-aligned silicide on bare silicon and RTA allow one to obtain junction about 0.15 μm deep with low contact and sheet resistivity and very good electrical characteristics.

Acknowledgments
The authors wish to thank A. Scorzoni for contact resistivity measurements, P.Castelli, E.Gabilli and S.Guerri for technical assistance. Work partially supported by CNR - Progetto Finalizzato MADESS.

References
1.	Murarka,S.P.: Microelectronic Materials and Processing, R.A.Leavy ed. Academic Press, New York 1988, 257-328.
2.	Wong, D.L.; Ku, Y.H.; Lee, S.K.; Louis, E.; Alvi, N.S.; Chu, P.: J.Appl.Phys. **61** (1987) 5084.
3.	Van Den Hove, L.; Wolters, R.; Maex, K.; De Keersmaecker, R.F.; Declerck, G.J.: IEEE Trans.Electron Devices 34 (1987) 554.
4.	Tsaur, B-Y.; Chen, C.K.; Anderson Jr., C.H.; Kwong, D.L.: J.Appl.Phys. **57** (1985) 1890.
5.	Laborde, O.; Thomas, O.; Senateur, J.P.; Madar, R.: J.Phys.F: Met.Phys. **16** (1986) 1745.
6.	Martin, T.L.; Mahan, J.E.: J.Mater.Res. 1 (1986) 493
7.	Gas, P.; Deline, V.; d'Heurle, F.M.; Michel, A.; Scilla, G.: J.Appl.Phys. **60** (1986) 1634.
8.	Gessner, J.; Rennau, M.; Schubert, S.; Vetter, E.: European Workshop on Refractory metals and Silicies, Heugelhoef, 20-22 March 1989.

A Study of Ultra Shallow Junctions by Diffusion from Self-aligned Silicides

H. Jiang[a,b], C.M. Osburn[a,c,], Z-G. Xiao[d], P. Smith[c], G. McGuire[c], and G.A. Rozgonyi[d]

[a]Dept. of Electrical and Computer Engineering, North Carolina State University, Raleigh, NC 27695-7911, USA
[b]The Royal Institute of Technology, Electrum, Solid State Electronics, S-16428, Kista, Sweden
[c]Microelectronics Center of North Carolina, Box 12889, Research Triangle Park, NC 27709, USA
[d]Dept. of Materials Science, North Carolina State University, Raleigh, NC 27695, USA

Shallow implantation into a thin silicide followed by a brief diffusion using Silicide As Diffusion Source (SADS), is a promising technique to achieve ultra shallow junctions. Since silicon consumption during silicide formation on source/drain areas of MOS transistors is a part of the total junction depth, both the silicide thickness and the diffusion of dopant beyond the silicide should be minimized. The present work investigates a number of trade-offs in SADS processing. In particular, the temperature dependence of silicide roughening, grain-boundary and bulk diffusion of dopants through the silicide, dopant evaporation, and diffusion of dopant in silicon must be understood in order to be optimized. For instance, the data indicate an enhanced diffusivity of dopants in silicon below a $CoSi_2$ SADS layer.

Shallow implantations were done with low-energy BF_2 (8 keV, 0.4-1.5E15) and As (15 keV, 1.5E15) into pre-formed $CoSi_2$ (50 nm, 300 nm), $TiSi_2$ (34 nm), NiSi (222 nm), Pd_2Si (105 nm), and PtSi (198 nm). Ge implantation at 25 keV to a dose of 4E14 was used prior to B introduction in an attempt to amorphize the polycrystalline silicides and thereby reduce the possible channeling of B in the silicides. Additional annealings at 600-1100 °C from 2 to 300 s were performed to diffuse the dopants into Si and to study the thermal stability of the silicides. SIMS was used for dopant profiling, RBS for As loss, and sheet resistivity and TEM for thermal stability analysis. Silicides were stripped off before profiling the dopant in the silicon substrate.

The implantation depths of B and As in various silicides (taken at 10^{19} and 10^{17} cm^{-3}) are listed in Table 1, together with corresponding values from TRIM calculations. It is seen that the difference between the SIMS and the TRIM values increases dramatically within the concentration range of 10^{19} -10^{17} cm^{-3}), indicating the possible presence of SIMS profile broadening. The effect of Ge-implantation on boron profiles (see Fig. 1), was found to be within the experimental error of SIMS for all silicides studied here. XTEM micrographs for $TiSi_2$ and $CoSi_2$ samples showed that the Ge-preamorphization gave rise to a heavily damaged region within 10-15 nm from the surface of both silicides rather than any amorphous layer, see Fig. 2. For $TiSi_2$, twins were found to extend throughout the 34 nm thick silicide layer. The implantation-induced damage was removed during annealing.

Boron was found to diffuse rapidly in $CoSi_2$ following a moderate 800 °C, 10 s annealing, see Fig. 3. However, the boron concentration seemed to plateau at about $4 \cdot 10^{19}$ cm^{-3} in the bulk of the silicide, leaving a large surface peak. This plateau is determined by the implantation dose rather than by the solid solubility of B in $CoSi_2$, since in a high-dose sample its level is raised beyond $1 \cdot 10^{20}$ cm^{-3}. The loss of boron to the ambient after 1000 °C, 10 s annealing was found to be 12%. The boron diffusion depths in Si (at 10^{17} cm^{-3}) were measured to be 38 nm for 10 s annealing at 1000 °C. For a limiting case with minimal diffusion, an initial slope of 3 nm/decade was measured and attributed to the presence of a surface Co-B-O-Si layer after silicide etching. Auger electron spectroscopy revealed that for a sample annealed at 900 °C for 300 s, this surface layer had SiO_2 as the major constituent, contained 0.1-1% B and less than 0.1% Co . Boron diffusion in $TiSi_2$ never gave an interface concentration exceeding $1 \cdot 10^{19}$ cm^{-3}, which was confirmed by substantial leakage of the diodes fabricated by this process. As diffuses readily after 900 °C for 10 s in $TiSi_2$, while grain boundary diffusion dominates for As motion in $CoSi_2$ during the same annealing cycle (Fig. 4). The slow motion of boron in $TiSi_2$ and of arsenic in $CoSi_2$ illustrate the difficulty associated with CMOS junction formation. Arsenic loss from both $CoSi_2$ and $TiSi_2$ is appreciable after annealing at 900 °C for 10 s, see Table 2.

In Ni, Pd and Pt silicides boron was found to diffuse by a grain boundary mechanism . For NiSi and Pd_2Si, the interface concentration of boron is approximately $1 \cdot 10^{19}$ cm^{-3} after annealing at 600 °C for 10 s, while boron diffusion is only marginal for PtSi even after 800 °C for 10 s (Fig. 5).

Thermal stability was evaluated by sheet resistivity measurements of the silicide after annealing at various temperatures for increasing durations, see Fig. 6. A degradation time constant is defined for each temperature as the time when the resistivity increases by 30% from its minimum value. XTEM micrographs in Fig. 7 show that this increase corresponds to the initial stage of agglomeration. Thermal stability lines derived thereby define the proper process time-temperature (t-T) region, and are plotted in Fig. 8 for 34 nm $TiSi_2$ (with 3.5 eV activation energy) and 50 nm $CoSi_2$ (E_a=6.0 eV) annealed in Ar. In Fig. 8 we also included calculated 40 nm constant boron diffusion depth (x_d) in Si, as well as extrapolated t-T data giving 80 nm x_d of B in Si, which was extracted from SIMS boron profiles shown in Fig. 9. It seems that the stability of 50 nm $CoSi_2$ would allow 40-80 nm B outdiffusion and therefore 90-130 nm junction formation. On the other hand, 34 nm $TiSi_2$ would barely withstand annealings giving 40 nm diffusion in Si. We also observed better stability for $CoSi_2$ after annealing in N_2 as compared to Ar. Palladium, Pt and Ni silicides all degrade at lower annealing temperatures than Co and Ti silicides, indicating that they are less interesting candidates for a SADS process.

Portions of this work were supported by the Microelectronics Center of North Carolina, the NSF Engineering Research Center at NCSU, and the John and Karin Engblom Scholarship Foundation in Sweden. The authors gratefully acknowledge B. Patnaik, M. Swanson, and N. Parikh at UNC-Chapel Hill for RBS analysis and D. Griffis, S. Corcoran and J. Hunter and S. Hofmeister at NCSU for SIMS and AES.

References

1) C.Y. Tsai, F.M. d'Heurle, S.S. Iyer and P.M. Fryer, Electrochem. Soc., Extended Abstracts 85-1, 387, 1985.
2) R.B. Fair, J. Electrochem. Soc., 122, 800 (1975)

Table 1. Implantation depth of B and As, by SIMS and from Gaussian profiles using TRIM

Silicide	Implantation depth (nm)							
	8 keV BF$_2$ (1.8 keV B)				15 keV As			
	SIMS (1E19)	TRIM (1E19)	SIMS (1E17)	TRIM (1E17)	SIMS (1E19)	TRIM (1E19)	SIMS (1E17)	TRIM (1E17)
TiSi$_2$	19	17	39	22	31	23	40	27
CoSi$_2$	18	16	40	21	24	22	33	26
NiSi	17	13	40	17		19		22
Pd$_2$Si	26	15	50	20		22		26
PtSi	17	15	30	19		21		25

Table 2. As and B loss in CoSi$_2$ and TiSi$_2$ after SADS processes

DOPANT LOSS AFTER SADS PROCESS WITH 50 nm CoSi$_2$

		Annealing			
	Implantation	900 °C, 10 s	1000 °C, 10 s	1000 °C, 30 s	950 °C, 300 s
SIMS	4E14 BF$_2$	0%	12%		
	1.5E15 BF$_2$		45%	67%	50%
	1.5E15 As			94%	
RBS	1.5E15 As	42%	90%	91%	93%

244

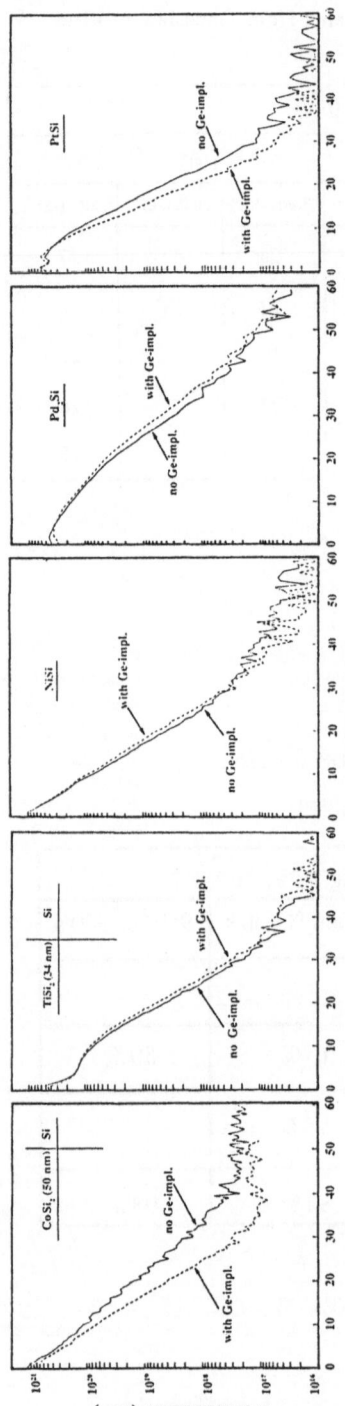

Fig.1 SIMS B Profile with and without Ge-preimplantation in CoSi$_2$, TiSi$_2$, NiSi, Pd$_2$Si and PtSi

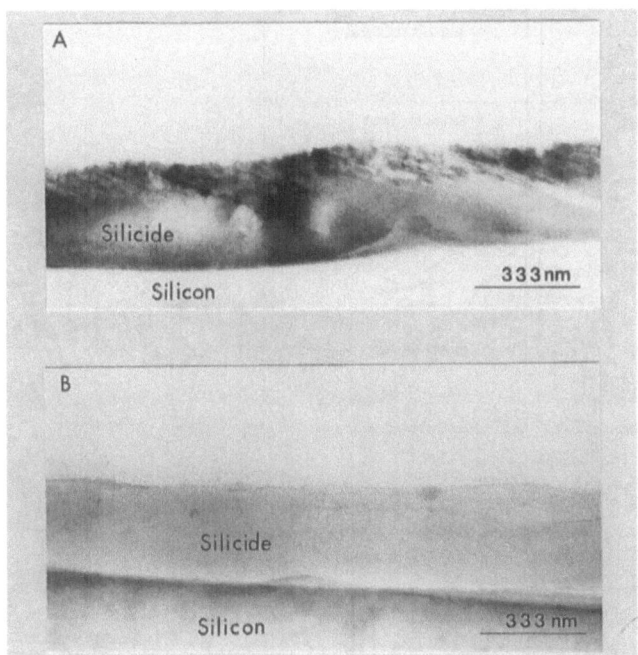

Fig. 2 XTEM micrographs of 34nm TiSi$_2$ implanted with 25 keV, 4E14 Ge and 8 keV, 4E14 BF$_2$, before and after annealing at 800°C for 10s

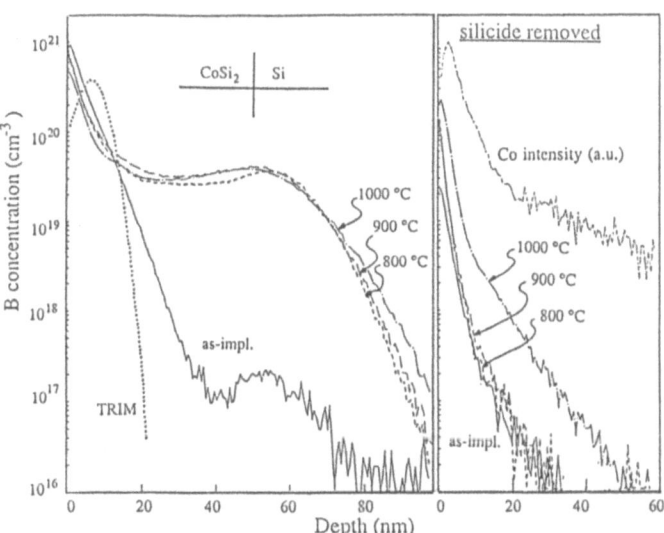

Fig.3.a) SIMS B profile for CoSi$_2$, implanted with Ge and with 8 keV, 4E14 BF$_2$, before and after annealing at 800-1000°C for 10s; b) with the silicide removed

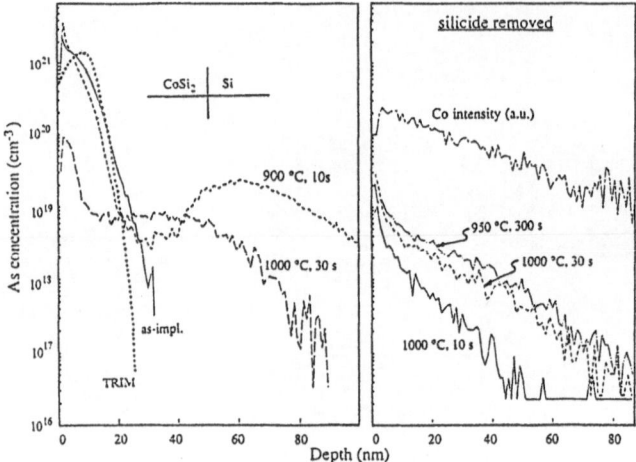

Fig.4 SIMS As profile in CoSi₂ before and after silicide removal

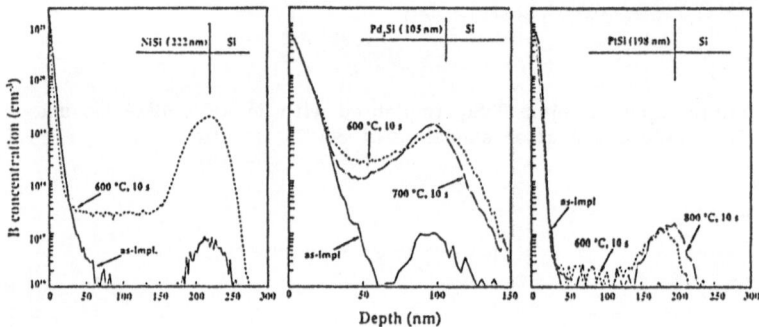

Fig.5 SIMS B profile in NiSi, Pd₂Si and PtSi before and after 10s annealing at 600–800°C

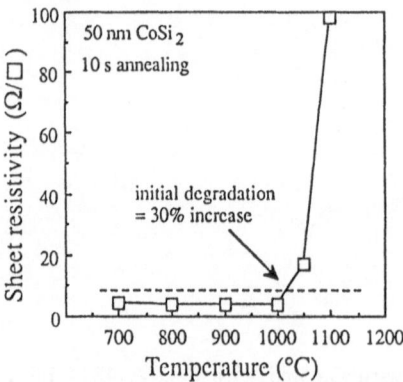

Fig.6 Sheet resistivity vs annealing temperature (time=10s) for 50nm CoSi₂

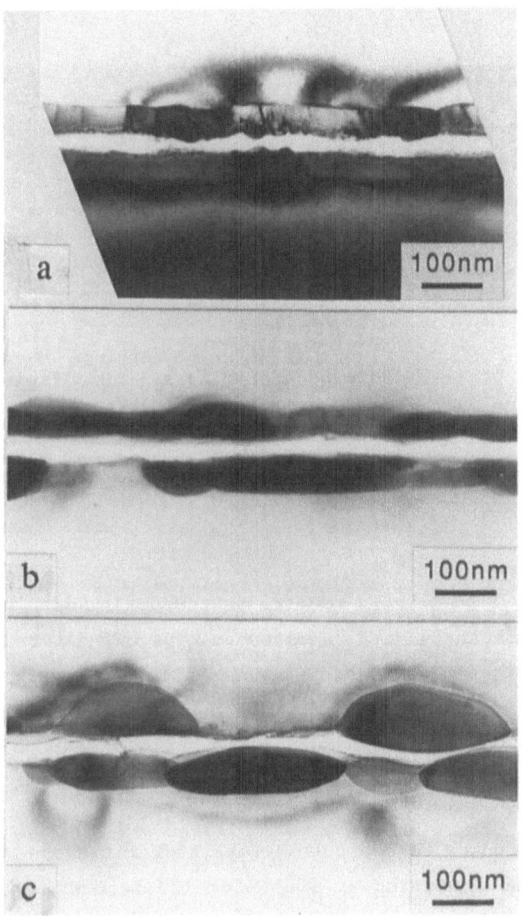

Fig.7 XTEM micrographs of 50nm SADS CoSi$_2$ a)as-implanted b)after 1000°C, 30s (30% increase in p) c)after 1000°C, 60s (250% increase in p)

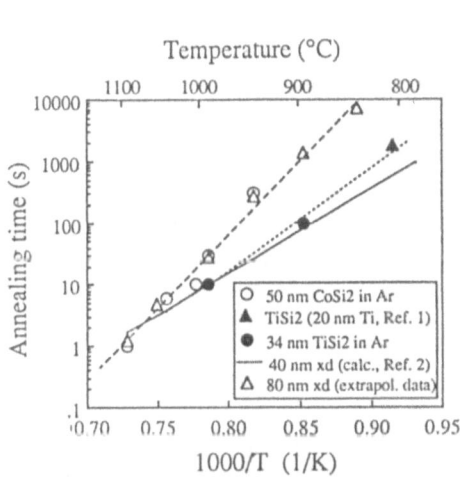

Fig.8 Arrhenius plot of degradation time constant for TiSi$_2$ and CoSi$_2$, and constant x_d data. The 40nm constant x_d line was calculated with C$_s$ = 1 · 10^{20}cm^{-3}

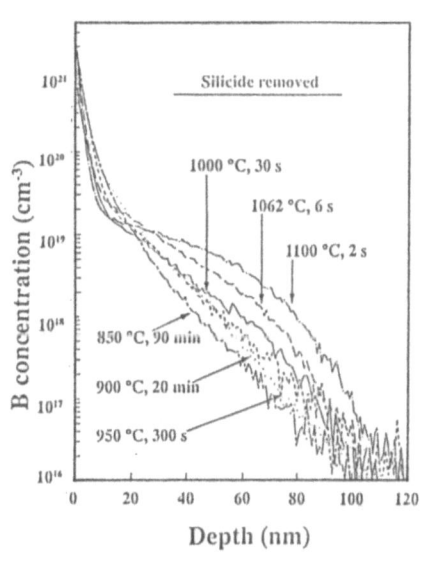

Fig 9 a) SIMS B profile for various annealings of CoSi$_2$, implanted with Ge and with 8 keV, 1.5E15 BF$_2$, after silicide removal.

The Outdiffusion of Boron and Arsenic from Pre-formed Cobalt Disilicide Layers

P. B. MOYNAGH, C. P. CHEW, K. B. AFFOLTER & P. J. ROSSER

STC Technology Ltd., London Rd., Harlow, Essex, England, CM17 9NA

Summary

Arsenic out-diffusing from Cobalt Disilicide into underlying silicon displays an enhanced activation level of up to 80%, and an enhanced diffusivity of greater than one order of magnitude. Boron, in contrast, displays a diffusivity as small as 0.25 times that expected. These observations are consistent with a considerable Si-vacancy injection and Si-interstitial depletion caused by the silicide layer in the silicon substrate.

Introduction

MOS devices with gate lengths of 0.5 μm or less require source/drain junctions with depths of less than 0.1 μm and much reduced resistances. The out-diffusion of dopants from pre-formed self-aligned refractory metal silicides has been identified as a means of meeting these requirements [1]. This paper describes the out-diffusion of B and As from $CoSi_2$ to form such junctions.

Experiments

Cobalt Disilicide layers were formed by three routes. Layers of thickness 0.5 μm were formed by depositing Co and annealing at 900°C for 120 seconds in an inert ambient. Layers of thickness 0.27 μm were formed by depositing a thinner layer of Co, ion-beam-mixing (i.b.m.) with a Si^{29} implant, and annealing at 900°C for 120 seconds. Layers of thickness 0.12 μm were formed by depositing Co, ion-beam-mixing, and performing a 450°C, 180 s anneal, a selective etch and a 700°C, 30 s anneal. Layers of LPCVD polysilicon 0.15 μm thick were deposited at 610°C on bare silicon wafers. These polysilicon layers were then processed in parallel with the silicide layers and therefore act as standards.

High doses of boron and arsenic (1E16 cm^{-2}) were implanted into the silicide and the polysilicon layers maintaining the implants within the layers. An oxide film of thickness 0.05 μm was deposited at 380°C on all samples to act as a barrier to the loss of dopant during any annealing stages. Rapid thermal anneals were then performed at 1000°C for times of 5 to 300 s thereby diffusing the dopant through the silicide or polysilicon layer and into the underlying silicon.

Results

Materials analysis has been undertaken to determine the process parameters for out-diffusion from $CoSi_2$. These include SIMS, SRA, AES, RBS, SEM, bevel & stain and delineation etching.

Figure 1a shows the dopant profile after As implantation into a 0.27μm CoSi$_2$ layer. The stopping power of CoSi$_2$ is high with B and As implant peaks occurring at 50 to 65 % of their peak depths in silicon.

Figure 1b shows dopant profile evolution after annealing the As implanted CoSi$_2$ layer at 1000°C for 30s. Upon annealing, grain growth rate within the silicide layer is very high, reaching film thickness in less than 5 s at 1000°C. As the grain boundaries move through the layer, the dopant segregates preferentially from the silicide grains to the grain boundaries. The dopant then diffuses very rapidly along the silicide grain boundaries to the silicide/silicon interface where the dopant segregates to and diffuses into the silicon substrate [2].

Figure 2 shows As profiles in the silicon substrate after out-diffusion from CoSi$_2$ and polysilicon sources at 1000°C for 30 s. Dopant profiles after out-diffusing from CoSi$_2$ are 50 to 80 % more active than from polysilicon as indicated by the peak SRA concentrations. This is of significant technological importance. The profiles also penetrate deeper into the substrate under CoSi$_2$ than under polysilicon. The dopant diffusivity enhancement factors have been extracted by fitting measured profiles using the process modelling program Suprem 3 [3]. These enhancements are plotted in figure 4. After 30 s at 1000°C, the As diffusivity enhancement factor is between 9 and 19.

Figure 3 shows B profiles in the substrate after out-diffusion from CoSi$_2$ and polysilicon sources at 1000°C for 30 s. Dopant profiles after out-diffusing from CoSi$_2$ are similar in activation level to that from polysilicon. For the 0.5 μm silicide layer, B displays no diffusivity modification. Under the ion-beam-mixed layers (0.12 μm & 0.27 μm), however, B is retarded with a diffusivity only 0.25 times that observed under polysilicon.

Analysis

The most probable mechanism for concurrent enhancement of As diffusivity, retardation of B diffusivity, and increased As activation level, is Si-vacancy injection and/or Si-interstitial depletion by the CoSi$_2$ layer. The increase in the concentration of vacancies in the substrate increases the diffusivity of As [4,5]. A decrease in the Si-interstitial concentration in the substrate would result from either surface depletion or bulk Frenkel recombination. This results in a reduction of B diffusivity [4,5]. A vacancy excess might also be expected to retard any dopant de-activation process, such as clustering, by providing unoccupied lattice sites on which As might remain substitutional and active. Increased activation levels would result.

RBS and AES analyses of the CoSi$_2$ before and after anneal at 1000°C show no change in the stoichiometry of the layer. It is possible however that a small percentage of unreacted Cobalt is available within the silicide layer during the out-diffusion anneal. The Cobalt may be available, a) because it did not react during the formation anneal, b) from Cobalt rich grain boundaries, c) through being 'knocked-on' during the dopant implantation process or, d) by being displaced at the silicide surface by a reaction with the annealing ambient. The free Cobalt would react with silicon during the 1000°C anneal to complete the $CoSi_2$ formation. The resulting removal of silicon atoms from the substrate results in an excess of Si-vacancies and a depletion of Si-interstitials and this in turn results in the diffusivity and activation modifications described above.

Since the available unreacted Cobalt is being consumed to cause point-defect perturbation, one would expect the injection to reduce with increasing time. This is observed in figure 4 in the form of a reduction in As diffusivity enhancement as anneal time increases.

Conclusion

When out-diffusing Arsenic from $CoSi_2$ one observes an enhanced activation level of up to 80%, and diffusivities which are enhanced by about one order of magnitude. Boron, in contrast, displays a diffusivity as small as 0.25 times that expected. These observations are consistent with a large Si-vacancy injection and Si-interstitial depletion caused by the silicide layer.

Acknowledgments

This work was partly supported by the UK Department of Trade & Industry through the Alvey Directorate under contract ALV/APP/VLSI/005.

References

1. F. C. Shone et. al., IEDM 85, p. 407, (1985)
2. P. B. Moynagh et. al., MRS Spring Meeting (1989), San Diego, to be published
3. Stanford University/ Technology Modelling Associates, SUPREM 3 process model
4. P. Fahey et. al., Semiconductor Silicon, p.571, ed. H. Huff et. al., (E.C.S., Pennington, N.J., 1986)
5. P. B. Moynagh et. al., ESSDERC 89 (1989), Berlin, to be published

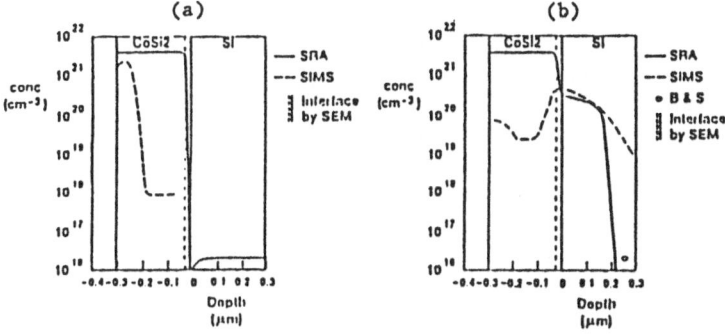

Implant: arsenic, 1E16cm^{-3} Anneal: 1000°C, 30s, N$_2$

Figure 1 Experimental data for cobalt disilicide + dopant implant + anneal

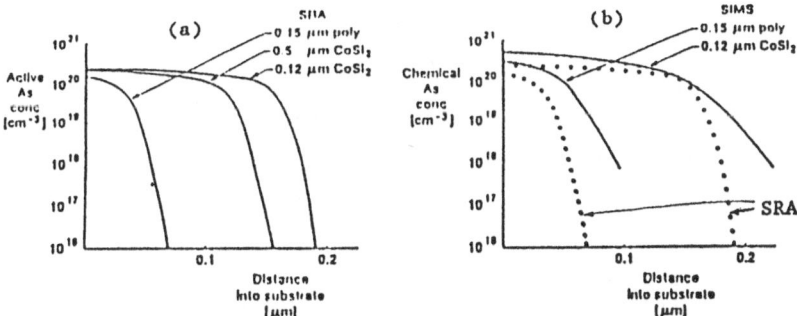

Arsenic implanted to 1E16 cm^{-2} into preformed CoSi$_2$ and polysilicon layers
and annealed, 1000°C, 30s, N$_2$

Figure 2 Substrate profiles after out-diffusion of As from CoSi$_2$ and polysilicon layers

252

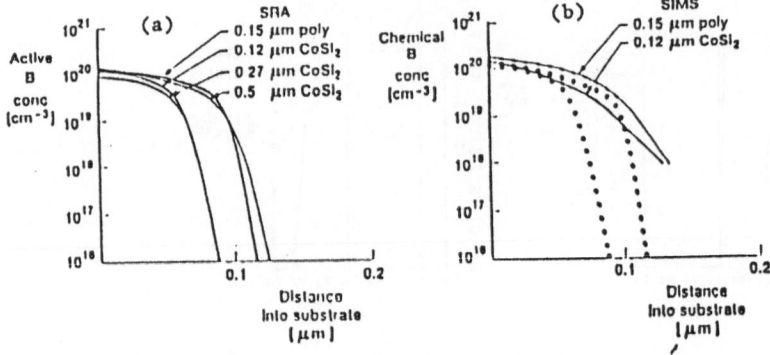

Boron implanted to 1E16-2 into preformed $CoSi_2$ and polysilicon layers
and annealed, 1000 °C, 30s, N_2

Figure 3   Substrate profiles after out-diffusion of B from $CoSi_2$ and poly layers

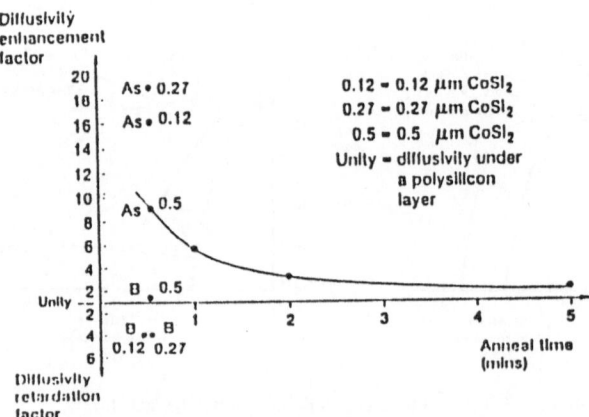

Figure 4   Diffusivity modification of Arsenic (As) and Boron (B) when out-diffusing
from $CoSi_2$ at 1000 °C

Ion-beam Mixed MoSi$_2$ Layers:
Formation and Contact Properties

C.Dehm, G.Vályi[1], J.Gyulai[1], and H.Ryssel

Fraunhofer-Arbeitsgruppe für Integrierte Schaltungen, Abteilung für Bauelementetechnologie
Artilleriestraße 12, 8520 Erlangen, FRG

ABSTRACT

n$^+$p-diodes with high conductivity MoSi$_2$-contacts were fabricated using self-aligned silicide technology and ion-beam mixing for silicidation. For optimizing contact properties, different ion-beam mixing processes were compared. The first mixing process investigated was a conventional one, using arsenic ions for silicide formation and doping of the underlying silicon substrate. After mixing, rapid thermal annealing was performed in a N$_2$ ambient to form stoichiometric MoSi$_2$ layers. The second ion-beam mixing process included a sequence of implantations: Silicon ions were used to cause mixing between the Mo-film and the underlying substrate. Rapid thermal annealing in N$_2$ ambient for silicidation was followed by a further silicon implantation to amorphize silicon substrate before junction doping by arsenic ions was performed. Electrical measurements were used to determine contact properties.

INTRODUCTION

The trend in VLSI-technology for faster and more complex microcircuits with structures of decreasing dimensions is causing short-channel effect or breakdown of MOS-devices. To overcome these deteriorating effects, the fabrication of shallow, good conductivity junctions is necessary. Shallow junction formation is well controlled by ion implantation and by rapid thermal annealing for dopant activation without considerable redistribution, but it causes some difficulties to achieve low-resistance contacts by conventional metallization methods. The increase of sheet resistance results from decreasing junction depths and widths with dopant concentrations reaching the upper limit of solubility in the silicon substrate. For formation of shallow and high conductivity junctions, self-aligned refractory silicides, like MoSi$_2$, TiSi$_2$, Pt$_2$Si, WSi$_2$, etc., are good candidates [1]. Silicide formation can be performed in several ways: (1) Thermal reaction of a thin metal layer with the underlying silicon substrate; (2) co-deposition of metal and silicon by sputtering or evaporation from a sintered target or from two separate targets and (3) silicidation by ion-beam mixing after metal deposition, etc.. Ion-beam mixing is especially advantageous for silicide formation, because the metal-silicon

---

[1] On leave from Central Research Institute,
H-1522 Budapest, P.O.Box 49, Hungary

reaction takes place at reduced temperatures and shorter times compared with (1) or (2), thus preventing the lateral growth over the isolation oxide or an appreciable redistribution of dopants. As a process, ion-beam mixing is very reproducible resulting in smooth surfaces and sharp interfaces. This latter is because mixing destroys contaminants like native oxide at the Si/silicide-interface, which allow only uncontrolled reactions at higher temperatures. Furthermore, ion-beam mixing fits into a self-aligned process technology. Phenomena during ion-beam mixing are rather complicated, because different effects like "ballistic mixing", "thermal spike effects" and "radiation enhanced diffusion" occur simultaneously during implantation [2]. It is obvious that these effects are leading to an improved reaction at reduced temperatures, but which one of the above mentioned effects dominates, is depending on the layers to be mixed, on temperature during mixing and on the mass of the incident ions.

The purpose of the present work is to suggest a mixing process new for shallow junction formation with $MoSi_2$ contact. To prove the advantages of this process, it was compared with a conventional mixing process using arsenic ions. $MoSi_2$ was selected because of its ease of processing and its high temperature stability. Implantation of arsenic ions was considered to be advantageous because mixing and doping occurs in one process step. Its application to shallow junction formation is, however, limited because appreciable dopant redistribution can occur during the following annealing step, necessary for formation of stoichiometric silicide and dopant activation. To prevent this, the second mixing process included a sequence of implantations. Silicidation was performed by interface mixing using silicon ions and rapid thermal annealing (RTA), followed by a second silicon implantation. This implantation was performed at an energy when the medium projected range was located at the silicide/silicon interface. Thus preamorphization of the silicon substrate was performed before junction doping in order to suppress dopant redistribution during annealing. It also helped to achieve better regrowth of the amorphous phase by preventing "end-of-range-defects". Compared to the conventional mixing, where $MoSi_2$-resistivities are of 150 $\mu\Omega$cm , the double-implanted silicides showed reduced values of 90 $\mu\Omega$cm as well as lower leakage currents (22 μA vs 90 μA for given geometry). In both cases, breakdown characteristics were hard with breakdown at 51 V.

EXPERIMENTAL

For the fabrication of the n^+p-diodes <100>-oriented, boron-doped Si-substrates with resistivities of 1-2 Ωcm were used. The process started by growing a field oxide of 400 nm thickness on the silicon. After contact hole opening, 25 nm DC-sputtered Mo was deposited. Silicon ions with 45 keV were then implanted to cause interface mixing ; the dose was $5 \cdot 10^{15}$ cm^{-2}. The mixing was followed by a 60 s RTA-step in a N_2 ambient at 700°C. Unreacted Mo on the isolation oxide was removed by selective etching and silicon was implanted again (90 keV, $5 \cdot 10^{15}$ cm^{-2}). High junction doping by arsenic ions (200 keV, $5 \cdot 10^{15}$ cm^{-2}) was then performed, followed by a further RTA-step. Process parameters were 1000°C and 10 s. In order to achieve good ohmic backside contacts, the wafers were doped with boron and 1 μm aluminium was deposited on both sides. At the end, Al was patterned and sintered in N_2/2%H_2-atmosphere. The wafers, where mixing was performed with arsenic only, were processed similarly, except that the arsenic ions used for mixing had energies of 180 keV to cause effective mixing of the 50 nm Mo. Arsenic dose was $1 \cdot 10^{16}$ cm^{-2}. The following RTA-step was again performed at 1000°C for 10s. All implantation energies were chosen to locate the peak concentration at the Mo/Si-interface or $MoSi_2$/Si-interface, respectively. The corresponding data were calculated by the simulation programm RAMM [3]. The implantations were carried out without intentional substrate heating, causing a substrate temperature rise not higher than 80°C, only induced by ion beam current. For electrical characterization sheet resistances and junction leakage currents were determined.

RESULTS

Sheet resistances (R_S) of the different fabricated $MoSi_2$ layers were determined by four-point probe measurements; the corresponding resistivity values were then calculated by multiplying with the $MoSi_2$ layer thickness. The mixing process by arsenic ions lead to $MoSi_2$-resistivity values of 150 $\mu\Omega cm$. $MoSi_2$-resistivity could, however, be reduced to 90 $\mu\Omega cm$ by using double silicon implantation. This value agrees very well with data determined by Murarka et al. [4].

For junction characterization, leakage currents and also breakdown characteristics were measured. Figure 1, shows the spread of junction leakage currents at a reverse bias of 5 V for the Mo-silicided n^+p-diodes formed by different processes. The area of the measured diodes was $9 \cdot 10^4$ μm^2. It can be seen from Figure 1, that the leakage current distribution of the doubly silicon mixed diodes is mainly concentrated in the 1 μA - 20 μA range, whereas one-step mixing lead to an enormous spread in leakage current distribution. The mean values for leakage currents are 22 μA and 90 μA for mixing by double-silicon implantation and mixing by arsenic ions, respectively. Typical reverse current characteristics for both types of diodes in logarithmical scale are shown in Figure 2. Breakdown characteristics are very hard, independent from the technology used, with breakdown voltages at 51 V (see Figure 3).

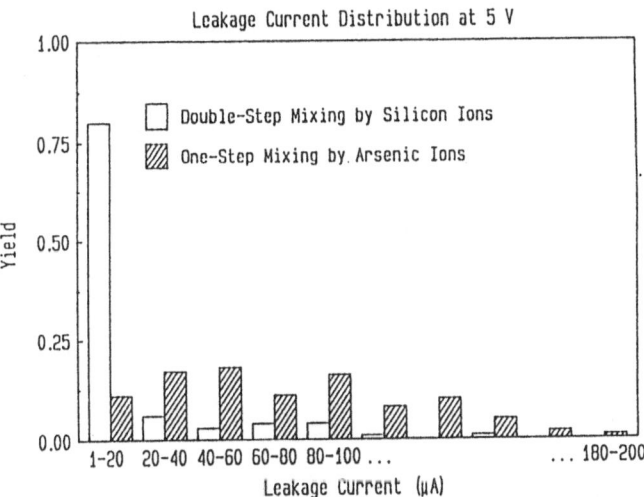

Fig.1 Histogram of leakage currents for 100 tested diodes at 5V

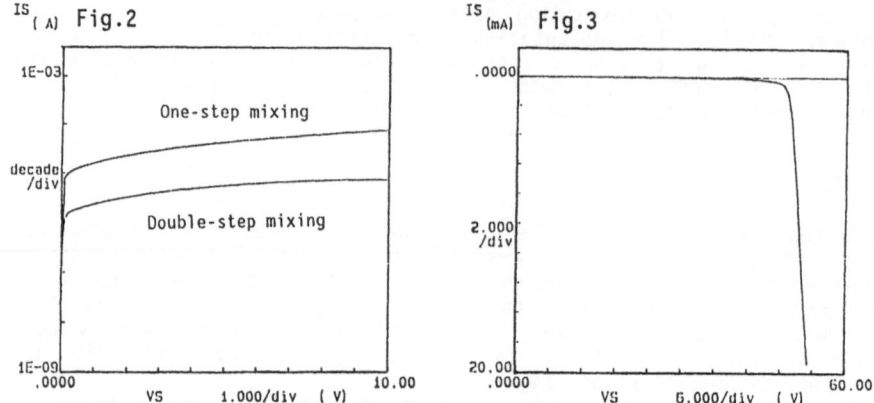

Fig.2 Reverse current characteristics of MoSi$_2$ n$^+$p-diodes fabricated by arsenic mixing and double silicon implantation, respectively.

Fig.3 Typical breakdown characteristics of Mo-silicided n$^+$p-diodes

DISCUSSION

A new mixing process using double silicon implantation has been described for fabrication of Mo-silicided n$^+$p-diodes. Silicon ions were used to cause silicide mixing and preamorphization before junction doping by arsenic ions. This process was compared with a one-step mixing process using arsenic ions. Resistivity measurements showed that after substrate amorphization by silicon implantation, reduced resistivity values of 90 $\mu\Omega$cm could be obtained compared to 150 $\mu\Omega$cm resulting from arsenic mixing. Reverse diode characteristics revealed also lower leakage currents (22 μA compared to 90 μA) when double silicon implantation was used. Independent from the technology performed, breakdown characteristics were hard with breakdown voltages of 51 V. Resistivities agreeing well with literature values and low leakage currents compared to arsenic mixing point out that the mixing process suggested [1] can improve silicide/silicon interface and is also appropriate for MoSi$_2$-based shallow junction formation. Cross-sectional TEM-investigations are in preparation to complete and prove electrical results.

ACKNOWLEDGEMENT

The authors would like to thank the technological crew of the Fraunhofer Institute for their appreciable help by diode processing.

REFERENCES

1. T.Yoshida, M.Fukumoto, T.Ohzone, J.Electrochem.Soc. 135, 481 (1988)
2. G.Vályi, C.Dehm, H.Ryssel, W.Möller, "Ion-Beam Mixed MoSi$_2$ Layers: Formation Kinetics and Contact Properties", unpublished
3. J.Lorenz, W.Krüger, A.Barthel, "Simulation of the Lateral Spread of Implanted Ions: Theory", NASECODE VI Conference, July 1989
4. S.P.Murarka, M.H.Read, C.J.Doherty, D.B.Fraser, J.Electrochem.Soc. 129, 293 (1982)

Compound
Semiconductor Devices

An Investigation into the Parasitic Effects Affecting the Operation of HEMT-based ICs

J.M. DUMAS[□], S. MOTTET[□], A. BELHADJ[□], A. CHRISTOU[•○], P. AUDREN[■]
and G.KIRIAKIDIS[•]

□ Centre National d'Etudes des Télécommunications
  BP 40, 22300 Lannion, France
• Fondation for Research and Technology - Hellas
  PO Box 1527, 711 10 Heraklio, Crete - Greece
○ Naval Research Lab.
  Washington D.C., USA
■ Institut Universitaire de Technologie
  BP 150, 22300 Lannion, France

Summary

A study of the parasitic effects penalizing the operation of two-dimensional electron gas field effect transistors (HEMTs) has been carried-out. In this paper, we present the results - experimental characterizations together with modelling - obtained for two major process-dependent parasitics: kink effect and backgating effect.

■ The kink effect penalizes the gate switching operation since it introduces a frequency-dependent threshold voltage [1].

■ The backgating effect is certainly the most studied parasitic until now in GaAs ICs because it drastically reduces the integration for both analog and digital circuits.

These two effects have been observed in HEMTs whatever the manufacturer and epilayer growth-processes (MBE or MOCVD). The results presented have been obtained on 0.5 μm recessed-gate depletion-mode HEMTs, of which MBE grown structure is ilustrated in fig. 1.

The kink effect corresponds to the extra drain-to-source current, I_{DS}, above a threshold voltage, V_{DS}^k, called *kink voltage*, as illustrated in fig. 2. After having established that V_{DS}^k presents a positive temperature coefficient between 100 K and 300 K, it is clear from fig. 2c that the kink effect is related to impact ionization mechanism.

Let us now consider the effect of a negative substrate-to-source voltage, V_{SUB}, as shown in fig 3. It illustrates the dramatic decrease on the drain current for $V_{DS} < V_{DS}^k$, due to the backgating effect, and the recovery of the drain current for $V_{DS} > V_{DS}^k$.

These experimental results can be explained as following : for $V_{SUB} < 0$ and $V_{DS} < V_{DS}^k$, the n AlGaAs/ p^- GaAs junction becomes reverse-biased, then the interface space charge region (SCR) extends, reducing I_{DS}. When $V_{DS} > V_{DS}^k$

impact ionization develops, holes are injected into the GaAs buffer layer and the SCR is minimized and a maximum of I_{DS} is reached independently of V_{SUB}. To confirm, it can be pointed-out that the kink effect is reduced when applying $V_{SUB} > 0$. As a first consequence it is obvious that kink effect is the *zero substrate-to-source voltage backgating effect*. Detailled experimental results with associated explanations can be found elsewhere [2].

It clearly appears that the epilayers and substrate equilibrium are responsible for these parasitic effects. A numerical model capable to describe this equilibrium as a function of the backgate-voltage has been set-up. The general set of equations used in the model an the numerical sheme have been previously reported [2,3]. Thus, the influences of the following technological parameters have been studied :

- GaAs buffer layer residual doping and thickness,
- AlGaAs donor layer doping level,
- semi-insulating material compensation ratio.

From our calculations it appears that the doping level of the AlGaAs donor layer has no influence on the parasitics. On the other hand, the residual doping levels of the undoped buffer layer and semi-insulating substrate have a major effect on the modulation of the two dimensional electron gas (2DEG) density versus drain-to-substrate voltage. These influences depend also on the thickness of the buffer layer. The figures 4-6 summarize these results. Moreover, let us report that the values of the residual doping levels introduced in the the model are typical of those we have experimentally measured on actual commercial devices and wafers.

From these numerical simulation results, it clearly appears that the backgating effect (which involves the kink effect) is closely related to the equilibrium between the GaAs buffer layer and the semi-insulating material.

This work has been partially supported by the European Economic Community under the ESPRIT contract n° 1270 "Advanced processing technology for GaAs FETs".

References

[1] Rocchi, M. : Status of the surface and bulk parasitic effects limiting the performances of GaAs ICs. J. P. Noblanc and J. Zimmermann (Eds.) Proceedings of the European Solid State Device Research Conf. Amsterdam : North-Holland *(1985) 119-138*

[2] ESPRIT *1270* reports over the year *1988*.

[3] Mottet, S. et al : Study of the parasitic effects affecting the operation of two-dimensional electron gas field effect transistors. To be published in the June *1989* issue of Revue de Physique Appliquée

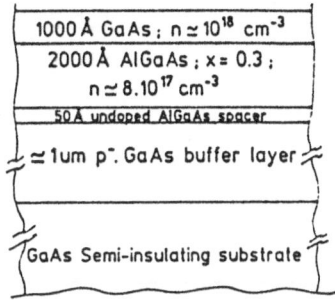

Fig. 1. HEMT structure grown by MBE.

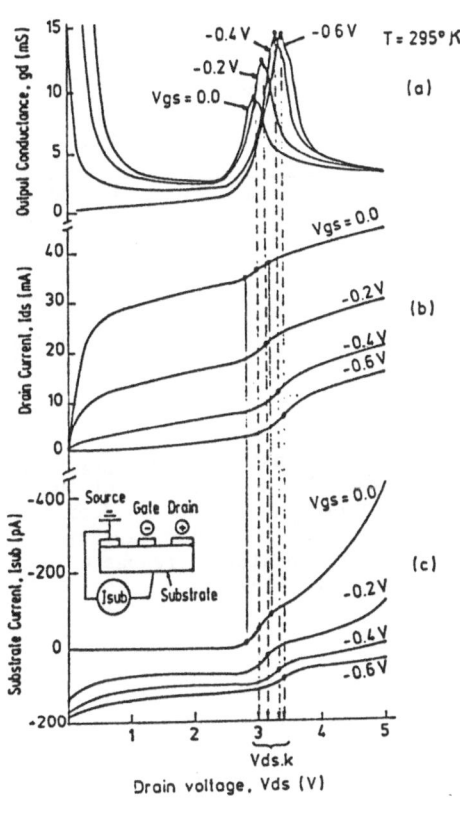

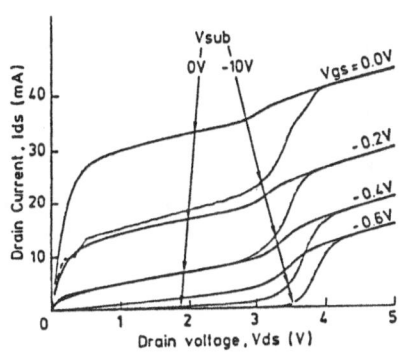

Fig. 3. Same characteristics as fig. 2b showing the backgating effect for $V_{DS} < V_{DS}^k$.

Fig. 2. (a) Output conductance gd versus V_{DS} and V_{GS} characteristics. The abrupt increase on gd easily visualize the kink voltage V_{DS}^k.
(b) I_{DS} versus V_{DS} and V_{GS} characteristics.
(c) I_{SUB} versus V_{DS} and V_{GS} characteristics. Substrate and source are short-circuited. The abrupt increase on I_{DS} is only indicated for $V_{GS} = 0$.

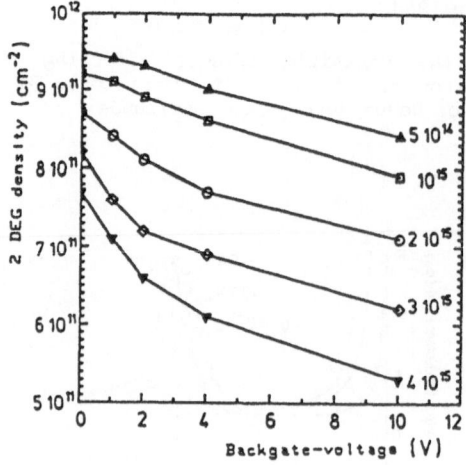

Fig. 4. 2DEG density versus V_{SUB} as a function of the p type residual doping of the GaAs buffer layer. One can note that for 3 to 4 10^{15}cm^{-3}, the variation of the 2DEG density is comparable to I_{DS} variation in fig. 3.

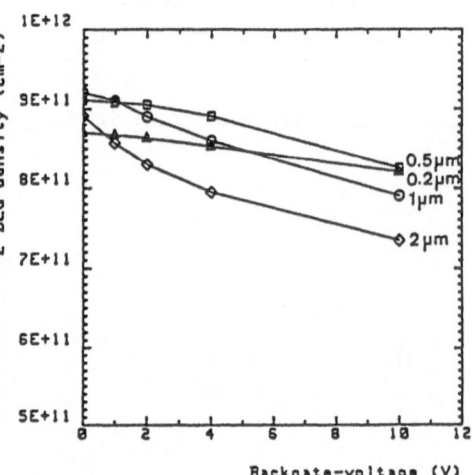

Fig. 5. 2DEG density versus V_{SUB} as a function of the thickness of the GaAs buffer layer for a doping density of 1 10^{15}cm^{-3}. The variation of the 2DEG density increases when increasing the thickness.

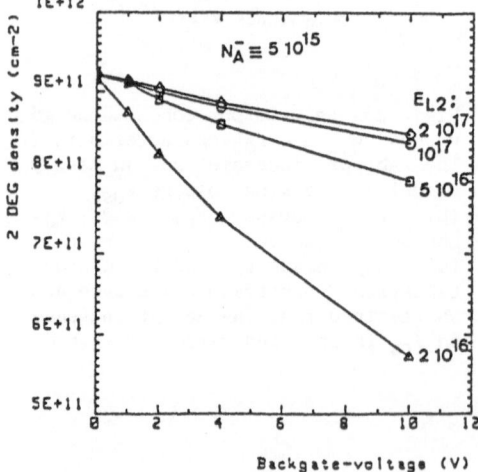

Fig. 6. 2DEG density versus V_{SUB} as a function of the compensation ratio, N_{EL2}/N_A^-, of the semi-insulating material. The possible variations of the E_{L2} concentration can induce large variations of the backgating effect sensitivity.

Automated Measurement of the Bias Dependence of Low Frequency Small-signal Parameter Dispersions in GaAs MESFETs

M. T. DE'FREITAS & J.G. SWANSON

Department of Electronic & Electrical Engineering, King's College London, Strand, LONDON WC2R 2LS, U.K.

Summary

An automated system has been developed which measures the small-signal g_m and g_o of FET's over a wide bias and frequency range. This behaviour has been investigated in GaAs MESFET's and several distinct mechanisms have been observed. Backgating has also been measured and related to the g_m dispersions of the same device.

Introduction

Low-frequency transconductance (g_m) and output-conductance (g_o) dispersions in GaAs MESFET's have been extensively reported [1-3], but usually only for a single gate and drain bias. During g_m tests V_{DS} is generally chosen to keep the device in the linear region and during g_o tests to maintain it in saturation. Bias dependent behaviour has not been considered and this limits the usefulness of any data obtained for model formulation.

A micro-computer controlled automated system has been developed by the authors which measures g_m , g_o and backgate transconductance (g_{mb}), and has been used to observe this behaviour with bias and frequency and present the data in a form that is easy to assimilate.

Measurement Technique

The bias stimuli are provided by two unity-gain operational summing amplifiers which add the d.c. bias to the small a.c. test signal. A two-phase lock-in amplifier is used to analyse the voltage produced by the resultant channel current across a 0·5 ohm resistor placed between the FET's source and ground. The bias error produced by this arrangement is corrected for by supplying the source voltage to an input of each of the summing amplifiers.

The gain and phase shift of the summing amplifiers vary significantly with load impedance. During g_o measurement this load impedance ($1/g_o$) varies widely with V_{DS} and V_{GS}. It is therefore necessary to use the signal at the amplifier output as a gain and phase reference. This precaution is necessary at each new bias and frequency.

Results

The typical results presented in Figure 1 show that $|g_m|$ and $|g_o|$ dispersions are strongly bias dependent. The percentage of $|g_m|$ dispersion was largest at low V_{DS}, decreased as V_{DS} was increased and became constant as the device entered saturation at ~1V. In the case of

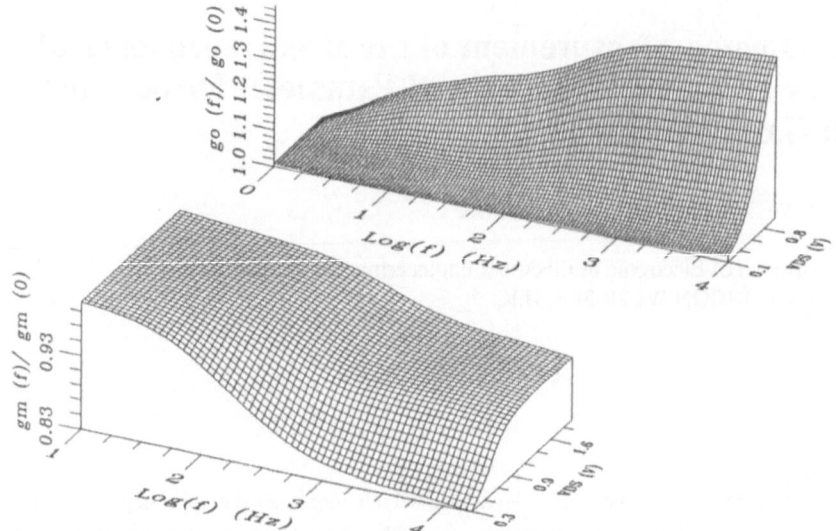

Figure 1   The effect of V_{DS} and frequency on the normalized magnitudes of g_m and g_o
for a commercial 1 micron GaAs MESFET (V_{GS} = 0V).

$|g_o|$ there was very little dispersion at low V_{DS} but almost 50% as saturation was reached.
The point of inflection in the $|g_m|$ versus frequency curve at constant V_{DS} corresponded
with a peak in phase shift and in imaginary component at a frequency f_{pk}. This suggests that
only one dispersive mechanism was operating and the locus of f_{pk} versus V_{DS} is shown in
Figure 2 (curve A). The corresponding g_o phase data indicated two distinct mechanisms, the
loci of their peaks are shown in Figure 2 (curves B and C). The variations of curves A and B
suggest a close relationship between these mechanisms.

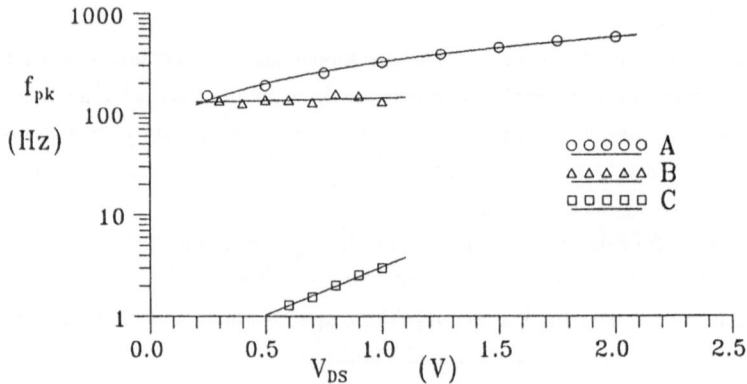

Figure 2   The V_{DS} dependence of f_{pk} for g_m and g_o (V_{GS} = 0V).

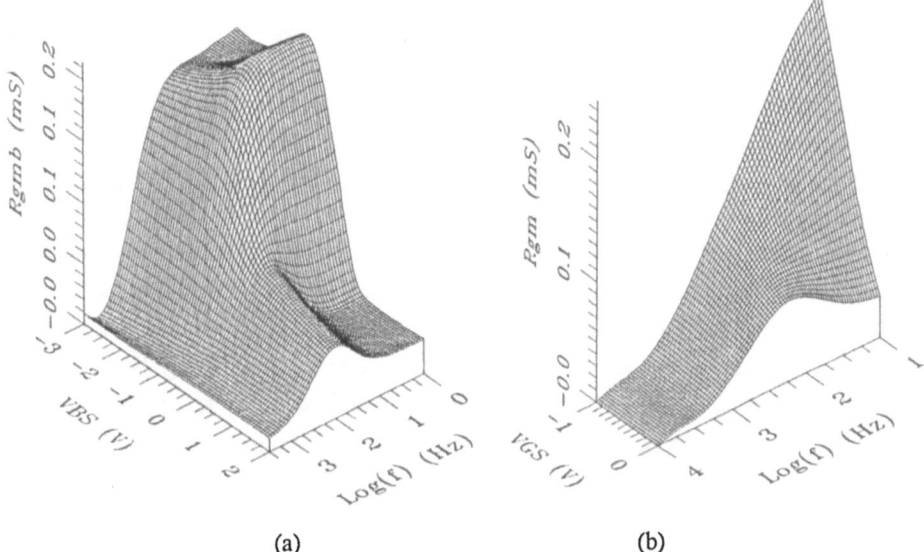

(a)   (b)

Figure 3   (a) The V_{BS} and frequency dependence of the Rg_{mb} and (b) the V_{GS} and

frequency dependence of Rg_m, for a 0·5 micron GaAs MESFET (V_{DS} = 0.5V).

The measurement system has been also used to explore the frequency and bias dependent behaviour of g_{mb} in ion-implanted proton-isolated MESFET's. The real component of g_{mb} (Rg_{mb}) is shown in Figure 3 (a). It is immediately obvious that at low frequencies Rg_{mb} was large when V_{BS} < 0V, falling abruptly at V_{BS} = 0V to a much smaller value when V_{BS} > 0V. Increasing the test frequency with V_{BS} > 0V revealed a resonance at around 100Hz [4] with a corresponding zero imaginary component at that point. This resonance was masked when

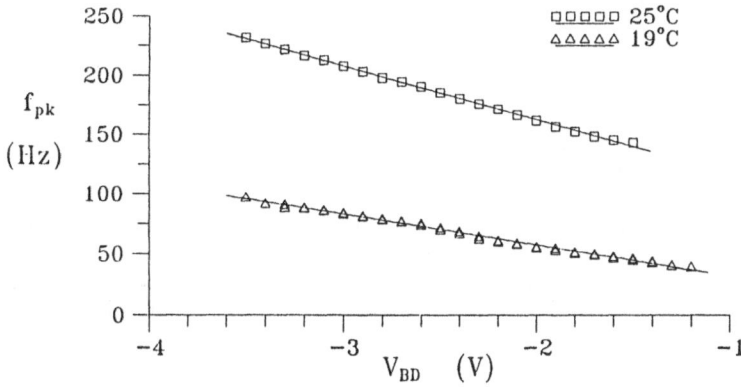

Figure 4   Bias and temperature dependent behaviour of backgate f_{pk} (V_{BS} < 0V).

V_{BS} < 0V by a relatively large dispersion of a different form. In this case the point of inflection in Rg_{mb} versus frequency corresponded with a single peak in the imaginary component, Ig_{mb}. This behaviour is reminiscent of an effect with a single time constant. The f_{pk} value showed a linear dependence when plotted against V_{BD} and was also strongly temperature dependent, as seen in Figure 4.

Neither the relaxation or the resonant behaviour was visible above 10kHz. In this frequency regime the g_m showed a V_{GS} dependence which was due to the ion-implanted profile. This effect was subtracted from the observed Rg_m in order to reveal its true low-frequency dependence for V_{GS} values in the range between pinch-off and forward gate conduction shown in Figure 3 (b). Over this bias range the excess Rg_m displayed a transition between resonance and relaxation which is remarkably similar to that seen in Rg_{mb}. This similarity suggests that the g_m dispersions contain effects which are due to backgating by the metal gate-pad of the device which was in contact with the proton-isolated material.

Conclusions

An automated system for observing the frequency and bias dependence of g_m, g_o and g_{mb} has demonstrated the importance of this type of measurement in GaAs MESFET's. The presence of distinguishable mechanisms which cause g_m and g_o dispersions has been confirmed and there appears to be a common mechanism linking these effects in some devices. In proton-isolated devices the low-frequency g_m dispersions display a close similarity with the behaviour of g_{mb} indicating that the gate-pad may act as a backgate.

Acknowledgments

This work has been performed at King's College London as part of a SERC CASE studentship in collaboration with GEC Hirst Research Centre. The authors would like to thank GEC who have provided materials and some of the test devices. Valuable design advice from J. Luck of King's College is greatfully acknowledged.

References

1. Ladbrooke, P.H.; Blight, S.R.: *Low-field low-frequency dispersions of transconductance in GaAs MESFET's with implications for other rate-dependent anomalies.* IEEE Trans. Electron Dev. ED-35 (1988) 257-267.

2. Kachwalla, Z.: *Characterizing traps in MESFET's using internal transconductance (g_m) frequency dispersions.* Solid State Electronics 31 (1988) 1315-1320.

3. Canfield, P.; Medinger, J.; Forbes, L.: *Buried-channel GaAs MESFET's with frequency-independent output conductance.* IEEE Electron Dev.Lett. EDL-8 (1987) 88-89.

4. Roach, J.W.; Wieder, H.H.: *Frequency dispersion of sidegating transconductance of GaAs junction field-effect transistors.* Appl.Phys.Lett. 47 (1985) 1285-1287.

High Speed $Ga_{0.47}In_{0.53}As$ MISFETs Grown by Metal Organic Vapor Phase Epitaxy

J. Splettstößer, F. Schulte, A.Trasser, D. Schmitz[*] and H. Beneking

Institute of Semiconductor Electronics, Technical University Aachen, Sommerfeldstrasse, D-5100 Aachen, FRG

* Aixtron, Jülicher Str. 336-338, D- 5100 Aachen

MISFETs with SiO_2 insulator have been successfully fabricated on MOVPE grown $Ga_{0.47}In_{0.53}As$ layers on s.i. InP substrate. The devices show a very high extrinsic transconductance of 250 mS/mm and 300 mS/mm for a gate length of 3 μm and 1.5 μm, respectively. The current gain cutoff frequency is 6 GHz and 16 GHz for a gate length of 3 um and 1,5 um, respectively. A corresponding response time of 130 ps and 70 ps is measured. Fabricated inverters in p- GaInAs show a voltage amplification of -7 for a supply bias of 5 V.

I. Introduction

$Ga_xIn_{1-x}As$ on InP exhibits higher electron mobility, higher peak electron velocity and a larger Γ -L intervalley separation than GaAs making this material very attractive for highest speed device applications which has been demonstrated by many authors[1,2]. In contrast to GaAs the semiconductor surface potential of $Ga_{0.47}In_{0.53}As$ can be varied between accumulation and inversion for n- [3] and p-type GaInAs [4] which allows the fabrication of MISFETs for high frequency and digital applications.

II. Passivation process

As the gate insulator SiO_2 has been chosen. The deposition of the SiO_2 has been performed by a conventional Chemical Vapour Deposition process (CVD) using silane and oxygen (SiH_4 to O_2 ratio about 1). The deposition temperature of the dielectric has been 250°C. Immediately before SiO_2 deposition the GaInAs surface has been oxidized in H_2O_2. After that the native oxide has been removed by buffered HF. A high temperature rapid thermal annealing (700°C - 800°C) has been used to improve the quality of the oxide and the interface. The interface state density has been reduced to about $10^{11}eV^{-1}cm^{-2}$. The C-V curves for low and high frequency show a well known behaviour in the accumulation and inversion mode and a small looping, see Fig. 1.

III. Device characteristics

On s.i. InP substrate a nominally undoped or a p- type InP buffer layer (d_{InP} =0.2um) and two n-type $Ga_{0.47}In_{0.53}As$ layers are deposited by MOVPE. The n^+ and n epilayers doping levels are $2x10^{18}$ cm^{-3} (d ~ 0.1um) and $1x10^{17}$ cm^{-3} (d ~ 0.1um), respectively. A schematic of the device is shown in Fig. 2. The device fabrication is described in /5/.

The I-V characteristic of an enhancement type MISFET (L_G=1.5μm, w=200um, d_{ox}=40nm) is shown in Fig. 3. The extrinsic transconductance is 60mS (300 mS/mm). For depletion mode MISFETs the extrinsic transconductance is 28mS (140mS/mm). S- parameter measurements have been performed with the Cascade microwave prober up to 20 GHz. From the fit of h_{21}^2 a cutoff frequency of 6 GHz and 16 GHz can be deduced for the 3 um and the 1.5 um gate MISFET, respectively, see Fig. 4. The corresponding maximum frequency of oszillation is determined to 12 GHz and 14 GHz. A significant improvement of f_{max} will be obtained by reducing the input resistance of 25Ω, see Fig. 4.

Time domain measurements have been made on mounted devices in a 50 system. Therefore, test pulse has been generated by a HP tunnel diode mount (t_r~20 ps). The response time for the 3 μm and 1.5 μm gate length devices are 130 ps and 70 ps, respectively, detected by a sampling oscilloscope with a bandwidth of 18 GHz, see Fig. 5.

IV. Inverter

The inverter is realized with an inversion mode /6/ switching FET and a depletion mode load. Therefore, an undoped buffer (d_{InP} = 0.5 μm) and a p-doped (p = 2 x 10^{16} cm^{-3}, d_{GaInAs} = 1.5 um) GaInAs epitaxial layer has been grown by MOVPE. Inverters have been fabricated using Si- ion implantation for n-type layer definition of source- drain regions and the channel region of the depletion type MISFETs. The transfer characteristics are shown in Fig. 6 for different supply voltages. The depletion type load is implanted in the channel region with Si (E=14 keV, D=$5x10^{11}cm^{-2}$). For a supply voltage of 5 V this inverter shows a output voltage swing of 4.7 V. The voltage magnification of this inverter is -7.

Acknowledgement

The authors wish to thank Ch. Steinberger for assistance in processing and M. Renaud (LEP, Philips) for fabrication of the masks.

This work is supported by the ESPRIT project 927 and 2518

References

/1/ U.K. Mishra, A.S. Brown and S.E. Rosenbaum, IEDM Tech. Digest, 1988, pp. 180-183

/2/ L.F. Lester, P.M. Smith, P. Ho, P.C. Chao, R.C. Tiberio, K.H.G. Duh and E.D. Wolf, IEDM Tech. Digest, 1988, pp.172-175

/3/ A.S.H. Liao, B. Tell, R.F. Leheny, R.E. Nahorny, J.C. DeWinter R.J. Martin, IEEE Electon Device Lett., vol. EDL-3, p. 158, 1982

/4/ R. Kaumanns, J. Selders, and H.Beneking, Inst. Phys. Conf. Ser., vol.63, p.329, 1982

/5/ J. Splettstößer and H. Beneking, IEEE Tans. on Electron Devices, vol. 36, No.4, 1989

/6/ J. Selders and H. Beneking, IEEE Electron Device Lett., vol. EDL-7, p. 434, 1986

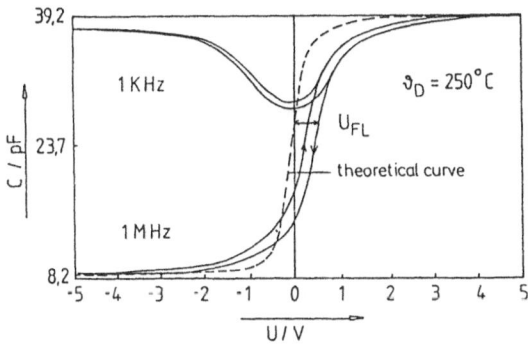

Fig.1: C/V- curves of a capacitor on n- GaInAs measured at 1 MHz and 1 kHz

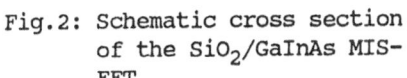

Fig.2: Schematic cross section of the SiO_2/GaInAs MISFET

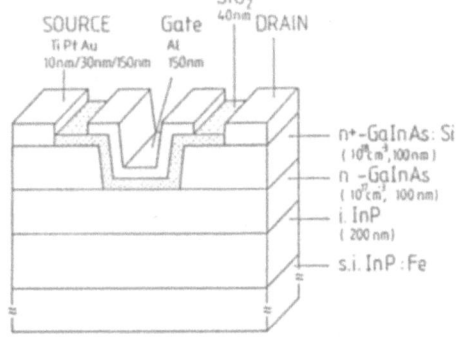

Fig.3: Output characteristic of a MISFET with a gate length of 1.5 μm (W=200μm, d_{SD}=5μm)

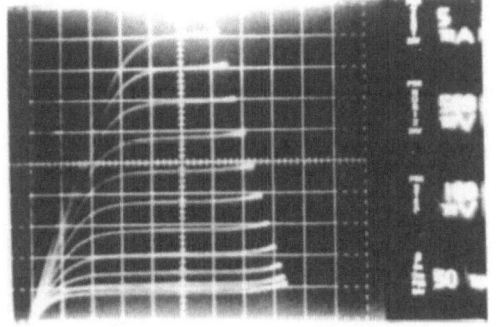

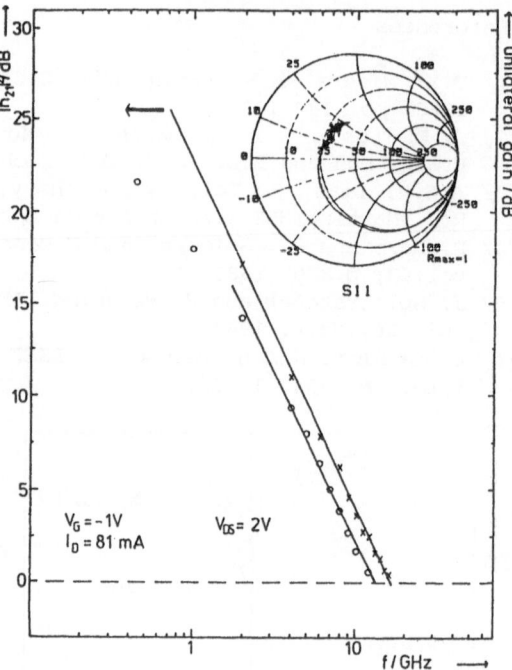

Fig.4: Current gain and
unilateral gain
for a MISFET with
a gate length of
1.5 μm (W=200μm,
d_{SG}=5μm)

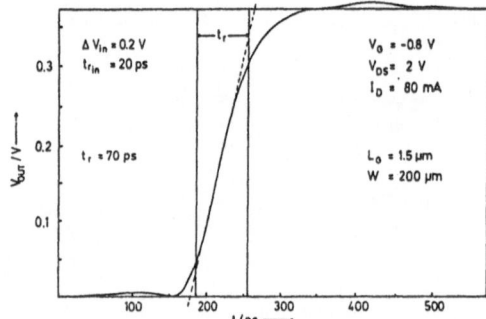

Fig.5: Switching behaviour of
the 1.5μm gate length
MISFET. The rise time
of the used input vol-
tage step was 20ps.

Fig.6: Transfer characteristics
of the inverter for dif-
ferent supply voltages

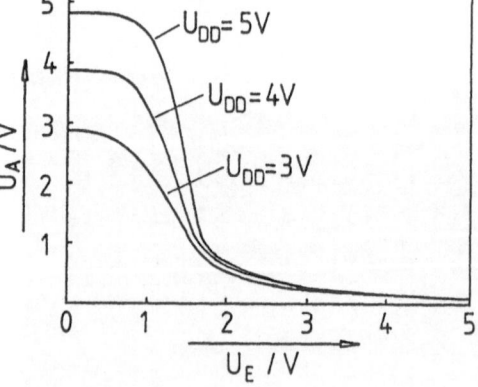

A Resonant Tunneling High Electron Mobility Transistor

M. Van Hove, C. Van Hoof, W. De Raedt, P. Jansen, I. Dobbelaere, J. Peeters, G. Borghs and M. Van Rossum

IMEC vzw, Kapeldreef 75, B-3030 Leuven, Belgium

Abstract
A double barrier resonant tunneling heterostructure has been combined with a pseudomorphic high electron mobility transistor. This three terminal device shows negative transconductance in addition to a clear negative differential resistance region which shifts with gate voltage. The device characteristics can be adequately explained by a simple device simulation model. Flip-flop and frequency doubling operations at room temperature are demonstrated.

1. Introduction

Double barrier resonant tunneling heterostructures are very promising for application in fast digital switches, high frequency oscillators and multiple valued logic devices. Although most of the current work has been concentrated on two terminal devices, several three terminal transistor structures have been proposed. Most of these concepts are based on the incorporation of the resonant tunneling double barrier in one terminal of a more conventional transistor, e.g. the series connection of a resonant tunneling structure (RTS) and a metal-semiconductor field-effect transistor (MESFET) [1,2]. A unipolar resonant tunneling hot electron transistor [3] with the quantum well in the emitter and a bipolar resonant tunneling transistor with a double barrier in the base [4,5] have also been reported. A different transistor concept is based on the direct control of the potential of the quantum well by the third terminal. A bipolar resonant tunneling transistor of this type was recently realised [6,7] and the same device concept for a unipolar structure was independently proposed [8,9].

The major disadvantage of the structure based on the series connection of a RTS with a MESFET is its speed limitation in the FET region. It is also subject to the same scaling limitations as a conventional MESFET. Better performance can be expected by integrating a RTS with a submicron high electron mobility transistor (HEMT), in particular when cooling down to 77K due to an increase of both the peak-to-valley ratio (PVR) of the RTS and the transconductance of the HEMT.

2. Experimental

In order to optimise the combined HEMT-RTS device characteristics both structures were fabricated separately and, depending on the application, the RTS was connected in series with the HEMT in different configurations. This is only a first step to the complete incorporation of the RTS in the source or drain region of the HEMT. The crucial parameters for high frequency operation, such as dimensions, parasitic resistance and noise figures can be considerably reduced by integrating both structures into a single device.

The layers were grown by molecular beam epitaxy. Processing was done using standard wet etching, metallisation and optical lithography techniques. For the HEMT layers a pseudo-morphic GaAs/InGaAs/AlGaAs layer sequence was used. Depending on the application, the layer structure and the processing were adapted to reach a specific threshold voltage, maximum transconductance or maximum channel current. As an example a maximum transconductance and current density of 475 mS/mm and 325 mA/mm respectively were obtained at 77K for a 1 μm gatelength. Hall effect measurements on this layer showed a sheet carrier concentration of $2.0 \cdot 10^{12}$ cm^{-2} and a mobility of 21750 cm^2/Vs. For the double barrier AlAs/GaAs/AlAs tunneling diode layer structure a compromise had to be made between an optimal PVR and a peak current compatible with the channel currents in the HEMT. This implied that, in order to reduce the peak current, the barrier thickness had to be increased at the expense of the PVR. For example a RTS structure with a 5.7 nm wide well and a 3 nm thick barrier yielded 2.8 as PVR and 5.7 kA/cm^2 as peak current density at room temperature.

3. RTS-HEMT operation and applications

Room temperature dc characteristics for the RTS in the source and drain of the HEMT are shown respectively in fig. 1a and b. For both configurations the negative differential resistance (NDR) shifts to lower source-drain voltages by increasing the bias at the gate terminal. The device operation is easily understood. The total drain bias is divided between the tunneling diode and the HEMT channel. As the resistance of the channel is changed a different total drain voltage must be applied in order to satisfy the tunneling voltage of the RTS. If the RTS is in

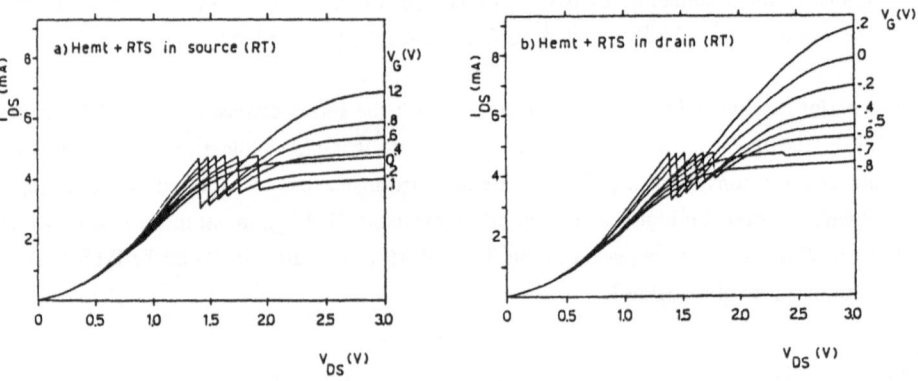

Fig.1. Experimental room temperature source-drain I-V curves with the gate voltage as a parameter for a RTS structure in source (a) and drain (b) of a pseudomorphic HEMT.

the source region, an appreciable part of the applied gate voltage drops over the double barrier. This explains the negative transconductance observed for this geometry and why NDR is obtained at higher external gate voltages. After modeling the dc characteristics of the HEMT

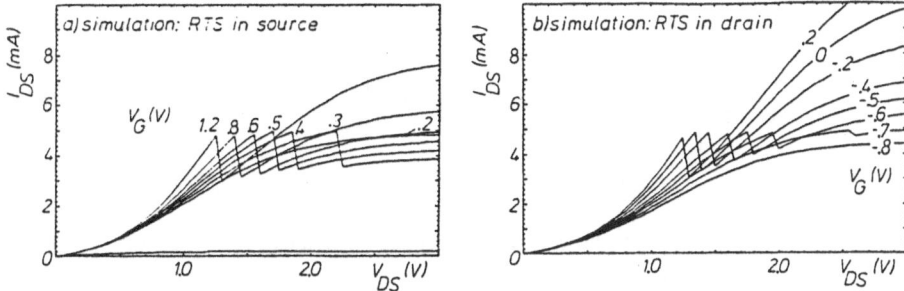

Fig.2. ELDO simulation of the source-drain I-V curves for the RTS structure in source (a) and drain (b) of a HEMT.

and the RTS separately, the circuit simulation program ELDO [10] was used to describe the I-V curves of the combined device. An excellent quantitative agreement between the experimental curves (fig. 1) and the simulation (fig. 2) was obtained. Device behaviour at 77K is illustrated in fig. 3. Both the NDR effect and the performance of the transistor were considerably increased. For these particular structures the transconductance of the HEMT was increased by 35 % and the PVR of the tunneling diode was multiplied by a factor 3.2. As an application of the negative transconductance, observed with the RTS in the source, the response of a voltage modulation at the gate input shows frequency doubling in the source-drain current output (fig. 4). Tristability was realised with two tunneling diodes in series used as load for the HEMT. The circuit configuration and flip-flop operation are shown in fig. 5.

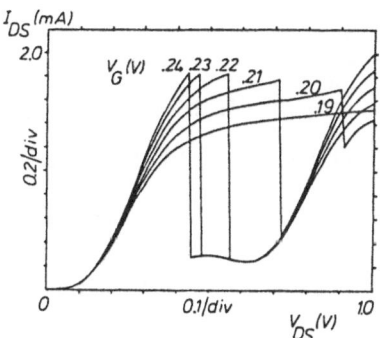

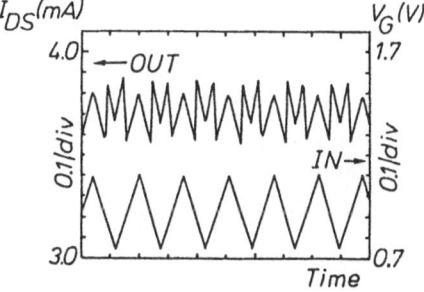

Fig.3. I-V curves at 77K for a pseudomorphic HEMT with a tunneling diode in the drain.

Fig. 4. Room temperature frequency multiplier input (gate voltage) output (source-drain) current diagrams.

274

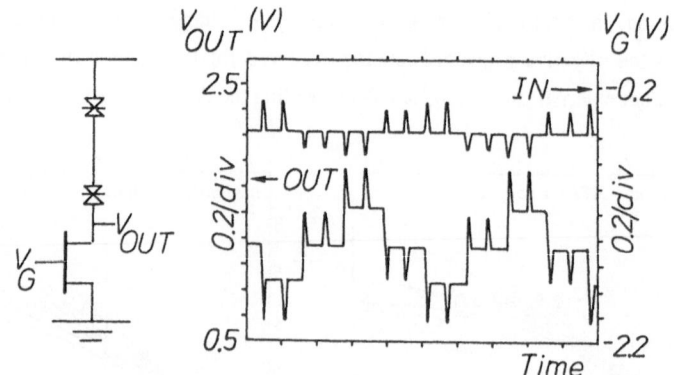

Fig.5. Room temperature tristability obtained with the circuit schematic shown in the inset.

4. Conclusion

As a first step to monolithic integration, the electronic behaviour of a resonant tunneling diode in series with a high electron mobility transistor was reported. Pronounced negative differential resistance in the combined device was obtained at room temperature. As potential applications flip-flop and frequency doubling operation were demonstrated.

Acknowledgements

We would like to acknowledge W. van de Graaf for support during MBE growth and L. Lauwers for his help with the simulation.

References
1. Bonnefoi, A.R.; McGill, T.C.; Burnham, R.D.; IEEE Electron Device Lett., EDL-6, (1985) 636-638
2. Woodward, T.K.; McGill, T.C.; Chung, H.F.; Burnham, R.D.; Appl. Phys. Lett. 51 (1987) 1542-1544
3. Yokoyama, N.; Imamura, K.; Muto, S.; Hiyamizu, S.; Nishi, H.; Jap. J. Appl. Phys., 24 (1985) L853-L854 and Gallium Arsenide and Related Compounds 1985, 739-740
4. Capasso, F.; Kiehl, R.A.; J. Appl. Phys. 58 (1985) 1366-1368
5. Capasso, F.; Sen, S.; Gossard, A.; Hutchinson, A.L.; English, J.H., IEEE Electron Device Lett., EDL-7 (1986) 573-576
6. Seabaugh. A.C.; Reed, M.A.; Frensley, W.R.; Randall, J.N.; Matyi, R.J.; IEDM Tech. Digest 1988, 900-901
7. Reed, M.A.; Frensley, W.R.; Matyi, R.J.; Randall, J.N.; Seabaugh, A.C.;Appl. Phys. Lett. 54 (1989) 1034-1036
8. Schulman, J.N.; Waldner, M., J. Appl. Phys. 63 (1988) 2859
9. Haddad, G.I.; Mains, R.K.; Reddy, U.K.; East, J.R.; Superlattices and Microstructures, 5 (1989) 437-441
10. ELDO is trademark of CNET Meylan France

Implanted-collector InGaAsP/InP Heterojunction Bipolar Transistor

L.M. SU, N. GROTE, P. SCHUMACHER, D. FRANKE

Heinrich-Hertz-Institut für Nachrichtentechnik Berlin GmbH, Einsteinufer 37, D-1000 Berlin 10 (FRG)

Summary

A novel structure for an InP-based heterojunction bipolar transistor is proposed and demonstrated which not only provides technological advantages over conventional structures but is also particularly suited for high-frequency applications due to the inherent reduction of parasitic time constants. The main structural feature of this transistor is an embedded rather than a mesa type collector which was created by ion implantation. Large-area devices showed well-behaved static I/V characteristics with a current gain of 250 and emitter-collector breakdown voltages between 3 V and 6 V.

Introduction

InP-based heterojunction bipolar transistors (HBT) are considered to represent very attractive devices for electronic and optoelectronic applications the performance of which may be superior even to that of their GaAs/GaAlAs counterparts. An InGaAsP/InP HBT exhibiting a cut-off frequency (f_T) of 110 GHz has been reported recently [1]. Until now, such devices have almost exclusively been designed with a mesa configuration for emitter, base, and collector, which not only entails technological difficulties but also introduces parasitic elements which may limit the high-frequency performance.

Device structure

The device structure which is shown in Fig. 1 is characterized by an embedded instead of a mesa etched collector made by ion implantation in semi-insulating InP:Fe. The major merits of this structure are:

a)   reduction of device height due to elimination of the collector mesa thereby facilitating device processing

b)   extremely low parasitic base/collector capacitance enhancing the high-frequency performance

c)   less critical demands upon base contact in terms of size and fabrication (e.g. no risk of damaging base/collector junction especially with very thin base layers)

d)   efficient symmetrization of emitter and collector junction areas allowing to achieve equivalent current gain behaviour with bilateral (i.e. emitter-up/collector-up) transistor operation

With respect to the base a p^+p^- doping structure has been adopted in that a highly doped region (InGaAsP(λ_g = 1.3 μm); p^+ = 2$\bullet$10^{18} cm^{-3}) is separated from the implanted collector by a weakly doped transition layer (p^- = 5$\bullet$10^{16} cm^{-3}). The intention of this design is, firstly, to guarantee a high collector efficiency due to the avoidance of an electron repelling effect at the base/collector heterojunction (refer to [2]). Secondly, by exploiting the velocity overshoot effect in the (depleted) p^- region the collector transit time [3] may be reduced thus increasing the ultimate intrinsic limits of high-frequence operation.

Device fabrication

For device fabrication the collector region was produced first by selective implantation of Si^{29} ions into a MOVPE-grown semi-insulating InP:Fe layer on (100) InP:Fe substrate. An epitaxial s.i. InP was preferred over an InP:Fe substrate because of the lower density of defects and the well-controlled Fe concentration [4]. A triple implant was applied using the following sequence of ion energies and doses: 100 keV, 5$\bullet$10^{12} cm^{-2}; 350 keV, 2$\bullet$10^{13} cm^{-2}; 700 keV, 4$\bullet$10^{13} cm^{-2}. Subsequently the implanted wafers were capless annealed in a MOVPE reactor in a PH$_3$/H$_2$ ambient (PH$_3$ mole fraction: 0.5%) at a temperature of 750°C for 20 min yielding a mirror-like surface. Fig. 2 depicts a typical, almost flat depth profile of the free electron concentration in the collector layer which was approximately 1 μm thick. The Fe concentration in the InP:Fe layer was adjusted to be 3$\bullet$10^{16} cm^{-3} giving a specific resistivity of some 3$\bullet$10^8 Ωcm [4]. The overgrowth of the epitaxial layers forming the base and emitter was carried out by LPE so far, but could certainly also be performed by more advanced techniques (MOVPE, MBE). Prior to epitaxy the wafers were slightly wet-etched to create a fresh surface. The base and emitter mesas were formed by wet etching utilizing RIE-etched grooves in the InP:Fe layer for photolithographic alignment purposes. For all ohmic contacts a Ti-Au metallization was used.

Results

Representative common-emitter characteristics (Fig. 3) measured on large-area devices ($A_E = A_c = 0.64 \cdot 10^{-4} cm^2$) showed current gains of up to 250 ($V_{CE} = 3$ V, $I_c = 3$ mA) proving the good quality of the base/collector hete-rointerface. The absence of a noticeable turn-on voltage confirms the effectiveness of the p^- layer in the base with regard to a high electron collection efficiency even at low bias voltages. The emitter-collector breakdown voltage was between 3 V and 6 V depending on the carrier concentration of the implanted collector at the heterojunction. Higher values, with concurrently decreasing the base/collector capacitance, could be easily attained by lowering the collector doping. Applying a more sophisticated implantation profile, a buried collector access layer providing a low series resistance can, nevertheless, be accomplished.

The h_{fe}-I_c characteristic is shown in Fig. 4 over the collector current (I_c) range from 10^{-6} A to 10^{-3} A. In this particular case a relationship $h_{fe} \approx I_c^{0.4}$ was found. Operation of the inverted transistor (collector-up) showed the reverse current gain to be much smaller than the forward gain, typically being in the range of 1...5. The reason for this behaviour is not yet understood.

Conclusion

To conclude we have proposed a novel HBT structure on InP which relies on an implanted collector layer. Functionality of such transistors with very promising I/V properties was demonstrated on large-area devices. The achievability of very high frequency operation is expected with small-size devices due to the inherent elimination of frequency determining parasitic capacitances.

278

References

[1] Nottenberg, R.N.: Chen, Y.K.; Panish, M.B.; Humphrey, D.A.; and Hamm, R.: IEEE Electron Device Lett. EDL-10, No. 1 (1989), pp. 30-32

[2] Su, L.M.; Grote, N.; Kaumanns, R.; and Schroeter-Janßen, H.: Appl. Phys. Lett. 47 (1) (1985), pp. 28-30

[3] Morizuka, K. Katoh, R.; Asaka, M.; Iizuka, N.; Tsuda, K.; and Obara, M.: IEEE Electron Device Lett. EDL-9, No. 11 (1988), pp. 585-587

[4] Grote, N.; Bach, H.G.; Franke, D.; Harde, P.; Sartorius, B.; Wolfram, P.: this conference, paper 1C4

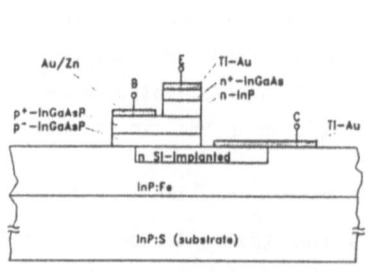

Fig. 1: Schematic cross-section of implanted-collector InGaAsP/InP HBT

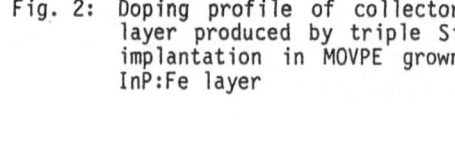

Fig. 2: Doping profile of collector layer produced by triple Si implantation in MOVPE grown InP:Fe layer

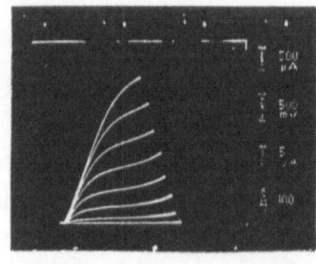

Fig. 3: Common-emitter I/V-characteristics of large-area HBT (emitter-up)

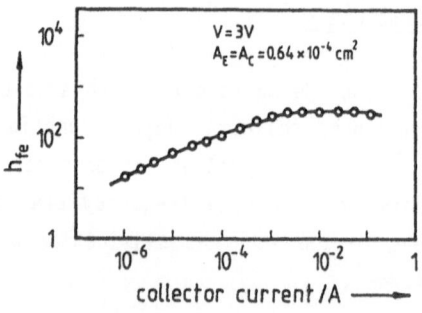

Fig. 4: Current gain as a function of collector current

0.25 μm All Level e-Beam Pseudomorphic AlGaAs/InGaAs MODFET with f_t over 65 GHz

E. Lopez[+][*], A. Marten[+], A. Forchel[+], J.L. Caceres[*]
H. Nickel[#], W. Schlapp[#], R. Lösch[#], D. Briggmann[#]
+ IV. Phys. Inst. Univ. Stuttgart, FRG.
* E.T.S.I.T., Univ. Polyt. Madrid, Spain
# FTZ, Darmstadt, FRG.

Abstract

This contribution describes the fabrication of 0.25 μm gatelength MODFETs on pseudomorphic $Al_{0.2}Ga_{0.8}As/In_{0.2}Ga_{0.8}As$ material. An e-beam writer was employed in all lithographic levels. We have obtained a DC extrinsic transconductance $g_m > 440$ mS/mm. From microwave measurements and the modeled equivalent circuit f_t and f_{max} values of over 65 GHz and 110 GHz are obtained.

Introduction

HEMTs provide better frequency and noise performance than MESFETs because of the spatial separation between the ionized impurities and the electrons induced by the heterojunction band offset. Compared with a more conventional AlGaAs/GaAs heterostructure, the use of pseudomorphic InGaAs channels results in a similar conduction band discontinuity between the doped AlGaAs layer and the channel with a lower Al mole fraction, or a larger discontinuity for the same Al percentage. In this way it is possible to increase the carrier concentration in the channel region and to avoid problems with DX centers. Even more important, the transport properties of InGaAs are superior to those of GaAs. Therefore the transistor performance improves with increasing In percentage [1].

Material structure and device fabrication

The pseudomorphic AlGaAs/InGaAs structures were MBE-grown on semi-insulating <100> GaAs substrates . First a 3.4 μm GaAs buffer was grown, followed by the undoped $In_{0.2}Ga_{0.8}As$ channel. The channel width is a critical design parameter in pseudomorphic InGaAs quantum well HEMTs. With decreasing channel width the effective conduction band discontinuity is lowered by increasing the quantized electronic states in the quantum well. This again is in direct contradiction to the desired electron confinement in the well. On the other hand the transport properties of a pseudomorphic

InGaAs channel degrade drastically if a critical layer thickness is exceeded. The decrease of the electron mobility can be explained by the strain-induced increase of the dislocation density, as reported in [2]. As a compromise value, we used a channel layer thickness of 130 $\overset{\circ}{A}$

The spacer thickness was 20 $\overset{\circ}{A}$ in our structures, consisting of undoped $Al_{0.2}Ga_{0.8}As$. The next layer is 200 $\overset{\circ}{A}$ thick Si-doped $(1.8 \times 10^{18}cm^{-3})$ $Al_{0.2}Ga_{0.8}As$, followed by a graded AlGaAs to GaAs transition, also Si-doped. Finally, as a contact and cap layer, 500 $\overset{\circ}{A}$ thick Si-doped $(2.0 \times 10^{18}cm^{-3})$ GaAs was grown on top of the structure. Hall effect measurements give a sheet carrier concentration of 1.6×10^{12} cm^{-2} and an electron mobility of 5030 cm^2/Vs at room temperature.

An all level e-beam fabrication process was employed . Electrical isolation of the devices was obtained by mesas, which were defined with CMS negative resist and dry etched down to 120 nm. The source and drain ohmic contacts were formed using a Au:Ge/Ni metallization, which was alloyed at 470°C for 90 s. The ohmic contact formation typically results in contact resistances of 0.1 Ωmm. In the next step contact pads were established, which allow the macroscopic access to the devices for DC and HF measurements. Finally gates were defined on 1 µm thick, 950 k molecular weight PMMA positive electron-beam resist. The adhesion of the PMMA was strongly improved by a dehydration step prior to the spin-coating of the resist. The width of the recess is determined by the lithographically defined gate width and the etch characteristic of the recess etch. Due to the dehydration step (a one hour bake at 250°C) no perceptible underetching of the PMMA edges occurs. The recess was accomplished with an etch solution consisting of citric acid : H_2O_2 : H_2O with volume ratios of 20:3:200. This leads to an etch rate of 3-4 A/s. The gate recess was controlled by monitoring the ungated drain-to-source current. The gate was formed by evaporating 100 $\overset{\circ}{A}$ Cr and 3000 $\overset{\circ}{A}$ Au and a lift-off step in acetone. The depletion-mode devices have gate lengths between 0.25 µm and 5 µm. For the short gate transistors the source-drain spacing amounts to 1.8 µm. The gate electrode is located asymmetrically between drain and source at a distance of 0.6 µm from the source. Fig. 1 shows a SEM photograph of the gate region of a 0.25 µm gate length HEMT.

DC and RF performance

The transistors exhibit a typical pinch-off voltage of -0.7 V. The maximum drain current before saturation is 320 mA/mm. This current is corresponds to an electron velocity of approximately 1.3×10^5 m/s, implying that no velocity overshoot occurs in this device [3], [4]. Fig. 2 shows the $g_m(V_g)$ and $I_{ds}(V_g)$ characteristics of a 50 µm wide, 0.25 µm gate length HEMT, that provides a maximum current of 400 mA/mm and a peak extrinsic DC transconductance g_m of 440 mS/mm. The ratio of the maximum transconductance and the

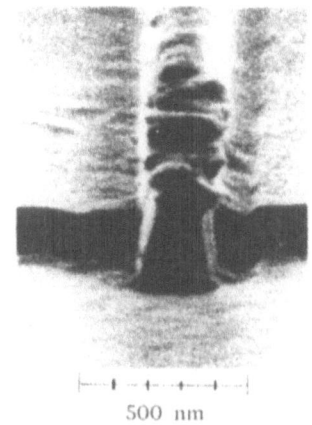

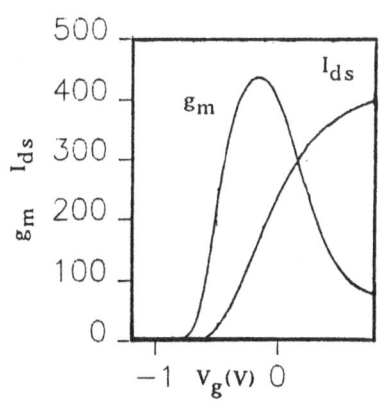

500 nm

Fig. 1 SEM photograph of the gate region.

Fig. 2 DC characteristics of a 0.25 μm gate length InGaAs/GaAlAs HEMT. g_m is in mS/mm, I_{ds} in mA/mm.

output conductance g_m/g_o is as high as 20. For gate lengths of 1 μm and 0.5 μm we obtained transconductances of 290 mS/mm and 310 mS/mm, respectively. The transistors were measured with a Cascade probe station from 1 to 15 GHz. Microwave measurements were made on 0.25 x 100 μm^2 and 0.25 x 50 μm^2 gate area devices. The s-parameters and the modeled equivalent circuit of a 0.25 x 100 μm^2 gate area HEMT are presented in figure 3 and 4. The RF transconductance is about 12% larger than the DC

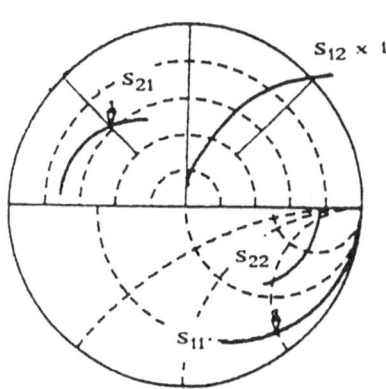

Start 1 GHz
Stop 15 GHz
Fig. 3 Measured HF s-parameters

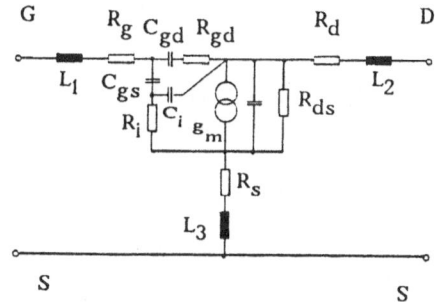

Fig. 4 The equivalent circuit to model the HEMT. The values for C_{gs} and g_m are 0.12 pF and 49 mS, respectively. They refer to a gatewidth of 100 μm.

value. From the equivalent circuit of the devices we obtain a unity current gain cutoff frequency of 66 GHz and a maximum oscillation frequency of 115 GHz.

Conclusions

We have developed submicrometer InGaAs/AlGaAs HEMTs . The process technology is based on an all-level e-beam pattern transfer that allows design changes quite easily. Transistors with 0.25 µm gate length were fabricated with a f_t over 65 GHz.

Acknowledgement

The authors are very grateful to M. H. Pilkuhn, H. Niederbach, B. Maile, M. Korn and J. Schäfer for their discussions and technical assistance. The financial support of this work by Stiftung Volkswagenwerk and the Stabstelle für internationale Beziehungen, Kernforschungszentrum Karlsruhe is gratefully appreciated.

References

[1] Kohki Hikosaka, S. Saga, N. Harada, Shigeru Kuroda : Current Gain Cutoff Frequency Comparison of InGaAs HEMTs, IEEE Electron Dev. Letters Vol. 9 No. 5 (1988) pp. 241-243

[2] J. M. Ballingall, P. Ho, G. J. Tessmer, P. A. Martin, N. Lewis, E. L. Hall : Novel Pseudomorphic High Electron Mobility Transistor Structures With $GaAs\text{-}Ga_{0.3}In_{0.7}As$ Thin Strained Superlattice Active Layers, Appl. Phys. Lett. 54 (21), 1989, pp. 2121-2123

[3] P. C. Chao, M. S. Shur, R. C. Tiberio, K. H. G. Duh, P. M. Smith, J. M. Ballingall, P. Ho, A. A. Jabra : DC and Microwaves Characteristics of Sub 0.1 µm Gate-Length Planar Doped Pseudomorphic HEMTs, IEEE Trans. El. Dev. Vol. 36 No. 3 (1989) pp. 461-473

[4] A. Cappy, B. Carnez, R. Fauquembergues, G. Salmer, E. Constant : Comperative Potential Performance of Si, GaAs, GaInAs, InAs Submicrometer-Gate FETs. IEEE Trans. El. Dev. Vol. 27 No. 11 (1980), pp. 2158-2160

Phase Noise in HBT Oscillator

H.Wang, C.Algani, E.Caquot, K.Iraki*, G.Alquié*, J.Tasselli**
Centre National d'Études des Télécommunications
Laboratoire de Bagneux, 196 Avenue Henri Ravéra
92220 Bagneux

* : Université de P. Et M.CURIE, Laboratoire de Dispositifs Infrarouges et Microondes - 4, Place Jussieu - 75005 PARIS
** : LAAS-CNRS, 7, Avenue Du Colonel Fabien - 31077 TOULOUSE cedex

High frequency low phase noise oscillator is one of the key element in modern communication systems. The origin of the phase noise in solid-state oscillators has been identified to be the low frequency excess noise (or 1/f noise). As a bipolar device, the HBT also presents inherent low 1/f noise characteristic, which makes HBT a very promising candidate, compared to GaAs MESFET or Silicon BJT, for the high frequency low phase noise oscillator applications [1,2].

This paper presents the preliminary results on the phase noise of HBT oscillators. The HBT low frequency excess noise has been characterized, its frequency dependance has been found to be $1/f^{1.2}$. The conversion factor of low frequency noise to the phase noise of HBT oscillator of 2.2E11 Hz/A has been obtained.

1/f NOISE ON HBT :

The source of the 1/f noise in a bipolar transistor was generally accepted as being located in the emitter-base region. In an HBT, in addition to the classical noise sources found in silicon bipolar transistors such as surface effects and generation-recombinations, the influence of the heterojunction emitter-base interface has to be taken into account. In reference [3], the low frequency excess noise in a double heterostructure HBT has been investigated at room temperature. We have investigated the low frequency excess noise on our single heterostructure high speed HBT. As in the silicon BJT, the noise is modelled by a current generator between the base and the emitter with a dependance of $1/f^1$:

$$i_{1/f}^2 = K J_B^a . \frac{\Delta f}{f^1}$$

Several HBTs low frequency noises, from 10Hz to 100kHz, have been measured.

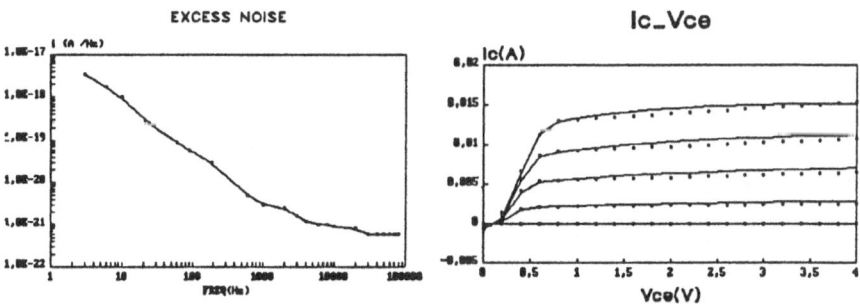

Figure 1 : low-frequency noise in HBT.   *Figure 2 : Measure and fitting static characteristics of HBT.*

The experimental results obtained on HBT show principally a noise spectral density of the form $f^{-1.2}$ (fig.1). Detailed measurement results and the caracterisations of the temperature and bias condition dependance will be presented.

A silicon transistor NEC has been compared. No significant improvement have been observed for HBT compared with NEC.

PHASE NOISE OF HBT OSCILLATOR :

A low Q HBT oscillator has been fabricated at 1.2GHz. It was realized using a hybrid technology.

The HBT has been measured from 1 to 18GHz. Non-linear modified Gummel-Poon model parameters of the transistor have been obtained by fitting static characteristics and scattering parameters (figure 2).

Oscillator has been simulated in linear and non-linear domains using ESOPE and SPICE softwares. Low frequency of 1.2 GHz and low Q were chosen in order to make a quick phase noise measurement.

The oscillator has been measured with an spectrum analyzer. The power efficiency of 25% has been obtained, as predicted by simulation. The oscillation frequency is at 1.28GHz.

The dominant noise contribution close to the center frequency is due to the phase noise. The single sideband noise spectrum is about -60dBc/Hz at 10kHz from carrier frequency.

The conversion factor defined as $(\Delta f_{rms}(Hz/\sqrt{Hz}))/(\sqrt{i^2}(A/\sqrt{Hz}))$ can be evaluated. It is close to 4.5E11 Hz/A. This result is similar to which of an MESFET $((\Delta f_{rms}(Hz/\sqrt{Hz}))/(\sqrt{V^2}(V/\sqrt{Hz})) = 2E8Hz/A)$ if the input impedance is $2.2k\Omega$. As the low frequency noise is lower in HBT, the final phase noise is lower for an HBT oscillator than an MESFET oscillator.

CONCLUSION :

The low frequency excess noise of HBT together with the phase noise of HBT oscillator have been investigated. The experimental results have shown a very promising future for HBT low phase noise application because of its low 1/f noise (less than $10^{-21}A^2/Hz$). The conversion factor of 4.5E11 Hz/A has been obtained in a free-running low-Q HBT oscillator.

The authors would like to thank S.Vuye for the hybrid circuit fabrication.

REFERENCES :

[1] - K.Agarwal, 'Dielectric Resonator oscillators using GaAs/GaAlAs heterojunction bipolar transistors ' Tech Digest 1986 MTT-S p 95

[2] - N.Hayama, et Al, ' Low-noise KU band AlGaAs/GaAs HBT oscillator ', MW symp. 1988 Vol 2 pp 679-682

[3] -X.N. ZHANG, A. VAN DER ZIEL, K.H. DUH and K. MORKOC : ' Burst and low noise generation-recombination noise in double-heterojonction bipolar transistors', IEEE Electron Device Letters, vol. 5, n°7, July , pp 277-279.

Process and Device Modelling II

The Influence of Point Defect Concentrations on the Diffusion of Gold in Silicon

H. ZIMMERMANN[*] and P. PICHLER[**]

* Lehrstuhl für Elektronische Bauelemente, Universität Erlangen-Nürnberg, Artilleriestrasse 12, D-8520 Erlangen

** Fraunhofer-Arbeitsgruppe für Integrierte Schaltungen, Artilleriestrasse 12, D-8520 Erlangen

Summary

The diffusion of gold in silicon is described by the kick-out and dissociative mechanism. The resulting set of four coupled partial differential equations is completely solved numerically. Splitting the self-diffusion coefficients, we find that the concentration of substitutional gold is strongly influenced by the equilibrium concentrations of self-interstitials and vacancies. As a result of our work, we can give an upper boundary for the value of the equilibrium concentration of vacancies and a lower boundary for the diffusivity of vacancies.

Theory

It is widely accepted that self-interstitials and vacancies are present at thermal equilibrium in silicon and that gold diffuses in silicon only via the so called kick-out and the dissociative mechanisms. This leads to the following three reactions:

$$Au_i \underset{k_{-1}}{\overset{k_{+1}}{\rightleftharpoons}} Au_s + I \quad (1); \qquad Au_i + V \underset{k_{-2}}{\overset{k_{+2}}{\rightleftharpoons}} Au_s \quad (2); \qquad I + V \underset{k_{-3}}{\overset{k_{+3}}{\rightleftharpoons}} 0 \quad (3);$$

where Au_i, Au_s, I, V, and 0 stand for interstitial gold, substitutional gold, silicon self-interstitials, vacancies, and the undisturbed lattice, respectively. The symbols k_{+1}, k_{-1}, k_{+2}, k_{-2}, k_{+3}, and k_{-3} are the reaction rates. In contrast to other approaches [1], the resulting equations are not simplified. The diffusion of gold in silicon is then described by the four following partial differential equations:

$$\frac{\partial C_s}{\partial t} = k_{+1}C_i - k_{-1}C_sC_I + k_{+2}C_iC_V - k_{-2}C_s \tag{4}$$

$$\frac{\partial C_i}{\partial t} = div \, D_i \, grad \, C_i - k_{+1}C_i + k_{-1}C_sC_I - k_{+2}C_iC_V + k_{-2}C_s \tag{5}$$

$$\frac{\partial C_I}{\partial t} = div \; D_I \; grad \; C_I + k_{+1}C_i - k_{-1}C_sC_I - K_B(C_IC_V - C_I{}^*C_V{}^*) \qquad (6)$$

$$\frac{\partial C_V}{\partial t} = div \; D_V \; grad \; C_V - k_{+2}C_iC_V + k_{-2}C_s - K_B(C_IC_V - C_I{}^*C_V{}^*) \qquad (7)$$

In these equations the characters C and D are used to denote concentrations and diffusion coefficients. The subindices s, i, I, and V denote substitutional gold, interstitial gold, silicon self-interstitials, and vacancies, respectively. $C_I{}^*$ and $C_V{}^*$ are the equilibrium concentrations of self-interstitials and vacancies. Table 1 shows the used parameters. Instead of k_{+3} and k_{-3}, $K_B = k_{+3}$ and $K_B C_I{}^* C_V{}^* = k_{-3}$ is used. According to Antoniadis [2], an activation energy of 1.4eV is assumed for the bulk reaction rate, whereas the prefactor is chosen so that at 1000°C the resulting K_B equals the value given by Budil et al. [3].

Results

The goal of our work was to find a splitting for the so called self-diffusion coefficients $C_I{}^*D_I$ and $C_V{}^*D_V$. The values of these self-diffusion coefficients have been taken from Morehead et al. [1].

The used values for the reaction rates k_{+1}, k_{-1}, k_{+2}, and k_{-2} are crude estimates. Their influence can be seen only in the vicinity of the surfaces and only for short diffusion times. The increase of the gold concentration in the bulk with respect to time was found to be determined mainly by the choice of $C_I{}^*$ and $C_V{}^*$. The values for $C_I{}^*$, D_I, $C_V{}^*$, and D_V given by Tan and Gösele [6] are a consistent splitting of the self-diffusion coefficients given by Morehead et al. [1]. With these parameters, however, the substitutional gold concentration in the bulk becomes too high (see Fig. 1 and 2).

The parameters have then been adapted to give a best fit for the gold diffusion profiles measured by Stolwijk et al. [4,5] (see Fig. 1 and 2). Based on these simulations, we can give the following upper boundaries for C_V and lower boundaries for D_V:

$$C_V < 2 \cdot 10^{22} \cdot exp(-2.0 \; eV \; / \; kT) \; cm^{-3}$$
$$D_V > 1 \cdot exp(-2.0 \; eV \; / \; kT) \; cm^2 s^{-1}$$

This means that C_V has to be a factor of ten lower than the values given by Tan and Gösele. Accordingly, D_V has to be a factor of ten greater than the values of [6].

Acknowledgement
This work has been supported in part by the Bundesministerium für Forschung und Technologie.

References

1. F.Morehead, N.A.Stolwijk, W.Meyberg, U.Gösele: Appl. Phys. Lett. 42(8), 690 (1983)

2. D.A.Antoniadis, I.Moskowitz: J. Appl. Phys. 53, 6788 (1982)

3. M.W.Budil, W.Jüngling, E.Guerrero, S.Selberherr, H.W.Pötzl:Simulation of semicond. Dev. and Processes: Proc. of the II. Int. Conf. in Swansea, 384 (1986)

4. N.A.Stolwijk, B.Schuster, J.Hölzl: Appl. Phys. A33, 133 (1984)

5. N.A.Stolwijk, J.Hölzl, W.Frank: Appl. Phys. A39, 37 (1986)

6. T.Y.Tan, U.Gösele: Appl. Phys. A37, 1 (1985)

Tab. 1. Values of the parameters for the best fits of Figs. 1 and 2

T[K]	1273	1371
k_{+1} [s_{-1}]	0.1	0.1
k_{-1} [$cm^3 s^{-1}$]	$2.8 \cdot 10^{-17}$	$4.8 \cdot 10^{-18}$
k_{+2} [$cm^3 s^{-1}$]	$4.9 \cdot 10^{-14}$	$8.8 \cdot 10^{-16}$
k_{-2} [s^{-1}]	$1.6 \cdot 10^{-3}$	$1.6 \cdot 10^{-3}$
C_i^* [cm^{-3}]	$8.0 \cdot 10^{13}$	$7.7 \cdot 10^{14}$
D_i [$cm^2 s^{-1}$]	$7.0 \cdot 10^{-6}$	$9.1 \cdot 10^{-6}$
C_I [cm^{-3}]	$1.9 \cdot 10^{13}$	$3.3 \cdot 10^{14}$
D_I [$cm^2 s^{-1}$]	$2.6 \cdot 10^{-7}$	$3.4 \cdot 10^{-7}$
C_V [cm^{-3}]	$2.4 \cdot 10^{13}$	$4.4 \cdot 10^{15}$
D_V [$cm^2 s^{-1}$]	$1.2 \cdot 10^{-7}$	$4.4 \cdot 10^{-7}$

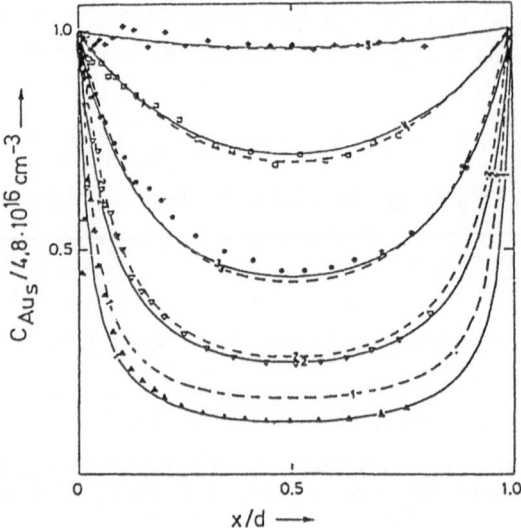

Fig. 1: —— Simulation with parameters of Tab. 1
       - - Simulation with parameters of [6]
       Measured profiles of gold for different annealing
       times at 1371 K [4] (▲ 0.428h, ▽ 1.97h, ● 6.16h,
       □ 19.8h, ◆ 70.1h), d = 500 μm

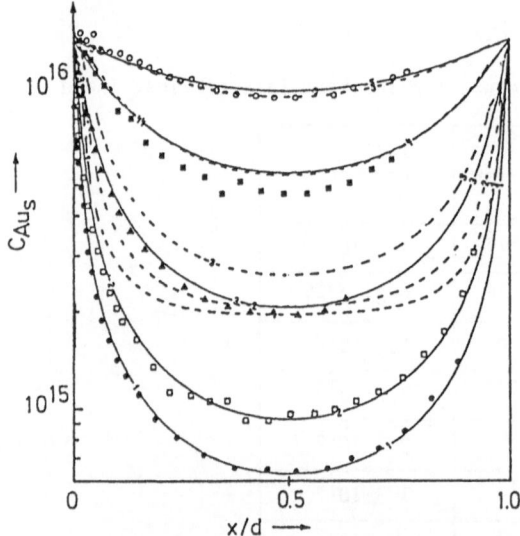

Fig. 2: —— Simulation with parameters of Tab. 1
       - - Simulation with parameters of [6]
       Measured profiles of gold for different annealing
       times at 1273 K [5] (● 0.467h, □ 1.03h, ▲ 4.27h,
       ■ 26.8h, ○ 100.6h), d = 500 μm

Quantification of Diffusion Mechanisms of Boron, Phosphorus, Arsenic, and Antimony in Silicon

P. B. MOYNAGH & P. J. ROSSER
STC Technology Ltd., London Rd., Harlow, Essex, England, CM17 9NA

Summary

This paper quantifies the Si-interstitial (I) and Si-vacancy (V) components of diffusion for B, P, As and Sb in silicon at 1100°C. It is determined that B and P diffuse predominantly by an I type mechanism and Sb predominantly by a V type mechanism. As displays an equal tendency to diffuse by I and V mechanisms. In the extrinsic diffusion regime the dopants maintain the same relative I and V components.

Introduction

Diffusion of the common dopants (A) in silicon takes place by interacting with point-defects, i.e. with Si-vacancies (V) and Si-interstitials (I) [1]. The diffusing species are therefore the dopant-defect pairs, AV and AI. In a non-faulted single crystal lattice with intrinsic concentrations of dopant, and with no surface reactions, the point-defect concentrations are in thermal equilibrium with the lattice and are approximately 1E15 cm^{-3} at 1100°C. Diffusivity of dopant A is described simply as

$$D_A^* = D_{AI}^* + D_{AV}^* \qquad (1)$$

where * indicates equilibrium diffusive conditions. If the equilibrium point-defect concentrations are perturbed then dopant diffusivity, D_A, will be modified.

$$\frac{D_A}{D_A^*} = f_{AI} \frac{C_{AI}}{C_{AI}^*} + (1-f_{AI}) \frac{C_{AV}}{C_{AV}^*} \qquad (2)$$

where $f_{AI} = \dfrac{D_{AI}^*}{D_A^*}$ and $C_{AI}, C_{AV} =$ dopant/defect pair concentrations

and f_{AI} is known as the fractional interstitialcy component of diffusion and is a fundamental property of each dopant. Since point-defect concentrations are perturbed by virtually every VLSI processing step (including oxidation, implantation, silicides and high dopant concentrations), it is important to determine the nature and magnitude of point-defect perturbations caused by each VLSI process, and the way in which the common dopants respond to these perturbations.

This paper quantifies the fractional interstitialcy component of diffusion for B, P, As and Sb in silicon at 1100°C using rapid thermal nitridation of Si and SiO$_2$ in an NH$_3$ ambient. The benifit of studying dopant diffusion in nitriding ambients derives from the fact that nitridation of Si results in a V supersaturation and I undersaturation in the underlying silicon, while nitridation of SiO$_2$ gives a V undersaturation and I supersaturation [2]. This is indicated in figure 1. This complementary behaviour provides an experimental situation where definitive statements can be made concerning the relative importance of the AI and AV diffusion mechanisms for dopants in silicon.

Experiments

Substrates used were p- or n-type Cz-(100) silicon with a doping concentration of 4 to 6×10^{14} cm^{-3}. Thermal oxides of 22 nm thickness were furnace grown at 900°C. Boron, Phosphorus, Arsenic or Antimony were then implanted to doses of 1 x 10^{13} cm^{-2} at 30, 80, 160 & 180 keV respectively. Implant damage was then annealed using an 800°C, 15 minute furnace treatment. Silicon nitride layers of 100 nm thickness were deposited and 200 μm lines and spaces photolithographically defined and wet etched giving the structure shown in figure 1. The structure was then annealed at 1100°C for 300s in an AG410 Rapid Anneal System. In region 1 where the substrate is protected by the nitride layer, no surface reaction occurs, and the equivalent of inert ambient diffusion takes place. In region 2 where silicon is exposed to the nitriding ambient, a V-supersaturation and I-undersaturation results [2], and dopant diffusivity is modified. In region 3 where oxide is exposed, a V-undersaturation and I-supersaturation results [2], and dopant diffusivity is modified accordingly.

After anneal, samples are bevelled at an angle of 7' and stained in a copper solution to delineate the n-type regions, thereby giving junction depths. Figure 2 shows a photomicrograph of such a sample. Diffusivity modification caused by the oxide nitridation (region 3) and silicon nitridation (region 2) is apparent by comparing the junction depths under those regions, with that under the nitride cap (region 1) where no nitridation induced point-defect perturbation takes place. Use of a monochromatic light source allows quantification of the distances diffused in each region. The resolution in determining differences in junction depths is better than 10 nm. Spreading Resistance Analysis (SRA) has been used to determine dopant profile shape and as a second measure of distance diffused. Dopant profiles measured on the experimental structures have been fitted using the process modelling program, Suprem 3 [3]. If diffusivity was enhanced relative to the control region (region 3 vs. 1 of figure 2) then a diffusivity enhancement factor was calculated as $D_{enhanced}/D_{equil}$. If retardation had taken place (region 2 vs. 1 of figure 2) then the retardation factor, $D_{equil}/D_{retarded}$, was calculated. The values determined are listed in figure 3.

Discussion

In figure 3a it is apparent that an intrinsic P profile is retarded by the silicon-nitridation induced V supersaturation and I undersaturation. Figure 3b shows that P is enhanced by the oxide-nitridation induced I supersaturation and V undersaturation. It can therefore be stated definitively that intrinsic P diffuses predominantly by an I type mechanism. Similarily one observes that B diffuses predominantly by an I type mechanism and Sb by a V type mechanism. As, on the other hand, exhibits enhancement under conditions of I and V excess and therefore diffuses by more evenly mixed I and V mechanisms.

It is reasonable to assume that the magnitude of the nitridation induced point-defect perturbations is independent of the dopant in the underlying silicon, particularily for intrinsic dopant concentrations. For the same anneal therefore, B, P, As and Sb will experience the same I and V perturbations. This consistency allows for manipulation of equation 2 above, as observed by Fahey et. al. [2]. Undertaking calculations based on these manipulations yields values for the fractional interstitialcy component of diffusion as listed in table 1.

Figure 5a shows junction depth variation in the 'diffusion assessment structure' when the intrinsic P profile is replaced with the extrinsic profile shown in figure 4. It is apparent that extrinsic P diffuses predominantly by an I mechanism, as it does in the intrinsic regime. Extrinsic concentrations of B also diffuse by a predominately I type mechanism. Extrinsic As displays a mixed behaviour but with an increased V component relative to the intrinsic regime.

The degenerate concentrations of P in figure 5b demonstrate no dependence on the surface reaction. The same is true for degenerate B and As. This indicates that, a) the concentration of surface generated point-defects is negligable compared to the concentrations in the degenerate profiles, b) the degenerate dopant regions efficiently consume excess point-defects or, c) the surface reactions are considerably modified by the high dopant concentrations.

Conclusion

Dopant diffusivity modification during rapid thermal nitridation has been examined and relative I and V components of diffusion determined.

References

1. P.B.Moynagh et. al., in 'Properties of Silicon', INSPEC, London, (1988)

2. P.Fahey et. al., Appl. Phys. Lett., 46, p.784, (1985)

3. Stanford University/Technology Modelling Associates, Suprem 3 process model

Acknowledgments

This work was part funded by the UK Department of Trade & Industry through the Alvey Directorate under contract ALV/APP/VLSI/005

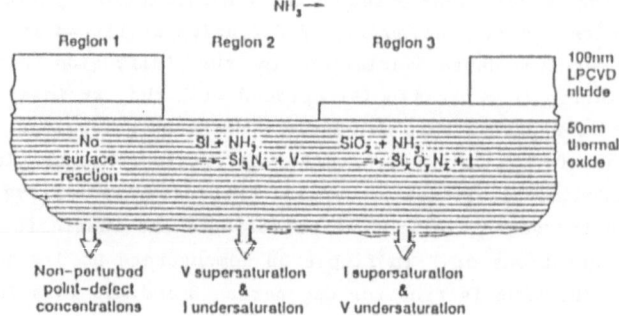

Figure 1:

The 'Diffusion assessment structure'.
Point-defect perturbations caused by thermal
nitridation in an NH_3 ambient are indicated.

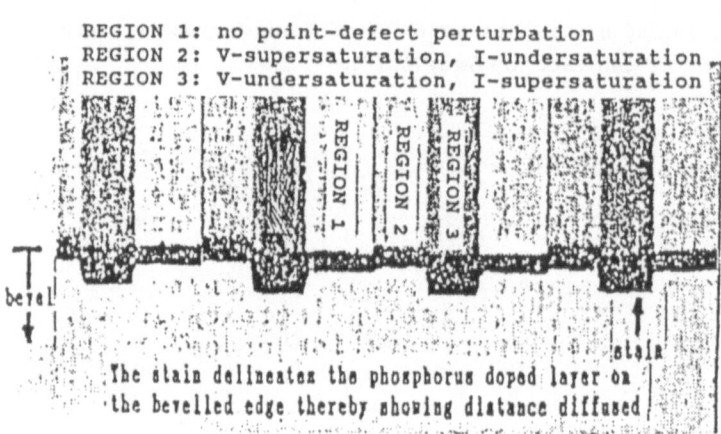

Figure 2:

Photomicrograph of a 'Diffusion assessment structure'
implanted with intrinsic phosphorus and annealed in
NH_3.

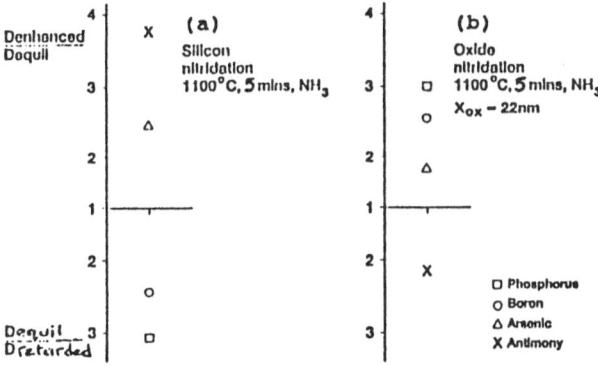

Figure 3:

Dopant diffusivity modification factors extracted from structures as shown in figures 1 & 2.

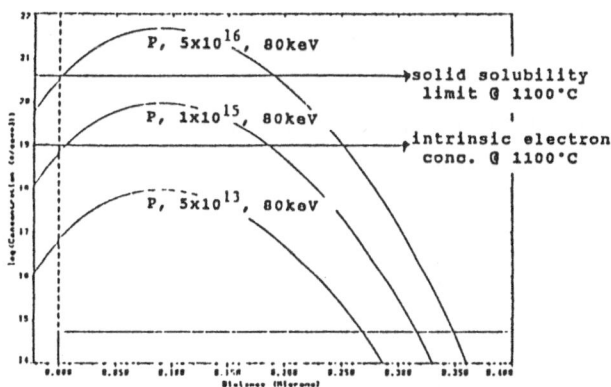

Figure 4:

Intrinsic (1×10^{13} cm^{-2} implant), extrinsic (1×10^{15} cm^{-2} implant) and degenerate (5×10^{16} cm^{-2} implant) phosphorus profiles after implantation. Also shown is the intrinsic electron concentration and the solid solubility limit at 1100°C.

296

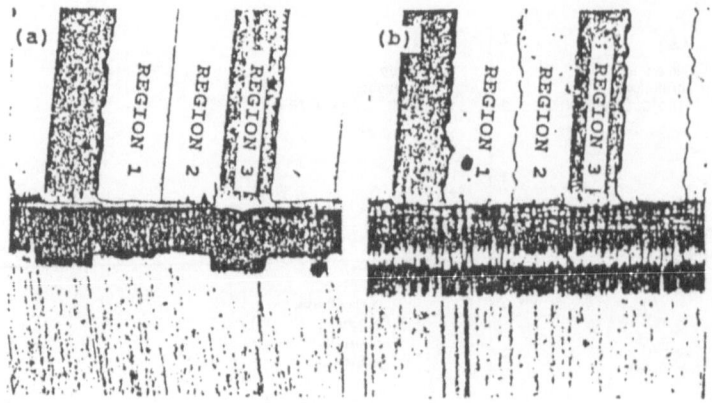

Figure 5:

Photomicrograph of a 'Diffusion assessment structure'.
a) Implanted with extrinsic phosphorus and annealed in
NH_3.
b) Implanted with degenerate phosphorus and annealed
in NH_3.

Table 1:

The fractional interstitialcy component of diffusion,
f_{AI},
as determined in this work and by others.

Reference	P	B	As	Sb
This work	) = 92%	88%	43–50%	(= 16%
Fahey, Dutton	) = 94%		30%	(= 29%
Antoniadis	38%	30%	35%	1.5%
Mathiot, Pfister	23%	10–30%	10–30%	10–30%

A Consistent Pair-diffusion Based Steady-state Model for Phosphorus Diffusion

R.Dürr[1] and P.Pichler[2]

[1]Lehrstuhl für Elektronische Bauelemente, Universität Erlangen-Nürnberg, Artilleriestrasse 12, D-8520 Erlangen, BRD

[2]Fraunhofer-Arbeitsgruppe für Integrierte Schaltungen, Abteilung für Bauelementetechnologie, Artilleriestrasse 12, D-8520 Erlangen, BRD

Summary
A simplified model for phosphorus diffusion has been developed. Assuming dopant diffusion to proceed by impurity-point defect pairs, the complete system of coupled partial differential equations describing both reaction kinetics and diffusion has been consistently simplified. The resulting equations are solved numerically. Important features of phosphorus diffusion arise naturally from the solution.

Theory

It is well known that high surface concentrations of phosphorus generate point defects. These point defects increase the diffusion coefficient in the bulk which leads to the phosphorus kink as well as to the emitter push effect [1]. Native point defects, silicon self-interstitials (I) and vacancies (V), are assumed to form mobile pairs with dopant atoms (A). The formation of dopant-point defect pairs (AI, AV) can be described by the following reactions

$$A + I \underset{k_2}{\overset{k_1}{\rightleftharpoons}} AI, \quad A + V \underset{k_4}{\overset{k_3}{\rightleftharpoons}} AV \tag{1}$$

where the k_1 to k_4 denote reaction constants. To simplify the following analysis, similar reactions involving charge states of point defects are neglected at the moment. A system of five coupled differential equations then has to be considered [2]:

$$\frac{\partial C}{\partial t} = - k_1 \cdot C \cdot C_i + k_2 \cdot C_{ai} - k_3 \cdot C \cdot C_v + k_4 \cdot C_{av} \tag{2a}$$

$$\frac{\partial C_{ai}}{\partial t} = \text{div } J_{ai} + k_1 \cdot C \cdot C_i - k_2 \cdot C_{ai} \tag{2b}$$

$$\frac{\partial C_{av}}{\partial t} = \text{div } J_{av} + k_3 \cdot C \cdot C_v - k_4 \cdot C_{av} \tag{2c}$$

$$\frac{\partial C_i}{\partial t} = \text{div } J_i - k_1 \cdot C \cdot C_i + k_2 \cdot C_{ai} - k_b \cdot (C_i \cdot C_v - C_i^* \cdot C_v^*) \tag{2d}$$

$$\frac{\partial C_v}{\partial t} = \text{div } J_v - k_3 \cdot C \cdot C_v + k_4 \cdot C_{av} - k_b \cdot (C_i \cdot C_v - C_i^* \cdot C_v^*) \tag{2e}$$

Here J_i, J_v, J_{ai}, and J_{av} are the fluxes of self interstitials, vacancies, dopant-interstitial pairs, and dopant-vacancy pairs. An additional reaction between interstitials and vacancies describing generation and recombination of Frenkel pairs is included in the respective continuity equations. Several assumptions can be made to reduce the number of equations and to simplify the system. The most convenient of these is the assumption of local equilibrium between point defects, dopants and dopant-point defect pairs, reducing the system to a set of three differential equations by adding equations (2a),(2b), and (2c). Assuming local equilibrium between interstitials and vacancies makes a further reduction of the number of equations possible. It should be pointed out, however, that this assumption might be invalid in some cases as has been discussed by several authors [3]. Finally, steady-state conditions for AI, AV, I and V are assumed, leading to a continuity equation for dopants and to a time-independent equation for interstitials

$$\frac{\partial C}{\partial t} = \text{div } J_{ai} + \text{div } J_{av} \tag{3a}$$

$$\text{div } J_i - \text{div } J_v + \text{div } J_{ai} - \text{div } J_{av} = 0 \tag{3b}$$

The above analysis can be easily extended to include different charge states of the point defects when local equilibrium between the charge states of the point defects and the carriers as well as between the dopant-point defect pairs and the carriers is assumed.

Equation (4) then describes the diffusion of an arbitrary dopant, where N denotes the mobile part of the concentration C, C_{i0} denotes the concentration of neutral interstitials, C_{i0}^* is the corresponding equilibrium value, z denotes the charge state of the dopant (+1 for singly ionized acceptors, -1 for singly charged donors), q the elementary charge, Ψ the electrostatic potential, k Boltzmann's constant, and T the temperature.

$$\frac{\partial C}{\partial t} = \frac{\partial}{\partial x}\left[(D_{av}\cdot\frac{C_{i0}^*}{C_{i0}} + D_{ai}\cdot\frac{C_{i0}}{C_{i0}^*})\cdot(\frac{\partial N}{\partial x} - \frac{z\cdot q}{k\cdot T}\cdot N\cdot\frac{\partial \Psi}{\partial x}) + \right.$$

$$\left. + N\cdot(D_{av}\cdot\frac{\partial}{\partial x}\frac{C_{i0}^*}{C_{i0}} + D_{ai}\cdot\frac{\partial}{\partial x}\frac{C_{i0}}{C_{i0}^*})\right] \tag{4}$$

D_{av} and D_{ai} are the diffusion coefficients via vacancies and interstitials, respectively. For, e.g., donors they can be described by:

$$D_{av} = D_{av0} + D_{av-}\cdot(n/n_i) + D_{av=}\cdot(n/n_i)^2 \tag{5a}$$

$$D_{ai} = D_{ai0} + D_{ai-}\cdot(n/n_i) + D_{ai=}\cdot(n/n_i)^2 \tag{5b}$$

where n is the electron concentration and n_i is the intrinsic concentration. D_{av0}, D_{av-} and $D_{av=}$ are the diffusion coefficients via neutral, singly charged and doubly charged vacancies, D_{ai0}, D_{ai-} and $D_{ai=}$ are the diffusion coefficients via neutral, singly charged and doubly charged interstitials. In the same way an expression describing the behavior of interstitials and vacancies in all charge states can be derived. Generally, this expression is valid for an arbitrary dopant interacting with point defects. In practice, however, only phosphorus seems to generate point defects so that the other dopants can be neglected. For the oversaturation of interstitials we obtain the following differential equation:

$$\left[[C_v^*\cdot D_v]\cdot(\frac{C_{i0}^*}{C_{i0}})^2 + [C_i^*\cdot D_i] + N_p\cdot D_{pv}\cdot(\frac{C_{i0}^*}{C_{i0}})^2 + N_p\cdot D_{pi} \right]\cdot\frac{\partial}{\partial x}\frac{C_{i0}}{C_{i0}^*} =$$

$$(D_{pi}\cdot\frac{C_{i0}}{C_{i0}^*} - D_{pv}\cdot\frac{C_{i0}^*}{C_{i0}})\cdot(\frac{\partial N_p}{\partial x} + \frac{q}{k\cdot T}\cdot N_p\cdot\frac{\partial \Psi}{\partial x}) \tag{6}$$

N_p is the concentration of phosphorus ions, D_{pv} and D_{pi} are their diffusion coefficients via vacancies and interstitials.
$[C_v^*\cdot D_v]$ and $[C_i^*\cdot D_i]$ are the Fermi-level dependent selfdiffusion coefficients of vacancies and interstitials, respectively. For the donor phosphorus, only neutral, singly, and doubly negatively charged point defects have to be taken into consideration and the so called selfdiffusion coefficients can be written in the form

$$[C_v^*\cdot D_v] = C_{v0}^*\cdot D_{v0} + C_{v-,i}^*\cdot D_{v-}\cdot(n/n_i) + C_{v=,i}^*\cdot D_{v=}\cdot(n/n_i)^2 \tag{7a}$$

$$[C_i^*\cdot D_i] = C_{i0}^*\cdot D_{i0} + C_{i-,i}^*\cdot D_{i-}\cdot(n/n_i) + C_{i=,i}^*\cdot D_{i=}\cdot(n/n_i)^2 \tag{7b}$$

where, e.g., $C^*_{V=,i} \cdot D_{V=}$ is the selfdiffusion coefficient of doubly negatively charged vacancies for intrinsic conditions ($n=n_i$).
Estimates for these products are given in literature [4] though the role
of the charged species in this context is not resolvable, so that presently
the selfdiffusion coefficients of the charged species have to be adapted to
give a best fit for measured profiles. In contrast to a previous steady-
state model [5] the contribution of vacancies is consistently included in
equation (6).

Results

The model described above has been implemented in the one-dimensional
process simulation program ICECREM [6], making the simulation of phosphorus
diffusion and of the emitter push effect possible with a low computing
time. Fig. 1 and 2 show examples for the simulation of the redistribution
of phosphorus during annealing at 900°C and 1000°C together with experimental results from SIMS measurements. The initial profiles were obtained
after ion implantation and furnace annealing at 900°C for 30min. All
simulations have been performed using the parameters listed in Table 1. The
selfdiffusion coefficients of the uncharged species have been taken from
[4] and the diffusion of phosphorus was assumed to proceed entirely by an
interstitial mechanism. As can be seen from Fig. 1 and 2 the calculated
profiles agree well with experimental data for both temperatures.

Table 1. Parameters used for the simulation with ICECREM.

	prefactor (μm^2/min)	activation energy (eV)		prefactor ($1/\mu m$/min)	activation energy (eV)
D_{pv}	0	0	$C^*_{v0} \cdot D_{v0}$	$1.8 \cdot 10^{20}$	4.03
D_{pi0}	$6.28 \cdot 10^{10}$	3.69	$C^*_{v-,i} \cdot D_{v-}$	$1.5 \cdot 10^{19}$	4.03
D_{pi-}	0	0	$C^*_{v=,i} \cdot D_{v=}$	0	0
$D_{pi=}$	$1.57 \cdot 10^8$	3.69	$C^*_{i0} \cdot D_{i0}$	$2.472 \cdot 10^{23}$	4.84
			$C^*_{i-,i} \cdot D_{i-}$	$1.2 \cdot 10^{22}$	4.84
			$C^*_{i=,i} \cdot D_{i=}$	0	0

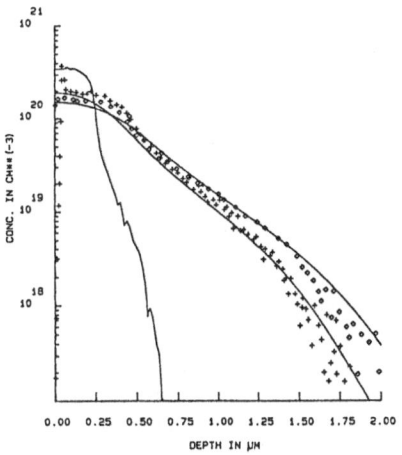 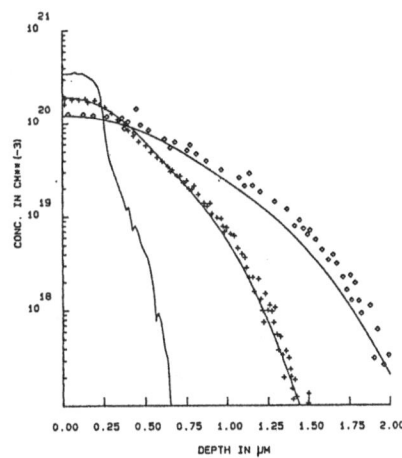

Fig.1 Initial phosphorus profile and profiles after diffusion for 600min (+) and 1200min (◊) at 900°C.

Fig.2 Initial phosphorus profile and profiles after diffusion for 60min (+) and 240min (◊) at 1000°C.

Conclusion

A simplified pair-diffusion based model for phosphorus diffusion has been presented. A continuity equation for dopants and a time-independent equation for point defects have been derived. The numerical solutions of these equations account for typical features of phosphorus diffusion.

Acknowledgment

This work has been supported in part by SIEMENS AG, München, BRD and by the Bayerisches Staatsministerium für Wirtschaft und Verkehr. The authors wish to thank H. Ryssel and J. Lorenz for helpful comments.

References

1. S.M. Hu, P. Fahey, and R.W. Dutton, J. Appl. Phys. 54, 6912, (1983)

2. W.B. Richardson and B.J. Mulvaney, Appl. Phys. Lett. 53, 1917, (1988)

3. D.A. Antoniadis and I. Moskowitz, J. Appl. Phys. 53, 6788, (1982)

4. T.Y. Tan and U. Gösele, Appl. Phys. A 37, 1, (1985)

5. F.F. Morehead and R.F. Lever, Appl. Phys. Lett. 48, 151, (1986)

6. P. Pichler, A. Barthel, R. Dürr, N. Holzer, J. Lorenz, H. Ryssel, K. Schott, NASECODE VI Software Forum Digest, (1989)

Numerical Simulation of Gas Flow and Temperature in a Diffusion Furnace

S. Howell, J.I. Ulacia F., Ch. Werner.
Siemens AG., Corporate Research and Development,
EL PT 33, 8000 Munich 83, West Germany.

Introduction:

Since the very beginning of semiconductor technology the diffusion furnace has been used as one of the workhorses in I.C. manufacture and has been employed for a number of different processes, such as dry and wet oxidation, LPCVD, high temperature annealing, etc. However, as wafer diameters increase from six to eight or ten inches, temperature homogeneity in the furnace becomes more and more of a problem, and optimization of the diffusion process is necessary for a high yield.

In this paper we present a novel simulation tool which can support furnace design as well as process optimization by calculating gas flow and heat transport both in the process gas and in the wafers. Using some simplifying assumptions for the wafer batch we can achieve full 3-dimensional solutions for the velocity and temperature distributions in the furnace, including the entire batch of wafers.

These simulations are a valuable help in furnace optimization because they provide visual feedback on gas flow patterns and temperature distributions in the furnace and can predict the effects that different operating conditions have on temperature homogeneity across the wafers.

Model Description:

The set of equations to be solved consists of the mass balance equation

$$\frac{\partial \rho}{\partial t} + \nabla \cdot (\rho \mathbf{v}) = 0 \tag{1}$$

the momentum balance or Navier-Stokes equation

$$\frac{\partial}{\partial t}(\rho v_i) + \nabla \cdot (\rho \mathbf{v} v_i - \mu \nabla v_i) = \rho g_i - \nabla_i P + S_{m,i} \tag{2}$$

the energy balance equation

$$\rho C_p \frac{\partial T}{\partial t} + \nabla \cdot (\rho C_p \mathbf{v} T - \lambda \nabla T) = S_h \tag{3}$$

and the Ideal gas law $\rho = PM/RT$ where ρ is the density of the medium, t is time, $\mathbf{v}$ is the velocity vector, ∇_i is the gradient in direction i, P is the pressure, v_i is the velocity resolute in direction i ($i = x, y, z$), μ is the viscosity of the medium, g_i is the gravity resolute in direction i, $S_{M,i}$ is the source of momentum in direction i, C_p is the specific heat capacity at constant pressure of the medium, T is the temperature, λ is the thermal conductivity of the medium, S_H is the heat source, M is the molecular weight of the gas, and R is the Universal gas constant. The physical parameters μ, C_p and λ are modelled as functions of temperature.

The equations 1 – 3 are solved for the 3-dimensional geometry of the diffusion tube. The boundary conditions applied are zero for the velocity at the furnace walls, a specified mass flux at the inlet and a fixed pressure at the outlet. For the temperature we used fixed values at the furnace sides in the region where the heating elements are placed and no heat flux through the walls outside this area, giving adiabatic wall boundary conditions.

In the batch, each wafer is not modelled individually but rather described as a whole block of porous material which allows gas transport only parallel to the wafer surfaces and sets the normal velocity component to zero.

The effective heat conductivity in this block is modelled differently normal and parallel to the wafer surfaces: parallel to the wafer surface we have a high conductivity λ_{Si} inside the wafers and very low values λ_{gas} in the gaps between the wafers. This makes an effective conductivity of

$$\lambda_{eff}{}^{\parallel} = \frac{\lambda_{Si}d_{Si} + \lambda_{gas}d_{gap}}{d_{Si} + d_{gap}} \approx \lambda_{Si}\frac{d_{Si}}{d_{Si} + d_{gap}} \tag{4}$$

where d_{Si} is the wafer thickness, and d_{gap} is the wafer spacing.

In the normal direction heat conduction is mostly by radiation between adjacent wafer surfaces, which makes a heat flux of

$$j_{rad}{}^{\perp} = \lambda_{eff}{}^{\perp}\frac{dT}{dz} \quad ; \quad \lambda_{eff}{}^{\perp} = 4C_w\sigma T^3(d_{Si} + d_{gap}) \tag{5}$$

where σ is the Stefan-Boltzmann constant $5.67 \times 10^{-8}\ Wm^{-2}K^{-4}$, and C_w is the exchange coefficient between the two surfaces.

A very important issue is the heat transfer between the heated furnace walls and the wafers. Since we solve the heat transport equation 3 numerically and care for the different heat conductivities in the wafers we automatically include the convective and conductive terms of heat transfer. The radiative part has been added artificially by considering special heat sources S_H on the wafer surfaces.

Equations 1 - 3 have been solved in 3-dimensions using the general-purpose fluid flow simulator PHOENICS [1] with models 4 and 5 added in appropriate subroutines. Typical calculation times are 1-3 hours on an Apollo 580T workstation.

Simulation Results:

According to Fig. 1 the general flow characteristics within the furnace are dominated by a large volume of recirculating gas between the gas inlet and the first wafer as a result of a large temperature gradient in the region. This is caused by the cold gas being fed into the furnace sinking because of its greater density, heating up as it approaches the wafers, expanding and rising due to the density decrease.

The basic effect of this recirculating gas is a strong cooling on the lower half of the first wafers on the inlet side. This can be seen in Figure 2 where we show the temperature profile for the hottest and coldest points on each wafer along the batch. We notice that the average temperature on the first wafers is significantly reduced and also that the maximum temperature difference ΔT_W within the first wafer is much larger than that for the rest of the batch. The use of multizone heating systems is a well established method to compensate for this cooling effect. We have simulated this system by setting a five-zone temperature profile as a boundary condition at the furnace walls. The temperature values of each zone have been chosen in such a way that the average temperature of each wafer is exactly the same in this simulation. However, as can be seen in Figure 2, this multizone heating system does not help to reduce the temperature non-uniformity ΔT_W within the first wafers, but rather gives approximately the same values of ΔT_W for both the one- and five-zone heating cases.

The simulation was performed for four different wafer diameters, four-, six-, eight- and ten-inch. As six-inch wafers are now standard for production, this simulation gives an idea of the scale of problems in the future. The four inch wafer simulation gave 1 degree

304

for the maximum value of ΔT_W. The simulated temperature uniformity gets progressively worse as the wafer diameter increases until, for a ten-inch wafer, ΔT_W was 30 degrees. This effect cannot be improved with five-zone heating. The values of ΔT_W for all four cases are shown in Fig. 3.

Since we have identified the density difference between hot and cold gas as the reason for the observed temperature non-uniformity, the simulations were performed for higher temperatures of the process gas on entering the furnace. This pre-heating improved ΔT_W for a six-inch wafer almost linearly as can be seen in Fig. 4. The force of the recirculation also slowly dies away as this is due to density, and hence temperature, differences. The recirculation finally dies in the region of 700 degrees Celsius inlet gas temperature.

An alternative way to improve temperature homogeneity is the case of the vertical diffusion furnace. Here, the directions of gravity and gas flow are parallel, so that recirculations, driven by natural convection, are efficiently suppressed. Figure 5 shows the maximum temperature differences, ΔT_W, on the wafers, which clearly indicate the advantages of the vertical furnace arrangement.

Acknowledgements:

The authors would like to thank Dr. Roesch and Dr. Reisinger for their cooperation and Dr C. Aldham of IKOSS for his assistance in the adaptation of the program PHOENICS to our needs.

References:

[1] Rosten H.I. and Spalding D.B., "The PHOENICS Reference Manual", CHAM TR/200, Wimbledon (1987).

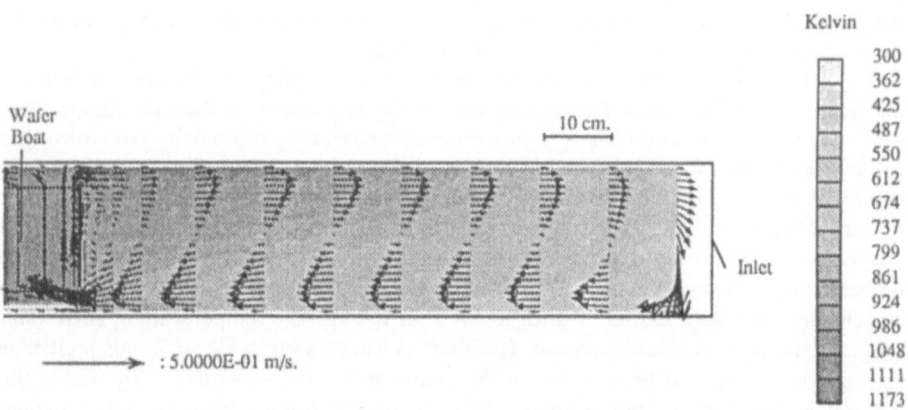

Figure 1: Gas flow vectors near the inlet port of a diffusion furnace showing a large volume of recirculating gas. The contours are of Temperature, showing how the cold gas flows along the bottom of the tube to the wafers.

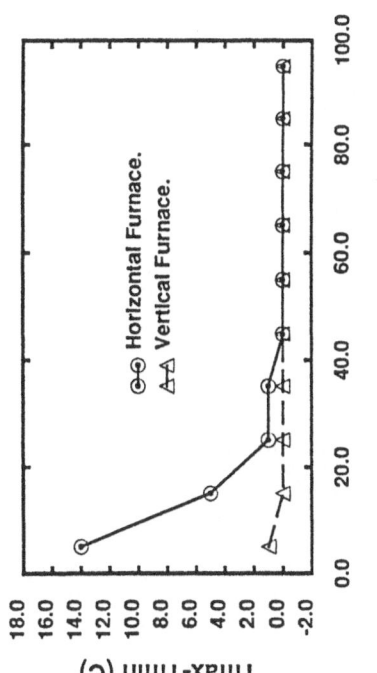

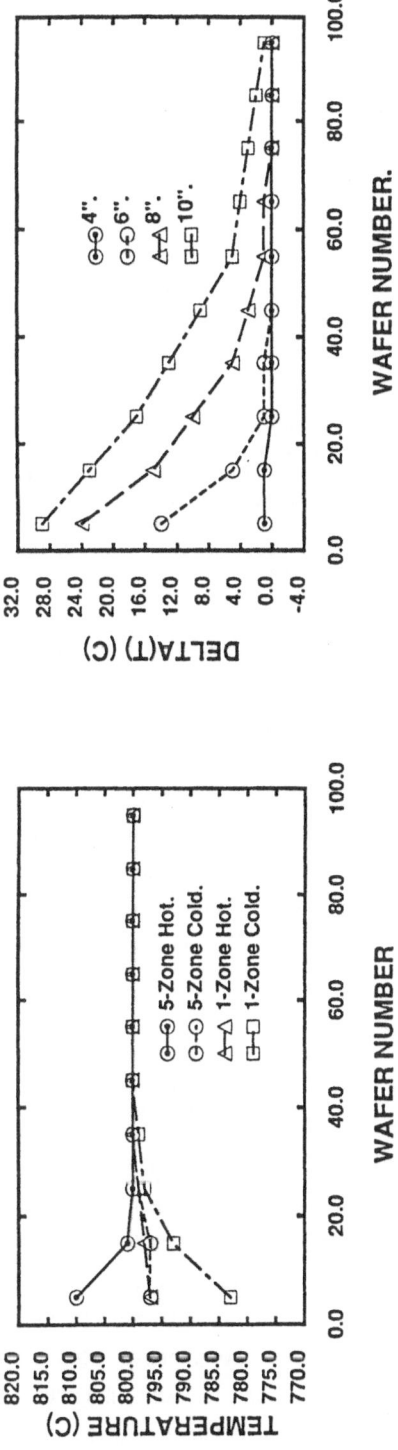

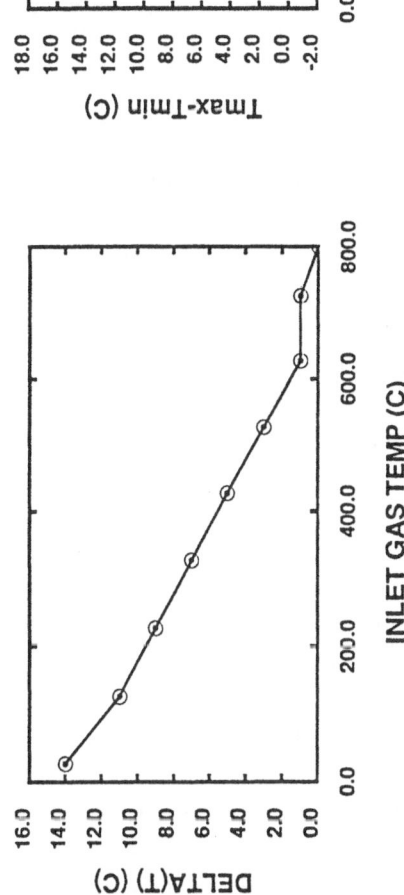

Figure 2: Comparison of the temperature profiles along the wafer batch with five- or one-zone heating.

Figure 3: Temperature uniformity in different-sized wafers.

Figure 4: Temperature uniformity in six-inch wafers for different temperatures of input gas.

Figure 5: Comparison of the temperature variation in wafers in Vertical and Horizontal furnaces.

The Impact of Metal Contamination on the Quality of Thermal SiO$_2$

H. Wendt and H. Cerva

Siemens AG, Otto-Hahn-Ring 6, 8000 München 83, FRG

Summary

To study the impact of a metal contamination on the quality of thermal oxides silicon wafers were intentionally contaminated by fast diffusing metals that tend to form silicides. Cu, Pd and Fe were found to be harmful, while Ni and Co did not cause any reductions of the breakdown field strengths of the oxides. Transmission electron microscopy studies showed that the reduction in gate oxide breakdown strength correlates with oxide thinning by metal silicide precipitates formed at the Si/SiO$_2$ interface.

One of the characteristics in the evolution of MOS technologies and products is a continuous increase of the gate oxide area (presently about 20-50mm^2 in 1M and 4M DRAMs). This trend results in a strongly increasing susceptibility of MOS devices to oxide quality and makes the dielectric breakdown one of the major factors determining the yield in integrated circuit production. Besides others, metal impurities have been found to deteriorate the quality of silicon dioxide layers by forming metal-rich silicides at the SiO$_2$/Si-interface [1,2,3]. In this paper we will focus on the silicide formation of different metals at the SiO$_2$/Si interface and on their influence on the gate oxide integrity.

Experimental

Boron doped 5 Ohm-cm (100) Czochralski-grown silicon wafers were used as starting material. Dry oxides with a thickness of 20nm were grown in a conventional furnace at a temperature of 900°C or in a rapid thermal annealer (RTA; Heatpulse 610) at 1200°C. The wafers were intentionally contaminated by scrapping with different metal wires (fast diffusing and silicide forming metals like Cu, Ni, Co, Pd and Fe) on the back surface of the wafers. The in-diffusion and precipitation of the different metals was performed by RTA (1200°C/30s or 900°C/60s; N$_2$). To find the process step in a device production process where a metal contamination is most harmful, intentional contaminations were performed (for Cu only) before, during and

after gate oxidation as well as after poly-Si (electrode material)
deposition. All other experiments correspond to a contamination occuring
after gate oxidation.
Transmission electron microscopy (TEM) combined with X-ray microanalysis on
cross-sections yields structural details and chemical composition of the
silicide phases formed at the interface. Electrical testing was either
performed in an electrolytic pinhole detector [4] or in an automatic tester
by a voltage ramp test.

Results and Discussion
During the routine control of oxide-monitor-wafers which were produced under
state-of-the-art processing conditions, a generel correlation between metal
contamination - as detected by shallow pits (S-pits) after Secco defect
etching - and the oxide quality of test structures could be established
(Fig.1).

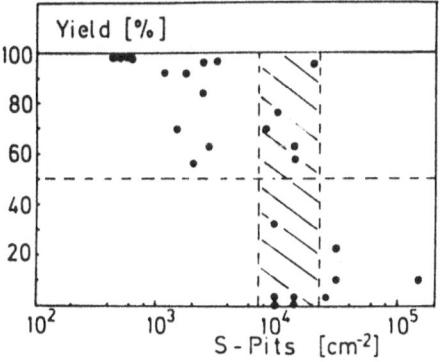

Fig.1. Yield of a gate oxide
test structure (ramped voltage test, electric breakdown field >8MV/cm) as a
function of the S-pit   density (Secco defect etching).

The yield decreases with increasing S-pit density. At defect densities
around $10^4/cm^2$ a transient region is observed with the yield varying between
0 and 95%, while at higher densities the yield stays below about 20%. From
this observation we infer that the probability for the occurence of a
critical, failure - causing defect increases with increasing defect density.
To find a correlation between a specific contamination, oxide failure and
the mechanism of the oxide breakdown, intentional contamination experiments
are necessary.
Reduced breakdown field strengths were observed in the pinhole detector in
Cu, Pd and Fe contaminated areas. Cu was found to be most effective in
reducing the oxide quality. High defect densities were observed at field
strengths as low as 2MV/cm and below. Ni and Co contaminations did not show
any influence on the oxide quality. It has to be mentioned that Fe-silicides

were difficult to obtain in the used Si-substrates and special treatments had to be performed (additionally low temperature "silicide growth" annealing) to observe an influence on the oxid quality.

Depending on the metal and the chosen supersaturation (as adjusted by the in-diffusion temperature) different failure mechanisms could be observed. In wafers with high Cu contamination levels (RTA at $1200^{o}C$) the observed silicide particles result in cracking and bending of the oxide as can be seen in Fig.2 and conduction paths to a later deposited layer (e.g. poly-Si) are possible. X-ray microanalysis in the TEM of these precipitates yields compositions of about 60%Cu and 40%Si (mole fractions).

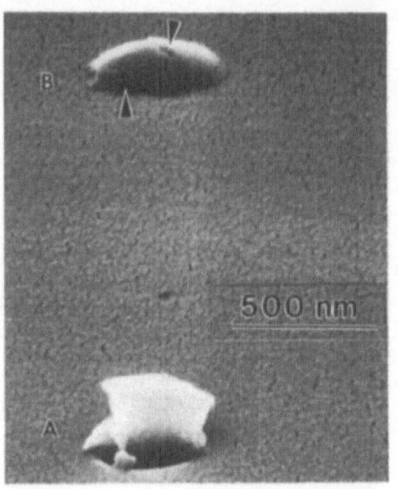

Fig.2. SEM-micrograph of Cu-precipitates formed at the SiO_2/Si-interface. The precipitates (A,B) penetrate the oxide or form cracks in the oxide.

At lower contamination levels (RTA at $900^{o}C$) and for Pd (even after RTA at $1200^{o}C$) only thinning of the oxide could be observed resulting in reduced breakdown fields. Cu-silicides show a rather strong effect with typical oxide thickness reductions of up to about 30%, while Pd-rich precipitates (Fig. 3) reduce the oxide thickness by about 10%.

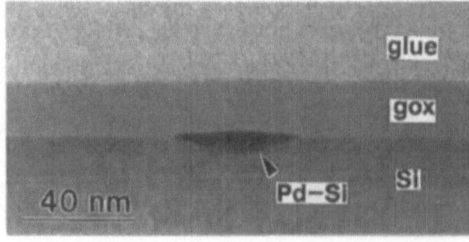

Fig.3. Pd-Si particle at the SiO_2/Si-interface reducing the oxide thickness. (TEM cross section)

Copper contamination before, during or after gate oxidation did not result in significantly different breakdown results. Only in the case of a contamination after poly-Si deposition the situation is slightly different. Even after RTA at 1200°C the silicide precipitates were not able to crack the SiO$_2$ layer because it is covered by the stiff poly-Si plate (Fig.4). An oxide thinnig of about 30% was detected.

In the case of Ni, 5-10µm long NiSi$_2$ platelets reach from the bulk to the Si/SiO$_2$ interface but thin the oxide only slightly (1nm) [5]. This effect is probably to small to be detected in the pin hole detector and should also be hardly detectible by other tests like the voltage ramp test.

Fig.4. Cu-rich silicides formed at the SiO$_2$/Si interface after an intentional contamination after poly-Si deposition. The oxide is thinned by about 30%. (TEM cross-section).

The experiments have shown that in order to attain lower gate oxide defect densities it is obviously necessary to control and reduce the metal contamination level present in the silicon semiconductor technology.

References
1. Honda, K.; Ohsawa, A.; Toyokura, N.; Appl. Phys. Lett. 46(6), (1985) 582.
2. Wendt, H.; Cerva, H.; Lehmann, V.; Pamler, W.; J. Appl. Phys. 65, (1989) 2402.
3. Cerva, H; Wendt, H; Mat. Res. Soc. Symp. Proc. 138 (1989), in press.
4. Eisenberg, P; Brion, K; Electronics 42 (1969) 45.
5. Cerva, H.; Wendt, H.; 6th Oxford Conf. Microscopy Semicond. Mater. 1989, to be published by Institute of Physics Conference Series.

A Novel Approach to Realistic Worst-case Simulations of CMOS Circuits

M. J. B. Bolt, J. Engel, M. Rocchi and A. van Steenwijk

Philips Research Laboratories,
P.O. Box 80000, 5600JA Eindhoven,
The Netherlands

Summary

To enhance the design of CMOS circuits a new method, named Gradient Analysis, has been proposed and verified. Gradient Analysis enables designers to realistically predict the standard deviation of the circuit performance from measured or guesstimated device parameter variations. It is realistic both in terms of its accuracy and in terms of the number of simulation runs required. With as few as one simulation run the approach is shown to accurately predict the variation of 64K SRAMs Read Access Time and Low Level Output Voltage. Gradient Analysis also provides designers with information on the sensitivity of the circuit performance variations to the device parameter variations.

1 INTRODUCTION

For the accurate modelling of the worst-case characteristics of a circuit a designer needs to relate the variations of the performance parameter to the variations of the device parameters and the correlations between the device parameters. In this paper a new approach to realistic worst-case CMOS simulations, named Gradient Analysis, is proposed and verified. Since, the Gradient Analysis approach uses the device parameters it can be applied by designers designing circuits in CMOS processes that are still in the development phase. This is an advantage over other proposed approaches [1,2], in which the device parameters must first be related to independent process parameters. This must be done, for example, by using regression analysis of measured data.

2 GRADIENT ANALYSIS

The Gradient Analysis approach consists in evaluating the standard deviation of the circuit performance and works on the following linear approximation. If a performance parameter P is a function of the device parameters A and B:

$$P = f(A, B) \tag{1}$$

the standard deviation σ_P of P due to the spreads of A, σ_A, and B, σ_B, can be calculated from:

$$\sigma_P = \sqrt{\left(\frac{\partial P}{\partial A}\right)^2 \sigma_A^2 + \left(\frac{\partial P}{\partial B}\right)^2 \sigma_B^2 + 2\left(\frac{\partial P}{\partial A}\right)\left(\frac{\partial P}{\partial B}\right)\rho_{AB}\sigma_A\sigma_B} \tag{2}$$

where, ρ_{AB} is the correlation coefficient between A and B.

In this simple example a designer would first calculate the gradients, $\frac{\partial P}{\partial A}$ and $\frac{\partial P}{\partial B}$ by circuit simulation. By knowing the device parameters σ_A, σ_B and ρ_{AB}, from process specifications, the designer would then be able to calculate the variation on the performance parameter P. Equation 2 can be extended to include all required parameters. For example, in reality the number of device parameters that are considered necessary to predict the variation of a CMOS circuit, i.e. N-channel and P-channel transistor parameters plus capacitance parameters, can be as many as 16. Since the calculation of the gradients is an internal option within the Philips circuit simulator PANACEA, only one simulation run using the nominal parameterset is needed to calculate all required gradients.

3   COMPARISON

The accuracy of the Gradient Analysis approach stems from the fact that the correct correlations between the device parameters are accounted for. In the traditional, and much used, approach to worst-case simulations [3] so-called FAST and SLOW parametersets are used to simulate maximum and minimum performance levels. In these parametersets the main device parameters are set to their $\pm3\sigma$ levels in such a way so as to result in maximum (FAST) and minimum (SLOW) circuit performances. In effect this approach assumes that the main device parameters are strongly correlated to each other. Since in reality this is not the case, it can result in very unrealistic simulations.

For example, figure 1 shows a histogram of the measured saturation current level from 250 N-channel transistors of channel length $1.2\mu m$ and width $20\mu m$. Superimposed on figure 1 are the predicted $\pm3\sigma$ current levels calculated using the FAST and SLOW parametersets compared with those calculated using the Gradient Analysis approach. It can be seen from this figure that the traditional approach overestimates the variation of the saturation current level while the Gradient Analysis approach results in a more accurate estimation of the $\pm3\sigma$ variation of the current level. In this example, by taking the $\pm3\sigma$ levels we have assumed a symmetrical distribution for the current.

4   RESULTS AND DISCUSSION

To verify the approach outlined the performance parameters of a 64K SRAM were extracted from a number of positions on 9 wafers from 3 different batches of a $1.2\mu m$ CMOS process. By extracting the device parameters from 29 positions on the same wafers the standard deviations and the correlations of the device parameters were obtained. Through simulations the gradients of the performance parameters to the device parameters were obtained. Substituting these gradients and the measured standard deviations from the device parameters plus the respective correlation coefficients into the Gradient Analysis formulation the standard deviation of the performance parameters were calculated.

Figures 2 and 3 compare the predicted performance parameter variations with the measured 64K SRAM data for, respectively, the Low Level Output Voltage, VOL, and the Read Access Time, τ_{acc}, performance parameters. In figure 2 the maximum and minimum levels for VOL were calculated, assuming a log-normal distribution, using the simulated mean value of VOL with the gradient analysis calculated σ value. For the clarity of figure 3 the predicted variation of the Access Time was calculated around the mean of the measured data. In figure 3 the gradient analysis predicted $\pm3\sigma$ levels are superimposed. It can be seen from figures 2 and 3 that the predicted performance parameter variations agree well with the measured variations.

312

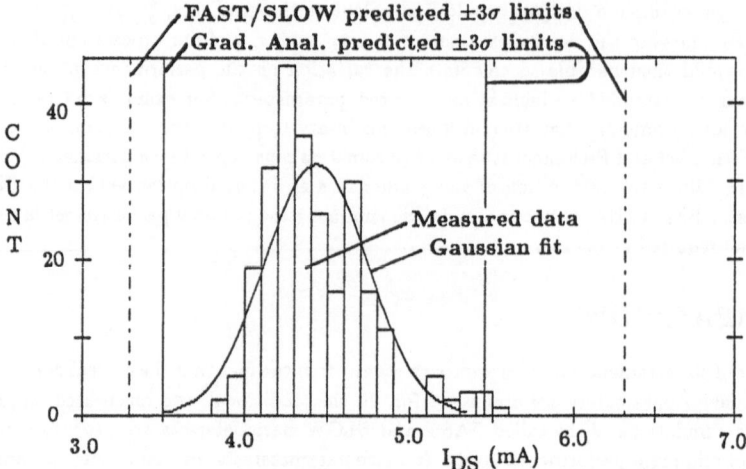

Figure 1: N-channel 20/1.2 measured saturation current levels. Superimposed on the histogram are the calculated ±3σ current levels using Gradient Analysis (solid) and the traditional FAST and SLOW parametersets (dashed).

In addition, the sensitivity information provided by Gradient Analysis indicates that the variation in τ_{acc} is caused nearly exclusively by the variation in the effective channel length of the transistors, LTOL. This is confirmed by seeing how the bimodality in the histogram for LTOL, shown in the inset of figure 3, result in bimodality in the measured data for τ_{acc}. This sensitivity information provides a useful way for designers to feedback to process engineers.

5   CONCLUSIONS

The Gradient Analysis approach is a method which provides IC designers with a means to accurately predict circuit performance variations from process specifications. It has been shown to result in an accurate prediction of 64K CMOS SRAM circuit performance variations. The approach also enables the main causes of the circuit performance variations to be pinpointed , i.e. which device parameter variations have the most influence on the performance parameter variations.

References

1. Yang P. & Chatterjee P., 1982 IEDM Technical Digest, p286.

2. Visvanathan V., 1986 ICCAD Technical Digest, p228.

3. Nassif S.R., Strojwas A.J. & Director S.W., IEEE Trans. CAD, vol.5, p104, 1986.

Acknowledgements

The authors would like to acknowledge P.Hens, B.Jongepier, J.Joosten and H.Tuinhout for their contributions to the work that has been described in this report.

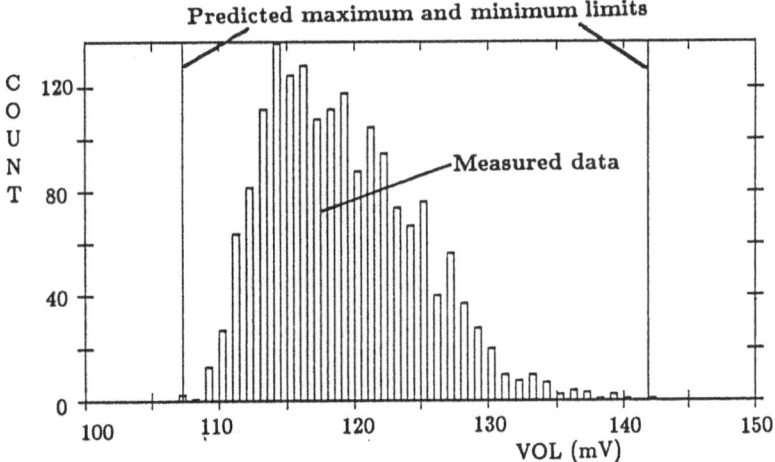

Figure 2: Measured Low Level Output Voltage, VOL, from the 64K SRAMs. Superimposed on the histogram are the calculated maximum and minimum levels using Gradient Analysis and assuming a log-normal distribution.

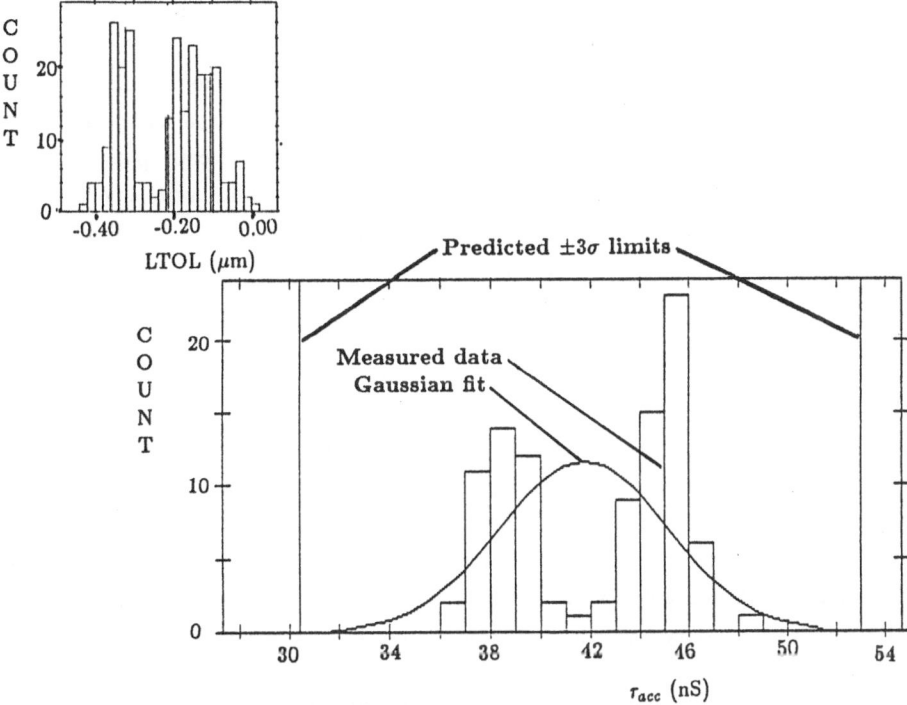

Figure 3: Measured Read Access Time, τ_{acc}, from the 64K SRAMs. Superimposed on the histogram are the calculated $\pm 3\sigma$ levels from the mean of the measured data using Gradient Analysis (solid) and the best gaussian fit to the measured data. The variation of the effective gate length, LTOL, measured on the same wafers is given in the inset.

Self-consistent Modelling of npn and pnp Transistors

Ann Kelleher, Denis J. F. Doyle, Colin Lyden, William A. Lane, Susan P. Edwards

National Microelectronics Research Centre,

University College, Cork, Ireland.

Abstract This paper examines recently published physical models for carrier transport, which distinguish between minority and majority carrier mobilities. The models have been implemented in a bipolar device simulator and comparisons have been made between predicted and measured results for both npn and pnp transistors. Discrepancies between measured and simulated device currents are discussed and further experiments to isolate sources of disagreement are outlined.

INTRODUCTION

The accuracy of numerical device simulators in predicting bipolar transistor currents depends on good physical models for carrier transport. In a bipolar transistor, holes and electrons contribute to device currents, both as majority and minority current carriers. Thus the set of physical models for mobilities, lifetimes, and bandgap narrowing for both holes and electrons must accurately represent their properties. Recently published models [1,2] indicate a significant increase in minority carrier mobility compared to majority carrier mobility in heavily doped silicon. Differences in lifetimes and bandgap narrowing compared to existing models were also reported. The set of models reported in [1] and [2] were derived in a self-consistent fashion and as a result are valid only when used together.

EXPERIMENTAL PROCEDURE

The models of Del Alamo et al [1] and Swirhun et al [2] were implemented into BIPOLE [3]. Separate triple diffused processes were used to fabricate npn and pnp transistors. Doping profiles for the devices were obtained from SIMS and SRP. The doping profile of the npn is shown in fig. 1. The fabricated transistors were measured over a well-controlled range of temperatures. The transistor characteristics were simulated using the Del Alamo and Swirhun models and also using the models recommended in BIPOLE. The measured and simulated results were then compared.

NPN RESULTS AND DISCUSSION

Fig. 2 shows measured and simulated Gummel plots for the npn transistor at 22°C. The models used in the simulation were those recommended by BIPOLE. These are the mobility models of Arora et al [4], the bandgap narrowing of Slotboom [5] and fitted lifetimes [6] which are referred to, collectively, as the BIPOLE models. Clearly there is good agreement for collector and base currents, confirming that this model set produces good results at 22°C. In fig. 3 the simulated currents using the models of Del Alamo et al and Swirhun et al are compared with the measured results. Again good agreement is obtained for the collector current. However, the base current is under-estimated by a factor of 3, giving an increase in the simulated gain.

The decrease in base current can be understood if we examine the models. Bandgap narrowing, from Del Alamo et al, at the peak doping level in the n+ emitter is less than that predicted by Slotboom's formula by 30 meV. This reduces the base current by a factor of 3.5. This decrease is offset by a triple increase of the minority carrier hole mobility in the emitter, which increases base current correspondingly. However the longer lifetimes described by Del Alamo et al bring about a net reduction in the base current. This is shown in fig. 4, where the lifetimes predicted by Del Alamo et al are compared to BIPOLE model lifetimes. At the emitter peak doping level there is a factor of 8 difference between these lifetimes.

In an effort to determine if one, or all, of the models are incorrect, we measured the fabricated npn transistor characteristics over a temperature range. Fig. 5 shows the measured gain over temperature and compares it to the simulated gain using the BIPOLE models. Extracting the bandgap narrowing from the simulated results indicate that the temperature dependence should be dominated by 120 meV of bandgap narrowing from Slotboom's equation. However, the measured gain showed less temperature dependance, thus producing a lower value of extracted bandgap narrowing, suggesting that the reduced bandgap formula of Del Alamo is more correct. This would suggest that either the minority hole mobility in the emitter is too small or the lifetimes are too long. There could be a difference in lifetimes between the samples used by Del Alamo and our samples, as his material was epitaxially grown [1] whereas the npn emitter was implanted. Another possible difference is that at Del Alamo's maximum doping level there is a large difference in the measured diffusion length. Therefore the curves fitted to his data may not extend to higher doping levels.

PNP RESULTS AND DISCUSSION

Fig. 6 shows the measured and simulated Gummel plots for the pnp transistor. The simulated currents in this plot use the BIPOLE models. Here, as with the npn, there is good agreement between the simulated and measured base currents at 22°C. There is some disagreement between collector currents. Fig. 7 shows the measured and simulated Gummel plots using the models of Swirhun et al and Del Alamo et al. In this case the collector currents are in better agreement. This is brought about by a slight increase in minority hole mobility in the n-type base which occurs even at the base doping level of $1e17\,cm^{-3}$.

There is a more interesting difference between measured base currents and the simulated base currents of Del Alamo et al and Swirhun et al. The base current depends on the heavily doped p+ emitter. This time, these models predict a higher base current than the measured value. According to these models this pnp transistor, with an emitter depth of 0.9μm and a doping level of 2.5E19 cm^{-3}, has a "short" emitter. This removes the lifetime dependancy of the base current. Since the bandgap narrowing formula is that of Slotboom in p+ material, the increase in base current is due to the increase in minority electron mobility. The conclusion is that the minority electron mobility of Swirhun et al is too large.

CONCLUSION

The carrier transport models of Arora et al and Slotboom used in BIPOLE predict measured currents for npn and pnp transistors at 22°C. However they do not agree with measured values over a range of temperatures. The set of models which include minority carrier mobility models do not predict base currents of our npn and pnp transistors. Our results suggest that this may be due to the minority hole mobility being too low, the lifetimes being too long in heavily doped n+ silicon and the minority electron mobility being too large in heavily doped p+ silicon.

To distinguish between mobility and lifetimes in the heavily doped n+ silicon we propose to fabricate shallower emitter NPN transistors to remove the lifetime dependancy. Also the emitter doping level will be reduced to the doping range used by Del Alamo et al.

[1] J. Del Alamo, R. M. Swanson, IEEE Trans. Electron Devices, ED-34, July 1987
[2] S. E. Swirhun et al. IEDM Tech. Digest, pp24, 1986
[3] BIPOLE User Manual, Waterloo Engineering Software
[4] N.D. Arora et al. IEEE Trans. Electron Devices, ED-29, pp292-295, Feb 1982
[5] J.W. Slotboom Solid State Electronics, 20, pp279-283, 1977
[6] D.J. Roulston et al. IEEE Trans. Electron Devices, ED-29, pp284-291, Feb 1982

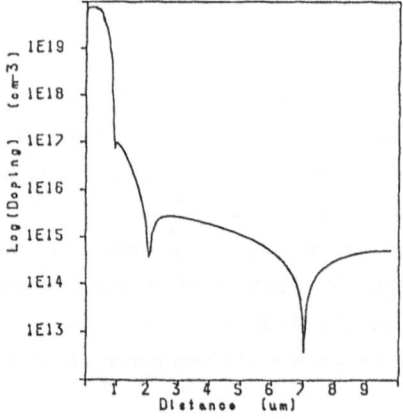

Fig. 1 NPN Doping Profile

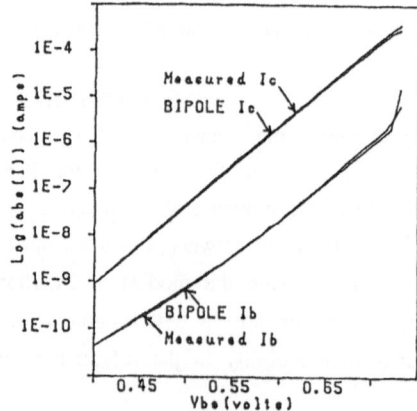

Fig. 2 Measured and predicted NPN current using BIPOLE recommended models.

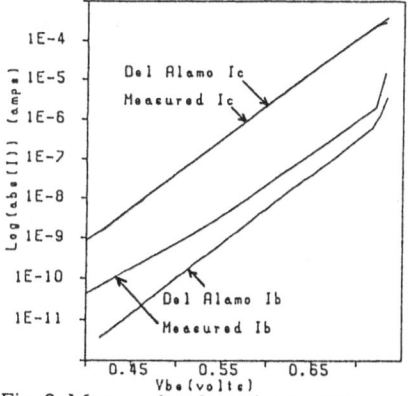

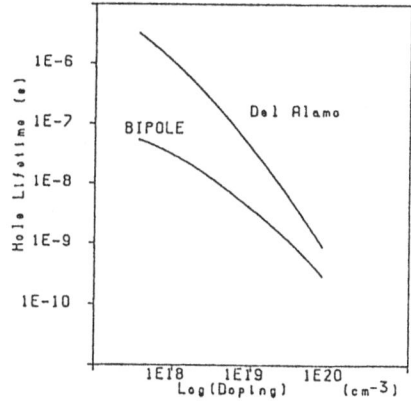

Fig. 3 Measured and predicted NPN currents using recently published models [1],[2].

Fig. 4 BIPOLE and Del Alamo hole lifetimes.

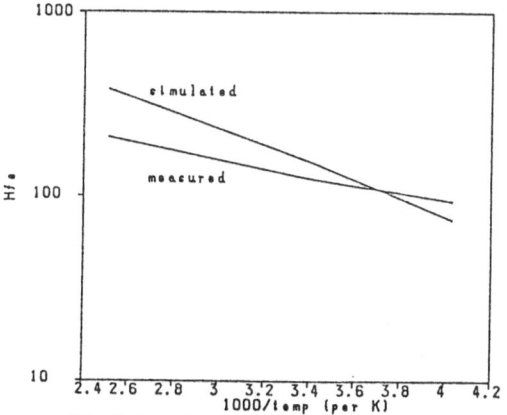

Fig. 5 Measured and BIPOLE predicted temperature dependence of gain

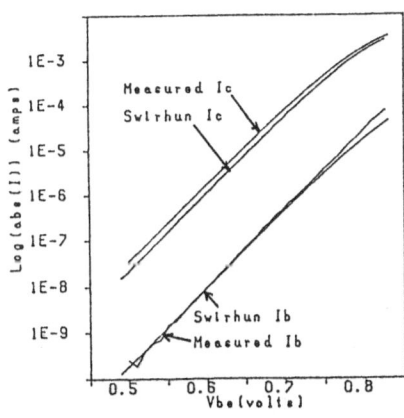

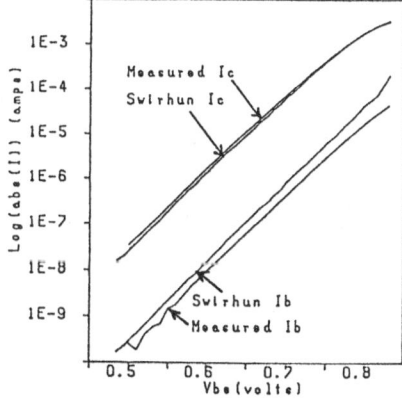

Fig. 6 Measured and predicted PNP currents using BIPOLE recommended models.

Fig. 7 Measured and predicted PNP currents using recently published models [1],[2].

Session 4

Invited Papers

Session 4

Technological Challenge of Artificial Neural Networks

Karl Goser

Bauelemente der Elektrotechnik
Universität Dortmund, FRG

Abstract

Artificial neural networks process data or information with new qualities, and offer new possibilities of developing IC processes and circuits. For the use of the full advantage of parallel processing in these networks we have to implement them into silicon. VLSI chips are shown for the digital technique, several concepts for analog versions are discussed. A short outlook about new technologies based on silicon, such as cooled circuits, optoelectronics, and molecular electronics closes the overview.

1. Introduction

From mathematics we are familiar with the fields of algebra and geometry. We are used to solve our problems by using algebraic methods on computers. The algorithm opens the solution to a problem. There are many problems, however, which belong to geometry, as pattern recognition or classification. In this case a computer which is different from the v.Neumann principle may be much more successful: this is the chance of neurocomputers and artificial neural networks. They show the interesting features of association, adaptation, optimization, self-organization, and fault-tolerance /1,2,3/.

2. The roots of artificial neural networks

In the fifties the neurologist Hebb and others have developed simple models of a neuron (Fig.1a): They characterised it by the synapses with the inhibitory and excitatory inputs, and a threshold stage (Fig.1b). We have to emphasize, however, that this equivalent circuit is only a very rough approximation. If we draw this circuitry in a symmetrical way, we get a matrix with every column representing one neuron, the crossovers with the connection elements are the synapses and the ends of the columns represent the threshold stages (Fig.1c).

Microelectronics uses the matrix as a fundamental structure of a neural network. This architecture distributes the data processing amongst the data storage so we get a low data movement and a high parallel processing. The input signals are weighted by the strength of the connection elements: The product of an input signal and an adaptive weight represents the output of a

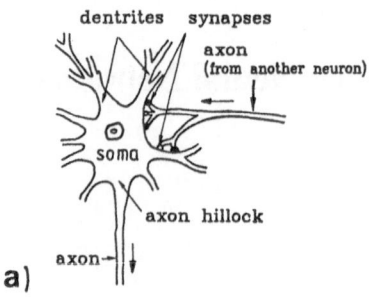

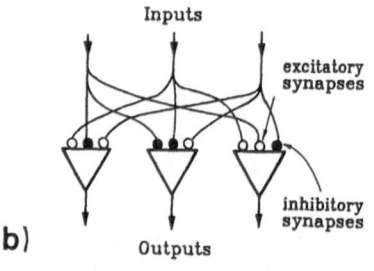

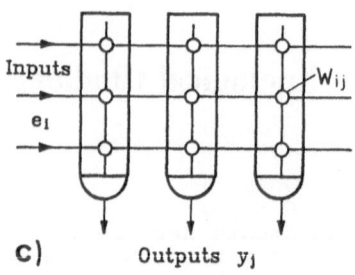

Weighted Sum of the Inputs:

$$S_j = \sum_{i=1}^{n+m} e_i \times W_{ij}$$

Transfer Function:

$$y_j = g_j(S_j)$$

Adaptation Function:

d) $\Delta W_{ij} = f(W_{ij}, e_i, y_j)$

Fig.1 The model of a neuron (a) from a biological neural
network can be a threshold stage (b) the inputs of which can
either excitatory or inhibitory. The threshold circuit (b) can
be redesigned as a matrix (c). The equations in (d) show one
of the algorithms of an artificial neural network.

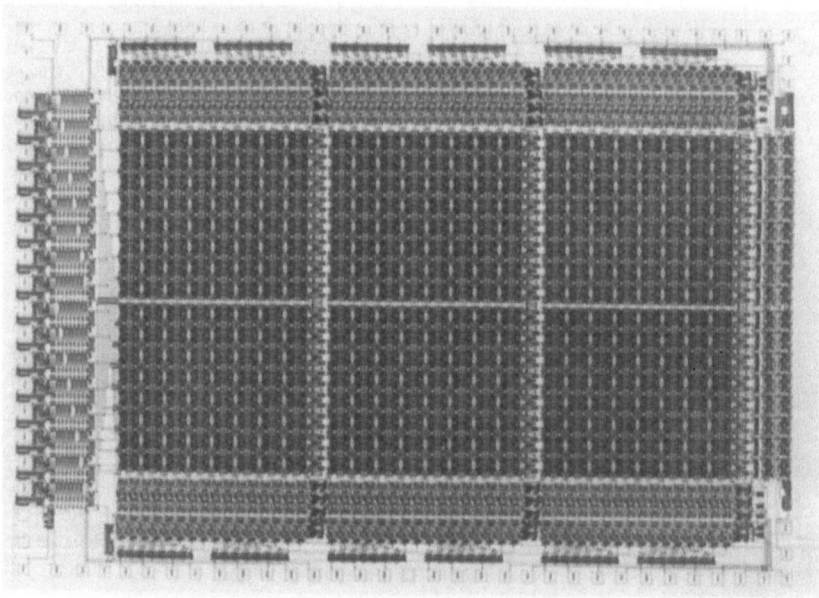

Fig.2 Microphotograph of a neural network matrix working
digitally (chip area 35 mm^2, 20,000 transistors, (96 x 16)
matrix, produced at the University of Dortmund).

synapse (Fig.1d). The sum of these output signals along the column is the input for the transfer function represented as a linear, sigmoid, or threshold function. This function yields the output of the neural network. The network should response quickly to an input information; this operation is the recall phase. During a longer time scale the adaption or learning phase takes place. This operation slowly alters the adaptive weights of the connection elements in accordance to an adaptation rule. The limitations of technology, however, restrict the adaptation rules to neighbouring neurons.

The theory of artificial neural networks is not fully developed. At the moment in physics they extend the theory of thermodynamics to such networks and in computer science they try to solve the open problems in the information space. It can not be expected that a complete theory is available in the next few years.

Neural networks can be subdivided into several classes: learning can be supervised or non-supervised; this means that the networks learn with or without a teacher or critique from outside. A further classification arises from the architecture: There are simple matrices, matrices in several layers, matrices with feedback, and processor arrays with some coupling between the neighbouring units. The last one shows the feature of self-organisation which can be applied to support the processing and the design of integrated devices and circuits /4/.

3. Silicon Technology

3.1 The basic problems of integration

The theory of neural networks reflects the highly parallel, regular and modular architecture of neural networks making them attractive for VLSI systems (Fig.1c). Such VLSI systems arise the following problems when being integrated:

- The processing units must be compact in order to accomodate many of them on a chip.
- The memory cells should store the adaptive weights as analog values which should be variable.
- The technology should offer solutions to the problem of the complex connectivity between the units.
- The learning algorithm must take into account the possibilities and restrictions of the technology, and should be integrated functionally in one device if possible.

In the following we discuss some basic integrated circuits for solving the above mentioned problems, as analog versus digital, connectivity, and storing adaptive weights.

3.2 Analog versus digital circuitry

Digital concepts are very important since they use the mainstream of technology. In addition they surprise by a quite good performance that is sufficient for many applications. One

324

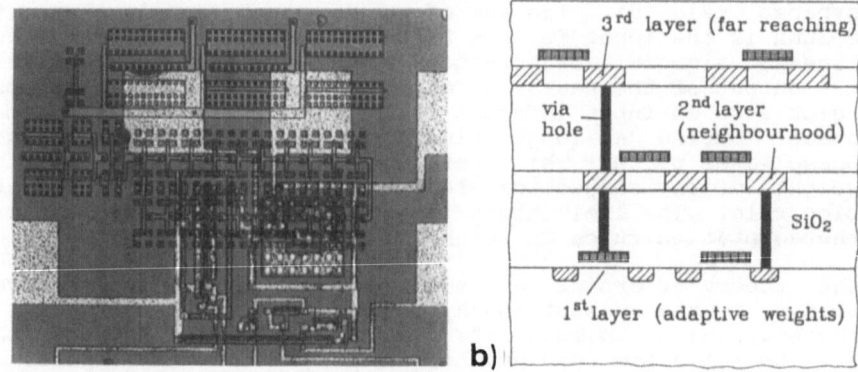

Fig.3 SOI technology for 3D integrated neural networks
a) Microphotograph of an SOI circuit /4/ (produced at the University of Dortmund).
b) Three layers of circuits for neural networks

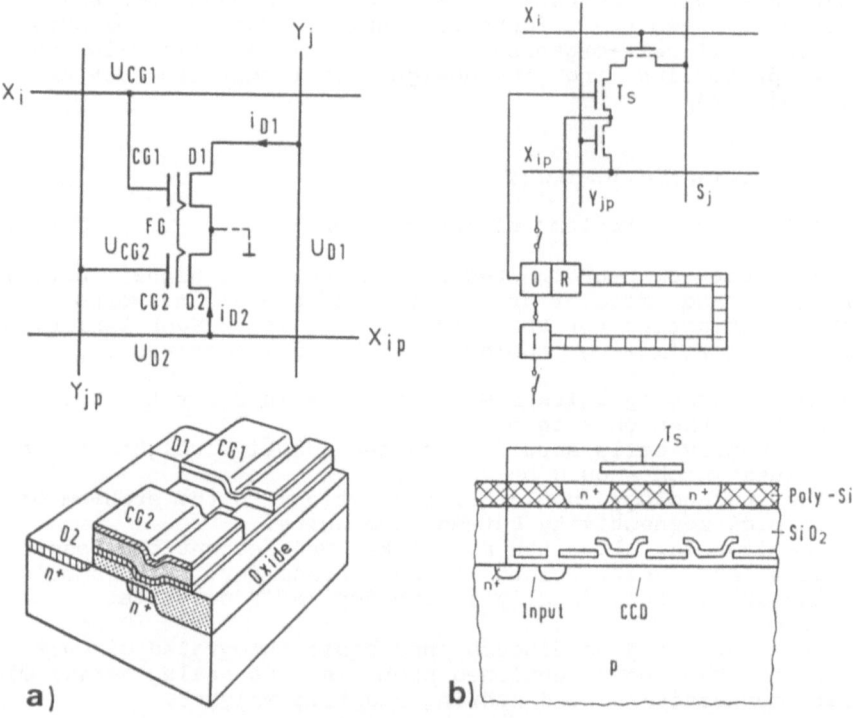

Fig.4 Integrated cells for storing adaptive weights in an analog way: a) A functionally integrated cell with a floating-gate transistor, and
b) a cell in SOI-technique with a CCD loop.

of several concepts is the simple associative matrix where only zero's and one's are used for weights /4/. We have designed and integrated the main building blocks for such a concept (Fig.2). The advantage of the concept is its modular structure. So we can achieve a quite large fully interconnected neural network of several hundred neurons by mounting only a few chips on a printed circuit board.

In our opinion, however, the analog version is more attractive because of size, power and speed /3/. The accuracy of analog circuits is not as high as for digital circuits, but more appropriate for highly parallel signal transfer operations. Therefore up to now almost all VLSI systems for neural networks make use of analog computation.

3.3 Connectivity

The usual way to solve the connectivity problem in electronics is the use of an additional grade of freedom, time or additional wiring layers. In the following, however, we will concentrate on technology to improve the data exchange in the networks. Besides using multilayers the three-dimensional integration offers interesting aspects for solving the problem of connectivity. One of the most interesting 3D techniques is the silicon on insulator (SOI-) technology which will play an important role for intelligent sensors in the future.

Such a concept is highly feasible, since the research on 3D integration is going on intensively. In our laboratory we have developed a method to recrystalline a polysilicon film by use of laser beam annealing (Fig.3a) /4/. The annealing temperature is high enough for integrating CMOS circuits into the silicon film and, on the other hand, it is low enough not to affect the integrated circuits in the bulk underneath.

For neural networks we should provide at least three circuit layers (Fig. 3b): The bulk at the bottom takes the circuitry for storing the adaptive weights. The circuits for data exchange between neurons and neighboured neurons are integrated in the middle layer, and the global interconnection network can be implemented on the top layer.

3.4 Adaptive weights

The most important basic circuits certainly are the elements for storing the adaptive weights. Therefore we propose two concepts in the following: a pure analog and a hybrid digital/analog concept.

The floating-gate transistor offers a good way for a functional integration of a storing element as already proposed in earlier publications /4/. Today the electrically erasable and programmable read only memories (EEPROM's) make use of this type of transistor. Based on it we developed a new storing element with dual drain and dual control gate, as depicted in Fig.4a /4/. Tunneling of electrons into the floating gate shifts the threshold voltage. The charge on the floating gate represents the adaptive weight. The device acts as a non-volatile storage cell where the electrical charge on the floating gate represents the information, in our case the adaptive weight. Since the charge can be varied continuously, the storage cell works analog. It holds the stored information independently of the power supply.

At the present state of the art some degradation effect due to charge trapping within the thin oxide limits the satisfactory longtime operation of this device. After around a million of modification cycles only a small threshold voltage shift is left so the cell becomes inoperable in a matrix. The technological results tend toward lower degradation by using better insulation layers. On the other hand today's cell characteristics are sufficient for learning the network once.

Charge coupled devices (CCD) or dynamic shift registers as storing elements overcome the disadvantages of low accuracy and longtime degradation of the floating gate transistor but need a larger chip area. The build up of such an element is more complex since the element requires a three dimensional integration with at least two layers of silicon (Fig.4b). In the first layer, in substrate, there is a CCD loop together with a regenerator and an input/output stage as well as the clock lines for the CCD. The number of CCD's in the loop yields the accuracy of the connection weight since the number of charge packets corresponds to the weight value, and only the area limits it. The cycle time for the CCD's depends on the leakage currents and can be relatively long in standby operation. In the second layer three MOS transistors represent the connection element itself, since one of the gate capacitances collects the charge packets of the CCD loop. The higher the electrical charge the lower the resistance will be or the higher the current through this transistor. The power supply lines and the input lines also are on top of this layer as suggested in Fig.3b.

A further advantage of this element is that the information stored into the loops can be read out by opening the loops and by switching all loops in series. In this way the output of the chip receives the charge packet strings of each storing element. An external memory, e.g. a magnetic disc, can store these data. Vice versa the adaptive weights can be written into the loops from outside; so the time consuming process of learning can be omitted.

3.5 The potential of sub-/μm technologies

In future the integration technology will probably proceed as quickly as in the past. In Europe the JESSI project will cause a progress from 1 Mega-bit memory chips to 64 Mega-bit chips, even 256 Mega bit chips are feasible /1/. These pilot products will use sub-micron-technologies down to 0.3 μm.

It is also a challenge for neural networks vice versa respectively. Today, we can realize 10 to 100 neurons on a chip. For complex digital concepts as the Boltzmann machine we can only expect one neuron on a chip; with a sub-μm-technology we have the possibilities to design chips up to 1,000 neurons. At the end of the JESSI program we can expect chips with about 10,000 neurons in analog technique if one weight can be stored in a single device; but this goal does not only require an adequate technology but also powerful design tools and test equipments. Even at such high-sophisticated technologies the number of neurons is quite small in comparison with that in biological systems. Therefore it will be a challenge to develop integration technologies beyond the scope of the JESSI program or to apply special technologies.

5. Aspects of future technologies

In the following we discuss some technologies which are all based on silicon and which will be available for other important applications in microelectronics in any case.

5.1 Low temperature circuits on silicon

The simplest way to high performance neural networks seems to be the cooling of the CMOS circuits. The results are not only a higher performance of digital logic but also a longer time constant for discharging a MOS capacitor. The leakage rate of a MOS capacitor is reduced by three orders of magnitude by cooling the circuits to -50°C and by five orders of magnitude by cooling to -100°C. Thus low temperature may solve the problem of storing the analog weights.

Moreover low temperatures offer the opportunity to introduce superconductors as lossless transmission lines so high data rates are practicable. The wiring with high temperature superconductors will be an important goal for future VLSI chips. In addition, it may be possible that superconductor rings store adaptive weights as a magnetic flux. Since a Josephson junction allows to change the flux step by step, the cell operates in an analog way. The summation of the signals in a neuron is the sum of the currents of all magnetic fluxes, the read operation can be reversible and lossless if special circuits are used.

5.2 Optoelectronics on silicon

Since silicon is a relatively cheap substrate, the research efforts have the goal to integrate light wave guides, photo diodes and light emitting sources into silicon. The first two components can be realized without difficulties as new results show, the last one is still open. On the one hand the molecular beam epitaxy may alter the crystal structure of silicon so light emitting devices are just ahead, on the other hand there are light emitting effects in silicon devices which we can probably develop further.

5.3 Molecular electronics on silicon

Molecular electronics is a vision which promises to solve all technological problems of neural nets /4/. Self-building and self-organizing in three dimensions offer the outlook to huge neural networks with high performance. In reality, however,

328

the biological molecular electronics will be very slow, and
the physical molecular electronics is just starting with first
test structures.

Delta films, that is a few atomic layers on silicon deposited
by molecular beam epitaxy, make possible new devices. Our
experiments show that capacitors with such films have a capacitance which is one magnitude higher than that of usual MOS
capacitors. The appearance of new devices as vertical switches
or tunnel switches impacts neural nets, since these devices
open new ways for functional integration. In addition the
quantum effects linked to such thin films are quite
interesting as regards storing adaptive weights.

6. Conclusion

The main stream of integration technology concentrates on
digital memories. This technology is not well suited for the
implementation of neural networks but offers a good opportunity. Since a broad application of microelectronics needs a
diversity of technology in any case, it is worthwhile to look
if such special technologies offer advantages for the integration of neural networks. Indeed, such advantages exist, as
shown above, and we should adapt the architecture and the
algorithms of neural networks to the facilities of these
technologies.

Acknowledgements

The author would like to thank his coworkers U. Rückert,
I. Kreuzer, K.M. Marks, U. Hilleringmann, A. Sönnecken and
V. Tryba for the cooperation, and he is grateful to the DFG
(Deutsche Forschungsgemeinschaft) and to the EC (project BRAIN
and ESPRIT) for the financial support.

References

1. Goser, K.: Mikroelektronik neuronaler Netze. Mikroelektronik 3 (1989), pp. 104-108.
2. Shriver, B. (Editor): Artificial neural systems. IEEE Computer 21 (1988), pp.8 - 117.
3. Mead, C.: Analog VLSI and neural systems. Addison-Wesley,
   Reading, Mas.,(1989).
4. Goser, K.; Hilleringmann, U.; Rückert, U.; Schumacher, K.:
   VLSI technologies for artificial neural networks. IEEE
   Micro 9 (Dec. 1989), in press.

Progress in Optoelectronic ICs

D. DECOSTER, J.P. VILCOT

Centre Hyperfréquences et Semiconducteurs
Université des Sciences et Techniques de Lille-Flandres-Artois
UA CNRS 287, Cité Scientifique
F-59655 VILLENEUVE D'ASCQ CEDEX

SUMMARY:

We present recent progress and discuss future developments of optoelectronic integrated circuits (OEIC's). Basic technologies (crystal growth, required components, IC structure and processing, packaging techniques) are reviewed. Examples of OEIC's (receivers, transmitters, integration of optical and optoelectronic functions) are given with their respective performance. Future trends are then proposed to stimulate the mass production of OEIC's.

Introduction

Owing to the advantages of an integrated optic system such as unsensitivity to electromagnetic interference, increased bandwith, small size, light weight and low loss signal transmission, optical data transmission and processing are becoming more and more important [1]. Optoelectronic devices such as lasers and detectors made on III-V materials are the key of the development of these optical systems, the speed has reach several gigabits and even the microwave range. But interconnection with electronic devices and circuits would degrade the performance of the systems because of the introduction of parasitic elements such as stray capacitances and conductances. Optoelectronic integrated circuits (OEIC's) would overcome these problems; more generally, OEIC's have the potential following advantages [2]:

- the speed and noise performance of optoelectronic devices could be improved significantly by the reduction of parasitic elements,

- the incorporation of complicated electronic signal processing circuits could expand the function of optoelectronic devices,

- the monolithic integration could simplify the component and system assembly, improving the compactness and reliability at much less cost.

The applicability of OEIC's to optical communication systems is clear either for digital or analogic (even microwave) signals. But other applications are also possible: they are expected to be useful in a variety of optical data processing systems. For example, they would be very attractive for the fabrication of optical interconnection between boards and chips, and even within the chip [2,3].

The technology of optoelectronic integration seems to be the basis of the future development of advanced optical communication and processing systems, as in the recent past, the marvelous developments of the electronic field is a consequence of the key technology of the monolithic integration of silicon devices. From an economic point of view, local area networks and integrated services digital networks over longer areas would create a vast demand for converting (optoelectronic) and processing chips. This demand is likely to create a $5 billion a year market within five years, according to the Japanese Optoelectronic Industry and Technology Development Association, Tokyo [4].

Because of all these aspects, many laboratories are involved in OEIC's research; among the most important, leaders include: the California Institute of Technology in Pasadena, the Massassuchets Institute of Technology as well as companies such as ATT, Honeywell, IBM and Rockwell in the United States, Fujitsu, Mitsubishi, Hitachi, NEC, NTT and Toshiba in Japan. In Europe most active research groups are from CGE (France), CNET (France), CSELT (Italy), GEC Marconi (U.K.), LEP (France), STC (U.K.), Thomson-CSF (France), HHI (Germany) and SEC (Germany) inside European Projects (ESPRIT and RACE).

The purpose of this paper is to present recent progress and to discuss future development of OEIC's.

I- Basic technologies

I.1) Material growth

Depending on the wavelength to be used, OEIC's are fabricated on GaAs or InP based III-V compounds. GaAs based optoelectronic devices are suitable for near infrared range applications ($\lambda \approx 0.8$ μm), and InP based components emit and receive in the infrared range (1.1 μm $< \lambda < 1.6$ μm).

Materials for IC's can be grown in a number of ways [5]: liquid phase epitaxy (LPE), vapor phase epitaxy (VPE), metal-organic chemical vapor deposition (MOCVD) or molecular beam epitaxy (MBE); alternative techniques have been proposed for the growth of InP based materials with MBE: gas source MBE (GSMBE), metal-organic MBE (MOMBE), ... Briefly, LPE is a mature technique that yields high quality epitaxial layers; however its application is limited to wafers of less than 1 square centimeter. VPE can produce very pure semiconductor layers; with this technique, however, the thickness of very thin films is difficult to control. On the other hand, with MOCVD and MBE, the thickness is easy to control and very high quality epitaxies have been obtained [6]. Such techniques allow us to fabricate components (lasers, transistors) for which electrons or light is confined into small dimensions providing very efficiency light to electricity conversion (lasers) or high speed (transistors) [7], showing their effectiveness in growing sophisticated heterojunction structures including superlattices with uniformity on a large diameter wafer with wafer to wafer reproductibility [8].

I.2) Required components

Thanks to the last improvements of growth techniques, high performance laser diodes [9,10], photodetectors [11-18] and transistors have been achieved. As examples, ultra low threshold, high bandwith and very low noise operation of 1.52 µm distributed feedback (DFB) laser diodes grown by MOCVD [19], 1.5 µm DFB laser source grown by GSMBE with low threshold current [20] and dual balanced PIN photodiodes with low leakage current and microwave frequency cut-off for 1.3-1.55µm wavelengths operations (grown by VPE) [21] have been recently fabricated; also Schottky photodiodes on GaAs material which exhibit 110GHz bandwidth [18] have been reported. Concerning the transistor fabrication, the most developped technology is on GaAs. For such component, high transconductance (800mS/mm) and high cut-off frequency (80 GHz) has been obtained [22]. A large variety of field effect transistor have been studied (high electron mobility transistor (HEMT) [23-25], pseudomorphic HEMT,...), generally on GaAs based materials for which it is easy to fabricate a Schottky contact; but some attempts have been made on InP based materials using generally solutions without Schottky contacts (MISFET, JFET,...) [26-28]. These components exhibit also high g_m and f_c, but this technology is not so mature than GaAs based one. Different transistor structures such as heterostructure bipolar transistor (HBT) have also been investigated, due to their high speed and high drive current capability.

A special mention has also to be made for optical devices; waveguides, modulators, multiplexers   since such components will be of great interest on future OEIC's. Very low loss optical waveguide have been obtained, either on GaAs (0.2 dB/cm) or InP based materials. Various modulators using electrooptic effect on III-V materials have been proposed [29], but because electrooptic effect is very low, and even if microwave operation has been experimentally demonstrated [30], up to now such type of modulator suffers from a high command power. Solutions based on multiquantum well structures, which are more easy to integrate with laser sources have been proposed [31].

I.3) Integrated circuits structure and processing

Considering an integrated circuit, the main problem is the compatibility between the epitaxial layers and technologies suitable for each component we want to integrate on the same substrate. It is clear that the epilayers suitable for a FET, for example, are not optimized for a PIN photodiode or a laser diode and vice versa. Such a problem increases the number of technological steps (selective etching, multiple or several growth,...) and, then, the difficulties for fabrication. Consequently, the efficiency of production and the performance of the integrated components are reduced. However, many efforts have been made, either for emitter (integration laser - driver) or for receivers (integration photodetector - preamplifier) and even for the integration of optoelectronic component with optical waveguides.

At last, if microelectronic is quiet matured on GaAs based materials, it is yet in its infancy on InP based materials. Such a problem increases the difficulties of fabrication for $1.3\mu m$-$1.55\mu m$ wavelengths OEIC's. Various solutions have been proposed (JFET, AlInAs/GaInAs heterojunctions, mismatch materials, heterojunction bipolar transistors, but up to now, no best way can be defined for this purpose.

I.4) Packaging technique

Packaging technique must follow the fabrication of OEIC's. The objective, here, is double: high coupling efficiency between light and optoelectronic device and high speed or microwave signal transport. Up to now, the conventional techniques used for discrete optoelectronic and microwave devices have been confirmed and applied. But they should be improved for applications where optical and/or electrical multichannel integration is involved, and special efforts have to be made in this way. The design of the OEIC, including pigtailed fibers and high performance connectors, is very important for the massive use of OEIC's.

II-Examples of OEIC's

II.1) Photoreceivers

The first and easier device to make is a GaAs based material photoreceiver because this material is ideal for Schottky contact. The compatibility with microelectronic is obvious, since GaAs is the MESFET and microwave monolithic integrated circuit (MMIC) material. Several OEIC's have been proposed, generally to benefit from the planar structure, associating photodetectors and MESFET preamplifiers. Photodetectors are MSM photodiodes [32,33], Schottky photodiodes [34,35], photoconductive detectors [36] or embedded PIN photodiodes [37]. Mesa structures have also been proposed, associating MESFET with PIN [38] or Schottky photodiode [39], but generally the photolithographic process suffers from the thickness of the optoelectronic device, up to several microns. For photoreceivers, it is well known, from Personick theoretical development [40], that at high bit rate the improvement of the minimum detectable power of a PIN FET is strongly dependent on the reduction of C_T^2/g_m where g_m is the transconductance of the FET and C_T is the total capacitance.

First attempts have been made using a submicron gate length MESFET and an interdigitated Schottky barrier photodiode [41]. Speed up to 5.2GHz have been recorded with this OEIC made on MBE epilayers, showing the advantages of the planar MMIC structure. It has also been experimentally shown [42] that the use of GaAlAs/GaAs HEMT grown by MBE instead of classical GaAs MESFET can improve the performance of photoreceiver in high bit rate applications. Monolithic integrated photoreceiver implemented with GaAs/GaAlAs hetero- junction bipolar phototransistor and transistors has also been proposed [43].

The technological problem is more difficult for long wavelength (1.3-1.6μm) photoreceivers, because MESFET cannot readily be fabricated on InP based materials. However, in view of the smaller transmission loss of optical fibers, OEIC's operating in the 1.3 to 1.6μm range would be more useful in fiber-optic communication systems. Compared to GaAs based photoreceivers, the increase of difficulty arises from the fabrication of transistors. In must cases, JFET grown by MOCVD are used for the implementation of the preamplifier [44-48] but sometimes, MESFET fabricated on AlInAs/GaInAs [49] or GaAs/InP heteroepitaxies [50] are used (a more complete discussion will be devoted to the use of mismatch heteroepitaxies in a next paragraph). InP MISFET's have also been used for the fabrication of the preamplifier [51,52]. Mesa structures are the most common but some attempts have been made to fabricate planar structures [49]. Integrated photodetector-amplifier circuit implemented with InP/GaInAs heterojunction bipolar transistors has also been fabricated [53]. We have also to mention the efforts made on coherent optical receivers. Recent twin PIN photodiodes show very low leakage current, high quantum efficiency and high frequency cut-off [21]. These last results are very promising for the development of such receivers.

II.2) Emitters

For emitters too, GaAs technology is more developped than the InP one. As an example, a transmitter OEIC designed for multichannel application has been fabricated on MBE GaAs epilayers [54]. LPE and MOVPE techniques have also been used leading to large scale integrated or high speed devices. For $1.3-1.6\mu m$ wavelength operation, the same problem occurs that for the photoreceivers fabrication. JFET or HBT has been monolithically integrated with laser diode on LPE grown epilayers; a 0.4 GBit/s speed limit has been obtained [55-57]. More recently, integration with MESFET, using GaAs on InP heteroepitaxies grown by LPE and MBE have been proposed, with a high speed limit (2.4GHz) and also MISFET on MOVPE epilayers [4]. Integration of 112 lasers in which the laser beams are emitted from the chip surface must also be mentioned [58].

II.3) Integration of optoelectronic and optical devices

The progress in the epitaxial growth techniques and process technologies make possible the integration of optoelectronic and optical devices. As an example, a first development is the integration of a photodetector and an optical waveguide. Several attempts have been made for such an association, mostly on InP substrate [60-63] but also on GaAs semi-insulating substrate. Recently GaAs MSM photodiode which absorbing layer is short (100 μm) and thin (0.2 μm), integrated an inverted rib waveguide has been reported [59].Using GaInAs/GaAs/AlGaAs heteroepitaxies grown on GaAs semi-insulating substrate by LPMOCVD, capability of short photoconductive detectors, to detect 90% of light with only $100\mu m$ have been theoreticaly and experimentally demonstrated [64-66].

Concerning the emitter, integration of laser and waveguide [67-69], even laser and modulator [70,71] has begun to appear for $0.8-1.3-1.5\mu m$ wavelength operations. The study of etched-cavity facet laser fabricated by dry-etching techniques (reactive ion etching) plays an important role for such integrated circuits and may contribute greatly to the scaling up of OEIC integration.

III-<u>Future trends</u>

Because of the technological difficulties inherent in their actual fabrication, OEIC's suffer from their cost which has to be reduced by a mass production. Moreover, up to now and even for their best results (CNET, STC, NEC), their performance is not as high as hybrid association. An ideal view, would be the possibility to combine independently, on the same substrate, the various epilayers and process needed for each function whatever the substrate is. Such a dream needs at least two important approachs:

- The mastery of growth techniques for heteroepitaxial layers with large lattice mismatch.

- The knowledge of a selective epitaxy method to deposite what we want, exactly where we want.

Concerning the first approach, it is now well known, since the growth of the first 1.3μm GaInAsP laser on GaAs substrate [72], that great progress have been made and excellent results have been reported. Many devices and integrated circuits have been fabricated with success on large lattice mismatch heteroepitaxies [73-77]: GaAs MESFET on silicon substrate [78], GaInAsPIN photodiode on GaAs substrates [79,80], integration of GaAs light emitting diode with silicon MOSFET on silicon substrate [81] ... For photoreceiver applications, in collaboration with Thomson CSF, we were the first to propose the use of heteroepitaxial layers to overcome the problem of the technological compatibility between the fabrication of a MESFET and the use of materials for 1.3-1.55μm photodetection [82]. Such a solution has also been proposed successfully by NEC for photoreceiver and transmitter applications. In this case, the substrate is an InP crystal, the optoelectronic device being fabricated on GaInAs matched to InP epilayers, and the MESFET being fabricated on GaAs/InP lattice mismatch epilayers. The results are to date among the best reported, and show very clearly that these ways are to be analysed very carefully.

Concerning the second approach, several groups in Japan and USA start to look for for selective epitaxy methods using either dielectric deposition previously etched before the growth of the active layers, or photochemistry assisted CVD selective deposition technique [83,84]. First results that we have obtained in collaboration with Thomson CSF show the possibility to grow by LPMOCVD GaInAs epilayers on a GaAs semi-insulating substrate previously covered with a SiO_2 layer and etched down to obtain a planar structure.

Conclusion

It is clear that, although successfull progress have been made, the performance and the cost of present OEIC's are still unsatisfactory to the customers of optical communication systems. No doubt that new technological development are needed to reach the goals of high performance, low cost and mass production. We can think that, combined with the use of heteroepitaxial layers, selective deposition technique would be one of the key for the OEIC mass production.

References

1. Shibata, J.; Kijiwara, T.: Optoelectronic integrated circuits using the InGaAsP/InP system. Opt. and Quant. Elect. 20 (1988) 363-371.

2. Wada, O.: Optoelectronic integration based on GaAs material. Opt. and Quant. Elect. 20 (1988) 441-474.

3. Hutcheson, L.D.; Hangen, P.; Optical interconnects replace hardwire. IEEE Spectrum (March 1987) 30-35.

4. Shibata, J.; Kajiwara, T.; Optics and electronics are living together. IEEE Spectrum (Feb. 1989) 34-38.

5. Olsen, G.H.; Nakajima, K.; Hirtz, J.P.; Razeghi, M.; Bonnet, M.; Duchemin, J.P.; Wood, C.E.C.; Pearsall, T.P. (eds.) GaInAsP alloy semiconductors. New York: John Wiley & Sons 1982.

6. Razeghi, M.; Tsang, W.T. (eds.) Lightwave technology for communications. New York: Academic 1985.

7. Razeghi, M.: The MOCVD challenge Vol.1: A survey of GaInAsP-InP for photonic and electronic applications. Bristol: Adam Hilger 1989.

8. Mircea, A; Daste, P.; Schiavini, G.; Couchaux, B.; Ougazzaden, A.: Extremely uniform GaInAsP growth by atmospheric pressure MOVPE. Proc. of the 172nd Meeting of the Electrochemical Society-SOTAPOCS VII, Oct. 1987

9. Yariv, A.; Lau, K.Y.; Ultra high speed semiconductor lasers. IEEE Journ. Quantum Elect. QE-21(2) (1985) 121-138.

10. Olhansky, R.; Hill, P.; Lanzivera, V.; Powazinik, W.: Frequency responses of 1.3 µm InGaAsP high speed semiconductor lasers. IEEE Journ. Quantum Elect. QE-23(9) (1987) 1410-1418.

11. Wang, S.Y.; Bloom, D.M.: 100 GHz bandwith planar GaAs Schottky photodiode. Elect. Lett. 19(14) (1983) 554-555.

12. Temkin, H.; Fram, R.E.; Olson, N.A.; Burrus, C.A.; McCoy R.J.; Very high speed operation of planar InGaAs/InP photodiode detectors. ELect. Lett. 22(23) (1986) 1267-1269.

13. Zebda, Y.; Bhattacharya, P.; Tobin, M.S.; Simpson, T.B.: Design and performance of very high speed InGaAs/InAlAs PIN photodiodes grown by molecular beam epitaxy. IEEE Elect. Dev. Lett. EDL-8(12) (1987) 579-581.

14. Miura, S.; Kuwatsuka, H.; Mikawa, T.; Wada, O.: Planar embedded InP/GaInAs PIN photodiode for very high speed operation. IEEE Journ. Light. Techn. LT-5(10) (1987) 1371-1376.

15. Wang, S.Y.; Carey, S.W.; Kolner, B.H.: A front side illumination InP/GaInAs/InP PIN photodiode with a 3 dB bandwith in excess of 18 GHz. IEEE Elect. Dev. ED-34(4) (1987) 938-940.

16. Parker, D.G.; Say, P.G.; Hamson, A.M.; Sibbett, W.: 110 GHz high efficiency photodiodes fabricated from indium tin oxide/GaAs. Elect. Lett. 23(10) (1987) 527-528.

17. Fair, C.; Yu, P.K.L.; Chen, P.C.: High speed, self passivated InGaAs PIN photodiode for microwave fibre links. Elect. Lett. 23(11) (1987) 571-572.

18. Wake, D.; Blank, L.C.;Walling, R.H.; Henning, I.D.: Top illuminated InGaAs/InP PIN photodiodes with a 3 dB bandwith in excess of 26 GHz. IEEE Elect. Dev. Lett. EDL-9(5) (1988) 226-228.

19. Krakowski, K.; Rondi, D.; Talneau, A.; Combemale, Y.; Deborgies, F.; Maillot, P.; Richin, P.; Blondeau, R.; De Cremoux, B.: Ultra low threshold, high-bandwith, very low noise operation of 1.52 μm GaInAsP/InP DFB buried-ridge structure laser diodes entirely grown by MOCVD. IEEE Journ. Quantum Elect. QE-25(6) (1989).

20. Fernier, B.; Artigue, C.; Bonnerie, D.; Goldstein, L.; Perales, A.; Benoit, J.: Low threshold 1.5 μm DFB laser grown by GSMBE. Elect. Lett. 25 (1989) 768-769.

21. Riglet, P.; Erman, M.; Chané, J.P.; Jarry, P.; Vingrieff, J.J.; Martin, B.G.; Decoster, D.; Gouy, J.P.: Low capacitance dual balanced detectors integrated for coherent communications. To be presented at the 16th International Symposium on Gallium Arsenide and Related Compounds, paper 145 (Sept. 1989), Karizawa, Japan.

22. Godts, P.; Vanbremeersch, J.; Constant, E.; Zimmerman, J.: Realization of very high transconductance GaAs MESFETs. Elect. Lett. 24(13) (1988).

23. Cirillo, N.C.; Shur, M.S.; Abrokwah, J.K.: Inverted GaAs/AlGaAs modulation- doped field effect transistors with extremely high transconductances. IEEE Elect. Dev. Lett. EDL-7 (1986) 71-74.

24. Fujishiro, H.T; Saito, T.; Nishi, S.; Seki, S.; Sano, Y.; Kaminishi, K.: Quartermicron gate inverted HEMT for high speed ICs. Gallium Arsenide and Related Compound 1987: ed.by Christou A. and Rupprecht H.S., Institute of Physics Conference Series nb.91, Bristol and Philadelphia.

25. Powell, A.L.; Mistry, P.; Roberts, J.S.; Rockett, P.J.: AlGaAs HEMTs grown by MOVPE exhibiting high transconductance. Elect. Lett. 23 (1987) 528-529.

26.Antreasyan, A.; Garbinski, P.A.; Mattera, V.D.; Temkin, H.: High speed enhancement mode InP metal insulator semiconductor field effect transistors exhibiting very high transconductance: Appl. Phys. Lett. 49 (1986) 513-515.

27.Raulin, J.Y.; Thorngren, E.; di Forte Poisson, M.A.; Razeghi, M.; Colomea, G.: Very high transconductance InGaAs/InP junction field effect transistor with submicron gate. Appl. Phys. Lett. 50 (1987) 535-536

28.Furutsu, M.; Sudo, H.; Soda, H.; Ishikowa, H.; Imai, H.: High speed and high power GaInAsP/InP junction field effect transistor with submicron gate. Elect. Lett.(24) (1988) 733-735.

29.Remiens, D.; Mallecot, F.; Vilcot, J.P.; Decoster, D.: Modulateur électro-optique à onde guidée sur GaAs. Rev. Phys. Appl. 22 (1987) 1581-1584.

30.Korothy, S.K.; Eisenstein, G.; Tucker, R.S.; Vesellera, J.J.; Reybon, G.: Optical intensity modulation to 40GHz using a waveguide electrooptic switch. Appl. Phys. Lett. 50(23) (1987) 83-85.

31.Wood, T.H.; Burrus, C.; Miller, D.A.B.; Chemla, D.S.; Damen, T.C.; Gossard, A.C.; Wiegmann,W.: High speed optical modulation with GaAs/GaAlAs quantum wells in a pin diode structure. Appl. Phys. Lett. 44 (1984) 16-18.

32.Rogers, D.: Monolithic integration of 3GHZ detector/amplifier using refractory gate ion-implanted MESFET process. IEEE Elect. Dev. Lett. EDL-7(11) (1986) 600-602.

33.Hamaguchi, H.; Makiuchi, M.; Kumai, T.; Wada, O.: GaAs optoelectronic integrated receiver with high output fast-response characteristics. IEEE Elect. Dev. Lett. EDL-8(1) (1987) 39-41.

34.Verriele, H.; Maricot, S.; Constant, M.; Ramdani, J.; Decoster, D.: Planar monolithic integration of a schottky photodiode and a GaAs field effect transistor for 0.8µm wavelength applications. Elect. Lett. 21 (1985) 878.

35.Verriele, H.; Remiens, D.; Ramdani, J.; Decoster, D.: A planar monolithic integrated photoreceiver: association of a GaAs Schottky photodiode with a GaAs FET. Sens. and Act. 11 (1987) 239-250.

36.Decoster, D.; Vilcot, J.P.; Constant, M.; Ramdani, J.; Verriele, H.; Vanbremeersch, J.: Planar monolithic integration of a GaAs photoconductor and a GaAs field effect transistor. Elect. Lett. 22(4) (1986) 193-195.

37.Miura, S.; Wada, O.; Nakai, K.: A novel planarization technique for optoelectronic integrated circuits and its application to a monolithic AlGaAs/GaAs pin FET. IEEE Trans. Elect. Dev. ED-34 (1987) 241-246.

38.Wada, O.; Hamaguchi, H.; Sakurai, T.; Nakai, K.; Iguchi, K.: Monolithic pin/preamplifier circuit integrated on a GaAs substrate. Elect. Lett. 19 (1983) 1031-1032.

39. Verriele, H.; Lorriaux, J.L.; Legry, P.; Gouy, J.P.; Vilcot, J.P.; Decoster, D.: GaAs monolithic integrated photoreceiver for 0.8 µm wavelength: association of Schottky photodiode and FET. IEE Proc. Part J 135(2) (1988) 92 95.

40. Personick, S.D.: Receiver design for digital fiber optic communication systems (Part I and II). Bell Syst. Techn. Journ. 52 (1973) 843-886.

41. Van Zeghbroeck, B.J.; Harder, C.; Halbart, J.M.; Jäckel, H.; Meier, H.; Patrick, W.; Vettiger, P.; Wolf, P.: 5.2 GHz monolithic GaAs optoelectronic receiver. Proc. IEDM (1987) 229-232.

42. Gouy, J.P.; Vilcot, J.P.; Lorriaux, J.L.; Raczy, L.; Decoster, D.: 0.8 µm wavelength integrated photoreceiver: improvements using a special hetero-epitaxy suitable for high electron mobility transistor fabrication. Thin Solid Films 172 (1989) L59-L63.

43. Wang, H.; Ankri, D.: Monolithic integrated photoreceiver implemented with GaAs/GaAlAs heterojunction bipolar phototransistors and transistors. Elect. Lett. 22 (1986) 391-393.

44. Wake, D.; Scott, E.G.; Henning, L.D.: Monolithically integrated InGaAs/InP PIN-JFET photoreceiver. Elect. Lett. 22 (1986) 719-721.

45. Renaud, J.C.; Nguyen, L.; Allovon, M.; Heliot, F.; Lugiez, F.; Scavennec, A.: Monolithic photoreceiver integrating GaInAs PIN/JFET with diffused junctions. Electron. Lett. 23 (1987) 1055-1056.

46. Akahari, Y.; Hata, S.; Ilseda, M.; Yuda, M.; Kawaguchi, Y.; Uehara, S.: Monolithic InP/GaInAs pin FET receiver using MOMBE grown crystal. Electron. Lett. 25 (1989) 37-38.

47. Spear, D.A.H.; Dawe, P.J.G; Antell, G.R.; Lee, W.S.; Bland, S.W. : New fabrication technology for long wavelength receiver OEICs. Electron. Lett. 25 (1989) 156-157.

48. Matsuda, S.; Kuno, M.; Ohnaka, K.; Shibata, J.: A monolithically integrated InGaAs/InP photoreceiver operating with a single 5V power supply. IEEE Trans. Elect. Dev. 35 (1988) 1284-1287.

49. Miura, S.; Mikawa, T.; Fujii, T.; Wada, O.: High speed monolithic GaInAs pin FET. Elect. Lett. 24 (1988) 394-395.

50. Suzuki, A.; Itoh, T.; Terakada, T.; Kasahara, K.; Asano, K.; Inomoto, Y.; Ishihara, H.; Torikai, T.; Fujita, S.: Long wavelength pin FET receiver OEIC on a GaAs on InP heterostructure. Elect. Lett. 23 (1987) 954-955.

51. Ohtsuka, K.; Sugimoto, H.; Abe, Y.; Matsui, T.; Ogata, H.: Monolithic integration of InGaAs/InP PIN PD with MISFET on stepless substrate. Elect. Lett. 22 (1986) 652-653.

52. Cheng, C.L.; Chang, R.P.H.; Tell, B.; Parker, S.M.Z.; Ota, Y.; Vella Colleiro, G.P.; Miller, R.C.; Zilko, J.L.; Kasper, B.L.; Brown-Goebeler, K.F.; Mattera, D.V.: Monolithically integrated receiver front end: InGaAs pin amplifier. IEEE Trans. Elect. Dev. ED-35 (1988) 1439-1443.

53. Chandrasekhar, S.; Campbell, J.C.; Dentai, A.G.; Joyner, C.H.; Qua, G.J.; Gnauck, A.H.; Feuer, M.D.: Integrated InP/GaInAs heterojonction bipolar photoreceiver. Elect. Lett. 24 (1988) 1443-1445.

54. Kuno, M.; Sanada, T.; Nobuhara, H.; Mulsiuchi, M.; Fujii, T.; Wada, O.; Sakurai, T.: Four-channel AlGaAs/GaAs optoelectronic integrated transmitter array.Appl. Phys. Lett. 49 (1986) 1575-1577.

55. Kasahara, K.; Suzuki, A.; Fujita, S.; Inomoto, Y.; Terakado, T.; Shikada, M.: InGaAsP/InP long wavelength transmitter and receiver OEIC 's for high speed optical transmission system. ECOC Barcelona (1986) 119.

56. Shibata, J.; Nalsao, I.; Sasai, Y.; Kimura, S.; Hase, N.; Serizawa, H.: Monolithic integration of an InGaAsP/InP laser diode with heterojonction bipolar transistors. Appl. Phys. Lett. 45 (1984) 191.

57. Tsuii, H.; Ohnalka, K.; Sasai, Y.; Shibata, J.: Monolithic integration of InGaAs/InP HBT's with a 1.3 µm laser diode for lightwave communication. IEEE Bipolar Circuits and Tech. Meet. (1987) Minneapolis 68.

58. Lian, Z.L.; Walpole, J.N.: Monolithic two-dimensional GaInAsP/InP laser arrays. IEDM Los Angeles (1986) 622.

59. Vinchant, J.F.; Vilcot, .P; Gouy, J.P.; Aboulhouda, S.; Decoster, D.: Thin and short GaAs MSM photodetector monolithically integrated on GaAlAs optical inverted rib waveguide on GaAs semi-insulating substrate. Eur. Conf. Opt. Int. Systems Amsterdam ECOISA 1989, paper D1, 25-29 Sept.

60. Bornhold, C.; Doldissen, W.; Fielder, F.; Kaiser, R.; Kowalski, W.: Waveguide integrated PIN photodiode on InP. Elect. Lett. 23 (1987) 2-4.

61. Cinguino, P.; Genova, F.; Rigo, C.; Cacciatore, C.; Stano, A.: Monolithic integrated InGaAlAs/InP ridge waveguide photodiodes for 1.55 µm operation grown by molecular beam epitaxy. Appl. Phys. Lett. 50 (1987) 1515-1517.

62. Chandrasekhar, S.; Campbell, J.C.; Dentai, A.G.; Qua, G.I.: Monolithic integrated waveguide photodetector. Elect. Lett. 23 (1987) 501-502.

63. Erman, M.; Jarry, P.; Gamonal, R.; Gentner, J.L.; Stephan, P.; Guedon, C.: Monolithic integration of a GaInAs PIN photodiode and an optical waveguide; modelling and realisation using chloride vapor phase epitaxy. IEEE Journ. Light. Techn. LT-6 (1987) 399-411.

64. Vinchant, J.F.; Mallecot, F.; Decoster, D.; Vilcot, J.P.: Effects of absorbing layers on the propagation constants: a four layer model on desk-top computer applied to photodetectors monolithically integrated with optical waveguides. Optics Communications 67 (1988) 266-270.

65. Mallecot, F.; Vinchant, J.F.; Razeghi, M.; Vandermoere, D.; Vilcot, J.P.; Decoster, D.: Monolithic integration of a short length GaInAs photoconductor with a GaAs/GaAlAs optical waveguide on a GaAs semi-insulating substrate. Appl. Phys. Lett. 53 (1988) 2522-2524.

66. Vinchant, J.F.; Mallecot, F.; Decoster, D.; Vilcot, J.P.: Photodetectors monolithically integrated with optical waveguides: theoretical and experimental study of absorbing layers. IEE Proceedings Pt.J 136 (1989) 72-75.

67. Liou, K.Y.; Koren, U.; Chandrasekhar, S.; Koch, T.L.; Shakar, A.; Burrus, C.A.; Gnall, R.P.: Monolithic integrated InGaAsP/InP distributed feedback laser with Y-branching waveguide and a monitoring photodetector grown by metal organic vapor deposition. Appl. Phys. Lett. 54 (1989) 114-116.

68. Ribot, H.; Sansonetti, P.; Brandon, J.; Carre, M.; Menigaux, L.; Azoulay, R.; Bouadma: Monolithic integration of GaAs/GaAlAs buried heterostructure orthogonal facet laser and optical waveguide. Appl. Phys. Lett. 54 (1989) 475-477.

69. Remiens, D.; Menigaux, L.; Dugrand, L.; Ben Assayag, G.; Gierak, J.; Sudrand, P.: GaAs/GaAlAs double-heterostructure laser with integrated passive waveguide. ECIO' 89, Paris.

70. Kawamura, Y.; Wakita, K.; Itaya, Y.; Yoshikuni, Y.; Asahi, H.: Monolithic integration of InGaAsP/InP DFB laser and InGaAS/InAlAS MQW optical modulators. Elect. Lett. 22 (1986) 242-243.

71. Suzuki, M.; Noda, Y.; Tanaka, H.; Akiba, S.; Kushiro, Y.; Isshiki, H.: Monolithic integration of InGaAsP/InP distributed feedback laser and electroabsorption modulator by vapor phase epitaxy. Journ. Light. Techn. LT5-9 (1987) 1277-1284.

72. Razeghi, M.: 1.3 μm laser on GaAs substrate. GaAs and Related compounds (1984) Biarritz.

73. Razeghi, M.: GaInAsP on Si Substrates and its device application. 20 th International Conference on Solid State Devices and Materials, Tokyo (1988) 363-366.

74. Razeghi, M.; Ramdani, J.; Legry, P.; Vilcot, J.P.; Decoster, D.: monolithic integration of a GaInAs/GaAs photoconduction with a GaAs FET for 1.3-1.5 μm wavelength applications. GaAs and Related Compounds (1987), Institute of Physics Confernces number 91 p 781-784.

75. Razeghi, M.; Hosseini Teherani; Vilcot, J.P.; Decoster, D.: Monolithic integration of a Scottky photodiode and a FET using a GaInP/GaInAs strained materials. GaAs and Related Compounds (1987), Institute of Physics Confernces number 91 p 625-628.

76. Hosseini Teherani, A.; Decoster, D.; Vilcot, J.P.; Razeghi, M.: Monolithic integration of a Schottky phtodiode and a field effect transistor on GaInP-GaInAs heteroepitaxy. Journ. Appl. Phys. 64 (1988) 2215-2218.

77. Ramdani, J.; Decoster, D.; Vilcot, J.P.; Gouy, J.P.; Razeghi, M.: 1.3-1.5 μm wavelength integrated photoreceiver using GaInAs-GaAs heteroepitaxy. IEE Proceedings Part J 136 (1989) 83-87.

78. Arch, D.K.; Morko, H.; Vold, P.J.; Longerbone, M.: High performance self aligned gate AlGaAs/GaAS MODFETs on MBE layers grown on (100) silicon substrates. IEEE Elect. Dev. Lett. EDL-7 (1986) 635-637.

79. Dentai, A.G.; Campbell, J.C.; Joyner, C.H.; Qua, G.J.: InGaAS PIN photodiodes grown on GaAs substrates by metal organic vapor phase epitaxy. Elect. Lett. 23 (1987) 38-39.

80. Hudson, P.D.; Wallis, R.H.; Davies, J.I.: Low leakage InGaAs photodiodes grown on GaAS substrates using a graded strained layer suppèrlattice. Elect. Lett. 23 (1987) 273-275.

81. Choi, H.K.; Mattia, J.P.; Turner, G.W.; Tsaur, B.Y.: monolithic integration of GaAs/GaAlAs LED and Si Driver Circuit. IEEE Elect. Dev. Lett. EDL-9 (1988) 512-514.

82. Razeghi, M.; Ramdani, J.; Verriele, H.; Decoster, D.; Constant, M.; Vambremeersch, J.: Planar monolithic integrated photoreceiver for 1.3-1.55 µm wavelength applications using GaInAs-GaAs heteroepitaxies. Appl. Phys.Lett. 49 (1986) 215-217.

83. Kamon, K.; Shimazu, M.; Kimura, K.; Mikara, M.; Ishii, M.: Selective growth of AlGaAs embedded in etched grooves on GaAs by low pressure OMVPE. Journ. of Cryst. growth 77 (1986) 297-302.

84. Jones, S.H.; Lou, M.: Selective area growth of high quality GaAs by OMCVD using native oxide masks. Journ. of Electrochemical Society, Solid State Science and Technology 134 (1987) 3149-3155.

Thin Dielectrics and Interfaces

The Effect of Rapid Thermal Annealing on the Quality of Thin Thermal Oxides

H. Wendt, A. Spitzer, W. Bensch and K. v. Sichart
Siemens AG, Otto-Hahn-Ring 6, D-8000 München 83, FRG

Summary
Rapid thermal processing (RTP) as a post oxidation annealing (POA) process is of considerable influence on the quality of thin thermal oxides: The number of defect related oxide failures is reduced whereas the intrinsic oxide quality is slightly reduced but to a degree that is of no importance with respect to reliability aspects. It is important to perform POA after poly-Si deposition and doping, while POA after poly-Si patterning was found to be a critical process. Infrared spectroscopy indicates stress relaxation due to the POA-treatment but no change of the SiO_2/Si interface roughness could be observed by high resolution electron microscopy. The improvement of the oxide quality is accompanied by the creation of electron traps in the oxide.

Rapid thermal processing has been suggested for different post gate processing steps like source/drain dopant activation, glass reflow, silicide formation etc.. Studies of the quality of gate oxides after RTP showed dramatic reductions of the oxide quality [1,2]. On the other hand, post oxidation annealing treatments (POA) using conventional furnaces were found to be successful in improving oxide quality [3]. In this paper we report on the observed quality increase of thin thermal oxides after rapid thermal treatments.

Experimental:
All the experiments were performed on 4" FZ-Si (20 Ohm-cm p-Si). A dry oxide with a thickness of 13nm was grown at a temperature of 900°C. Poly-Si was deposited on the oxide (300nm) and doped with phosphorus. The rapid thermal treatments were carried out in an EATON ROA 400. POA was performed after poly-Si deposition, after poly-Si doping, after poly-Si doping with the phosphorus containing glassy layer (PSG) still on top of the poly-Si or after poly-Si patterning. The temperature was between 1000 and 1150°C and the annealing times went up to 60 seconds. Constant current tests were used to characterize the oxide quality. The oxide yield was determined by stressing the oxide using the following conditions: Test 1: $60\mu A/cm^2$ for 70ms; Test 2: $1mA/cm^2$ for 500ms (area $8mm^2$). The charge to breakdown values Q_{bd} of the oxides were determined on small area capacitors by applying a constant current ($0.5A/cm^2$) until breakdown was detected.

Results and Discussion

The results of the tests can be summarized in the following way:

(a) The number of defect related breakdown events of the oxide decreases with increasing POA-temperature. Fig.1a shows the yield of about 500 test capacitors. The increase in yield due to POA is significant and reaches values of about 40% (Test 2).

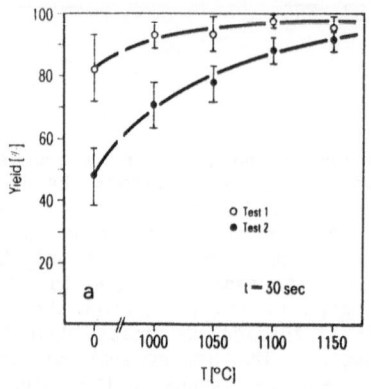 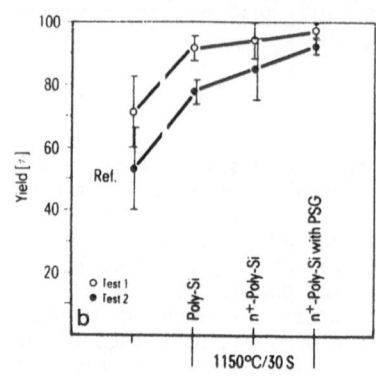

Fig.1a. Yield increase of test capacitors due to isochronal POA. The annealing time was 30s. Fig.1b. Yield increase due to POA (1150°C/30s). During POA the poly-Si electrode was undoped/ n^+-doped/ n^+-doped with phosphorous glass on top.

(b) Isothermal POA at 1100°C shows that annealing times longer than 30s do not improve the oxide quality any further.

(c) The best results are obtained if the phosphorus doped glassy layer on top of the poly-Si used for the n^+-doping is still present during POA (see Fig. 1b). Ranking the different process sequences, the yield increase is lowest for the undoped poly-Si, higher for the n^+-doped electrode, and highest if the PSG layer is still present during POA. The phosphorus has obviously a beneficial effect on the yield increase.

(d) The intrinsic quality of the POA treated oxides is reduced. The charge to breakdown values Q_{bd} decrease with increasing temperature from about 17C/cm^2 to 7C/cm^2 (Fig.2). However, this decrease in the intrinsic quality is not critical for device lifetimes.

(e) POA performed after poly-Si gate definition does also increase the yield but the yield improvement decreases as a function of the temperature in contrast to the preceding experiments. Additionally, the observed scattering of the test results increases significantly. It indicates that this POA-sequence is not a reliable process.

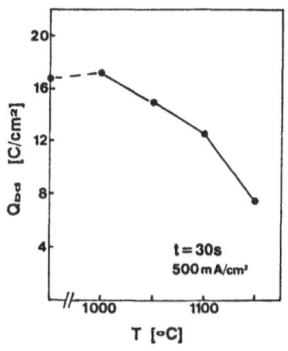

Fig.2. Intrinsic charge to breakdown (Q_{bd}) as a function of the POA temperature. The annealing time was 30s. and the poly-Si n^+-doped during the treatment.

Further experiments conducted to elucidate the observed oxide quality increase yield the following results:

(i) High resolution electron microscopy does not reveal any differences in the SiO_2/Si-interface roughness between POA-oxides and unannealed oxides. In all cases the interfaces were found to be very smooth showing step heights less than 2-3 (100) atomic layers.

(ii) Infrared spectroscopy (IR) indicates stress relaxation caused by the POA-treatment. The position of the asymmetric Si-O-Si stretching vibration (ASM) which depends on properties of the oxide, such as homogeneity, impurities, porosity and stress or bond strain, was found to be shifted to higher wave numbers (see Fig.3).

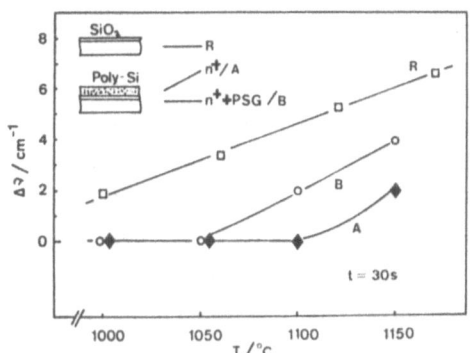

Fig.3. IR-shift of the absorption band at $1070 cm^{-1}$ as a function of the POA temperature. R: POA after oxidation. A: POA of n^+-doped poly-Si. B: POA of n^+-doped poly-Si with PSG layer.

This shift is due to a relaxation of internal stress/bondstrain caused by a rearrangement of the "constrained" amorphous network. Fig.3 shows that an uncovered SiO_2 shows a large shift of the ASM due to POA (curve R). The shifts for the n^+-doped poly-Si (A) are within the scatter of the method whereas POA of the n^+-poly-Si with the PSG layer present during POA shows a clear shift of the ASM. This observation is in agreement with the results shown in Fig.1b.

348

(iii) More insight into the reasons for the reduction of defect related oxide failures due to POA could be achieved by a constant current injection (gate negative; 10mA/cm^2). A voltage change to maintain the current indicates the build up of a space charge near the injecting electrode. The voltage change monitored for the different POA treated oxides is shown in Fig.4. The voltage has to be increased for all oxides and the change of the injection voltage increases with increasing POA temperature. Since electrons are the dominant charge carrier in Fowler-Nordheim injection experiments the increased voltage indicates that electrons are trapped in the oxides and a negative space charge has been built up. Particularly electrons trapped at

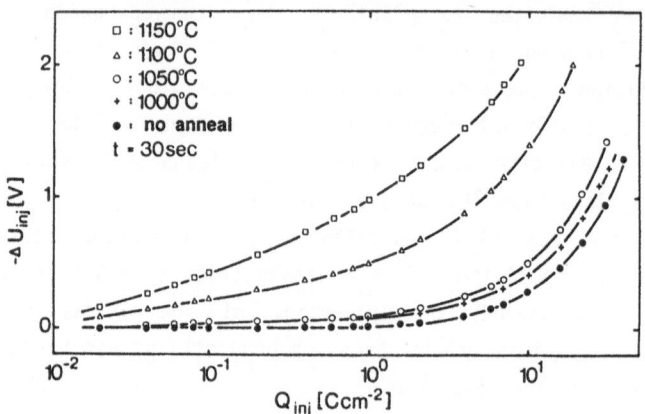

Fig.4. Voltage change during a constant current test for different POA treated capacitors (0.02mm^2, 10mA/cm^2, gate negative).

a defect site reduce the current density at this weak spot by a reduction of the electric field. By this mechanism defects are electrically shielded and defect related breakdown events are reduced. On the other hand, the reduced Q_{bd}-values for the intrinsic failures are a consequence of the enhanced electron trapping after POA. The negative space charge increases the electric field near the anode. For higher electric fields at the same current density the intrinsic Q_{bd}-values are reduced [4].

In summary, we showed that rapid thermal annealing of oxides with properly adjusted parameters can increase the oxide quality considerably.

References

1. McGruer, N.E.; Oikari, R.A.; IEEE Trans. ED-33 (1986) 929.
2. Cosway, R.G.; Hodel, M.W.; J. Electrochem. Soc. 135 (1988) 533.
3. Bhattacharyya, A.; Vorst, C.; Charim, A.H.; J. Electrochem. Soc. 132 (1985) 1900.
4. Hu, C.; IEDM Tech. Dig. (1985) 368.

Charge Trapping in Dry and Pyrogenic Gate Oxides

M. SEVERI, M. IMPRONTA, P. NEGRINI and S. VASSURA

CNR - Istituto LAMEL, Via Castagnoli 1, 40126 Bologna (Italy)

Summary

Hole traps and interface trap generation under avalanche hole injection have been studied in dry and wet (pyrogenic) gate oxides 30 nm thick. We found that pyrogenic oxides are characterized by a lower hole trap density only when they are grown under reaction-limited conditions. On the other hand, fast and slow states are more easily generated in pyrogenic oxides.

Introduction

Oxide and interface traps seriously degrade the reliability of silicon MOS integrated circuits. The properties of these traps are strongly dependent on the oxide growth conditions. Hydrogen present in various forms in the oxide plays an important role in determining the charge trapping behavior. It is well known that the presence of H_2O in the oxidizing ambient increases both the electron trap concentration and the interface trap generation under electron injection. On the other hand, it has been shown very recently that 20 nm pyrogenic oxides grown at 850°C are characterized by a lower hole trap density as compared to dry oxides grown at 1000°C [1]. The doubt might arise that the reduction of the hole traps is due to the lower oxidation temperature. Since hole traps play a dominant role for the hot carrier degradation process [2], it seems worthwhile to further investigate the hole trapping and interface trap generation in dry and wet (pyrogenic) oxides.

Experimental

MOS capacitors with Al gate electrodes were realized on n-type <100> silicon substrates of 0.1 Ω cm resistivity to allow laterally homogeneous avalanche injection. The pyrogenic oxidation was performed at 900°C for 12 min obtaining a SiO_2 film 33 nm thick. The dry oxide was thermally grown at the same temperature for 140 min to a 30 nm thickness. The oxidation was followed by a nitrogen anneal, also at 900°C for 10 min. The Al gate was deposited by evaporation. Electrodes with areas of 2.46×10^{-3} cm^2 were defined by standard photolithography or using a metal mask. A final postmetallization anneal for 20 min at 400°C in N_2 was carried out.

The trapping characterization was performed by injecting hot holes from the substrate using the avalanche technique. A 150 kHz sawtooth voltage was applied to the gate electrode to generate avalanche breakdown in the Si substrate. The average injected current density was kept constant at the value of 8×10^{-8} A/cm^2 by a feedback circuit. All measurements were done at room temperature. The oxide charge buildup was monitored by periodically interrupting the injection and determining the flatband voltage, V_{fb}, shift from the high frequency C-V curve. The density, D_{it}, and distribution of interface traps were determined immediately after injection using the quasi-static technique.

Results and Discussion

Fig.1 shows the negative V_{fb} shift as a function of the avalanche hole injection time for both the pyrogenic and the dry oxides. As can be seen, the pyrogenic oxides are characterized by

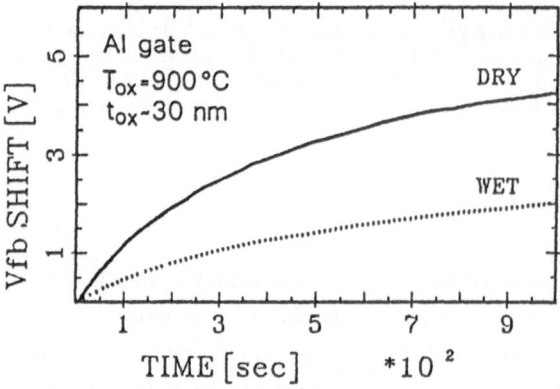

Fig.1. Negative V_{fb} shift by avalanche injection of holes in dry and pyrogenic SiO$_2$. The injection current density is 8×10^{-8} A/cm^2.

a lower shift due to accumulation of positive charge as compared to dry oxides. We suggest that this effect is not directly related to the presence of hydrogen, but it is due to different oxidation kinetics. Because of the large compressive stress in the oxide, we expect the growth of the dry oxide to follow a parabolic law [3]. On the contrary, the pyrogenic growth of 30 nm SiO$_2$ films at 900°C is a reaction-controlled process. In these conditions, there is a large number of oxygen species at the Si/SiO$_2$ interface available to reduce the oxygen vacancies or Si-Si stretched bonds which are precursors of hole traps and E' centers [4]. To check this hypothesis, we have performed some measurements on 100 nm pyrogenic and dry oxides grown at 1100°C, in such a way that the wet oxide too grows under diffusion-limited conditions. In this case, the negative V_{fb} shift in the pyrogenic oxides is larger than that in the dry oxides. The higher hole trapping may be related to an higher density of Si-H bonds, that have been suggested to act as hole traps [5].

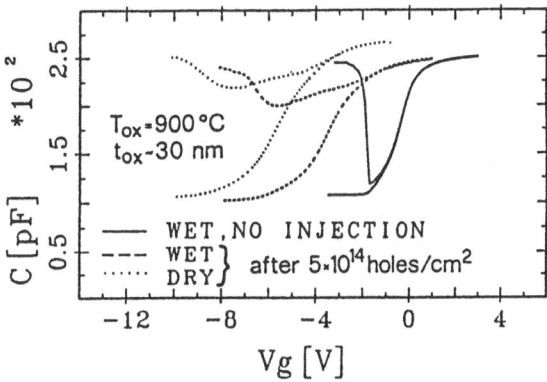

Fig. 2. High frequency and quasi-static C-V curves for both the dry and wet oxides after 1000 s of hole injection. C-V curves for the wet oxide before stress are also shown for comparison.

As far as the generation of interface traps during hole injection is concerned, we found that the interface trap buildup is almost the same in both oxides although hole traps are larger in the dry ones (Fig. 2). This means that higher hole trapping does not simply correlate with higher interface trap generation. In Fig. 3 the increase of the midgap interface trap density induced by the stress is reported vs the effective trapped hole density, as determined from the midgap voltage shift (where the interface trap charge is assumed as being zero). As can

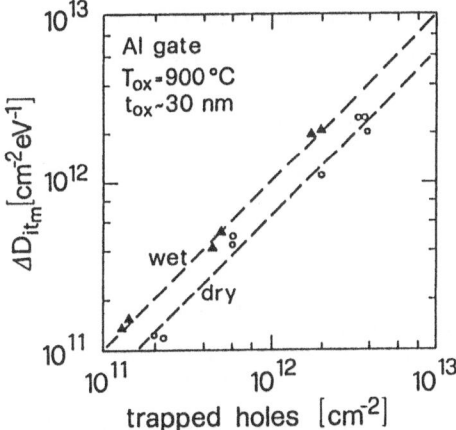

Fig.3. Density of interface traps generated at midgap vs density of trapped holes.

be seen, the ratio between the densities of the generated interface states and the trapped holes is roughly one in the pyrogenic oxides, while it is lower in the dry oxides. This suggests that the generation of fast interface traps during avalanche hole injection is favoured by the presence of water-related species. It has already been reported that radiation induced positive charges more easily convert into fast interface traps in presence of hydrogen [6]. Several models for the interface trap generation that depend on hydrogen being present in the oxide have been proposed [5,7,8]. The lower "yield" of the interface trap generation in dry oxides could also be due to the fact that the strained transition region between Si and SiO_2 is thicker in dry oxides [9]: therefore, carriers trapped relatively far from the Si/SiO_2 interface do not lead to the generation of interfacial defects.

We also found that slow states are more easily generated in pyrogenic oxides. The slow state density was determined by taking the difference between the midgap voltage after discharging and charging the slow states [10]. It is also interesting to observe that, during the annealing of the trapped holes (by a tunneling recombination process involving electrons from the Si valence band), a significant interface degradation occurs: in particular, there is an increase in the interface traps peaked at about 0.75 eV above the valence band edge. This is consistent with the Lai's model, according to which, when electrons are captured by trapped holes, a weak Si-O bond forms due to local relaxation of the oxide network [11]. This model does not depend directly on hydrogen being present; indeed, the generation is the same in both dry and pyrogenic oxides.

Conclusions

A comparison of hole trapping and interface trap generation in dry and pyrogenic gate oxides has been made with the aim of providing a better framework for understanding these phenomena. It is found that pyrogenic oxides are characterized by a lower hole trap density only when they are grown under reaction-limited conditions. On the other hand, the generation of fast and slow interface traps is more efficient in wet oxides. From our results it can be concluded that in order to diminish the degradation of the oxide during charge injection it is important to lower not only the density of the hole traps but also the hydrogen content.

References
1.   Miki, H.; Noguchi, M.; Yogokawa, K.; Kim, B.; Asada, K.; Sugano, T.: IEEE Trans. Electron Devices **ED-35** (1988) 2245-2251
2.   Weber, W.: IEEE Trans.Electron Devices **ED-35** (1988) 1476-1486.
3.   Sze, S.M.: Semiconductor Devices, Physics and Technology, John Wiley Ed. 1985, p.348
4.   Withan, H.S.; Lenahan, P.M.: Appl.Phys.Lett. **51** (1987) 1007-1009.
5.   Revesz, A.G.: IEEE Trans. Nucl.Sci. **NS-24** (1977) 2102-2107.
6.   Sabnis, A.G.; Nelson, J.T.; Billig, J.N.: IEDM 81 (1981) 244-247.
7.   McLean, F.B.: IEEE Trans. Nucl. Sci **NS-27** (1980) 1651-1657.
8.   Griscom, D.L.: J.Appl.Phys. **58** (1985) 2524-2533.
9.   Grunthaner, F.J.; Grunthaner, P.J.: Materials Science Reports 1 (1986) 65-160.
10.   Heyns, M.M.; De Keersmaeker, R.F.: Paper presented at "The Physics and Technology of amorphous SiO_2" Les Arcs (France), June 29-July 3, 1987.
11.   Lai, S.K.: J.Appl.Phys. **54** (1983) 2540-2545.

IC Process Compatible Preparation of Silicon Interfaces Using the Silicon-to-Silicon Direct Bonding Method

Stefan Bengtsson and Olof Engström
Department of Solid State Electronics
Chalmers University of Technology
S-412 96 Göteborg, Sweden

Abstract - Silicon/silicon interfaces were prepared by wafer bonding using the silicon-to-silicon direct bonding method. Silicon/silicon interfaces (n/n-type) with excellent electric properties were prepared, by using hydrophobic wafer surfaces, at temperatures in the range of 700° C to 1000° C. The influence of the bonded interface on device characteristics of bonded p^+n junctions prepared at low temperatures was seen as n-factors larger than 2 at forward bias. A correlation between n-factors and the density of states in the bandgap was found. In hydrophilic samples, the density of voids at the bonded interface was determined mainly in the contacting of the wafer surfaces at room temperature. Compared to hydrophilic samples, hydrophobic samples were held together at a smaller fraction of the area before heat treatment. After the heat treatments, no difference in the density of voids was found.

INTRODUCTION

The interfaces Si/Si, Si/SiO_2 and SiO_2/SiO_2, prepared by the silicon-to-silicon direct bonding method, have been utilized to attain useful structures and device geometries in a number of different applications such as SOI-materials [1-2], power devices [3-4] and sensors [5]. Recently, an energy barrier, influencing the current-voltage characteristic, was found at bonded Si/Si interfaces prepared from wafers with hydrophilic surface properties. By instead using wafers having hydrophobic properties, the density of interface charge was controlled and excellent electrical properties were found for interfaces prepared from moderately doped wafers [6-7]. However, in these studies temperatures at and above 1000° C were used during the preparation of the interfaces. In order to use wafer bonding in connection with standard types of IC processes, it is advantageous if the bonding process can be performed at temperatures lower than usual process temperatures. In this paper we report on electrical and mechanical properties of bonded Si/Si interfaces prepared at temperatures below 1000° C.

SAMPLE PREPARATION

Three inch silicon wafers were dipped in a 2% HF-solution and cleaned in H_2O_2:NH_4OH:H_2O (1:1:5) and H_2O_2:HCl:H_2O (1:1:6) mixtures at 75° C, followed by a water rinse. To give two groups of interfaces different properties, they were treated in warm HNO_3 and HF:H_2O (1:50), respectively. The treatment in HNO_3 maintained the *hydrophilic* properties and is known to increase the density of OH molecules present on the surfaces. The other group of wafers was given *hydrophobic* surface properties by treatment in HF. After the treatment in HF or HNO_3 the wafers were rinsed in water and dried. The wafers were brought together using a mechanical fixture. After being released from the fixture the wafers were annealed at temperatures in the range 500° to 1000° C for different periods of time using oxygen as an ambient.

EXPERIMENTAL RESULTS

Current-voltage characteristics were measured to investigate the influence of the bonded interface on the electrical properties of Si/Si n/n-type structures. Linear I-V curves were found for samples prepared from wafers having hydrophobic surface properties, indicating the absence of a potential barrier at the interface. No influence of temperature in the range 700° C to 1000° C was observed. Figure 1 shows the resistance for a number of samples of resistivity 3-7 Ωcm and area 1 cm^2, prepared at 700° C for 5 min. and 30 min. respectively. The spreading of series resistances (bulk, contacts, cables), as measured on reference samples without bonded interface, is given by two horizontal lines in Fig. 1. Attempts to reduce the bonding temperature to values considerably below 700° C were thwarted by the decreasing mechanical strength of the bonded interface. Samples prepared from wafers having hydrophilic surface properties exhibited a potential barrier at the bonded interface, in agreement with earlier reported results [6-7].

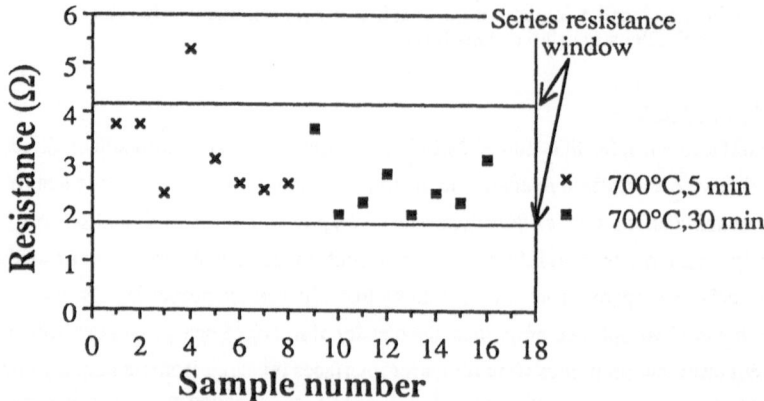

Fig. 1. Measured resistance of bonded Si/Si interfaces prepared at 700° C using wafers of resistivity 3-7 Ωcm and hydrophobic surface properties. The two horizontal lines represent the spreading of series resistances, as measured on reference samples.

The p$^+$n-junctions were prepared by direct bonding at 700° C for 5 min. using wafers having hydrophobic surface properties, and showed properties close to those of diffused junctions. Figure 2 shows the current-voltage characteristic for one such junction. At a forward bias of a few tenths of a volt, n-factors in the range of 2 to 5 were found for various samples. From DLTS measurements a correlation between high n-factors and a high density of states in the bandgap was found [8]. Preliminary results suggest that these high n-factors are due to an excess tunneling current via states in the bandgap. The current in the reverse direction can be described as due to charge carrier generation in the space charge region.

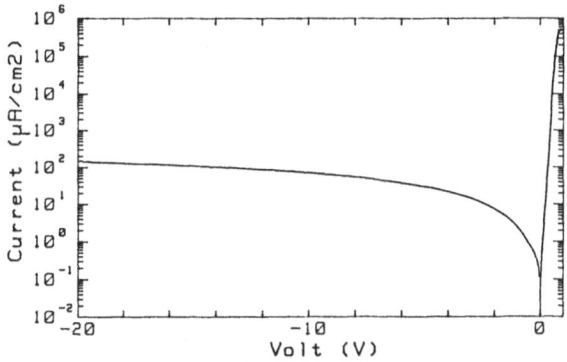

Fig. 2. Current-voltage characteristic of a p^+n-junction prepared with wafers having hydrophobic surface properties using a heat treatment at 700° C for 5 min.

To investigate the density of voids at bonded Si/Si interfaces, infrared light transmitted through the samples was converted to visible light and detected photographically. The density of voids in hydrophilic samples was mainly determined in the contacting of the wafer surfaces at room temperature (Fig. 3a). The influence of heat treatments was weak. Hydrophobic samples, which were brought together but not annealed, were held together at a smaller fraction of the area (Fig. 3b). A short heat treatment considerably changed the density and size distribution of voids for these samples. Figure 3c shows the sample in Fig. 3b after a heat treatment at 700° C for 5 min. The larger defective areas were split into smaller ones and point-shaped voids, probably due to enclosed gas, were generated. These voids disappeared when the annealing proceeded (Fig. 3d) and the required annealing time exhibited an exponential dependence on the temperature (Fig. 4).

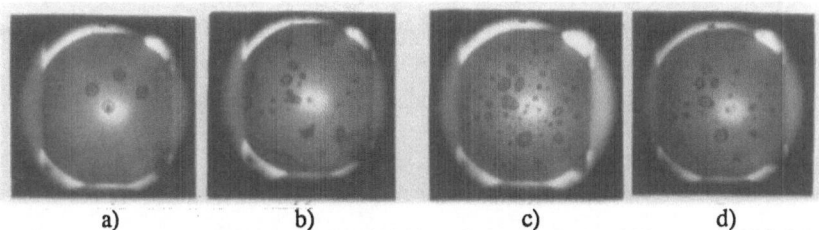

a)   b)   c)   d)

Fig. 3. Thermal images of directly bonded Si/Si interfaces: a) Hydrophilic sample, prior to heat treatment; b-d) Hydrophobic sample; b) prior to heat treatment; c) 700° C 5 min.; d) 700° C 24 h.

At the lower limit of the vertical bars in Fig. 4 no reduction in density was seen. At the upper limit the density of voids was reduced by at least 90 %. In Fig. 4, a straight line corresponding to an activation energy of 2.3 eV is shown. After the heat treatments were performed, no difference in the density of voids due to chemical pre-treatment was found. The small number of voids present after a long time at elevated temperature may be due to dust particles.

356

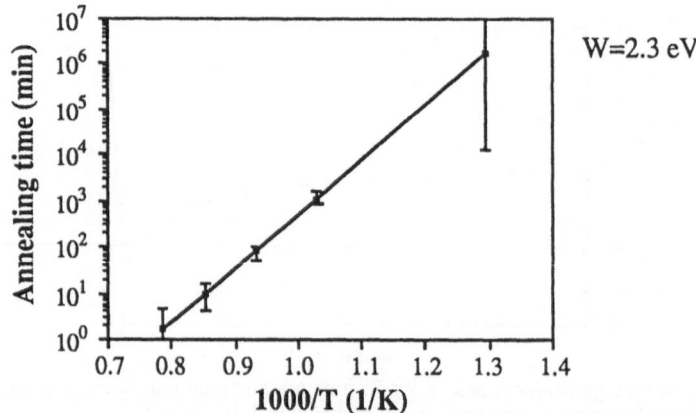

Fig. 4. Time at elevated temperature required to reduce the density of voids generated in hydrophobic samples during the first short heat treatment. The vertical bars represent the experimental periods of time during which the density of voids was reduced by more than 90 %.

DISCUSSION

The results of the present study demonstrate the feasibility of processing new geometrical structures in combination with standard IC technology. The Si/Si (n/n-type) interfaces which were prepared at 700° C have a negligible influence on the I-V characteristics of measured samples. The influence of the bonded interface was seen as n-factors larger than 2 in bonded p^+n junctions. These high n-factors were probably due to an excess tunneling current via states in the bandgap. Attempts to reduce the bonding temperature considerably below 700° C were thwarted by mechanical rather than electrical properties of the interface. Differences in the bonding mechanisms between hydrophilic and hydrophobic surfaces were found as well. A high density of OH molecules on the wafer surfaces prior to bonding was found not to be necessary for the preparation of bonded Si/Si interfaces. A thermally activated process was found to control the effect of reducing the density of voids in hydrophobic samples.

ACKNOWLEDGEMENT
This work was financed by the Swedish National Board for Technical Development.

REFERENCES
1. J.B. Lasky, Appl.Phys.Lett. **48** (1986) 78-80.
2. W.P. Maszara, G.Goets, A.Caviglia and J.B. McKitterick, J.Appl.Phys. **64** (1988) 4943-4950.
3. M. Shimbo, K. Furukawa, K. Fukuda and K. Tanzawa, J.Appl.Phys. **60** (1986) 2987-2989.
4. H. Ohashi, K.Furukawa, M. Atsuta, A. Nakagawa and K. Imamura, 1987 IEEE IEDM Tech.Dig. 678-681.
5. S. Shoji, T. Nisase, M. Esashi and T. Matsuo, Proc. TRANSDUCERS'87 (1987) 305-308.
6. S. Bengtsson and O. Engström, Proc. ESSDERC 88 (1988) C4-63-C4-66.
7. S. Bengtsson and O. Engström, Accepted for publication in J. Appl. Phys. 1 Aug. 1989.
8. S. Bengtsson, G.I. Andersson and O. Engström, unpublished.

Electrical Properties of Ultra Thin Multilayer Dielectrics on Polysilicon

J. Hirschler, L. Do Thanh, K.H. Küsters and K. v.Sichart

Siemens AG, Dept. HL T 111, 8000 München 83, Otto-Hahn-Ring 6, West Germany

Abstract: We demonstrate the feasibility of very thin ONO ($SiO_2/Si_3N_4/SiO_2$) dielectrics on poly-Si for 16M DRAM application. For $D_{ox\ eff.} = 8.5nm$, the leakage current density is lower than 10^{-8} A/cm^2, and the long term stability is more than 10 years at electric fields of 5.5MV/cm. The onset of tunneling current in ONO dielectrics depends on the thickness of the SiO_2 entry potential barrier. The high field current characteristic depends on the thickness of the SiO_2 exit potential barrier.

1. Introduction

Dielectrics with effective thicknesses ($D_{ox\ eff.}$) of 10nm and below, especially deposited on poly-Si, are necessary for storage capacitors in 16 Megabit DRAMs. In this paper we present a study of the properties of stacked films consisting of ultra thin oxide and nitride layers.

2. Experimental

The planar capacitors used in this study were prepared by thermally oxidizing a smooth, amorphous deposited, n$^+$-doped poly-Si layer. This was followed by a LPCVD nitride film deposition. The top oxide layer was grown by dry thermal oxidation of the nitride at 900 °C. A n$^+$-doped poly-Si layer was used as the gate electrode. Fig.1 shows schematically the configuration of the ONO multilayer dielectrics. In some cases ON- (without top oxide) and NO- (without bottom oxide) doublelayer dielectrics have also been prepared. The effective thicknesses $D_{ox\ eff.}$ of the films range from 7.1 to 12.4nm.

The defect density was evaluated using time zero breakdown measurements. The effective thickness $D_{ox\ eff.}$ was determined by TEM cross sections and CV measurements. Leakage current measurements were carried out using the IV ramp technique up to field strengths of 12MV/cm. The long term reliability of the films was investigated by stressing the samples at 150 °C at different field strengths until dielectric breakdown.

3. Results and discussion

3.1. TEM cross section and defect density

Fig.2 shows a typical TEM cross section of an ONO triple dielectric. The bottom poly-Si surface is smooth and well defined. Roughening at grain boundaries is not a serious problem with amorphous deposited poly-Si layers. Also of note is the homogeneous thickness of the triple films.

Defect densities extracted from time zero breakdown measurements are generally less than 0.1 defect/cm^2.

3.2. Current-field characteristics of ONO-, ON- and NO- multilayer dielectrics

Fig.3 shows the current-field characteristics of a 12.4nm ONO sample. For positive gate voltage the typical trapping behaviour of ONO samples is seen in the step like behaviour of the IV curve. Electrons which are trapped in the nitride reduce the electric field at the injecting electrode (Fig.4), resulting in a slower increase of the current density with increasing field. For negative gate voltage the current density is orders of magnitude lower at the same field strength. This is due to the thicker SiO_2 exit/ejection potential barrier and the larger trapping efficiency.

ONO-, ON- and NO- multilayers with different top oxide and bottom oxide thicknesses were investigated (Fig.5). Although the conductivity of sub-10nm ONO films is larger compared to 12nm ONO layers, the current density at electric fields below 2 MV/cm is still $\leq 10^{-8}$ A/cm ^{2}, low enough for their application in DRAM storage capacitors.

The electrical behaviour of ONO films can be highlighted by relating the J vs. E curves of these samples (Figs.6 and 7) to the band model. Samples with different bottom oxides differ at negative V_g only in the high field range and at positive V_g only in the low field range (Fig.6). In the first case electrons have to penetrate nearly the same injection barrier thicknesses (top oxide) of the upper gate, but different SiO_2 ejection barriers, ranging from 1.5nm native oxide up to 3 and 6nm bottom oxide. A thicker ejection barrier results in a higher trapping efficiency and a lower current density at the same field strength. Hence the different thicknesses of the SiO_2 ejection/exit barriers correlate with the current in the high field range. On the other hand, at positive V_g the J vs. E curves differ only at low fields. The onset of tunneling depends only on the thickness of the SiO_2 injection barrier. The results of samples with different top oxide thicknesses ranging from 0nm to 2nm can be similarly interpreted. In this case the IV curves at high negative fields converge, but diverge at large positive V_g (Fig.7).

3.3. Long term reliability

Fig.8 shows the time dependent breakdown characteristics (Weibull plot) obtained from constant voltage stress measurements. ONO samples with $D_{ox\ eff.} = 8.5$nm were stressed with positive V_g at different field strengths at 150 °C. Typical times to 63% cumulative breakdown were $4*10^3$s at 8MV/cm. Fig.9 shows a field acceleration plot according to the model of Chen, Holland and Hu /2/. In this plot the intrinsic breakdown behaviour is extrapolated to lower fields and longer times. In the case of the 8.5nm ONO samples lifetimes of 10 years can be attained at field strengths of 5.5MV/cm.

4. Summary

ONO multilayer dielectrics with effective thicknesses ranging from 7nm to 12.4nm have been investigated. The defect density is lower than 0.1 defect/cm ^{2}. The leakage current density is less than 10^{-8} A/cm ^{2}. Time-dependent breakdown results achieved by constant field stressing and corresponding extrapolations show that the integrity of $D_{ox\ eff.} = 8.5$nm ONO dielectrics can be preserved for more than 10 years at electric fields of 5MV/cm. The onset of tunneling current in ONO dielectrics depends on the SiO_2 injection/entry potential barrier, whereas the high field current varies with the thickness of the SiO_2 ejection/exit potential barrier.

References:

/1/ N.C. Lu, IEEE Circuits and Dev. Magazine, **27** (1989)

/2/ I. Chen et.al., IEEE Trans. on Solid State Circuits **SC-20**, 333 (Feb 1985)

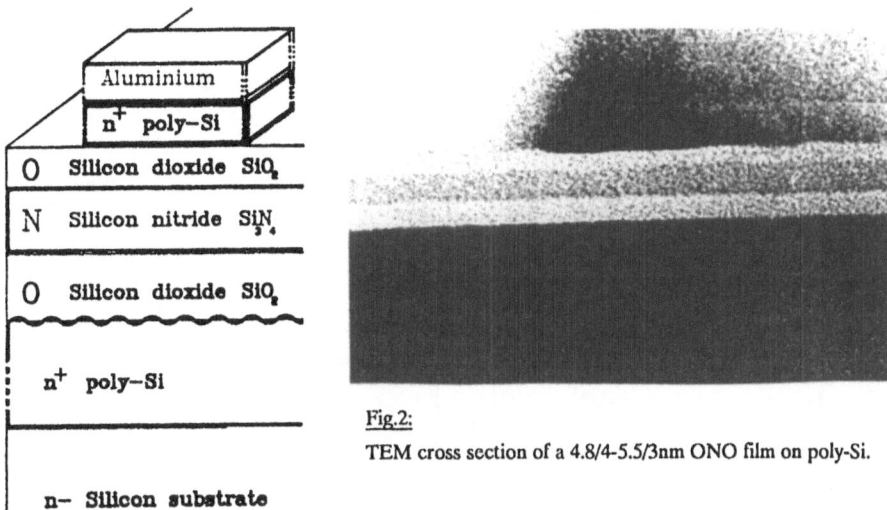

Fig.2:

TEM cross section of a 4.8/4-5.5/3nm ONO film on poly-Si.

Fig.1: Schematic of ONO layers on poly-Si.

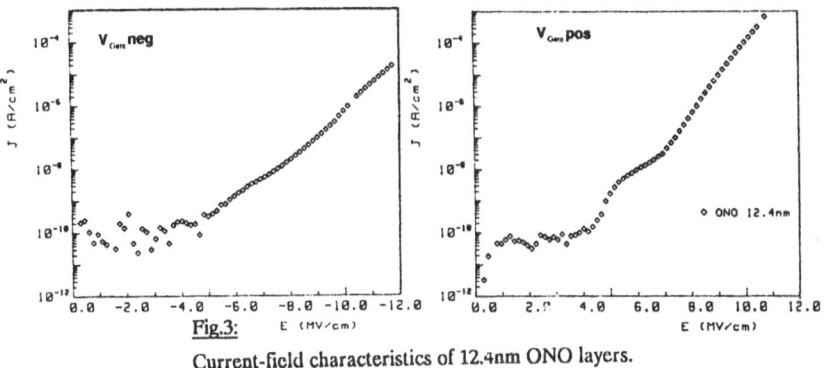

Fig.3:

Current-field characteristics of 12.4nm ONO layers.

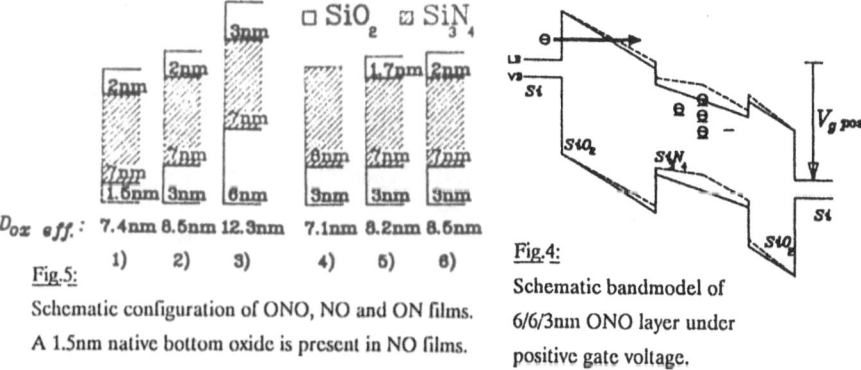

Fig.5:

Schematic configuration of ONO, NO and ON films.
A 1.5nm native bottom oxide is present in NO films.

Fig.4:

Schematic bandmodel of
6/6/3nm ONO layer under
positive gate voltage.

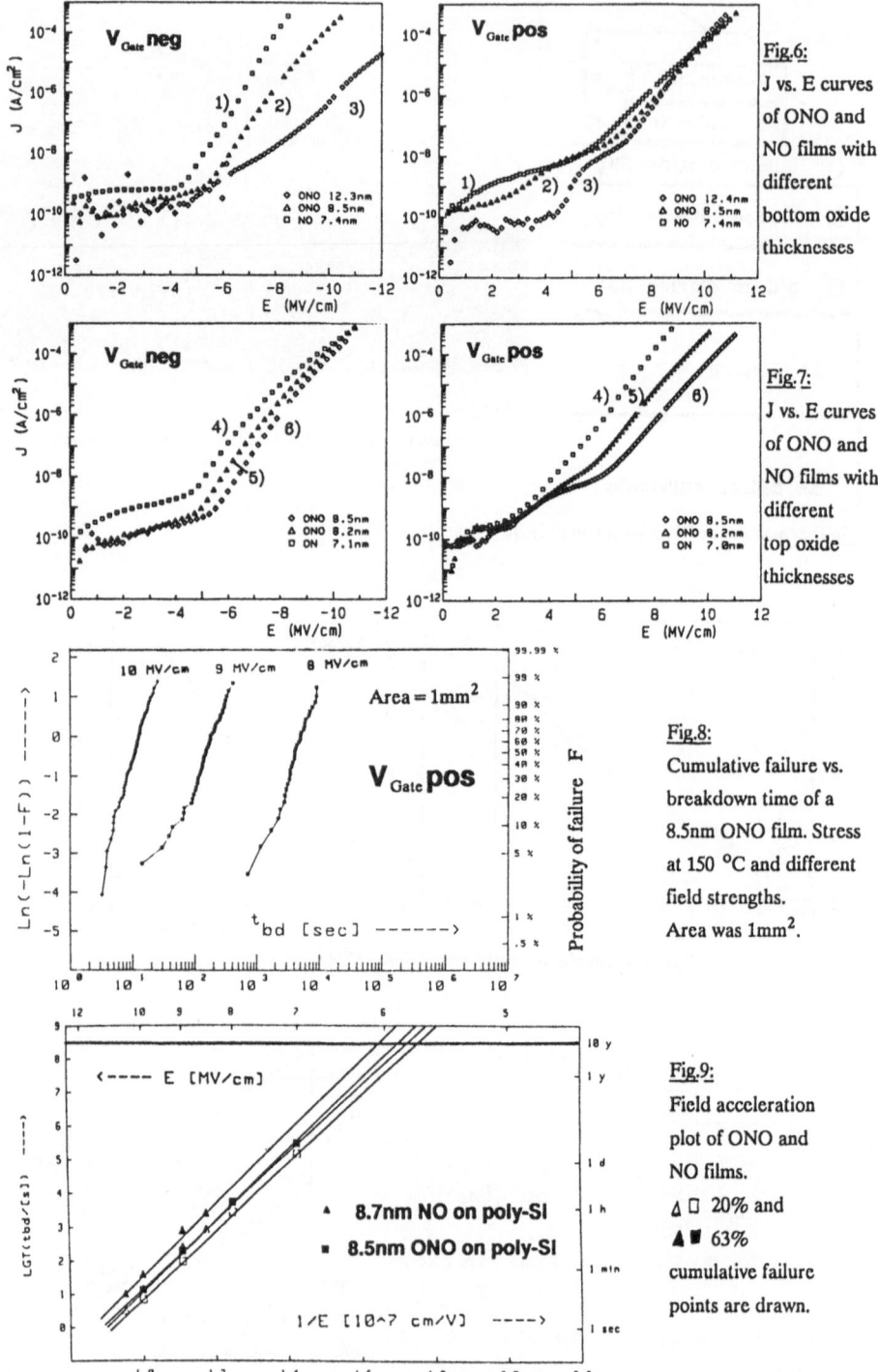

Fig.6:
J vs. E curves
of ONO and
NO films with
different
bottom oxide
thicknesses

Fig.7:
J vs. E curves
of ONO and
NO films with
different
top oxide
thicknesses

Fig.8:
Cumulative failure vs.
breakdown time of a
8.5nm ONO film. Stress
at 150 °C and different
field strengths.
Area was 1mm².

Fig.9:
Field acceleration
plot of ONO and
NO films.
△ □ 20% and
▲ ■ 63%
cumulative failure
points are drawn.

Gate Oxide Reliability in a Sealed Interface Local Oxidation Scheme

I. J. VOORS, K. OSINSKI, F. H. A. VOLLEBREGT and C. A. SEAMS

Advanced MOS Process Development Group
Philips Research Laboratories, P.O. Box 80.000, 5600 JA Eindhoven
The Netherlands

Abstract

The influence of process steps, preceding the gate oxidation, on gate oxide integrity in a Sealed Interface Local Oxidation (SILO) scheme, is studied. Bird's beak etchback procedure and double sacrificial oxidation appeared to have great effect on defect densities, measured by constant current (Q_{bd}) technique. By means of ramped field (E_{bd}) measurements no influence of field oxidation temperature, bird's beak etchback and sacrificial oxidation procedure was seen.

1 Introduction

Together with the introduction of selective field oxidation or LOCOS techniques for the definition of active areas in MOS processes problems with the gate oxide integrity (GOI) were observed. Probably the first paper on this is the one by Kooi et. al. on the so called "white ribbon" effect [1]. Since then many investigations on this phenomenon and the impact of processing steps regarding were reported. These process steps include field oxidation growth temperature [2], oxide etchants [3] and the use of sacrificial oxides [2]. It all concerned here standard LOCOS techniques i.e. with a thermally grown pad oxide layer as a stress relief layer underneath the oxidation blocking nitride layer.

The ongoing miniaturization required the development of new field isolation techniques in order to recess the lateral encroachment (bird's beak). Data on GOI when using recessed field oxidation techniques are scarce. Due to the inherent nature of these techniques it is expected that effects on GOI will become worse compared to the standard techniques [4]. In this paper we will report on the effect of processing steps preceding the gate oxidation on the quality of the gate oxide in a Sealed Interface Local Oxidation (SILO) scheme [5]. In short, the SILO scheme used here comes to the replacement of the thermally grown pad oxide layer as used in standard LOCOS techniques by an LPCVD silicon oxynitride (SiO_xN_y) layer. This results in a recessed bird's beak growth. The SILO scheme results in an abrupt transition region ("sharp edge" point) between the bird's beak and the edge of the field oxide. This allows for a further widening of the active regions by an etchback of the bird's beak. The nominal etchback consists of a complete removal of the bird's beak i.e. until this "sharp edge" point is reached. The etchback is fully integrated in the removal of the sacrificial oxide(s). More details on the used SILO scheme can be found in [6].

Details on the process flow, experimental splits and measurement methods are given in section 2, in section 3 the results are presented and discussed. Finally, conclusions are stated in section 4.

2 Experimental

MOS capacitors were prepared on p-type (100) Si wafers. The process flow is presented in table 1. In this descriptive flow, the main areas around which the experiments on gate oxide integrity concentrate, are also shown. An overview of the experimental splits is given here.

Step	Process	
1	48 nm LPCVD oxynitride deposition	
2	110 nm LPCVD nitride deposition	
3	Definition of oxidation mask	
4	Field oxidation 90% H2O at 900 C	(exp. 1)
5	Oxidation mask removal in hot H3PO4	
6	Bird's beak nominal etchback in 1% HF	(exp. 2)
7	Sacrificial oxidation 55 nm dry O2 at 950 C	(exp. 3)
8	Sacrificial oxide removal in 1% HF	
9	20 nm gate oxidation, dry O2 at 950 C	
10	500 nm poly-Si deposition	
11	Poly-Si dope and definition	
12	Anneal, N2 at 450 C	

Table 1. Descriptive process flow; areas at which experimental splits are performed are also presented (see text).

Exp. 1. Field oxidation temperature : 900°C; 950°C; 1000°C.

Exp. 2a. Etchant of bird's beak etchback : 1% HF; buffered oxide etch (BOE).

Exp. 2b. Bird's beak etchback procedure : 100 nm underetch; nominal etchback; 120 nm overetch.

Exp. 3. Sacrificial oxidation : 55 nm dry O_2 at 950°C; 55 nm dry O_2 at 1000°C; 55 nm H_2O at 950°C; double sacrificial oxidation : 100 nm, strip, 55 nm, both dry O_2 at 950°C.

The gate oxide is characterized by means of ramped field (E_{bd}) and constant current (Q_{bd}) measurements [7,8]. Two different capacitor structures were measured. Module C (plate capacitor) has an area of 0.006 cm^2 and 0.4 cm perimeter of the gate/gate oxide/field oxide boundary. Module D (field edge sensitive capacitor) has an area of 0.0015 cm^2 and 12 cm perimeter.

Using the E_{bd} technique, a negative voltage was applied to the top electrode. Electron injection occured from the gate causing the Si-SiO$_2$ interface to be in accumulation. The field ramp used was 20 MV/cm/sec and the current limit was 1 mA/cm^2. The sample size of every experimental split consisted of at least 6 wafers. On each wafer 234 capacitors of each module type were measured. With the constant current Q_{bd} measurements a negative voltage was applied to the gate. The used current density was 100 mA/cm^2 for module C and 50 mA/cm^2 for module D. Q_{bd} results are based on measurements of 2 wafers, each containing 72 capacitors, per experimental split.

3 Results and discussion

The intrinsic field at breakdown ($E_{bd,intr}$) and the defect density at intrinsic breakdown are derived from extreme value distribution plots for both module C and D. The results for the different experiments are presented in tables 2 to 4. The obtained defect levels range from 1 - 5 per cm^2 for module C and from 1 - 7 per cm^2 for module D. Statistical evaluation pointed out that, with the used sample size, the 99% confidence intervals for the minimum and maximum values of obtained defect densities overlap. This is true for both modules. Hence it cannot be concluded that the defect density distributions between the investigated procedures are different.

Differences were found for the intrinsic field breakdown $E_{bd,intr}$ level between the two modules. Table 2 shows lower levels for module D for the nominal etched and the overetched samples. The samples with an incomplete etchback have the same level for both modules. These results indicate that the presence of a small remaining bird's beak (i.e. the absence of a sharp edge) has a benificial effect on the $E_{bd,intr}$ level. Gate oxide thinning effects around corners may play a role here. Apparently, the samples with nominal etchback were slightly overetched. All other experimental splits, of which the $E_{bd,intr}$ values are presented in tables 3 and 4, have been etched back nominally. For these samples bird's beak absence is expected and lower $E_{bd,intr}$ values were found. However, for the samples processed with a double sacrificial oxidation, no decreased values were observed for the edge sensitive module. It is likely that, using the double oxidation procedure, a bird's beak remains. The integration of the double sacrificial oxidation in the bird's beak etchback procedure causes, especially during the thick sacrificial oxidation, the bird's beak to grow. This is not compensated for during subsequent bird's beak etching.

Overall good $Q_{bd,max}$ values were reached. Defect densities at Q_{bd} levels of 0.1 C/cm^2 and 1.0 C/cm^2 are calculated from extreme value Q_{bd} plots. The results for module C for all experiments are presented in fig. 1 to 3. From fig. 1 can be seen that for all three field oxidation temperatures the samples etched back with BOE have lower defect densities than the samples etched back using 1% HF. An interaction between field oxidation temperature and etchant is present. The samples etched with 1% HF showed decreasing defect densities at increasing field oxidation temperature while this is insignificant for the BOE etchant.

The results for various sacrificial oxidations are shown in fig. 2. The highest defect level is obtained using a single wet oxidation at 950°C. Increasing the oxidation temperature from 950°C to 1000°C in O_2 results in lower defect density. The lowest defect densities are obtained with application of a double sacrificial oxidation. This result can be explained by the remaining of a small bird's beak. For the same reason, the incomplete backetched samples yielded a lower defect level (see fig. 3). For the field oxide etchback split, the edge sensitive module was also measured. It always exhibited a higher defect level.

4 Conclusion

It is shown that reliable gate oxides can be achieved using a sealed interface local oxidation scheme. Various field oxidation temperatures, field oxide etchback procedures and sacrificial oxidations all resulted in a low intrinsic field breakdown defect level of 1 - 7 per cm^2. Q_{bd} testing showed that field oxide etchback in BOE is preferred over 1% HF since lower defect levels were obtained. The presence of a small bird's beak during gate oxide growth has a benificial effect on gate oxide quality. This appeared to occur by use of an incomplete bird's beak etchback, as well as, by application of a second sacrificial oxidation integrated in the bird's beak etchback.

References

[1] E. Kooi, J.G. van Lierop and J.A. Appels, J. Electrochem. Soc., 123 (71), 1976, p. 1117

[2] M. Itsumi and F. Kiyosumi, ibid., 129 (4), 1982, p. 800

[3] T.A. Shankoff et. al., ibid., 127 (1), 1980, p. 216

[4] L.C. Parillo et. al., IEDM Tech. Dig., 1986, pp. 244-247.

[5] J. Hui, T.Y. Chiu, S. Wong and W.G. Oldham in IEDM, 1982, p. 220

[6] K. Osinski et. al., to be published in Philips Journal of Research, 44, 1989

[7] J.J. van der Schoot and D.R. Wolters, Insulating films and Semiconductors, Amsterdam, North Holland Publishing Co, eds. J.F.Verwey and D.R. Wolters, 1983, p. 270.

[8] D.R. Wolters and A.T.A. Zegers-v. Duynhoven, J. Vac. Sci. Technol., A5 (1985), p. 1563.

364

Table 2
Intrinsic field breakdown and defect density for various field
oxidation temperatures and - etchants

Procedure	900°C HF	900°C BOE	950°C HF	950°C BOE	1000°C HF	1000°C BOE
Mod.C; Ebd,intr (MV/cm) Def.dens (per cm2)	9.6 2	9.5 1	9.7 1	9.6 3	9.6 4	9.6 2
Mod.D; Ebd,intr (MV/cm) Def.dens (per cm2)	8.8 4	9.2 1	9.0 1	9.2 3	9.0 1	9.2 1

Table 3
Intrinsic field breakdown and defect density for various field
oxide etchback procedures

Procedure	overetch 120 nm	etchback nominal	underetch 100 nm
Mod.C; Ebd,intr (MV/cm) Def.dens (per cm2)	9.5 1	9.5 2	9.5 1
Mod.D; Ebd,intr (MV/cm) Def.dens (per cm2)	9.0 1	8.8 2	9.5 1

Table 4
Intrinsic field breakdown and defect density for various
sacrificial oxidation procedures

Procedure	55 nm 950°C H2O2	55 nm 950°C O2	55 nm 1000°C O2	100+55 nm 950°C O2
Mod.C; Ebd,intr (MV/cm) Def.dens (per cm2)	9.2 3	9.4 3	9.4 4	9.4 5
Mod.D; Ebd,intr (MV/cm) Def.dens (per cm2)	8.4 6	8.7 1	8.8 6	9.5 7

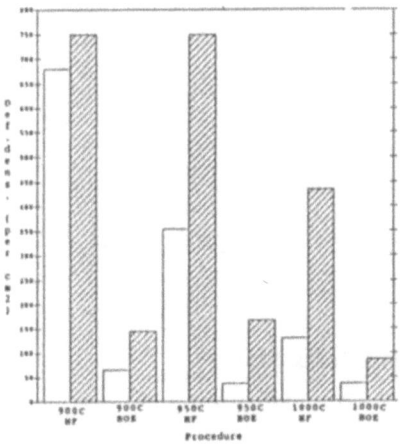

Fig. 1. Defect density of 0.006 cm2 capacitors
for various field oxidation temperatures and - etchants

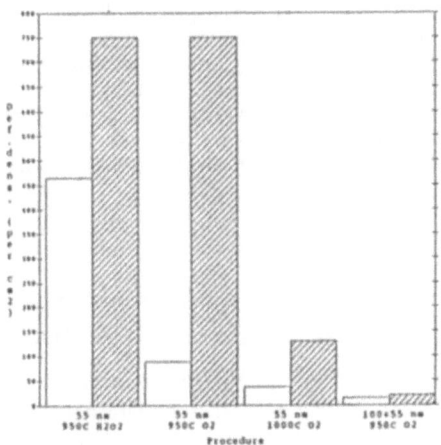

Fig. 2. Defect density of 0.006 cm2 capacitors
for various sacrificial oxidation procedures

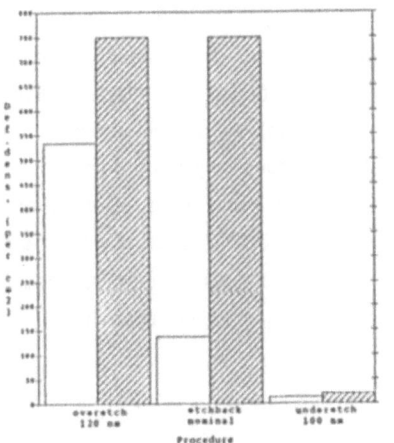

Fig. 3. Defect density of 0.006 cm2 capacitors
for various field oxide etchback procedures

Qbd = 0.1 C/cm2
Qbd = 1.0 C/cm2

Interface Properties and Channel Mobility
of Plasma Nitrided Devices

B. PIOT, A. STRABONI, B. VUILLERMOZ, K. BARLA, M. BERENGUER,
J.F. PORTAILLER

FRANCE TELECOM, CNET , BP 98, F-38243 Meylan Cedex, France.

Abstract : Electrical properties of gate oxides nitrided in plasma at 950 °C and 70 µbar NH3 pressure are studied for thicknesses ranging from 10 nm to 60 nm, and various RF powers and nitridation times. No interface states are generated during plasma nitridation, but fixed charges density rises with increasing nitridation times and/or RF power. Plasma nitridation (950 °C, 3 hours, 70 µbar pNH3, 0.6 kW RF power) of 15 nm gate oxide MOSFET achieves an improvement of channel mobility and current drivability under high normal field (Vg-Vt > 2.5 Volt), whereas the peak transconductance remains comparable to that of a thermal oxide . Finally, hot channel stressing experiments on N-mosfet (tox=15 nm, Leff ≈ 1 .5 µm) show that plasma nitridation reduces the transconductance degradation rate by a factor 2 .

INTRODUCTION

Recent works on nitridation of thin thermal oxides report about their good barrier properties (1)(2), as well as their better endurance under stress conditions (3). However, oxide nitridation induces fixed charges near the Si/SiO2 interface which are usually attributed to the presence of nitrogen. The major effect of these charges is to shift the threshold voltage towards negative values and are reported to be responsible for surface mobility degradation up to 50% in N-channel transistors (3)(4).

In this paper, electrical properties of MIS structures prepared by means of plasma enhanced nitridation at 950 °C of dry thermal oxides have been studied. The effects of different nitridation parameters (time, RF power, oxide thickness) were examined through cartographic C-V hf & quasi static measurements on Al or N doped polysilicon gate capacitors (88 points on 4" wafers). Finally, mobility measurements and long term reliability experiments were performed on devices using a standard 2 µm CMOS technology with a 15 nm, 3 hours nitrided gate oxide. The ammonia plasma treatments were performed in an industrial set-up operating at 13,56 MHz, which has become commercially available recently (5).

RESULTS

a) Process dependance of electrical properties

In order to determine the best nitridation conditions, which preserve the properties of the Si/SiO2 interface, thermal oxides with thicknesses ranging from 10 nm to 60 nm were nitrided for various RF powers, NH3 pressures and nitridation times. **Fig 1** shows that the fixed charge density (Nf) introduced by the nitridation process decreases as a function of the initial oxide thickness, for the 0.7 and 1.0 KW RF powers. These values are calculated from the shift in Vfb after 3 hours nitridation; (this measures the position-weighted charge distribution through the film and not only the fixed interface charge).

The midgap interface state density Nit, calculated from quasi-static C-V measurements remains in the low 10^{10} cm^{-2} ev^{-1} range. Therefore, as no interface states are generated during nitridation , the Vfb shift is mainly due to positive fixed charge generation near the Si/SiO2 interface .

When reducing the thickness of the oxides submitted to a 3 hours nitridation at 0.7 KW RF power and 100 μbar ammonia pressure, Nf is shown to increase from 1.5 10^{11} cm^{-2} for 60 nm to 6.0 10^{11} cm^{-2} for 10 nm. The values obtained for the 1 KW nitrided oxides are always higher, especially for smaller thicknesses. These results are to be related to already reported AES analyses of plasma nitrided oxides (2), which show 3 distinct nitrided regions : the surface oxide which concentrates the major part of the nitrogen content, a nitrogen-rich interface monolayer and the bulk region with a very low nitrogen content. The nitrogen content increases in every region with longer nitridation time. Using high RF power, mainly enhances the interface nitrogen concentration while the surface concentration is slightly affected. The fixed charge generation could be related to the amount of nitrogen incorporated at the interface, but other phenomena resulting from the interactions with the NH3 plasma might play a role, especially in the case of thin oxides nitrided at very high RF power. However the origin of this positive fixed charge generation is still not solved and needs further investigation.

By varying RF power and ammonia pressure at a fixed time (3 h) for 15 nm oxides, we determine the best nitridation conditions at 950 °C to be about 70 μbar NH3 pressure and 0.6 kW RF power. Under these conditions, Nf was ≈ 2.10^{11} cm^{-2}. **Fig 2** shows the nitridation time effect on the Vfb shift. The flat band continously shifts towards negative values from 60 mV after 1 hour up to 260 mV for a 6 hours nitridation. No turnaround effect is observed . (7)

Nitrided transistor characteristics.

For the nitridation of the 15 nm gate oxide of the transistors, we chose an intermediate duration of 3 hours at 950°C, 70μbar, 0.6 kW. That was the strongest nitridation conditions leading to a substantial nitrogen incorporation without producing too high Vfb shift and disporsion.

Id-Vg characteristics of both n and p channel MOSFETS were measured at 300°K for devices with effective channel length (width) of 4.6 μm (48.6 μm). The insulator capacitance Ci was evaluated by C-V measurements on adjacent capacitors. The change in Ci was about +4 % after nitridation. The extrapolated threshold voltage shift (-150 mV) quantitatively corresponds to the shift in Vfb previously evaluated (fig 2) and is equivalent to a ΔNf of+2.2 10^{11} cm^{-2}.

Table 1 summarizes the peak values of the field effect mobility (μFE) (6) for electrons and holes . As compared to thermal nitridation results (3)(4)(7), for which a severe reduction of electron mobility up to 50% and up to 20% for holes have been reported, our values obtained for a 3 hours 950 °C plasma nitridation are about 90% of those measured on the as grown oxide for both N and P channel devices. This could be explained by the low degree of nitridation of the Si/SiO2 interface , i.e. a very slight modification of the initial thermal oxide interface, that has been shown to be characteristic of plasma nitridation (2).

Furthermore,the **fig 3** clearly indicates that plasma nitridation hardly reduces the transconductance peak value and achieves improvement of Gm under high Eeff, leading thus to a significant increase of current drivability in the high Vg-Vt region. Similar results for very light R.T.N. have been reported by Hori (8).

Channel hot carriers stressing .

Fig 4 shows the time dependence of the transconductance degradation rate ΔGm/Gm0 (peak values at Vd=50mV) for N devices (nitrided and not nitrided) with effective channel length (width) of 1.6 μm (48.6 μm) and a 15 nm gate oxide. The gate bias conditions were defined at the maximum of the Ibulk-Vg characteristic for Vd=8 V ; under these stress conditions the bulk current was about 1.6 mA .

It appears that the plasma nitridation of the gate oxide reduces by a factor 2 the degradation rate of Gm as compared to thermal oxides with similar initial characteristics .

CONCLUSION

It has been shown that, when using low RF power, a 3 hours plasma nitridation of thin gate oxides does not degrade the initial properties of the interface. There is no interface states generation, and lower fixed charge densities are obtained as compared to thermal NH3 annealing processes. These properties are mainly due to the characteristics of the plasma process, that induces a preferential nitridation of the surface oxide, while both the bulk and the interface are slightly affected. Furthermore, the nitridation hardly reduces the channel mobility and achieves improvement of Gm and current drivability under high normal fields; finally plasma nitrided devices show a better endurance when submitted to hot electron channel stressing.

References .
(1) S.S.Wong, C.G.Sodini, T.W.Ekstedt, H.R.Grinolds, K.H.Jackson, S.H.Kwan, W.G.Oldham, J. Electrochem. Soc., Vol 130, No 5, p 1139, 1983
(2) Ph.Debenest, K.Barla, A.Straboni, B.Vuillermoz, Appl.Surface Sci., 36, p196,1989
(3) W.Yang, R.Jayaraman, C.G.Sodini, IEEE Trans. Electron Device, Vol. 35, p 935, 1988
(4) T.Kusaka, A.Hiraiwa, K.Mukai, J. Electrochem. Soc., Vol 135, No 1, p 166,1988
(5) ATEA, BP 32, 44472 Carquefou Cedex FRANCE
(6) S.C.Sun, J.D. Plummer, IEEE Trans. Electron Device,Vol 27, p 1497, 1980
(7) M.A.Schmidt,F.L.Terry,B.P.Mathur,S.D.Senturia, IEEE Trans. Electron Device,Vol 35, p 1627,1988
(8) T.Hori, H.Iwasaki,IEEE Electron Devices Lett.,Vol 10, No 5, p 195,May 1989

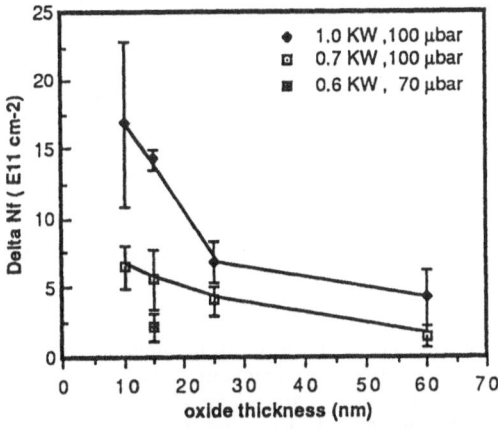

Fig1: Fixed charge density variation vs oxide thickness
Plasma nitridation : 950 °C, 3 hours

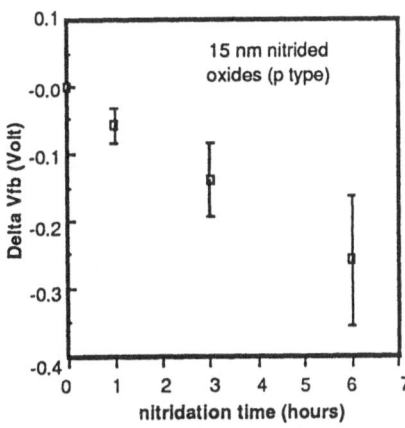

Fig2: Flat band voltage shift vs nitridation time
at 950 °C,pNH3:70 µbar;RF power:600 watt

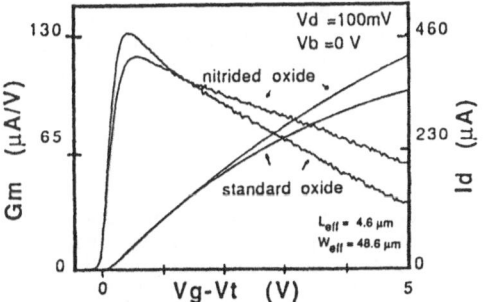

Fig3:Transconductance & drain current
vs gate voltage for 15 nm standard oxide
and 3 hours, 950°C, 70 µbar, 600 watt
plasma nitrided oxide.

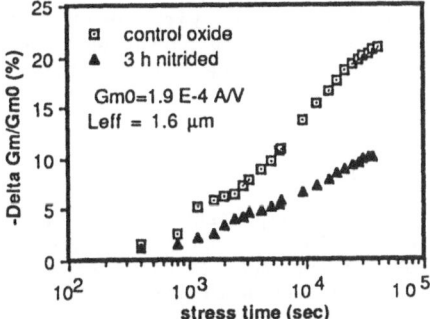

Fig4: Peak transconductance degradation
Stress conditions: Vg=3 V ; Vd=8 V,Vb=0V
Measurements at Vd= 50 mV

Table 1: Peak effective channel mobility :
μFE = Gmmax/(Cox . Vd . Weff/Leff) ; Vd = 50 mV

μFE (cm2/V.s)	standard oxide	nitrided oxide	nitridation 950°C, NH3
electrons	535 ± 10	500 ± 40	3 h plasma 600 Watt
holes	189 ± 3	175 ± 5	
μn (ref 4)	540	400	1 h thermal

The Influence of Cleaning on SiO$_2$ Growth

R. Wiget[1], H. Ryssel[1][2] and W. Aderhold[2]

1) Fraunhofer-Arbeitsgruppe für Integrierte Schaltungen, Artilleriestrasse 12, 8520 Erlangen, FRG
2) Lehrstuhl für Elektronische Bauelemente, Artilleriestrasse 12, 8520 Erlangen, FRG

Abstract - The oxidation rate of single crystalline silicon depends on the preoxidation cleaning procedure. Various cleaning procedures were compared using aqueous solutions of $NH_4OH-H_2O_2$, $HCl-H_2O_2$ and $H_2SO_4 \cdot H_2O_2$. A dip in 10% HF was used either at the beginning or at the end of the cleaning procedure. Thermal oxidation was carried out at 885°C, 985°C and 1085°C in a conventional diffusion furnace. The oxidation rates are maximum for the samples treated with HF as a final step in the cleaning procedure whereas a final treatment with NH_4OH leads to a minimum oxidation rate. ESCA measurements were performed to investigate the chemical state of the SiO_2-Si interface. The oxide contains more silicon after HF etching than after NH_4OH etching.

INTRODUCTION

A very important step in VLSI technology is wafer cleaning. Especially before high temperature processes, cleaning steps are used in order to obtain high quality oxide layers and to avoid an unintended doping of the bulk silicon with life time killers such as gold or iron. Without proper cleaning procedures, high yield and reliability, which are mandatory for the increasing integration density of advanced devices, cannot be achieved.

It has been reported that preoxidation cleaning influences the oxidation kinetics (as well as crystal orientation, doping level, etc. do) [1-6]. In this study, the oxidation kinetics of ⟨100⟩-oriented silicon were investigated for oxidation temperatures of 885, 985, and 1085°C and oxidation periods up to 300 min. After different surface treatments the kinetic parameters were extracted from the experimental data using the model of Deal and Grove [7].

To determine the reason for different growth rates as a function of the cleaning process, ESCA (Electron spectroscopy for chemical analysis) and TXRF (total reflection x-ray fluorescence) measurements were performed.

EXPERIMENTAL

The silicon wafers used in this investigation were CZ, ⟨100⟩ oriented, phosphorus doped with resistivity in the range of 1 to 10 Ωcm from Wacker.

The chemicals used for cleaning were MOS grade Selectipur H_2SO_4 (96%), HCl (32%) and NH_4OH (29%) mixed with peroxide (30%), optionally followed by an HF dip. In Table 1, details are given concerning etching period, sequence and temperature. For comparison, wafers were used as received from the manufacturer, i. e. without cleaning.

In several cases, oxidized wafers were used as starting material. In these cases, the oxide was removed with 10 % HF prior to the above cleaning sequence. But no difference in oxidation kinetic was observed in comparison to silicon wafers used as delivered.

Thermal oxidations were performed at temperatures of 885, 985 and 1085°C with a temperature stability of ±0.5°C in dry oxygen under atmospheric pressure. At 1085°C, also experiments with wet oxidation were performed. The duration of the oxidation was chosen to result in thicknesses between 40 and 210 nm. The samples were slowly pushed into the furnace with 25 cm/min and the temperature was ramped up from 800°C with 8°C/min. Both steps were performed in a dry nitrogen atmosphere. The cool-down after oxidation was again performed in nitrogen with a rate of 5°C/min.

Wafers with different pre-oxidation cleaning were oxidized together in the same run. An automated ellipsometer was used to determine the oxide thicknesses. The oxidation data are the average of at least five measurements per wafer and the average from two to four runs. After cleaning, oxidations were performed within 4 days. Prior to oxidation, the native oxide was measured in several cases using dummy wafers. Without HF dip, it was found to be 1.25 and 1.4 nm, after HF dip it was found to be irrespective of the cleaning 0.6 nm.

ESCA and TXRF measurements were performed after cleaning # 1 with and without HF dip and # 2 and # 3 only without HF dip.

RESULTS

In Fig. 1, the oxide thickness is given as a function of the oxidation time for cleaning etchants # 1 to # 3 and the non-cleaned reference. With etch # 3 (NH_4OH : H_2O_2), the oxide thickness is the smallest, with etch # 1 (H_2SO_4 : H_2O_2) it is always the largest. For the samples which were dip etched in HF after the cleaning step, however, the oxide thicknesses are identical for all the different procedures and still larger than in case of etch # 1. The samples which underwent no particular cleaning step have an oxide thickness close to that obtained after cleaning with etch # 3. These results agree well with those of Kao and Gould [3,5]. The samples cleaned with etchants 1 and 2 as well as the uncleaned sample always lie between the limits set by etch # 3 and all cleaning steps followed by an HF dip, the results, however, tend to depend slightly on doping concentration and the batch number of the wafers. Only with etch # 3 or HF dips, reproducible results are possible.

Fitting the linear-parabolic model by Deal and Grove to the experimental data, the parabolic and linear rate constants were determined, see Fig. 3.

In contrast to Kao and Deal [5] as well as Schwettmann et al. [2], we found that the parabolic rate is influenced by the cleaning procedure, whereas the linear rate constant remains essentially unchanged. This agrees with the findings of Ruzyllo [4], that oxide thicknesses remain unchanged for layers thinner than 25 nm. In Fig. 2, the measured difference between the oxide thickness of wafers cleaned with procedures # 1 to # 3 with an additional HF dip (which always resulted in the same oxide thickness) and wafers cleaned with procedure # 3 without HF dip is shown. The solid line was calculated by using our rate constants. The broken line was calculated by using the same parabolic rate constant, the linear rate constant was arbitrarily multiplied by a factor of 2. It can be seen that the thickness difference cannot be described satisfactory by using a different linear rate constant, whereas for different parabolic rate constants the agreement with the data points is good. Like other authors [5,6], we found that in the case of NH_4OH cleaning without HF dip there is a transport limiting surface layer that stays on top of the oxide during oxidation. After etching off this layer with HF, the dependence of the oxide growth on the cleaning procedures disappeared.

Measurements of the refractive index by ellipsometry on wafers with 65 nm of oxide showed a difference between samples that underwent different preoxidation cleanings. For NH_4OH cleaned samples, we found 1.465 and for samples with an HF dip 1.457 for the refractive index. Applying the Lorentz-Lorenz relationship, densities of 2.224 and 2.192 g/cm^3 can be calculated. Other authors reported similar values [3]. It can be assumed that in the early stage of dry oxidation, an oxide film of different density grows depending on the cleaning procedure. Since for dry oxidation very little exchange of the oxygen with the silica lattice takes place [11], this denser layer remains on top of the oxide forming a barrier for the oxygen diffusion. Fig. 2 also shows that for an oxide thickness larger than 200 nm, the difference of oxide thicknesses does not follow the calculated solid line. This can be explained if a two-layer linear-parabolic model is applied [5]. In this model, two parabolic rate constants are used, one for the initial oxide and one for the oxide that grows after approx. 200 nm. For the oxide thicknesses of interest in dry oxidation, however, the simple linear parabolic law describes the growth sufficiently well.

We also carried out wet oxidations (1085°C) for wafers cleaned with the different etchants and did not measure any significant difference in oxide thickness. An explanation for this might be that in wet oxidation, a reaction of water molecules with the silica lattice takes place determining the oxide structure. So neither a denser top layer can form nor remain.

Using ESCA measurements, it is possible to investigate the surfaces of cleaned wafers. In Table 2 results are given for wafers with etch # 1, # 2, and # 3 as well as etch # 1 with HF dip. The measurements revealed a thin layer of SiO_x with an x between 1.45 and 2 at the Si/SiO_2 - interface. Such layers are well described in several articles [8,9,10]. In the spectra, all oxidation states of silicon are present. However, the distribution of the intermediate states of silicon and, therefore, the stoichiometry is different. Low oxidation states are dominant for the HF dip and etch # 1, whereas the high oxidation state (Si^{3+}) is dominant for silicon wafers cleaned with etch # 3. These results show that the retarded oxide diffusion in case of etch # 3 is caused by the relatively high concentration of Si^{3+} states.

In addition, TXRF - measurements were performed to investigate a possible influence of surface contamination. The spectra of silicon wafers with an final HF dip show actually no detectable contamination with transition metal, whereas all other silicon surfaces were contaminated. But no correlation with the oxidation data could be found.

REFERENCES

1. F. J. Grunthaner, J. Masserjian, IEEE Trans. Nucl. Sci., NS-24, 2108 (1977)
2. F. N. Schwettmann, K. L. Chiang, and W. A. Brown, Electrochem. Soc., Meeting, May 1987
3. G. Gould and E. A. Irene, J. Electrochem. Soc., July 1987
4. Jerzy Ruzyllo, J. Electrochem. Soc., July 1987
5. D. B. Kao and B. E. Deal, Ext. Abstr. 88-1, No. 255, Electrochem. Soc., 173th Meeting, 1988
6. J. M. Delarios, C. R. Helms, D. B. Kao, B. E. Deal, Infos 87, Proceedings of the 5th intern. Conf. on Isolating Films on Semiconductors, North Holland 1987, p. 17
7. B. E. Deal and A. S. Grove, J. Appl. Phys., 36 (12), 3770, (1965)
8. F. J. Himpsel, Appl. Phys. Lett., Vol 44, No. 1, 1. January 1984
9. M. Grundner and H. Jacob, Appl. Phys. A 39, 73 - 82, (1986)
10. R. Flitsch and S. I. Raider, J. Vac. Sci. Technol., Vol 12, No.1, Jan/Feb. 1975
11. G. Barbottin, A. Vapaille "Instabilities in Silicon Devices", Vol. 1, North Holland,1986

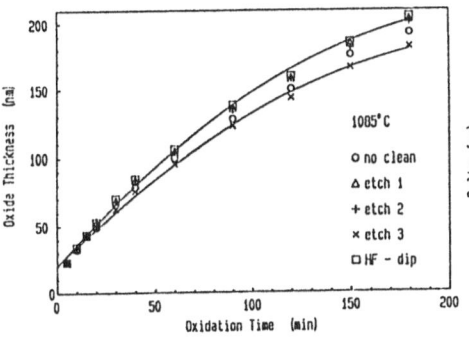

Fig.1: Measured oxide tickness after different cleaning for dry oxidation at 1085°C.
solid line: least square fit to the experimental data.

Fig.2: Difference in thickness of silicon wafers oxidized at 1085°C, precleaned with a final HF dip and etch # 3 without HF dip (x) compared to theory.

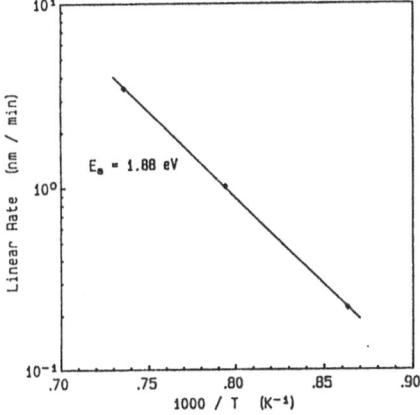

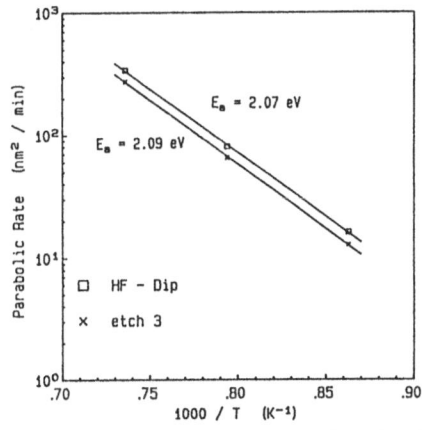

Fig.3: Arrhenius plots of the linear and parabolic rate constants for ⟨100⟩ silicon oxidized in dry oxygen (cleaning procedures as indicated).

cleaning sequence	1
a) cleaning etch # 1, 2, or 3	$5H_2SO_4 : 2H_2O$
b) DI rinse to 18 MΩcm	5 min, 130 °C
c) dip - etching in 10% HF	
d) DI rinse to 18 MΩcm	2
e) dry spinning in nitrogen	$HCl : H_2O_2 : 6H_2O$
c + d optinal	10 min, 80 °C
no clean	3
as received	$NH_4OH : H_2O_2 : 5H_2O$
	10 min, 80 °C

Cleaning Procedure	etch 1 + HF dip	etch 1	etch 2	etch 3
Oxide Thickness	0.62 nm	1.03 nm	1.4 nm	1.25 nm
Si^{1+}	6.4 %	4.4 %	5.1 %	1.3 %
Si^{2+}	2.9 %	3.1 %	0.9 %	1.1 %
Si^{3+}	1.4 %	1.2 %	5.1 %	6.5 %
Stoichiometry of the SiO_2/ Si Interface	$SiO_{1.48}$	$SiO_{1.58}$	$SiO_{1.96}$	$SiO_{1.75}$
High Oxidation-Rate Low Oxide Density	⟵		⟶	Low Oxidation-Rate High Oxide Density

Table 1: Pre-Oxidation cleaning procedure

Table 2: Distribution of the intermidiate states Si^{1+}, Si^{2+} and Si^{3+} of ⟨100⟩ Si/SiO_2 - interfaces after different chemical cleanings

Optoelectronic Devices

Assessment of Pulse-to-Pulse Timing Jitter in Periodically Gain-switched Semiconductor Lasers

E. H. Böttcher and D. Bimberg

Institut für Festkörperphysik I der Technischen Universität Berlin, Sekr. PN 5-2 , Hardenbergstr. 36 , D-1000 Berlin 12, F.R.G.

SUMMARY

Methods to assess pulse-to-pulse timing jitter in periodically gain-switched semiconductor lasers are investigated. The power spectrum of the optical pulse train for mutually incoherent fluctuations of the turn-on delay time and the pulse amplitude is calculated. The result shows that a measurement of the power spectrum alone is not sufficient for the characterization of the inherent timing jitter. Additional knowledge of the optical waveform and the transfer characteristics of the detection system is necessary.

In this investigation, we focus on the characteristics of the source of noise in opical fiber communications which has attracted considerable attention most recently: the pulse-to-pulse timing jitter of gain-switched semiconductor lasers due to random fluctuations of the turn-on delay time [1-4]. The origin of this type of timing jitter is attributed to transient fluctuations of the optical power during the turn-on process of the laser. With respect to long range optical fiber communications, the experimental observation that the timing jitter in single-mode lasers is considerable larger than in multi-mode ones [1,4] makes a detailed treatment even more important.

Basically, two different experimental approaches were followed to investigate the timing jitter of optical pulse trains converted to electrical pulse trains by fast photodetectors: sequential sampling by fast sampling scopes [1,3,4] and spectral analysis by microwave spectral analysers [2]. We compare these two techniques and demonstrate that the latter one is only of limited practical value for the characterization of the inherent timing jitter.

In order to model the pulse-to-pulse jitter we use the following mathematical description for the derivation of its power spectrum. We consider a train of optical pulses generated by periodical current modulation of a semiconductor laser. In the absence of noise the temporal waveform of the optical signal $y_\Delta(t)$ is given by a sum of single pulses separated by the time period T of the excitation

$$y_\Delta(t) = \sum_{n=-N}^{N} x \; f(t-\tau-nT) \tag{1}$$

x denotes the amplitude, $f(t)$ the temporal pulse shape, and τ the turn-on delay time of a single pulse. Δ is the time interval under consideration and equals 2NT. In the following we omit the summation limits. Summation is always meant from $-N$ to N unless it is explicitly stated otherwise. Now we include turn-on as well as pulse amplitude fluctuations

$$y_{\Delta'}(t) = \sum_{n} x_n \; f(t-\tau_n-nT) \tag{2}$$

where x_n and τ_n are random variables which refer to the n-th excitation event of the laser. Their statistics is defined by the respective probability density functions (PDF's) given by $p(x_n)$ and $p(\tau_n)$. We assume that x_n and τ_n are independent random variables with equal PDF's from pulse to pulse. Thus

$$p(x_n) = p(x_m) \quad \text{and} \quad p(\tau_n) = p(\tau_m) \quad \text{for all n,m} \tag{3}$$

The model for the random process formed refers to timing and amplitude fluctuations from pulse to pulse caused by the inherent noise during the turn-on of the laser. Extrinsic noise contributions like phase and amplitude jitter of the electrical driving signal and temperature variations of the laser which can affect or even dominate the amount of the pulse-to-pulse fluctuations are not included. In these circumstances, the random signal $y_\Delta(t)$ corresponds to a periodical stochastic process with period T for $\Delta \to \infty$. In order to determine the power spectrum of the signal we use a representation of the Wiener-Khinchine theorem which establishes the relation between the power spectrum $G_y(\omega)$ and the expectation value of the square modulus of the Fourier transform of the truncated random signal

$$G_y(\omega) = \lim_{\Delta \to \infty} \frac{E\left[\left|Y_\Delta(\omega)\right|^2\right]}{\Delta} \tag{4}$$

$E[\ ]$ is the expection value operator and means averaging over all realizations of the random signal $\left|Y_\Delta(\omega)\right|^2$ where $Y_\Delta(\omega)$ is the Fourier transform of $y(t)$ which is given by

$$Y_\Delta(\omega) = \sum_n x_n \, F(\omega) \, e^{-i\omega(nT-\tau_n)} \tag{5}$$

$F(\omega)$ is the Fourier transformation of $f(t)$. Then taking the square modulus and performing the averaging operation with respect to the statistically independent random variables τ_n and x_n we obtain the power spectrum defined by Eq. (4)

$$G_y(\omega) = \lim_{N \to \infty} \frac{(2N+1)\left|F(\omega)\right|^2\left(\overline{x^2} + \overline{x}^2 \left|P_\tau(\omega)\right|^2 \sum_n e^{-i\omega n T}\right)}{2 N T} \tag{6}$$

$\overline{x^2}$ is the mean square value and $\overline{x}$ is the mean value of x_n which are equal for all n. $P_\tau(\omega)$ is the Fourier transform of $p(\tau_n)$. Hence, taking the limit $N \to \infty$ we obtain

$$G_y(\omega) = \frac{\left|F(\omega)\right|^2}{T} \left[\overline{x^2} + \frac{2\pi \, \overline{x}^2 \, \left|P_\tau(\omega)\right|^2}{T} \sum_{n=-\infty}^{+\infty} \delta\left(\omega - \frac{2\pi n}{T}\right)\right] \tag{7}$$

Hence the power spectrum is essentially a train of δ-functions with period T, the envelope of which is given by the product $\left|F(\omega)\right|^2 \left|P_\tau(\omega)\right|^2$.

The switch-on process from a level below the laser threshold to a state above threshold is nonstationary. Due to the periodical excitation, the process we are investigating is termed periodically nonstationary. Therefore in general, nonstationary processes have to be decribed by nonstationary statistical parameters. The technique applied here circumvents this problem. The result of Eq. (7) is a time independent power spectrum because it contains a transformation of the nonstationary process into a stationary one by implicite phase randomizing with the consequence that the statistical parameters are averaged over one period T.

Now we wish to examine how information on the timing jitter can be extracted from a power spectrum in the form of Eq.(7). Since we have to take into account the transfer characteristics of the detection system the experimentally observed power spectrum $G_z(\omega)$ can be expressed by

$$G_z(\omega) = 2\pi \ |H(\omega)|^2 \ G_y(\omega) \tag{8}$$

where $H(\omega)$ is the tranfer function of the detection system which is assumed to be linear. It is evident from Eqs.(7) and (8) that for the extraction of timing noise from experimental data both the knowledge of the waveform $f(t)$ of the light pulses and of the tranfer function $H(\omega)$ is required. Consequently, the sole measurement of the microwave spectra of rapidly current modulated semiconductor lasers is not sufficient for the characterization of the inherent timing jitter.

We believe that our results can contribute to a better understanding of the relatively large differences of timing jitter values reported so far. Though different types of injection lasers and driving conditions were used the following tendency is observed. Considerable larger values of timing jitter were detected by sampling techniques [1,3,4] compared to spectral analysis [4]. In view of the results presented here, this is evident, because inherent timing jitter can not be measured directly by power spectra analysis. It yields useful information only if the pulse-to-pulse fluctuations are mutually coherent. Therefore, sampling techniques should be prefered for the assessment of these fluctuations.

REFERENCES

1. M. M. Choy, P. L. Liu, P. W. Shumate, T. P. Lee,and S. Tsuji, Appl. Phys. Lett. 47, 448 (1985).

2. A. J. Taylor, J. M. Wiesenfeld, G. Eisenstein, and R. S. Tucker, Appl. Phys. Lett. 49, 681 (1986).

3. E. H. Böttcher, K. Ketterer, and D. Bimberg, J. Appl. Phys. 63, 2469 (1988).

4. P. Spano, A. D'Ottavi, A. Mecozzi, and B. Daino, Appl. Phys. Lett. 52, 2203 (1988).

Monolithic $Pb_{1-x}Sn_xSe$ on Si Infrared Sensor Array for the 8-12 µm Range

C. Maissen, J. Masek, H. Zogg, S. Blunier
AFIF (Arbeitsgemeinschaft für industrielle Forschung)
at Swiss Federal Institute of Technology,
ETH-Hönggerberg, CH-8093 Zürich, Switzerland

A. Lambrecht, M. Tacke
Fraunhofer-Institut für Physikalische Messtechnik,
Heidenhofstr. 8, D-7800 Freiburg, FRG

Abstract
   Linear photovoltaic IR-sensor arrays for the 8-12 µm wavelength range have been fabricated for the first time in narrow gap $Pb_{1-x}Sn_xSe$ layers grown heteroepitaxially on Si(111) substrates. Heteroepitaxy was achieved using stacked intermediate CaF_2-BaF_2 buffer layers. Both, the 2000 Å thick fluoride buffer layer and the $Pb_{1-x}Sn_xSe$ were grown by molecular beam epitaxy (MBE). Differential resistance times area products (R_0A) are up to 0.3 Ωcm^2 at 77 K. This corresponds to junction noise limited detectivities of $2.4 \cdot 10^{10}$ cm$\sqrt{Hz}$/W which are only 5-10 times lower than those of state of the art $Hg_{1-x}Cd_xTe$ sensors on non Si substrates with the same cut-off wavelength, but still above the 300 K background noise limit.

INTRODUCTION

Thermal imaging is not only confined to military applications but also finds more and more use in medicine, earth observation, biology and meteorology. To observe objects having a temperature around 300 K the best spectral wavelength range is 8-12 µm. In this range the atmosphere is relatively transparent and the maximum power emitted by a 300 K blackbody is around 10 µm, according to Wien's displacement law.

Highest detectivity at a given operation temperature is obtained by using a fully staring array of intrinsic narrow gap semiconductor (NGS) sensors. The NGS sensors have the advantages that their cut-off wavelength can be tailored to the desired value by using a suitable composition of ternary alloys and their quantum efficiencies are very high.

Unfortunately, fabrication of signal processing circuitries in narrow gap semiconductors is very difficult and not developed. Therefore the only practical way is to fabricate the read-out electronics in a silicon chip, and then interconnect each individual sensor element with the read-out electronics. The usual procedure is to solder every detector element with an indium bump to the read-out electronics [1]. This technology is rather complicated and expensive for a large 2-dimensional array, and moreover the pixel size is limited by the indium bump. As a consequence of the increasing number of pixels the interconnections between the detectors and the signal processing unit becomes more and more difficult.

A much more elegant approach would be to grow the NGS layers heteroepitaxially on the Si-Chip which already contains the read-out electronics, and then fabricate the detectors in the NGS thin film. To this end we have grown epitaxial $Pb_{1-x}Sn_xSe$ layers on Si(111) by using an intermediate stacked CaF_2-BaF_2 buffer layer to overcome the large lattice mismatch of up to 20%, as well as the thermal expansion mismatch between $Pb_{1-x}Sn_xSe$ and the silicon substrate, which differs by a factor of 7. The thermal strain relaxation capability on temperature changes is due to movements of misfit dislocations. In the NGS layers we have fabricated a fully working IR-sensor array with a cut-off wavelength of 11.6 μm at 77 K.

We have used $Pb_{1-x}Sn_xSe$ instead of $Hg_{1-x}Cd_xTe$, the most used NGS for the 8-12 μm wavelength range. The main reasons are: the technology for growing epitaxial layers and the fabrication of sensors is much simpler for $Pb_{1-x}Sn_xSe$ than for $Hg_{1-x}Cd_xTe$ and we have already fabricated linear arrays in PbTe on Si [2] whose best sensors reach detectivities comparable to $Hg_{1-x}Cd_xTe$ on *non* Si substrates with the same cut-off wavelength of 5.7 μm at 77K [3].

SENSOR FABRICATION

We have used Si(111) substrates for sensor fabrication, which were cleaned by a modified Shiraki method [4]. This method needs a high temperature (up to 1000° C) rapid thermal anneal. However, another low temperature procedure [5] can be used instead. The stacked 2000 Å thick CaF_2-BaF_2 buffer layer is grown on Si(111) by MBE [6]. The growth temperature for CaF_2 was about 750° C and was then decreased to 550° C for BaF_2 growth. On the fluoride covered substrates the $Pb_{1-x}Sn_xSe$ was deposited at 380° C in a separate MBE system. We use PbSe and SnSe as effusion material. The appropriate composition was determined by the flux ratio of the PbSe and the SnSe. Coevaporation of Se leads to the desired p-type carrier concentration of about $4 \cdot 10^{17}$ cm^{-3}.

A common ohmic contact was formed by Pt. Our array consists of 66 Schottky diodes each of 50×100 μm^2 area, which were delineated by depositing Pb on the p-type $Pb_{1-x}Sn_xSe$. The diodes are arranged in a staggered way with a pitch of 100 μm [7]. No antireflection coating and no passivation layer was used.

RESULTS

Fig. 1 shows the current-voltage characteristics of a sensor at 77 K. The differential

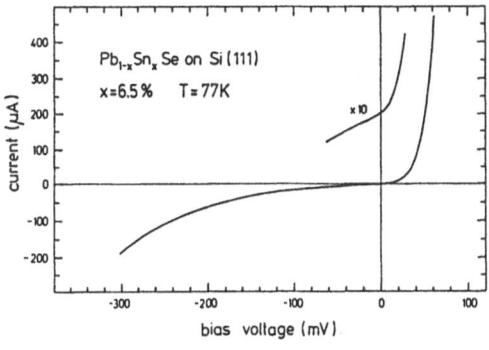

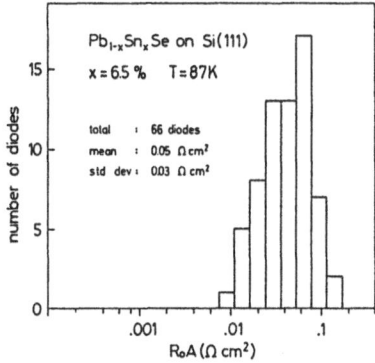

Fig. 1. Current-voltage characteristics of an epitaxial $Pb_{1-x}Sn_xSe$ on fluoride covered Si photovoltaic sensor array at 77 K.

Fig. 2. Distribution of resistance-area products R_0A of a 66 element linear photovoltaic IR-sensor array at 87 K.

resistance at zero bias is 6 kΩ and the R_0A product is 0.3 Ωcm^2. Assuming 30% external quantum efficiency this corresponds to a junction noise limited detectivity $D^*=2.4 \cdot 10^{10}$ cm$\sqrt{Hz}$/W, which is only 5-10 times lower than the best published [8] values for $Hg_{1-x}Cd_xTe$ sensors with the same cut-off wavelength. A histogram of the R_0A products for one array with 11.3 μm cut-off wavelength at 87 K is shown in fig. 2. All the 66 elements are functional and the mean R_0A value is about 0.05 Ωcm^2. The temperature dependence of the R_0A product indicates that the noise behaviour is diffusion limited for temperatures down to 100 K and depletion limited for temperatures between 100 K and 77 K [9].

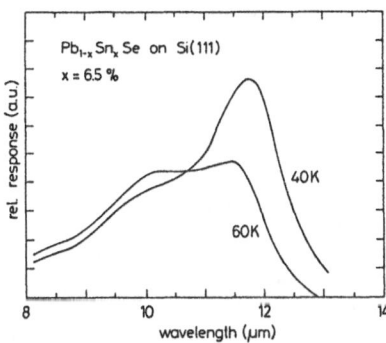

Fig. 3. Spectral response (Photocurrent per Watt incident power) for a photovoltaic $Pb_{1-x}Sn_xSe$ on Si IR-sensor at 40 K and 60 K.

The measured quantum efficiencies were up to 30% without antireflection coating. In fig. 3 the spectral response is shown at 40 and 60 K. At 60 K the cut-off wavelength is about 12 µm which increases to 12.3 µm at 40K due to the positive temperature coefficient of the band gap energy.

CONCLUSION

The first IR-sensor array with a cut-off wavelength as high as 12 µm at 60 K has been fabricated in narrow gap $Pb_{1-x}Sn_xSe$ grown heteroepitaxially on Si(111). We achieved junction noise limited detectivities D^* up to $2.4 \cdot 10^{10}$ cm$\sqrt{Hz}$/W at 77 K, which is still above the 300K background noise limit. In regard of our crude technology and the progresses we achieved with PbTe further improvements to detectivities comparable with state of the art $Hg_{1-x}Cd_xTe$ sensors seem feasible. Growth of the $Pb_{1-x}Sn_xSe$ films and sensor fabrication are compatible with Si device processing. Therefore we have opened the way to a heteroepitaxial and fully monolithic approach for making IR focal plane arrays on Si which already include read-out electronics.

Acknowledgment: The work is sponsored by the Swiss Defense Technology and Procurement Agency, the Swiss National Science Foundation, and the European Space Agency.

REFERENCES

[1] K. Vural, *Mercury cadmium telluride short- and medium-wavelength infrared staring focal plane arrays,* Optical Engineering, **26** (1987), p. 201.

[2] C. Maissen, J. Masek, H. Zogg, S. Blunier, *Photovoltaic infrared sensors in heteroepitaxial PbTe on Si,* Appl. Phys. Lett. **53**,(1988), p. 1608.

[3] C. Maissen, J. Masek, S. Blunier, H. Zogg, *Photovoltaische Infrarotsensoren auf Silizium Substraten,* Helv. Phys. Acta **62** (1989), p. 270.

[4] A. Ishizaka, Y. Shiraki, *Low Temperature Surface Cleaning of Silicon and Its Application to Silicon MBE,* J. Electrochem. Soc. **133** (1986), p. 666.

[5] P. J. Grunthaner, F. J. Grunthaner, R. W. Fathauer, T. L. Lin, F. D. Schowengerdt, B. Pate, J. H. Mazur, *Low temperature substrate cleaning technology for Si MBE,* 2nd Int. Symp. on Si MBE, Honolulu, Oct. 1987, Electrochem. Soc. Proc. **88-8,** p. 375

[6] H. Zogg, S. Blunier, J. Masek, *Progress in Compound-Semiconductor-on-Silicon-Heteroepitaxy with Fluoride Buffer Layers,* J. Electrochem. Soc. Vol. **136,** No. 3 (1989), p. 775.

[7] J. Masek, C. Maissen, H. Zogg, *Photovoltaic infrared sensor array in heteroepitaxial narrow gap lead-chalcogenides on silicon,* J. de Physique, Colloque C4, supplément au n°9, Tome **49,** (1988), p. 697.

[8] J. Ameurlaine, A. Rousseau, T. Nguyen-Duy, R. Triboulet, *(HgZn)Te infrared detectors performance,* Proc. SPIE **865** (1987), p. 30.

[9] H. Holloway, in Physics of Thin Films, Academic Press, G. Haas, M. H. Francombe ed., Vol. **11,** 1980, p. 105.

InP Based Integrated Laser Driver Circuit

A. PARASKEVOPOULOS, H.G. BACH, G. MEKONNEN, H. SCHROETER-JANβEN,
F. FIEDLER*, N. GROTE

Heinrich-Hertz-Institut für Nachrichtentechnik Berlin GmbH, Einsteinufer
37, D-1000 Berlin (FRG)
*Present address: Forschungsinstitut im Fernmeldetechnischen Zentralamt
der Deutschen Bundespost, Am Kavalleriesand, 6100 Darmstadt (FRG)

Abstract

An integrated laser driver circuit representing a step towards a monolithic optical transmitter was fabricated on InGaAsP/InP employing invertible double-heterojunction bipolar transistors. Static transconductance
of up to 200 mS was attained. Well-behaved eye pattern diagrams have been
demonstrated at modulation rates of as high as 1.12 Gbit/s.

Introduction

Optical transmitters comprising a laser eventually coupled to an output
waveguide, a monitor-diode and an electronic driver circuitry represent
key components for optical transmission systems. Monolithic integration of
such opto-electronic circuits offers the potential advantage of compactness, superior performance, high reliability and low cost. In this paper
we report on the development of a laser driver circuit based on
InGaAsP/InP invertible double heterojunction bipolar transistors (DHBT) to
form part of a monolithic "long-wavelength" transmitter.

Device structure

The laser driver circuit, which is depicted in fig. 1, comprises a
differential amplifier (transistors T_1 and T_2) and a common current source
(transistor T_3), similar to what has been implemented in related MeSFET-
based GaAs OEICs (e.g. [1]). In InP technology, circuits of this kind employing heterostructure bipolar transistors have already been integrated
with laser diodes by using conventional metal interconnects [2, 3]. Here
we have pursued a different scheme in that the transistors are coupled via
a common subcollector layer. This was accomplished by exploiting the invertibility of DHBTs [4]. As a result, a compact and fairly planar structure is achieved, the cross-sectional view of which is sketched in fig. 2.

The outer elements act as the switching transistors (T_1 and T_3) and operate in the "emitter-down" configuration. The current source transistor (T_2) has been designed with a split emitter and an inside base contact in order to attain minimum lateral resistances to the neighbouring emitter regions of T_1 and T_3.

Fabrication and results

The LPE grown epitaxial double heterostructure consists of a p-InGaAsP (λ_g = 1.3 μm) base layer (p = 3-5•10^{17} cm^{-3}) sandwiched between the n-InP emitter and collector layers (n = 1-3•10^{17} cm^{-3}). Localised Zn diffusion has been used in order to provide access to the base layer and to ensure bilateral transistor operation [4]. Au-Zn alloy has been used for the p-type base contact, while Ti-Au metallisation on an n$^+$-InGaAs layer has been utilised for the n-type emitter contact the area of which was 16•150 μm^2. Although the integrated transistor structure can be fabricated on substrates of any type of doping, n$^+$-doped material has been chosen here to allow separate testing of the individual transistors.

Depending on the thickness of the base layer high current gains in the range of 100-1000 (max ≈ 5000) were easily attainable with these transistors. The gain values of the inverted structure ("emitter-down") generally tended to be lower than those for the "emitter-up" configuration but were still as high as 200.

Laser driver ICs were mounted and bonded into 16-lead ceramic chip carriers (fig. 3) and evaluated using a test circuit which was realized in thick film board technology. In fig. 4 the measured static transfer characteristic of a representative IC is shown at a total supply current of 62 mA. In this particular case a load resistor of 27 Ω was used. Typical values of transconductance and IC current amplification of 150 mS and >100, respectively, have been measured on these devices.

High-frequency operation of the ICs was tested using a differential driving scheme, which allows for symmetrical switch-on/switch-off delay. To simulate a driven laser diode a 10 Ω load resistor was chosen for the current modulation tests. Well-behaved eye pattern diagrams could be achieved at modulation rates up to 1.12 Gbit/s (fig. 5), which is regarded as a quite satisfactory figure in view of the still relatively large transistors involved. Reduction of the crucial geometrical dimensions and parasitic elements of the transistor devices will lead to a considerable enhancement of the high-frequency capability.

Summary

In conclusion, a compact differential amplifier-type bipolar laser driver circuit was fabricated on InGaAsP/InP, exploiting the invertibility of DHBTs. Transconductance up to 200 mS, current amplification in excess of 100 and current modulation rates of as high as 1.12 Gbit/s have been demonstrated. Further improvement of the high-frequency performance will be readily achievable by reducing the relevant geometrical dimensions of the transistors, which the current work is focused on.

This work was conducted under the ESPRIT-programme (project 263B).

References

[1]   Nakano, H.; Yamashita, S.; Tanaka, T.; Hirao, M.; Maeda, M.: J. Lightwave Technol. LT-4, (1986) p. 574

[2]   Kasahara, K.; Suzuki, A.; Fujita, S.; Inomoto, Y.; Terakado, T.; Shikada, M.: 12th Europ. Conf., Optical Comm. (ECOC), Techn. Dig., Vol. 1, (1986) p. 119

[3]   Shibata, J.; Natao, I.; Sasai, Y.; Kimura, S.; Hase, N.; Serizawa, H.: Appl. Phys. Lett. 45, (1984) p. 191

[4]   Bach, H.G.; Grote, N.; Fiedler, F.: Conf. Proc. 17th ESSDERC, Sept. 1987, Bologna, Elsevier Science Publishers B.V. (North Holland), edited by G. Soncini, (1987) pp. 883-886

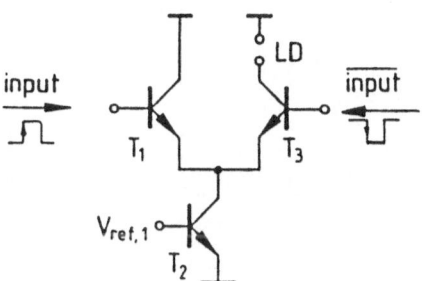

Fig. 1:   Equivalent circuit of bipolar laser driver

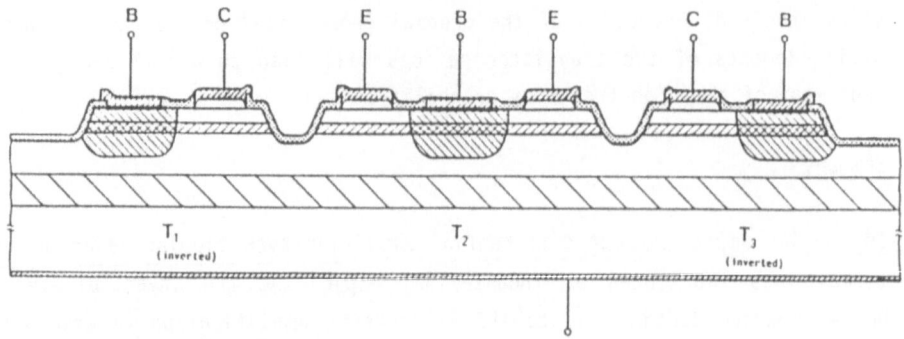

Fig. 2: Schematic cross-section of the transistor array

Fig. 3: Mounted laser driver IC

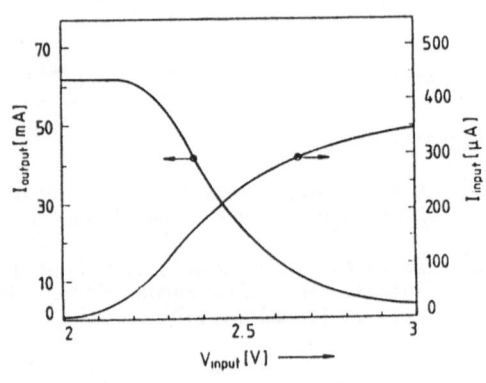

Fig. 4: Static characteristics

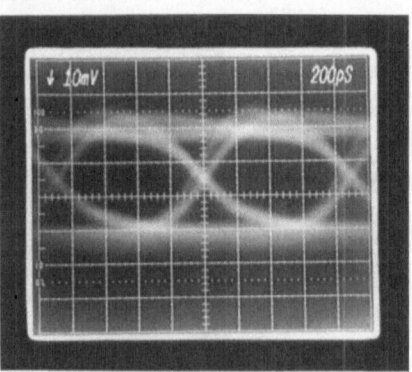

Fig. 5: Eye diagramme at 1.12 Gbit/s current modulation

High-sensitivity, Polysilicon-emitter Phototransistors

N.S. Nam and I. Zólomy

Technical University of Budapest, Department of Electron
Devices H-1521

J. Berkecz and Gy. Pásztor

MEV (Microelectronic Co.), Budapest

Abstract

Phototransistors were prepared and investigated with polysilicon emitter. The high current gain (till 10^4) ensured a very high light sensitivity as compared to normal phototransistors. The spectral sensitivity was also more favorable. Transistors with an interfacial tunnel oxide layer show a much better performance than the transistors without it. Also higher current amplifications were obtained in the case of <100> oriented substrate.

Device fabrication

The polysilicon emitter ensures a very high current gain [1...3]. Thus polysilicon emitter was used for preparing high-sensitivity phototransistors. Devices were prepared with and without interfacial oxide layer between the polysilicon and the substrate. The polysilicon layer was deposited by LPCVD process at 630°C and at 0.32 torr during 50 min., resulting in a thickness of 640 nm.

OXIDE	yes				no	
ORIENTATION	100		111		100	
DOSE(cm^{-2})	10^{13}	5×10^{13}	10^{13}	5×10^{13}	10^{13}	5×10^{13}
$\beta(I_c = 10mA)$	5500	3500	2400	2000	130	100
$V_{CEO}(V)$	22	24	27	30	2	3

Table 1. Current amplifications and maximum collector-emitter voltages of the polysilicon emitter phototransistors

The average grain size was 167 nm after deposition and 250 nm after diffusion step. The n-type doping of the polysilicon was made after the deposition by diffusion of phosphorus at 900°C during 30 or 60 min.. If there were no tunnel oxide layer between the substrate and the polysilicon, the diffusion of phosphorus took place also in the monocrystalline silicon resulting in a junction depth between 0.1 and 0.2 µm, as measured by spreading resistance meter. In the presence of a tunnel oxide , diffused n-type layer in the substrate could not be measured. The p-type base was prepared before the polysilicon deposition by ion implantation of boron atoms through an oxide layer of 100 nm thickness, (40 keV... 100 keV, doses $10^{13}...5 \times 10^{13} cm^{-2}$). The implanted layer was annealed at 900°C in dry oxigen during 30 min.. The resistivity of the substrate was 1-2 Ωcm, the orientations were <100> and <111>. The 3 nm thick tunnel oxides were grown at 550°C in dry oxigen during 25 min.. The base area is 470x470 µm, the emitter area is 100x100 µm. Standard phototransistors with traditional structure were also fabricated with the same masks, but with different doping profiles (base-collector junction depth 6 µm, base thickness 2 µm) for the sake of comparison.

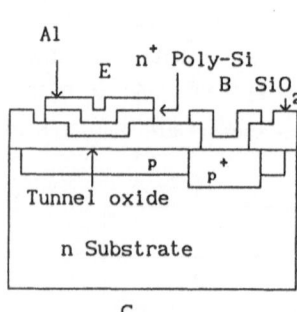

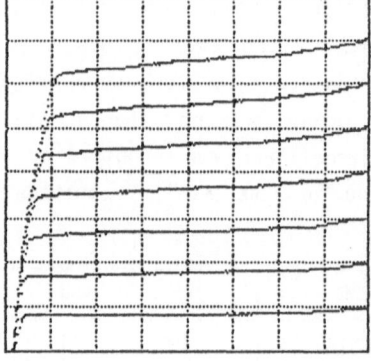

Fig.1. Transistor cross-section

Fig.3. Output characteristics of the <100> polysilicon emitter transistor with tunnel oxide and $10^{13} cm^{-2}$ base-dose.
Hor. 1V/div; vert. 2mA/div; base-step: 0.2µA/div.

Results

The cross section of the device is shown in Fig.1. The polysilicon-emitter transistor has a very shallow base, thus the light sensitive part of the transistor is near to the surface. Therefore the absorption of light, before reaching the light-sensitive region, took place at a much lower degree, than in the case of the traditional phototransistors. This difference was more significant at shorter wave-lengths, as it can be seen in Fig.2., where the spectral sensitivities of the traditional (a) and of the polysilicon emitter (b) phototransistors are shown. The measured current gains (β) and maximum collector-emitter voltages (V_{CEO}) are given in Table 1.. The interfacial tunnel oxide increases very much the current amplification. The <100> orientation is more favorable for the electron tunnel current, resulting in a higher current gain. This fact also shows that tunnelling play an important role in the operation of the polysilicon emitter phototransistor. Decreasing base dose results in increasing current gain. The tunnel oxide prevented also the diffusion of phosphorus into the substrate, therefore the maximum voltage of the devices prepared with tunnel oxide is much higher, than the maximum voltage of the devices prepared without it, where it was limited by punch-through. The I_c-V_{CE} characteristics of such a transistor with tunnel oxide and $10^{13} cm^{-2}$ base dose is shown in Fig.3., as it was measured and printed with a computer-controlled measuring instrument.

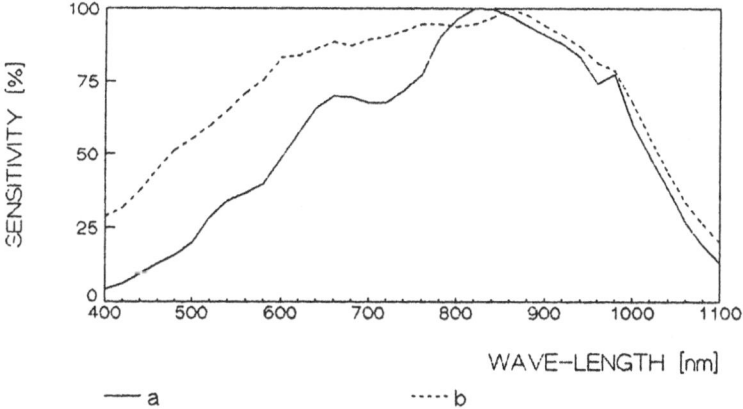

Fig.2. Spectral sensitivity of the traditional phototransistors (a) and of the polysilicon emitter phototransistors (b)

The current amplification versus the collector current is shown in Fig.4. for the <100> (1) and <111> (2) orientations with tunnel oxide (dose $10^{13} cm^{-2}$). The maximum amplification reaches 10000. The collector current versus illumination intensity is shown in Fig.5 for <100> (1) and for <111> (2) orientations. A traditional phototransistor (3) is also shown with a current amplification of about 300. The sensitivity of the <100> phototransistor is about one order of magnitude higher than the sensitivity of the traditional phototransistor.

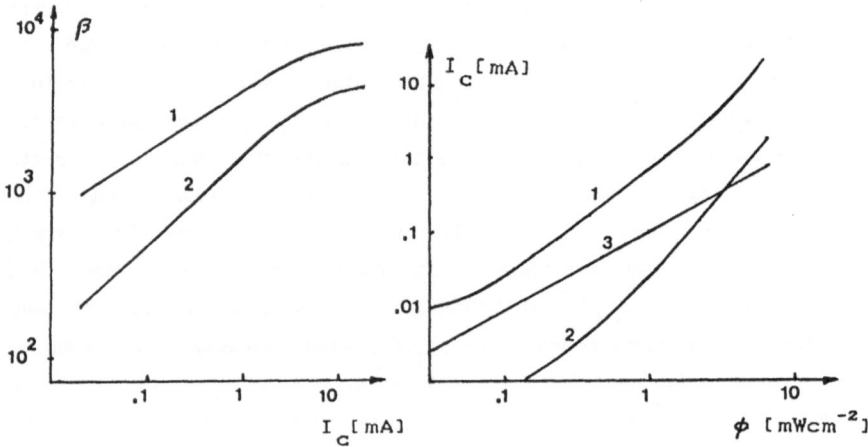

Fig.4. Current amplification versus collector current with (1) <100> and (2) <111> orientation substrates

Fig.5. Photo-sensitivity of <100> (1) and of <111> (2) polysilicon emitter transistors with tunnel oxide and $10^{13} cm^{-2}$ base-dose. (3) conventional phototransistor

References

1. Cuthbertson, A.; Ashburn, P.: Self-aligned transistor with polysilicon emitters for bipolar VLSI. IEEE Trans.on El.Dev., ED-32, 342 (1985)

2. Halen, P.V.; Pulfrey, D.L.: High gain bipolar transistor with polysilicon tunnel junction emitter contacts. IEEE Trans. on El. Dev., ED-32, 1307 (1985)

3. Keyes, E.P.; Tarr, N.G.: Superbeta polysilicon emitter transistors. IEEE El. Dev. Lett., EDL-8, 312 (1987)

$Ga_{0.96}Al_{0.04}Sb$ Implanted Avalanche Photodiode; Perspective for a 2.55 µm SAM APD Photodetector

M.PEROTIN, H. LUQUET, L. GOUSKOV, P. ABIALE-ABI, H. ARCHIDI, M. LAHBABI, B. MBOW.

Centre d'Electronique de Montpellier (UA 391),
Université des Sciences et Techniques du Languedoc,
Place Eugène Bataillon, 34060 MONTPELLIER Cedex 1, FRANCE.

A. PEREZ.

Département de Physique des Matériaux,
Université Claude Bernard, Lyon I, 69100 VILLEURBANNE, FRANCE.

Summary

Be^+ implantation into Liquid Phase Epitaxial n $Ga_{0.96}Al_{0.04}Sb$ has been used in order to study the photodetection characteristics of the resulting p^+/n diode,in the multiplication region.The great difference observed between the multiplication in the respective cases of pure electron and hole photoinjections confirms that the room temperature ionization coefficients ratio is very different from unity,making this material very promising for an antimonide based Separated Absorption Multiplication Avalanche Photodiode. (SAM APD).

Introduction.

In the perspective of long distance, large capacity optical communication systems, there is a need for emiter and detector components at a wavelength of 2.55 µm. The couple of antimonide compounds $Ga_{1-x}In_xAs_ySb_{1-y}$ and $Ga_{0.96}Al_{0.04}Sb$ appear as good candidate in this range of wavelength and a device based on the absorption in GaInAsSb and multiplication in GaAlSb (SAM APD) is particularly attractive for photodetection, due to the high value of the ratio of the impact ionization coefficients for holes: k_p , and electrons: k_n[1,2]

In this perspective,the object of this paper is to study the photoelectrical properties of Be implanted $Ga_{0.96}Al_{0.04}Sb$ diodes.

Results and discussion

The junction is realized by a double implantation (50 keV and 100 keV) of the same dose of Be ions (this dose was varied between 1 and 4×10^{13} cm^{-2}) into Liquid Phase Epitaxially grown Te doped $Ga_{0.96}Al_{0.04}Sb$ layers ($n = 2 \times 10^{16}$ cm^{-3}), (3). The implantation is followed by a thermal annealing under H_2 flux (T= 450°C, t = 15 mn). Calculated and measured Be profiles presented in Figure 1 are in rather good agreement, giving a junction depth of about 0.6 µm. Hall effect and resistivity measurements have been made on two samples implanted respectively with the doses 1 and 3×10^{13} cm^{-2}. The p surface density and the mean values of the p concentration and mobility μ_p values in the implanted layers are:

Dose (cm^{-2})	P_{surf} (cm^{-2})	p (cm^{-3})	μ_p (cm^2/V.s)
2 x 1x10^{13}	2x10^{13}	3.6x10^{17}	512
2 x 3x10^{13}	6x10^{13}	1.0x10^{18}	422

These values indicate that implanted Be ions are electrically active and that the annealing has been efficient to restore a carrier mobility equivalent to bulk doped single crystal's one. Mesa photodiodes have been realized by photolithography on these implanted layers.

* The Capacity-Voltage variations are quite typical of abrupt junctions; the effective doping level is equal to that of the starting layer.

* The dark reverse I-V characteristics at 300 and 77 K are shown in figure 2. At 300 K the current is dominated at low voltages by generation of carriers with the time constant t_{G-R} = 2.2x10^{-10} s; tunnel current is dominant at V > 15 V and impact multiplication leads to a breakdown voltage of 20 V. At 77 K breakdown appears below 15 V. This behavior is common to GaAlSb diodes. The t_{G-R} value is equivalent to those observed in diffused junctions (3, 4) but lower than the best value observed in epitaxially grown junctions: 10^{-9} s (5). Figure 2 is relative to a device which does not exhibit any excedentary current at 300 K, some other devices have presented leakages in the medium voltage range.

* Spectral responses are typical of shallow junctions with rather low front diffusion length L_n in the implanted layer. Figure 3 illustrates the spectral response of an implanted diode with a 3x10^{13} cm^{-2} dose. From the fitted calculated response, L_n= 0.4 μm has been derived; L_n value reaches 1.2 μm in implanted layers with a lower Be dose. In LPE p layers this parameter value was 5 μm. The photoresponse due to the GaSb substrate is apparent on every diode. The external efficiency at a wavelength of 1.3μm is 0.5.

* The multiplication of the photocurrent at 1.7, 1.3 and 0.78 μm is presented in Figure 4. The topography of the photocurrent at 1.3 μm under multiplication condition indicates spatial variations corresponding to electric field fluctuations lower than 3%.

At 1.7 μm $Ga_{0.96}Al_{0.04}Sb$ is transparent (absorption coefficient a <10 cm^{-1}) and the multiplication is initiated only by the photoholes created in the GaSb substrate : a =2x10^3 cm^{-1}. At 0.78 μm, a = 5X10^4 cm^{-1} in $Ga_{0.96}Al_{0.04}Sb$, the photocurrent is principally due to the electrons created in the front layer.

The strong difference between $M_{1.7 \, \mu m}$ and $M_{0.78 \, \mu m}$ indicates that the hole ionization coefficient k_p is much more greater than k_n. The variations of k as a function of the electric field F allowing to account for the three experimental M variations are:

$$k_p = 340 \exp(-112/F)$$

$$k_n = 24 \exp(-122/F) \quad (k: \mu m^{-1}, F: V/\mu m)$$

in the electric field range 18 < F < 32.

The figure 5 compares our results to those of T. Mikawa et al. (2). Our k_p values coincide with the extrapolation of their results, but our k_n values are lower leading to a k_p / k_n ratio of about 20 in our field variation range.

Conclusion

Avalanche diode have been obtained by Be^+ ions implantation into $Ga_{0.96}Al_{0.04}Sb$ LPE layers. The multiplication over the surface of these diodes appear homogeneous. The high value of the k_p / k_n ratio is confirmed in this material. These conditions had to be fulfilled prior to the elaboration of planar SAM APD devices.

References

1. J. Benoit, M. Boulou, G. Soulage, A. Joullié, H. Mani
   J. Opt. Com. 9-2, (1988) 55.

2. T. Mikawa, S. Miura, H. Kuwatsuka, N. Yasuoka, T. Tanahashi, O. Wada
   Proc. of the I.E.D.M. San Francisco 11-14/11/1988. I.E.E.E. Catalog Nb 88 CH 2528-8 p. 487.

3. H. Luquet, L. Gouskov, M. Pérotin, A. Jean, A. Rjeb, T. Zarouri, G. Bougnot.
   J. Appl. Phys. 60, (1986) 3582.

4. I.A.Andreev, A.N.Baranov, M.Z. Zingharev, V.I. Korolkov, M.P. Mikhailova, Y.P. Yakovlev.
   Soviet. Phys. S.C. 19 , (1985) 987.

5. M. Pérotin, L. Gouskov, H. Luquet, P. Silvestre, P. Abiale Abi, D. Magallon, G. Bougnot.
   J. of Crystal Growth (1989) to be published.

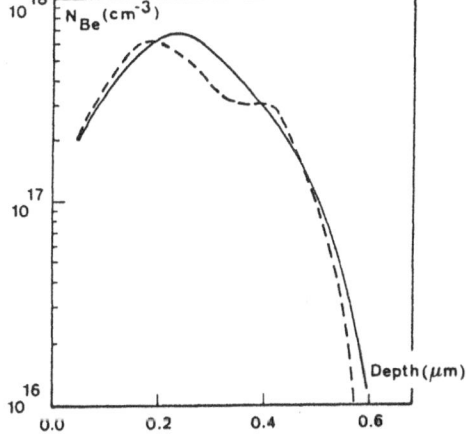

Figure 1

Calculated: ---- and SIMS: —— Be profiles in the sample A 43-1 (dose: 1 x 10^{13} cm^{-2})

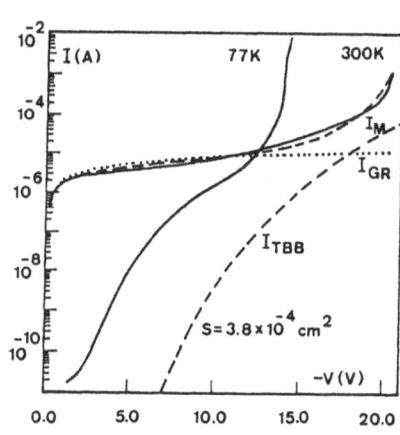

Figure 2

Reverse dark current in implanted $Ga_{0.96}Al_{0.04}Sb$ diode:
Solid line: experimental; other lines : calculated G-R,tunnel, multiplied currents.

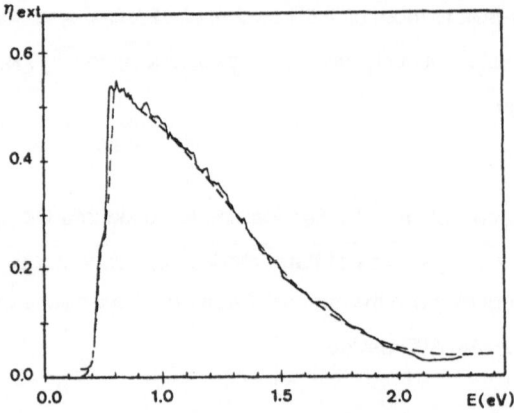

Figure 3

Experimental: —— and calculated: ――― spectral responses of the diode 46-4 (dose= 4x 10[13] cm[-2])

Parameters of the calculated curve: Junction depth: 0.65 µm, L_n: 0.4 µm, $Z_n = s.t_n/L_n = 10$ (s_n: surface recombination velocity)

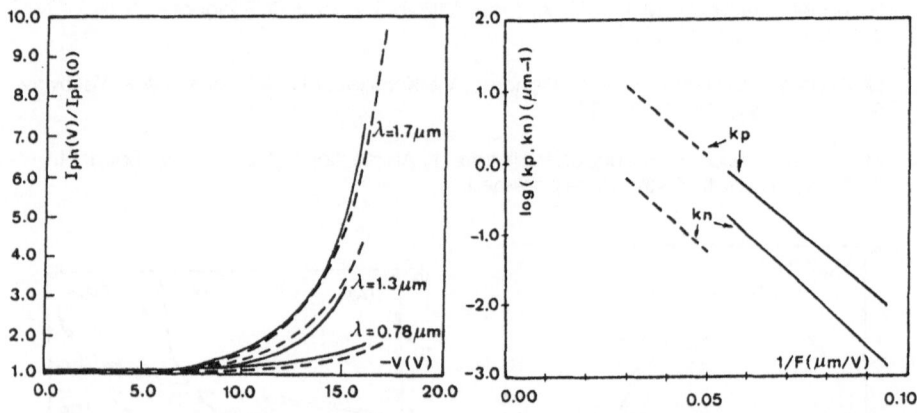

Figure 4
Photocurrent multiplication:
――― : calculated,
—— : experimental.

Figure 5
Variations of the ionization coefficients as a function of the inverse of the electric field: F.
――― : our results
—— : T. Mikawa et al (2) results.

Crack Formation and Selective Growth in MOVPE-GaAs on Si and its Application to OEICs

A. ACKAERT, L. BUYDENS, D. LOOTENS, P. VAN DAELE and P. DEMEESTER.
University of Gent, (L.E.A.-IMEC), St. Pietersnieuwstraat 41,
B-9000 Gent, Belgium

The potential technological advantages of growing GaAs on Si, such as the development of a new generation of OEIC's, has attracted a large research effort during recent years. Many problems encountered in the hetero-epitaxial growth such as lattice mismatch, anti-phase boundaries, pre-growth cleaning and difference in thermal expansion coefficients, have been extensively studied. In particular the thermal mismatch between the two materials involved is still a severe problem, since it induces a high tensile stress within the GaAs layer on cooling down from the epitaxial growth temperature. These stresses cause the wafer to bend, which hinders lithography and further processing. Above a certain strain level, this means thicker epilayers, microcracks will appear in the GaAs along its cleaving directions, reducing the material stresses. These arbitrarely occuring microcracks form isolated regions in the layerstructure which may cause improper device operation.

In this paper selective growth has been used in an original way to solve the problem of random crack formation in thick GaAs on Si layers [1]. The hetero-epitaxial growth was carried out in a small MOVPE system working at atmospheric pressure. The sources used are TMG and a 5% mixture of AsH_3 in H_2, and the conventional two-step hetero-epitaxial growth procedure was followed [2]. For the selective growth off-oriented Si substrates, partially covered with a PECVD-deposited SiO_2 layer were used. A special mask design allowed us to investigate the influence of so-called "wedges" defined in the SiO_2 mask and directed towards the open growth zones, on the formation of micro-cracks in thermally strained GaAs on Si layers. The principle of wedges defined on the sides or in the middle of the growth zones is shown in figure 1, together with the different investigated shapes of the wedges. The topology of the mask used gives us the possibility to examine the influence of the dimensions and the shape of the growth window, the interdistance of the wedges and the occurance of wedges on perpendicular and/or opposite sides of a growth window.

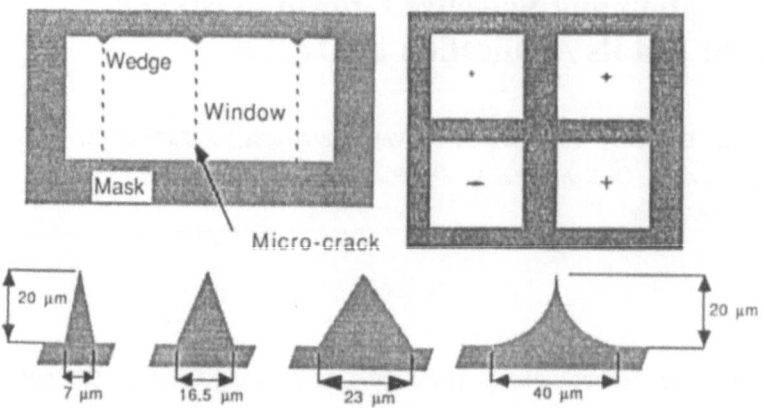

Figure 1 : Principle of wedge-mask technique.

In the selectively grown GaAs on Si epilayers with a thickness exceeding approximately 3 μm thickness the thermal stress induced during cooling down from the growth temperature, seems to relax precisely at the extremities of the predefined wedges. Figure 2 gives a cross-sectional SEM view of a 6 μm GaAs on Si layer. Monocrystalline growth occurs in the open growth window, while an inhomogeneous, low density polycrystalline deposition is visible on the mask regions. Due to the bad nucleation behaviour of the GaAs on the SiO_2 mask material [3] there is a slight increase in growth velocity at the sides of the growth window. The eventual micro-cracks formed are made clearly visible through a slight etching in a H_2SO_4:H_2O_2:H_2O mixture. Over the entire substrate surface the occuring microcracks run exactly parallel or exactly perpendicular to each other, this in accordance with the fact that the cracks run in the [011] and the [01$\bar{1}$] cleaving directions of the monocrystalline GaAs material. Some of the cracks stop spontanously in the GaAs window, while others stop on crossing a perpendicular microcrack. The average distance over which a crack propagate and the overal density of occuring cracks increase with increasing layerthickness. Microcrack formation seems to be independent of the exact wedge or growth-window shape. Figure 3 shows an SEM top view of microcracks induced at isolated SiO_2 "wedge"-islands of different shape in a 6 μm GaAs on Si layer. Important to notice is also that even under extreme conditions (dipping in liquid nitrogen, pressurising through vacuum contact) only few of the already nucleated microcracks propagate further in the monocrystalline GaAs and this in the predicted direction.

The advantage of precise control of the location of microcrack-formation has been investigated for the fabrication of opto-electronic devices in GaAs

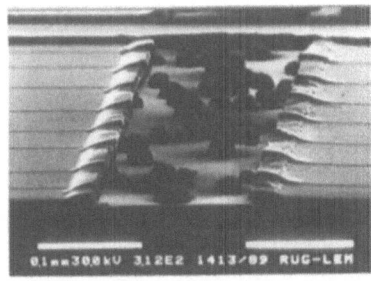

Figure 2 : Cross Section

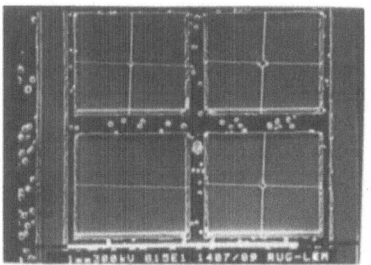

Figure 3 : Top view

on Si. Therefore Double Heterostructure Infra Red GaAs LED structures have been realised on highly doped n^+ Si substrates, using "broad area" (± 1 cm^2) and selective MOVPE growth. In the latter case the lithography was carried out in such a way that the complete LED-devices were fabricated in a crack-free monocrystalline zone, defined by the location of the wedges on the growth mask (see figure 4). The total thickness of the structure amounts to 4.6 μm, including the use of a 3 μm GaAs buffer-layer, two 0.5 μm $Al_{25}Ga_{75}As$ cladding layers, an 0.5 μm p-type GaAs active layer and a thin $Al_{10}Ga_{90}As$ upper contact layer. The use of an $Al_{25}Ga_{75}As$ bufferlayer instead of GaAs was introduced to improve the homogeneity of the deposition on the masked SiO_2-zones in the case of selective growth. For identical structures no degradation of optical and electrical properties was measured between the LED structures realised in the selectively grown monocrystalline zones and the LEDs realised in the broad-area grown material. The output power P of several LED structures (a, b : "broad area" growth with respectively a GaAs and a AlGaAs bufferlayer, c, d : "selective" growth) is shown in figure 5 as a function of the device-current I.

The different spectral characteristics obtained with an optical spectrum analyzer, at room temperature and for a device current of 30 mA, also seem to give comparable results regarding the central wavelength λ_c, the FWHM value and the emitted peak power at λ_c (see table 1).

Yield-performances however are better in the case of selective growth as the LED-devices were defined in completely crack-free zones. Finally figure 6 illustrates a completed "broad area" LED device with a microcrack running through the active window area. Such devices were found to have very short life time when working at high currents.

First results obtained with this wedge-mask technique in controlling the location of microcrack-formation have shown the strength of this technique. In thermally strained GaAs on Si layers it is possible to precisely control

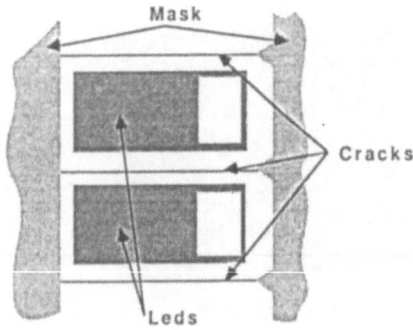
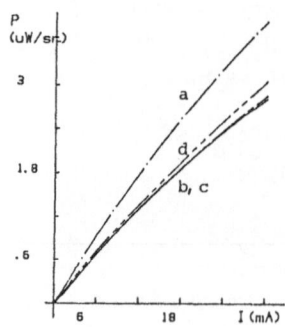

Figure 4 : Selectively grown LED   Figure 5 : P/I curve LEDs

LED	growth techn.	buffer	λ_c (nm)	FWHM (nm)	peak power (mV)
a	broad area	GaAs	887-889	47-49	160
b	broad area	AlGaAs	887-891	47-53	125
c	selective	AlGaAs	888-889	44-47	110
d	selective	AlGaAs	886-891	44-50	120

Table 1 : Spectral characteristics LED-structures.

Figure 6 : "Broad Area" GaAs/Si LED with microcrack.

the location of microcrack formation through the use of small area SiO_2-wedges defined in a selective growth mask. This gives the possiblity to define the active devices in the GaAs within completely crack-free zones. Further investigation of eventual improvement of optical layer quality through this technique is under investigation.

[1] patent pending, EU 31.1.1989, nr 89200205.6
[2] M. Akiyama, Y. Kawarada and K. Kaminishi, J. Cryst. Growth 64, 21, 1984
[3] P. Demeester, P. Van Daele, A. Ackaert, R. Baets, Int. Phys. Conf. Ser. 91, 183, 1987

Integration of GaAs LEDs on Si by Epi-lift-off

I. POLLENTIER, P. DE DOBBELAERE, F. DE PESTEL, P. VAN DAELE, P. DEMEESTER

University of Gent (LEA) - IMEC
St. Pietersnieuwstraat 41, B-9000, Gent, Belgium

In recent years, there has been a persistent interest in the integration of III-V semiconductors with Si. This material combination finds many applications in fast signal processing and optical interconnect, ranging from on chip to long distance communication. Up to now, this integration was mostly realised in combining discrete III-V components, interconnected to eachother by means of wire bonding or flip-chip technique. Further on, there is an ever increasing interest in the monolithic integration, in which III-V components are combined with Si circuits on chip scale. In this area, most research is done on heteroepitaxial growth of III-V semiconductors on Si with MOVPE or MBE. In this paper, we present an alternative quasi-monolithic integration method, which makes use of an epitaxial layer lift-off procedure (in brief epi-lift-off), introduced by E. Yablonovitch [1]. Here, a GaAs component is, lattice matched, grown and processed on a GaAs substrate. By means of the epi-lift-off (ELO) procedure, the GaAs component is removed from the GaAs substrate and placed on an appropriate place on the Si substrate and finally interconnected to the Si circuits with photolithographically patterned metallisation. The mentioned epi-lift-off mainly consists of a very selective chemical etch of an AlAs layer grown beneath the device structure.

The ELO approach has different important advantages in this integration, e.g. high quality III-V components with low dislocation density can be integrated, and the processing of Si-circuits can be completed including the Al-metallisation. Furthermore, the ELO procedure has the advantage that GaAs components can be integrated with arbitrary materials, e.g. glass or LiNbO$_3$. However, there are still several problems in the ELO approach, e.g. the presence of cracks or strain in the thin layers, the handling and the adhesion of the thin films, the alignment of ELO components to Si-circuits, etc. Nevertheless, we used this technique recently with succes for the transplantation of GaAs MESFETs to Si and InP [2] and MQW modulators to glass.

402

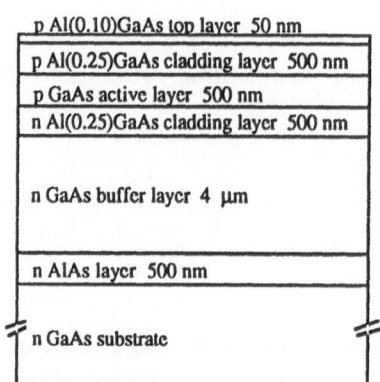

p Al(0.10)GaAs top layer 50 nm

p Al(0.25)GaAs cladding layer 500 nm

p GaAs active layer 500 nm

n Al(0.25)GaAs cladding layer 500 nm

n GaAs buffer layer 4 μm

n AlAs layer 500 nm

n GaAs substrate

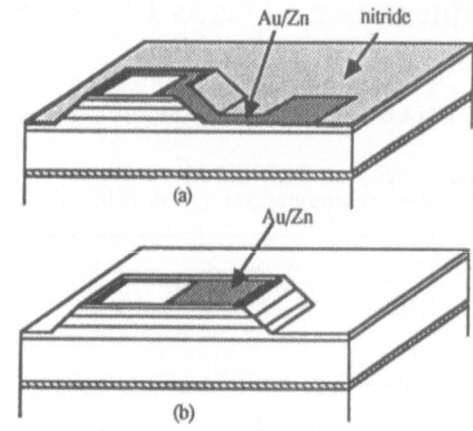

Fig. 1: LED layer structure

Fig. 2 : LED structure prior ELO
(a) Type I LEDs
(b) Type II LEDs

In this paper we present for the first time the succesful transfer of GaAs LEDs from a GaAs substrate to a Si substrate. In this integration we used several ELO strategies, as will be described below. These strategies mainly concern about the problem of cracks, manipulations of ELO samples and the adhesive and electrical contact to Si.

Metal Organic Vapour Phase Epitaxy (MOVPE) was used to grow the LED structure on the GaAs substrate (Figure 1). A 500 nm AlAs layer, which is selectivily etched in a hydrofluoric acid (HF) solution, is used for the lift-off. A 4 μm GaAs buffer layer is incorporated to give mechanical strength to the ELO layer. The following processing steps were completed prior to the ELO in two ways. In a first set (type I LEDs) a Si_3N_4 layer was deposited after the 1.4 μm mesa etch. This Si_3N_4 layer was used to insulate the Au/Zn metallisation from the bottommost cladding layers (Figure 2a). A second set (type II LEDs) was processed without the use of a Si_3N_4 isolation, resulting in a LED structure, in which the bonding pad is on mesa next to the LED window (Figure 2b).

Following these processing steps, we used two ELO strategies on both types. In a first strategy [2], the GaAs samples were cleaved into small strips (smallest size less than 1 mm). They were attached upside down on a (GaAs) carrier substrate with resist. This configuration was placed into a HF:DI (1:5) solution and after a few hours the original substrate came off, leaving the 5.5 μm ELO film on the resist of the intermediate carrier. On this configuration and on a separate Si substrate a AuGe/Ni metallisation was deposited. Then the thin film was removed from the carrier with aceton and placed on the metallised Si substrate. The n and p contacts were alloyed at 420 °C in forming gas, resulting in a good adhesive and electrical contact between the ELO layer and Si, provided that the ELO films are sufficiently small (order 500 x 500 μm^2). This final result is schematically shown in figure 3.

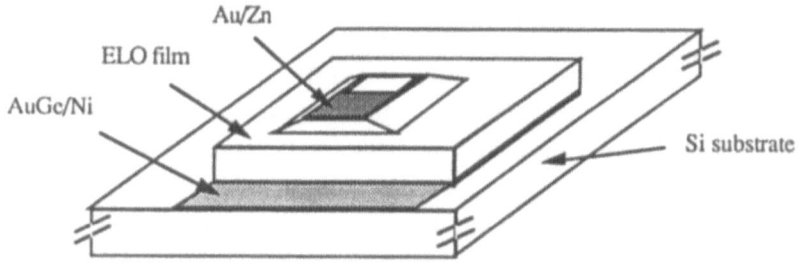

Fig. 3 : Schematical representation of LED integration after ELO

In a second way we used the ELO procedure proposed by E. Yablonovitch [2]. Here, a supporting wax coating (Apiezon W diluted in trichloroethylene) was deposited on the top surface of the original substrate. After drying, this coated substrates were placed in the HF-solution. After a few hours the ELO layers floated off and could be handled, since they were still supported by the wax. A AuGe/Ni backside contact was deposited and next the wax was washed away with trichloroetylene. Contact to the metallised Si substrate was done in two ways. For sufficiently small ELO films we used the fast alloying, as described above. However, due to the concave curvature of the ELO layer, adhesive contact to Si after alloying became impossible for large area films. This curvature, caused by a difference in thermal expansions between the GaAs ELO layer and the AuGe/Ni, was quite impressive after alloying : we observed in crack-free ELO films a radius of curvature of about 1 cm! Those large films were alloyed separately and finally mounted on the metallised Si with silver epoxy. Hereafter, the resulting structure is more or less similar to Figure 3.

Because of the very fragile nature of the thin films, the strain and the crack problem are very important in the ELO approach. In the case of type II LEDs, we obtained a rather high crack density using the carrier strategy, which may be caused by a non-uniformity in resist layer due to the corrosive HF etch. Crack-free ELO layers could be obtained only if the smallest size after cleaving was less than 1 mm. Using the wax strategy however, large area crack-free ELO films were obtained in a relative short time due to tensile forces in the wax [1]. Besides the aspect of stress, the wax coating also provided a better protection of the processed LEDs during the corrosive ELO etch. However, the handling of the ELO films supported by the wax was not as easy as in the carrier strategy. In the case of type I LEDs , ELO samples showed in both strategies a rather high crack density. These cracks were probably induced by strain effects between the Si_3N_4 and the GaAs ELO layer. Through these cracks Si_3N_4 was etched, creating short circuits between the Au/Zn metallisation and the bottommost cladding.

Operation characteristics of type I LEDs showed a significant deterioration due to this strain problem. However, in the case of type II LEDs, I/V and output power measurements (Figure 4) showed no degradation of the LED operation by the ELO. Furthermore, there was no significant difference in operation characteristics between ELO LEDs attached by the alloying and by silver epoxy.

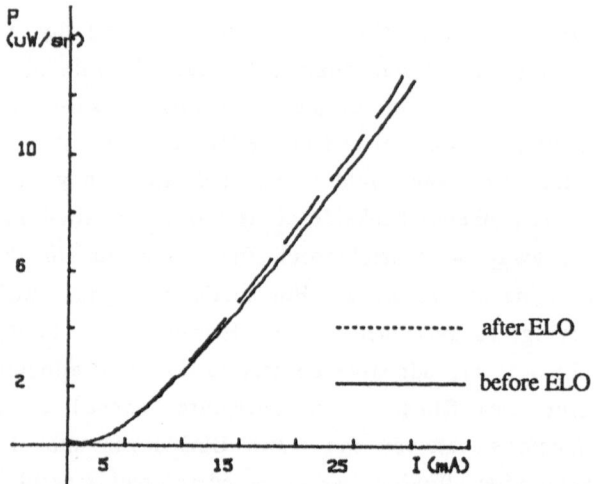

Fig. 4 : Power vs. current characteristic of a type II LED before and after ELO.

In conclusion, we report for the first time the succesful transfer of GaAs LEDs to Si using the ELO technique. In this integration, we used several ELO strategies to overcome problems concerning strain, cracks, handling and contact to Si. Problems which are still present, e.g. strain effects of dielectrics and metallisation, and the alignment to Si-circuits, are under investigation.

References :
[1] Yablonovitch E. et al., Appl. Phys. Lett. 51 (26), 2222 (1987)
[2] P. Demeester et al., to be presented at ECOC Conference 1989

Analytical Device Models

Submicron CMOS Circuit Simulation Model Accurate from Liquid Nitrogen to Room Temperature

H.I. Hanafi, T.J.Bucelot, D.S. Zicherman, R.W. Weiss, P.J. Restle,

IBM Thomas J. Watson Research Center
Yorktown Heights, New York 10598

Abstract

A drain-current device model for circuits operating over the temperature range from liquid nitrogen to room temperature has been developed. The model is accurate for short (> 0.3 μm) surface channel CMOS devices and has been used to predict circuit performance and switching characteristics at varying temperatures. The model includes temperature dependance of important device parameters such as "apparent" mobility μ, saturation velocity and threshold voltage. Model parameters are designed for extraction at room temperature; except for device threshold which is measured at the operating temperature. This allows simple extraction of model parameters and results in an accurate description of device currents. The model is incorporated in the ASTAP circuit simulation program and is used to project improvement in delay of optimized low temperature CMOS over room temperature CMOS. The model is also used to design temperature insensitive off-chip driver/receiver circuits for maximum noise immunity between 77°K and 125°K using a 0.5μm liquid-nitrogen CMOS technology.

1. Introduction

There are significant performance and reliability advantages in operating MOSFET's at liquid nitrogen temperature. At the device level, the advantages include increased carrier mobility, decreased junction leakage, sharper subthreshold turn-off transition [1], and the sharper transition of the gate-to-channel capacitance [2]. At the circuit and system level, the advantages include lower interconnect line resistance and decreased electromigration and thermal wearout [1]. Low temperature CMOS (LT-CMOS) is also unconditionally latch-up free because of the decreased bipolar gains of the parasitic transistors. It is more extendable to VLSI's with channel lengths smaller than 0.5 μm because of more scalable threshold voltage and reduced leakage.

Proper design of circuits in LT-CMOS technology requires a drain-current model accurate over the range of operating temperature. In this paper, the main features of a short (> 0.3 μm) surface channel CMOS model for circuits operating over the temperature range from liquid nitrogen to room temperature are presented. The primary emphasis is on model accuracy and simplicity of formulation which make it suitable for implementation in circuit analysis programs.

2. Model Formulation

The model formulation is based on a room temperature version [3] except for the redefinition of the NFET device threshold to account for impurity scatter effects in this device at liquid nitrogen for 77°K operation. The new n-channel threshold is taken to be the gate voltage required to draw 3 μA / W/L drain current. To clarify this point further, it is necessary to discuss in some detail the approach taken in modeling device "apparent" mobility and threshold voltage

2.1. device "apparent" mobility

At small drain voltage in the linear mode of operation, the current of a long FET device is modeled as:

$$I_{ds} = \mu\, C_{ox}\, \frac{W}{L} \left[V_{gs} - V_T - \frac{V_{ds}}{2} \right] V_{ds} \qquad (1)$$

$$\frac{1}{\mu} = \frac{1}{\mu_o} + \frac{\theta_o}{\mu_o} \left[V_{gs} - V_T - \frac{V_{ds}}{2} \right] \qquad (2)$$

where C_{ox} is the oxide capacitance per unit area, W is the channel width, L is the channel Length, V_T is the device threshold voltage, μ_o is the low field carrier "apparent" mobility, and θ_o is the surface scatter factor.

Equation (2) represents the best linear fit for the $1/\mu$ versus $V_{gs} - V_T - V_{ds}/2$ (gate drive) relationship. This linear form enables us to include the device source and drain series resistance R_s in the mobility degradation expression as follows:

$$\frac{1}{\mu} = \frac{1}{\mu_0} + \frac{\theta}{\mu_0}\left[V_{gs} - V_T - \frac{V_{ds}}{2}\right] + C_{ox}\frac{W}{L}R_s\left[V_{gs} - V_T - \frac{V_{ds}}{2}\right]$$

and not as separate resistance elements in series with the device [3]. This in turn reduces the number of elements modeled in the circuit simulation program resulting in much faster simulation time.

2.2. device threshold voltage

V_T can be measured by extrapolating a line tangent to I_{ds} versus V_{gs} characteristics from the point of maximum inflection. This line intercepts $I_{ds} = 0$ at $V_{gs} = V_T + V_{ds}/2$. This technique becomes sensitive to small amounts of noise in the data, however, especially when the inflection point moves to high I_{ds} as it does in NFETs at 77°K. An alternate technique, which is less sensitive to measurement noise, is to define the threshold as the gate voltage required to allow a given drain current density to flow. 50nA x W/L has been shown to be the appropriate current density at threshold for the NFET at room temperature. Using this definition of V_T, the reciprocal of device mobility at room temperature is plotted against gate drive in Fig. 1. The best linear fit to this relation is also shown in the figure resulting in μ_0 and θ_0 values of 542 cm^2 /Vsec and 0.15 /V respectively.

Using the above current density definition for an NFET operated at liquid nitrogen results in $1/\mu$ versus gate drive as shown in Fig. 2. Clearly there is no simple linear fit that can be obtained over the voltage range of interest. However the accuracy of modeling the drain current in the device, I_{ds}, depends on the accuracy of determining the product of the "apparent" mobility μ and the gate drive and not on the accuracy of determining each separately. Hence, by redefining the device threshold, it is possible to obtain a mobility versus gate drive relationship which can be linearized. Figure 3 shows a plot of $1/\mu$ versus gate drive with device threshold defined at different current densities as a parameter. It is clear from this plot that an n-channel V_T definition taken to be the gate voltage required to draw 3 μA /·W/L drain current results in the best linear fit for $1/\mu$ versus gate drive over the voltage range of interest. This linear fit results in a μ_0 and θ_0 values of 3200 cm^2 /Vsec and 0.1 /V respectively at 77°K.

For a p-channel device; impurity scattering effects on mobility at liquid nitrogen temperature are very small and device threshold at this temperature may be defined at the same current density used for room temperature operation, namely, 50 nA x W/L. This results in an approximately linear relation between $1/\mu$ and gate drive at liquid nitrogen temperature as shown in Fig. 4.

For any intermediate temperature T_i between 300°K and 77°K, μ_0 and θ_0 can be determined by: (1) assuming a linear relation between device threshold and temperature in the range 300°K to 77°K, V_{Ti} is determined at T_i using measured V_T values at 300°K and 77°K. (2) using V_{Ti} determined in (1) and following similar steps as outlined in Fig. 3 but at temperature T_i, μ_{oi} and θ_{oi} are determined. A plot of μ_0 and θ_0 versus temperature is shown in Fig. 5.

3. Model Verification

Using the above definitions of V_T and the corresponding μ_0 and θ_0 values, the model has been correlated with hardware at different temperatures. Device channel length and width were extracted at room temperature while device threshold was measured at the operating temperature according to model design. Figure 6 show the DC model verification results indicating model inaccuracy less than $\pm$ 10%.

A.C. model verification has also been carried out using CMOS ring oscillators. Figure 7 shows model to hardware correlation of a CMOS NAND with fan-in = fan-out = 3 and 1 μm lithography layout rules. Very good correlation is obtained at all temperatures.

4. Modeled Circuit Delay

The model has been used to project logic delay of a 3 input CMOS NAND with 0.2 fF load, 15 μm device width and 0.7 μm layout rules when designed in an optimized liquid nitrogen process and in an optimized room temperature process. It was found that for the liquid nitrogen case with NFET and PFET channel lengths = 0.4 μm, worst case power supply (2.25 volts) and worst case temperature (85°K), the delay is 230 ps. For room temperature design: L_n = 0.4 μm, L_p = 0.5 μm (due to buried channel operation of PMOS), worst case power supply (3V) and worst case temperature (85°C), the gate delay is 590 ps. This represents a factor of 2.57 improvement in gate delay.

5. Design of Off-Chip Driver/Receiver Circuits

The model has also been used to design temperature insensitive off-chip driver/receiver circuits (OCD/OCR) for maximum noise immunity between 77°K and 125°K in a 0.5µm LT-CMOS technology [4]. Fig. 8 shows hardware and simulation results of the OCD/OCR waveforms at 77°K. Modeled and measured OCR noise immunity at different temperatures are shown in Fig. 9. This figure is obtained by applying a fixed dc offset to the OCR input then superimposing a noise pulse to it. The width and magnitude of the pulse was adjusted until the OCR output changed state thereby detecting the point where the noise immunity was exceeded. Very good model to hardware correlation is obtained in Figs. 8 and 9.

6. Conclusion

A device model for circuits operating over the temperature range from liquid nitrogen to room temperature has been developed. The model has been correlated with hardware at different temperatures. Model inaccuracy is less than ± 10%. The model is used to predict a 2.5X improvement in circuit performance for optimized LT-CMOS over optimized RT-CMOS. The model has also been used to design temperature insensitive OCD/OCR for maximum noise immunity between 77°K and 125°K in a 0.5µm LT-CMOS technology.

References

1. F. H. Gaensslen, et al., IEEE Trans. Electron Dev., vol. ED-24, P. 218, 1978.

2. G. Baccarani, et al., IEEE Trans. Electron Dev., vol. ED-31, P. 452, 1984.

3. H. I. Hanafi, "Current Modeling for MOSFET," Circuit Analysis, Simulation and Design, North-Holland (A. E. Ruehli, Editor), 1986, Chapter 3.

4. J. Y.-C. Sun et al., 1988 ESSDERC Conference Abstracts, P. C4-25.

410

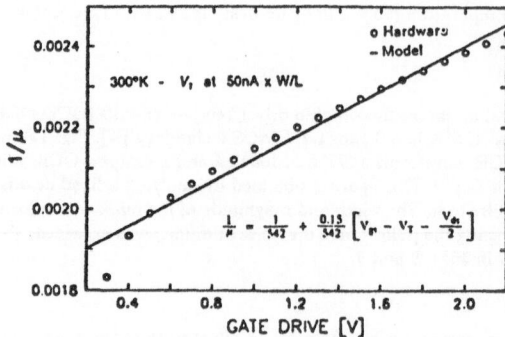

Fig. 1 NFET 1/apparent mobility vs. gate drive

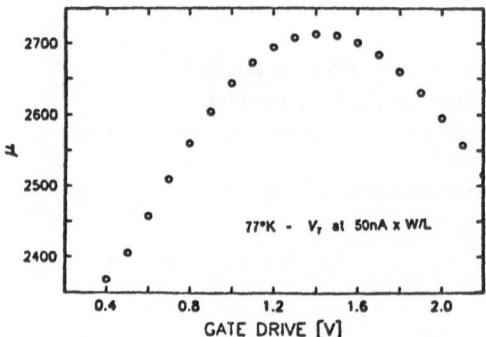

Fig. 2 NFET apparent mobility vs. gate drive

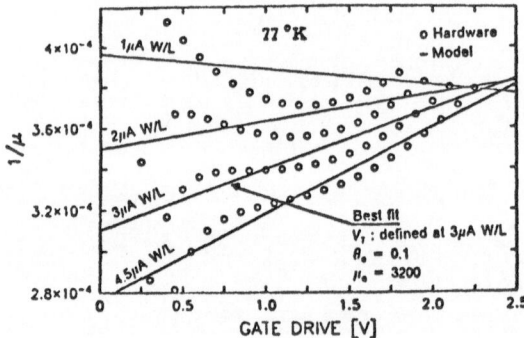

Fig. 3 Linear fit for NFET 1/apparent mobility vs. gate drive

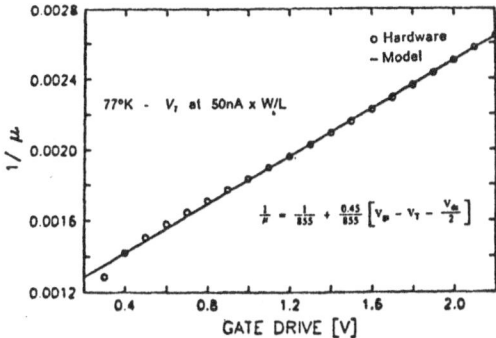

Fig. 4 PFET 1/apparent mobility vs. gate drive

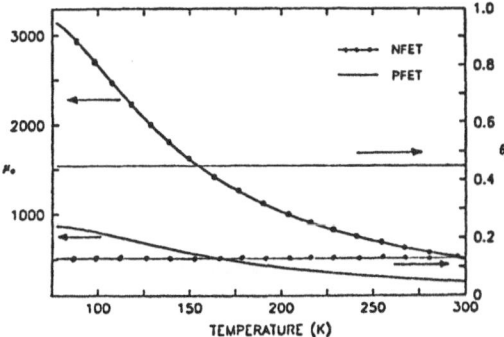

Fig. 5 Model parameters versus temperature

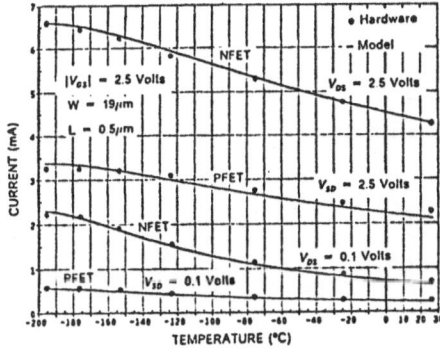

Fig. 6 DC model verification.

412

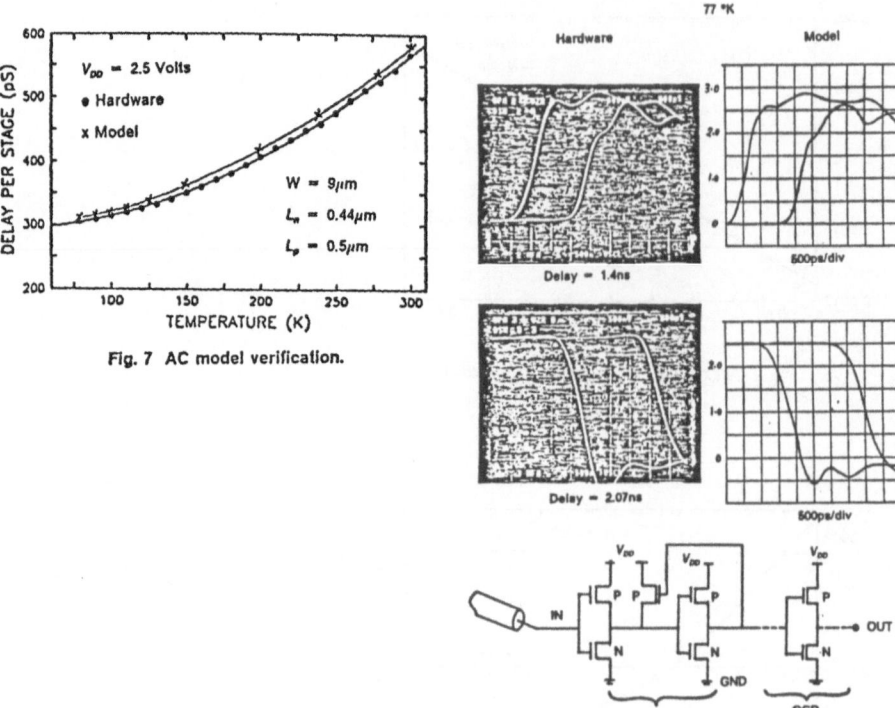

Fig. 7 AC model verification.

Fig. 8 OCD/OCR waveforms

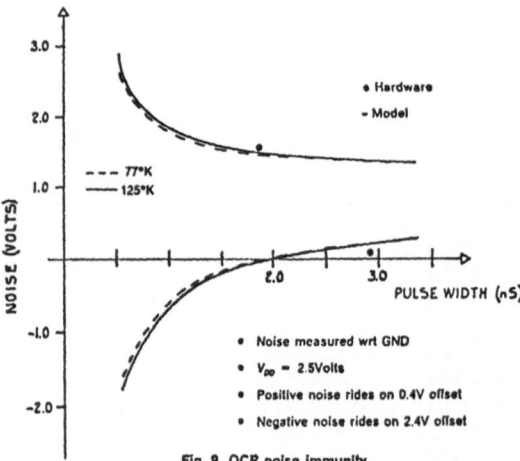

Fig. 9 OCR noise immunity

A New Methodology to Build-up Accurate Empirical Models for VLSI MOSFETs

M.Conti (*), C.Turchetti (*), G.Masetti (**)

(*) Dept. of Electronics, University of Ancona, v. Brecce Bianche -60131 Ancona - ITALY
(**) DEIS - University of Bologna, v.le Risorgimento 2, 40136 Bologna -ITALY.

ABSTRACT
A procedure to derive a complete set of piecewise continuous orthogonal functions able to represent wide ranges of experimental data for VLSI MOSFETs by using simple analytical device models, is presented. Based on this procedure and starting from zero-order models, a new methodology to build-up accurate empirical models which are continuous functions of all bias voltages and parameters is proposed. As an example, an advanced HCMOS process is considered.

1. INTRODUCTION
Circuit simulators such as SPICE are widely adopted tools for the analysis and design of MOS ICs. However, to account for two or three-dimensional effects and to save CPU run time, several empirical parameters have to be introduced in the analytical MOSFET models implemented in these simulators. Besides, the complex dependence of the drain current on device geometry and bias voltages is often empirically modeled by introducing in the formulation of the device model several parameters whose values are determined by a fitting procedure[1-4].
The main lacks of these completely empirical approches are:
i) they are based on the " natural assumption " that more are the parameters in the model more accurate is the fitting with experiments. Unfortunately, however, the extraction procedure may cause unphysical parameter values without any improvement in the fitting of experimental data; in fact, it has been found that a reduction of the fitting parameter set, discarding the parameters which have larger uncertainty, can result in a more accurate fitting [4];
ii) when the parameters in the model are numerous, a large amount of CPU time is required for the extraction of the optimal set of parameters through a "global" fitting procedure.
Purpose of this work is to develop a new general methodology which enables to formulate accurate empirical models for VLSI MOSFETs starting from a zero order formulation, and to save a significative amount of CPU time in the parameter extraction procedure.

2. A PROCEDURE TO DERIVE A COMPLETE SET OF ORTHOGONAL FUNCTIONS TO REPRESENT EXPERIMENTAL DATA

Let us suppose that the DC characteristics of MOS transistors fabricated with several channel length (L) and width (W) have been measured in all regions of operation and for a wide range of applied voltages (V_{DS}, V_{GS}, V_{BS}). The set S of the available experimental data can thus be defined as

$$S = \{ (y, x) : y \in \mathcal{A}, \ x \in \mathcal{D} \} = \mathcal{A} \times \mathcal{D} \tag{1}$$

where x is the vector of independent variables of the device and y is the vector of dependent variables. For the DC operation y=y represents the measured values of the drain current, while
$$x = [x_j]^T = [V_{GS}, V_{DS}, V_{BS}, W, L]^T . \tag{2}$$
If $\mathcal{D}$ is partitioned in a set of N mutually exclusive subdomains $\mathcal{D}_j$ whose union equals $\mathcal{D}$

$$\mathcal{D}_1 \cup \mathcal{D}_2 \ldots \cup \mathcal{D}_N = \mathcal{D} \qquad\qquad (3)$$

one can find, through a suitable best-fit procedure applied to each of the subdomains $\mathcal{D}_j$, a set of N values for the parameters $\mathbf{p}$. (In this work the SOBS parameter extractor [6] was used to achieve all the parameter values in each subdomain pertaining to the model considered).

In such a way a piecewise continuous approximating function $\tilde{y}$ (x) can be assumed as the sum of N functions $\phi_j(x,p^j)$, which are equal to the functions $f (x, p^j)$ approximating the y data in each of the subdomain and zero outside; in formulae

$$\tilde{y} (x) = \sum_{j=1}^{N} f(x,p^j)\ \delta_j(x) = \sum_{j=1}^{N} \phi_j(x, p^j) \qquad\qquad (4)$$

where $\delta_j(x)$ is defined as

$$\delta_j(x) = \begin{cases} 1 & x \in \mathcal{D}_j \\ 0 & x \notin \mathcal{D}_j \end{cases} \qquad\qquad (5)$$

Two are the main features of such an approach: i) it is quite general because it can be applied to model the physical behaviour of all electronic devices characterized through a set of measured data, ii) it is very cost effective since the extraction procedure is applied to small sets of experimental data only. Conversely, its main lack is that the device model is expressed in terms of large numbers of distinct values of the parameters and the approximating function (4) does not result a continuous function of $\mathbf{p}$. However, if the values of the parameters obtained by the extraction procedure described above are located in a small interval, only small variations of the parameters are needed to represent the model.

3 - A METHODOLOGY TO BUILD-UP ACCURATE EMPIRICAL MODELS

In this section we wish to develop a methodology to built-up accurate empirical models starting from zero-order models, as continuous functions of x. From eq. (4) it follows that in order to represent the set of experimental data by the function f(x, p), that is through a zero-order model with a given error ε, the vector p has to be a function of x such that

$$p(x^j) = p^j \qquad x \in \mathcal{D}_j \qquad j = 1, \ldots, N \qquad\qquad (6)$$

that is a step map.

By indicating with $p (x, \vartheta)$ a continuous function of x and of a vector of parameters ϑ, the approximating function f for $x \in \mathcal{D}_j$ can be written as

$$f(x, p(x, \vartheta)) = f(x, (p(x, \vartheta) - p^j) + p^j) = f(x, \sigma^j(x, \vartheta) + p^j)$$

which can be expanded in Taylor's series around the point $\sigma^j = 0$

$$f(x, \sigma^j + p^j) = f(x, p^j) + \sigma^{jT} (\partial f / \partial \sigma^j)_{\sigma'^j}, \qquad \text{where } \sigma'^j \in (0, \sigma^j).$$

Now let us evaluate the average error between this function and the piecewise continuous function. If the derivative $\partial f / \partial \sigma^j$ is bounded, that is $\| \partial f / \partial \sigma^j \| < M$, it can be shown that it results

$$\| f(x, p(x, \vartheta)) - \Sigma_j f(x, p^j) \delta_j \|^2 \leq M^2 \int_{\mathcal{D}} [p(x, \vartheta)) - \Sigma_j p^j \delta_j]^2 \, dx$$

where $\Sigma_j p^j \delta_j$ represents a step map.

From this relationship it follows that the more the function $p(x, \vartheta)$ is close to the step map the more the error tends to the error of the piecewise continuous function. To evaluate this functional dependence an empirical fitting procedure has been used. The choice of the function $p(x, \vartheta)$ is quite easy in that only few values of parameters have to be fitted in comparison with the total measured data.

4. RESULTS

As examples of the goodness of the proposed methodology a new accurate first-order empirical model was built-up for MOS transistors starting from a zero-order model and by assuming 2000 values of experimental data of an advanced HCMOS process, obtained for a wide range of W, L, V_{GS}, V_{DS} and V_{BS} values. The model, whose relationships for drain current are reported in Appendix, results a continuous model which holds in all regions of operation of the device, in particular in moderate and in weak inversion. Starting from the five constant parameters μ_0, δ, V_{MAX}, ϑ and λ, a more accurate models has been built-up by adopting the methodology developed in Sect.3 to the complete set of measured data partitioned in 168 subdomains.

For each parameter a dependence on x has been derived, in formulae:

$$p (x) = [\; \mu_0(x), \; V_{T0}(x), \; \vartheta(x), \; V_{MAX}(x), \; \lambda(x) \;]$$

The resulting second order models are expressed as continuous functions of x with 51 fitting parameters. In Figs. 1-8 some of the results achieved with the accurate model are reported as continuous lines and compared with the measured data (dots).

As an example, the maximum relative error ε between models and experiments was 2.6 % for a 1.5 μm n-channel MOSFET and 4.3% for a 1.5 μm p-channel MOSFET in the whole I_D - V_{DS} plane (see Figs. 1-4). Besides, the transfer curves for n-channel devices with several channel lengths shown in Figs. 5-8 evidence an average relative error between model and measurements of 1.9 % over a current range of about 5 decades.

ACKNOWLEDGEMENTS

This work has been supported by the Consiglio Nazionale delle Ricerche under the Project: "Materiali e Componenti per l'Elettronica a Stato Solido". Support from SGS-THOMSON Microelectronics is gratefully acknowledged. The authors would like to thank Dr. G. Giaccaglini, Dr. M. Lissoni and Dr. A. Saporito for their valuable contribution in characterizing MOS devices and supporting the SOBS program.

APPENDIX
Zero-order Model

In this model, which follows the analysis presented in [7], the drain current, expressed as a function of the surface potential at the borders of the channel, results a continuous function of all the voltages applied to the device. Moreover, since it includes the diffusion component of the current, it is also valid in weak inversion.

The DC drain current is given by the following expression:

$$I_{DS} = W / L_{eff} \; \mu_{eff} \; C_{ox} \, (kT/q)^2 [\, P_2 (U_L, U_D) \, - \, P_2 (U_0, U_S) \,] * (1 + \lambda \, V_{DS}) \qquad (B.8)$$

where the symbols have the same meaning of those reported in [7]. In addition, the channel modulation effect have been modeled through the parameter λ.

REFERENCES

[1] - A. Vladimirescu and S. Liu, "The simulation of MOS integrated circuits using SPICE2", Mem. No. UCB / ERLM80 / 7, Feb. 1980, Univ. of California, Berkeley.

[2] - T.J.Krutsick, M.White, H.Wang, R.Booth, "An improved method of MOSFET modeling and parameter extraction", IEEE Trans.on Electron Devices, vol. ED-34, No.8, pp. 1676-1680, Aug. 1987.

[3] - K.Doganis, D. L. Scharfetter, "General optimization and extraction of IC device model parameters", IEEE Trans. on Electron Devices, vol. ED-30, No. 9, pp. 1219, Sept. 1983.

[4] - S. W. Lee and R. C. Rennick, "A compact IGFET Model - ASIM", IEEE Trans. on CAD, Vol. 7, No. 9, pp. 952-975, Sept. 1988.

[5] - C.F.Machala, P.C.Pattnaick and P.Yang, "An efficient algorithm for the extraction of parameters with high confidence from non linear models", IEEE Electron Device Letters, vol. EDL-7, No.4, pp.214, April 1986.

[6] - A.Saporito et al., ICCAD Conf., Santa Clara, Nov. 1984.

[7] - C. Turchetti and G. Masetti, "A CAD-oriented analytical MOSFET model for high-accurancy applications", IEEE Trans. on CAD, Vol. CAD-3, No. 2, pp. 117-122, April 1984.

416

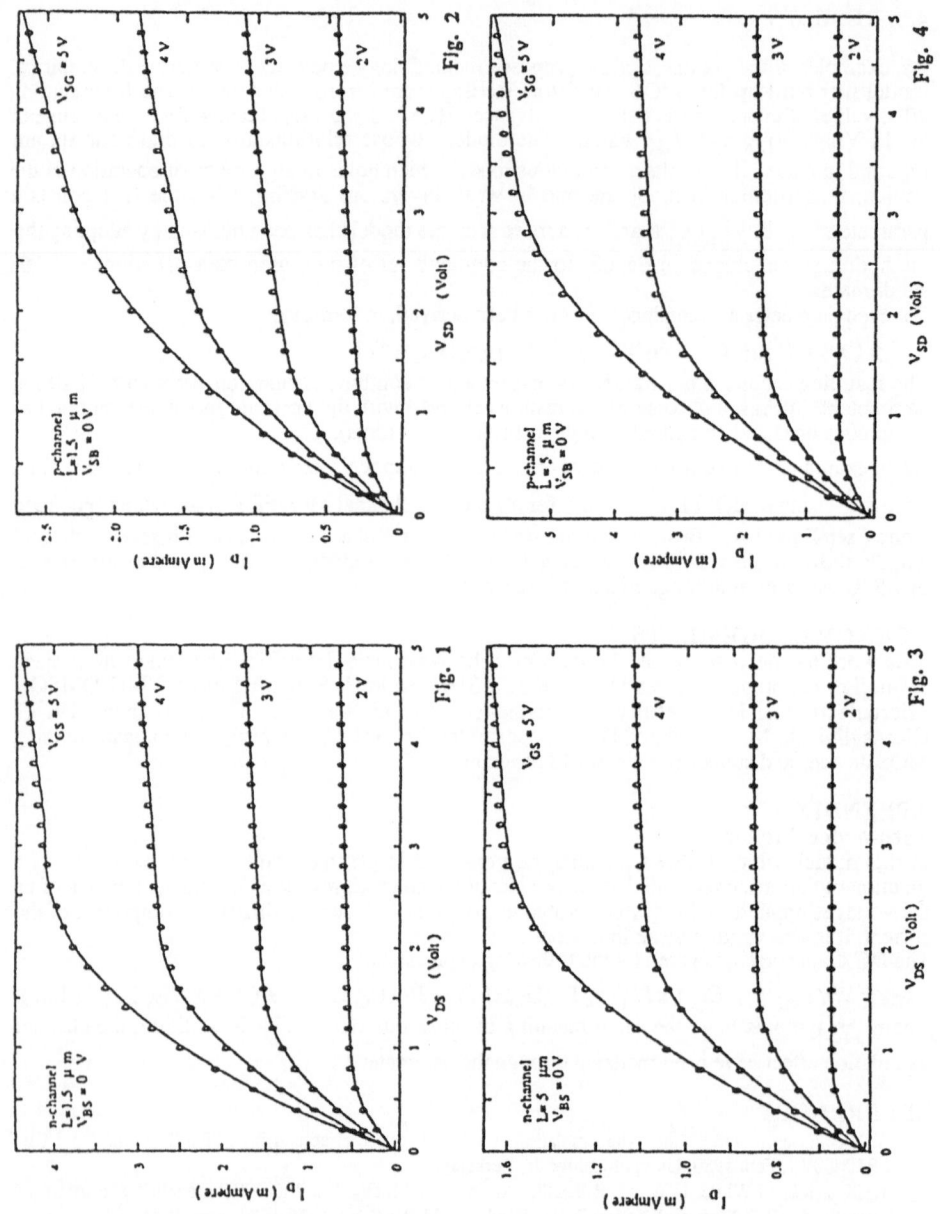

417

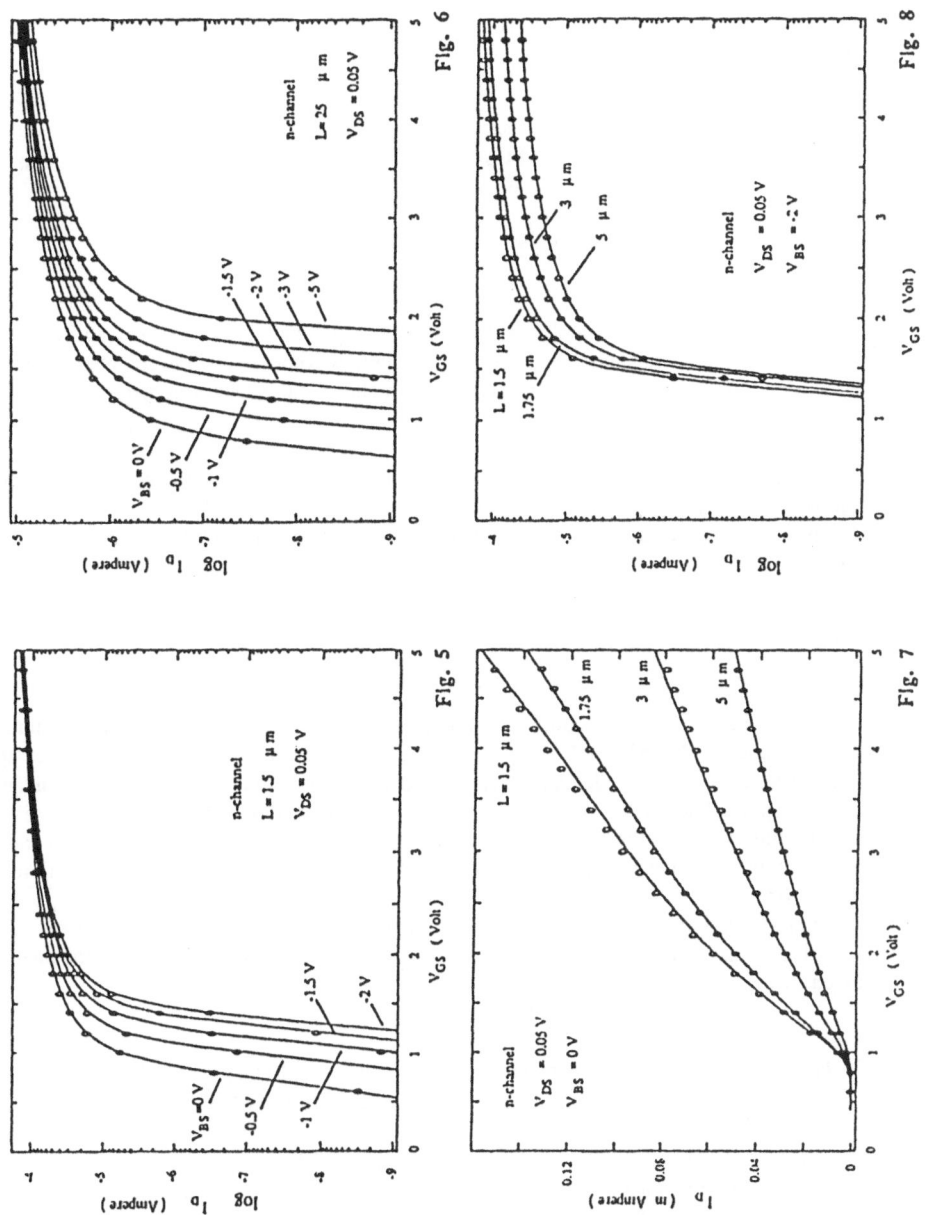

Compact Modelling of the MOSFET Drain Conductance

F.M. Klaassen and R.M.D. Velghe

Philips Research Laboratories, P.O.Box 80000, 5600 JA Eindhoven, The Netherlands

Abstract

A new compact MOSFET model for analog design applications is presented. In particular results for the drain conductance are discussed in detail. A comparison between measured and modelled data for a wide variety of CMOS devices shows an excellent agreement.

Introduction

For analog design applications, even after recent improvements [1-4], MOSFET compact models generally fail to describe the following features:
- the drain conductance in a wide range of bias voltages and device dimensions,
- the subthreshold and weak inversion region of short-channel devices,
- the saturated drain current with a threshold voltage characterized by two values of the body coefficient,
- the effect of a source and drain series resistance,
- the charges and capacitances of short-channel devices.

As an example in fig. 1 the measured (dots) and calculated (fully drawn lines) drain conductance of an n-channel MOSFET (W/L = 20 μm/2.2 μm) is given. Two major deficiencies appear. The saturated drain conductance has a large deviation (almost 100%) at gate voltages close to threshold (V_T = 0.80 Volts). Furthermore even larger deviations are observed in the transition region between the linear and the saturated mode of operation. Both defects cause a calculation of the voltage gain of a CMOS difference amplifier by circuit simulation to become erroneous.

Recently by taking into account the relevant physical mechanisms, a model has been conceived, which satisfies the requirements mentioned above. Here only the drain conductance is discussed in detail.

A new compact model

In fact the new model is an extension of a previous model [5,6] in which several terms, which cause the deficiencies mentioned are corrected. Successively, the following corrections have been implemented. The dependence of the threshold voltage on back bias in the case of an implanted substrate is described by two distinct values of the body coefficient. Correspondingly, the saturation voltage at zero back bias also depends on the smaller value of the body coefficient. Next, the effect of drain-induced barrier lowering on the subthreshold characteristics has been newly formulated. In addition, a smooth transition between subthreshold and strong inversion regime has been adopted [2]. The effect of a source and drain series resistance on the current expression has been taken into account implicitly [7]. A much improved description of drain conductance is obtained by applying a new formulation of static feedback and channel length modulation in the strong inversion regime. Finally, a new description of the four intrinsic charges has been conceived [8].

The result of all above corrections on the modelled characteristics is shown in figs. 3, 4 and 5. These results apply to an n-channel device (W/L = 20 μm/1.4 μm), which is subject to all known short-channel effects. From the comparison between the calculated (fully-drawn lines) and measured results (dots) it can be concluded that the accuracy of the model is satisfactory. For figs. 3, 4 and 5 the average deviation amounts to 3%, 15% and 8%, respectively. Furthermore the largest deviations occur in regions of less importance.

The drain conductance

Since channel length modulation by an increase of drain bias can only be calculated via a 2-D solution of Poisson's equation, including mobile carrier space charge (compare fig. 2), very few published analytical models sustain a comparison with experimental results of the drain conductance. Using a slight adaption of one model [9], we have obtained satisfactory results. According to this, the effective channel length is expressed by

$$L_{eff} = \left[1 + \alpha \ln\left(1 + \frac{V_{DS} - V_{DSS}}{\alpha V_p} \right) \right]^{-1} , \qquad (1)$$

where V_{DSS} is the saturation voltage and α, V_p are parameters.

Although static feedback has been taken into account in earlier models [5] to correct the gate drive, it appears that the drain bias dependence is a little more complicated. Here we use the formulation

$$V_{GT} = V_{GS} - V_T + f(V_{DS}) . \qquad (2)$$

When the saturation conductance is calculated using eqs. (1) and (2), it is easily shown [10] that the first term causes the drain conductance to vary proportionally with the saturation current I_{DSS}, while the second term yields $g_{DSS} \sim V_{DSS}$.

Although accurate results for the drain conductance in the linear and saturation regions, are found after taking into account the above expressions in the drain current, no solution is given for the defect observed in the transition region. Generally, the latter is caused by a discontinuity of the second derivative of I_{DS} at $V_{DS} = V_{DSS}$. Unfortunately, no satisfactory physical criterion is known to enforce continuity. Therefore, a simple smoothing procedure for the transition between V_{DS} and V_{DSS} is applied.

Figs. 5 and 6 give a comparison between modelled and measured results. The agreement is good. For long-channel devices, always the first mechanism dominates causing $g_{DSS} \sim (V_{GS} - V_T)^2$. On the other hand, in LDD type, short-channel devices (fig. 5) the second mechanism is dominant causing $g_{DSS} \sim (V_{GS} - V_T)$. In conventional, not too short p-type devices (fig. 6) both mechanisms are present. This explains the qualitatively large difference between figs. 5 and 6. Finally, the geometry dependence of the characteristic parameters (fig. 7) corresponds to physical intuition.

References

1. B.J. Sheu et al., IEEE JSSC-22, p.558 (1987).
2. G.T. Wright, IEEE Trans. ED-34, p.823 (1987).
3. L. Lauwers et al., ESSDERC 1988, p.249.
4. S.W. Lee. et al., IEEE Trans. CAD-7, p.952 (1988).
5. F.M. Klaassen et al., Solid State Electr. 23, p.237 (1980).
6. F.M. Klaassen, Ch.12 in Process and Device Modelling, Elsevier (1986).
7. F.M. Klaassen et al., ESSDERC 1988, p.257.
8. T. Smedes et al., to be published.
9. P.K. Ko et al., Techn. Digest IEDM 81, p.600 (1981).
10. H.C. de Graaff and F.M. Klaassen, Ch.7 in Compact transistor models for circuit simulation, Springer (1989).

420

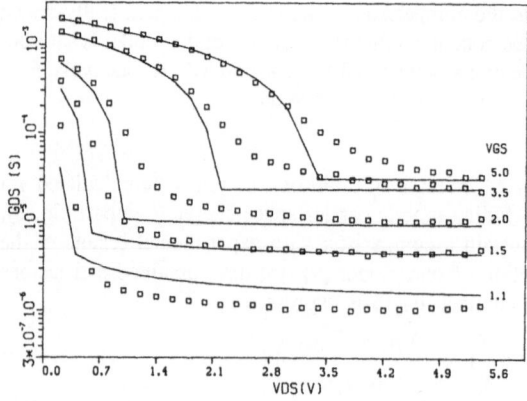

Fig. 1. Deviations between measured (dots) and calculated drain conductance ($Z/L = 20 \, \mu\text{m}/2.2. \, \mu\text{m}$) typical for state of the art models.

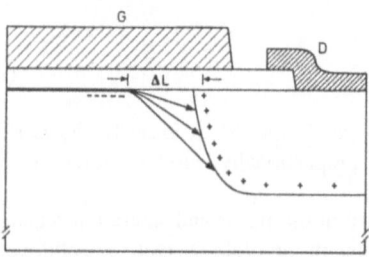

Fig. 2. Illustration of physical mechanisms affecting the value of the drain conductance.

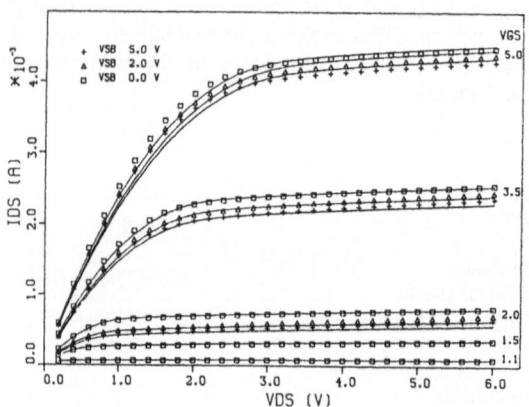

Fig. 3. Measured (dots) and modelled characteristics of the drain current (n-channel device, $Z/L = 20 \, \mu\text{m}/1.4 \, \mu\text{m}$).

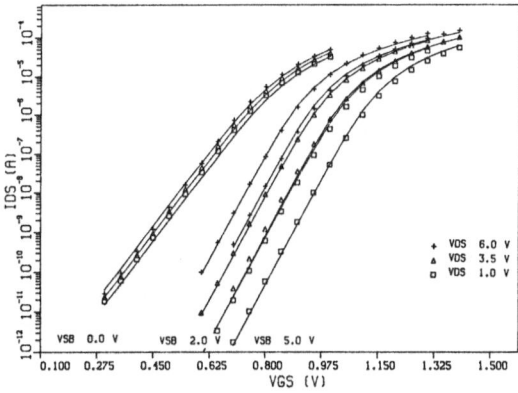

Fig. 4. Measured (dots) and modelled substhreshold characteristics corresponding to fig. 3.

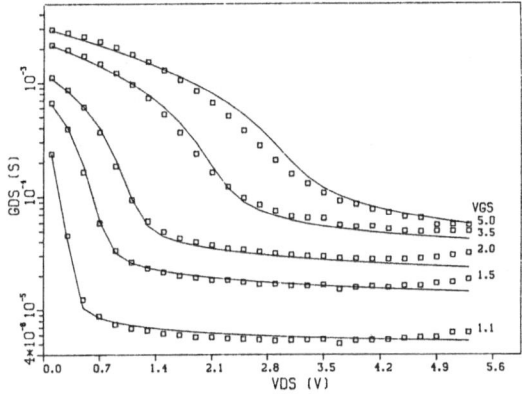

Fig. 5. Measured (dots) and modelled drain conductance corresponding to fig. 3.

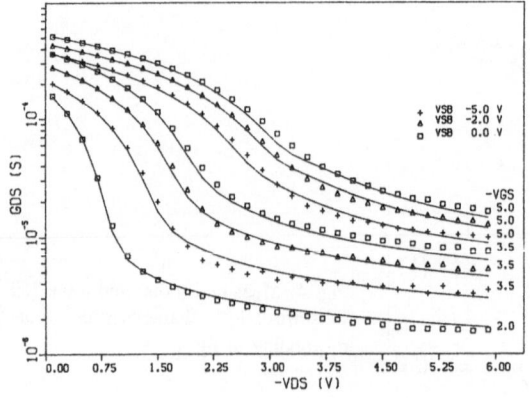

Fig. 6. Measured (dots) and modelled drain conductance of p-channel device (Z/L = 20 μm/2.4 μm).

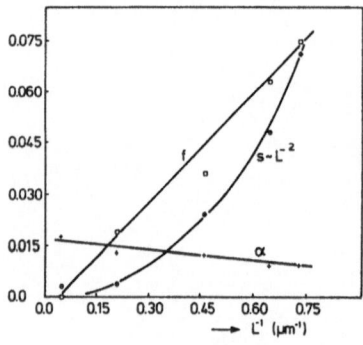

Fig. 7. Dependence of channel shortening (α), static feedback (f) and DIBL parameter (s) on channel length.

A New Analytical Model of CMOS Latch-up

C. LEROUX , J. GAUTIER , J.P. CHANTE (*)

D.LETI,CENG , 85X av. des Martyrs , 38041 Grenoble Cédex
(*) LPCA, INSA, 20 av. Albert Einstein, 69621 Villeurbanne Cédex

Summary
Classical analytical models of CMOS latch-up assume that the latch-up phenomenon is the result of the interaction of two or more bipolar transistors. With such a model, sustaining of latch-up can be only explained by a sufficiently high value for the product of bipolar gains (classically ßn.ßp≥1 or ßn.ßp≥ f(Iddmax)[1]). Such an explanation fails with the reduction of gain at high injection current levels[2]. We can also ask, whether we can even speak of bipolar transistors when the base-collector junctions have disappeared. This effect can be observed in figure 1[3]. Our model of latch-up includes a novel, four terminal, bipolar transistor model. With it we can explain the sustaining of latch-up at high injection levels, voltage drops through the base can be accounted for, and the disappearance of the base-collector junction can be represented. Our model is compared with experimental results, and good correlation was seen between measured and simulated values of holding voltage.

Classical models

Considering that classical models consist of two or more bipolar transistors and resistors between the different nodes, we distinguish two major effects for modelling latch-up: injection phenomena and conduction phenomena.

Our specific model for conduction and injection phenomena

For conduction phenomena, we have two kinds of resistances: well and substrate resistances. In the well conduction has been modelled by evaluation of the sheet resistance plus lateral edge conduction. Our model of resistance for the epitaxial substrate is based on the transmission line model of Troutman [4].

The injection phenomena are correctly modelled by the classical bipolar models (Webster[2], Gummel&Poon [5],...),but they are not suited for latchup modelling. Their model of high injection is based on the hypothesis that majority conduction can be neglected in the bipolar transistor base in the direction from emitter to collector. It is no longer valid in latch-up structures and SCR`s where we have a non negligible current of majority carriers which is the collector current of the other transistor.

We have reconsidered the current continuity equations in the case of latch-up. It leads to a new bipolar model represented in figure 4. We have taken into account majority carrier current, electrical field induced by this current, and minority carrier current driven by this electrical field. Instead of three nodes, as shown in figure 3, we have four nodes because we have separated the base into two nodes: one node for the point near the emitter-base junction, and another for the point near the collector-base junction. This model is used in a general model of latch-up which is shown in figure 5.

With this new model we have a more accurate representation of the latchup phenomenon, because each node corresponds to a physical point. This is not the case of the model represented in fig.2, where the point Bn and Bp cannot be exactly located in the structure. In our case, the points Bn, BIn, Bp, BIp correspond to the four different points of the two bases near the junctions. In our model the disappearance of the collector-base junctions is explained by an equipotential for the points BIn and BIp. The voltage drops through the base of the transistors can be represented by voltage difference between Bp and BIp, Bn and BIn.

Comparison between model and experimental values

Experimental values are obtained on inverter structures 50 μm wide which were processed using an 0.8 μm CMOS technology. In figures 6 and 7, we have compared experimental to simulated values for holding voltage and triggering current. We have tested two epitaxial thicknesses, 8 and 5 μm. In each case, measurements have been done with and without backside contact. There is good agreement between simulated results and experimental values. In this model we have used no fitting parameters, thus all parameters can be related to the technology except for the bipolar gains which are measured values.

Conclusion
With such a model, the latchup phenomenon can be related to technology. The main physical effects such as high injection and its consequences are taken into account, and latchup parameters(IH,VH,Itrig) can be predicted.

Acknowledgements
The authors would like to thank the different teams of D.LETI/ SMSC for device processing and testing.

References
1.ESTREICH (D.B.) : Ph. D. dis. , Stanford University, 302 p.
2.WEBSTER (W.M.) :Proc. of IRE,1954, vol. 42, n° 6, p.914-920
3.WIEDER (A.W.) WERNER(C.) HARTER(J.) : IEEE ED-30,n° 3, p.240-5
4.TROUTMAN (R.R.) HARGROVE (M.J.) : IEEE ED-33, n° 7, p. 945-954
5.GUMMEL (H.K.) POON (H.C.): Bell syst. tech. j., 1970,p.826-54

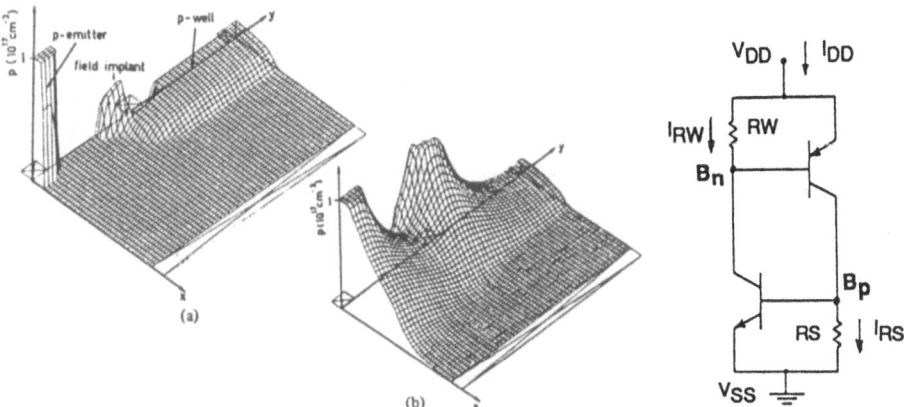

Fig.1. Hole level in CMOS inverter a) before and b) after latchup [3]

Fig.2. Classical model for CMOS latch-up

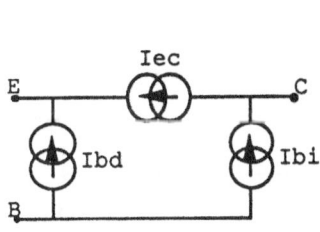

Fig.3. classical bipolar model

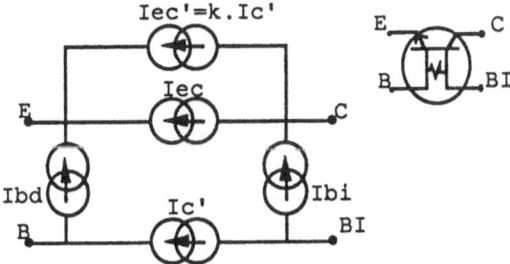

Fig.4. a) New bipolar model used for CMOS latch-up b) Notation

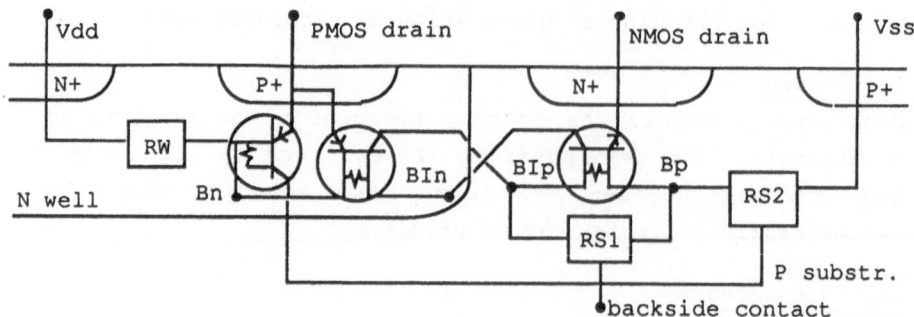

Fig.5. Novel model for CMOS latch-up

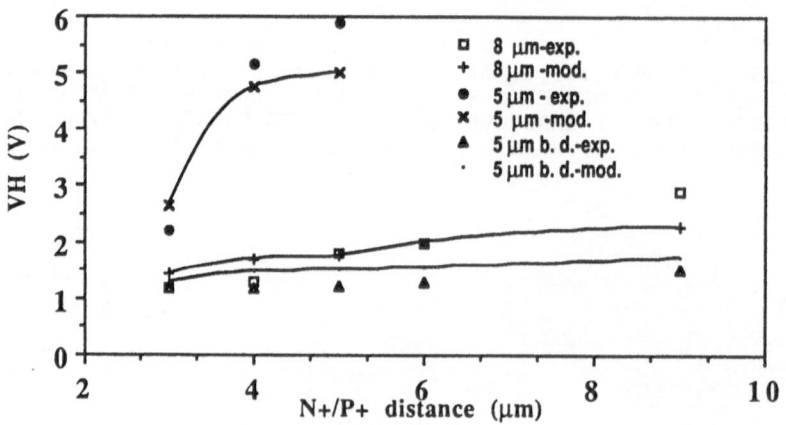

Fig.6. Holding voltage versus N+/P+ distance (exp.= measured values, mod.= modelled values, b. d. = backside disconnected)

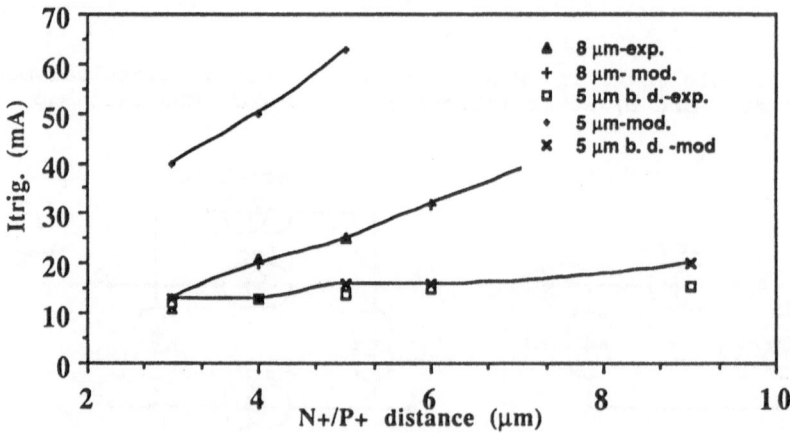

Fig.7. Triggering current versus N+/P+ distance (exp.= measured values, mod.= modelled values, b. d. = backside disconnected)

Physical Modelling of Bipolar Mode Field Effect Transistor (BMFET) for Circuit Simulation

G.BUSATTO

IRECE-CNR
Naples, Italy

G.F.VITALE

Dept. Electronic Engineering
University of Naples, Italy

Summary

A charge-control model of the BMFET, based on a physical analysis of its operation is presented and discussed. This model is used to derive an equivalent circuit of BMFET capable of describing, with continuity, both bipolar and unipolar regions of its characteristics. The equivalent circuit is presented in a form suitable to be easily incorporated in a circuit simulator such as SPICE.

The charge control model

The BMFET is a multicellular device where each elementary cell has the structure of a vertical-JFET with the conduction channel confined by gate P^+ lateral diffusion [1,3]. If the combination of epilayer doping and aspect ratio of the conduction channel is such that the depletion regions of N^-P^+ lateral junctions do not overlay, the BMFET results in a normally-on device. For this device unipolar and bipolar regions of output characteristics are separated by a zero-bias "ohmic" characteristic and correspond, respectively, to a reverse and a forward gate-source biassing [4].

The theoretical models presently available to describe steady state bipolar characteristics [2], unipolar characteristics [3], and transient behavior [4] of the BMFET were intended at the understanding of the basic physics of device operations. In

---

(*) Work supported by National Research Council of Italy under the project "MADESS".

this paper a new formulation of the charge control model is presented and a large-signal equivalent circuit, suitable for circuit simulators is developed. The circuit is obtained by merging the model of a conventional JFET with the one appropriate to the bipolar operation. The latter is described in the following Section.

The bipolar region: Into the bipolar region, the conductivity modulation effect takes place and the minority-carriers charge is virtually stored only within the low-doped epilayer being it surrounded by heavily doped regions. The integration of the continuity equation within the epilayer domain results in the following charge control equation:

$$I_G = \frac{dQ}{dt} + \frac{Q}{\tau} + I_{RG} + I_{RS} + I_{RD} + \frac{dQ_{GS}}{dt} + \frac{dQ_{GD}}{dt} \qquad (1)$$

where: Q is the minority-carrier charge stored within the plasma region, τ is the minority carrier lifetime into the epilayer, I_{RG} and I_{RS} are the recombination currents within source and gate heavily doped regions, I_{RD} is the recombination current within the heavily doped substrate (drain), dQ/dt accounts for the displacement current due to the variation of the charge stored within the plasma region. dQ_{GS}/dt and dQ_{GD}/dt are the displacement currents of the gate-source and gate-drain depletion capacitance.

The charge control equation for the drain current is:

$$I_D = \frac{Q}{\tau_T} - \frac{dQ_{GD}}{dt} - I_{RG} - I_{RD} \qquad (2)$$

where: τ_T is the transit time.

Because of the very low epilayer doping, most of BMFET bipolar operation is in high injection, therefore: $\tau_T = x_1^2/4D_n$, x_1 being the boundary of the conductivity modulated region which is a function of the drain voltage and drain current.

In the portion of the BMFET characteristics where the transport of carriers can be considered ohmic, x_1 can be written as [4]:

$$x_1 = w \left(1 - \frac{V_{DS}}{R_{dr}I_D} \right) \qquad (3)$$

where w is the source to drain distance and R_{dr} is the equivalent resistance associated to the drift region of the epilayer.

In this case the steady state gate current becomes:

$$I_G = I_{GC} \left(1 - \frac{V_{DS}}{I_D R_{dr}}\right) + I_{RG} + I_{RS} \qquad (4)$$

where: I_{GC} is the epilayer recombination current of the diode whose thickness is w, the term in parenthesis is the correction due to the modulation of the recombination volume.

The steady state drain current takes the following expression:

$$I_D = \beta_F I_{GC} + \frac{V_{DS}}{R_{dr}} - I_{RG} - I_{RD} \qquad (5)$$

where: $\beta_F = 4\tau D_n/w^2$ represents the gain of a transistor having a base thickness w, and the term V_{DS}/R_{dr} is the leakage current associated with the ohmic transport between source and drain.

Equations (1) through (5) have the circuit representation reported in Fig. 1 where also the JFET portion has been added. It is worth to note that only the V_{GS} controlled current generator is required to introduce the unipolar branch. In fact the depletion capacitances and the reverse saturation currents of gate-drain and gate-source junctions are already accounted for by C_{jgs}, C_{jgd}, I_{RGS} and I_{RGD}, of the bipolar model, moreover both unipolar and bipolar branches present an output resistance R_{dr}.

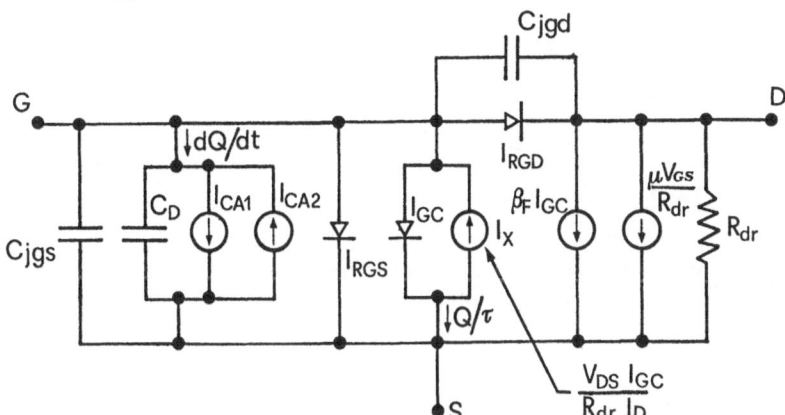

Fig.1. Equivalent circuit describing the BMFET in the region of the characteristics where the unipolar transport can by considered ohmic.

In Fig. 1 the recombination currents due to the heavily doped regions of gate-source and gate-drain structures are described by two diodes, I_{RGS} and I_{RGD} respectively, which have an emission coefficient of 1. This result comes out from the calculation of I_{RG}, I_{RS} and I_{RD} according to [2] by considering equal minority carriers concentrations at the low doped sides of both the p^+n and n^+n transitions.

In the equivalent circuit of Fig. 1, the variation of the gate recombination current Q/τ with x_1, as follows from equation (4), is accounted by the current generator I_x which is controlled by I_{GC}, V_{DS} and I_D. Moreover the controlling diode I_{GC} has an emission coefficient of 2, due to the high injection.

The difference between the emission factors of the controlling diode and of the recombination diode is responsible of the β falloff of the BMFET at higher current levels.

In the circuit model, the displacement current, dQ/dt, due to the diffusion capacitance is described by a normal diffusion capacitance, C_D, and by two current generators, I_{CA1} and I_{CA2}, controlled by the time derivatives of V_{DS} and I_D respectively, which account for the effect of the motion of the boundary x_1.

References

1. Y.Nakamura, H.Tadano, M.Takigawa, I.Igarashi, J.Nishizawa: "Experimental study on current gain of BSIT," IEEE Trans. Electron Devices, Vol.ED-33, pp.810-815, 1986.

2. S.Bellone, A.Caruso, P.Spirito, G.Vitale, "A quasi-one-dimensional analysis of vertical JFET devices operated in the bipolar mode," Solid State Electronics, Vol. 26, No. 5, pp. 403-413, 1983.

3. C.Bulucea, and A.Rusu, "A first-order theory of the static induction transistor", Solid-State Electronics, Vol. 30, No. 12, pp.1227-1242, 1987.

4. G.Vitale, G.Busatto: "The Turnoff Transient of the Bipolar-Mode Field-Effect Transistor" IEEE Trans. Electron Devices, Vol. ED-35, pp. 1676-1682, 1988.

An Analytical Model for the Vertical Insulated Gate Bipolar Transistor

K. GRAHN and S. ERÄNEN

Semiconductor Laboratory
Technical Research Centre of Finland

Summary

An analytical steady-state model has been developed for the vertical IGBT. Effects like conductivity modulation, collector bias variations, emitter injection efficiency, recombination and channel conductance are included. Simulated results are in good agreement with measured characteristics for a 1000V IGBT device.

Introduction

The cross-section of a vertical insulated gate bipolar transistor (IGBT) with an additional buffer layer is drawn in Fig. 1a. Its equivalent circuit model is shown in Fig. 1b. The IGBT behaves as a vertical bipolar transistor (pnp) driven by a n-channel DMOS-transistor. The IGBT contains a parasitic thyristor structure which is sensitive to latch-up at higher current levels /1/. The conductivity of the epitaxial region is modulated by carrier injection from the anode. The introduction of the buffer n^+-layer increases the punch-through voltage and speeds up the turn-off process . The forward voltage of the device is then slightly increased /2/.

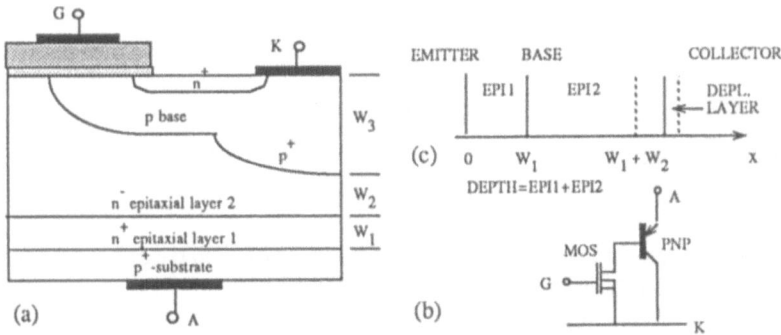

Fig. 1. (a) The cross-section of the DMOS portion of a 1000V IGBT structure. The total depth $(W_1 + W_2 + W_3)$ is equal to 107 μm.

(b) The equivalent circuit of the device. Parasitic elements are not shown.

(c) Coordinate system used in the model.

432

In this paper, a new steady-state model based on analytical expressions has been developed. The model is based on the ambipolar transport equation /4/. Collector bias variations, emitter injection efficiency, high injection, base transport factor and channel conductance are included. In deriving our model the procedure has been the following:

(1) solve the steady-state ambipolar transport equation in the double epilayer,
(2) relate emitter hole current density to the ambipolar transport equation,
(3) obtain the transport factor from derivatives of carrier density,
(4) give depletion width as a function of base-collector voltage,
(5) relate base-collector (drain-source) voltage to DMOSFET current,
(6) express emitter current as a function of carrier density at emitter-base junction.

Finally, the predictions of the model are compared with measured data from experimental devices.

Theory

To describe the low-gain, high-level injection characteristics the ambipolar transport model must be used.. The solution of the steady-state ambipolar transport equation can be found to be

$$n_1(x) \approx p_1(x) = A_1 \sinh((x-W_1)/L_1) + B_1 \cosh((x-W_1)/L_1), \quad 0 \le x \le W_1 \quad (1a)$$

$$n_2(x) \approx p_2(x) = A_2 \sinh((x-W_1)/L_2) + B_2 \cosh((x-W_1)/L_2), \quad W_1 \le x \le W_2 \quad (1b)$$

with boundary conditions

$$p_1(x) = p_0 \qquad\qquad , \text{when } x = 0 \quad ; \; p_1(x) = p_2(x) \;, \text{when } x = W_1$$

$$(2)$$

$$\partial p_1(x)/\partial x = \partial p_2(x)/\partial x \;, \text{when } x = W_1 \; ; \; p_2(x) = 0 \qquad , \text{when } x = W_1 + W_2$$

where p_0 is the hole density at the edge of the emitter-base junction, L_1 and L_2 are the ambipolar diffusion lengths, and W_1 is the width of epitaxial layer 1. Shown in Fig. 1c is the used coordinate system.

By relating the emitter hole current at the emitter-base junction to the ambipolar equation for carrier transport /2/, we get

$$qD_a \partial p_1(x)/\partial x = (b_1(\gamma + 1) - 1) \cdot J_E \qquad , \text{when } x = 0 \quad (3)$$

where γ is the emitter injection efficiency, $b_1 = \mu_{n1}/\mu_{p1}$ is the electron/hole mobility ratio in epitaxial layer 1, D_a is the ambipolar diffusion coefficient, q is the electron charge and J_E is the emitter current density. Electron and hole mobilities are functions of the total dopant concentration /4/. The solution to eqn (3) is

$$A_1 \cosh(W_1/L_1) - B_1 \sinh(W_1/L_1) = -S \cdot L_1 \cdot J_E \tag{4}$$

where $S = (\gamma(b_1 + 1) - 1)/2qD_{n1}$. The diffusion coefficient for electrons in the buffer layer is given by D_{n1}.

By definition, the base transportation factor α_T will be given by,

$$\alpha_T = L_1/(L_1 \cosh(W_1/L_1)\cosh(W_2/L_2) + L_2 \sinh(W_1/L_1)\sinh(W_2/L_2)) \tag{5}$$

The collector-base junction is reverse-biased. By using the depletion approximation the junction depletion region can be written in the form

$$W_d = \text{DEPTH} - W_1 - W_2 = (2\varepsilon_{Si}(\Phi_0 + V_{bc})/qN_{epi2})^{1/2} \tag{6}$$

For the DMOS-current we have used the simple expression

$$I_{ds} = k'(W/L)((V_{gs} - V_t)V_{ds} - V_{ds}^2/2)) \tag{7}$$

At the emitter-base junction, small injection on p^+-side and high injection on n-side is considered. This gives

$$(1 - \gamma)J_E = (qD_n/L_n N_A)p_0^2 \tag{8}$$

The resulting equation system in Fig. 2 has been solved using Brent's method /5/.

$$
\begin{aligned}
&\text{(1)} \quad p_0 = -A_1 \sinh(W_1/L_1) + B_1 \cosh(W_1/L_1)\\
&\text{(2)} \quad B_1 = B_2\\
&\text{(3)} \quad A_1 = L_1 A_2/L_2\\
&\text{(4)} \quad A_2 \sinh(W_2/L_2) + B_2 \cosh(W_2/L_2) = 0\\
&\text{(5)} \quad A_1 \cosh(W_1/L_1) - B_1 \sinh(W_1/L_1) = -L_1 J_E(\gamma(b_1+1)-1)/(2\,q\,D_n)\\
&\text{(6)} \quad \alpha_T = L_1/(L_1 \cosh(W_1/L_1) \cosh(W_2/L_2) + L_2 \sinh(W_1/L_1) \sinh(W_2/L_2))\\
&\text{(7)} \quad \text{DEPTH} - W_1 - W_2 = (2\varepsilon_{Si}(\Phi_0 + V_{bc})/(q\,N_{EPI2}))^{1/2}\\
&\text{(8)} \quad I_{ds} = k'(W/L)((V_{gs} - V_t)V_{ds} - V_{ds}^2/2)\\
&\text{(9)} \quad (1 - \gamma)J_E = (q\,D_n/L_n N_A)p_0^2
\end{aligned}
$$

Fig. 2. The resulting equation system.

Experimental Verification.

The experimental devices were made on p-type, <100> Si-wafers with two n-type epitaxial layers as shown in Fig. 1a. In the first process step the field oxide was opened for the p⁺-implantation over the p-well and guard ring regions. Then the n-type channel steps were formed followed by the double diffusion of the NMOS p-body and the shallow As-source. After the growth of the gate oxide the n⁺-poly gate was formed. Finally, the 1μm non-doped low temperature oxide was opened for the contact windows, and in our case the front side metallization is 5μm thick Al. The back side anode metallization is alternatively 1μm Al or three level Ti/Ni/Ag.

In the simulations, the MOSFET transconductance parameter is used as a fitting variable at different gate voltages. Measured and simulated I/V-characteristics are shown in Fig. 3. Used parameters are given in Table 1. Measured and simulated results are in good agreement.

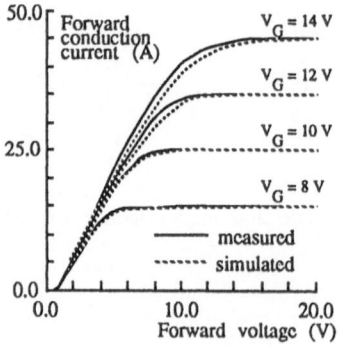

Table 1. Device parameters.

$W1 = 12\ \mu m$
$W2 = 80\ \mu m$
$W3 = 15\ \mu m$
$N_{EPI1} = 1 \cdot 10^{16}\ cm^{-3}$
$N_{EPI2} = 5 \cdot 10^{13}\ cm^{-3}$
base
lifetime $= 2\ \mu s$

Fig. 3. Measured and simulated I/V-characteristics. The threshold voltage is 4.2 V.

References

1. Baliga, B.J. et al.: Suppressing Latchup in IGTs, IEEE Electron Device Letters. 5 (1984) 323-325 .

2. Kuo, D-S.; Hu, C.: An Analytical Model for the Power Bipolar-MOS Transistor, Solid-State Electronics. 29 (1986) 1129-1237.

3. Ghandi, S.: Semiconductor Power Devices. New York: John Wiley 1977.

4. Muller, R.S.; Kamins, T.I.: Device Electronics for Integrated Circuits. New York: John   Wiley 1977

5. Press, W.H. et al: Numerical Recipes. Cambridge: Cambridge University Press 1988.

Effect of Velocity Saturation on Small Signal Behaviour of Submicron MOSFETs: Analytical Modelling and 2-D Simulations

T. SMEDES

Eindhoven University of Technology, Faculty of Electrical Engineering, P.O. Box 513, 5600 MB Eindhoven, the Netherlands

Summary
In short channel MOSFETs the effect of carrier velocity saturation becomes increasingly important. This paper concentrates on its effect on the small signal behaviour. An analytical model, which is compared with 2D simulations for a wide frequency range and quasi-static measurements, shows an important influence on the lf and hf behaviour.

0. Introduction

Although non-quasi-static (NQS) small signal models for long channel MOSFETs have been published [1,2], such models become questionable for submicron transistors. In this paper an analytical NQS small signal model for (short channel) MOSFETs will be presented. The model includes the short channel effect of carrier velocity saturation.

1. Analysis

For a MOSFET, in general an admittance matrix can be defined with (complex) elements $Y_{kl} = \frac{\partial I_k}{\partial V_l}$, where k and l denote terminals of the device. For the sake of simplicity we will first neglect the bulk of the transistor, later this will be reviewed.

To be able to find closed form expressions, the following field dependent mobility model was used:

$$\mu = \mu_0(1 + E_x/E_c)^{-1} \tag{1}$$

Here μ_0 is the low-field mobility, E_x the lateral electrical field and E_c the critical field for velocity saturation. Starting from the continuity equation and the transport equation and using the small signal approximation, we get a coupled set of DC equations, leading to the well-known model [3], and a similar set of AC equations. This set can be rewritten as the second order partial differential equation:

$$\frac{\partial^2 i}{\partial V^2} - jDVi = 0, \qquad \text{with } D = \omega\mu_0\frac{C_{ox}^2}{I_{DC}^2}, \tag{2}$$

where i is the small signal current and V the DC potential in the channel, ω the angular frequency of the signal and μ_0 the low-field mobility. I_{DC} is the current expression from the DC model, containing a velocity saturation component. This equation can be solved in terms of Bessel functions [4]:

$$i = V^{1/2}\{AJ_{1/3}(\frac{2}{3}D^{3/2}e^{\frac{3}{4}\pi j}) + BJ_{-1/3}(\frac{2}{3}D^{3/2}e^{\frac{3}{4}\pi j})\} \tag{3}$$

A and B are constants defined by the boundary conditions. A similar expression can be derived for the small signal potential. The gate current is found from charge neutrality, assuming an ideal gate. The admittance parameters Y_{kl} can be found as $\frac{i_k}{v_l}$, where v is the small signal voltage.

Whereas 2 terms of the Bessel series expansions are sufficient for long channel transistors [2], the more complex form of the short channel expressions requires many terms for a correct model. Because this is impractical for the formulation of a compact model, an asymptotic approximation for the Bessel functions [4] is chosen. Figure 1 shows the effect of the number of series terms in the calculation of Y_{SG} vs. frequency for a MOSFET biased in saturation. The asymptotic approximation is also shown. The effect of non-convergence when too few terms for the Bessel series are used can be seen clearly. The alternative approximation shows acceptable results. Analogous effects are found for other admittance parameters.

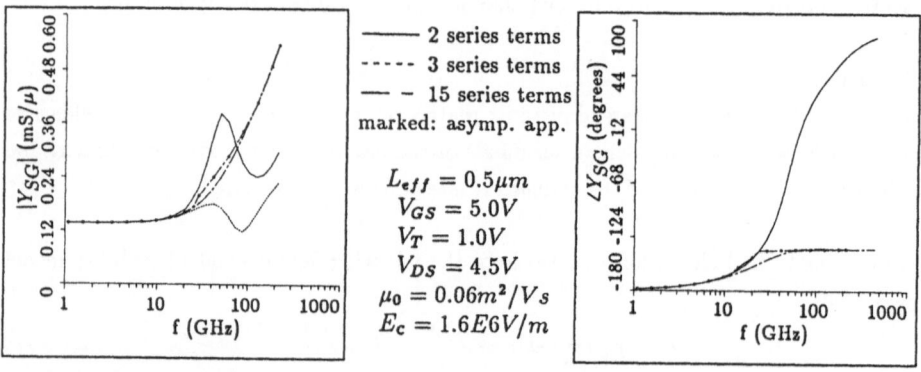

Figure 1: Effect of truncation of Bessel series expansion on calculated Y_{SG}.

2. Simulations

Because it is very difficult, due to several parasitic effects (eg. contacting, packaging), to measure the behaviour of the intrinsic device, a 2-D device simulator [5] was used to carry out numerical calculations for a realistic LDD MOSFET. These simulations included the bulk of the device. Figure 2 shows the magnitude of the simulated gate-related parameters, ie. Y_{kG}. It is clear that the magnitude of the bulk parameter is several orders smaller than the other parameters. This is independent of the mobility model used in the

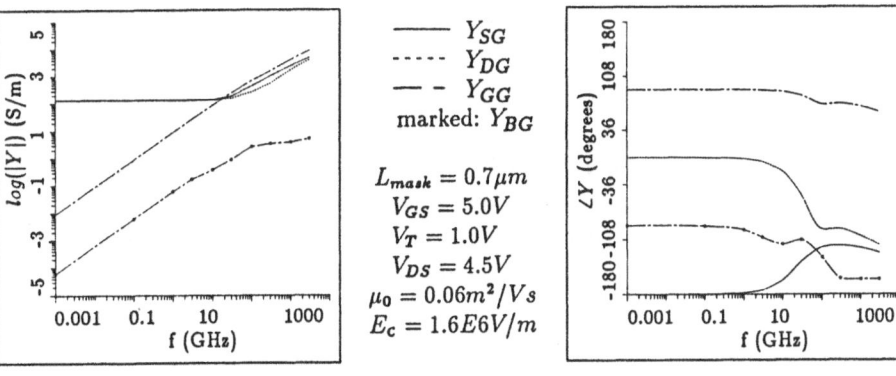

Figure 2: Simulated gate-related parameters vs. frequency.

simulations, thus the bulk can be neglected except for very specific applications.

3. Discussion

To investigate the influence of velocity saturation on the other admittance parameters, simulations were carried out with several mobility models. Figure 3 shows the results of simulations of Y_{SG} for three models: constant (low-field) mobility, velocity saturation and a complete mobility model, including the transverse field dependence. Clearly, one observes a shift of the point where hf behaviour starts to lower frequencies, if velocity saturation is taken into account. This occurs in both the magnitude and the phase. The use of the constant mobility gives inaccurate results, even at low frequencies. Analogous observations can be found for the other parameters. Furthermore, it is apparent that the velocity saturation is a reasonable approximation for the more complete model.

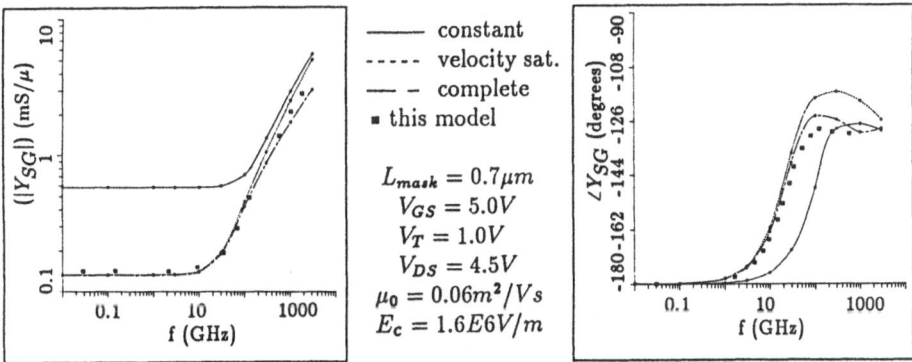

Figure 3: Simulated Y_{SG} vs. frequency for 3 mobility models.

Figure 3 also shows the analytically calculated results. The differences between these and the numerical results are due to the approximation of the Bessel functions and to the

438

parasitics within the intrinsic device, such as overlap capacitances and series resistances, which are not taken into account in the model.

For low frequencies the parameters reduce to their quasi-static equivalents. Figure 4 shows the transconductance, Y_{DG}, and drain conductance, Y_{DD}, of a 0.7 μm device. The difference in slope of Y_{DG} is due to the power used in the mobility model. The saturated part of Y_{DD} is caused by effects not implemented in the model, eg. static feedback. Note that a constant mobility would result in linear $G - V_{DS}$ curves.

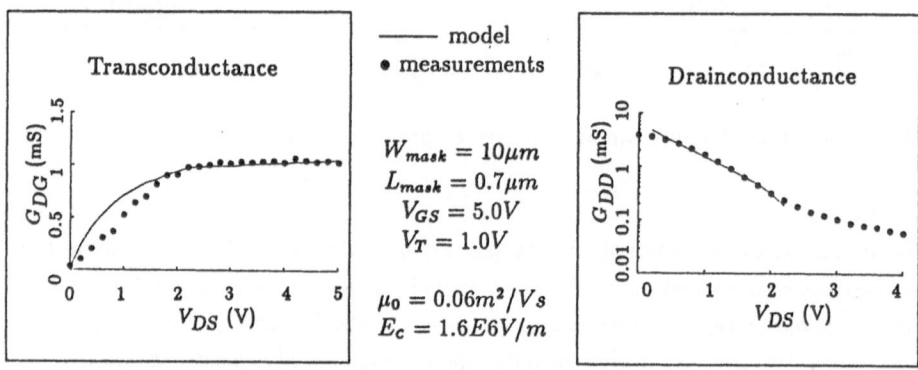

Figure 4: Quasi-static transconductance and drain conductance vs. V_{DS}.

4. Conclusions

An analysis, which was not based on a quasi-static assumption, for a compact small signal model was presented. The resulting model is consistent with 2D device simulations up to high frequencies and with quasistatic analysis, modelling and measurements. It is shown that inclusion of carrier velocity saturation has an important influence on the admittance parameters in both the low and high frequency region.

Acknowledgement

The author would like to thank Dr. F.M. Klaassen for offering the opportunity to work with CURRY at the Philips Research Laboratories in Eindhoven.

References
1. Haslett, J.W.; ea.: Small-signal, high-frequency equivalent circuit for the metal-oxide-semiconductor field-effect transistor. Proc. IEE. 116 (1969) 699-224.
2. Paulos, J.J.; ea.: Limitations of quasi-static capacitance models for the MOS transistor. IEEE El.Dev.Let. 4 (1983) 221-224.
3. Klaassen, F.M.: Compact MOSFET modelling. Advances in CAD for VLSI, vol.1 North Holland 1986.
4. Abramowitz, M.; ea.(eds.): Handbook of mathematical functions. Dover, New York.
5. CURRY user reference manual, Philips propriety.

Session 5

Invited Paper

Chemical Sensors

A. D'AMICO, M. FACCIO, G. FERRI, I. GIANNINI [#]

University of L'Aquila - Department of Electrical Engineering
Monteluco di Roio - 67100 L'AQUILA - Italy

Eniricerche Spa - 00015 Monterotondo (ROMA) - Italy

SUMMARY

This paper takes into consideration some basic devices which are utilised for the realization of chemical sensors. In particular metal-oxide capacitors (MOS), Field Effect Transistors (MOSFET), pyroelectric systems, surface acoustic wave devices (SAW), ion selective FET, immunosensors, electrochemical devices and thermopiles are presented and commented.

INTRODUCTION

In recent years there has been a strong interest in particular solid state devices capable of interfacing the chemical ambient with analog and digital microelectronics.

By the words chemical ambient (C.A.) we shall intend its broadest meaning, i.e. either the surronding atmosphere or any closed or open space where organic and/or inorganic chemical species are present, and eventually time evolving in either or both the gas or liquid phase under equilibrium or non-equilibrium conditions.

Thus the C.A. is characterized by physical parameters such as pressure, temperature, density and chemical parameters: concentrations of chemical substances, chemical potentials, activity, etc.

Chemical sensors (C.S.) are beeing studied and developed for a very broad variety of applications which have a direct impact in the following areas:

biomedics, microbiology, neuropharmacology, industrial control (combustion, biogas systems, wastes etc.), atmospheric monitoring, security systems etc.....

Generally speaking these devices are rather complex because they require a basic multidisciplinary knowledge (at both design and fabrication process levels), ranging from solid state physics and microelectronics, to electrochemistry, biochemistry and membrane science.

Furthermore these devices have direct implication with the following problems: technological compatibility and ambient compatibility. The first has direct consequences on the overall cost of the C.S., the second concerns the possibility of either beeing contaminated or producing contamination.

This article aims to give the reader on overview of the most important basic devices employed for the fabrication of C.S..

In particular for each of them the operating principle will be illustrated and some data on the best performances so far obtained will be commented.

To semplify the discussion the topics will concern the following categories:

1. Pd based Metal-Oxide-Semiconductors sensors,
2. Thermally based C.S. using pyroelectric systems,
3. Surface acoustic wave (SAW) based C.S.,
4. ISFET,
5. Immunosensors,
6. Electrochemical sensors,
7. Thermopiles based H_2 sensors.

Some of these kind of C.S. will be wiewed as an example in terms of their sensitivity to hydrogen and for this reason the palladium in thin films form, as catalytic material for H_2, will be considered. On the other hand it is worth noting that the following considerations are in principle valid also for C.S. of different chemical species, providing that the membrane which is utilised, is suitable for the purpose.

C-MOS H_2 SENSORS

Structures like Metal-Oxide-Semiconductor capacitors, (C-MOS) [1] and (MOSFET) [2], using Pd as a metal have been largely studied as H_2 sensors.

Fig. 1 schematically shows this kind of device whose working principle is the following:

adsorption and diffusion of H atoms

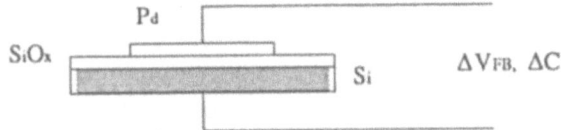

$$H_2 \longrightarrow P_d \xrightarrow{\ H\ } P_d \ SiO_2 \longrightarrow \begin{vmatrix} - \text{ dipole formation} \\ - \text{ work function change} \\ - \text{ interface states variation} \end{vmatrix}$$

H_2 desorption from P_d by O_2

$$O_2 \longrightarrow P_d \longrightarrow \text{formation of } OH, H_2O \text{ groups} \begin{vmatrix} -\text{dipole distruction} \\ -\text{work function change} \\ -\text{inter. states variat.} \end{vmatrix}$$

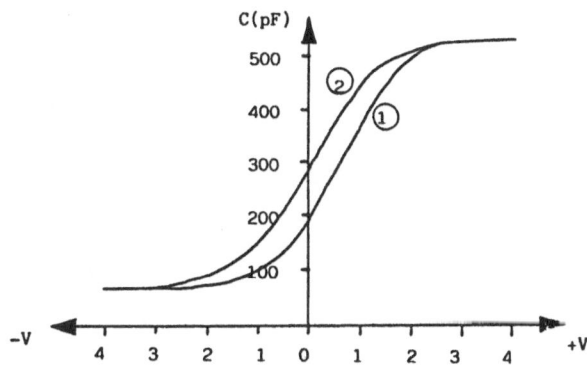

Fig. 1 : Schematic of a P_d MOS H_2 sensor

After the catalytic reaction has taken place at the P_d outer surface, in presence of H_2, it is possible at the P_d-SiO_2 interface to have contribution to the flat band voltage variation (ΔV_{FB}) or to capacitance variation ΔC from: a dipole layer formation or a change in the P_d work function (also possible could be a certain role played by surface state density variations) [3]. In practice the shift of the C/V curves under adsorption and desorption is measured as ΔV_{FB} and is then correlated to the H_2 concentration (see fig. 2).

Fig. 2 : Typical C/V measurement at 1 MHz of P_d MOS capacitor in air (1) and in a 0.5% of H_2 in N_2 (2). Oxide thickness to 1000 $^\circ$A, n-type S_i

PYROELECTRIC SENSORS

These devices are based on pyroelectric materials which shows pyroelectricity, a phenomenon equivalent to piezoelectricity.

In a piezoelectric crystal a strain can induce a net polarization along suitable crystallographic orientations, while in a pyroelectric material a change in temperature due to a change of absorbed thermal energy, causes a net polarization which is measurable by suitable electronic circuits.

A typical noise equivalent power for a commonly used pyroelectric materials such as $LiTaO_3$ is about 10^{-9} watts/cm^2 which corresponds to a thermal input of $0.24*10^{-15}$ cal/cm^2 sec or $6*10^{10}$ eV/cm^2 sec.

This indicates, at least in theory, the possibility of studying thermal processes of chemical interaction of small amount of chemically active substances.

It is worth pointing out that under appropriate calorimetric conditions, temperature changes as small as 10^{-5} C$^\circ$ can be measured by this device.

The working principle of a pyroelectric based C.S. specialised for the analysis of the heat developed by the Pd-H_2 (adsorption) and Pd-O_2 (desorption) processes is indicated in the following flow diagram:

$$\Delta Q \longrightarrow \Delta T \longrightarrow \Delta p \longrightarrow \Delta q \longrightarrow \Delta V, \ \Delta I$$

A change in thermal energy ΔQ produces a temperature variation ΔT which induces a polarization change Δp and as a consequence a charge variation Δq on the two surfaces. Then a voltage or current variation can be measured [4].

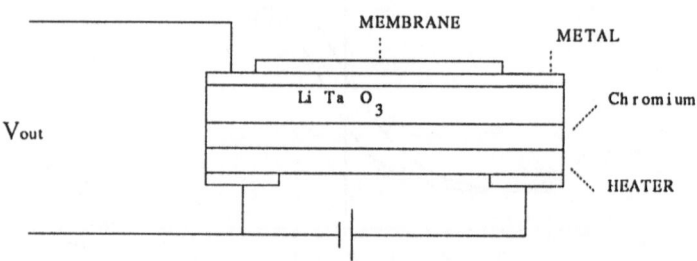

Fig. 3 : Schematic of a pyro-sensor

S.A.W. DEVICES

Fig. 4 shows a Surface Acoustic Wave (SAW) delay-line whose propagation path is coated with a selective sorbent membrane. By using interdigital transducers (I.T.) it is possible to launch and detect SAW's.

Absorption and desorption of a given gas causes a change in the physical properties of the membrane thus effecting the propagation velocity of the acoustic modes. A widely used technique for the detection of SAW's velocity changes is rappresented by an oscillator made by the SAW delay-line and an amplifier (Fig. 4). Frequency changes can be measured whith high accurancy so that small change in the velocity can be detected. Then the output frequency behavior can be associated and correlated to the adsorbed chemical specie.

At present SAW based devices are also fabricated on silicon making use of ZnO thin films as piezotransducers for the interdigital fingers [5]

It is worth mentioning that also vapor [6] and liquid-phase sensing SAW devices have been recently realised with success.

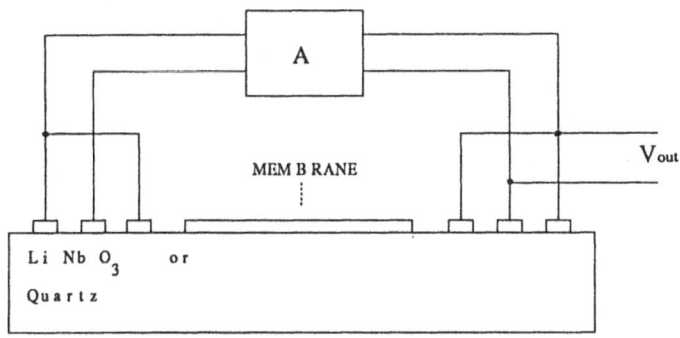

Fig. 4 : Schematic of a SAW delay-line oscillator

ISFET

In recent works on ion selective field effect transistors (ISFET's) many interesting new applications should be mentioned.

Stable membranes selective to heavy metal ions have been reported, that may work for more than a month in solutions [7]: this findings should allow a good control of metal effluents (on surface waters) for environmental protection. A differential measurement using a double gate structure has been proposed by [8] for the detection of alcaline ions in critical care, using a very small blood flow taken from the patient "ex vivo".

In general the most exciting field of development appears to be that of biosensors, where so different knowledges converge for the realization of specialized solid state devices. When an ISFET structure is employed enzymatic electrodes may be realized [9] bat also membranes containing antibodies, or receptors [10].

A classical enzymatic membrane contains glucose-oxidase and is aimed to the detection of glucose in blood. This is perhaps the most heavily studied among the biosensors: the goal is the realization of an artificial pancreas or, at least a better control on diabetic subjects.

Glucose-oxidase is a very stable enzyme, such as urease, so that many of the enzymatic ISFET realized so far are dealing with the measurements of glucose or urea.

A recent realization useful for enviromental protection utilizes the enzymatic approch to quantify the amount of pesticides, like organophosphates and carbamates, in ground waters: the enzyme molecule that is used is acetylcholine-esterase, which is strongly inhibited by the pesticides of these classes.[11]

Immunochemical assays have also been developed using a stable membrane chemically linked to an ISFET in order to detect a substance present in the ambient; an antibody for this substance is usually produced and linked to the insulator of the FET.

An enzymatic technique is then used in order to generate and amplify the electrical signal related to the binding of substances to be detected. The sensitivity of this assay may be very high; in fact 0.01 ppb of a low molecules weight substance can be revealed.

ELECTROCHEMICAL SENSORS

Electrochemical (EC) cells can work as sensors when the chemical species to be measured reacts at the electrodes, giving rise to voltage or current variations.

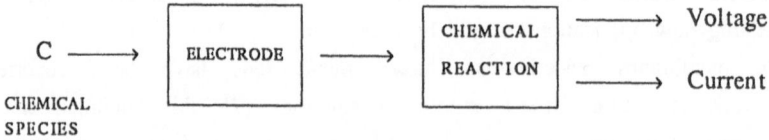

Many of such sensors are now currently in use as macroscopic devices.

This method quite easily may evolve toward smaller microfabricated structures allowing the selective detection of more than one species. This can be made especially by using solid electrolytes, and by forming electrodes through conventional deposition techniques.

Solid electrolytes working at both ambient and at high temperatures are available, so that it is possible design many different structures for a variety of applications. Zirconia oxide or doped β-alumina are useful ionic conductors for high temperature applications. For ambient temperatures, there is availability of some polymeric electrolytes, or, in alternative, proton inorganic conductors. Zirconia exploitation for the development of oxygen miniaturized probes is well reported in ref. [12].

A good solution was suggested [13] for the detection of toxic gases at high temperatures using a couple of solid state ionic conductors: the first forming a thick layer of the principal electrolyte of a potentiometric cell; the second one is deposided at the external surface of the first, forming a sensitive structure coupled to the outer electrode.

This type of structure has been realized using Ag doped β-alumina coupled to $AgCl$ to detect chlorine, or coupled to Ag_2SO_4 to detect sulfur oxides, or NASICON and Na_2CO_3 for the detection of CO and CO_2.

By this technique, when some problems due to the stability of external layers would be solved, an array of EC sensor with different gas sensitivity may be easily assembled for the realization of a smart gas sensor.

In lower temperature range some attempt for the construction of a miniaturized device has been made using NAFION as proton conductor. NAFION could be a good choice because of his high conductivity, however it shows problems in working in dry atmospheres, and its operation would be satisfactory only at room temperature in a reasonable humidity range.

A good solution for the realization of a hydrogen sensor, working between 0 and 200 °C is related to the utilization of α-Zr-phosphate as electrolyte. [14]

THERMOPILES BASED H2 SENSOR

Recently new kinds of thermocuples sensitive to H2 made by thin films of Cu/Pd have been fabricated and tested.[15]

With a temperature difference (ΔT) between the cold and hot junction of 10 °C the responsivity was found to be 25 nV/ppm. The noise equivalent

Hydrogen concentration (NEC_{H_2}) which is the H_2 concentration which gives rise to a signal to noise ratio equal to 1 is about 10^{-2} ppm, with the following operative conditions : flux of 500 ppm of H_2 in H_2 + N_2 mixture equal to 20 cc/min, $\Delta T=100$ °C.

Finally it is worth pointing out the positive role played by the semiconductor oxide such as Z_nO and in particular S_nO_2 which are beeing promising candidate for the detection, at moderately high temperature (200 - 400 °C), of gases such as H_2, NO_x, SO_2, CH_4 etc. [16,17].

An up-date collection of results on C.S.'s can be seen on the recent TRANSDUCERS '89 Proceeding (Montreux June 25-30, 1989) relative to : the 5th Int. Conference on Solid-State Sensor and Actuator.

REFERENCES

1. Lundström K. I. : Hydrogen sensitive MOS-structures. Part 1 : Principles and Applications. Sensors and Actuators, 1 (1981) 403-426.

2. Lundström K.I.; Shivaraman M.S.; Svensson C.M. : A hydrogen-sensitive P_d-gate MOS transistor. J. Appl. Phys. 46 (1975) 3876-3881.

3. Fortunato G.; Bearzotti A.; Caliendo C.; D'Amico A. : Hydrogen sensitivity of $P_d/S_iO_2/S_i$ structure : a correlation with the hydrogen induced modification on optical and transport properties of α-phase P_d films. Sensors and Actuators, 16 (1989) 43-54.

4. D'Amico A; Zemel J.N. : Pyroelectric enthalpymetric detection. J. Appl. Phys. 57 (1985) 2460.

5. D'Amico A.; Verona E. : SAW sensors. Sensors and Actuators, 17 (1985) 15.

6. Wenzel S.W.; White R.M. : Flexural plate wave chemical vapor sensor Transducers '89, Montreux June 1989, 214.

7. Battilotti M.; Mercuri R.; Mazzamurro G.; Giannini I.; Giongo M. : Lead ion-sensitive membrane for ISFETs. Transducers '89, Montreux June 1989, 74.

8. Gumbrecht W.; Schelter W.; Montag B. : A chemFET microcell system for medical and biotechnological online electrolyte monitoring. Eurosensor I, Cambridge Sept. 1983, 198.

9. van der Schoot B.H.; Bergveld P. : ISFET based enzyme sensors. Biosensors 3 (1987/88) 161-186.

10. Tedesco J.L.; Krull U.J.; Thompson M : Molecular receptors and their potential for artificial transduction. Biosensors 4 (1989) 135-167.

11. Colapicchioni C.; Barbaro A.; Giannini I.; Mercuri R. : Pesticide determination by a FET-linked acetylcholinesterase device. SEB Meeting, Edinburgh Apr. 1989, 63.

12. Pribat D.; Velasco G. : Microionic gas sensor for pollution and energy control in the consumer market. Sensors and Actuators 13 (1988), 173-194.

13. Hötzel G.; Wepper W. : Application of fast ionic conductors in solid state galvanic cells for gas sensors. Solid State Ionics (1986) 1223-1227.

14. De Angelis L.; Maimone A.; Modica L.; Alberti G.; Palombari R. : A new hydrogen sensor with pellicular Z_r phosphate as proton conductor.Transducers '89, Montrex June'89, 20.

15. Bearzotti A.; Gentili M.; Lucchesini A.; Caliendo C.; Verona E.; D'Amico A. : Cu/Pd thin film thermopile as a temperature and hydrogen sensor. Transducers '89, Montreux June 1989, 202

16. Göpel W.; Shierbaum K.D.; Schmeisser D.; Wiemhöfer H.D. : Prototype chemical sensors for the selective detection of O_2 and NO_2 in gases. Sensors and Actuators, 17 (1989) 377-384.

17. Huck R.; Böttger U.; Kohl D.; Heiland G. : Spillover effects in the detection of H_2 and CH_4 by sputtered SnO_2 films with Pd and PdO deposits. Sensors and Actuators, 17 (1989) 355-359.

Bipolar/BICMOS Technology

Fabrication and Characterisation of pnp Polysilicon Emitter Bipolar Transistors

I.R.C. POST and P. ASHBURN

Dept. of Electronics and Computer Science,
The University, Southampton, SO9 5NH, England.

Abstract
The fabrication of *pnp* polysilicon emitter bipolar transistors is described for devices both with and without a deliberately grown interfacial oxide layer. SIMS is used as an aid to determine the optimum emitter drive-in conditions, and this is compared with electrical results from actual devices. Precipitation of boron in the polysilicon is shown to occur for devices with high implant doses and low drive-in temperatures. Preliminary modelling of devices with an interfacial oxide layer has been undertaken in terms of the barrier heights to majority and minority carriers.

1. Introduction

Polysilicon is now widely used as a contact to the emitter of *npn* bipolar transistors. This results in a suppression of base current, thereby allowing a trade-off between current gain and device speed. If an interfacial oxide layer is deliberately grown at the polysilicon/silicon interface [1] a larger gain improvement is obtained, when compared to polysilicon contacts with no interfacial oxide [2], which has been attributed to minority carriers being forced to tunnel through the oxide.

Up until the present time the study of polysilicon emitter transistors has been wholly confined to *npn* structures. For high speed applications this choice is quite natural since the electron mobility in silicon is over 3 times that of holes in silicon. On the other hand *pnp* polysilicon emitter transistors are important, not only in their own right for analogue and BICMOS applications, but in that they allow the physics of the polysilicon/silicon interface to be studied through its interaction with an opposite type charge carrier.

2. Fabrication of devices

Fabrication of the *pnp* polysilicon emitter transistor was along conventional lines up to and including the base drive-in. Emitter windows were opened and, immediately before polysilicon deposition, the interfacial layer was characterised either by a HF dip etch (to remove any native oxide) or an RCA clean (to chemically grow a thin oxide of nominal thickness around 14Å). The polysilicon was then implanted with BF_2 ions of 70keV energy and doses in the range 2.5×10^{15} to 1×10^{16} cm^{-2}. In order to determine the optimum emitter drive-in conditions, experiments were performed in which the time and temperature of the drive-in were varied.

3. Doping profiles

Secondary ion mass spectroscopy (SIMS) was used as an aid to analyse the doping profile of boron in the polysilicon and single-crystal silicon, and the results are shown in figs. 1 to 3 for implant doses of 2.5×10^{15}, 5×10^{15}, and 1×10^{16} cm^{-2} respectively. The dopant peaks at the surface indicate boron precipitation, and similar studies in single-crystal silicon have shown the boron in these peaks to be immobile [3] and electrically inactive [4]. TEM analysis of the polysilicon in this region shows that the grains are small and heavily defected, contrasting greatly with the 'plateau' region of the SIMS profile, which the TEM analysis shows to consist of large grained ($\sim$0.2μm) polysilicon. The maximum concentration of boron, before precipitation occurs, can be defined to exist at the shoulder of the peak [3], and this analysis has been carried out on the SIMS profiles, with the results summarised in table 1. The maximum concentrations in table 1 are about a factor of 3 higher when compared to single-crystal silicon [3] (6×10^{19} cm^{-3} for a 900°C drive-in), although this may not be surprising since the diffusion properties of polysilicon are quite different from that of single-crystal silicon. For the highest dose implant (fig. 3), which is typical of the implant given to devices, the peak remains very prominent even for the highest drive-in temperature of 950°C. This indicates that for the drive-in temperatures studied, boron precipitation limits the boron concentration close to the interface to around 2.2×10^{20} cm^{-3}, with higher dose implants merely adding to the concentration of precipitated boron. In single-crystal silicon, the boron peak can only be removed by high temperature anneals (1000-1100°C), which suggests that rapid thermal annealing (RTA) could be a very viable solution to produce heavily boron doped polysilicon, with shallow emitter junctions.

4. Electrical characteristics of devices

Typical electrical results for the devices, in the form of Gummel plots, are shown in fig. 4 (for devices given a 60 minute drive-in at 850°C) and in fig. 5 (for devices given a 180 minute drive-in at 850°C). The base currents in fig. 4 are non-ideal, although referring to the SIMS profile (fig. 2) it can be seen that the emitter/base junction is extremely shallow ($\leq$100Å). The depletion region will therefore reside close to, or even at the interface, thus causing non-ideal characteristics. A gain enhancement of over a factor of 10 is obtained for the RCA device when compared to the HF device, indicating that a substantial tunnelling barrier is present at the polysilicon/silicon interface for the RCA device. On the other hand the base currents for the longer drive-in devices (fig. 5) are ideal for over 5 decades of current. A gain enhancement is also obtained for RCA devices over HF devices, although this is only by a factor of 3.5, which suggests that recombination in the single-crystal emitter is appreciable.

In *pnp* devices, the electron barrier height controls the base current, and hence it should be possible to model the RCA characteristic in fig. 5 using a value of 0.4 eV for χ_e [5]. Fig. 6 compares predicted and measured base characteristics for the case where recombination in the single-crystal emitter is suppressed. Although this assumption is not valid in reality, it does allow a minimum value of χ_e to be deduced. It is clear from fig. 6 that it is just possible to model the base current using a value of 0.4 eV, however the inclusion of realistic levels of recombination in the single-crystal emitter would increase the base current by a factor of about 1.2, and hence make a good fit impossible.

Studies of the base current of *npn* devices have yielded a minimum value for the hole barrier height, χ_h of around 1.2 eV [5]. In *pnp* devices, χ_h controls the emitter resistance, and therefore influences the collector current at high forward bias, as illustrated in fig. 6. These results highlight a large discrepancy in the hole barrier height as obtained from *npn* and *pnp* devices.

References
1. H.C. de Graaff et al, IEEE Trans. Electron Devices, vol.ED-26, p.1771, 1979.
2. T.H. Ning et al, IEEE Trans. Electron Devices, vol.ED-27, p.2051, 1980.
3. H. Ryssel, et al, Applied Physics, vol.22, p.35, 1980.
4. W.K. Hofker et al, Applied Physics, vol.4, p.125, 1974.
5. P. Ashburn et al, IEEE Trans. Electron Devices, vol.ED-34, p.1346, 1987.

Temperature [°C]	850	900	950
Maximum boron concentration [cm^{-3}]	1.5-2.0x10^{20}	2x10^{20}	2.2x10^{20}

Table 1. Maximum boron concentration in polysilicon before precipitation occurs, as a function of temperature, for a drive-in of 60 minutes in dry N$_2$

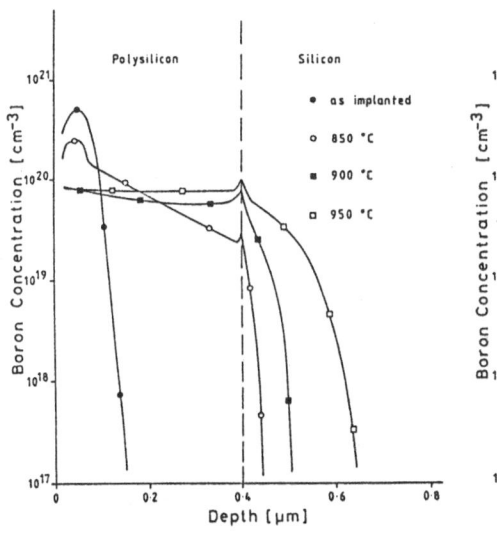

Fig. 1. Boron profile for implanted BF$_2$ (dose 2.5x10^{15} cm^{-2}) driven-in for 60 mins in dry N$_2$ at 850, 900 and 950°C.

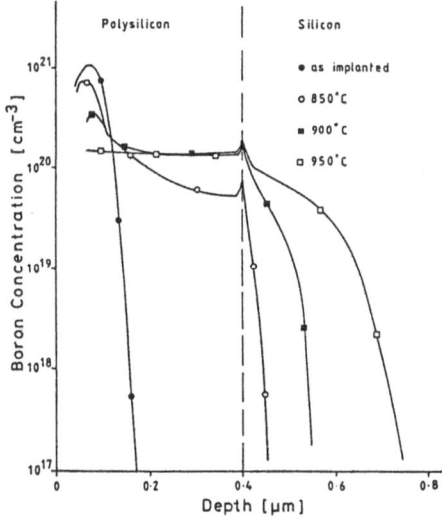

Fig. 2. Boron profile for implanted BF$_2$ (dose 5x10^{15} cm^{-2}) driven-in for 60 mins in dry N$_2$ at 850, 900 and 950°C

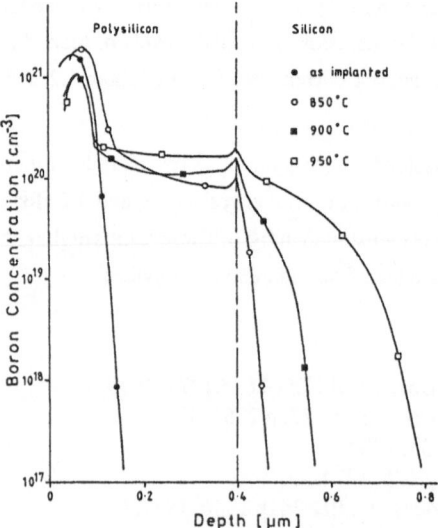

Fig. 3. Boron profile for implanted BF_2 (dose 1×10^{16} cm^{-2}) driven-in for 60 mins in dry N_2 at 850, 900 and 950°C

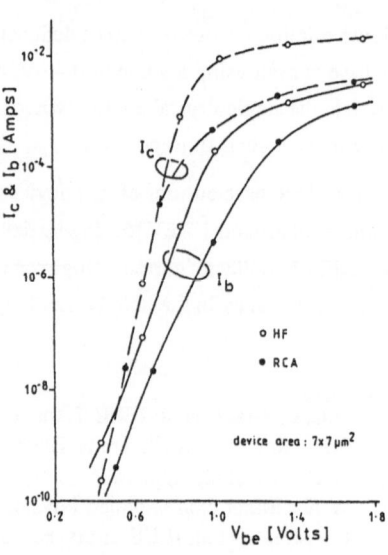

Fig. 4. Gummel plot of devices with an emitter drive-in of 60 mins at 850°C in dry N_2 (implant dose 5×10^{15} cm^{-2})

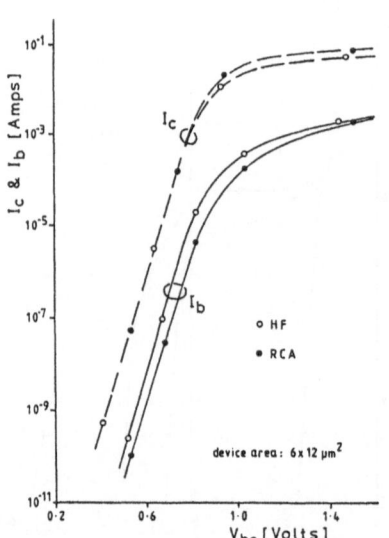

Fig. 5. Gummel plot of devices with an emitter drive-in of 180 mins at 850°C in dry N_2 (implant dose 1×10^{16} cm^{-2})

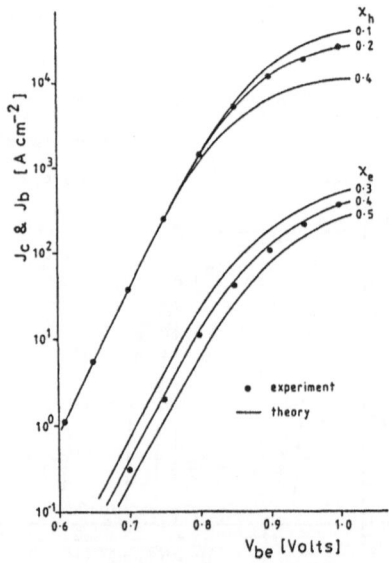

Fig. 6. Gummel plot of RCA device from fig. 5, modelled in terms of a hole (χ_h) and electron (χ_e) oxide barrier height

BICMOSG³ Cell – A Novel High-Speed DRAM Cell Taking Full Advantage of BICMOS Technology

R. Richter , W. Winkler , W.-E. Matzke , K. E. Ehwald , B. Heinemann

Institute for Physics of Semiconductors
Academy of Sciences of the GDR
Walter-Korsing-Str. 2, 1200 Frankfurt(Oder), GDR

Summary

The BiCMOSG³ cell, a novel high density memory cell using BiCMOS technology is described. The cell is based on merging four transistors and occupies approximately the area of a single MOS transistor. The implementation of three gain mechanisms and a bipolar output driver transistor into the cell ensures both a very high operation speed and a bit line read out signal of about 1 volt using only operation voltages between 0 and 5 volts. The BiCMOSG³ cell allows a completely selective access to single cells or groups of them. Therefore, the recharging of unselected bit lines is not necessary and the power dissipation of an array based on BiCMOSG³ cells is reduced compared to one-transistor cells.

1. Introduction

Up to now, alternative dRAM cells based on the principle of internal charge gain, e.g., [1], [2], could not prevail against the one transistor cell and its three-dimensional successors. This is mainly due to the fact, that the peripheric circuitry and the user's environment are very closely adapted to the one transistor cell concept, and the advantages of the alternative candidates are not sufficient to break into a well established line of products. The novel BiCMOSG³ memory cell concept overcomes the major drawbacks of previous gain cells by exploiting the advantages of BiCMOS technology even in the memory cell array itself.

2. Device Structure

A configuration of the BiCMOSG³ cell and its lumped component model are shown in fig. 1 a and b, respectively. From these figures, the following devices are identified: 1) A P-channel JFET Q2: The P⁺-region (connected with the word line) forms the source. The P-region below the N⁺-region (connected with the read bit line) forms the drain, while the storage region and the N-well form the topgate and the backgate, respectively. The intermediate P-layer forms the channel. The N-well is connected to V_{DD}. 2) A first bipolar junction transistor (BJT) Q1: The N-well forms the collector, the channel and the topgate of the JFET form the base and the emitter, respectively. 3) A second BJT Q3: The N⁺-region forms the emitter, the drain of the JFET is the base and the N-well is its collector. 4) A third lateral BJT Q4: The emitter and the base are common with Q3. The collector is formed by the topgate of the JFET (storage region). 5) A MOS-capacitor formed by a polysilicon plate (write bit line), the gate oxide and the N-storage region, stores the "0" and "1" states. We define "1" as having the storage region filled with electrons, whereas "0" corresponds to a depleted storage region.
WRITING: In the writing operation, a positive voltage V_{WL} is applied to the word line WL. Writing "0" or "1" is determined by the bit line voltages.

When the write bit line WBL is grounded and the read bit line RBL is biased at V_{RBL}^+, "0" is written, as shown in fig.2. With the grounding of WBL, the potential of the storage region becomes more negative resulting from the capacitive coupling. In consequence of the positive word line voltage Q1 turns on, injecting the electrons of the emitter, i.e., the storage region into the base. While the write bit line voltage V_{WBL} returns to V_{DD}, the storage region potential ψ_{SR} is increased w.r.t. the starting point (usually the "1" state represented by $\psi_{SR}-$) as a result of the electron deficit. After "0" writing the potential of the storage region is given by $\psi_{SR}^+ = \psi_{SR}- + Q_{WR}/(C_{ox}+C_{PN})$, where Q_{WR} is the charge written out as emitter current of Q1. C_{ox} and C_{PN} are the oxide capacity and the junction capacity between the storage region and the channel, respectively. When WBL is held at V_{DD} and RBL is precharged at about 0 V, "1" is written. As a result of the high word line voltage, the gate source voltage V_{GS} of Q2 is lower than its pinch off voltage V_P. Q2 charges the base of Q3 and Q4, turning on both transistors. Q3, operating in the grounded-collector configuration, charges RBL, while the lateral BJT Q4 discharges its collector, i.e., the storage region. Due to the collected electrons, the potential of the storage region falls, until Q1 turns on, removing excess electrons to the N-well. The storage region potential is given by $\psi_{SR}- = V_{WLWR} - V_{DBE}Q1$. V_{WLWR} is the word line voltage during writing and $V_{DBE}Q1$ is the diffusion voltage of the base emitter diode of Q1.

READING: As the result of the writing operations ψ_{SR}^+ and $\psi_{SR}-$ were built up in the storage region representing the "0" and "1", respectively. In the reading operation an intermediate voltage V_{WLRD} ($V_{WLRD} < V_{WLWR}$) is applied to WL, while RBL is precharged at about 0 V and WBL is held at V_{DD}. If a logic "1" is stored, the gate source voltage of Q2, which is given by $V_{GS}RD = \psi_{SR} - V_{WLRD}$ is lower than V_P and Q2 turns on. The cell amplifies the stored charge three times. The stored charge of the MOS capacity modulates the drain current of Q2 (first gain). Simultaneously, the JFET drain current represents the base current of Q3 and Q4. Q3 charges RBL (second gain), and Q4 discharges the storage region increasing the stored charge and causing a positive feedback to the JFET drain current (third gain). In the case of reading "0", $V_{GS}RD$ is greater than V_P and the JFET drain current is suppressed. Therefore, the RBL voltage remains at its precharge level. Assuming a RBL precharge voltage of 0 V, the obtainable read out signal is given by $V_S = V_{WLRD} - V_{DBE}Q3$, where $V_{DBE}Q3$ is the diffusion voltage of the base emitter diode of Q3.

3. Device Design

To determine the external voltages on the control lines of the cell and the pinch off voltage of Q2, one has to meet the following requirements : $V_P < \psi_{SR}^+ - V_{WLRD}$ (read "0" condition), $V_P > \psi_{SR}- - V_{WLRD}$ (read "1" cond.), $V_P > \psi_{SR}^+ - V_{WLWR}$ (write "0"->"1" cond.). Furthermore, it is necessary, that V_{WLRD} has to be greater than $V_{BE}Q3$ to obtain a detectable read out signal. Fig. 3 shows a vertical doping concentration profile of the stratified charge memory structure as well as the corresponding calculated potential profiles, which satisfies the above mentioned conditions. If V_{WLRD} and V_{WLWR} are chosen to 2.2 V and 3.5 V, respectively, ψ_{SR}^+ and $\psi_{SR}-$ were calculated to 6 V and 2.8 V, respectively. The presented doping profile determines V_P to about 3.3 V. The current gain of Q1 follows from the design of Q2 and is not open to influence. However, the base gummel number is lower than 5.10^{11} s/cm^4, resulting in a current gain of about 500 (measured value) for relevant collector current densities. Because the collector emitter voltage is always lower than 3 V, there is no danger of punch through for Q1. Q3 is the conventional vertical npn-transistor of the used BiCMOS technology.

4. Experimental Results and Discussion

Experimental test devices were fabricated using a 1.5 um BiCMOS technology [3]. One added masking step is used to realize the test structure. This covers the peripheric transistors from the ion implantations of the storage region and the buried channel. As is shown in fig. 3 the stratified charge memory structure has the inverse doping configuration as previous gain cells [1], [2]. The storage region and the buried channel of the BiCMOSG3 cell are doped with As and B, respectively. Hence, it follows as a result of the very abrupt pn-junction a more efficient control of the JFET drain current by the potential of the storage region and a better α-particle immunity compared with [1],[2]. For masking of the emitter implantation, electron beam lithography was used to ensure a space of about 0.3 μm between the gate (WBL) and the N+-region defining the lateral base width of Q4. The electron beam lithography step is not required, if an oxide spacer at the edge of the poly silicon gate is used to displace the emitter w.r.t. the base implantation. As especially important for the cell operation the following technology and device parameters are presented :

$$
\begin{array}{lll}
x_{je} = 0.3 \ \mu m & t_{ox} = 45 \ nm & R_s \ pinch\text{-}base = 10 \ k\Omega/\square \\
x_{jb} = 0.6 \ \mu m & BV_{CBO} = 20 \ V & f_{Tnpn} = 3 \ GHz \\
x_{jP+} = 0.35 \ \mu m & BV_{EBO} = 7 \ V & \beta_{npn} = 80 \\
t_{epi} = 1.5 \ \mu m & BV_{CEO} = 9 \ V & I_{CEO}/cm^2 < 100 \ nA/cm^2
\end{array}
$$

Fig. 4 shows an oscillograph of the operation voltage waveforms according fig. 2 (Note the large read out voltage of about 1 V with a load resistance of 1 kΩ). The same method of measurement was used to get signal retention times. At 85°C values between 0.15 s and 3.5 s were obtained for a charge leakage of 20 % and a cell area of 150 μm². Fig. 5 illustates the influence of the "0" write operation to selected and unselected cells connected with WBL. V_{WL}^* denotes the word line voltage applied at the time when WBL is active (low). $V_{WL}"1"$ is the minimum word line voltage required to turn on Q2, that means, to read or write an "1" state, respectively. In the storage mode V_{WL}^* has to be lower than $V_{WL}^{ST}max$ in order to inhibit "0" writing for unselected cells. If V_{WL}^* is greater than $V_{WL}^{ST}max$ and lower than $V_{WL}^{WR}min$ a WBL pulse generates undefined states between "0" and "1" in the cell. A reduction of the stored charge caused by V_{RBL} pulses is prevented by the reverse biased base emitter diode of Q3. The existence of a passive state for both bit line voltages (high levels) allows a completely selective access to single cells. The high driving capability of the cell is illustrated in fig. 6 by static I-V characteristics. The high scaling capability of the cell is pointed out by a cell variant suggestion suitable for a 16 Mbit integration, as shown in fig. 7. Assuming a minimum feature size of 0.7 μm for trench isolation and emitter length a cell area of about 5 μm² was estimated.

References

1. P.K. Chatterjee, G.W. Taylor, J.E. Leiss; Proc. of the 11th Conference on Solid State Devices, Tokyo 1979 pp. 209-212
2. A.G. Eldin, M.I. Elmasry; IEEE Journal of Solid State Circuits, Vol. SC-20, No. 3, June 1985, pp. 715-723
3. D. Temmler, K.E. Ehwald, W. Winkler, R. Barth, P. Schley ; 17th Yugoslav Conf. on Microelectronics, Miel, May 1989

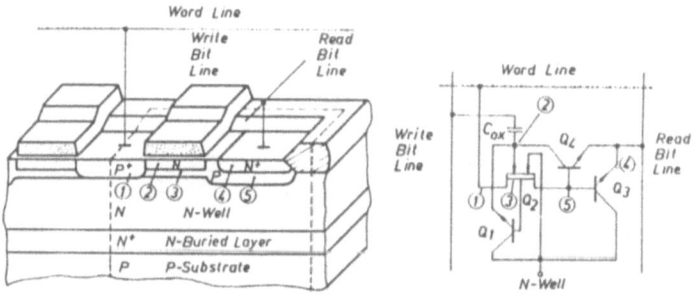

① P^+-Source ② N-Storage Region ③ P-Buried Channel ④ N^+-Emitter ⑤ P-Base, Drain

(a)   (b)

Fig. 1. BiCMOSC³ cell structure (a) Cross section (b) Lumped component model

460

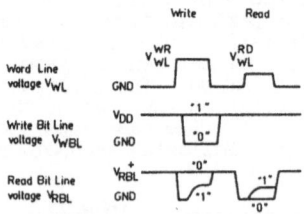

Fig. 2. Operation voltage waveforms

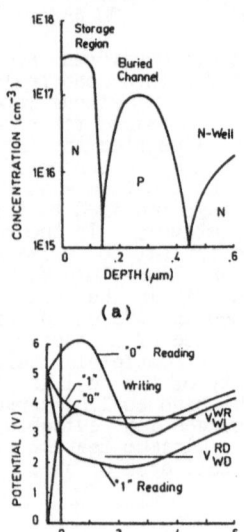

(a)

(b)

Fig. 3. (a) Simulated concentration profiles of the storage region and the buried channel (b) Corresponding potential profils illustrating the operation of write and read functions

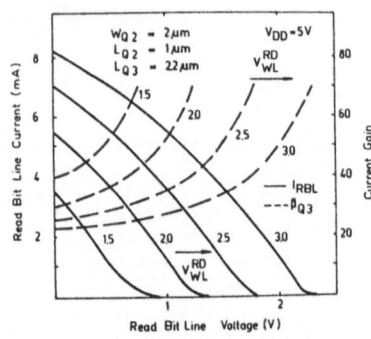

Fig. 6. Static I-V characteristics: Read bit line current and current gain of Q3 versus RBL voltage

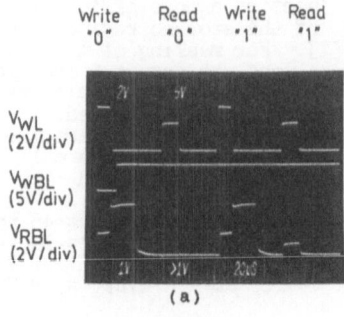

(a)

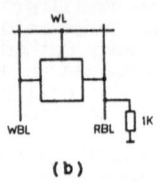

(b)

Fig. 4. (a) Oscillograph illustrating the operation of the write and read functions (b) Circuit configuration used to generate oscillograph

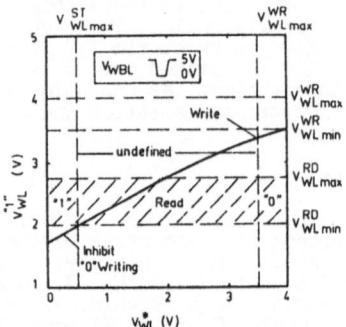

Fig. 5. Minimum word line voltage for reading and writing "1", respectively versus word line voltage applied at the time when V_{BWL} is low

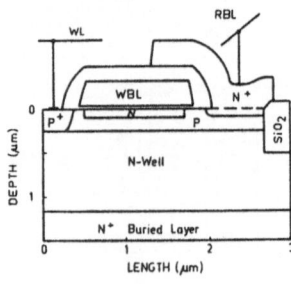

Fig. 7. 16-Mbit cell suggestion with 5 μm^2 cell area using polyemitter and trench isolation

1.2 μm BICMOS Technology for Mixed Analog-digital Applications

C.MALLARDEAU,P.KEEN,A.MONROY,J.C.MARIN,D.CELI,P.A. BRUNEL, M.ROCHE

SGS-THOMSON MICROELECTRONICS,B.P.217,38019 GRENOBLE Ced. FRANCE

Summary

A 1.2μm mixed analog-digital BICMOS technology has been developed and characterized. The objective of this technology is to achieve simultaneously high density CMOS logic (5V) and high performance bipolar transistors for analog applications (12V). In this paper we particulary discuss the results of the optimization of the bipolar device parameters, in order to combine high voltage analog requirements and high frequency performance.

1- INTRODUCTION

The use of both CMOS and bipolar structures on the same chip is regarded as an ideal technology to exploit the specific advantages of each process: high speed components, high driving capability, high precision analog circuitry for bipolar, and high integration density, low power consumption for the CMOS. However, the implementation of such a complex technology can only be successful if the improved performance can counterbalance the higher costs of fabrication. One optimal use of BICMOS technology is the one-chip integration of complex analog-digital systems, where complex digital functions are achieved in 5 Volts CMOS, with added NPN bipolar for speed, drive capability and analog functions [1],[2],[3].

In this paper, we present the characteristics of a 1.2μm BICMOS technology for mixed analog digital applications, with 5V/12V voltage supply specifications. The process starting point is a 1.2μm CMOS technology, to which five extra masking steps have been added for bipolar transistors. The emitter-base structure of the NPN transistor has been optimized in order to achieve both high voltage analog requirements (12V) and high speed performance. Static and dynamic measurements have been performed to characterize the technology.

2- MAIN FEATURES OF THE PROCESS

The process is set up using 1.2μm design rules with photolithography implemented on a CANON FPA1550 5X stepper. Fig.1 shows a cross section of a 1.2μm BICMOS structure. The available components of the technology are

NMOS, PMOS and high speed NPN bipolar. The process starts from a 1.2μm CMOS technology, with five additionnal levels for bipolar process. The outline of the process is given in fig.2.

The main features of the 1.2μm CMOS starting point technology are the following: P substrate, twin well structure for high integration, mixed isolation with conventionnal LOCOS and junction, 1.2μm polygate with silicide ($TaSi_2$), LDD structure for NMOS, with spacer to prevent hot electron degradation, conventionnal structure for PMOS, BPSG deposition and reflow to smooth down the topography. The double level metal structure consists in Ti/TiN/AlCu for metal 1 to achieve good contact resistance to silicon and to provide a good diffusion barrier between Si and AlCu, a sandwich of PECVD oxide and SOG for planarized mineral interdielectric, and AlCu for metal 2.

To add the bipolar components to the technology, the following steps have been introduced: N+ buried layer combined with N+ deep collector sink is used for low collector acces resistance, P+ buried layer is added for isolation between components, a 3μm thick N-type epilayer, with higher resistivity than the Nwell, is required for the NPN collector to achieve high performance analog functions, the P- base of the NPN requires a specific photolithography and implantation step.

3-EMITTER-BASE STRUCTURE`OF THE NPN BIPOLAR

Emitter and base contacts are implanted at the same time as the source/drain of NMOS and PMOS transistors respectively. The emitter is formed by As direct implantation into monosilicon. The base uses an additionnal mask and is implanted in two steps: a shallow BF_2 implantation to reduce the base sheet resistance between emitter and base contact; and a deeper Boron implantation for intrinsic base doping. The emitter-base structure is formed at the end of the process and only sees a final 950°C drive-in , also used as BPSG reflow.

Process and electrical simulations have been performed to determine the emitter base implantation and drive-in conditions, in order to achieve the optimized cut-off frequency while guarantying specific analog requirements: high Early voltage, high BVCEo, low noise figure at low and high frequency, low offset voltage. Fig.3 gives the bipolar transistor impurity profile measured by spreading resistance and compared to process simulations obtained with SUPREM 3. 2-D device simulations were performed with IMPACT and HFIELDS to predict current gain and cut-off frequency of the NPN transistor.

4-DEVICE CHARACTERISTICS

Table1 shows the main device parameters. The main electrical parameters required for an analog BICMOS process are high Early voltage combined with high gain (Hfe=120), high f_T, short delay times and low collector resistance. As a high Early voltage and high f_T have conflicting collector requirements, the f_T optimization relied mostly on the base width reduction. An Early voltage of 50V is obtained for the minimal size transistor. Fig.4 shows the Ic dependence of f_T for $S_E=3.6\mu mx3.6\mu m$. F_T was obtained from S parameters measurements performed directly on wafer using Cascade Microtech microwave probe head [4]. A 7.5GHz f_{Tmax} is obtained at VCE=5V.

High frequency and speed of bipolar transistors were tested on packaged devices, demonstrating performances of f_{max}=890MHz on D flip-flop, and 170ps propagation delay per stage for 21-stage CML ring oscillators (table 2). Measurements were in good agreement with the simulations.

5-CONCLUSION

A 1.2μm BICMOS process for mixed analog digital applications has been developed. The choice of the right epilayer thickness and resistivity, and the implementation of very shallow emitter-base junctions for the bipolar device allowed the combination of performant analog features (high Early voltage, high breakdown voltages, low noise figure...) together with high speed characteristics for the NPN transistor. The high performance of the technology is demonstrated on ring oscillators and D flip-flop.

ACKNOWLEDGEMENTS
The authors would like to thank G.TROILLARD and J.MOURIER for their contribution to process set up and device characterisation. Special thanks to all the people of the silicon process facility (prototype line).

This work is supported by the EEC program ESPRIT under contract n°2268

REFERENCES
1.H.Momose et al , "1μm N -Well CMOS/Bipolar Device Technology " . IEEE Trans. ED , Vol 32 , 1985 p 217 - 223
2.P.A.H Hart et al , " BICMOS , ESPRIT 86 : Results and achievement " p 221
3.T.Yamauchi et al , " 20V BICMOS Technology with polysilicon Emitter structure" Electrochem . Soc. Spring Meeting . 1987 , Philadelphia - Abst. n° 286 .
4. D. CELI - " Method for accurate determination of the intrinsic cut-off frequency of IC Bipolar transistors " ICRTS , p 200 - 203 , february 1988 .

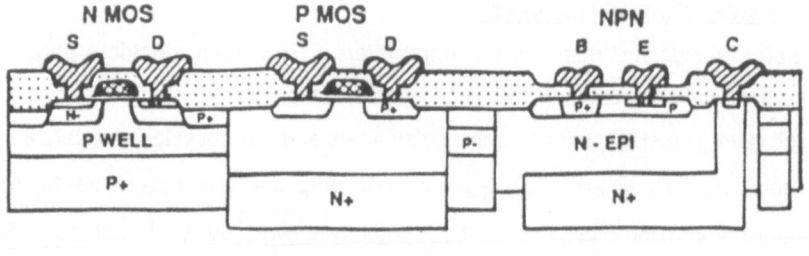

Fig.1 - BICMOS Process cross section

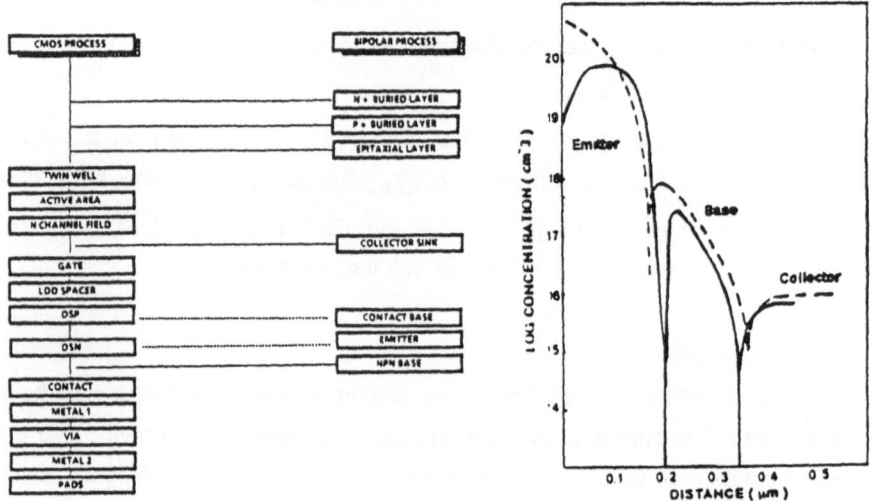

Fig.2 - Process architecture

Fig.3 - Bipolar transistor profile
— spreading resistance measurement
..... simulation results

Bipolar (NPN)	CMOS
S emit. : 3.6 x 3.6 µm2	NMOS (50/50) Vth : 0.7 V
hfe : 120	PMOS (50/50) Vth : 1.0 V
BVCE0: 14 V	BVDSS : > 7 V
BVCB0 : 25 V	td (fin=fout) : 300 ps
BVEB0 : 5 V	
VEA : 50 V	
Ceb : 39 pF	
Ccb : 40 pF	
Ccs : 119 pF	
rbb' : 370 Ω	
ft max : 7.5 GHz	

Table 1 : Main device parameters

Supply voltage	: VEE = - 5 V
Internal logic swing	: ∆V = 250 mV
Current per stage	: 500 µA
Propagation delay time	: tpd = 170 ps (fi=fo=1)

Table 2 : 21-stage CML ring oscillator characteristics
(Se = 3.6 x 3.6 µm2).

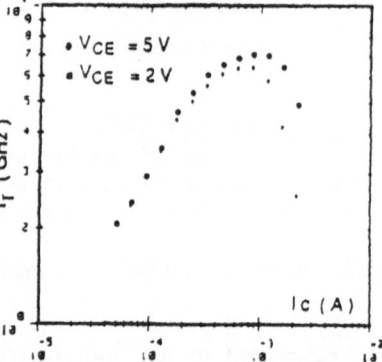

Fig.4 -Cut-off frequency f_T versus
collector current Ic .
S_E: 3.6 µm x 3.6 µm

Study of a Polycide (WSi$_2$/Polysilicon) Emitter for a CMOS Compatible Self-aligned Bipolar Transistor

G.GIROULT and A.NOUAILHAT

CNET/CNS - BP 98 - Chemin du Vieux Chene - 38240 Meylan Cedex - France
and M.GAUNEAU

CNET/Lannion - BP40 - Route de Tregastel - 22301 Lannion Cedex - France

Abstract

A WSi2/polysilicon/Si bulk system has been studied in the framework of a CMOS compatible self-aligned bipolar transistor technology. The WSi2 silicide is implanted with Arsenic and used as a dopant source for the formation of the polysilicon-bulk emitter. Arsenic and boron profiles have been studied as a single dopant and in the transistor configuration. Experimental results give dopant diffusion and segregation coefficients and are used to adjust the parameters of our process simulation program on the technological process. WSi2 polycide appears to be particularly well adapted as a dopant source for self-aligned bipolar emitter in a large temperature range (1030 C - 1130 C).

Introduction

It is now well known that BICMOS technology allows a large improvement in circuit performances in comparison with pure CMOS technology. Nevertheless the drawback of BICMOS is the cost of the technology if it is too complicated. In this context, we have developed [1] a very simple CMOS-compatible bipolar transistor, the stucture of which being similar to the PMOS and requiring a single additional mask for the base implantation and the gate oxide removal. The emitter corresponds to the As doped MOS gate : the arsenic diffuses from the polycide into the mono-silicon in the presence of the boron coming from the extrinsic base implantation (PMOS source/drain) in the self-aligned process. In this framework, we have studied, on plain wafers for SIMS analysis conveniency, a WSi2/polysilicon bilayer as a dopant source, dopants being implanted in the silicide (Fig.1). The sample process (Fig.2) is the same as the transistor process without the lithography and etch steps. Taking

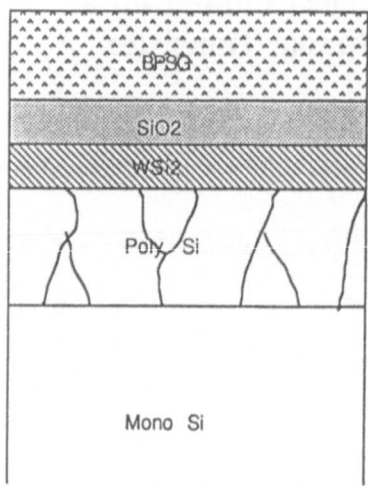

Fig.1 Sample cross-section

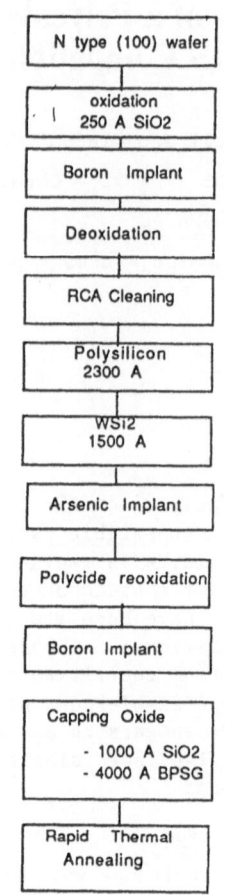

Fig.2 Process flow chart

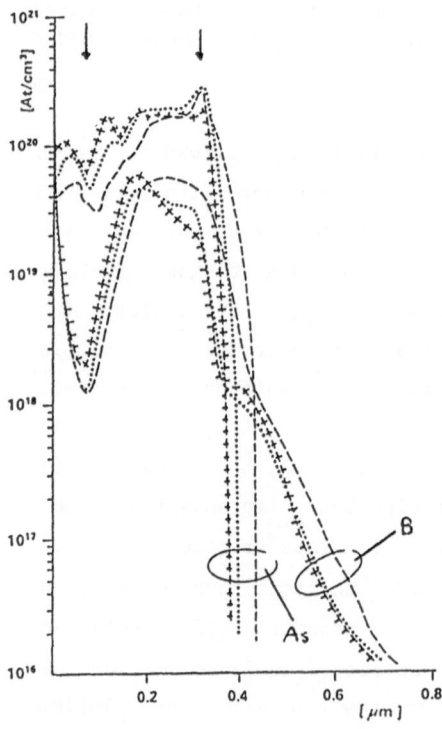

Fig.3 SIMS profiles of arsenic
(1e16at/cm2,100KeV) and boron
(2e13at/cm2 at 25KeV in Si-bulk
2e15at/cm2 at 30KeV in polycide)
in transistor configuration,
for several RTA conditions:
+ + + 1030C, 20s
. . .   1080C, 20s
— —   1130C, 20s
↓ estimated position of the
WSi2/poly-Si and poly-Si/Si-bulk
interfaces

into account the complexity of the involved diffusion mechanisms, single dopant diffusion is also analysed in the same conditions. The RCA cleaning procedure prior to polysilicon deposition, and the polycide annealing (corresponding to the polycide reoxidation in the process flow chart), ensure a good reproducibility of the dopant profiles in the temperature range of the final rapid thermal annealing (RTA) (1030 C - 1130 C).

Experimental results

Fig.3 shows the emitter-base junction SIMS profiles in the transistor process case : the silicon wafer was implanted with 2e13/cm2 boron at 25KeV (intrinsic base implant), the polycide with 1e16/cm2 Arsenic at 100KeV (emitter implant) and then with 2e15/cm2 boron at 30KeV which corresponds to the extrinsic base implant in the self-aligned emitter bipolar transistor process. The final RTA was performed during 20s with the above temperature range.

The segregation coefficient between polysilicon and WSi2 defined as (dopant concentration in polysilicon / dopant concentration in WSi2) at the interface is about 3 to 6 for arsenic and 33 to 40 for boron. Arsenic and Boron quickly re-distribute from silicide to polysilicon, the diffusivity coefficients in WSi2 being at least of the same order of magnitude as in polysilicon, and then diffuse into monosilicon.

The computed diffusivity values in monosilicon for As and B are the following :

$$D(Boron) = 2*exp(-3.46eV/kT)*[1+0.1(p/ni)]/(1+0.1) \ cm2/s$$
$$D(Arsenic) = 22.43*exp(4.1eV/kT)*[1+100(n/ni)]/(1+100) \ cm2/s$$

and are similar to those given in experiments with conventional annealing [2] . As RTA has been performed after the polycide annealing , which eliminates the defects created in monosilicon after the (boron) intrinsic base implantation , the transient enhanced diffusion , generally observed with RTA [3] does not occur. This is confirmed by A.Martinez et al. in recent experiments (*).

Emitter widths in monosilicon have been obtained from 400 A at T=1030 C to 1000 A at T=1130 C with a base width of 1500 A to 1300 A (at boron concentration = 1e17 at/cm3) respectively.

Conclusions

At the WSi2/polysilicon interface, segregation coefficients values

	Rapid Thermal Annealing		
	1030 C 20s	1080 C 20s	1130 C 20s
Poly-Si/WSi2 Arsenic segregation Cp/Cw			
Arsenic as a single dopant	10	10	12
Arsenic with Boron (transistor case)	2.85	4	5.45
Poly-Si/WSi2 Boron segregation Cp/Cw			
Boron as a single dopant	4.5	3.3	0.83
Boron with Arsenic (transistor case)	33	33	40

Table.1 Boron and Arsenic segregation : Cp/Cw (Cp,Cw: dopant concentration at the poly-Si/WSi2 interface in polysilicon and in WSi2 respectively)

- single dopant case:
  *Arsenic =(1e16 at/cm2 100KeV in WSi2)
  *Boron =(2e15 at/cm2 30 KeV in WSi2)

- transistor case :
  *Arsenic =(1e16 at/cm2 100KeV in WSi2)
  *Boron =(2e15 at/cm2 30 KeV in WSi2) (2e13 at/cm2 25 Kev in Si-bulk)

found when arsenic and boron codiffuse are different from the ones obtained when dopants are studied separately (table 1) and suitable for transistor processing as WSi2 acts in this case as a perfect dopant source.

In polysilicon the diffusivity coefficients used for boron and arsenic are the following:

$$D(boron) = 50*exp(-3.42eV/kT) cm2/s$$

$$D(Arsenic) = 70000*exp(-22.43eV/kT) cm2/s$$

With respect to these values, the pre-exponential factor in WSi2 was increased by a factor 30 for boron and unchanged for arsenic so that the dopant concentration in the silicide remains constant in our process conditions.

WSi2, convenient as a dopant souurce and as an emitter contact providing low access resistance, appears to be particularly well adapted for self-aligned bipolar transistors.

References
[1] A.Nouailhat, G.Giroult, P.Delpech, A.Gerodolle
Electronics letters, 1988, Vol.24, No.25, pp. 1581-1583
[2] V.Probst, H.J.Bohm, H.Schaber, H.Oppolzer and I.Weitzel
Journal of electrochemical society, 1988, 135,3, pp.671-676
[3] R.B.Fair, J.J.Wortman and J.Liu
Journal of electrochemical society, 1984,Vol.131,No.10, pp.2387-2394
* Private communication

Physical Analysis of Peripheral Base Currents in an Advanced Polysilicon Self-aligned Bipolar Transistor Structure

A. CHANTRE, G. GIROULT and A. NOUAILHAT
CNET/CNS, B.P. 98, Chemin du Vieux Chêne, F-38243 MEYLAN CEDEX, FRANCE

SUMMARY

The emitter/base junction properties of a CMOS compatible bipolar transistor structure have been studied. Dry etching induced damage occuring at the polysilicon emitter patterning level is found to account for the observed poor low current gain behaviour. Simple modifications of the device design are described and shown to result in high performance DC characteristics.

1. INTRODUCTION

BICMOS is emerging as a promising technology for the fabrication of high speed analog–digital VLSI circuits, such as required for telecommunication applications. In this context, we have recently (1) developed an advanced self–aligned bipolar transistor structure, which is fully compatible with our 1 micron CMOS technology (2). The device structure is very similar to the pMOS transistor and requires a single additional mask for the (boron) implantation of the intrinsic base (B_i) region, and for the removal of the gate oxide prior to polysilicon deposition. The gate arsenic–doped polysilicon acts as a contact and dopant source for the emitter (E). The extrinsic base (B_e) regions are formed during source/drain boron implantation. The PECVD oxide–sidewall spacers of the pMOS provide direct self–alignment of the bipolar devices. Finally, the n well acts as the collector (3). We report here detailed electrical characterization of the sidewall region of this transistor (Fig. 1). Good control of peripheral emitter/base (E/B) junction properties is indeed required for high performance, small geometry device fabrication.

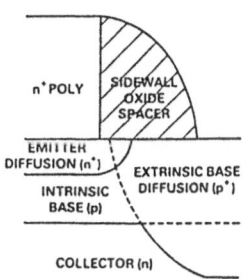

Figure 1 : Sidewall region blow-up of the device structure

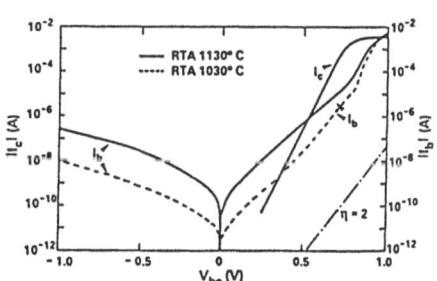

Figure 2 : Nonideal base currents in typical Gummel plot (emitter area : 335μm^2)

470

2. RESULTS AND DISCUSSION

Figure 2 shows typical Gummel plots for devices fabricated using our stabilized 1 micron CMOS process (i.e. without process parameter adjustment). Large abnormal base currents (are observed at both forward and reverse biases. From their dependence on device perimeter – to – area ratio, we conclude that these currents originate from the E/B_e junction under the PECVD sidewall oxide. The base current temperature dependence is shown in Fig. 3. The ln I_b α $-Eg(T)$ (forward bias) and ln I_b α $-Eg(T)^{3/2}$ (reverse bias) variations, where Eg(T) is the bandgap at temperature T, are indicative of an excess (i.e. gap state assisted) tunneling current mechanism. The bias voltage dependence of the current is fully consistent with such mechanism.

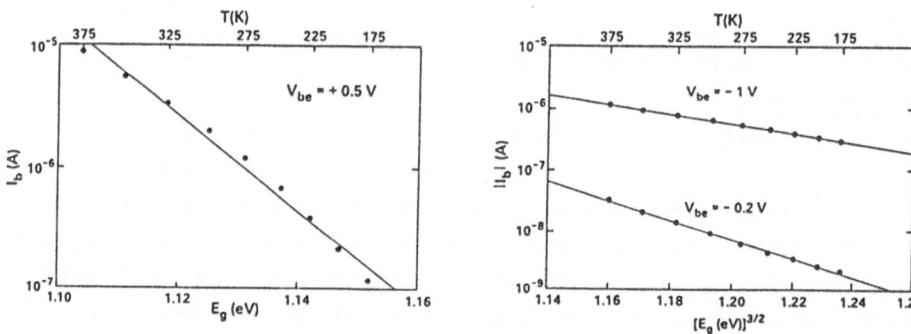

<u>Figure 3</u> : Temperature dependence of forward bias and reverse bias base currents.

The above results demonstrate that non ideal base currents in this device structure are caused by a defective region crossing an n+/p+ junction under the sidewall oxide spacer (4). Tunnel junction formation can be avoided by adjusting (within the range imposed by CMOS compatibility constraints) the characteristics of the final rapid thermal anneal (RTA) of the structures, to fine – tune the E/B_e profile overlap. This is illustrated in Fig. 2, where a lowering of the RTA temperature from 1130° to 1030°C is seen to result in significant (about two orders of magnitude) tunnelling current reduction. However, emergence of a generation – recombination component in the foward base characteristics (ideality factor η~2) sets a limit to such simple CMOS process adjustment.

Further improvement of the base characteristics requires electrical defect removal from the E/B junction region.

Two potential sources of electronic defects have been considered : (i) the PECVD oxide/silicon interface, and (ii) the reactive ion etching (RIE) of the polysilicon emitter.

A polysilicon reoxidation step was inserted in the process prior to PECVD oxide deposition in an attempt to suppress these defects. Devices incorporating a 280 Å thick thermal oxide under the PECVD oxide showed only limited base current reduction, pointing to a buried (depth > 130 Å) RIE–induced damage. Increasing the thickness of the pad thermal oxide to consume the defective region can be expected to significantly reduce peripheral base currents. However, this may require unacceptably thick oxide films, as can be anticipated from recent literature (5). This issue is under current investigation in our laboratory.

An alternative approach is depicted in Fig. 4. Here, the polysilicon etch step is so adjusted to result in a "walled" emitter configuration (3). As a result, the E/B junction no longer intersects the defective region underneath the sidewall oxide spacer. The lateral Si/SiO$_2$ interface region now crossing the junction is expected to be of much higher quality due to the anisotropic properties of the RIE etch.

This concept was tested using the poly/mono n+/p junction structure shown in the insert of Fig. 4. This device can be fabricated using a single photolithographic step, allowing for fast process optimization. As can be seen in the figure, near–ideal current–voltage characteristics are thus obtained ($\eta \sim 1.01$). The data also demonstrate the beneficial effect of a 280 Å pad thermal oxide on the reverse current (about two orders of magnitude current reduction at 5V reverse bias). Such sidewall oxide structure is also likely to prove more reliable with respect to stress–induced E/B junction degradation (6). The rapid increase in the current above 2–3V (Fig. 4) can be ascribed to tunneling in the n+/p junction (peak boron concentration ~ 5 x 10^{17} cm^{-3}) (7).

Figure 5 shows a typical Gummel plot for bipolar transistors incorporating this optimized E/B junction structure (peak base doping ~ 5 x 10^{18} cm^{-3}). A dramatic improvement in the base characteristics is observed, compared to the data shown in Fig. 2 : Constant common–emitter current gain is obtained over five decades of collector currents. This result demonstrates the potential capabilities of this CMOS compatible bipolar process.

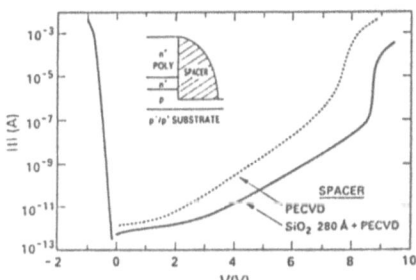

Figure 4 : I-V characteristics of "walled" poly/mono n+/p junctions (area : 100 x 100µm^2)

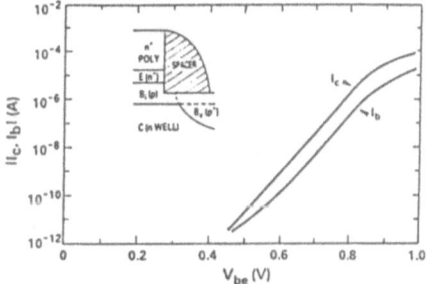

Figure 5 : Typical Gummel plot for "walled" bipolar transistors (emitter area 1 x 2µm^2)

472

3. CONCLUSION

We have investigated the origin of non – ideal base currents in the bipolar transistor structure sketched in Fig. 1. An improved device design, featuring a "walled" emitter configuration and a stacked SiO_2 / PECVD spacer structure, was shown to result in excellent low current gain behaviour.
Future work will be directed towards reducing emitter and collector resistances to improve high current characteristics.

ACKNOWLEDGEMENTS

The authors are indebted to G. Festes for assistance in electrical characterization, and wish to thank the staff of the CNET pilot line for device processing.

REFERENCES

1. A. Nouailhat, G. Giroult, P. Delpech and A. Gerodolle,
Electronics Letters 24 (1988) 1581 – 1582
2. D. Bois, B. Minghetti, P. Normandon, G. Goltz, J. Galvier, G. Gimine and
J.M. Dumant,
Int. Conf. on Solid State and Integrated Circuit Technology ; edited by Wang Xiuying and Mo Bangxian
(World Scientific, Singapore, 1986) ; p.p. 223 – 225
3. M.P. Brassington, M.H. El – Diwary, R.R. Razouk, M.E. Thomas and P.T. Tuntasood,
A related BICMOS approach is described, IEEE Trans. Electron Devices, vol 36 (1989) 712 – 719
4. G.P. Li, E. Hackbarth and T.C. Chen,
IEEE Trans. Electron Devices, vol 35 (1988) 89 – 95
5. S.W. Pang, D.D. Rathman, D.J. Silversmith, R.W. Mountain and P.D. De Graff,
J. Appl. Phys. 54 (1983) 3272 – 3277
6. E. Hackbarth and D. Duan – Lee Tang,
IEEE Trans. Electron Devices, vol 35 (1988) 2108 – 2118
7. J.M.C. Stork and R.D. Isaac,
IEEE Trans. Electron Devices, vol 30 (1983) 1527 – 1534

Sidewall Contact for npn/pnp Transistors by Selective Oxidation (SELOX)

Horst H. Berger, Bernt Müller, Erich Biermann,
Peter Linke and Andreas Gauckler

Institute of Microelectronics, J13
Technical University Berlin West
Jebensstr. 1
D-1000 Berlin 12

Summary

The feasibility of a selfaligned sidewall base contact npn-transistor for high-speed integrated circuits using selective oxidation (Selox) for the definition of the polysilicon sidewall contact region has been shown. The selective oxidation is based on the increased low-temperature oxidation rate of n-type monosilicon for dopings above $\approx 10^{19} cm^{-3}$. Besides npn transistors with reduced collector-base capacitance, lateral pnp transistors with p+ polysilicon emitters have been realized. Due to the underlying oxide, these do not exhibit the usual excessive charge storage in and below the emitter, so that $f_T = 130$ MHz has been achieved with a basewidth of 2.5 µm.

Introduction

An important improvement of bipolar transistor structures has been the p^+-polysilicon sidewall access to the intrinsic base [1,2]. It substantially reduces junction capacitances like C_{bc} as well as charge storages in upward operated npn transistors or in lateral pnp transistors. Schemes realized so far have been named "SICOS" or "SELECT" [2,3].

The approach to be reported here ("SELOX") [1] offers a simpler way of obtaining a selfalignment of the poly-silicon contact to the intrinsic base region.

Process scheme

As in the known sidewall contact schemes, a recess is etched, saving the region of the intrinsic transistor (fig.1). Then, however, oxidation conditions are used that enhance the oxide growth at the bottom of the recess relatively to the sidewall. Hence, the latter need not be covered by nitride during oxidation (fig.2). A final dip etch removes the thinner oxide there to define the polysilicon contact location (fig.3).

474

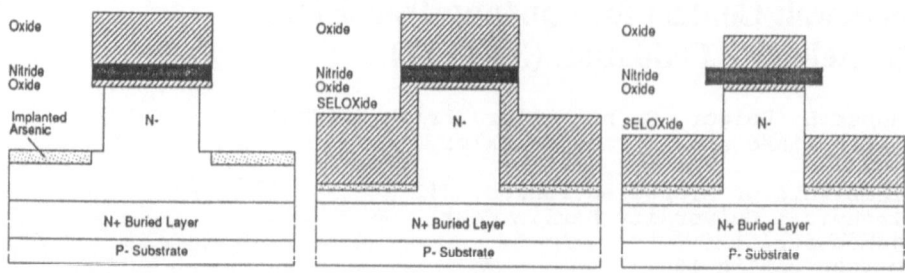

Fig.1 Recess for Poly-   Fig.2 Seloxidized   Fig.3 Seloxidized
silicon Sidewall   recess before etch   recess after etch

This Selective Oxidation (**SELOX**) relies on the higher oxidation rate of highly n^+ doped surfaces at lower temperatures [4]. Ref. [4] as well as our own experiments on As doped surfaces (fig.4) show that concentrations well above the 10^{19} cm^{-3} level are needed to obtain a sufficient acceleration of the oxidation at the favourable temperature of 800°C (for more details see [5]). Such dopings are naturally found only in the peak doping region of the buried layer.

To avoid the disadvantages of having to etch the recess so far down, the horizontal surface doping was increased by an As implantation with a dose of $3 * 10^{15}$ cm^{-2} at an acceleration voltage of 75 keV. For the preceding recess etch, RIE with $SiCl_4$ was used. To protect the sidewall of the mesa (80°angle) from the As implantation, the overhanging resist was kept as a mask.

After a 20-minutes anneal at 800°C in nitrogen atmosphere, the seloxidation was performed in a steam atmosphere at 800°C for 4 hours. The oxide thickness over the implanted surface was 554 nm, while at the mesa sidewall (rho = 0,4 Ω cm) only 140 nm were grown. A 2-minutes buffered etch removed the sidewall oxide and left 360 nm at the bottom of the recess.

Npn and pnp transistors

Both vertical npn and lateral pnp transistors have been fabricated using SELOX. The lateral pnp is obtained by leaving out the upper part of the npn structure, i.e., the intrinsic base implantation and n^+-polyemitter deposition

Fig. 5 shows the top view of a npn transistor having a 5 * 5 μm² emitter, while fig. 6 is a SEM microphotograph of the critical sidewall/bottom region.

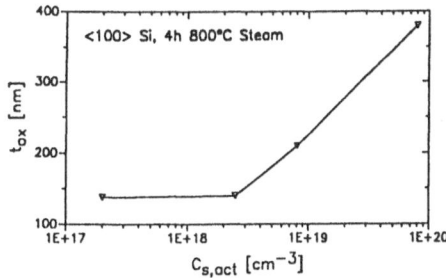

Fig.4 Oxide thickness vs. active surface doping (Arsenic)

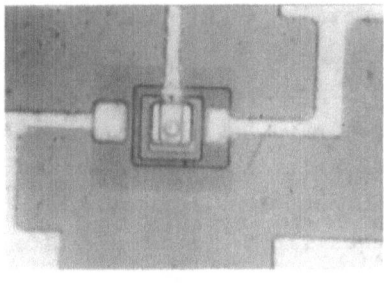

Fig.5 NPN transistor with 5μm*5μm Emitter

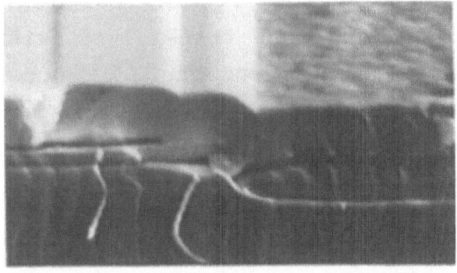

Fig.6a SEM picture of polysilicon/ monosilicon contact

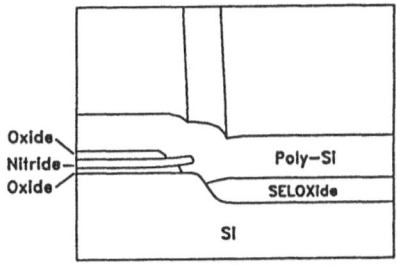

Fig.6b Schematic diagram of SEM picture

Grounded emitter output characteristics of a pnp and a npn transistor are shown in figs. 7 and 8 . Current gains are 14 for the pnp, but, due to excessive base current, only 2 for the npn . The effect is still under investigation. It occurs less severe in inverse operation ($ß_R≈15$)

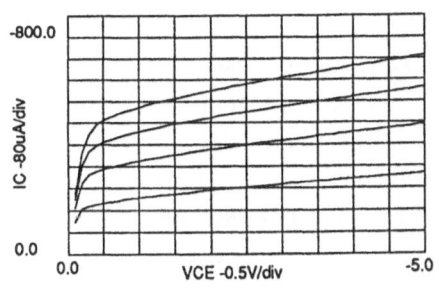

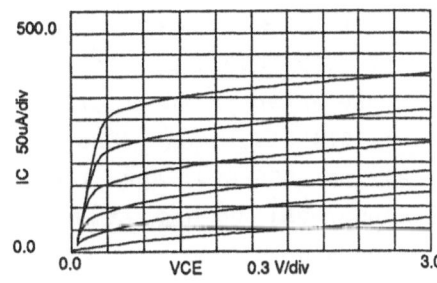

Fig.7 Output characteristic of pnp transistor

Fig.8 Output characteristic of npn transistor

In contrast to conventional pnp's., with a base width of $W_b \approx 2.5 \, \mu m$, the lateral pnp has a high $f_T = 130$ MHz. This confirms, that the pnp's emitter bottom area does not contribute anymore to charge storage. From this we may also assume an ideal behavior according to $f_T \sim W_b^{-2}$. Considering the 0.5 μm base width for which a 3 GHz transit frequency was reported in [6], our result scales well. The f_t of the npn transistors could not be measured yet due to a metalization defect that consistently caused a broken connection line and thus impeded a proper packaging of devices.

Conclusion

The feasibility of SELOX as a precise tool for selfalignment has been proven by producing pnp and npn transistors with selfaligned p^+-polysilicon sidewall contacts. Process parameters for SELOX are being further investigated to establish optimum conditions and to assess its full potential for integrated circuit fabrication.

Acknowledgement

The partial financial support by the Commission of the European Communities is gratefully acknowledged (ESPRIT 243, sub-contract by Thomson CSF France). Valuable technical support was received from M. Depey and M. Roche (Thomson). Also, the ESPRIT 243 partners Plessey (England) and Telefunken electronic (Germany) provided helpful technological information .

References

[1] **Horst H. Berger et al.**,IBM Techn. Discl. Bulletin, Vol.23, pp. 1089-1090 & pp. 1487-1488, Aug./Sept.1980

[2] **Tohru Nakamura et al.**,ISSCC 81, Digest of Technical Papers, pp. 214-215, and IEEE Trans. Electron Dev., Vol. ED-29, pp. 596-600, Apr. 1982

[3] **Katsuyoshi Washio et al.**, ISSCC 87, Digest of Technical Papers,pp. 58-59

[4] **C. P. Ho et al.**, J. Electrochem. Soc. Vol.125, pp. 665-671, 1978

[5] **E. Biermann**, this conference, paper 1B7, 1989

[6] **Kazuo Nakazuto et al.**, IEDM 86 Digest pp.416-417

A 10 GHz High Performance BICMOS Technology for Mixed CMOS/ECL ICs

B. Hoffmann, H. Klose, T. Meister, I. Kerner, R. Schreiter
Siemens AG, Corporate Research, Otto-Hahn-Ring 6, 8000 Munich 83, FRG

Abstract
A 1.2 μm BICMOS process is presented for the realization of high complexity CMOS- circuits together with high performance bipolar transistors on the same chip. n^+ -/ p- buried layers, a p- well CMOS-process and a double - polysilicon selfaligned bipolar process are the main technology features. A cut - off frequency of 10 GHz as well as a CML gate delay time of 65 ps are the results obtained with this process.

1. Introduction

The trend in the development of modern ICs is heading towards ever faster circuits and chips with ever higher complexity (≥100 k equivalent gates at a data rate of 600 Mbit/s and above). The only way to reach this goal is to combine high density CMOS circuits with high speed bipolar devices on the same chip.

The aim of this paper is to present the characteristics of a high performance BICMOS process, which is based on a 1.2 um CMOS-technology. The eminent device features are reported and the performance capabilities are demonstrated by basic circuits.

2. Technology

In fig. 1 the schematic flowchart of the overall BICMOS process is sketched.

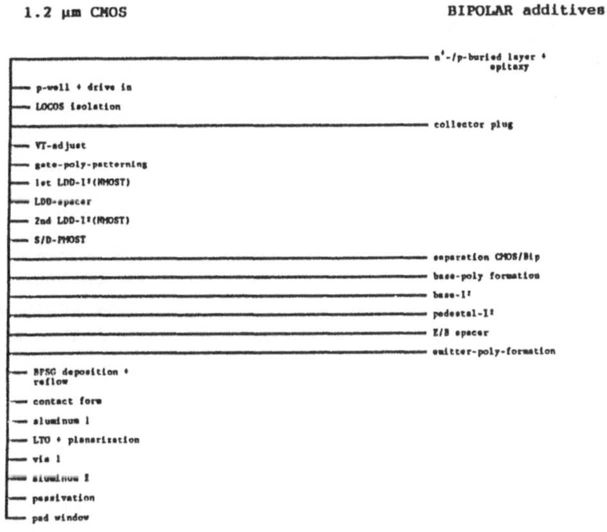

Fig.1: Schematic flowchart

Since the main advantages of CMOS are high packing density and low power dissipation the majority of devices are designed in CMOS technology. For this reason the presented process is based on an existing 1.2 µm CMOS technology, for which optimization and yield maximization is already done. The main steps of this process are outlined on the left hand side of the diagram in fig. 1. To the right the additional process blocks for the bipolar devices are shown. They are modularly plugged into the CMOS-core process. This modular concept allows to decouple the most sensitive parts of both device types, namely source- / drain-regions and emitter- / base- junctions.

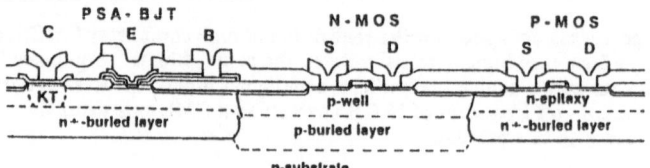

Fig.2: Cross - section of BICMOS high performance devices

A cross - section of the bipolar transistor is shown in fig.2 (Polysilicon Selfaligned - Bipolar Junction Transistor: PSA - BJT). The n^+- buried layer is mandatory for low collector resistance of this device. Beneath the p- MOS transistor it prevents punch through between p- drain and substrate. Moreover, due to the n^+- buried layer the latch up immunity is greatly improved. The most striking advantage of a p-buried layer is the reduction of n^+ / n^+- buried layer spacing. The risk of punch through of n- MOS source / drain is decreased, too. Therefore a selfaligned n^+ / p- buried layer process is implemented. The additional technological effort is worth-while.

The growth of a 1.5 µm arsenic doped expitaxial layer, p- well definition /1/ and LOCOS isolation are done before the next bipolar specific step is inserted into the CMOS process: the implantation of the collector plug (fig. 1). After threshold and antipunch implantation, the gate definition mask is used to cover all bipolar devices with a gate-polysilicon cap. Such the emitter- / base- regions are protected during source- / drain- processing. For the n- MOS devices a LDD- structure is implemented in order to minimize hot carrier effects.

With the bipolar definition mask a protective TEOS layer is left on all CMOS-devices, the bipolar section is opened. By this double protection concept bipolar and MOS device parameters can be optimized independently, i.e. hot carrier hardness or base- / emitter-breakdown.

For this BICMOS version a double polysilicon selfaligned process is implemented for the bipolar transistor /2/. Current gain degradation due to shallow emitter junctions is prevented by the use of a polysilicon emitter. By the application of the selfalignment scheme the external base resistance as well as the base collector capacitance are kept low /3/. A summary of the main process features is given in fig. 3. Fig. 4 shows a micrograph of a n- MOST, a p- MOST and a bipolar npn - transistor.

- selfaligned n+ / p- buried layer
- n- type epitaxy

- collector plug

- 1.2 µm p- well CMOS process
- polysilicon selfaligned
  base / emitter structure
- 2 layers of metal

Fig.3: Main features of the high performance BICMOS process

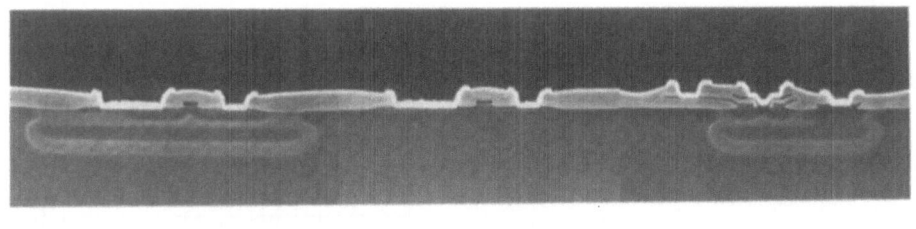

PMOST   NMOST   BJT

Fig.4: Micrograph of the cross-section of the BICMOS devices

3. Device Performance

Due to the lightly doped epitaxy the dopant concentration in the region between base and n^+-buried layer is to low to obtain reasonable current densities for a high performance bipolar transistor. Therefore a high energy implantation of phosphorus through the emitter window is introduced. With this selfaligned pedestal implantation the collector doping is sufficiently increased within the active transistor region. The sub-collector below the inactive base contact areas remains lightly doped. Good high current behaviour coincides with low inactive-base-collector-capacitances . In fig. 5 the dopant profile of the pedestal collector and the n^+-buried layer is depicted, for simulated data as well as for measurement. Obviously the experimental results obtained by C(V)-measurement are in excellent agreement with the simulations done with SUPREM 3.

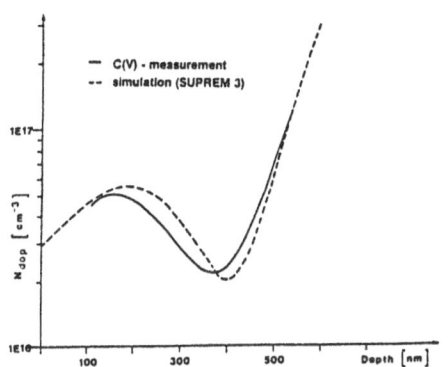

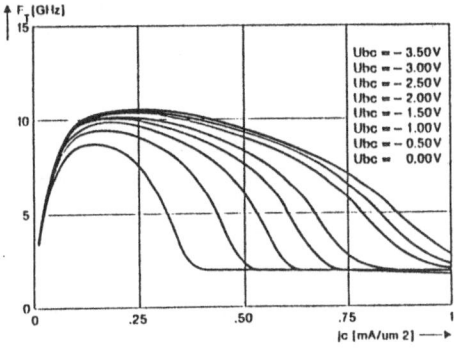

Fig.5: Doping profile of the sub-base region: Phosphorus (pedestal implantation) and antimony (buried layer)

Fig.6: Cut-off frequency of the bipolar (npn) device vs.collector current density

The current density which marks the 10% roll off point for the cut-off frequency, which is 10 GHz (see fig. 6), is as high as 0.25 mA / μm^2 and 0.50 mA / μm^2 for V_{BC} values of 0 V and -3 V respectively. Simultaneously the base collector capacitance is as low as 18 fF.To keep the up-diffusion of the buried layers low, the time of well-drive-in has to be reduced. The CMOS process parameters concerned were adapted. The width specific drain current received at a drain- / gate- voltage of 5 V is 0.23 mA / μm for the n- MOST and 0.1 mA / μm for the p- MOST. The bodyfactor is 0.9 $V^{1/2}$ and -0.6 $V^{1/2}$ for the n- MOST and the p- MOST respectively. A summary of the electrical data is given in fig. 7.

Device	Quantity	Value	Unit
NMOST (LG=1.2 μm)	VT	0.82	V
		0.9	V1/2
	IDC (VGS=VDC=5V)	0.23	mA/μm
PMOST (LG=1.2 μm)	VT	0.95	V
		0.6	V1/2
	IDC (VGS=VDC=5V)	0.1	mA/μm
BJT (1.0x4.5 μm2)	BF	100	1
	RB	225	Ω
	CBE	30	fF
	CBC	18	fF
	CCs	85	fF
	fT (VBC=3V)	10	GHz

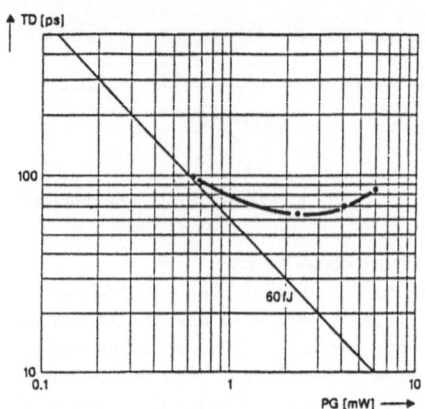

Fig.7: Electrical device data   Fig.8: ECL - Gate delay times

4. Circuit Performance

In order to demonstrate ECL / CML capability of our technology CML ringoscillators with 200 mV voltage swing were realized and evaluated. The gate delay time vs. power dissipation plot is depicted in fig. 8. The minimal gate delay time is 65 ps, which is the shortest CML gate delay time of a BICMOS process published so far /4/. The power delay product for low power applications is 60 fJ.

5. Conclusion

The results achieved with the presented process indicate that ECL - performance with CMOS complexity can be realized on the same chip. As soon as high packing density, low power dissipation and high speed are desired simultaneously, i.e. for analogue applications or mixed ECL / CMOS ICs, BICMOS is inevitable in the future. These circuit features impossible to get with pure bipolar or CMOS technologies compensate higher production costs.

The work presented is part of the ESPRIT 412 / 2430 project.

/1/ H.Klose, T.Meister, B.Hoffmann, H.Kabza, J.Weng, ESSDERC, 1988

/2/ H. Kabza et al., to be published by IEEE EDL in Aug. 1989

/3/ H. Schaber, J. Bieger, T.F. Meister, K. Ehinger and R. Kakoschke,
    IEEE IEDM Techn. Digest (1987), pp.170

/4/ T.Chin, Proceedings IEDM, San Francisco 1988, pp.752

Quantum Well Devices and Optoelectronic Devices

Capacitance-voltage Investigation of Rechargeable Traps in Isotype Laser Heterojunctions

H.-G. Bach[1] and G. Beister[2]

[1] Heinrich-Hertz-Institut für Nachrichtentechnik Berlin GmbH, D-1000 Berlin 10, FRG
[2] Zentralinstitut für Elektronenphysik der Akademie der Wissenschaften der DDR, Berlin

Summary

A new analytical C-V modelling procedure is presented which comprises deep level recharging processes at the isotype Pp heterojunction in conventional PpN DH laser structures. Experimental results on AlGaAs laser diodes can be explained by this approach.

Introduction

Investigations of degradation phenomena in AlGaAs/GaAs double-heterostructure lasers usually comprise capacitance-voltage (C-V) profiling. Sometimes laser diodes, exhibiting good lifetime expectations, show besides the well-known interface bending /1/ additional (sharp) peaks in the effective free carrier profile (Fig. 1), or even loopings (Fig. 2), which are believed to be caused by rechargeable interface states. The existence of such states was further confirmed by DLTS and admittance measurements, the latter of which is shown for zero bias in Fig. 3. Here in section II. two traps have been identified by applying a series equivalent RC network for each trap /2/. The DLTS detected a corresponding trap at about room temperature.

A corresponding influence of rechargeable traps was already proposed by Kazmierski et al. /3/ as observed in asymmetric p^+n diodes depleting isotype GaInAs/InP heterojunctions. In our case in double-heterostructure laser diodes with nearly symmetric doping we had to consider two different space charge regions.

C-V Modelling

We implemented a new analytical C-V modelling procedure comprising the influence of rechargeable traps at the isotype Pp-heterointerface. The important part of the structure used for calculation is given in Fig. 4. The low frequency capacitance as measured at the contacts is calculated from the change of electric field strength inside the structure with applied bias. This field strength has to be taken in the depleted region between the two space charge layers, where the conductive part j_{cond} of the total ac current j_{tot}

$$j_{tot} = j_{cond} + \frac{\partial \vec{D}}{\partial t}$$

(1)

is negligible: $j_{cond} \ll \delta \vec{D}/\delta t$.

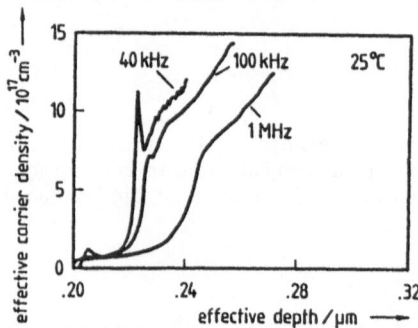

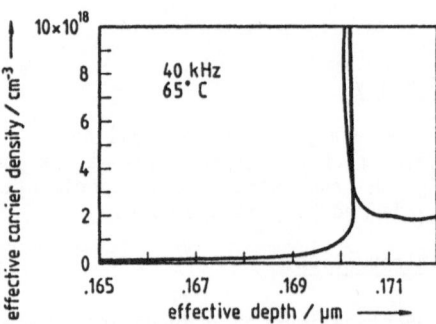

Fig. 1 Effective carrier profile of an AlGaAs/GaAs laser diode measured at different frequencies

Fig. 2 Effective carrier profile of the sample from Fig. 1 for 65° C and 40 kHz

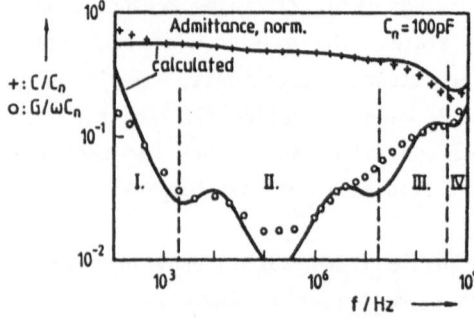

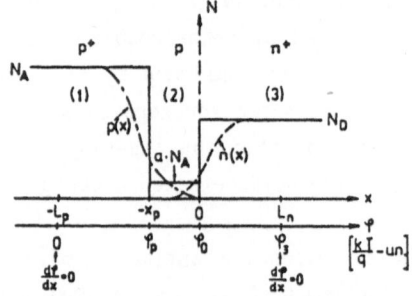

Fig. 3 Admittance vs. frequency at 25° C of the sample in Fig. 1

Fig. 4 Scheme of structure used for C-V calculations, discontinuities omitted for simplicity, a:doping ratio region (2)/(1)

We used

$$c = \varepsilon_2 \varepsilon_0 \frac{d}{d\varphi_3} \left[(\frac{d\varphi}{dx})_{(2)} \Big/ \right]_{x \to 0}$$

(2)

with the electrostatic potential φ (in kT/q units), assumed to be continuous through the whole structure, and φ_3 the total potential drop over the space charge regions. In the analytical solution, equation (2) contains the derivatives of φ_p and φ_o with respect to φ_3. Both can be derived from the condition of displacement continuity at x = 0 and from the $x_p(\varphi_p, \varphi_o)$ expression :

$$|x_p| = \frac{L_{D2}}{\sqrt{2}} \int_{\varphi_p}^{\varphi_o} \frac{d\varphi}{(\frac{d\varphi}{dx})} \cdot \tag{3}$$

The interface states (here donor-like) were introduced via the corresponding discontinuity of the displacement at x = - x_p:

$$\vec{D}_{(1)} \Big|_{-x_p} + q r_p N_t = \vec{D}_{(2)} \Big|_{-x_p} \cdot \tag{4}$$

N_t denotes their concentration and r_p the occupation rate for holes. The latter can be deduced from the balance equation for the trap occupation to

$$r_p = \frac{\alpha_p P_p + e_n}{\alpha_p P_p + e_n + e_p} \cdot \tag{5}$$

(P_p : hole concentration at x = - x_p, α_p : specific hole capture rate $[cm^3 s^{-1}]$ and e_n, e_p : emission rates $[s^{-1}]$ for electrons and holes, respectively).

The calculated C-φ_3 dependence was evaluated like a measured C-V curve for determination of the apparent free carrier profile N^x with respect to an effective depth x = $\varepsilon_0 \varepsilon_2 / C$ and

$$N^x = \frac{2kT}{q^2 \varepsilon_2 \varepsilon_0} \cdot \frac{1}{\frac{d(C^{-2})}{d\varphi_3}} \tag{6}$$

Results

Fig. 5 demonstrates an example. It shows accumulation and depletion of free carriers caused by the discontinuities ($N_A = 5 \cdot 10^{17}$ cm^{-3}; a = 0.1; $N_D = 3 \cdot 10^{17}$ cm^{-3}) in section (I) and in section (II) recharging of an interface trap (concentration $N_t = 2 \cdot 10^{11}$ cm^{-2}, activation energy $E_t - E_V(2) = 0.5$ eV, capture cross section ratio (h/e) $\sigma_p / \sigma_n = 10^{-3}$), which compares well with the looping in Fig. 2.

486

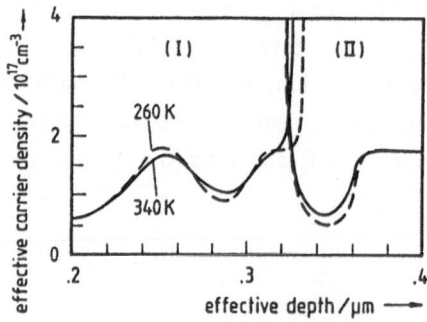

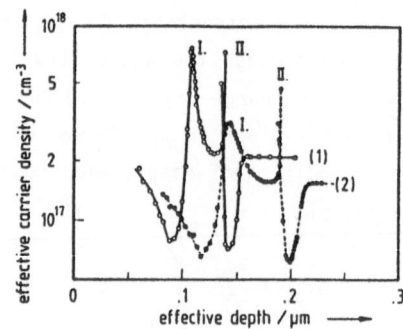

Fig. 5 Calculated apparent free carrier concentration of a PpN structure according to Fig. 4

Fig. 6 Carrier concentration of an pPn structure:
$N_A = 2 \cdot 10^{17}$ cm^{-3} and
$N_D = 1 \cdot 10^{22}$ cm^{-3} (1), or
$N_D = 5 \cdot 10^{17}$ cm^{-3} (2)

Our modelling can also be applied to situations, where a displacement of the pn-junction due to p-dopant outdiffusion into the N-region (3) has to be considered. In this case depletion starts from the wide-gap material (3) --> (2). In Fig. 6 a corresponding result is shown for the "inverse" structure pPn with a comparison between an asymmetric Pn$^+$-junction ("Schottky barrier diode")- curve 1 - and a "normal" Pn structure (curve 2). Firstly, the "smoothing" of the discontinuity peak in nearly symmetric pn-structures is obvious (section I). Secondly, the looping in section II caused by interface trap recharging ($N_t = 4 \cdot 10^{10}$ cm^{-2}; $E_t - E_v(2) = 0.3$ eV; $\sigma_p/\sigma_n = 10^{-3}$) is larger in the asymmetric structure (curve 1) than in the normal one. This result supports the explanation of the observed sharp peaks (Fig. 1) by rechargeable interface traps instead by discontinuities alone.

References

/1/ H. Kroemer, W. Y. Chien, J. S. Harris, and D. D. Edwall, Appl. Phys. Lett. 36, 295 (1980)

/2/ S. R. Forrest and O. K. Kim, J. Appl. Phys. 53, 5738 (1982)

/3/ K. Kazmierski, P. Philippe, P. Poulain, and B. de Cremoux, J. Appl. Phys. 61, 1941 (1987).

Optical Properties of GaInAs/InP Multi-quantum Wells Grown by Low Pressure MOVPE

J.P. LAURENTI[*], B. REYNES[*], J. CAMASSEL[*], D. GRÜTZMACHER[**], K. WOLTER[**],
H. KURZ[**], P. BALK[**]

* Groupe d'Etude des Semiconducteurs, Université des Sciences et Techniques, 34060 - MONTPELLIER-Cédex, FRANCE.
** Institut of Semiconductor Electronics, AAachen Technical University, 5100 Aachen, F.R.G.

Summary

Absorption, derivative absorption and reflectivity measurements have been performed on a series of GaInAs/InP multi-quantum well structures (MQWs) grown by low pressure (LP) metal organic vapor phase epitaxy (MOVPE). Excitonic transitions up to n=4 between confined states have been resolved and, in some cases, reveal clearly at room temperature. Comparison with previous photoluminescence (PL) data shows Stokes shifts as weak as 4 meV, indicating minor potential fluctuations. Each intrinsic excitonic transition energy is carefully determined by a series of theoretical fits in the light of the two-dimensional (2D) exciton model. The results reasonably agree with those expected from recent calculations without any adjustable parameter. Some departs of the alloy compositions or QW widths with respect to the nominal ones have been found. They result in a sizeable shift of the series of excitonic transitions with respect to their calculated positions.

GaInAs/InP multi-quantum well structures (MQWs) are of increasing interest for achieving optoelectronic devices acting in the spectral range of optical fiber communications (1.3-1.55 μm). Previously, high quality lattice-matched GaInAs/InP MQWs have been grown by low pressure (LP) metal organic vapor phase epitaxy (MOVPE) across 2" wafers [1]. Under a total pressure of 20 mbar and high flow rates (~ 140 cm/s), a good homogeneity in thickness, composition and residual doping could be achieved. The photoluminescence (PL) line shifts at 2K have been already described [1] and were found in excellent agreement with the best results reported for MOVPE [2,3] and molecular beam epitaxy (MBE) [4,5]. The small linewidths indicate abrupt interfaces on the monolayer level.

In optoelectronic device applications, a systematic check of the well compositions and thicknesses is needed. This is done here by performing absorption and derivative absorption as well as reflectivity measurements on the same series of MQWs. The nominal well thicknesses ranged from 50 to 200 Å. Typical transmission spectra are shown in Fig.1. EHn (ELn) designate the transitions between confined electrons and heavy (light) holes. The excitonic character of most EHn transitions is evidenced by sharp structures and can be observed even at room temperature. For larger well widths, transitions up to n = 4 (not shown here) have been also resolved. All these results correlate with the high quality of our MQWs, previously evidenced by PL. This high quality is still confirmed by the sharpness of the reflectivity features (Fig.2) and by the weak PL Stokes shift. In Fig.2 the experimental value is 4 meV. It is very small compared with most previous reports (see e.g. ref. 6) and indicates small exciton trapping effects, i.e. very weak potential fluctuations.

Precise values for each intrinsic excitonic transition energy have been obtained from the absorption and derivative absorption spectra, by performing a series of theoretical fits in the light of the two-dimensionnal (2D) theory of direct-allowed excitonic transitions [7]. An example of fit is shown in Fig.3. The exciton binding energies E_B are free parameters which compare well with recent variational calculations using anisotropic wave functions [8]. The best fits are obtained using broadening parameters Γ between 5 and 30 meV. It should be noticed that attempts to fit with the three-dimensional (3D) model have been unsuccessfull. This agrees with the 2D character of excitons reported in Ref.8 for QW widths ranging from 10 to 200 Å.

In order to check quantitatively the optical properties of our MQWs, we have compared these transition energies to the theoretical results :

$$E_{ex} = E_g + E_{nh} + E_{ne} - E_B \qquad (1)$$

where E_g is the 3D band gap of lattice-matched GaInAs ; E_{nh} and E_{ne} are the confinement energies of holes and electrons, respectively [6], and E_B the exciton bending energy [8]. These expected excitonic transition energies are drawn in Fig. 4 as a function of the well width for confinements levels $1 \leq n \leq 4$, and for transitions involving heavy (full lines) and light (dashed lines) holes. The measured transition energies are plotted as open symbols. Most experimental data are in reasonable agreement with the expected values. However, in some cases, a better agreement is obtained after readjusting the QW width with respect to the nominal one : 59 Å instead of 50 Å ; 56 Å instead of 65 Å. This is shown as full symbols in Fig. 4. In other cases, a departure from the x = 0.47 alloy composition for lattice matched $Ga_xIn_{1-x}As$ is observed. This results in a sizeable shift of the series of excitonic transitions. For instance, for the 150 and 200 Å wide QWs, a correction of the alloy composition, i.e. the 3D band gap, is needed. This gives a best agreement indicated by crosses in Fig. 4. The actual compositions deduced in this way are, respectively, x = 0.48 (E_g = 0.823 eV) and x = 0.445 (E_g = 0.785 eV), instead of x = 0.47 (E_g = 0.812 eV). Similar discrepancies appear when comparing the results of this computation with previous works (see e.g. Refs. 6,9) and emphasize the difficulty of the GaInAs/InP system with respect to GaAs/GaAlAs : not only the QW width but also the alloy composition must be perfectly controled for optoelectronic device production.

The results reported here confirm that LP-MOVPE is an attractive approch for growing sophisticated structures over large substrates, for integrated optoelectronic device production. They establish also excitonic absorption and/or reflectivity measurements as a quantitative and non-destructive tool for controling the wells compositions and thicknesses.

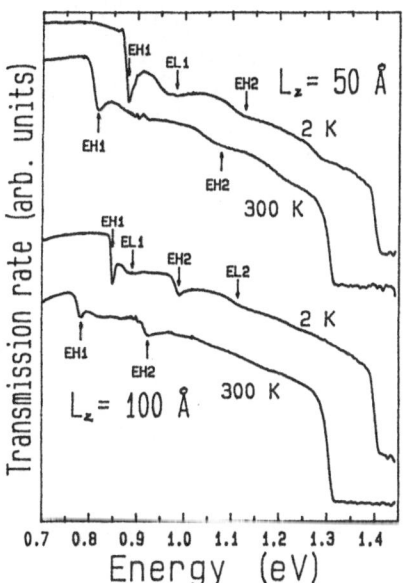

Fig.1. Transmission spectra of two GaInAs/InP MQWs recorded at 2 and 300K. All spectra have been normalized for the system response.

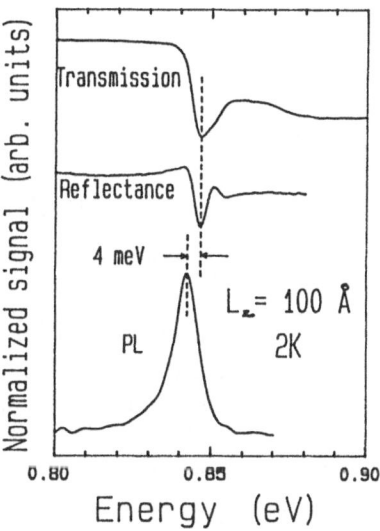

Fig.2. Comparative plots of transmission, reflectance and PL for 100 Å well width.

490

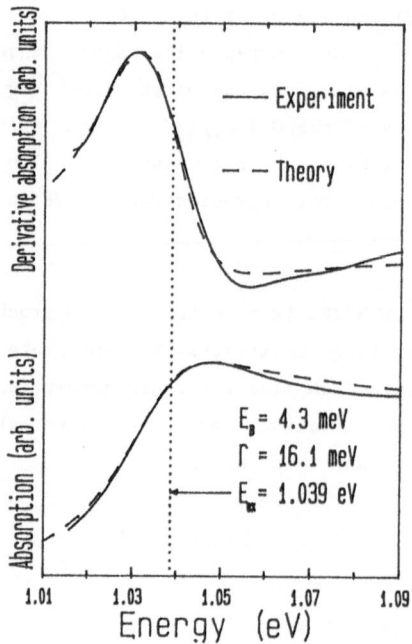

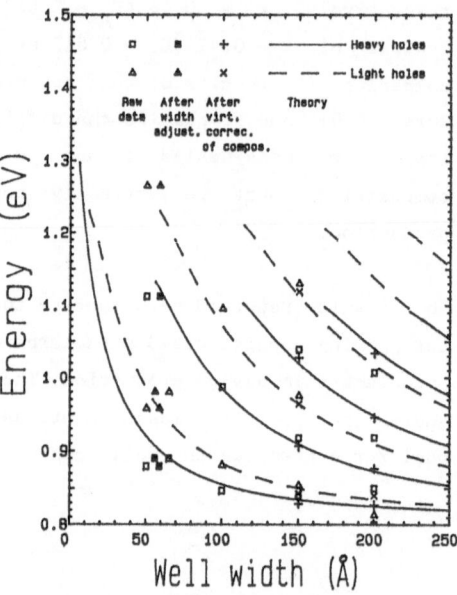

Fig.3. Fits of absorption and derivative absorption spectra for 2D direct-allowed excitonic transition model (see text).

Fig.4. Comparaison of intrinsic excitonic transition energies deduced from transmission measurements, with the results of recent calculations as a function of the QW width. For the two thinner QWs, the agreement is improved by readjusting the QW width with respect to the nominal one. For the two thicker QWs, a better agreement is found after a virtual shift of the 3D GaInAs band gap from the actual value to the one of the lattice-matched alloy.

References

1. D. Grützmacher, K. Wolter, M. Zachau, H. Jürgensen, H. Kurz, P. Balk; Inst. Phys. Conf. Ser. n° 91, IOP Publishing Ltd (1988), p. 613.

2. M. Razeghi, J. Nagle and C. Weisbuch ; Inst. Phys. Conf. Ser. n° 74, Adam Hilger Ltd (1985), p. 379.

3. B.I. Miller, E.F. Schubert, U. Koren, A. Ourmazd, A.H. Dayem and R.J. Capik ; Appl. Phys. Lett. 49, 1384 (1986).

4. W.T. Tsang and E.F. Schubert ; Appl. Phys. Lett. 49, 220 (1986).

5. M.B. Panish, H. Temkin, R.A. Hamon and S.N.G. CHU ; Appl. Phys. Lett. 49, 164 (1986).

6. D. Gershoni, H. Temkin and M.B. Panish ; Phys. Rev. B.38, 7870 (1988).

7. J. Camassel, P. Merle and H. Mathieu ; Physica 99B, 309 (1980) and therein refs.

8. M. Grundmann and D. Bimberg ; Phys. Rev. B38, 13486 (1988).

9. M.S. Skolnick, L.L. Taylor, S.J. Bass, A.D. Pitt, D.J. Mowbray, A.G. Cullis and N.G. Chew ; Appl. Phys. Lett. 51, 24 (1987).

Technology and Characterization
of a Photoconductive Device on InP

H.-H. Wehmann and A. Schlachetzki

Institut für Halbleitertechnik, Technische Universität
Postfach 3329, D-3300 Braunschweig, Federal Republic of Germany

Summary
We have fabricated highly sensitive, planar, I-bar-shaped photoconductive devices in unintentionally doped LPE-grown InGaAs
layers on semi-insulating InP substrates. The gain is measured
in dependence on the optical power and the temperature, showing
a maximum above 10^6 at low temperature and intensity. This
behavior can be described by a model based on the Shockley-
Hall-Read formalism. Further results on the wavelength-dependent
gain and the speed of the devices demonstrate their suitability
for operation in optical communication systems.

Introduction

Photoconductive devices (PCDs) are simple and provide considerable gain at low voltages. Therefore they are attractive for
integrated optics and optoelectronics applications in general.
Since the introduction of glass-fiber technology a special
demand exists for the wavelength region from 1.3 to 1.6 μm,
which can be met by InP-based alloys.
We report on the fabrication and characterization of PCDs in
$In_{0.53}Ga_{0.47}As$ lattice-matched to (100)-InP. For better understanding a simple model is presented describing the measured
temperature and intensity dependent optical gain.

Technology

The liquid-phase-epitaxially grown InGaAs layers show an electron concentration $n = 1.3 \times 10^{15}$ cm^{-3} and a mobility
$\mu_n = 8700$ cm^2/Vs at room temperature (RT). The average thickness t was microscopically determined to be 1 μm.
After Au/Ge evaporation, ohmic contacts were formed by rapid
thermal alloying (20 s; H_2-atmosphere at 330 $^\circ$C). The I-bar-
shaped mesa structures with $l = 42$ μm channel length and a width
$w = 56$ μm were wet-chemically etched. A selective etchant for
InGaAs was employed consisting of hydrogen peroxide (33 % by
vol. H_2O_2) and citric acid (50 % by vol. $C_6H_8O_7$). The dependence
of the etching rate r on the volume fraction x of hydrogen

peroxide is shown in Fig. 1 where the temperature T is a parameter. The low etching rate and the change from isotropic to anisotropic behavior near $x = 0.2$ (see insert) make this etchant very versatile for device fabrication.

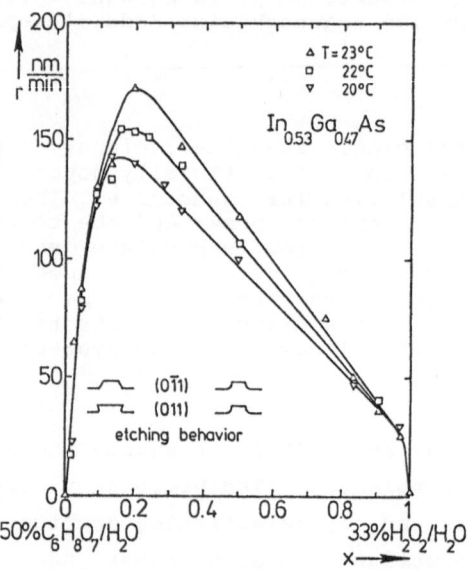

Fig. 1: InGaAs etching rate r versus volume fraction x of hydrogen peroxide/citric acid with the temperature T as a parameter. The insert shows the corresponding mesa ramps.

Static $I(V)$-characteristics

The measured current-voltage characteristics ($I(V)$) of the devices show a linear region from 0 to 4 V obeying Ohm's law with a 4.1 kΩ gradient, then leveling off to a saturation region between 5 and 10 V. A possible cause for this is a high-field domain coinciding with the saturation drift velocity of the electrons. Above 11 V the current increases again due to carrier injection effects. The following optical measurements were performed in the linear region, mainly at 1.3 V.

Gain-measurements

To determine the gain

$$\Gamma = (I_{ph}/q)/(P_{opt}/\hbar\omega) \tag{1}$$

of the detectors, we used the chopped light of a 1.3 μm laser-diode, or a halogen lamp with monochromator. The optical power P_{opt} was measured by calibrated beamsplitter and Ge-photodiode. The resulting Γ in dependence on P_{opt} is shown in Fig. 2 (a for RT and b for $T = -92$ °C) with the wavelength λ as a parameter.

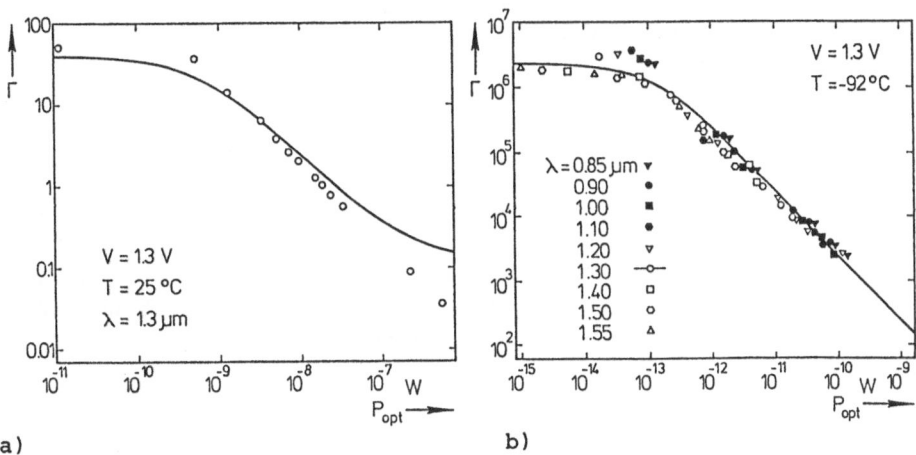

a)   b)

Fig.2: Gain Γ of a 42 μm long InGaAs-PCD as a function of optical power a) at RT and 1.3 μm wavelength; b) at -92 °C and different wavelengths. The curves represent the results calculated according to our model.

Theoretical model

In the linear region of the $I(V)$-characteristics, the photocurrent due to the optically generated free carriers Δn and Δp can be easily expressed using Ohm's law

$$I_{ph} = qV \, (\mu_n \Delta n + \mu_p \Delta p) \, wt/l \; . \tag{2}$$

The Shockley-Hall-Read formalism [1] of Fig. 3 with the densities N_T of traps at the energy W_T and N_R of recombination centers at W_R gives the optical generation rate

$$G = (1-r) \, (1-exp(-\alpha t)) \, (P_{opt}/\hbar\omega)/(wtl) \; , \tag{3}$$

assumed to be homogeneous. r and α are the wavelength-dependent reflectivity and absorbtion coefficient, resp. [2]. Electrons are captured by optically active traps at a rate

$$G_T = 1/N_0 \, G \, (1-f_T)N_T \tag{4}$$

proportional $(1/N_0)$ to G and the density of empty traps. The arrows in Fig. 3 show the transitions possible for the electrons and characterized by the transition constants C_n, C_p, and C_T.

Assuming quasi-neutrality and stationary operation, we numerically solve the continuity equations leading to Δn and Δp and resulting in the curves of Fig. 2. We used four constants for fitting ($W_R = 0.73$ eV, $W_T = 0.374$ eV, $N_R = 1.4\times10^{12}$ cm^{-3}, $N_T = 8\times10^{15}$ cm^{-3}) and three parameters slightly varying with temperature ($C_n = C_p = 0.15$ cm^3/s, $C_T N_0 = 8\times10^5$ s^{-1} for RT and $C_n = C_p = 0.12$ cm^3/s, $C_T N_0 = 1.9\times10^5$ s^{-1} for -92 °C).

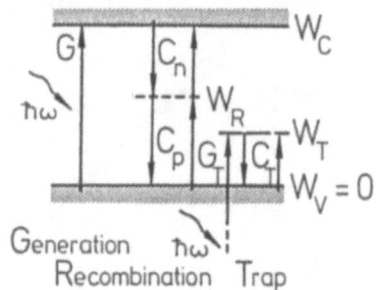

Fig. 3: Energy band diagram used for the calculations with optical generation, recombination centers and optically active traps. W_C and W_V are the energies of the conduction and the valence band edge, respectively.

We attribute the slight difference between simulation and measurement at RT and high optical powers to surface and contact effects. They might also be responsible for the relatively small dependence of the gain on the voltage. In general, we find good qualitative agreement with [3].

Dynamic measurements

The velocity of the PCDs was determined by modulating the laser with a digital signal and recording the amplified detector signal on an oscilloscope. The 3 dB roll-off frequencies depend on the optical power and the temperature, and range from 10 MHz up to more than 200 MHz (modulation limit of the laser diode).

Acknowledgement
We thank S. Aytaç for comprehensive measurements and W. Schultz for enlightening discussions.

References
[1] Sze, S.M.:Physics of Semiconductor Devices, 2nd ed. New York: John Wiley & Sons 1981.
[2] Kowalsky, W.; Wehmann, H.-H.; Fiedler, F.; Schlachetzki, A.: Optical Absorption and Refractive Index near the Bandgap for InGaAsP. phys. stat. sol. (a) 77 (1983) K75-K80.
[3] Antreasyan, A.; Tsang, W.T.: High performance Ga$_{0.47}$In$_{0.53}$As photoconductive detectors grown by chemical beam epitaxy. Appl. Phys. Lett. 49 (1986) 322-324.

MOMBE Growth of GaInAs/InP Structures:
Quantum Wells and Selective Epitaxy

M. GAILHANOU, L. GOLDSTEIN, M. LAMBERT, M. BOULOU, C. STARCK and L. LE GOUEZIGOU

Laboratoires de Marcoussis, CR-CGE, Route de Nozay
91460 Marcoussis (France)

Abstract: High quality InP and GaInAs have been grown by MOMBE. 77K mobilities of $57000 cm^2/Vs$ for InP and $50000 cm^2/Vs$ for GaInAs were obtained. We have prepared GaInAs/InP multi single quantum well structure showing sharp photoluminescence peaks with linewidths as low as 7 meV for 1.7nm well. We have also studied selective epitaxy on different shapes of mesas and performed the planarization of a RIE etched ridge.

1-Introduction

Metal Organic Molecular Beam Epitaxy (MOMBE) is a powerful technique for the epitaxial growth of high quality GaInAsP/InP materials [1,2]. Combined with its ability to achieve very good interfaces, resulting in the realization of high optical quality Quantum Wells (QWs) [3], this makes MOMBE very attractive for optoelectronic devices fabrication. Another feature of MOMBE concerns selective growth [4,5]. This characteristic enhances the interest of MOMBE, as compared to conventional MBE or Gas Source MBE (GSMBE), for the complete device elaboration. In this paper we report on the growth of high quality InP and GaInAs as well as GaInAs/InP QWs by MOMBE. We have also studied selective epitaxy on etched substrate.

2-Experimental-InP-GaInAs

The growth experiments were performed in a home made MOMBE system with non rotative substrate holder, equipped with two organometallic (OM) lines (Triethyl Gallium: TEG, Trimethyl Indium: TMI) and two hydride lines (AsH_3,PH_3). The control of stable OM fluxes is achieved by using an hydrogen carrier flow at reduced pressure (25 to 50 torr in the OM bottle). OM are mixed and introduced in the growth chamber through a low temperature (60°C) cell. A low pressure cell heated at 1000°C provides for the cracking of hydrides to As_2 andP_2.

InP as well as GaInAs layers grown with this equipment exhibited excellent electrical properties. We obtained $n=3.10^{14}$ cm^{-3} with $\mu=57000$ cm^2/Vs for InP and $n=5.10^{15}$ cm^{-3} with $\mu=50000$ cm^2/Vs for GaInAs.

4-GaInAs/InP QWs

We have grown on a InP:Fe substrate a structure consisting of an InP buffer layer (130nm), a GaInAs thick layer and four wells of different thicknesses separated by 130nm InP barriers. The thicknesses of the layers have been determined by Wedge Transmission Electron Microscopy

(WTEM) [6]. InP was grown at low temperature (495°C) in order to obtain smooth surfaces. Interfaces and GaInAs layers and wells were made 20°C higher. Growth interruptions between GaInAs and InP layers were about 10s.

Low temperature (4.2K) photoluminescence measurements were performed on the multilayer structure by using the 488nm line of an Ar laser. A typical spectrum is shown on fig 1. For this spectrum, the pumping power was $0.5W/cm^2$. Peaks corresponding to the different well are extremely sharp. For instance linewidth as low as 7meV for the 1.7nm well is observed. This is among the best reported values for such narrow wells [3,7].

5-Selective epitaxy

We have studied InP selective MOMBE on (001) InP patterned substrate with Si_3N_4 or SiO_2 masked mesas along $\alpha[1\bar{1}0]$ and $\beta[110]$ directions. The mesas were fabricated either by wet etching or by Reactive Ion Etching (RIE). Etching depth was approximatively 1mm for both technique. The growth profile is independant of the nature of the dielectric mask but strongly influenced by the etching technique especially at the edge of the mesa.

a-Selective growth on wet etched masked mesas

The etch used for defining the mesas was $HCl:HNO_3$ (1:1).Two kinds of profiles were obtained, depending on the direction of the mesa. Alternate 0.3 μm InP layers and 0.03μm GaAs marker layers were deposit to show how the growth proceeds.

α orientation

The mesa shows inverted V shape. Fig 2a is a micrograph of (110) cleaved plane. We can notice on a detail (fig 2b) that one family of planes, identified as (116) grows faster than (001) planes. In fact, the etched substrate didn't show any (116) plane and the GaAs markers enable the observation of the development of this plane during the growth.

β orientation

Vertical sidewalls and a signicative undercuting of the dielectric are the main characteristics of $HCl:HNO_3$ (1:1) etched β mesa. We can remark on fig 2c that no growth occurs at the edge of the mesa and that the growth stops on (111)B type planes. We assume that a shadow effect due to the undercutting of the mask arises at the beginning of the growth. However this can not explain the shape of the structure and in particular the emergence of (111)B type non growing planes.The outcome of this, and it is a drawback for device applications, is the absence of planarization of the structure.

b-Selective growth on a dry etched InP substrate

We etched masked InP with a CH_4/H_2 mixture in a RIE chamber. The walls of the mesas were vertical, presenting no undercutting of the dielectric. a and b direction oriented mesas showed the same etching profile. Two kinds of experiments were performed on these contoured substrates. First we grew InP with the substrate manipulator in the standard position of epitaxy (OM cell axis close to the direction perpendicular to the substrate surface). Cleavage of the grown structure exhibited a good planarization of the structure. However voids were remaining at one edge of the mesa (fig 3a). This problem was overcomed by tilting periodically the substrate manipulator with an angle of $\pm\theta$ from normal position of epitaxy. Fig 3b shows the planarization obtained with this process for $\theta= 30°$ and the continuity of the material at the edge of the mesa.Despite the absence of rotation of the substrate during epitaxy, we have succeed in obtaining a good planarization of RIE etched mesas.

5-Conclusion

We have grown high quality InP and GaInAs materials. Some qualitative aspects of their growth have also been evidenced. Availibility of MOMBE for the growth of low dimensional structures (QWs) as well as for planarization of mesas has been demonstrated, making this technique powerful for future devices fabrication.

AcknowledgementsWe would like to thank R. Vergnaud and J.P. Chardon for their assistance on MOMBE, Y. Louis and F. Poingt for providing the masked substrates.We are also greatly endebted to J.D. Ganière from Ecole Polytechnique Fédérale de Lausanne for WTEM experiments.

References
1. W.T.Tsang, A.H.Dayem, T.H. Chiu, J.E. Cunningham, E.F. Schubert,, J.A. Ditzenberger, J. Shah, J.L. Zyskind, N. Tabatabaie, Appl.Phys.Lett. 49,170(1986)

2. Y.Kawaguchi, H. Asahi, H. Nagai, Proceedings of 12th International Symposium on GaAs and related Compounds, Karuizawa, Japan.

3. W.T.Tsang,E.F. Schubert, Appl.Phys.Lett. 49,220(1986)

4. W.T.Tsang, J.Electron.Mater. 15,235(1986)

5.D.A.Andrews,M.A.Z. Rejman-Greene, B. Wakefield, G.J. Davies, J.Cryst.Growth 95,167(1989)

6. P.A. Buffat, P. Stadelmann, J.D. Ganière, D. Martin and F.K. Reinhart, Inst. Phys. Conf. Ser. No. 87: Section 3 p207

7.P.J.A. Thijs,E.A. Montie, H.W.van Kesteren, G.W. 't Hooft , Appl.Phys.Lett. 53,971(1988)

498

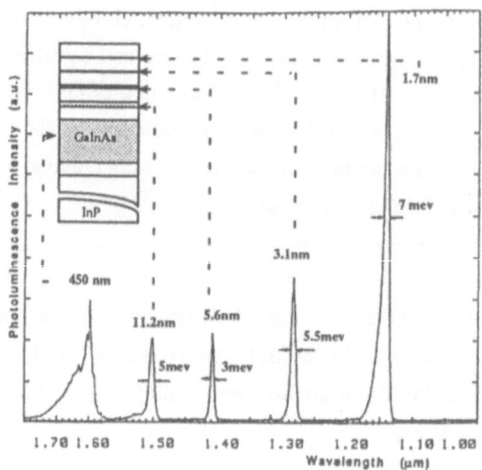

Fig1-Photoluminescence spectrum (4.2K) of
GaInAs/InP structure

Fig2- InP selective MOMBE on wet etched mesa

2a

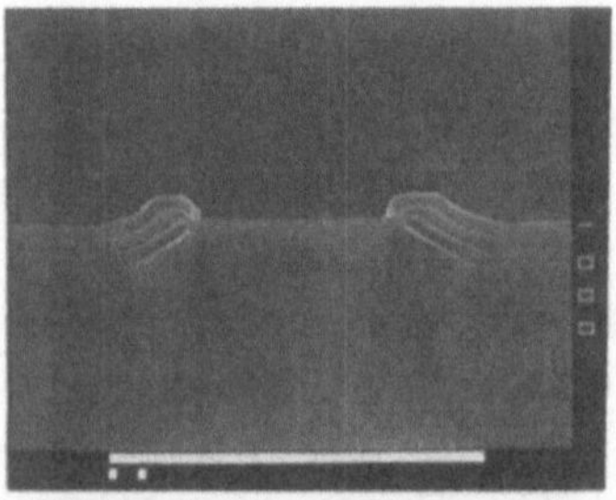

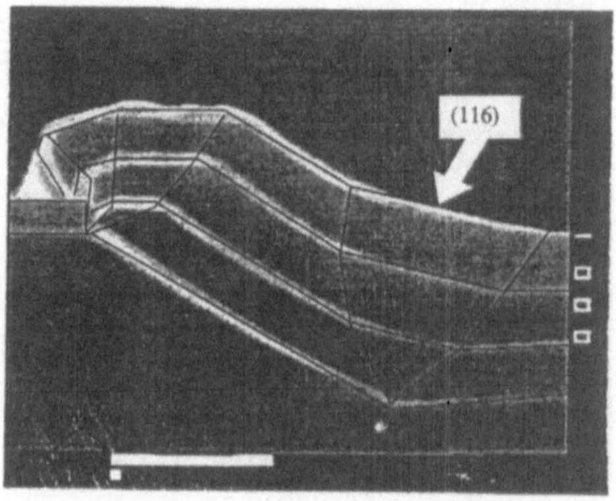

2c

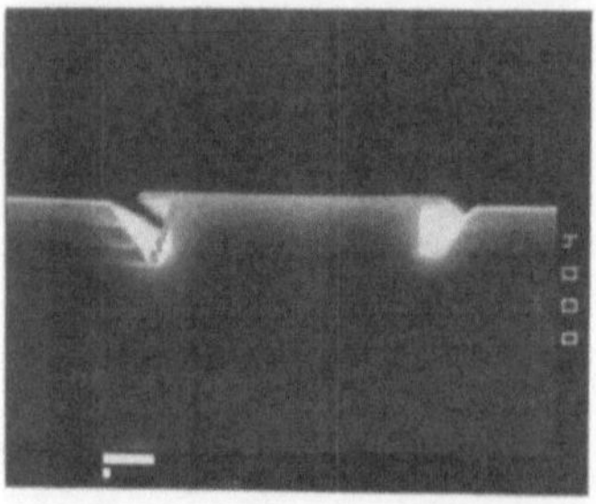

2b

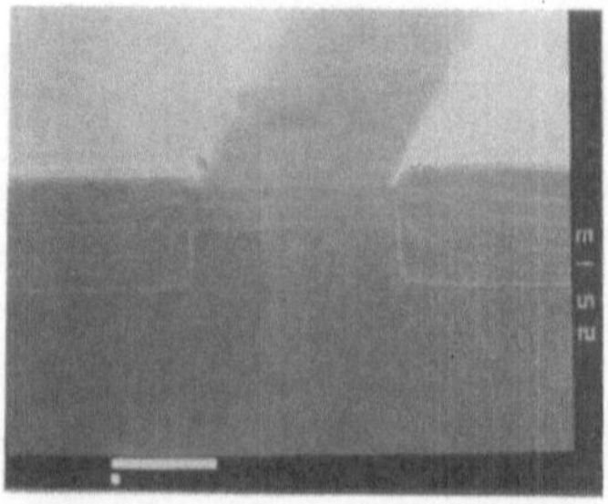

(116)

Fig3- InP selective MOMBE on RIE etched mesa

3a

3b

Coupling Induced Enhancement of Interface Recombination in GaAs Multiple Quantum Well Structures

M. Krahl[a], D. Bimberg[a], R.K. Bauer[a], D.E. Mars[b], J.N. Miller[b]

[a]Technische Universität Berlin, Institut für Festkörperphysik,
Hardenbergstr. 36, 1000 Berlin 12, Federal Republic of Germany

[b]Hewlett Packard Laboratories, 3500 Deer Creek Road,
Palo Alto, CA 94304, USA

Abstract

The variation of luminescence transients of GaAs multiple quantum wells as a function of barrier widths is studied in the range between 0.87nm (superlattice) and 18.1 nm (uncoupled wells) by means of cathodo- and photoluminescence. With decreasing barrier width the nonradiative recombination rate is found to be drastically enhanced whereas the radiative recombination probability decreases. Thus a pronounced decrease of the quantum efficiency results. The controlled variation of the barrier width is found to be decisive for an unambigous identification of the origin of the traps: Comparison with a theoretical calculation shows that they are localized at the heterointerfaces and **not** in the barriers.

I. Introduction

The superiority of quantum well (QW) lasers as compared to conventional double hetero-structure lasers is a direct consequence of the carrier confinement in two dimensions[1]. Recently a strong enhancement of the radiative mono- and bimolecular recombination coefficients was discovered and accounted for by an increased overlap of the electron and hole wave functions[2]. Multiple quantum well (MQW) lasers with narrow barriers in contrast to SQW-lasers have a dimensionality somewhere between two and three, depending on the width of the barriers. Coupling of wells leads to a loss of the localization induced pecularities of twodimensional structures[3]. Here we will demonstrate the impact of coupling and thus delocalization on the dynamic properties of MQW structures, in particular on the radiative and nonradiative time constants τ_r and τ_{nr}. The knowledge of both parameters and a physical insight into the origin of their variations is essential for any modeling of a number of most fundamental laser charac-teristics like the threshold current j_{th}, the gain or the cut-off frequency.

II. Radiative and Nonradiative Recombination in QWs

Several authors reported observations of a well width dependent enhancement of the radiative recombination rate e.g. [1, 2, 4] in QW structures. Other groups observed that formation of QWs might also favorize nonradiative recombination [5]. Obviously, the potential advantage of

QW structures turns into an actual one only if the localization induced enhancement of radiative recombination can overcompensate the technology dependent enhancement of the nonradiative one. Traps causing the latter effect are generally assumed to be localized either at the interface or in the barrier region. An "effective" surface recombination velocity is usually introduced to describe the trapping ignoring the two possible completely different trap localizations. This recombination velocity can be derived in terms of the wave function ψ^{SQW}, the usual solution of a single quantum well. In this model the trapping rates at the interfaces and in the barriers are assumed to be proportional to the probability of finding a carrier at the interface or in the barrier region, respectively. Thus the interface trapping rate is proportional to $|\psi^{SQW}|^2$ at the interface. For the description of the barrier trapping $|\psi^{SQW}|^2$ has to be integrated over the whole barrier region. Numerical evaluation of both terms yields a dependence on L_z, which is qualitatively the same for well widths of practical use, i.e. $L_z \geq 2.5$ nm: S_{eff} increases for both trapping mechanisms with decreasing L_z. Thus L_z-dependent time resolved experiments cannot distinguish between interface and barrier trapping and only speculations about the real origin of the nonradiative processes and the "optical quality" of the barrier material exist [e.g. 4, 6]. Here we will show that controlled variation of the barrier width L_B does provide a tool to distinguish unambiguously between the two possible localizations of the dominant traps. In order to calculate the trapping probability at different regions outside the QW in the case of small L_Bs we have to consider the coupling induced modification of the wavefunction: Coupling of wells allows movement of carriers perpendicular to the layers and the QW wavefunction has to be replaced by a superlattice wavefunction ψ^{SL}, accounting for the energy dispersion in z- (growth-) direction. Details of it's derivation are given in Ref. 3.

III. Samples and Experiment

All samples studied were successively grown by molecular beam epitaxy as described in Ref. 7. The well width $L_z = 4.7$ nm was kept constant, but the barrier width L_B was varied from 18.1 nm (limit of completely decoupled wells) to 0.87 nm (true SL, strongly coupled wells). The Al-content of the barriers is 40 %. Time-resolved cathodo- and photoluminescence (CL and PL) experiments were performed.(For details see Refs. 8 and 9).

IV. Results and discussion

Fig. 1 shows typical room temperature CL-transients of three samples with well widths $L_z = 4.7$ nm and barrier widths of $L_B = 18.1$ nm (uncoupled wells), $L_B = 4.5$ nm and $L_B = 0.87$ nm (strongly coupled wells) on a semilogarithmic scale. Obviously the carrier decay is drastically increased upon coupling of wells. For predominant radiative recombination we would expect the opposite: the radiative recombination rate should decrease upon an decreased overlap of the carrier wavefunctions. Increased total recombination rate with increasing coupling strength therefore can only be caused by the dominance of nonradiative recombination: The corresponding shortest time constants in the decay of each sample derived from a χ^2-

analysis are strongly correlated with the barrier width L_B: Upon increasing coupling strength τ_{nr} is lowered from 2.5 ns (L_B = 18.1 nm) down to 0.9 ns (L_B = 0.87 nm).

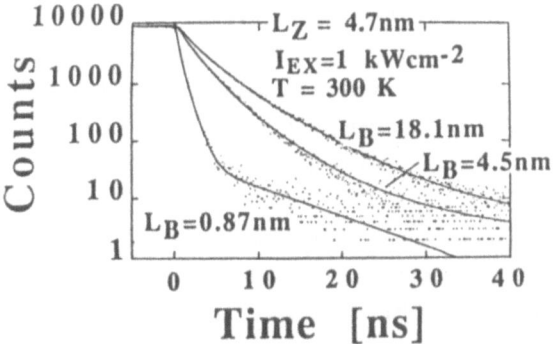

Fig. 1: CL-transients of three samples with different coupling strength (dots). Full lines give fits based on a χ^2-analysis. With decreasing barrier width the decay becomes more rapid.

In order to explain this enhancement and to identify the location of the dominant trap Fig. 2 shows a comparison of the experimental data with our theoretical predictions based on the SL-wave-functions. Obviously the data can **not** be explained if the nonradiative recombination is assumed to occur in the barrier since the probability of finding an electron in the barrier is decreased upon coupling of wells. On the other hand the assumption that the dominant traps are localized at the interface yields excellent agreement with experiment. Thus we have unambigously shown that the nonradiative recombination in our MQW structures is dominated by the interfaces.

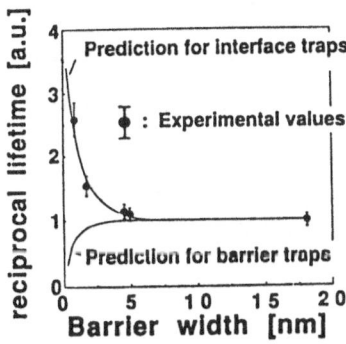

Fig. 2: Comparison of experimentally observed recombination rates with theoretical predictions based either on pure interface or barrier trapping.

502

Finally we found that the radiative excitonic lifetime also depends on the coupling strength: If τ_r were independent of the coupling, $I_{int} = f(\tau)$ would be a linear function of τ with constant slope proportional to $1/\tau_r$. In reality I_{int} depends nonlinearly on τ. The slope increases with increasing τ and thus with decreasing L_B. Apparently, the radiative lifetime increases with decreasing barrier width, as expected from theory[10]. The enhancement of the radiative recombination rate is a typical localization effect, which should disappear with increasing coupling and subsequent delocalization of the carriers. Coupling of the wells initiates a transition from two-dimensional to three-dimensional properties of carriers and excitons.

In conclusion, our timeresolved luminescence experiments show, that coupling of wells in GaAs MQW structures results in a strong enhancement of the nonradiative recombination probability. Comparison with a theoretical model based on the SL wavefunctions gives clear evidence that the traps are localized at the interfaces and not in the barriers. The decrease of nonradiative lifetime together with the increase of the radiative lifetime results in a drastically lowered quantum efficiency with increased coupling.

References

1) D. Bimberg, J. Christen and A. Steckenborn; in "Two-Dimensional Systems, Heterostructures, and Superlattices", (ed. by G. Bauer, F. Kuchar and H. Heinrich, Springer, Berlin, 1984), pp. 136 - 146 and refs. therein.

2) E.H. Böttcher, K. Ketterer, D. Bimberg, G. Weimann, and W. Schlapp; Appl. Phys. Lett. 50 (16), 1074 (1987)

3) M. Krahl, J. Christen, D. Bimberg, D.E. Mars, and J.N. Miller; IEEE J. Quant. Electron., in print and
M. Krahl, J. Christen, D. Bimberg, G. Weimann, and W. Schlapp; Appl. Phys. Lett. 52, 798 (1988)

4) K. Fujiwara , A.Nakamura, Y.Tokuda, T.Nakayama, and M.Hirai; Appl. Phys. Lett. 49 (18), 1198 (1986)

5) see the review by D. Bimberg; in Solid State Devices 1985, ed. by P. Balk and O.G. Folberth, Elsevier Science Publishers B.V., 1986, P. 101

6) B.Sermage, M.F.Pereira, F.Alexandre, J.Beerens, R.Azoulay, and N.Kobayashi; Inst. Phys. Conf. Ser. 91, 605 (1988)

7) D. Bimberg, D. Mars, J.N. Miller, R.K. Bauer, and D. Oertel; J. of Vac. Sci. and Technol. B4 (4), 1014 (1986)

8) D. Bimberg, H. Münzel, A. Steckenborn, and J. Christen; Phys. Rev. B31, 7788 (1985) and
J. Christen, and D. Bimberg; Oyo Buturi 57, 69 (1988)

9) M. Engel, R.K. Bauer, D. Bimberg, D. Grützmacher, and H. Jürgensen; J. of Crystal Growth, 93, 359 (1988)

10) M. Grundmann, and D. Bimberg; Phys. Rev. B 38 , 13486 (1988)

Scanning Photoluminescence Assessment MOCVD InGaAs/InP Lattice Mismatched Heterostructures During the Fabrication of Photodiode Arrays

K. Schohe, J. Y. Longère, S. Krawczyk*
Laboratoire d'Electronique, UA CNRS 848, Ecole Centrale de Lyon,
36, avenue Guy de collongue, 69131 Ecully Cedex, France

B. Vilotitch, C. Lenoble, M. Villard and X. Hugon
Thomson Composants Militaires et Spatiaux, avenue de Rocheplaine
38521 Saint-Egrève, France

Summary
We report on scanning photoluminescence (SPL) measurements carried out on MOCVD lattice mismatched InGaAs/InP heterostructures during the fabrication of planar PIN photodiode arrays. Electrical characterization was performed on completed devices. We show that SPL measurements reveal a large number of defects and non-uniformities in the deposited films. The correlation of SPL measurements and the reverse current of individual diodes is presented.

Introduction

Lattice mismatched InGaAs/InP heterostructures are good candidates for the realization of photodetectors and photodiode arrays with extended spectral response up to 2.5µm. Such components are required for space applications (e.g. the SPOT-satellite program) and optical fibre communication.

Thus, our motivation is to develop high quality lattice mismatched InGaAs heterostructures and appropriate qualification techniques to control the reproducibility of the fabricated layers.

In this contribution we report on the assessment of defects and nonuniformities of lattice mismatched InGaAs/InP heterostructures using scanning photoluminescence measurements and electrical characterization of the fabricated photodiodes.

Experimental

Linear arrays of PIN photodiodes were realized by planar technology, using Zn diffusion into InP/InGaAs/InP/InP heterostructures grown by Metal Organic Chemical Vapor Deposition (MOCVD) in a horizontal reactor. The dimensions of the elementary photodiode cells are 30µm x 30µm. Scanning PL measurements (room temperature, 5µm resolution) were performed

---

* currently with the Institut für Hochfrequenztechnik, Technische Hochschule Darmstadt, West-Germany under an Alexander-von-Humbold fellowship

with the SCAT Imageur (Scantek, France) on the InP substrates and on the epitaxial layers after the MOCVD process. Electrical characterization (I-V measurements, spectral response) was carried out on the finished devices. The lattice misfit $\Delta a/a$ between the InGaAs layer and the InP substrate was found to be in the range of $-2 \cdot 10^{-3}$ to $+2 \cdot 10^{-3}$, as measured by X-ray diffraction. The bandgaps of the deposited layers were determined by infrared transmission measurement. The obtained values correspond to wavelengths of 1.65 to 1.75 μm.

Results and discussion

SPL measurements revealed the following defects and nonuniformities in the InGaAs layers:

a) isolated dislocations propagating from the substrate; as in the case of InP substrates, they appear as local depressions of the PL intensity (Fig. 3)

b) mismatch dislocations which appear as parallel lines in the case of InGaAs/InP heterostructures (Fig. 1)

c) mismatch dislocations which appear as cross hatched lines in the case of InP/InGaAs/InP double heterostructures (Fig. 2)

d) short range nonuniformities (5...10μm, density $> 10^5/cm^2$) which result in small fluctuations of the PL signal (Fig. 4)

e) long range nonuniformities (several millimeters wide) which result in a smooth variation of the PL signal across the wafer (Fig. 5)

In most of the samples we observed the simultaneous presence of several types of the above defects. This concerns in particular defects due to the mismatch and dislocations propagating from the substrate (Fig. 6).

The SPL images correlate with typical X-ray topographs of these epitaxial layers (1) and with SPL measurements on the substrates prior to the epitaxy. However, the physico-chemical nature and origin of the short range fluctuations of the PL intensity and of the long range nonuniformities remains unknown.

We found that generally even very small varations of the process conditions strongly affect the density and the type of the defects in the layer. This indicates, that a tight control of the epitaxial layer quality is indispensable during device processing.

In order to estimate the influence of different types of defects on the performance of the photodiodes, we compared SPL images with results of electrical characterization. In Fig. 7 we show the variation of the reverse current of photodiodes (Io) along two lines across the sample. Their position is indicated in the PL image at the right of the figure. It is apparent, that the long range nonuniformities are correlated with the variation of Io, the central part exhibiting both high PL intensity and an anomalous high reverse current (> 500 pA) of the photodiodes. Outside this region, the reverse current of most of the diodes is lower than 10 pA and for only 1 - 2% of the diodes Io exceeds 500pA.

High resolution SPL measurements of this part of the wafer indicated the presence of dislocations with a density of about $2 \cdot 10^4 cm^2$ and the short range fluctuations, but no mismatch defects. If dislocations propagating from the substrate were responsible for high Io values, about 10% of the diodes of this area would be affected. But since their number is about ten times lower, the presence of dislocations propagating from the substrate is not directly correlated to intolerable high dark current values. A similar conclusion in the case of avalanche diodes is reported in (2,3). However, the presence of dislocations can be harmful to the breakdown voltage of the devices and their long-term stability (2,3).

Fig. 8 shows a comparison of Io of a line of photodiodes with the dispersion of the PL intensity measured on the InGaAs layer prior to the lithography. The respective area of the sample showed strongly pronounced misfit dislocations. It appears, that the presence of this type of defect is correlated to spatial fluctuations of the dark current.

Conclusion

A large number of defects and nonuniformities of mismatched InGaAs/InP heterostructures can be revealed by scanning photoluminescence measurements. Some of these defects apparently are related to the reverse current of the photodiode arrays. The defect creation in the epitaxial layers is extremely sensitive to small variations of the process conditions. Since SPL measurements are nondestructive, contactless and compatible with compound semiconductor technology, they are well adapted for the evaluation of the quality of epitaxial layers before ulterior steps in the device processing.

Acknowledgements

This work is supported by the french Ministry of Research and Technology (S1127) and by the Commission of the European Communities (ST2*265). The autors gratefully express their appreciation.

References

1. Yamazaki, S.; Nakajima, K.; Komiya, S.; Kishi, Y.; Akita, K. : Misfit dislocation in (111) A InP/InGaAs/InP double heterostructure wafers grown by LPE. J. Appl. Phys. 10 (1984) 3478 - 3884.

2. Matsushima, Y.; Sakai, K.; Akiba, S.; Yamamoto, T.: Zn-diffused InGaAs/InP avalanche photodetector. Appl. Phys. Lett. 35 (1979) 466 - 468.

3. Susa, N.; Yamauchi, Y.; Ando, H.: Effect of imperfections in InP avalanche photodiodes with vapor phase epitaxially grown p-n junctions. J. Appl. Phys. 53 (1982) 7044 - 7050.

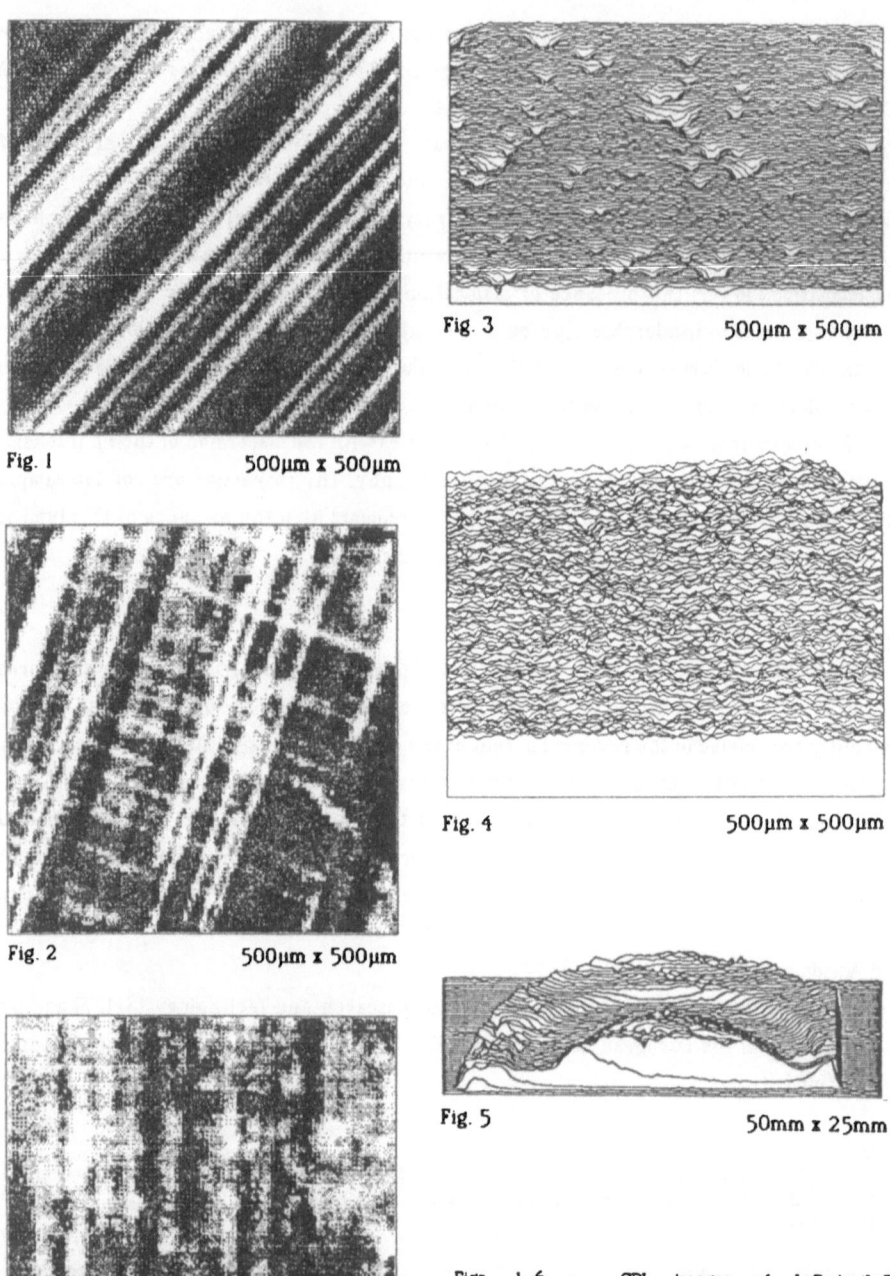

Fig. 1   500µm x 500µm

Fig. 2   500µm x 500µm

Fig. 6   500µm x 500µm

Fig. 3   500µm x 500µm

Fig. 4   500µm x 500µm

Fig. 5   50mm x 25mm

Figs. 1-6 : SPL images of InGaAs/InP heterostructures. They correspond to the most typical defects and nonuniformities observed in those structures. (Refer to the text for the description.) In the grey mapping the black pixels correspond to high PL intensities.

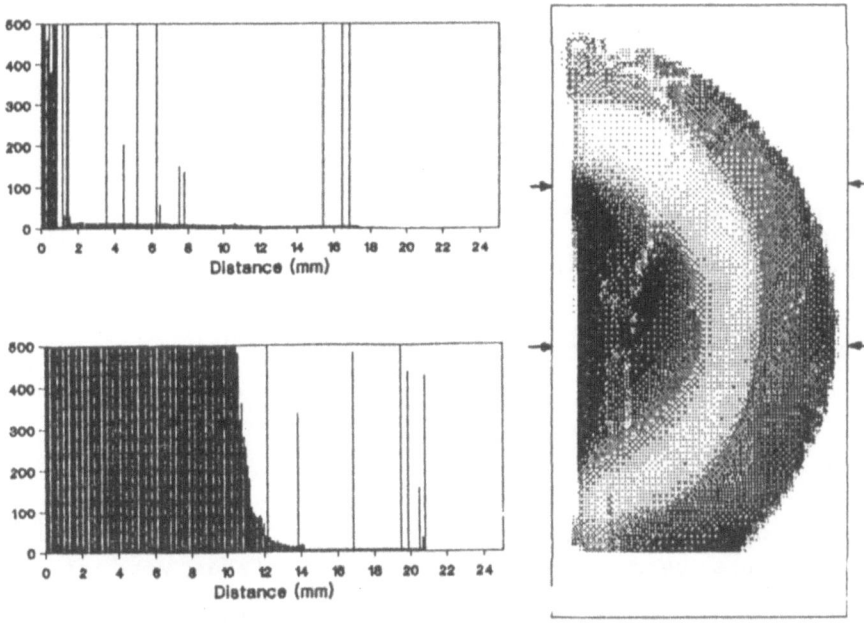

Fig. 7 : Variations of the reverse current of photodiodes along two lines across an InGaAs/InP heterostructure. The corresponding SPL image is presented at the right.

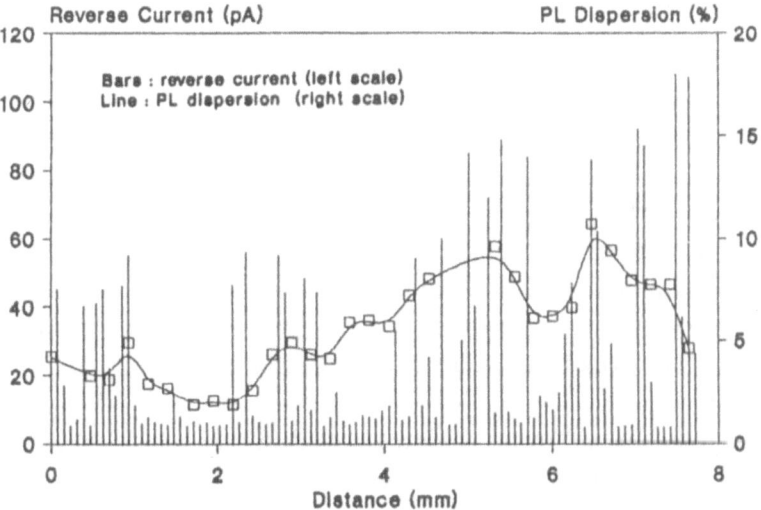

Fig. 8 : Spatial variation of the reverse current of a photodiode array and the standart deviation of the PL intensity at the same area

Investigation of GaAs/AlGaAs Quantum Well Lasers by Micro Raman Spectroscopy

S. Beeck, T. Egeler, G. Abstreiter
Walter Schottky Institut, TUM, D-8046 Garching, FRG

H. Brugger*, P.W. Epperlein, D.J. Webb
IBM Zürich Research Laboratory, CH-8803 Rüschlikon, Switzerland
*Present address: Daimler Benz AG Research Inst. D-7900 Ulm, FRG

C. Hanke, C. Hoyler, L. Korte
Siemens AG Research Lab., D-8000 München, FRG

Abstract:

GaAs/AlGaAs quantum well lasers are investigated with non destructive micro Raman spectroscopy. Electronic properties and
temperature behaviour at the mirror surfaces were studied by
probe lasers with high spatial resolution. Electric field induced Raman scattering (EFIRS) is suitable for observing the
band bending, caused by surface states. Temperature behaviour of
coated and uncoated mirrors and the spatial temperature profile
are investigated by measuring the Stokes- and anti-Stokes intensities of the GaAs TO-phonon. Resonant electronic Raman scattering leads to spatial distribution of carrier densities.

One of the main problems in the achievement of high power
GaAs semiconductor lasers is the catastrophic self destruction
of the facet mirrors during operation. In the past years several
efforts have been made in changing the laser structure like
mirror coatings, different materials and new topological design
to increase the lifetime and maximum power of semiconductor
lasers. In this contribution we present optical studies of the
mirror surfaces of semiconductor lasers using non destructive
micro Raman spectroscopy. A special microscope set up with high
spatial resolution (FWHM of laser spot <1.0 μm), xy-translation
stages with piezo drivers (positioning accuracy <0.1 μm) and a
cryostat for microscopy (T>5ºK) have been used. The Raman spectra were measured with a triple grating spectrometer and a multichannel detector system (DILOR). This configuration enables us
to examine electronic properties, temperature behaviour of laser
functions and rapid damaging effects on submicron scale during
laser operation. Three aspects of Raman scattering have been

used to investigate various local properties of GaAs/AlGaAs
GRINSCH MQW lasers close to their mirror surfaces.

Electric field induced "forbidden" Raman scattering (EFIRS)
leads to direct information on the surface barrier height and
consequently on the width of the depletion region at the (110)
orientated laser mirrors. The intensity of the LO phonon mode of
the GaAs multi quantum wells depends roughly on the square of
the surface electric field. The field independent allowed TO
mode can be used as an intensity calibration. Details of this
method are described for example in [1]. The LO/TO phonon intensity ratio measured at the mirror of a GaAs/AlGaAs MQW
GRINSCH laser is displayed versus injected current through the
laser in Fig. 1. The geometry is shown in the inset. The steep
decrease of the LO phonon intensity at small currents demonstrates the strong reduction of the depletion region close to the
mirror cleavage plane, already far below the threshold current
of the laser. The reduction of the barrier due to the injected
current leads to a high electron and hole concentration close to
the surface which might be responsible for strong non radiative
recombination via surface states. These are first results of
this type which support the assumption that non radiative decay
can be a major source for the strong temperature increase at the
active mirror area, concomitant with degradation for increasing
power densities.

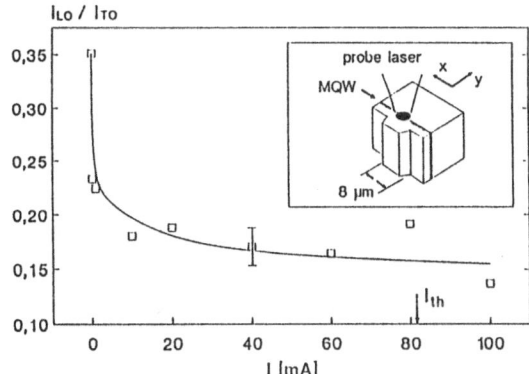

Fig. 1. LO/TO-phonon intensity ratio as a function of laser current; threshold current is indicated. The inset shows the Raman
scattering geometry.

The actual temperature distribution can be studied as well by micro Raman spectroscopy. Both the vibrational frequency and the intensity ratio of Stokes and anti-Stokes scattering of the TO phonons of GaAs are temperature sensitive quantities, especially at room temperature and above [2,3]. The microscope set-up allows temperature mapping of the mirror planes around the active region. The temperature distribution of a cleaved GaAs GRINSCH laser is shown in Fig.2 [3]. For current densities below the threshold no temperature increase is detected. Above threshold a "hot spot" is observed at the position of the active area. After degradation the laser shows a much stronger mirror heating even at much lower laser power density. The observed temperature profiles reflect the near field pattern of the laser.

We have also studied the influence of mirror coatings on the temperature distribution. The heating of the "hot spot" is drastically different for different mirror technologies as shown in Fig. 3, where the temperature increase is plotted versus laser power for four different lasers. The lowest temperature rise was observed with an anti-reflection coating, which also yields the highest laser output power. Raman spectra of degraded lasers exhibit in addition a new, so-called disorder activated mode indicating a strong crystallographic damage of the mirror.

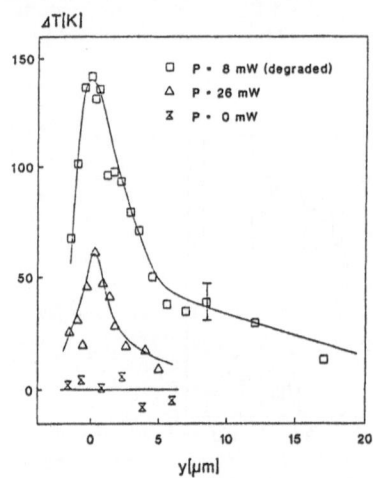

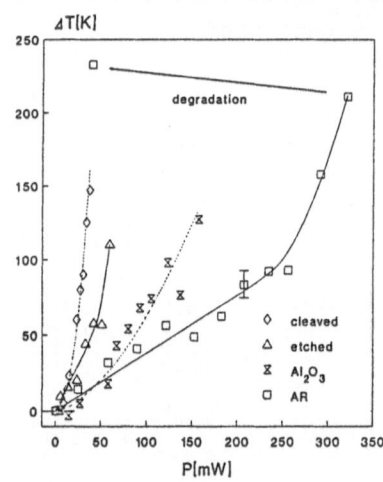

Fig. 2. Temperature profile in growth direction of cleaved mirror surface in dependance on power and laser degradation.

Fig. 3. Heating of laser mirror as a function of laser power in dependance on mirror properties.

Under resonant conditions, further information can be obtained from electronic light scattering [4,5]. Single particle
and plasmon excitations are sensitve to electron density, -tem-
perature, and -velocity distribution. As a first result we show
the strength of the single particle excitations across a GaAs
MQW laser without and with laser operation (Fig. 4).In case of
laser operation the maximum signal of the injected carriers is
shifted towards the active layer. From a lineshape analysis we
can deduce an increased electron density and an increased
electron temperature in the quantum wells during laser operation. Further work, which allows a quantitative separation of
the lattice- and electron temperatures and details on the spatial and velocity distribution of the carriers is in progress.

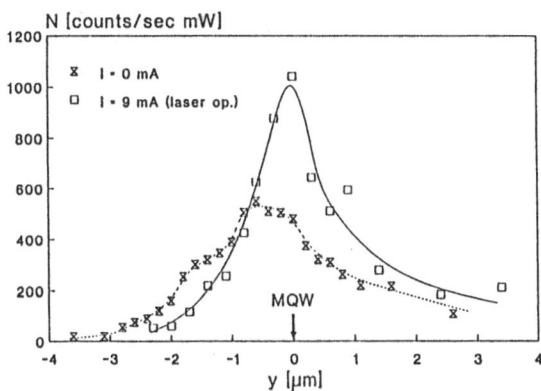

Fig. 4. Integral intensity of single particle excitation as a
function of probe laser position in dependance on laser operation. The position of the multi quantum well (MQW) is indicated.

References:

1. Schäffler, F.; Abstreiter, G.; Electric-field-induced Raman
    scattering: Resonance, temperature, and screening effects.
    Phys. Rev. B 34, 4017 (1986).
2. Compaan, A.; Trodahl, H.J.; Resonance Raman scattering in Si
    at elevated temperatures. Phys. Rev. B 29, 793 (1984).
3. Brugger, H.; Epperlein, P.W.; Mapping of local temperatures
    on mirrors of GRIN-SCH GaAs quantum well laser diodes. Sub-
    mitted for publication.
4. Mooradian, A.; Light scattering from single-particle elec-
    tron excitations in semiconductors. Phys. Rev. Letters 20,
    1102 (1968).
5. Cardona, M.; Güntherodt, G. (eds.); Light scattering in so-
    lids, 4th ed.; Topics in Appl. Phys.; vol. 54; p. 5;
    Springer Berlin, Heidelberg, New York, Tokio (1984)

Analogue and
High Voltage Devices

Optimizing the Epilayer Doping Concentration with Respect to Bipolar Transistor Performance in a Low-Power UHF-Process

J. Nagel

Philips GmbH, Röhren- und Halbleiterwerke
D-2000 Hamburg 54

Abstract
The performance of bipolar devices (NPN-transistors, lateral PNP-transistors and IIL-gates) in a low-power UHF-process at varying epilayer doping concentration is discussed on the basis of both two-dimensional device simulation and experimental data. While IIL-gates and lateral PNP-transistors would benefit by an increased doping, the observed strong decrease of the NPNs' forward Early Voltage Veaf imposes a severe restriction on a choice towards raised doping concentrations, since Veaf is an important electrical device parameter in linear analog bipolar processes. The presented results help to find a reasonable, application orientated trade-off between the various electrical device parameters.

Introduction

Many important parameters of bipolar transistors, such as cut-off frequency, forward Early voltage, breakdown voltages and current gain, as well as the performance of IIL-gates (propagation delay time, upward current gain) are mainly determined -or at least affected- by the choice of epilayer doping concentration. In order to find the optimum doping concentration with respect to bipolar transistor performance in a low-power UHF-process, a thorough investigation of NPN-transistors, lateral PNP-transistors and IIL-gates on the basis of both numerical device simulation and experimental data has been made. The devices were fabricated in Philips' standard V480 process, using a special component mask set. This article reviews and discusses the main dependencies between epitaxial parameters and resulting electrical device performance.

NPN-Transistors

Since in most circuit designs the vertical NPN-transistor plays the dominant role, the effects of a change in epilayer doping on the NPN-transistor performance has to be considered carefully. For each realized resistivity - ranging from ρ_{epi}=1.5Ωcm to 0.3Ωcm - all relevant dc- and ac-parameters of the standard NPN-transistor have been measured on-wafer. The cut-off frequencies, illustrated in **fig. 1**, show a pronounced increase with higher epilayer doping levels at collector currents exceeding 100µA; doubling the doping concentration N_D (i.e. a change from ρ_{epi}=1.5Ωcm to 0.7Ωcm) brings about a gain in maximum cut-off frequency f_{Tmax} of approx. 30%. Simultaneously the maximum is being shifted towards higher collector currents due to the retarded onset of

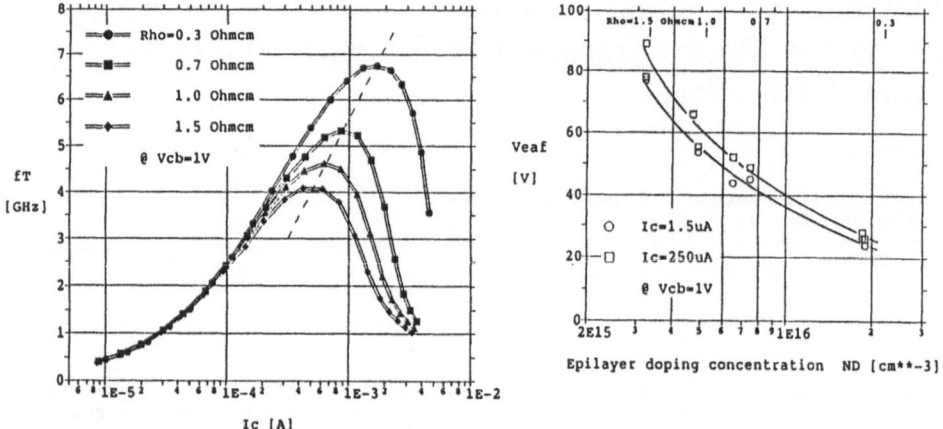

Fig. 1: F_T-Characteristics.   Fig. 2: Forward Early Voltages.

high injection. Besides this also the narrower base width and hence shorter
base transit time contributes to the observed increase of f_{Tmax}. The **forward
Early voltages** plotted in **fig. 2** for two collector current levels exhibit a
strong deterioration with increased doping concentration. If the doping is
changed from ρ_{epi}=1.5Ωcm to 1.0Ωcm then V_{eaf} drops from approx. 80V to 60V or
even to 45V with ρ_{epi}=0.7Ωcm. This can partly be explained by the observed
rise of the pinched base resistor, which amounts to about 10% in the range between 1.5Ωcm and 0.7Ωcm. Measurements of the **common-emitter current gain** h_{FE}
finally show an increase with epilayer doping of the order of 20% if ρ_{epi}=1.5
Ωcm is compared with 0.7Ωcm. The higher values for h_{FE} are accompanied with a
slightly improved linearity and with a shift of the falling slope in the
$h_{FE}(I_c)$-characteristic towards larger currents. The increase of both h_{FE} and
f_T with epilayer doping concentration correlates with the already mentioned
rise of the pinched base resistor.

Lateral PNP-Transistors

The two-dimensional device simulation package CURRY was employed to study the
dynamical behaviour of lateral PNP-transistors with different geometrical variations and epilayer doping concentrations. **Fig. 3** summarizes predicted and
measured results for the **maximum cut-off frequency** as a function of epilayer
doping concentration; the emitter area of the device is 8x8μm². The increase of
f_{Tmax} with decreasing base width w_B is clearly to be seen. To even further reduce the charge stored in the epitaxial base (which is mainly responsible for
f_{Tmax}), the emitter can be scaled down. As measurements show, a reduction from
A_{em}=8x8μm² to 7x7μm² brings about an increase of approx. 10%. The slope of the
$f_{Tmax}(N_D)$ curves can be explained as follows: A higher epilayer doping reduces

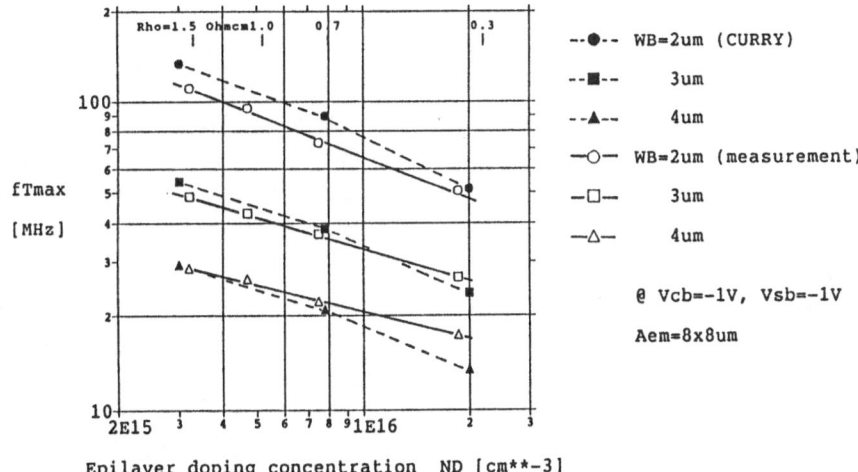

Fig. 3: Maximum cut-off frequencies of lateral PNP-transistors.

the outdiffusion of both the buried layer and the emitter/collector implant.
This in turn leads to an increased minority charge storage in an enlarged epitaxial base. CURRY simulations show that charge storage in the emitter-base
depletion capacitance at current levels under consideration only marginally
contributes to the total transit time and therefore cannot account for the observed slope. In **fig.4** two **breakdown mechanisms**, avalanche and punch-thru, can
be discerned by different dependencies of their respective breakdown voltages
on epilayer doping variation. For low doping concentrations and 2μm base width
punch-thru prevails, recognizable from an upward directed $BV_{ceo}(N_D)$-character-

Fig. 4: Breakdown voltages of lateral PNP-transistors.

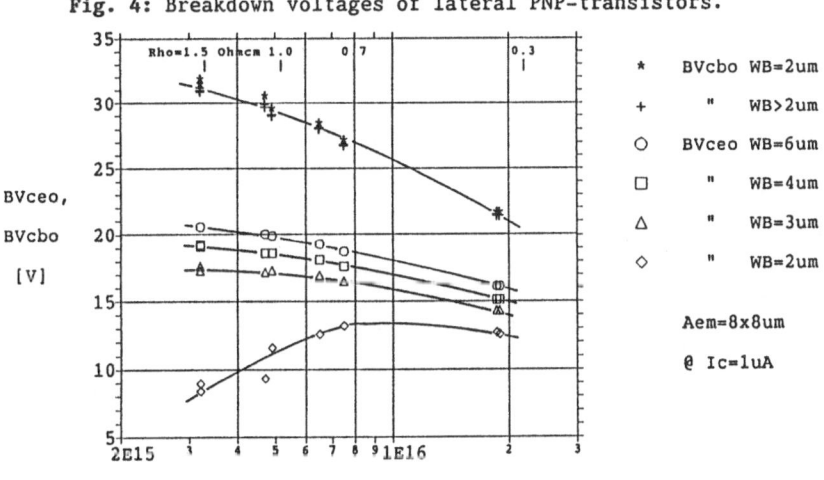

istic until a maximum is reached at approx. $8*10^{15}cm^{-3}$. At higher doping concentrations avalanche will occur before the collector-base depletion region can reach the emitter to cause punch-thru. **Fig. 4** therefore confirmes the known fact that an increase in epilayer doping concentration will improve the transistor blocking capabilities in regions - and only there - where punch-thru is the limiting factor; as long as avalanche prevails, an increase of N_D brings about a deterioration of breakdown voltages.

IIL-Gates

The effects of a change in epilayer doping on the static and dynamic performance of IIL-gates at two different current levels can be read from fig. 5.The size of all 4 collectors is 3x8µm², the injector distance being 10µm. The increase of both current gain and speed with doping level can be explained partly by a decreased base width, partly by an raised emitter majority charge.

Conclusions

The dynamic performance of all discussed bipolar devices improves with increased epilayer doping concentration (in the case of lateral PNPs a higher N_D allows for minimum base widths at $BV_{ceo}>10V$), the strong decrease of the NPNs' Early Voltage, however,imposes a severe restriction on a choice towards high doping concentrations, since V_{eaf} is one of the most important electrical parameters in analog bipolar processes. Depending on the specific application field for the considered UHF-process, a reasonable trade-off between Early Voltage and cut-off frequency has to be sought. Where compact circuit design and the premise of optimum analog performance dictate high Early Voltages, low doping levels will be chosen. Medium N_D will lead to an optimum with respect to the breakdown voltage and cut-off frequency of lateral PNP-transistors.

Fig. 5: Static and dynamic performance of IIL-gates (4th collector).

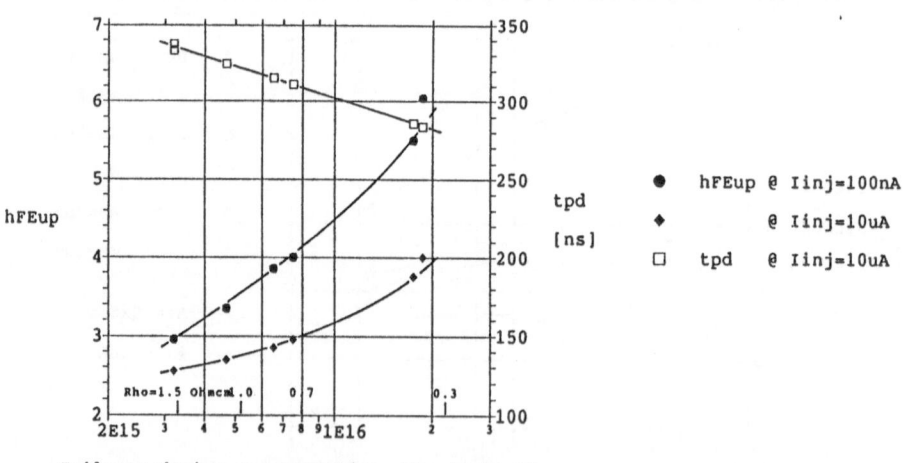

Epilayer doping concentration ND [cm**-3]

An 1 GHz All-implanted Vertical pnp Transistor

F.W. Ragay, A.A.I. Aarnink & J. Middelhoek

University of Twente, IC-Technology & Electronics group
P.O. Box 217, 7500 AE Enschede (NL)

Abstract
Vertical all implanted PNP transistors have been fabricated using high energy ion-implantation. The PNP transistor process can be implemented in standard NPN buried collector processes, with epitaxial layers larger than 2.5 μm, to achieve a high performance complementary bipolar process.
The collector is formed by implantation of doubly charged boron ions with an energy of 500 keV. Base and emitter regions are also implanted. Independent change of the base concentration is possible.
The base and collector currents are ideal over 5 decades. The current gain is $\cong$ 35 and constant over 4 decades. Cut off frequencies of the PNP transistors of over 1 GHz have been measured.

Introduction

The most important device for both digital and analog bipolar circuits is the NPN transistor. A PNP transistor is often also available on chip, but the process is optimized for the NPN transistor.

In general the electrical characteristics of the NPN transistor are superior to the characteristics of the PNP transistor. Not only the higher electron mobility is responsible for the better performance, but also the layout of the devices. Usually the NPN transistor is made vertical and the PNP transistor lateral. The lithography determines the minimum distance between emitter and collector and therefore the base width of a lateral transistor . Base widths of some microns can be achieved resulting in low current gain and cut off frequency. The current gain is generally a strong function of the biasing conditions and high injection effects show up at relatively low currents.

Vertical PNP processes [1] and complementary bipolar processes with vertical pnp transistors have been developed [2,3]. The buried collector is made by epitaxy or diffusion.

Use of a complementary bipolar process leads to simpler designs for analog circuits. The PNP transistor is not only used as a load but also for signal processing.

520

Device fabrication

Vertical PNP transistors are fabricated in phosphorus doped <100> CZ wafers. No epitaxial layer is used here. The transistors are made by ion implantation with an ion accelerator of High Voltage Engineering. Because of the low ion current, < 1 μA, photoresist can be used as masking layer. After each implantation, resulting in a junction, a damage anneal at 600 °C is performed. The electrical activation is achieved during the flow of the BPSG layer at 900 °C.

After the formation of LOCOS with 900 nm thick field oxide, the collector and collector contact regions are implanted. Doubly charged boron ions are implanted with an energy of 500 keV. The emitter area is formed by a high dose boron implantation of 40 keV. To deactivate the surface states at the silicondioxide-silicon interface the wafers are annealed in a NH_3 plasma at 350 °C [4].

A schematic cross section of the vertical PNP transistor is shown in figure 1. The solid lines represent the junctions and dashed lines additional implantations to contact the deep layers.

Figure 2 shows the Spreading Resistance Profile of the PNP transistor. The total depth of the transistor is 2 μm.

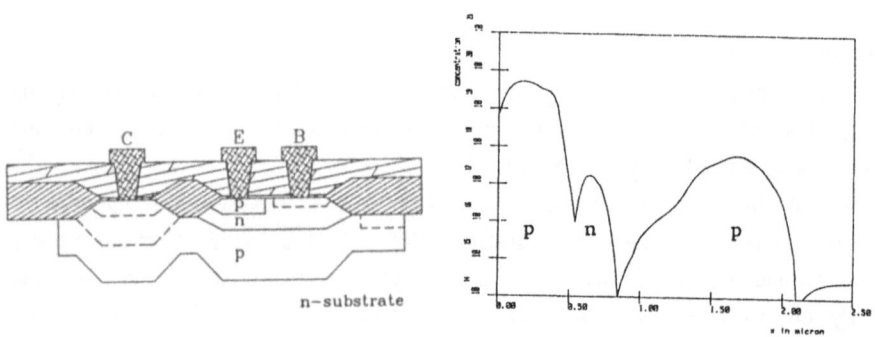

Fig.1. A schematic cross section of   Fig. 2. Spreading Resistance Profile
      the vertical PNP transistor

Electrical characteristics

The Gummel plot of a PNP transistor, figure 3, shows ideal collector and base currents over more than 5 decades. The current gain is not high, ≅ 35, but almost constant over 4 decades. Ion implantation provides easy changes of the doping profiles. Transistors with different base and collector concentration have been made. The collector current is in first order proportional to the base sheet resistance R_{sb} according to eq. (1) [5]. The measured data are shown in figure 4.

$$I_c = q^2 A_e \left[\overline{D_p \, \mu_n \, n_{ieff}^2} \right] R_{sb} \, e^{qVbe/kT} \tag{1}$$

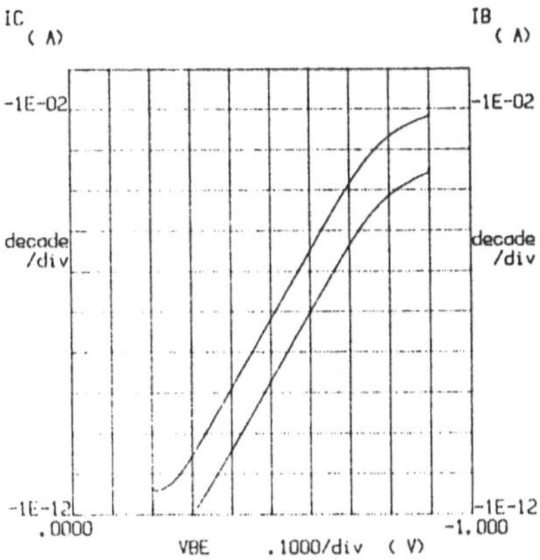

Fig. 3. Gummel plot

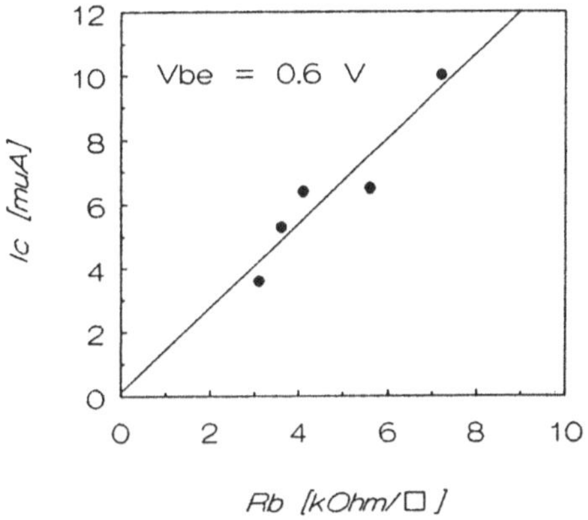

Fig. 4. I_c versus R_{sb}

Due to the implantation of the collector the process is simpler than conventional buried collector processes. However some device parameters suffer from the relatively high collector concentration at the base-collector junction. The Early voltage is only 15 V, however output characteristics can be improved by cascode circuits.

The breakdown voltage of the base-collector junction is 23 V, in active mode V_{bc} is however limited to about 10 V. Due to impact ionization in the bc-junction generated holes drift into the base. In a voltage driven device I_b decreases.

The cut off frequency F_t of the PNP transistors is measured on chip by determining the s-parameters. Figure 5 shows the F_t of two different transistors as a function of I_c for different bias conditions. Cut off frequencies of over 1 GHz have been measured. The transistors with a lower collector concentration have lower F_t due to a larger collector resistance. F_t is given by the eq. (2) [6] and shown in figure 6. The y-axis intercept $w_b^2/2D_p + r_c C_{bc}$ is consistent with data from DC measurements. The base transit time is estimated from device simulation.

$$\frac{1}{2\pi F_t} = \frac{(C_{eb} + C_{bc} + C_{cs})}{I_c / V_t} + \frac{w_b^2}{2 D_p} + r_c C_{bc} \tag{2}$$

	$w_b^2/2D_p$	DC : $r_c\,C_{bc}$	AC : $w_b^2/2D_p + r_c\,C_{bc}$
$R_c = 260\ \Omega$	30 ps	140 ps	170 ps
$R_c = 575\ \Omega$	30 ps	260 ps	240 ps

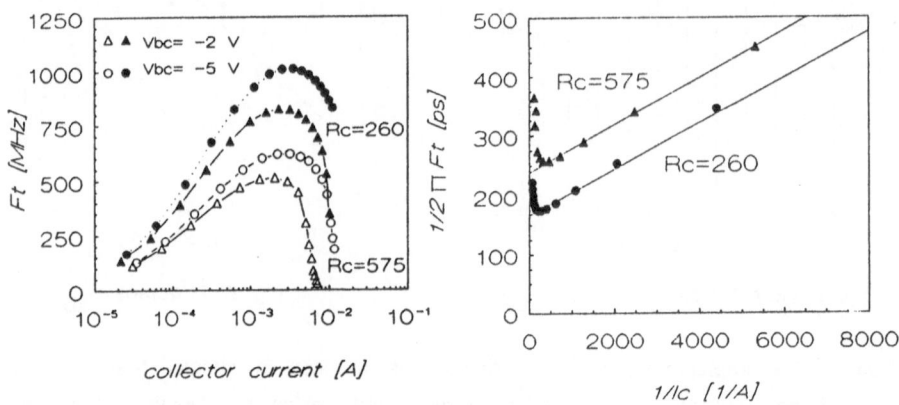

Fig. 5. F_t versus I_c

Fig 6. $1/F_t$ versus $1/I_c$

Conclusion

All implanted vertical PNP transistors have been realized using high energy ion implantation, on a wafer without epitaxial layer. The current gain is constant over 4 decades and can be easily tuned by the dose of the base implantation. Cut off frequencies of over 1 GHz have been measured. These PNP transistors can be implemented in a standard NPN buried collector process.

Acknowledgements
The authors like to thank Dr. van der Vlist (Philips Nijmegen) for the high frequency measurements, Dr. Vandervorst (IMEC Leuven, Belgium) for the SRP measurements and R.C.M. Wijburg, L.P.F. Warmerdam, and K. Lippe for their encouragement. This research is financially supported by the Foundation for Fundamental Research on Matter (FOM).

References
1. I. Magdo, IEEE Trans. Electr. Dev. ED-27 (1980) 1394,
2. T. Kikkawa et al., Technical Digest IEDM 1980 p. 65
3. M. Inoue et al., IEEE Trans. Electr. Dev. ED-34 (1987) 2146,
4. K. Aite et al., P.P. NO 35961
   K. Aite & F. W. Ragay, submitted for publication to Appl. Phys. Lett.
5. J.M.C. Stork et al., Technical Digest IEDM 1987, p. 405
6. S.M. Sze, Physics of semiconductor devices

High Voltage IC with Vertical Current Flow and Junction Isolation

R. Zambrano, G. Fallico and G. Ferla

SGS - Thomson Microelectronics
Stradale Primosole, 50
95121 Catania (Italy)

Operation in the high voltage, high current range (above 1 KVA) at reasonable costs requires junction isolated ICs with vertical current flow for the power stage.
In this paper VIPower M1 technology is presented: we first introduce the structure and its main features, then discuss both the power stage and the low voltage components main performances.

The ever increasing demand for low cost, compact, power drive systems with high reliability and full control and protection features, has concentrated research in the semiconductor field into producing intelligent devices on a single silicon chip, that is, smart power.

It has been demonstrated that the most effective technologies in this field use junction isolation. Vertical current flow for the power stage allows the best exploitation of Si area, since it maximizes the current density. The power stage must be built on lightly doped, n-type layers, so that it is possible to integrate either a single transistor with an insulated collector (drain), or many transistors sharing a common collector (drain.)

For the above reasons, we call these technologies VIPower, i. e. Vertical Intelligent Power. In this paper we present the structure and the main characteristics of both low and high voltage bipolar components of VIPower M1 technology.

VIPower M1 (figure 1) uses a totally bipolar manufacturing process with 9 basics masks and several options.

After the first, thick, epitaxial layer growth, a p-type buried layer is used to insulate the low voltage components from each other and from the power part (while a p-type substrate is usually used for the standard ICs.) The p-type buried layer also serves as the base of the power transistor (different doping levels can be set for the two regions, if needed.) On the other hand, the same n-type buried layer is used both in the control part to provide the low resistance collector regions, and in the power stage, as emitter of the power transistor. After the second epilayer growth, isolation and collector sink regions are diffused, and eventually the low voltage npn transistors base and emitter regions are formed.

Due to the flexibility of this approach it is possible to utilize almost all known LV bipolar processes. As a matter of fact, both the thickness of the second epilayer and the heat cycles needed to make the LV components do not affect significantly neither the isolation structure nor the power stage.

The concentration of the two epilayers is set mainly in accordance with the power stage voltage rating. The resistivity of the second epilayer can anyway be adjusted with a selective phosphorous implant in order to mantain compatibility with the choosen LV process.

The electrical equivalent of the isolation structure is shown in figure 2. The breakdown between the two buried layers sets an upper limit for the operation voltage of the control circuit, while there are no significant differences between the power stage and the isolation to substrate breakdowns.

The main parasitic transistors which have to be controlled are shown in figure 3: their DC current gain has to be as low as possible, so that no SCR latching can occur. A summary of the main techniques used can be found in [1].

The DC characteristics of a 25.8 sq.mm power stage are displayed in figures 4 and 5. Proper termination techniques are used in order to have breakdown (BVces) voltages as high as 1200 Volts, while the output characteristics in the saturation region show collector currents in excess of 15 Amps, thus allowing operation in the 10 KVA range. The AC performances (storage and fall times) are plotted as a function of collector current in figure 6.

Minimum layout rules of 4 μm are used for the LV components.

The doping profiles of the LV components have been designed in order to have open base breakdown of the npn transistor in excess of 28 Volts: the junction depths for the emitter and the base are 2 and 4 μm, respectively, while the junction between the two buried layers lies at 15 μm.

The DC current gain vs. collector current is shown in figure 7 for three LV npn transistors with different emitter areas.

Both the DC current gain and the collector to emitter (BVces) breakdown of lateral LV pnp transistors depend upon their basewidth. Figures 8 and 9 show, respectively, how the inverse of gain and the square root of breakdown are linearly dependant from the basewidth (as defined on the mask.)

Working devices have already been made using VIPower M1 technology. A layout of the device which optimizes the grounding network and the power stage geometry allows to drive a DC motor, switching 10 Amps at 600 Volts, with dV/dt as high as 1000 Volts/μsec.

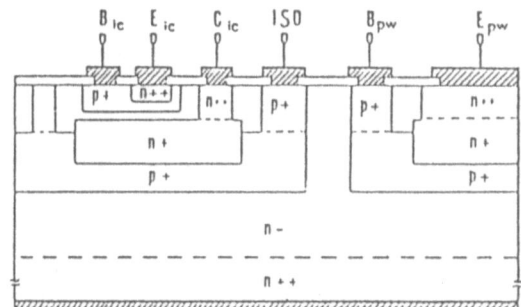

Figure 1: VIPower M1 structure

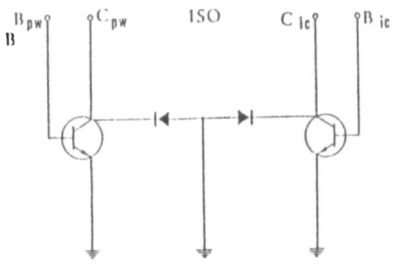

Figure 2: Electrical equivalent of the isolation structure

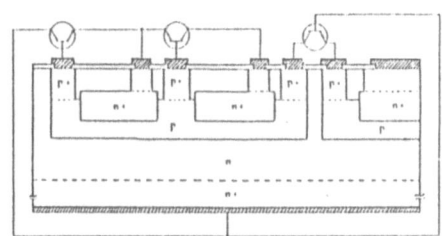

Figure 3: Parasitic transistors in VIPower M1 structure

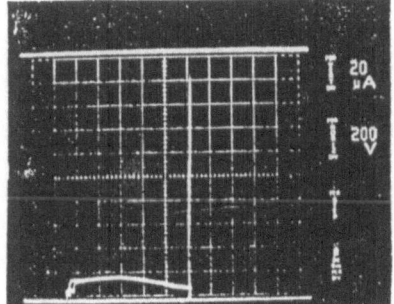

Figure 4: Power stage breakdown

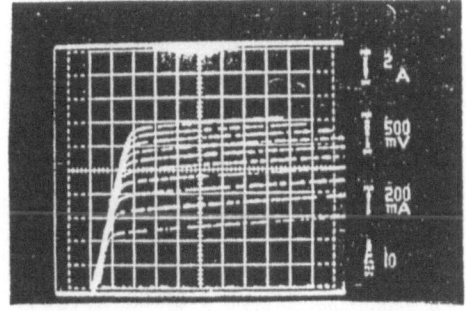

Figure 5: Power stage output characteristics

526

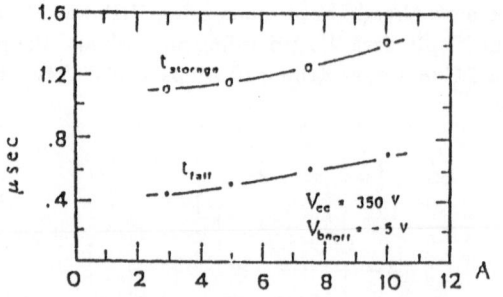

Figure 6: Power stage switching performances

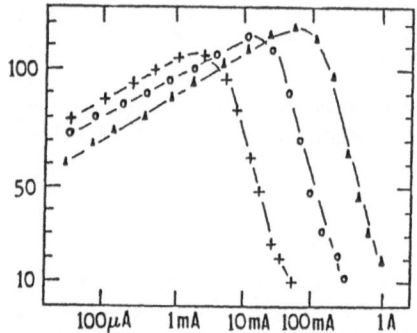

Figure 7: LV npn transistors DC current gain

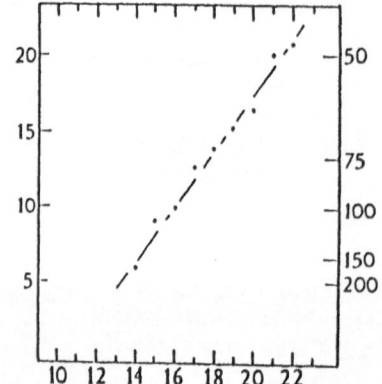

Figure 8: Inverse of DC current gain of
pnp transistors vs. basewidth (μm)

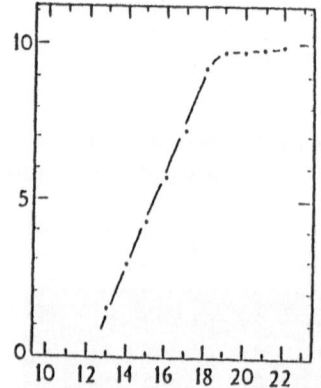

Figure 9: BVces square root ($V^{1/2}$) of
pnp transistors vs. basewidth (μm)

Reference

[1] R. Zambrano, G. Ferla, S. Musumeci and M. Paparo
Isolation Techniques in Power ICs with Vertical Current Flow
17th European Solid State Device Research Conference
Bologna, September 1987

Characterisation and Performance of Short-channel Polysilicon Thin-film Transistors

Alan Lewis, I-Wei Wu, Tiao Huang, Richard Bruce and Anne Chiang

Electronic and Imaging Laboratory, Xerox Palo Alto Research Center
3333 Coyote Hill Rd.
Palo Alto. CA94304 USA

Abstract

The influence of a range of device parameters on short-channel threshold shifts in n- and p-channel polysilicon TFTs is reported. Reducing the gate dielectric thickness reduces the shifts at low drain bias, but is less effective at high drain bias where channel avalanche multiplication is significant. The polysilicon island thickness has little influence, but passivation of the grain boundaries by hydrogenation does improve the short channel performance.

Introduction

The fabrication of thin-film transistors (TFTs) on large area substrates offers the possibility of integrating addressing or multiplexing circuitry onto, for example, flat panel displays or page-width scan bars. In this respect, polysilicon TFTs are particularly attractive due to their relatively high drive currents and their ability to form CMOS circuits. Reducing the geometry of polysilicon TFTs will clearly allow improvements in performance and packing density; however, short-channel effects in these devices have been shown to be very severe [1], and present a major problem to such scaling. In this paper, we examine both appropriate characterisation techniques for small geometry TFTs, and the effect of device parameters such as gate dielectric and island thickness on short-channel effects.

The effect of gate dielectric thickness

Figs. 1 and 2 show the threshold voltages of n- and p-channel polysilicon TFTs as a function of gate length for devices with 1000Å and 500Å gate oxide thicknesses. Here, the thresholds are defined at a fixed drain current scaled by the device geometry (that is at a drain current of $I_{DN}.W/L$ where I_{DN} is fixed for all devices). This definition is used since the strong dependence of effective channel mobility on gate bias makes an unambiguous extrapolated threshold (such as is used with single crystal transistors) very difficult to define [1]. The strong dependence of threshold on gate length is clear, especially at high drain bias. Fig. 3 shows

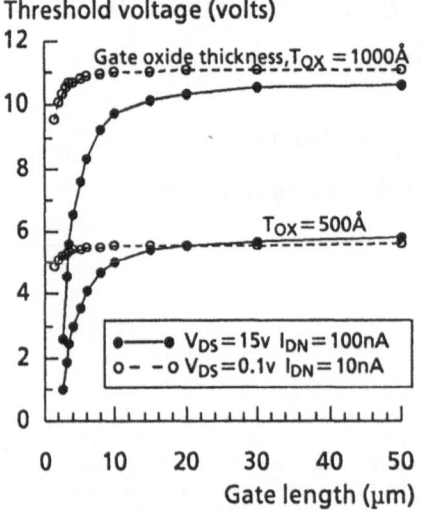

Fig.1 Unhydrogenated n-channel TFT thresholds. Polysilicon island thickness = 1000Å.

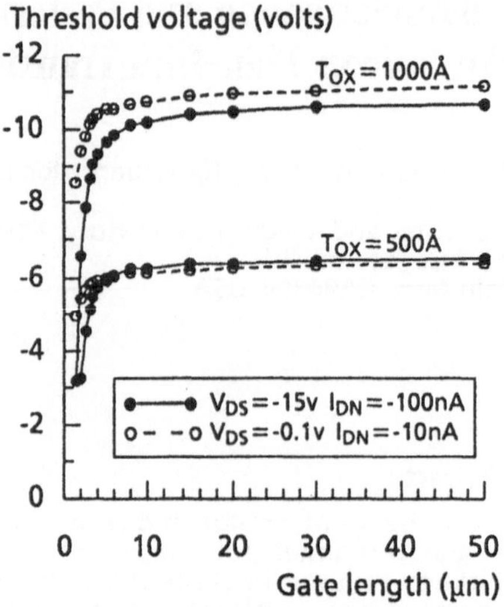

Fig.2 Unhydrogenated p-channel TFT thresholds. Polysilicon island thickness = 1000Å.

subthreshold characteristics of the TFTs; the roughly parallel shift in the curves with gate length indicates that the results shown in figs. 1 and 2 represent true threshold shifts. A small increase in the sub-threshold slope can be seen for the shorter n-channel TFTs, but this is due to channel avalanche effects [2].

The results plotted in figs. 1 and 2 also show that reducing the gate oxide thickness reduces both the long-channel threshold voltage, as expected, and also improves the short-channel threshold shifts. This latter effect is characteristic of threshold shifts caused by charge supported in the channel by the source and drain regions. Such a model also predicts threshold shifts which are inversely proportional to channel length. Fig. 4 shows that such a relationship holds well at low drain bias. At high drain bias, the curves are less linear and there is also a lower proportionate difference between the two gate dielectric thicknesses. The mechanism responsible for the threshold shifts here is channel avalanche multiplication; electron-hole pairs are generated by impact ionization in the high field region near the drain, and while the electrons are gathered by the drain, the holes charge the floating device island, lowering the threshold voltage. P-channel devices are less susceptible to this effect due to the lower ionization rates for holes. The thin gate oxide TFTs might be expected to be worse in this respect due

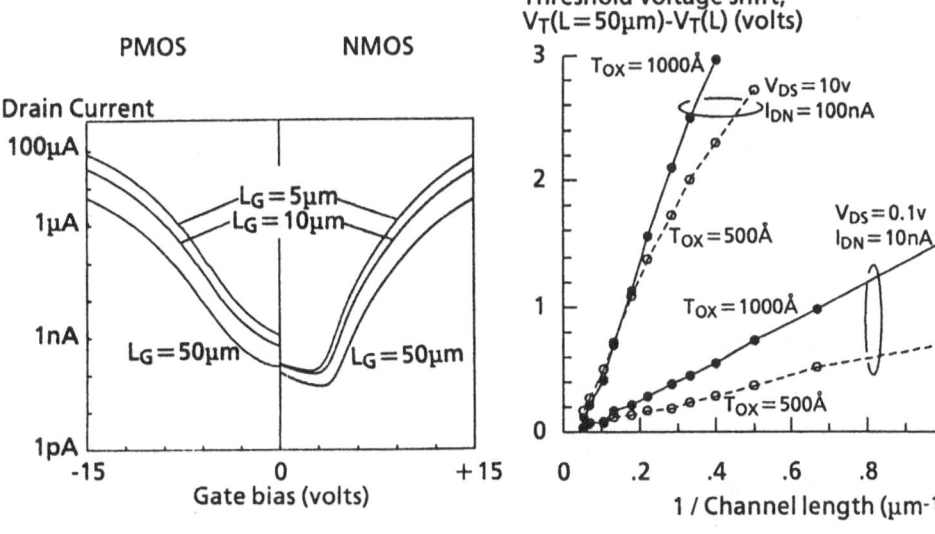

Fig.3 Unhydrogenated n- and p-
channel TFT sub-threshold
characteristics. V$_{DS}$=10v.

Fig.4 Unhydrogenated n-channel TFT
threshold shifts. Polysilicon
island thickness=1000Å.

to the higher drain field in saturation, but the reduced gate oxide thickness also lowers the sensitivity to island charging, so there is relatively little change with gate dielectric.

The effect of polysilicon island thickness

Fig. 5 shows the influence of island thickness on p-channel thresholds. The shifts are about the same for each thickness at both low and high drain bias. This is in contrast to single crystal silicon-on-insulator transistors, where reduced island thickness reduces avalanche effects and short-channel phenomena [3], and may be accounted for by the high density of trapped charge.

The effect of hydrogenation

Finally, the threshold voltages of n-channel TFTs after grain boundary passivation by hydrogenation are shown in fig. 6.   As expected, the long-channel thresholds are greatly reduced. The short-channel shifts are also improved at both low and high drain bias. At high drain bias, the increased drain field expected due to the higher drain to gate bias seems to be more than compensated by the reduction in trapped charge in the channel. The thresholds are also almost independent of

530

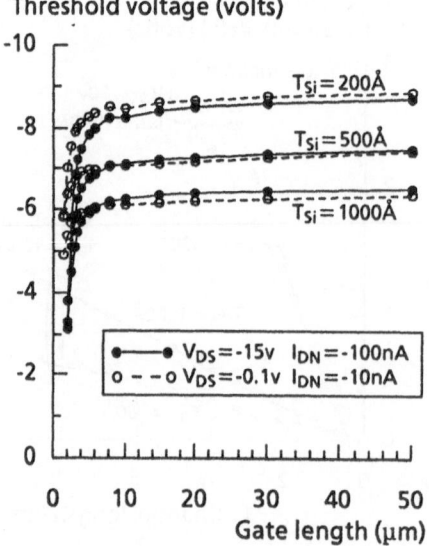

Fig.5 Unhydrogenated p-channel thresholds. Gate oxide thickness = 500Å.

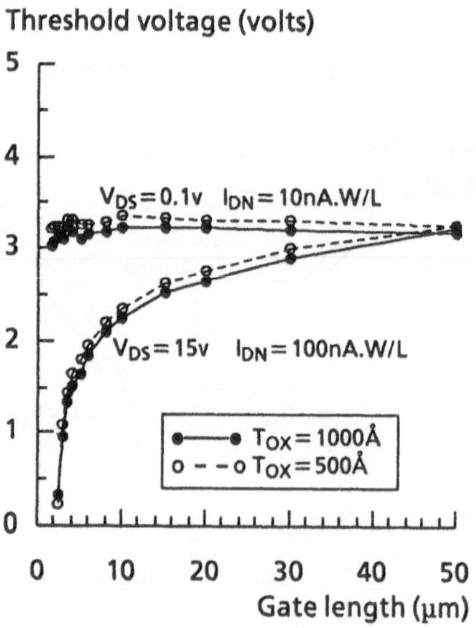

Fig.6 Hydrogenated n-channel TFT thresholds. Polysilicon island thickness = 1000Å

gate oxide thickness, which is characteristic of a fully depleted, lightly doped island.

Conclusion

This paper demonstrates that short-channel effects in polysilicon TFTs are generally similar in origin to equivalent phenomena observed in single crystal devices. However, they are much more severe and their alleviation by adjusting such device parameters as island thickness is more difficult.

[1] A.G.Lewis et al. 1988 IEDM Technical Digest, pp260-263.

[2] J.R.Davis et al. IEE Electron Device Letters, vol EDL-7, no.10, 570-572, 1986.

[3] J.-P.Collinge. Electronics Letters, vol.22, pp187-188, 1986.

Analysis of the Specific ON-resistance in Conventional VDMOS Transistors

J. FERNANDEZ, J. PAREDES, S. HIDALGO, C. DOMINGUEZ, F. BERTA, J. REBOLLO
and J. MILLAN.
Centro Nacional de Microelectrónica (CNM), CSIC-UAB, 08193 Bellaterra.
Barcelona. Spain.

Abstract

A distributed modellization of the specific ON-resistance is experimentally corroborated for VDMOS transistors with different cellular designs. The results for low voltage capability devices show differences for the optimum percentage of the cell area outside the body at which the minimum of the specific ON-resistance occurs and for this minimum value compared with previous formulations. In addition, a new VDMOS design with a waved gate electrode is reported, which seems to be attractive for medium voltage applications.

Specific ON-resistance

VDMOS transistors have been widely used as discrete power devices as well as drivers in Power Integrated Circuits [1]. One of the most important aspects in this field is the device optimization in order to minimize its specific ON-resistance, $R_{ON}xS$. In this sense, the use of cellular designs for low voltage capability is recommended in order to increase the effective channel width per unit area and, therefore, decrease the channel resistance component [2,3].

Up to the present, an ON-resistance modellization in terms of four components [3] has been used to obtain the optimum cell spacing which minimizes $R_{ON}xS$. However, in order to take into account the direction change of the current flow in the epitaxial layer, the two-dimensional character of the access resistance below the gate electrode has been considered by means of a distributed network. This fact alters both the minimum value of $R_{ON}xS$ and the cell spacing at which this minimum occurs, and has been already confirmed experimentally in the case of interdigitated designs [4].

Several cellular VDMOS structures with different layouts have been integrated on the same wafer in order to corroborate this distributed formulation. These devices, with a breakdown voltage in the range of 80 V, have been fabricated using a conventional process and with different geometrical dimensions for the elementary device cell; i.e., body width and cell spacing. Some of the results obtained are plotted on Fig. 1, showing differences between both modellizations for the optimum percentage of the cell area outside the body at which the minimum of $R_{ON}xS$ occurs (around 25%) and for the $R_{ON}xS$ minimum value (around 9%). Moreover, the experimental results agree with the distributed model corroborating its validity.

Waved gate VDMOS device

In addition, a new VDMOS design with a waved gate electrode (its layout is pictured in Fig. 2) is reported. In this design, given a body width, the cell spacing is limited to a maximum value above which the neighbouring gates overlap and the design becomes cellular. The analytical $R_{ON}xS$ distributed modellization for this design also agrees with the experimental results from the fabricated structures, as shown in Fig. 3.

The results obtained from this modellization show that for medium and high voltage capability devices ($V_{BR} > 200$ V), this design exhibits a minimum $R_{ON}xS$ comparable with that of the cellular designs (Fig. 4). Besides, note that for a given body width the optimum percentage of the active area outside the body is greater in the case of the waved gate design (provided it can be fabricated) than that in the cellular VDMOS transistors, thus indicating a lower current limitation due to the quasi-saturation effect for this new design. This fact is shown in Fig.5 where the current density at which the device enters the quasi-saturation regime [5] is plotted versus the breakdown voltage. Consequently, this new waved gate VDMOS design could be interesting within view of the constrains that both the optimization of the specific ON-resistance and the quasi-saturation effect impose on the cell spacing.

Conclusions

In this work, a distributed formulation of the specific ON-resistance is experimentally corroborated for cellular VDMOS transistors. This modellization differs from the previous ones in the $R_{ON}xS$ minimum value and in the optimum percentage of the cell area outside the body. These differences are around 9% and 25%, respectively, for devices with a breakdown voltage of 80 V.

Besides, a new waved gate VDMOS design is considered. Its experimental $R_{ON}xS$ also agrees with the distributed model. This design seems to be attractive for medium voltage applications bearing in mind the constraints that both the minimization of $R_{ON}xS$ and the quasi-saturation effect impose on the cell spacing.

References

1. Baliga, B.J.:Power Integrated Circuits - A Brief Overview IEEE Trans. Electron Dev., ED-33, 1986, 1936-1939.

2. Ilower, P.L.and Geisler, M.J.: Comparison of various source-gate geometries for power MOSFET's IEEE Trans. Electron Dev., ED-28, 1981, 1098-1101.

3. Hu, C.; Chi, M-H. and Patel, V.M.: Optimum Design of Power MOSFET's IEEE Trans. Electron Dev., ED-31, 1984, 1693-1700.

4. Fernández, J.; Hidalgo, S.; Paredes, J.; Berta, F.; Rebollo, J.; Millán, J. and Serra-Mestres, F.: An ON-Resistance closed form for VDMOS Devices. IEEE Electron Dev. Lett., EDL-10, 1989, 212-215.

5. Darwish, M.N.: Study of the Quasi-Saturation effect in VDMOS transistors IEEE Trans. Electron Dev., ED-33, 1986, 1710-1716.

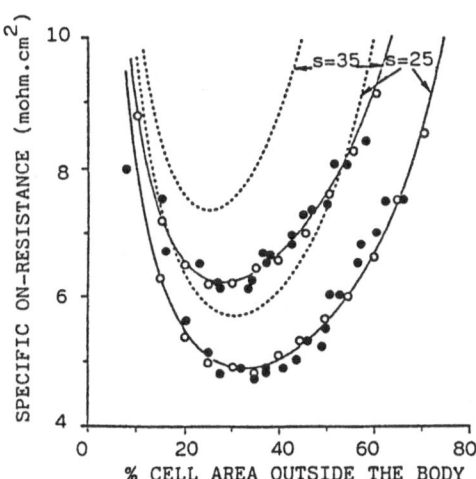

Fig.1. $R_{ON}xS$ versus the percentage of the cell area outside the body for cellular designs.
───── distributed model. ------- four component model
● experimental points. ○ two-dimensional simulations
s: body width

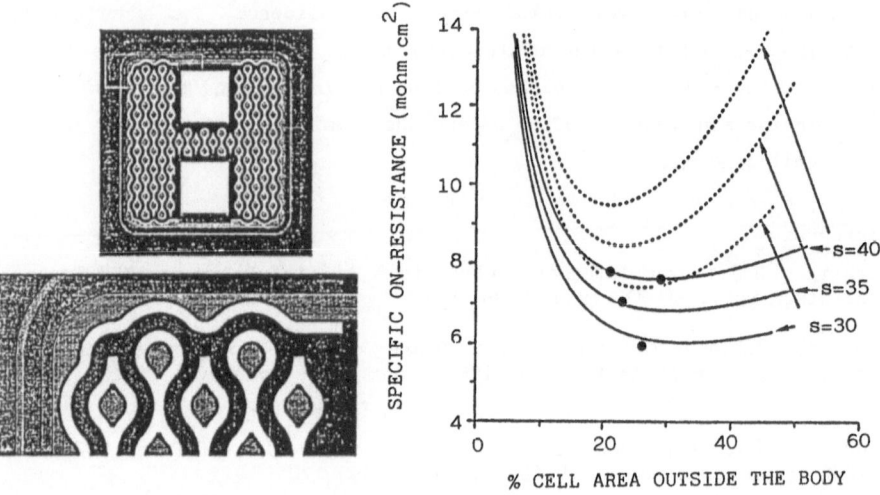

Fig.2. Layout of the waved gate VDMOS transistor.

Fig.3. $R_{ON}xS$ versus the percentage of the cell area outside the body for the waved gate design.
———— distributed model. - - - - four component model. ● experimental points.

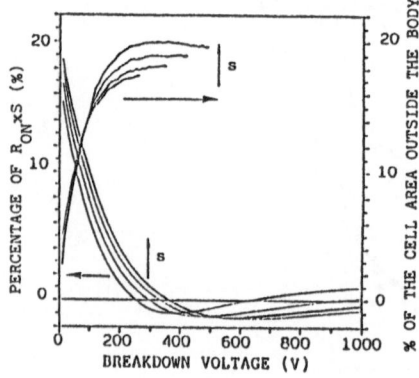

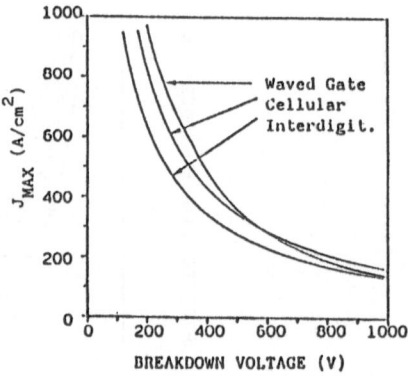

Fig.4. Percentage of the minimum of $R_{ON}xS$ for the waved gate design in respect to the cellular, versus V_{BR} with s as a parameter s=30,35,40,45 um. The percentage of the cell area outside the body for the minimum of $R_{ON}xS$ is also indicated.

Fig.5. Current density at which the device with minimum $R_{ON}xS$ enters the quasi-saturation regime versus V_{BR}, for interdigitated, cellular and waved gate VDMOS transistors. s=35 um.

CMOS-compatible High-voltage Complementary LDMOS Devices

M.R.Duncan, J.M.Robertson, R.J.Holwill and R.Rodrigues † .

EMF, Dept. of Electrical Engineering,   † Seagate Microelectronics Ltd.,
Edinburgh University,   MacIntosh Road,
The King's Buildings,   Kirkton Campus,
Edinburgh,   Livingston,
EH9 3JL.   EH54 7BW.

ABSTRACT

The design and implementation of a CMOS-compatible high-voltage process is described. It is shown that small changes can be made in an established n-well process to produce both high-voltage p- and n- channel power LDMOS transistors. These changes do not affect the performance of the low-voltage devices, and result in breakdown voltages of 50 volts for the p-channel, and 120 volts for the n-channel power transistors.

1. INTRODUCTION

Many circuit applications such as dc-dc converters, motor controllers and automotive switches need an interface between high-voltage devices and low voltage control circuitry. If a power device can be integrated with a control circuit on a single die, there are substantial savings in performance, at lower costs than when logic devices and power ICs are packaged separately and then connected. These savings come from eliminating the many packages to house individual chips, abolishing the interface circuits between power ICs and control ICs and shrinking the overall system size.

The usual approach to producing so-called 'smart power' chips, is to take an optimised power process and simply add it to an optimised logic process [1,2]. This method gives the best overall performance, but the resulting complex process is only suitable for high-volume applications. If a smart power chip is to be used for ASIC applications then a less-expensive compromise approach is necessary.

In this paper, high-voltage transistors fabricated within a CMOS process are presented. These have been realised by analysing the CMOS and LDMOS processes and finding the common processing steps that will lead to a smart power solution with a minimum mask set.

A benefit of this approach is that the low voltage transistors are not adversely affected by the compromises. This means that the existing logic cell libraries can be used with the new high-voltage transistors to create smart power designs.

2. PROCESS DESIGN

This work was based on the EMF's scaleable 5 μm CMOS/NMOS process. This flexible process can be scaled down to 1.5 μm design rules to enable high packing density of the logic elements. The design rules were kept at 5 μm to be compatible with existing designs. The devices are fabricated on 14 - 20 Ω p-type (100) silicon substrate. Isolation between the devices is provided by a 1.5 μm LOCOS field oxide. The process uses 850 Å gate oxide and n+ polysilicon gates.

One additional masking step is required to produce the n-drift region implant for the p-channel LDMOS transistors. It was also necessary to alter the doping profile of the n-well. This well has to be driven deeper, to 8 μm, to prevent vertical punchthrough when the drain voltages of the p-channel power transistors are increased. The well also has to keep the original surface concentration so that the logic elements are not adversely affected. These goals were achieved by increasing the implant dose and the well drive-in time.

3. DEVICE DESIGN

High-voltage operation within a standard process can be achieved by circuit techniques [3] or by device techniques [4]. The approach here uses device techniques to achieve high-voltage operation. In order to support high voltages the basic MOSFET structure must be altered. The power device chosen is the LDMOS, where the highly doped drain diffusion is separated from the channel by a lightly doped drift region. This region serves to reduce the electric field at the gate-drain interface resulting in higher breakdown voltages.

Cross sections of the n- and p-channel transistors are shown in figures 1 and 2. The n-well is used conventionally in both the low and high-voltage p-channel transistors, but it is also used to form the lightly doped drift region in the high-voltage n-channel devices. In the p-channel LDMOS, the n- drift region is formed by the extra masking step discussed earlier.

For this type of process there is a trade-off between breakdown voltage and on-resistance. The optimum doses for the drift regions were found using SUPREM3 and SUPRA in conjunction with the device simulator ,CANDE.

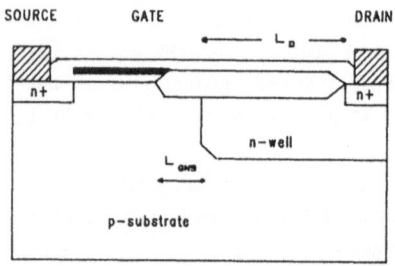

Figure 1. N-channel LDMOS Transistor

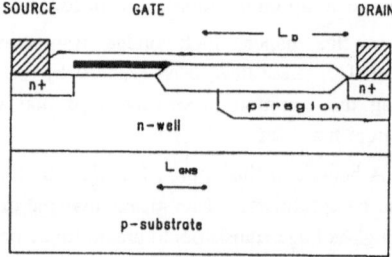

Figure 2. P-channel LDMOS Transistor

4. RESULTS

A set of n- and p-channel transistors were fabricated with different geometrical parameters. The drift region length L_D, and the gate - well separation L_{GNS} were the two parameters varied for both types of power transistors. The resulting breakdown values for the n- and p-channel transistors are given in tables 1 and 2 respectively. A set of n-channel transistors were also fabricated using the conventional n-well CMOS process with a 5 μm deep well. The results of this are shown in table 3.

N-Channel Transistors		
L_D (μm)	L_{GNS} (μm)	BV_{DSS} (Volts)
5.0	2.5	-
7.5	5.0	90
10.0	5.0	120
12.5	7.5	120
15.0	10.0	120

P-Channel Transistors		
L_D (μm)	L_{GNS} (μm)	BV_{DSS} (Volts)
5.0	2.5	-20
7.5	5.0	-40
10.0	5.0	-55
12.5	7.5	-55
15.0	10.0	-55

Table 1. N-channel Power Transistor Breakdown Voltages

Table 2. P-channel Power Transistor Breakdown Voltages

N-Channel Transistors		
L_D (μm)	L_{GNS} (μm)	BV_{DSS} (Volts)
5.0	2.5	20
7.5	5.0	120
10.0	5.0	125
12.5	7.5	125
15.0	10.0	125

Table 3. Standard Well N-channel Power Transistor Breakdown Voltages

For the n-channel power transistors it can be seen that above critical values of L_D and L_{GNS} the breakdown voltage is constant. These values of L_D and L_{GNS} are the drawn mask dimensions, not the actual dimensions in silicon. When the actual dopings are such that n-well does not overlap the drain, and that the drift region is 10 μm (on the mask), then the maximum BV_{DSS} is obtained. Further increases in the drift region length result in an increase of on-resistance only.

Similarly for the p-channel transistors, the maximum breakdown voltage occurs when the p-drift region does not overlap the gate. This should happen for a shorter drift region than the n-channel LDMOSs, but the step size on the test chip was too coarse to detect this.

538

For the n-channel transistors fabricated within the standard CMOS process with the shallower n-well, the breakdown voltages are similar to those of the complementary power n-channel transistors. This demonstrates that the extra well depth does not positively affect performance and should only be used if both n- and p-channel power transistors are needed in the design.

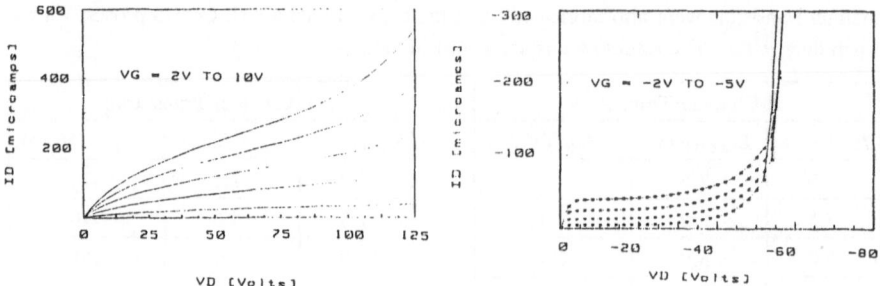

Figure 3. N-channel LDMOS Drain Characteristics

Figure 4. P-channel LDMOS Drain Characteristics

The drain current - voltage characteristics of the n- and p-channel transistors are shown in figures 3 and 4 respectively.

5. CONCLUSIONS

This paper has demonstrated that high-voltage elements can be fabricated within a CMOS process without compromising the logic devices. It has lead to a minimum mask count smart power process. Breakdown voltages of 120 volts and -50 volts for the n- and p-channel transistors respectively have been reported.

6. ACKNOWLEDGEMENTS

This work has been supported by the Science and Engineering Research Council of Great Britain. Additionally, M.Duncan would like to acknowledge the financial support of Seagate.

7. REFERENCES

[1] Power devices are in the chips, V.Rumennik, IEEE Spectrum, pp 42-48, July 1985.

[2] Multipower BCD - A versatile technology from 60 to 250V that integrates Bipolar, CMOS and DMOS devices, C.Contiero, A.Andreini and P.Galbiati, IEE Colloquium on Integrated Intelligent Power Devices, March 1987.

[3] High voltage circuits in standard CMOS processes, C.Petersen and A.R.Barlow , IEDM Tech. Dig., pp 77-80, 1982.

[4] Modelling and Characterization of CMOS-compatible High-Voltage Device Structures, Z.Parpia, C.A.T.Salama and R.Hadaway, IEEE Trans. Electron Devices, vol. ED-34, no.11, pp 2335-2343, November 1987.

Analysis of the Modifications Induced by Electron Irradiation on the Electrical Characteristics of High Power GTOs

P.G. Fuochi[1], B. Passerini[2], F. Fasce[3], M. Zambelli[4]

1) Istituto FRAE-CNR, Via de' Castagnoli 1, 40126 Bologna, Italy
2) International Rectifier Corporation Italiana S.p.A., C.so Novara 17, 10078 Venaria, Torino, Italy
3) Ansaldo Trasporti, Semiconductor Dept., Via N. Lorenzi 8, 16152 Genova, Italy

Summary
Electron irradiation has been used to modify the electrical characteristics and to enhance the switching speed of high power GTO thyristors. The defects introduced by the impinging electrons have been annealed through several thermal steps, carried out at three different temperatures, after which the main electrical characteristics have been tested. The trade-off between static and dynamic characteristics, obtained with this method of lifetime control, has been compared with that obtained in standard gold doped devices. Special attention is paid to the control of power losses during the turn-off phase.

Introduction

Electron irradiation has been widely used for many years to modify the switching characteristics of semiconductor power devices, like diodes and thyristors, by means of deep levels introduced into the silicon crystal [1]. In recent years this technique has been successfully applied to Gate Turn-off thyristors (GTO) [2], a relatively new power semiconductor device that offers the advantages of the conventional silicon controlled rectifiers (SCRs), namely high blocking voltage and surge current capabilities together with the ability to be switched off by the application of a gate control signal. Similarly to the Au-doping, electron irradiation has beneficial effects on the dynamic characteristics, while it affects adversely the static ones. Thus a trade-off between these characteristics which behave in opposite way must be reached. The aim of this work is to asses this trade-off in electron irradiated high power GTOs. A comparison of these characteristics with those of standard Au-doped GTOs is presented.

This work has been supported by the National Research Council of Italy within the Finalized Project "Materiali e dispositivi per l'Elettronica a stato solido".

Experimental Section

The devices used for this study were high power GTOs manufactured on 38 mm silicon wafers, alloyed on tungsten discs and alluminum metallized. Details of the fabrication process are reported elsewhere [3]. The resistivity of the starting material was 100 Ohm.cm leading to a blocking voltage in excess of 2000 V. The electron irradiation was performed on finished devices, at room temperature, using pulses of electrons from a 12 MeV linear accelerator. The accelerator was used in repetitive mode with pulses of 50 ns duration and repetition rate of 600 p.p.s. giving fluxes of a few 10^{11} e^-/cm^2 s at the sample position. The devices were irradiated with a dose of 2.3 10^{13} e^-/cm^2. Further details on the irradiation process and dosimetry have been reported elsewhere [2]. After irradiation the devices underwent a mixed isothermal isochronal annealing under inert atmosphere in a furnace controlled with an accuracy of +/- 1 °C. Several annealing time steps at 250, 300 and 360 °C were carried out on each sample. The device characteristics were measured after each annealing step. Temperature was increased whenever no further appreciable variation in the characteristics was observed.

Results and Discussion

In order to evaluate the effect of the electron irradiation on the behaviour of the devices, static characteristics (such as: threshold voltage (VTO), slope resistance (rT), trigger voltage (Vgt) and current (Igt), forward (VDRM) and reverse blocking voltage (VRRM)) and dynamic characteristics have been tested. As far as the dynamic characteristics are concerned we have focused our attention on the turn-off process and evaluated thoroughly the variation of the related parameters: storage time ts, fall time tf, peak reverse gate current IRGM and reverse gate recovery charge QRG. The characteristics are defined as reported in reference [3].

Static Characteristics

The effect of electron irradiation on the on state and trigger characteristics is an overall worsening. VTO and Igt show a large increase after irradiation, while a slight increase is observed for Vgt. All these parameters are decreasing functions of the annealing time and temperature. At the end of the annealing process the values prior irradiation were reached [2].

Dynamic Characteristics

IRGM, QRG and ts are increasing functions of annealing time and temperature [2]. This behaviour is related to the increase of the minority carrier lifetime [1]. As the annealing proceeds and ts and IRGM increase, the value of the anode current that can be switched off without endangering the device increases accordingly. The fall time tf shows no significant variation during all the annealing process. Nevertheless, the fall phase has turned out to last for a sufficient time to avoid spikes of the reapplied anode voltage. This means an intrinsically safe commutation since these spikes could cause an excessive instantaneous power dissipation leading to the failure of the device.

Au-doping versus B-irradiation

The comparison between the B-irradiated and the gold doped GTOs has been carried out considering the forward voltage drop characteristics and the turn-off parameters. The forward V-I characteristics of the two groups of devices are quite similar: the B-irradiated GTOs show a voltage drop slightly lower than the gold doped ones. This fact results in a higher current carrying capability during the conduction period for the B-irradiated GTOs. In table 1 are reported the values of the main turn-off parameters of the B-irradiated and the gold doped GTOs when switching off an anode current of 800 A at the maximum junction temperature with a reverse gate current slope of 40 A/µs. The turn-off time tgq is shorter for the B-irradiated device that are therefore showing higher frequency operation performances. Analizing the turn-off time it is remarkable that while the storage time is shorter for the B-irradiated devices, the fall time is shorter for the gold doped ones. The shorter the storage time, the higher the turn-off gain and, therefore, the gold doped devices present a better behaviour with respect to the maximum interruptible current. On the other hand the
shorter the fall time, the higher the reapplied voltage spike during this pediod. As a consequence the gold doped GTO presents a higher peak of power loss during tf. The tail current is higher for the B-irradiated GTOs. As far as the energy loss are concerned we can say that they are quite similar during the fall

DEVICE PARAMETER	Gold doped	β-Irradiated
ts [us]	5.3	5.02
tf [us]	0.54	0.64
tgq [us]	5.84	5.66
IRGM [A]	180	158
QRG [uC]	701	609
Vs [V]	396	352
ITtail [A]	24	28
Pmax [kW]	132	121
Pmax(tail)[kW]	13.6	17.9
Eoff(tf) [mJ]	51.7	55.4
Eoff(tail) [mJ]	30.9	48.1

Table 1. Comparison between some turn off parameters of the electron irradiated GTOs at the end of the annealing process, and the gold doped GTOs. Turn off conditions:
Tj   = 125 °C
ITC   = 800 A
diRG/dt = 40 A/μs
VD   = 800 V
Cs   = 4   μF

period, while during the tail phase the β-irradiated devices show a higher dissipation due to the higher tail current. The total turn-off energy loss is then lower for the gold doped GTO and, as a consequence, this device shows a better behaviour with respect to the frequency operation.

Conclusions

The results presented in this paper show that the same performances are obtained by controlling the dynamic characteristics of GTOs either with gold-doping or with β-irradiation, but this latter technique has the following unique advantages:
1) a better control of the static and dynamic characteristics,
2) the possibility of annealing the defects created within the crystal,
3) the enhanced maximum junction temperature.

References
1. Fuochi, P.G.; Di Marco, P.G.; Bisio, G.M.; Di Zitti, E.; Passerini, B.; Tenconi, S.; Zambelli, M.: Alta Frequenza 55 (1986) 47-55 and references therein.
2. Fasce, F.; Fuochi, P.G.; Mauri, A.; Passerini, B.; Tenconi, S.; Zambelli, M.: Proc. 2nd PEVSD Conference, Birmingham (U.K.) (1986) 10-14.
3. Fasce, F.; Zambelli, M.: Proc. 1st EPE Conference, Brussels (B) (1985).

Session 6

Technology (Miscellaneous)

Influence of Top Oxide Thickness on Electrical Properties of ONO Stacked Insulators

H. Reisinger and A. Spitzer
Siemens AG, Otto Hahn Ring 6, D-8000 München 83, Federal Republic of Germany

Summary
Electrical properties of stacked oxide-nitride-oxide insulators as a function of the thickness of the thermally oxidized top oxide and the oxidation agent have been investigated. Though currents through the films increase for a given electric field and decreasing top oxide thickness very thin top oxides are still effective as a barrier against tunnelling of charge carriers into the nitride. The intrinsic lifetime decreases with decreasing top oxide thickness but it is still above 10^{10}sec at 4MV/cm.

In silicon technology thin oxide-nitride-oxide (ONO) films are used as a dielectric in DRAM storage capacitors. For future DRAM generations the thickness of ONO layers has to be scaled down below 10nm equivalent oxide thickness. In this paper we studied the influence of the thermally oxidized top oxide thickness on the intrinsic electrical properties of ONO films.

Experimental

MIS capacitors were fabricated on p-Si substrate. A 6nm bottom oxide was grown in dry O_2 at 900°C and a 9nm LPCVD-nitride was deposited at 800°C. The nitride was then thermally oxidized in wet and dry ambients at 900°C except for the 3.6nm dry top oxide which was oxidized at 1000°C. The obtained top oxide thicknesses can be seen in Fig.1. As gate material n^+-poly-Si was used. All thicknesses were determined by ellipsometry and CV-measurements.

Results

Current densities at low fields for samples with different top oxides are given in Fig.1. These current densities are extremely low for fields below 5MV/cm but they could be measured conventionally using an electrometer and large gates with an area of several mm^2. The sensitivity achieved this way is actually one order of magnitude higher than the one measured in ref. [1] employing long-time measurement of flatband shifts.

For current densities below $10^{-8}A/cm^2$ the amount of charge trapped in the nitride during the time of measurement is negligible. Therefore the curves in Fig.1 are not affected by charge trapping. For positive gate voltages (V_g) and fields above 6MV/cm, the currents are dominated by Fowler-Nordheim tunnelling of electrons through the relatively thick bottom oxide into the

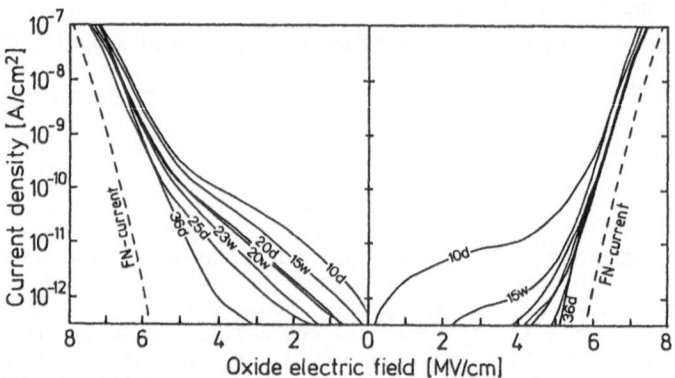

<u>Fig.1.:</u> Current densities as a function of the electric field in the oxide layers. The curves are labelled with the thickness of the top oxide in Angstroms; a "d" denotes dry, a "w" wet oxidation of the top oxide. The calculated Fowler-Nordheim current is plotted for comparison.

nitride. Below 5MV/cm a current increase with decreasing top oxide thickness can be observed. Flatband voltage shifts to negative values as a function of time have shown that the currents at these electric fields are due to hole tunnelling from the gate across the top oxide.

For negative gate voltages the electron tunnelling current depends strongly on the top oxide thickness. Even for the 1nm top oxide the current is orders of magnitude lower than for a nitride layer without a top oxide.

In Fig.2 some of the measured IV-curves from Fig.1 are compared with calculated currents. Electron tunnelling currents -from Si-conduction band across a SiO_2-barrier into states in the bandgap of the nitride- are calculated in

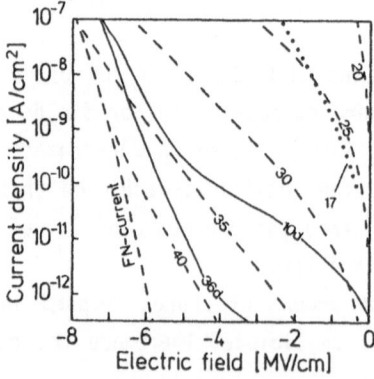

<u>Fig.2:</u> Comparison of measured current densities (full lines) with calculated current densities (broken lines). Calculated was the electron tunnelling current across a thin oxide barrier. The dotted line shows the current of an ONO layer with a deposited top oxide. The curves are labelled with the thickness of the top oxide in Angstroms.

WKB-approximation . All calculated currents in Fig.2 above the Fowler-Nordheim current result from direct tunnelling across the oxide. The energy levels of the final states were assumed to be 2eV below the conduction band edge of the nitride [2] but are not really known. Therefore the calculated curves in Fig.2 have to be understood as a more qualitative than quantitative result. The 4nm curve is insensitive to the energy of the final states and roughly agrees with the experimental data. For the thinner top oxides, especially the 1nm top oxide the calculation yields currents far too high. For comparison, the current density of an ONO layer with a deposited top oxide is drawn in Fig.2. It is closer to the currents expected by the calculation and shows a behaviour significantly different from the thermally grown top oxides. The conclusion is that the model of an oxide with a sharp boundary is not adequate for thermally oxidized top oxides on nitrides. Most likely the thermal oxidation produces a graded transition between oxide and nitride as well as changes of the trap density in the nitride . With respect to the scaling of ONO insulators this means that the top oxide can be much thinner than the bottom oxide in order to be an equally good barrier against tunnelling of carriers.

In Fig.3 intrinsic charge to breakdown values (Q_{bd}) are plotted together with steady-state high field currents. For the shown polarity electrons were injected from the substrate and the top oxide is at the anode. In contrast to the data in Fig.1 the currents in Fig.3 were measured with a saturation of trapped charge in the nitride.

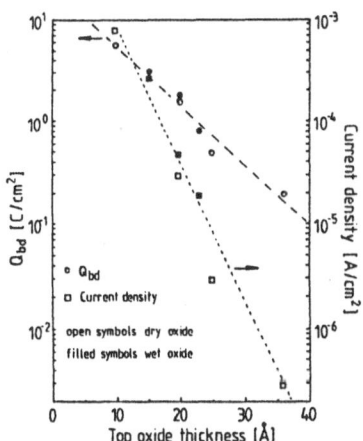

Fig.3: Values of charge to breakdown and steady state currents at an average electric field of 10MV/cm. Average electric field is calculated from the voltage drop across the ONO film and its equivalent oxide thickness.

There is a strong dependence of the current density on the top oxide thickness which can be explained in the following way: The large amount of negative trapped charge in the nitride causes the electric field in the oxide at the cathode to be lower than the field in the oxide at the anode. For a given amount of trapped charge in the nitride and a given gate voltage the low field at the cathode decreases when the oxide thickness at the anode is increased since the total potential drop across the film has to be constant. The field at the cathode controls the electron tunneling current. Therefore for the case of the top oxide being at the anode the current for a given average field (i.e. V_g divided by total film thickness) decreases with increasing top oxide thickness.

From the Q_{bd} dependence on top oxide thickness in Fig.3 it can be seen that a breakdown of the ONO films is associated with the breakdown of the oxide at the anode. The charge to breakdown values of this oxide increase with decreasing thickness similar to single layer oxides.

Using the data in Fig.3 time to breakdown (t_{bd}) was calculated as well as directly measured at 10MV/cm; t_{bd} increases from around $3*10^3$sec to around 10^5sec for top oxide thicknesses from 1nm to 3.6nm. Values calculated for a field of 4MV/cm assuming the Q_{bd}'s to be the same as for 10MV/cm give t_{bd}'s from 10^{10}sec to 10^{11}sec. This intrinsic lifetime is well above a time of $3*10^8$sec required as the lifetime for DRAMs. Since Q_{bd} is expected to increase with decreasing field [3] these values can be considered to be a lower bound.

Data for negative V_g are not shown in Fig.3. They are close to the data of the thickest top oxide for positive V_g. Since in this case the top oxide is at the cathode the influence of its thickness is - as one would expect - not drastic.

There is no major difference between wet and dry top oxides, except for a larger decrease of the gate voltage during the Q_{bd}-measurements (done at constant current) for the wet oxides. This effect is assumed to be due to an increased trapping of positive charge in the wet top oxides.

Our results give insight into charge transport properties and wearout in ONO layers with scaled top oxides. For application to DRAMs furthermore investigations of defect related breakdowns on large areas are important and will be performed.

References:
1. S. Manzini a. G. Queirolo, Solid-State Electronics 30, 587 (1987)
2. V.J. Kapoor a. S.B. Bibyk, in: Physics of MOS Insulators, ed. G. Lucovsky et. al., (Pergamon Press, New York, 1980), p.117
3. C. Hu, IEDM Tech. Dig. 368 (1985)

Electrical and Optical Spectroscopic Characterization of Oxide Traps Induced by Hot Hole Injection

M.Bourcerie, J.-C.Marchetaux, A.Boudou, BULL S.A, Direction des technologies Avancées, 78340 Les Clayes sous bois, FRANCE.
D.Vuillaume, Laboratoire d'Etude des surfaces et Interfaces, Equipe de Physique des Solides, URA D0253 CNRS, ISEN 41 Bd Vauban 59046 Lille Cedex, FRANCE.

ABSTRACT
For a better understanding of defects induced by hot carrier stressing in NMOS transistors, the effect of emission enhanced by electric field (Poole-Frenkel effect) and the photodepopulation spectroscopy are used to characterize the traps in the thin gate oxide.

Recent studies have shown that hole injection ($V_g = V_d/4$) was responsible for the creation of oxide traps [1,2] in NMOS. After stressing, the (Id,Vg) curve is weakly changed, but a subsequent short injection of electrons at Vg = Vd induces a large shift of (Id,Vg) curve towards positive gate voltages (Fig 1). Furthermore, a brief injection of holes (Vg = Vd/4) induces a return to the previous (Id,Vg) curve. This phenomenon has been interpreted as due to the creation of electron traps by hole injection, which are then filled by electrons and emptied by holes. The change of the charge state of these defects is directly observed through the reversible changes of the (Id,Vg) curve. It was supposed that these traps are localized in SiO2 above the Si conduction band level and near the interface.

The transistors studied have conventional junctions (no LDD) with an oxide thickness of 40 nm and a 2μm gate length.

The devices are aged at $V_g = V_d/4 = 2V$ for 6.10^4 sec. The created oxide traps are then filled by electrons during a 100 sec injection period at $V_G = V_d = 8V$. We monitor the detrapping of electrons from these defects to the oxide conduction band by the evolution of the threshold voltage (V_T) measured on the $I_d - V_g$ curves (fig. 1).

It has been found that three methods can be used to empty these traps:

1) Injecting hot holes (Fig 1). 2) Applying a field between drain and gate (Fig 2). 3) Illuminating the device (Fig 3).

The filling of these traps is only achieved by hot electron injection.

Photodepopulation spectroscopy, with photons in the 1-4 eV range, was used to characterize the photoionization energy of the defect. Detrapping is induced by the illumination, and the photo-emitted electrons are collected at the drain by applying a small voltage between gate and drain. The use of photons with an energy lower than 4 eV allows to avoid photoinjection from the valence band of the poly-Si gate into the oxide, and of holes from the Si-p substrate. We have followed the photodetrapping kinetics by recording the time evolution of the threshold voltage shift ΔV_T . The collected data are interpreted using a first order kinetics (no retrapping of the photoemitted electrons) from which the photoionization cross-section $\sigma^o(h\upsilon)$ is obtained (Fig 2). The optical threshold between 2.5 and 3.5 eV is the photoionization energy of an electron from the defect level to the SiO_2 conduction band.

The electric field enhanced-emission measurements have been performed to determine the thermal activation energy of this defect. The difference between this energy and the photoionization energy is an indication of the lattice relaxation (or distorsion) amplitude [3]. In these experiments detrapping is observed when a large oxide electric field (in the range 2.5 to $3.5 \times 10^6 V/$cm), but smaller than the threshold for a Fowler-Nordheim injection, is applied between gate and drain (Source, substrate, gate grounded). This means that the oxide traps are located above the drain . It was found that the threshold voltage V_T decreases with time and follows an exponential law. Moreover, the assymptotic value of V_T is this given by curve 2 on fig. 1. This means that the emptying of the defect by both the methods is complete, and that we have observed all the defects introduced by hot hole injection. Figure 3 shows the emptying of the traps by electric field, for various values of Vd.

Figure 4 shows that the emission rate seems to follow an exponential dependence on the square root of the field. This is in agreement with a Poole-Frenkel model for a defect with an attractive large range potential (like a Coulomb potential) [4]. From data in fig. 4, we conclude that E_T lies between 2.8 and 3.1 eV if we assume that the capture cross-section ranges from 10^{-19} to 10^{-15} cm^2 respectively. However, in the limits of the electic field probed in this study, the data are also well fitted by an exponential dependence on the field. This would indicate a defect with a repulsive shorter range potential [5,6]. In this case, we extract an activation energy between 1.6 and 1.8 eV below the SiO_2 conduction band (with the same range for the capture cross-sections). The deconvolution between a F or $\sqrt{F}$ dependence need the measurements of the detrapping kinetics under a larger range of electric field. This is not easy at room temperature and needs a study as a function of temperature.

For the $\sqrt{F}$ dependence of the emission rate occurs, the defect must acquire a net charge opposite to that of the carrier emitted from the defect. Thus, detecting this dependence establishes that the defect has a donor electronic character upon the emission of an electron (attractive large range potential). On the contrary, the F dependence of the emission rate means that the defect has an acceptor character (repulsive shorter range potential). On one hand, if the oxide trap created by hot hole injection is a donor state with a small lattice relaxation, its energy level in the SiO_2 band gap is located at 3 eV below the conduction band. On the other hand, if this oxide trap is an acceptor state with a large lattice relaxation and its energy level is at 1.7 eV from the conduction band. In the present status we cannot assert which conclusion is the right one.

In summary, we have shown that an electron trap is created in the oxide by hot hole injection in NMOS transistors. This trap has an optical ionization energy of 3 eV from the SiO_2 conduction band edge. The donor or acceptor nature of this defect is still under question. However, we have shown that it may have a small lattice relaxation if it is a donor, or a large lattice relaxation if it is an acceptor. In the former case, this oxide trap induced an energy level at 3 eV below the SiO_2 conduction band, and in the latter case, at 1.7 eV below it. This trap is localized in the overlap region between the gate and the drain.

We thank A. Zylbersztejn for a critical reading of the manuscript.

1. B.S. Doyle, M. Bourcerie, C. Bergonzoni, R. Benecci,A. Bravaix and A.Boudou, IEEE Trans. Electron devices, submitted for publication.
2. P. Heremans, R. Bellens, G. Groseneken and H.E. Maes, IEEE Trans. Electron Devices 35, 2194, 1988.
3. J. Bourgoin and M. Lannoo, in "Point defects in semiconductors" (Springer-Verlag, Berlin, 1983), vol. 2, pp. 88-121.
4. J. Bourgoin and M. Lannoo, "Point defects in semiconductors", (Spriger-Verlag, Berlin, 1983), vol. 2, p. 199.
5. A. Chantre, G. Vincent, D. Bois, Phys. Rev. 15, 23, 5335 (1981).
6. M. Lannoo, private communication.

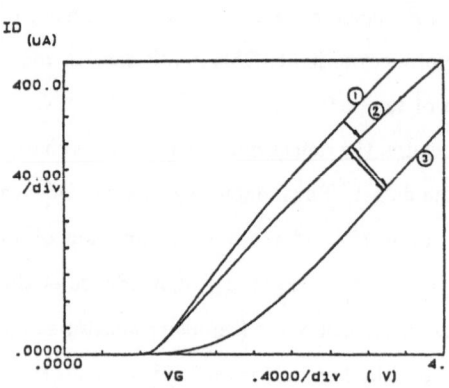

Fig 1 : I_d-V_g characteristics : 1) Before aging
2) After aging at $V_g = V_d/4 = 2$ volts during
6.10^4 s 3) After a short electron injection
at $V_g = V_d = 8$ volts during 100 seconds.

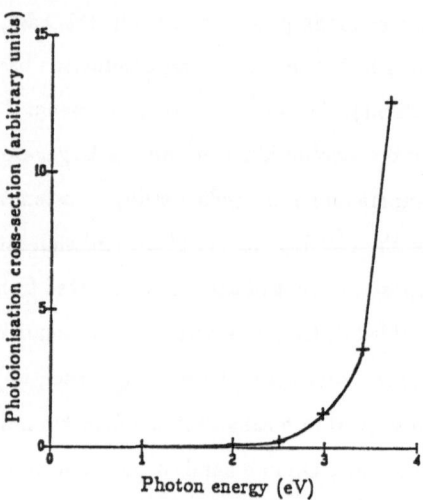

Fig 2 : Photoionization cross section
versus photon energy measured by
photodepopulation spectroscopy.

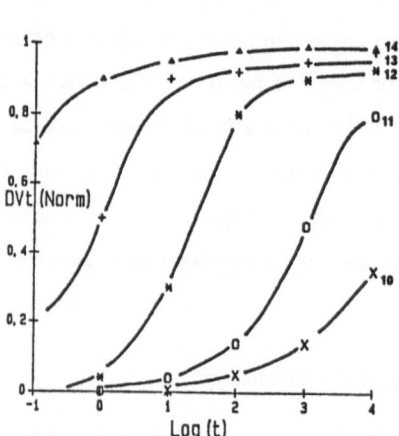

Fig 3 : Emptying by field $(10 < V_d < 14)$
between 0.1 second and 10^4 seconds.

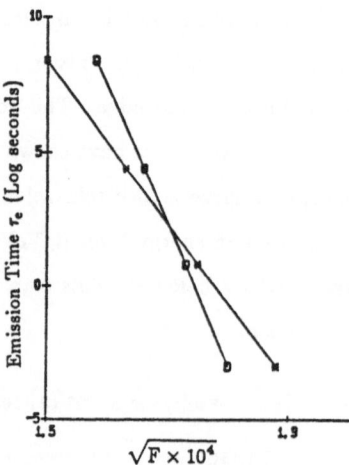

Fig 4 : Dependance of the emission
time constant with the oxide electric
field F (∗) and with the square root
of the field (o).

Well Technologies for Half-micron CMOS Processes

H.-M. Mühlhoff, F. Lau, P. Küpper, W.U. Kellner, and S. Röhl
Siemens AG, Semiconductor Division,
Otto-Hahn-Ring 6, München, West Germany

Summary

Scaling CMOS processes to sub-µm dimensions requires a reduction of lateral well extensions to be able to shrink the n^+-p^+ spacing. This paper describes concepts for reducing the lateral extension of n-wells on p-substrate. Three well types have been compared: (1) a deep n-well fabricated by long drive-in, (2) a shallow n-well made by very short drive-in, (3) an n-well superimposed on the p-well by counter doping the p-well in the n-well areas. Whereas device performance of thin oxide PMOSFETs is identical for all three wells, considerable difference has been observed for lateral well isolation and latchup.

Well process flow

Wafers with deep n-wells were taken from a twin well process in which first p-wells are patterned, implanted and driven in before n-wells are patterned and implanted with Phosphorus ($2 \times 10^{13} cm^{-2}$) followed by a drive-in of 10h at 1150°C. This is a two mask well process. The shallow n-wells were made on p-doped 4µm thick 5Ohm-cm EPI on 20mOhm-cm substrate with no p-well. N-well dose was $1.5 \times 10^{13} cm^{-2}$. The drive-in at 1150°C was optimized such that maximum well depth could be achieved. Too long drive-in times cause the well-depth to be reduced due to the outdiffusion of Boron from the highly doped substrate. A well depth of 1.5µm was realized. The superimposed n-wells were made by counter doping p-wells with a high dose ($3.5 \times 10^{13} cm^{-2}$) and using the same drive-in conditions as for the deep n-well. Process flow of this twin-well structure is inherently simple because only one mask is used. Process complexity is much reduced compared to more conventional process flows for self-aligned wells, which use selective oxidation to achieve covering of one well during the implantation of the other well [1]. Fig.1 shows the simulated 2d doping profiles of the n-wells and Table 1 lists their dimensions. For the counter doped n-well, the distance between the pn-junctions and the location of $10^{16} cm^{-3}$ doping concentration is a minimum in both the vertical and the horizontal direction. The well profile at the well edges resembles that of retrograde n-wells [2,3].

PMOS Device Performance

N-well implant doses were adjusted to achieve identical surface concentrations for all three well types. This value was optimized for good short channel behavior and punchthrough resistance. P-channel active devices were formed with $t_{ox}=18nm$ and n-doped poly-gates requiring buried channel implants at 25keV. Fig.2 shows that subthreshold behavior is identical for all three well types and PMOSFETs with 0.5 μm L_{eff} (0.7μm gate length) can be realized. The saturation currents are independent of well type as well (Fig.3). The high impurity concentration of the counter doped n-well has no negative effect on carrier mobility.

Lateral Isolation

Results of measurements on field oxide transistors between p-substrate, n-well and p^+ diffusion are shown in Fig.4. In these measurements, the position of the well-edge was varied with respect to the p^+ diffusion. Locations of pn-junctions were taken from simulation. The data represents the threshold-voltages of the poly-gate field oxide transistor at drain voltages of 0.25V and 10V. At very small distances, punchthrough occurs. But at 1μm distance between diffusion and well-edge, sufficient isolation is obtained for the counter doped n-well whereas more than 2μm are needed for the other well types. This data shows that using counter doping for well formation in a twin well process yields a much more compact isolation structure in which virtually no lateral space is wasted due to lowly doped transition areas. An abrupt lateral pn junction can be formed between highly doped p- and n-wells and smaller n^+-p^+ spacings than in conventional well processes can be realized.

Latchup

Latchup sensitivity is a critical factor when optimizing wells. In order to better distinguish between latchup problems related to the n-well and problems related to the p-well and p-substrate, current ex citation on the emitter terminals was used as characterization tool [4,5].

Sourcing current into the parasitic npn causes the parasitic pnp to trigger latchup. The basewidth of the pnp, which is dependent on well depth and lateral well extension, determines the current amplification of the pnp and has to be kept large for good latchup resistivity. Also the well resistance which shunts the base of the pnp has to be kept low. These goals can be met with a deep and highly doped n-well with large lateral extension. Fig.5 shows currents required to trigger latchup via well injection. As expected, the shallow well performs much worse than the deep well. The counter doped well, however, exhibits only slightly degraded trigger currents.

Current ex citation on the emitter of the parasitic pnp causes the parasitic npn to trigger latchup. Substrate resistivity is critical for this kind of trigger mode. N-well parameters are not very important in this case and no difference between the deep n-well and the counter doped n-well are observed. Keeping substrate or p-well resistance low is crucial for achieving good latchup protection for substrate triggering. Fig.6 shows that using EPI increases trigger currents substantially.

Conclusion

Two well configurations were evaluated as alternatives for conventional deep n-well processes. Active device performance is not dependent on well type but on net surface concentration of n-dopant. Using shallow n-wells on thin EPI gives superior latchup protection for substrate triggering but degraded protection for well triggering. Retrograde well implants for this process may be required. Counter doping the p-well in n-well areas yields a twin well process in one masking step an with low complexity. It permits the reduction of n^+-p^+ spacing because of its improved isolation properties. Latchup performance is comparable to conventional deep well processes. This process does not require retrograde well implants.

References

1. L.C. Parillo et al.,
   IEDM Techn. Digest, pp706-709 (1980)

2. Y. Taur et al.,
   IEEE Trans. Electron Dev., pp203-209 (1985)

3. R.A. Martin et al.,
   IEDM Techn. Digest, pp403-406 (1985)

4. Latchup in CMOS Technology,
   R.R. Troutman, Kluwer Academic Pub., p135 (1986)

5. C. Mazure, W. Reczek, D. Takacs and W. Winnerl,
   IEEE Trans. Electron Dev., pp1609-1615 (1988)

Table 1

	deep n-well	shallow n-well	counter doped n-well
well depth	4.0μm	1.5μm	3.0μm
mask-xj	3.5μm	2.0μm	0.5μm
ver. 1E16-xj	2.0μm	0.9μm	0.6μm
hor. 1E16-xj	1.4μm	0.8μm	0.5μm

556

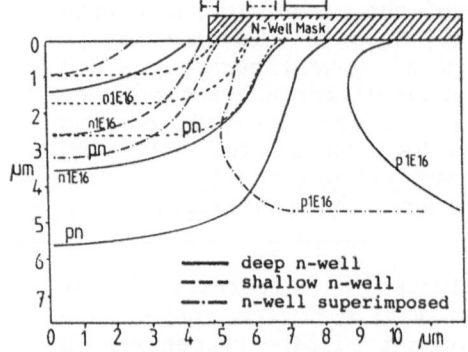

Fig.1: Simulated 2d doping profiles.

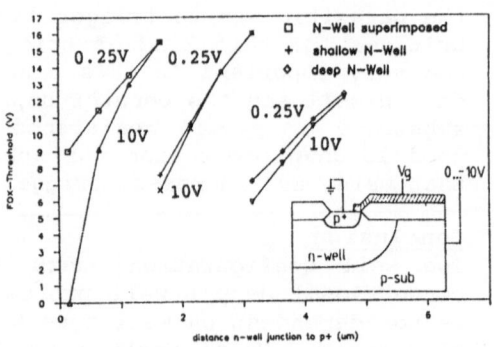

Fig.4: Lateral Isolation of N-Well. Threshold voltages of FOX-transistor for Vds=0.25 and Vds=10V at 1pA/μm. D_{ox}=400nm.

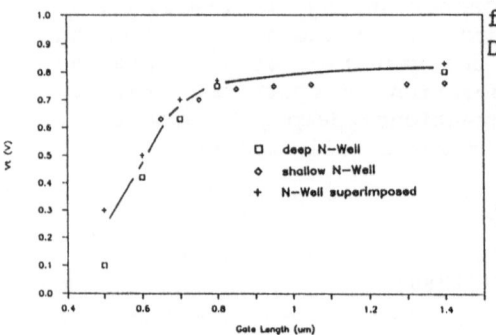

Fig.2: PMOS short channel effect. Vt(L) for Vds=5V.

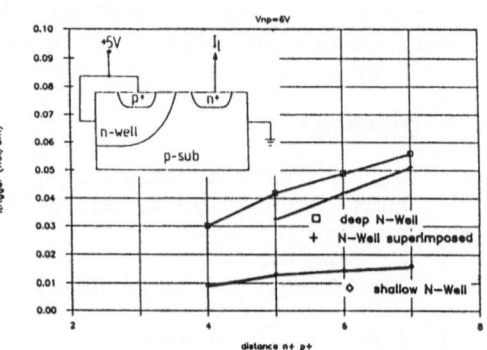

Fig.5: Latchup - Well Triggering. Trigger currents vs. n^+-p^+spacing.

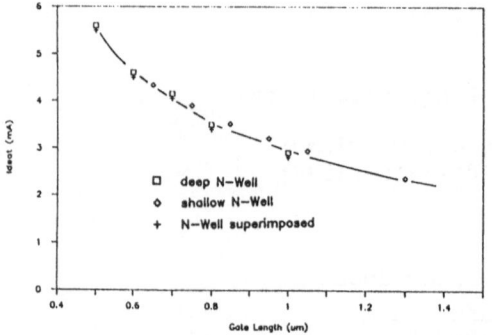

Fig.3: PMOS saturation current. Id(L) for Vds=Vgs=5V. W=20μm.

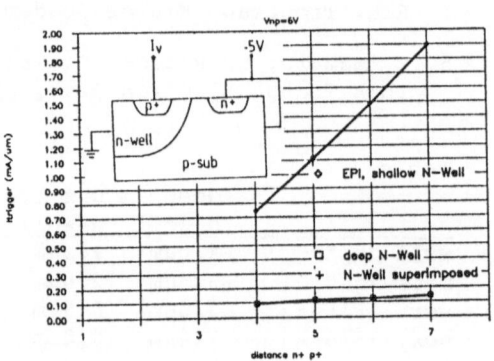

Fig.6: Latchup - Substrate Triggering. Trigger currents vs. n^+-p^+spacing.

An Investigation into the Effects of RTA Processing on Low Frequency Noise and Other Characteristics of CMOS FETs

D.C.Murray, J.C.Carter, A.G.R.Evans, A.Gougam[†] and J.L Altrip

Dept. of Electronics and Computer Science,
The University, Southampton, SO9 5NH, England.

† INELEC, Boumerdes, 3500, Algeria.

Abstract

CMOS devices fabricated with a rapid thermal anneal source and drain activation have been compared with similar devices fabricated with a standard furnace anneal. Measurements have been made of low frequency noise, average interface state density, channel mobility, threshold voltage and bulk states using DLTS. Results show that in some instances RTA can have a deleterious effect on device performance that appears to be due to the rapid rates of cooling that can occur when using this techology.

Introduction

Rapid thermal annealing is now replacing conventional furnace annealing as a method for implantation damage removal and dopant activation due, mainly, to the minimal dopant redistribution that results. However, possible side-effects may arise when using this technology because of the high rates of temperature change that are possible [1,2]. An attempt has been made to quantify this damage by comparing n and p-channel CMOS devices fabricated with an RTA stage with similar devices fabricated using a standard furnace anneal.

Fabrication of devices

CMOS devices were fabricated using an n-well process that includes a boron implant threshold adjust for both n and p-channel devices through a sacrificial oxide. The 40nm gate oxide was grown in dry oxygen at 950°C with a 15min anneal in nitrogen at the same temperature. Polysilicon gates were solid source doped at 1000°C and dry etched. N-channel devices had 5E15 As^+ 80keV source and drain implants and p-channel devices had 5E15 BF_2^+ at 50keV. Activation either occured in a furnace at 1000°C for 15mins followed by 35mins at 950°C or in a 30keV electron beam annealer [3]. The peak temperature used was 1100°C for times at this peak ranging from 0-50s. The RTA cycles ended with either a fast cool(FC) (200°Cs^{-1}) or a controlled cool (CC) (10°Cs^{-1}). Samples were placed face down in the electron beam annealer to avoid electron damage. After metallization wafers were given a 450°C anneal in forming gas.

Measurement methods

Low frequency noise was measured using a system known as ACUMENS [4] developed at AERE Harwell Laboratories on 100x8 μm^2 MOSFETs operating in strong inversion and drain current saturation. Noise is quoted in this paper as the equivalent input noise voltage $E_n(f)$. This is the square root of the input referred spectral density $S_v(f)$. MOSFETs have a characteristic 1/f noise spectrum due to defects in the oxide close to the silicon-silicon dioxide interface [5,6]. If bulk defects are present the shape becomes $1/f^a$ where 1<a<2 - a is known as the noise index. Average interface state densities were measured using a charge pumping technique [7] and channel mobilities and threshold voltages were obtained from I_d-V_g curves. DLTS was performed on large area diodes (0.86mm²) using the Biorad Polaron DL4600 system.

Results

Figure 1 shows the noise value at 10Hz as a function of different anneal cycles for n and p-channel devices. Figure 2 shows the noise index for the same devices. It can be seen that the 0s FC devices have a noise figure an order of magnitude higher than the furnace annealed devices. The noise figures for 10s and 50s RTA devices are similar. The addition of a slow cool gave a marked improvement in noise, levels similar to furnace annealed samples were seen in n-channel devices. The frequency index is close to unity for n-channel devices for all anneals except the 0s FC. For p-channel devices the frequency index is around 1.1 for all RTA devices; the furnace samples have a noise index of 0.95. Table 1 shows average interface state density (for n-channel only) channel mobility and threshold voltage. Lowest interface state densities are seen in furnace annealed samples RTA device levels being two to three times this level. Mobilities are all similar for p-channel devices but decrease with increased RTA duration for n-channel devices. Threshold voltages are also similar for all p-channel devices but become more negative as RTA duration is increased in the case of n-channel devices.

DLTS studies did not detect any high defect densities for the majority of the anneals. However, an exception was found for short duration RTAs (0s FC and 0s CC) after As+ implantation.

Discussion

The increase in noise of n-channel RTA devices is thought to be due to an increase in oxide defects which appear to be related to the rapid cooling rates at the end of the anneal cycle. This is consisitent with the the low noise levels in slow cooled devices and the duration of RTA having little effect on noise levels. The increased noise for short duration RTA is perhaps due to bulk defects that have been left unannealed. These are indicated by the frequency index shift to 1.1 for these devices and directly seen as peaks in noise temperature plots and in DLTS.

The analysis of the noise results for p-channel devices is complicated by their partial buried channel mode of operation. This results in the peak carrier concentration being away from the silicon-silcon dioxide interface. This means that noise contributed by 1/f oxide defects will be reduced and noise due to bulk defects increased [8]. An increase in bulk noise was demonstrated by the shift in frequency index for all RTA devices to near 1.1. This bulk noise increase masks a small increase in oxide noise. As for n-channel devices this increase in bulk noise is especially associated with short duration RTAs.

For p-channel devices RTA does not appear to influence the channel mobility or threshold voltage to any significant degree. This is not true for n-channel devices where there is a deterioration in interface state density, channel mobility and threshold voltage as the length of RTA anneal is increased. There appear to be two effects causing device damage during RTA; the first is the cooling rate which, if extreme, causes an increase in defects responsible for low requency noise; the second is an unknown factor which depends on the length of time of the anneal. This deterioration might be accounted for by damage from x-rays that result from electron impact in the e-beam annealer - the use of lamp annealing for comparison would be useful.

Conclusions

It has been shown that RTA significantly increases low frequency noise in both n and p-channel CMOS devices compared to furnace annealed devices. This degradation appears to be caused by rapid cooling rates as it is not a function of anneal time. The deterioration in low frequency noise after RTA would be serious in applications where MOSFETs are used at the input of first stage amplification in on-chip transducer technology for example. The noise can be reduced by giving devices a slow cool. Changes in interface state density, channel mobility and threshold voltage are also observed in RTA devices however these changes are in most cases small.

References
1.   J.Y.C Sun et al., Proc. 18th ESSDERC Conf., Journal de Physique, ppC4_401-404,1988
2.   D.C.Murray et al., Semicond. Sci. Technol., **4**, 393-398, 1989
3.   M.J.Hart and A.G.R.Evans, Semicond. Sci. Technol, **3**, 421-36, 1988
4.   C.Cox, AERE **R10642** (HMSO), 1982
5.   A.L.M°Whorter, Semiconductor Surface Physics (Univ. Pennsylvania Press),207-228, 1956
6.   M.J.Uren, Appl. Phys. Lett. **47**, 1195-1197, 1985
7.   G. Groeseneken, IEEE Trans. Electron. Devices, **ED-31**, 42-53, 1984
8.   K. Nakamura, Jap. J. Appl. Phys., **46**, 3189-3193, 1975

Anneal	n-channel					p-channel				
	FA	0s FC	0s CC	10s FC	50s FC	FA	0s FC	0s CC	10s FC	50s FC
D_{it}	1.0	2.9	2.4	2.5	2.7	-	-	-	-	-
μ	538	572	596	538	528	197	199	194	195	199
Vt	+0.34	+0.44	+0.44	0.00	-0.20	-0.89	-0.84	-0.83	-0.82	-0.70

Table 1. Average interface state density (10^{10} eV^{-1}cm^{-2}), channel mobility (cm^{-2}V^{-1}s^{-1}) and threshold voltage as a function of source and drain anneal.

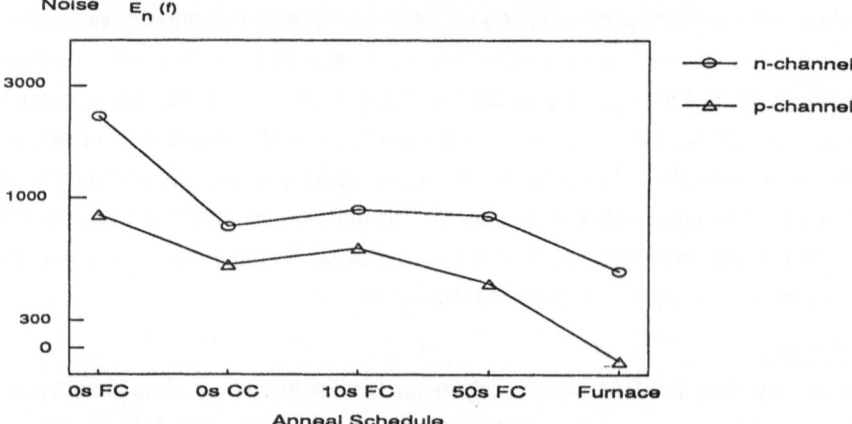

Fig. 1. Input referred noise $E_n(f)$ as a function or source and drain anneal.

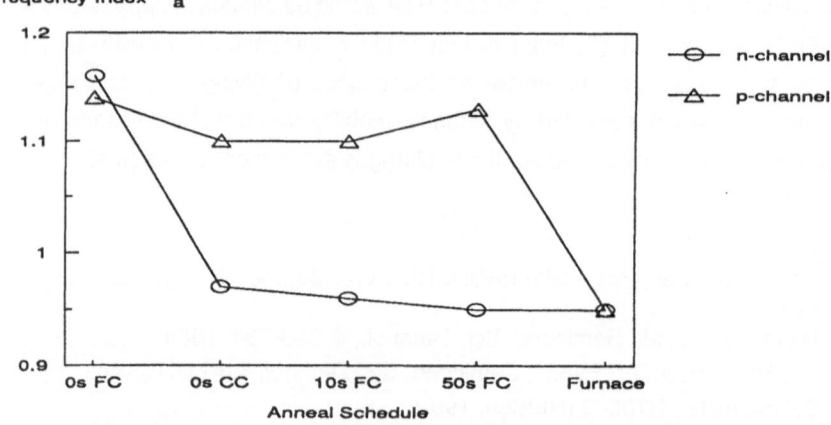

Fig. 2. Frequency index a as a function of source and drain anneal.

Avoiding Lateral Diffusion of Dopants in n$^+$/p$^+$ Polycide Gates

D.T.Amm[1]), D.Lévy[2]), M.Paoli, P.Delpech, T.Ternisien d'Ouville, H.Mingam and G.Göltz.

Centre National d'Etudes des Télécommunications,
Chemin du Vieux Chêne, BP 98, 38243 MEYLAN Cedex, FRANCE.

Summary: The lateral diffusion of dopants in a TiSi$_2$ salicide process was studied for various BPSG reflow RTP anneals. Lateral diffusion of arsenic was detected for anneals above 1000°C (20 sec.) whereas no boron diffusion could be observed for anneals up to 1060°C. A novel "weakly doped gate" test structure proved to be a sensitive detector of lateral diffusion.

Introduction: The dual-type poly gate MOS structure (N+ poly gate for NMOS and P+ poly gate for PMOS) is becoming widely accepted in submicron CMOS technologies [1-3]. However, when used in conjunction with a silicide gate shunting layer, the dual-type poly can be problematic if the N and P type dopants diffuse laterally via the silicide. This lateral diffusion can result in dopant mixing and uncontrollable threshold voltages which vary with MOSFET layout and interconnection lengths [4,5].

Earlier work has shown that lateral diffusion can be severe when dopants were implanted directly into the WSi$_2$ of a polycide gate structure [6]. However, the use of a titanium salicide process, where the implantation and activation of dopants are performed before silicide formation, appears to be less sensitive to this problem. This paper concentrates the effect of the BPSG reflow anneal on lateral diffusion in a TiSi$_2$ salicide process. The results of a novel test structure which is more sensitive to lateral diffusion than standard structures, are also presented.

Process: The principle steps of the TiSi$_2$ process that relate to lateral diffusion are listed in Table 1. It is important to note that in this process, the N+ and P+ junctions are implanted and annealed before the titanium deposition.

---

[1]) Present address : Dept. of Physics, Queen's University,
Kingston, Ontario, CANADA K7L 3N6
[2]) Permanent address: Bull, Rue Jean Jaurès,
78340 Les-Clayes-sous-Bois, FRANCE.

After silicide formation, the gate consists of about 800Å of
TiSi2 and 3000Å of polysilicon. RTA has been used in most of
the process steps as indicated in Table 1. In order to study
temperature effects, wafers were subjected to BPSG reflow
anneals of 950, 1000, 1030 and 1060°C. The integrity of the
TiSi$_2$ was monitored closely at high temperatures.

---

* deposition undoped polysilicon 3800Å
* Drain/source/gate implantation NMOS - As 4×10^{15} 1/cm²
  PMOS - B 2×10^{15} 1/cm²
* activation anneal 1060°C, 20 s
* deposition titanium 800Å
* anneal for silicide formation, N$_2$, 700°C, 50 s
* BPSG reflow anneal, variable temperature, 20 s

Table 1. Principle Steps in the TiSi$_2$ Salicide Process

Test Structures: The first type of device is a "worst case"
test structure shown in Figure 1(a). An isolated MOSFET is
connected via polycide to large source/sink pads of polycide
over field oxide. This represents the "worst case" geometry
envisioned for lateral diffusion. The source/sink pads are
either undoped, or are implanted N+ or P+. The implantation
area starts 1±0.5 μm away from the transistor active area. The
combination of N+ (NMOS) and P+ (PMOS) gates surrounded by N+,
intrinsic or P+ doped pads gives six test transistors per
module.

The second type of test structure consists of a "chain" of
transistors which is used to determine the distance of dopant
lateral diffusion (Figure 1(b)). Transistors are connected with
a common gate at distances of 1, 6, 11, 26, 51, and 91 μm from a
source pad. A "PMOS chain" of devices (P+ gate) is connected
to a N+ implanted pad, and an "NMOS chain" (N+ gate) is
connected to a P+ pad.

A third and novel test structure consists similarly of an
NMOS chain and connected polycide pad, but the NMOS gates are
masked during the N+ implantation and thus only see the much
lower concentration LDD implant (As, 1×10^{13} 1/cm²). With this
lower dopant concentration in the NMOS gates, the MOSFET's are
much more sensitive to the lateral diffusion of impurities.

The standard transistor size used in all the about
structures has a gate length of 2μm and a width of 3μm.
Assuming all the implanted dopants are initially uniformly

distributed in the polysilicon layer, the concentrations are: $N+ = 1\times10^{20}$, $P+ = 5\times10^{19}$, and $LDD(N) = 3\times10^{17}$ atom/cm^3.

Fig. 1

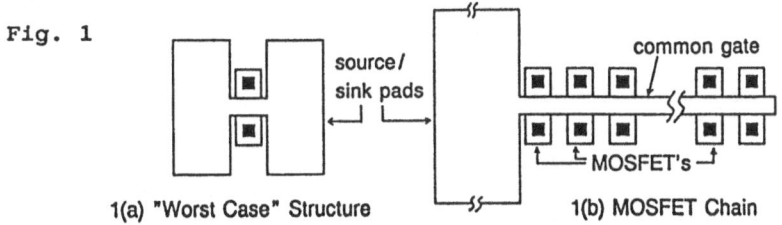

source/sink pads

common gate

MOSFET's

1(a) "Worst Case" Structure   1(b) MOSFET Chain

<u>Results:</u>   Isolated MOSFET's, where the gate dopants can diffuse only vertically between the poly and TiSi$_2$, show no threshold variation whatsoever as a function of final anneal temperature.   Omitting the anneal, or annealing from 950°C to 1060°C all gave a PMOS Vt of -0.69±0.03 V and an NMOS Vt of 0.81±0.02 V.   This shows that there is no strong segregation of dopants into the silicide layer which is in contrast to results of some other silicides particularly TaSi$_2$ [7].   This result also indicates that, even though a 20% increase of TiSi$_2$ resistance was noted at higher temperatures, the eventual degradation of the silicide does not affect Vt measurements.

The results for the PMOS "worst case" test structures are given in Figure 2.   PMOS devices surrounded by non-doped polycide gave the same Vt's as those surrounded by P+ doped polycide.   This reveals that there is no significant out-diffusion of boron from the device.   In contrast, the PMOS devices surrounded by N+ polycide show a decrease in Vt at 1030 and 1060°C indicating that substantial amounts of arsenic are diffusing into the PMOS gate.   The NMOS "worst case" devices showed no Vt variations for any combination of temperature or pad doping.

Similarly, the "NMOS chain" devices showed no Vt variation as a function of distance from the P+ doped pad.   For the "PMOS chain" test structure, the device 1μm from the N+ doped pad had a Vt shift of only 70 mV at 1060°C.   PMOS devices at 6μm or further away, where unaffected by the arsenic diffusion.   Thus the lateral diffusion of boron was found not to affect standard devices whereas the arsenic diffusion for temperatures above 1000°C may affect PMOS devices up to a distance of 6μm.

The "NMOS LDD chain" devices were, as expected, much more sensitive to lateral diffusion.   Approximately, the LDD transistors should be sensitive to concentrations 1000 times lower than standard NMOS devices.   The results given in Figure 3 indicate that the arsenic diffuses about 30 and 50 μm at 1030 and 1060°C.   Although there is some threshold variation for the LDD NMOS with a P+ doped polycide pad, the relative change can

564

be explained almost entirely by the out-diffusion of the LDD arsenic into the polycide pad.

Conclusion: The "NMOS LDD chain" test structure is confirmed to be a sensitive detector for lateral diffusion of dopants. Nevertheless, in order to measure convenient diffusion lengths, the BPSG reflow temperature was intentionally high (up to 1060°C, RTA). Based on our data, we can conclude that dopant lateral diffusion should not affect device performance when a reflow temperature compatible with TiSi$_2$ salicide process (about 950 to 980°C, RTA) is used. Furthermore, an extrapolation of these results to lower temperatures and dimensions will be applicable to the problem of lateral dopant diffusion in "butting contacts".

Figure 2.   Figure 3.

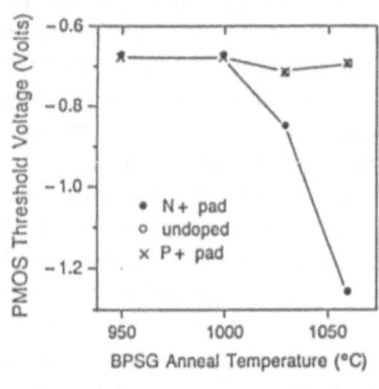

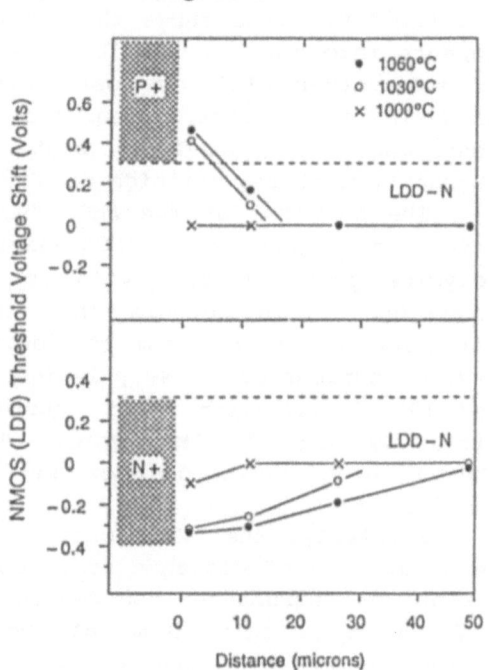

Figure 3.   Dotted Line = Estimated Vt of NMOSFET with intrinsic poly gate

REFERENCES:

[1] R.A. Chapman et al, IEDM Tech. Dig., p.52 (1988)
[2] B. Davari et al, IEDM Tech. Dig., p.56 (1988)
[3] J-Y.C. Sun et al, IEDM Tech. Dig., p.236 (1986)
[4] S.J.Hillenius et al, IEDM Tech. Dig., p.252 (1986)
[5] L.C.Parillo et al, IEDM Tech. Dig., p.418 (1984)
[6] D.Lévy et al, Refractory Metal and Silicide Workshop, Houten, Belgium, Appl.Surf.Science, in press (1989)
[7] C.Mazure et al, J. de Physique, vol. C4-49, p.405 (1988)

Electrical Properties and Sputter-etched Induced Defects in Ti-W/Si Schottky Diodes

D. Bauza*, C. Mallardeau**, Y. Morand**,

*L.P.C.S., E.N.S.E.R.G., B.P. 257, 38016 Grenoble Cedex FRANCE.
**SGS- THOMSON MICROELECTRONICS, B.P. 217, 38019 Grenoble Cedex FRANCE.

Abstract

Ti-W/Si (n type) Schottky diodes which have been back-sputter etched prior to metal deposition are studied. It is found that the barrier height strongly depends on the etching conditions. Defects levels are measured using DLTS. Their characteristics are studied as function of the etching conditions.

INTRODUCTION

Ti-W/Si Schottky diodes are used with Pt-Si/Si ones in fast VLSI STL logic circuits. This logic requires manufacturing of both low and high Schottky barrier diodes on the same substrate. The high and low levels are given by the difference between the threshold voltages of the two types of diodes. Pt-Si/Si diodes, with 0.8 Volt Schottky barrier height value have been widely studied and exhibit a satisfactory behaviour. On the contrary, there are only few studies on Ti-W/Si diodes which have electrical behaviour far from ideality.

EXPERIMENTAL DETAILS

The samples (supplied by SGS-Thomson) were made using $<100>$ Czochralski grown silicon wafer (p doped, 10-20 Ω.cm). An N^+ buried layer was realised with 4.10^{15} cm^{-2} As implantation at 50 keV. After annealing at 1160°C during 90 mn, an epi-layer of 1.4 μm thickness (0.4 Ω.cm) was grown. Then, the Pt-Si/Si diodes were processed. Prior to Ti-W sputter deposition, the substrate was Ar etched (RF diode principle). The Ti-W diodes were fabricated by sputter-deposition of 1500 Å Ti-W from a composite target. The concentration of Ti and W in atoms in the target were 30 and 70 % respectively. 8000 Å thick Al-Si-Ti was then deposited and patterned for interconnect. Finally, the samples were annealed at 420°C for 30 minutes.

SAMPLES CHARACTERISATION

Measurements were carried out on samples wich area were 1600 μm^2. The quality of the Schottky barrier was evaluated by I-V measurements. For this , an HP 4145B analyser was used. A DLTS system using a double boxcar averager (2) and operating at 1 Mhz was used (Biorad Polaron S4600 model spectrometer). Spectra were recorded in the temperature range from 90 to 330K. The measured emission rates were between 1000 s^{-1} and 80 s^{-1} and the filling pulse width was set equal to 100μs. The energies E_t and capture cross sections σ of the traps were calculated from the Arrhenius plot $e_n/T^2 = f(1/T)$. Defect profiling was achieved by using constant filling voltage and variable reverse bias method (3). The profiles were corrected to take into account the filling factor due to non abrupt space charge region (4).

RESULTS

The I-V characteristics of diodes back-sputter etched using different conditions are depicted in fig. 1. The etching conditions were chosen so that a layer of 50 $\overset{\circ}{A}$ to 75 $\overset{\circ}{A}$ was removed. It can be seen that the highest barrier is obtained for samples etched at 15 $\overset{\circ}{A}$/min (etched depth equal to 60 $\overset{\circ}{A}$). The barrier height value of this sample is 0.602 Volt. This value is near the optimum value for this type of diodes (5).

Compared to the previous sample, the samples etched at 8 $\overset{\circ}{A}$/min and 45 $\overset{\circ}{A}$/min show a barrier height lowering. Such a lowering is usually observed after sputtering processes and is due to donnor-like defects created near the metal-silicon interface (6,7). Because of the large energy of the Ar$^+$ ions in the case of the samples etched at high etching rate, the strong lowering observed for the sample etched at 45 $\overset{\circ}{A}$/min is not a surprising result. On the other hand, the lowering observed on the sample etched at 8 $\overset{\circ}{A}$/min means that a large amount of defects can be created even at very low power. In figure 2 is shown DLTS spectrum recorded from such a sample. It exhibits four trap E(0.26), E(0.28), E(0.36) and E(0.49) located respectively at 0.26, 0.28, 0.36, and 0.49 eV from the conduction band edge .Their capture cross sections are written in the same figure. Depth concentration profiles of these traps are shown in figure 3. As found generally in the case of sputter-induced defects, the profiles decrease nearly exponentially in the depth (8,9), except for trap E(0.28). For clarity data point of defects E(0.26), E(0.36) and E(0.49) have been least square fitted and their fitting have been extrapolated toward the interface.

DLTS spectra recorded from samples etched under the other conditions are displayed in figure 4. From this figure it can be seen that the samples etched at 15 $\overset{\circ}{A}$/min exhibit the same four main traps. The possibility that defects could be due to platinum contamination during the Pt-Si/Si diodes preparation have been hypothesised. So the sample etched at 15 $\overset{\circ}{A}$/min with an etched depth equal to 60 $\overset{\circ}{A}$ have been processed without Pt-Si/Si diodes realisation. The same traps are still present. The sample etched at 45 A/min shows the trap E(0.36) plus two new traps at lower temperatures. The trap E(0.36) have been found for all etching conditions and always exhibited the largest density. So the defect concentration profiles of these traps have been recorded to compare their permeation an their density as functions of the etching conditions. These profiles are shown in figure 5.

From this figure it can be seen that the highest defect concentration is obtained from the sample etched using the lowest etching rate. This is in agreement with the evolution of the barrier height found previously and comfirms that a large amount of defects can be created at low etching rate. The largest permeation (lowest slope) is obtained from the sample etched with the highest etching rate. This can be explained by the highest power used for this sample but the defect density (the lowest) cannot explain the largest Schottky barrier height lowering observed. To explain this, a large amount of defects close to the metal-silicon interface and wich cannot be sampled by DLTS must be assumed.

To resume, DLTS measurements shows that the largest defect concentration between 0.1 μm and 0.3 μm from the interface are obtained from the sample etched at the lowest etching rate whereas the highest defect concentration is measured in the sample etched at the highest etching rate.

As the results reported in ref. 8 these results could be explained by a change in the type of radiation and/or particles responsible of the defects when the etching conditions are varied.

CONCLUSION

The sputter-etching conditions can have a strong effect on the electrical characteristics of Ti-W/Si Schottky diodes. The maximum of the barrier height value is found to be near 0.6 Volts. The evolution of the barrier height with the measured defects concentrations and with the etching conditions are not those found previously but are in good agreement with those reported in ref. 8. Finally, good quality Ti-W/Si Schottky diodes can be realised using sputter etching provided that the etching conditions are properly chosen.

References

(1) S. M. Sze, Physics of semiconductor devices, Wiley Interscience Publication, New York 1981.

(2) D. V. Lang, J. Appl. Phys. 45, 7, 3023 (1974).

(3) K. Yamasaki, M. Yoshida and T. Sugano, Jap. J. of Appl. Phys. 18, 1, 113 (1979).

(4) J. Bourgoin and M. Lannoo in Point defects in semiconductors II, Springer-Verlag publication, 1983.

(5) M. O. Aboelfotoh, J. Appl. Phys. 61, 7, 2558 (1987).

(6) S. J. Fonash, S. Ashok and R. Singh, Appl. Phys. Lett. 39, 5, 423, (1981).

(7) N. A. Bojarczuk and O. Paz, J. Vac. Sci. Technol. A1 2, 615 (1983)

(8) F. D. Auret, M. Nel and N. A. Bojarczuk, J. Vac. Sci. Technol. B4, 5, 1168 (1986).

(9) F. D. Auret, O. Paz and N. A. Bojarczuk, J. Appl. Phys. 55, 6, 1581 (1984).

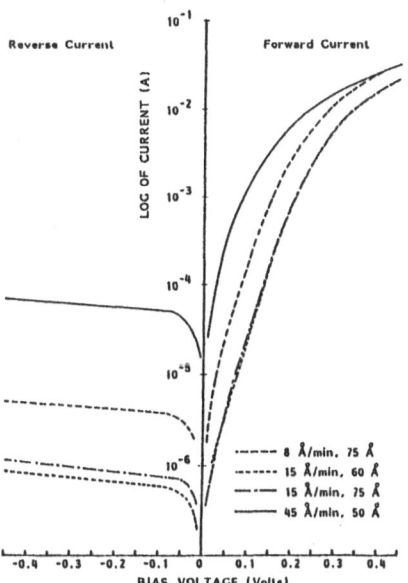

Fig. 1 : I-V characteristics of Ti-W/Si Schottky diodes for different etching conditions : etching rate and etched depth

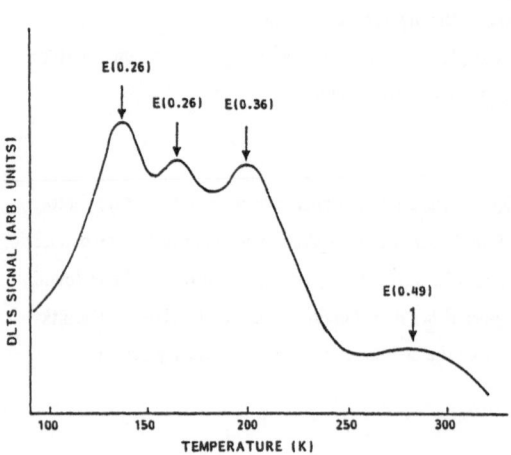

Fig. 2 : DLTS spectrum recorded from a sample etched at 8 Å/mn (etched depth : 75 Å ; Quescient bias : -0.3 V, pulse height : 0.6 V)

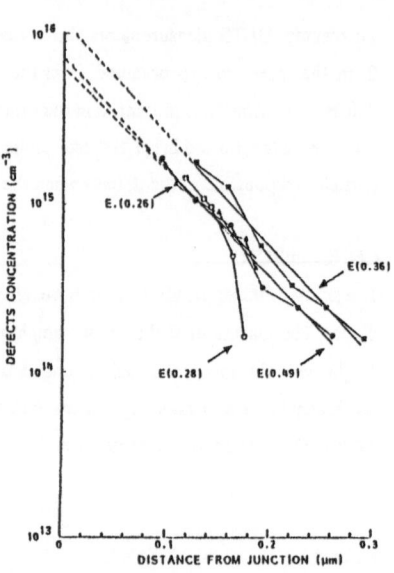

Fig. 3 : Depth concentration profiles of traps observed in Ti-W/Si Schottky diodes etched at 8 Å/mn (etched depth : 75 Å)

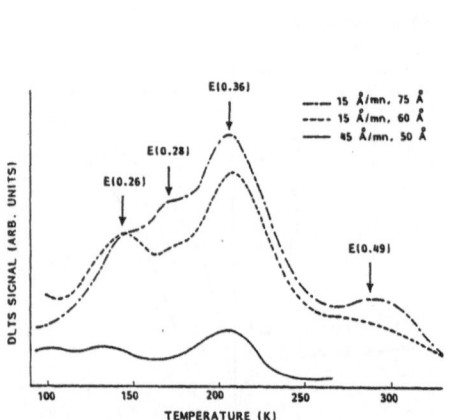

Fig. 4 : Typical DLTS spectra recorded from samples etched at 15 Å/mn and 45 Å/mn (Quescient bias : -0.3 V, pulse height : 0.6 V)

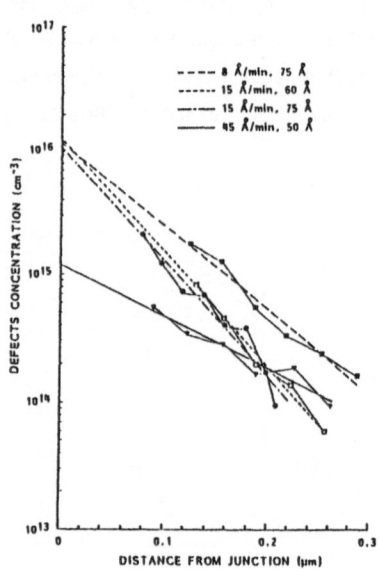

Fig. 5 : Depth concentration profiles of traps E(0.36) recorded from samples etched under the different etching conditions

Si/Si$_{0.8}$Ge$_{0.2}$ Heterojunction Bipolar Transistors with Ion Implanted Device Contacts

H.-U. SCHREIBER

Ruhr-Universität Bochum, Mikroelektronik-Zentrum (A),
4630 Bochum, West Germany

Summary

High performance npn Si/SiGe heterojunction bipolar transistors were fabricated with As ad BF$_2$ implanted emitter and base contacts, respectively. Annealing was carried out in a conventional diffusion furnace at 900°C for 2 up to 8 minutes. The DC output characteristcs were clearly improved. A maximum current gain of more than 1000 was achieved and the typical V_{CE} offset was negligible. A serious problem is the SiGe base sheet resistance, which is higher in the outer as compared with the inner base. A 900°C annealing step may reduce the outer base resistance. The first high frequency measurement resulted in a transit frequency of 8.5 GHz at V_{CE}=1V.

Introduction

The Si/SiGe heterojunction bipolar transistor (HBT) is an appropiate candidate for improving the switching speed of future silicon based integrated circuits. The basic idea is to provide an additional energy barrier ΔE_V to holes injected from the base into the emitter. This energy barrier permits an increased base doping concentration and a reduced base width as compared with present self-aligned Si bipolar transistors. The decreasing base width and base resistance are the conditions for increasing the switching speed.

Device Fabrication

The complete vertical semiconductor structure of the Si/SiGe-HBT was grown by molecular beam epitaxy (MBE) on a highly doped (100)-n$^+$-Si-substrate. B and Sb were used as epitaxial dopants for p- and n-layers,respectively. The Ge fraction of the commensurate, strained base layer was 20%. This low Ge percentage yields an increased critical SiGe thickness. For the HBT a double mesa structure was chosen. A cross-section of the transistor is given in Fig. 1. The emitter area is 5x35 μm^2. The cap and emitter layers were etched selectively, using a solution, which stops etching at the SiGe surface. The second mesa was formed by reactive ion etching and wet chemical etching [1]. After depositing a sputtered SiO$_2$ film, the contact windows were opened. For improving the device contacts BF$_2$ was implanted into the base contact window and As into the emitter. The implantation dose was

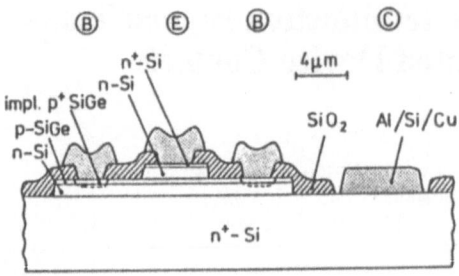

Fig.1. Cross-section of the heterojunction bipolar transistor

$10^{15}/cm^2$ or more. Contrary to reported HBT fabrication [2], annealing was
carried out in a conventional diffusion furnace for 2 up to 8 minutes at
900°C. Finally the Al/Si/Cu-metallization was sputtered, which was followed
by an annealing step at 400°C.

Results and Discussion

The contact resistance is a serious problem during HBT fabrication. An
improper base contact or a high total base resistance may produce base
current dependent V_{CE} offset voltages and reduced current gains or may
suppress the complete DC output characteristics, as was the case for a non-
implanted sample with the layer sructure given in Fig. 2. A base contact
implant revealed a current gain of 800 (Fig.2) or more than 1000 for a
square emitter transistor. This is up to now the highest value for a
Si/SiGe-HBT, which has been reported. The output characteristics of Fig.2
is marked by the generally observed collector current non-linearities at
low currents and by a poor current rise at low V_{CE}, which partly comes from
a high emitter contact resistance. An emitter implant and proper annealing
conditions improve the transistor, as it is demonstrated in Fig.3 for a

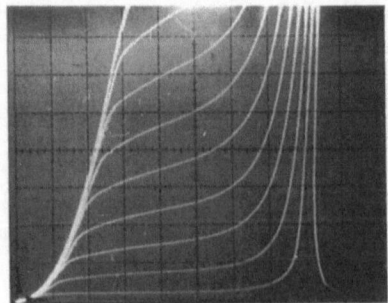

Layer	$N(cm^{-3})$	t(nm)
Cap	5×10^{19}	90
Emitter	1×10^{18}	180
Base	1×10^{18}	80

I_B=10 μA/step
I_C= 5 mA/div
U_{CE}=0.5 V/div

Fig.2. Measured DC output characteristics and HBT layer structure. The
base contact is implanted and annealed 2 min at 900°C. $\beta_{max} \approx 800$

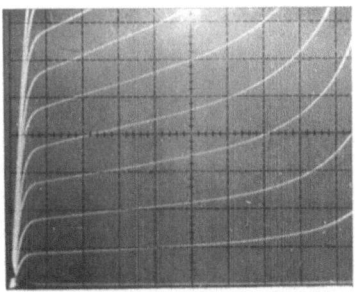

Layer	N(cm^{-3})	t(nm)
Cap	impl. 2x10^{20}	⎫
Emitter	5x10^{17}	⎬ 270
Base	1x10^{19}	80

I_B= 0.1 mA/step
I_C= 1 mA/div
U_{CE}= 0.5 V/div

Fig.3. Measured DC output characteristics and HBT layer structure.
Emitter and base contacts are implanted and annealed 6 min at 900°C.

different sample, which was annealed 6 min at 900°C after the emitter and
base implants. Parasitic resistances were reduced to such a kind that a
V_{CE}-offset is negligible. Moreover, the annealing process annihilated recombination centers in such a way that non-linearities disappeared.

A high temperature anneal is beneficial for the HBT, even if no contact
implantation is performed (Fig.4). However, it must be presumed that the
prevailing doping concentrations yield working contacts. For the samples in
Fig.4 the HBT layer structure of Fig.3 was used except that the emitter of
the non-implanted transistor was covered with a 5x10^{19}/cm^3 doped cap layer.
If annealing the HBT for 8 minutes at 900°C, the current gain is abruptly
reduced. The reason is the dynamic, time- and temperature-dependent critical SiGe base thickness, which is surpassed under these conditions and
consequently, the strained SiGe layer is partly destroyed by stress re-
lease. Using thinner comparable SiGe base layers, the current gain does not
change after annealing 8 minutes at 900°C.

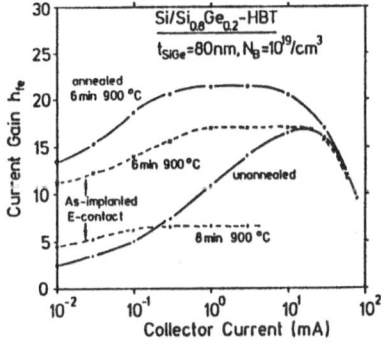

Fig.4. Small signal current gain versus collector current for a
non-implanted (with a cap layer) and an As-implanted HBT.

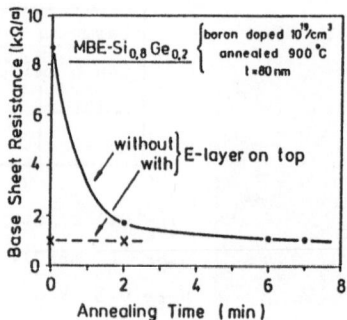

Fig.5. Base sheet resistance versus annealing time

For a high speed transistor the base sheet resistance has to be low. However, there is an unexpected difference between inner and outer base sheet resistance (Fig.5). The inner resistance under the active emitter nearly yields the Si value for the layer structure of Fig.3, but this behaviour strongly depends on the epitaxial growth conditions. The outer uncovered or oxide covered base is marked by an increased sheet resistance, which is reduced by annealing at 900°C.

The first transit frequency measurement resulted in 8.5 GHz for V_{CE}=1V. This value seems to be low [3], but it should be noted that the measured transistor with a layer structure different from Fig.3 was characterized by a collector doping concentration of $10^{17}/cm^3$ and by an enlarged outer base sheet resistance. It may be concluded that the transit frequenccy will drastically improve, if the epitaxial doping concentrations and the base sheet resistance are conform to a high frequency transistor.

Acknowledgement

The author would like to thank E. Kasper from AEG Research Center Ulm for depositing the MBE layers.

References

1. Schreiber,H.-U.; Bosch,B.G.; Kasper,E; Kibbel,H.: Si/SiGe heterojunction bipolar transistor with base doping highly exceeding emitter doping concentration. Electronics Lett.25 (1989) 185-186.

2. King,C.A. et al.: Si/Si$_{1-x}$Ge$_x$ heterojunction bipolar transistors produced by limited reaction processing. IEEE Electron Device Lett.10 (1989) 52-54.

3. Harame,D.L. et al.: High performance Si and SiGe-base pnp transistors. IEDM Technical Digest (1988) 889-891.

New Devices

Cryoelectronic Application of a Hybrid Device Concept Based on Semiconducting and Superconducting Components

J. PARISI, W. CLAUSS, U. RAU, A. KITTEL, M. BAYERBACH, J. PEINKE

Physikalisches Institut, Lehrstuhl Experimentalphysik II, Universität Tübingen, Morgenstelle 14, D-7400 Tübingen, F. R. Germany

Abstract

We have experimentally realized a miniaturized approach to a cryoelectronic device family, the operating principle of which looks highly promising for both analog and digital circuit applications.

Introductory Remarks

The realization of faster electronic devices with the potential of high packing density and correspondingly low power dissipation requires new concepts. So far, cryoelectronic devices based on superconducting phenomena such as the Josephson effect have played an important role in research and development. On the other hand, new semiconducting heterostructures based on hot electron effects are being extensively investigated at present. It appears that new concepts utilizing the attractive features of both superconducting and semiconducting components in a novel hybrid configuration are discussed very little up to now.

Our contribution deals with the first experimental realization of a cryoelectronic device family including magnetic field effect transistors, logic gates, and memory circuits. The devices we have in mind utilize the magnetic control of the filamentary current flow during impurity impact ionization breakdown in a semiconductor at low temperatures (for an overview of magnetoresistance and current filamentation see /1-8/). Here the magnetic control field is generated by superconducting lines attached to the surface of the semiconductor. We have modeled and characterized miniaturized device configurations using extrinsic germanium, silicon, and gallium arsenide semiconductors together with superconducting niobium lines for generating the magnetic control field and for all interconnections.

Experimental

The samples investigated were prepared from homogeneously p-doped
Ge (Si) single-crystal slices with an In (B) acceptor concentra-
tion in the range of $(1-3) \times 10^{14}$ cm^{-3} $(7 \times 10^{14} - 1 \times 10^{16}$ cm$^{-3})$.
The compensation ratio was definitely smaller than 5×10^{-3} (1×10^{-2}). Moreover, high-purity n-type GaAs epitaxial layers with a
Si donor concentration ranging from 6×10^{14} to 5×10^{15} cm^{-3} at
about 50 percent compensation ratio were deposited on semi-insu-
lating GaAs substrates. The semiconductor samples (of about 0.1
mm thickness) carried properly arranged ohmic contacts placed on
one of the two broad surfaces. The distance between the contacts
varied from 0.5 mm to 10 µm. The ohmic contacts were fabricated
either by alloying or by ion implantation /9-12/.

To provide these contacts with an electric field, a voltage bias
was applied to a series combination of the sample and an 1 Ω load
resistor. The sample current was obtained from the voltage drop
at the load resistor. An external magnetic field perpendicular to
the broad sample surfaces could be generated either by a superconducting coil surrounding the semiconductor sample or by a superconducting loop attached directly to the surface of the sample.
Shaped in the form of a long hairpin, the superconducting Nb lines were fabricated by standard thin-film technology, and electric
separation from the underlying semiconductor was achieved by a
thin-film insulator. During the experiments, the samples investigated were kept in direct contact with the liquid-helium bath at
4.2 K and were protected against external visible and far-infrared
irradiation by an appropriate Cu shield.

Results and Conclusions

A first experimental demonstration of the basic operating principle underlying this novel cryoelectronic transistor scheme has
been given recently /13/. Based on the hybrid concept outlined a-
bove, a three-terminal device was fabricated acting as a magnetic
field effect transistor. Adequate magnetic control could be achieved already with field strengths as small as 1 - 10 mT. This extremely sensitive response to the magnetic control field is a result of the high carrier mobility at low temperatures and the correspondingly large Lorentz force acting upon the moving charge
carriers /14/. Beyond analog operation as magnetic field effect

transistor, the magnetoresistive effect further enables digital cryoelectronic device applications including logic gates and data storage /15/. The digital operation modes are based on the magnetically induced switching behavior of current filaments between two stable states of different conductance (logic gates) or between two stable channels of high conductance (memory device).

In order to evaluate the limitations of the most important performance characteristics, we have experimentally performed a miniaturized approach to both analog and digital cryoelectronic device concepts using different sample geometries (contact distances) and semiconductor materials (Ge, Si, GaAs). From the measured electric transport properties (carrier density, mobility, and drift velocity) in the post-breakdown regime of the materials analyzed, we derive values of about 50 ps/μm for the time of flight per unit length of the charge carriers. Taking further into account the typical electric breakdown fields of a few V/cm and the relating current flow of some mA, we obtain values of less than 1 μW/μm for the power losses per distance between the ohmic contacts. Thus, nearly independent of the semiconductor material employed, the miniaturized sample geometry enables operation up to frequencies in the GHz regime. This estimate agrees reasonably well with the experimentally observed frequency limits of about 10 MHz up to a few GHz for samples with lengths ranging from 0.5 mm to 10 μm, respectively.

Particularly attractive features of such a hybrid configuration are its simplicity, the utilization of well-established semiconductor technology, and its robust nature promising a high production yield. Most importantly, the power losses of these devices are similar to those in a superconducting Josephson junction, representing the high-speed circuit with the lowest power consumption up to now. We take note of the fact that superconducting wiring can preserve extremely fast pulse rise times required in ultrafast electronic systems. In this way, combining semiconducting (Ge, Si, or GaAs) device components operating at low temperature with superconducting (Nb) lines for magnetic control and for all interconnections discloses an interesting new direction which has been neglected so far and which definitely deserves much more consideration.

578

References

1. Mannhart, J.; Huebener, R.P.; Parisi, J.; Peinke, J.: Magnetic
   control and switching of current filaments in a semiconductor.
   Solid State Commun. 58 (1986) 323-325.

2. Mayer, K.M.; Gross, R.; Parisi, J.; Peinke, J.; Huebener, R.P.:
   Spatially resolved observation of current filament dynamics in
   semiconductors. Solid State Commun. 63 (1987) 55-59.

3. Mayer, K.M.; Peinke, J.; Röhricht, B.; Parisi, J.; Huebener,
   R.P.: Spatial and temporal current instabilities in germanium.
   Physica Scripta T 19 (1987) 505-510.

4. Mayer, K.M.; Parisi, J.; Huebener, R.P.: Imaging of self-gen-
   erated multifilamentary current patterns in GaAs. Z. Phys. B -
   Condensed Matter 71 (1988) 171-178.

5. Rau, U.; Peinke, J.; Parisi, J.; Huebener, R.P.: Switching behavior of current filaments in p-germanium connected in parallel. Z. Phys. B - Condensed Matter 71 (1988) 305-310.

6. Mayer, K.M.; Parisi, J.; Peinke, J.; Huebener, R.P.: Resonance
   imaging of dynamical filamentary current structures in a semiconductor. Physica D 32 (1988) 306-317.

7. Peinke, J.; Parisi, J.; Röhricht, B.; Mayer, K.M.; Rau, U.;
   Huebener, R.P.: Spatio-temporal instabilities in the electric
   breakdown of p-germanium. Solid State Electron. 31 (1988) 817-
   820.

8. Mayer, K.M.; Parisi, J.; Rau, U.; Peinke, J.; Huebener, R.P.:
   Nascent states of current filamentation in semiconductors governed by negative differential resistance. Phys. Rev. B (to be
   published).

9. Peinke, J.; Parisi, J.; Röhricht, B.; Mayer, K.M.; Rau, U.;
   Clauß, W.; Huebener, R.P.; Jungwirt, G.; Prettl, W.: Classification of current instabilities during low-temperature breakdown in germanium. Appl. Phys. A 48 (1989) 155-160.

10. Beck, S.: Tieftemperatur-Dunkeldurchbruch in einkristallinem
    p-Silizium. Diploma Thesis (Tübingen, 1988).

11. Mayer, K.M.: Stromfilamente und deren raum-zeitliche Dynamik
    in Halbleitern. PhD Thesis (Tübingen, 1989).

12. Ion-implanted contacts were fabricated at the Microstructure
    Laboratory of the University of Stuttgart.

13. Clauß, W.; Parisi, J.; Peinke, J.; Rau, U.; Huebener, R.P.:
    A miniaturized approach to the cryoelectronic magnetic field
    effect transistor. J. Phys. (Paris) 49 (1988) C4-637-639.

14. Parisi, J.; Rau, U.; Peinke, J.; Mayer, K.M.: Determination
    of electric transport properties in the pre- and post-break-
    down regime of p-germanium. Z. Phys. B - Condensed Matter 72
    (1988) 225-233.

15. Mannhart, J.; Parisi, J.; Mayer, K.M.; Huebener, R.P.: A new
    cryoelectronic device family. IEEE Trans. Electron Devices ED
    -34 (1987) 1802-1806.

Si Permeable Based Transistor Realization Using a MOS Compatible Technology

P. LETOURNEAU, G. VINCENT, P. PERRET, P.A. BADOZ and E. ROSENCHER *

CNET-CNS, BP 98, 38243 Meylan Cedex, FRANCE
* Present address: Thomson-CSF, BP 10, 91401 Orsay Cedex, FRANCE

Abstract: We have fabricated silicon permeable base transistors (PBT) using a MOS technology process. The critical lithography defining the metal base grating is a standard optical lithography, which allows submicronic periodicity of the base grating, and therefore permits the realisation of PBT with either normally on or normally off transistor characteristics. Thanks to the high current density achievable in these devices and the high degree of compatibility with MOS technologies, PBT clearly appears as a possible alternative for bipolar metal oxide semiconductor (BICMOS) applications.

I INTRODUCTION:

The permeable base transistor is considered as a very promising device. Indeed PBT's simulations (1) predict high power millimeter wave amplification and GaAs (2) as well as Si (3) realizations exhibit a f_{max} in the few tens Gigahertz range. These high performances can be accounted for by the structure of the PBT which looks like a vertical field effect transistor with a very short gate length.

Until now, two different technologies have mainly been used:

i) In the first one, the metallic base is embedded in the semiconductor lattice by epitaxial regrowth of the semiconductor (4). This method involves complex epitaxial techniques and is therefore not compatible with MOS technology. ii) In the second one, the metal base is evaporated in the bottom of semiconductor etched grooves (3,5); this avoids the critical epitaxial step. But two important problems remain uncompletely solved i.e. the passivation of the sidewalls of the Si grooves and the electrical connection to the top emitter.

We present here a fabrication technology which can be entirely integrated in a MOS process, using:

i) an optical lithography in order to define the metal grid,

ii) the passivation of the Si sidewalls with a thin thermally grown oxide

iii) the filling of the openings using a thermohardening dielectric before taking the emitter and base contacts.

II TECHNOLOGICAL PROCESS:

The starting material is a $1\Omega.cm$, n-type Si, phosphorus doped, 4 in substrate. Device active areas are first defined by an optical lithography and chemical etching of a thermal 500 nm field oxide.

A first arsenic implantation is performed on the front side in order to get a good emitter ohmic contact while a second one improves the back collector contact. Then a 150 nm thermal oxide is grown and covered by a 30 nm tungsten film sputter deposited. A second optical lithography defines a submicrometer grating in a positive resist (fig.1a). This grating is transfered successively by anisotropic reactive ion etching (RIE) in W using SF_6, in SiO_2 using CHF_3, and finally in Si using a mixture of SF_6 and O_2 (fig.1b). This then leads to 600 nm deep Si grooves with abrupt sidewalls (about 85 degrees).

580

The SF_6 RIE plasma selectivity between silicon and tungsten is small and therefore, the 30 nm thick W film is completely removed during etching of the silicon grooves while the oxide film acts as countermask; the remaining oxide film is subsequently removed by a dip in a HF solution.

The sidewall passivation is obtained using a third grown thermal oxide etched using anisotropic RIE. The result is a few tens of nm oxide on the sidewall whereas the top and the bottom of the trenches are oxide free (f ig.1c). A 10 nm thick Pt film is then electron gun evaporated on the structure. Platinum silicide has been chosen for its good Schottky barrier height on n type silicon, around 0.84 eV. A thermal anneal at 420°C leads to the formation of PtSi on the top and bottom of the grooves, while the oxide prevents the reaction of Pt on the sidewalls. Non reacted Pt is then selectively etched with aqua regia (fig.1d). On the sidewalls, the absence of Pt is confirmed by electrical insulation between the base and the emitter contacts.

The last point concerns the planarisation of the structure.This critical step is achieved by covering the whole structure using a polyimide (PPQ200). Because of both the different heights between emitter and base levels, and the different thicknesses of polyimide on these levels, two lithographics and two different polyimide etching steps are needed in order to open successively the emitter and base contacts.

At last, a 300 nm film of aluminium is evaporated. A fifth lithography defines emitter and base pads. Aluminium is chemically etched with orthophosphoric acid solution (fig.1e).

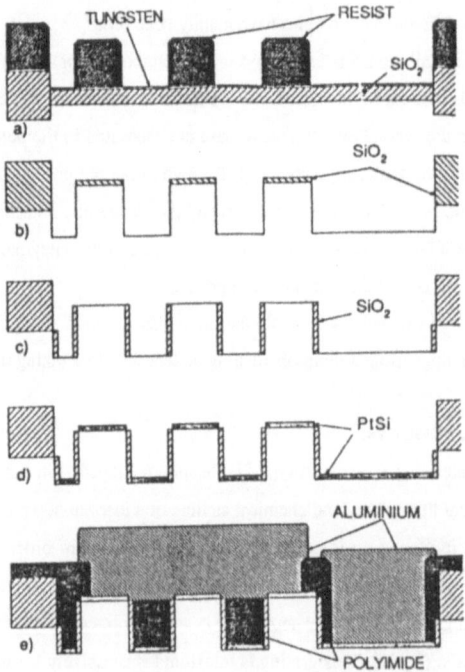

Fig.1 Si permeable base transistor fabrication sequence.

We have succesfully realized PBT of different emitter areas, i.e. 4×10^{-6} cm^2 and 4×10^{-4} cm^2 with various metal grid spacings: 0.4 µm, 0.7 µm and 1.0 µm. Furthermore, we shall stress that this technology makes it possible to test the PBT devices just after the chemical etching of the non reacted Pt: indeed, two tungsten probes carefully laid down on respectively the top of the grooves and the base pad take respectively the emitter and base contacts

III ELECTRICAL RESULTS:

We now present the characteristics of fully completed PBT, i.e. after the planarization process. We shall note first that in these devices, both base-collector and base-emitter Schottky diodes exhibit excellent rectifying properties with leakage current density less than 0.15 A/cm^2 at 5 V reverse bias. Static transistor characteristics, both normally on and off, have been obtained on devices with different grid periodicity; we shall note that these various electrical behaviors are obtained on devices realized on the same wafer, with a good reproducibility from wafer to wafer. Devices with a 2 µm grid periodicity (1 µm base width and 1 µm emitter width) exhibit a normally on behavior, while devices with a 1 µm grid periodicity (0.6 µm base width and 0.4 µm emitter width) present a normally off behavior. Fig. 2a shows common base transistor characteristics of such a device. It is clearly seen on the figure that the transistor is normally off, and turned on with a negative emitter-base bias V_{BE}. Fig. 2b shows common base characteristics of PBT with intermediate 1.4 µm grid periodicity (0.7 µm PtSi and 0.7 µm Si). This transistor, which is only partially open, can be either turned off (switch-off bias of 1.5 V), or turned on, depending on the sign of emitter-base bias.

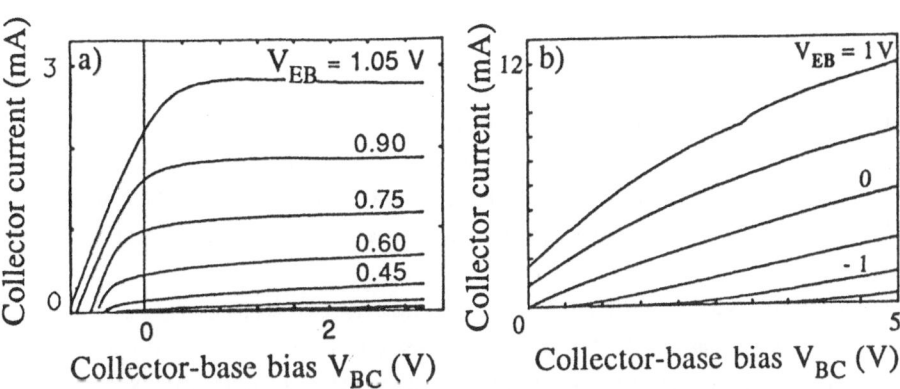

Fig.2 a) Common base characteristics of a 1 µm periodicity Si PBT, i.e. 0.6 µm PtSi base and 0.4 µm Si channel. This transistor is normally off and turned on with a negative emitter-base bias.

b) Common base characteristics of a Si PBT with a 1.4 µm periodicity, i.e. 0.7 µm PtSi base and 0.7 µm Si channel. This transistor which is only partially open, can be either turned off (switch-off bias of 1.5 V), or turned on, depending on the sign of emitter-base bias.

It clearly appears that it is necessary to apply high emitter-base bias to switch off or open the device. This effect is probably coming from a high emitter series resistance, due to a non optimum emitter implantation which is critical for the ohmic PtSi contact. Furthermore, the transconductance is clearly damaged after the planarisation process. This degradation is believed to come from parasitic MOS phenomena which occur close to the emitter.

IV CONCLUSION:

We have presented a fabrication process of etched groove Si PBT aiming at full compatibility with MOS technology. In that respect, the following problems have been addressed and solved:

i) use of classical optical lithography and optimized anisotropic etching,

ii) thermal oxide passivation of Si sidewalls, and

iii) emitter-base self alignment using this thermal oxide.

Si PBT with excellent electrical characteristics have thus been fabricated, while some technological steps have still to be optimized, in particular the implantation profile in order to minimise access resistance. This study clearly shows that this Si PBT technology is a very attractive alternative to the highly complex BICMOS technology.

ACKNOWLEDGEMENT

The authors are grateful to C. Puissant for her skillfull technical assistance during the etching process. They also wish to express thanks to Pr. J.C. Pfister for many enlightening discussions and critical reading of the manuscript.

REFERENCES

(1) C.O. BOZLER and G.D. HALLEY:" Fabrication and numerical simulation of the permeable base transistor". I.E.E.E Transaction on Electronic Devices, 1980, ED - 27, 6, p 1128.

(2) M.A. HOLLIS, K.B. NICHOLS, R.A. MURPHY, R.P. GAL, S. RABE, W.J. PIACENTINI and C.O. BOZLER: " Reproducible fabrication of high performance AsGa permeable base transistor". I.E.E.E, IEDM 85, p 102.

(3) D.D. RATHMAN and W.K. NIBLACK : "Silicon permeable base transistor for low phase noise oscillator applications up to 20 Ghz". I.E.E.E, MTT International Microwave Symposium Digest, 1988, vol.1 p. 537.

(4) G. GLASTRE, E. ROSENCHER, F. ARNAUD d'AVITAYA, C. PUISSANT, M. PONS, G. VINCENT, and J.C. PFISTER: "CoSi$_2$ and Si epitaxial growth in <111> Si submicron lines with application to permeable base transistor", Appl. Phys. Letters, 1988, 52, p. 898.

(5) A. GRUHLE, H. BENEKING: "Silicon permeable base transistors fabricated by selective epitaxial growth", Electronics Letters, 1989, vol. 25, N° 1.

Enhanced Transconductance in Deep Submicron MOSFET

W. HÄNSCH and H. JACOBS

Siemens AG, Corporate Research and Development, Microelectronics, EL PT

Otto Hahn Ring 6, 8000 Munich 83, FRG

ABSTRACT: We will demonstrate that the enhanced transconductance which is observed in deep submicron MOSFET at 77K and 300K can be modeled using an extended drift-diffusion approximation with mobility parameters taken from measurements on MOSFETs of the 4M DRAM generation. We can show that for the terminal characteristics of these devices velocity overshoot is of little importance at least for devices with channel lengths larger than 60nm.

I. INTRODUCTION

It is of considerable concern for technology development whether the current/voltage characteristics of a deep submicron ($L < 0.3\mu m$) MOSFET can still be reliably predicted with a drift-diffusion type approach for carrier transport modeling. Terminal characteristics are integral quantities and therefore should only be little sensitive to the exact form of the distributions in the device. It is, however, of interest to find features of the internal distributions that can be detected in the terminal currents. Recently, a series of papers addressing this subject were published from the IBM group [1,2,3] in which Monte Carlo techniques were used to evaluate the electrical characteristics of deep submicron MOSFETs. An important conclusion drawn from their investigation is the observation of a considerable amount of velocity overshoot that leads to an enhanced transconductance for the ultra short gate length devices. We will show that this enhanced transconductance is also reproduced using the extended drift-diffusion approximation.

II. MODELING DEEP SUBMICRON MOSFET

For this investigation we aimed at modeling the MOSFETs considered by the IBM group, using process and device simulation tools. The doping profiles were reconstructed from processing data provided in the literature [4,5] using a 2-D process simulator. In Fig. 1 we show the $I_D(V_D)$ characteristics of our reconstructed devices as compared to the measured data provided in [2]. The simulations were performed with a modified version of MINIMOS which includes the enhanced drift-diffusion model by Hänsch and Selberherr [6] as extended to cover low temperature effects [7] such as freeze out and mobility variation with temperature that includes a temperature dependent saturation velocity v_{sat}. In addition we developed a model for spatial velocity overshoot in the realm of the extended drift-diffusion model. This can not be modeled within the conventional drift-diffusion approach where v_{sat} is the limiting velocity. We showed [8] that the mobility is primarily a functional of the carrier temperature and not of the driving force for the current density. Only if an expansion in this driving force is performed in lowest order a mobility is obtained that only depends on that driving force and accounts for velocity saturation in the usual way [6]. To account for velocity overshoot we have to carry that expansion one step further and collect contributions proportional to the spatial variation of U_T. The result of such a calculation is Eq. (1).

$$\mu_{LISFO} = \frac{\mu_{LISF}}{1 + \alpha \dfrac{\mu_{LI}}{v_{sat}} \dfrac{U_0}{U_T} e_j * grad\ (U_T)} \tag{1}$$

Here e_j is the unit vector in the direction of the current density and U_T is the self consistently calculated carrier temperature divided by the elementary charge $U_T = k_B T_c / q$ [6]. In Equ. (1) μ_{LISF} accounts for lattice (L), impurity (I), surface (S) scattering, and velocity saturation (F) [7]. α is a dimensionless number of order one and is proportional to the ratio of the high and low field diffusion constant. If in a region the particle velocity vector coincides with the direction of a large variation of U_T the denominator is reduced and a velocity larger than v_{sat} is possible. In a constant carrier temperature configuration $(grad(U_T) = 0)$ there is no velocity overshoot.

In any phenomenological approach to carrier transport modeling the parameters put into the mobility model are of crucial importance. A considerable spread of the parameter values can be found in the literature. The mobility model of MINIMOS was matched to experimental data from several sources for temperatures of 300 and 77 K as reported in [7]. This set of parameters also gave the best fit to our typical 4M DRAM MOSFET generation at room temperature. For our investigation that part of the mobility that accounts of velocity saturation is most crucial, whereas the temperature dependence of the low field mobility is not important. A comparison of v_{sat}^{300} and v_{sat}^{77} with that proposed by the IBM group [2] for longer channels indicates that our values are too large. However, if we use the values suggested by the IBM group we are not able to obtain a good fit for our 4M DRAM devices. In Fig. 1 we show the calculated MOSFET characteristics using the mobility parameters that have been adjusted to our 4M DRAM technology. No additional fitting is required to obtain the results at 77K. In the simulation we have included a series resistance of 100 $\Omega\mu$m at both source and drain as reported in Ref. [1].

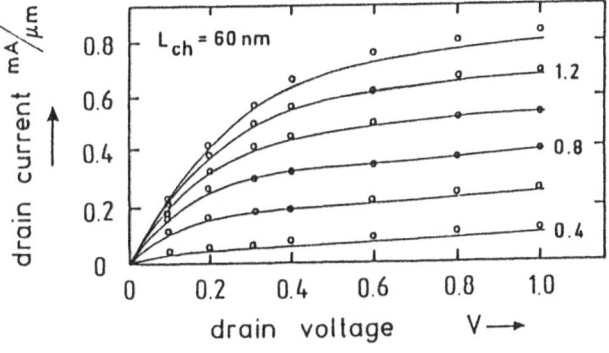

FIGURE 1: Comparison of measured and calculated I-V characteristic of a MOSFET with 60nm channel length at 77K. The measurements were taken from Ref. 2. The simulation (open circles) accounts for an series resistance of 100 $\Omega\mu$m each at the source and drain terminals .

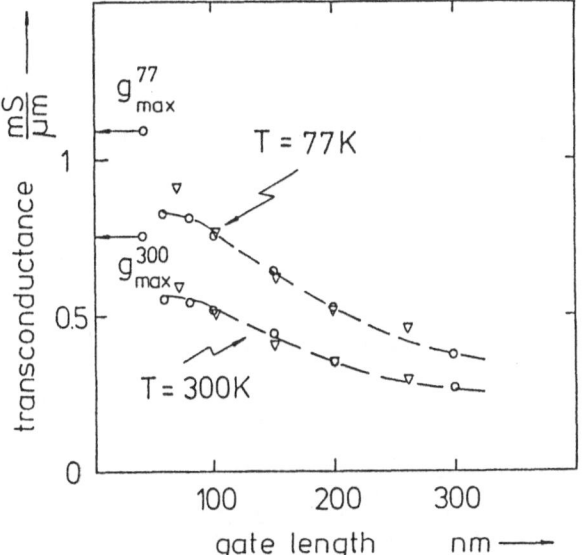

FIGURE 2: Measured (Ref. [2], triangles) and calculated (circles) transconductance for different channel lengths. The calculation was done at $V_G = V_{th} + 0.6V$ and $V_D = 0.8V$; A series restance of 100 $\Omega\mu$m and 200 $\Omega\mu$m was included on source and drain for T=77K and T=300K, respectively; we include the the upper limit $g_{max} = C_{ox}v_{sat}$ for both temperatures.

In Fig. 2 we present the transconductance as a function of gate length and temperature which we obtain in our investigation. The transconductance was calculated with a small signal analysis postprocessor to the MINIMOS program. Fig. 2 shows that the enhanced drift-diffusion model indeed reproduces the measured transconductances very well without any further adjustments. In contrast to the results published in [2] the effect of the velocity overshoot as modeled in Eqn. (1) is small, never exceeding about 10% variation in transconductance. Although we can reproduce the trend of the measured data very well down to a gate length of 100nm there is a noticeable deviation of our calculations from the measured data at the extreme end of gate lengths (smaller than 100nm). At these channel lengths the departure of our results may be due to small deviations in the doping profiles as well as uncertainties in the determination of the gate length which become increasingly important.

III. CONCLUSIONS

Our results demonstrate that extended versions of the drift-diffusion equations can indeed be helpful in evaluating a technology even in the deep submicron region. Provided that the doping profiles can be calculated with sufficient accuracy, the terminal characteristics can still be successfully predicted using mobility parameters which were validated for today's technologies. However, the physical description in the extended drift-diffusion equation might no longer be adequate for the correct determination of the distributions inside the device. Nevertheless the advantage of the phenomenological approach will become obvious in the following: The evaluation of the 42 bias points for the $I_D(V_D)$ characteristics in Fig. 1 took about 1h CPU time on a VP200 vector processor using the most sophisticated MOS model that MINIMOS provides (Model = HOT). The evaluation of the transconductance (6 bias points for various channel lengths) took another 12 min. The memory requirement for the program including the small signal analysis postprocessor is 15MByte. A further reduction of computational cost is achieved by using a less sophisticated MOS model (Model = AVAL), however, at the cost of lower accuracy. These numbers should be compared with the CPU and memory requirements for the Monte Carlo code [2].

In conclusion we would like to emphasize that although the need for more sophisticated modeling is indisputable there is still room for useful applications of extended drift-diffusion models even in the deep submicron region.

References:

[1] S. E. Laux and M. V. Fischetti, "Monte-Carlo Simulation of Submicrometer Si-n MOSFET at 77 and 300K", IEEE EDL9 467 (1988)

[2] G. A. Sai-Halasz, M. R. Wordeman, D. P. Kern, S. Rishton, and E. Ganin, "High Transconductance and Velocity Overshoot in NMOS Devices at the 0.1 μm Gate Length Level", IEEE EDL9 464 (1988)

[3] M. V. Fischetti and S. E. Laux, "Monte Carlo analysis of electron transport in small semiconductor devices including band-structure and space-charge effects", Phys. Rev. B38 Nov.15th (1988)

[4] G. A. Sai-Halasz, M. R. Wordeman, D. P. Kern, E. Ganin, S. Rishton, D. S. Zicherman, H. Schmid, M. R. Polcari, H. Y. NG, P. J. Restle, T. H. P. Chang, and R. H. Dennard, " Design and Experimental Technology for 0.1 μm Gate-Length Low-Temperature Operation FET's", IEEE EDL8 463 (1987)

[5] G. A. Sai-Halasz and H. B. Harrison, "Device-Grade Ultra-Shallow Junctions Fabricated with Antimony", IEEE EDL7 534 (1986)

[6] W. Hänsch and S. Selberherr, " MINIMOS3: A MOSFET Simulator that includes Energy Balance", IEEE Trans. Electron. Dev. ED34 1074 (1987)

[7] S. Selberherr, " MOS Device Modeling at 77K", to be published in IEEE Trans. Electron. Dev. special issue on low temperature electronics

[8] W. Hänsch, M. Orlowski, and W. Weber, "The hot electron problem in submicron MOSFET", Proceedings of the ESSDERC'88, Montpellier France 1988, pp 597

A Planar Reach-through p^+np^+ Device for the Detection of Surface Acoustic Waves

J.C. HAARTSEN and A. VENEMA

Electrical Engineering Department
Delft University of Technology, the Netherlands

Summary

A novel SAW detector is presented. It basically consists of a planar reach-through p^+np^+ structure covered by a piezoelectric ZnO layer. Mechanical displacements in the ZnO layer will induce electric fields that penetrate into the depleted gap between the two p^+ junctions. There they will modulate the potential distribution, which results in a modulation of the injected current. Experiments at 80 MHz demonstrated its operation.

Introduction

Reach-through devices have excellent HF properties. The current mainly consists of drifting charge carriers in a high electric field crossing a depleted area. This results in a short transit time and very little charge storage.

In the past, reach-through diodes used as microwave devices were studied extensively [1]. The considered BARITT diodes consist of MnM or p^+np^+ sandwich structures. Recently, planar MnM reach-through devices on GaAs have been developed to operate as high-speed photodetectors [2]. In these structures the current flows laterally, as opposed to the vertical current in common BARITT diodes. In the current paper a planar p^+np^+ device in silicon is presented as an ultrasonic detector for the detection of Surface Acoustic Waves (SAW).

Theory

The basic planar structure is shown in Figure 1. Two implanted p^+ junctions in a lightly-doped n^- region form the source and drain electrodes of the detector. After a thermal oxidation, the surface is covered with a piezoelectric ZnO layer. Metal contacts to source and drain are placed at the ends of the electrodes to avoid discontinuities in the SAW propagation path, which would cause reflections.

When the drain is reverse biased, the device acts as a bipolar transistor with floating base. If the reversed drain voltage is increased, the drain depletion region will expand and eventually reach the source depletion region. At this reach-through point the gap between source and drain is completely depleted. Majority carriers (holes) are injected by the source across a potential barrier and pulled away by the large drift field of the drain.

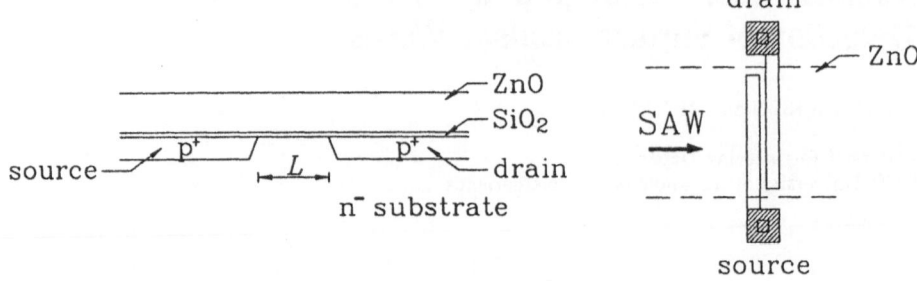

Fig.1. Cross-section and topview of the ultrasonic detector. Source and drain terminals are placed outside the propagation path.

To examine the two-dimensional potential distribution in the gap beyond reach-through, a computer program was developed. The Poisson equation was discretized using a finite-difference scheme, and solved by using a point-by-point over-relaxation method. For small current levels the injected charge density of the holes can be ignored and the space-charge distribution completely determined by the doping level N_d of the n^- bulk region. Figure 2 shows an example for $V_{ds}=-12V$, $V_{bs}=+5V$, $N_d=5.10^{14}$ cm^{-3} and a gap length $L=6\mu m$. In the gap between source and drain a potential barrier is observed. The height ϕ_b of this barrier is defined with respect to the source potential.

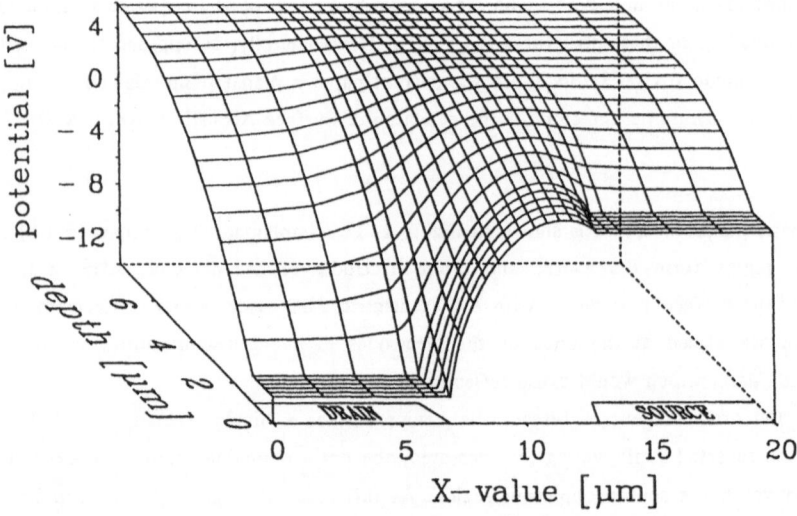

Fig.2. Two-dimensional potential distribution around the gap. The depth is measured from the SiO$_2$-Si interface.

In the gap three regions can be distinguished: 1) the *injection* or *barrier* region between the source junction and the barrier maximum, where diffusion mechanisms dominate; 2) the *ohmic* region just after the barrier, in which the injected holes are accelerated by the rising electric field ($v=\mu E$); 3) the *drift* region, in which the holes have reached the scattering velocity ($v=v_s$). By increasing the drain reverse voltage the drift region will expand at the expense of the ohmic and injection regions. At high reverse voltages the current is dominated by drift.

The thermionically injected current density is controlled by the barrier height ϕ_b:

$$J_{p^+\to n} = A^* T^2 exp(\frac{q\phi_b}{kT}) \qquad [A/m^2] \qquad (1)$$

in which A^* is the Richardson constant for holes [3]. If the reverse voltage is increased, the barrier will be reduced which will give an exponential rise in the injected current density. As can be seen in Figure 2, ϕ_b is minimal at the silicon surface and rapidly increases with increasing depth. As a consequence, the lateral current will be concentrated at the silicon surface.

For larger current densities the injected space-charge density must be taken into account. The effective charge density in the gap will increase, which will retard the exponential rise in the current. At high current densities the potential distribution is completely determined by the injected charge and the current becomes space-charge-limited (SCL) [3]. Due to the current concentration at the silicon surface, SCL effects will already occur in this planar structure at relatively low current levels.

Mechanical displacements in the piezoelectric ZnO layer caused by a SAW will induce electric fields that will penetrate into the depleted gap and change the potential distribution. This will cause a modulation of ϕ_b and thus of the injected current. This device has been called the Barrier-Modulated Tap [4]. The sensitivity can be obtained with:

$$g_e = \frac{\partial I}{\partial \phi_b} = -\frac{qI_0}{kT} \qquad [S/m^2] \qquad (2)$$

and is linearly propartional to the bias current I_0.

Results

Monolithic test devices have been realized in a ZnO-SiO_2-Si layered structure with a 9 μm thick epilayer of 10 ohm.cm, a 0.1 μm thick oxide layer and a 10 μm thick sputtered ZnO layer. To compare the experimental results with the theory an equivalent circuit model in SPICE has been developed [4]. In Figure 3a the measured DC response is depicted, which shows the injection current vs. V_{ds}. After reach-through an exponential rise in the current is observed, which retards at higher current levels due to SCL effects. In the equivalent circuit the SCL effect is modelled by a series resistance.

590

To obtain the AC characteristic, interdigital transducers with an operation frequency of 80 MHz have been placed on the test chip to generate SAWs. In Figure 3b the normalized sensitivity vs. the bias current I_0 is shown. In accordance with equation (2) the sensitivity is proportional to the bias current. The non-linearity is caused by a feedback effect: the drain-induced barrier modulation which is inversely proportional to I_0 [4]. In the equivalent circuit this feedback effect is modelled by a shunt resistance between source and drain. At large bias current, SCL limits the feedback effect, which explains the linear response.

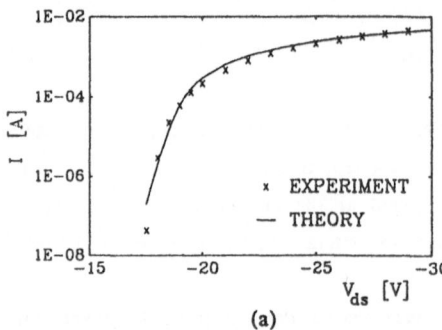 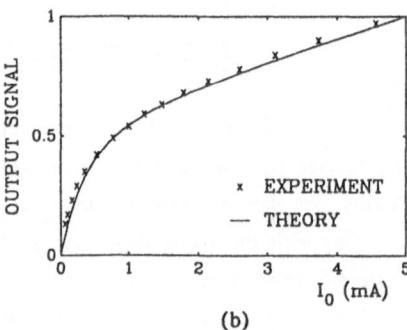

(a)   (b)

Fig.3. Experimental and theoretical results: a. DC response; b. normalized AC response at 80 MHz

Conclusions

A novel reach-through detector for the detection of ultrasonic waves has been presented. The sensitivity is proprotional to the bias current. The measured DC and AC response at 80 MHz agree well with theoretical results obtained with an equivalent circuit model. The measurement of the detector bandwidth was limited by the frequency response of the applied interdigital transducers, but is expected to extend to several hundreds of MHz.

References
1. Coleman, D.J.: Transit-Time Oscillations in BARITT Diodes, J. Appl. Phys. 43 (1972) 1812-1818
2. Ito, M.; Wada, O: Low Dark Current GaAs Metal-Semiconductor-Metal (MSM) Photodiodes Using WSi$_x$ Contacts, IEEE J. Quant. Electron. QE-22 (1986) 1073-1077
3. Chu, J.L; Persky, G; Sze, S.M.: Thermionic Injection and Space-Charge-Limited Current in Reach-Through p^+np^+ Structures, J. appl. Phys. 43 (1972) 3510-3515
4. Haartsen, J.C.; Venema,A: The Barrier-Modulated Tap: A New Detection Method in Silicon, IEEE Ultrasonics Symp. Proc. (1988) 159-163

A Monolithic Integrated Balanced Mixer Circuit for 9 to 12 GHz

A. Schüppen, L. Vescan, C. Werres*, H. Beneking**

 Institute of Thin Film and Ion Technology(ISI), KFA-Jülich
 5170 Jülich, W-Germany
* present address: Telefunken electronic, Heilbronn, W-Germany
**Institute of Semiconductor Electronic, RWTH Aachen, W-Germany

Abstract
Selectively grown silicon camel diodes were integrated into a
microstrip-line balanced mixer circuit using a rat-race
coupler. The cutoff frequency of the single diodes lies in
the range 50 GHz to 330 GHz depending on diode area. This
monolithic integrated mixer circuit exhibits a minimum of
conversion loss of 7 dB and a SSB noise figure of 7.5 dB
(P_{LO}=14 dBm) at 10.5 GHz.

Silicon Camel Diodes

The advantage of bulk unipolar diodes ($n^+n^-p^+n^+$) over Schottky
diodes consists in the possibility to tune the barrier height
by mainly changing thickness and doping concentration of the
p^+-layer /1/. Low pressure vapour phase epitaxy (LPVPE) at low
temperature allows the growth of the required thin and highly
doped multilayer structure /2/. Si-camel diodes prepared from
material grown by LPVPE were reported to have a conversion loss
of 7 dB at 12 GHz/3,4/. Here, the monolithic integration of two
camel diodes into a balanced mixer is examined. Starting with a
high resistivity p-type Si-substrate (2000Ωcm) n^+-regions were
formed by ion implantation using a SiO_2 mask. The diodes were
selectively grown by LPVPE in dry etchted stripes in SiO_2
(1x6 to 2x60 μm^2) to achieve low series resistances and
capacitances. The fabrication process has been developed
earlier/5/. Fig.1 shows a cross section of a complete camel
diode.

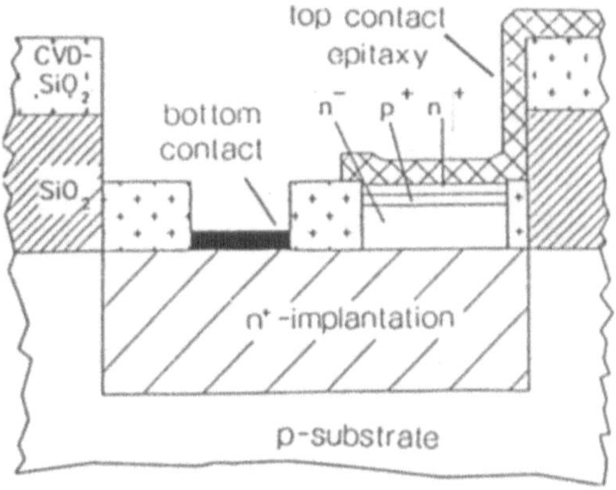

Fig.1. Cross-sectional
view of a selectively
grown silicon cameldiode

DC-measurements for finding pairs of diodes in a special position for the balanced mixer with nearly identical I-V characteristics revealed the good uniformity of the epitaxial layers. The cutoff frequency obtained from S-parameter measurements are in the range 50 GHz to 330 GHz depending on diode area. For mixer applications the cutoff frequency must be 4 to 6 times higher than the operating frequency /6/.

Microstripline Components

Each passive component of a microstrip network can be realized with different microstriplines by changing its length and width which for a given impedance Z_L depends on the height of the substrate, the permittivity, the thickness of metallization and the operation frequency /7/. This mixer circuit consists of a directional coupler, two high-pass filters, two $\lambda/4$ matching lines, a pair of diodes, two low-pass filters and an "air bridge". A photograph of the balanced mixer is shown in fig.2.

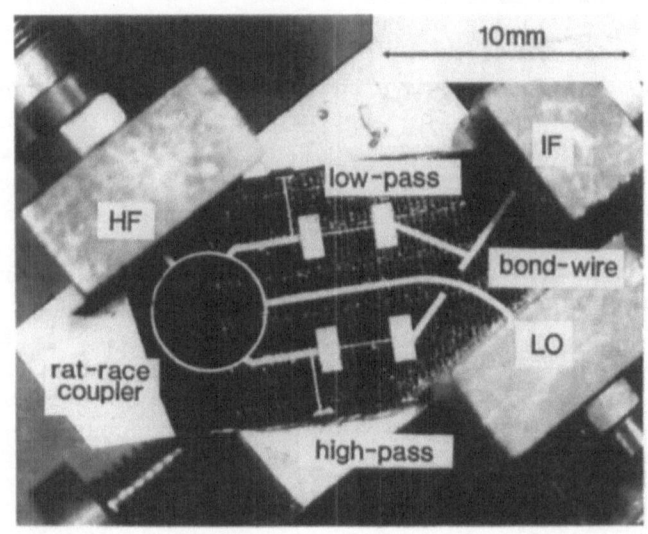

Fig.2 Photograph of the monolithic integrated mixer using two camel diodes

The rat-race coupler sums up both input signals (Fig.3: branch 1,3), i.e. the high frequency signal HF and local oscillator signal LO and divides them symmetrically to the two output branches (2,4). The advantage of this hybrid coupler arises from its high isolation D of HF and LO at operation point (Fig.3a), its good symmetry of the coupled output signals (Fig.3b) and its low voltage standing wave ratio . The isolation from branch 3 to 1 is defined as /7/:

$$D_1 = 20 \log(|S_{12}|/|S_{13}|) \tag{1}$$

where S_{12} and S_{13} are the transmisson coefficients from port 2 to 1 and from port 3 to 1, respectively. The coupling of the input branches 1 and 4 is: $\quad C_{14} = 20 \log(1/|S_{14}|)$ (2)

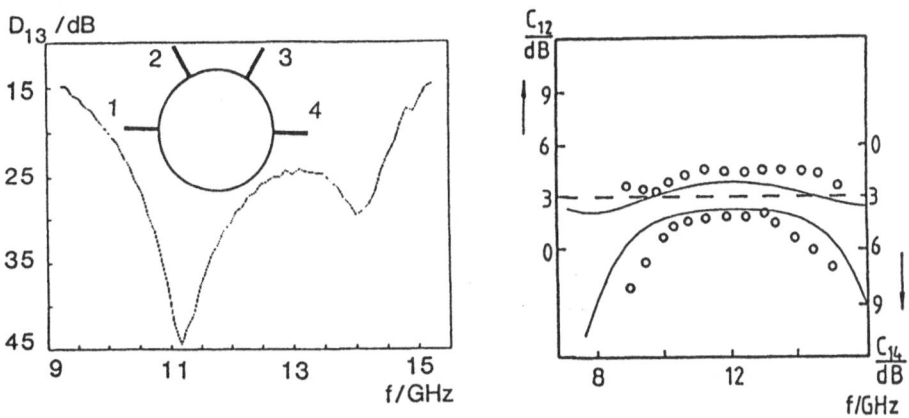

Fig.3 Isolation of the rat-race coupler versus frequency
a) D_{13} and b) C_{12} and C_{14}, -- simulated, oo measured

The high-pass filters (Fig.4) consisting of a 50Ω-line and a short-circuit terminated 80Ω- λ/4-line should suppress the IF-current through the LO resistance. This reduces the conversion loss considerably. The camel diodes which lie between the high-pass and the low pass filters (Fig.2) generate the IF signal. The low-pass filters (Fig.5) connected in series with the diodes suppress HF and LO signals at the IF output. Because a rat-race coupler was used it isn't possible to have a crossing free microstrip design, but measurements at a special test circuit with a bond wire yielded a transmission lower than -40 dB from the HF to the IF side.

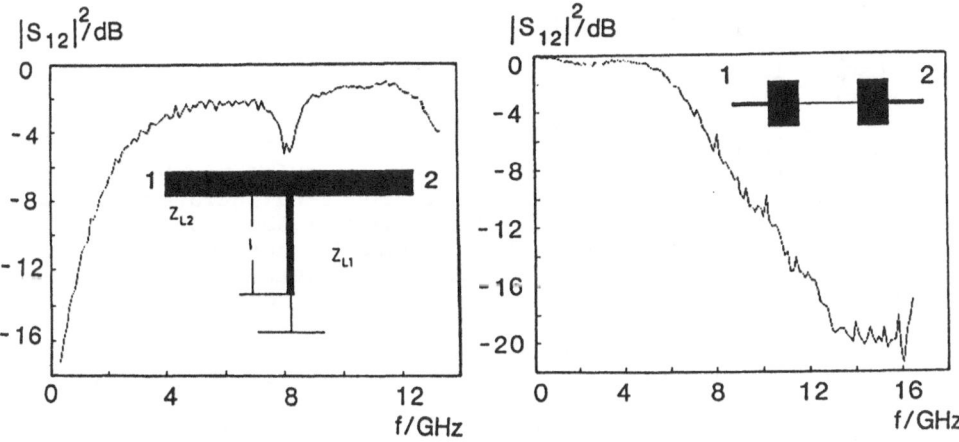

Fig.4. Transmission of a
high-pass filter

Fig.5. Transmission of a
low-pass filter

594

Conversion Loss L and Noise Figure F_{SSB}

The conversion loss of a broad band mixer can be written as a product of four factors/8/: $L = P_{HF}/P_{IF} = L_{intr} \times L_{IF} \times L_{HF} \times L_e$, where L_{intr} are conversion losses of each intrinsic camel diode, which arises from additve mixing principle. L_{IF} is the loss at the IF port due to series resistance of the mixer diodes. L_{HF} takes into consideration loss of mismatching at the HF port and L_e is the insertion loss of the microstripline circuit, which reaches 0.4 dB/cm for a 50Ω-microstripline at 12 GHz on a high resistivity p-substrate (2000Ωcm)/9/.The conversion loss measurement was done without additional matching elements and the single sideband noise figure F_{SSB} was measured by y-factor method/3/. The results (Fig.6) of conversion loss and noise measurement show a relative good behaviour of the monolithic balanced mixer circuit in the frequency range of 9.6 GHz to 11.8 GHz.

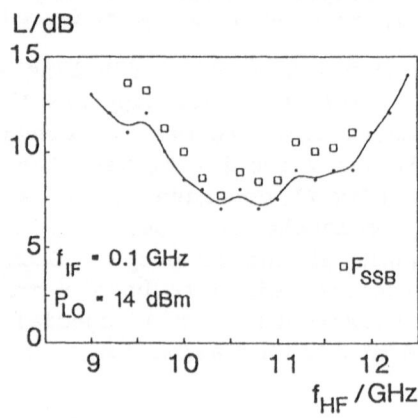

Fig.6. Conversion loss and single sideband noise figure of a balanced camel diode mixer for $f_{HF} > f_{LO}$ and $P_{HF} = -20dBm$
Diode area $1\times20\mu m^2$

References
1. Shannon, J.M.: A Majority-Carrier Camel Diode. Appl.Phys.Lett. 35 (1979) 63-65
2. Vescan, L.; Beneking, H.: LPVPE-Grown Silicon Camel Diodes. Electron.Lett. 22 (1986) 994-996
3. Werres, C.; Vescan, L.; Beneking H.: Noise in Epitaxial Silicon Bulk Unipolar Diodes. Electron.Lett. 23 (1987) 613-614
4. Vescan, L. and Beneking, H.: Effectiveness of Si-LPVPE for Narrow, Highly Doped Multilayers. Proc.2-nd Int.Symp. Si-MBE eds.Bean,J., Schowalter,L., J.Electrochem.Soc. 88-8 549-558
5. Schüppen, A.:Integrierter Cameldioden-Mischer. Master Thesis Inst.of Semicond.Electr. RWTH Aachen, W-Germany (1988)
6. Dragone, C.: Performance and Stability of Schottky Barrier Mixers. Bell.Syst.Techn.J. 51 (1972) 2169-2192
7. Mehran, R.: Grundelemente des Rechnergestützten Entwurfs von Mikrostreifenleitungs-Schaltungen, ed. Wiss.Techn. Bücher H. Wolf, Aachen 1983, W-Germany
8. Held, D.N.; Kerr, A.R.: Conversion Loss and Noise of Microwave and Millimeter Wave Mixers. IEEE-MTT-26 (1978) 49-60
9. Bahl, I.J.; Trivedi, D.K.: A Designer's Guide to Microstrip Line. Microwave May (1977) 174-180

High Temperature Superconducting Ceramic RF SQUID

Ilyichyov E.V., Tulin V.A., Zakosarenko V.M.

Institute of Microelectronics Technology and High Purity
Materials, USSR Academy of Sciences, 142432 Chernogolovka,
Moscow District, USSR

The discovery of high temperature superconductivity of metal oxide compounds is of particular interest from the viewpoint of practical application of these materials. The simplest and most promising device is the superconducting quantum interferometer (SQUID). We have fabricated several versions of Zimmerman-type RF SQUIDs from bulk ceramic $YBa_2Cu_3O_x$ samples and studied their characteristics. The investigations have revealed that high temperature superconducting ceramic SQUIDs exhibit the properties similar to those of conventional niobium SQUIDs. Yet, they display some distinctions that are primarily related to the superconducting properties of the ceramics and the macroscopic size of the weak link $(d > \lambda, \xi$, where λ and ξ are magnetic field penetration depth and coherence length, respectively). This refers both to bulk ceramic devices and film samples.

It has been found that the operation distinctions of ceramic SQUIDs can be described by the relation between the size of the magnetic vortex in the ceramics (in the intergrain volume) and the size of the weak link (junction). If the size of the vortex is less than that of the junction, the SQUID operates as a current concentrator with the junction as a "bulk" SQUID. In the reverse case the high temperature superconducting ceramic SQUID is analogous in operation to conventional nonhysteretic SQUIDs.

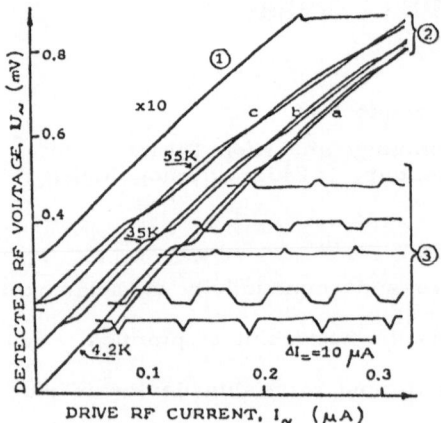

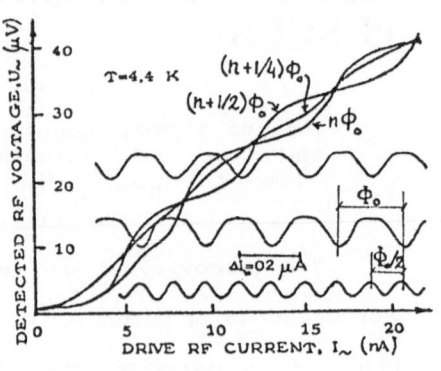

Fig.1 Characteristics of the SQUID with the size of the junction exceeding that of the vortex in ceramics (with a single weak link). The RF V–I characteristic of the SQUID (1), the initial section of the RF V–I characteristic at different temperatures (2) and a family of signal characteristics.(3).

Fig.2 Characteristics of the SQUID with the size of the junction less than the vortex size.

A Monolithic Integrated Active Thin Film Head for Magnetic Recording – An Example for a Combination Technology Suitable for Smart Sensors

H. Rudolph, K. Vernie, D. Seitzer
Lehrstuhl für Technische Elektronik
Universität Erlangen - Nürnberg
Cauerstraße 9, 8520 Erlangen, FRG

0. Introduction

Since the procedures of the semiconductor technology have been used for fabricating sensors, it is possible to scale the sensors down to extremely small sizes. This means lower mass, shorter response time and lower manufacturing costs. Unfortunately, this often implies lower output signals. Hence it would be desirable if a preamplifier could be integrated monolithically together with the sensor.

An example is a magnetic head for reading digital data from a hard or a flexible disk. For achieving high recording density thin film magnetic heads (TFMHs) are necessary. A TFMH developed at our institute makes recording densities up to 2000 fluxchanges/mm on a flexible disk possible /1,2/, but then the readout signal is less then 50 µV. So we have developed a process, which has enabled us to integrate sensors realized in thin films monolithically with the semiconductor circuit (SCC). /3/

1. Problem: Combination of Semiconductor and Thin Film Processes

Although closely related to semiconductor technologies most sensors require a different sequence of processing steps. As such the sensor must be fabricated separately from the SCC, either beforehand or afterwards. For two reasons we have decided to realize first the SCC and then the sensor (TFMH):

* Not all of the materials used for the TFMH can withstand the temperatures which occur during diffusion or oxidation processes.

* Moreover these high temperature processes could induce thermally generated stresses which could destroy the thin films.

On the other hand fabricating the semiconductor circuit first means that the circuit has to be protected from mechanical and chemical influences during processing of the TFMH.

2. Solution: Choice Of Processing Steps

The concept of the combination technology consists of three further fabrication steps, and these are:

1) Applying adjustment marks which enable accurate and correct positioning of the sensor beside the SCC (adjustment mask).

2) Protecting of the semiconductor from the thin film processes. The SCC is coated by two sputtered protective layers. The first layer is a SiO_2 film, which acts as a passivation coating directly over the SCC. It soon became evident, that SiO_2 is very brittle and so has many small cracks. Some etchants and strippers can attack the SCC through these cracks and damage it. Thus a second protective layer was applied, a soft magnetic alloy, which acts simultaneously in this case as the first layer of the TFMH. Because the alloy is rather soft there are no cracks and the semiconductor is shielded. The sensor can now be realized. Following this the protective layer can be removed using the same mask, but with negative photoresist (protection mask).

3) Depositing and structuring of the metal layer needed to connect both the head and the amplifier (connection mask).

Fig. 1 shows a photograph of the complete active TFMH and fig. 2 a schematic cross section illustrating the combination technology in the case of a TFMH with a bipolar transistor.

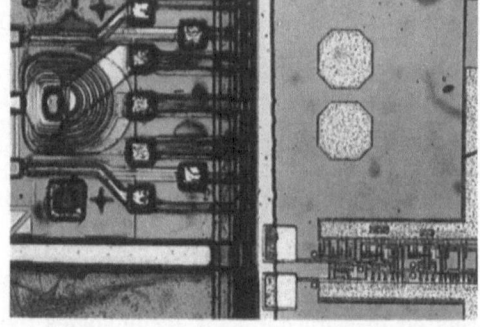

Fig.1: Photograph of the active thin film magnetic head.

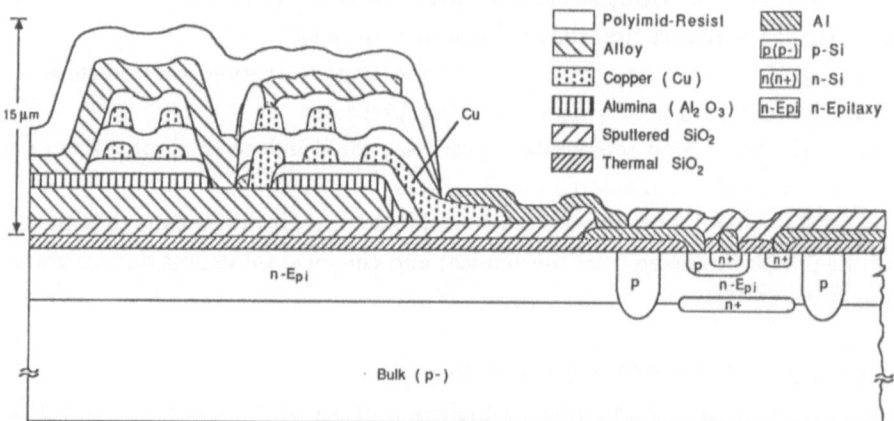

Fig.2: Schematic cross section of a monolithic integration of a TFMH and bipolar devices

3. Results

To verify our considerations about this combination technology we realized a TFMH integrated with two different SCCs.

First, the TFMH described below and a simple bipolar analog array fabricated with our own semiconductor process line were integrated on the same chip.

The TFMH is an inductive head with 12 windings in two layers suitable for both reading and writing. It was realized with the standard thin film processes. The films are applied by sputtering, evaporation and spincoating. Afterwards they are structured by the usual photolithographic techniques. Most of the etching was done with an ion milling system, but wet etching processes were also employed. The TFMH is composed of nine layers. Its fabrication requires fifteen photolithographic steps and in total one hundred process steps were necessary.

The comparsion of the transistor output characteristics before and after applying the thin film processes showed no remarkable difference (Fig. 3).

Fig.3: Output characteristics of a bipolar transistor <u>before</u> (left) and <u>after</u> (right) applying the thin film process. Vert.: $I_C = 0.2$ mA/Div, $I_B = 5$ µV/Step, Horiz.: $U_{CE} = 1$ V/Div.

Second, we combined the TFMH with an amplifier, fabricated with an industrial bipolar process ("CIT1", Intermetall, Freiburg).

The SCC is a two stage, differential cascode amplifier, developed for lowest noise and high bandwidth. The circuit has a voltage gain of 54 dB and an input voltage noise density of 5.4 nV/√Hz as simulated by SPICE. Measurements of the selected chips show a typical gain of 56 dB and a typical noise density of 6 nV/√Hz. A bandwidth of 60 MHz was both simulated and measured for capsulated chips (Fig. 4).

600

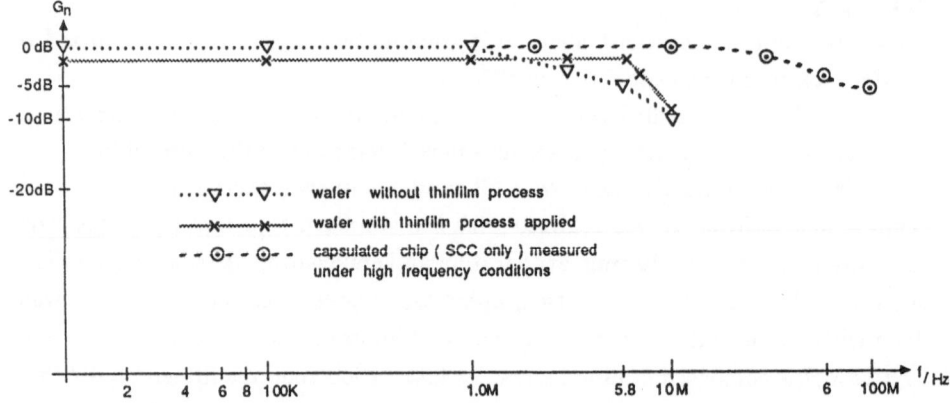

Fig.4: Normalized gain G_n vs. frequency for the amplifier mentioned above, <u>before</u> and <u>after</u> applying the thin film processes. The low corner frequency is due to the measurement method with a manual wafer prober direct on wafer. The dotted line shows the result for a capsulated chip measured under high frequency conditions. This result agrees very well with the simulated data.

4. Conclusions

A thin film sensor was realized on a silicon wafer on which an analog circuit was integrated beforehand. By suitable protective layers we have achieved, that the performance of the SCC was not affected by the ca. hundred process steps necessary for fabricating the sensor. Thus, a combination technology has been developed which makes it possible to integrate a wide span of thin film sensors with a complete SCC, and this juxtaposition of technologies allows to realize real "smart sensors".

5. Acknowledgements

Thanks are due to INTERMETALL, Freiburg, for the friendly cooperation in carrying out the mask layout for the amplifier and the bipolar fabrication process. This project was partly supported by the Bundesministerium für Forschung und Technologie and the Deutsche Forschungsgemeinschaft.

6. References

/1/ J. Kollmann, K. P. Frohmader: "A Thin Film Head For High Density Storage On Flexible Disks" IEEE Transaction on Magnetics mag 23, vol. 5 (Sept. 1987), pp. 2937

/2/ F. Oehme, R. Plankenbühler, J. Bäsig: "Hochdichte Magnetaufzeichnung digitaler Information" ntz Nachr.-techn. Zeitung 40 (1987) Heft 4, S. 268-272

/3/ H. Rudolph, K. Vernie: "An Active Thin Film Magnetic Head By Monolithic Integration With A Preamplifier" poster presented at the "Sensor 88", Nürnberg, 1988

Devices
for Special Applications

The Effect of the Metal, Interface, and Semiconductor Parameters on the Electrical Behaviour of Schottky Junctions

Zs. J. Horváth

Research Institute for Technical Physics of the Hungarian Academy of Sciences, Budapest, P.O.Box 76, H-1325, Hungary

Abstract
    The effect of the metal, interface and semiconductor parameters on the barrier height of n-type Schottky junctions is analysed by means of a general expression.

It is known that the current-voltage (I-V) and capacitance-voltage (C-V) characteristics of Schottky junctions are mainly determined by the barrier height and the free carrier concentration in the semiconductor. Lately great efforts have been made to understand the origin and the effect of the energy distribution of interface states on the Fermi-level pinning and thus on the barrier height. However the energy distribution of the interface states is only one aspect of the problem, because their depth distribution has at least as important role in the Schottky barrier formation as the energy distribution.

Recently an evaluation method based on the interfacial layer model of Schottky junctions, has been proposed for the extraction of the interface state energy distribution and of the relative interfacial layer thickness (the ratio of the thickness of the interfacial layer to its relative dielectric constant: δ/ϵ_i) from the Schottky I-V characteristics [1]. (The thickness of the interfacial layer may be considered as the depth of the interface charge centroid from the metal-semiconductor interface.)

Further on a general interfacial layer expression for the equilibrium (zero bias) Schottky barrier height has also been derived [2]. This expression has no initial assumptions concerning the energy distribution of the interface states and the equilibrium interface charge, in contrast to the previous expressions [3-5]. Using this expression and the above evaluation the first relations between the experimental barrier height, relative interfacial layer thickness and interface charge values, and the interface state distribution spectra has been obtained in Au/GaAs junctions as a function of the annealing temperature [2].

In this paper the effect of the metal, interface and semiconductor parameters on the barrier height of n-type Schottky junctions is analysed by means of the above expression.

The charge neutrality of the Schottky junctions may be expressed as

$$Q_{M0} + Q_{i0} + Q_{s0} = 0 \qquad (1)$$

where Q_{M0} is the equilibrium charge on the metal, Q_{i0} the equilibrium interface charge including that of the interface states and the fixed charge, and Q_{s0} is the equilibrium space charge in the semiconductor. This expression may be rewritten for n-type semiconductors as [2]

$$Q_{i0}(\phi_{b0}) - \epsilon_0 \epsilon_i (\phi_M - X - \phi_{b0})/\delta + [2\epsilon_0 \epsilon_s qN(\phi_{b0} - \phi_n)]^{1/2} = 0 \qquad (2)$$

where $Q_{i0}(\phi_{b0})$ is a single-valued function of the equilibrium barrier height ϕ_{b0}, ϵ_0 is the dielectric constant of the vacuum, ϵ_i and ϵ_s the relative dielectric constants of the interfacial layer and the semiconductor respectively, ϕ_M the metal work function, X the electron affinity of the semiconductor, δ the interfacial layer thickness, q the electron charge, N the free electron concentration in the semiconductor and ϕ_n the energy difference between the conduction band edge and the Fermi-level in the semiconductor. (ϕ_{b0}, ϕ_M, ϕ_n and X are considered in volts.)

For each actual relative interfacial layer thickness value and $Q_{i0}(\phi_{b0})$ function (i.e. interface state energy distribution spectrum and fixed interface charge) the only variable of this expression is the equilibrium barrier height. Thus the equilibrium barrier height of the Schottky contacts is determined by Eq. (2). For some special cases of $Q_{i0}(\phi_{b0})$ (as e.g. treated in the above works [3-5]) ϕ_{b0} may be expressed in explicit form.

The effect of the metal, interface and semiconductor parameters on the Schottky barrier height may be analysed by the differentiation of Eq. (2) according to the different parameters. This yields the following expressions:

$$\partial\phi_{bo}/\partial\phi_M = C_{ii}/(C_{is}+C_{ii}+C_{so}) \tag{3}$$

$$\partial\phi_{bo}/\partial X = -C_{ii}/(C_{is}+C_{ii}+C_{so}) \tag{4}$$

$$\partial\phi_{bo}/\partial(\delta/\epsilon_i) = -(\phi_M - X - \phi_{bo})/(\delta/\epsilon_i)*C_{ii}/(C_{is}+C_{ii}+C_{so}) \tag{5}$$

$$\partial\phi_{bo}/\partial N = -(\phi_{bo}-\phi_n+kT/q)/N*C_{so}/(C_{is}+C_{ii}+C_{so}) \tag{6}$$

$$\partial\phi_{bo}/\partial\epsilon_s = -(\phi_{bo}-\phi_n)/\epsilon_s*C_{so}/(C_{is}+C_{ii}+C_{so}) \tag{7}$$

$$\partial\phi_{bo}/\partial E_g = 1/(2q)*C_{so}/(C_{is}+C_{ii}+C_{so}) \tag{8}$$

$$\partial\phi_{bo}/\partial Q_{io} = -1/(C_{ii}+C_{so}) \tag{9}$$

where E_g is the forbidden gap, and C_{is}, C_{ii} and C_{so} are the interface state, interfacial layer and zero bias depletion layer capacitances per unit area:

$$C_{is} = dQ_{io}/d\phi_{bo} = q^2 D_s \tag{10}$$

(D_s is the density of interface states.)

$$C_{ii} = \epsilon_0 \epsilon_i /\delta \tag{11}$$

$$C_{so} = \{\epsilon_0 \epsilon_s qN/[2(\phi_{bo}-\phi_n)]\}^{1/2} \tag{12}$$

On the basis of Eqs. (3-9) (see signs) one can conclude that for n-type semiconductors the Schottky barrier height increases with increasing metal work function and increasing forbidden gap, while it decreases with increasing free electron concentration, with increasing electron affinity and dielectric constant of the semiconductor, and with increasing equilibrium interface charge. If $\phi_M - X > \phi_{bo}$ [2], then the barrier height decreases with increasing relative interfacial layer thickness, in the opposite case this dependence is reversed.

The denominator in Eqs. (3-8) is the same, i.e. the sum of the interface state, interfacial layer and zero bias depletion layer capacitances. As the zero bias depletion layer capacitance is usually much lower, than that of the interface states and interfacial layer, the value of the denominator is dominated by the interface state density and the relative interfacial layer thickness. Due to the same fact the effect of the metal work function, the electron affinity and the relative interfacial layer thickness must be much higher than that of the free electron concentration, relative dielectric constant and forbidden gap, as in the numerator in Eqs. (3-5) the interfacial layer capacitance figures, while in the numerator in Eqs. (6-8) the zero bias depletion layer capacitance.

If the equilibrium interface charge does not depend on the barrier height, then the derivatives according to the metal work function and electron affinity are close to +1 and -1, respectively [Eqs. (3-4)] (Schottky limit for ϕ_M [3]).

The above effects with the exception of those of the forbidden gap and relative dielectric constant may be demonstrated by a great amount of the experimental results [2,3,6-15].

It should be mentioned that the near-interface concentration change also affect the barrier height. This effect was mainly studied by using the I-V characteristics. Recently a general expression for the C-V characteristics of such structures has also been derived [16].

References:

1. Horváth, Zs.J.; J.Appl.Phys., 63 (1988) 976.
2. Horváth, Zs.J.; Appl.Phys.Lett., 54 (1989) 931.
3. Cowley, A.M.; Sze, S.M.; J.Appl.Phys. 36 (1965) 3212.
4. Wu, C.-Y.; J.Appl.Phys. 51 (1980) 3786.
5. Szatkowski, J.; Sieranski, K.; Solid-State Electron. 31 (1988) 257.
6. Kurtin, S.; McGill, T.C.; Mead, C.A.; Phys.Rev.Lett. 22 (1969) 1433.
7. Schmidt, M.T.; Podlesnik, D.V., Yu C.F.; Wu X.; Osgood R.M.; Yang E.S.; J.Vac.Sci.Technol.B 6 (1988) 1436.
8. Weber, E.R.; Spicer, W.E.; Newman, N.; Liliental-Weber, Z.; Kendelewicz, T.; Proc. 19th Int. Conf. Physics of Semiconductors, Warsaw, Poland, Aug.15-19, 1988, p.705.
9. Chiaradia, P.; Fanfoni, M.; Nataletti, P.; De Padova, P.; Viturro, R.E.; Brillson, L.J.; J.Vac.Sci.Technol.B 7 (1989) 195.
10. Crowell, C.R.; Sze, S.M.; Spitzer, W.G.; Appl.Phys.Lett. 4 (1964) 91.
11. Srivastava, A.K.; Arora, B.M., Guha, S.; Solid-State Electron. 24 (1981) 185.
12. Aboelfotoh, M.O.; Phys.Rev.B 39 (1989) 5070.
13. Shan, W.; Li, M.F.; Yu, P.Y.; Hansen, W.L.; Walukiewicz, W.; Appl.Phys.Lett. 53 (1988) 974.
14. Sobolewski, M.A.; Helms, C.R.; Appl.Phys.Lett. 54 (1989) 638.
15. Horváth, Zs.J.; Gyúró, I.; Németh-Sallay, M.; Szentpáli, B.; Kazi, K.; Phys.Stat.Sol.(A), 94 (1986) 719.
16. Horváth, Zs.J.; J.Appl.Phys., 64 (1988) 443.

A New Silicon Diaphragm Pressure Sensor and its Stress Analysis

Lian – Zhong Yu, Min – Hang Bao, Xian – Ping Wu

Department of Electronic Engineering
FUDAN University, Shanhai, P.R.China

Abstract

A conventional flat silicon diaphragm does not perform well in a low pressure range sensor that requires high output and high accuracy because the large deflection of diaphragm causes severe nonlinearity when the diaphragm thickness is small for high output. This phenominon is known as the balloon effect. In order to solve this problem, an improved structure with two bulk masses on the back of a square or rectangular diaphragm has been proposed for silicon pressure sensors. The diffused piezoresistive gauges are formed on the diaphragm between the two bulk masses and between either bulk mass and diaphragm edge, where the stress is concentrated. The two dimensional stress analysis of this structure using finite element method are given, The results are helpful for the optimization design. The experimental results show that the sensitivity of this kind of sensor is about two times or more as large as that of conventional square diaphragm pressure sensor with same diaphragm thickness and size, which agrees well with the analytical results.

I. The Structure of New Diaphragm Pressure Sensor

Silicon diaphragm pressure sensors are widely used because of its excellent electrical and mechanical performances. Typically, the sensitivity of these sensor can be as high as $50 \sim 150 \mu V/mmHg$ at $V_o = 5V$, which is, however, still not satisfied for many low pressure range applications because of accuracy limit, that is the balloon effect[1].

In oder to solve the contradiction between sensitivity and accuracy of pressure sensor, first, a silicon pressure sensor with a concentric bulk mass on the back of a circular diaphragm was reported [2]. Then a structure with two bulk masses on the back of a basically circular diaphragm was proposed [3]. Here a square silicon diaphragm pressure sensor with two rectangular bulk masses on the back is obtained by using compensating undercutting technology [4], as shown in Fig.1. When external pressure is applied, the diaphragm will deflect and the displacement of two bulk masses followed. Meanwhile, the stress in the narrow diaphragm regions between thick regions is much greater than ususal. It is called as stress concentration effect. The piezoresistive gauges are diffused on the stress concentrated regions. The stress distribution in the central and edge regions of the diaphragm is more uniform than that in the structure with only one bulk mass[4]. Therefore it is easy to locate the piezoresistive gauges in the required regions without decline of sensitivity when its location deviates a little.

II. Stress Analysis Using Finite Element Method

In order to know the stress distribution of this new diaphragm structure so that we can obtain an optimum design, finite element method has been used to make stress analysis for this structure.

Finite element method is a kind of direct method based on the veriational principle. By partitioning the continum, that has an infinite number of nodal points, this method has been applied to continuous material as well as discrete ones. The balance differential equation describing characteristics of elastic continum can approximately be replaced by a group of algebraic balance equations.

According to the minimum potential energy principle [5][6] we have:

$$\Delta\Pi = \Delta(U - V) = 0 \tag{1}$$

where Δ means veriation, Π is the whole potential energy, U is the strain energy and V is the potential energy of external force.

The potential energy veriation is:

$$\Delta\Pi = \sum_1^n \int \int_A (M_x\Delta\epsilon_x + M_y\Delta\epsilon_y + M_{xy}\Delta\epsilon_{xy})dxdy - \sum_1^n \int \int_A (P)dxdy = 0 \tag{2}$$

where n is the number of equal-thickness elements that form the whole plate; A is the surface area of every element and M represents the moment of force; ϵ,P are the strain and applied loading respectively.

According to elastic mechanics, through transformations we can obtain:

$$\sum_1^n \{\Delta\delta\}_e^T([K]_e\{\delta\}_e - \{Q\}_e) = 0 \tag{3}$$

$\{\delta\}_e$ is the matrix of displacement. $\{Q\}_e$, the "nodal force" matrix of element caused by initial strain ϵ_0 and transverse loading P. Subscript e represents single element.

Summing up equation (3) to describe the whole plate, we can obtain the distribution of displacement $\{\delta\}$, so that the stress distribution also can be calculated.

For the new diaphragm structure, shown in Fig.1 , stress profile is analyzed by using finite element method. The diaphragm is a square with width of $1600\mu m$ so that it can be compared with the square flat diaphragm of same size at same condition, i.e. applied pressure $P = 0.4Kg/Cm^2$ and diaphragm thickness h=$30\mu m$. First, we calculate the stress at x/a=0.8 (where a strain gauge is located) for square flat diaphragm under above conditions. The result of $\sigma_x = 1.39 \times 10^8 dyn/Cm^2$ is obtained and chosen as a reference to be compared later with that of new structure to demonstrate the stress concentration effect.

It can be found in Fig.2 that when the width of bulk mass W_3 is less than $400\mu m$, the stress in diaphragm center of new structure is obviously larger than above-mentioned reference value of square flat diaphragm. Stress two times as large as that of square flat diaphragm can be obtained as long as appropriate parameters W_1 and W_2 are selected. Therefore the effetc of stress concentration is clearly shown.

Also from Fig.2 it can be seen that with the increasing of W_3, both the stresses at center and edge of diaphragm are decreasing. It can be explained that when the width W_3 is increased, the whole diaphragm area is decreased, therefore the decreased bending in Y direction greatly impede the bending in X direction which influence the above stresses. Fig.3 shows the influence of center width W_1 on the effect of stress concentration. With the increasing of center width W_1, the stress at center decreases sharply at first and then increases a little, but the stress at edge increases continuously. It also can be explained that when the center width W_1 increases, the center diaphragm area increases which causes the increasing displacement of two bulk masses and larger bending of edge diaphragm so that the stress at edge increases. Meanwhile the relative displacement at center is smaller, therefore the stress at center decreases.

In addition, it can be found that when center width W_1 equals to about 170μm and edge width W_2 about 80μm (that the center width is as 2.13 times as that of edge width), the stress at center is equal to that at edge. This result is useful for designing a balanced piezoresistive bridge.

Fig.4 shows that with the increasing of edge width W_2, the stress at center grows gradualy and then decreases, but the stress at edge decreases sharply. This can also be explained by the above mentioned reasons. Another result is that when the center width is equal to the edge width, the stress at center is two times as large as that at edge.

III. Experiment and Conclusion

The sensor with new structure has been fabricated and the effect of stress concentration has been investigated. Two p-type four-terminal gauges are diffused in center and edge of diaphragm of (100) n-type silicon substrate. The diaphragm with two bulk masses is anisotropically etched by EPW. Also the pressure sensor with square flat diaphragm is measured. For both structure, the geometric parameters are: $2a=1640\mu$m, $W_1 = W_2 = 120\mu$m and $W_3 = 920\mu$m. The experimental results are shown in Table I.

TABLE I

Sensitivity (μV/mmHg), $V_s = 5V$					
No.	1		2		3
square diaphragm	181.9		144.4		160.0
New diaphragm	#1	#2	#1	#2	#1	#2
	283.7	141.8	277.8	179.4	281.3	161.2
Diaphragm Thickness	18μm		24μm		21μm

It can be obviously seen that the sensitivity of device #1 (center gauge) in new structure is 1.5 to 2 times as large as that of square flat diaphragm. Also the sensitivity of device #1 (center gauge) is 1.5 to 2 times as large as that of device #2 (edge gauge). All the results agrees well with the analytical results. Although exist the errors exist due to the diaphragm etching, geometric parametres and lithography, the agreement between theory and experiments is very satisfactory.

The calculation shows that the new diaphragm structure can increase the output of pressure sensor two times as large as that of conventional pressure sensor and can be used in low pressure range measurement. The analytical results and computer programm are helpful for designing this new diaphragm and other structure pressure sensor which is being developed now in solid state sensor laboratory.

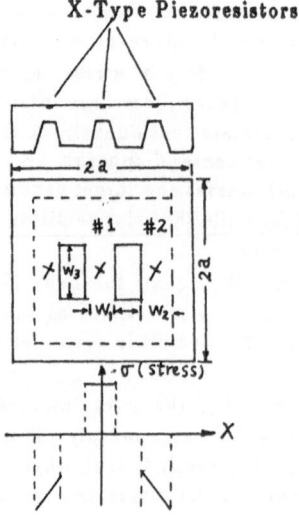

Fig.1 Cross sectional and top view of new diaphragm pressure sensor and schematic diagram for its stress distribution

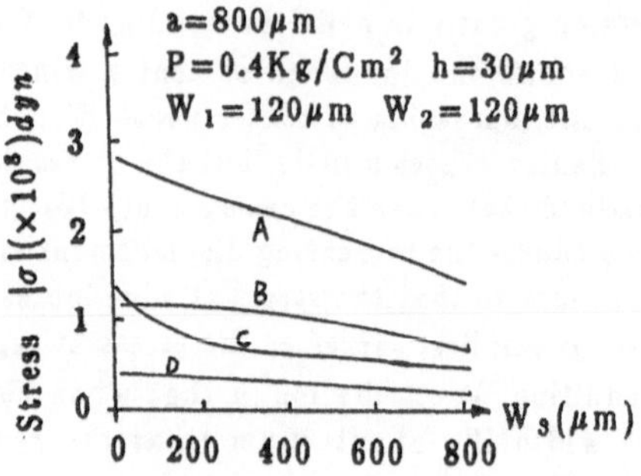

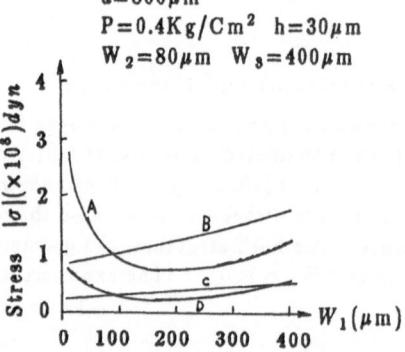

Fig.2 Stress in center and edge of diaphragm with changing of longitudinal width of bulk mass W_3.

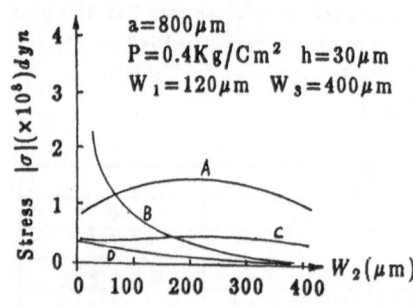

Fig.3 Stress in center and edge of diaphragm with changing of center width of diaphragm W_1.

Fig.4 Stress in center and edge of diaphragm with changing of edge width of diaphragm W_2.

N.B.:in Fig.2,Fig.3,Fig.4, A:σ_x in center; B:σ_x in edge; C:σ_y in center; D:σ_y in edge.

REFERENCES

1. O.N.Tufte *et al*, J.Appl.Phys.33,(1962)3322-3327
2. M.Shimamazoe *et al*, Sensors and Actuators,2(1982)275-282
3. Basic Advantages of the Anisotropic Etched, Transverse Gauge Pressure Transducer, Endeveco Tech. Paper
4. Xian-ping Wu and Wen H.Ko, Proceeding of Transducers'87,p126. Tokey.
5. S.P.Timoshenko *et al*, Theory of Plates and Shells. McGraw- Hill Book Company, Inc.New York,1959.
6. L.J.Segerlind, Applied Finite Element Analysis, New York:Wiley,1976.

A Distribution Function of Injected Electrons in the Planar Doped Barrier Transistors

N. A. BANNOV and A. A. SVYATCHENKO

Institute of Microelectronics,
Academy of Sciences of the USSR, Yaroslavl

Summary

The injection properties of the planar doped barrier transistors (PDBT) at low temperature are investigated by means of numerical simulation. It is shown that due to the random distribution of the charged impurities electric field potential fluctuations may result in essential deviation of hot electrons injected from emitter into the base region from the direction normal to the layers of PDBT.

After prior reports by Shannon, Malic, etc. [1]-[2] the semiconductor structures with planar doped layers are the subject of a detailed research (see the review [3]). It is connected with the possibility of development of very high speed transistors [4] as well as the possibility of using them as hot electron spectrometer [5].

In the present report the injection characteristics of the planar doped barrier transistors (PDBT) are investigated. The structure under discussion (see Fig. 1) consists of the series of semiconductor layers n^{+}-i-p^{+}-i-n^{+} forming potential barrier for electrons (see Fig. 1; pay attention to the notation, it will be used in a text). Let us assume the barrier to be strongly assymetric and the acceptor layer to be thin ($l_1 \gg l_2, d_1$). The barrier height $\varepsilon_b \simeq e^2 N_a d_1 l_2$ (e is the electron charge) is supposed to be much more than the energy T corresponding to the lattice temperature but less than half-bandgap ε_g (the effect of holes is ignored). This is true under the following conditions of the surface concentration of the acceptors

$$\frac{T}{e^2 l_2} \ll N_a d_1 < \frac{\varepsilon_g}{2e^2 l_2}.$$

Such a diod structure serves as a high energetic electrons injector into n^+- base region of PDBT. The distribution function of injected electrons is usually considered (see, for example, [3], [5]-[7]) to be close to that of monoenergetic beam and electron momenta to be normal to the layers of PDBT. However, simple estimates demonstrate that discrete nature of impurities distribution and related with it potential fluctuations can result in significant deviation of injected electrons from the normal to the layers of structure. In fact under typical values of doping concentrations $N_d \propto N_a \propto 10^{18} cm^{-3}$ the interimpurity distance l_I is comparable with a dimension of a region where an electron gains energy at forward biased structure (see Fig. 1). Therefore potential energy fluctuations can strongly affect the injected electrons distribution function form.

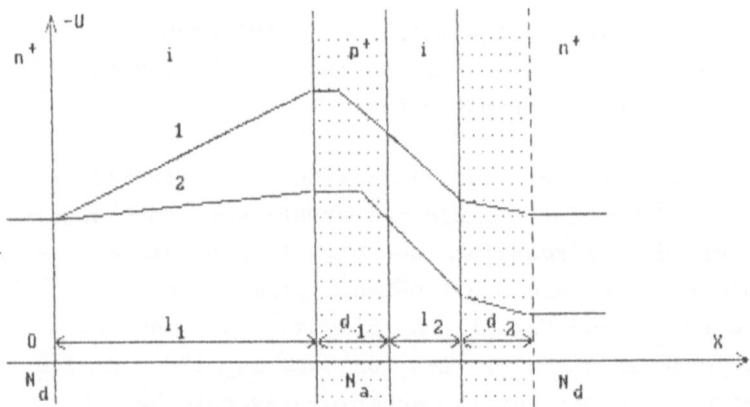

Fig. 1. Potential energy of electron versus position at zero (1) and forward (2) bias.

Angular dependence of distribution function of electrons passing p^+-layer, i-layer and n^+-layer depleted region was calculated. Our calculations were carried out on the basis of Newton equation for electron transport. While numerical simulation charged layers were considered to be finite with radius $R_0 \gg d_1 + l_2 + d_2$. Acceptors and donors distribution in these layers was assumed to be non-correlated and determined by means of uniformly distributed random number generator. Fig. 2 demonstrates the distribution function F of electrons injected into the base region of PDBT for the structure on the basis of GaAs with $N_a = N_d = 10^{18}$ cm^{-3}, $d_1 = l_1 = d_2 = 10^{-6}$cm. Here the average energy of electrons, enfering the quasi-neutral region of n^+-layer was 0.25 eV.

Considerable electron deviations from the normal to layers result in significant spread of longitudinal rates of electrons, injected into n^+-layer. So angular spread of injected electrons as well as electron scattering in the base region of PDBT accounts for smooth character of hot electron spectra, observed experimentally [3], [5], [6].

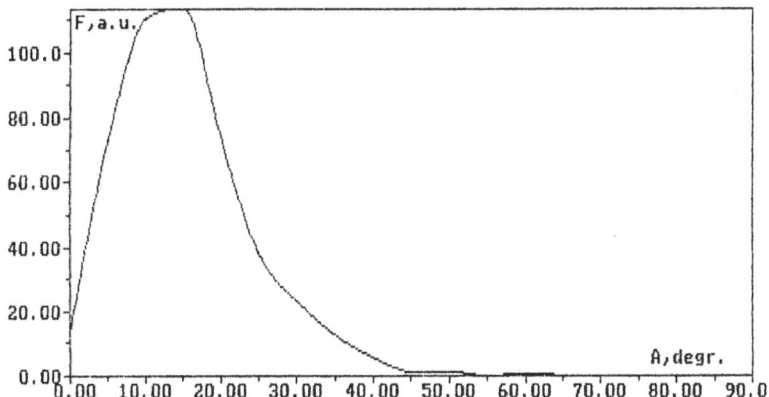

Fig. 2. Angular dependence of distribution function of electrons injected into quasi-neutral n^+-region. (A is the angle between electron momentum and normal to the layers of PDBT).

Let us discuss the validity of the approach mentioned. Its main restriction consists in comparability of Debroil electron wave length λ with a distance between impurities. That consequently results in necessity of quantum description of the electron transport in PDBT. However, when moving in the electric field between the charged layers, an electron accumulates energy to a great extent $\varepsilon \gg T$ so in the considerable part of the simulated region the condition $\lambda < l_I$ is satisfied. Taking into account that injected electrons rates component parallel to the structure surfaces is comparable with the entire rate (see Fig. 2) we can come to a conclusion that investigated structure regions where electron energy is considerable contribute mainly in electron scattering so quantum effects can be neglected.

References

1. Shannon, J.M.: A majority carrier camel diode. Appl. Phys. Lett. 35 (1979) 63-65.
2. Malic, R.J.; AnCoin, T.R.: Ross, R.L.; Board, K.; Wood, C.E.; Eastman, L.F.: Planar doped barrier in GaAs by molecule beam epitaxy. Electron. Lett. 16 (1980) 836-838.
3. Hayes, J.R.; Levi, A.F.J.: Dynamics of extremly nonequilibrium electron transport in GaAs. J. of Quantum Electronics. 22 (1986) 1744-1752.
4. Woodcock, J.M.; Harris, J.J.; Shannon, J.M.: Control growth of GaAs layers for monolithic photoelectron transistors. J. Vac. Sci. & Technol. B4 (1986) 609-611.
5. Levi, A.F.J.; Hayes, J.R.; Platzman, P.M.; Wiegmann, W.: Injected hot electron transport in GaAs. Phys. Rev. Lett. 55 (1985) 2071-2073.
6. Beton, P.H.; Long, A.P.; Kelly, M.J.: Hot-electron transport in GaAs in the presense of a magnetic field. Appl. Phys. Lett. 51 (1987) 1425-1427.
7. Gruzinskis, V.; Kersulis, S.; Mickevicius, R.; Pozela, J.; Reclaitis, A.: Collective electron interaction in double-barrier transistors. Sol. St. Electron. 31 (1988) 345-347.

Dependence on Gate Length of Electrical Properties of Self-aligned AlGaAs/GaAs HEMTs Studied by Monte Carlo Technique

G. U. JENSEN[1], B. LUND[1], T. A. FJELDLY[1] and M. S. SHUR[2]

[1]Department of Electrical Engineering and Computer Science, Norwegian Institute of Technology, University of Trondheim, N-7034, Trondheim, Norway.

[2]Department of Electrical Engineering and Supercomputer Institute, University of Minnesota, Minneapolis, Minnesota 55416, USA.

Summary

Various electrical properties of self–aligned AlGaAs/GaAs heterostructure field–effect transistors (HEMTs) have been studied by self–consistent ensemble Monte Carlo simulations for gatelengths between 0.1 and 1.0 μm, and for different depths of ohmic contacts. The maximum transconductance is found to be only weakly dependent on gate length and contact depth. The output drain conductance and the threshold voltage show pronounced short–channel effects for gate lengths at or below 0.2 μm and shallow contacts. With deeper contacts, the short–channel effects appear at longer gate lengths. The switching times vary with gate length from 1.0 ps at 0.1 μm to 5.0 ps at 1.0 μm. The observed effects are discussed in terms of simple models.

Simulation of modern semiconductor devices by self–consistent Monte Carlo technique has become a valuable tool for extracting information on macroscopic electrical behaviour, and for gaining insight into the complicated transport physics of such devices. By comparing simulation results with experimental data it may also be possible to identify and, hopefully, eliminate effects that tend to degrade device performance. In addition, Monte Carlo results, in conjunction with experiments, are an ideal basis for the development of practical and accurate device models.

The technology of heterostructure field effect transistors (HFETs or HEMTs) is rapidly maturing to a point where widespread application in high–performance digital and microwave systems appears imminent. However, questions still remain concerning details of transport properties of these devices, particularly for transistors with very short gate lengths in the sub–μm range and beyond. In this work, we have addressed the issues of short–channel effects and switching transients in HEMTs with gate lengths between 1 μm and 0.1 μm, by means of self–consistent Monte Carlo simulatons.

Fig. 1 shows the geometry of the simulated samples. The devices are designed for high–speed applications by assuming self–aligned gates and planar doping of the AlGaAs

616

layer. The gate lengths are 1.0, 0.5, 0.3, 0.2, and 0.1 μm. The doped plane has a sheet doping density of 4.8×10^{12} cm^{-2}, and is located 200 Å below the gate electrode. The AlGaAs layer is 250 Å thick. Two different depths of ohmic contacts are used, 0.05 and 0.28 μm, and the doping of the contact regions is 4.3×10^{17} cm^{-3}. The simulated samples have a total depth of 0.44 μm. The artificial source and drain electrodes are given a vertical potential distribution like that expected deep inside the ohmic regions if real top electrodes were used. This construction simplifies the charge exhange at the contacts.

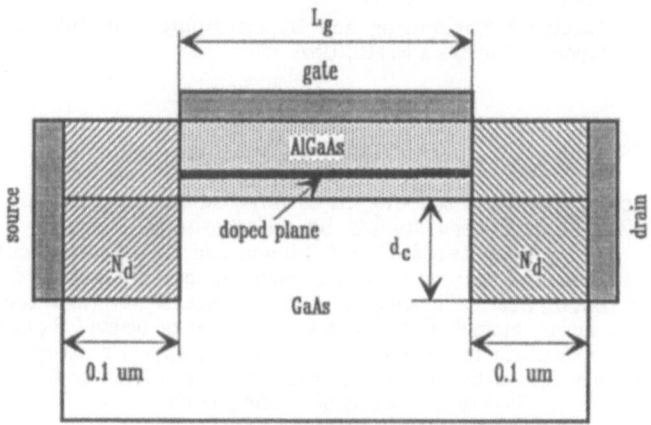

Fig.1. Device geometry

The scattering mechanisms used in the Monte Carlo simulation are intravalley polar optic and deformation potential acoustic phonons, equivalent and non—equivalent intervalley phonons, and ionized impurities. The semiconductor conduction bands are described by a three—valley (Γ,L,X)—model using Kane—type non—parabolicity. Electrons impinging on the hetereointerface are reflected or undergo real—space transfer according to classical transmission laws. The two—dimensional electrical field is updated every 9 fs by solving Poisson's equation self—consistently on a rectangular mesh with typical cell size 70x33 Å. The number of electrons simulated is in the range 30,—50,000. Transconductances g_m and output drain conductances in saturation g_d are derived from computed stationary current—voltage characteristics. The switching transients are calculated separately with the inclusion of displacement current terms.

Fig. 2 shows g_m vs. gate voltage V_g at different gate lengths L_g, for drain voltages V_d well into saturation. The contact depth d_c was 0.05 μm. The maximum transconductance g_{mo} is seen to be quite high, in the range 1100 − 1300 mS/mm, but appears to be relatively weakly dependent on L_g. This is to be expected from simplified models which show that, for submicrometer devices, g_{mo} depends only indirectly on gate length through the effective saturation velocity v_s [1]. The simulations also show that g_{mo} is quite insensitive to contact depth, as could be expected.

Estimates of the threshold voltage V_t can be obtained from plots as shown in Fig. 2 by extrapolating the linear part of g_m vs. Vg to $g_m = 0$ at low values of V_g. The results for V_t, corresponding to $V_d = 1.5$ V, and the computed values of g_d are presented in Fig. 3 for the various devices simulated.

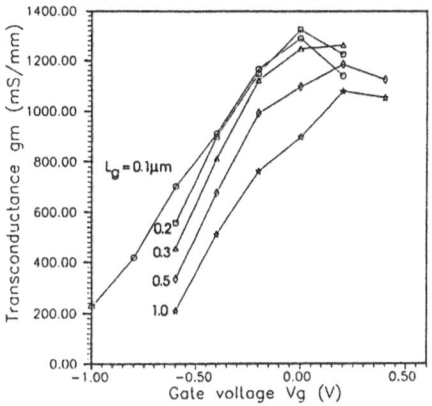

Fig. 2. Computed transconductance vs. gate voltage at different gate lengths.

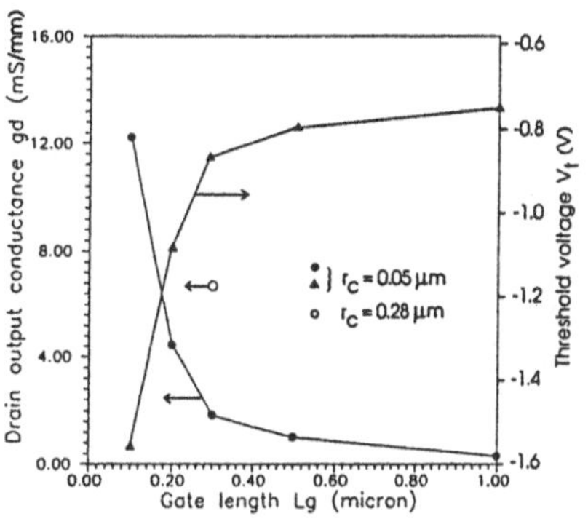

Fig. 3. Threshold voltage and output drain conductance vs. gate length.

V_t and g_d show clear indications of short–channel behaviour. The effect is particularly apparent in the sharp change in V_t and g_d for gate lengths at or below 0.2 μm and shallow ohmic contacts (0.05 μm). With deeper contacts, similar short–channel effects appear at even larger gate lengths. These phenomena are primarily associated with space charge injection into the region between the ohmic contacts, as evidenced by the dependence on contact depth, and further ascertained by microscopic details in the Monte Carlo results. According to this mechanism, g_d and the deviation in V_t should be roughly proportional to the inverse square of the effective gate length [2], which is in good overall agreement with the simulations.

The space charge injection gives rise to a drain current contribution which is relatively insensitive to V_g, but quite linear in V_d. At short gate lengths, this manifests itself as i) a high subthreshold current, ii) a large g_d (but independent of V_d) and iii) a large deviation in V_t. These are all short–channel phenomena which signify reduced gate control and, ultimately, loss of transistor action.

Examples of *on* and *off* transients for the simulated devices in saturation are presented in Fig. 4. For simplicity, *on* and *off* states in all simulations are defined with reference

618

to $V_g = 0$ and $V_g = -1.0$ V, respectively, in spite of the presence of significant subthreshold current in the *off* state of the smallest devices. The switching time τ is defined as the point where 90 percent of the total change in the drain current is completed. The results for the various devices are shown in Fig. 5 (for $r_c=0.05$ μm).

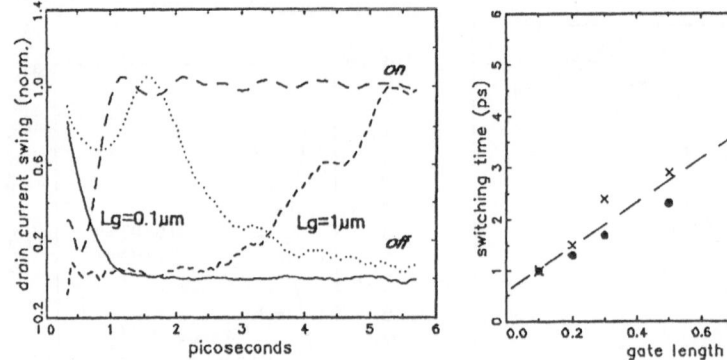

Fig. 4 Examples of switching transients, $r_c=0.05$ μm.

Fig. 5. *On* and *off* switching times vs gate length.

The computed transients show a reduction in switching time with decreasing gate length, as expected. The very simplest model indicates that τ is equal to kL_g/v_s, where k is a constant of order unity which depends on the definition used for τ. However, instead of following this model, the simulated switching times extrapolate to a finite value $\tau_0=0.6$ ps at zero gate length. This behaviour can be explained in terms of a residual parasitic RC time constant associated with the contacts.

This work was supported by the Royal Norwegian Council for Scientific and Industrial Research, the Supercomputer Center at SINTEF/the Norwegian Institute of Technology, the Supercomputer Institute at the University of Minnesota, and the NATO Scientific Affairs Division.

References

1. Chao, P–C.; Shur, M.S.; Tibero, R.C.; Duh, K.H.G.; Smith, P.M.; Ballingall, J.M.; Ho, P.; Jabra, A.A. IEEE Trans. Electron Devices, ED 36 (1989) 461–472.

2. Han, C.J.; Ruden, P.P.; Grider, D.; Fraasch, A.; Newstrom, K.; Joslyn, P. Shur, M. IEDM Technical Digest, San Fransisco, (1988) 696–699.

Frequency Dependent CV Measurements of GaAs/AlGaAs Heterostructures

M. Berroth, R. Bosch and V. Hurm

Fraunhofer Institut fuer Angewandte Festkoerperphysik

Eckerstr. 4, 7800 Freiburg, FRG

Summary

A procedure is described to determine the carrier density profile in the channel of a FET by evaluating S-parameters measured over a broad frequency range. Applying this method to GaAs/AlGaAs heterostructures, a frequency dispersion of the gate capacitance has been found, which is attributed to a parasitic conducting channel.

Introduction

The capacitance measurement of a Schottky contact versus the applied bias voltage is a commonly used technique to determine the carrier concentration profile of a device. Due to the operating frequency of about 1 MHz, which is very low compared to applications of field effect transistors far in the GHz range, relatively large structures are required, so that the carrier profile can not be measured at the transistors used in microwave circuits. Further the junction resistance becomes rapidly smaller than the impedance of the junction capacitance at low frequencies, if an increasing forward bias is applied thus making CV measurements difficult. In addition, some of these transisors like MESFETs and MODFETs are showing low frequency effects; CV measurement at high frequencies would therefore be of great interest.

Measurements

The measurements are performed using a HP 8510 network analyzer with a Cascade prober for on-wafer testing in a 50 ohms environment. Different samples have been measured using a standard microwave transistor with 1 gate length and 250 μm width. The drain to source voltage is set to zero and the scattering parameter S_{11} is measured at various gate voltages from beyond pinch-off up to far positive gate voltages. The S-parameters are then converted to the Z-parameters to simplify the deembedding.

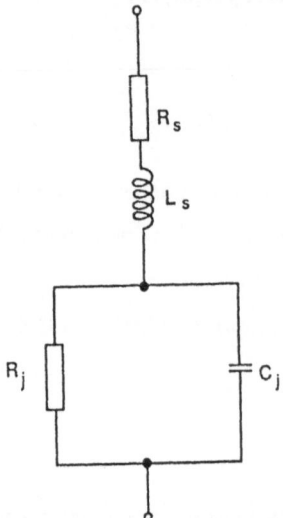

Fig. 1: Equivalent circuit of a Schottky contact

Deembedding Procedure

As shown in fig. 1, the equivalent circuit of a Schottky contact at high frequencies consists of four elements: a parasistic ohmic resistance in series with an inductance and the intrinsic Schottky contact, described as a junction resistance and a junction capacitance in parallel. The impedance of the Schottky contact is given by the following equation:

$$Z_{11} = R_s + j\omega L_s + \frac{R_j}{1 + j\omega R_j C_j} \tag{1}$$

Assuming that the parasitic ohmic resistance and inductance are independent of the bias, both can be determined precisely if the gate is strongly positive biased, because at these high current levels the junction resistance is negligeable and shunts the junction capacitance. The real part of the Z-parameters describes the ohmic series resistance, while the imaginary part determines the inductance at strongly positive gate voltage. Using this procedure to determine the parasitic series resistance and inductance, the measured Z-parameters at all bias points and frequencies can now simply be deembedded by subtracting Rs from the real part and ωL_s from the imaginary part of Z_{11}. The resultant Z-parameters are then converted to Y-parameters, the real part of which represents the juntion conductance, and the imaginary part the junction capacitance.

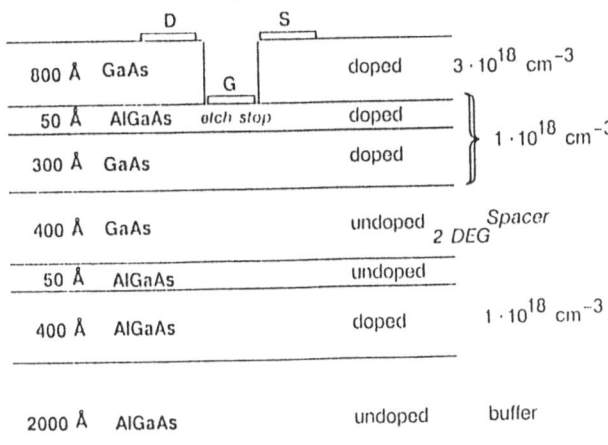

Fig. 2: Cross-sectional view of the heterostructure transistor

Results

We have measured MESFETs grown by MBE on an AlGaAs buffer layer. The CV curves determined were very close to those measured at 1 MHz for the whole range from 50 MHz up to 25 GHz. However a modulation doped field effect transistor with a vertical structure, as shown in fig. 2, exhibited a strong frequency dependence of the junction capacitance, as presented in fig. 3. The proposed explanation for this behaviour is that there is a parasitic conductive channel in the doped AlGaAs layer. The diminishing of this path at higher frequencies is thought to be due to the increased contact resistance

622

to this layer and the reduced mobility of the carriers, establishing a low pass response.

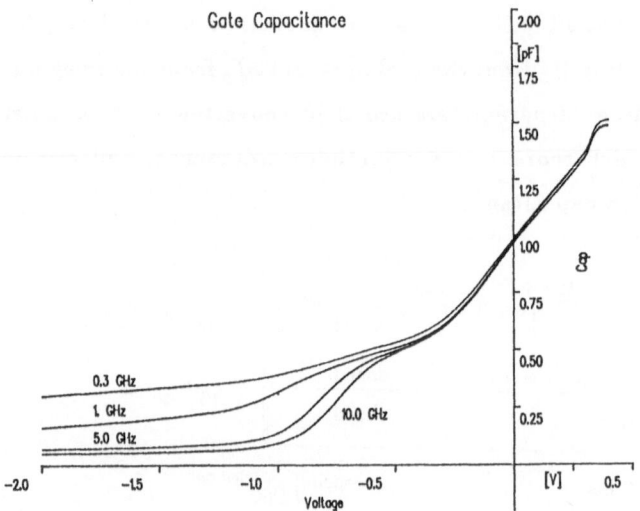

Fig. 3: Gate capacitance extracted from S-parameter data for various frequencies

Conclusion

A method to determine the gate capacitance of even submicrometer gate transistors is proposed. Due to the broad band S-parameter measurement this procedure is of great interest for the characterization of heterostructure FETs, e.g. GaAs/AlGaAs MODFETs. Unlike other methods (1), no fitting procedure is necessary. Especially for the characterization of thin layers needed in enhancement FETs can this procedure be used to determine the junction capacitance of a Schottky contact up to moderately positive bias voltages.

References

1. Y. K. Fukai and M. Muraguchi, "New Method for Measuring Carrier Concentration Profile Near a GaAs Surface Through a Network Analyzer," IEEE Electron Device Letters, vol. 9, pp. 74-76, Feb. 1988.

Nuclear Reaction Analysis of Electronic Materials – Part II: Instability Analysis of a-Si: H Solar Cell Material

M. A. Briere[1] and H. C. Neitzert[2]

Hahn-Meitner-Institute Berlin GmbH,
[1] Dept. Dataprocessing and Electronics,
[2] Dept. Radiation Chemistry,
D- 1000 Berlin 39, F. R. G.

Summary

The hydrogen content is important to the performance of glow discharge amorphous silicon solar cells. In fact, a-Si:H was one of the first materials studied using the ^{15}N NRA method [1]. However, there are some outstanding issues which must be addressed in order to understand the stability of this material in a stress environment (i. e. thermal annealing, radiation). The low temperature (20 - 400°C) diffusion of hydrogen in this material is also still of interest. Further, an investigation into the effects of doping on the concentration and spatial distribution of hydrogen in a-Si:H solar cell material is required.

Experimental Results

Figure 1 shows the effect of annealing treatments on the distribution of hydrogen in a two layer structure formed by subsequent depositions of a-Si:H at two different temperatures (200 °C / 300 °C).Two of three identical samples were additionally annealed for 1 h at 300 °C, resp. for 2 h at 400 °C. It can be seen that the hydrogen profile at the interface region is very stable with respect to these annealing conditions, though there is understandable diffusion into the quartz substrate during the 400 °C annealing.The accumulation of hydrogen at the quartz interface, as well as that between the two layers is found to be a common feature and is under investigation.

In Figure 2 we show the profile of a 3-layer structure, also including the effects of electron irradiation. The layer are grown at different temperatures (250 °C/ 120 °C/ 250°C) which is known to lead to different hydrogen incorporation into the films [2].We see again a slight increase of the hydrogen content at the substrate interface. Electron irradiation did not change the hydrogen distribution in this structure.

Figure 3 shows the hydrogen profile for a commercially available p-i-n-solar cell before and after irradiation by 2.5 MeV electrons for a dose of 5 Mrad. After irradiation, there is a further accumulation of hydrogen in the n+-region and a slight depletion in the i-region as well as a small increase of the hydrogen concentration in the p+-region. The general shape of the original curve is in agreement with that shown by Müller et al. [3], though the accumulation of hydrogen in the n+-region is significant greater than expected. The contactless technique of microwave reflectance [4] was used to evaluate the effect of the electron irradiation on the carrier dynamics. Results indicate that, as expected, the lifetimes of the carriers are dramatically decreased by the the radiation treatment.

Figure 4 shows the profile for a thick p-i-n structure made in our lab. Here, the accumulation of hydrogen in the highly doped ([PH3] / [SiH4] > 10^{-3}) p+-layer is opposite to the results Müller et al. [3] found for highly Boron doped samples.The reason for this discrepancy is not yet clear. It is clear, however, that the presence of Boron enhances surface denution in agreement with the results of Figure 3 and enhanced hydrogen diffusion found in Boron doped samples [4].

Conclusions

The results indicate that the hydrogen distribution in a-Si:H materials are generally stable with respect to thermal annealing up to 300ºC as well as electron irradiation. Accumulation of hydrogen at substrate and layer interfaces is found in some cases. Phosphorous as well as Boron doping enhances the hydrogen concentration in bulk material. Absence of the denuded region at the substrate interface as well as the magni-tude of the doping effects calls into question the general applicability of the previous results [3].

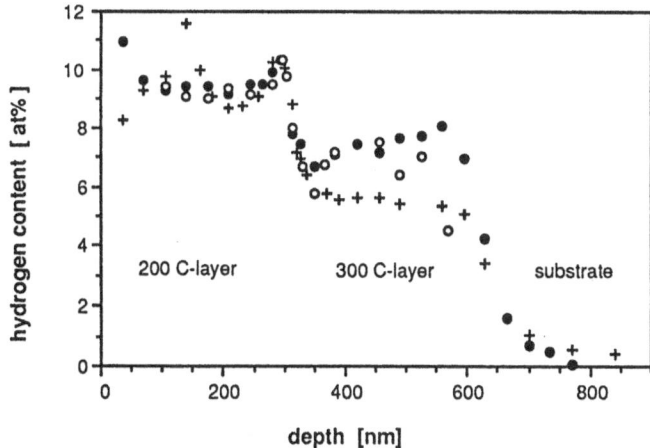

Fig. 1 Hydrogen profiles of a 2- layer a-Si:H structure
deposited at different temperatures (300°C / 200°C)
● as grown   ○ 1h anneal at 300°C   + 2h anneal at 400°C

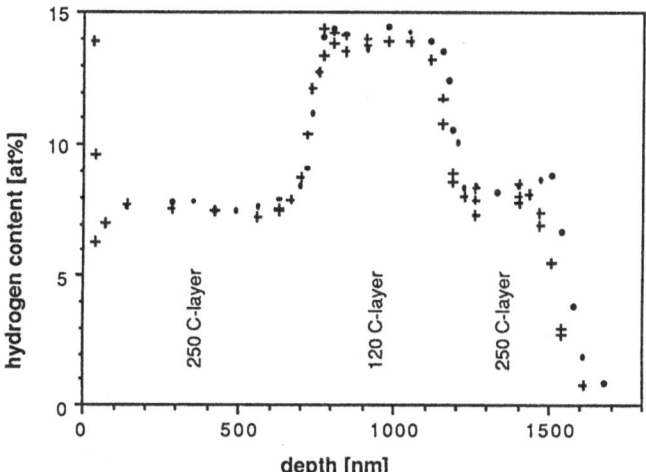

Fig. 2 Influence of electron irradiation on hydrogen profile of a 3-layer a-Si:H
structure deposited at different temperatures (250°C,120°C,250°C)
● before e-irradiation   + after e-irradiation (2.5MeV, 5Mrad)

626

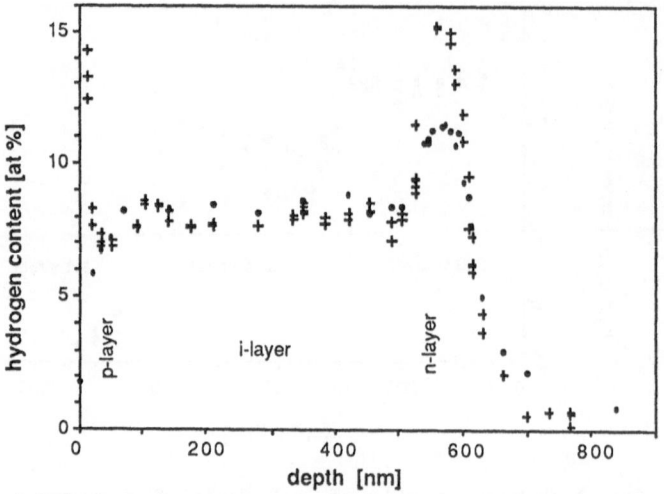

Fig. 3 Influence of electron irradiation on the hydrogen profile of a p-i-n a-Si:H solar cell.
• before e-irradiation   + after e-irradiation (2.5MeV, 5Mrad)

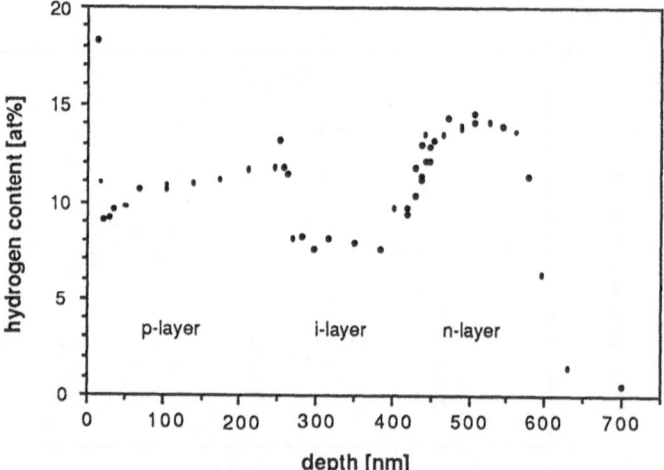

Fig. 4 Hydrogen profile of a p-i-n a-Si:H structure

References

[1]   M. H. Brodsky et al., Appl.Phys.Lett. 30 (1977) 561.

[2]   D. E. Carlson et al., Phil. Mag. B 45 (1982) 51.

[3]   G. Müller et al., Phil.Mag.B 41 (1980) 571.

[4]   M. Kunst and G. Beck, J. Appl. Phys. 63, (1988) 1093.

[5]   R. A. Street et al., Phil.Mag.B 56 (1987) 305.

Miniaturized Accelerometers Based on Silicon Micromachining Techniques

W. Benecke, W. Riethmüller, U. Schnakenberg, B. Wagner

Fraunhofer-Institut für Mikrostrukturtechnik
Dillenburger Str. 53
D-1000 Berlin 33, Federal Republic of Germany

Abstract

A miniaturized acceleration sensor has been fabricated using silicon micromachining techniques. Integrated monocrystalline silicon piezoresistors are used to detect the bending of a beam attached to a seismic mass. An outline is presented for improvement of the sensor performance with respect to temperature sensitivity and overrange protection.

Introduction

Silicon micromachining for three dimensional structuring of the substrate, combined with standard planar IC processes allows the realization of new classes of sensors and actuators /1,2,3/.

Most of the improvements with respect to size, weight, geometrical accuracy, reproducibility, as well as reliability of micromechanical devices are a direct consequence of the batch processing technique employed in the usual silicon IC manufacturing process. The use of silicon as a base material allows the monolithic integration of the mechanical structures with electronic circuitries on the same chip /4/. This renders possible the fabrication of sophisticated systems with superior performance for a wide variety of new applications.

Silicon accelerometers generally contain a seismic mass, suspended by one or more flexible beams, which moves relative to the substrate while under the influence of an acceleration. Primarily integrated strain sensitive piezoresistors /5,6/, or mechanical resonators /7/ are used for signal conversion. The mechanical deflection can also be picked up by capacitive techniques./8,9/

Sensor Design and Fabrication

A schematic view of a piezoresistive accelerometer is shown in Fig. 1.

The seismic mass consists of a silicon plate attached to three bending beams. To increase the mass, electrodeposition of heavy metals, as well as anisotropically structured substrate silicon can be used. Single crystalline piezoresistors are integrated on the central beam for strain detection. A fully active Wheatstone bridge is formed by a combination of two longitudinally and two transversely active strain gauges. The piezoresistive gauge factor $K = (\Delta R/R)/(\Delta L/L)$ for silicon depends on the dopant type, doping concentration and temperature /10/. For single crystalline silicon the K-factor is anisotropic and can have values well above 100. The two side supports of the mass plate prevent torsion modes of the structure. To maximize the sensitivity and resonant frequency simultaneously, short bending beams, mass plates with the largest possible weight and a center of mass close to the suspension point should be used /11/.

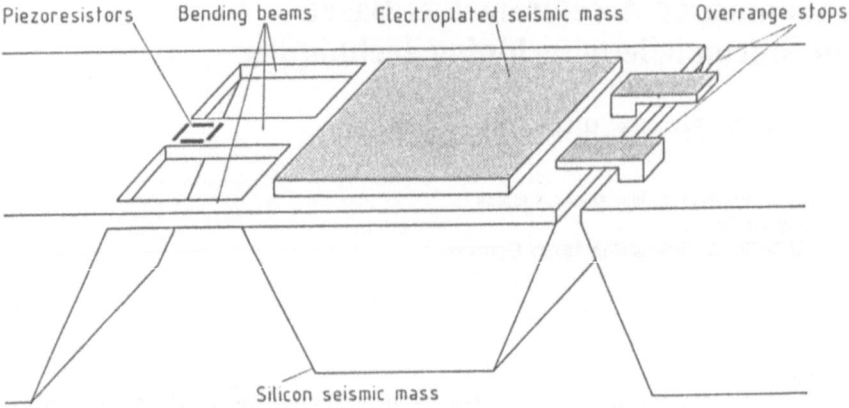

Fig. 1: Schematic view of a piezoresistive cantilever-type accelerometer with bidirectional over-range protection. The mass can be realized by electroplating and/or monolithically by silicon.

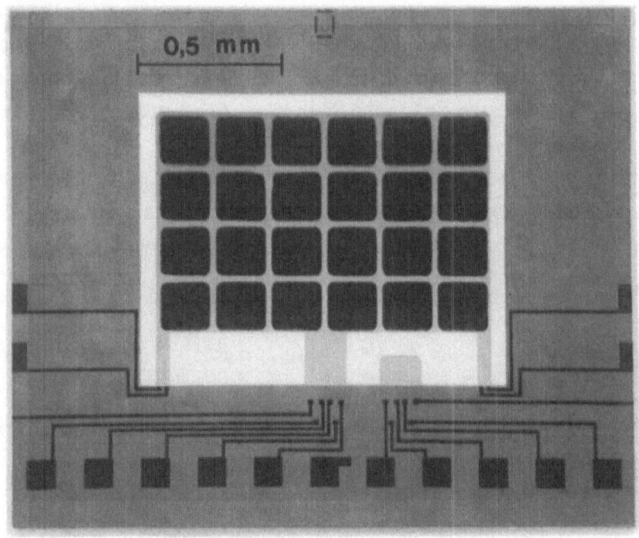

Fig. 2: Top view of the fabricated accelerometer showing the 24 separate gold bumps on the mass plate

Cantilever-type accelerometers, as discussed here, need mechanical overrange protection in order to be resistant to shock damage. As shown in Fig. 1, bidirectional overrange stops can be integrated monolithically. These structures could be produced by electroplating techniques with a two-level resist process /12/.

An optical micrograph of the fabricated sensor is shown in Fig. 2. The bending beams have a thickness of 6 μm. The seismic mass was deposited by electroplating of separate gold bumps with a thickness of 20 μm. Piezoresistors located on the outer suspending beams are for break detection. The short cantilever not attached to the mass plate can be used to compensate cross sensitivities.

The processing sequence starts with an n-epitaxial layer on the p-type, (100)-oriented substrate. The piezoresistors, as well as their connection to the gold interconnection lines, are produced by boron implantation and diffusion.

The anisotropic etching of the substrate is performed in a KOH-water solution from the back of the wafer. During etching, a positive voltage is applied to the n-epitaxial layer with respect to a counter electrode in the solution. The etching process stops when the pn-junction is reached /13/. The accelerometer structure is released by anisotropic etching from the front of the wafer. Instead of KOH, etchants based on NH_4OH can be used. They can be applied under clean room conditions within standard IC fabrication lines /14/.

Results

The experimental results are summarized in the following table:

Constant voltage supply	5 V
Piezoresistors	10 kΩ
Sensitivity	2 mV/g
Temperature coefficient of sensitivity (TCS)	- 0.3 %/°C
Offset voltage	± 15 mV
Temperature coefficient of offset (TCO)	0.2 mV/°C
Nonlinearity	< 1% (to 50 g)
Resonant frequency	750 Hz

Discussion

The temperature dependence of the sensitivity is basically due to the temperature coefficient of the piezoresistive K-factor (TCK) of silicon. By choosing an appropriate doping concentration it is possible to obtain a temperature coefficient of resistance (TCR) which is equal but of the opposite sign to the TCK. In this case, the TCK can be compensated with a constant current excitation of the Wheatstone bridge.

An important improvement in the temperature coefficient of the offset voltage (TCO) was achieved by forming separate gold bumps on the mass plate instead of a solid gold mass. It could be demonstrated that the TCO could be reduced by a factor of ten by this measure. This can be explained by the bimetallic effect, causing a deformation of the gold/silicon sandwiched plate, which results in a large temperature coefficient of the dc-offset of the Wheatstone bridge. A further reduction of the TCO can be expected if the seismic mass is achieved monolithically from silicon. This requires additional space and increases the overall chip size.

Conclusions

Silicon micromachining allows the fabrication of very small accelerometers. The piezoresistive signal conversion is very convenient, but special attention has to be given to cross sensitivities to temperature. Technological and design considerations have been given to improve the sensor performance. Moreover, to create exchangeable sensors, laser-trimmed resistor networks could be added for compensation of offset voltage, temperature coefficients and for adjustment of the sensitivity.

References

/1/ A. Heuberger (ed.), Mikromechanik, Springer 1989.

/2/ W. Benecke, Micromechanical sensors, Proc. Comp. Euro., 3-39, Hamburg 1989

/3/ J.C. Greenwood, Silicon in mechanical sensors, J. Phys. E.: Sci. Instrum. 21 (1988) 1114

/4/ W. Benecke, A. Heuberger, W. Riethmüller, Technologien zur monolithischen Integration von mikromechanischen und mikroelektronischen Funktionsgruppen, GME-Fachtagung Mikroelektronik, Baden-Baden 1989

/5/ L. Roylance, J. Angell, A batch-fabricated silicon accelerometer, IEEE Trans. Electr. Dev., ED-26 (1979) 1911

/6/ H. Sandmaier, K. Kühl, E. Obermeier, A silicon based micromechanical accelerometer with cross sensitivity compensation, Proc. Transducers'87, 399, Tokio 1987

/7/ S. Chang, M. Putty, D. Hicks, C. Li, R. Howe, Resonant-bridge two axis microaccelerometer, Proc. Transducers '89, Montreux 1989

/8/ F. Rudolf, A. Jornod, J. Bergqvist, H. Leuthold, Precision accelerometers with μ-g resolution, Proc. Transducers '89, Montreux 1989

/9/ H. Seidel, H. Riedel, R. Kolbeck, G. Mück, M. Königer, Capacitive silicon accelerometer with highly symmetrical design, Proc. Transducers '89, Montreux 1989

/10/ Y. Kanda, A graphical representation of piezoresistance coefficients in silicon, IEEE Trans. Electr. Dev., ED-29 (1982) 64

/11/ H. Seidel, L. Csepregi, Design optimization for cantilever-type accelerometers, Sensors and actuators, 6(1984) 81

/12/ B. Wagner, W. Benecke, W. Riethmüller, U. Schnakenberg, patent pending

/13/ T.N. Jackson, M.A. Tischler, K.D. Wise, An electrochemical p-n junction etchstop for the formation of silicon microstructures, IEEE Electron Dev. Letts, EDL-2 (1981)44

/14/ U. Schnakenberg, W. Benecke, B. Löchel, NH_4OH based etchants for silicon micromachining, Transducers 89, Montreux 1989, patent pending

Session 7

Invited Papers

Sub-0.1 µm Silicon MOSFETs

D.P. KERN

IBM Research Division
T. J. Watson Research Center
Yorktown Heights, New York 10598

Summary

Work aimed at demonstrating feasibility of silicon FET technology at the 0.1µm gate length level and below has produced results which indicate that scaling of FETs to smaller dimensions well beyond what is practiced presently is a trend worth while pursuing. Extrinsic transconductances of over 940µS/µm at 0.07µm gate length and 13ps delay per stage in 0.1µm gate length ring oscillators have been achieved with self-aligned, n-channel, polysilicon-gate FETs fabricated with direct write electron beam lithography of all levels. There is not only clear evidence for velocity overshoot but at the same time it is shown that devices can be built in such a way that such effects translate into improved extrinsic performance.

Introduction

Field effect transistors (FETs) made from silicon are the most common components of integrated electronics. Integration has reached its highest levels employing Si-FET devices and circuits due to their simple fabrication, superior material quality and relatively low power consumption. The most striking feature in Si-FET device and circuit development has been the continuous progress in scaling[1] of devices to ever-smaller dimensions. However, there are questions regarding the limits to which this miniaturization can be carried. A variety of non-scaling parameters and associated detrimental effects, such as mobility degradation[2], inversion-layer broadening[3], tunneling through the gate oxide, and the onset of velocity saturation will reduce the performance to be gained by scaling, not to mention such practical difficulties as linewidth and overlay control, thin insulator reliability, shallow junction fabrication, the limitations of contact and spreading resistivities, and many others. Low temperature operation, however, will help with voltage scaling and small dimensions together with low voltages may have advantageous effects as well, such as the possibility of velocity overshoot[4]in very short devices and the weakening of detrimental effects associated with energy thresholds such as avalanche breakdown and hot carrier injection into insulators.

Although theoretical treatments abounded on the perceived limits of silicon FET technology, prior to our efforts there was relatively little experimental work addressing these issues in the deep-submicron regime. The objectives of the project fall into two categories: a desire to understand the limits of traditionally scaled FETs, of established notions of FET behavior and classical transport, the emergence of new effects, if any; on a more practical side, to explore how far FET technology is extendable using as few deviations from standard processing as possible, and what performance can be achieved.

Design Considerations

As a first feasibility study at the 0.1µm gate length level our efforts[5-8] were directed exclusively towards n-channel devices (NMOS). Following classical scaling theory, all horizontal and vertical dimensions, and voltages as well, were attempted to be reduced in proportion with the gate length. However, for funda-

634

mental or practical reasons, scaling had to be modified in several instances. Voltages cannot be decreased indefinitely with dimensions due to non-scalable parameters and noise margins. Due to velocity saturation in the high fields associated with small dimensions, performance will saturate below the operating voltage, and therefore the high operating voltage causes higher power consumption not accompanied by a corresponding gain in performance. Design for the lowest power supply voltage feasible is therefore a key issue. This inevitably leads to low temperature (LT), 77°K, operation which increases the turn-off rate below threshold and reduces threshold voltage changes caused by temperature variations. Since these two effects are the main obstacles to lowering the threshold, and thus the operating voltage, LT enables a design point which is not reachable at room temperature (RT). In addition, LT operation leads to improved performance[9], mainly due to lower interconnect resistance, and results in better punch through behavior. A threshold voltage of 150mV was considered a lower practical limit, lending itself to a 0.6V - 0.8V power supply (V_{DD}). Another exception was made with the gate oxide. For reliability reasons the gate oxide was grown to 4.5nm in most cases, rather than the ~2.5nm called for by full scaling, while all other vertical dimensions were scaled with the 0.1μm gate length design. In the horizontal dimensions the diffusion level was not fully scaled for practical reasons, since we decided to stay with a simple semi-recessed oxide isolation, compatible with 0.25μm ground rules.

In order to reduce geometry effects in short channel devices it is necessary to reduce depletion layer widths. At LT it is possible to forward bias the substrate by 0.6V without experiencing significant leakage currents[10]. To minimize the impact of impurity scattering on inversion carrier mobility at 77°K, the acceptor doping in the channel is peaked away from the surface. A source drain (S/D) extension structure[11] is used providing a shallow junction near the channel and a deeper junction displaced from the gate by a spacer. In order to assure low S/D resistance, the single most important factor in determining the extrinsic transconductances, antimony was used in some of the S/D extensions, which has advantages over arsenic for very shallow junctions[12]. This structure provides reduced geometry effects, low parasitic resistance and allows the formation of self-aligned metal or silicide on the diffusions. A schematic cross section of the device together with design voltage levels is shown in Fig. 1.

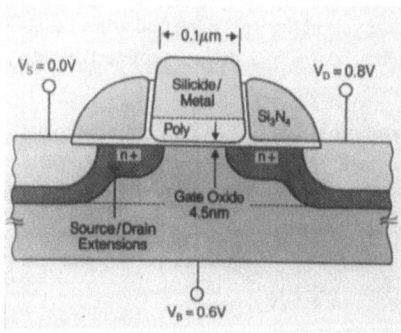

Figure 1: Schematic cross section of typical 100nm gate length device and bias levels.

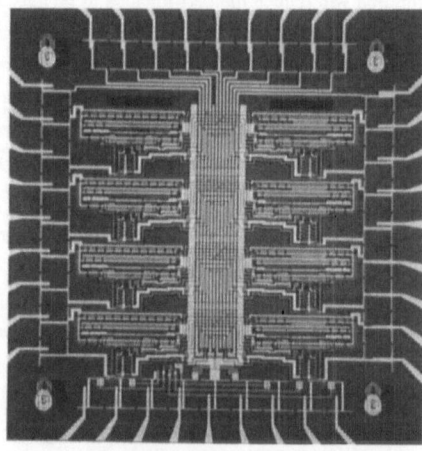

Figure 2: Micrograph of chip containing inverter chains of varying gate length.

Seven different chips have been assembled with a variety of test structures ranging from simple parametric test sites to inverter chains and ring oscillator circuits. Most of the designs were repeated on the chips with different gate lengths, with actually fabricated gate lengths varying from 0.27μm to 0.07μm, as measured by SEM on hardware. Fig. 2 shows the picture of a chip containing inverter chains and associated support circuitry.

Fabrication

Lithographic capability is clearly an important issue in this work[13]. Not only is high resolution required for fabrication of the short gates, but besides the use of low-resistance materials such as silicides and of appropriate dopant densities, miniaturization of all device elements, in particular reduction of the distance between S/D contacts and optimization of the contact size and geometry are key factors in reducing parasitic effects. Increased demands on level-to-level overlay and dimensional control in the contact process are direct consequences.

All five lithography levels, including one for high resolution alignment marks were performed in a Gaussian electron beam system, with a high resolution field size of 250μm $\times$ 250μm. The system was specifically designed for sub-100nm lithography [14] with a minimum probe size of 8nm. It could be operated at up to 80kV, with 25kV being the typical value used for this work. Electronic noise, vibrations and other interferences have been carefully reduced to a level consistent with the nanometer lithography requirements, and methods have been developed to verify these conditions routinely [15, 16].

The highest resolution step was the gate fabrication. It was performed by lift-off of a metal mask using double layer PMMA, followed by reactive ion etching (RIE) to form the polysilicon gates down to 70nm minimum dimensions. Lithography for the contact holes, where below 0.2μm resolution was required, was achieved using a novolac-based resist. This resist exhibited excellent behavior during contact hole RIE. The level-to-level overlay was found to be consistently better than 30nm. This was achieved using tantalum silicide alignment marks with 0.25μm minimum features, lifted off at the first lithography level. Further details of the lithography have been described previously [6, 13, 15].

Processing was carried out on a batch of wafers in parallel, with different process variations from wafer to wafer. Except for lithography related work the processing was carried out in the silicon processing facilities of our laboratory. The main processing steps together with some of the reasons behind them and significant experimental variations are summarized below:

1. Implant boron into the backside of 2Ωcm p-type substrates to provide good contact at 77°K.

2. Liftoff TaSi$_2$ alignment marks with minimum dimensions of 0.25μm.

3. Grow 10nm pad oxide and deposit 100nm nitride, pattern the diffusion regions, implant channel stop boron, grow 160nm dry, semi-recessed field oxide at 950°C.

4. Strip the nitride-oxide stack from diffusion regions, implant boron for threshold control and punch-through stop. Doses ranged from 2.5 to 5.5×10^{12}/cm^2. In most cases two implant energies, 15keV and 30keV, were used.

5. Grow thermal SiO$_2$ gate insulator ranging in thickness from 3.3nm to 4.5nm, at 800°C in dry oxygen with HCl. Thickness was measured during processing by ellipsometry and upon completion by capacitance.

6. Deposit 100nm thick undoped polysilicon for the gate.

7. Write the gate pattern in double layer PMMA resist, lift-off metal mask for a two-step RIE process to form the poly gates.

8. Grow 7.5nm oxide on polysilicon at 850°C in dry oxygen with HCl.

9. Implant S/D extensions[11]. Both arsenic and antimony were used as dopant species. Doses were in the low $10^{14}/cm^2$ range, with energy down to 10keV. Multiple-energy implants were used in some cases to achieve more 'box-like' profiles.

10. Deposit and RIE 100nm nitride to from an insulating gate-sidewall spacer.

11. Implant arsenic S/D. Doses were 2 to $5x10^{15}/cm^2$, at 20keV.

12. Activate S/D. Furnace thermal budgets ranged from 900°C, 30 minutes to 850°C, 20 minutes. Some wafers received rapid thermal anneals of 1050°C and 1100°C for 10sec. The final depth was estimated to be ~50nm for the S/D extensions and 100nm for the 'deep' junctions.

13. Form self-aligned titanium-silicide ($TiSi_2$) on selected wafers. Standard treatment at 800°C for 30 minutes resulted in gate shorts to the S/D and/or channel, caused by excessive Ti diffusion along grain boundaries. Rapid thermal processing in the 10 to 30 second range solved this problem. However, sheet resistance of $TiSi_2$ was ~30Ω/□ instead of <10Ω/□ Furthermore, a resistance increase of ~30% was observed for 0.1μm wide lines. Overall, the silicide resistivity is not considered to be a fundamental problem, there are alternatives. In the present work, however, it limited the performance of the fastest circuits.

14. Deposit 100nm low-temperature conformable oxide overlay.

15. Pattern the contact hole level and RIE contact openings.

16. Form metal level via lift-off of 300nm Ti/Al and anneal in a partial hydrogen ambient.

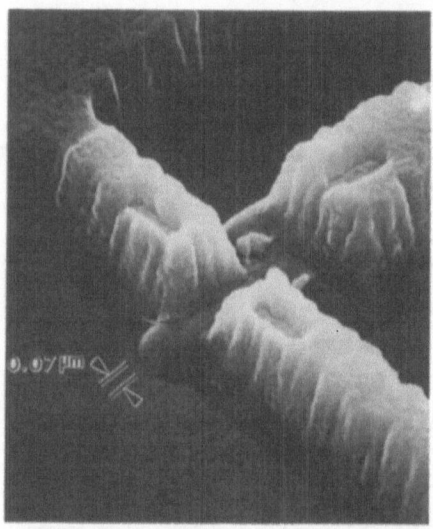

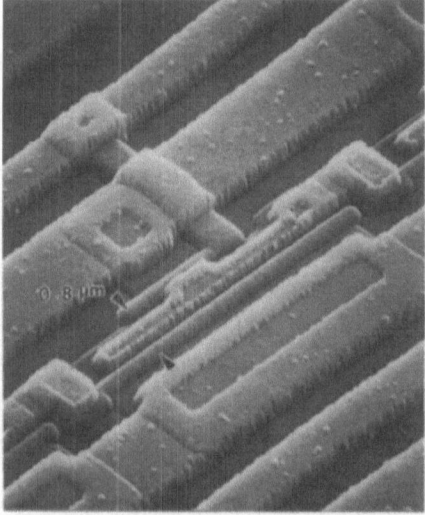

Figure 3: SEM micrograph of a 0.5μm wide FET with 0.07μm gate.

Figure 4: Section of an inverter chain with the 5μm wide active device in the center and the 1.25μm wide load device to the upper left.

To illustrate the precision in the fabrication process, Fig. 3 shows a SEM micrograph of an FET with a gate 0.07μm long and 0.5μm wide, probably the smallest FET ever made. Fig. 4 shows a section of an inverter chain with 5μm wide active devices, while Fig. 5 shows a similar region with the metal and oxide layers removed, exhibiting the 0.07μm wide gates with line smoothness better than 0.01μm.

Results and Discussion

Overall ~75% of the tested structures were operational. This included sites which depended on the operation of many devices, such as inverter chains and ring oscillators. Defects were mostly catastrophic, caused by contamination and independent of the gate length. This yield is quite satisfactory, considering the experimental nature of the work.

Device Performance: In terms of basic device function, the design expectations have been well met. Fig. 6 shows the measured change in threshold voltage as a function of gate length. While similar behavior is seen at both RT and LT, the utility of the forward bias at LT is clearly seen. Also the deep channel implant is quite effective in improving the punch through behavior at LT for even the shortest devices. Overall these characteristics show the benefit of reduced oxide thickness and increased channel doping in controlling geometry effects down to 0.1μm dimensions. The threshold fall-off at 0.1μm gate length, however, will require tight linewidth tolerances when large numbers of these devices are to be fabricated, although further optimization in terms of reduced oxide thickness below 4.5nm, increase of the dose of the punch through control implant, and finally a ~50-100mV higher threshold seem possible. The subthreshold slope at RT and LT also behaved as expected[17].

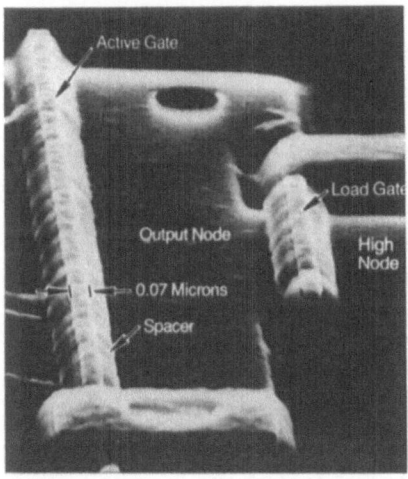

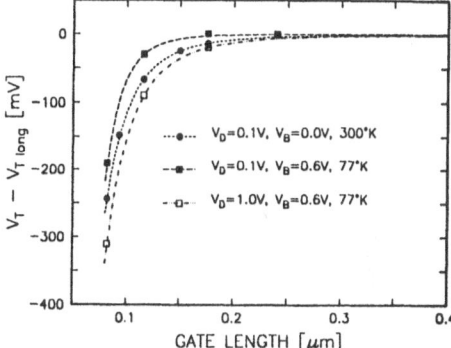

Figure 5: Section of an inverter chain after stripping of metal and oxide layers, exhibiting 0.07μm long gates.

Figure 6: Threshold voltage change as a function of gate length.

As expected, the extrinsic characteristics depended strongly on the processing variations. In order to highlight meaningful trends rather than incidental, process related ones, the results presented from now on will all be from a single wafer, unless specifically pointed out otherwise. The wafer had 4.5nm gate oxide, antimony S/D extensions, self-aligned $TiSi_2$ on S/D and gate, and a single threshold adjust implant

638

of $5.5 \times 10^{12}/cm^2$ boron at 30keV. The resulting threshold of 0.3V for the 0.1μm devices at 0.6V substrate bias, was at the high end of the range we explored. However, this wafer was chosen, because it had the best combination of low S/D resistance and good quality TiSi$_2$.

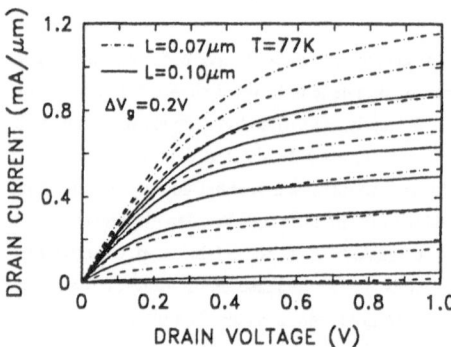

Figure 7: Device terminal characteristics at 77°K. Maximum V_G is 1.5V, substrate is biased to 0.6V.

Fig. 7 shows the LT terminal characteristics of the 0.07μm and 0.1μm gate-length devices. As can be seen, the device characteristics are excellent. The 0.07μm device at a drain voltage V_D of 1V has a maximum transconductance (g_m) of over 940μS/μm, while the 0.1μm device has 770μS/μm. The room temperature g_m of these devices with 0V substrate bias is 590μS.μm and 505μS/μm, respectively. Both the LT and RT transconductances are the highest ones measured to date in silicon FETs, and compare well even with values attained in compound semiconductors [18]. The same very high transconductances were also measured on another of our wafers[5, 17]. The common processing feature of the two wafers was the antimony S/D extensions. The sharp lateral S/D junction edge achieved with antimony was necessary for lowering the parasitic resistances. For a detailed discussion of device performance as a function of gate length at both RT and LT we refer to previous publications [5, 6, 17].

It is instructive to assess the performance of such devices in a dense circuit environment where wiring capacitances have to be driven over full voltage-signal swings. It can be seen from Fig. 7 or from device characteristics presented earlier[5, 17], that a ~2.5μm wide 0.1μm gate-length device, with an input capacitance of <3fF, can switch more than 0.5mA of current upon a 0.5V input swing. Such current and voltage values are typical of the highest performance emitter coupled logic (ECL) bipolar circuits. Thus LT 0.1μm NMOS devices are in a domain so far reserved only to bipolar transistors. Such a performance in small FETs is to a considerable extent due to velocity overshoot, as will become clear from the following.

Velocity Overshoot: First predicted from Monte Carlo simulations[4], this is an effect that can only be observed in devices that are small enough for very strong field gradients to be tolerated over major portions of the device. In a steady state condition, i.e. when carriers are exposed only to slowly varying fields in both space and time, to any given field strength there belongs a unique carrier temperature. This temperature, or equivalent energy, determines the possible scattering mechanisms and therefore the carrier mobility. Velocity saturation comes about because in stronger fields the carrier energy is higher, the scattering rate increases and therefore the mobility decreases. The phenomenon of velocity overshoot arises when either the temporal or the spatial scale (in practice both) is such that the one to one correspondence between field strength and carrier energy breaks down: when carriers experience a field increase, e.g. from F_0 to F_1, then for that time period before they are heated up to the steady state energy corresponding to

F_1 they can exceed, or overshoot, the velocity they would reach under steady state conditions in a field F_1.

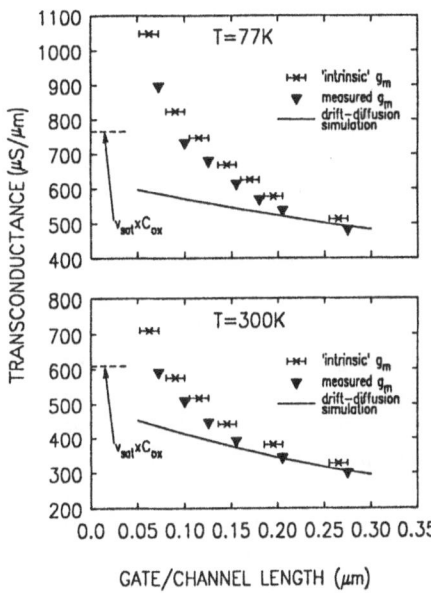

Figure 8: Plot of measured transconductance (solid triangles) vs. gate length, intrinsic transconductance (crosses) vs. channel length with error bars indicating uncertainty in effective channel length, and result of two-dimensional finite element drift-diffusion transport simulation (solid line), for two temperatures. Limits on the intrinsic transconductance are marked as $v_{sat} \times C_{ox}$.

Drift times, field strengths and spatial scales in deep-submicron NMOS devices are such that one can expect to observe velocity overshoot and the effect has been reported as responsible for the intrinsic behavior of short silicon inversion layers[19, 20], although S/D resistance dominated the terminal characteristics in those investigations. Nevertheless, our results are in qualitative agreement with these earlier reports. For a rigorous theoretical treatment one has to resort to an array of powerful tools in band structure theory and Monte Carlo techniques[21].

Fig. 8 shows the measured (extrinsic) transconductance as a function of gate length and intrinsic g_m as a function of channel length. The plot was made for $V_D = 0.8V$, and for each gate length at the gate voltage (V_g) where the transconductance peaked. This typically happened at a gate drive ($V_g - V_t$)~0.6V, where V_t is the threshold voltage. The transconductances are plotted without fitting or adjustment except for width-normalization. The 'intrinsic' transconductances, on the other hand, and the channel lengths against which they are plotted were the result of fitting, and therefore do contain uncertainties as indicated by the error bars in Fig. 8.

The evidence for velocity overshoot is observed in two ways. First there is the sheer magnitude of the transconductance in the shortest devices. If carriers were not capable of exceeding a saturation velocity v_{sat} there would be a theoretical upper limit for the intrinsic transconductance of $v_{sat} \times C_{ox}$ where C_{ox} is the gate oxide capacitance per unit area. For saturation velocities of $8.0 \times 10^6 cm/sec$ and $1.0 \times 10^7 cm/sec$ at RT and LT, respectively, these limits are marked on Fig. 8 as $610 \mu S/\mu m$ and $770 \mu S/\mu m$. It can be seen that even the measured, extrinsic transconductances exceed these limits. The second manifestation of velocity overshoot lies in the trend of transconductance increase with decreasing gate or channel length. Extensive finite element simulations have been performed[7, 22], using FIELDAY, a two-dimensional simulator containing all the physics of local, drift-diffusion transport. On the one hand, the effect of device details such as S/D resistance, S/D junction profiles, contact and wiring resistance on device performance have been investigated, but also transconductance has been modeled under the experimental conditions, using mobilities of $720 cm^2/Vsec$ and $390 cm^2/Vsec$ at LT and RT, respectively, and the before mentioned saturation velocities. Values for v_{sat} scatter widely in the literature [23, 24], with the values used here being among the higher ones. The solid curve in Fig. 8 therefore shows how the transconductance would have behaved in the absence of velocity overshoot. While the longer gate devices are matched quite well, the increasing discrepancy between the modeled curve and the experimental data for shorter gates can only mean that the electron velocity exceeds v_{sat}. Detailed Monte Carlo simulations [21, 25] show indeed that the carriers reached speeds of more than double that of v_{sat}.

Circuit Performance Three of the seven chips were designed for measurement of delay times. They contain 21-stage unloaded ring oscillators and open ended inverter chains. An enhancement mode device of the same gate length, with its gate tied to an independent power supply, served as the pull-up element, under these conditions equivalent to a resistor. The width of the active devices was 5μm, that of the load devices 1.25μm. The output of each ring oscillator was buffered into either a push-pull circuit or a source-follower for driving off chip, this support circuitry having 0.25μm gate lengths. Of the ring oscillators and inverter chains tested on 33 chips from 5 different wafers, ~70% were operational, and functionality was not correlated with gate length.

Fig. 9 shows an oscilloscope trace of a 0.1μm gate length ring oscillator output at LT. The period of 550ps translates into a delay per stage of 13.1ps, the shortest switching time obtained to date in any kind of silicon device at both RT and 77°K[26], including bipolars[27]. The amplitude does not represent the internal signal of the ring oscillator; it is only the output of the on-chip driver into a low impedance load, limited by the circumstance that the power to the driver was supplied via a long, relatively high resistance silicided polysilicon line. The consistency of all the delay measurements and the excellent characteristics of the individual inverter stages indicated that the ring oscillators internally had the full voltage swing. As mentioned before, in this regime of transconductance saturation the switching does not depend strongly on the power. In fact, a power increase by a factor of 4 resulted only in 15% delay improvement[8, 22]. The power-delay product of the 0.1μm devices at 0.9V V_{DD} is 1.1fJ per 1μm of channel width.

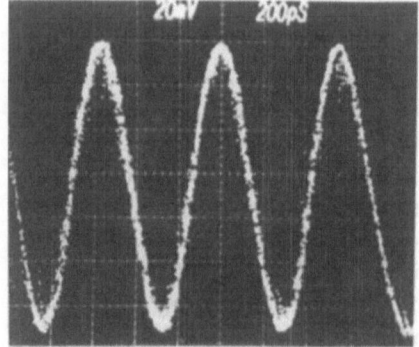

Figure 9: Performance of a 21-stage 0.1μm gate length ring oscillator at 77°K.

The fastest switching delays measured at selected gate lengths, always on the same wafer, are as follows[8]: 19.5ps at 0.20μm, 17.8ps at 0.16μm, 16ps at 0.13μm, and 13.2ps at 0.10μm. For the 0.1μm gate length ring oscillators the shortest RT delay was found to be 17.1ps at $V_{DD} = 1.3$V with 0.12mW of power consumption per 1μm of channel width. Detailed circuit simulations with measured device characteristics as input[22] revealed that the measured switching times were significantly deteriorated (~6ps) by the high sheet resistivity of the $TiSi_2$. This also explained why the 0.07μm gate length circuits were actually the slowest of all. The simulations indicated, that without the RC-constant constraints associated with the gate resistance, the 0.1μm gate length ring oscillators should have reached a delay per stage of 7ps and with horizontally fully scaled devices employing shallow trenches for instance, delays could be below 5ps per stage.

Conclusions

Experimental NMOS-FET devices and circuits with gate lengths down to $0.07\mu m$ (estimated $\leq 0.05\mu m$ channel length) have been successfully built using existing design and processing approaches. Their performance with a transconductance of $940\mu S/\mu m$ and 13ps delay per stage in ring oscillators is the best ever for silicon devices at room temperature or $77°K$. Ultra-high resolution lithography and accurate overlay - as provided by our nanolithography electron beam system - proved essential for the work reported here, improvements in throughput, however, still pose one of the challenges for large scale application of such deep-submicron technology. Selective applications at the $0.1\mu m$ level, on the other hand, could be practically within reach. The ability to utilize non-steady-state transport in ultra short channel devices and the associated performance gain provides additional incentive to continue miniaturization of silicon FETs, even below the $0.1\mu m$ level.

Acknowledgements

The work reported here is the result of a highly collaborative effort, in particular with G.A. Sai-Halasz, M.R. Wordeman, S. Rishton and E. Ganin. Gratefully acknowledged are the contributions of the members of the Nanostructure Technology group, the Silicon Facility and the Exploratory Materials group in our laboratory, in particular H. Ng and D. Moy. Illuminating discussions with S.E. Laux, M.V. Fischetti, R.H. Dennard and T.H.P. Chang were of great benefit. D.S. Zicherman and M.D. Rodriguez are acknowledged for help in characterization.

References

1. Dennard, R. H., Gaensslen, F. H., Kuhn, L., and Yu, R. H. Design of Micron MOS Switching Devices. *IEEE IEDM Tech. Dig.*, page 344, 1972.

2. Baccarani, G., and Wordeman, M. R. Transconductance Degradation in Thin-oxide MOSFETs. *IEEE Trans. Electron Devices*, ED-30:1295, 1983.

3. Leburton, J. P., and Dorda, G. Effect of the Electron Temperature on the Gate-Induced Charge in Small Size MOS Transistors. *Solid-State Elec.*, 26:611, 1983.

4. Ruch, J. G. Electron Dynamics in Short Channel Field-Effect Transistors. *IEEE Trans. Electron Devices*, ED-19:652, 1972.

5. Sai-Halasz, G. A., Wordeman, M. R., Kern, D. P., Ganin, E., Rishton, S., Zicherman, D. S., Schmid, H., Polcari, M. R., Ng, P. J., Restle, P. J., Chang, T. H., and Dennard, R. H. Design and Experimental Technology for 0.1μm-Gate-Length Low-Temperature Operation FETs. *IEEE Electron Device Lett.*, EDL-8:463, 1987.

6. Rishton, S. A., Schmid, H., Kern, D. P., Luhn, H. E., Chang, T. H., Sai-Halasz, G. A., Wordeman, M. R., Ganin, E., and Polcari, M. Lithography for Ultrashort Channel Silicon Field Effect Transistor Circuits. *J. Vac. Sci. Technol.*, B5:140, 1988.

7. Sai-Halasz, G. A., Wordeman, M. R., Kern, D. P., Rishton, S., and Ganin, E. High Transconductance and Velocity Overshoot in NMOS Devices at the 0.1μm-Gate-Length Level. *IEEE Electron Device Lett.*, EDL-9:464, 1988.

8. Sai-Halasz, G. A., Wordeman, M. R., Kern, D. P., Rishton, S., Ganin, E., Ng, D., Moy, D., Chang, T. H., and Dennard, R. H. Inverter Performance of Deep Submicron MOSFETs. *IEEE Electron Device Lett.*, EDL-9:633, 1988.

9. Gaensslen, F. H., Rideout, V. L., Walker, E. J., and Walker, J. J. Very Small MOSFETs for Low Temperature Operation. *IEEE Trans. Electron Devices*, ED-24:218, 1977.

10. Baccarani, G., Wordeman, M. R., and Dennard, R. H. Generalized Scaling Theory and its Application to a 1/4 Micrometer MOSFET Design. *IEEE Trans. Electron Devices*, ED-31:452, 1984.

642

11. Wordeman, M. R., Schweighart, A. M., Dennard, R. H., Sai-Halasz, G. A., and Molzen, W. W. A Fully Scaled Submicrometer NMOS Technology Using Direct-Write E-Beam Lithography. *IEEE Trans. Electron Devices*, ED-32:2214, 1985.

12. Sai-Halasz, G. A., and Harrison, H. B. Device Grade, Ultra Shallow Junctions Fabricated with Antimony. *IEEE Electron Device Lett.*, EDL-7:534, 1986.

13. Kern, D. P., Rishton, S. A., Chang, T. H., Sai-Halasz, G. A., Wordeman, M. R., and Ganin, E. Lithography Issues in Fabricating High-Performance sub-100nm Channel Metal-Oxide Semiconductor Field Effect Transistors. *J. Vac. Sci. Technol.*, B6:1836, 1988.

14. Coane, P. J., Kern, D. P., Speth, A. J., and Chang, T. H. An Electron Beam Microfabrication System for Lithography Below 1000 Å. in R. Bakish, editor, *Proceedings of the 10th Conference on Electron and Ion Beam Science and Technology*, page 2, Electrochemical Society, Montreal, 1983.

15. Rishton, S. A., Kern, D. P., Kratschmer, E., and Chang, T. H. Electron Beam Lithography of sub-0.1 μm Circuits. *Microelectronic Engineering*, 1989. in press.

16. Kratschmer, E., Rishton, S. A., Kern, D. P., and Chang, T. H. Quantitative Analysis of Resolution and Stability in Nanometer Electron Beam Lithography. *J. Vac. Sci. Technol.*, B6:2074, 1988.

17. Sai-Halasz, G. A., Wordeman, M. R., Kern, D. P., Ganin, E., Rishton, S., Ng, D. S., Zicherman, D. S., Moy, D., Chang, T. H., and Dennard, R. H. Experimental Technology and Characterizations of Self-aligned 0.1 μm-Gate-Length Low-Temperature Operation NMOS Devices. *IEEE IEDM Tech. Dig.*, page 397, 1987.

18. Chao, P. C., Smith, P. M., Duh, K. H., Ballingall, J. M., Lester, L. F., Lee, B. R., Jabra, A. A., and Tiberio, R. C. High Performance 0.1 μm Gate-Length Planar-Doped HEMTs. *IEEE IEDM Tech. Dig.*, page 410, 1987.

19. Chou, S. Y., Antoniadis, D. A., and Smith, H. I. Observation of Electron Velocity Overshoot in Sub-100nm-channel MOSFETs in Silicon. *IEEE Electron Device Lett.*, EDL-6:665, 1985.

20. Shahidi, G. G., Antoniadis, D. A., and Smith, H. I. Electron Velocity Overshoot at Room and Liquid Nitrogen Temperatures in Silicon Inversion Layers. *IEEE Electron Device Lett.*, EDL-9:94, 1988.

21. Fischetti, M. V., and Laux, S. E. Monte Carlo Analysis of Electron Transport in Small Semiconductor Devices Including Band-Structure and Space-Charge Effects. *Phys. Rev. B*, 38:9721, 1988.

22. Sai-Halasz, G. A., Wordeman, M. R., Kern, D. P., Rishton, S., Ganin, E., Chang, T. H., and Dennard, R. H. Experimental Technology and Performance of 0.1 μm-Gate-Length FETs Operated at Liquid-Nitrogen Temperature. *IBM J. Res. Develop.*, 1989. in press.

23. Modelli, A., and Manzini, S. High-Field Drift Velocity of Electrons in Silicon Inversion Layers. *Solid-State Electron.*, 31:99, 1988.

24. Fang, F. F., and Fowler, A. B. Hot Electron Effects and Saturation Velocities in Silicon Inversion Layers. *J. Appl. Phys.*, 41:1825, 1970.

25. Laux, S. E., and Fischetti, M. V. Monte Carlo Simulation of Submicron silicon MOSFETs at 77 and 300K. *IEEE Electron Device Lett.*, EDL-9:467, 1988.

26. Kobayashi, T., Miyake, M., Okazaki, Y., Sato, M., Defuchi, D., Ohki, S., and Oda, M. 8.6ps/Gate Chilled Si E/E JNMOS Integrated Circuit. *IEEE IEDM Tech. Dig.*, page 881, 1988.

27. Toh, K. Y., Chuang, C. T., Chen, T. C., Warnok, J., Li, K., Chin, K., and Ning, T. H. A 23ps/2.1mW ECL Gate. *IEEE ISSCC Digest*, page 224, 1989.

Interconnect Technology

A TiN/TiSi$_2$ Interconnect Structure Durable for High Temperature Processing

D. Gloesener, G. Rivas, B. Goffin*, P. Verlinden and F. Van de Wiele

Laboratoire de Microélectronique, Université Catholique de Louvain
Place du Levant, 3   1348 Louvain-la Neuve, Belgium
*now with ACEC Space Defense Telecommunication Division, SDT 42.1,BP 4, 6000 Charleroi, Belgium

Summary

A simple method has been developed to enhance the thermal stability of TiSi$_2$. It allows a RTA reflow step at temperatures higher than 1000°c without any increase in TiSi$_2$ resistivity nor degradation of the contact structure.Its efficiency has been separately demonstrated on a polycide process and on diffusion areas.

Introduction

The use of TiSi$_2$ for self aligned applications has many advantages but has its main drawback is that the titanium easily reacts with SiO$_2$, limiting the TiSi$_2$ formation temperature to below 700°c. Also TiSi$_2$ is readily etched in HF or its mixtures and exhibits thermal instability problems when submitted to high temperature anneal as a PSG reflow step.It has been shown that TiSi$_2$ agglomerates under these conditions resulting in a dramatic increase in sheet resistance[1].Both shallow junctions and MOS structure integrity can be affected by this mechanism when TiSi$_2$ is formed on single crystal silicon or polysilicon : an important outdiffusion of dopants and a roughening of the silicide/silicon interface have been oberved in both cases[2].

The present work investigates the thermal behaviour of a TiSi$_2$ /TiN bilayer interconnect structure on the gate and diffusion areas. The nitride layer is believed to slow down the agglomeration reaction, by restraining the motion of titanium and silicon atoms, and also to prevent an excessive outdiffusion of dopants. In addition, TiN can withstand a cleaning step in HF solutions[3] and acts as a barrier against aluminum/silicide interdiffusion.

The use of rapid thermal processing to limit the silicide/silicon or polysilicon interactions is also investigated. In particular, the silicide resistivity lowering, the dopants activation and the glass reflow are simultaneously carried out by a one-step rapid thermal annealing (RTA).

Experimental

TiSi$_2$ was formed by evaporating 500Å of titanium on a 4000Å n$^+$ polysilicon layer (on 250Å gate oxide) as well as on n$^+$ and p$^+$ implanted junctions. A RTA step at 650°C was then carried out for 100 sec. in N$_2$. During this anneal, a nitrided layer was formed on the surface of the silicide. AES analysis revealed that it effectively consisted in a TiN film with an oxidized extreme surface (about 30Å).The thicknesses of the layers were about 300Å for the TiN film, 1100Å for the silicide and

3000Å for the polysilicon. On half of the wafers, the TiN film was selectively removed in a $NH_4OH:H_2O_2:H_2O$ (1:1:4) solution to obtain reference samples. The wafers were capped with 5000Å of PSG, deposited by APCVD and then subjected to RTA at 1100°C for 30 seconds in order to simultaneously reduce the resistivity of the titanium silicide, to activate the dopants and to flow the oxide layer. After holes opening and metallization (Al-Si 1%), we measured sheet resistance and contact resistance on the gate (Al/silicide) and diffusion areas (silicide/silicon). Unpatterned wafers were also processed following a similar sequence. Changes in the films characteristics were observed after each process step by combining RBS, AES, SEM, TEM, and sheet resistance measurements. The activation and diffusion of dopants were verified by Spreading Resistance technique.

Results and discussion

I. $TiSi_2$ on polysilicon

The electrical results are summarized in table I.

From these measurements, two important phenomena can be pointed out :

- when the TiN layer is kept on top of the silicide, the resisitivity lowering of $TiSi_2$ is achieved after the high temperature step. In the other case, a severe degradation of the resistivity is observed.

- the contact resistance $Al/TiSi_2$ for a classical polycide process exhibits a large spread of high values indicating important damages in the contact while a specific contact resistance of about 2E-7 Ω cm2 is obtained for the $Al/TiN/TiSi_2$ structure.

Two facts can explain this better behaviour [4]. First, as shown on the SEM micrographs of figure 1, the silicide film undergoes severe agglomeration during annealing when the nitride layer is selectively removed before the high temperature treatment. Agglomeration causes $TiSi_2$ to change from a uniform continuous film to a semicontinuous film and finally to a discontinuous film made up of isolated islands, leading to sheet resistance increase. Similarly, such irregular surface explains the large spread of high values obtained for contact resistances. The nitride film slows down the kinetics of $TiSi_2$ multiple islands formation by restraining the movement of titanium and silicon atoms. The alteration of the thin films stresses is mainly responsible for this slower movement.

The nitride layer also prevents outdiffusion of dopants: the sheet resistance of polysilicon after silicide removal is 20 to 50 percent higher than the sheet resistance of polysilicon when the TiN layer has been etched, indicating that, in this case, a more important part of dopants is consumed. Such dopants outdiffusion has been pointed out by Chow et al.[5] and results in polysilicon grain boundaries migration and therefore in grain growth enhancement. TEM analysis was performed on the polysilicon films after silicide removal. It showed that, in the case of a classical polycide structure (i.e. without TiN) - which exhibits the greatest movement of dopants - the final grain size is higher. It is suggested that, in this case, a greater number of excess vacancies is generated in the polysilicon film since the outdiffusion of dopants is more important; the grain boundaries migration is therefore favoured and the final grain size enhanced. This phenomenon can introduce additionnal constraints in the silicide film and accelerate the islands formation.

Finally, RBS analysis showed that the use of RTA for the high temperature step prevents silicide/polysilicon interdiffusion and guarantees the MOS structure integrity.

II.TiSi$_2$ on single crystal silicon

TiSi$_2$ on single crystal silicon was also studied. Its behaviour is roughly similar to that on polysilicon. Table II presents the electrical results on diffusion areas. First, it appears clearly that the sheet resistance changes of TiSi$_2$ without the TiN layer are smaller -though present- in the case of single crystal silicon than for polysilicon. The silicide film undergoes here less severe agglomeration during annealing. This different behaviour can be attributed to the additional constraints introduced in the TiSi$_2$ film by the polysilicon grain growth and to the initial roughness of the polysilicon surface.

The contact resistivities between silicide and metal, measured using several contact chains as test devices, are typically 10^{-7} to $10^{-8} \Omega cm^2$ for both Al/TiSi$_2$ and Al/TiN/TiSi$_2$ systems, which adds negligible resistance even for micronic size contact holes. Therefore, the main feature appeared to be the contact resistance between silicon and silicide. A specific test structure for the evaluation of contact resistance was used. This was proposed by Taur et al.[6] and enables the current to flow through the TiSi$_2$-Si interface.

While good values are obtained for the Al/TiN/TiSi$_2$ system, we obtain a considerable degradation of the contact resistance when the nitride film has been etched. The TiN layer acts as a barrier and prevents an excessive outdiffusion of dopants as verified by spreading resistance measurements. Nevertheless, the silicide formation process has to be controlled since both low silicide sheet resistance and shallow junction integrity are required at the same time. This necessarily implies a careful optimization of the junction parameters, the initial thickness of the titanium layer and the annealing conditions.

Conclusions

The use of a TiN/TiSi$_2$ bilayer, allows RTA treatments, at temperatures higher than 1000°c, without any resistivity nor contact resistance degradation and without excessive silicide/silicon interdiffusion. The efficiency of the method has been demonstrated on a polycide process and on diffusion areas. Since TiSi2 can be converted into TiN by direct thermal nitridation[4],the method can be easily implemented in a self-aligned process providing a careful optimisation of all the junction parameters.

References

1.C.Y.Ting, F.M.d'Heurle S.S.Iyer and P.M.Fryer,J.Electrochem.Soc.,vol.133, n°12, 2621 (1986)

2.K.B. Affolter, A.A. Brown, S.R. Jennings, P.J. Rosser and L. Van Den Hove, 1987 Proceedings of the IEEE VLSI Multilevel Interconnection Conference, June 15-16, 1987,Santa Clara CA, pp. 138-147.

3.H. Kaneko, M. Koyanachi, S. Shimizu, Y Kubota and S. Kishino, IEEE Trans. Electron Devices, vol ED-33 n°11, 1702 (1986)

4.D. Gloesener, G. Rivas, B. Goffin, J.L. Coppée and F. Van de Wiele, 1988 Proceedings of the IEEE VLSI Multilevel Interconnection Conference, June 13-14, pp. 43-50

5. T.C. Chow, C.Y.Wong and K.N. Tu, J. Appl. Phys., 62 (7), 2722 (1987)

6.Y. Taur, J.Y.C. Sun, D. Moy, L.K. Wang, B. Dawari, S.P. Klepner and C.Y. Ting, IEEE Trans. Electron Devices, vol ED-34 n°3, 575 (1987).

TABLE I : Electrical Measurements on Polysilicon

	SHEET RESISTANCE* (Van der Pauw structure) R (Ω/sq.)	SHEET RESISTANCE (four-point measurement)	CONTACT RESISTANCE* Al/gate contact Rc (Ω) for 3x3 μm²	CONTACT RESISTANCE* Al/gate contact Rc (Ω) R (Ω/sq.) for 1.5x1.5 μm²
n+ polysilicon gate	30.48 ± 2.11	31.3	6.9 ± 2.7	-
TiSi₂/n+ poly. gate (1100°C, 30 sec.)	32.01 ± 5.02	36.4	43 to 86.8	49 to 140
TiN/TiSi₂/n+ poly. gate (1100°C, 30 sec.)	1.81 ± 0.14	1.9	2.02 ± 0.67	5.79 ± 2.91

* mean value and standard deviation related to 18 test structures

TABLE II : Contact resistance and sheet resistance of TiN/TiSi2/diff. and TiSi2/diff. structures after RTA at 1100°C for 30 sec.

		contact resist. Rc (Ω) 3μmx3μm.	Sheet resistance VdP strut. (Ω/ sq.)
TiSi₂ / n+ diffusion As 5e15 cm⁻² 140 keV (1100°C 30 sec.)	with TiN	1.6 +/- 0.2	1.48 +/- 0.12
	without TiN	135 +/- 12.5	1.89 +/- 0.15
TiSi₂ / p+ diffusion BF₂ 3e15 cm⁻² 45 keV (1100°C 30 sec.)	with TiN	4.7 +/- 0.6	1.61 +/- 0.1
	without TiN	> 5K	1.77 +/- 0.19

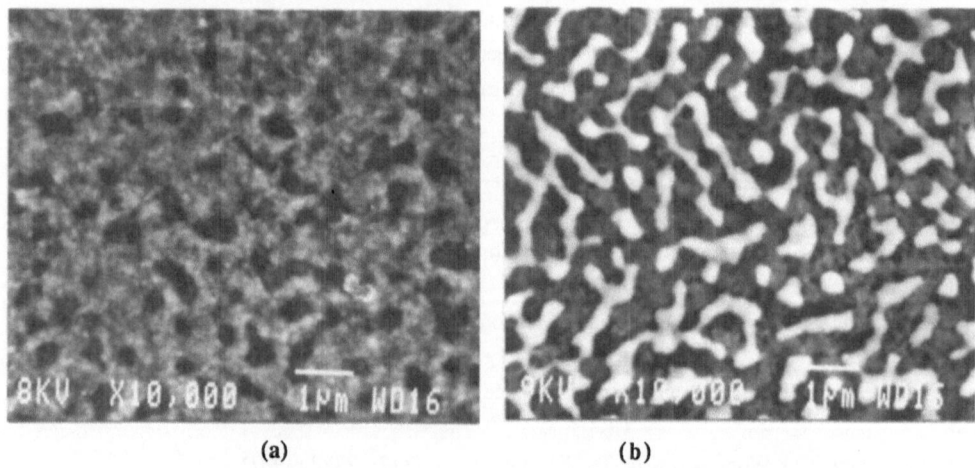

(a)   (b)

Fig. I : SEM micrographs of (a) TiN/TiSi2/polySi (b) TiSi2/polySi structures after RTA at 1100°C for 30 sec.

Formation of Reliable Al-Si/Si Contacts by Chemical Oxidation of the Contact Area

F. T. AGRICOLA, W. G. M. SUIJKERBUIJK and M. A. SPROKEL

Advanced MOS Process Development Group
Philips Research Laboratories, P.O. Box 80.000, 5600 JA Eindhoven, The Netherlands

Abstract

As feature sizes become smaller than 2 μm, and also contact sizes cross this limit, the resistance and reliability of AlSi to Si contacts are limited by silicon recrystallization from the AlSi into the contacts. This problem, often referred to as Si Solid Phase Epitaxy, Si-SPE, can be both a yield limiter and a reliability hazard. This can be prevented by applying a simple chemical oxidation just prior to the metallization. Such a cheap, simple, reliable and reproducible step completely inhibits preferential Si recrystallization in contacts to Si by modification of the surface. Contact resistances can be kept in the range of diffusion sheet resistances, with a very small spread, and VLSI device yield and performance can be made more reproducible and reliable.

1 Introduction

The use of AlSi as a metallization material is common in the production of intergrated circuits. The addition of Si to the aluminum, usually around 1%, is done to prevent spiking in the contacts by dissolution of Si into the aluminum, caused by the solid solubility being around 0.5% of Si in aluminum at normal annealing temperatures [1]. However, the excess Si in the AlSi may preferentially recrystallize at the interface to both n^+ and p^+ diffusion areas and thereby cause highly resistive contacts for contact sizes smaller than around 2 μm [2,3]. This phenomenon, called Si Solid Phase Epitaxy, Si-SPE, can be prevented by several means. The most effective way is by adding an extra diffusion barrier between the silicon and the aluminum, like silicides, metal nitrides or refractory metals [4]. Another solution is the use of rapid thermal sintering to reduce the heat treatment time [5]. Both solutions are not easily implementable in conventional double metal processes.

In this study a new, simple and cheap solution to Si-SPE is presented. Chemical modification of the contact area by formation of a thin chemical oxide using fuming nitric acid, has proven to be very effective in eliminating Si-SPE.

2 Experimental

Experiments were carried out in a 1.2 μm CMOS process on 150 mm wafers. Junctions were formed by a B^+ implant for the p-doped areas and a combined P^+ and As^+ implant for the n-doped areas. The wafers contained both SRAM devices as well as several test structures, including both n^+ and p^+ contact strings and Kelvin contacts with 1.4 μm contact sizes. The double metal process was carried out with conventional metallization using Al-Si(1%) and a highest post metallization temperature step of 30 minutes at 420°C at the end of the process. The process and SRAM have been described earlier [6]. Physical analysis of high resistive contacts was done by selective removal of passivation, metal layers and intermetal oxide, and some of the oxide below the metal. After this the contacts were investigated with a SEM.

3 Contact resistance of Kelvin structures and SRAMs

For the evaluation of the oxidation by fuming nitric acid three lots of 20 wafers were processed including a 10 min. fuming nitric treatment at room temperature (Type B). This was done after an HF dip to remove remaining oxide in the contact holes to diffusion, and before first aluminum deposition. Also three lots were processed without the fuming nitric treatment (Type A). On both types of wafers Kelvin structures (ten per wafer) of contacts to n^+ and p^+ diffusion were measured and the SRAMs were tested for functionality. The results of the contact resistance to p^+ are given in Fig. 1. As can be seen the median of the contact resistance is somewhat lower for type A wafers, but the spread is considerably higher. This can be explained by the fact that type B contacts give an in situ higher contact resistance because of the presence of oxide at the aluminum-silicon interface. This oxide prevents the formation of Si-SPE, and thus reduces the spread in the contact resistance. Physical analysis of the highly resistive contacts of type A material indeed showed clear Si-SPE over the whole contact area.

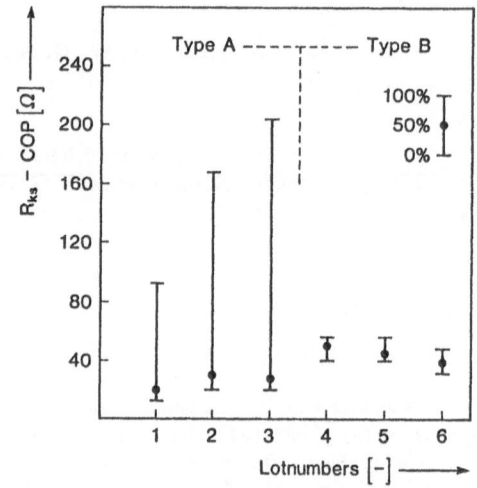

Figure 1: Plot of contact resistance to p^+ active area with and without chemical oxidation.

An example of the difference of Si-SPE between type A and type B material can clearly be seen in the contact holes to p^+ shown in Fig.2.

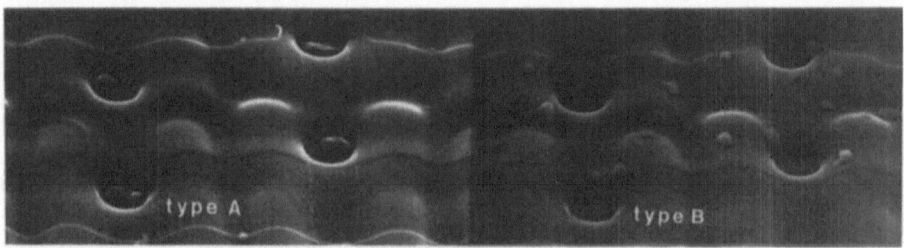

Figure 2: Comparision of type A (not oxidized) and type B (oxidized) contacts to p^+ active area. Note the complete absence of Si-SPE on type B material.

Degradation of contacts to n^+ due to Si-SPE occurs much less frequently, and therefore could not be detected on Kelvin structures. However, when occuring in our Static RAM, they give a specific failure mode causing two failing bits in one row, socalled row pair failures. This is caused by high resistance of a ground contact shared by two cells. Analysis of this specific failure mode on both type A and B wafers revealed that no Si-SPE was present in the contacts to n^+ of type B material, with a chemical oxidation. Failures due Si-SPE ranging from 1 contact out of 1000 to 1 out of a million were observed in contacts to n^+ silicon of type A, without oxidation. The cell layout and an example of a cell failing due to Si-SPE in the n^+ ground contact can be seen in Fig. 3.

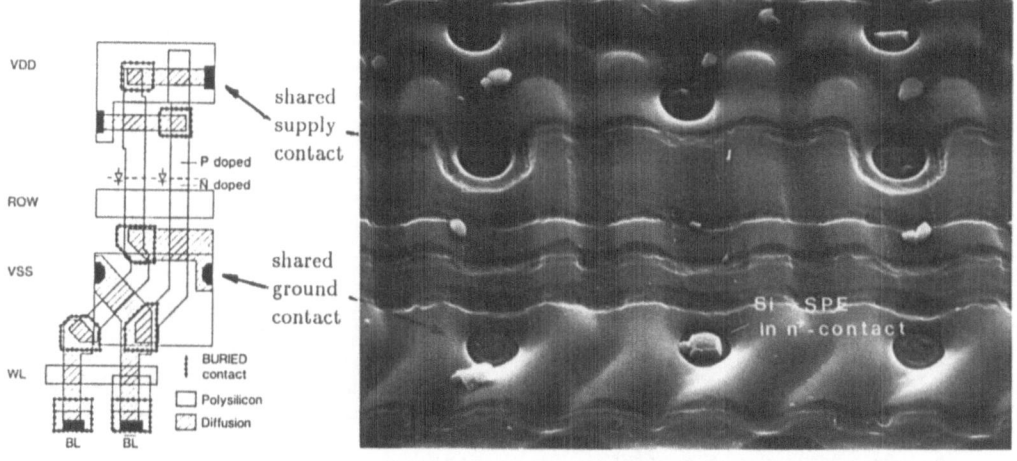

Figure 3: Cell layout showing the shared supply voltage and ground contacts, and a SEM picture showing a high resistive contact to ground due to Si-SPE, causing two failing cells.

4 Reliability investigations

To study the influence of extra temperature steps an additional anneal of 30 min. at 450°C was carried out on six wafers, out of three lots, of both type A and B, after which Kelvin structures and SRAMs were remeasured. The difference of the contact resistance to p^+ before and after anneal on the wafers of type A and B are given in Fig. 4 and 5: Both contact resistance to p^+ and row pair failures increase strongly on type A material, but are very stable on type B material. Physical analysis confirmed that Si-SPE was the cause of the higher contact resistances and the horizontal pair failures on wafers of type A, without the oxidation.

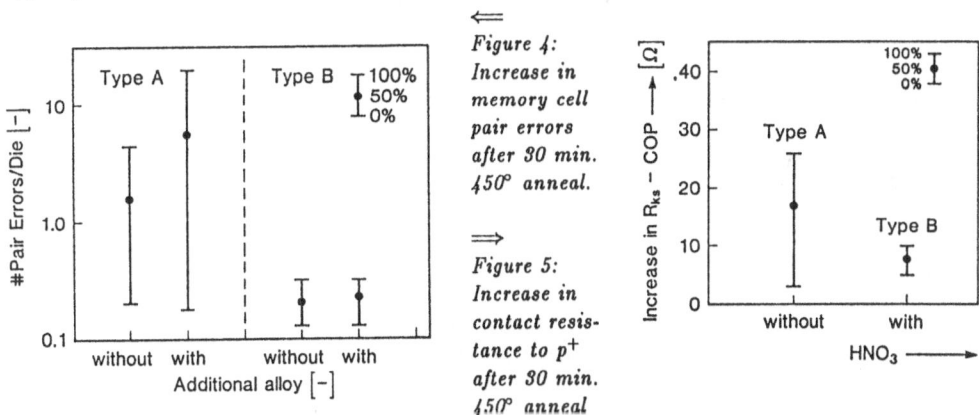

⟸

Figure 4:
Increase in
memory cell
pair errors
after 30 min.
450° anneal.

⟹

Figure 5:
Increase in
contact resis-
tance to p^+
after 30 min.
450° anneal

To study the reliability of the contacts formed on the type B wafers, contact strings for both CO to n^+ and CO to p^+ of these wafers were stressed. Strings containing 18 1.4 μm contacts were tested at 150°C with a current density of 3mA/contact. No failures were observed after 1500 hours, showing exellent contact migration resistivity behaviour.

For specific analysis of contacts in the SRAM matrix, 450 fully functional mounted SRAM devices from both type A and B were subjected to a dynamic burn-in test at 7 volt and 150°C for 168 hours. During this test of devices of type A, 17 typical row pair failures occured, which is unacceptably high. Physical analysis revealed clearly that this was due to Si-SPE. The devices of type B did not show any horizontal pair failure related to Si-SPE at all.

652

5   Additional physical analysis

XTEM pictures of blanket wafers that recieved a treatment of fuming nitric acid, metal deposition and anneal were made to study the aluminum-silicon interface. Fig. 6 shows the Si-AlSi interface: Direct contact of AlSi to Si is observed because the contact area is not continuous and smooth, but contains areas with oxide of about 2.5 nm thick. This way Si-SPE is limited to a very small area, and does not cause complete coverage of contacts.

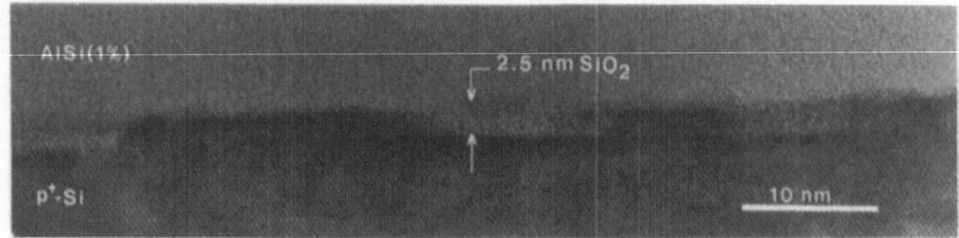

Figure 6: XTEM picture showing local presence of SiO_2 areas on the AlSi-Si interface.

To examine the amount of oxide formed during the treatment with fuming nitric acid, wafers that received this treatment, as well as wafers that received the treatment combined with aluminum deposition, anneal and aluminum removal, were measured with RBS. This showed that both before and after anneal an average oxide thickness of 2.0 nm was present at the interface.

6   Conclusion

It is clearly demonstrated that a chemical oxidation step before metal deposition can eliminate Si-SPE and can thus improve the quality and reliability of small AlSi to Si contacts. Contact resistance spreads were strongly reduced, yield loss and reliability problems on SRAM devices due to Si-SPE were eliminated, and excellent contact migration was obtained. The formation a thin oxide film, which is not uniform after anneal, is responsible for this.

Acknowledgements

The authors would like to thank everybody that contributed to this work, with special thanks to S.Bastiaens and R.Wolters for encouraging discussions, and the Process Integration team for their processing and analysis support.

References

[1] M.Hansen, Constitution of Binary Alloys, McGraw-Hill, New York, 1958

[2] S.M. Sze, VLSI Technology, p368, McGraw-Hill, New York, 1983

[3] T.Maeda c.s., IEDM 1985, pp 610-613.

[4] Y.Pauleau, Solid State Technology, vol.30(4), p155, april 1987.

[5] E.Umemura c.s., Proc. 26th IRPS, p.230, 1988.

[6] W.Gubbels c.s., IEEE Journal of Solid State Circuits, Vol.SC-22,5, october 1987.

The Influence of Tungsten Contact Filling on Junction Quality and Contact Resistance

W.L.T.M. RAMSELAAR, F.T. AGRICOLA and C.A. SEAMS

Advanced MOS Process Development Group
Philips Research Laboratories
P.O. Box 80.000, 5600 JA Eindhoven, The Netherlands

In this paper the benificial effect of tungsten contact filling is described for two aspects of modern VLSI technology. First, aluminium spiking through thin TiW barrier layers is effectively prevented by W-plugs. This results in lower diode leakage currents, both for N^+ and P^+ junctions. Second, the contact resistance to N^+ and P^+ junctions is lower when W contact filling is used. This effect becomes more pronounced as the diameter of the contacts decreases. It is concluded that W contact filling is needed to guarantee low leakage currents and low contact resistances.

1 Introduction

To take full advantage of transistor scaling it is also necessary to scale the metal system. This implies agressive design rules for metal line and spacing as well as for contact and via openings. To meet these requirements many of today's and tomorrow's processes feature planarization at LOCOS, polysilicon and metal levels. Together with the need for submicron contacts this has stimulated the interest in W contact filling [1]. The studies presented so far however, were mainly limited to the manufacturing of so-called W-plugs and the resistance of contacts with a minimum diameter of 0.9 μm. In this paper the influence of W contact filling on the diode leakage current of 0.9 μm contact holes is evaluated. In addition, a comparison is made of the contact resistance with and without W-plugs for contacts with diameters ranging from 0.7 μm to 1.5 μm. From the data presented it is will be concluded that W contact filling results in reduced diode leakage currents and guarantees low contact resistances for contacts smaller than 0.9 μm.

2 Experimental

The results described in this paper were obtained from material which had been fully processed in the Philips' 0.7 μm CMOS technology [2], which is used to fabricate 256K and 1M SRAMs. The process flow of this technology has been described in detail elsewhere [3,4]. Therefore, only the relevant process steps after contact etch are summarized in table 1, for material with and without W-plugs.

To evaluate the influence of the W contact filling, strings consisting of 746 contacts to N^+ or P^+ regions were electrically tested on a Keithley 350 system. Diode leakage currents of complete strings were measured at a voltage difference of 5.5 Volts between both string terminals and the Nwell (P^+contacts) or the substrate (N^+contacts). The resistance per contact was determined from the current through the string at a voltage difference between the two string terminals of 100 mV.

Table 1: process flow after contact etch

with W	without W
TiW deposition - 100nm	TiW deposition - 100nm
W deposition - 800nm	
W/TiW etchback	
TiW deposition - 100nm	
AlSiCu deposition - 500nm	AlSiCu deposition - 500 nm

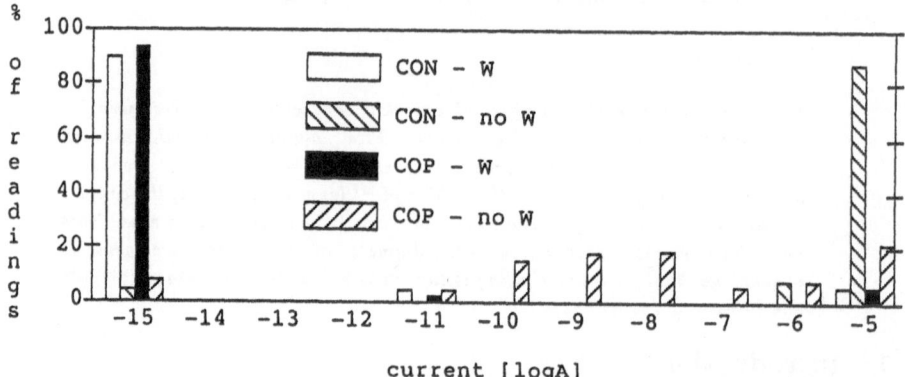

Figure 1: *diode leakage current for strings consisting of 746 contacts to N^+ (CON) and P^+ (COP)*

3   Results and discussion

In Figure 1 diode leakage currents of strings consisting of 746 0.9 μm contacts of P^+ to Nwell and N^+ to substrate are shown. It is obvious from this figure that there is a difference in diode leakage between material with and without W. When W contact filling is used there is hardly any leakage (logA = -15 means that no leakage could be detected). On the other hand, when no W-plugs are present considerable leakage currents are measured. To investigate the origin of the difference in diode leakage current, cross-section SEM pictures were taken. These pictures revealed that whereas without W-plugs Al-spiking occurs, this does not happen in the presence of W-plugs.

Another difference between material with and without W contact filling was seen in operation of the SRAM cells. It appeared that in the material without W, several SRAM cells did not work properly. The origin of this phenomenon was investigated by electrical characterization of isolated failing cells. It was found that close to unfilled contacts, the driver transistors suffered from asymmetry. In figure 2 a SEM picture of such a failing cell is shown. It can be seen from this picture that Si suck-out has occured of the gate region. With W-plugs, no such electrical or physical effects are observed.

So, from the data presented it can be concluded that W-plugs prevent Si suck-out. The fact that Al-spiking does occur on the material without W is in agreement with results obtained by Chang et al. [5]. These authors demonstrated that a 0.1 um thick TiW barrier layer , as is used in our process flow without W (see Table 1), is insufficient to prevent Al-spiking.

Low contact resistances are of great importance in VLSI circuits. In order to evaluate the influence of W contact filling, resistances of contacts with diameters ranging from 0.7 μm to

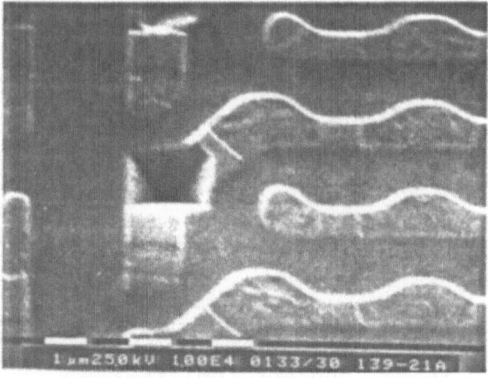

Figure 2: *SEM picture of failing memory cell due to Si suck-out of the transistor gate*

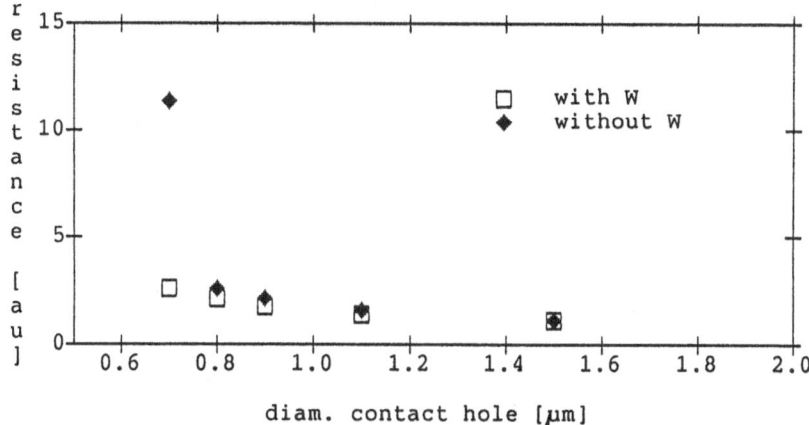

diam. contact hole [μm]

Figure 3: *Geometry dependence of contact resistance of TiW/W/TiW/AlSiCu and TiW/AlSiCu to N^+ regions*

1.5 μm were measured. In figure 3 the geometry dependence of contacts with and without W filling to N^+ are shown.Similar results are obtained for contacts to P^+. It can be observed that for contacts with a diameter of 0.9 μm and larger W contact filling has no influence on the resistance. This implies that for present day processes (minimum contact size 0.9 μm) W-plugs have no benificial effect on the contact resistance. On the other hand, for contacts with a diameter smaller than 0.9 μm W contact filling clearly leads to a lower resistance. This effect becomes more pronounced as the contact size decreases. This result establishes that for contacts with a diameter smaller than 0.9 μm, which will be used in future processes, W contact filling is necessary to guarantee as low a contact resistance as possible.

The difference in resistance between contacts with and without W-plugs is most likely due to poor step coverage of TiW and AlSiCu in high aspect ratio contacts with a diameter smaller than 0.9 μm. This shown by plotting the contact resistance as a function of the inverse square contact diameter, see figure 4. It can be observed from this graph that the datapoints of the W filled contacts lie on a straight line, indicating that the resistance is influenced predominantly by the interfacial resistance, which is inversely proportional to the contact area. The plot of contacts without W-plug is not linear. For this material, the datapoints for contacts with a

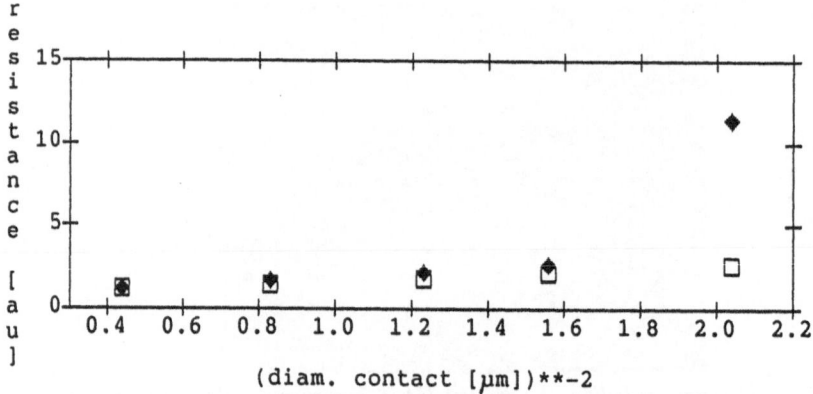

Figure 4: *contact resistance of TiW/W/TiW/AlSiCu and TiW/AlSiCu to N^+ regions as a function of inverse square contact diameter*

diameter below 0.9 μm lie above the straight line which can be drawn through the datapoint at larger contact diameters. This implies a lower step coverage for contact with a diameter below 0.9 μm. Hence, it is concluded that W contact filling guarantees minimum contact resistances, limited only by interface resistance. This holds in particular for contact diameters smaller than 0.9 μm, where poor stepcoverage of traditional metals in high aspect ratio contacts leads to a severe increase in the contact resistance without W contact filling.

4 Conclusion

The present paper clearly demonstrates the beneficial effect of W contact filling for two aspects of VLSI technology. In the first place W-plugs prevent Al spiking through thin TiW barrier layers. In the second place, W contact filling secures as low a contact resistance as possible. Whereas the latter effect only plays a minor role in present day processes, it will become increasingly important with further decreasing contact sizes.

Acknowledgements

The authors would like to thank all members of the Philips' MEGA Technology Team for their pleasant cooperation, and in particular A. Stolmeijer and P. Zandveld for useful discussions.

References

1. Proceedings of the 18th ESSDERC, J. Phys. Colloq. C4, Suppl.9, Tome 49 p. 171-201 and p. 489-525 plus references cited therein

2. R. de Werdt et. al., 1987 IEDM Technical Digest, p. 532.

3. T. Doan et. al., Proceedings of the 5th international IEEE VMIC (1988), p. 13

4. A. Stolmeijer et. al, Proceedings 1989 international symposium on VLSI technology, systems and applications, p. 317

5. P.-H. Chang, R. Hawkins, T.D. Bonifield and L.A. Melton, Appl. Phys. Lett., 52(4) p. 272

AlSi/Ti/AlSi Multisandwich Systems on TiN/Ti for VLSI Interconnect Applications

G.Röska[1], P.Kücher[1], B.Vollmer[2], H.Oppolzer[2], G.Zorn[2], H.Cichy[2]

Siemens AG, [1]Semiconductor Division and
[2]Corporate Research and Development, Microelectronics,
Otto-Hahn-Ring 6, D-8000 München 83, F.R.G.

Summary

The influence of the Ti thickness of the AlSi/Ti/AlSi system on Ti/TiN was investigated by transmission electron microscopy (TEM), x-ray diffraction measurements, scanning electron microscopy (SEM) and energy dispersive x-ray analysis (EDX). The results are discussed and correlated to the data of accelerated electromigration stress tests.

Introduction

Low resistivity and electromigration hardness are required for the minimization of the interconnect width and for the improvement of the speed performance of VLSI circuits. We investigated AlSi(1%)/Ti/AlSi(1%) multisandwich systems with Ti interlayer thicknesses between 10 and 80 nm. They were fabricated on a Ti/TiN underlayer which is suitable as diffusion barrier between the silicon substrate and an Al based metallization /1/.

Experimental

Ti (20 nm) and TiN (150 nm) were sputter deposited on an oxidized silicon substrate. TiN was reactively sputtered in an Ar/N_2 mixture. The AlSi/Ti/AlSi multisandwich systems were deposited without breaking the vacuum at a base pressure below 10^{-5} pa. Each AlSi layer was 400 nm thick. After patterning, the wafers were annealed 15 min at 450°C in forming gas. For reliability tests, 500 μm long and 3.0 μm wide metal interconnects were used as test structures . The samples were Al wedge bonded and encapsulated in ceramic packages.

Results

The multisandwich systems were analyzed by TEM cross sections, which indicate zones of chemical reaction between Al and Ti (Fig.1a-d). Interlayers up to 40 nm are totally transformed into Al_3Ti (Fig.1b) and form non continuous layers. This is confirmed by SEM top views after chemical etch back of the upper AlSi layer showing pores in the interlayers of Fig 2a,b. For the 60 nm Ti (Fig. 2c) and 80 nm Ti (Fig. 2d) interlayers the reaction is not complete. Energy-dispersive x-ray (EDX) analysis in the TEM showed that there is indeed unreacted, fine grained (5nm) Ti inside the reaction zones of Al_3Ti /2/.

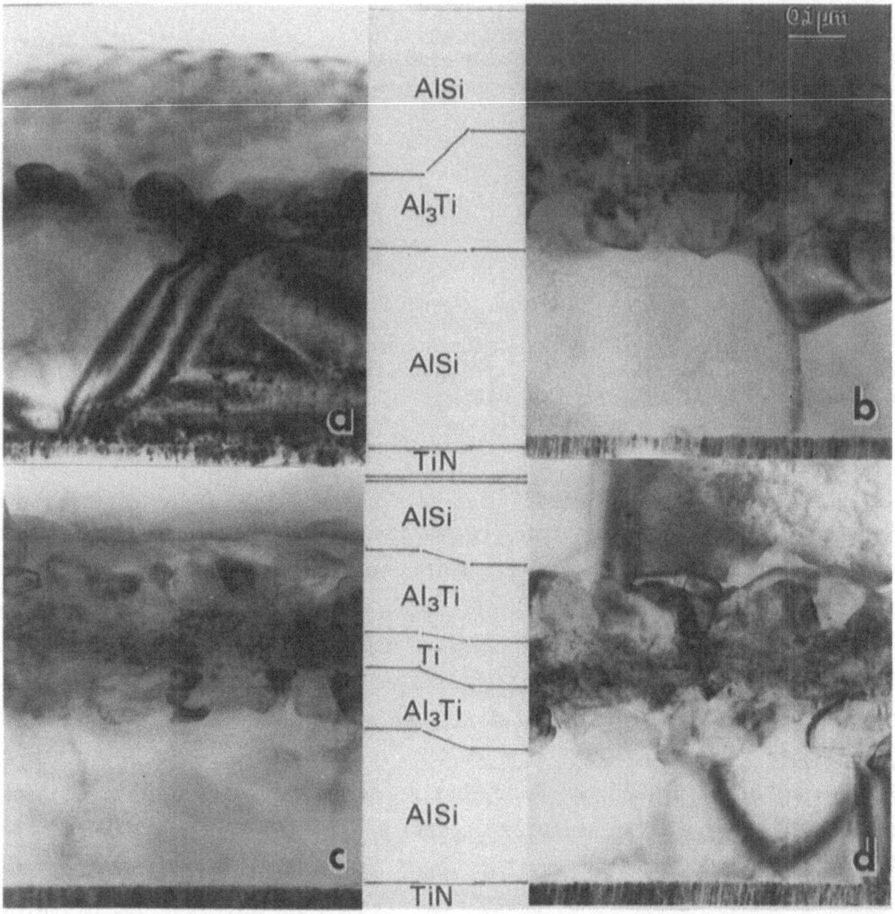

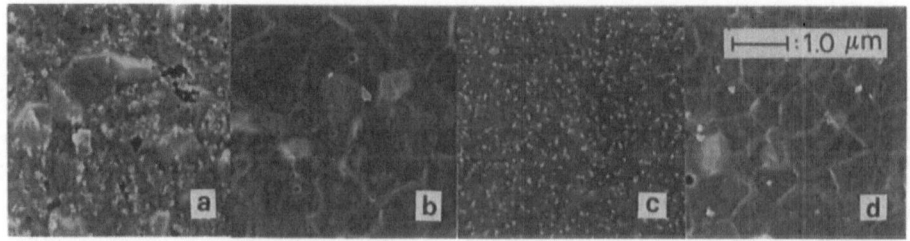

Fig. 1: TEM cross sections of layered AlSi/Ti/AlSi sandwich interconnects with Ti thicknesses of a) 20 nm, b) 40 nm, c) 60 nm and d) 80 nm

Fig. 2: Surfaces of the interlayers after etch back of the upper AlSi for Ti thicknesses of a) 20 nm, 40 nm, 60 nm and 80 nm.

X-ray diffraction at small incidence angles (1°) using a parallel beam geometry also proves the formation of Al₃Ti (Fig. 3a). The intensity of the Al (111) peak is strongly suppressed using an asymmetric geometry because most Al crystals are oriented with their (111) lattice planes parallel to the sample surface. In Fig. 3b in-situ high temperature x-ray diffraction measurements in symmetric geometry of samples with 20 nm (solid line) and 80 nm (dotted line) Ti are compared. In contrast to Fig. 3a the most prominent features origin from the oriented crystals of TiN and Al. For 80 nm Ti the substantial increase of the Al₃Ti peak between 450°C and 470°C shows that Ti is not totally transformed during 15 min annealing at 450 °C. Reaction in the 20nm Ti interlayer was complete, as is indicated by constant intensity of the Al₃Ti peak.

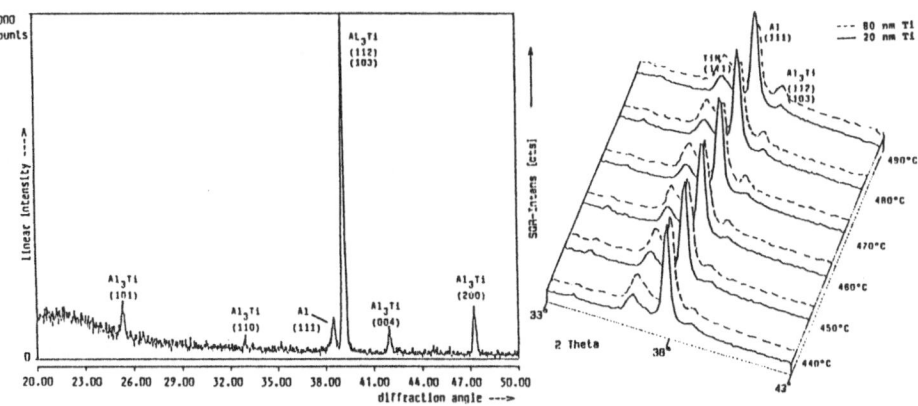

Fig. 3: (a) X-ray diffraction measurement of AlSi/80 nm Ti/AlSi on TiN with
            1° incidence angle and parallel beam geometry.
        (b) High-temperature X-ray diffraction measurements of 20 nm and
            80 nm Ti interlayer.

In microelectronics the formation of hillocks is a risk because it can lead to electrical shorts. Photographs of the metal surfaces (Fig. 4) indicate that Ti layers thicker than 20 nm suppress hillock growth significantly. This correlate with the formation of a stable interlayer (Fig. 2).

Fig. 4: Influence of Ti thickness on hillock formation
(a) no Ti, (b) 10 nm, (c) 20 nm, (d) 40 nm, (e) 60 nm ,(f) 80 nm.

660

Electromigration hardness was examined by accelerated testing for an interconnect width of 3 μm (Fig. 5) because for this width a minimum in the lifetime of AlSi exists. The stress conditions were $4*10^6$ A/cm^2 and ambient temperatures of 150°C and 130°C. The additional line heating was between 80°C and 150°C. For a correct comparison the mean time to failure (MTTF) was temperature standardized to 250°C using Black's formula /3/. The current density exponent was determined as 2 and the activation energies E_a as (0.4±0.1) eV for a sample without Ti interlayer, (0.7±0.1) eV for 20 nm Ti and (1.3±0.1) eV for 80 nm Ti (Fig. 5a). For 40 nm and 60 nm Ti, E_a was interpolated. As failure criterion for MTTF, total metal open or the formation of high resistance regions at the cathode /4/ were used. The MTTF maximum is at a sputtered Ti thickness of about 60 nm.

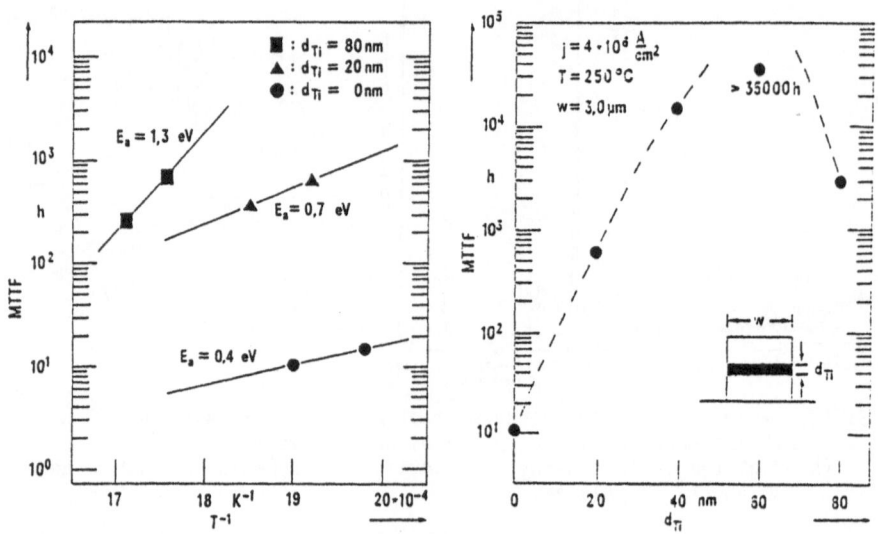

Fig. 5: (a) 1/T dependence of the MTTF
(b) MTTF versus the Ti interlayer thickness d_{Ti}

Conclusions

The increase of interconnect reliability hardness is coupled with achieving continuous interlayers. The decrease of MTTF with sputtered Ti thickness is correlated to the increase of the unreacted Ti layer between the Al_3Ti zones.

References
/1/ J.Stimmell, J. Vac. Sci. Technol. B4(6), p. 1377-82, 1986
/2/ D.Gardner,T.Michalka,K.Saraswat,T.Barbee,J.McVittie,J.Meindl,IEEE
    J. Solid-State Circuits, vol. sc-20, No. 1, p.94, 1985
/3/ J.Black, Proc 6th Ann. Rel. Phys. Symp. IEEE, p. 148, 1967
/4/ P.Kücher,G.Röska, Proc. Int. Symp. on Trends and New Applications in
    Thin Films (Soc. Franc. du Vide, Paris 1987) vol.1, p. 139

A Three-micron Pitch, Three Layer Metallization and Dielectric Scheme with Application to a One Micron CMOS Process

N.F.STOGDALE, P.G.HUGGETT, B.MARTIN and I.SNOWDEN.

Plessey Research Caswell Ltd.
Caswell, Towcester.
Northants. U.K.

ABSTRACT.

An interlayer dielectric scheme and three layer metal process which has been implemented as part of a one micron CMOS process for ASIC products is described. The interlayer dielectric scheme is outlined with particular reference to step coverage and planarisation issues. The advantages of the use of Anti-Reflective Coatings (ARC) to improve control at contact and metal prints are described. Good control of the etched features has also been achieved. The process has been proven by its implementation in a one micron CMOS process schedule which has yielded working circuitry of high complexity, including a 38K SRAM cell.

INTRODUCTION.

A three layer metal process has been developed for a one micron ASIC CMOS process. To maximise the packing density the metal pitch is set to three microns for all layers; two micron track and one micron gap, with vias and contacts at 1.2 and 1 micron respectively.

1. INTERLAYER DIELECTRIC AND CONTACTS PHOTOLTHOGRAPHY.

The interlayer dielectric was chosen to smooth over the polysilicon tracks, to give good metal step coverage and to provide a good basis for the addition of a further three layers of metal. Two main schemes were evaluated; doped and undoped CVD deposited LTO layers and LTO combined with spin-on-glass. The LTO only structures tended to form cusps or grooves between minimum spacing (1μm) polysilicon lines which leads to poor metal step coverage or even broken tracks. The final choice was a combination of doped and undoped LTO films and a spin-on-glass smoothing layer. A 1000Å layer of undoped LTO is deposited, this acts as a barrier between the doped glass layers above and the source-drain regions below. A second LTO layer is deposited; 3000Å of phosphorus (4%) doped glass. A layer of SOG is applied of thickness 1000Å (on a planar surface, but thicker at polysilicon edges where the glass acts to smooth the corners) this is given an anneal to drive out the solvents and to cure the SOG to its final glassy structure. The structure is then capped with a final layer of 2000Å doped (4%) LTO.

2. CONTACTS PHOTOLITHOGRAPHY.

The printing of one-micron contacts with adequate control was found to be difficult over the interlayer surface even with the improved interlayer scheme. This was due to the variation in glass thickness over the highly topographic surface and a consequent variation in resist thickness. There is also the effect of lensing in the films over the edges of the polysilicon gates. To combat this an ARC (anti-reflective coat) was spun-on under the resist, this film is opaque at the exposure wavelength and hence removes any effect of the substrate on the print, the ARC/resist process has a greater latitude than the resist alone leading to greater control of the photolithographic process. A contrast enhancement solution was also evaluated and, although this improved the contact definition, it did not reduce the effects of the substrate on the print and control was not improved. Contacts are etched in a CHF_3/O_2 chemistry with high anisotropy. Printed and etched contacts are shown in figure 1. The sidewall coverage of the metal over the contacts was seen to be good with the printing and etching schemes used.

3. METALLISATION SCHEME.

The metal scheme is based on Al/Cu (4%) alloys throughout. At metal one a barrier/adhesion layer of 1500Å TiW is used in combination with 0.5μm Al/Cu. Metals two and three are both of 0.5μm thickness to aid planarisation and are used without barrier layers. Intermetal dielectrics and passivation layers are 1.3μm spun on and cured Polyimide films.

4. METAL PHOTOLITHOGRAPHY.

Control of the metal print at all layers is achieved by using an ARC/resist combination [1]. This removes reflective notching from the substrate, lessens the effect of resist thickness variations over topography and eases photolithographic control on highly reflective surfaces. The resist process latitude is much improved with the use of an ARC as illustrated in figure 2. ARC allows the as-drawn 2μm track, 1μm gap features to be reproduced at the print.

5. METAL ETCHING.

Metal etching is carried out in a multi-chamber etching system (Electrotech Omega-4) using a $SiCl_4$ based chemistry [2]. The resist is deep-UV hardened prior to etch to give resistance to the aggressive metal etch chemistry. The Al/CU layer and the TiW layer are etched in different chambers of the same machine. Excellent sidewall profiles are achieved with no TiW barrier undercut. Etched metal features are illustrated in figures 3 and 4. No residual copper residues were seen following the etch; the etching of the 4% Cu film was seen to be as good as the etching of 1% Cu alloys which were also evaluated.

6. VIA FORMATION.

The two via layers are resist printed and etched in an oxygen plasma in the polyimide layers. Improvements in anisotropy of the via etch system lead to via size increase being maintained at 0.2µm total increase on drawn size leading to a good coverage of metal over vias at all layers.

CONCLUSION.

The salient features of a three layer metal and interlayer dielectric scheme at three micron pitch has been described. Good control of the print and etch at all layers has been achieved by the use of anti-reflective coatings and highly anisotropic etch processes. The process has been proven by its inclusion in a one micron CMOS process schedule which has yielded working circuitry of high complexity.

ACKNOWLEDGEMENTS.

The support of the UK Alvey Directorate and the Integrated Circuits Division, Plessey Research is acknowledged

REFERENCES.

[1] B.Martin and J.E.Lamb III Proceedings SPIE. 1989
[2] A.Marsh and I.M.Snowden. Procs. SEMI Technical and Educational programme. Oct 1989. Brussels. (to be published)

FIG 1. PRINTED AND ETCHED 1um CONTACT FEATURES.

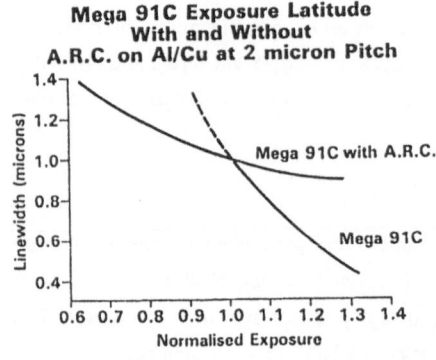

Mega 91C Exposure Latitude With and Without A.R.C. on Al/Cu at 2 micron Pitch

Linewidth (microns) vs Normalised Exposure

Mega 91C with A.R.C.

Mega 91C

Mega 91C Focus Latitude With and Without A.R.C. on Al/Cu at 2 micron Pitch
Without A.R.C. pattern only resolved between focus = −1 and +2 microns

Linewidth (microns) vs Stepper Focus (microns)

Mega 91C

Mega 91C with A.R.C.

FIG 2. IMPROVED PRINT LATITUDE WITH ARC.

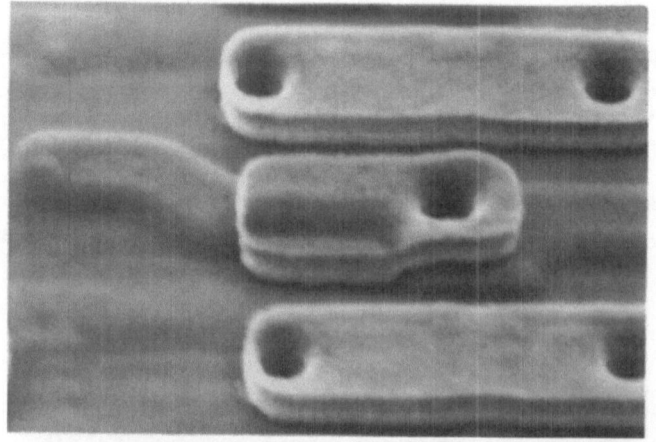

FIG 3. ETCHED METAL 1 FEATURES. (3UM PITCH)

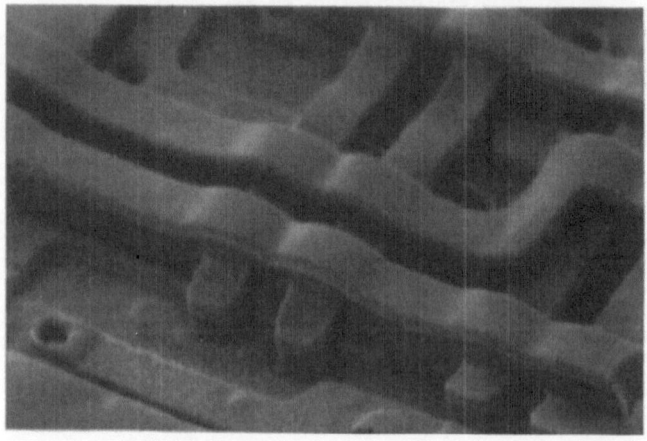

FIG 4. THREE LAYERS OF METAL PRINTED AND ETCHED. (POLYIMIDE REMOVED).

Comparison Between Different Intermetallic Dielectric Processes and Consequences on Field Transistor Behaviour

S.Deleonibus,C.Arena,M.Heitzmann,F.Martin,
J.Lajzerowicz,F.Vinet

LETI   CENG   Avenue des Martyrs
38041 Grenoble Cedex France.

INTRODUCTION In this paper,we report on the leakage current of metal 2 gate NMOS field transistors in double aluminium level process.Important consumption is observed on 0.8um 16K SRAM if permanent SOG is used in the metal2/metal1 insulator. We identify the origin of the problems as SOG related and compare different solutions in terms of process.

PROCESS CONDITIONS

Three types of processes have been investigated :

1-A sandwich of intrinsic PECVD oxide with permanent spin-on-glass (SOG).PECVD oxide was deposited on GOUPYL (SEMY ENGINEERING) or ELECTROTECH equipments .The ACP5 (silicate type) or AC211(siloxane type) SOGs are spinned and baked on GCA track.

2-The former sandwich with partial etch-back of the SOG. Etch-back was performed in an ALCATEL RGV 100 high pressure RIE reactor in CHF3 + C2F6 + Ar mixture.

3-Totally sacrificial SOG for planarization scheme with a first layer of APCVD borosilicate glass(BSG).

In the three cases, tapered vias were etched in a mixed wet (buffered HF) etch process finished by an anisotropic step in RGV ALCATEL machine.Metal 1 is AlCu (0.5%) on TiN/Ti barrier.Metal2 is also AlCu (0.5%).SOG was cured after coating at 450°C during 60mn under nitrogen ambient .

APCVD BSG has been chosen for its good conformality properties and its low temperature deposition process compatible with aluminium metallization suitable for a total etch-back solution.

RESULTS,DISCUSSION

From the three scenarii characterization we found that :

1) Etch-back of SOG is a solution to solve crack problems and improve via resistance (figures 1a&b).

2) The metal2 gate NMOS 50um/1.6um field transistor leakage current is also largely improved by etch-back of SOG.We show(figure 2) the VGOFF dependance on designed field transistor length (VGOFF is defined as a leakage limit of 10pA/um width) and that etching back the SOG drastically improves VGOFF for metal 2 gate NMOS field transistors .

The Log Id (Vg) characteristics show large leakage at Vg= 0V for permanent SOG process when etch-back SOG process gives a good off-current at 5V (figure 3a&b).PMOS field transistors, on the other hand, show no degradation.

We can explain this phenomenon by the large contribution of the fixed charges contained in SOG as it can be seen through the variations of flat-band voltage VFB on plane NMOS capacitors as a function of the insulator thickness (figure 4) : the insulator is the addition of the different layers of the interlevel dielectric through the different phases of process .A stress cycle on plane capacitors at 200°C with +90 V polarization for 60 mn does not evidence any mobile ions contribution to VFB.No difference of behaviour between both types of SOGs (siloxane or silicate) has been evidenced on our devices .D.Pramanik et al.[1],published on hydrogen capturing into the siloxane type SOGs [1]:in their case,the hydrogen source was a nitride passivation layer.Triangular Voltage Sweep (TVS) has been applied to SOG by N.Lifschitz et al.[2] :great hydrogen retention ability via water capturing especially in siloxane type materials is invoked in their work.

In our case,the best situation is obtained for the APCVD BSG interlevel dielectric with total etch-back of SOG , as the final VFB is the lowest (see figure 4).Partial etch-back process is an intermediate solution and the permanent SOG case is the worst case.That confirms that the fixed charges are trapped in SOG.

0.8um 16K SRAM (6 transistors CMOS memory cell) has been processed with the three processes.The functional dies show the same dynamic performances of 10 ns address access time.We clearly observe a main difference ,though, on the CMOS standby current(figure 5).In the total etch-back solution this current is lowered by a factor of 10 as compared to the permanent SOG solution. A good correlation is observed between standby current , VFB and field threshold values.

ACKNOWLEDGEMENTS Special thanks to the silicon facility (SAME) and the silicon research laboratory of LETI(MSC) people for their contribution in this work.

[1]   D.Pramanik,S.Nariani,G.Spadini   Proceedings   VMIC Conference 1989,11-14 June 1989 ,Santa Clara, p.454-462 .
[2] N.Lifschitz,G.Smolinsky,J.M.Andrews J.Electrochem.Soc., Vol.136 ,No.5,May 1989.p.1440-1446.

Via Kelvin resistance(Ω/via)   Vias string resistance yield (%)

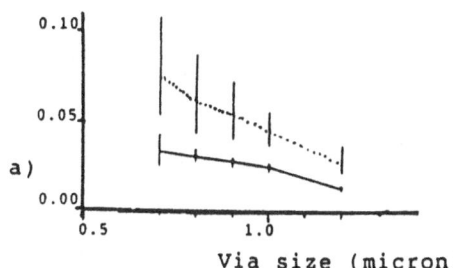
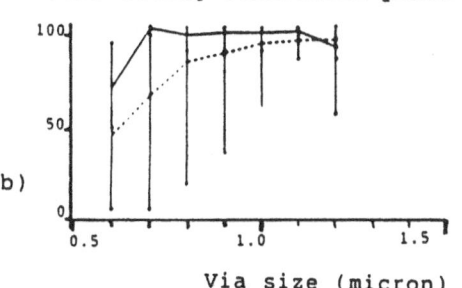

a)   b)

Via size (micron)   Via size (micron)

Figure 1 Via characteristics as a function of via size:
    a)Kelvin resistance value   b)100 vias string yield

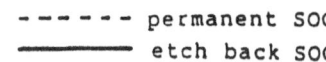

- - - - - - permanent SOG
─────── etch back SOG

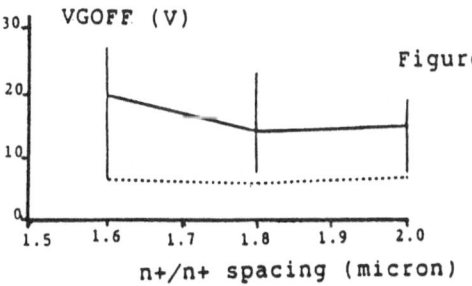

Figure 2 Metal 2 gate NMOS field transistor VGOFF as a function of n+/n+ spacing.

n+/n+ spacing (micron)

Drain current (A)

Drain current (A)

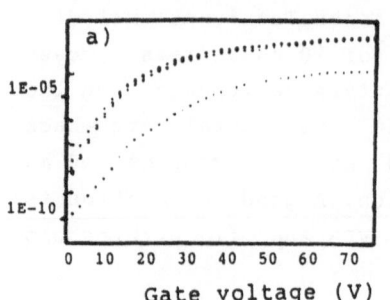

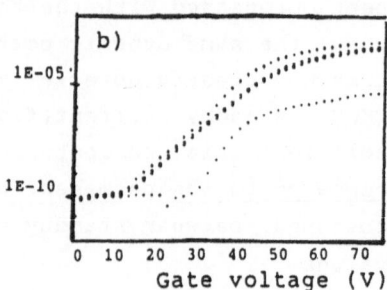

Figure 3 50/1.6 Metal 2 gate NMOS field transistor Ids(Vgs)
characteristics for :
  a) Permanent SOG process.   b) Etch-back SOG process
                                  (partial or total).
Drain voltage values are 0.1V,5V,10V

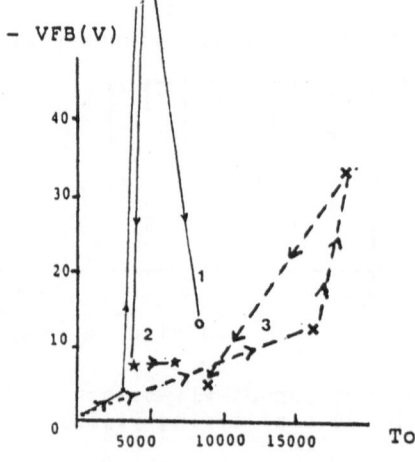

Figure 4 Flat band voltage - VFB
of plane capacitors as a function
of oxide thickness (Tox) through
the different steps of process.
Parameter is the type of process.

1o— Permanent SOG process
2*— Partial SOG etch-back process
3x--Total SOG etch-back process
      (APCVD BSG insulator)

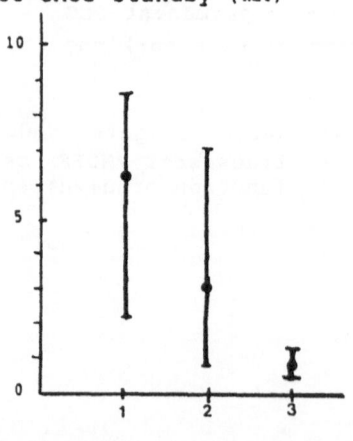

Figure 5 CMOS standby current of
0.8 micron design rules
6 transistors 16 K SRAM for
three different processes.

  1 Permanent SOG process
  2 Partial SOG etch-back process
  3 Total SOG etch-back process
      (APCVD BSG insulator)

Selective Tungsten Metallization for
0.5 µm MOS Processes

D. Friedrich, P. Staudt-Fischbach, D. Wagenaar, W. Windbracke

Fraunhofer-Institut für Mikrostrukturtechnik
Dillenburger Str. 53
1000 Berlin 33, Federal Republic of Germany

Abstract

Up to now conventional aluminum sputtering is a standard method for metallization in VLSI processes. With decreasing contact hole sizes down to the 0.5 µm range and aspect ratios > 1, sputtering becomes more and more critical due to step coverage and reliability problems. A promising technique for future metallization processes is the contact hole filling by selective tungsten deposition. This metallization method was examined in a partially scaled 0.5 µm NMOS and PMOS process. Different contact schemes $Si/TiSi_2/W/Al$, $Si/W/Al$ and Si/Al were compared with regard to leakage current and contact resistance measurements, performed on diodes and Kelvin structures, respectively. Resistivity values $2 \cdot 10^{-7}$ - $5 \cdot 10^{-8} \Omega \; cm^2$ were evaluated for $Si/TiSi_2/W/Al$ contact systems.

Contact hole structures, designed for scanning electron microscopy (SEM) analysis with dimensions down to 0.5 µm have been filled by selective W deposition.

Introduction

In recent years intensive work has been done in the field of selective tungsten chemical vapour deposition (W-CVD) by many researchers. For the selective W deposition with WF_6 chemistry two different deposition processes have to be distinguished, depending on the reduction gas H_2 or SiH_4 used /1,2/. The H_2 reduction of WF_6 has been found to be not appropriate for VSLI application because of several drawbacks like lateral encroachment, wormholes effects, low depositon rate and non compatibility to Ti silicided areas /3,4/. Alternatively, the SiH_4 reduction of WF_6 is a method, avoiding most of the drawbacks mentioned above. However, the small process window of the SiH_4 process requires an extremely accurate control of the process parameters especially the temperature to achieve reproducible and wafer surface type independent results, which can not be sufficiently achieved by now /5,6/. Nevertheless, the H_2 and the SiH_4 W processes were applied for the fabrication of NMOS and PMOS test devices with $Si/W/Al$ and $Si/TiSi_2/W_{SiH4}/Al$ metallization schemes for an electrical characterization of the test devices with regard to junction leakage and contact resistance measurements. The aim of this paper is to compare the electrical parameters of identically processed device wafers with exception of the metallization. Functioning Sub-µm transistors with selective W-metallization have been fabricated.

Experimental

The pattern transfer for the fabrication of NMOS and PMOS test devices was done by X-ray lithography at all 4 levels which allows a structuring of feature sizes down to 0.3 µm /7/. The process except the metallization has been described in a recent paper /8/. A LOCOS isolation technique is been applied for definition of the active areas with a field oxide thickness of 400 nm. Diffusion areas and Poly-Si were implanted simultaneously with As (80 keV, $1 \cdot 10^{16} cm^{-2}$) and BF_2 (25 keV, $5 \cdot 10^{15} cm^{-2}$) for NMOS and PMOS devices, respectively. The surface concentration, determined by spreading resistance measurement, was found to be $5 \cdot 10^{19}$ atoms/cm^3. A subsequent activation and diffusion was carried out at 900 °C for 30 min in N_2 ambient, followed by a 4 step Ti-salizidation sequence. The junction depth for BF_2 and As implanted areas was about 0.2 µm in both cases. A 50 nm sputter deposited Ti layer was converted by thermal Ti-Si reaction into TiSi at 600 °C in a 30 sec N_2 rapid thermal anneal (RTA). Wet chemical selective etch in $H_2O_2:H_2SO_4$ (1:4) for removal of unreacted Ti and following a second RTA in N_2 atmosphere at $800\,^{\circ}C$ for 30 sec results in a low resistance 5 $\Omega/\square$ $TiSi_2$ layer within the active diffusion and Poly-Si areas. After contact hole definition into 500 nm TEOS by reactive ion etching (RIE), a 3 step cleaning procedure was carried out just before the selective W deposition was performed. For removal of organics and Ti particles, generated during contact hole RIE, the wafers were etched in $H_2O_2{:}H_2SO_4$ (1:4) for 10 min. A subsequent 10 sec dip in $HNO_3{:}HF{:}H_2O$ was employed to clean the contact hole sidewalls of sputtered Si, preventing excessive sidewall nucleation during the W deposition process. The sample preparation for W-CVD was finished with a 20 sec HF (3%) dip for native oxide removal.

Selective W-CVD was carried out in a cold wall, low pressure, single wafer reactor. Both processes, H_2 and SiH_4 reduction of WF_6 have been performed by pyrometer temperature control at approximately $450\,^{\circ}C$ and $250\,^{\circ}C$, respectively. The flow ratios of H_2/WF_6 and SiH_4/WF_6 have been adjusted to 20 and 0.7. Before Al/1%Si/0.2%Ti sputtering, an in situ Ar sputter cleaning was carried out in order to prepare the tungsten surface. The process was completed by an $450\,^{\circ}C$, 20 min anneal step in forming gas.

Results and Discussions

Diode leakage currents of different metallization schemes were measured on p^+n and n^+p diodes 250 μm^2 in size to compare the junction integrity of different contact systems. The diagrams shown in Fig. 1 represent typical reverse current characteristics of measurements within a 4 inch wafer. The $Si/W_{SiH4}/Al$ contact system of p^+ and n^+ diodes exhibits similar leakage current behaviour below $10^{-12}A$, with current evidently lower than the current of salicided diodes. This can be explained by reduction of the junction depth due to 100 nm $TiSi_2$ formation. The hydrogen W_{H2} deposition process in a $n^+Si/W_{H2}/Al$ contact structure results in one order of magnitude higher leakage current compared to the silane process, probably due to a higher rate of Si consumption during the initial deposition phase with related encroachment and wormhole effects. Contact resistances were measured on Kelvin structures with different contact hole dimensions from $0.8 \cdot 0.8$ μm^2 to $2.9 \cdot 2.9$ μm^2 on $Si/TiSi_2/W_{SiH4}/Al$ and Si/Al contact systems. The results, as shown in Fig. 2, are mean values averaged over a 4 inch wafer. Resistance values less than 5 Ω have been obtained with contact sizes down to $1.3 \cdot 1.3$ μm^2 for the silicided contacts, whereas the reference Si/Al contacts exhibit significant higher values. For the 0.8 μm n^+ Kelvin structure a resistance increase up to 40 Ω was observed, which is not completely understood and could be caused by the W deposition process. Resistivities for different contact systems, evaluated on 1.3 μm n^+ and p^+ Kelvin structures are summarized in Table 1. The lowest resistivity values were achieved with the $Si/TiSi_2/W_{SiH4}/Al$ system, due to the silicided interface and W planarized contacts. Functioning MOS transistors were fabricated with effective channel length down to 0.4 μm. As shown in Fig. 3 the drain leakage current of the silicided device is increased by one decade compared to the $Si/W_{SiH4}/Al$ - contacted transistor. A cross section of a 0.5 μm, selective tungsten filled contact structure, designed for SEM analysis, is shown in Fig. 4, which demonstrates the ability of this method for devices scaled down to 0.5 μm design rules.

Conclusions

Different metallization schemes have been examined in partially scaled NMOS and PMOS processes with main interest on the $Si/TiSi_2/W_{SiH4}/Al$ contact structure. This metallization scheme exhibits the lowest contact resistance compared to the other contact systems. An increased leakage current behaviour was observed due to $TiSi_2$ formation on shallow junctions. Further investigations on this contact structure seem to be promising, with regard to shrinked contacts down to the sub 0.5 μm range. Improvements of the selective W_{SiH4} - CVD process in terms of reproducibility and better process control have to be done.

References

/1/ R.F. Foster, D.L. Brors, S. Tseng.
Tungsten and other refractory metals for VLSI application II, 1987 MRC.

/2/ M.L. Tu, B.N. Eldridge, R.V. Joshi.
Tungsten and other refractory metals for VLSI application IV, 1989 MRC.

/3/ R.C. Ellwanger, J.E.J. Schmitz, A.J.M. van Dijk.
Tungsten and other refractory metals for VLSI application III, 1988 MRC.

/4/ S.S. Chen, S. Sivaram, R.K. Shukla.
J. Vac. Sci. Technol., B5(6) Nov/Dec 1987

/5/ K.Y. Ahn, P.M. Fryer, J.M.E. Harper, R.V. Joshi, C.W. Miller
Tungsten and other refactory metals for VLSI application IV, 1989 MRC

/6/ J.E.J. Schmitz, M.J. Buiting, R.C. Ellwanger
Tungsten and other refractory metals for VLSI application IV, 1989 MRC

/7/ W. Windbracke, H.L. Huber, P. Staudt, G. Zwicker
Microelectronic Engineering, Vienna 1988.

/8/ G. Zwicker, W. Windbracke, H. Bernt, D. Friedrich, H.L. Huber, E. Krullmann, M. Pelka, P. Lange, P. Hemicker, P. Staudt
EIPB, Monterey 1989, to be published.

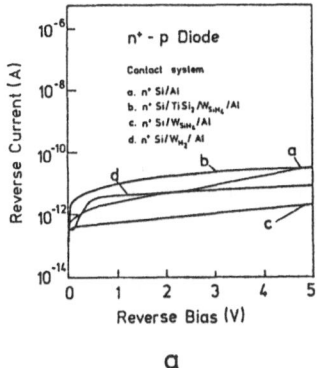

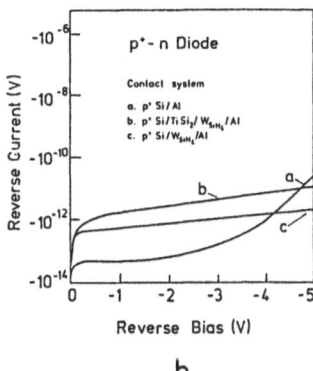

a   b

Fig.1 Reverse current characteristics for a) n^+diodes, b) p^+diodes of different metallization schemes

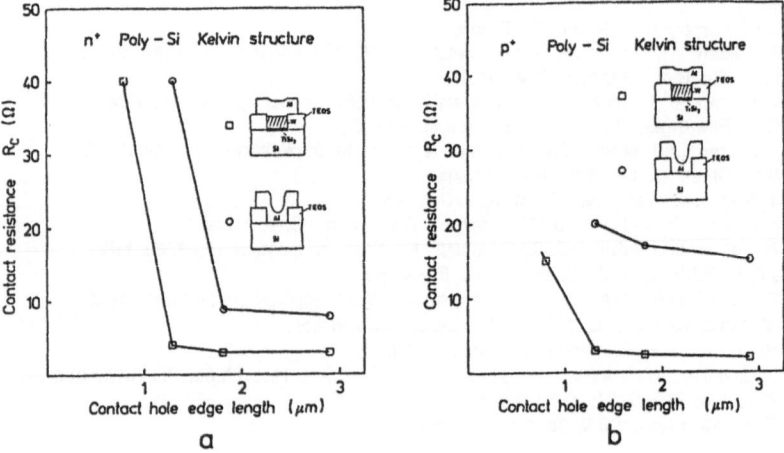

Fig.2 Contact resistance of a) n$^+$ Poly-Si, b) p$^+$ Poly-Si Kelvin structures for Si/TiSi$_2$/W$_{SiH_4}$/Al and Si/Al contact schemes

Table 1 Contact resistivities for different contact systems evaluated on 1.3 μm Kelvin structures

Contact system	n$^+$		p$^+$	
	Poly-Si (Ωcm^2)	Si (Ωcm^2)	Poly-Si (Ωcm^2)	Si (Ωcm^2)
Si / Al	$7 \cdot 10^{-7}$	$2 \cdot 10^{-6}$	$4 \cdot 10^{-7}$	$8 \cdot 10^{-7}$
Si/ W$_{H_2}$ /Al	$3 \cdot 10^{-7}$	$5 \cdot 10^{-7}$	—	—
Si/TiSi$_2$/W$_{SiH_4}$ /Al	$7 \cdot 10^{-8}$	$2 \cdot 10^{-7}$	$5 \cdot 10^{-8}$	$8 \cdot 10^{-7}$

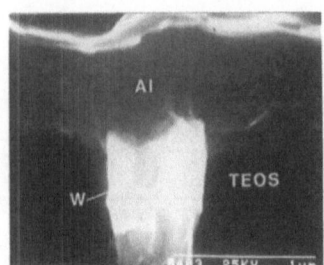

Fig.4 0.5 μm W filled contact hole by selective W-CVD

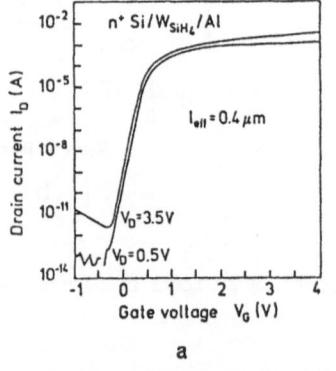

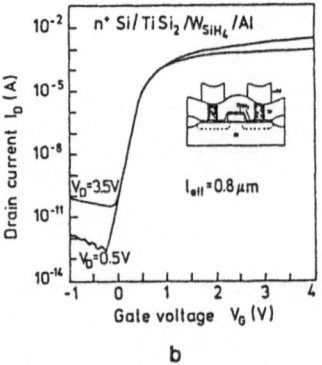

Fig.3 Subthreshold characteristics for NMOS transistors with a) n$^+$Si/W$_{SiH_4}$/Al, b) Si/TiSi$_2$/W$_{SiH_4}$/Al metallization

VLSI MOS

Self-heating and Temperature Measurement in Sub-µm-MOSFETs

P. G. Mautrý and J. Trager

Siemens AG, HLT 131, Otto-Hahn-Ring 6, D-8000 Munich 83, FRG

Summary

For the first time we report the observation of lattice self-heating effects by temperature measurements inside sub-µm MOSFETs and by dynamic drain current measurements with a time resolution of 3 ns. The temperature dependent electrical resistance of a polysilicon gate has been used as the temperature sensor. With decreasing channel length L the temperature increase rises faster than 1/L. For a n-channel MOSFET with design dimensions W/L = 10/0.6 µm the temperature rise approach 60 K at a supply voltage of 5.0 V.

In 1972 Sesnic and Craig [1] reported the self-heating effect in MOSFETs at cryogenic temperatures. In this temperature region the effect is very strong due to the fact that silicon has a low thermal conductivity at low temperatures. As for past technologies the power density in digital MOSFET circuits has been relatively moderate and operation has taken place at room temperature, there has been no need to regard this effect. However, as design dimensions approach the sub-µm region the effect is enhanced and must be taken into account.

A relatively simple measurement technique using a special transistor test structure enables a fairly good characterisation of the self-heating effect. Since gate and channel of a MOSFET are seperated only by the thin gate oxide, the gate material has a good thermal coupling to the channel. The average gate temperature can therfore be considered as the average temperature inside the transistor. This temperature is obtained by measuring the temperature-dependent ohmic resistance of the gate stripe. Controlling the ambient temperature by a thermo-chuck allows calibration of the measurements. Fig. 1 shows the measurement configuration. The voltage sources V_d ang V_g define the operation point of the transistor. Another voltage source, not shown, supplies the bulk bias. The ohmic resistance of the gate stripe is obtained by forcing a small current I_t through the connections I and II and measuring the voltage drop V_t at connections III and IV.

As an example, in Fig. 2(a) the average gate temperature rise of a n-channel MOSFET (design dimensions W/L = 10/0.6 µm) is shown for various gate and drain voltages at constant ambient temperature. At $V_d = V_g = 5.0$ V, where a power of $P = 46$ mW is dissipated within the transistor, the temperature rise reaches 59 K.

676

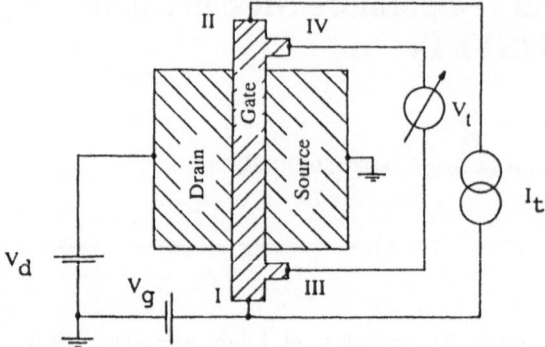

Fig. 1. Electrical configuration for measurement of the gate resistance during MOSFET operation.

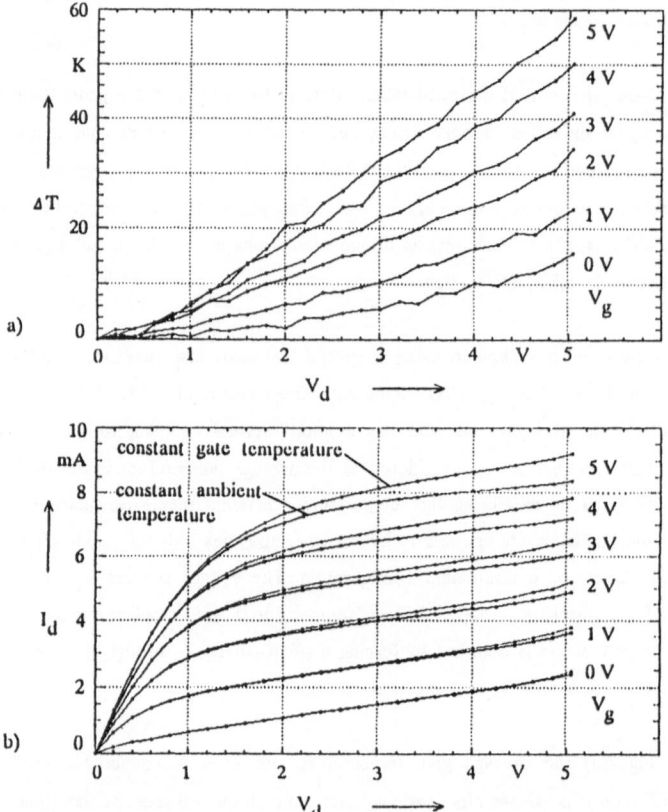

Fig. 2. Measurement results for a n-channel MOSFET, $W/L = 10/0.6$ μm a) Increase of gate temperature above ambient temperature at various operating points and constant ambient temperature $T = 80^{\circ}C$ b) Comparison of output characteristics measured at constant gate temperature or constant ambient temperature, both $T = 80^{\circ}C$.

In order to measure the output characteristics of a transistor at constant channel temperature, the gate temperature was kept constant. This was done by lowering the thermo-chuck temperature to achieve the same electrical resistance of the gate stripe at each operating point. Fig. 2(b) compares data measured in this way with data of a conventional measurement technique, where the channel temperature depends on the dissipated power in the transistor. As carrier mobility decreases with rising lattice temperature, the self-heating imposes a transconductance reduction in comparison to the cold transistor of up to 10%. Measurements of transistors with channel lengths ranging from 0.6 μm to 7.0 μm show that with decreasing channel length L the temperature increase rises faster than 1/L (Fig. 3). This is due to the higher power density in short transistors.

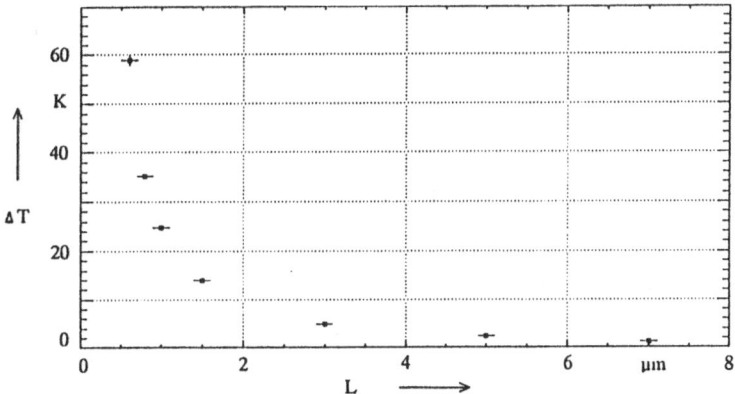

Fig. 3. Dependence of the temperature increase upon channel length. All measurements for n-MOSFETs : $W = 10$ μm, $V_g = V_d = 5.0$ V, bulk voltage $V_b = -2.5$ V.

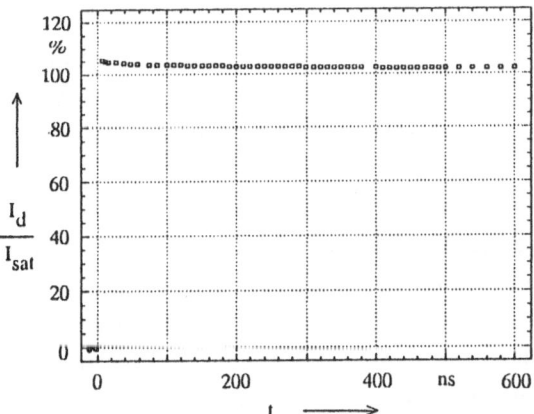

Fig. 4. Measured variation of drain current with time for a n-channel MOSFET, $W/L = 10/0.6$ μm, dissipated power 36 mW. At $t = 0$ a rectangular pulse with rise time 2 ns has been applied to the drain. $I_{sat} = I_d(t \rightarrow \infty)$

678

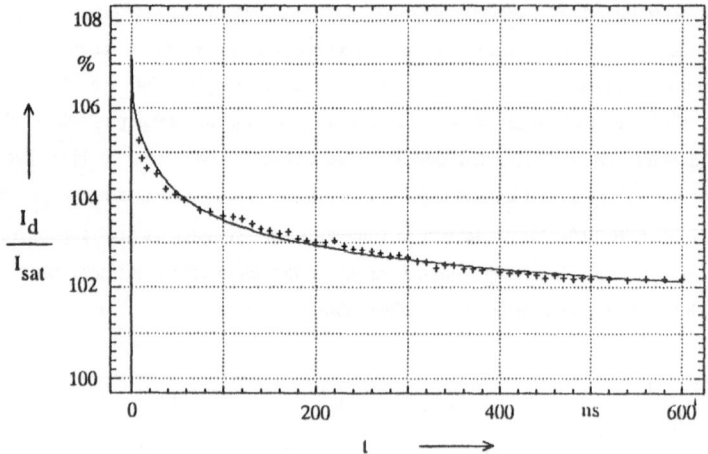

Fig. 5. Enlarged top section of Fig. 4: measured variation of drain current with time (crosses), compared to simulation (solid line). $I_{sat} = I_d(\ t \to \infty)$

In order to investigate the time evolution of the temperature rise after switching, the reduction of the transconductance due to self-heating has been observed using a fast pulse-generator, sampling oscilloscope and a micro-stripline assembly. In Fig. 4 the transient reduction of the drain current due to the self-heating is shown, in Fig. 5 the enlarged top section of Fig. 4. From a state of high electron mobility at $t = 0$ in the cold transistor, heating-up leads to a fast reduction of the drain current within the first 100 ns. The solid line is the result of a 3D-simulation of the temperature rise in combination with a calculation of the drain current reduction based on data from the static measurements. It emphasizes the fast reduction of the drain current during the first few nanoseconds. Steady-state of lattice temperature is reached a few microseconds later. In real circuits switching happens within the first nanosecond after pulse-raising. Standard static measurement methods therfore underestimate their transconductance because they are always taken in the self-heated state. This may cause errors in circuit modeling with DC-measured model parameters.

In conclusion, the two independent measurment methods described above clearly show that self-heating takes place in both VLSI and ULSI devices. Drain current reductions of up to 10% and increased transistor temperatures of up to 60 K due to self-heating were detected. Errors in circuit modeling may arise. Furthermore, the interpretation of physical models fitted to DC-measured data can be misleading if the models omit the self-heating effect. As feature sizes further scale down for future IC-generations, the observed effect becomes increasingly significant and must be considered.

References
[1] Sesnic, S.S.; Craig, G.R. : Thermal effects in JFET and MOSFET devices at cryogenic temperatures. IEEE Trans. Electron Devices vol. ED-19, 8 (1972) 933-942
[2] Takacs, D.; Trager, J. : Temperature increase by self-heating in VLSI CMOS. ESSDERC 1987, Bologna, pp. 59-62

Latch-up Free CMOS Using Buried Polysilicon Diodes

W. Reczek, J. Winnerl*, F. Bonner, B. Murphy

Siemens AG, Components Group
*Siemens AG, Corporate Research and Development
Otto-Hahn-Ring 6, D-8000 Munich 83, FRG

Summary

Latch-up free CMOS circuit operation is achieved through the use of buried polysilicon diodes instead of conventional (ohmic) well contacts. In a DRAM technology with polysilicon bit line a buried polysilicon diode can be realized with no additional process step and no additional die area is required. No degradation of MOS device parameters occurs. The basis for experiment is a 4M DRAM 0.9µm n-well CMOS technology with substrate bias generator [1].

Introduction

In CMOS circuits negative injection is prevented through the use of a substrate bias (VBB) generator. For positive overvoltages, however, a significant latch-up risk exists due to the PMOS device in the n-well. To achieve latch-up free CMOS output buffers injection of the p+/n-well junction has to be prevented.

Buried polysilicon diode

The use of a non-linear element to supply the n-well bias [2] combines the good static latch-up hardness of floating well concepts [3] with the low latch-up susceptibility to supply voltage transients of conventional CMOS. The n-well contact of a PMOS device is replaced by a buried polysilicon diode (Fig. 1b). Majority carriers can only flow out of the well. Proper biasing of the well is therefore guaranteed even when an electron current is collected by the well. On the other hand no electron current can be supplied to the well in the event of a positive overvoltage at the p+ diffusion. At forward bias the voltage drop has to be low to maintain the shunt efficiency of the well contact, while at reverse bias the leakage current has to be small (Fig. 2). A SEM picture and the I/V characteristic of a buried polysilicon diode are shown in Fig 3. The diode is realized by the outdiffusion of arsenic from the polysilicon bit line (100 nm deep) into the p-channel source region (600 nm deep). For a DRAM technology with an As doped polysilicon bit line no additional process step is required for the buried polysilicon diode and the area requirement is comparable to a conventional well contact. In Fig.4 the forward characteristic of the buried polysilicon diode is compared to that of the p+/n-well junction. At low forward bias the poly diode forward current is four orders of magnitude higher.

Latch-up characterization

Fig.5 shows the four different CMOS output buffers under consideration (summary of results Tab.1).

Static latch-up behaviour was characterized by forcing a current into the output of the circuit (DO) while both the PMOS and NMOS devices are in the "OFF" state. At positive injection (Fig.6a, V_Trig > VDD) the circuits with conventional (ohmic) well contacts latch at VDD + 0.7V. With non-linear well contacts the devices are latch-up free even for trigger voltages greater than VDD + 3V. The poly diode, however requires less space and its latch-up trigger characteristic is comparable to the MOS diode. At higher trigger voltages the PMOS is turned on additionally, thereby increasing the trigger current. At negative injection (Fig.6b, V_Trig < VSS) the CMOS buffers with on-chip VBB generator are latch-up free for trigger currents of up to 100mA independent of the type of the well contact used. Without a VBB generator the trigger voltage is lower but up to 100mA no latch-up occurs.

Power-on latch-up hardness was investigated with a 7V/2ns ramp rate and a DC voltage source at the DO pin (measurement setup Fig.7, results Tab.1). While the conventional approach is very susceptible to power-on latch-up with externally applied voltages at the DO pin, the new buried device concept guarantees latch-up free circuits at power-on even for externally applied voltages at the DO pin in the range of -1V to 6V.

buffer design	static latch-up trigger current/voltage	power-on latch-up voltage VDO
CMOS with VBB-generator (b)	+25mA / 5.7V no LU at -100mA/-3.4V	> 1.7V < -3V
buried poly Si diode (c)	no LU at VDO <8V no LU at -100mA/-3.4V	> 6V < -1V

Tab. 1: Latch-up hardness of different CMOS buffers (see Fig. 5). W/L = 700/1.2 and W/L = 300/1.2 for PMOS and NMOS devices, respectively. VDO = 5V; ramp rate at power-on = 7V/2ns. LU = latch-up.

Conclusions

The use of a non-linear well contact provides static and power-on latch-up free output buffers for CMOS circuits with VBB generator. Such buffers are suitable for DRAM application and avoid the performance drawback of pure NMOS output buffers. In a DRAM technology with polysilicon bit line a buried poly diode can be realized with no additional process step and no additional die area is required.

Acknowledgement

The authors would like to thank their colleagues in design, technology, testing and packaging for their support and contributions. This report is based on a project which has been supported by the Minister of Research and Technology of the Federal Republic of Germany under the support-no. NT 2696. For the contents the authors alone are responsible.

References

[1] K.H. Küsters et al., Proc. of the 1987 Symposium on VLSI Technology, Nagano, Japan, pp. 93-94.
[2] W. Pribyl et al., IEEE JSSC, Vol. 23, No.3, June 1988, pp. 816-819.
[3] H. P. Zappe et al., IEDM Tech. Dig. pp. 517-520, 1985.

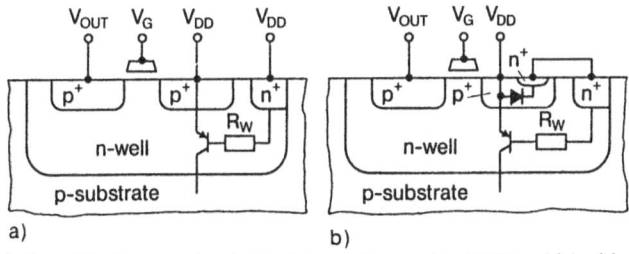

Fig.1: PMOS device with a.) conventional (ohmic) n-well contact to VDD and b.) with a non-linear element (buried poly silicon diode).

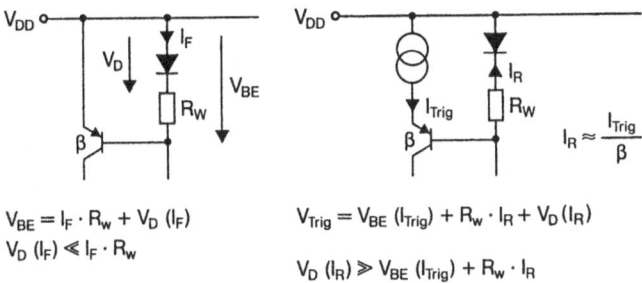

Fig.2: Equivalent lumped element model for the parasitic elements of the PMOS device using a non-linear element instead of an ohmic n-well contact to VDD. a.) Normal operating condition with diode forward biased and b.) overvoltage trigger condition with diode reverse biased.

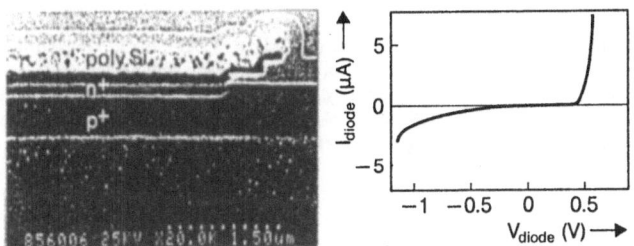

Fig.3: SEM microphotograph (detail of Fig.1) and I/V characteristic of a buried poly diode (area $1 \times 4 \mu m^2$).

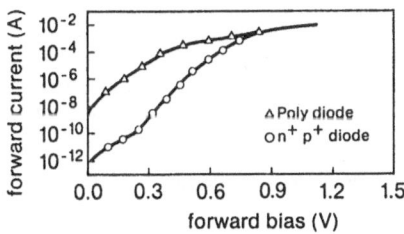

Fig.4: Forward I/V characteristic of the buried polysilicon diode compared to a p+/n junction.

682

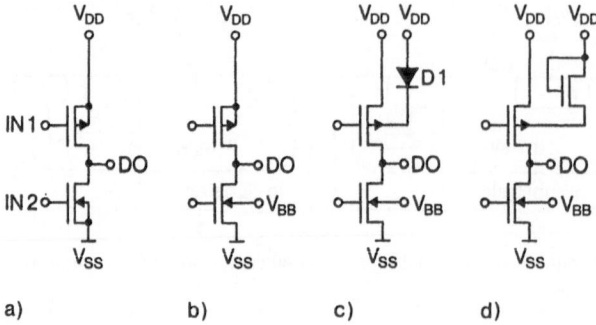

Fig.5: CMOS output stages: a.) conventional, b.) with VBB generator, c.) with VBB generator and n-well connected by an additional diode, d.) with VBB generator and n-well connected by a MOS diode.

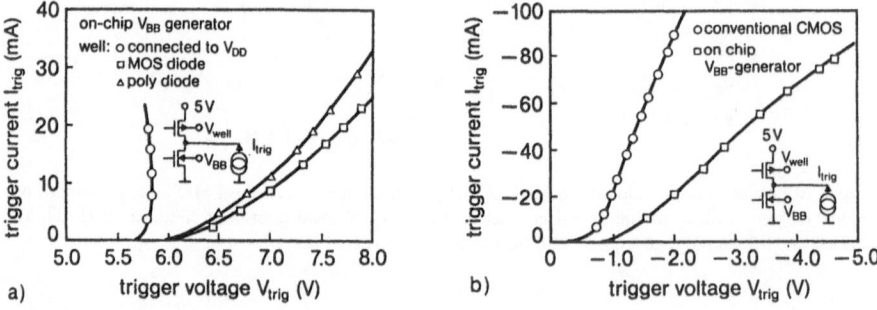

Fig.6: Static latch-up behaviour. a.) Positive injection (V_Trig > VDD) at the DO pin. Comparison of n-well CMOS with VBB generator and ohmic well contact, well connected with a NMOS diode (W/L = 100/1.2) and well connected with a buried polysilicon diode. b.) Negative injection (V_Trig < VSS) at the DO pin. Results for n-well CMOS with on-chip VBB generator and the three different well connections (Fig.5, case b, c and d) to VDD compared to conventional CMOS.

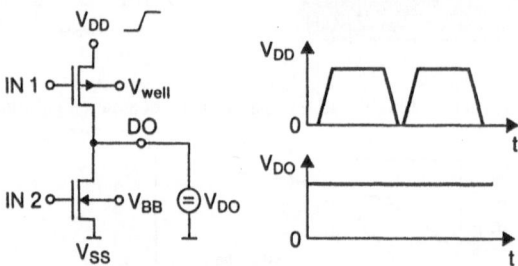

Fig.7: Measurement setup for power-on latch-up characterization.

Anomalous Punchthrough in ULSI Buried-channel P-MOSFETs

Tomasz SKOTNICKI, Gérard MERCKEL and Thierry PEDRON

Centre National d'Etude des Télécommunications (CNET-CNS)
B.P. 98, Chemin du Vieux Chêne 38243 Meylan, FRANCE

Buried-channel (BC) P-MOSFET is more susceptible on punchthrough (pt) than surface-channel (SC) N-MOSFET and thereby it contributes at least equally as the latter, to a total stand-by power dissipation in CMOS circuits. Cham *et. al.* [1] have found that the channel thickness t_C is the most crucial parameter for pt conditions in BC-P-MOSFETs. They showed that when making the channel thinner than $0.1\mu m$ it is possible to achieve pt immune BC-P-MOSFETs with electrical channel length L_{el} as small as $0.6\mu m$. However, below $0.5\mu m$ of channel length BC-P-MOSFET always shows punchthrough,which is commonly attributed to the difficulty in achieving channel thicknesses sufficiently small.

We have shown that the above reasoning is wrong because the mechanism of punchthrough in extremly short BC-P-MOSFETs is different. Using the 2-dim process simulator TITAN [2] in conjunction with the 2-dim device electrical simulator JUPIN [3] we have shown that punchthrough in a very short, $L_{el}=0.4\mu m$, BC-P-MOSFET with the typical well doping N_D of $1.3\times10^{16}cm^{-3}$ is almost independent of t_C. The numerical results shown in Fig. 1 demonstrate that already at $L_{el}=0.4\mu m$ no choice of t_C is able to suppress punchthrough. Note that Fig. 1c) shows a limiting case $t_C=0$ (no counterdoping implant is performed) and even then the transistor shows punchthrough.

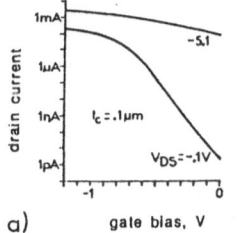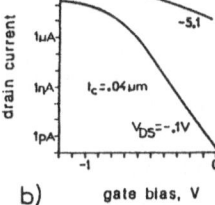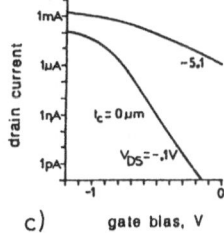

Fig.1. Simulated subthreshold characteristics for a series of BC-P-MOSFETs of $L_{el}=0.4\mu m$ and $t_j=0.21\mu m$ differing by the channel thickness t_C which reads: a) $0.1\mu m$, b) $0.04\mu m$ and c) $0\mu m$ - no counterdoping implant.

An explanation of this anomalous punchthrough is found on the basis of the voltage-doping transformation (VDT) [4] and [5], which as applied to a BC-P-MOSFET shows that the influence of the drain bias V_{DS} on the potential barrier height is equivalent to and can be replaced by the increase in the

684

effective p-channel doping N^*_A and the decrease in the effective n-well doping N^*_D according to the following formulae:

$$N^*_A = N_A - \frac{2\varepsilon_s}{q\,l^2_0}\, V^*_{DS}\, \exp\left(\frac{-L_{el}}{2\,l_0}\right) \tag{1a}$$

$$N^*_D = N_D + \frac{2\varepsilon_s}{q\,l^2_0}\, V^*_{DS}\, \exp\left(\frac{-L}{2\,l_0}\right) \tag{1b}$$

where N_A and N_D are the real channel and well dopings, respectively, L_{el} is the electrical channel length, L is the mean length of a current flow line within a well region and $V^*_{DS} \propto (V_{DS}, V_{SB}) < 0$. As results from (1b), N^*_D must inevitably change sign in the case of short channel and/or large enough negative V_{DS} bias. Electrically it means that a part of the n-well, adjacent to the channel, is transformed, under an influence of the drain field, from a n- to a p-type thus enlarging the p-channel and giving rise to punchthrough. It is purely this effect, hereafter called the drain-induced channel enlargement (DICE), which makes punchthrough to become almost independent of channel thickness.

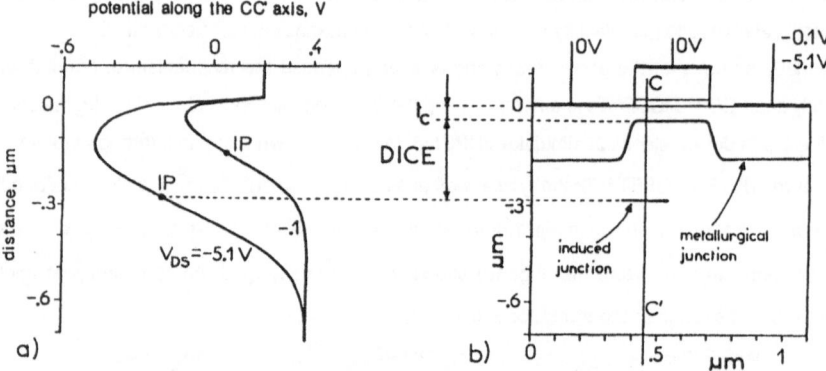

Fig. 2. An illustration of the DICE effect. The inflection point (IP) on the plot of the potential distribution across the channel (in a given cross-section CC') distinguishes between the p-channel and n-well regions. The numerical results demonstrate that the IP is shifted down into the bulk by a negative drain bias which purely means the electrical enlargement of the p-channel region.

The DICE effect and its physical interpretation are both confirmed by results of numerical simulation. Fig. 2a) shows the transversal potential distribution in the cross-section CC', depicted in Fig. 2b). The inflection point (IP) obviously corresponds to the border-line between the p-channel and n-well regions. As seen in Fig. 2 the inflection point is shifted towards the bulk with increasing negative drain bias. This purely means that the p-channel is electrically enlarged by the drain field. Indeed, at $V_{DS} = -0.1V$ the pt current path is confined to the metallurgical p-channel, see Fig. 3a), whereas at $V_{DS} = -5.1V$ it goes distinctly below the metallurgical channel passing through the n-well region, and thus indicating the DICE effect.

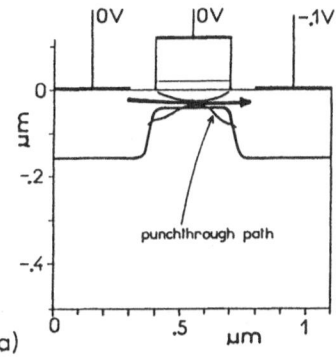 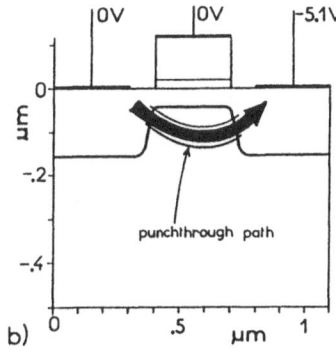

Fig. 3. At V_{DS}=-0.1V the punchthrough current path (numerical simulation) is confined to the metallurgical channel - a), however at V_{DS}=-5.1V it goes distinctly below the channel across the n-well region - b).

This anomalous pt mechanism can be suppressed by any action leading to an increase in N_D or/and L which both, according to (1b) prevent N^*_D from changing sign, thereby impeding the DICE effect. Fig. 4 shows the characteristics of a pt resistant BC-P-MOSFET of L_{el}=0.3μm which has the same channel thickness t_c and channel doping N_A as that one from Fig. 1b) which shows a significant punchthrough already at L_{el}=0.4μm. This suppression of punchthrough is achieved by a use of the LDD (lightly doped drain) structure in conjunction with an heavy dose deep Arsenic implant, hereafter called the retrograde implant (RI). The LDD structure gives rise to L by about 0.4μm and the RI is aimed at a local increase in N_D within a part of the well, adjacent to the p-channel. It is to be noted that the RI is different from the deep Arsenic implant proposed by Cham and Chiang in [1]. Their Arsenic implant is aimed only at better channel tailoring, whereby it is too weak to suppress the DICE effect. The number one reason for the RI is rather the local increase in the well doping, although a simultanous channel tailoring can also be performed. As a result, the RI must be deeper and much heavier than the implant proposed in [1].

Once the retrograde implant is performed the pt conditions become almost independent of the source/drain junction depth t_j - see Fig. 5a), although without the retrograde implant an exponential dependence is observed - see Fig. 5b), the latter being in agreement with [1]. An explanation of the anomaly is given by the VDT which shows that the dependence on t_j is exclusively via the DICE effect; t_j intervenes (through L) in (1b), describing the DICE effect but does not intervene in (1a) describing the DIBL effect. As a result, once the DICE effect is suppressed by the RI the dependence on t_j vanishes.

In conclusion, it has been shown that punchthrough in very short BC-P-MOSFETs can not be suppressed by making the p-channel thinner. This is because of the drain-induced channel enlargement (DICE) effect which becomes a dominant pt mechanism at very short channels. A new implant step, called the retrograde implant, and the LDD structure of a transistor are shown to be able to suppress the DICE effect and thereby to shift the limit of pt immune BC-P-MOSFETs from 0.5μm down to 0.3μm. As an additional advantage, it has been found that the RI renders punchthrough independent of tj thus relaxing the needs for shallow junctions. A BC-P-MOSFET of $L_{el}=0.4\mu m$ has been shown to be pt immune up to $t_j=0.33\mu m$. Deeper junctions have not been tried but the VDT suggests that even then pt remains suppressed. This is a very favorable finding from the stand-point of hot carrier injection into a gate oxide, s/d sheet resistance and drain break-down voltage, which all suffer from junction shallowing.

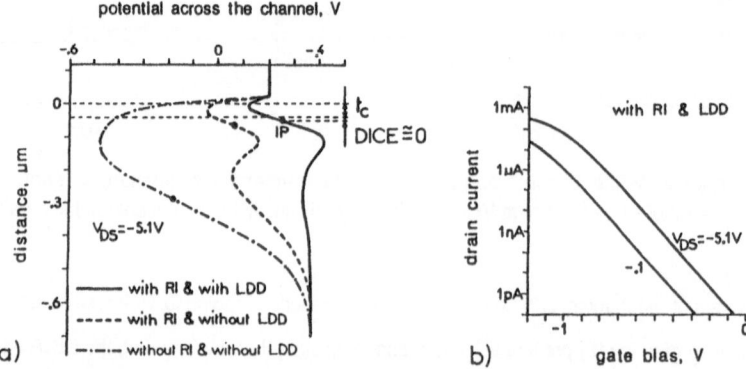

Fig. 4. A comparison of the potential distribution across the channel - a) reveals that the DICE effect is greatly reduced by the RI although its complete elimination needs both the RI and LDD structure. The joint use of the RI and LDD enables the BC-P-MOSFET to be free of pt, see b), down to L_{el} as short as $0.3\mu m$.

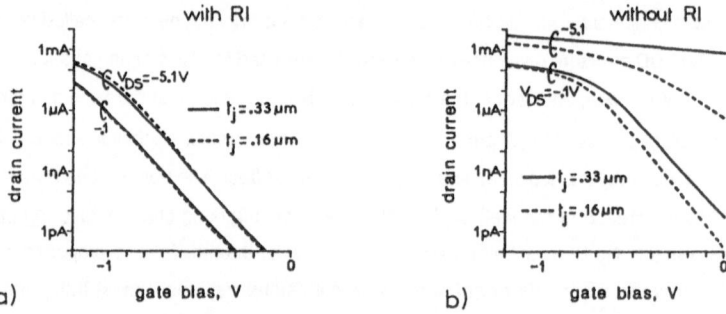

Fig. 5. An anomaly in the pt dependence on tj. With the RI the subthreshold characteristics of the two simulated BC-P-MOSFETs overlap - see a), in spite of one of them having the s/d junctions twice as deep as the other. The characteristics of the same two transistors but without the RI split out - see b), showing much more of pt for the transistor of $t_j=0.33\mu m$ than for that of $t_j=0.16\mu m$. The simulated transistors are identical, except for tj, and have L_{el} of $0.4\mu m$.

References

1. K.M. Cham and S-Y. Chiang. IEEE Trans. Electron Dev., ED-31, 964-968, July 1984.
2. A. Gerodolle and S. Martin: TITAN 4 - 2d process simulation. Technical report, CNET-CNS, Meylan, France, May 1987.
3. H. Belhaddad, C. Corbex and A. Poncet. Proceedings NASECODE IV, 193-198, June 1985.
4. T. Skotnicki, G. Merckel and T. Pedron. IEEE Trans. Electron Dev., ED-35, 1076-1086, July 1988.
5. T. Skotnicki, G. Merckel and T. Pedron. IEEE Trans. Electron Dev., ED-36, 690-705, April 1989.

Charge Transport Near the Si/SiO$_2$ Interface in MOSFET Devices

Th. Vogelsang, W. Hänsch, and R. Kircher
Siemens AG, Corporate Research and Development, ZFE ELPT

SUMMARY: The influence of quantum mechanical corrections of particle density near the Si/SiO$_2$ interface on the channel mobility of a MOSFET is investigated by an approximation of the quantum statistical expectation values and by an exact solution of Schrödinger's equation perpendicular to the interface. A new model for mobility reduction at the interface is derived. The model is implemented in the device simulator MINIMOS, the results show good agreement with n- and p-channel devices of the 4M-DRAM generation.

Modeling the channel mobility of a MOSFET is a cornerstone in the calculation of the output characteristics of a MOSFET device. Most commonly used are empirical models that account by a number of adjustable parameters for different physical effects, as there are: surface scattering, gate field reduction, quantum mechanical effects in the inversion channel, and velocity saturation [1]. In this work we present a unified treatment of an improved description of carrier transport near the Si/SiO$_2$ interface that includes quantum mechanical corrections of the carrier distributions near the surface [2] and a consistent mobility model which holds for the linear regime. Velocity saturation is included in the usual way [3].

The basic idea of our work is to expand the quantum statistical expectation value by separation of the slowly varying electric potential from the sharp band edge. The band edge is accounted for by the choice of the wave functions that are used in the calculation of the expectation values [4]. Together with an exact solution of Boltzmann's equation on a semi-infinite space [5], which is appropriate for the MOSFET geometry, we derive a current density equation that is modified by an effective potential in the vicinity of the interface and a mobility model that takes heed of the symmetry breaking interface by one parameter p only. Moreover this parameter is directly related to the scattering properties of the interface. It ranges from one to zero which covers total elastic to complete inelastic scattering respectively. An interesting feature of the channel mobility is that it does not depend on the local transverse electric field but rather on the shape of the electric potential in an extended vicinity of the interface.

We compare our approximation with the exact quantum mechanical solution which is only possible close to equilibrium (low current level), therefore it is restricted to small drain voltages only. Here we do not, as it is usually done, solve Schrödinger's equation together with Poisson's equation. Instead we couple Schrödinger's equation directly with the current density equation by the introduction of an effective potential. This automatically gives a density distribution that is consistent with the continuity equation. The objective for this is found in [6].

$$\vec{J} = -q\,\mu\,\nabla V^* - k_s T\,\mu\,\nabla n \tag{1}$$

$$V^* = V - \frac{k_s T}{q}\,\ln\gamma \tag{2}$$

Here $\vec{J}$ is the current density, V the electric potential, μ the mobility and n the particle density.

We get the classical drift diffusion equations with $\gamma = 1$.

The approximation we derive in [6] gives

$$\gamma = 1 - \exp\left(-\frac{y^2}{\lambda_{th}^2}\right) \tag{3}$$

where λ_{th} is the thermal wavelength of the electrons. For the mobility we get:

$$\frac{\mu}{\mu_{Bulk}} = 1 - \frac{1-p}{\sqrt{\pi}\,\gamma}\int_0^{\hat{V}^{\frac{1}{2}}} d\xi\, e^{-\xi^2}\left(1-\cos\left(2\frac{y}{\lambda_{th}}\xi\right)\right)\cdot\exp\left(\frac{\hat{V}_0\,\arctan\left[\left(\frac{\hat{V}_0}{\hat{V}-\xi^2}-1\right)^{\frac{1}{2}}\right]}{c(\hat{V}-\xi^2)^{\frac{1}{2}}}\right)\cdot$$

$$\cdot\frac{\cosh\left(\frac{\hat{V}_0}{c\,(\hat{V}-\xi^2)^{\frac{1}{2}}}\arctan\left[\left(\frac{\hat{V}}{\hat{V}-\xi^2}-1\right)^{\frac{1}{2}}\right]\right)}{1-p\exp\left(\frac{2\,\hat{V}_0}{c\,(\hat{V}-\xi^2)^{\frac{1}{2}}}\arctan\left[\left(\frac{\hat{V}_0}{\hat{V}-\xi^2}-1\right)^{\frac{1}{2}}\right]\right)}\;;\quad \hat{V} = \frac{q\,V_{Bulk}-q\,V}{k_s T} \tag{4}$$

This expression is very effectively evaluated on a vector processor with no additional computational costs compared to conventional models.

The coupling of Schrödinger's equation gives a self consistent determination of γ. The procedure is outlined in Fig. 1. In this way we achieve a self consistent solution that is correct not only for the equilibrium (no drain voltage) but also for small currents parallel to the Si/SiO$_2$ interface. The more rapid decrease of particle density in a greater distance from the interface with the exact quantum mechanical solution that has been reported [7], compared to the conventional drift diffusion results, is no true physical effect. It is the result of taking into account only the first few bound states. With a complete spectrum the particle density reaches the equilibrium density in the bulk and the differences between the drift diffusion equations and Schrödinger's equation are largest near the interface, as expected. Therefore it is important to take into account the total number of bound states in the inversion channel (some hundred) together with the continuum states as well. (Fig. 2)

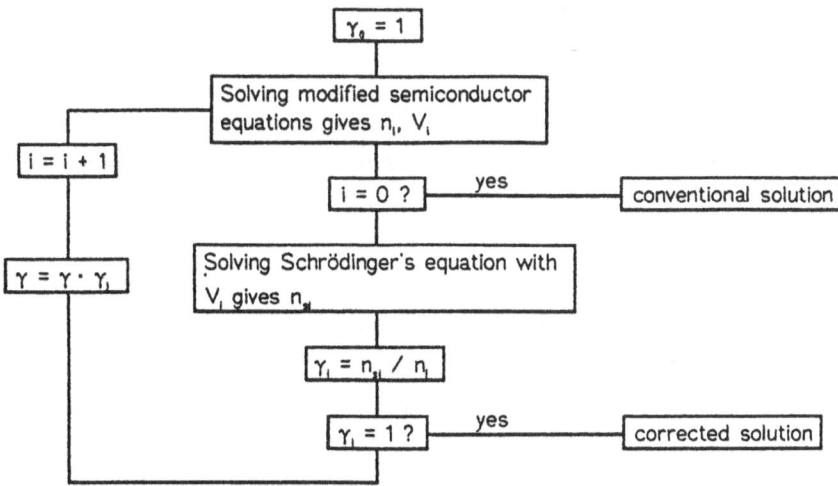

Fig. 1: Coupling of Schrödinger's equation with drift diffusion equations

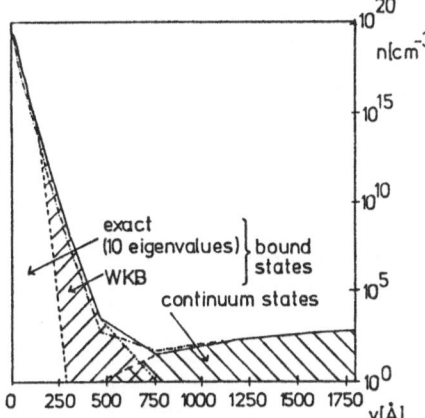

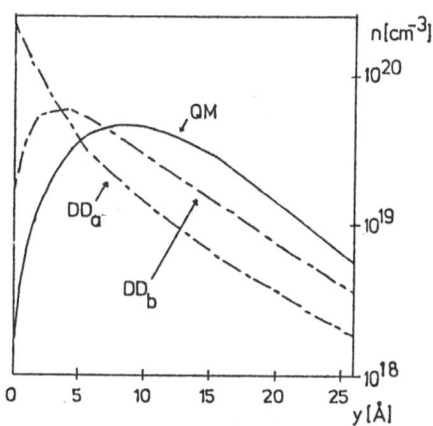

Fig. 2: particle density near equilibrium
a) self consistent solution of Schrödinger's equation together with Poisson's equation and current density equation: small dash-dash line - the first 10 exact bound states; dot-dot line - all other bound states within WKB-approximation; large dash-dash line - continuum states; solid line - complete solution of Schrödinger's equation;

b) Blowup of the region near the surface: solid line: QM - Schrödinger's equation; large dash-dot line: DD_a - conventional drift diffusion equation; small dash-dot line: DD_b - modified current equation (equation (1))

The new transport model was implemented in the device simulator MINIMOS [3]. The I-V-characteristics for devices of the 4M-DRAM generation were calculated. We could verify the characteristics for n- and p-channel devices covering the gate lengths L_G from 0.8 to 3.0µm and oxide thicknesses t_{ox} 16nm, 18nm, and 20nm (Fig. 3,4).

Both kinds of devices require a surface scattering parameter p around 0.9 which means that we have an almost perfect interface. We also compared our new model with the conventional mobility model provided by MINIMOS where the in-

690

fluence of the interface is accounted for by four parameters. We find that the mobility model provided by MINIMOS represents the features of our new model, especially the explicit dependence from the interface, very well. Furthermore we are able to conclude that for the currently developed MOSFET devices quantum mechanical effects near the Si/SiO$_2$ interface do not significantly change integral quantities like terminal currents.

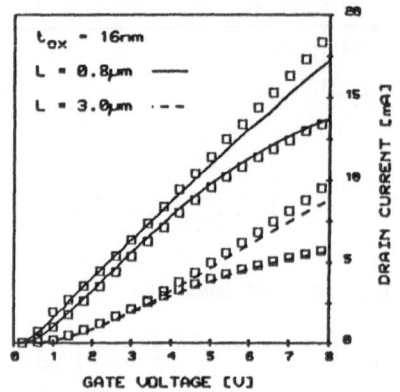 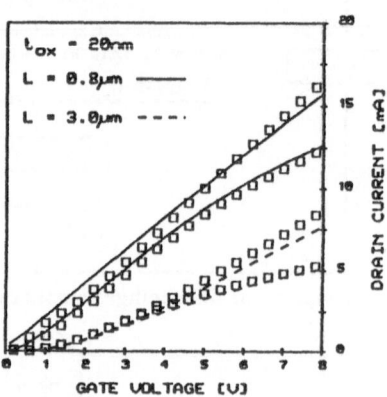

a) t_{ox} = 16nm;   b) t_{ox} = 20nm;

Fig. 3: 4M-DRAM n-channel device with gate length L_0 = 0.8µm (upper curves) and L_0 = 3.0µm (lower curves); V_d = 2V and 6V; scattering parameter p = 0.95; measurement - solid lines; simulation squares;

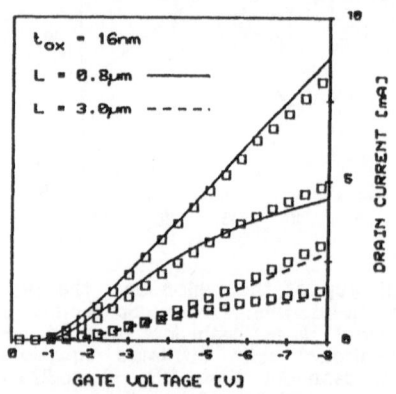 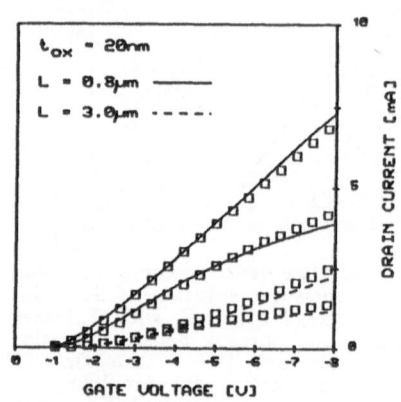

a) t_{ox} = 16nm;   b) t_{ox} = 20nm;

Fig. 4: 4M-DRAM p-channel device with gate length L_0 = 0.8µm (upper curves) and L_0 = 3.0µm (lower curves); V_d = -2V and -7V; scattering parameter p = 0.85; measurement - solid lines; simulation squares;

[1] S. Selberherr; Analysis and Simulation of Semiconductor Devices, (Springer Wien New York 1984) chap. 4
[2] F. Stern and W. E. Howard; Phy. Rev. 163 816 (1967)
[3] W. Hänsch and S. Selberherr; IEEE Trans. Electr. Dev. ED34 1074 (1987)
[4] W. Macke and P. Rennert; Ann. Phys. (Leipzig GDR) 12 32 (1964); Ann. Phys. (Leipzig GDR) 12 84 (1964)
[5] R. F. Greene, D. R. Frankel and Joy Zemel; Phys. Rev. 118 967 (1960)
[6] W. Hänsch, Th. Vogelsang, R. Kircher, M. Orlowski; accepted for publication in solid state electronics
[7] C. T. Hsing et. al.; Phys. stat. sol. (a) 56 129 (1979)

Characterization of MOSFET Gate Oxides by Injection of Controlled Quantities of Electrons

W.H. Krautschneider

Siemens AG, Central Research and Development
Otto-Hahn-Ring 6, D-8000 Munich 83, West Germany

Abstract

A simple technique is presented for uniformly injecting a controlled quantity of electrons in a range from 10^{10} cm^{-2} to 10^{18} cm^{-2} into the gate oxide of small-geometry MOS transistors. It can be used for probing oxide characteristics and oxide stability under hot electron stress. In conjunction with a refined evaluation of charge pumping experiments, the influence of hot electron injection on spatial and energetic trap distribution is studied.

Introduction

When electrons are injected into the gate oxide of a MOS transistor, traps in the silicon dioxide located close to the Si-SiO$_2$ interface can be filled. The capture of electrons by these traps leads to a shift of the threshold voltage, which saturates when all these states are occupied [1]. When the injection is continued, neutral traps located in the bulk of the oxide possessing a smaller capture cross section are filled.

Electron Injection

The injection was performed by a short voltage pulse applied to the substrate which forward-biases the source, drain and inversion layer of the MOS transistor (Fig. 1). This is followed by a reverse-bias pulse which provides the carrier heating field [2]. We modified this pulse train to a three-level pulse with constant heating and injection pulse widths which were small compared to the pulse period, so that the injected charge per cycle is independent of the pulse frequency. The gate current, from which the injected charge is calculated, is consequently proportional to the frequency of the substrate pulse train. This allows the undetectably small gate currents at very low injection levels to be estimated by extrapolation.

Electron injection was carried out alternately with measurements of the threshold voltage V_{th}. This is shifted by the charge build-up due to the trapped electrons, which can be modelled by the following equation [3].

$$\Delta V_{th} = \sum_{i=1}^{n}(q \cdot N_{iti} \cdot x_{ci}/(\epsilon_0 \cdot \epsilon_{ox}) \cdot (1-\exp(-\sigma_i \cdot N_{inj})))$$

N_{it} is the number of trap centers, x_c the centroid of the trap distribution, σ the capture cross-section of the traps, N_{inj} is the number of injected electrons, and n denotes the number of different trap types. The other symbols have the usual meanings.

A fit with the experimental curve (Fig. 2) leads to two different trap types with comparable trap densities but different capture cross-sections ($\sigma_1 = 4 \cdot 10^{-16} cm^2$ and $\sigma_2 = 2 \cdot 10^{-18} cm^2$) distinguished by the plateau at $N_{inj} \approx 5 \cdot 10^{16} cm^{-2}$.

Energetic and Spatial Distribution of Interface States

Further insight into gate oxide behavior under hot electron stress is given by the distribution of oxide traps in a direction perpendicular to the interface measured at different injection levels. This distribution can be determined by charge pumping experiments [4] using a data evaluation based on the non-constant frequency behavior of the pumped charge per cycle, $Q_{it}=i_p/f$, which is the pumping current normalized to the frequency f (Fig. 3). Traps located deeper in the oxide have a longer time constant [5] and cannot respond to higher frequencies.

The number of traps per area N_{it} located up to the distance x from the interface is

$$N_{it} = \int_{0}^{x} N_{itd}(x)dx \qquad (1)$$

The concentration of traps $N_{itd}(x)$ as a function of the distance from the interface is obtained from (1) by differentation

$$N_{itd}(x) = \frac{dN_{it}}{dx} = \frac{1}{q \cdot A} \cdot \frac{dQ_{it}}{dx} \approx \frac{1}{q \cdot A} \cdot \frac{\Delta Q_{it}}{\Delta x}$$

The distance x, in which the traps can follow the gate signal, is obtained from the trap time constant $\tau=1/2f$ by [5]

$$x = \frac{1}{a} \cdot \ln\left(\frac{\sigma \cdot v_{th} \cdot n_s}{2 \cdot f}\right) \qquad (2)$$

where $a \approx 10^8 cm^{-1}$, σ is the capture cross-section, v_{th} the thermal velocity, and n_s denotes the surface concentration of the electrons. Fig. 4 shows the concentration of the oxide traps as a function of the distance from the Si-SiO$_2$ interface for a fresh device and after hot electron injections.

The oxide characterization can be completed by measuring the energetic trap distribution using a modified charge pumping technique [6], as depicted in Fig. 5. The electron injection is reflected in an increase of the energetic trap distribution D_{it} and the trap concentration N_{itd}. On a relative scale, the latter is increased predominantly at a greater distance from the interface (Fig. 4).

Conclusion

In conclusion, the uniform injection of hot electrons into the gate oxide of MOS transistors allows a determination of the oxide parameters and a quantitative assessment of oxide stability under hot electron stress.

Acknowledgment

The author is very grateful to J. Graf and M. Fellermeier for their careful measurements and H. Mulatz for setting up the test devices.

References

[1] J.M. Aitken, IEEE J. Sol.-State Circ. SC-14, p. 294, 1979.

[2] T.H. Ning, J. Appl. Phys., vol. 49, p. 5997, 1978.

[3] T.H. Ning, J. Appl. Phys., vol. 47, p. 3203, 1976.

[4] J.S. Brugler and G.A. Jespers, IEEE Trans. Electr. Dev., ED-16, p. 297, 1969.

[5] F.P. Heiman and G. Warfield, IEEE Trans. Electr. Dev., ED-12, p. 167, 1965.

[6] F. Hofmann and W.H. Krautschneider, J. Appl. Phys., vol. 65, p. 1358, 1989.

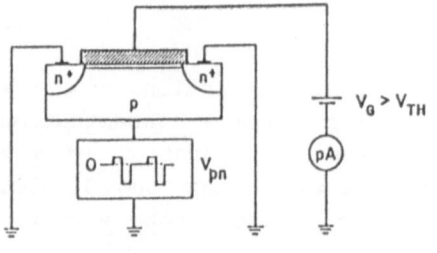

Fig. 1 Basic setup for double pulsed injection technique.

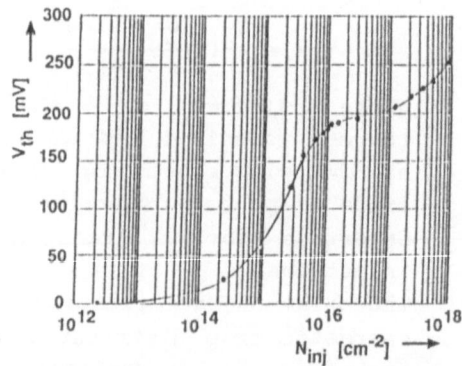

Fig. 2 Threshold voltage shift of a MOS transistor versus number of injected electrons. Channel length: 1.2μm, channel width: 10μm, oxide thickness: 20nm. Dots: Experimental values. Solid line: Model curve. Density of interface traps: $2 \cdot 10^{11} cm^{-2}$, capture cross section: $4 \cdot 10^{-16} cm^2$. Density of bulk traps: $1.5 \cdot 10^{11} cm^{-2}$, capture cross section: $2 \cdot 10^{-18} cm^2$.

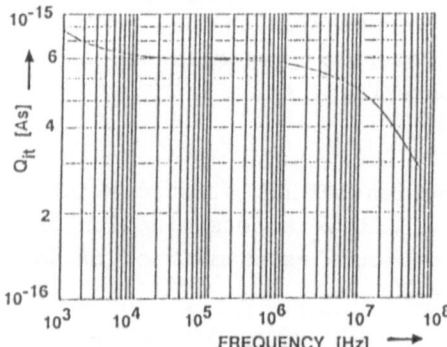

Fig. 3 Pumped charge per cycle of a MOS transistor as a function of pumping frequency. Channel length: 0.9μm, channel width: 10μm, oxide thickness: 20nm.

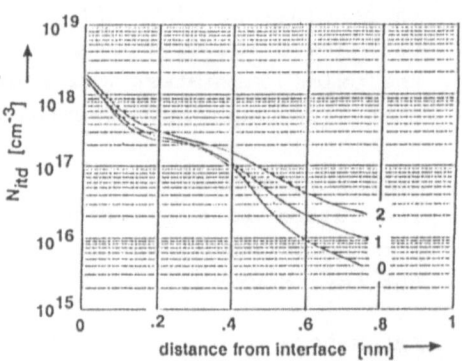

Fig. 4 Trap concentration within a MOS transistor gate oxide versus the distance from the Si-SiO₂ interface. Device data as in Fig. 3
0: fresh device
1: after hot electron injection of $3 \cdot 10^{17} cm^{-2}$.
2: after hot electron injection of $9 \cdot 10^{17} cm^{-2}$.

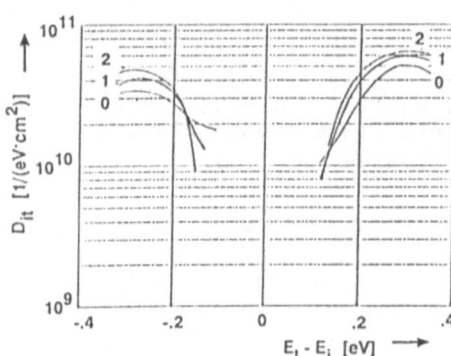

Fig. 5 Trap distribution as a function of energetic position within the bandgap.
Device data and stress parameter as in Fig. 4.

The Influence of an Externally Induced Lateral Electric Field on Bipolar Transistors

L. Deferm, C. Claeys and G. Declerck

IMEC, Kapeldreef 75, B-3030 Leuven, Belgium

1. Introduction

The presence of an electric field in the base region of a bipolar transistor is influencing the different current components and the minority carrier charge built-up in this region. This electric field can be caused by either an external applied voltage over the substrate in which the bipolar transistor is present (figure 1), or by the presence of parasitic current sources in CMOS circuits.

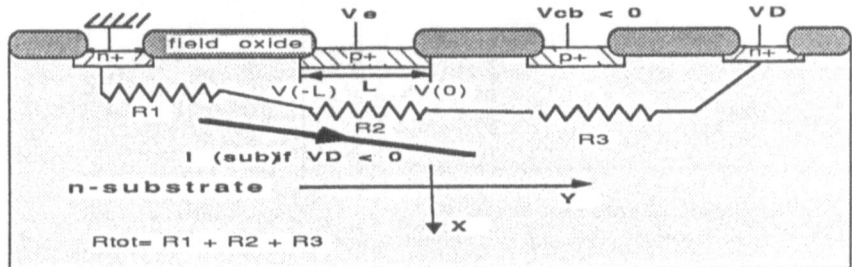

Figure 1: Cross-section of a lateral bipolar transistor

Especially in the latter case the change of the characteristics of the parasitic bipolar devices is certainly not intended. This last situation is occurring in a SCR structure, where the collector current of the vertical parasitic bipolar transistor is causing an electric field in the substrate, enhancing the beta of the lateral bipolar device and leading to an increased sensitivity towards latch-up (figure 2). Additionally the collector current of the lateral transistor will also create a lateral field in the base region of the vertical bipolar transistor. In order to explain the influence of the electric field on the bipolar behaviour, first the fundamental equations are solved, then these equations will be used for investigating the limitations regarding the improvement of lateral bipolar transistors and finally the influence on the latch-up behaviour will be studied.

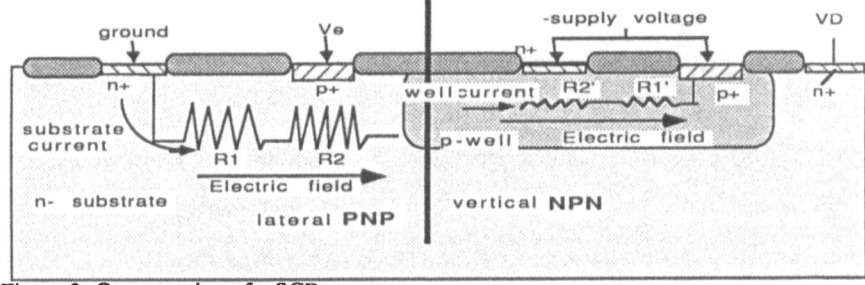

Figure 2: Cross-section of a SCR structure.

2. Derivation of the equations.

For calculating the influence of the electric field(E_{ad}), the continuity equation implementing both the external electric field and the electric field resulting from the majority carrier built-up is solved. The resulting equations show an increase of both the injected current and the collector current (figure 3) for higher external induced electric fields. The increase in collector current can very well be explained by the increase of the slope of the minority carrier distribution at the collector edge as shown in figure 4.

In order to correlate the induced electric field and the base-emitter voltage with the externally applied voltage over the substrate, the potential drops in the substrate, presented as lumped current dependent resistors in figure 1, must be taken into account. At low level injection these resistances can be considered to be constant, but at high level injection the majority carrier concentration becomes larger than the impurity concentration (N_B), resulting in a resistance decrease for both the intrinsic and extrinsic base. This decrease leads to a reduction of the externally induced electric field as can be seen in figure 5.

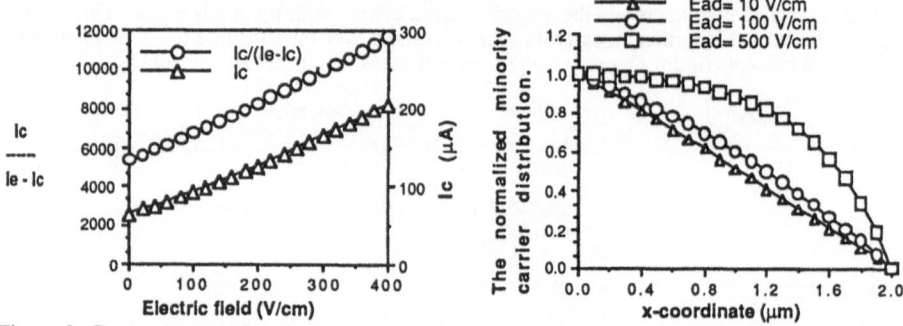

Figure 3: Current versus electric field   Figure 4: Minority carrier distribution versus field.

A second important phenomenon for investigating the influence of the lateral electric field on the bipolar characteristics is the reduction of the effective emitter area for vertical injection due to the presence of the electric field under the emitter as shown in figure 6.

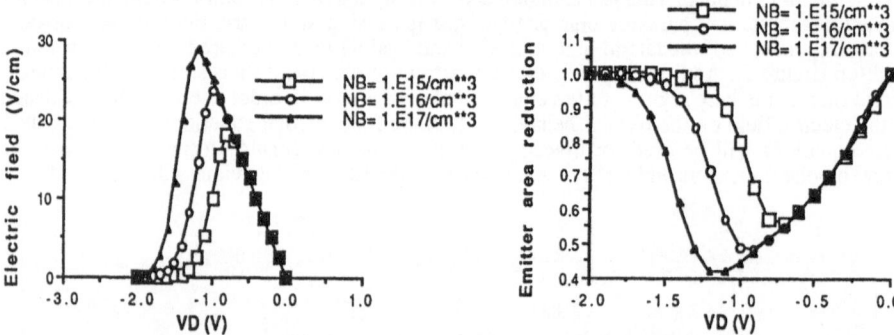

Figure 5: The behaviour of the induced electric field.   Figure 6: The emitter area reduction factor

The relationship between the external voltage, necessary for creating the electric field, and the area reduction is calculated using the next equations derived from the continuity equation and the relationship between resistances and the majority carrier distribution [1]:

$$\text{Area factor} = (-\frac{N_B}{2} + \frac{G(0)}{L} - \frac{G(-L)}{L})/(A(0) - \frac{N_B}{2})$$

$$G(y) = \frac{2V_t}{E_{ad}}(A(y) + \frac{N_B}{4}\ln(\frac{A(y) - \frac{N_B}{2}}{A(y) + \frac{N_B}{2}}))$$

$$A(y) = \sqrt{(\frac{N_B}{2})^2 + n_i^2 \exp((V_{be\ ref} + E_{ad}\ y)/V_t)}$$

The $V_{be\ ref}$ is the base-emitter voltage at the intrinsic base side and V_t is the thermal voltage.

For increasing electric fields the effective area decreases in a similar way as calculated by Estreich [2]. However due to the occurrence of high level injection both the electric field and the area reduction will disappear as shown in figures 5 and 6.

3. The influence of the electric field on the bipolar behaviour.

The most important current components in lateral bipolar transistors are the collector current, the current injected into the emitter and the bulk recombination current. The bulk recombination current, which is strongly dependent on the bulk parameters, and which is the dominant base current component in low level injection [3], must be reduced in order to increase the current amplification factor. One possibility is to implement a vertical electric field opposing the carrier injection, which can be done if epi is used on a highly doped substrate. Additionally the amount of injection current can be reduced if the effective area for vertical injection is reduced, which can be accomplished if a lateral electric field is generated. This lateral electric field will cause an increase of the current amplification factor as shown in figure 7.

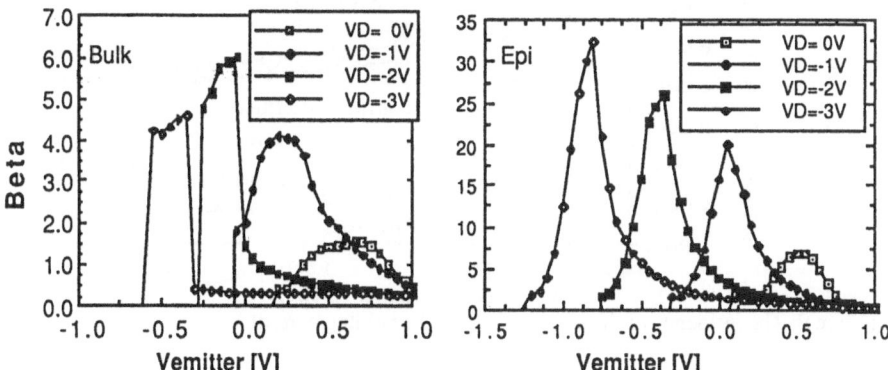

Figure 7: The beta behaviour for epi and bulk wafers in the presence of an electric field.

However due to the occurrence of high level injection the beta improvement is limited as shown in figure 8.

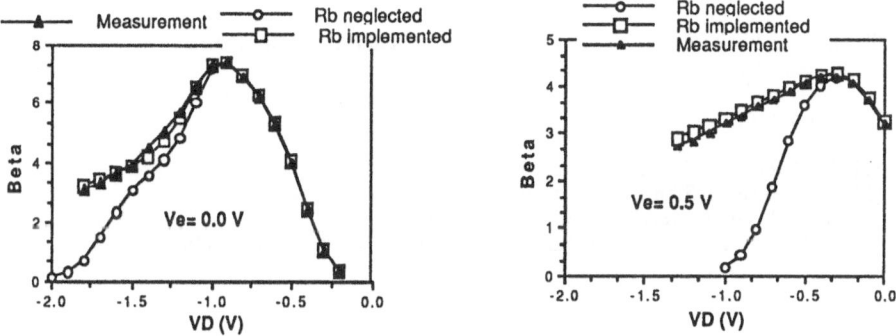

Figure 8: The influence of the external voltage on the beta behaviour.

698

The influence of the external voltage on the current amplification factor can be very well calculated using the previous equations extended with the base resistance influence.

The explained limitation of the beta improvement in the presence of an electric field due to the occurrence of high injection limits the applicability for active bipolar transistors. However, as mentioned before, the parasitic bipolar transistors present in every CMOS technology are also influenced by a lateral electric field and in this case the increase of the beta leads to an increased latch-up sensitivity.

4. Latch-up and the lateral electric field.

In the case of a parasitic thyristor, which causes latch-up when fired and which can be approximated by a two transistor model, the collector current of one type of bipolar transistor causes a lateral electric field in the base regions of the other bipolar transistor as shown before in figure 2. In the substrate this leads to an increase of the current amplification factor of the parasitic lateral transistor as explained in the previous section and results in an increased sensitivity to latch-up (figure 9).

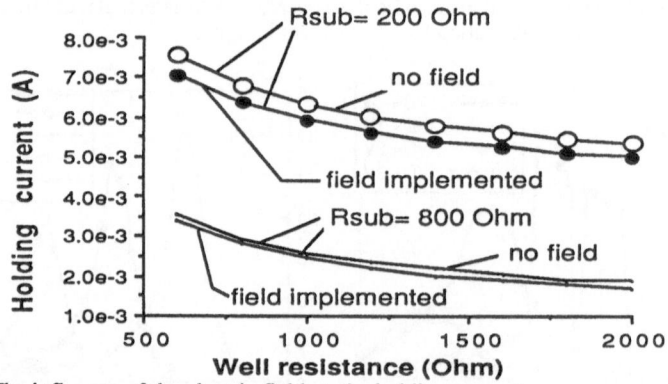

Figure 9: The influence of the electric field on the holding current.

5. Conclusions

The presence of a lateral field increases the beta until high level injection occurs, what limits the applicability for active bipolar devices.
Additionally the latch-up sensitivity increases for higher lateral electric fields.

6. Acknowledgement

The authors like to thank M. Van Dievel and G. Vanhorebeek for their assistance during the measurements.

7. References

[1] L. Deferm, PhD. Dissertation, May 1989
[2] D.B. Estreich, Stanford Electronics Laboratories, Technical report
     no G-201-9, November 1980
[3] L. Deferm et al., Solid St. Electron, Vol. 32, no. 2, pp. 103, 1989.

Current Crowding in Nested and Self-aligned Micrometric Contacts

M. DELLAGIOVANNA, G. DE SANTI and N. CIRCELLI

SGS-Thomson Microelectronics R.& D. Central Dep., 20041 Agrate Brianza (Mi) - Italy

A. SCORZONI

CNR - Istituto LAMEL, via Castagnoli 1, 40126, Bologna - Italy

Abstract

A comparison between interfacial contact resistance (R_i) and End Resistance (R_e) on n^+ nested and self-aligned contacts metallized with two different schemes: Al-1%Si and Ti/TiN/Al-Si is presented. The quite different values of R_i and R_e on the contacts with Ti/TiN/Al-Si have been explained according to the considerable current crowding phenomenon which occurs when the tranfer length $"l_t"$ of the contact (defined by the transmission line model) is much shorter than the contact linear dimension.

Introduction

Self-aligned contacts (SAC, see fig.1a) are designed larger than the active area and are widely used as cell drain contacts in VLSI memories because they provide a denser device integration. The shorting of the active area junction is avoided by an ion implantation doping performed after the contact opening. Nested contacts, instead, are surrounded with an active area square head (fig.1b) that covers the photolitographic misalignment tolerances. The progressive shrinkage of the geometries and contacts has suggested to modify the metallurgical scheme using Al-Si with a Ti/TiN barrier layer. This solution provides a lower contact resistance but a current crowding phenomenon is experimentally detectable.

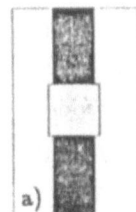 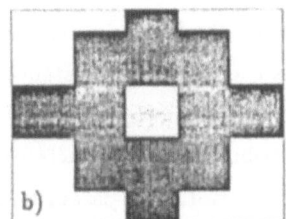

Figure 1 : a) self-aligned contact, b) nested contact

Experimental

Contacts of $1 \times 1 \mu m^2$ were used. The active area head for nested contacts was $3 \times 3 \mu m^2$. For both types of contacts (nested and self-aligned) the active area was doped with an Arsenic implant dose of 5×10^{15} cm^{-2} at $60 keV$. After the RIE contacts opening, a Phosphoros implant dose of 1×10^{15} cm^{-2} at $80 keV$ was performed. Two kinds of metallurgical schemes were used: Al-1%Si and Ti/TiN/Al-1%Si. Metal layers were obtained by sputtering (TiN was reactively sputtered).

The six terminal CKR (Cross Kelvin Resistor)[1] (fig.2a) was used for interfacial resistance (R_i) measurements on n^+ nested contacts; the terminal n°1 and 4 were employed for current injection ($0.5 mA$) while the terminals n°2 and 5 were used for voltage drop measurement. If the voltage drop is taken between the terminals n°3 and 5 the structure is used as CER (Contact End Resistor)[2] and the R_e measurement is obtained. According to the nature of self-aligned contacts, only the CER structure is feasible (fig. 2b) and therefore only the R_e measurement is possible. The End Resistance of self-aligned contacts will be indicated as R_{e-sa}.

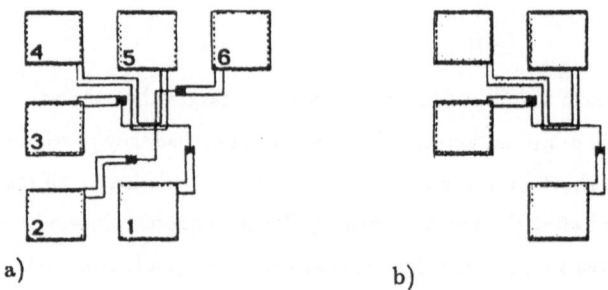

Figure 2 : a) Cross Kelvin Resistor, b) Contact End Resistor

In fig.3 the R_i, R_e and R_{e-sa} distributions for contacts with Al-1%Si metallization are presented. All of them show both a "low-resistance" contribution centered at 100 Ω and a considerable "high-resistance" contribution around 2 $k\Omega$. Tab.1 collects the mean values $\overline{R}$ and dispersions σ of the "low-resistance" contribution ($R < 500\Omega$). The three distributions are quite similar, both in shape and in statistical parameters.

R_i, R_e and R_{e-sa} distributions for contacts with Ti/TiN/Al- 1%Si metallization are reported in fig.4 and their statistical parameters in Tab.1. With this kind of metallization the "high-resistance" component disappears and $\overline{R_i}$ is much lower than that one detectable on contacts without barrier. Besides that, $\overline{R_e}$ is about 50% of $\overline{R_i}$ and $\overline{R_{e-sa}}$ is one order of magnitude lower than $\overline{R_i}$.

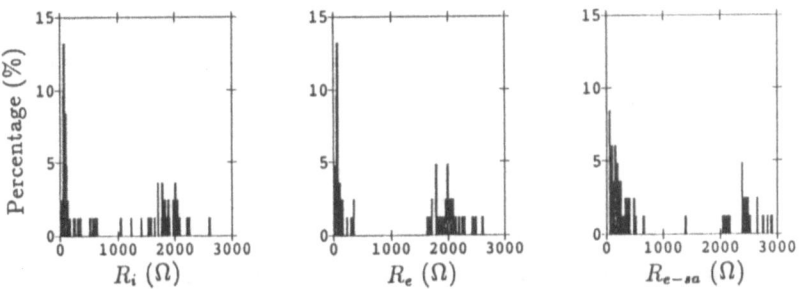

Figure 3 : R_i, R_e and R_{e-sa} distributions with Al-1%Si metallization

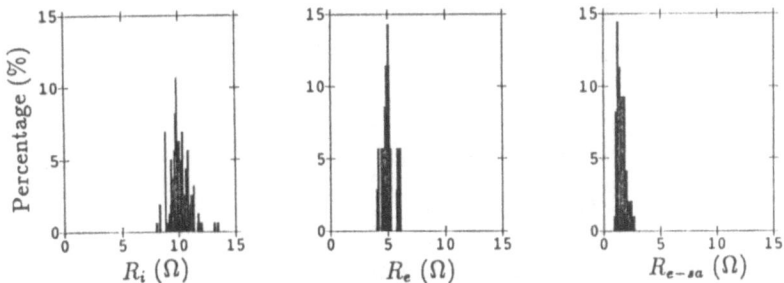

Figure 4 : R_i, R_e and R_{e-sa} distributions with Ti/TiN/Al-1%Si metallization

	$\overline{R_i}$ (Ω)	σ_i (%)	$\overline{R_e}$ (Ω)	σ_e (%)	$\overline{R_{e-sa}}$ (Ω)	σ_{e-sa} (%)
Al-1%Si	100	72	98	82	189	63
Ti/TiN/Al-1%Si	10	7	5	9.8	1.5	24

Tab.1: statistical parameters for the above distributions

Discussion

The lower R_i values for contacts with barrier are due to Ti/Si interface, while, the complete lack of a high-resistance component in their statistical distribution, can be mainly justified by the absence of preferential silicon segregation inside the contacts themselves as it was indicated by SEM observations.

The low $\overline{R_e}$ and $\overline{R_{e-sa}}$ compared with $\overline{R_i}$, observed only for the contacts with a barrier layer, can be explained invoking the current crowding phenomenon.

For present discussion we will refer to the contact one-dimensional transmission line model TLM [3] according to which current crowding can occur only if the following condition is satisfied:

$$\sqrt{\frac{\rho_c}{R_{sk}}} \ll d \tag{1}$$

where ρ_c is the contact resistivity, R_{sk} the active area sheet resistance under the contact and d the contact linear dimension. The quantity $\ell_t = \sqrt{\rho_c/R_{sk}}$ is the contact transfer length. In Tab.2 ρ_c and ℓ_t values are reported as calculated using a computer model based on a pseudo-three-dimensional resistor network that takes into account the current crowding effects [4,5]. In the calculation R_i data from Tab.1 were used and R_{sk} was assumed equal to the sheet resistance of the active area outside the contact ($45\Omega/\square$).

	ρ_c (Ωcm^2)	ℓ_t (μm)
Al-1%Si	9.5×10^{-7}	1.46
Ti/TiN/Al-1%Si	5.4×10^{-8}	0.34

Tab.2: calculated ρ_c and ℓ_t

From these results current crowding should affect only the contacts with barrier metallization. This prediction is experimentally confirmed by the R_e data compared with the R_i ones. R_e, defined by the voltage drop at the end edge of the contact respect to the total current flowing through the contact, is expected to be comparable with R_i if the current, coming from the active area arm, flows uniformly through the contact area up to the metal line; actually, this is the case of the contacts without Ti/TiN. On the contrary, whenever the current enters the metal layer preferentially at the leading edge of the contact (i.e. a current crowding takes place) an R_e much lower than R_i must be expected. This electrical behaviour is actually observed in contacts with a Ti/TiN barrier layer.

Conclusion

$1\times1\mu m^2$ contacts metallized with Al-1%Si and Ti/TiN/Al-1%Si have been compared as far as Interfacial and End Resistances are concerned. The contacts with Ti/TiN have shown a very low ρ_c which meets VLSI requirements. However, starting from the evaluation of contact transfer length with respect to contact linear dimension a remarkable current crowding must be expected on these contacts rather than on the others. This foresight has been experimentally confirmed by a comparison of Interfacial Contact Resistances and End Resistances on both nested and self-aligned micrometric contacts.

References

[1] A.Scorzoni and M.Finetti, *Metal/Semiconductor contact resistivity and its determination from contact resistance measurements*, Material Science Reports, vol.3 n.2, pp.123.

[2] ref.[1], pp.120.

[3] H.H.Berger, Solid State Electron.15 (1972) 148.

[4] A.Scorzoni, M.Finetti, K.Grahn, I.Suni,and P. Cappelletti, IEEE Trans. Electron Devices, ED-34 (1987) 525.

[5] P.Cappelletti, M. Finetti, A.Scorzoni, I.Suni, N.Circelli, and G. Dalla Libera, IEEE Trans. Electron devices, ED-34 (1987) 532.

Hot Carriers

Transient Behaviour of UV-induced Interface States in Au/SiO$_2$/n-Si Tunnel Structures

P.E. Bagnoli

Istituto di Elettronica e Telecomunicazioni, Universita' di Pisa,
Via Diotisalvi 2, 56100 Pisa, Italy

Summary

The generation and subsequent spontaneous decay of interface states in semi-transparent metal / tunnel oxide / silicon diodes after exposure to UV irradiation were studied by means of analysis of capacitance-voltage characteristics. The densities of the two groups of interface states, in communication with metal or semiconductor, were evaluated as a function of time using a recently proposed method which makes use of the slope and the voltage axis intercept of the $1/C^2$ curve.

Introduction

It is well known that ionizing radiations (X rays, UV photons) provoke changes in the interface state density of silicon based MOS devices [1]. In this communication, the effects of UV radiations on the interface states of Au / tunnel SiO$_2$ / n-Si structures are reported. The phenomenon was studied by means of high frequency capacitance-voltage measurements.

The $1/C^2$-V plot of an MIS tunnel device has been extensively examined in two recent papers [2,3] which take into account the effects of two groups of interface states whose occupation functions are governed by the metal Fermi level (E_{fm}) or the semiconductor Fermi level (E_{fs}) respectively. The main conclusions are that both the slope (P) and the voltage axis intercept (V_{int}) of the $1/C^2$-V plot differ from those obtained from an oxide-free Schottky diode with the same donor density. By assuming, as a first approximation, that both densities of state are uniformly distributed in energy, the following relationship holds at the thermal equilibrium:

$$\phi_M - \phi_S = V_{bo} + \frac{Q_{sc}(0)}{C_i} - \frac{1}{C_i} \left[(qD_{SS} + qD_{SM})(E_G - V_n - V_{bo} - \phi_o) \right] \tag{1}$$

where ϕ_M and ϕ_S are the metal and semiconductor work functions, V_n is the difference in energy between the semiconductor Fermi level and the bottom of the conduction band, E_G is the semiconductor energy gap, V_{bo} is the equilibrium silicon band bending, C_i is the oxide layer capacitance, D_{SM} and D_{SS} are the densities (cm^{-2} eV^{-1}) of the states in communication with metal and semiconductor respectively and Q_{sc} is the semiconductor space charge . In both groups ϕ_0 is the neutral level which separates the acceptor states (above) and the donor states (below). The term within the square brackets in (1) is the total charge stored in the interface states.

If the oxide thickness lies in the range 10 - 30 Å, the two parameters of the $1/C^2$ plot can be expressed by the following relationships [3] :

$$P = P_o \frac{C_i + q\, D_{SM}}{C_i + q\, D_{SS} + q\, D_{SM}} \quad , \quad V_{int} = \left(V_{bo} - \frac{KT}{q} \right) \frac{C_i + q\, D_{SS} + q\, D_{SM}}{C_i + q\, D_{SM}} \tag{2}$$

where KT/q is the thermal voltage and Po, which depends on donor density only, is the slope of the curve of the corresponding oxide-free Schottky contact. From (1) and (2) , the semiconductor band bending V_{bo} and the average values of the densities D_{SM} and D_{SS} can be obtained [3].

Experiment

The MIS devices were fabricated on phosphorus doped silicon by oxidizing the semiconductor surface at 400 °C for 30 minutes in a wet oxygen atmosphere. The oxide thickness is in the range 20-30 Å. Circular dots (diameter 1 mm) were obtained by depositing a semi-transparent (150 Å) gold film through a metal mask. MS junctions were also fabricated using the same silicon wafer and with anodes of the same thickness. A standard mercury lamp (6 W), principal emission wave length 2518 Å, was used for device irradiation. High frequency (1 MHz) capacitance and d.c. current measurements were performed every ten minutes before and after 30 minutes of irradiation for a period of several hours.

The ideality factor of the forward I-V characteristic of the MS diodes proved very close to unit (n = 1.02) confirming the absence of an interposed oxide layer; the true value of the donor density can therefore be obtained directly from the slope of their $1/C^2$-V curve ($N_d = 2.5\ 10^{15}$ cm^{-3}).

While the capacitance plots of the MS diodes do not change after irradiation, the curves of the MIS devices show large modifications both in slope and in voltage axis intercept. Fig. 1 shows the curves of an MIS diode measured at given times before and after irradiation after switching off the lamp.

In fig. 2 the kinetics of D_{SS} and D_{SM} are reported as a function of time. The values at each time were calculated using the slope and the voltage intercept measured on the corresponding $1/C^2$ curve. The following values were used: oxide thickness 25 Å, $\phi_0 = 0.23$ eV, $\phi_M = 4.7$ eV.

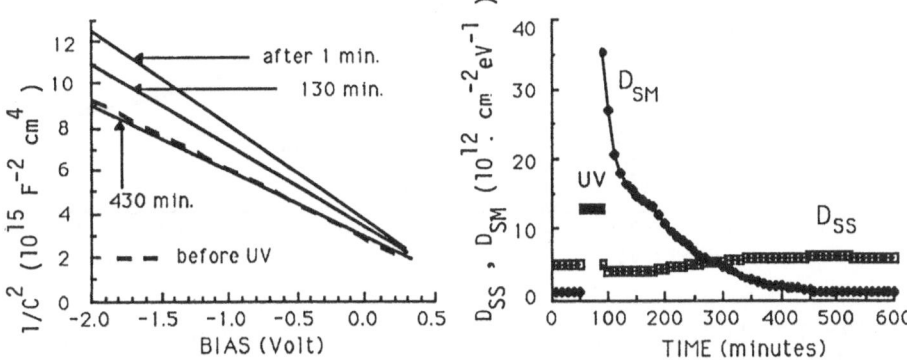

FIG. 1 : $1/C^2$ plot of an MIS diode before and after irradiation.

FIG. 2 : Evolution in time of the two interface state densities.

As can be seen, after the irradiation the states in communication with the metal undergo a large increase followed by a slow decay; the steady-state value was close to the initial one. The generation and spontaneous annealing of the interface states are also evident from the I-V characteristics shown in fig. 3. The tunnel current via interface states is in fact a large part of the total current across MIS junctions, specially under reverse bias. Therefore, the transient behavior of the reverse current in fig. 4 can be directly related to the kinetics of the densities of interface states.

Discussion

The different behavior of the two groups of states shown in fig. 2 can be explained as follows. Over the whole bias range explored, E_{fm} is always above the neutral level whereas the energy range spanned by E_{fs} is mainly located below ϕ_0. This implies that the values for D_{SM} calculated from (1) and (2) may refer to the acceptor-like states, whereas the values obtained for D_{SS} refer mainly to the donor-like states. It could therefore be said that the UV photons generate acceptor states located in energy above ϕ_0, whereas they have a weak influence on donor states below ϕ_0. This statement is also supported by the fact

710

that the total equilibrium charge stored in the interface states is negative and shows the same kinetics as D_{SM} states before and after irradiation.

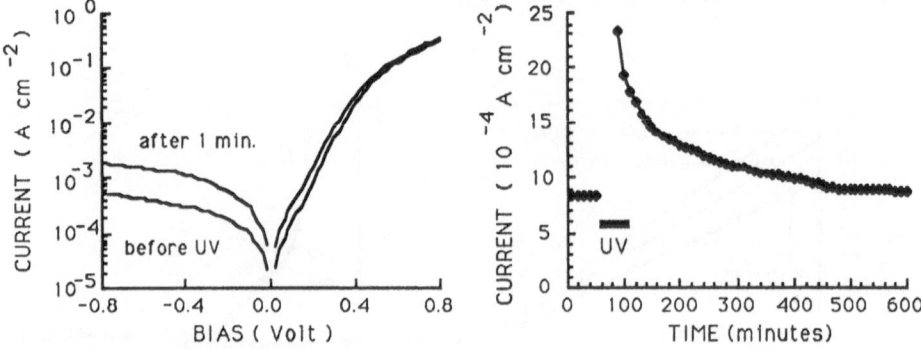

FIG. 3 : I-V characteristics of an MIS diode before and just after irradiation

FIG. 4 : Evolution in time of the diode current at -0.8 V of reverse bias.

The acceptor-like nature of the UV-induced traps in the Si/SiO$_2$ interface, deduced from the experiment, agrees with experimental and theoretical results concerning thick oxide MOS devices irradiated by UV photons [4]. The authors ascribe the generation of traps to the breaking of the Si-OH bonds producing interstitial hydrogen atoms. On the other hand, the slow decrease of the generated states observed in our devices is also consistent with a recapturing of free hydrogen atoms by the dangling bonds in the insulator [5].

References

[1]   E.H.Snow,A.S.Grove, D.J.Fitzgerald, *Proc. IEEE* **55**, 1168 (1967)

[2]   S.J.Fonash, *J. appl. Phys.* **54**, 1966 (1983)

[3]   P.E.Bagnoli, A.Nannini, *Solid-St. Electron.* **30**, 1005 (1987)

[4]   J.Kassabov, D.Dimitrov, A.Grueva, *Solid-St. Electron.* **31**, 49 (1988)

[5]   W.Wan-Li Lin, C.T. Sah, *Solid-St. Electron.* **31**, 1451 (1988)

Advanced Simulation for Reliability Optimization of Submicron LDD MOSFETs

M. Orlowski, C. Mazuré, A. Lill, H.-M. Mühlhoff, W. Hänsch,
A. v. Schwerin and F. Neppl

Siemens AG, Corporate Research and Development,
Otto-Hahn-Ring 6, 8000 München 83, FRG

Abstract

We present an advanced technique for optimization of source/drain structures of submicron MOSFETs with respect to hot carrier degradation. This technique is based on the simulation and analysis of the distribution of hot electrones and holes injected into the gate oxide obtained from a consistent, temperature-dependent gate and substrate current model. This method allows to judge the reliability performance of S/D structures for which the analysis of the electric fields *fails* to provide the necessary insight. The applicability of the method is demonstrated by comparison with experimental data from subμm-logic and 16M DRAM MOSFETs.

1 Introduction

Despite advanced simulation capabilities and increased insight into degradation mechanisms there are still no safe guidelines to achieve an optimal LDD S/D structure for given technology and transistor requirements with respect to reliability performance. Since degradation by hot carriers is closely related to the existence of electric field peaks at or near the gate edge, it was customary hitherto to guide the optimization effort by an analysis of the lateral and vertical field distributions for given channel and S/D profiles [1,2]. While this procedure provides the first clues as to which S/D structure is more preferable than other, it is too crude to allow for refined optimization. An increase of the lateral field maximum or of substrate current does not necessarily turn out in a lifetime reduction. Consideration of the magnitude and finite depth of the overlap between the current path and the electric field are important as well. Therefore it is necessary to evaluate the magnitude and position of injection of both carrier types that are able to surmount SiO_2 barrier and eventually to create the detrimental oxide damage.

2 Results and Discussion

In Ref.[3] an analytical expression for the distribution function for high energetic carriers in an arbitrary 2D or 3D electric field utilizing Boltzmann's equation has been derived. The distribution functions is of Maxwell-Boltzmann type if the system is in thermal equilibrium. We apply this new model, which has been implemented in the device simulator MINIMOS 4 [4], to calculate the substrate and gate current in MOSFET LDD-transistors. We first demonstrate that the simulation can explain experimental data for which substrate current is not a sufficient criterion for device lifetime τ. Stress measurements at three different operating points but for equal substrate current as shown in Fig.1 were carried out. The simulation describes correctly the substrate current dependence on V_G and V_{DS}. As can be seen from Table I the lifetime for the three operation points does not correlate with the maxima of the simulated lateral fields, see case 2 versus case 3. Consider now the corresponding injected hole and electron distribution at the $Si - SiO_2$ interface shown in Fig.2. Although case 3 exhibits the largest hot electron injection, it has the longest lifetime. Conversely, case 1 exhibits the smallest electron injection and has the shortest lifetime. Consequently the oxide damage cannot be determined by electron injection alone. Rather, simulation suggests that τ correlates with the overlap between the distributions of injected holes and electrons. This finding is in surprisingly good agreement with the two-step-degradation model by Hoffman et al [5] in which holes play the initial and key role in the final formation of interface states. In Fig.3 and 4 the experimental data for I_{sub}/I_{DS} and τ are shown for three otherwise identic transistors with $L_G = 1\mu m$ and $Tox = 20nm$ for phosphorus implantation with A): $4e13cm^{-2}$, 60 keV; B): $3E13cm^{-2}$, 60 keV; C): $3E13cm^{-2}$, 80 keV. Again, transistor with the highest I_{sub}/I_{DS} exhibits the lowest lifetime, and the transistor with lowest I_{sub}/I_{DS} exhibits the shortest lifetime. The experimental data are analyzed in terms of simulated quantities displayed in Table II. One observes that simulation correctly describes the relation between I_{sub}/I_{DS}-curves. From the maximum concentration of injected carriers in SiO_2 it is readily explained why case B has the shortest lifetime. However, at first glance simulation suggests that A should degrade stronger than C, in contradiction to experiment. However, it is now crucial to consider the lateral position of the injection maximum which is for both carrier types the same and is indicated in the last row of Table II. In contrast to the case A the injection of hot carriers for the cases B and C takes place (just) outside of the gate electrode, i.e. into spacer oxide. Because of the lower quality of the spacer oxide than the gate oxide smaller hot carrier injection causes larger oxide damage than injection underneath the gate causing, in turn, stronger degradation. These examples demonstrate that the advanced simulation can effectively guide and accelerate design efforts for optimum LDD S/D structures.

Applying the insights gained above to transistors with extremely shallow S/D regions (ESSD) having a junction depth χ_j smaller than $0.1\mu m$ we find a 5 to 6 orders of magnitude higher level of electron and hole injection into gate oxide, see Table III, than for conventional LDD MOSFET. Even at reduced drain voltage of 3V the electron injection level of the ESSD MOSFET is still a factor of 100 higher than that of the conventional LDD MOSFET at $V_{DS} = 5V$ although the substrate current and maximum lateral field of the ESSD are lower than those of conventional LDD MOSFET. We conclude that, in contrast to the usual criteria a successfull reliability optimization must include: 1) distributions of injected hot carriers (both types are important), 2) kinetics of traps, fixed charges and interface states determined by the injected species, oxide quality, and field distribution and 3) feedback of the damage on the transistor performance.

References

[1] M. Orlowski et al, VLSI TECH. SYMP. (1987), p.57

[2] K. Mayaram et al, IEEE ED-34 (1987), p.1509

[3] W. Hänsch and A. v. Schwerin, J. Appl. Phys. 1989, in print

[4] W. Hänsch and S. Selberherr, IEEE ED-34 (1987), p.1087

[5] K.R. Hoffmann et al, IEEE ED-32 (1985), p.691

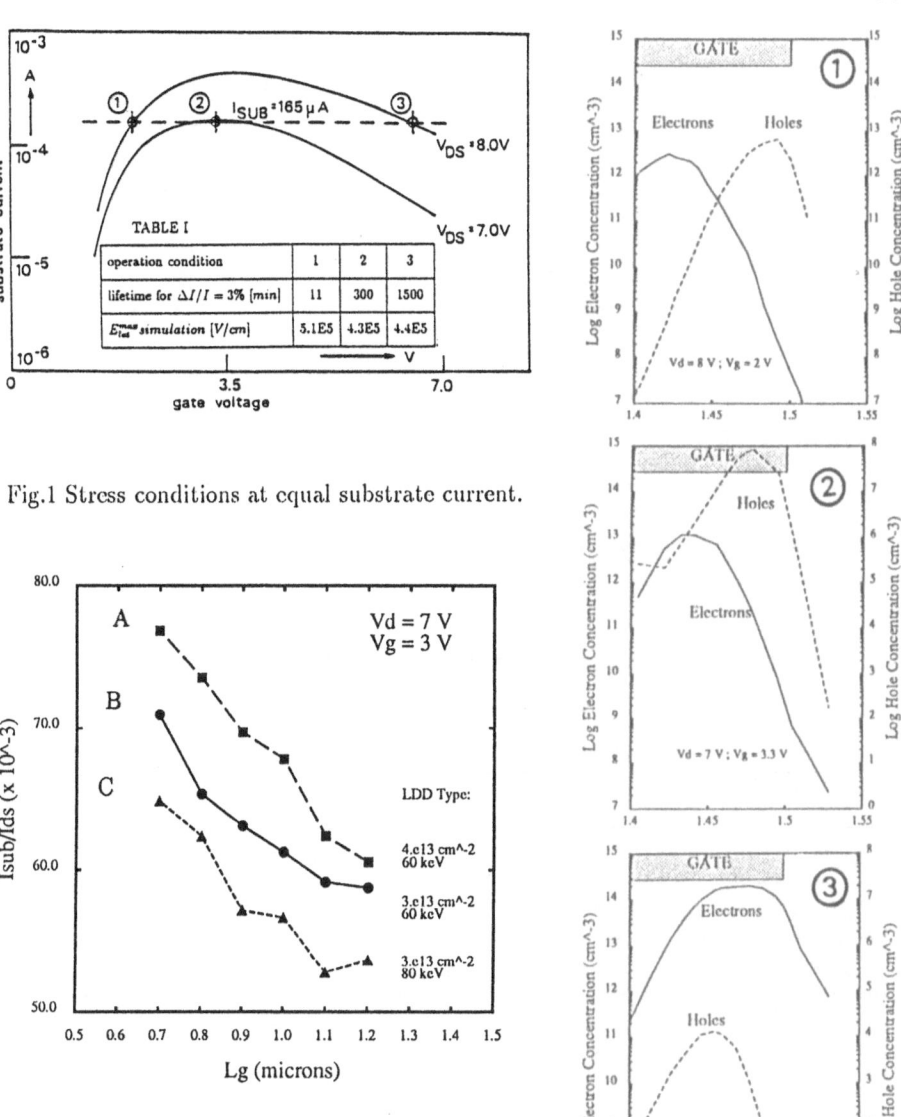

Fig.1 Stress conditions at equal substrate current.

Fig.3 I_{sub}/I_{DS} experimental curves for three LDD transistors A, B, and C.

Fig.2 Simulated distribution of injected hot holes and electron at $Si - SiO_2$ interface for the stress conditions shown in Fig.1 Note that the scale for hot holes for case 1 is different than in case 2 and 3.

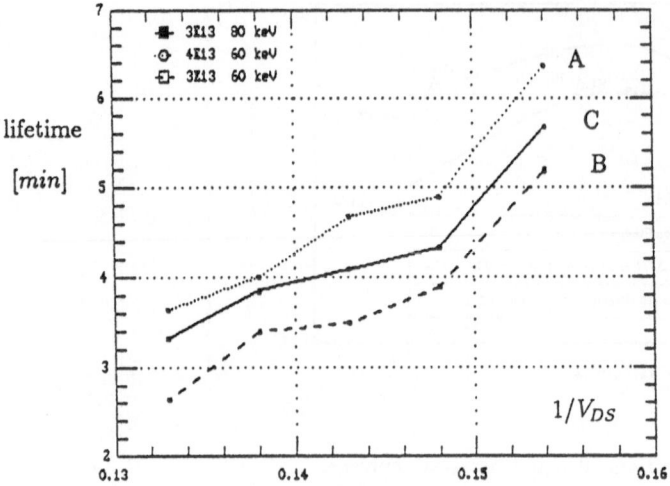

Fig.4 Lifetime τ in [min] for transistors A, B, and C as a function of the inverse drain voltage.

TABLE II: Simulation $V_{DS} = 6V$, $V_G = 3V$, $L_G = 1\mu m$, $Tox = 20nm$

LDD Type	A $4E13cm^{-2}, 60keV$	B $3E13cm^{-2}, 60keV$	C $3E13cm^{-2}, 80keV$
$E_{lat}^{max}[V/cm]$	3.6E5	3.5E5	3.4E5
$I_{sub}/W[A]$	4.2E-6	3.2E-6	1.8E-6
$C_{elec}^{max}[cm^{-3}]$	2.4E6	2.7E7	2.4E6
$C_{hole}^{max}[cm^{-3}]$	1.8E9	1.4E11	5.7E8
pos. of C^{max} underneath gate?	yes	no	no

TABLE III

	conv. n-LDD $V_D = 5V, V_G = 2.5V$	ESSD $V_D = 5V, V_G = 2.5V$	ESSD $V_D = 3V, V_G = 1.5V$
$I_{sub}[A]$	1.8E-6	1.9E-5	2.9E-7
$E_{lat}^{max}[V/cm]$	3.9E5	5.9E5	3.9E5
$C_{elec}^{max}[cm^{-3}]$	4.4E11	2.5E16	2.6E13
$C_{hole}^{max}[cm^{-3}]$	1.0E3	7.1E12	1.1E4

Annealing of Hot Carrier Damaged Double Metal MOSFET

R.Annunziata, G.Dalla Libera, E.Ghio, A.Maggis

S.T. Microelectronics, Central R & D, Via C.Olivetti 2
20041 Agrate Brianza (MI) – Italy

Abstract *The annealing of damage induced by static hot carrier aging on double metal MOS devices was investigated. To get more insight about the nature of hot carrier damage a new experimental approach is proposed: devices were subjected to cycles of hot carrier stress, followed by annealing at fixed temperature. The activation energy for the bake recovery of parameters was evaluated. The method allows to distinguish between permanent and recoverable hot carrier effects. The recovery kinetics suggests that the prevailing degradation source is electron trapping into shallow interface states induced by double metal processing.*

1 Introduction

One of the key problems in hot carrier aging of MOS devices is to understand the nature of the damage caused by hot carrier injection, in fact it is not easy to separate the phenomena causing device degradation (bulk oxide trapping, interface state trapping, mobility effects) basing upon electrical measurements alone [1] .

The main purpose of this presentation is to raise the following question: is hot carrier induced degradation a reversible phenomenon or does it really involve a permanent modification of device properties ?

In order to find an answer, two new experiments are proposed:

1. Cycles of static hot carrier stressing followed by annealing at fixed temperature.

2. Changes in transistor parameters at three different temperature were monitored to evaluate the recovery kinetics of hot carrier degraded MOSFET.

The method was applied to MOSFET fabricated with a two layer metal process, which are known to show much more degradation than single metal devices [2] .

2 Experimental

Double metal N–MOSFET with effective length of about 1.2 μm and gate oxide thickness of 330 Å, PSG capped and assembled in standard ceramic package were evaluated. Aging tests were carried out at room temperature under the following bias conditions:

$$V_{GS} = V_{DS}/2 \qquad V_{GS} = V_{DS} \qquad (V_{DS} = 7\,V)$$

Bake recovery experiments were performed at a temperature $T = 200$ °C for the test 1. (stress/bake cycle) and $T = 150, 200, 250$ °C for the test 2. (recovery kinetic).

Changes in device characteristics induced by both electrical stress and bake treatment were monitored by taking transfer curve in the linear zone ($V_{DS} = 100$ mV), and bulk current $I_B(V_{GS})$ at $V_{DS} = 3.5V$.

3 Results

1. Aging/bake cycle

The change in maximum transconductance as a function of stress time is reported in fig.1. Very similar aging behaviors are observed at both stress conditions, despite what is usually reported for single metal devices [3] .

The transfer curves taken at fixed time steps are shown in fig.2 in the case $V_{GS} = V_{DS}/2$..

A strong deformation in the transconductance curve is observed after prolonged stress .

The behavior of bulk current during stress is shown in fig.3.

The modification of transfer curves (fig.4) and bulk current (fig.5) during bake shows just the opposite behavior compared with the aging phase; the characteristics are fully recovered after a long bake time (6×10^5 sec.).

Aging/recovery cycles were iterated three times; the aging rate is slightly reduced after the first cycle and not further affected by the next one (fig.6).

2. Recovery kinetics

The change of maximum transconductance $vs.$ bake time at three different temperatures ($T = 150, 200, 250$ °C) for devices formerly stressed for 6×10^5 seconds at $V_{GS} = V_{DS}/2$, is reported in fig.7. Assuming a first order kinetics [4], data were fitted by an exponential law:

$$\frac{\Delta g_{mR}}{\Delta g_{mA}} = (1 - exp-\frac{t}{\tau(T)})^\alpha$$

where Δg_{mR} and Δg_{mA} are the change in maximum transconductance during bake and aging respectively, and α is a fitting parameter.

From the Arrhenius plot of calculated recovery time constant $\tau(T)$, we estimated activation energy E_a in the range 0.6 – 0.8 eV (fig.8) for both aging conditions.

To be sure that baking actually anneals only the damage induced by hot carrier stressing, some units were baked for 6×10^5 seconds at $T = 200$ °C before aging. No difference in aging behavior was found with devices not subjected to bake treatment before stress.

4 Discussion and Conclusions

Stress/bake experiments evidenced that the hot carrier induced damage is almost totally recovered by bake treatment, even at low temperature.

The slight reduction in aging rate after the first cycle is related to some permanent interface modification induced by the first hot carrier aging.

The obtained activation energy values suggest that the main cause of degradation in this experiment is electron trapping into shallow acceptor type interface states.

It has been demonstrated from experiments on capacitors [5] and directly verified on aged MOSFET by charge pumping measurements [6], that a bake treatment in temperature range $150 - 250\,^{\circ}C$ can anneal hot carrier induced interface states.

This means that the recovery of degraded characteristics can be ascribed to both electron detrapping and interface state annealing. Our aging/recovery experiments showed very similar behavior at $V_{GS} = V_{DS}/2$ (maximum interface state generation) [7] and $V_{GS} = V_{DS}$ (maximum electron injection), indicating that in our samples the contribution of hot carrier generated interface traps is negligible. Thus we can conclude that the prevailing cause of degradation is electron trapping into interface states induced by double level metallization, and not annealed by the cold processing involved.

References

[1] H.Hansch, M.Orlowski, W.Weber, Proc.ESSDERC p.596, 1988

[2] M.L. Chen, C.W. Leung, W.T. Cochran, S.Jain, H.P.W.Hey, H.Chew, C.Dziuba, IEDM Tech.Dig. p.55, 1987

[3] B.S. Doyle, M.Bourcerie, J.C.Marchetaux, A.Boudou, Proc.ESSDERC p.155, 1987

[4] R.Mahnkopf, G.Przyrembel, H.G.Wagemann, Proc. ESSDERC p.771, 1988

[5] M.V. Fischetti, R.Gastaldi, F.Maggioni, A.Modelli, J.Appl.Phys. 53, p.3136, 1982

[6] R. Annunziata, R.Benecchi, C.Bergonzoni, G.Dalla Libera, workshop "Physic of hot carrier injection", Munchen 1989

[7] P. Heremans, R.Bellens, G.Groeseneken, H.E.Maes, IEEE Trans. El. Dev. ED-35, p.2194, Dec. 1988

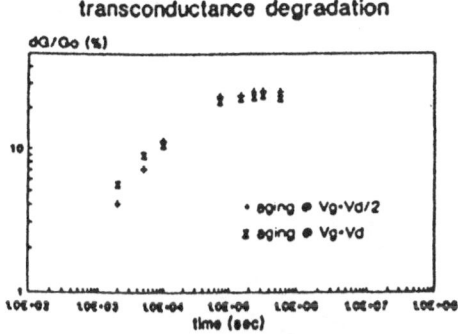

Fig.1 Comparison between transconductance degradation at different stress biases.

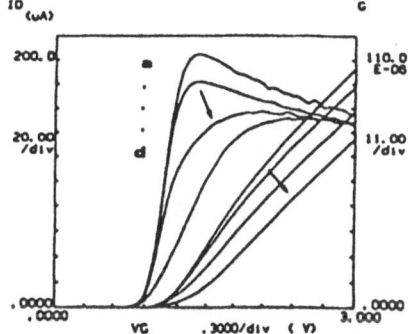

Fig.2 Effects of stress ($V_{GS} = V_{DS}/2$) on transfer curve. Stress time (sec): a. 0 b. 10^4 c. 7.5×10^4 d. 6×10^5

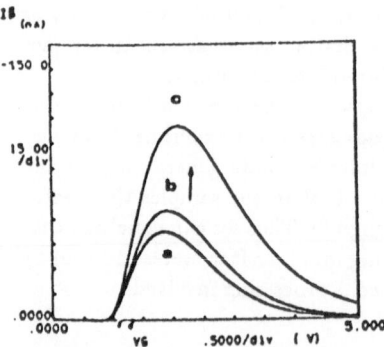

Fig.3 Effects of stress ($V_{GS} = V_{DS}/2$) on bulk current. Stress time (sec): a. 0 b. 6×10^3 c. 6×10^5

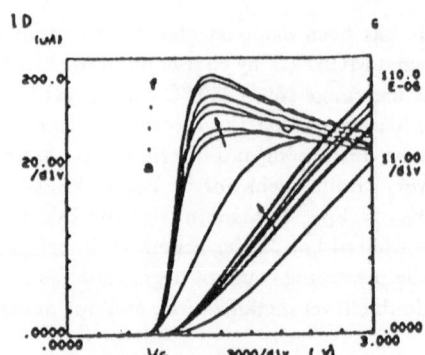

Fig.4 Bake induced recovery of transfer curve. Bake at $T = 200°C$, bake time (sec):
a. 600   b. 1200   c. 6000
d. 3×10^4 e. 6×10^4 f. 6×10^5

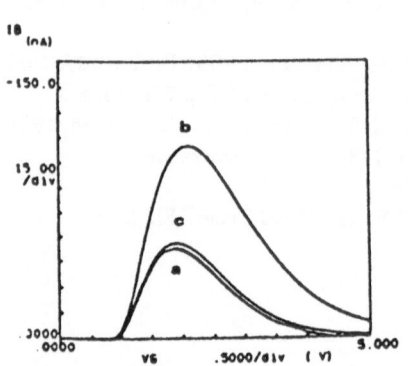

Fig.5 Bake induced recovery of bulk current. Bake at $T = 200°C$.
Device were stressed and baked for 6×10^5 sec.
a. virgin b. aged c. recovered

transconductance degradation

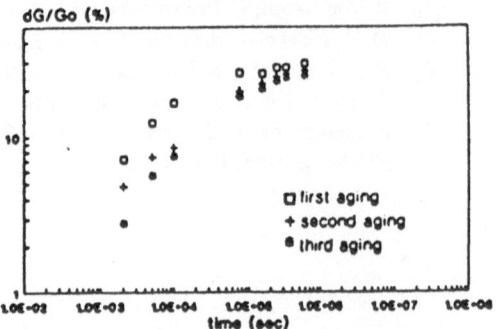

Fig.6 Comparison between transconductance degradation during the three stress cycles.

transconductance recovery

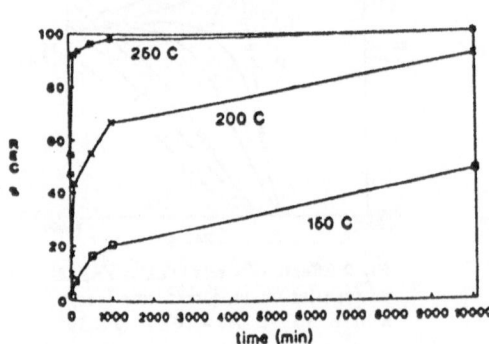

Fig.7 Change in maximum transconductance vs. bake time at $T = 150, 200, 250 °C$.

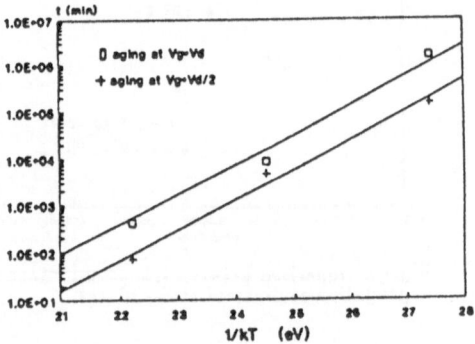

Fig.8 Arrhenius plot of bake recovery time constant $\tau(T)$ vs. bake temperature.

Hot-hole and Electron Effects in Dynamically Stressed n-MOSFETs

W. Weber and I. Borchert

Siemens, Central Research & Development, Microelectronics
Otto-Hahn-Ring 6, D 8000 München 83, FRG

Abstract
Dynamic hot-carrier stresses have been performed with frequency and stress time as parameters. The resulting data show a steeper time dependence than any single static stress. By frequency variations and combinations of static stresses it was found that no transient effects are responsible for it. The results can be explained qualitatively by the subsequent injection of holes and electrons in the dynamic case. This understanding is of great importance for the quality assurance of devices operating in real circuits.

Results

The devices used in this investigation were single n-MOSFET's with a gate length of 0.9 µm, an oxide thickness of 20nm, and a lightly doped drain (LDD) structure with n^-=$4x10^{13}cm^{-2}$ P-doping from a 4MBit DRAM product development.

In Fig. 1 the current change is plotted versus the stress time of two single transistors, one stressed statically under maximum substrate current conditions, the other stressed dynamically under inverter-like conditions (see insert in Fig.2). We observe that their time behaviors are different, manifesting in the fact that the dynamic degradation shows a somewhat stronger time dependence than the static degradation [1].

Can this result be explained by a duty cycle effect ? The assumption of duty cycle behavior means that for equal numbers of injected carriers, equal degradations result no matter what current of injected carriers is chosen. This leads to a shift of the degradation curve on the logarithmic time scale (as an example, see the 8V and 9V curves in Fig. 1) for different injection currents but not to a deformation. The results cannot therefore be explained by a duty cycle effect.

Another idea plausible at first sight is that of transient effects in the dynamic degradation, which are frequently reported in the literature (see e.g. [2-4]). To check on this, we performed frequency-dependent measurements under constant

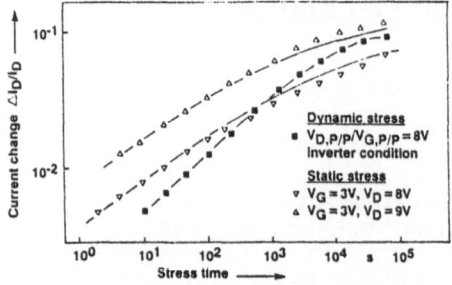

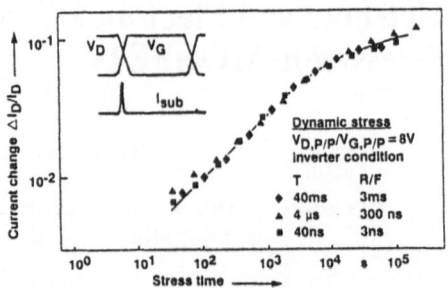

Fig. 1 Dynamic degradation shows a stronger stress time dependence than static degradation.

Fig.2 The strong dynamic stress time dependence is not a function of frequency.

duty cycles (Fig. 2, the time scale has been adjusted somewhat, using a substrate current criterion in order to compensate for fluctuations in the device channel lengths) showing that even for pulse periods as long as 40ms the same time dependence appears. The effect seems clearly not to originate from a transient phenomenon as no frequency dependent effect occurs.

The next step in elucidating this problem is to compare the static and the dynamic stressing *conditions*. The gate voltage of 3V used in the static experiments is possibly not the relevant one to compare with dynamic results where other (gate and drain) voltage conditions could produce more significant degradation contributions. We thus performed static stress experiments with V_G=1V, V_D=8V, but obtained the same weak time dependence as for the V_G=3V experiment (Fig. 3).

Now the question arises whether a static stress with a *single fixed* stressing condition can be directly correlated with the inverter-type dynamic condition which contains stresses at different gate and drain voltages (see insert in Fig. 2) corresponding to variations of the injection of holes and electrons. Thus we investigated quasi-static stresses with two different gate voltages applied alternately on a 2s time base. The result is also shown in Fig. 3 and shows a clear coincidence with the time dependence of the dynamic stress results.

Discussion

From the above results we can conclude that the differences in the time dependences after static and dynamic stress can be explained by a sequence of different static stressing conditions proving the absence of transient degradation effects. We must, however, clearly state that this result is

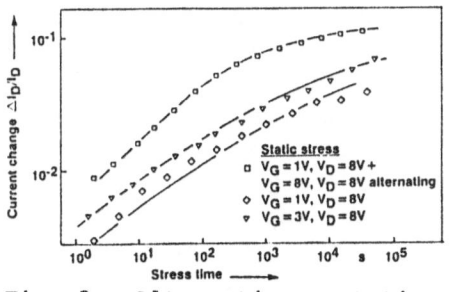

Fig. 3 Alternating static stress shows the strong stress time dependence originally found after dynamic stress.

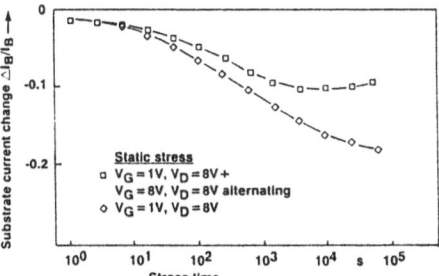

Fig. 4 The different time dependences can be explained with the aid of substrate current measurements.

not obtained for every technology. In some technologies, like conventional nitride passivated devices, we observe strong transient effects [5]! However, in the modern LDD-type devices from 1M-Bit and 4M-Bit DRAM product processes the above result is found to be reproducible.

Obviously the combination of two different static stressing conditions leads to a stronger time dependent degradation than any single one. How can this outcome be explained ? From the literature it is well known that at low gate voltages hole-effects are important, and combinations of hole- and electron effects are also discussed [6,7]. Thus the following appears to be a possible mechanism: at low gate voltages holes (and probably a small number of electrons) are injected into the oxide. Interface states and positive fixed charges are formed [7]. In the single static case with $V_G<V_D/2$ this process leads to a certain amount of net positive fixed charge after a certain stress time. The interface states (which are predominantly of acceptor-type for the positions accessible to the Fermi level [4]) are essentially uncharged at the stress conditions with $V_G<V_D/2$ due to a low position of the Fermi level beyond pinch-off. Thus the positive charges dominate, which causes a reduction of the electric field and thus of the injection of holes for increasing stress times. In the alternating stress experiment, injected electrons at the $V_G=8V$ condition can neutralize part of the fixed positive charges, thus leading to a smaller electric field reduction and to a smaller reduction of the injection of hot holes. This leads, in essence, to a stronger time dependence of the degradation in this stress case.

This model was checked by substrate current measurements (Fig. 4, the measurements are performed during stressing at

V_G=1V, V_D=8V) which show a weaker substrate current reduction after stress in the alternating stress experiment. Positive charges located at the drain are known to reduce the electric field in an n-MOSFET, and as the substrate current is a measure of the electric field, the results in Fig. 4 show that in both stress cases net positive charges are effective. However, in the alternating stress case the substrate current reduction is weaker because a certain proportion of the positive charges are neutralized by the hot-electron injection. This represents a strong indication for the validity of the model presented above.

Conclusion

In conclusion, further strong indications for combined electron-hole effects in n-channel MOSFET degradation are found. Furthermore, these results allow the definition of a new stress method to provide more realistic statements about the stability of a CMOS-process: by alternating dc-stresses, dynamic stressing conditions can be simulated experimentally. In combination with proper duty cycle calculations [5] the lifetimes of MOSFET's in a circuit can be specified.

References
[1] W. Weber, L. Risch, W. Krautschneider, and Q. Wang, "Hot-carrier degradation of CMOS-inverters", IEDM Tech. Dig., pp. 208-211, 1988.

[2] W. Weber, C. Werner, and G. Dorda, "Degradation of n-MOS-transistors after pulsed stress", IEEE EDL-5, pp. 518-520, 1984.

[3] J. Y. Choi, P. K. Ko, and C. Hu, "Cause of enhanced degradation under pulse stress", Proc. VLSI Symp. pp. 45-46, 1987.

[4] W. Weber, "Dynamic stress experiments for understanding degradation phenomena", IEEE TED-35, pp. 1476-1486, 1988.

[5] W. Weber, C. Werner, and A. v. Schwerin, "Lifetimes and substrate currents in static and dynamic hot-carrier degradation", IEDM Tech. Dig., pp. 390-393, 1986.

[6] K. R. Hofmann, C. Werner, W. Weber, and G. Dorda, "Hot-electron and hole emission effects in short n-channel MOSFET's", IEEE TED-32, pp. 691-699, 1986.

[7] P. Heremans, R. Bellens, G. Groeseneken, and H. E. Maes, "Consistent model for the hot carrier degradation in n-channel and p-channel MOSFET's", IEEE TED-35, pp. 2194-2209, 1988.

Characterization of the Hot Carrier Related MOS Parameters Using Negative Feedback Circuits in VLSI CMOS

D.Takacs, M.Steger

Siemens AG, Corporate Research and Development, Microelectronics,
Otto Hahn Ring,6 8000 München 83, FRG

Abstract:

An external resistor in the collector circuit of the vertical bipolar transistor in CMOS was used to realize negative feedback into the MOST in the well. Using these feedback circuits a characterization technique for impact ionization related currents in VLSI CMOS was developed. The operating mode- and the operating point-dependent current gains and their influence on the measurable impact ionization current are discussed.

1. Introduction

Hot carrier effects in MOSTs are usually characterized by measuring the substrate currents. The measured substrate current, however, can differ from the total impact ionization current, if parasitic bipolar action takes place. The difference between these currents is a strong function of the current gain of the parasitic bipolar transistor. This effect has been described for short channel NMOS devices in the bulk [1], where only a lateral bipolar source-bulk-drain transistor exists.

In CMOS, an additional vertical bipolar transistor exists, which leads to an even more complicated bipolar action. The various base current components of the existing bipolar transistors reduce the measurable substrate current. The current reduction is a strong function of the bipolar current gains. The measurement of the current gains in normal NMOS or in CMOS· operating conditions, however, is difficult. Current gain values obtained from measurements in the active bipolar mode are not relevant, because the required changes of the external bias affect its magnitude. In this paper, we describe a characterization technique for the impact ionization related bipolar action in VLSI CMOS using negative feedback circuits. Experiments were carried out on samples fabricated in p-well based twin-tub CMOS processes with feature sizes down to 0.5 μm and gate oxide thickness of 20 nm.

2. Negative feedback circuits

The cross section of a p-well CMOS structure with the nMOS- and bipolar-transistors is shown in Fig.1a. The nMOST is operated in deep saturation, leading to an impact ionization current I_{ion}. If an injection from the source takes place, the minority carriers are effectively collected by the well/substrate junction. Therefore the equivalent circuit of the nMOST in the p-well can be given as shown in Fig.1b. On the sensitivity of the nMOST to $V_{CC} = V_{DD}$ was reported in [2]. R_C is an external resistor in the collector circuit of the vertical bipolar transistor. The voltage drop $V_x = I_C \cdot R_C$ caused by the collector current I_C is an indicator for the impact ionization induced bipolar action. It can be used to realize negative feedback to the impact ionization of the nMOST. It can be applied to the well-, drain- or the gate-terminal of the nMOST, realizing negative feedback to the well, drain or gate, respectively. The required polarity of the control-voltage V_x can be easily set with the help of buffer operational amplifiers.

The effect of the negative feedback $V_{BB} = -I_C \cdot R_C$ on the impact ionization is demonstrated in Fig.2. The collector current with and without feedback and the feedback voltage V_{BB} vs. V_{GS} are shown. The bipolar action is effectively suppressed.

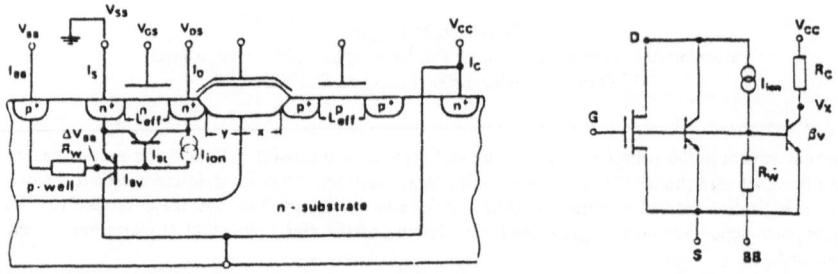

Fig.1: (a) Cross sectional view of the nMOST in p-well CMOS, and (b) its equivalent circuit with an external resistor R_C.

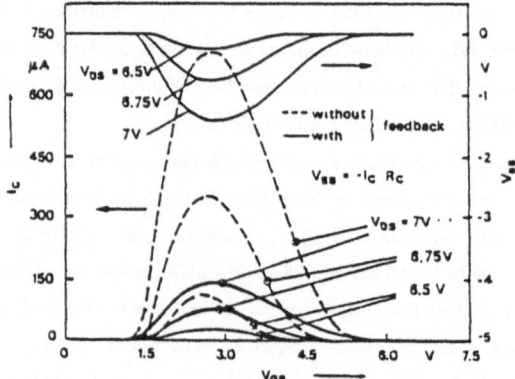

Fig.2: Effect of the negative feedback $V_{BB} = -I_C \cdot \dot{R}_C$ on the impact ionization rate in the nMOST as demonstrated by the collector current of the vertical transistor I_C and the feedback voltage V_{BB}.

The effect of a negative feedback into the drain terminal is shown in Fig.3. The feedback voltage is $V_{DS} = V_{DSo} - I_C \cdot R_C$. The onset of the feedback is evident: the impact ionization is strongly reduced by decreasing the drain voltage. The third possible nMOST terminal with which to realize negative feedback is the gate.

The influence of the feedback $V_{GS} = V_{GSo} - I_C \cdot R_C$ on the drain current I_D is shown in Fig.4a for two different well voltages V_{BB}. Increasing V_{DS} for any given nominal gate voltage, the drain current is decreasing due to the decreased effective gate voltage $V_{GS} = V_{GSo} - I_C \cdot R_C$.

Measuring the hot carrier related CMOS currents I_{BB}, I_C or I_D with and without negative feedback and from a detailed theoretical analysis of the circuit of Fig.1b, the current gain of the vertical bipolar transistor can be determined in operating conditions relevant for CMOS.

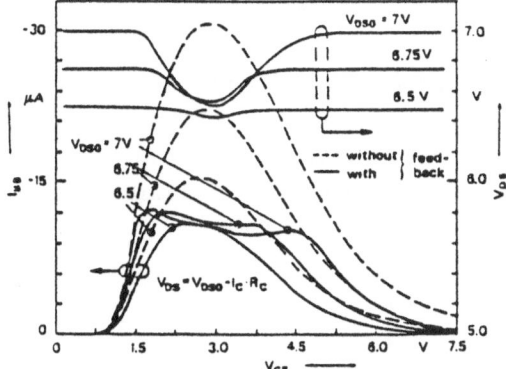

Fig.3: Effect of the negative feedback $V_{DS} = V_{DSO} - I_C \cdot R_C$ on the impact ionization rate as demonstrated by the well current I_{BB} and the feedback voltage V_{DS}.

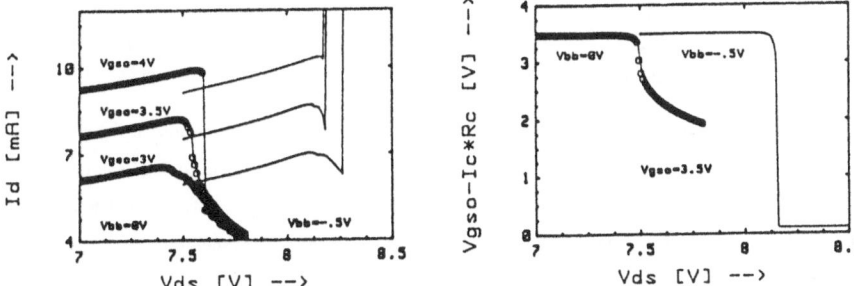

Fig.4: (a) Effect of the negative fedback $V_{GS} = V_{GSO} - I_C \cdot R_C$ on the impact ionization rate as demonstrated by the drain current and (b) the feedback voltage $V_{GS} - I_C \cdot R_C$.

Using one of the three possible negative feedback circuits, the one into the gate terminal for example, the total output conductance of the nMOST can be approximated:

$$dI_D/dV_{DS} \approx g_{da}(1 - g_{da} \cdot \beta_v \cdot R_C/(\partial V_{GS}/\partial V_{DS})) \quad (1)$$

where g_{da} is the output conductance caused by avalanche multiplication, β_v is the common emitter current gain of the vertical transistor in normal nMOST operation. The actual gate voltage V_{GS} is an implicit function of the drain voltage, as shown in Fig.4b. From the condition $dI_D/dV_{DS} = 0$, the current gain β_v can be determined experimentally:

$$\beta_v = (\partial V_{GS}/\partial V_{DS})/(g_{da} \cdot R_C) \quad (2)$$

The current gain β_v is shown for two different well voltages in Fig.5a. The value is $\beta_v \approx 1$. Measuring β_v

in the externally biased bipolar mode, its magnitude is about two orders of magnitude higher for all operating points (Fig.5b).

This result shows the effect of the parasitic bipolar action on the well current: the measured well current corresponds to the impact ionization current only if $I_C = 0$. With vertical bipolar action, a significant portion of the impact ionization current is consumed by the base current of the bipolar transistor. In this case the measured well current I_{BB} underestimates the impact ionization rate and thus can lead to incorrect predictions of the corresponding hot carrier effects.

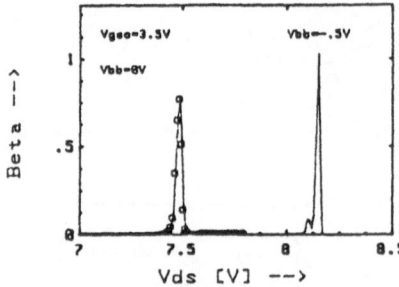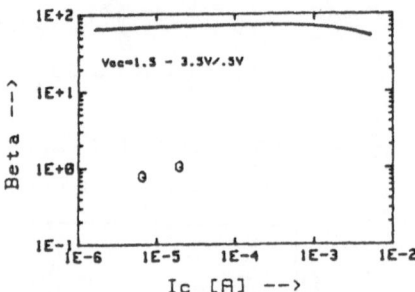

Fig.5: (a) Common emitter current gain of the vertical transistor Beta, measured in normal nMOST operation mode using negative feedback circuit $V_{GS} = V_{GSo}-I_C \cdot R_C$, as a function of the drain voltage V_{DS} and (b) in externally biased bipolar mode vs. collector current I_C for different collectior voltages V_{CC}. The β_v-values from Fig.5a are shown in the corresponding operating point too.

3. Conclusions

An additional resistor in the collector circuit of the vertical bipolar transistor in CMOS was used to realize negative feedback into the MOST in the well.

There are three possibilities to control the impact ionization rate of the MOST using negative feedback into the drain-, well- and gate-terminals, respectively.

The impact ionization related MOS- and bipolar-currents with the respective feedbacks have been used to determine the common emitter current gain of the vertical transistor in MOST relevant operating points.

It has been shown that the common emitter current gain of the vertical transistor in normal nMOST operaton mode is about two orders of magnitude lower than in externally biased bipolar operation mode. This fact indicates that the measurement of the impact ionization current at the well terminal may lead to incorrect predictions of the device degradation.

References:

[1] G.Krieger et al,IEEE EDL,Vol.9,No.1,p26,1988

[2] D.Takacs et al,ESSDERC 88,J.de Physique,Vol49,Suppl.9,pC4-387

The Dependence of Channel Hot-carrier Degradation on Temperature in the Range 77 K to 300 K

P. Heremans[*], G. Van den bosch, R. Bellens, G. Groeseneken, H. E. Maes

IMEC v.z.w. - Kapeldreef 75 - B3030 Leuven - Belgium

Abstract
The temperature dependence of the hot-carrier degradation behavior of n- and p-channel transistors is investigated. It is found that the efficiency of hot-electrons and hot-holes to generate interface traps is *reduced* at low temperatures, and that this generation can therefore be characterized by a positive activation energy. The much larger degradation of current characteristics commonly observed for n-channel MOSFETs at 77 K is shown to originate from the larger influence of any local damage in the transistor channel at low temperatures. For p-channel transistors, however, the influence of trapped electrons near the drain on I_d-V_g curves is not significantly dependent on temperature.

Introduction

The problem of hot-carrier reliability is generally accepted to be aggravated at lower temperatures in n-channel transistors [1]-[3]. The origin of this effect is analyzed in the present study. The charge pumping current I_{cp} [4] was checked to show no freeze-out effects for temperatures as low as 77 K, and was therefore used for the determination of the increase of the interface trap density (D_{it}) after stress. The transistor lifetime τ was defined in two different ways : firstly, as the time necessary to obtain a certain threshold voltage shift (ΔV_t) defined at a certain current level with source and drain reversed and $V_d = 5$ V, and secondly, as the time necessary to obtain a certain arbitrary increase of the maximal charge pumping current (ΔI_{cp}).

Experimental results and discussion

Fig. 1 shows the lifetime plots (i.e., the plots of $\tau^* I_s$ versus I_{sub}/I_s, where I_s and I_{sub} are the source and substrate current, respectively) for n-channel transistors, measured with charge pumping. Curve '295K' of Fig. 1 is measured at 295 K, after stresses at this temperature. Curve '77K' gives the lifetime plot measured at 77 K after stresses at liquid nitrogen temperature. These curves cannot be readily compared, because at 77 K a larger portion of the bandgap is measured than at 295 K. To correct for this, the increases of I_{cp} after the stresses at 77 K were also measured at 295 K. The lifetime curve then obtained is labeled 'S=77K M=295K'. Two striking features are observed on the curves of Fig. 1:

1. Both the curves for 77 K and the one for 295 K show two slopes. The smaller slope at low I_{sub}/I_s can be associated with a generation process of fast interface traps by hot-electrons [5], while the larger slope at large I_{sub}/I_s, is believed to be due to D_{it}-generation

---

[*] P. Heremans is a Research Assistant of the Belgian Fund for Scientific Research

by hot-holes [6]. It should be noted that this more efficient hot-hole degradation mechanism is activated whenever the product $\lambda.E$ is large: at low V_g and high V_d (large E), and at lower temperatures (large carrier mean free path λ). From the slopes of Fig. 1, the critical energies for formation of interface traps by hot-electrons and hot-holes are evaluated: $\phi_{it,e}$ = 3.7 V [5] and $\phi_{it,h}$ = 4.2 V [6],[7]. The fraction of electrons and holes that acquire the energies $\phi_{it,e}$ and $\phi_{it,h}$, respectively, in the field E can be evaluated using the following equations [5]-[7]:

$$I_{HE}/I_S = C \cdot \exp(-\phi_{it,e}/q.\lambda_e.E) \tag{1}$$

$$I_{HH}/I_{sub} = C \cdot \exp(-\phi_{it,h} \cdot \lambda_h/q.\lambda_e.E) \tag{2}$$

Here, λ_e and λ_h are the mean free paths for electrons and holes, respectively.

To eliminate the field E, the substrate current can be expressed as [5]:

$$I_{sub}/I_S = C \cdot \exp(-\phi_i/q.\lambda_e.E) \tag{3}$$

where ϕ_i ($\approx$ 1.3 V) is the impact ionization threshold [5].

This last expression can be fitted to measurements, to find the constant C. At both 295 K and 77 K, it was found that C $\approx$ 1.9. The equations (1)-(3) can now be plotted, as in Fig. 2. From Fig. 1, it follows that the hot-electron-induced D_{it}-generation process is taken over by a hot-hole assisted process at the value $I_{sub}/I_S \approx 8 \times 10^{-2}$. From Fig. 2, it can be seen that at this value, there are about 2000 times more hot-electrons present in the lateral field than hot-holes, independent of temperature. It is therefore concluded that hot-holes are 3 to 4 orders of magnitude more efficient to form interface traps than hot-electrons, independent of temperature. This result extends the conclusion obtained in [7] to the range 77 K to 295 K.

2. From comparison of the curves '295K' and 'S=77K M=295K' of Fig. 1, it follows that the lifetimes after stresses at 295 K are shorter than after stresses at 77 K. This clearly indicates that *the increase of D_{it} is smaller after low temperature stress* than after room temperature stress, when compared on the basis of equal I_{sub}/I_S. From eq. (3), I_{sub}/I_S is an approximate measure for the mean carrier energy, given by the product $\lambda.E$. The equations for the lifetime plots in the regions of hot-electron-induced and hot-hole-induced D_{it} are, respectively [5]-[7]:

$$\tau.I_S/W = C_e \cdot (I_{sub}/I_S)^{-\phi_{it,e}/\phi_i} \tag{4}$$

$$\tau.I_S/W = C_h \cdot (I_{sub}/I_S)^{-1-\phi_{it,h}.\lambda_e/\phi_i.\lambda_h} \tag{5}$$

From Fig. 1, C_e and C_h are found to be dependent on temperature. Taking the general expression:

$$C = C^0 \cdot \exp (E_a/k.T) \tag{6}$$

and fitting this to experimental values of C_e and C_h in the range 77 to 295 K, the activation energy for the generation of D_{it} by both types of hot-carriers was found to be $E_a \approx$ +15 meV.

Curves '295 K' and '77 K' of Fig. 3 show lifetime curves for n-channel transistors, now measured with ΔV_t. As can be seen, the temperature dependence of the curves is opposite to the one of Fig. 1: at 77 K, the lifetime is reduced by a factor of 70 as compared to 295 K. To investigate the origin of this effect, the devices stressed at 77 K were warmed up to 295

K, and their lifetimes were evaluated at this temperature, yielding curve 'S=77K M=295K'. These lifetimes are comparable to the ones of curve '295K'. However, when the devices are cooled back to 77 K, they again show a much reduced τ (curve '77K + WU'). This shows that the dramatic shortening of the lifetime at 77 K as compared to 295 K is due to the large impact of a given damage at low temperatures, and can *not be explained by the enhanced electron trapping* in the gate oxide at low temperatures. Indeed, when devices *stressed at 295 K* are measured at 77 K, they show a much reduced lifetime too: curve 'S=295K M=77K' ! This phenomenon can be explained by assuming that the interface traps formed in the upper-half of the bandgap during hot-carrier degradation are of acceptor-type (i.e. negative when filled with electrons), and narrowly distributed at the position of the peak lateral electric field, i.e. right inside the drain junction. Since they are located inside the junction, they form a potential barrier that is more or less independent of the gate voltage. Simulations of I_d-V_g curves were therefore performed with the MINIMOS 4 simulator [8], using a fixed potential barrier, as shown on Fig. 4a. The forward and reverse characteristics simulated at V_d = 5 V with such a barrier are shown in Fig. 4b (295 K) and Fig. 4c (77 K). It can be seen that the reverse mode current is much more affected at 77 K. Indeed, from analysis of the simulations, it is found that even if at the position of the charge the current is forced to flow at a certain depth below the surface, the carriers still encounter an extra potential barrier induced by that charge (40 mV in the case of Fig. 4 in reverse mode at V_g = 1.5 V). They have to surmount this barrier by thermal emission, and therefore a considerable voltage drop can occur if the barrier is large as compared to kT/q. This means that this barrier can significantly reduce the channel current. The only case when this charge-induced barrier at the source does *not* influence the channel current is when the 'natural' potential barrier, determined by V_g and always present at the source, is larger than the barrier induced by the charge, i.e. in subthreshold. Finally, when the transistor is operated in forward mode at V_d = 5 V, the current is not affected by the potential barrier for the following reasons: firstly, the barrier is now located at the position where most of the voltage drop occurs, and secondly in saturation the carriers at the drain flow deeper below the surface, where the barrier is much smaller. These features completely explain the measured current characteristics, in forward and reverse mode, at all temperatures.

In the case of p-channel transistors, the degradation is due to the trapping of hot-electrons above the drain junction. Where this trapping occurs, the $|V_t|$ is lowered, and therefore this region is no longer part of the channel that controls the current I_s. As a result, the channel current increases due to channel shortening. This shortening does not depend on the temperature at which it is measured, as is experimentally verified on Fig. 5.

Conclusions
In the complete range of 77 K to 295 K, the hot-hole-induced D_{it} generation is 3 to 4 orders of magnitude more effective than the hot-electron-induced D_{it} generation. At low temperatures, the generation efficiency of D_{it} by both types of carriers is *reduced*. The much smaller lifetimes observed for n-MOSFETs in reverse mode at low temperatures with ΔV_t are

730

due to the dramatic influence that the potential barrier at the source, present after stress, induces on the channel current at low temperatures. In p-channel MOSFETs, the reduction of the channel length is not significantly dependent on the measurement temperature.

References
[1] E. Takeda, proceedings VLSI Symposium, p. 2, 1985.
[2] J. Tzou et. al., IEEE Electron Device Lett., EDL-6, p. 450, 1985.
[3] T.-C. Ong et al., IEEE Trans. Electron Devices, ED-34, p. 2129, 1987.
[4] G. Groeseneken et. al., IEEE Trans. Electron Devices, ED-31, p. 42, 1984.
[5] C. Hu et. al., IEEE Trans. Electron Devices, ED-32, p. 375, 1985.
[6] R. Bellens et. al, proceedings of IRPS, p. 8, 1988.
[7] P. Heremans et al.,IEEE Trans. Electron Devices, ED-35, p. 2194, 1988.
[8] S. Selberherr, IEDM Tech. Dig., p. 496, 1988.

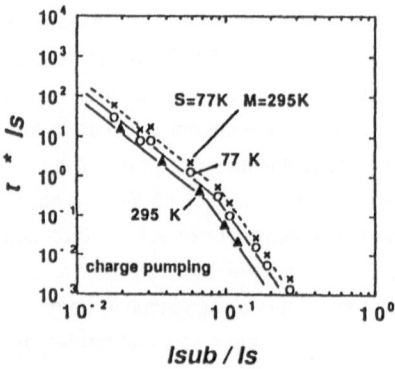

Figure 1: Lifetime plots for conven tional nMOSFETs (L=1.7µm, Tox=30nm) measured with charge pumping at 77K and 295K. See text for discussion.

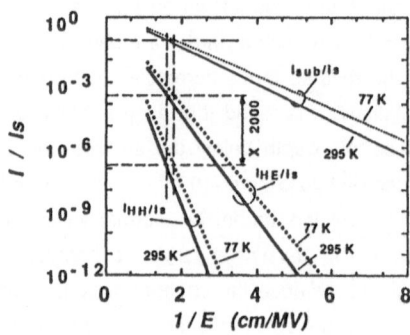

Figure 2: Calculated I_{sub}/I_s, I_{HE}/I_s and I_{HH}/I_s as a function of the lateral field, for n-channel MOSFETs. The curve of I_{sub}/I_s was fitted to experiments, using the model described in ref. [5].

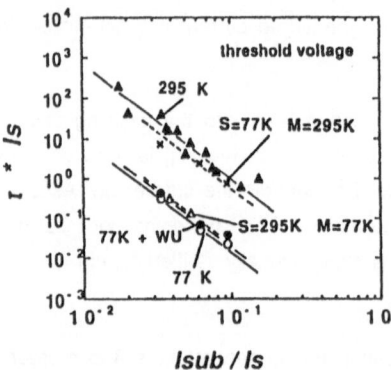

Figure 3: Lifetime plots for nMOSFETs measured with threshold voltage shifts. See text for discussion.

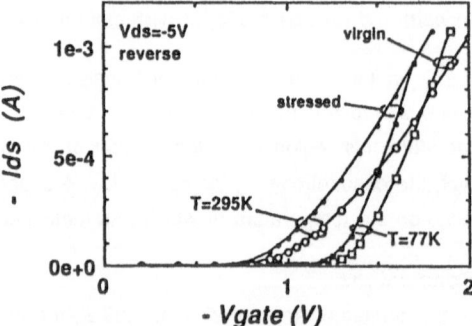

Figure 5: Measured I_d-V_g characteristics for p-channel MOSFETs, at 295 K and 77 K, before stress and after a stress

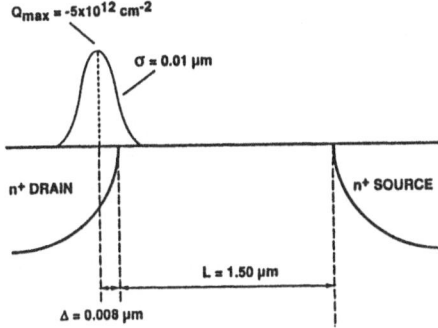

Figure 4 a: Fixed charge profile used in MINIMOS simulations for n-channel MOSFETs.

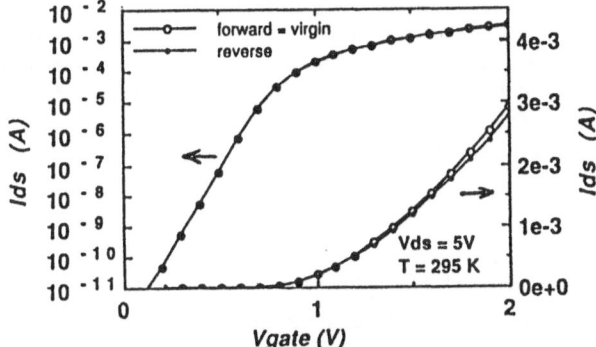

Figure 4 b: Simulated I_d-V_g characteristics at 295 K and V_d=5V, before and after introduction of the profile at the source (reverse) and at the drain (forward).

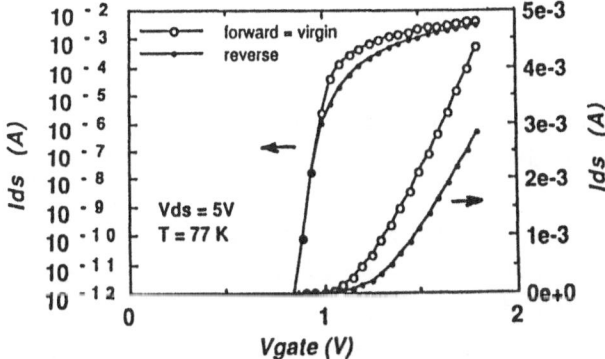

Figure 4 c: Analogous simulations as Fig. 4a, but now at 77 K.

Gate Oxide Thickness Dependence of Hot Carrier Induced Degradation on PMOSFETs

Y. Hiruta, H. Oyamatsu, H. S. Momose*, H. Iwai*, and K. Maeguchi

Semiconductor Device Enginnering Laboratory, Toshiba Corp.
*Toshiba ULSI Research Center.
1, Komukai-Toshiba-cho, Saiwai-ku, Kawasaki, 210, Japan,   Phone: 44-549-2210

ABSTRACT

Gate oxide thickness dependences of pMOSFET hot carrier degradation characteristics were studied at 300K and 77K. It was found that a thin gate oxide pMOSFET shows superior characteristics regarding hot carrier degradation. Threshold voltage shift becomes very small, with reduction in the gate oxide thickness, due to the tunneling effect. Interface state generation increases with the gate oxide thickness reduction. Fortunately, however, it is not enough to affect device characteristics. It was found that the interface state increase is described by the gate current uniquely, independently from the gate oxide thickness, bias condition, and temperature.

INTRODUCTION

Hot carrier induced degradation becomes a serious problem, not only for an nMOSFET but also for a pMOSFET with the scaling of device feature size [1,2]. There have been a few reports for an nMOSFET on the gate oxide thickness dependences of hot carrier induced degradation characteristics for a scaled-down thin gate oxide device [3,4]. However, there have been very few reports on that for a pMOSFET. This paper reports results of an investigation made on the gate oxide thickness dependences of pMOSFET hot carrier degradation characteristics at room and liquid nitrogen temperatures.

EXPERIMENT AND RESULTS

P^+ poly gate pMOSFETs, with 3 to 10 nm oxide thicknesses were fabricated by the 800°C process. Hot carrier stress was applied under various bias conditions at 300K and 77K for 1000 seconds.

Figure 1 shows the I-V characteristics before and after stress for 10 nm and 3 nm gate oxide devices. It was found that the thinner gate oxide device shows very little degradation. This is because electrons are easily detrapped by tunneling.

Figure 2 shows the stress gate bias dependences of degradation characteristics for various gate oxide thickness pMOSFETs. With a reduction in the gate oxide thickness, the threshold voltage shift ΔV_{TH} reduces drastically as shown in Fig.2(a). On the contrary, interface state generation, which corresponds to the charge pumping current increase, ΔI_{cp} [5], is large for a thinner gate oxide. Fortunately, however, the interface state density increase is not large enough to affect the device characteristics, as shown in Fig.1(b).

Figure 3(a) shows the substrate current dependence of ΔI_{cp}. In this figure, the ΔI_{cp} curves have a clear gate oxide thickness dependence. The ΔI_{cp} curves are quite similar to the substrate current dependence of the gate current shown in Fig.3(b). This means that hot carrier degradation is strongly correlated to the gate current rather than to the substrate current. The correlation between ΔI_{cp} and the gate current is shown in Fig.4. It was found from the figure that interface state generation is defined uniquely by the amount of gate current throughout a wide range, independently from the bias condition and gate oxide thickness. It should be noted that it is also defined uniquely, independently from the temperature. The threshold voltage shift is not independent from the gate oxide thickness, as shown in Fig.5. However, they are defined uniquely by the gate current for devices with the same gate oxide thickness.

Figure 6(a) summarizes the gate oxide thickness dependences of degradation characteristics at 77K and 300K. With a reduction in the gate oxide thickness, ΔV_{TH}s become small due to the tunneling effect. However, ΔI_{cp} values become large with reduction in the gate oxide thickness, because the gate current is larger in the thin gate oxide device, as shown in Fig.6(b). ΔV_{TH} and ΔI_{cp} at 77K are larger than those at 300K. One of the reasons is that the gate current at 77K is larger than that at 300K, as shown in Fig.6(b).

CONCLUSION

Gate oxide thickness dependences of pMOSFET hot carrier degradation characteristics were studied at 77K and 300K.

A thin gate oxide pMOSFET has superior characteristics regarding hot carrier degradation. By reducing the gate oxide thickness, the threshold voltage shift becomes very small, due to the tunneling effect. Interface state generation increases with gate oxide thickness reduction. Fortunately, however, the degree is not large enough to affect device characteristics.

It was found that the interface state generation at 77K has the same gate current dependence as that at 300K. The interface state generation is defined uniquely by the amount of gate current throughout a wide range, independently from the bias condition, gate oxide thickness, and temperature.

REFERENCES

[1] M. Koyanagi, A. G. Lewis, R. A. Martin, T. Huang, and J-Y. Chen, "Hot Electron Induced Punchthrough in Submicron PMOSFETs", *Extended Abstracts Conference on Solid State Devices and Materials*, pp.475-478, 1986.

[2] Y. Hiruta, K. Maeguchi, and K. Kanzaki, "Impact of Hot Electron Trapping on Half Micron PMOSFETs with P$^+$ Poly Si Gate", *IEDM Tech. Dig.*, pp.718-721, 1986.

[3] Tong-Chern Ong, Ping K. Ko, and Chenming Hu, "50-A gate-oxide MOSFET's at 77K," *IEEE Trans. Electron Devices*, vol. ED-34, no.10, pp.2129-2135, 1987.

734

[4] Y.Toyoshima, F.Matsuoka, H.Hayashida, H.Iwai, and K.Kanzaki, "A study on gate oxide thickness dependence of hot carrier induced degradation for n-MOSFETs," *Digest of Technical Papers, VLSI Symposium on Technology*, San Diego, pp.39-40, May, 1988.

[5] D. Schmitt, and G. Dorda, "Interface States in MOSFETs due to Hot-Electron Injection Determined by the Charge Pumping Technique", *Electron Lett.* vol.17, No.20, pp.761-763, 1981.

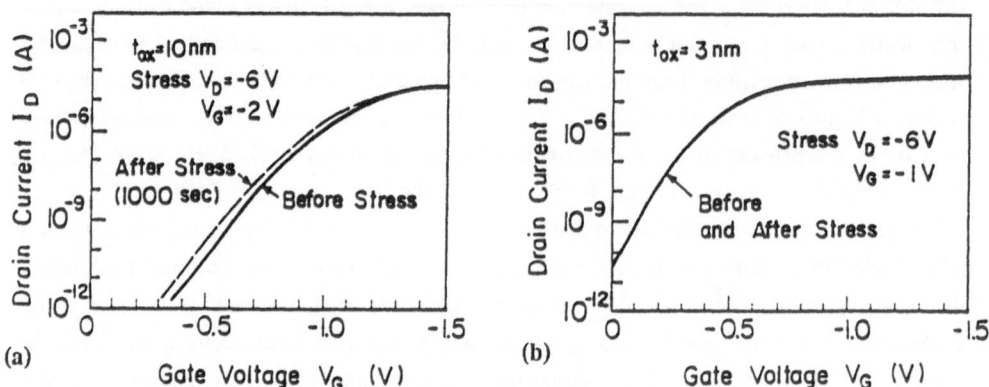

Fig.1 I-V characteristics change by hot carrier stress, $L_{poly} = 0.6\mu m$.
(a) 10nm gate oxide MOSFET, and (b) 3nm gate oxide MOSFET.

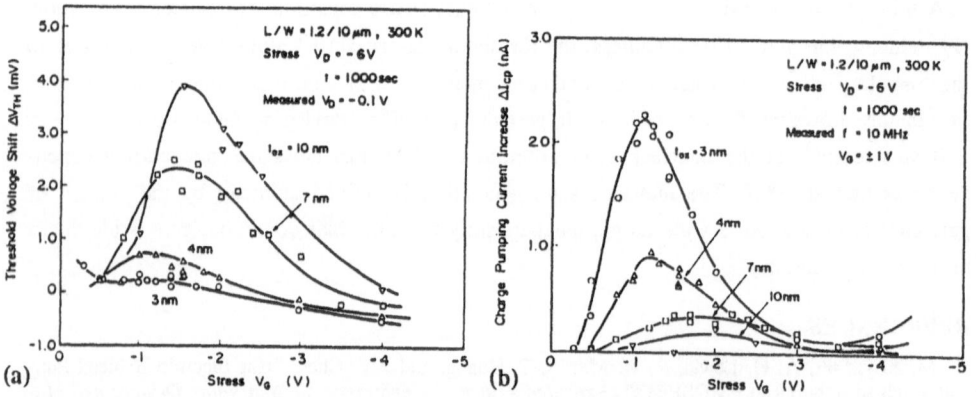

Fig.2 Stress V_G dependence of degradation characteristics, $L_{poly} = 1.2\mu m$.
(a) ΔV_{TH}, and (b) ΔI_{cp}, which corresponds to the interface state generation.

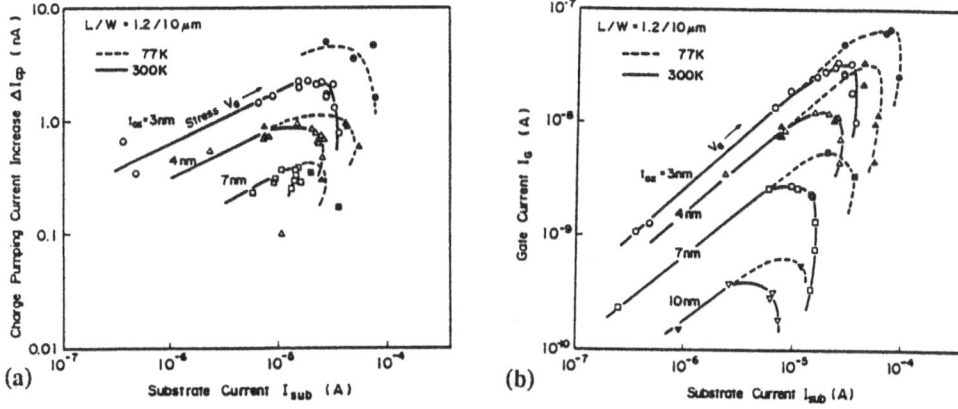

Fig.3 Substrate current dependences of ΔI$_{cp}$ and gate current.
(a) ΔI$_{cp}$, and (b) gate current.

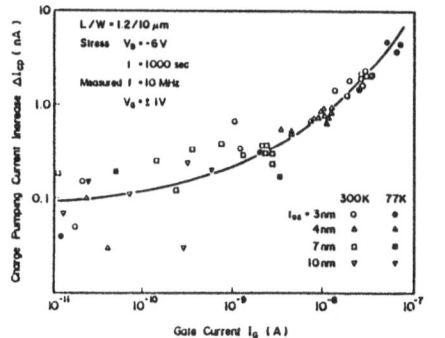

Fig.4 Gate current dependence of ΔI$_{cp}$.

Fig.5 Gate current dependence of ΔV$_{TH}$.

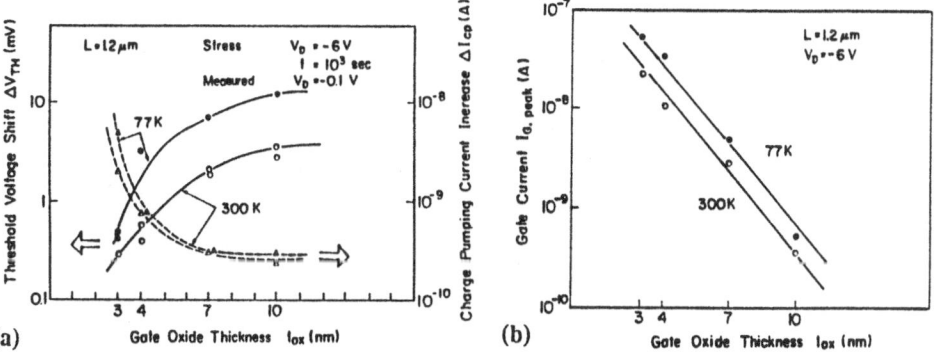

Fig.6 t$_{ox}$ dependence of the degradation at 300K and 77K.
(a) t$_{ox}$ dependence for ΔV$_{TH}$ and ΔI$_{cp}$, and (b) t$_{ox}$ dependence for gate current.

Session 8

Invited Papers

Ultimate Speed of Bipolar Devices: Will the Gap Between CMOS and Bipolar be Maintained?

A. Wieder, H. Klose

Siemens AG, München, Otto Hahn Ring 6, FRG

Summary of the talk

The history of integrated circuits started with bipolar technology. During the 70's the complementary MOS technology (CMOS) arose. This silicon technology was restricted first to slow but very low power watch circuits. Since this time the market share of CMOS products increased a lot. Today this technology can be regarded as the mainstream technology worldwide for digital integrated circuits. For quite a long time it was thought that bipolar even would be pushed out of its designated market segments namely high speed IC products. In recent years, however, a renaissance of bipolar technology occured due to a lot of technology innovations just to name the polysilicon emitter and novel self-alignment schemes. Thus today the question is again open for discussion where the borderline of application areas has to be drawn for bipolar and CMOS.

The presentation gives a state of the art review of bipolar technology and proves its unique speed advantage by demonstrator circuits and products. It is shown that already today 10 Gbit/s systems can be done in silicon bipolar. The perspectives of this technology are summarized next. Bipolar and CMOS performances are compared then with respect to speed and power dissipation in a general way. Finally the application areas for CMOS and bipolar are split up based on this reasoning.

CMOS-SOI Technologies for High Speed and Radiation Hard Circuits

A.J. AUBERTON-HERVE

C.E.A -IRDI- LETI Centre d'Etudes Nucléaires de Grenoble
B.P. 85X 38041 GRENOBLE CEDEX - FRANCE -

ABSTRACT

SOI is a very promising technology for CMOS-VLSI applications. The use of thin silicon film drastically improves the device characteristics. In this paper we will discuss the main trends to perform a CMOS-SOI technology for high-speed and rad-hard applications.

INTRODUCTION

The advantages of SOI have been demonstrated in a wide range of activities : JFET [1], bipolar [2], MESFET [3], or 3D integration [4] [5]. However VLSI-CMOS technologies are the driving force of SOI, mainly for military or space applications.

All the SOI advantages compared to bulk CMOS technologies are due to the complete dielectric isolation between devices. So, the latch-up is suppressed, allowing a reduced N^+/P^+ distance and a higher packing density; the speed is increased due to junction capacitance reduction; the radiation behavior by the collection volume reduction is improved mainly for transient doses and heavy ions (SEU immunity). Some recent results [6] have demonstrated high temperature (over 300°C) SOI circuits. Interesting performances at low temperature have also been demonstrated [7].

More advanced research on SOI aims to offer most of the previous advantages and mainly high-speed and rad-hard circuits. After a brief discussion of the SOI fabrication method, we will discuss what kind of SOI structure is adapted to these applications and where the limitations lie.

WHAT STRUCTURE FOR WHAT APPLICATION ?

There are two kinds of methods of obtaining a silicon layer on top of the SiO_2 with physical characteristics good enough to achieve high-density circuits:
- The "in-line" methods. The circuit design has to take into account the SOI fabrication technique. In this group the most advanced technique is the FIPOS (Full Isolation by Porous Oxidized Silicon) [8].
- The "out-line" methods. The SOI material is completely independent of its use. In this group, we find all the emerging techniques : SIMOX (Separation by IMplanted OXygen) Wafer bonding and ZMR (Zone Melted Recrystallization) [9] [10] [11]. SIMOX is certainly the most widely used technique.

The choice for VLSI-CMOS of one of these techniques is determined by the film thicknesses required. We have summarized in Table(1) the key parameters of this choice.

	BULK	THICK SOI E>0,3um	THIN SOI 0,1um<E<0,3um	ULTRA THIN SOI E<0,1um
DEVICES :				
SHORT CHANNEL EFFECTS	.	. =BULK	++	++
SUBTHRESHOLD SLOPE	+	+	++	++
FLOATING BODY EFFECTS	++	.	+	++
BREAKDOWN VOLTAGE	++	.	+	++
RADIATION HARDENING	.	. TRANSIENT DOSE	+	. CUMULATIVE DOSE
COMPLEX CIRCUIT :				HIGH SPEED CASE
PACKING DENSITY	+	+	++	++
PARASITIC ELEMENTS	.	+	++	++
PROCESS FACILITY	.	+	++	+
LATERAL ISOLATION	.	+	++	++
JUNCTION METALLISATION	.	+	++	?
MATERIAL	++	.	+	.
++ VERY GOOD + GOOD - LESS GOOD				

TABLE (1) : TRENDS IN THE SILICON FILM THICKNESS CHOICE AND COMPARISON WITH THE BULK TECHNOLOGIES.

For high-speed applications it has been demonstrated that reducing the silicon film thickness improves the MOS devices behavior in submicron region. The limit between a bulk equivalent behavior and a thin film behavior is determined by the maximum depletion depth controled by the gate. In case of a film thickness in the range of the maximum depletion depth or twice this value, full depletion of the film can occur. In this case the experimental results on fully depleted devices demonstrate a reduction of short channel [12] and floating body effects [13], an improved subthreshold slope (better drivability) [14] and a better punchthrough behavior [15]. The models proposed on SOI [16] demonstrate an equivalence between the film thickness and the channel dopant concentration. Therefore, for a given back interface potential and a given dopant concentration, reducing the film thickness is equivalent to increasing the dopant concentration. So, for the same device characteristics and the same channel length the dopant concentration on SOI could be reduced with respect to the bulk, and the channel mobility is thus maintained even for short channels by the use of thin film transistors. The underlying oxide thickness has to be high enough to reduce the junction capacitance.

The rad-hard SOI structure is quite different. The effect of cumulative dose irradiation is mainly to create positive fixed charges in the oxides. As the radiation-induced threshold voltage shift of the back parasitic transistor is proportional to the square of the oxide thickness, it is obvious that the underlying oxide thickness has to be reduced. If the film is fully depleted, the gate oxide threshold shift depends on the back oxide threshold shift, so the film thickness has to be high enough to decorrelate both interfaces. But for transient dose the active silicon volume has to be reduced. The film thickness therefore has to be around the maximum depletion depth to optimize both the transient and cumulative dose behavior.

The film thickness ranges adapted to the different targets are shown in Fig(1).

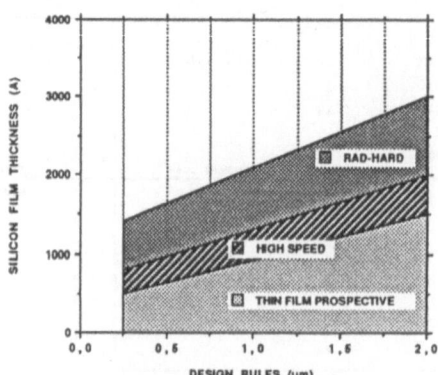

Fig(1): Choice of the film thickness best suited
to the different applications

For VLSI applications, SIMOX is the most suitable method because it provides thin silicon films on top of a medium thick SiO_2 film, with low thickness dipersion (around $\pm 5\%$). ZMR or Wafer-Bonding methods provide an underlying oxide several microns thick, with a wide range of silicon thicknesses and a good homogeneity over $0.5 \mu m$ thickness. So ZMR and Wafer-Bonding are more adapted to power or bipolar devices.

However, the SIMOX ability to provide thin silicon films has to be completed by the ability to provide high quality crystalline silicon films. It is only the case for high temperature post-implantation annealing, around 1300°C. For lower temperature, an epitaxial silicon layer is often grown on the initial SOI structure to increase the silicon thickness to $0.3-0.5 \mu m$, and to improve the active silicon quality [17]. In this case, ZMR or Wafer-Bonding are lower cost challengers.

A new but high cost approach aims to improve the dislocation density and the buried oxide integrity obtained in SIMOX. Multi-implantation and annealing are performed with low doses each time [18] [19]. There are no electrical results which demonstrate the advantages of this technique.

The SIMOX wafers used at LETI were obtained by an O+ implantation at 600°C (dose: $1.8 \times 10^{18}/cm^2$ energy : 200keV) and a 1300°C post-implantation annealing. The dislocation density is less than $10^5/cm^2$ [20]. Starting with this material, we have processed complex circuits (16K SRAM μProcessors) with a film thickness under the gate of 150nm in the as-obtained SOI structure, without an epitaxy [21] [22].

CMOS-SOI FOR RAD-HARD APPLICATION

SOI is mainly developed for radiation hardening to compete with SOS(Silicon On Sapphire) in VLSI technologies. A comparison of the main challenger in military and space technologies in Table(2) clearly demonstrates the advantage of SOI in transient dose. The key parameters to reach high cumulative doses and also the limitation in transient doses will be discussed in this paper.

	APPLICATIONS		TECHNOLOGY		
	SPATIAL	MILITARY	CMOS BULK	SOI	GaAs
Cumulative dose (rads)	10^6	$>10^5$	>100K tolerant > 1Mrads optimized	>100K tolerant > 1Mrads optimized	>1M
Transient dose (rads/s)	<0,1	10^{10}	10^6 LATCH-UP	$>10^{10}$	$>10^9$
SEU (errors/bit/day)	10^{-8}	10^{-8}	$<10^{-5}$	$<10^{-8}$	10^{-6}
NEUTRONS (1 MeV eq) (neutrons/cm²)	-	$>10^{13}$	$>10^{14}$	$>10^{14}$	10^{15}

TABLE (2) : RADIATION-HARDENING REQUIREMENTS

- CUMULATIVE DOSE

For cumulative dose, the main MOSFET limitation is the threshold voltage shift induced by the oxide charge trapping. In SOI there are three oxide involved in the radiation hardening : gate, underlying, and the lateral isolation. Techniques to perform hardened gate oxides are the same on SOI as on bulk technology. The underlying oxide has a 1Mrads(Si) threshold shift of 12-15V. It is possible to compensate the back channel threshold shift by a back interface threshold adjustment around 30V. The main limitation corresponds to the lateral isolation because of the silicon thinning near the edges, which correlates both the lateral and back interfaces near the edges Fig(2). So to increase the edge parasitic MOS hardening level both the lateral isolation and the buried oxide have to be hardened.

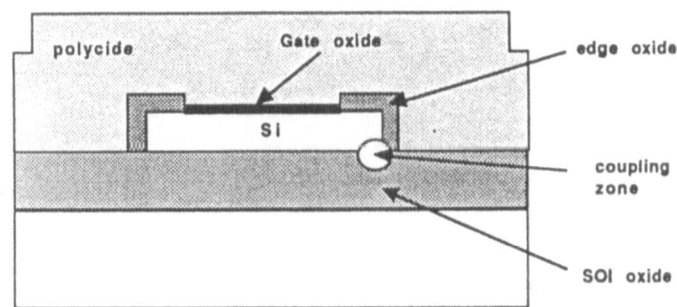

Fig (2) : INTERFACE COUPLING AT THE MOS EDGES

Buried oxide layer hardening.

Nitride or oxynitride buried layers are the most promising for buried SIMOX isolation hardening. The published results [23] on oxygen combined with nitrogen implantations, demonstrate that the 1Mrads(Si) back channel threshold shift is better than a single oxygen implantation, for a negative bias at the back interface. The minimum shift value obtained is 3.7V for a -3V applied substrate voltage during irradiation. The implantation conditions are $2x10^{18}/cm^2$ [O+] and $5x10^{16}/cm^2$ [N+]. The use of a different supply voltage on the substrate and the circuit is a limitation of this method.

Another way is to create interface states at the back Si/SiO$_2$ interface. These interface states have a screening effect on the buried oxide threshold shift by fixing the surface potential. The main difficulty is to find a compromise between leakage current and radiation hardening. To create such interfaces the easier way is to decrease the post-implantation temperature. It has been demonstrated, by creating a polycrystalline silicon near the buried oxide, that the 2Mrads(Si) level could be reached [24].

Lateral isolation hardening.

The lateral isolation proposed for radiation hardening in litterature is formed by bilayer structures in bulk technology. The bilayer could be nitride on oxide [25], doped oxide on thin undoped oxide [26], or polysilicon on thin oxide [27]. The principle is always to obtain the hardening level of the thin underlying oxide and the isolation equivalent to the thick deposited layer. In SOI a polysilicon spacer Fig(3) formed on the edges of the MOS has been proposed to reach the previously mentioned 2Mrad(Si) level. The polysilicon layer acts as a potential screen and is connected to the grounded substrate.

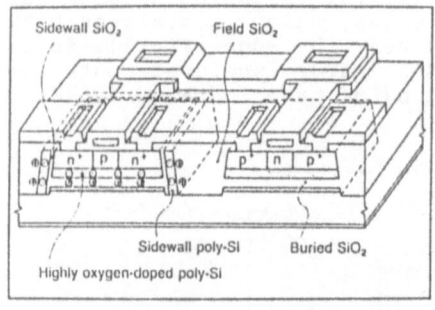
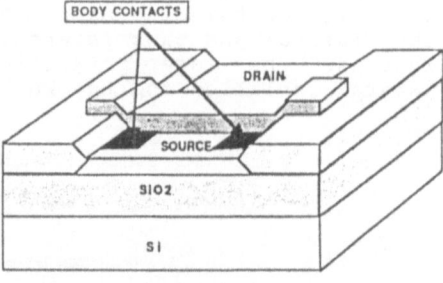

Fig (3) : SCHEMATIC CROSS SECTION OF A RADIATION -HARDENED CMOS-SOI INVERTER (24)

Fig (4) : BODY CONTACTS USED AS LATERAL ISOLATION IMPLANTATION

Junction isolation.

The last way to obtain a hardened lateral isolation is to use a junction isolation Fig(4). This method consists in a conventional lateral isolation (LOCOS or Trenches) which is not hardened, and a junction at the edges of the devices which prevents the edge MOS from turning on. The hardening level is thus obtained by a specific design which creates the junctions [28]. The density lost is around 15% compared to a bulk design. We have obtained a 3Mrads(Si) immunity on a 15000-transistor circuit without major changes in the characteristics.

- TRANSIENT DOSE

Transient doses create photocurrents proportional to the depletion volume. In thin SOI films the junctions reach the underlying oxide, reducing the collection volume to the junction perimeter and no longer to the junction area. But, the MOSFET structure is formed by a parasitic bipolar in parallel with the MOSFET. This parasitic bipolar in SOI has a floating base. Therefore, the SOI transient dose behavior will be determined by the bipolar parasitic gain [29] [30]. Fig(5) clearly demonstrates the influence of the bipolar gain on the transient radiation behavior. So, to improve the transient dose immunity, the bipolar gain has to be reduced by a degradation of the silicon filmlife time, or by decreasing the injection efficiency which is predominant for short channel. In this way the LDD (Low Doped Drain) which also forms a LDS (Low Doped Source) reduces the emitter efficiency, because it reduces the dopant ratio between the bipolar emitter (or MOS source) and the bipolar base (or MOS channel). The floating body can also be connected to the source to obtain a common base parasitic bipolar. But for high photocurrent the body contact efficiency is reduced by the access resistance. Some designs have been proposed with alternate body contact pitch in the source region as low as $2\mu m$ [31].

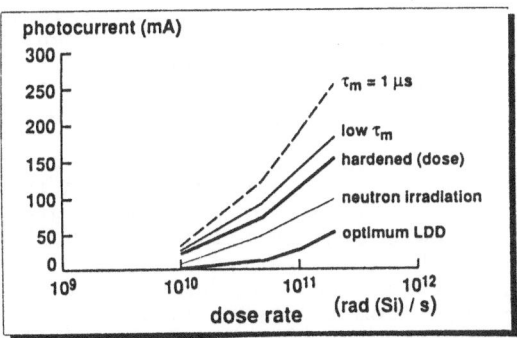

Fig (5) : RADIATION INDUCED PHOTOCURRENT IN A 29101 PROCESSOR FOR DIFFERENT PARASITIC BIPOLAR DEGRADATIONS

In transient dose radiations, memory cells or latches are the most sensitive. It is also possible to increase the hardening level by a design optimization [32]. Resistive decoupling has been proposed [33] which slows the latch cell response but the speed is also degraded.

The main SOI optimization efforts are focused on these two points: lateral isolation and parasitic bipolar. The main paradox of radiation hardening is that increasing the material quality decreases the radiation hardening. Specific design approaches enable high radiation performances to be reached on both cumulative and transient approaches even with a high material quality. However to increase the intrinsic SOI hardening level, the SIMOX wafer itself has to be modified.

CMOS-SOI FOR HIGH SPEED APPLICATIONS

The first demonstration of a 16K SRAM on SIMOX substrate was published by TI in 1986 [34]. A bulk design was used in this case, with an address access time of 30-35 ns. Some new results were published in 1988 by TI on 16K SRAM and 64K SRAM with SOI specific design Table(3). All these circuits used an SOI structure formed by an epitaxial silicon layer on SIMOX, to increase the silicon on insulator thickness.

COMPANY	ACCES TIME (ns)	POWER DISSIPATION (mW)	SUPPLY VOLTAGE(V)	MATERIAL	DESIGN RULES(um)	CIRCUIT
NTT 1982	12	45	5	SIMOX	1,5	1K
TI 1984	55	85	5	SIMOX	2,5	4K
NTT 1985	20	520	5.2	FIPOS	1,5	64K
TI 1987	30	.	5	SIMOX	1,5	16K
NTT 1987	14	45	5	SIMOX	1,3	4K
HARRIS 1987	130	30	5.5	SIMOX	3	4K
LETI 1988-89	11	200	5	SIMOX	1,2	16K
TI 1988	20	60	5	SIMOX	1,25	16K
TI 1988	25	.	5	SIMOX	<1	64K

specific SOI design optimization

TABLE (3) : SRAM RESULTS ON SOI

The 16K SRAM we have achieved at LETI is processed in the SIMOX thin SOI layer obtained by using a 1300°C furnace annealing without epitaxy. To compare the speed with a bulk CMOS technology, this circuit was processed with the same design on bulk by using a conventional 1.2μm CMOS technology. Address access time as low as 6ns at 77K was obtained on SOI.

For a 5V power voltage and a temperature of 300K the address access time is 10 to 12ns on SOI and 16 to 18ns on bulk. For 1MHz the active current is 43mA Fig(6) on SOI and 45mA on bulk. The access time on SOI is reduced by around 30% compared to the bulk. A comparison of the active current versus frequency for bulk and SOI demonstrates the reduction of the slope for SOI from 1.4mA/MHz for bulk to 1mA/MHz for SOI, due to the reduction of the total capacitance.

The temperature characterization demonstrates a good behavior from 77K to 400K . At 400K, there is a change of the active power dissipation on SOI, which is due to the parasitic bipolar effect which increases the standby power dissipation [35]. The access time at 400K is around 14ns for SOI and 21ns for bulk so the speed gain of 30% is true for the whole military range of temperature and the access time of SOI at 400K is equivalent to that of bulk at 218K.

These results have clearly demonstrated the speed improvement obtained by the SOI structure. This is mainly due to the junction capacitance reduction. The drain forms the gate of a MOS capacitor. So the drain capacitance is determined by both the oxide thickness and the depletion depth in the underlying substrate. By the use of high resistivity P-type substrate, the junction capacitance is drastically reduced.

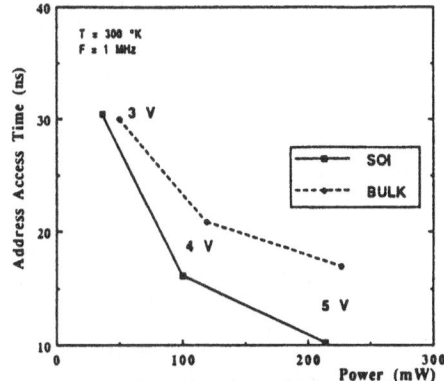

FIG. (6) : Comparison of SOI and BULK 16k SRAM
Address Access Time vs. Active Power dissipation
for different Supply voltages

CONCLUSION

Radiation survivability will be an increasing demand for electronic systems. It is the case for military systems, space systems but also for instrumentation, industrial robotics and medical applications which are increasingly required to operate in man-made radiation environments. High-speed circuits in a large temperature range are also the major market in VLSI application. BICMOS circuits are a first answer to the speed improvement with also a higher power consumption. CMOS-SOI is a versatile answer on both the increasing demand in radiation hardening and high-speed application, without major change in process complexity and power consumption. It could be a good challenger for bulk CMOS technology.

SIMOX is the most widely used material approach to achieve high complexity SOI circuits. The SOI wafer cost is today the main limitation to compete with conventional technologies. In a bulk CMOS technology the wafer cost is around 10% of the process cost. The wafer price is between 30$ and 50$. SIMOX prospective wafer price could be around 100$ in 1991 and the price reduction under 50$ is foreseen after 1992 . The crossing point in production cost between CMOS-SOI and CMOS conventional technologies could be around $0.5\mu m$ in ULSI circuits [36]. This is only possible if $1.2\mu m$ and $0.8\mu m$ technologies are developed and justify a wafer volume high enough to reduce the wafer cost.

Therefore SOI is today of major interest and a very promising future exists. Ultra-thin film devices offer a very wide research and application interest. The maximum development of SOI in a wider market than specific military applications should occur in the mid-nineties.

748

[1] D.G.PLATTETER, T.F.CHEEK IEEE Trans.on Nucl.Science 1988 N.6 p.1350
[2] HONG K.CHOI, BOR-YEU TSAUR, CK.CHEN IEEE Electron Device Letter Vol.EDL-8 N.3 March 1987.
[3] J.P.COLINGE, S.-Y CHIANG IEEE Elec.Dev.Letter Vol.EDL-7 N.12 1986.
[4] Y.AKASAKA Proceedings of the IEEE,Vol.74,N°12,1986 p1703-1714
[5] H.ACHARD,et al,A.MONROY,et al,C.G.CAHILL 1988 IEEE SOS/SOI Technology Workshop p66
[6] W.A.KRULL, J.C.LEE 1988 IEEE SOS/SOI Technology Workshop
[7] T.ELEWA,F.BALESTRA,S.CRISTOLOVEANU,A.J.AUBERTON-HERVE,J.DAVIS 1988 IEEE SOS/SOI Technology Workshop
[8] N.J.THOMAS, J.R.DAVIS, J.M.KEEN, and al IEEE Electron Device Letters, Vol.10,N°3,1989
[9] M.BRUEL, J.MARGAIL, A.J.AUBERTON-HERVE, J.STOEMINOS Microelectronic Engeneering 8 1988 149-161 North Holland
[10] J.M. McNAMARA, J.S. RABY 1988 IEEE SOS/SOI Technology Workshop
[11] P.ZAVRACKY,D.P.VU,L.ALLEN,1988 IEEE SOS/SOI Technology Workshop
[12] S.VEERARAGHAVAN,J.G.FOSSUM IEEE Trans.on Elec.Dev.Vol.36 N°3,1989p522
[13] J.P.COLINGE IEEE Electron device letter Vol 9 N°2 p97 1988
[14] J.P.COLINGE IEEE Electron device letter Vol EDL7 N°4 p244 1986
[15] M.YOSHIMI,H.HAZAMA,M.TAKAHASHI,S.KAMBAYASHI,T.WADA,K.KATO, H.TANGO IEEE trans. on Electron Devices Vol 36 ,N°3 March 1989
[16] K.K.YOUNG IEEE Trans. on Electron Dev. Vol.36 N°2, 1989 p.399
[17] M.BRUEL, J.MARGAIL, C.JAUSSAUD, A.J.AUBERTON-HERVE, J.STOEMENOS Microelectronic Engineering 8 (1988) 149-161 North Holland
[18] J.MARGAIL,J.STOEMENOS,C.JAUSSAUD,M.BRUEL,European SOI Workshop,1988
[19] D.HILL,P.FRAUNDORF,G.FRAUNDORF,Journal of Applied Phys. 63 1988 p4933
[20] J.STOEMENOS, J.MARGAIL, Thin Solid Films 135 1986 p115
[21] A.J.AUBERTON-HERVE, M.BRUEL, et al, W.D'HESPEL,J.F.PERE, et al., 1988 Symposium on VLSI Technology SAN DIEGO, CA
[22] J.L.LERAY and al. A.J.AUBERTON-HERVE and al. IEEE Trans. on Nuclear Science 1988 N.6 p.1355
[23] B.Y. MAO,C.E. CHEN,G.POLLACK,H.L.HUGHES, G.E. DAVIS IEEE Trans. on Nuclear Science 1987 N.6 p.1692
[24] T.OHNO, K.IZUMI Elec. Letters 12 February 1987 Vol.23 N.4 p141-143
[25] F.L.TERRY, R.J.AUCOIN, M.L. NAIMAN, S.D.SENTURIA, IEEE Eectron Dev. Letters Vol. EDL4 p191 1983
[26] K.KASAMA, F.TOYOKAWA, M.SAKAMOTO, K.KOBAYASHI, IEEE Trans. on Nuclear Sciences Vol. NS32 N°6 December1985
[27] L.MANCHANDA,S.J.HILLENIUS,W.T.LYNCH,HONG-HI CONG,K.K.NG,R.L.FIELD IEEE Trans. on Electron Dev. Vol36 N°4 1989 p651-657
[28] J.TIHANIL,H.SCHLOTTERER IEEE Trans. On Electron Dev ED22 1975
[29] G.E. DAVIS, L.M.HITE, T.G.BLAKE, C.E.CHEN, H.W.LAM, R.DEMOYER IEEE Trans. On Nuclear Science Vol 6 December 1985
[30] A.J.AUBERTON-HERVE 1988 IEEE SOS/SOI Technology Workshop
[31] Y.OMURA K.IZUMI IEEE Trans. on Elec. Dev. Vol.35. N.8. August 1988
[32] T.M.MNINCH, S.E.DIEHL, B.D.SHAFER, R.KOGA, W.A.KOLASINSKI, A.OCHOA IEEE Trans. Nuclear Science Vol. NS30 p4620, 1983
[33] A.OCHOA,JR,C.L.AXNESS,HARRY T. WEAVER,J.S.FU IEEE Electron Device Letter EDL-8,N°11, November 1987 p.537-539
[34] C.E.CHEN,M.MATLOUBIAN,B.Y.MAO,S.SUNDARESAN,C.SLAWINSKI,H.W.LAM, T.G.W.BLAKE,L.R.HITE,R.K.HESTER IEEE Trans. Elec. Dev.,ED33,1840 (1986)
[35] A.-J.AUBERTON-HERVE 1988 SOS/SOI Technology Workshop p.55
[36] J.-P.COLINGE 5th International Workshop On Future Electron Devices Three Dimensional Integration Miyagi,Zao, JAPAN 30 May-1June 1988

SOI 1

Novel Electrical Characterization of Edge Effects in SIMOX Transistors

T. ELEWA, B. KLEVELAND, B. BOUKRISS, T. OUISSE, A. CHOVET AND S. CRISTOLOVEANU[*]

Lab.PCS (UA CNRS), INPG ENSERG, B.P. 257, 38016 Grenoble Cedex, France

[*] On sabbatical leave at the University of Maryland, College Park, USA

Abstract - We demonstrate that in order to get a better understanding of the parasitic edge conduction in silicon on insulator transistors and to evaluate the technological solutions proposed to avoid it, charge pumping, dynamic transconductance and noise measurements in addition to the static characteristics should be performed. The advantages of these complementary electrical characterization methods are illustrated by extracting new results related to the edge interface in SIMOX devices.

1. Introduction

It is well known that parasitic transistors activated at the sidewalls of the channel can dominate the subthreshold characteristics, cause high leakage currents and lead to substantial threshold voltage shifts in both bulk and SOI narrow devices. The conventional method of suppressing this parasitic conduction is to heavily implant the sidewalls in order to increase the threshold voltage of the edge transistors. Several new isolation techniques (SILO, BOX, modified LOCOS and MESA) have been compared by studying the static characteristics and threshold voltages[1,2]. We present new results concerning the edge properties (threshold V_{th} and flatband V_{fb} voltages, densities of fast D_{it} and slow N_t interface traps) which are obtained using complementary electrical characterization methods.

2. Experiment

The SIMOX wafers used in the experiment were produced by deep oxygen implantation (1.8×10^{18} O_2/cm^2 at 400keV,55μA,550°C and annealing at 1300°C for 5 hours in N_2). LOCOS oxides were used to isolate the silicon islands.

2.1. Static characteristics

The static $I_D(V_G)$ characteristics of three different transistors are compared in fig.1. The first slope (for negative V_G) is not due to the back interface but to the edges because devices with same channel lengths and different widths exhibit identical subthreshold characteristics. Moreover, the edgeless transistor has only one slope. Fig.2 shows the progressive suppression of the edge conduction with increasing negative substrate bias, V_{SUB}.

2.2. Charge pumping measurements

Results of charge pumping measurements for the transistor of fig.2 are shown in fig.3. The charge pumping current (I_{cp}) as a function of the base voltage (V_B) has normally a rectangular shape with V_{th} and V_{fb} being extracted from the turn on/turn off transitions of the I_{cp} current[3]. The shape obtained in fig.3 for $V_{SUB}=0V$ is not a simple superposition of two rectangles as could be expected from the parallell combination of two different transistors suggested by fig.2. The reason for this is that the edge conduction is not due to a single edge transistor but to a continuum of transistors with different V_{th} and V_{fb} (i.e. different oxide thicknesses, densities of interface states and fixed oxide charges[4]). The different edge components thus constitute a continuum of charge pumping current rectangles and a smooth transition is observed between the edge and front transistors.

The results obtained for the three different substrate biases (fig.3) can be explained by a nearly constant rectangle due to the front transistor which is not much affected by the increasing negative substrate bias and the edge contributions which are shifted towards higher values of V_G (see also fig.2). However, for $V_{SUB} \leqslant -4V$ the edge effects can not be detected by the static characteristics of fig.2 whereas they are easily revealed by the charge pumping measurements. The average density of interface traps deduced from these measurements is $D_{it} = 2\times10^{10}$ cm^{-2} eV^{-1} at the front interface and only slightly higher at the edges which confirms other observations that the LOCOC isolation does not generate a high density of interface traps[1]. It is here illustrated that the edge effects can not be suppressed merely by increasing the threshold voltage (unless it becomes much higher than that of the front transistor), and will constitute a significant frequency dependent parasitic component.

A similar qualitative behaviour is found for the back interface. The edge and back interface components are separated by comparing V_{th} and V_{fb} voltages and by measuring diodes and transistors with different dimensions. By pulsing the substrate instead of the gate, a better determination of D_{it} is achieved at the edges because the large swing applied (24V) allows measuring the whole edge contribution. The average density at the edges ($D_{it} = 6\times10^{11}$ cm^{-2} eV^{-1} at f=100kHz) is now more than one order of magnitude higher than the value deduced from front gate experiments. This suggests that D_{it} at the edges increases towards the back interface. As far as the back interface is concerned, we find about the same density of fast states (100kHz) as at the front interface but at low frequencies (<1kHz) we detect slow traps with a density two orders of magnitude (10^{12}cm^{-2} eV^{-1}) higher than that of the fast states. Other components of I_{cp} have been detected and attributed to localized traps.

2.3. Noise measurements

The 1/f noise levels of several front channel transistors are shown as a function of gate voltage in fig.4. At negative gate voltage, the front channel is turned off and only the edge-transistor noise is measured. In strong inversion, both the edgeless and conventional transistors (with $V_{SUB} \leqslant 0$) exhibit similar $S_I(V_G)$-characteristics which are dominated by the front channel only. The S_I/I_D^2 noise level for the edge transistors is higher than that of the front transistor. As the noise is inversely proportional to the "gate" area, this is explained by the smaller surface of the edges. In the intermediate range ($0V < V_G < 0.5V$) there is a coupling between the edge and front transistors, and the S_I/I_D^2 characteristics are very much dependent on the transistor width. The noise spectra are very useful as being directly related to the nature of slow traps. In fig.5, for $V_G = -0.7V$, only the edge fluctuations are measured. The plateau at low frequencies can be attributed to the existence of very slow traps. For $V_G = 0.4V$ a contribution of several types of traps is obtained; both localized traps ($1/f^2$) and flicker noise ($1/f$) is observed[6].

2.4. Dynamic transconductance measurements

This method gives the density of fast states by measuring the imaginary part of the transconductance as a function of frequency[7]. The peaks of the curves of fig.6 correspond to D_{it} and their position to the time constant. It is observed that the edge traps (negative gate voltage in fig.6) are slower than the traps at the front interface. In the intermediate range, the decrease of G_p/ω can be explained by the coupling of the edge and front transistors. At gate voltages above 0.5V, the slow traps at the edges do not interact any more and the behaviour is dominated by the front interface. We therefore observe a "normal" shift of the peak of G_p/ω to higher frequencies with increasing V_G.

3. Conclusion

The static characteristics of the front channel SIMOX transistors show the existence of a significant edge conduction. Charge pumping measurements reveal that the parasitic effects are not suppressed by a negative voltage but are only shifted to higher gate voltage. The charge pumping method is an elegant tool to understand the edge effects and measure the effictiveness of the additional edge doping and the frequency dependence of the parasitic effects. It was also shown that more information of the edge interface can be extracted by measuring the back interface behaviour. The noise spectra can give very useful information about low current parasitic effects and the time constants of different traps can be separated in the frequency spectrum. The dynamic transconductance measurements complement the charge pumping and noise measurements in the whole spectrum except for very low frequencies. All these measurements are necessary to get a thourough understanding of the edge effects in SIMOX transistors.

Acknowledgements - Drs. J.R. Davis and N.J. Thomas (British Telecom) are sincerely thanked for providing the devices and encouraging this work.

References

1. S.W.Sun and K.C.Weng, Solid State Electron 32, 333(1989)
2. J.L.Coppee,E.Figueras,B.Goffin,D.Gloesener,F.Wiele, ESSDERC Paris, J.Phys., sup.n°9, 749(1988)
3. N.G.Groeseneken, H.E.Maes,N.Beltran,R.F.Dekeormaeker, IEEE Trans.El.Dev.,31, 42(1984)
4. J.-C.Marchetaux,B.S.Doyle,A.Boudou, Solid State Electron 30, 745(1987)
5. T.Elewa,H.Haddara,S.Cristoloveanu,M.Bruel, ESSDERC Paris, J.Phys., sup.n°9, 137(1988)
6. A.Chovet,B.Boukriss,T.Elewa,S.Cristoloveanu, Int.Conf.Noise, Montreal,World Sc.Publ. 390(1987)
7. H.Haddara and G.Ghibaudo, Solid State Electron 31, 1077(1988)

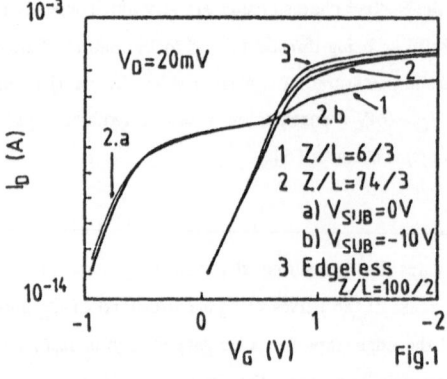

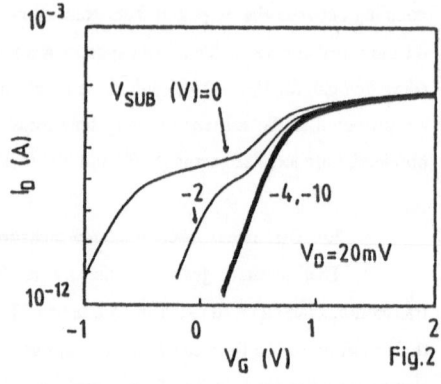

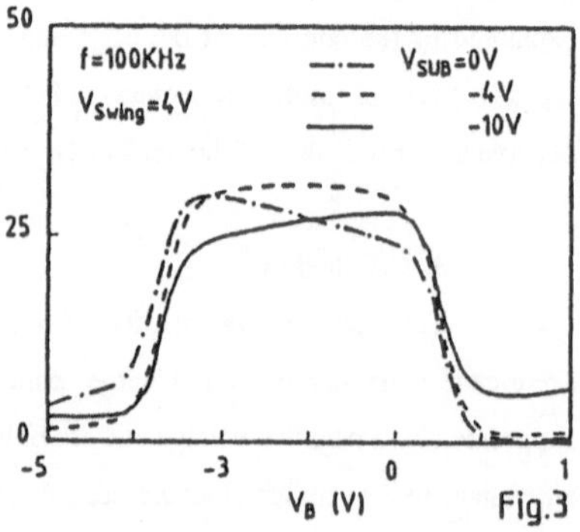

FIG.1. Static characteristics.
Front channel enhancement transistors.

FIG.2. Static characteristics.
Front channel enhancement transistor.

FIG.3. Front channel charge pumping current versus base voltage.

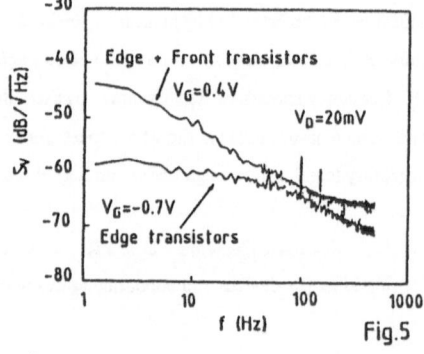

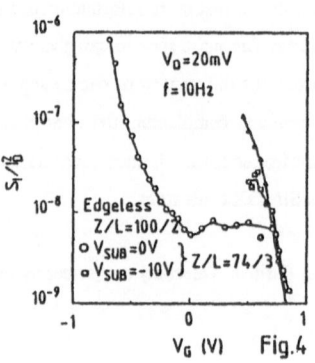

FIG.4. Normalized channel current fluctuations versus gate voltage.

FIG.5. Channel fluctuations versus frequency.

FIG.6. Dynamic transconductance versus frequency.

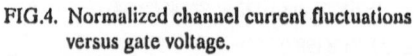

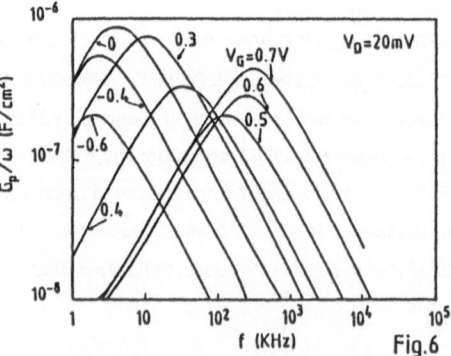

A Physical Model for the Characterization of SOI MOSFETs in Linear Operation

D. FLANDRE and F. VAN DE WIELE

Laboratoire de Microélectronique,
Université Catholique de Louvain,
Place du Levant 3, 1348 Louvain-la-Neuve (Belgium)
*D. Flandre is sponsored by IRSIA.

Summary

The present paper deals with the electrical characterization of long-channel SOI MOSFET's in linear operation. Limitations of previous analytical analyses are emphasized and a new physical model with a larger scope of application is presented. The model is shown to be in good agreement with I-V measurements performed on SOI MOSFET's of various thicknesses, in strong inversion as well as in subthreshold operation.

Introduction

The purpose of our work was to develop a physical model, very efficient for the characterization of SOI MOSFET's in linear operation (fig.1). Indeed, previous analytical models [1] are only valid in limited ranges of SOI film thicknesses and of gate voltages and are inefficient in analyzing the back substrate doping effects recently highlighted [2,3,4]. These basic limitations result from first-order approximations: full depletion of the SOI film and consideration of the underlying Si substrate as a metallic back gate. Moreover, a comparison of I-V characteristics predicted by a one-dimensional (1D) numerical device simulator [5] and by a first-order analytical model shows the latter's inefficiency for reliable electrical characterization, even in the case of thin-film SOI MOSFET's (fig.2). In response to those needs, we have developed a second-order analytical model for long-channel SOI MOSFET's in linear operation, which incorporates the physical effects related to the modulation of the depletion charge in the SOI film and the influence of the underlying semiconductor Si substrate. Our paper briefly describes that analysis, then discusses its validity and usefulness.

Presentation of the model

The uniform channel charge density of a long-channel SOI MOSFET in linear operation may be accurately modeled by means of a charge-sheet based 1D analytical analysis near the source contact. Our analysis results from the extension of a simplified model presented in [3], to include the influences of the source contact, of work-function differences, of fixed interface charges and of fast surface-state densities. In particular, the impact of the source contact is to fix the value of the floating SOI film quasi-Fermi level at equilibrium, Φ_O. We assume uniform doping densities in the SOI film and in the back substrate. 1D Poisson's equation is solved rigorously in the underlying Si substrate by assuming that it is of infinite length.

Such a rigorous analytical solution is not feasible in the SOI film, due to its finite thickness. Therefore, 1D Poisson's equation is solved quasi-rigorously by approximating the widths of the front and back space charge regions and assuming that their sum is equal to the film thickness. In the case of a P-type SOI film (doping level Na), four realistic types of front-back coupling are to be considered and are modeled by imposing physical boundary conditions on the vertical electric field (E_O), the carrier concentrations (p_O, n_O) and the potential (V_O) at the coupling point:

- case 1 : the front and back space-charge regions are separated by a quasi-neutral region : $p_O \cong Na$, $E_O = 0$, $V_O = \Phi_O$.

- case 2 : an accumulation region and a depletion one interact at some "quasi-neutral" coupling point where the electric field has a non-zero minimum value : $p_O \cong Na$, $E_O \neq 0$, $V_O = \Phi_O$.

- case 3 : two depletion regions interact at a coupling point with a zero electric field : $p_O \cong Na.e^{-\beta.Vo}$, $E_O = 0$, $V_O \neq \Phi_O$.

- case 4 : a single depletion region extends over the SOI film, so that the coupling point is confounded with one of the film Si/SiO$_2$ interfaces.

We finally obtain a non-linear system of seven analytical equations, five of which are common to all four cases of operation, the last two being specific to each case. This system can be solved by means of a Newton-Raphson iterative method, the appropriate case of operation being selected at each iteration through a simple examination of the surface potentials. The system solution is then used to compute the front and back channel charge densities by means of the charge-sheet approximation [6] and finally to derive the drain current considering constant effective carrier mobilities. A set of 12 parameters can be adjusted to match experimental measurements, instead of 11 in previous first-order analyses.

Discussion

1° Very narrow differences were found between the model results and 1D numerical simulations, which demonstrates the validity of our analysis. Moreover, our model sheds light on the electrical behavior of SOI MOSFET's. The four modeled cases of operation actually occur for different values of front and back gate voltages and are continuously linked (figs.3,4,5). In particular, it can be shown that a thin-film SOI MOSFET does not remain fully depleted for its whole range of operation, a physical feature that previous analytical models are unable to take into account.

2° To demonstrate the usefulness of our model, its results were compared to I-V measurements performed on thin as well as thick-film SOI MOSFET's (figs.3,4). Apart from the mobility degradation effect, which is complex to model analytically in thin-film SOI MOSFET's [7], excellent agreement between experimental and theoretical curves has been obtained, using sets of parameters very close to the expected technological values. Moreover, the subthreshold characteristics are also well fitted (fig.5). On the contrary, previous first-order models can not accurately match the experimental curves for the whole range of Vgb; the "best" fitting being obtained with model parameters up to 10% different from their expected values (fig.6).

3° Our model is able to reveal the influence of parameters unaccounted in previous analyses, for example, the variations of the subthreshold current with the charge condition in the underlying Si substrate (fig.5).

Conclusion

A physical model for the SOI MOSFET in linear operation has been developed by incorporating important effects unaccounted in previous analyses. Our model is proved to be valid for a wide range of technological parameters and of operating regimes and is in good agreement with measurements performed on long-channel SOI MOSFET's of various film thicknesses. Therefore, the model is efficient for the characterization of long-channel SOI MOSFET's and can also serve as an evaluating tool in the development of other - more approximated - models for CAD applications.

References

1. Lim, H.-K.; Fossum, J.G.: Current-voltage characteristics of thin-film SOI MOSFET's in strong inversion. IEEE T. Elec. Dev. 31 (1984) 401-408.
2. Davis, J.R. et al.: Measurement and modeling of circuit speed of CMOS on oxygen-implanted SOI. IEEE T. Elec. Dev. 34 (1987) 1713-1718.
3. Flandre, D.; Van de Wiele, F.: A new analytical model for the two-terminal MOS capacitor on SOI substrate. IEEE Elec. Dev. Lett. 9 (1988) 296-299.
4. Paelinck, P. et al.: Theoretical analysis of the two-terminal MOS capacitor on SOI substrate. Proc. 18th ESSDERC (1988) C4 67-70.
5. Van de Wiele, F.; Paelinck, P.: A general 1D-model for SOI structures - Numerical analysis by a fixed function integration method (FIXFUN). To be published in Solid State Electr.
6. Van de Wiele, F.: A long-channel MOSFET model. Solid State Electr. 22 (1979) 991-997.
7. Sturm, J.C.: Performance advantages of submicron Silicon-on-Insulator devices for ULSI. in MRS Symp. Proc. 107 (1987) 295-307.

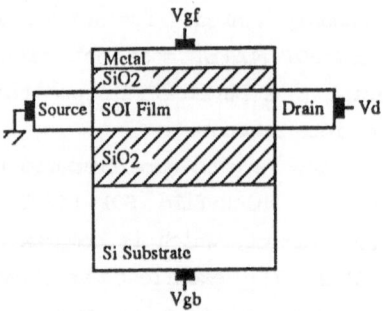

Fig.1: Cross-sectional schematic of a SOI MOSFET.

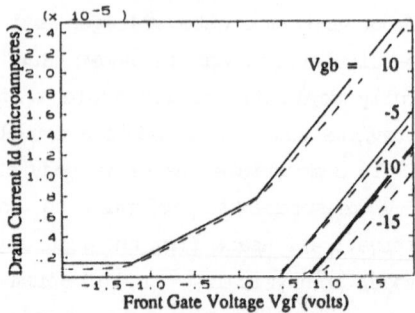

Fig.2: I-V curves predicted by 1D numerical simulations (– – –) and by 1st-order analytical analyses (———). (n-channel MOSFET; film thickness = 100nm; W/L = 50/10μm; Vd=.02V; fast surface-state densities=0; P-type back substrate).

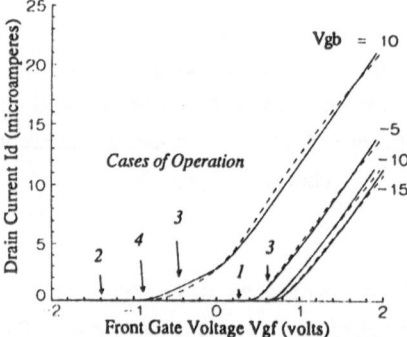

Fig.3: Measured (– – – –) and 2nd-order modeled (———) I-V curves of a thin-film SOI MOSFET. (same parameters as in fig.2 except for the fast surface-state densities)

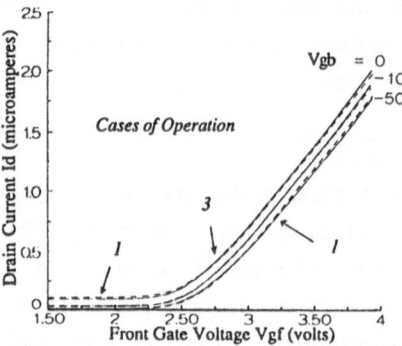

Fig.4: Measured (– – – – –) and 2nd-order modeled (———) I-V curves of a thick-film SOI MOSFET (film thickness = 415nm; W/L = 18/6μm; Vd = .02V).

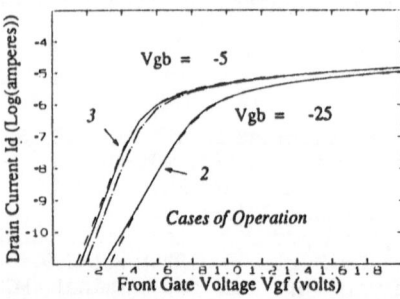

Fig.5: Measured (– – – –) and 2nd-order modeled (———) subthreshold characteristics of the thin-film SOI MOSFET of fig.3. A simplified 2nd-order analysis on a metallic back-gate equivalent device (–·–) shows the influence of a depleted underlying substrate.

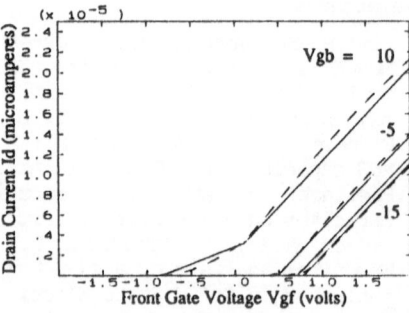

Fig.6: "Best" fitting obtained with a 1st-order model on the measurements of fig.3.

The Origin of the Anomalous Off-current in SOI-Transistors

[*]L.J. MCDAID, [*]S. HALL, and [*]W. ECCLESTON, [+]J.C. ALDERMAN

[*]The University of Liverpool, Department of Electrical Engineering and Electronics, Brownlow Hill, P.O. Box 147, Liverpool, L69 3BX.
[+]Allen Clark Research Centre, Plessey Research (Caswell), Towcester, Northants, NN12 8EQ.

Abstract

The leakage currents exhibited by silicon-on-insulator (SOI) transistors show a strong dependence on gate and drain voltage. Measurements on lateral p[+]-n junction diodes fabricated using separation by implanted oxide (SIMOX) technology suggest that reverse leakage currents originate from field enhanced generation in the depletion region. A model is presented which supports this process. In SOI-transistors field intensification occurs in the drain depletion region for increasing drain voltage and increasing gate voltage in the "off" direction. This mechanism can therefore explain the strong dependence of leakage currents on terminal voltage.

Introduction

CMOS integrated circuits are a primary application of SOI as latch-up is eliminated with this technology. However, it is imperative that off-state leakage currents, which determine the stand-by power, are low. Although thin-film SOI-transistors show near ideal subthreshold slopes [1] the off-state leakage current is unacceptably high at large drain voltages. Furthermore, these currents are strongly dependent on gate voltage for the "off" condition, such behaviour cannot be explained by thermal generation alone.

Fig.1. shows typical off-currents for an n-channel SIMOX transistor with drain voltage as a parameter. These currents are constrained to flow to the source where they can be amplified to an extent that depends on the carrier lifetime within the film and the channel length. If the gain of the parasitic bipolar times the multiplication factor is ≥ 1 then the drain current is no longer influenced by the gate voltage and the transistor will remain on. Hence, the presence of this effective base current sets a limit on scaling and drain voltage. Before design steps can be taken to eliminate this current it is necessary to identify its origin. To this end the following work was carried out.

Experimental results

Measurements were performed on p[+]-n lateral junctions fabricated using SIMOX technology (1405[o]C anneal). This enables a one-dimensional analysis of the data as the diodes are sandwiched between the buried oxide and an upper thick deposited oxide. Presented in Fig.2. is a typical diode leakage current characteristic as a function of reverse voltage and Fig.3. shows this current plotted against $1/kT$, where k is the Boltzmann constant and T is temperature. Two conclusions can be drawn from this data. Firstly, the leakage currents are strongly voltage and temperature dependent; secondly, the slope of the graphs in Fig.3. reduces with applied voltage.

Several mechanisms can give rise to leakage currents in reverse biased p^+-n junctions. These are thermal generation, weak multiplication, direct band-to-band tunnelling or tunnelling via a localised state, thermal generation enhanced by the presence of an electric field.

Tunnelling is unlikely as this process is weakly dependent on temperature. However, with increasing voltage and thin gate oxides, tunnelling can dominate [2]. Weak multiplication has a negative temperature coefficient inconsistent with the results presented in Fig.3. Thermal generation is strongly temperature dependent, through the intrinsic carrier concentration, with almost constant activation energy. This again is inconsistent with the results shown in Fig.3. As a consequence of the qualitative discussion above we are left with mechanism (4); that is thermal generation enhanced by the presence of an electric field. The equation that describes this mechanism in a region of varying electric field is now derived.

Theory

If the field dependence of thermal generation is ignored then the current density dJ, resulting from thermal generation over a small distance dx, of the diode depletion region is given by

$$dJ = -\frac{qn_i}{2\tau_g} dx \tag{1}$$

where n_i is the intrinsic carrier concentration, τ_g is the generation lifetime and q is the electronic charge. In this case the activiation energy E_a for the generation process is $E_g/2$, where E_g is the energy gap of silicon. If we now assume that the traps are Coulombic and wellspaced, then the presence of an electric field leads to a reduced activation energy which is given by [3]

$$E_a = \frac{E_g}{2} - \left(\frac{qE(x)}{\pi\varepsilon_o\varepsilon_s}\right)^{\frac{1}{2}} \tag{2}$$

where ε_o and ε_s is the permittivity of free space and the relative permittivity of silicon respectively and E(x) is the field at distance x, From (1) and (2) the total current density is given by

$$J_t = -\frac{qn_i}{2\tau_g} \int_o^w \exp\left[\frac{q}{kT}\left(\frac{qE(x)}{\pi\varepsilon_o\varepsilon_s}\right)^{\frac{1}{2}}\right] dx \tag{3}$$

where W is the depletion width. Writing the variables in (3) in terms of potential ϕ, yields

$$J_t = \frac{qn_i}{2\alpha\tau_g} \int_o^{V_r}\left[\frac{\exp(\beta\phi)}{\phi^{\frac{1}{2}}}\right] d\phi \tag{4}$$

where the constants β and α are given by

$$\beta = \frac{q}{kT} \left[\frac{2q^3 N_d}{\pi^2 \varepsilon_o^3 \varepsilon_s^3} \right]^{1/4} \tag{5}$$

$$\alpha = \left[\frac{2qN_d}{\varepsilon_o \varepsilon_s} \right]^{1/2} \tag{6}$$

V_r is the applied voltage, N_d is the doping concentration on the low doped side and we have assumed a one-sided step junction approximation. Carrying out this integral, using integration by parts, gives an expression for the leakage current of the form

$$I_r = \frac{2A_j n_i}{\alpha\beta\tau_g} \left\{ \frac{1}{\beta^2} (\exp(\beta V_r^{1/4})-1) + V_r^{1/2}\exp(\beta V_r^{1/4}) \right\} \tag{7}$$

where A_j is the junction area. For $V_r > 1$ equation (7) approximates to

$$I_r \simeq KV_r^{1/4} \exp(\beta V_r^{1/4}) \tag{8}$$

result for the proposed mechanism to be valid. Fig.4. shows a plot of $Ln(I_r/V_r^{0.25})$ vs $V_r^{0.25}$ and indicates a good straight line fit with a linear regression coefficient >0.99. Note that at high voltages the current no longer obeys equation (8). This region was found to be consistent with band-to-band tunnelling. The doping level, calculated from the slope of the graph in Fig.4. through use of equation (5), was 9 x 10^{18} cm^{-3}. This level of doping is that expected on the high doped side of the junction, the p$^+$ region, and this therefore suggests that a fraction of reverse voltage falls across the high doped region probably due to field intensification in the thin-film [4].

Conclusion
Experimental evidence and supporting theory presented in this paper suggest that the anomalous leakage currents observed in SIMOX p$^+$-n diodes result from field enhanced generation in the depletion region. The temperature dependence of this leakage current indicates barrier lowering resulting from high fields in the depletion region due to thin-film effects. Because field intensification occurs in the drain region of SOI-transistors with both increasing drain and gate voltages, this mechanism can therefore explain the off-currents, which act as additional base current for the parasitic bipolar transistor.

Acknowledgements
The authors would like to thank the University of Surrey for the SIMOX material and G. Celler of AT & T Bell Labs. for the 1405°C anneal. This work has been partially supported by the procurement executive MOD (RSRE) under contract RP009/335.

References
[1] J.P. Collinge, IEEE Electron Dev. Lett, Vol.EDL-7, No.4, pp.244-246, April 1986.
[2] J. Chen, et al, IEEE Electron Dev. Lett, Vol. EDL-8, No.11, November 1987.
[3] J. Frenkel, Tech. phys. USSR 5, 685 (1938); phys. Rev. 54, 647 (1938).
[4] M. Yoshimi et al, VLSI Tech. Conference 1080.

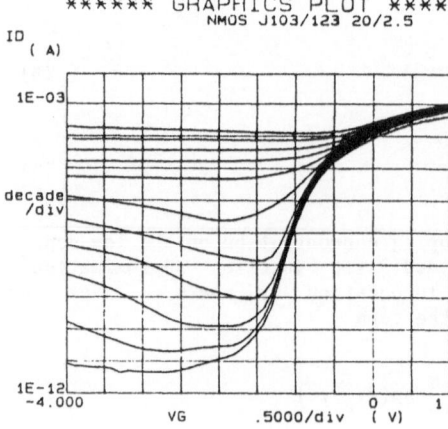

Fig.1. Typical off-currents
for an n-channel SIMOX transistor
as a function of gate voltage with
drain voltage varying from 0.5 to
5.5 volts in steps of 0.5 volts.

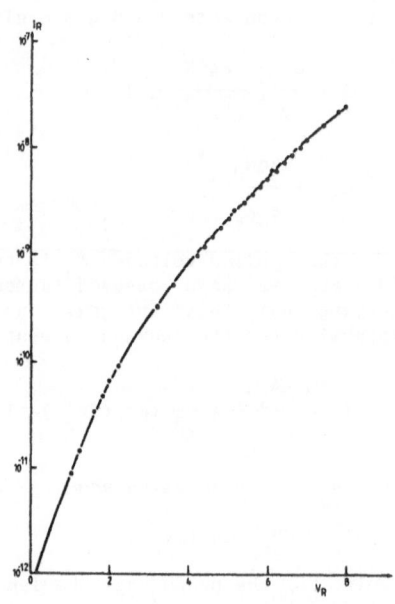

Fig.2. Reverse leakage current
from p^+-n diode.

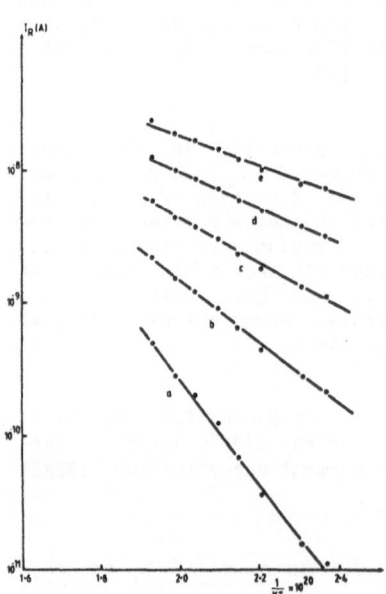

Fig.3. Temperature dependence
of the leakage current with reverse
voltage as a parameter. V_r = 1 to
5 volts in steps of 1 volt (a-e)

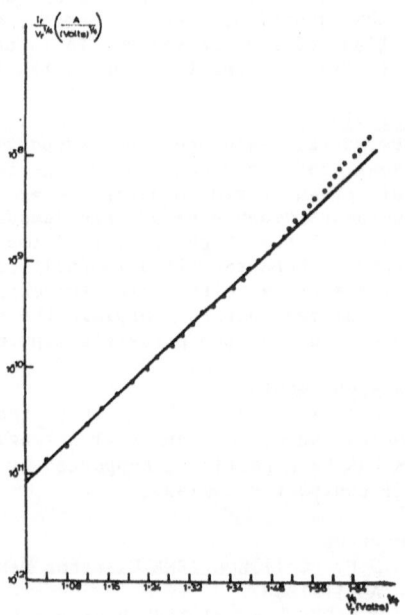

Fig.4. Data in Fig.2. plotted
according to equation (8).

Two-dimensional Simulation of SOI MOSFETs

H. Belhaddad (*), A. Poncet (**), G. Merckel (**)

(*)   Nucletudes S.A. Avenue du Hoggar, Z.A. Courtaboeuf BP 117 –
      F – 91944 LES ULIS Cedex
(**) CNET–Grenoble Chemin du vieux Chene, BP 98 F – 38243 MEYLAN
      Cedex

I - INTRODUCTION :

SOI MOSFET is an attractive device for realizing three-dimensionnal integrated circuits. It offers many advantages including fast switching speeds and increased radiation hardness , and provides an alternative to silicon-on-saphire. The main drawback of SOI, for both design and fabrication of SOI IC's, is the "Kink" effect. Several methods can be used to control or to reduce this undesirable effect : applying of a suitable back gate bias, controlling properly the lifetime of silicon material, using thin films... This latter solution has the major advantages.

This work presents simulation and experimental results on SOI MOSFET, laying stress on the floating body. The analysis of the sensitive parameters which affect the operating of such structures has been done : influence of lifetime, of film thickness and substrate concentration on both subthreshold and strong inversion behavior.

Avalanche generation due to impact ionization is modeled using an original model (/1/), named "Pnstat". Simulation results are compared with both numerical results using standard Chynoweth model, and measurements.

II - NUMERICAL MODELLING USING TITAN-V PROCESS/DEVICE SIMULATOR :

Computer simulations have been performed with **TITAN-V** process/device simulator (/2/); the device part of the simulator solves coupled Poisson's and carrier continuity equations, and is designed to deal with both steady-state and transient conditions. All the major physical effects are taken into account,i.e. :

* Mobility versus electric field, doping profiles and carrier concentrations ; * Shockley-Read-Hall
  recombination including doping dependent lifetime model ; * Auger recombination;
* Impact ionisation : the new model for avalanche simulation "Pnstat" is based on an explicit expression
  of ionization coefficients including "quasi-non-stationnary" aspects and a lucky electron concept.
* Hot carriers (/3/).

III - SIMULATION RESULTS - COMPARISON WITH MEASUREMENTS :

A measured ID(VDS) characteristic, for a 0.25 µm thick silicon film device (Fig.1), and a fixed gate bias of 3 V, is compared with both simulated results obtained when using standard Chynoweth or

Pnstat models (Fig.2-a). A best fit is obtained with the new impact ionization model. In fact, Chynoweth model, for which ionization coefficients are those calculated by Moll and Van Overstraeten, overestimates the multiplication hole current, either in the case of bulk MOS devices (/2/,/4/) or SOI (/5/). This propensity is clearly illustrated in Fig.2-b where the potential within the floating region (VBS) is plotted versus VDS for the two avalanche models.

Fig.3-a shows the output characteristic ID(VDS) for two different lifetimes. It can be seen that both threshold of Kink effect and abruptness of breakdown mode are influenced by this parameter (/5/). Such a result is easily explained looking at Fig.3-b : as lifetime increases, the floating body potential increases, which leads to a stronger enhancement of the channel current.

The simulated characteristics in the subthreshold region are in good agreement with measurements (Fig.4-a). The simulation describes accurately the rapid increase of VBS with VDS in the subthreshold region. Fig.4-b represents VBS(VGS) curves when VDS is stepped from 0.1 to 5 V and for two different lifetimes.

The effects of hole accumulation in the floating region are shown on Fig.6 : the potential profiles, at a distance of 0.22 μm from the surface, are plotted for VDS = 2, 3.7, 6 V , and VGS = 3 V ; it can be seen that the source junction becomes more and more forward biased. Fig.5 exhibits the strong injection mechanism related to the substrate-source junction.

A cross section of the electric field, near the front Si-SiO2 interface, is plotted on Fig.7 , for the same gate and drain voltages. The large increase in the electric field peak, resulting from increasing VDS, enhances drastically the generated carrier concentration; this is due to the exponential relationship between electrical field and ionization rate.

Fig. 8a-8b show the computed generation and recombination distributions of carriers at VDS = 6 V and VGS = 3 V . The peak near the substrate-drain junction (Fig. 8-a) corresponds to carrier generation originated from the avalanche process. The recombination, near the substrate-source junction (Fig. 8-b), increases with the drain voltage.

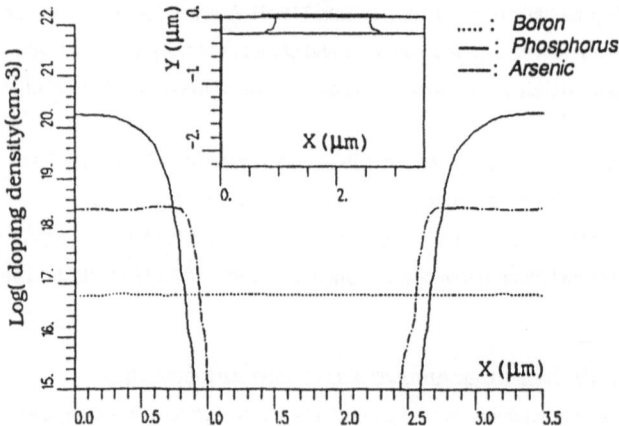

Fig. 1. Impurity profile accross substrate near the front SI-SiO2 interface . The simulated structure is a 0.25 μm film device .

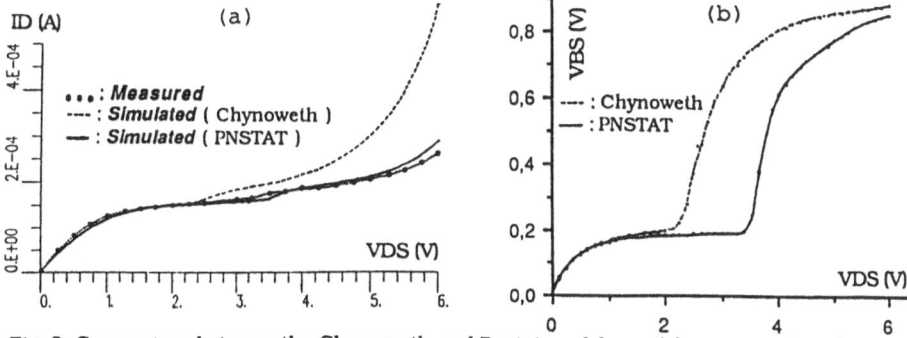

Fig. 2. Comparison between the Chynoweth and Pnstat models ; (a) measured and simulated ID versus VDS characteristics (b) simulated VBS versus VDS characteristics . In both cases VGS=3. V and τ=5.e-8 s

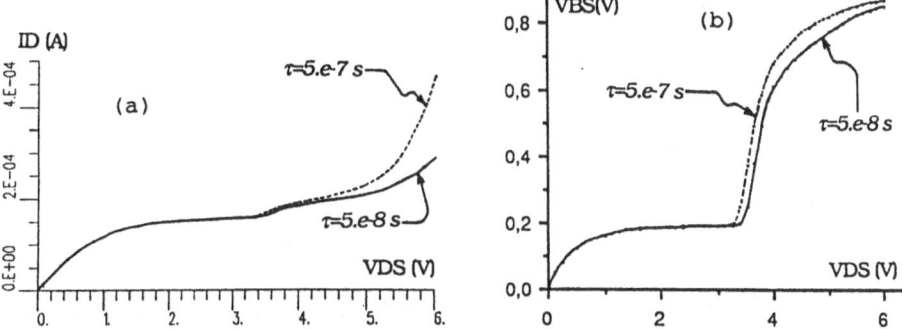

Fig. 3. Simulated (a) ID(VDS) and (b) VBS(VDS) characteristics for two different lifetimes : 5.e-8 s and 5.e-7 s . In both cases VGS =3. V

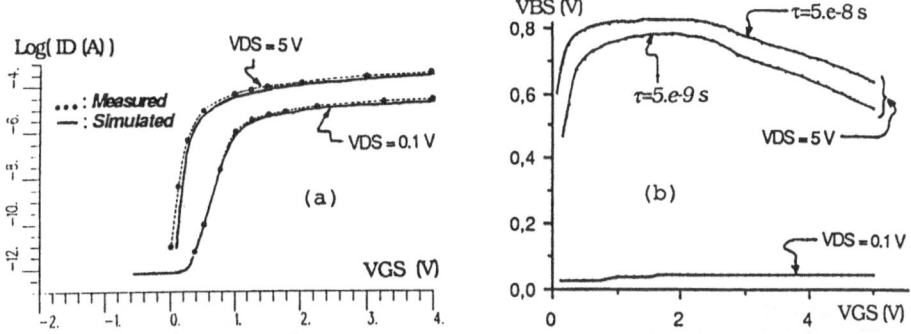

Fig. 4. (a) Measured and simulated ID(VGS) subthreshold characteristics for VDS=0.1 and 5 V (b) Simulated VBS(VGS) characteristics for two different lifetimes : 5.e-8 s and 5.e-9 s

IV- ACKNOWLEDGEMENTS :

The authors wish to express their gratitude to M. Haond and T. Pedron,from CNET, for assistance with the measurements and for fruitful discussions.

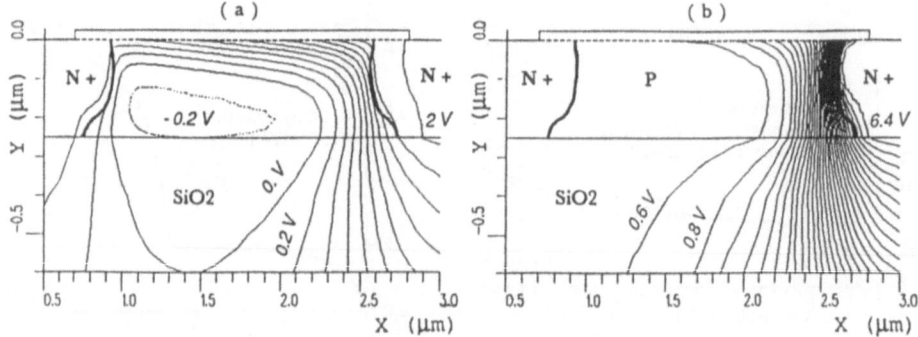

Fig. 5 . Equipotential lines (VGS=3. V) for : (a) VDS=1.5 V (b) VDS=6. V

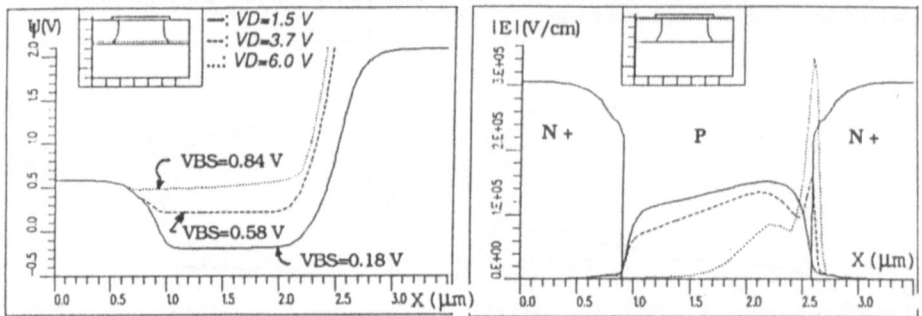

Fig. 6 . Potential profile accross substrate , near the back Si-SiO2 interface

Fig. 7 . Electrical field profile accross substrate, near the front Si-SiO2 interface

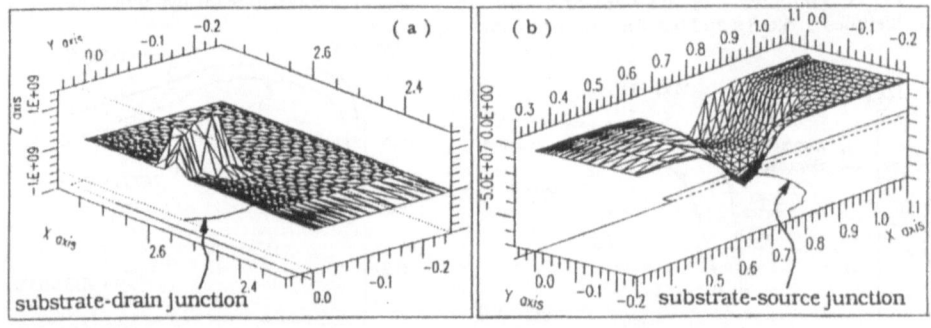

Fig. 8. Generation and recombination distributions in the silicon layer near :   (a) the substrate-drain junction (b) the substrate-source junction for VDS=6.V and VGS=3. V

V - REFERENCES :

/1/ M. Garrigues, A. Alexandre, T. Pedron - To be published

/2/ A. Gerodolle et al. "TITAN-V process/device simulator" NASECODE VI ,Short course, Boole Press 1989

/3/ M. Garrigues et al "Two dimensionnal computer simulation of hot carrier degradation in NMOSFETs" ESSDERC'88 Montpellier -Journal de Physique - Sept. 1988

/4/ J.W. Slotboom et al. "Surface impact ionization in silicon devices" IEEE IEDM 87 p.494

/5/ S.P. Edwards et al. "Two-dimensionnal numerical analysis of the floating region in SOI MOSFET's" IEEE Trans. El. Dev. Vol.35 N° 7 July 1988 p.1012

Characterization of Different SOI-MOS Technologies at Cryogenic Temperatures

M. TACK, M. H. GAO, C. CLAEYS and G. DECLERCK
IMEC, Kapeldreef 75, B-3030 Leuven, Belgium

Abstract:

The operation of SOI nMOS and pMOS transistors working in both inversion mode and buried channel/accumulation mode is investigated at liquid helium temperature, and compared to conventional inversion mode bulk MOS transistors. The pronounced hysteresis and kink behaviour seen in bulk MOS inversion mode transistors is strongly reduced in SOI MOS inversion mode transistors, and fully eliminated in SOI MOS buried channel/accumulation mode transistors. A general charge based model for both bulk and SOI MOS transistors is presented and discussed. It is concluded that SOI offers some substantial advantages over bulk for operation at cryogenic temperatures.

1. - INTRODUCTION

Conventional inversion mode nMOS transistors, made in bulk silicon, exhibit a pronounced kink and hysteresis behaviour at temperatures in the 4 Kelvin (liquid He) range (Fig.1 and [1-2]). This seriously affects their use for a number of practical applications.

In our work, we investigate the use of both inversion mode and buried channel/accumulation mode transistors made in a SOI technology, as an alternative for cryogenic electronics.

2. - EXPERIMENTS

Both nMOS and pMOS transistors are made using a modified 3 μm CMOS process in a ZMR-SOI technology [3]. The thickness of the silicon film, the front oxide and the underlying oxide is respectively 450 nm, 60 nm and 500 nm. A N+ polysilicon gate is used and the silicon film doping level is about $2 \cdot 10^{16}$ cm^{-3}. An important feature is the use of junctions reaching through the Si film. Also, devices made in bulk silicon are added for reference. Transistors made in bulk and in SOI material show a very comparable behaviour at room temperature, confirming the good SOI material quality, except for the well known 'kink effect' in the SOI devices due to the floating Si film [4]. The cryogenic measurements are performed by inserting the DIL mounted devices directly in a liquid helium storage dewar.

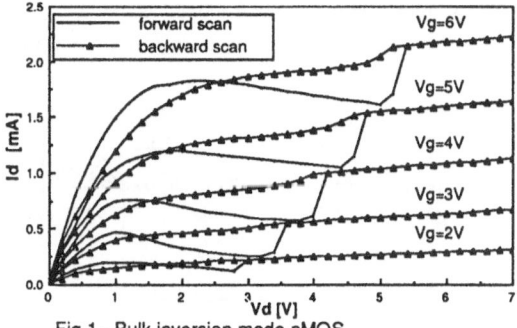

Fig.1 - Bulk inversion mode nMOS

3. - INVERSION MODE TRANSISTORS

3a. Bulk inversion mode transistors

When immersed in liquid helium (4.2 Kelvin), the bulk inversion mode nMOS (N+P-N+) transistors exhibit a very pronounced hysteresis and kink behaviour (Fig.1). These effects have been explained in terms of a 'kink effect' model similar to that of room temperature SOI transistors [1]. Recently this theory is partly contradicted and extended by the *'Forced Depletion Layer Formation'* model [2], which especially accounts for the hysteresis.

3b. SOI inversion mode transistors

To investigate both models, inversion mode SOI transistors are characterized at 4.2 K. As is clear in Fig.2, the 'after kink' behaviour of SOI nMOS transistors is very similar to the bulk case. However, both the kink magnitude and the hysteresis are much reduced for the SOI devices. A very similar behaviour is also obtained for SOI inversion mode pMOS transistors (P+N-P+)(Fig.3).

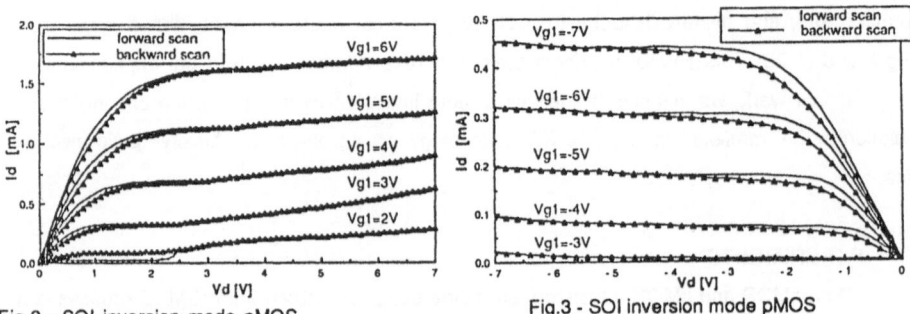

Fig.2 - SOI inversion mode nMOS   Fig.3 - SOI inversion mode pMOS

3c. Discussion - a charge based model

The main difference between SOI and bulk inversion mode transistors is the presence of an underlying oxide in the SOI case, which results in a floating Si film. For the bulk case, this Si film is connected to the substrate or well contact through a substrate or well resistance. At higher drain voltages impact ionization occurs in the pinch-off region near the drain, driving a hole current I_f (for the case of a nMOS) into the Si film. In this regime the SOI and

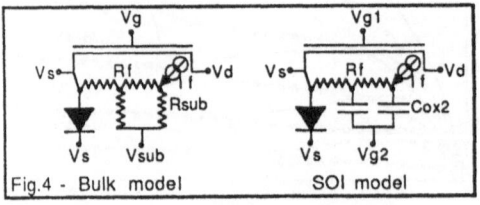

Fig.4 - Bulk model   SOI model

bulk transistors can schematically be represented as shown in Fig.4. The intrinsic current source I_f is dependent on Vg1, Vd and Vf (local Si film voltage). Cox2 represents the capacitance of the underlying oxide (SOI), Rf the series resistance under the channel, Rsub the substrate or well resistance (bulk) and the diode the source junction.

At 4.2 K Rsub should be regarded as a non-linear resistance caused by charge build-up Qf in the Si film (*'space charge limited current'* effect). This Qf will cause an increase in Vf, thus a decrease in Vt through the body-effect and originates the first kink in the bulk devices. The second, flattening, kink is caused by the turn on of the source diode at Vf ≈1.1V, ie. the turn-on voltage of the diode at 4.2 K. This model is recently discussed in more detail by Simoen et al. [5].

In SOI a very similar mechanism occurs: charge Qf is built up in the 'floating' Si film, thus increasing Vf and decreasing Vt, until the source diode is turned on and Vf is pinned at the diode turn-on voltage.

However, the main difference between SOI and bulk transistors is caused by the very different mechanisms by which Qf is reduced to its equilibrium condition at low Vd, ie. when the current source If is switched off. For the bulk case, these accumulated or trapped (eg. freeze-out) holes will be removed mainly through the body contact; the time constants involved will lead to the hysteresis behaviour below the kink. In SOI transistors however, Qf can only flow away through the source/drain junctions. Because of the very abrupt diode beha -viour at 4.2 K (thus very small currents below the turn-on voltage) and the small junction area in SOI devices with reach through junctions, this will be an extremely slow process (at least several hours). Consequently, in SOI transistors operating at 4.2 K Vf will always be pinned at the diode turn-on voltage in practical applications. This explains the absence of both a pronounced hysteresis and kink in SOI inversion mode transistors, with respect to bulk inversion mode transistors. It should be noted that SOI inversion mode transistors operating at cryogenic temperatures thus always behave in a stable non-equilibrium condition [6].

The further increase in output current at higher drain voltages can be explained by a continuous adjustment of Vf (thus Vt and Id) at increasing If and is determined mainly by the shape of the diode on-characteristic and by Rf. As this should be very similar for SOI and bulk devices no big difference is expected in this region, as is confirmed above.

4. BURIED CHANNEL/ACCUMULATION MODE TRANSISTORS

Although it is shown above that inversion mode transistors made in SOI behave much better than their bulk counterparts, they still exhibit some hysteresis and kink.

To overcome this, an alternative SOI CMOS process is presented, based on buried channel/accumulation mode transistors (N+N-N+ and P+P-P+ structures). As these devices have much reduced fields (no junctions, as opposed to inversion mode transistors, N+P-N+ and

P+N-P+), and so a much reduced impact ionization effect, and as they don't have a floating body, they are expected to show a strongly suppressed hysteresis and kink behaviour. Furthermore, by constructing these devices in a rather thin SOI layer, it is possible to suppress the parallel leakage current seen in the bulk case [7]. Figure 5 displays the characteristics of a buried channel/accumulation mode SOI pMOS transistor at 4.2 K, as opposed to an inversion mode SOI pMOS device in Fig.3. No kink or hysteresis is seen for the former case. Furthermore, this device also has a small output conductance, which makes it attractive for analog applications, and an enhanced current drive capability. Similar results are obtained for a buried channel/accumulation mode SOI nMOS transistor, as shown in Fig.6.

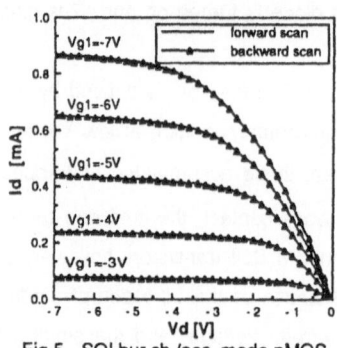
Fig.5 - SOI bur.ch./acc. mode pMOS

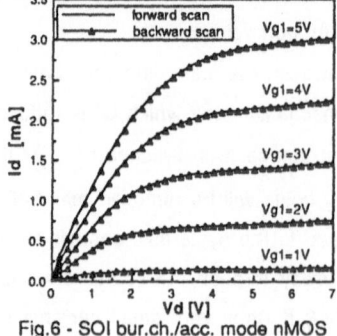
Fig.6 - SOI bur.ch./acc. mode nMOS

5. CONCLUSIONS

The operation of different SOI MOS technologies at cryogenic temperatures is studied and compared to conventional bulk MOS technologies. A general charge based model is presented to explain the hysteresis and kink behaviour at 4.2 K in both SOI and bulk inversion mode transistors. It is shown that both inversion mode and buried channel/accumulation mode transistors made in SOI behave superior over their bulk counterparts. SOI thus offers some substantial advantages over bulk for operation at cryogenic temperatures.

References
[1] F.Balestra et al., Solid State Electron. 30, p.321 1987
[2] B.Dierickx et al., IEEE Trans.Electron Devices ED-35, p.1120 1988
[3] D.Wouters et al., Solid State Devices, p.549 1988
[4] S.Eaton et al., IEEE Trans.Electron Devices ED-25, p.907 1978
[5] E.Simoen et al., IEEE Workshop on Low Temp Electronics, Burlington Aug.1989
[6] M.Tack et al., IEEE Workshop on Low Temp Electronics, Burlington Aug.1989
[7] B.Dierickx et al., Journ. de Physique 49, C4 p.741 1988

Analysis of Charge Conservation
in Isolated Silicon Regions

R. Kircher and W. Bergner

Siemens AG, Corporate Research and Development,
Otto-Hahn-Ring 6, 8000 München 83, FRG

Abstract

A simple formalism for charge conservation is presented, which allows the physically correct prediction of the electrical behavior of isolated silicon regions in two or three dimensions. It is shown that an integral boundary condition can be applied to the semiconductor equations to obtain a physical solution. The charge conservation formalism has been used to study the physics of isolated silicon regions. The influence of charge conservation on the electrical coupling of multi-layered structures is seen in the C-V curve of a two-terminal SOI capacitor. This formalism does not only allow the correct modeling of the electrical characteristic of a trench isolation with polysilicon as trench filling and a capping oxide, it does also allow the prediction of the electrical behavior influenced by a mobile charge confined in the trench filling.

1 Charge Conservation Formalism

The correct modeling of isolated silicon regions e.g. SOI layers or trench fillings represents a non-trivial problem for the device simulation in more than one dimension, especially if these regions are surrounded by an isolating oxide. In this case it is not possible to apply Dirichlet boundary conditions to the variables which make the solution of the semiconductor equations unique. By physical considerations an integral boundary condition can be formulated [1] which allows to find a physical and unique solution of the equations

$$\nabla \cdot \epsilon \nabla \Psi = -q(p - n + N) - Q_{it} \tag{1}$$

$$\frac{\partial n}{\partial t} = G - R - \nabla \cdot \vec{j}_n \tag{2}$$

$$\frac{\partial p}{\partial t} = G - R - \nabla \cdot \vec{j}_p \tag{3}$$

for any bias condition. The charge conservation can be expressed by the free carrier densities p and n, the ionized impurity concentration N and the fixed interface and oxide charges Q_{it}

$$Q = q \int_{V_i} (p - n + N) dv + Q_{it} = const. \tag{4}$$

A relaxation procedure between the semiconductor equations and charge conservation is used to take into account a coupling between the electrostatic potential and the charge confined in the isolated region V_i. By an iterative method the physically correct value of the quasi-Fermi levels is found. The method has been implemented into the two-dimensional device simulator GALENE II [2, 3] for arbitrary device structures. The schematic diagram of the algorithm is shown in Fig. 1.

2 Examples

The charge conservation formalism has been applied to a two-terminal SOI capacitor, consisting of a 250nm thick silicon film sandwiched between two metal gates, as shown in Fig. 2. For the doping profile a homogeneous p-type doping concentration of $1 \cdot 10^{15} cm^{-3}$ has been assumed. The gate oxide thickness is $100 nm$ for both sides. For simplicity the work function difference Φ_{ms} is set to $0V$, and interface and oxide charges are neglected because simulations have shown that a non-vanishing work function difference and oxide charges only slightly modify the results and are not essential for the understanding of charge conservation.

The terminal charge has been calculated as a function of the applied voltage. The upper gate voltage was varied in the range $-15V \leq V_{Gu} \leq 15V$, whereas the lower gate is fixed on ground potential. The capacitance is then obtained by the derivative $\partial Q / \partial \Psi$. This procedure was carried out with application of the integral boundary condition of charge conservation, and for the case that no boundary condition is given for the SOI film. The result of this comparison is shown in Fig. 3. The straight line represents the result with, and the dotted line the result without integral boundary condition. They differ not only in magnitude but also in shape. The curve with charge conservation exhibits two minima at $V_{Gu} = \pm 2V$. This is because the special symmetry of this device causes different charge conditions at the interfaces. For positive gate bias we have depletion at the upper and accumulation at the lower interface and vice versa for negative gate voltages.

Another device structure which is of a more practical interest is the trench isolation with polysilicon filling and a capping oxide. Such a structure is illustrated in Fig. 4. It is used to isolate neighboring active devices. For the trench a depth of $3\mu m$ and a width of $1\mu m$ has been assumed. To improve the isolation behavior a channel stopper profile has been implanted at the trench bottom. The doping profiles are taken from SUPREM 3 [4] simulations.

To investigate then influence of charge carriers confined in the trench filling, the leakage current has been calculated for the fixed bias condition of $V_S = 0V, V_D = -5V$ and $V_B = -5V$. The results of the simulations are shown in Fig. 5 for two different values of the oxide thickness.

If the polysilicon filling is connected to an ohmic contact, the quasi Fermi potentials are fixed. However, in the absence of the contact they are determined by the charge conservation formalism. Our simulations have shown that for the case $Q = 0C$, their value is $\Phi = -2.7V$. Considering the device as a MOSFET this means that the gate voltage is positive relative to bulk, and the transistor may be in on-state. To guarantee a proper isolation the threshold voltage of this parasitic transistor should be sufficiently high. This explains the strong dependence of leakage current on the oxide thickness.

References

[1] Kircher, R.; Bergner, W.: Modeling of Charge Conservation in Isolated Silicon Regions. Submitted to JJAP.

[2] Engl, W.L.; Kircher, R.; Bach, K.H.; Götzlich, J.; VLSI Process/Device Modeling Workshop (1987), Tokyo.

[3] GALENE II User's Guide (1988). RWTH Aachen, West-Germany.

[4] Ho, C.P.; Hansen, S.E.: SUPREM 3 - Program for integrated circuits process modeling and simulation. Stanford University Technical Report 1983.

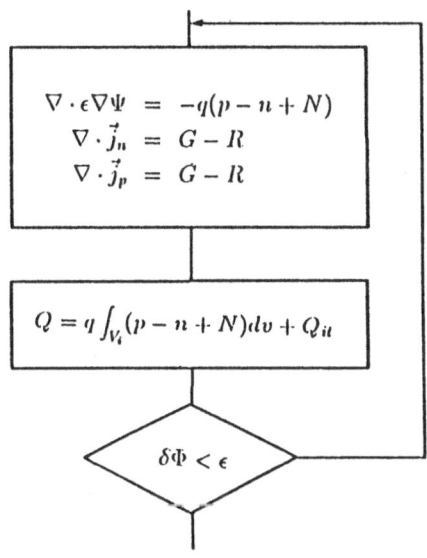

Fig. 1. Schematic diagram of the relaxation algorithm for charge conservation.

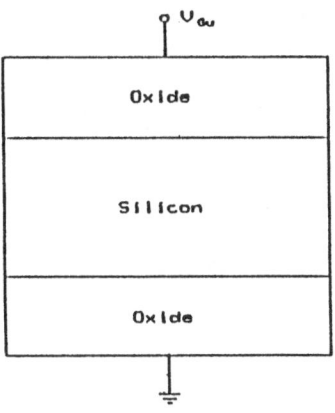

Fig. 2. Cross section of the simulated structure of an isolated silicon film sandwiched between two metal gates.

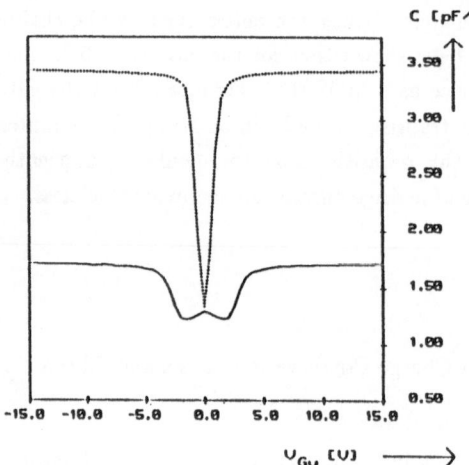

Fig. 3. Comparison of the C-V curves of the sandwiched film. The solid line represents the result with and the dotted line that without charge conservation.

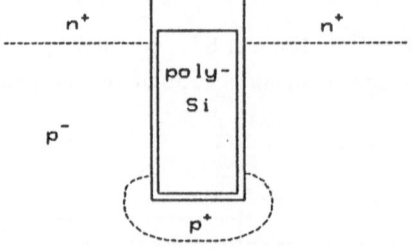

Fig. 4. Illustration of the trench isolation structure.

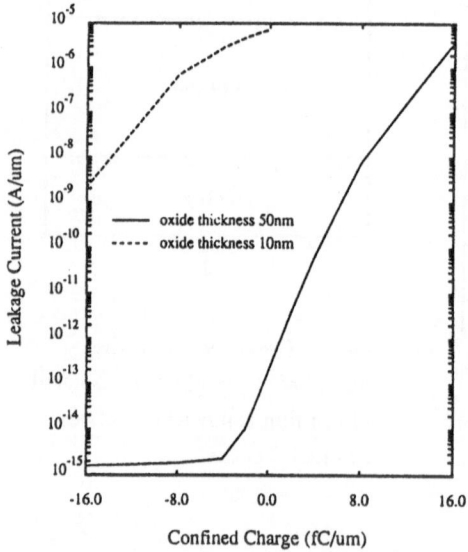

Fig. 5. Influence of a confined charge on the leakage current.

An Advanced Fabrication Process for 3D-CMOS Devices

R. Buchner, K. Haberger, S. Seitz, J. Weber, W. van der Wel*,P. Seegebrecht

Fraunhofer-Institut für Festkörpertechnologie
  Paul-Gerhardt-Allee 42, D-8000 München 60, FRG
* Philips Research Laboratories, Hamburg, FRG

Abstract

An advanced 2 µm 3D-CMOS process was developed which allows the fabrication
of NMOS devices in the substrate and CMOS devices in a thin laser recrystallized polysilicon layer. The processing parameters were determined carefully in order to obtain a high-quality SOI layer and to avoid any degradation of underlying substrate devices. The fabricated devices in both layers
show customary bulk device quality.

Introduction

Integrated circuit semiconductor development is characterized by constantly
increasing device packing density. Apart from a reduction of the lateral
device dimensions, this can also be achieved by using three dimensional
(3D) integration techniques /1/. This approach also offers most promising
opportunities for the realization of multifunctional circuits: mixed circuit technologies (e.g. digital / analog), mixed process technologies (e.g.
CMOS / bipolar) and the combination of different semiconductor materials.
As one step towards this aim, we developed a 2 µm 3D-CMOS process.

Process

The process allows the fabrication of MOS devices in the silicon substrate
and in a thin overlying recrystallized polysilicon layer. Because of the
thermal stability of arsenic, we fabricated NMOS devices in the substrate.
In the recrystallized polysilicon layer we realized CMOS devices.

Starting material for the process are 4" p-type (100) silicon wafers. The
first phase of the process is the utilization of a polysilicon gate NMOS
process to fabricate the n-channel devices using a 500 A thick gate oxide

and a minimum feature size of 2 µm. In order to simplify the contacting of the devices in the first layer, buried contacts are used. At the completion of this phase the surface exhibits significant topography, which will disturb the recrystallization process. Therefore, a planarization process is applied, which serves simultaneously to insulate the first device layer.

For obtaining a recrystallized silicon film of a definite crystal orientation it is necessary to use seeds [1]. Therefore, seed windows are opened in the insulating oxide. Then a 0.5 µm thick polysilicon film is deposited and covered with a 1700 Å thick LPCVD oxide capping layer and a 590 Å thick Si_3N_4 layer. The nitride film, which acts as antireflective layer, is patterned into 6 µm wide stripes with a spacing of 12 µm. These integrated absorbers modify the melt zone temperature profile for obtaining a grain boundary entrainment. The polysilicon is recrystallized by means of an argon laser system. Thereby care has to be exercised with respect to the occurrence of substrate damage, which will degrade substrate devices. Therefore we have optimized the recrystallization process for obtaining a high-quality SOI layer and a defect-free substrate simultaneously [2]. In this process window the laser power, beam width, scan speed and substrate temperature are 14 W, 120 µm x 50 µm, 10 cm/s and 500°C, respectively.

Following recrystallization, the capping layers are removed and the silicon film is divided into individual islands to realize dielectric insulation. Then a polysilicon gate CMOS process is used to fabricate devices in the recrystallized layer with its active areas lying in defect-free regions. A schematic cross-section of the fabricated structure is shown in Fig. 1.

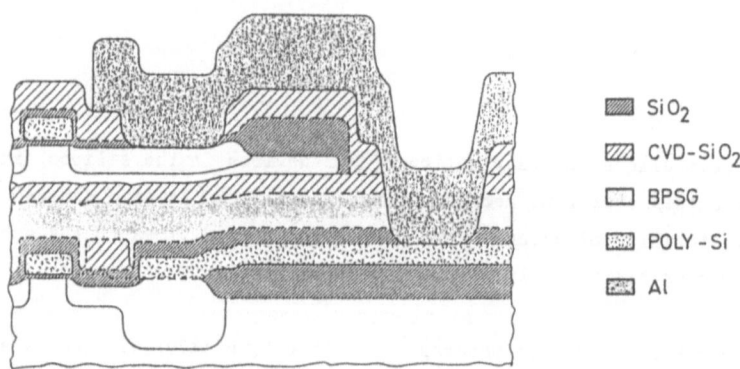

Fig. 1: Schematic cross-section of a fabricated 3D device.

Results

For the fabrication of 3D devices it is very important to avoid any damage
to existing substrate devices, especially by the laser recrystallization.
Our investigations have shown that the electrical characteristics of NMOS
substrate devices have not been influenced by our recrystallization process. Threshold voltage and channel mobility remain the same and are 0.9 V
and 770 cm²/Vs respectively. After completion of the second active layer,
the device parameters have been changed slightly, but reproducibly to
0.7 V and 800 cm²/Vs. Since MOS devices are not so sensitive to substrate
damage, we looked at the leakage current to examine the quality of the
substrate after laser recrystallization. Fig. 2 shows the subthreshold
characteristics of the bulk devices before and after recrystallization and
it proves that there is no substrate damage, the leakage current remains
the same and this is consistent with crystal investigations.

We have seen that the recrystallization process does not affect the electrical behaviour of the underlying devices. However, the quality of the
CMOS devices fabricated in the thin recrystallized polysilicon layer, too,
is not of minor importance. The n-channel devices exhibit a threshold voltage of 0.8 V and a channel mobility of about 850 cm²/Vs, whereas the values
of the p-channel devices are -1.3 V and 300 cm²/Vs respectively. These data
illustrate that the devices fabricated in recrystallized polysilicon
achieve customary bulk device quality. Additionally we have fabricated CMOS

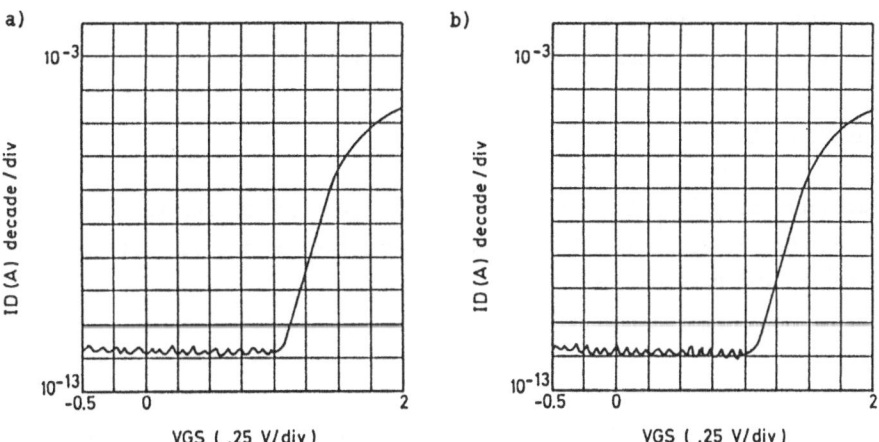

Fig. 2: Subthreshold characteristic of a bulk NMOS transistor before (a)
and after (b) laser recrystallization.

inverters and oscillators in the second layer. The output signal of a ring oscillator at a supplying voltage of 5 V is shown in Fig. 3. The measurements also demonstrate that the oscillator frequency also shows the well-known linear dependence on the supplying voltage.

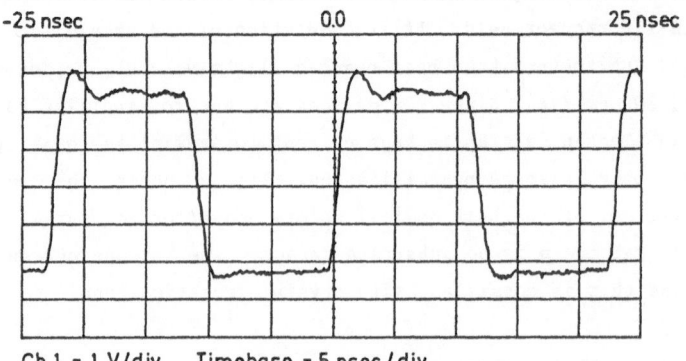

Fig. 3: Output signal of a 23 stage buffered SOI ring oscillator.

Summary

An advanced 3D-CMOS process has been developed which allows the fabrication of CMOS devices in recrystallized polysilicon with electrical characteristics similar to those of devices in monocrystalline material. The investigation has also shown that the fabrication process does not influence the performance of underlying MOS devices already existing.

Acknowledgements

The authors are grateful for the technical support of M. Forster, A. Heidenreich and B. Schmiedt. This work was supported by the Federal Ministry of Research and Development.

References

1. Y. Akasaka: Proc. IEEE, Vol. 74, p. 1703 (1986)
2. W. van der Wel, R. Buchner, K. Haberger, P. Seegebrecht: Extended Abstracts 175th Electrochem. Soc. Meeting, p. 285, May 1989

VLSI Bipolar

Predicted Propagation Delay of Si/Si$_{1-x}$Ge$_x$ Heterojunction Bipolar Circuits

Z. A. SHAFI, P. ASHBURN and G. J. PARKER

Dept. of Electronics and Computer Science,
The University, Southampton, SO9 5NH, England.

Abstract
A comparison is made between silicon homojunction and Si/SiGe heterojunction technology by means of ECL gate delay calculations. The calculations are based on state of the art 1μm technology and basewidths of initially 0.1 μm and then scaled basewidths of 0.025 μm. Results of 42.5 ps for silicon homojunction circuits and 35.4 ps for HBT circuits demonstrate improvement in gate delays for HBT silicon heterojunction technology at 0.1 μm basewidths. Further improvements are predicted for basewidths scaled to 0.025 μm, with delays of 33.7 ps for homojunction and 28.3 ps for HBT circuits. In addition, sub 20 ps delays are predicted for appropriately designed devices.

1. Introduction

The use of Si/Si$_{1-x}$Ge$_x$ heterojunctions for high speed silicon based technology has recently generated considerable interest. HBTs utilizing SiGe as the base material have successfully been fabricated by MBE [1] and CVD [2] and have exhibited current gains of up to 400 [2]. To date no study has been reported which indicates the expected improvement in logic gate delay.

In this paper, a quasi-analytical expression for the gate delay of an ECL circuit [3] is used to predict the performance of Si/Si$_{1-x}$Ge$_x$ heterojunction bipolar circuits. This expression allows the transistor design options such as emitter and base doping levels, emitter/base capacitance and intrinsic base resistance to be investigated for a given transistor geometry. The geometry used for this work was a 1 μm, self-aligned transistor. Initial calculations are based around a 100nm basewidth which is typical for ion implanted, high-speed bipolar transistors, and later calculations around scaled basewidths as narrow as 10nm.

2. Procedure

Experimental results on Si/Si$_{0.88}$Ge$_{0.12}$ heterojunction bipolar transistors [1] indicate a 10 times improvement in gain is obtainable, which is equivalent to a valence band offset of 59 meV. This increased gain can be traded for an increase in base doping and a decrease in emitter doping. Experimental results also indicate that a reduction in mobility is obtained compared with silicon due to the presence of the germanium [1,2]. These findings are in contrast to theoretical predictions that the mobility in the direction of growth is expected to increase for SiGe grown on a (100) silicon substrate. We therefore base the ECL gate delay calculations for Si/Si$_{0.88}$Ge$_{0.12}$ devices on two models. Model 1 uses the 25% mobility reduction observed by Patton et al [1] and model 2 is based on the expected 50% mobility enhancement in the direction of growth predicted by Smith et al. [5]. Silicon bandgap narrowing [6] has been used in the absence of any reliable data on SiGe. The ECL gate output voltage swing used here is 0.4V. The operating collector current is chosen to correspond to the maximum current before base widening effects start to degrade the high frequency performance of the devices [7]. In optimising the device design for minimum gate delays, the constraint of $\beta \geq 50$ has been applied.

3. Predicted ECL gate delay results and discussion

In order to identify a suitable starting point for the calculations, the gate delay of a state of the art silicon homojunction technology with a 0.1 μm basewidth [4] has been calculated. An ECL gate delay of 42.5 ps is predicted which is in good agreement with the reported value of sub 40 ps [4] and 40.0 ps obtained from direct SPICE simulation. A breakdown of the time constants is given in table 1, which indicates that the base resistance (R_B) and load resistance (R_L) terms are dominating.

The gate delay of an ECL circuit consisting of an equivalent 0.1 μm basewidth $Si/Si_{0.88}Ge_{0.12}$ HBTs is next calculated. Optimisation of the transistor design leads to a base doping of 10^{20} cm^{-3} and emitter doping of 4×10^{19} cm^{-3} for model 1 and a base doping of 10^{20} cm^{-3} and emitter doping of 3×10^{18} cm^{-3} for model 2. Gate delays of 43.9 ps for model 1 and 36.2 ps for model 2 are predicted compared with 42.5 ps for the corresponding homojunction technology. For model 2, the reduced delay clearly shows the advantage of the HBT circuit in allowing a reduction in base resistance through an increase in base doping. The trade-off between base doping and base transit time is shown in Fig. 1.

The above results clearly demonstrate the desirability of scaling the basewidth for the heterojunction transistors in order to reduce the base transit time. For a basewidth of 25 nm [2], an emitter doping of 2×10^{18} cm^{-3} and a base doping of 10^{20} cm^{-3} can be achieved for both mobility models. In fact, at this basewidth, the optimum doping is achieved at $\beta > 50$. This indicates that gain is no longer a constraint. Calculated gate delays of $\tau_D = 29.3$ ps for model 1 and 28.3 ps for model 2 are obtained compared with 33.7 ps for the comparable homojunction circuit. Fig. 2 demonstrates the variation of gate delay with base doping for 25 nm basewidth devices.

Figs. 3 and 4 show how the gate delay varies with basewidth for homojunction and heterojunction circuits. It should be noticed that a minimum value of gate delay is achieved for a given base doping contour. This minimum represents a compromise between high base transit time at relatively wide basewidths and high base resistance at narrower basewidths. The maximum base doping is in practice constrained by the solid solubility limit at the growth temperature.

Table 1 shows a breakdown of the gate delay components for the HBT circuit incorporating transistors with a 25 nm basewidth (model 2). The load resistance terms now dominate the overall delay (64%). The load resistance is constrained by the ECL gate output voltage swing and the collector current. A reduction in load resistance could therefore be achieved by choosing a smaller logic swing. Ichino et al. [8] predict higher driving currents for scaled basewidth HBTs, allowing a smaller load resistance to be used. The effect of operating collector current and voltage swing and hence load resistance on overall gate delay is shown in Fig. 5 indicating further improvements in delay may be obtainable by optimising the collector current at which the peak cut-off frequency occurs or by utilising a 0.2 V logic swing.

References
[1] G.L. Patton, S.S. Iyer, S.L. Delage, S. Tiwari and J.M.C. Stork, IEEE Electron Dev. Letts., vol. 9, p.165, 1988.
[2] C.A. King, J.L. Hoyt, C.M. Gronet, J.F. Gibbons, M.P. Scott and J.Turner, IEEE Electron Dev. Letts., vol. 10, p.52, 1989.
[3] W. Fang, A. Brunnschwelier and P. Ashburn. To be published.
[4] M.C. Wilson, P.C. Hunt, S. Duncan and D.J. Bazley, Electronics Letters, Vol.24, p.920, 1988.
[5] C. Smith and A.D. Welbourn, IEEE BCTM., p.57, 1987.
[6] P. Ashburn, 'Design and Realization of Bipolar Transistors', John Wiley, New York, 1988.
[7] D.J. Roulston, S.G. Chamberlain, J. Sehgal, IEEE Trans. Electron. Devices, p.809, 1972.
[8] H. Ichino, M. Suzuki, K. Konaka, T. Wakimoto and T. Sakai, IEEE BCTM., p.15, 1988.

Time	Delay Component (ps)	
Constant	Homojunction $W_B=0.1\mu m$	Heterojunction $W_B=0.025\mu m$
τ_F	8.4	2.02
$R_{BI}C_{JCI}$	0.68	0.10
$R_{BI}C_{JCX}$	0.20	0.03
$R_{BI}C_{JE}$	0.96	0.16
$R_{BI}C_D$	2.21	0.08
$R_{BI}C_{JS}$	0.23	0.03
$R_{BX}C_{JCI}$	0.73	0.15
$R_{BX}C_{JCX}$	0.22	0.04
$R_{BX}C_{JE}$	1.04	0.24
$R_{BX}C_D$	2.39	0.12
$R_{BX}C_{JS}$	0.24	0.05
$R_L C_{JCI}$	2.18	2.18
$R_L C_{JCX}$	9.86	9.86
$R_L C_{JE}$	0.29	0.34
$R_L C_{JS}$	3.16	3.16
$R_L C_L$	2.48	2.48
$R_C C_{JCI}$	0.60	0.60
$R_C C_{JCX}$	2.43	2.42
$R_C C_D$	0.70	0.17
$R_C C_{JS}$	2.32	2.32
$R_E C_{JCI}$	0.12	0.19
$R_E C_{JCX}$	0.51	0.77
$R_E C_{JE}$	0.06	0.10
$R_E C_D$	0.05	0.02
$R_E C_{JS}$	0.44	0.66
τ_D	42.5	28.3

Table 1 - ECL gate delay components for homo- and heterojunction circuits.

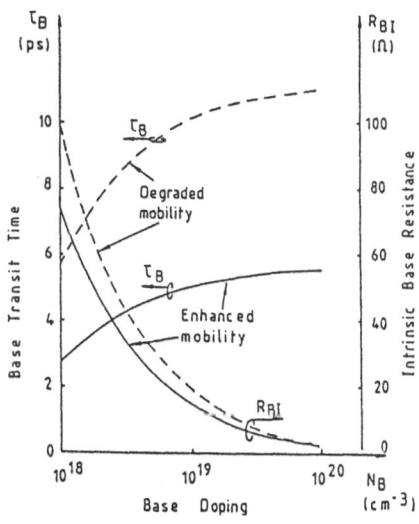

Fig 1.- Base transit time and intrinsic base resistance vs base doping for heterojunction transistors with a 0.1 μm basewidth.

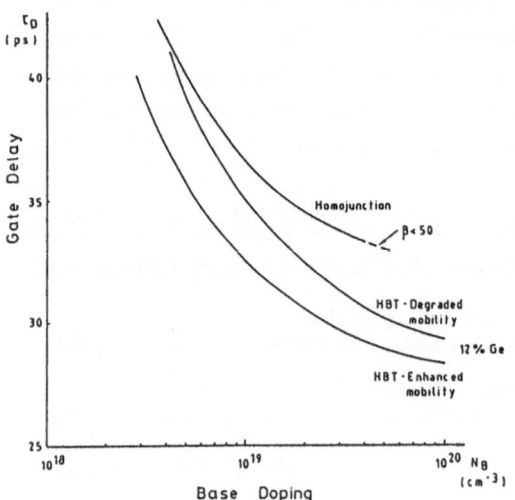

Fig 2 - ECL gate delay vs. base doping
for 25 nm basewidth homo- and heterojunction technology.

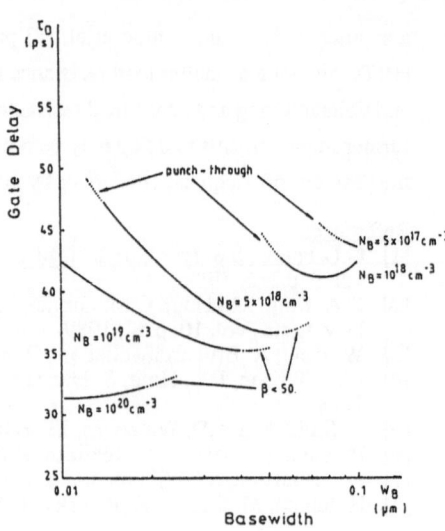

Fig 3 - ECL gate delay vs. basewidth
for base doping contours using
homojunction technology.

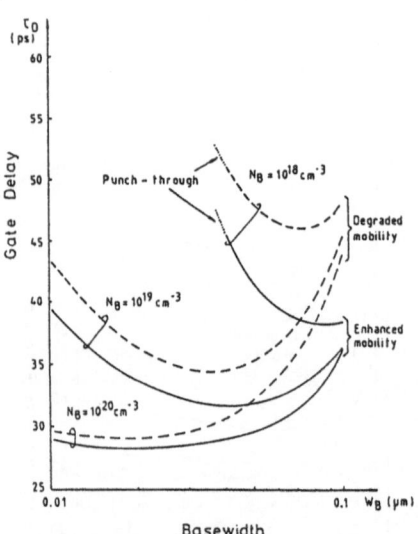

Fig 4 - ECL gate delay vs. basewidth
for base doping contours using
heterojunction technology.

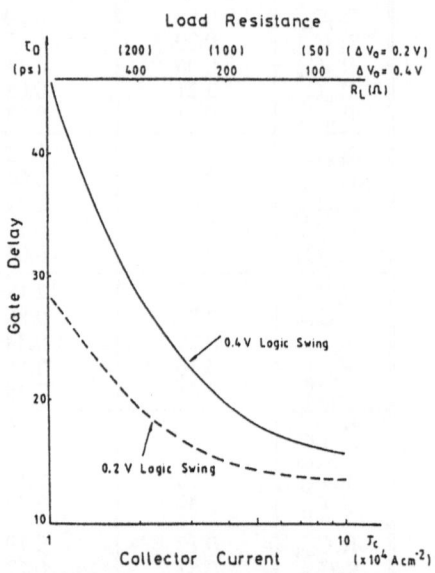

Fig 5 - ECL gate delay vs. collector
operating current and load resistance
using 25 nm heterojunction technology.

Study of the Effective Valence Band Offset of Si/SiGe Heterojunctions with a Doping Interface Dipole

Stephen P. Wilson and **Duncan W.E. Allsopp**

Department of Electronics, University of York,

Heslington, YORK, YO1 5DD, ENGLAND.

The modulation of heterojunction band offsets by the addition of ultra thin p^+ and n^+ layers either side of the hetero-interface has been extensively modelled. It has been found that the resulting structure, the Doping Interface Dipole (DID), has a voltage tunable pseudo-offset dependent on structural parameters which can be readily controlled during epitaxial growth. The results have enabled a full understanding of the basic physics of DID's and elucidate their possible device applications.

Introduction

The heterojunction offset is an important feature in electron devices based on layered structures. It is ideally dependent only on material parameters, but a voltage dependence of the offset would be desirable, as it provides greater versatility in the design of heterostructure devices. This can be achieved via a doping interface dipole (DID), which consists of ultra-thin n^+ and p^+ layers on either side of the interface, as shown in Figure 1. This structure was first proposed and demonstrated by Capasso and co-workers using AlGaAs/GaAs heterostructures grown by molecular beam epitaxy (MBE) [1].

In addition to the diffusion induced field of the heterojunction a DID contains a dipole field which is easily modified by an applied bias. As a consequence, by defining the effective heterojunction offset as the difference in potential between the two dipole layers (Figure 1), a bias dependent offset is achieved.

This paper reports the results of a theoretical study of the performance of DIDs. In particular the effects of variations in the doping dipole parameters and background doping on the bias dependence of the effective offset of a DID are described in detail. Realistic device structures which can be readily fabricated using state of the art epitaxial techniques have been modelled, enabling future comparisons between experiment and the results presented here.

Model

In this work a strained layer $Si/Si_{0.75}Ge_{0.25}$ single heterojunction with p-type background doping was chosen as the host for a doping dipole, consisting of a layer of shallow acceptors in the strained $Si_{0.75}Ge_{0.25}$ and a layer of shallow donors in the unstrained Si. In this hetero-epitaxial system almost all the band gap difference is taken up in the upper valence band discontinuity when lattice matched to Si [2], giving rise to a strong diffusion induced dipole in p-isotype structures. In the case described here the upper valence band discontinuity in the basic $Si/Si_{0.75}Ge_{0.25}$ heterojunction was assumed to be 0.21 eV [2]. The thickness of the alloy layer was taken to be 50 nm, which is just within the thickness boundary for commensurate growth as determined by transmission electron microscopy studies [3].

A one-dimensional model was used in which current flow was assumed to be negligible. Fermi-Dirac statistics were straightforwardly utilised to describe the carrier population and dopant ionisation. A strain induced splitting of the heavy and light hole valence bands of the lattice mismatched $Si_{0.75}Ge_{0.25}$ layer at k = 0 of Es = 41.5 meV was incorporated into the model by multiplying the light hole contribution to the effective mass density of states by a Boltzmann factor of exp(-Es/kT). Ionisation energies of 45 meV for acceptors and 33 meV for donors were used in the calculations, values which yielded almost total ionisation of all shallow impurities. Possible quantum size effects at the hetero-interface and in the narrow, highly doped regions [5] have been neglected in determining the net charge distribution. The Poisson equation thus derived was solved self-consistently, using Dirichlet boundary conditions in the bulk and at the surface. Asymmetric three point differences were used to impose the discontinuity in the electric field at both of the dipole layers and at the hetero-interface, under the assumption that the charge layers did not exceed two mesh spacings in width. The effective valence band offset was then calculated for each value of surface bias.

Results and discussion

Typical potential profiles for a range of surface biases, for a given set of device parameters, are shown in Figure 2. It can be seen that the potential difference measured between the poles (marked A and B in Figure 2) of the DID is strongly bias dependent. This sensitivity to applied bias was in turn found to be a function of doping density in the layers forming the poles of the

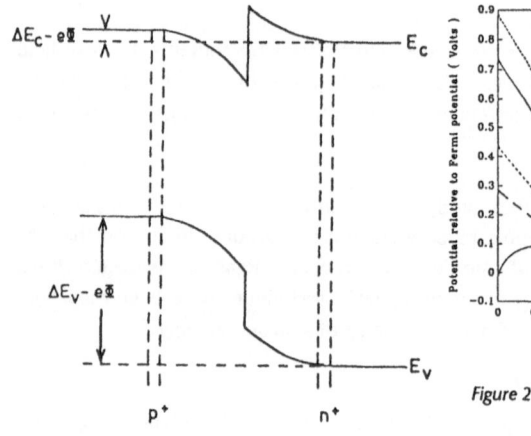

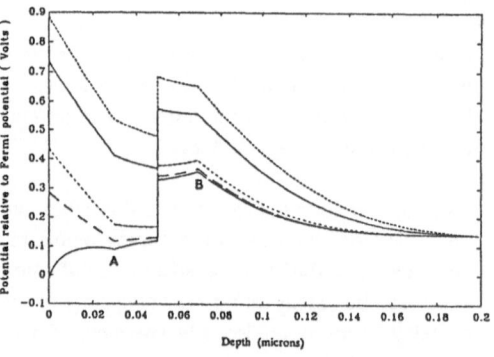

Figure 2   Valence band edge potential versus depth for a typical DID with 1 nm thick layers of 5 x 10^{18} cm^{-3} doping level 40 nm apart. The solution is shown for several values of applied bias.

Figure 1   Schematic potential diagram of the DID structure proposed by Capasso et al (ref. 1).

DID, the thickness and the separation of these layers and of the background doping concentration. As an example the effect of changing the pole doping density on the voltage modulation of the effective offset is shown in Figure 3. As expected the effective offset increases with increasing pole doping level since this enhances the strength of the dipole moment. However, at higher values of reverse bias the applied field has, in all cases, effectively counteracted any enhancement of the overall offset induced by the doping dipole.

Increasing the width of the poles, or widening their separation has essentially the same effect on the effective offset as increasing the doping in poles. This follows directly from the contribution of the doping dipole to the overall potential drop across the structure. In the absence of any screening effects this additional dipole potential is given by

$$\Phi = qNtd/\epsilon \qquad (1)$$

where N is the space charge density in each pole, t is the thickness of the layers comprising the poles and d is the pole separation.

However, the change in the effective offset (an enhancement in the structures considered here) will rarely equal the ideal of the dipole potential given by equation (1). This is because in any practical device structure the charge contained in the poles will be partially screened by the heterojunction's own diffusion induced dipole, the strength of which is related to the background density of ionised impurities. Here the diffusion induced dipole field of the discrete heterojunction opposes that of the doping dipole. Hence, the enhancement of the heterojunction offset can be significantly less than that predicted by equation (1) for the ideal case. This is shown in Figure 4 in which the calculated effective offsets for two values of applied surface bias are plotted against the p-type background doping in the $Si/Si_{0.75}Ge_{0.25}$ layers. These results show that the internal screening in a DID by the diffusion induced dipole can still be substantial for background doping densities as low as 1×10^{15} cm^{-3}. For the DID considered in Figure 4 the enhancement of the offset at roughly zero bias was 0.16eV for a background doping of 1×10^{15} cm^{-3}, a value barely half the 0.3eV predicted from equation (1) for the DID parameters used.

One other feature common to all the DID structures described here was that the effective offset was found to be strongly dependent on reverse bias. This sensitivity to bias arises from the way in which the different parts of the DID respond to the surface charge from which the applied electric field emanates. The pole nearest this surface charge partially screens the rest of the DID, leaving the potential distribution through the rest of the device relatively unperturbed at low applied fields. The DID potential distributions shown in Figure 2 clearly indicate this shielding of the surface charge at x = 0 by the nearest pole, marked A. Comparisons between the potential distributions for a small forward bias (lowest curve in Figure 2) and those for small reverse bias (the next lowest two curves) demonstrate that the potential of the more remote pole, B, varies only slightly compared with that of pole A. However, at high reverse bias the screening becomes ineffective and the near parabolic potential distribution associated with surface depletion becomes apparent (top curve in Figure 4).

788

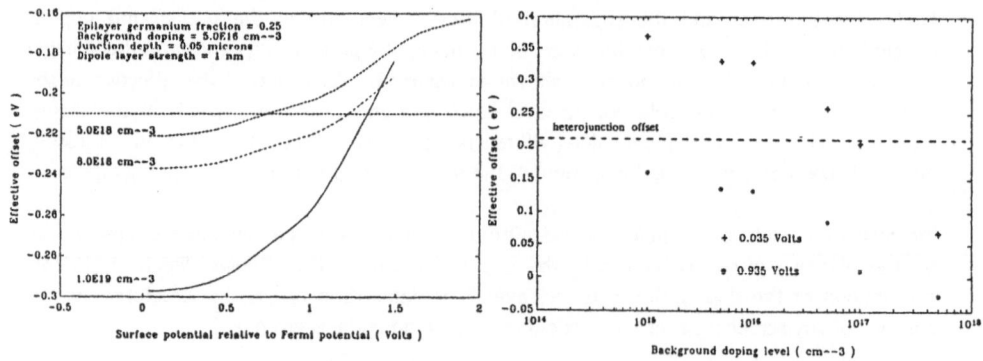

| Figure 3 | The effect on the effective offset of varying the doping level in the dipole layers, while maintaining the other dipole parameters. These values are indicated. | Figure 4 | The effect of varying the background doping level on the effective offset for surface bias values of 0.035 and 0.935 eV. The doping dipole has 1 nm thick layers and a 5 x 10^{18} cm^{-3} doping level. The separation is 40 nm. |

Further evidence of this is shown in Figure 3, in which the DID with the most heavily doped poles (N = 1 x 10^{19} cm^{-3}) has the most bias dependent effective offset. Initially the effective offset is almost independent of applied voltage, as the DID is almost completely screened by its pole nearest the surface charge. However, once the applied field is sufficiently strong to penetrate this pole, its effect on the band bending within the DID and, ultimately beyond, is dramatic giving rise to a rapid reduction in the effective offset until there is a net decrement, rather than an enhancement, of the overall potential difference compared with the heterojunction band offset. (See the upper 2 curves in Figure 2 and Figure 3.)

In conclusion, the performance of doping interface dipoles as a function of its design and growth parameters has been investigated by extensive device modelling. It has been found that these devices exhibit a voltage controlled modulation of their effective band offsets at voltage levels compatible with integrated circuit technology, thus extending further the design flexibility of heterostructure devices.

Acknowledgements

The authors wish to acknowledge conversations with R J Malik of A.T. and T. Bell Laboratories during ESSDERC 88. One of us (SPW) also wishes to thank the Science and Engineering Research Council (UK) for providing a studentship.

References

1. Capasso F., Cho A.Y., Mohammed K. and Foy P.Y., Appl Phys Lett 46 (1969), 664-666.
2. Van der Walle C.G., Phys Rev B 34 (1986), 5621-5634.
3. Kohama Y., Fukada Y. and Seiki M., Appl Phys Lett 52 (1988), 380-382.
4. People R., Phys Rev B 32 (1985), 1405-1408.
5. Cunningham J.E., Tsang W.T., Timp G., Schubert E.F., Chang A.M. and Owusu-Sekyere K., Phys Rev B 37 (1987), 4317-4320

Analysis of the Punchthrough Effect in Walled Emitter Bipolar Transistors

Emmanuel DUBOIS , Bruno BACCUS , Dominique COLLARD

Institut Supérieur d'Electronique du Nord
41, Boulevard Vauban 59046 LILLE Cedex, France

SUMMARY

A detailed study of the punchthrough formation mechanism induced by the walled emitter configuration is proposed. Efficient process and device simulation tools -*IMPACT*- have been used and results have been successfully compared to measured characteristics.

I-INTRODUCTION

High packing density in bipolar integrated circuits is greatly improved by the systematic use of walled emitter transistors [1]. However, both vertical and lateral scaling require a severe control of the process steps, specially when isolation is assumed by a fully-recessed LOCal Oxidation of Silicon. Indeed, the major drawback related to the buried LOCOS technique is its extreme sensitivity to emitter/collector leakage currents caused by a local punchthrough effect [2].

This paper proposes a detailed study of the punchthrough formation mechanism induced by the walled emitter configuration in order to clarify the following unresolved points:

I- What are the key processing steps critical for the punchthrough control ?

II- Is the punchthrough current path located into the silicon bulk or along the Si/SiO$_2$ interface ?

For this aim, accurate process and device simulation tools have been extensively used to investigate the typical two-dimensional effects involved in base thinning and emitter/collector shorts. Simulation results have been checked by measured current characteristics.

II-SIMULATION TOOLS DESCRIPTION

The complete doping profiles have been calculated with the two-dimensional process simulator IMPACT4 [3] which has proved its capabilities in dealing with complex bipolar structures. The process simulation results (impurity concentration and oxide shape) have been introduced in the device simulator IMPACT3.2 [4] where the dicretization grid was generated with strong refinements to ensure an accurate topology and doping profile description. Since the punchthrough phenomenon relates closely to the generation current in the depletion region, the Auger, thermal and impact generation/recombination processes have been carefully taken into account.

III-PROCESS DESCRIPTION

A simplified processing steps flow chart is given in table I. The main characteristics of the studied NPN transistors are as following:
- Cut-off frequency: 7.5 GHz (V_{cb}= 2 V).
- All implanted process.
- Buried LOCOS isolation technique.
- Walled emitter.

IV-BASE THINNING FORMATION

A representation of the device under study is given in figure 1: the corresponding emitter/base and base/collector junctions location issued from process simulation are also provided. Vertical cross-sections of the impurity profiles in the intrinsic region on one hand and in the vicinity of the bird's beak region on the other hand clearly indicate the difference in base profiles caused by the walled emitter configuration (fig. 2). The base narrowing effect is mainly due to a geometrical effect related to the non planar two-layers structure, and to the difference between arsenic and boron implantation ranges. The opposed segregation effect of these two impurities at the Si/SiO_2 interface is not responsible for base thinning, in the present case.

V-SENSITIVITY TO OXIDE OVERETCH

Numerous simulations based on processing steps variations demonstrate that an unsufficient control of the oxide etching step realized between the base drive-in and the emitter ion implantation contributes significantly to a more pronounced base thinning effect. In figure 3 are shown the final junctions locations for 3 different oxide overetch thicknesses. The more the overetched oxide thickness is important, the more the emitter/base junction penetrates locally deeper in the bulk whereas the base and collector regions are kept unaltered by the overetching step. Moreover, the compensation concentration at the emitter/base junction proves that the punchthrough current path will be initiated into the silicon bulk and not along the Si/SiO_2 interface (fig. 4).

Device simulations have been performed for two devices with an 0 and 200 Å oxide overetch, to determine the I_c-V_{ce} characteristics, emitter and base being grounded. The large discrepancies of the punchthrough characteristics observed on measured transistors (fig. 5) are fully coherent with the computed currents (fig. 6). The electron and hole concentration profiles for a 200 Å overetched device at V_{ce}= 17 volts are displayed in figure 7. These profiles clearly demonstrate that the punchthrough current path corresponds to the base thinning region.

VI-CONCLUSION

Punchthrough effect in walled emitter transistors has been investigated by process/device simulations and experimental validation. The presented results have clearly shown that the current path is located into the silicon bulk. Moreover, a quantitative analysis has proved that a slight uncontrolled oxide overetch (200 Å) at the bird's beak edge induces a drastic reduction of the punchthrough voltage, so that an extreme control of this step is required.

REFERENCES

[1]: W.C. KO, T.C. GWO, P.H. YEUNG, S.J. RADIGAN , IEEE Trans. Electron. Devices, pp 236-239, March 1983

[2]: H. STORK , Symposium on VLSI Technology , pp 97-116 , May 1986

[3]: B. BACCUS, D. COLLARD, E. DUBOIS, D.MOREL , Bipolar circuits and technology meeting, pp 164-167, Sept. 1988

[4]: E. DUBOIS, J.L. COPPEE, B. BACCUS, D. COLLARD, pp 151-161 , SISDEP-88, Sept. 1988

This work was partially supported by RTC-PHILIPS COMPONENTS, CAEN FRANCE

TABLE I : PROCESS FLOW CHART

1- BURIED LAYER I/I AND DIFFUSION (SB)

2- EPITAXIAL GROWTH (AS)

3- CHANNEL STOP I/I (B)

4- BURIED LOCOS OXIDATION

5- DEEP N I/I AND DIFFUSION (P)
FOR COLLECTOR CONTACT

6- DRY OXIDATION AND BASE I/I (B)

7- INERT BASE DRIVE-IN

8- OXIDE ETCHING
(POSSIBLE OVERETCHING)

9- EMITTER I/I (AS)

10- FINAL INERT EMITTER DRIVE-IN

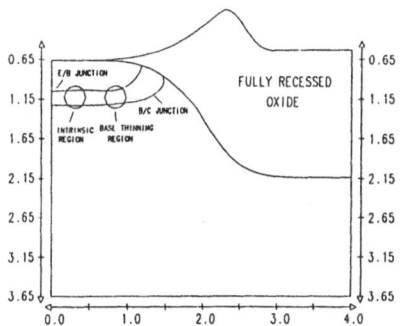

Fig. 1 : Base thinning effect and junctions locations.

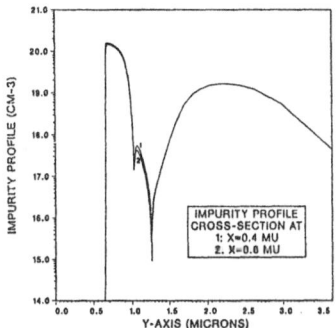

Fig. 2 : Doping profile cross-sections in the intrinsic region and in the base thinning region.

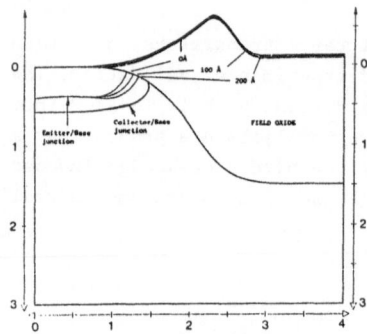

Fig. 3 : Junctions locations for
3 different etching steps.

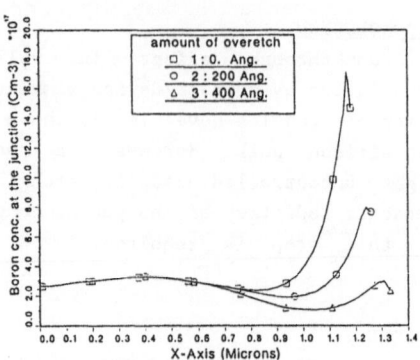

Fig. 4 : Compensation concentration
along the Emitter/Base junction
for three different etching steps

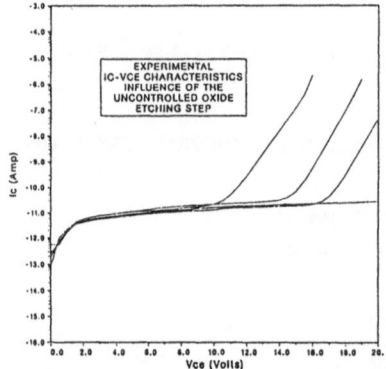

Fig. 5 : Measured Ic= f(Vce) characteristics
revealing drastic reductions of the punchthrough
voltage due to the uncontrolled oxide overetch.

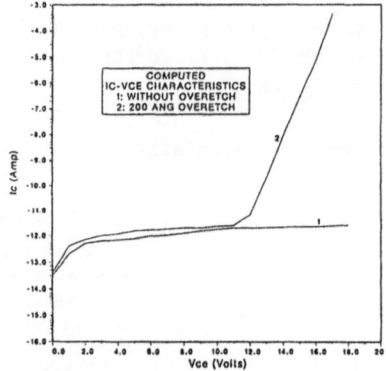

Fig. 6 : Ic= f(Vce) characteristics for 2
devices with and without oxide overetch.
Vemitter=Vbase=0 Volt.

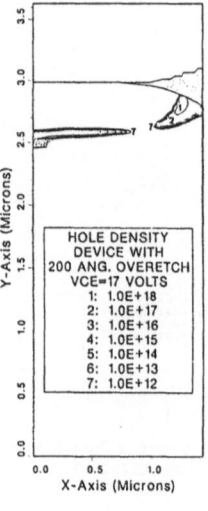

Fig. 7 : Hole and electron
concentration distributions
in punchthrough mode.
Device with 200 ang
overetch at
Vce=17 Volts.

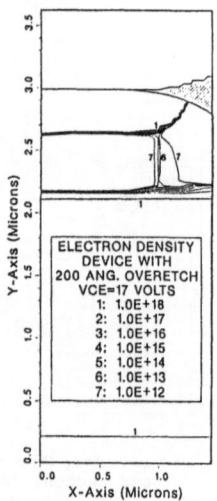

Modelling Forward-Biased Tunneling

G.A.M. Hurkx, F.G. O'Hara and M.P.G. Knuvers

Philips Research Laboratories, 5600 JA Eindhoven, The Netherlands

ABSTRACT

A recombination model is presented that incorporates trap-assisted tunnelling in a Shockley-Read-Hall type formulation. This model is based on a simple quantum-mechanical treatment of the p-n junction, using the envelope-function approach. By comparison with experiments it is shown that the model agrees well with the measured reduced temperature dependence and large non-ideality factor associated with tunnelling currents in forward-biased junctions.

1. INTRODUCTION

In downscaled bipolar devices where the doping concentration at the emitter-base junction is in the order of $10^{18}cm^{-3}$ or more, forward-bias tunnelling causes a severe reduction of the low-bias current gain [1,2,3]. This is due to tunnelling via impurity states (see fig. 1) which makes the non-ideal base current considerably larger than that predicted by the conventional Shockley-Read-Hall model [4]. Therefore the inclusion of tunnelling in modelling the non-ideal base current is essential.

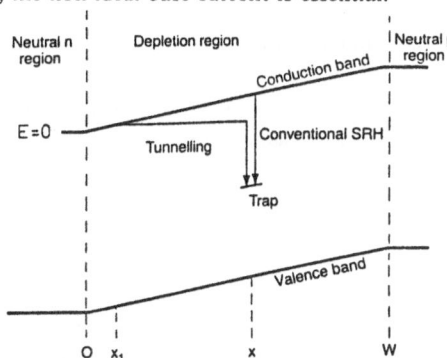

Fig. 1. Schematic energy-band diagram of a forward-biased pn junction. Both the conventional SRH process and the trap-assisted tunnelling process are denoted.

Existing models for forward-biased tunnelling [1,4] give a semi-empirical relation between the current density and a certain function $h(V_j,F)$ of the applied voltage V_j and the average electric field F , i.e.

$$J_{tun} = c \, e^{h(V_j, F)} \tag{1}$$

These models, however, suffer from the following drawbacks:
- The magnitude of the proportionality constant c cannot be obtained from the model but must be determined by fitting the experimental I-V curves.
- For these models it is assumed that only electrons at $x = 0$ (and similarly holes at $x = W$) contribute to the tunnelling current and, therefore, they neglect the possibility of injected electrons in the depletion layer (e.g. at x_i in fig. 1) tunnelling to a trap.

- A model that describes the tunnelling effect in the form of a current density is not suitable for implementation into a numerical device simulator. To this end the tunnelling contribution must be given in the form of a recombination term in the continuity equations.

In this paper we present a recombination model that includes both SRH recombination and trap-assisted tunnelling. As will be shown, this model does not suffer from the above-mentioned drawbacks.

2. THE MODEL

Basically, SRH recombination and trap-assisted tunnelling are the same mechanism, i.e. recombination via impurity states. The difference lies in the character of the state from which the electron makes a transition to a trap state. In the conventional SRH process the initial state is a non-localized (Bloch) state in the conduction band and in the tunnelling process the initial state is a tunnelling state in the gap. In our model the recombination rate is given by a SRH-like expression

$$R = \frac{np - n_1 p_1}{\tau_p (n + n_1) + \tau_n (p + p_1)} \tag{2}$$

In the above expression the electron density $n(x)$ contains both the conventional injected electron density (i.e. from states in the conduction band) and a tunnelling term originating from the tunnelling states. The same holds for the hole density. τ_p and τ_n are the recombination lifetimes while n_1 and p_1 are the carrier densities at equilibrium, assuming midgap states. The expression for the electron density is obtained by application of the so-called envelope-function formalism, first developed by Slater [5] and frequently used, for instance, in heterojunction problems and to obtain impurity energy levels [6]. In this formalism, $n(x)$ is given by

$$n(x) = \int_0^\infty N(E) f(E) \phi_E^2(x) \, dE \tag{3}$$

where $N(E)$ is the density of states, $f(E)$ is the Fermi-Dirac function and $\phi_E(x)$ is the envelope wave function which is the eigenfunction of the effective-mass Schrodinger equation:

$$\left[-\frac{\hbar^2}{2m^\times} \frac{d^2}{dx^2} + U(x) \right] \phi_E(x) = E \phi_E(x). \tag{4}$$

For a linearly varying potential in the depletion layer as sketched in fig.1 ($U(x) = qFx$) the solutions of (4) are Airy functions, i.e. $\phi_E(x) \sim Ai[\gamma(x - x_1)]$ with $x_1 = E/(qF)$ and $\gamma = (2qFm^\times \hbar^{-2})^{1/3}$, $m^\times$ being the tunnelling effective mass. The exact value of $m^\times$ to be used is not clear and, moreover, depends on the crystal orientation. Theoretical treatments on Zener tunnelling suggest a value between 0.1 and 0.3. For $x > x_1$ the Airy function is decaying and represents the tunnelling state while for $x < x_1$ the Airy function oscillates. In order to make the result suitable for implementation into a conventional numerical device simulator we express $n(x)$ as a function of the conventional injected electron density $n_{inj}(x)$. After some manipulation, putting $N(E) f(E) = dn_{inj}/dE$ and subsequently replacing $(dn_{inj}/dE).dE$ by $(dn_{inj}/dx_1).dx_1$, we obtain

$$n(x) = n_{inj}(x) + Ai^{-2}(0) \int_0^x \left(-\frac{d n_{inj}(x)}{dx} \right)_{x=x_1} Ai^2 [\gamma(x - x_1)] \, dx_1 \tag{5}$$

The second term on the right-hand side of the above expression is the tunnelling contribution. The physical meaning of $Ai^2[\gamma(x - x_1)]$ is the probability that an electron at x_1 will tunnel to a trap at x. It should be noted that, since only electrons at an energy level above the trap level contribute to the recombination, in the case of shallow impurities the lower bound of the integral in (5) should be replaced by $x - E_T/(qF)$, where E_T is the trap level measured from the conduction band. It is interesting to note that for a reverse-biased junction, where the tunnelling distance is $E_g/(qF)$, the asymptotic expansion of the Airy function yields the well-known field-dependence of Zener tunnelling, viz.

$$Ai^2 \left(\gamma E_g/qF \right) \sim \exp\left(-\frac{4\sqrt{2m^x}\, E_g^{3/2}}{3qhF} \right) \tag{6}$$

3. COMPARISON WITH EXPERIMENTS

In order to test the model, we have compared calculations with measurements on diodes having linearly-graded junctions. Because eq. (5) applies only to a linear potential we have approximated the potential in the depletion layer by a straight line and used the dependence of W on V_j as obtained from capacitance measurements. We have used the average electric field and not the field at the electrical junction because, unlike the the conventional SRH process, with tunnelling the whole depletion region contributes to the values of n and p at the electrical junction. A standard model for the recombination lifetime is used [7], together with the Slotboom-De Graaff model for bandgap-narrowing [8]. The current density is obtained by numerical integration of eq. (2). In fig. 2 the measured and calculated I-V characteristics are plotted for a diode with a 200Å zero-bias depletion width. From the calculations it follows that the ratio of the current obtained using the proposed model and that obtained with the conventional SRH model, depends strongly on the zero-bias depletion width or, equivalently, on the electric field. This is shown in fig. 3. Furthermore, from this figure is apparent that, in agreement with experiments [1], the tunnelling effect increases dramatically when the zero-bias depletion width is less than about 300Å.

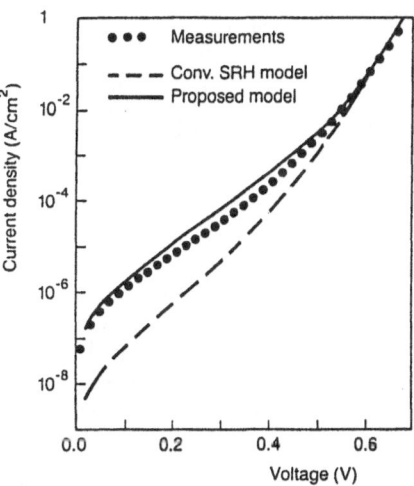

Fig. 2. Measured and calculated I-V characteristics for a diode with a linearly-graded junction. $m^x = 0.2$ and T = 294 K.

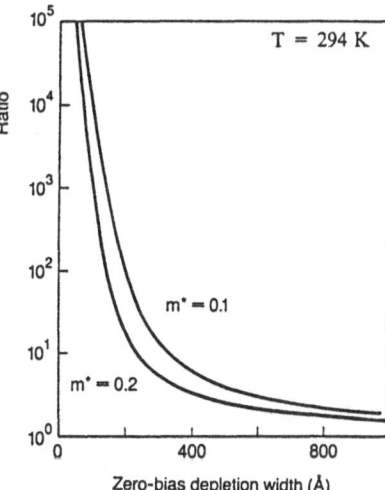

Fig. 3. Ratio of the calculated current obtained using the proposed model and that obtained with the conventional SRH model as a function of the zero-bias depletion width.

Notice that the difference between the conventionally calculated current and that obtained by the proposed model is about one decade at low bias. Figs. 4 and 5 show two characteristic features associated with trap-assisted tunnelling which are described well by the proposed model, viz.

- The non-ideality factor m in $J \sim \exp(qV_j/mkT)$ is larger than two whereas the conventional SRH model predicts a value less than two (fig. 4).
- The temperature dependence is much less than that predicted by the conventional SRH model (fig. 5). As can be observed from fig. 5, the difference between the current obtained with the proposed model and that obtained with the conventional SRH model can be as much as six decades at 150K. This, of course, is due to the small temperature dependence of the tunnelling contribution to the carrier density.

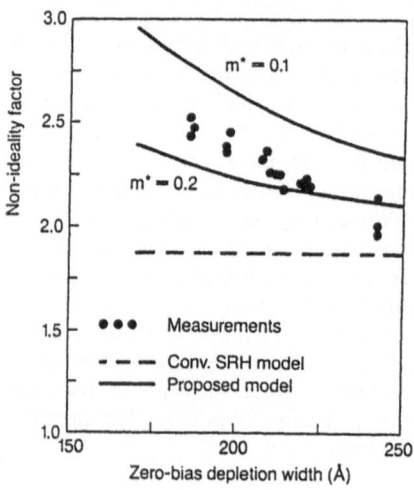

Fig. 4. Measured as well as calculated non-ideality factor as a function of the zero-bias depletion width. The calculations are performed for two values of the effective mass. $T = 294K$.

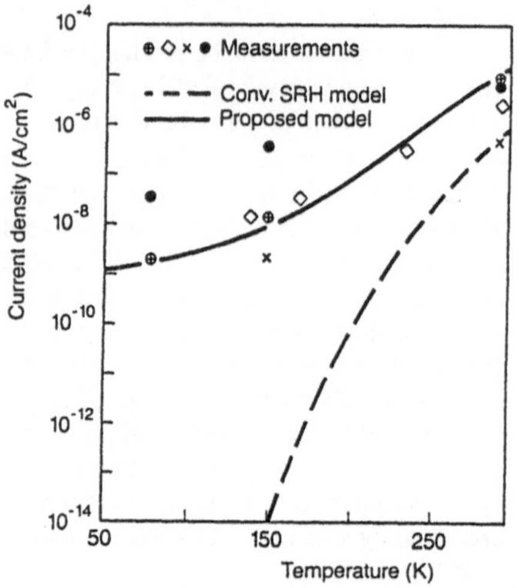

Fig. 5. Measured and calculated ($m^x = 0.2$, zero-bias depletion width $= 200\text{Å}$) non-ideal diode current as a function of temperature at $0.2V$ junction voltage. Each symbol denotes a diode with a zero-bias depletion width in the range $200 - 250\text{Å}$ (the diamond symbols are from [1]).

4. CONCLUSIONS

The proposed model agrees well with the experimentally observed characteristic features associated with trap-assisted tunnelling. From comparison with experiments it is found that an effective mass of $0.1 - 0.2$ has to be used, in rough agreement with Zener tunnelling. Furthermore, from our calculations it is seen that, in agreement with experiments [1] , the tunnelling effect increases dramatically when the zero-bias depletion width is less than about 300Å , which is equivalent to a doping level of $3 \ 10^{18} cm^{-3}$ for a two-sided abrupt junction.

REFERENCES

1 J.A. Del Alamo and R.M. Swanson, IEEE Electron Device Letters, EDL-7,1986, p. 629
2 G.P. Li, E. Hackbarth and T.C. Chen, IEEE Trans. Electron Devices, ED-35, 1988, p. 89
3 H. Schaber, J. Bieger, B. Benna and T Meister, Proc. ESSDERC '87, p. 365
4 A.G. Chynoweth, W.L. Feldman and R.A. Logan, Phys. Rev. 121, 1961, p. 684
5 J.C. Slater, Phys. Rev. 76, 1949, p.1592
6 R.A. Smith, "Wave Mechanics of Chrystalline Solids", Chapman and Hall Ltd, London 1969, p. 403
7 S. Selberherr, " Analysis and Simulation of Semiconductor Devices", Springer, Wien 1984, p. 106
8 J.W. Slotboom and H.C. de Graaff, Solid-State Electr., 19, 1976, p. 857

Pedestal Collector in Advanced Bipolar Technology for Improved Speed Power Performance

K. Ehinger, M. Reisch, H.W. Meul, D. Hartwig, E. Pfoser,
R. Köpl, J. Weng

Siemens AG, Corporate Research and Technology, Otto-Hahn-Ring 6,
D-8000 München 83

Summary

A double poly-Si self-aligning bipolar process employing 1 μm lithography has been developed for very high speed/low power circuit applications. Shallow base-emitter profiles were obtained by combining low energy boron implantation and rapid thermal annealing for the emitter drive-in. Cut-off frequencies of 14 GHz at $V_{BC} = -1$ V, ECL gate delay times of 43 ps and minimum power delay products of 30 fJ were achieved with conventional epitaxial collector configurations. Further improvements in cut-off frequency up to 18 GHz, gate delay and minimum power delay product as low as 36 ps and 23 fJ, respectively, were attained by an additional implantation of doubly ionized phosphorus into the active device area forming a so-called pedestal collector.

1. Introduction

The introduction of self-alignment schemes, smaller minimum feature sizes and scaling of vertical doping profiles significantly improved speed-power performance of bipolar devices [1,2]. Further progress is possible and should focus in particular on the reduction of the base collector capacitance C_{BC} which strongly influences the power delay product and gate delay time. Distinguishing between internal and external contributions to C_{BC}, associated with the areas A_{int}, A_{ext} (cf fig.1) and specific capacitances C_{int}, C_{ext} respectively, the C_{BC} related contributions to the ECL gate delay time Δt_D can be approximated by an expression of the functional form [3] $\Delta t_D = k_1 V_1 C_c / I_c = k_1 V_1 (C_{int}/j_c)\{1 + (A_{ext}/A_{int})(C_{ext}/C_{int})\}$ where j_c denotes the current density, k_1 an appropriate weighting factor and V_1 the voltage swing. Since high-performance ECL circuits usually are operated near the limit to the high-current regime, there is not much tolerance towards lower values of C_{int} by simply reducing dopant concentration of the epi layer concentration N_{epi} if j_c is kept constant because of base pushout. Constant current scaling even requires to increase the doping concentration in the active collector region resulting in an enhanced C_{int}. Reduction of

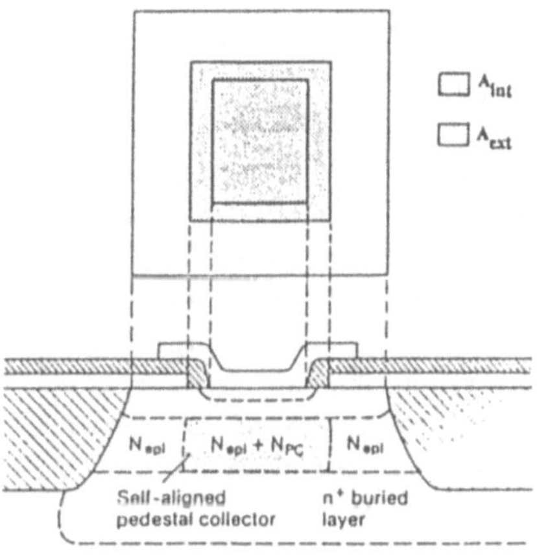

Fig.1. Layout and schematic cross-section of a bipolar transistor. The self-aligned pedestal collector (SAPC) is formed by implantation into the dotted area.

the term $(A_{ext}/A_{int})(C_{ext}/C_{int})$ requires optimization of A_{ext}/A_{int} and benefits from a reduction of C_{ext}/C_{int}. This is possible by a reduction of N_{epi} and by an additional self-aligned implantation of a pedestal collector (SAPC) into the active device area. In contrast to previously published investigations [4] which attempted to realize extremely narrow base widths by cutting off the channeling tail of the implanted base profile, we focus on the reduction of external C_{BC}. The concept presented here provides an additional degree of freedom for gate delay optimization; its implementation in a double poly-Si self-aligned process is outlined in the following section.

2. Technology

Besides two modifications the process was as outlined in [5] using an Sb doped n^+-buried layer, n^--epitaxy and planar recessed field oxide regions for device isolation. A double poly-Si self-aligned E/B configuration with shallow emitter/base profiles (fig 2.) was obtained by low energy (10 keV) boron implantation and rapid thermal annealing for emitter drive-in.

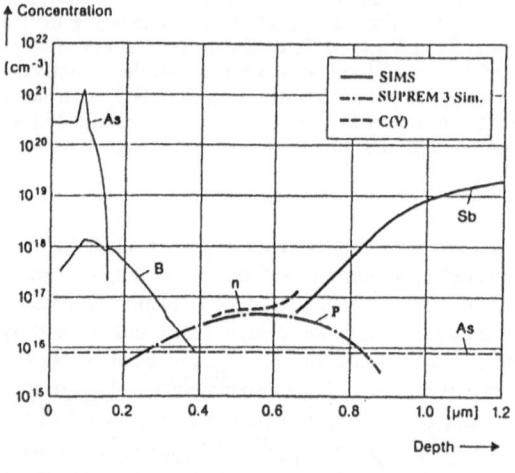

Fig.2. Concentration profile as obtained by SIMS, C(V)-measurement and SUPREM 3 simulation

In contrast to the processing scheme described in [5] N_{epi} was reduced to $8 \cdot 10^{15} cm^{-3}$ to optimize the external base collector capacitance C_{ext}. Base stretching at unacceptable low j_c values is avoided by an additional implantation of doubly ionized phosphorus to realize a pedestal collector [6]. This leads to a collector doping profile with slightly increasing phosphorus concentration towards the buried layer (fig. 2). Since this implantation is performed subsequently to the base implant and therefore uses the emitter opening as a mask, the pedestal collector is self-aligned to the active base layer, i.e. no changes occur in the doping concentration in the inactive part of the collector (fig. 1). Besides the reduction of external C_{BC} this process sequence allows excellent control of the vertical doping profile in the active base collector diode due to the use of the same reference plane for the base and collector implantation. Thus the electrical parameters like C_{BC}, breakdown voltages BU_{CB0} and BU_{CE0}, as well as Kirk current are almost insensitive with respect to variations of epi layer thickness. The benefits due to the self-aligned pedestal collector (SAPC) are discussed below by a comparison with devices without SAPC but comparable values of the doping concentration at the base/collector junction.

Table 1. Data of devices with and without SAPC
(self-aligned pedestal collector)

	unit	no SAPC	SAPC
$C_{int}(0V)$	$[fF/\mu m^2]$	0.46	0.50
$C_{ext}(0V)$	$[fF/\mu m^2]$	0.56	0.32
t_0	$[ps]$	8.2	7.0

3. Results

Table 1 compiles the specific capacitances C_{int} and C_{ext} of both internal and external base collector junctions (determined from large area test structures) for devices with and without SAPC. Due to a comparable dopant distribution in the active collector region both types of transistors show approximately the same value of C_{int}. Reduction of N_{epi} and implantation of the SAPC, however, lowers the value of C_{ext} by more than 40 %.

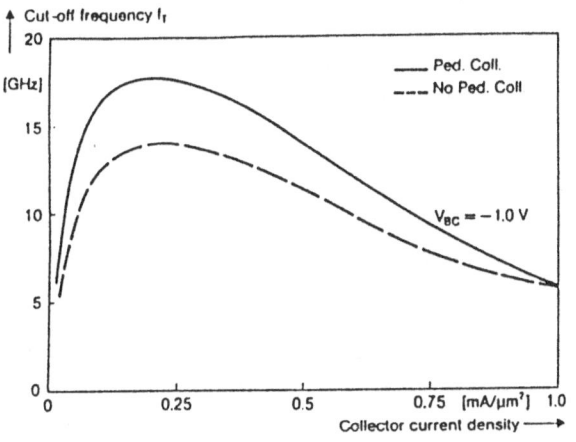

Fig.3. Cut-off frequency f_T-vs. collector current density at $V_{BC}=-1V$.

In comparison with conventionally processed devices cut-off frequency f_T is enhanced from 14 to 18 GHz at $V_{BC}=-1$ V for the devices with SAPC implant (cf fig. 3). From these f_T-curves the values of the transit time t_F were extracted in the conventional way [7]. For $V_{BC}=-1$ V $t_0 = t_F (j_c \rightarrow 0)$ of the devices with SAPC implant is 7.0 ps, i.e. 1.2 ps below the value of t_0 for conventional transistors (cf table 1). A steeper increase in the $t_F(j_c)$ curve is observed for the devices with SAPC which is attributed to reduced current spreading [8]. In contrast to laterally uniformly doped epi layers, current flow is more confined to the volume defined by the SAPC.

Improvement in circuit performance is demonstrated by a comparison of the gate delay time t_D vs. power consumption per gate curves for ECL ringoscillators (1.8 V supply voltage, 200 mV voltage swing) with transistors of two different emitter sizes, respectively (fig. 5). The benefits by the SAPC show up in a reduction in t_D from 43 to 36 ps and - even more important - in a reduction of minimum power delay product by more than 20 %.

800

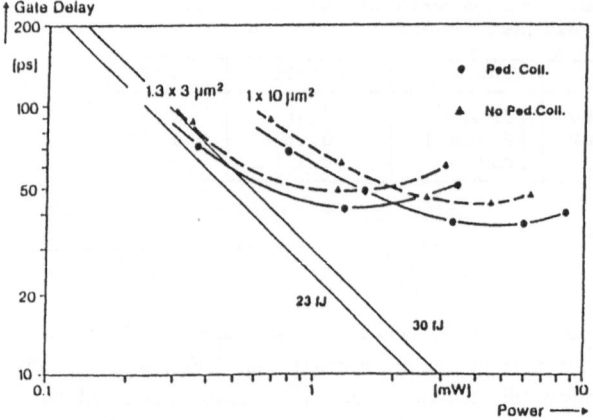

Fig.4. Gate delay time vs. power consumption per gate of two ECL ringoscillators having different emitter sizes, respectively.

4. Conclusions

Implantation of a self-aligned pedestal collector (SAPC) opens an additional degree of freedom for the optimization of double poly-Si self-aligned (DPSA) bipolar transistors. As this implantation is performed through the emitter openings, only the collector doping of the active transistor area is enhanced. Therefore it is possible to attain lower C_{ext} values by reducing N_{epi} without compromising t_F at j_c values relevant for applications. The benefits of the SAPC were demonstrated experimentally: in comparison with conventionally processed [5] transistors we obtained a higher f_T, an improved ringoscillator speed with a gate delay time as low as 36 ps and a reduced minimum power delay product. We conclude that the implanted SAPC provides a valuable and easily implementable tool for the optimization of digital circuits built with DPSA transistors. Since C_{int} is comparable for both types of devices, the Early voltage is only slightly affected by the SAPC implant. Proper adjustment of the SAPC implantation dose therefore should qualify this concept for analogue circuits as well. As further reductions in minimal feature size of DPSA transistors are expected to decrease A_{int}/A_{ext}, advantages due to the SAPC will become even more pronounced.

We would like to thank the personnel in the silicon facilities for wafer processing, R. Schreiter for electrical measurements and R. Zeininger for SIMS.

References

1. for a recent review see "Ultra-Fast Silicon Bipolar Technology", Springer Verlag Berlin, L. Treitinger and M. Miuara-Mattausch (1989)
2. K. Ehinger et al., ESSDERC 88, Journal de Physique C4, Tome 49, C4-109-112
3. T. Gormi et al., IEDM Techn. Digest, (1988) 744-742
4. D.D. Tang, P.M. Solomon, IEEE J. Solid-State Circuits, SC-14, (1979) 679-684
5. H. Kabza et al., to be published in IEEE-EDL (Aug. 1989)
6. B. Hoffmann et al., this conference
7. P. Ashburn, "Design and Realization of Bipolar Transistors", John Wiley & Sons, Chichester (1988)
8. M. Schröter, PhD-thesis, University of Bochum (FRG) 1987

Dependence of the Propagation Delay of an ECL Gate on the Transistor and Circuit Parameters

M. Y. Ghannam*, R. P. Mertens and R. Van Overstraeten
Interuniversity MicroElectronics Center IMEC
Kapeldreef 75, Leuven, Belgium
*Also with the Electronics and Communications Dept., Cairo University, Guiza, Egypt

Summary

The propagation delay T of an Emitter-Coupled-Logic (ECL) gate is studied as a function of the different device and circuit parameters involved. It is found that the delay in the response of the switching transistor is the dominant delay component. The propagation delay is highly sensitive to the change in the cut-off frequency and in the magnitude of the current source. This sensitivity is, however, a strong function of the range in which these parameters are considered. The delay is also highly but monotically sensitive to the product of the base resistance with the collector junction capacitance. The results show a weaker sensitivity of the propagation delay to the voltage swing, to the collector to substrate capacitance and to the emitter junction capacitance. Finally, the delay is almost insensitive to the emitter resistance and to the current gain.

Introduction

In the last few years a great attention has been given to bipolar technology aiming at the realization of scaled-down super-fast bipolar transistors with a base region thinner than 0.1 μm and with a cut-off frequency greater than 20 GHz. Emitter-Coupled-Logic gates with sub-50 ps propagation delays is now readily achievable [1-3]. The devices are, however, operating near the fundamental physical limits (e.g. tunneling, punchthrough, quantum mechanical, etc..), and trade-offs are imposed. A higher speed of operation can be achieved by more optimization. Therefore, new technologies (self-aligned, double-poly, etc..) are being continuously introduced and new materials (e.g. SiGe alloys [4,5]) are under investigation. This optimization has to be guided, however, by the accurate knowledge of the dependence of the propagation delay on the different parameters involved. Such a dependence can be obtained using standard computer simulations (e.g. SPICE). Developing an analytical expression which is able to substitute the lengthy process of computer simulations is very advantageous. This complex exercise has been the subject of only very few papers [6-11]. We proposed an analytical method [11] by which the full transient response of the ECL gate can be determined. In the present paper we apply this method and evaluate quantitatively the dependence of the propagation delay of the ECL gate and of its principal components on the main device and circuit parameters.

Components of the Propagation Delay of the ECL gate

The basic configuration of an Emitter-Coupled-Logic gate is shown in Fig.1. When the input voltage is changed abruptly from $V_R - V_s/2$ to $V_R + V_s/2$, where V_R is the reference voltage and V_s is the voltage swing, the output voltage (at the emitter of the emitter follower stage) will respond accordingly and changes its logic state. Since one of the tasks of the emitter follower is to insure the input/output logic level compatibility, the output voltage swing is also V_s. The propagation delay of the gate is defined as the period between the instant at which the input voltage changes and the instant at which the change in the output voltage reaches half the voltage swing. This delay T is equal to the sum of three principal components: 1) the delay t_t due to the carrier transit time in the base of the switching transistor and to the time constant of its load (collector) circuit, 2) the delay t_{jc} caused by the time constant related to the capacitance of the junction space charge regions, and 3) the delay t_{ef} due to the emitter follower.

1) Switching transistor delay t_t

In the proposed method [11] the base circuit of the switching transistor is decoupled from its load (collector) circuit. The charge control model and Kirchhoff law are applied to the base circuit in order to determine the time response of the collector current I_c to an input voltage step V_s. A time constant T_b is defined and describes the collector current (base circuit) time response. Knowing $I_c(t)$, the voltage at the load $V_L(t)$ is determined from the response of the RC load circuit characterized by a time constant T_c. The component t_t of the delay T is evaluated using conditioned analytical expressions involving T_b and T_c. It is shown here that this component is the dominant component and contributes by more than 65% to the propagation delay.

2) Junction capacitance delay t_{jc}

Before the switching transistor can respond naturally to the input voltage change, the emitter junction capacitance and the collector (Miller) junction capacitance must be charged. This charging period introduces a delay to the collector current response especially at low current levels. This capacitive effect is considered in the charge control model and results in the delay component t_{jc}. This component represents about 20% of the propagation delay for the typical conditions investigated here.

3) Emitter follower delay t_{ef}

As mentioned earlier, an emitter follower is used as the output stage of the ECL gate in order to insure input/output logic level compatibility. It also acts as a buffer between the stages. The delay t_{ef} caused by the emitter follower is evaluated by considering the emitter follower as a double pole RC circuit. The results obtained here show that the contribution of t_{ef} to T lies around 15%.

<u>Analysis and Numerical Results</u>

<u>*A) Dependence of T on the device parameters*</u>

The device parameters which are considered here are: the collector junction capacitance C_c, the emitter junction capacitance C_e, the collector to substrate capacitance C_{cs}, the cut-off frequency f_T, the base resistance R_b, the emitter resistance R_e, and the current gain β. The circuit parameters are the voltage swing V_s and the current source I_s. The load resistance R_L of the switching transistor, the source resistance R_s and the emitter resistance R_{ef} of the emitter follower are not taken as independent parameters but $R_L = V_s/\alpha I_s$ where α is the common base current gain, and $R_s = R_{ef} = V_s/I_s$. In this study, all the parameters are fixed at a chosen quiescent value, except one at a time. This free parameter is allowed to vary and the effect of such variation on T and on the three delay components, t_t, t_{jc}, and t_{ef} is recorded. By this way the effect of each parameter on the gate delay is studied independently. The quiescent values are chosen to represent an average for typical state of the art advanced silicon self-aligned bipolar technology and are listed in table I. The limits within which the free parameter is varied represent typical values recently reported by different laboratories and are also listed in table I.

As shown in Fig.2a, T is highly sensitive to any change in f_T in the low frequency range (f_T <20 GHz). Under such conditions, t_t is the dominant delay component and is mainly controlled by the base transit time, i.e. the base circuit time constant T_b. In the high frequency range, t_t is relatively less dominant and is controlled to a lesser extent by the base transit time, which explains the saturation tendency of T versus f_T.

On the other hand, T as well as its three components t_t, t_{jc} and t_{ef} increase almost linearly with R_b and with C_c as shown in Fig.2b and Fig.2c, respectively. The delay t_t is again the dominant component in both cases. For values of R_b smaller than 150 Ω t_t is mainly controlled by T_c, whereas the contribution of T_c to t_t is increasing for values of C_c greater than 10 fF. Note that if the scale of the horizontal axis in Fig.2b is multiplied by the quiescent value C_c and the scale of the horizontal axis in Fig.2c is multiplied by the quiescent value R_b, the resulting curves in either situation are almost identical. A slight difference might, however, exist and is attributed to the contribution of R_b to the emitter follower delay. Therefore, the optimization of the transistor for minimum delay should consider the product $R_b C_c$ rather than R_b and C_c separately.

The collector substrate capacitance C_{cs} has a moderate influence on T through the component t_t due to the delay in the response of the load RC circuit. As shown in Fig.2d, this capacitance has no effect on the emitter follower delay t_{ef} because in this case the collector is directly connected to the power supply. The emitter junction capacitance C_e has also a moderate influence on T, as shown in Fig.2e mainly through the component t_{jc}. Because the emitter follower is always on t_{ef} is almost unaffected by C_e due to the domination of the diffusion capacitance $\tau_F g_m$.

In the range of values considered in this study, the emitter resistance and the current gain have almost no influence on the delay T.

B) Dependence of T on the circuit parameters

At low I_s values, the load resistance R_L should be relatively large in order to satisfy the voltage swing limitation. Consequently, the time constant of the load circuit T_c will also be large and results in a long delay t_t. This explains the fast decrease of t_t and of T with increasing I_s at a constant value of V_s as shown in Fig.3a. For larger values of I_s the load time constant T_c is reduced and the time constant of the base circuit T_b starts to contribute effectively to t_t. For values of I_s greater than 1 mA T_b becomes dominant which results in a much less sensitivity of t_t and of T to any change in I_s. It should be mentioned that the model in [11] is not valid if the current I_s drives the transistor into high level injection.

Since R_L and consequently T_c increase with increasing V_s, t_t should also follow the same behavior. A moderate increase of T and t_t with increasing V_s is confirmed in Fig.3b.

References

1. K. Ueno, H. Goto, E. Sugiyama, and H. Tsunoi, "A Sub-40 ps ECL Circuit at a Switching Current of 1.28 mA", IEDM Technical Digest, 1987, 371.

2. T. C. Chen, D. D. Tang, C. T. Chuang, J. D. Cressler, J. Warnock, G. P. Li, P. E. Biolsi, D. A. Danner, M. R. Polcari, and T. H. Ning, "A Sub-50 ps Single Poly Planar Bipolar Technology", IEDM Technical Digest, 1988, 740.

3. T. Gomi, H. miwa, H. Sasaki, H. Yamamoto, M. Nakamura, and A. Kayanuma, "A Sub-30 ps Si Bipolar LSI Technology", IEDM Technical Digest, 1988, 744.

4. P. Narzony, M. Hamacher, H. Dambkes, H. Kibbel and E. Kasper, "Si/SiGe heterojunction Bipolar Transistor with Graded GaP-SiGe Base Made by Molecular Beam Epitaxy", IEDM Technical Digest, 1988, 562.

5. J. .F. Gibbons, C. A. King, J. L. Hoyt, D. B. Noble, C. M. Gronet, M. P. Scott, S. J. Rosner, G. Reid, S. Laderman, K. Nauka, J. Turner, and T. I. Kamins, "Si/Si$_{1-x}$Ge$_x$ Heterojunction Bipolar Transistors Fabricated by Limited Reaction Processing", IEDM Technical Digest, 1988, 566.

6. D. D. Tang and P. M. Solomon, "Bipolar Transistor Design for Optimized Power-Delay Logic Circuits", IEEE Trans. Electron Devices, ED-19, 1979, 809-820.

7. A. Barna, "Analytic Approximations for Propagation Delays in Current Mode Switching Circuits Including Collector-Base Capacitances", IEEE J. Solid-State Circuits, SC-16, 1981, 597.

8. R. Ranfft and H.-M. Rein, "A Simple Optimization Procedure for Bipolar Subnanosecond ICs with Low power Dissipation", Microelectron. J. 13, 1982, 23-28.

9. E. F. Chor, A. Brunnschweiler, and P. Ashburn, "A Propagation Delay Expression and its Application to the Optimization of Polysilicon Emitter ECL Processes", IEEE J. Solid-State Circuits, SC-23, 1988, 251.

10. J. M. C. Stork, "Bipolar Transistor Scaling For Minimum Switching Delay and Energy Dissipation", IEDM Technical Digest, 1988, 550.

11. M. Y. Ghannam, R. P. Mertens and R. Van Overstraeten, "An Analytical Model For the Determination of the Transient Response of CML and ECL Gates", accepted for publication in IEEE Trans. Electron devices.

Parameter	Quiescent Value	Minimum Value	Maximum Value
f_T (GHz)	20	10	30
R_b (Ω)	250	50	500
C_c (fF)	8	4	12
C_{cs} (fF)	20	10	30
C_e (fF)	10	4	16
R_e (Ω)	25	0	100
β	100	20	200
I_s (mA)	1	0.2	1.8
V_s (mV)	500	300	700

Table I

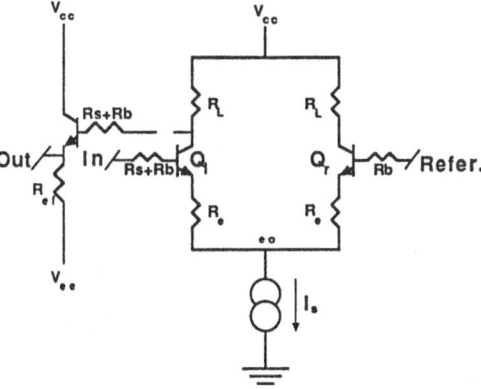

Fig.1. Configuration of a Standard ECL gate

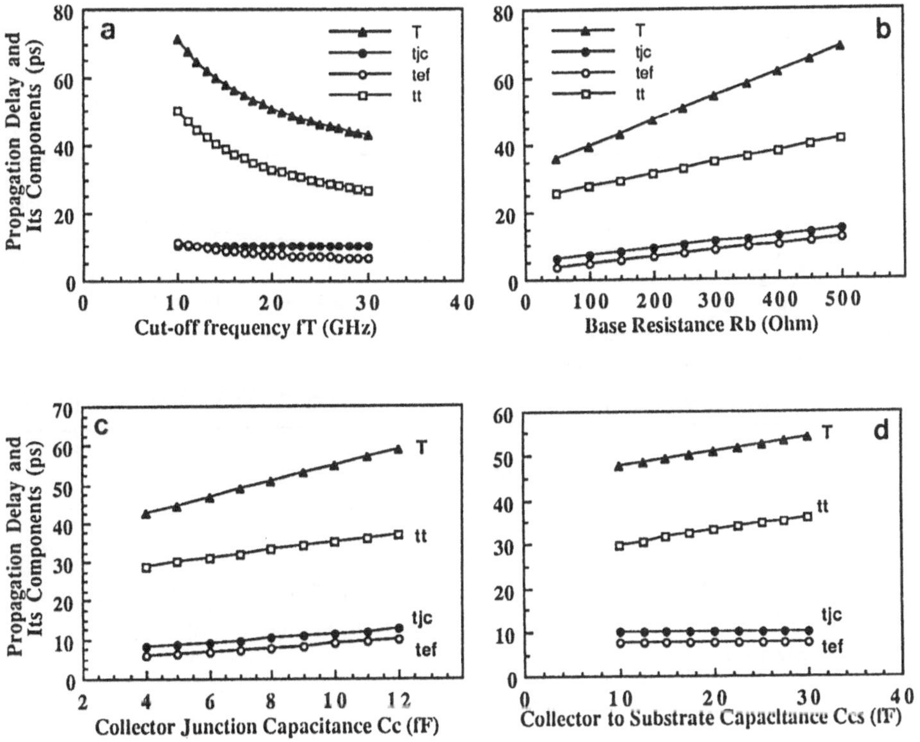

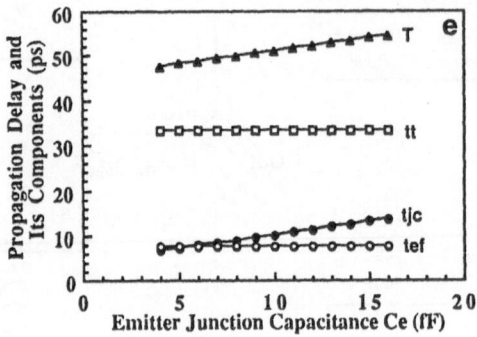

Fig.2: Propagation Delay T and its components t_t, t_{jc}, t_{ef} as a function of: a) cut-off frequency f_T, b) Base resistance R_b, c) Collector junction capacitance C_c, d) collector to substrate capacitance C_{cs}, e) emitter junction capacitance C_e.

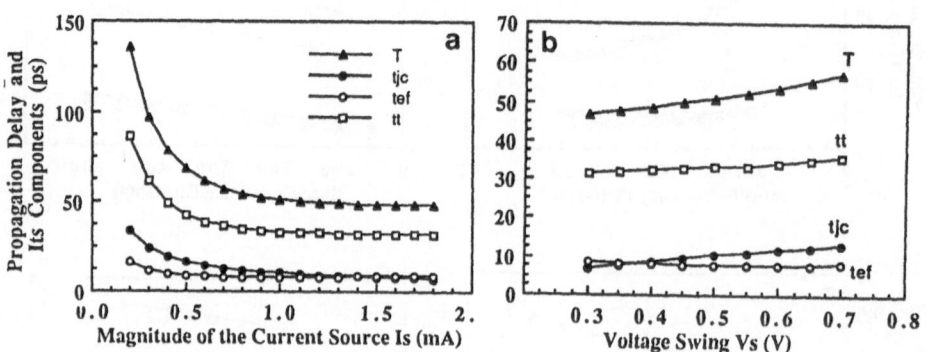

Fig.3: Propagation Delay T and its components t_t, t_{jc}, t_{ef} as a function of: a) Magnitude of the current source I_s, b) Voltage Swing V_s.

Bandgap Narrowing Due to Heavy Doping in $Si_{1-x}Ge_x$ Layers

J.Poortmans, R.P. Mertens, S.C.Jain*, J.Nijs,
R.Van Overstraeten

IMEC, Kapeldreef 75, B3030 Leuven,Belgium
*Present Address Clarendon Laboratory, Oxford, U.K.

Summary

In this work the bandgap narrowing due to heavy doping in SiGe alloys is calculated using first-order models at T=0K. A detailed overview of the way we have proceeded is given as well as a clarification of the different assumptions. Results are presented for n and p type doping, indicating that the values for the bandgap narrowing are approximately the same. We have also tried to extend this model for the case of strained SiGe layers. We found that the influence of the strain is rather small for p type layers, but is substantial for n type layers.

Calculations

The calculations are based on the work of Mahan[1] and Berggren and Sernelius[2]. Basically, we followed the approach of Mahan, but corrected it for the self-energy due to ionized impurity scattering.

As a result we have three terms in the total bandgap narrowing. The **Hartree-Fock exchange energy** is given by ,

$$exx = 2e^2 kf\Lambda/(\pi\varepsilon); \quad (e=1.6e-19) \quad (SI-units) \ [2]$$

where kf is the wavevector at the Fermi-level, ε the background dielectric constant and Λ takes into account the effect of the mass anisotropy. This overestimates the self-energy in the majority carrier band due to electron-electron (hole-hole) interaction by 15 percent when compared with values,given by Berggren.

When applying the Thomas-fermi approximation the total **bandgap narrowing due to impurity scattering**, is given by,

$$eix = 4\pi Ne^2/(a\varepsilon ks^3)(1+(mhh+mlh)/(2md)); \quad (SI-units) \ [2]$$

where N is the dopant concentration and ks the inverse

Thomas-Fermi screening length and a the effective Bohr-radius. This result overestimates the effect of impurity scattering because the frequency dependence of the dielectric screening is not included. md is the effective mass in the conduction band, while mhh and mlh are the masses for the heavy- and light-hole subbands in the valence band. For p-type material we have to substract the shifts due to impurity scattering of valence and conduction band.

The **correlation energy of a free carrier in the minority band** was found by Mahan,

$$ec=2e^2/(\pi\epsilon)(m.wp/(2h))^{0.5}J(so); \quad (\text{SI-units}) \quad [1]$$

where m is the mass in the minority carrier band, wp the plasma frequency of the degenerate free carrier gas and h Planck's constant. The value of J(so) is 0.8 for the doping levels studied in this paper.

Before going to the calculation we have to make some assumptions concerning the background dielectric constant and the masses in the valence band as a function of the alloy composition. For both we take a linear interpolation between the values of pure Si and pure Ge. This is a crude approximation because experimental results do not confirm such a simple behaviour for the masses in the valence band of strained $Si_{1-x}Ge_x$ layers [3]. As far as the conduction band is concerned, there is less uncertainty, because it is known that the conduction band structure changes rather abruptly from 'Si-like' (six equivalent conduction band minima along the <100>-axes) to 'Ge-like' (four equivalent conduction band minima along the <111>-axes) [4]. Furthermore it is stated by Krishnamurthy that the transverse and longitudinal masses do not change much as a function of concentration [5]. Therefore the values of the silicon effective masses for the conduction band are used up to a concentration of germanium of 85%.

Furtheron we suppose the conduction band and the valence band to be parabolic. To obtain ks we take the general formula which states that

$$ks=(4\pi e^2/\epsilon)D(Ef); \quad (SI\text{-units}) \quad [6]$$

where D(Ef) is the density of states at the Fermi-level.

Results for unstrained n-type material (fig.1)

For the unstrained material we obtain values for the bandgap narrowing, which correspond well to the values,experimentally determined by Wagner [7]. The discontinuity at x=0.85 comes from the rather abrupt change of the degeneracy and the effective masses from the values for Si to those of germanium. We did not try to find the exact shape of the curve around this point, because the critical thickness for layers with such high concentrations of germanium is so small (3 nm) [8] that they cannot be used as base material in a heterojunction bipolar transistor.

Results for unstrained p-type material (fig.2)

For p-type material we obtain almost the same values as for n-type material. This corresponds to the values, measured by Wagner [7]. This result is rather surprising, because of the important simplifying assumptions we have made for the valence band.

Results for strained n- and p-type layers (fig.1 and fig.2)

The influence of bandgap narrowing in strained $Si_{1-x}Ge_x$ layers has been estimated by calculating the split in the majority carrier band on the basis of the deformation potentials of Si and Ge [9]. Knowing this split we calculate the value of the Fermi-level as to accomodate all the free carriers. Then we use the above mentioned formulas where kf is now approximated by a weighted mean over the different kf of the split valleys. We define the weight factors as the ratio of particles in a valley to the total number of free carriers.

For strained n-type material the bandgap narrowing increases substantially when compared to unstrained material. This is caused by the degeneracy change which causes a change in kf and ks. For p-type material the difference is small. This is explained by the fact that, in unstrained material, the properties of the heavy-hole band already dominate.

810

Conclusions

We have calculated the relation between the Ge-content and the
bandgap narrowing due to heavy doping effects. These results
should be included in the interpretation of the bandgap
offset that one finds in the heterojunction bipolar
transistors, because the bandgap narrowing due to heavy doping
is comparable to that, caused by strain [10,11,12,13].

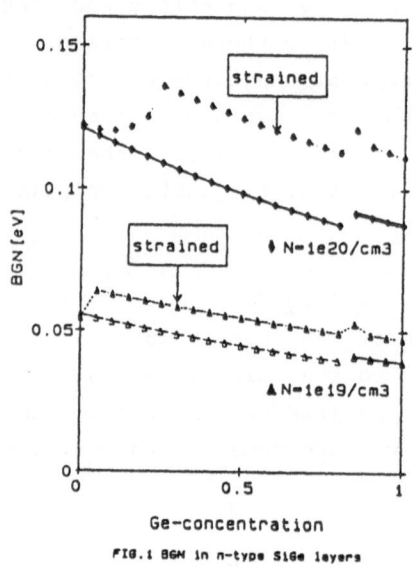

FIG.1 BGN in n-type SiGe layers

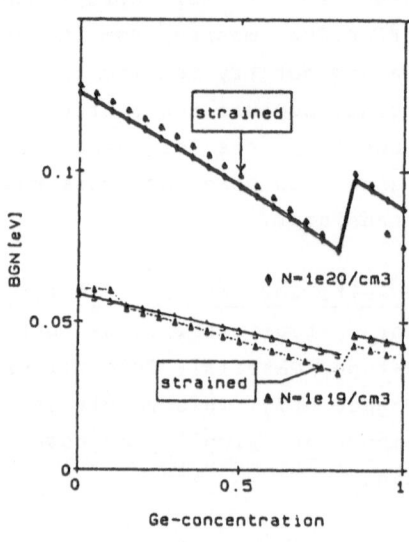

FIG.2 BGN in p-type SiGe layers

References

1. G.D. Mahan, J. Appl. Phys., 51, 2634(1980).
2. K.F. Berggren and B.E. Sernelius, Phys. Rev. B24,1971
   (1981).
3. R. People et al. Appl. Phys. Lett. dec. 1984.
4. C. Smith et al. 1987 bipolar circuits and Technol. Meeting
   Dig. p.57.
5. S. Krishnamurthy and A.Sher, Appl. Phys. Lett.,47,160(1985)
6. C. Kittel, Introduction to Solid State Physics 6th ed. p.265
7. J. Wagner and J.A. del Alamo, J. Appl. Phys.63, 425(1985)
8. J.C. Bean, J. Vac. Sci.Technol.B 1427(1986)
9. Landolt-Bornstein tables, Springer-Verlag Vol.17a (1982)
10. S.S. Iyer et al., IEDM Techn. Dig., 874(1987)
11. T.Tatsumi et al., Appl. Phys. Lett. 52, 895(1988)
12. D.X Xu et al., Appl. Phys. Lett. 52, 2239(1988)
13. G.L. Patton et al., IEEE Elec. Dev. Lett. 9, 165(1988)

Reliability

Thermally Activated Failure Modes and Mechanisms of High Electron Mobility Transistors

Claudio Canali(*), Giuseppe Castellaneta (**), Fabrizio Magistrali (***), Marco Sangalli (***), Carlo Tedesco (*) and Enrico Zanoni (*)

(*) Universita' di Padova, Dipartimento di Elettronica ed Informatica, Via Gradenigo 6A, I-35131 Padova, Italy.
(**) Tecnopolis- CSATA, I-70010 Valenzano, Bari, Italy
(***) Telettra S.p.A., Quality and Reliability Dept., Via Trento 30, 20059 Vimercate, Milano, Italy.

In recent years, High Electron Mobility Transistor (HEMT) has emerged to be a very promising device for both high-speed digital circuits and high frequency low-noise amplifiers, becoming a possible replacement for conventional GaAs MESFETs. These devices take advantage of the enhanced mobility and velocity of electrons in the two dimensional electron gas formed at selectively doped GaAs-AlGaAs heterojunction and show a high transconductance with best results in the range 300/400 ms/mm [1].

Up to now, the reliability of this device has not been thoroughly studied and very few data are available concerning failure mechanisms [2].

To evaluate failure modes and mechanisms of low-noise HEMTs, devices of five different suppliers were submitted to different high temperature storages and life tests, together with low-noise GaAs MESFETs for comparison. We will report here results concerning the first 2500 hours of the thermal storage test at 250 °C with no bias applied. Parametric degradation has been followed by monitoring Idss, gm, Vp and the gate breakdown voltage. Parasitic resistances Rs, Rd and Rg were evaluated by means of "end-resistance" [3] measurements.

Technological characteristics of the five suppliers are summarized in Table I. Devices mainly differ in gate metallization: some suppliers (A,D,) adopt pure Al, which can not be directly deposited on AlGaAs, because this gives raise to a leaky junction and therefore requires growing a thin n^+-GaAs layer on top of the AlGaAs. On the contrary, refractory gates (C,E) can be directly deposited on the ternary phase.

Degradation appears to be dominated for shorter times (<500 hours) by interdiffusion/interfacial effects which affect the Schottky contact and/or cause variations in the 2DEG

Table I: Technological characteristics of HEMT devices submitted to storage test at 250°C.

DEVICES	A	B	C	D	E
GATE	Al	Al/Ni	WSi/Ti/Pt/Au	Al	Al/Ti
W/L (um)	.35/200	.70/200	.25/220	.60/200	.40/190
OHMIC CONTACTS	AuGeNi	AuGeNi	AuGeNi	AuGeNi	AuGeNi

(#) Work partially supported by CNR Progetto Finalizzato MADESS.

814

concentration, with consequent degradation of Idss and Vp.
    For longer times (>500 hours) increase in Rs and Rd takes
place, which gives raise to an increase in RL and is possibly
due to ohmic contacts degradation.
    Figures 1 and 2 report as a reference percentage variation
of Idss and RL respectively (averaged on all tested devices) as
a function of time. RL is defined as Vds/Ids for Vds=100 mV and
Vgs=0 V and comprises obviously both the contribution of the
channel resistance and of parasitic resistances Rs and Rd.

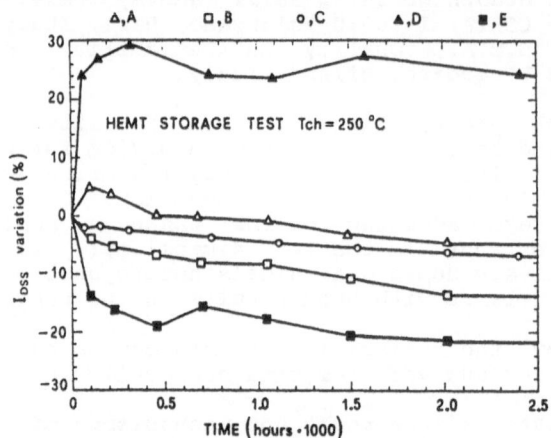

Fig. 1   Idss percentage variation during storage test at 250
°C of HEMT devices of five different suppliers.

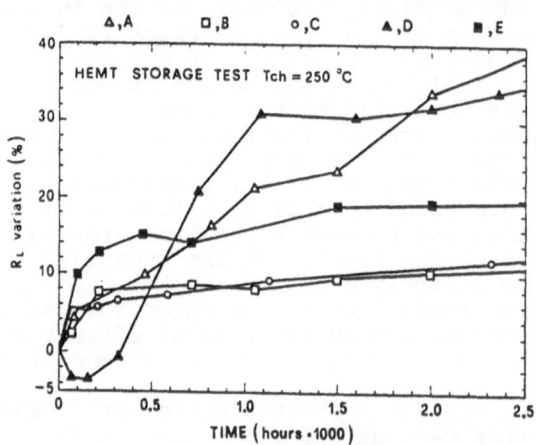

Fig. 2 RL percentage variation during HEMT storage test at 250
°C. Results are averaged on all treated samples.

    Four different effects are observed in failed devices:
(a) some suppliers (D and E) show a noticeable change in Idss
in the first 500 hours. Changes in Idss seem to be correlated
with RL behaviour. In fact in devices D, with pure Al gate
metallization, Idss increases (+30% in the first 500 hours)
while RL decreases (-5%). On the other hand, the I-V

characteristics of the gate Schottky diode remain virtually unchanged, suggesting that the observed degradation is due to variations in the 2DEG concentration, possibly caused by channel deconfinement and/or interdiffusion effects.

(b) devices from supplier B, with Al/Ni gate metallization show a slight decrease of Idss accompanied by RL increase. This effect is correlated with an increase of knee voltage of the gate diode, Fig. 3. Since the behaviour of the gate diode at low forward voltages (<0.8 V) is dominated by the metal/AlGaAs Schottky diode [4], the observed change in the I-V characteristics is attributed to an increase in the Schottky barrier height due to metal-semiconductor interdiffusion. This effect causes both a rigid shift of the Id(Vgs) and gm(Vgs) characteristics, Fig. 4, and the observed Idss degradation.

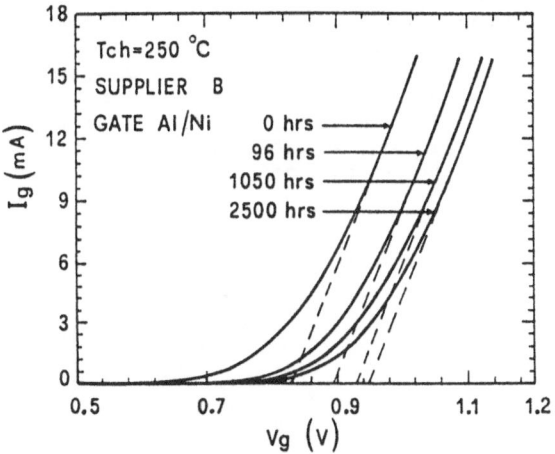

Fig. 3 Parallel shift of linear Ig(Vg) characteristic due to change of gate diode barrier height.

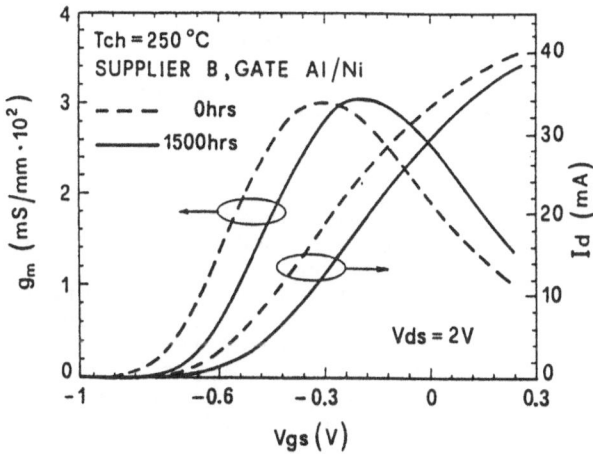

Fig.4 Shift in Id(Vgs) and gm(Vgs) characteristics due to thermal storage in one device from supplier B.

816

(c) after approximately 500 hours of test RL begins to increase markedly in some devices (D and A), but no effect is measured on Idss. Complementary results suggest that this increase in RL is not due to an increase in channel resistance, but should attributed to ohmic contacts degradation. RL increases up to 30% in devices D; it should be noted that much larger increases in Rs and Rd should take place in these devices to have an observable effect on Idss.

(d) Finally, supplier E presents another failure mode, which consists in a drastic increase (+300% in 1000 hours) in the gate series resistance Rg, Fig.5. This effect does not influence the dc Idss value, but can lead to degradation in device rf performances. The increase in Rg is possibly due to interaction phenomena.

In conclusion, results on thermal storage of HEMTs have evidentiated both interdiffusion/interfacial effects in gate active area (at shorter times) and ohmic contacts degradation (at longer times). By comparing Idss percentage variation, however, only suppliers D and E show significant changes in electrical characteristics, while other devices show excellent stability up to 2500 hours of test at 250 °C.

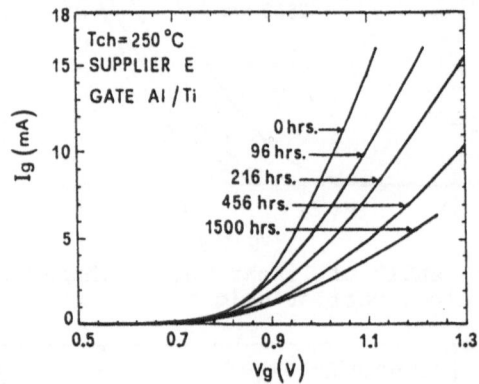

Fig.5 Degradation of I-V characteristics of gate diode (supplier E) due to increase of Rg.

REFERENCES

[1] A. Eskandarian, "Determination of the small-signal parameters of an AlGaAs/GaAs Modfet", IEEE Trans. on El. Dev., ED 35, pp.1793-1801, 1988.
[2] A. Christou, "Reliability problems in state of the art GaAs devices and circuits", Quality and Reliability Engineering International, pp.37-46, 1989.
[3] K. Lee, M.S. Shur, A.J. Valois, G.Y. Robinson, X.C. Zhu and A. Ziel, "A new technique for characterization of the "end" resistance in modulation doped FETs", IEEE Trans. El. Dev., pp. 1394-1398, 1984.
[4] C. Chen, S.M. Baier, D.K. Arch and M.S. Shur, "A new and simple model for GaAs Heterojunction FET gate characteristics", IEEE Trans. on El. Dev., ED 35, pp.570-577, 1988.

Computer Simulation of Transient Charge Collection from Ionizing Radiation in Silicon Junctions

H.Belhaddad,R.Gaillard,G.Poirault (*),A.Poncet (**)

(*)   Nucletudes S.A. Avenue du Hoggar, Z.A. Courtaboeuf BP 117 -
      F - 91944 LES ULIS Cedex
(**) CNET-Grenoble Chemin du vieux Chene, BP 98 F - 38243 MEYLAN
      Cedex

I - INTRODUCTION :

The collection of charge from ion tracks can produce logic upset and memory change in integrated circuits. A good understanding of the fundamental aspects of the charge collection phenomena is necessary to determine the perturbation threshold of the device struck. To this respect, the shape and the amplitude of the current pulse must be accurately modellized. However, due to very high density of electron-hole pairs along the ion track and the field-enhanced funneling phenomenon, the analysis of charge collection requires numerical resolution of device equations.

First order analytical solutions have been developed (/1/,/2/,/3/), which are based on a significant number of assumptions on electrical behaviour and are limited to very simple devices, mainly diodes; however, the space distributions of carrier densities and electrical potential require accurate computer simulations in order to validate these models before use in circuit simulators.

The purpose of this work is to show to what extend a 2-D simulator can provide a better understanding of the charge collection mechanism, and to exhibit the dependence of electrical characteristics, like photocurrent pulse, versus both technological parameters and physical models.

II - NUMERICAL MODELLING USING A 2-D DEVICE SIMULATOR :

Strictly speaking, the accurate modelling of charge collection process requires general three-dimensionnal device simulation. It can be brought back to 2-D when the device structure is axi-symmetrical and the ion track follows the axis of symmetry; therefore, using a 2-D simulator for modelling realistic situations cannot provide more than qualitative results. In the present work, computer simulations were performed using **JUPIN** 2-D device simulator (/4/), which is designed to analyze general transport problems in semiconductor devices under both steady-state and transient conditions.

Ionizing radiation is modellized through a specific expression in the generation term.

The coupled resolution of Poisson's and carrier continuity equations provide a robust treatment of these additional non-linearities, and the flexibility of triangular meshes allows drastic mesh refinements in the neighborhood of the ion track.

III - SIMULATION RESULTS :

From simulations performed on two very simple device structures, P+N and P+NN+ junctions (Fig.1), the influence of main parameters have been exhibited, i.e. :

* Ion track thickness r (while keeping constant the implanted dose) : the maximum current decreases with r and stabilizes for r ≤ 0.1 μm. The time required to establish this maximum current is, for r=0.1 μm, very close to the value commonly suggested ;

* Substrate concentration : The maximum current increases with the doping concentration and the collection time decreases ;

* Mobility models : The Figs.2,3 show that all the parameters which usually affect the mobility must be taken into account : electric field , doping concentration, carrier concentrations . Indeed, both current characteristic shape and collection time constant of the junction are highly dependent on the degree of accuracy of the mobility model (/1/) ;

* Funneling phenomenon : the funneling process is due to the electrical field wich is created in the quasi-neutral substrate region (Fig.5) . Fig.4 illustrates the radial ambipolar diffusion mechanism : the carriers move from the ion track, where the carrier concentrations strongly exceed the background doping, to the rest of the substrate.

* Angle of incidence (Fig.6) : The charge collection time is greater than the time at normal incidence.

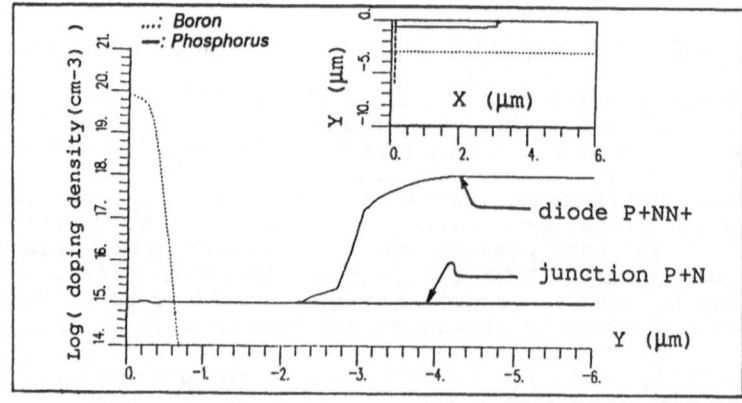

Fig.1.   Cross-sectionnal views of impurity profile for the two simulated devices (P+N - P+NN+)

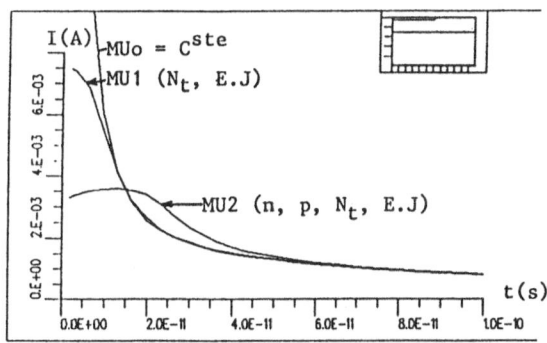

Fig.2. Current pulse obtained with three different
mobility models :
MU0 = constant mobility
MU1 = mobility depending on electric field and doping
concentration
MU2 = mobility depending on electric field and doping
and carrier concentrations

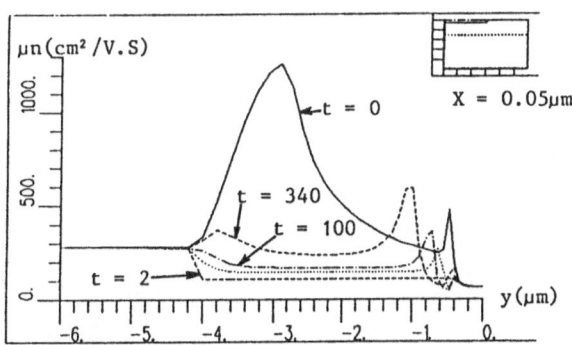

Fig.3. Mobility along the ion track at different time
steps (t = 0, 2, 50, 100, 340 ps)

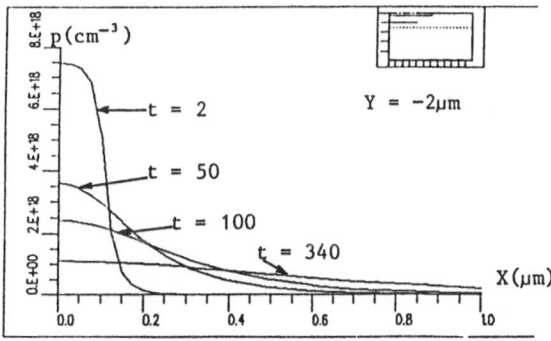

Fig.4. Carrier concentration along the radial X axis at
different time steps (t = 2, 50, 100, 340 ps)

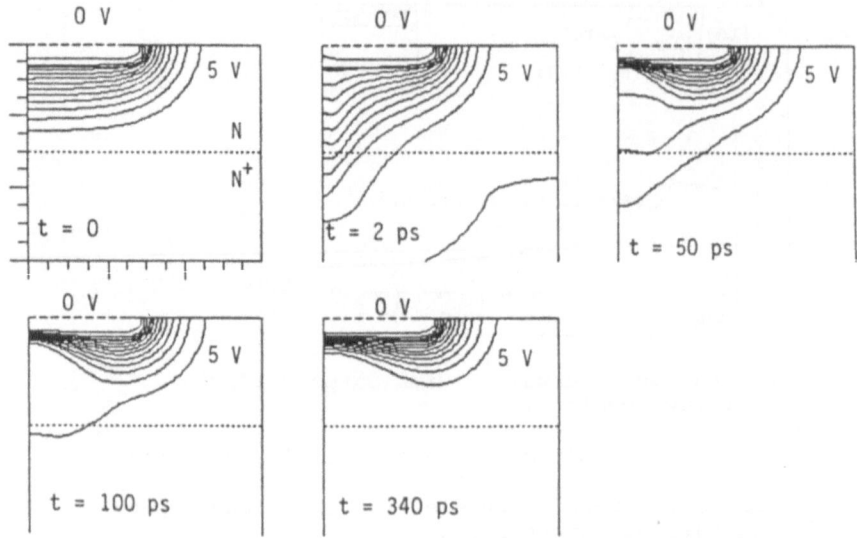

Fig.5.   Equipotential lines at t = 0, 2, 50, 100, 340 ps

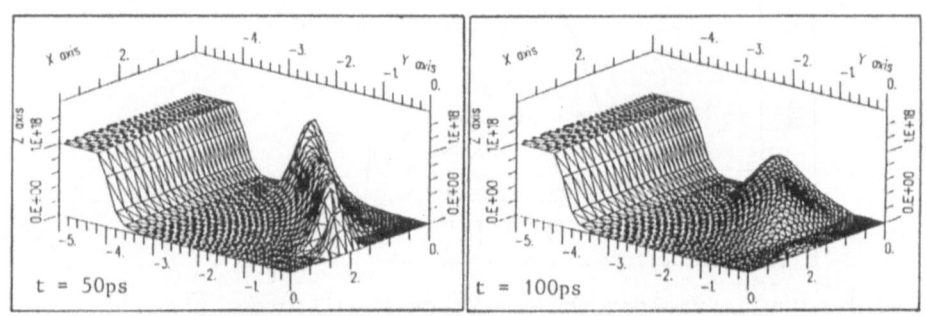

Fig.6.   Electron concentration at 50, 100 ps under oblic
incidence (50°)

III - REFERENCES :

/1/ G.C. Messenger IEEE/NS 29 No.6 December 1982
/2/ F.B. McLean & T.R. Oldham IEEE/NS 26 No.6 1982
/3/ M.Shur IEEE/NS 33 No.5 1986
/4/ H.Belhaddad, C. Corbex & A. Poncet Nasecode-IV Proc. Boole
    Press June 1985

Model for Simulation of Soft Error Rate of Megabit DRAMs

W.H. Krautschneider and Th. Künemund

Siemens, Central Research and Development, Microelectronics
D-8000 Munich 83, West Germany

Abstract
A major constraint in designing dynamic memories of the 4 or 16 megabit level are soft errors caused by natural alpha radiation. This dictates a minimum capacitance for meeting the soft error rate (SER) specifications. A self-consistent engineering model has been developed for simulating the SER in dependence on memory cell and circuit parameters. SERs with different memory technologies will be presented and discussed.

Introduction
Electronic circuits such as memory chips are exposed to alpha radiation which is emitted from traces of radioactive elements (mainly U and Th) contained in the packaging material and in the aluminum metallization. The flux from the packaging material can be completely shielded by a protective layer, in contrast to the flux from the aluminum, which represents the main part of the alpha particles impinging on today's circuits. There is an additional flux from alpha particles originating from the decay of silicon atoms after capturing neutrons or protons stemming from cosmic rays [1].

Charge Generation and Collection
When an alpha particle strikes a silicon device and its track traverses a pn-junction, electron-hole pairs are generated with an energy-dependent ionization density N_α ranging from 30 to 95 pairs/nm [2]. The charge generated in the space-charge region (section 1 in Fig. 1) is completely collected by drift. This charge component is given by the length of the track section in the space-charge region and by the ionization density N_α. As long as the concentration of free carriers in the space-charge region is high, the electric field can penetrate the substrate and give rise to an additional component of collected charge known as the funneling charge [3,4]. Because of their higher mobility, electrons within a distance of $\mu_n/\mu_p \cdot w_p$ (section 2) can

reach the boundary of the pn-space charge region during the time in which the space-charge region is emptied of free carriers. Since this carrier flow is along the field across the space-charge region, it is independent of the angle of the alpha particle track [5]. Thus, the charge collected by drift Q_f is given by

$$Q_f = q \cdot N_\alpha \cdot w_p \cdot \left(\frac{1}{\cos\beta} + \frac{\mu_n}{\mu_p}\right) \tag{1}$$

with w_p, the width of the space-charge region

$$w_p = \left(\frac{2 \cdot \epsilon_0 \cdot \epsilon_{si}}{q \cdot N_B} \cdot V_{pn}\right)^{1/2}.$$

The other symbols have the usual meanings. Eq. (1) shows that for higher dopant concentrations N_B the collected charge decreases significantly.

The charge collection by diffusion (section 3) does not amount to more than a few femtocoulombs for collector areas in the submicrometer range [6], so it can be neglected compared with the drift charge.

Calculation of the Soft Error Rate

A soft error is induced when an alpha particle strike causes charge collection of more than the critical charge Q_c within the time span or shortly before the hit node can change its logical state. The contribution of a node to the SER is given by

$$SER_n = \Phi_\alpha \cdot A \cdot s \cdot P(Q \geq Q_c). \tag{2}$$

where Φ_α is the flux (alpha particles per time and area), A the area of the space-charge region, s the quotient of the time during which the node is sensitive to alpha particles to the operation time. The factor s causes a dependence of the SER on the memory cycle time. $P(Q)$ denotes the probability of the occurence of an alpha particle strike that transfers a charge greater or equal to Q. If Q equals or exceeds Q_c, a soft error will be triggered.

The $P(Q)$-function can be obtained by direct measurement [7] or by simulation of alpha particle strikes allowing for their energy and angle spectrum. It is a sensitive function of the dopant concentration (Fig. 2), and in case of a trench type capacitor, of the depth of the trench (Fig. 3). A very flat trench with a depth of 0.2µm, which is equivalent to a planar structure, shows the lowest charge collection (Fig. 3). Cells with deeper trenches gather more charge because of the increased length of the alpha particle track within the space-charge region.

For applying eq. (2), the critical charge Q_c has to be known. It is given mainly by the storage cell, the bitline capacitance and the noise level during signal sensing. Using the simulation program SPICE, the response of the memory circuit to alpha particle strikes was investigated for varying amounts of collected charge, which were represented by transient current injections at the nodes sensitive to alpha particles, i.e. the cell, bitline and sense amplifier nodes. By this means the critical charge was determined for different cell capacitances (Fig. 4). The bitline capacitance in our simulations was 330 fF and the substrate dopant concentration $4 \cdot 10^{16}$ cm^{-3}.

The SER of a memory chip is then calculated by summing up the contributions of cell, bitline and sense amplifier nodes,

$$SER = \Phi_\alpha \cdot (\Sigma A_c \cdot P(Q_c) + s \cdot (\Sigma A_b \cdot P_b(Q_c) + \Sigma A_{sa} \cdot P(Q_c))).$$

In Fig. 5 the SERs of DRAMs with stacked and trench storage cells are compared, the capacitance of the stacked cell was 25 fF. It exhibits a clear superiority of the stacked cells in respect of alpha particle sensitivity.

Using a ^{226}Ra-source, the model has been verified by accelerated soft error tests of 4Mbit memory chips with 4.5μm deep trench cells (Fig. 6).

Conclusion

In summary, a self-consistent engineering model has been developed and experimentally verified as a tool for simulating the SER of dynamic memories with different cell concepts.

Acknowledgment
The help of K. Lau in computer processing is gratefully acknowledged.

References
[1] J.F. Ziegler and W.A. Lanford, Science, vol. 206, p. 776, 1979.
[2] T.C. May and M.H. Woods, IEEE Trans. Electr. Dev., vol. ED-26, p. 2, 1979.
[3] C.M. Hsieh, P.C. Murley, and R.R. O'Brien, IEEE Elec-tr. Dev. Lett., vol. EDL-2, p. 103, 1981.
[4] C. Hu, IEEE Electr. Dev. Lett., vol. EDL-3, p. 31, 1982.
[5] E. Takeda, K. Takeuchi, T. Toyabe, K. Ohshima, and K. Itoh, IEDM Dig. Techn. Pap., p. 542, 1986.
[6] K. Terrill, C. Hu, and A. Neureuther, Solid-State Electr., vol. 27, p. 42, 1984.
[7] W.H. Krautschneider and W. Meyberg, Proc. 17th ESSDERC, Bologna, Italy, p. 709, 1987.

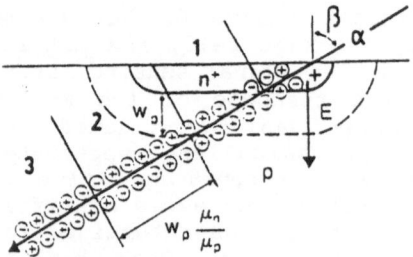

Fig. 1 Charge collection after alpha particle strike. Section 1: Space charge region.
Section 2: Funneling region
Section 3: Charge collection by diffusion.

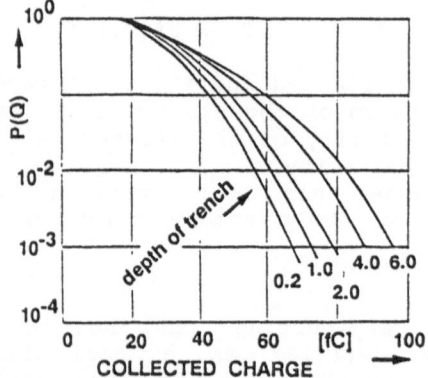

Fig.3 P(Q)-function for different trench depths (in μm). $N_B = 10^{16}$cm^{-3}; node voltage: 7.5V; node area: 5μm^2

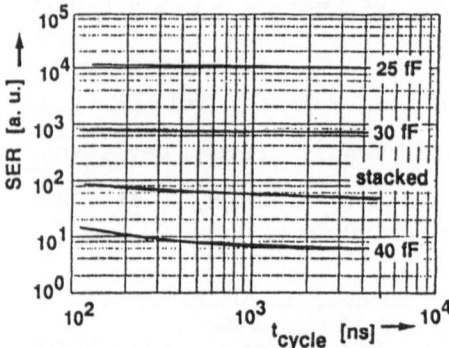

Fig. 5 Soft error rate of dynamic memories with trench cells (cell capacitances are 25, 30 and 40 fF) and with a stacked cell of 25 fF.
$V_{DD} = 4.5$ V. $N_B = 4 \cdot 10^{16}$cm^{-3}.

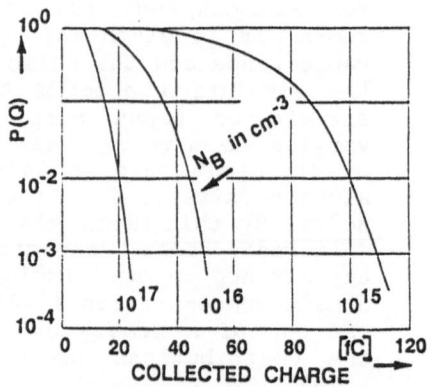

Fig. 2 Probability for charge collection P(Q) caused by alpha particle strikes for different substrate dopant concentration N_B. Node voltage is 5V. node area is 1μm^2.

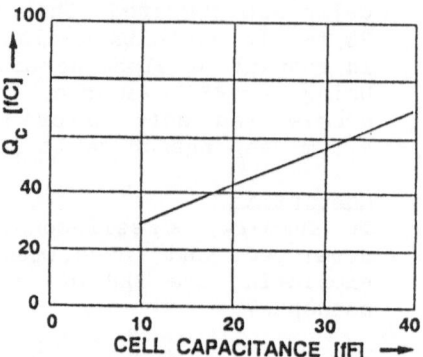

Fig. 4 Critical charge Q_c as a function of the cell capacitance.

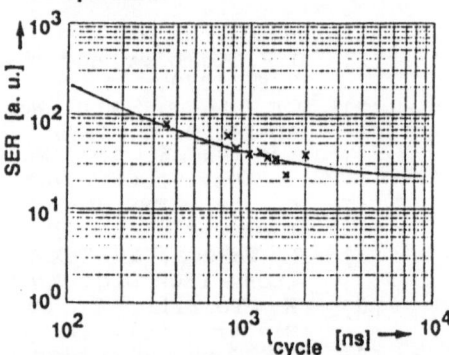

Fig. 6 Soft error rate of a 4Mbit dynamic memory Solid line: Simulation. Crosses: Experimental data.

Optimization of Submicron n-channel Devices for Performance and Reliability

S. J. S. NAGALINGAM, P. DERKS, M. J. B. BOLT, K. OSINSKI

Philips Research Laboratories,
P.O. Box 80000, 5600JA Eindhoven,
The Netherlands

Abstract

The DC performance of silicided N-channel devices is characterised primarily by the current drive capability. However, this is adversely influenced by the series resistance of the device which arises primarily from the lightly doped region located between the channel and the highly doped source/drain regions. This region is introduced to meet the lifetime specification of the device by reducing the hot carrier effects at the drain edge. The goal of this investigation was to achieve maximum current drive capability while satisfying the lifetime specifications. Initially N-channel devices with different drain structures were fabricated and their current drive capabilities, electric field distributions and lifetimes were characterised. Simulations with MINIMOS-4 were performed and the results were correlated with the measured data. Based on these verified simulations, improved N channel drain structures were then defined and devices with these drain structures were fabricated. These devices proved to have both higher current drive and improved lifetime. However, as expected, adverse short channel effects were seen on these structures. However, a compromise can be achieved that gives better performance and improved lifetime while minimising short channel effects.

1   Introduction

In scaling of technology, one critical measure of DC N-channel performance improvement is the drive current, ie the saturated drain current. Silicidation of source/drain regions is used to further increase the drive current. However in submicron technology lightly doped drain regions are necessary to achieve acceptable device lifetimes. But these lightly doped regions manifest themselves as an increase in series resistance which limits the drain current. The goal of this investigation was to design an N-channel device structure that maximises the current drive capability while still meeting the lifetime requirements.

2   Preliminary Experiment

Initially simple drain structures were defined based on SUPREM-3 simulations and these devices were fabricated together with the conventional lightly doped drain (surface LDD-Phosphorous doped) structure. These structures, which included a buried LDD (S1), a graded buried LDD (S2) and a hard buried LDD (S3) [1] were designed with specific combinations of Arsenic and Phosphorous implants. Comparison of the doping profiles of a graded buried structure and conventional surface LDD structure is shown in figure 1.

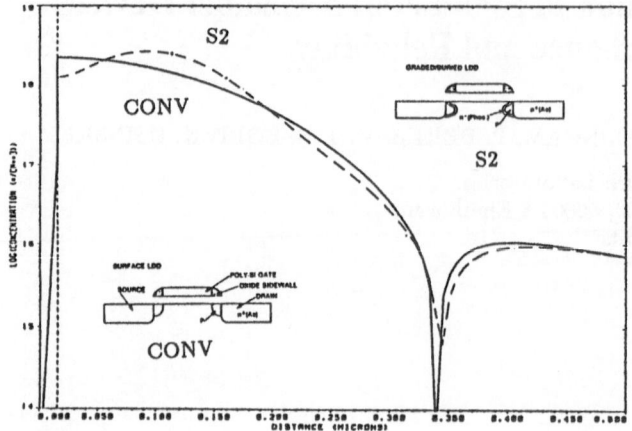

Figure 1: *Doping profiles of structures conventional CONV and S2.*

Accelerated hot electron life testing was done in addition to standard device characterisation. For the effective channel length of 0.80 um, the dependence of lifetime on Ib/Id is shown in Figure 2. The dual slope behaviour is consistent with the results reported in the literature [2], except that the slope in the higher Ib/Id region is steeper than reported (7.7 cf with 5.5). Based on these results we used Ib/Id as a measure of the device Hot carrier lifetime.

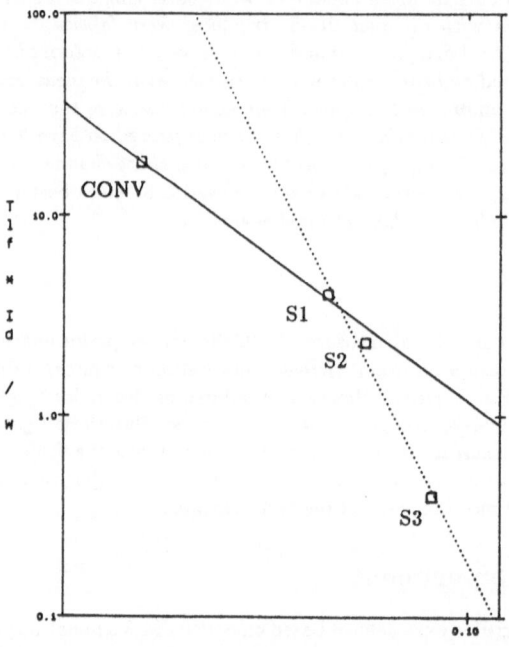

Ib/Id

Figure 2: *The dependence of lifetime (Tlf) on Ib/Id, for the effective channel length of 0.8μm.*

3 Simulation

At this point, the electrical behaviour of these structures using the device simulator MINIMOS-4 was determined. Drain currents and substrate currents for the accelerated test conditions used to measure lifetimes were determined from the simulations. Using the experimentally determined results in figure 2, lifetimes were calculated. The saturated drain currents were also simulated. The comparison of these simulated values, lifetime and drive current, with the experimentally determined values are shown in figure 3. The excellent agreement obtained gave us confidence to use our simulations to predict the Phosphorous and Arsenic dose and energy levels which would result in an optimised drain structure.

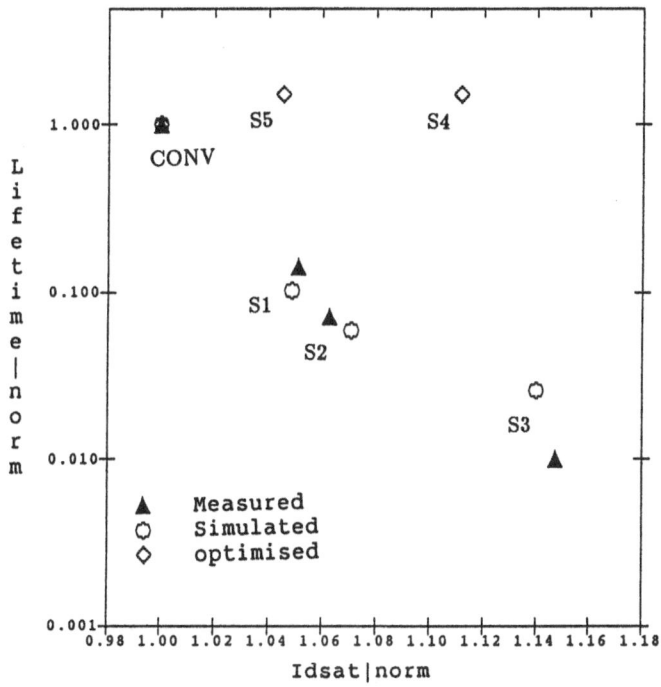

Figure 3: *Lifetime versus I_{DS} normalised to the conventional structure.*

4 Results and Discussions

Devices with this optimised drain structure (S4) and with a simple variation (S5) were fabricated together with the conventional surface LDD structure. The lifetimes of these structures evaluated from the Ib/Id value and the Idsat are also plotted in figure 3. The improvement in lifetime for these structures was obtained by moving the peak current further into the bulk. However, the optimised structure (S4) suffers from short channel effects which is illustrated by the shift of the threshold voltage with increase in drain to source voltage (figure 4). A higher value for the shift is caused by an increased drain induced barrier lowering and results in higher leakage current. As evidenced in figure 4, a compromise has to be reached between the saturated drain current, hot carrier lifetime and short channel effects. A simple variations of the optimised structure (S5) illustrate this compromise (figures 3 & 4).

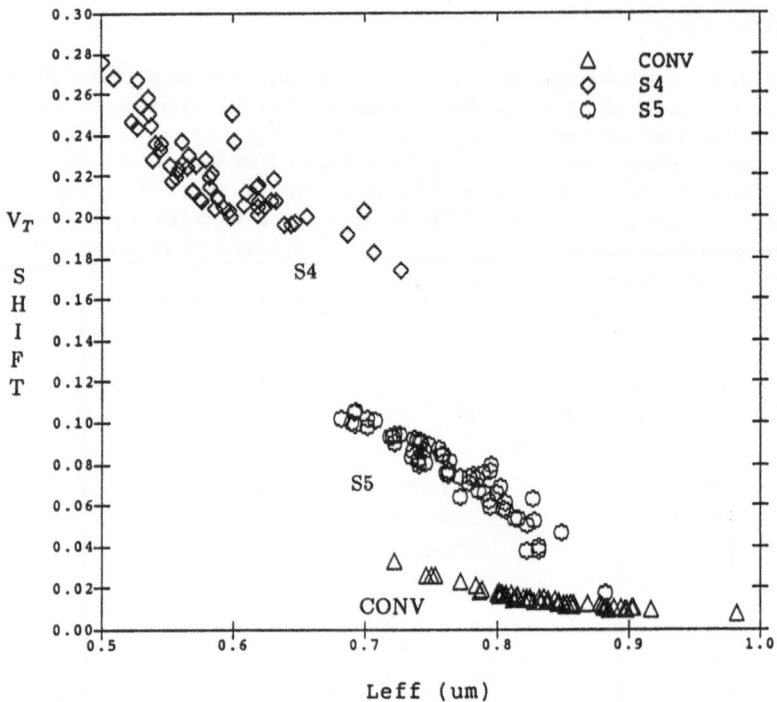

Figure 4: *Threshold voltage shift with drain to source voltage versus L_{eff}.*

5 Conclusion

In conclusion we have developed, with the help of process and device simulators, the capability to design device structures optimised for hot carrier lifetime, current drive and short channel effects. This technique was used to define a structure which maximised the drive capability while meeting the lifetime requirements and achieving acceptable short channel effects.

Acknowledgements

The authors wish to thank Joost Van Hurk for support in fabrication of these structures, J. Joosten for characterisation, H. Verhagen for design of the experiment and D. Guidry for reviewing the article.

References

[1] C.Y. Wei, J. M. Pimbley and Y. Nissan-Cohen,
"Buried and Graded/Buried LDD structures for improved Hot-Electron reliability",
IEEE Electron Device Letters, vol EDL-7, No 6, June 1986.
[2] P.Heremans, R.Bellens, G.Groeseneken and H.E.Maes,
" Consistent Model for the Hot Carrier Degradation in n-Channel and p-Channel MOSFET's ",
IEEE Trans. on Electron Devices, vol ED-35, no 12, pp 2194 - 2209, 1988.

Performance and Reliability of Sub-micron PMOS Devices Formed by Implantation into Silicide

K.J. BARLOW AND N.F. STOGDALE

Plessey Research Caswell Ltd
Caswell, Towcester, Northants.
NN12 8EQ, UK.

Abstract

Dopant implantation into TiS_{12} and low temperature annealing have been
used to fabricate submicron PMOS transistors. Spreading resistance
measurements show that very shallow junctions can be formed by the
technique. Dopant penetration through the silicide during implantation
is found to occur in some cases. Electrical measurement show that good
quality devices can be formed, whilst electrically stressing the devices
indicates that no enhanced device degradation is observed as compared to
conventionally formed PMOS devices.

Introduction

The use of dopant implantation into silicide can offer advantages such as
shallow junction formation, minimal ion damage in the silicon substrate,
no anisotropy or shadowing effects and consequent low temperature
processing [1-3]. In this paper the performance and reliability of PMOS
transistors formed by the implantation of boron or boron fluoride into
TiS_{12} will be presented.

Experiment and Discussion

The devices were fabricated using a CMOS process that employs a 20nm gate
oxide with silicided source, drain and gate. The TiS_{12} layer was around
68nm thick. PMOS devices formed by the more conventional method of
junction implantation and then silicidation were also fabricated for
comparison. Fig. 1 shows a schematic of the fabrication process. A more
comprehensive description of the device fabrication sequence is given
elsewhere [3].

In order to fabricate the implant into silicide devices both boron and
boron fluoride were used, this enabled us to obtain very low energy boron
ion implants (⩽5Kv) to assess the effects of boron channelling during the
implant. This was done by implanting the dopant into the silicide using
a range of energies, then the silicide was removed by chemical etching,
the wafers were then covered in undoped LTO and annealed at 800°C for 1
hour in N_2. Following this spreading resistance measurements were
performed to determine if any dopant penetrated the silicide. The
results showed that for boron ion implant energies greater than around
9Kv channelling of dopant into the silicon occurred. Further spreading

resistance measurements indicate that outdiffusion of boron from the TiSi$_2$ layer is inhibited and this can lead to a high contact resistance and resistance problems due to the dopant not diffusing under the sidewall oxide spacer of the transistor. The diffusion characteristics of boron from TiSi$_2$ observed are consistent with data presented elsewhere (4,5) which shows that outdiffusion of boron from TiSi$_2$ is inhibited by the formation of metal-dopant compounds leading to lower boron surface concentrations in the silicon than would be expected and consequently to high values of contact resistance. In the work presented in this paper some dopant penetration through the silicide has occurred. This factor still allows the formation of shallow junctions, shallower than conventional P$^+$ junctions formed by the same implant conditions into silicon. Also low temperature processing (~800°C) can still be used because any implant damage in the silicon does not encroach into the junction depletion region, and thus low junction leakage can be maintained. Fig. 2 shows a spreading resistance profile of a P$^+$ junction formed by boron implantation into silicide and then annealed at 800°C.

For a given energy of 20KV the effect of implant dose on the electrical characteristics of PMOS transistors was assessed. Fig. 3 shows the linear transconductance measured as a function of the boron dose, for a fixed channel length. For doses around 3E15/cm^2 and greater the transconductance increases only at a slow rate, as the dose is reduced below 3E15/cm^2 the transconductance falls. This is due to a reducing boron surface concentration and thus an increase in contact resistance. The reduced active boron concentration as the dose is reduced is confirmed by spreading resistance measurements. The influence of the boron dose on the saturated drive current, again for a given channel length, is shown in Fig. 4 and a similar plot to that shown for the transconductance in Fig. 3 is seen. For doses around 3E15/cm^2 and greater the current is essentially constant, and below 3E15/cm^2 it falls due to increasing contact resistance. Comparison with conventional PMOS devices of the same channel length shows that for doses of 3E15/cm^2 and higher the implant into silicide devices have a slightly higher drive current. The implant dose was found to have no noticeable effect on the subthreshold characteristics, whilst junction breakdown was around -11.5V for a range of doses.

An initial assessment of the influence of implantation into silicide on device reliability has been performed by comparing the devices fabricated to conventional PMOS devices. The ratio I_{sub}/I_{D} was measured under conditions of peak substrate current and was $<10^{-4}$, for devices with channel lengths of ~0.6μm. A similar value was obtained for conventionally fabricated devices. Electrical stressing showed no major difference between the two types of devices. The shift in the interface trap density, measured by charge pumping, and the gain were about the same for both types of device 2% and 2.5% respectively under the conditions used [V_{D}= -7v, V_{G}= -3.5v, 14 hour stress] Fig. 5 shows the threshold voltage shift as a function of time, the implant into silicide device has a channel length of 0.58μm as opposed to 0.7μm for the conventional device. The plot indicates that there is little difference in the two types of device, with electron trapping causing the threshold shift and both saturating after an ~20mV shift.

Conclusion

PMOS transistors have been fabricated by dopant implantation into TiSi$_2$. Spreading resistance data and electrical measurements indicate that outdiffusion of dopant from the silicide is inhibited and this can lead to high contact and series resistance effects causing degradation of PMOS device charactersitics.

Characteristics comparable to conventionally fabricated PMOS devices are obtained when the dopant penetrates the silicide during implantation, this still allows the formation of shallow, low resistance junctions. Electrical stressing indicates the devices formed are no less reliable than conventional PMOS devices of similar dimensions.

References

1. R. Liu, D.S. Williams and W.T. Lynch. Ext. abs. IEDM 1986, p58
2. D.L. Kwong, Y Hu, S.K. Lee, E. Louis, N. Alvi and P. Chu. J. App. Phys 61, 11, p5084 July 1987.
3. K.J. Barlow. Proc. ESSDERC 88, p507.
4. V. Probst, H. Schaber, P. Lippens, L. Van Den Hove and R. De Keersmaecker. Appl. Phys. lett 52, 21, p1803 May 1988.
5. L. Van Den Hove, P. Lippens, K. Maex, L. Holdos, R. De Keersmaecker V. Probst and H. Schaber. Eur. w/shop on Refractory Metals + Silicides abstract A2-2.

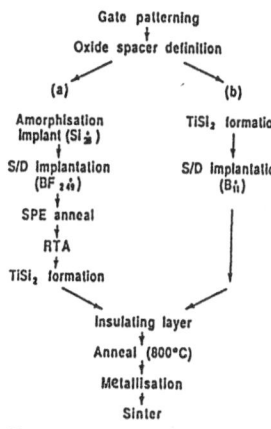

Fig. 1

Schematic of the silicided source/drain Schedule.

(a) Conventional first form junction then silicidation.
(b) Silicidation then implant into silicide and activation at 800°C.

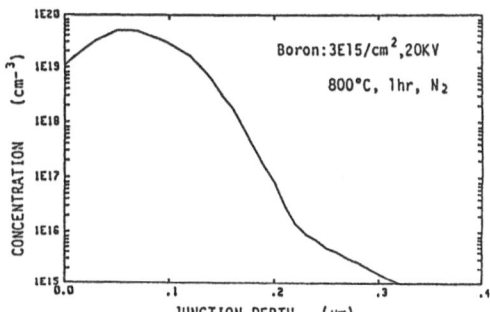

Fig. 2

Spreading resistance profile for p+ junction formed by implantation into silicide.

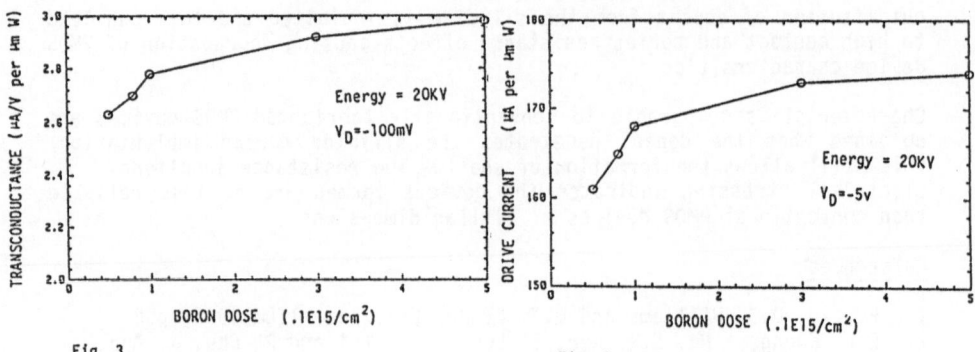

Fig. 3.

Linear transconductance as a
function of boron dose.

Fig. 4.

Saturated drain current as a
function of boron dose.

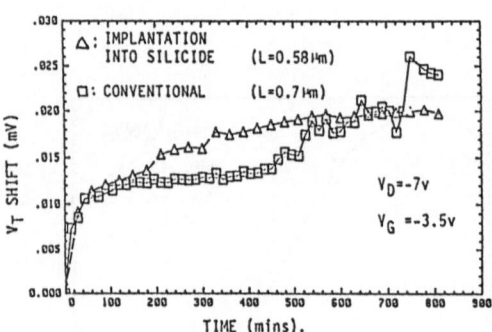

Fig. 5.

Threshold voltage shift as a
function stress time.

Performance Drift Mechanisms of 1.3 µm Lasers During Aging

A. CHANGENET[■] S. MOTTET[□] C. CARRIERE[●] E. VETU[●]

Centre National d'Etudes des Télécommunications.
[■] 3, avenue de la République 92131 ISSY les MOULINEAUX FRANCE
[□] Route de Trégastel 22300 LANNION FRANCE
[●] ALCATEL C.S.O. Centre de Villarceaux, Route de NOZAY
91620 La VILLE du BOIS FRANCE

ABSTRACT.
To understand the aging mechanisms of lasers, we have developed a semi analytical electrical model giving access to the physical parameters of the quaternary active layer of 1.3 µm laser diodes. Coupled with the simulation of the leakage paths, this model is used to analyse the aging behaviour of a particular 1.3 µm B.H. laser diode's structure.

INTRODUCTION

In previous papers we have described a semi-analytical, simplified model able to fit the current-voltage characteristic of the active part of 1.3µm laser diodes(1-2-3). Two terms contribute to the total current flowing in a device (3):
- the active branch, which produces the laser effect and is simulated by the model. This contribution is not technology dependent.
- the leakage current, due to the blocking layers and the lateral surface effects, which must be studied for a particular structure.
We show here how to isolate these two terms of current on a semi-logarithmic plot of the current voltage characteristic for a particular 1.3µm buried heterostructure laser diode, what is the contribution of the leakage current to the threshold current, and how these two currents are affected by screening stresses.

THE SEMI-ANALYTICAL MODEL

$$V_{(\nu)} = \frac{k.T}{q} \left[\overline{\mathcal{F}_{\frac{1}{2}}} \left(\frac{\nu}{Nc} \right) + \overline{\mathcal{F}_{\frac{1}{2}}} \left(\frac{\nu}{Nv} \right) \right] + \frac{Eg}{q} + Rs.I_{(\nu)}$$

$$I_{(\nu)} = \nu^2 \left[\frac{1}{\tau_0 (2.\nu + n_1)} + Bo + 2.C_A.\nu \right] . Vol$$

This model involves three recombination terms: Auger, radiative spontaneous and thermal assisted by deep centers. The parameters are τ_0, the lifetime and n_1 which is related to the location of the deep centers in the bandgap).

LASER STRUCTURE
The figure 1 shows the structure of the 1.3μm buried heterostructure ALCATEL C.S.O. laser analysed here. The arrows represent the different current paths.

Two paths of leakage current can exist in such a structure (figure 1) :
1) the current flows through the blocking layers.
2) the current flows through the p access layer, the p blocking layer and the n layers.
This branch usually can be modeled by a single serial resistor-diode cell according to $I = I_s \cdot (\exp \dfrac{q(V - R \cdot I)}{\eta kT} - 1)$ where the ideality factor η can vary from 1 to 10 or more.

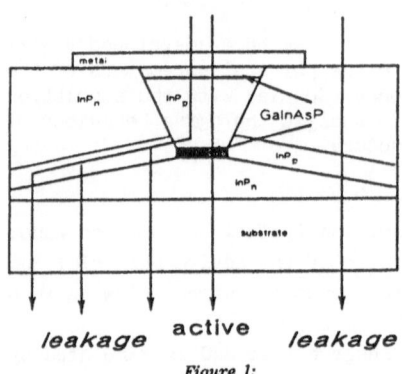

Figure 1:

Alcatel C.S.O. B.H. laser diode structure

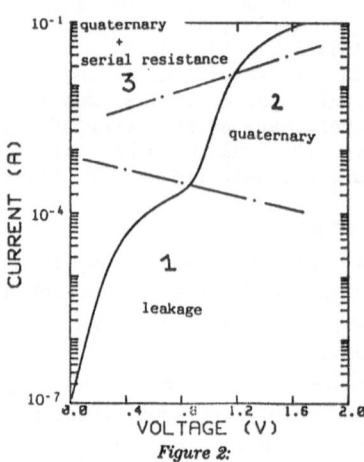

Figure 2:

The 3 parts of the I(V) characteristic.

ANALYSIS OF THE CURRENT-VOLTAGE CHARACTERISTIC
The current-voltage characteristic of such a laser diode exhibits three particular parts (figure 2):
- part 1 is due to the leakage behaviour.
- part 2 is the contribution of the quaternary active layer.
- part 3 includes the quaternary active layer affected by the serial resistance involving the material resistance and the contact resistance.

Parts 2 and 3 are fitted by the semi-analytical model described above by adjusting the three parameters (τ_0, n_1, R) and part 1 by the resistor-diode cell by adjusting η, Is, Rs.

EFFECT OF AGING
We subjected components representative of the technology to screening type stresses. An increase of the threshold current was observed on all the

components. This increase is due to two effects : the degradation of the quaternary active layer and the increase of the leakage.

Degradation of the active layer
The electrical measurements show a decrease of the lifetime of the thermal recombination (figure 3) correlated to an increase of the threshold current (figure 4). This degradation is due to a reorganisation of the deep centers located in the bandgap of the quaternary material because of the thermal and high current stresses. This effect was observed on all the components.

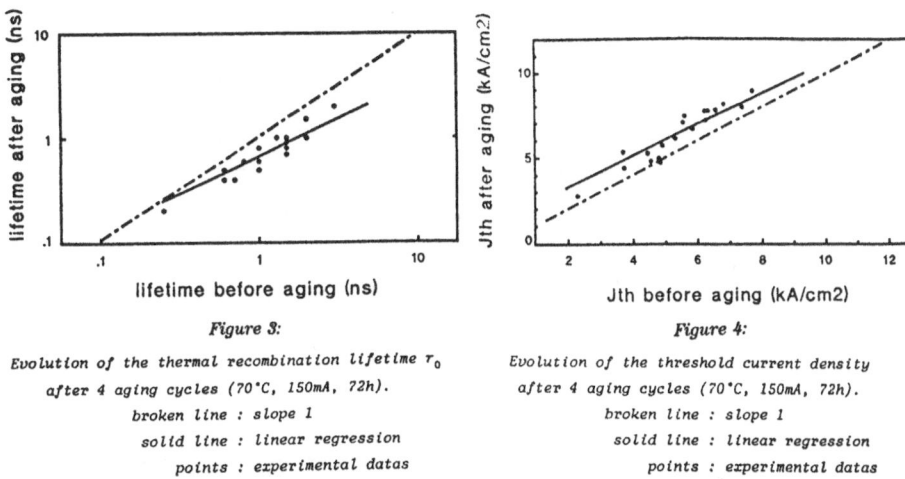

Figure 3:
Evolution of the thermal recombination lifetime τ_0 after 4 aging cycles (70°C, 150mA, 72h). broken line : slope 1 solid line : linear regression points : experimental datas

Figure 4:
Evolution of the threshold current density after 4 aging cycles (70°C, 150mA, 72h). broken line : slope 1 solid line : linear regression points : experimental datas

The evolution of the thermal recombination lifetime value after aging stresses is a function of its initial value. Figure 5 shows that a large value of τ_0 before aging moves more than a small one. We assume that if the value is small, the material has not yet reached its steady state. The decrease of the lifetime value is correlated with the increase of the threshold current density.

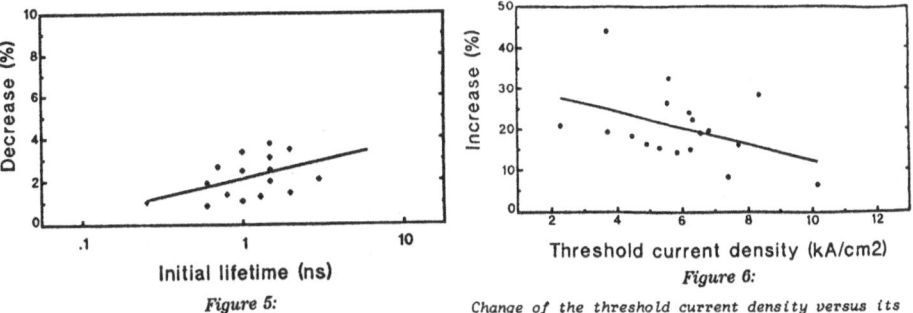

Figure 5:
Change of the τ_0 thermal recombination lifetime versus its initial value

Figure 6:
Change of the threshold current density versus its initial value. Leakage current at threshold has been subtracted.

836

Figure 6 shows that a small threshold current density before aging increases more than a big one. It is necessary to make measurements after long term aging to evaluate the degradation law.

Contrary to the behaviour of the quaternary active layer, it is impossible to correlate the evolution of the leakage current as a function of its initial value (figure 7).

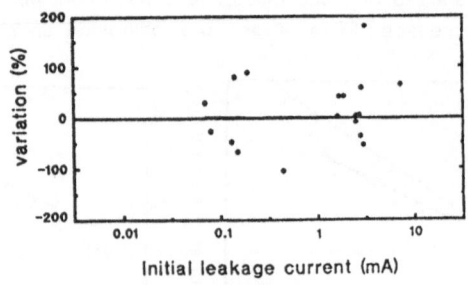

Initial leakage current (mA)

Figure 7:

Change of the leakage current value at the
threshold voltage.

CONCLUSION

We have shown that two mechanisms are involved in the evolution of the behaviour of laser diodes. A degradation of the quaternary active layer is observed: the thermal recombination becomes more and more important, this implies a modification of the threshold current. We observe a modification of the leakage current but no rule can be pointed out.

Components which before aging exhibit low leakages and small threshold currents may show more evolution than components with initially higher threshold and leakage currents. This is due to the fact that these lasers are already stabilized.

These components were choosen to illustrate typical behaviour of a particular technology. The results shown here are not representative of commercially available ALCATEL C.S.O. laser diodes.

References:

(1) S. MOTTET et Al."Conduction mechanisms analysis and simulation of 1.3μm laser diodes", Proc. 17thESSDERC, 1987, 1003,1006

(2) S. MOTTET et Al."Analysis and modelling of 1.3 μm laser diodes", Proc; Int. Symp. GaAs and related compounds, Inst. Phys. Conf. Ser. Ser. n°91:chap 7, 1988, 621-627

(3) A. CHANGENET "Contribution à l'analyse des diodes laser à 1,3μm" Thèse de Doctorat, Université de RENNES 1, 1988.

Reliability and Resistance Minimization Studies of the Laser Diffused Diode Links in Wafer-scale-integration

P. FANG , G.H. MASSIHA , T.M. CHEN and J.G. COTTLE

Electrical Engineering Department, University Of South Florida, Tampa, Florida 33620

SUMMARY:
The reliability of the Laser-diffused Diode Links (LDL) has been studied. For low resistance devices the electromigration damages were found at the oxide steps as expected. However, for the high resistance devices, failures occurred around the device links. our study shows that the high resistance links failure can be related to the high joule heating enhanced Si electromigration.

The LDL technology enables a direct connection between devices at the substrate level over a wafer scale area [1]. The linking is formed by laser induced carrier redistribution between two adjacent diodes. The major advantage of this linking technology, compare to double metal link, is its CMOS process compatibility. However, the resistance of LDL links are a few hundreds of ohms higher than the resistance of metal links. Additional reliability problems could be confronted as high joule heating causes an increase in the temperature of LDL links. In this paper we will show electromigration may make LDL the weak point of the interconnect in Wafer Scale Integration (WSI).

The test structure before it is linked by laser is illustrated in Figure 1. The linking process is done after all other processing steps have been completed. The link comes about as a result of dopant diffusion from the p junction regions into the gap that separates the diodes. The fast diffusion is made possible by the creation of a small melten zone when laser beam is applied which encompasses parts of the diode regions and the gap between them.

Two kinds of LDL links were subjected to the reliability test. For the low resistance links, which were zapped twice by laser, the link resistances are about 200 ohms; and for the high resistance links, which were zapped once by laser, the link resistances are about 300 ohms. All of the LDL samples used in this study are p doped on n-type Si wafer. The gap separated the p regions is 4 μm and p region depth is about 2 μm.

The accelerated life time test was performed at the substrate temperature ranging from 135 °C to 160 °C and the bias current density of 5×10^5 A/cm^2. The increase in the link resistance was recorded by monitoring the voltage across the device. The link actual temperature due to combined ambient temperature and joule heating was estimated to be around 230 °C to 260 °C using temperature versus resistance plots [2]. Stress process was halted when 10% increase in resistance was reached. Scanning Electron Microscopy (SEM) showed no visible change for both link area and aluminum contacts (Figure 2a). Samples were placed under initial stressing conditions until total failure was reached. A typical SEM picture of a failed link (Figure 2b) shows bright spots around the area where laser beam was applied but no visible damage was found on Al contact area.

Energy Dispersion X-ray Spectroscopy (EDS) test on the bright spots revealed that the spots contain 74% wt Al, 25% wt Si, and 1% wt P. This is quite a change from a prestressed link EDS test which showed 5% wt Al, 94% wt Si, and and 1% wt P at the same spots. Then selective etching was performed on stressed samples to take off the glass and metal (aluminum) by using diluted HF. SEM and EDS tests on etched samples (Figure 3) showed Si build up at the negative contact and void at the positive contact. All twelve high resistance samples tested showed the same result.

Accelerated life time test on low resistance LDL samples showed conventional electromigration of Al ions through the Al matrix which resulted in formation of voids and hillocks and eventually causes the device failure. These kind of failure were observed around the oxide steps.

Previous study showed that Electromigration of Si through Al matrix at temperatures above 230 °C is an important device failure mechanism [3]. The growth of pits in Si at negative contacts and the build up of resistive Si at positive contacts would result in device failure. The driving force of the Si migration is the momentum exchange between the conducting carriers and the Si atoms. The direction of Si migration is the same as the moving direction of holes for p-type links which agrees with our experimental results on the high resistance links. The failure process can be interpreted as follows. During electromigration process Si ions migrate into Al at the negative contact and vacancies created by the departure of Si ions will then be occupied by Al ions. So the Si ions and Al ions are moving in opposite directions. In this manner, The Al ions migrate from the the negative contacts to the positive contacts. When Al ions reach the laser zapped spot, they flow out of the laser zapped holes as seen in Figure 2b. Eventually an open circuit will be created at the negative end where the Si and Al contacted.

Our study has shown that reducing link resistance is an important task for developing more reliable links. Theoretical study[4] has indicated that the uniform distribution of the carriers in the gap gives the minimum resistance ;and which could be obtained by controlling laser pulse duration time.

The authors would like to acknowledge Alicia Sue Slater for her Electron Microscopy work. This work was supported by DARPA under Grant No. MDA 972-88-J-1006.

References:

[1] Simon S. Cohen, et al,IEEE trans. Ed-35,(9), p1533,1988.

[2] G.H.Massiha, T.M.Chen and G.J. Scott, IEEE EDL Vol.10, No.2. Feb. 1989,p58.

[3] J.R. Black, Proc. of Reliability Physics, 1978, p233.

[4] P.Fang and T.M.Chen,Proc. of IEEE Southeastcon,1989,p1421.

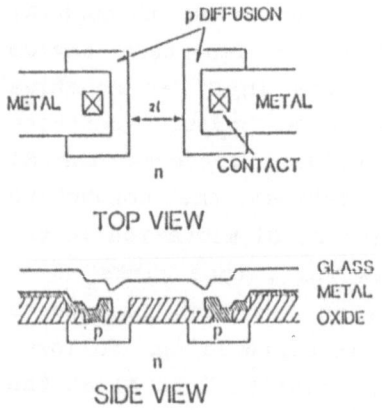

Fig.1 The LDL before it is
linked by laser.

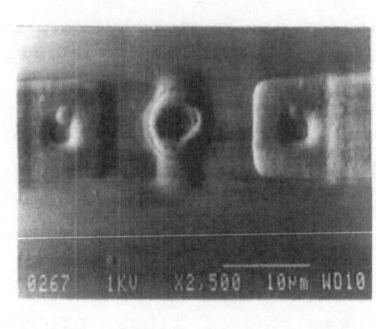

Fig.2 a) Top view of an LDL.

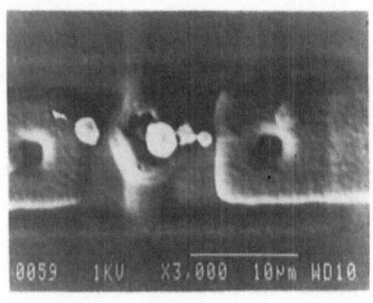

Fig.2 b) A failed LDL, Al
comes out from the
laser zapped spot.

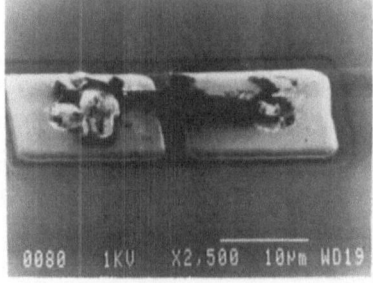

Fig.3 A failed p-type LDL
after selective etching.

Session 9

Invited Papers

Limitations and Use of Analytical Techniques for ULSI

W.Vandervorst and H.Bender, Imec, Belgium

Abstract

The applications and use of analytical techniques in ULSI are discussed in this paper. Emphasis is on the use of RBS,AES, ESCA, SIMS, SRP with respect to routine analysis requirements in semiconductor processing. The techniques are reviewed through recent applications whereby the emphasis is placed on new developments and limitations as encountered when applying the techniques to small area or 2D-analysis.

Introduction

Characterization of semiconductor materials and structures has been the key process towards the understanding of the technological steps used during integrated circuit processing leading to the development of various techniques for this purpose [1]. Whereas in the past the demands could be satisfied by one-dimensional analysis on uniform structures, the evolution into ULSI has outlined the necessity to perform analysis not only with better depth resolution and higher sensitivity but also to probe in small areas and eventually in two- and three-dimensions. The problem raised in small area analysis is achieving a very high spatial resolution while maintaining sufficient sensitivity to perform the analysis adequately, and providing a quantitative interpretation of the results even in the case of a multilayer structure. A point of concern in these analysis is certainly the interaction of the probing beam with the sample itself and the eventual modifications this might introduce.

Basically the characterization of semiconductor structures can be subdivided into the following areas :
1) **layer composition analysis** dealing with thickness measurements and the determination of the composition and distribution of the matrix elements

2) **impurity analysis** providing the detection and distribution of elements (dopants or contaminants) at very low levels (<ppm)

3) **chemical analysis** of layers and surfaces identifying the chemical state of the various elements

4) **crystallographic analysis** identifying the presence of defects, the determination of the crystal structure and phases

5) **electrical analysis** providing information regarding electrical activity of the various layers and interfaces

Although many techniques are available which probe some of the previous points, we will limit ourselves in the present paper to those which can be considered to be of very general use and we will emphasize some of the recent developments of interest for ULSI-applications and in particular in relation to quantitative small area analysis. Techniques discussed further are Nuclear techniques RBS and ERD (items 1,4), AES (item 1,3), ESCA (item 1,3), SIMS (1,2) and SRP (item 5). TEM (item 1, 4) is also a very important technique but this is not discussed in the present paper since it will be dealt with extensively in another paper at this conference.

2. Nuclear techniques

2.a Rutherford backscattering spectrometry (RBS)
As a nuclear technique based on the detection of backscattered high energy (Mev) particles, RBS is a very powerfull technique which is fairly well understood. [2,3]. It lends its capability to characterize materials quantitatively on its element specificity (through the kinematic factor), its depth perception (through the energy loss which is very well tabulated [4]) and the predictable scattering cross-section allowing the conversion of intensities to concentration profiles. Since the energy-loss provides the depth scale in principle no sputtering is required which eliminates the problems this might introduce (preferential sputtering, atomic mixing, topography etc). Analysis of the interaction of metals with silicon during thermal processing leading to silicide formation is a typical example where RBS has proven its potential.

Since the energy of the backscattered particle contains the depth information as well as the mass information problems can arise due to overlap of various signals especially for deeper lying layers. An elegant solution for this problem has been introduced by Wittmaack [5] by combining sputtering with RBS. As the toplayers are removed by sputtering the signal information from deeper lying elements becomes background free and accessible for interpretation. In this way Wittmaack et al have been capable of deriving the segregation coefficient of As during the oxidation of silicon. The interesting fact is here that only minor sputtering was required such that not the layer of interest itself needed to be sputtered but only the top layer was removed. Hence any artifacts due to sputtering process are neglegible or minor. The attractiveness of this approach is sustained by the problems encountered when using other techniques like SIMS and AES where ion- and electron beam induced segregation makes it impossible to derive a correct segregation coefficient. [6]

The use of MBE-grown material has spurred the need for very high depth resolution measurements. The depth resolution in RBS is a complex function of detector resolution, energy straggling, multiple scattering, angular acceptance, angle of incidence and detector angle. [1,7] The major problem is that there is no unique answer to the optimum configuration since this depends on the analysis depth. Although minimum depth resolution (3-4 nm) can be obtained for near surface analysis using a very small exit angle, the advantage is lost for larger depths.

The use of ions heavier than He might theoretically offer an advantage because of the larger stopping power and increased mas resolution. [2] Unfortunately detector resolution is degraded at the same time such that finally no better mass and depth resolution is obtained at medium energies (2-10 MeV). [8] The most interesting aspect however for using a heavy ion beam is the greatly enhanced sensitivity since the scattering cross-section scales with the square of the atomic number of the primary particle. Due to this effect Knapp et al.[9] have been capable of extending the detection limit for Au on Si down to $1 \cdot 10^{10}$ at/cm^2 using a 400 kev C-beam as compared to a limit of $1.3 \cdot 10^{13}$ at/cm^2 using a 1.5 Mev He -beam.

In relation to ULSI the use of RBS to perform localized analysis is directly related with the possibility to create and use small He-beams. The situation becomes particularly difficult if one wants to maintain a collimated beam for channeling measurements also since in that case a strong focussing lens cannot be used. Several facilities [10-14] are now available which are capable of producing beams with a diameter of 1-3 um with He-currents ranging from 500 pA to a few nA. By storing the energy of every detected particle in combination with the beam position it is possible to produce a three-dimensional multi-element distribution of the sample under investigation. [11,14]. Localization of the point of impact becomes now a point of concern but can usually be performed using a He-beam induced secondary electron image. [10] Although technically available the use of (sub)micron He -beams does create a problem because of the high damage density introduced in the sample. Especially in channeling analysis it was found that unscanned highly focussed beams do introduce severe crystal damage such that chi-min values become much higher. A typical example is the measurement of the substutional fraction of As implanted in Silicon (fig 1) which is easily displaced from its lattice site. The data contained in fig 1 indicate that scanned areas of 1 x 1 mm2 are required to measure damage free the substitutionality of the As. [15]

2b Elastic recoil detection

Elements lighter than Si represent a problem in RBS since their signal is superimposed on a large background. Elastic recoil detection (ERD) is the inverse of RBS in that now the sample is bombarded with heavy ions and the emitted light particles are detected. [16-18]. It has grown now into a usefull technique for the characterization of H,O, C and N in oxy(nitri)des and other semiconductor structures. The use of Time-of -flight detectors in this area [17,18] has extended the applicability of the technique by providing it independent mass and energy resolution to allow even the composition analysis (Al, As, Ga) of III-V materials something which is impossible with RBS because of the large signal overlap.[17] The main limitation of this technique is instrumental since high energy (20 -30 Mev) beams of heavy ions are required which are not always available on many accelerators. In principle small area ERD would be possible by focussing the heavy ion beam similar to the He beam however since the experimental ERD-set up requires small impact angles with respect to the sample surface the spot size will always be larger than in a RBS-experiment. Beam damage from such a focussed, high energetic heavy ion beam would certainly be a limiting factor for small area analysis. [19]

3. Auger electron spectroscopy

For quantitative multilayer depth profiling AES in combination with sputtering has proven to be a very powerfull technique. (fig 2). The main reason for its quantification capability are a reasonable understanding of the physics involved (background generation, cross section), the very moderate matrix dependence of the important mechanisms and elaborate measurements of the most important parameters (sensistivity factors, inelastic mean free path) [20-22]

Since the attenuation length of the AES electron is quite small, AES offers excellent depth resolution (typically 1-2 nm). However during depth profiling with ion beam sputtering the final depth resolution can deteriorate because of ion beam induced displacements. Fig 3 illustrates this effect on the quantification of the In-content of a thin InGaAs layer embedded in a GaAs substrate. The apparent In concentration for very thin layers is much smaller than anticipated from the growth conditions and only reaches its normal value for layers thicker than 20 nm. Of course the smearing effect does depend on the experimental sputtering conditions used but this effect will always be present to some degree.

Detailed studies of multilayer profiling (III-V, silicides) [23,24] have shown that the matrix correction factor cannot yet be predicted properly and requires experimental verification. It was also found that the major limiting factor in depth profiling compound materials is the preferential removal of one component as compared to the other one(s) such that even after applying the correct matrix corrections the results still are different as compared to non-destructive techniques. [23,24]

Another limitation in obtaining quantitative analysis has been the overlap of different AES-peaks. A typical example encountered during the study of TiN on $TiSi_2$ has been the overlap between the nitrogen and the titanium peak . This problem can be resolved by relying on factor analysis to deconvolute the overlapping peaks. [25,26]. Although quite powerfull this technique becomes more difficult if oxygen and carbon are also present at the same time.

Since the excitation source is now an electron beam, very fine focussed beams can be obtained. Commercially instruments available claim spot sizes down to 10-20 nm. Although usefull for SEM-imaging and localizing the spot of interest, normal point AES-analysis will require more current and hence a larger spot size (50-100 nm). The instrumental set-up also has a strong influence on the final resolution obtained. Depending on the angle of incidence and the detector position, edge effects, shadowing and backscattering may generate apparent broader lines and show the unreal presence of non-existing indiffusions. [27] A major limitation in acquiring quantitative AES-images is the drift of the set-up during the experiment [28]. In a detailed study of Au-contacts on GaAs we have tried to image a series of Au-gates with a length ranging from 50 nm to 1 um. Due to the long acquisition times required for a complete image (1-2 hours), the gates never appeared as straight lines but were distorted by drift despite the fact that the instrument was located in a clean room facility with tight temperature control. The average displacement of the sample amounted to 0.2-0.5 um/hour which is however unacceptable in high resolution work. A further problem is that the small currents available limit the signal to noise ratio in high resolution work. Prutton et al. [29-31] have tried to circumvent this problem by using a multi-detection scheme (peak and background) and by adapting an interpretation procedure to enhance the information contained in the images based on constructing scatter diagrams and Hotelling transformations. This approach essentially uses the fact that through the image processing (anti)-correlations between the signals from various elements can be used as a contrast enhancement criteria to improve the image quality and to pinpoint areas with otherwise undetected phases.

A final concern in the analysis of small areas is that the focussed beam can induce severe damage to the sample itself. Recent work by Bender [32] shows that under focussed conditions oxynitrides become irreversily depleted from nitrogen. Earlier work by Remmerie [33] and Hoesler [34] also indicates that the depth profiling analysis of a nitride/oxide/silicon multilayer always leads to the apparerent detection of nitrogen at the SiO_2/Si-interface although this cannot be detected in a ESCA-depth profile or with AES after chemically removing the top nitride before the depth profiling. The prime artifact is here electron-beam induced migration of the nitrogen which is enhanced under strong focussed conditions.

4. Electron Spectroscopy for Chemical Analysis

ESCA or XPS is a very appealing technique since it provides information regarding the chemical state of the element under investigation. Typical applications have been the analysis of the wafer surface after various cleaning procedures [35] and the characterization of the bonding state of Si at the interface of thin oxides and the relation of the presence of suboxide states to electrical behavior of the oxide and the understanding of the growth mechanisms [36,37]. Again the depth resolution is intrinsically quite good (2-3 nm) but deteriorates in the depth profiling mode due to ion beam sputtering. Although hence apparently similar to AES, it is possible to extract more information from an ESCA-depth profile. Fig 4 shows a three-dimensional representation of a Si-depth profile through a silicide layer which has been formed on top of SiO_2-layer. The Si-peak positions as a function of depth indicate that just before the SiO_2-interface a silicide is formed. From the AES-depth profile this information is not obtainable since the total Si-signal is increasing very steeply in that region as the SiO_2 is approached. [38]

One of the more recent applications has been the characterization of sililyated photoresists. By measuring the sample under several exit angles with respect to the spectrometer (angular resolved XPS) it is possible to obtain a non-destructive chemical depth profile typically from a depth of 5-6 nm. [39,40]. This technique allows for the characterization of the SiO_2-layer (thickness, stoechiometry) formed on the photoresist during the oxygen plasma exposure.

Due to the use of x-rays fine focussed beams as in AES are not feasible. Full XPS-analysis has therefore been limited to rather large areas (150 um and more). Recently a new concept has been introduced whereby for the energy selected photoelectrons an direct image from the sample is formed on a position sensitive detector. Under these conditions it becomes possible to map out the different chemical states of an element with a resolution slightly better than 10 um. Applications assessing non-destructively the effectiveness of etch procedures (for instance removing SiO_2 in small holes) are now becoming feasible with this technique.

5.Secondary Ion Mass Spectrometry

In terms of semiconductor profiling Sims has certainly been the most heavily used one, primarily due to its unchallenged sensitivity (<ppm), high dynamic range and good depth resolution. [41] Dopant profiling (B, As, P) in simple silicon structures has matured now into a straightforward quantitative application and certainly represents a large part of the Sims applications. Additionally its usefullness also originates from the capability to measure elements like hydrogen, oxygen, nitrogen in trace amounts (1-10 ppm) [42,43,44]

Again the intrinsic depth resolution is quite good since the escape depth of a sputtered particle is only a few atom layers but during depth profiling the effect of the ion beam can introduce atomic displacements. Detailed studies on this subject [45-49] have eluded the role of the primary energy (as low as possible), primary mass (heavy preferably) and angle of incidence (between 40-60 degrees). It should however be realized that in most cases the beam induced disturbances only introduce a degradation of the depth resolution in the order of 10-20 nm. Longer tails on recorded profiles have to be considered real or induced by topography which can develop during the sputtering of metallic systems.

The main Sims requirements originating from ULSI-analysis can be correlated with the following requests :

    a) lighter doping levels and tighter contamination control requiring higher sensitivity

    b) shallower profiles implying near surface quantification, increased depth resolution and undistorted profiles

    c) Steeper profiles like MBE-grown layers which emphasize the need for very high depth resolution

    d) multilayer structures (heterostructures, silicide/silicon interfaces, poly/insulator/si) necessitating multilayer quantification as well as interface quantification

    e) small area analysis which invokes high spatial resolution (area definition), incresed sensitivity as well as means to localize the area of interest

    f) 2D- and 3D-analysis implying high spatial and depth resolution simultaneously concurrent with very high sensitivity

    g) On-chip analysis which further complicates the measurements by the presence of surrounding layers which can be insulating or can contain very high amounts of the species of interest, as well as the presence of high aspect ratios.

The problems can be summarized as a need for higher sensitivity, better quantification properties and increased lateral resolution. Recent studies into these problems have clarified the aspects of near-surface transients [50,51] and shown the means to overcome the quantification problems in this region by using special molecular signals [52] or oxygen flooding [45]. Non-linearities in the signal with concentration of for instance BF_2-profiles have been observed for high F-concentrations [53] which implies that this need to checked always very carefully. Multilayer quantification is possible using elaborate calibration procedures [54-56] whereby even the variation of the dopant across the interface can be obtained quantitatively. [6] A potentially more universal solution for this problem has arisen from the incorporation of post-ionization into Sims -equipment which decouples the process of sputtering and ionization. By post-ionizing the sputtered neutral cloud (which constitutes the majority of the atoms ,>80 %) one not only obtains an enhanced sensitivity by a more effcient ionization but when element-selective processes like resonant excitation are used, also better detection limits by the elimination of mass interferences.[57] Detection limits of for instance 2 ppb for Fe in Si have been reported using this technique [58]. The decoupling of the ionization and the sputter process also eliminates (or at least strongly reduces) the matrix dependence of the ionization yield which up till now has been the major drawback of Sims-quantification. Preliminary experiments on B and Si sputtered from oxides, silicides and GaAs support this expectation [59].

Sims imaging as performed in the microscope mode forms a direct image of the sample surface on a position sensitive detector. In principle this method provides the highest sensitivity [60] but it is limited to 0.5-1 um resolution. In the Ion probe mode where the image is obtained from the synchronization of the detector system with the position of the rastered beam the spatial resolution is limited by the spot size achievable which depending on the primary ion used ranges from 0.5-1 um (oxygen) down to 50 nm (Ga, Cs). The limited amount of atoms available in a pixel of 50x 50 x 20 nm (2.5×10^5 atoms for Si) does require however extremely high ionization (which is not possible with a Ga source) and collection efficiency to obtain reasonable detection limits. Again post-ionization with its increased ionization yield (a Ga-source now has no negative effect) will inevitable play a role in future high resolution work. At the moment depth profiles in small areas (3 x 3 um) have been reported by several authors [61,62] with reasonable detection limits (10^{17} at/cm^3) However it was found also that in a number of cases the effect of the surrounding material (i.e a BPSG-layer surrounding a contact hole where a B-profile needed to be made) is so dramatic that usefull data could not be obtained [64,65].

A problem requiring serious attention for ULSI-applications is the measurement of 2D-diffusions. In view of the sensitivity and the resolution required point analysis of the lateral diffusion is not directly possible. Recently special sample preparation techniques have been proposed (special masks, sample beveling) which do provide a geometrical magnification of the area of interest such that more atoms become available for the analysis and that the spatial resolution requirements at the same time are alleviated. [66,67] The preliminary results are very promising in that lateral resolutions of a few nm can obtained with still reasonable detection limits (1-10 ppm). Along the same lines using special sample preparations it also has been possible to analyse the doping profile in a trench wall which constitutes a major problem because of the difficult aspect ratios. [61] Further progress for small area analysis and 2D-profiling is however required to meet the technology requirements and this inevitably will imply more work on the the use of sample preparation, LMIS and post-ionization.

6.Spreading resistance probe

All the previous techniques do provide physical characterization but not much information on the electrical performance of the structure. The spreading resistance probe (SRP) is one of the only techniques available with nearly unlimited dynamic range and the capabilty to probe across several junctions with good depth resolution. As such it has been in use for a long time in process evaluation. Moving into the ULSI-applications it has however become clear that the technique itself needed to be developed further also. One of the major results of recent studies [68] has been the developement of new software for correcting the raw data in a more consistent manner [69] The application of the technique to more abrupt profiles has outlined the importance of carrier spilling on the junction position . Fig 5 shows a comparison between a Sims, SRP profile of an annealed ion implant. The apparent discrepancy in the tail has in the past quite often been interpreted as incomplete activation of the dopant, interstitial diffusion or as a mixing effect during the Sims measurements.[70,71] Neither of them is true however but the junction displacements is entirely caused by the redistribution of the mobile carriers in the bevelled sample . This effect is not solely limited to SRP but is present in every technique which uses material removal either by bevelling (SRP, staining), stripping in combination with anodic oxidation or chemical etching (Electrochemical C-V). This is also illustrated in fig 5 where the profile obtained from resistivity measurements (VanderPauw structures) in combination with anodic oxidation and stripping closely agrees with the SRP-profile. A full treatment of the carrier spilling problem can be found in ref [74, 75, 72] The importance of the effect can be derived from fig 6 which gives the junction displacement for a B-implant in n-type material as compared to the position derived from Sims and/or an implant in p-type substrate. From a comparison with a theoretical carrier spilling model it was found that the slope of the profile in the neighbourhood of the junction together with the substrate concentration are the determining parameters. [73,76]. Slowly varying profiles and lowly doped substrates will show the largest effect. It should be realized also that for structures which do not contain a junction the displacement is much smaller and hence this kind of structure should be used in comparative SIMS-SRP studies to analyse the activation of dopants.

The technology of the probe points has evolved such that now routinely probe separations less than 20 um can be used. In an attempt to evalutate the potential of SRP for small area analysis we have measured the speading resistance on a given structure whereby the boundaries (formed by a junction) were brought closer together. Fig 7 gives the increase of the resistance with the distance to the boundary. It is clear that once the boundaries come closer than twice the probe separation they start to influence the resistance value. The latter can only be accounted for by a three-dimensional treatment which is not incorporated in the correction programs presently available. Hence this implies that SRP-data should be taken from areas which are at least twice the probe separation. Small area analysis will therefore imply further development of closely spaced tips and/or a full 3D-treatment of the current distribution in the sample.

Conclusions

Characterization of semiconductor structures with the various techniques described here, has proven the potential and the benefit of the individual techniques. In order to accommodate the ever increasing demands of technology, developments have been necessary in instrumentation as well as in data interpretation. Despite some succes in achieving the required spatial resolutions for ULSI-applications one is faced with the problem that signal intensities from these small areas are strongly reduced and that modification of the sample by the probing beam becomes more likely. As the individual techniques are being pushed towards their limits a stronger synergism between them will be required to live to the ULSI-expectations. A number of improvements is still possible but they will require an increased theoretical and practical effort in instrumentation development and data assessment.

References

1. H.W.Werner : "Diagnostic techniques for micro-electronic materials, processes and devices", Proc. NATO-advanced institute (1986)
2. W.K.Chu, J.M.Mayer and M.A.Nicolet : "Backscattering Spectrometry", Academic Press, New York (1978)
3. L.C.Feldman, J.M.Mayer and S.T.Picraux : "Materials analysis by ion Channeling", Academic Press, New York (1982)
4. J.F.Ziegler : "Handbook of stopping cross sections for energetic ions in all elements", Pergamon Press, New York (1980)
5. K.Wittmaack and N.Menzel : Appl. Phys. Lett 53,1708 (1988)
6. A.E. Morgan and P.Maillot : Proc. Sims-VI, Ed A.Benninghoven, A.M.Huber and H.W.Werner, Wiley (1987), p 706

854

7. J.S.Williams : Nucl. Instr. Meth. 126, 205 (1975)
8. J.A.Leavitt, L.C. McIntyre, P.Stoss, M.D.Ashbaugh, B.Dezouly-Arjomandy,
   M.F.Hinedi and G.Van Zijl : Nucl. Instr. Meth. B35, 333 (1988)
9. J.Knapp and B.Doyle : Proc. Ion Beam Analysis 89, Kingston (to be
   published)
10. G.J.F Legge and I.Hammond : J. Microsc. 117, 209 (1979)
11. B.L.Doyle : Nucl. Instr. Meth. B15, 654 (1986)
12. W.G.Morris, H.Bakhru and A.W.Haberl : Nucl.Instr. Meth. B15, 661 (1986)
13. P.M.Read, J.A.Cookson and G.D.Alton : Nucl. Instr. Meth. B24/25, 627
    (1987)
14. A.Kinomura, M.Takai, K.Inoue, K.Matsunnaga, M.Izumi, T.Matsuo, K.Gamo,
    S.Namba and M.Satou : Nucl. Instr. Meth B33, 862 (1988)
15. J.S.Williams, J.C.McCallum and R.A.Brown : in "Principles and
    applications of High-Energy Ion microbeams" eds F.Watt and
    G.W.Grime (1987), chapter 9.
16. C.P.M.Dunselman, W.M.Arnold Bik, F.H.P.M.Habraken and W.F.Van der Weg :
    MRS Bulletin, august 16 (1987), 35
17. H.J.Whitlow, G.Possnert and C.S.Petersson : Nucl. Instr. Meth. B27, 448
    (1987)
18. R.Groleau, S.C.Gujrathi and J.P.Martin : Nucl. Instr. Meth. 218, 11 (1983)
19. W.Vandervorst and H.Maes : Spectroch. Acta 40B,781 (1985)
20. E.N.Sickafus : Phys. Rev. B16,1448 (1977)
21. M.P.Seah : SEM 1983 (SEM Inc.,), p521 (1983)
22. J.A.D. Matthew, M.Prutton, M.M. El Gomati and D.C.Peacock : Surf. Interf.
    Anal. 11, 173 (1988)
23. W.D.Chen, H.Bender, A.Demesmaeker, W.Vandervorst and H.E.Maes : Surf.
    Interf. Anal. 12,156 (1988)
24. W.D.Chen, H.Bender, W.Vandervorst and H.E.Maes : Surf. Interf. anal.
    12,151 (1988)
25. H.Bender, J.Portillo and W.Vandervorst : Surf. Interf. Anal (accepted for
    publication) (1989)
26. S.Gaarenstroom : Appl. surf. Sci. 7,7 (1981)
27. M.M. El Gomati, M.Prutton, B.Lamb and C.G.Tuppen : Surf. Interf. Anal.
    11,251 (1988)
28. C.G.H.Walker, D.C.Peacock, M.Prutton and M.M.El Gomati : Surf. Interf.
    Anal. 11,266 (1988)
29. M.Prutton, M.M.El Gomati and C.G.Walker : Inst. Phys. Conf.Ser. No 90,
    Chap 1, (1987)
30. M.M.El Gomati, D.C.Peacock, M.Prutton and C.G.Walker : Journ. Microsc.
    147,149 (1987)

31. M.Prutton, M.M.El Gomati and P.G.Kenny : J. Elctron. Spec. and Rel. Phen. (accepted for publication, 1989)
32. H.Bender and W.D.Chen : Surf. Interf. Anal. (submitted, 1989)
33. F.Fransen, R.Vanderberge, R. Vlaeminck, M.Hinoul, J. Remmerie and H.E.Maes : Surf. Interf. Anal. 7,79 (1985)
34. W.Hoessler : Proc. Imec summer course on Characterization techniques for VLSI and advanced semiconductor devices. (1987)
35. S.Matteson and R.A.Bowling : MRS-proceedings Vol48, 215 (1985)
36. F.J.Grunthaner and P.J.Grunthaner : Materials Science reports 1,65 (1986)
37. F.J.Grunthaner, P.J.Grunthaner, R.P.Vasquez, B.F.Lewis, J.Maserjian and A.Madhukar : J. Vac. sci. Techn. 16,1443 (1979)
38. H.Bender, W.D.Chen, J.Portillo, L.Vandenhove and W.Vandervorst : Appl.Surf. Sci. (accepted 1989)
39. M.A.Hartney, J.N.Chiang, D.S.Soane, D.W.Hess and R.D.Allen : SPIE (1989)
40. J.M.Hill, C.S.Fadley, L.F.Wager and F.J.Grunthaner : Chem. Phys. Lett. 44,225 (1976)
41. A.Benninghoven, F.G.Rudenauer and H.W.Werner : "Secondary ion mass spectrometry" , Wiley, Chemical series Vol 86 (1987)
42. P.Mertens, W.Vandervorst and J.Leclair : Proc Ion Beam analysis (1989)
43. J.Kilner, R. Chater and K.Reeson : Proc. Ion Beam analysis (1989)
44. M.Meuris, G.Borghs and W.vandervorst : J. Vac Sci. Techn. (accepted 1989)
45. W.Vandervorst and F.R.Shepherd : J. Vac. sci. Techn.. A5 (3),313 (1987)
46. W.Vandervorst, H.E.Maes and R. De Keersmaecker : J. Appl. Phys. 56,1425 47. M.Meuris, W.Vandervorst, P.Debisschop and D.Avau : Appl.Phys.Lett.   54,1531 (1989)
48. G.Horcher, A.forchel, S.Steiner, R.Germann, G.Weimann and W.Schlapp : Proc Sims-VI, Ed A.Benninghoven, A.M.Huber amd H.W.Werner, Wiley (1987),457
49. E.Frenzel : Nucl.Instr. Meth. B15, 183 (1985)
50. W.Vandervorst, F.R.S.Shepherd, B.Philips, J.Newman and J.Remmerie : J. Vac. Sci. Techn A,1359 (1985)
51. W.Vandervorst and F.R.Shepherd : Appl. surf. Sci. 21,230 (1985)
52. D.Avau, W.Vandervorst and H.E.Maes : Surf. Interf. Anal. (1988)
53. M.Anderle, R.Canteri, D.robba and G.Quierolo : Proc Sims-VI, Ed A.Benninghoven, A.M.Huber amd H.W.Werner, Wiley (1987),747
54. F.A.Stevie, P.M.Kahora, S.Singh and L.Kroko : Proc Sims-VI, Ed A.Benninghoven, A.M.Huber amd H.W.Werner, Wiley (1987),319
55. R.G.Wilson and S.W.Novak : Proc Sims-VI, Ed A.Benninghoven, A.M.Huber

    and H.W.Werner, Wiley (1987),57
56. D.Avau, W.Vandervorst and H.E.Maes : Fresenius Zeitschrift fur Analytische Chemie 329,220 (1987)
57. W.Reuter : Proc Sims-5, eds A.Beningoven, R.J.Colton, D.S.simons and H.W.Werner, Springer (1986), 84
58. M.J.Pellin, C.E.Young, W.F.Callaway, J.W.Burnett and D.M.Gruen : Proc. ECS-symposium on "Diagnostic techniques for semiconductor materials and devices" Vol 88-20,73 (1988)
59. P.Debisschop and W.Vandervorst : Proc Sims-VI, Ed A.Benninghoven, A.M.Huber amd H.W.Werner, Wiley (1987),809
60. G.slodzian : Proc Sims-VI, Ed A.Benninghoven, A.M.Huber amd H.W.Werner, Wiley (1987),3
61. R.von Criegern, H.Zeininger and S.Rohl : Proc Sims-VI, Ed A.Benninghoven, A.M.Huber amd H.W.Werner, Wiley (1987),419
62. M.Gauneau, R.Chaplain, J.M.Dumas and M.Schumacher : Proc 4th int colloquiem on quality in electronic components (bordeaux 1989), 86
63. H.M.Migeon, M.Schumacher, J.J.Le Goux and B.Rasser : Proc. Angewandte Oberflachenanalytik, Julich 1988.
64. W.Vandervorst and H.W.Werner : Proc Sims-VI, Ed A.Benninghoven, A.M.Huber amd H.W.Werner, Wiley (1987),409
65. M.Meuris, P.Debisschop and W.Vandervorst : Proc sims-7, to be published
66. M.Dowsett : paper presented at QSA'88, London.
67. W.Vandervorst and J.Leclair : unpublished
68. This work is performed under ESPRIT-519 contract.
69. The new software package is available through Imec.
70. P.Picco and M.L.Polignano : J. Elec. Soc. 128,2034 (1981)
71. S.Al-Maryati, K.Shenai, N.Lewis, B.J.Baliga : Proc. ECS-symposium Fall 1988.
72. A.Casel and H.Jorke : Appl. Phys. Lett 50,989 (1987)
73. W.Vandervorst and T.Clarysse : Proc. ECS-symposium on "Diagnostic techniques for semiconductor materials and devices" Vol 88-20,267 (1988)
74. S.M.Hu : J. Appl. Phys. 53, 1499 (1982)
75. J.Albers : ASTM STP 960 eds G.C.gupta and P.H.Langer (1986)

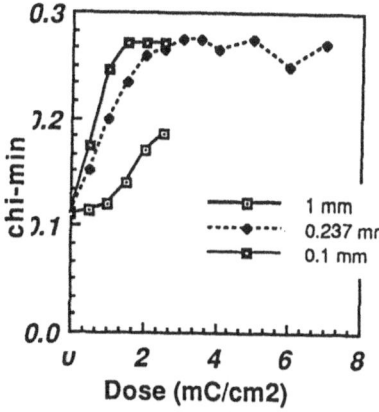

Fig 1 : Degradation of the substitutionality of As in Si as a function of He-dose. Parameter is the raster size.

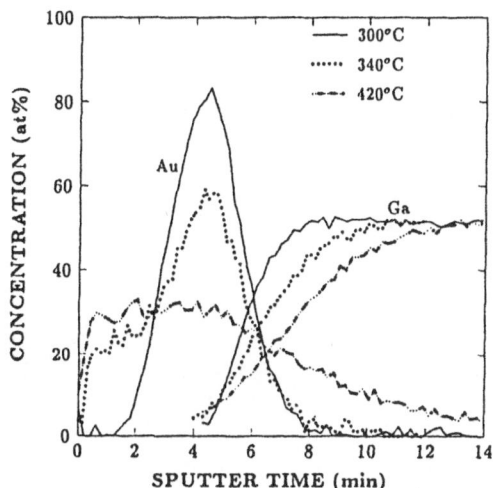

Fig 2 : Quantitative AES-profiles from a Ni/Au(Ge)-layer on GaAs as a function of anneal temperature.

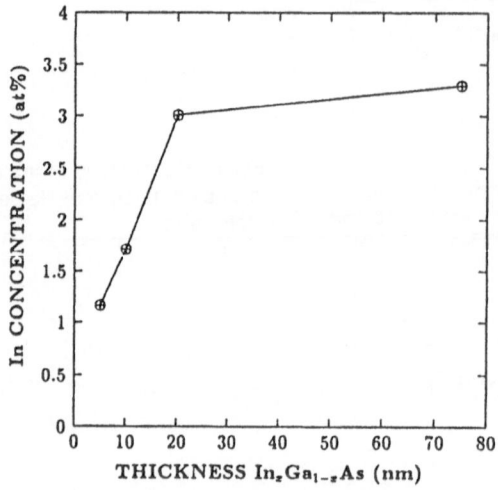

Fig 3 : Apparent In-concentration as derived from AES-depth profiles with 1,5 kev Ar sputtering of layers with identical In concentrations.

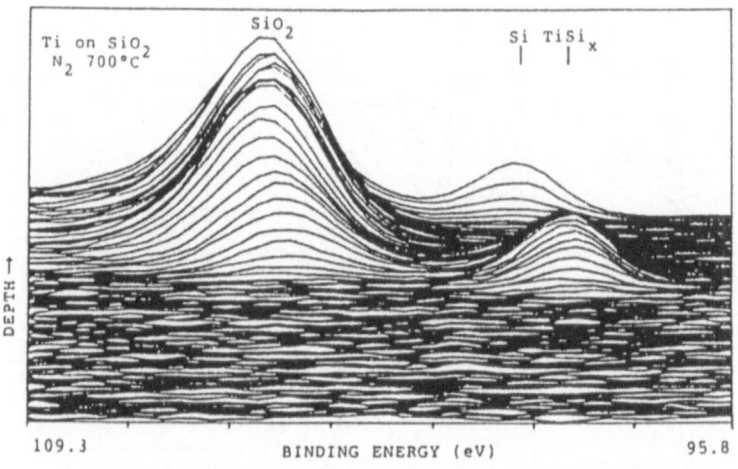

Fig 4 : 3D-plot of a ESCA-depth profile through a Ti/SiO2/Si-structure annealed at 700 C in N. Shown is the shift of the Si-peak as a function of depth (the front is the surface). The Si peak clearly indicate the three different phases (Si, oxide and silicide).

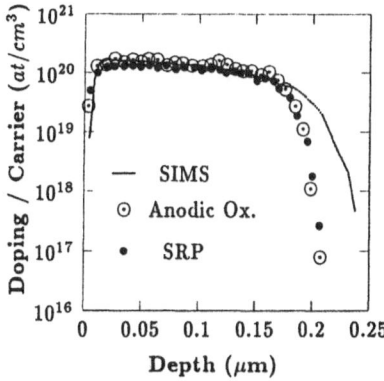

Fig 5 : Comparison between Sims, SRP and sheet resistivity measurements combined with anodic oxidation. The difference in tail is entirely due to carrier spilling.

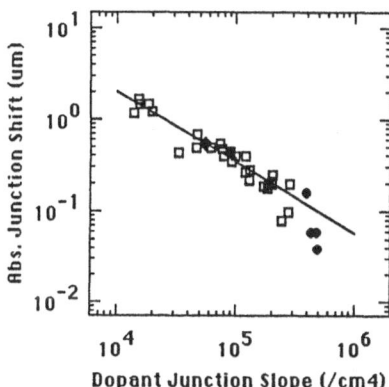

Fig 6 : Absolute junction shift for a p-type profile in n-type substrate determined from the same profile in p-type substrates and comparative SIMS/SRP-measurements. The determining parameter is the slope of the profile in the proximity of the junction. Substrate concentration 1 10 [15] at/cm3

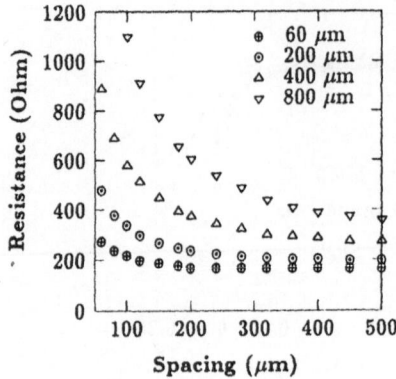

Fig 7 : Variation of the spreading resistance on the surface of a B-implant in n-type substrate. Perpendicular to the probe tips boundaries are formed by junctions also. The spacing relates to the distance between the boundaries, parameter is the probe spacing.

Is There Anything Beyond the 64 MDRAM

Kyoichi Shibayama

LSI R&D Lab., Mitsubishi Electric Corp.*
Mizuhara, Itami 664 JAPAN

Abstract
    Physical and economical constraints which provides obsta-
cles for the improvement of VLSI technology beyond 64MDRAM or
0.3um, and the approach and physically new device concepts how
to prevent these issues are discussed quantitatively as far as
possible. Fundamentally, a MOS transistor can work at 0.1um
gate length probably on the chip level with 1V power supply at
77K. SOI structure or precisely controlled channel engineering
under the low temperature heat cycle is essentially important.
New materials or new device concepts are substantially needed
for a DRAM cell in future. Other physically new concepts or
attempt which provides larger operational margin or better
performance of a miniaturized device will also be discussed.

Introduction

    Semiconductor industry is now entering deep sub-micron era,
where further miniaturization is confronting physical limitations
and processings become quite complicated to realize such devices
in higher integration. Therefore are arising questions; "Is there
anything beyond 64MDRAM?", or "Is it profitable for semiconductor
industry to go beyond 64MDRAM?" In this paper, discussions will
be made on the physical limitations of ULSI and new concepts for
device structure, higher density DRAM, and super integrations as
a proposed answer to these questions.

Limitations for ULSI

    Several limitations have to be solved to increase the pack-
ing density further by scaling down into the deep submicron region. A deep submicron MOS transistor as a fundamental component

*Present adress; Ryouden Kasei Company,
                 2-6-1 Miwa, Sanda, 669-13 JAPAN

of ULSI will encounter the operation limit due to short channel effect, hot carrier degradation, breakdown and punchthrough voltage lowering, velocity saturation, and direct tunneling in thin gate oxide. Many attempts have been in progress to cope with these limits including drain engineering, channel engineering, low voltage operation, cryogenic operation, and so on. Velocity overshoot below 0.1 um at the cryogenic operation is encouraging for further scaling.

Lithography technologies have continued to be developed to advance the integration density. A g-line (436nm) optical stepper has been improving the resolution down to 0.8 um by increasing the numerical aperture and used for the mass production of 4M DRAM. To improve the resolution and the depth of focus, the wavelength is shortened as presented in Fig.1. An i-line (365nm) stepper will play an important role in 16M DRAM production with the 0.5um design rule. For further decrease in the wave length, the excimer laser steppers using KrF(248 nm) and ArF(193nm) have been under development to improve the resolution down to 0.3 um for 64M DRAM production. New resist materials have to be developed for the shortened wavelength. A SOR or point source X-ray stepper will be needed beyond 64M DRAM.

The remaining problem to be solved is the improvement of lithographic alignment. Device areas will be limited by the alignment accuracy of lithographic tools, therefore self-aligned structures will be further required.

The chip size of DRAM increases through the successive generation as presented in the Fig.3. Due to this trend, the image field of an optical stepper has to be increased. In addition package size will be required to be increased.

Physically New Devices

New transistor structures are necessary to increase the integration density further by reducing the device area. Several attempts have been made such as the surrounding gate MOS transistor (SGT) and the polysilicon source/drain transistor (PSD)[1] . Compared with the conventional MOS transistor, the new MOS transistor PSD employs the polysilicon source/drain with the buried insulator between the polysilicons for the isolation. The buried isolation can realize the deep submicron isolation and suppress the narrow channel effect in contrast to the conventional LOCOS isolation. In addition, the contact, the source/drain, and the interconnect are merged and formed by the self-aligned method, resulting in the reduction of device area. This new structure can also reduce junction capacitance and realize high speed operation.

In scaling down the transistor size into the deep submicron region, hot carrier degradation is one of the limiting factors. To cope with this difficulty, drain engineering will be increasingly important to optimize the impurity profile at the drain edge. The LDD structure has been used to reduce the drain electric field strength. However the problem for the LDD is that current drivability reduces due to the high parasitic resistance of lightly doped regions and that thereby it becomes difficult to achieve high speed operation even though the transistor size is scaled down into deep submicron region. Recent drain engineering has produced the gate/N^- overlapped structure[2] to overcome the drawback and increase the current drivability as presented in Fig.2. The normal electric field from the gate electrode into the N^- regions forms the accumulation layers and reduce the parasitic resistance.

The short channel effects are another serious limiting factors. Channel engineering will definitely be important to cope with this difficulty. For example, the delta doping[3] in Fig.3 is a new promising concept. Major features of these structures are (1) heavily doped layers to suppress punch-through, (2) insensitivity to the statistical fluctuation of channel impurities, and (3) high carrier mobility and small junction capacitance due to the low impurity concentration in the substrate.

Cryogenic CMOS operation will open the way to low supply voltage operation and high circuit speed below quarter micron. Thin SOI transistor will also be promising as well as cryogenic CMOS for low voltage operation and high speed due to the steep subthreshold slope and the small parasitic capacitance.

High Density DRAM

There was an expectation that the gain cell might be better candidate for high density DRAMs because of elimination of capacitor which needs large area to keep capacitance sufficiently large. However, as a matter of fact, most people believe that one transistor-one capacitor cell would further be utilized for future DRAMs. Because, as technology advances, the capacitors can be made more smaller by using three dimensional extension of the capacitance area combined with the highly reliable ultra thin dielectric films.

Fig.4 shows a technology trend of DRAM memory cell. The capacitor area can be kept almost constant in spite of decreasing occupied area by using stacked and/or trench structure. Actually mass-production of 4MDRAM has started with stacked or trench cell. It was also demonstrated that 16M and even 64MDRAMs will be realized with improved three dimensional capacitor structures.[4,5]

However, it is also true that there is a limit in enlargement of the effective capacitor area. The thickness of dielectric film has also a lower limit. Fig.5 shows the capacitance value as a function of the occupied capacitor area. At the bottom of the figure, the expected maximum capacitor area, and the enlargement factor which is defined as a ratio of the actual capacitor area to the cell area, are plotted for each DRAM generation. The realistic value of the enlargement factor would be limited to below five. Therefore it is quite difficult to realize 256MDRAM and beyond even if dielectric film thickness can be reduced to 3nm oxide equivalent. Tantalum oxide dose not change the situation remarkably. The most hopeful candidate to realize DRAMs beyond 64M would be ferroelectric thin films, such as PZT, having quite high dielectric constant. For example, if PZT films of 0.3um can be utilized, a planar type 64M and three dimensional type 256M and 1GDRAM would be realized.

Increase of DRAM integration and high speed operation cause an increase of power consumption. Fig.6 shows the DRAM technology trend, actual and anticipated RAS access time, and power dissipation dependence on the DRAM generations. The progress in the MPU (micro-processor-unit) operating speed requests the DRAM to operate at higher speed. At the transition from 256K to 1M, CMOS technology was unavoidably adopted to realize high speed and low power dissipation. And the multidivided bit-lines were also adopted to increase the S/N ratio and to reduce operating power with the partial array activation technique.

From 64K through 4M, 5V has been used in spite of the scaling down of the design rule. If the power supply of 5V is continuously used even in the 16MDRAM, the operating power at a maximum cycle rate of 120ns (60ns access version) will reach 700mW. The reliability problems concerned with high electric field have been solved by the innovations of the drain engineering such as LDD. But the condition may change in the 16MDRAM. The drain voltage reduction to around 3.3V must be certainly necessary for reducing the power and insuring the same reliability as 1M and 4M. Using 3.3V and partial activation design of 1/16 can realize reasonable operating current of 80mA at maximum cycle time of 120ns. For the sake of user's convenience, an on-chip regulator will be adopted to regulate the voltage from external 5V to internal 3.3V, so that the operating power dissipation becomes about 400mW. In the generation of 64MDRAM, the external supply voltage should be reduced to 3.3V. Moreover internal voltage reduction to below 3V may be necessary to get high reliability in the range of 0.3um dimension.

The extremely low operating voltage will bring another difficulty of maintaining adequate stored signal charge for the stable operation. It is easy to forecast that the situation will become exceedingly severe in 256M era. Can enough signal charge be obtained using so low voltage? Although lowering the external supply voltage is very effective to reduce the power dissipation, will the users accept the lowering of the supply voltage twice in succession from 64M DRAM?

Cooling the chip by liquid nitrogen may become "must" in order to achieve stable operation of the RAM with 0.2um transistor, sufficiently long retention time and also low power. So, technological "revolution" in circuit and process will be indispensably required to attain the Gb era beyond 64M.

Super Integration

3-D structure or wafer scale integration is helpful for realizing a very high integration and superior system operation performance in future especially beyond 64MDRAM. Monolithic approach of 3-D IC is now under development very actively in Japan. As shown in Fig.7, active 3 IC layers are stacked monolithically and are verified to work successfully as an image sensing and processing device. 2 layer stacked CMOS, which can be called inter-CMOS is expected to realize world first practical VLSI SRAM chip.

Hybrid 3-D approach which is represented by wafer attachment technology is also very promising. Instead of tough work of laser recrystallization in monolithic 3-D, residual thin bulk crystalline layer can be used, although making the crystalline layer very thin down to 0.1um is also another tough process technology. By using these technologies, i.e. wafer attachment and thining, a new type device system might be realized. A wafer scale integration may be rather a present technology not in future, but will be applied more successfully by introducing new wiring material, wiring technology and a sophisticated testing method.

Manufacturing

For developing new devices, simulation is very effective for understanding the fabrication processes and device operations without actual fabrication. Fabrication process becomes long and complicated so that experimental approach for ULSI development takes enormous period. Therefore, the developments using simulation will become a main approach.

As the device structure of ULSIs become miniaturized and complicated, three dimensional simulation with high precision is required. Several 3D-simulators for process[6] and devices[7] are reported.

Monte-Carlo method is recently required to understand hot carrier phenomena and behavior of high energy particles such as the decrease of impact ionization rate in 0.1um device and ion trajectory in implantation process.

It is well known that USLIs have become key components of all products. Customers of semiconductor devices require chips with high performance, superior function and special features. In order to provide the necessary functions, circuit of chips become more complicated and the number of masks also increase. The device structure of ULSIs becomes more complex in order to realize high performances. As a result, the number of process steps has increased with increasing the packing density of chips. It is predicted that the number of process steps of ULSIs beyond 64MDRAM exceeds 500 steps, and the number of critical processes increases and process window becomes narrow. Thereforer the average yield of each process step of ULSIs beyond 64MDRAM should be greater than 99.9% for the stable fabrication.

Process automation including process optimization will be essential to fabricate ULSIs which requires a long and sophisticated processes. Therefore, feed-forward fabrication should be developed. In the feed-forward system, wafer treatment is mechanized, and lot/wafer tracking and management are systematized. Subsequent processes are dynamically optimized using the results of data analysis and simulation in the midway of the process.

References
1) M.Shimizu et al.; Technical Digest of IEDM 1988, pp.96
2) M.Inuishi et al.; 1989 Symposium on VLSI Technology, pp.33
3) K.Yamaguchi et al.; Proceedings of 20th Solid State Devices and Materials, Tokyo, 1989 pp.17
4) W.F.Richardson et al.; 1989 Symposium on VLSI Technology, pp.65
5) W.Wakamiya et al.; ibid., pp.69
6) M.Fujinaga et al.; Technical Digest of IEDM 1989, pp.332
7) H.Umimoto et al.; 1989 Symposium on VLSI Technology, pp.47
8) H.Masuda et al; ISCAS 1984, pp1163

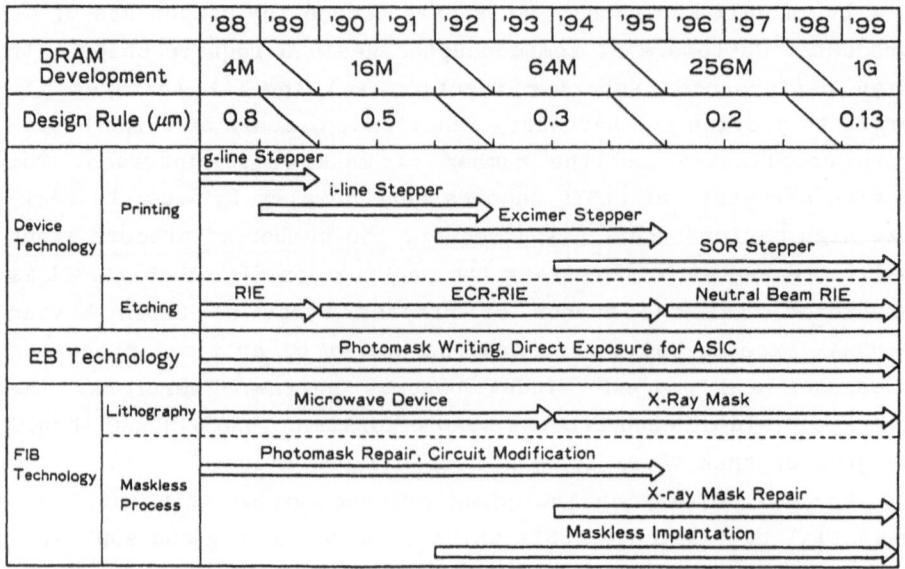

Fig.1   Trend of lithography technology with the DRAM
         development.

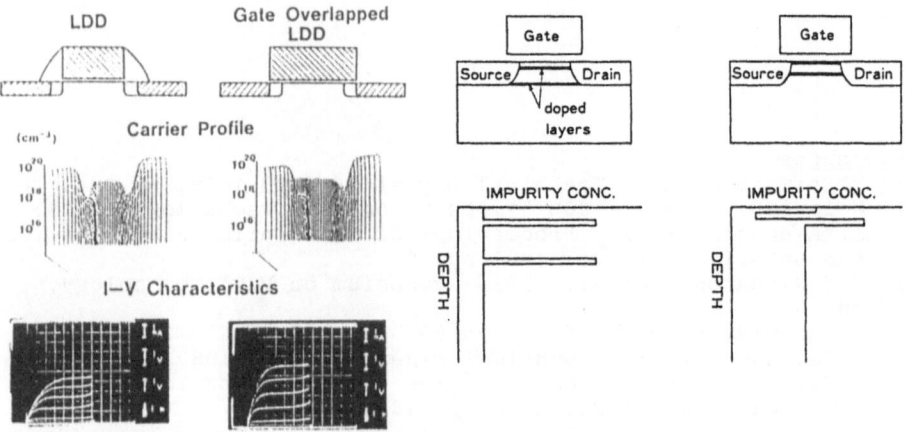

Fig.2   Drain engineering.
        (gate overlapped LDD)

Fig.3   Channel engineering.
        (delta doping)

Year	'90	'91	'92	'93	'94	'95	'96	'97	'98	'99	2000
DRAM Genera-tion	R & D										1G
	Produc-tion	4M			16M			64M			256M
Design Rule		0.8μm			0.5μm			0.25μm			0.13～0.1μm
Memory Cell Cross Section 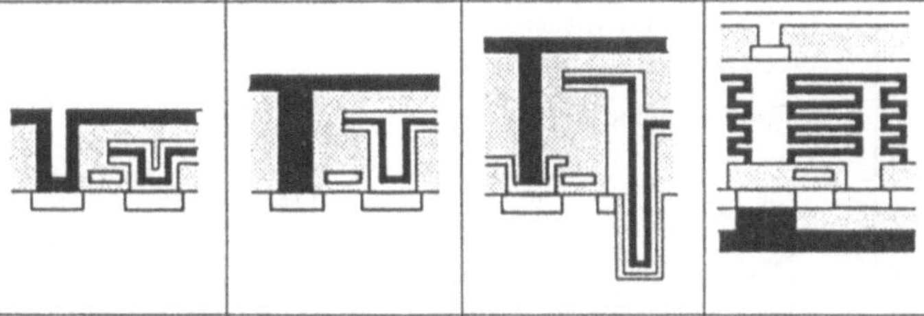											
Structure		Stacked Capacitor			T−Shape Stacked Capacitor			Stacked Trench Capacitor			SOI Capacitor
Lithography		G−line Stepper			G−line Stepper			Excimer Laser Stepper			X−ray(SOR) Stepper
Material		Al−Based Alloy (AlSiCu)			Contact Filled Wiring (CVD−W)			Ferroelectric Film			Super Conductive Wiring

Fig.4   Technology trend of DRAM memory cell structures.

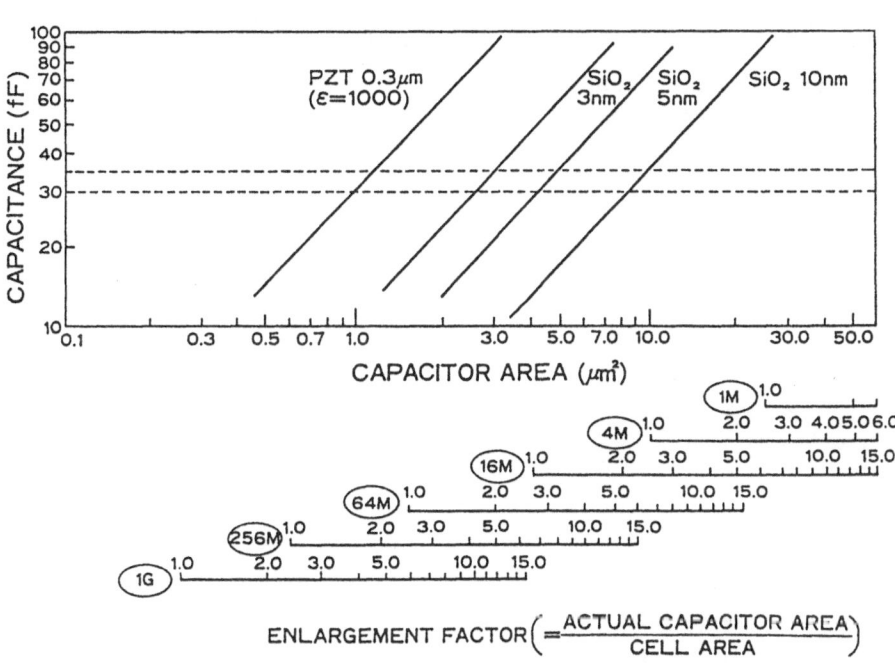

ENLARGEMENT FACTOR $\left(=\dfrac{\text{ACTUAL CAPACITOR AREA}}{\text{CELL AREA}}\right)$

Fig.5   Capacitance as a function of capacitor area and enlargement factor (actual capacitor area/cell area) to keep sufficiently large capacitance.

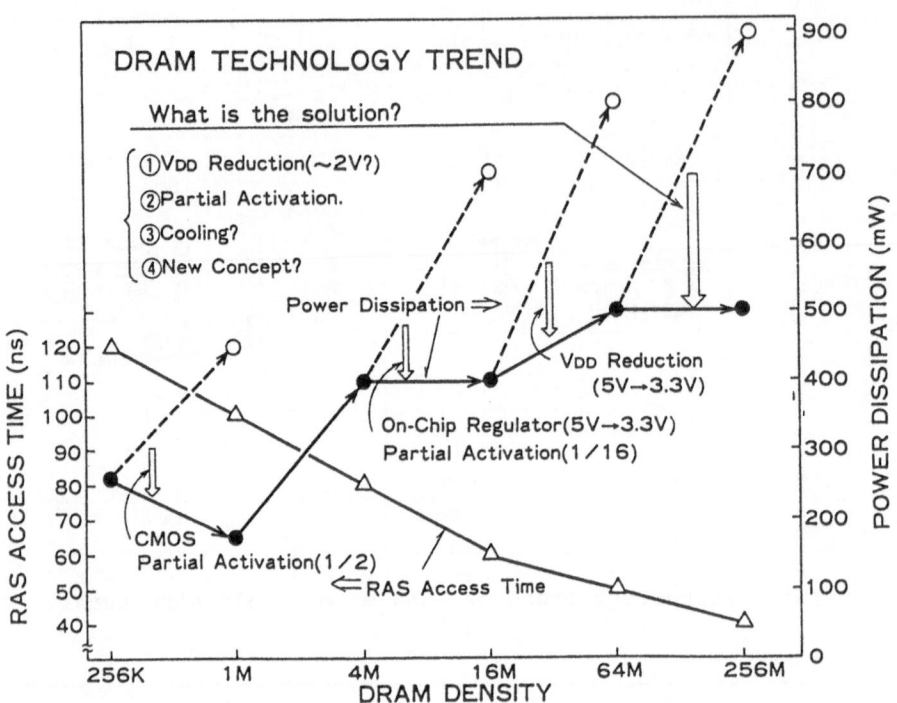

Fig.6 DRAM technology trend for RAS access time and power dissipation.

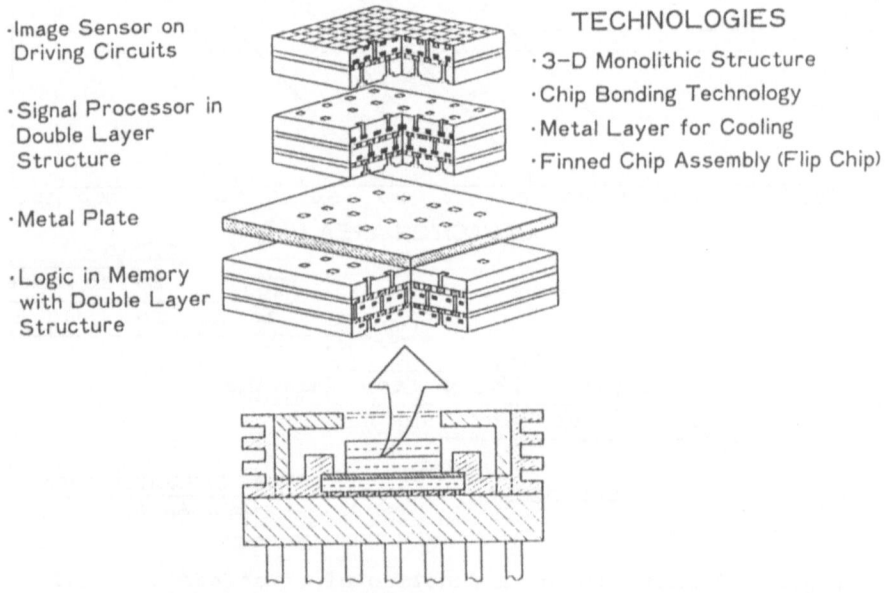

Fig.7 Super three dimensional integration using a chip bonding technology.

SOI 2

Characterization and Simulation of SOI-CMOS Devices for 3D-integration

J.GÖTZLICH, R.KIRCHER, K.GIESEN[*] and G.PÖSCHL

Siemens AG
ZFE EL PT / ZPL TW 61 [*]
Otto-Hahn-Ring 6
8000 München 83,FRG

Summary

SOI MOSFETs were fabricated in argon laser recrystallized silicon layers. The SOI film thickness was varied between 100nm and 350 nm using a planarizing etch back technique. The electrical characteristics of the devices were compared with two-dimensional device simulations.

Introduction

Three-dimensional integration offers new possibilities for further increase of circuit performance and for realization of new integration concepts, e.g. multifunctional circuits. Up to now argon laser and electron beam recrystallization are the best techniques for realization of SOI layers in 3D circuits. By these two methods SOI layers in chip dimensions can be achieved without degradation of underlaying device structures in the bulk silicon wafer [1,2]. However, the lateral temperature gradients in the liquid silicon film usually lead to a mass transport in the SOI film. Therefore planarization of the recrystallized layers is necessary, especially when thin SOI devices are fabricated.

In this paper, firstly the technology of a SOI CMOS process is reported, which can be used for processing 3D ICs. This technology is based on argon laser recrystallization including seeding and planarization of the SOI layer. The electrical characteristics of n- and p-channel MOSFETs fabricated in these SOI layers are presented and compared with the predictions made by two-dimensional device simulation.

Experimental

Argon laser recrystallization of 500 nm thick LPCVD deposited polysilicon layers was performed by a 25 W laser beam, which was focussed into an elliptical spot (50 x 150 μm^2). The scanning velocity was 10 cm/sec and the substrate temperature 500°C. To obtain large single SOI crystals with a defined orientation, seeding holes and antireflecting stripes on top of the SOI film were used. After recrystallization the SOI layer

874

was patterned into individual mesa islands by etching. Some SOI films were thinned before device processing by back-etching using an organic resist for planarization. The sidewall isolation was performed by LOCOS oxidation and the doping of the islands as well as the threshold voltage adjust of the MOSFET by one single ion implantation. Then a conventional n+-polysilicon gate CMOS process was applied. To investigate the influence of a fixed bias on the body of SOI MOSFETs, some of the transistor structures had an external body contact.

Results

In Fig. 1 the distribution of threshold voltage of a n-channel MOSFET (W/L = 21/um/3/um) on one wafer is shown, which were placed either 15/um or 415/um from a seeding window. Two effects can be observed : 1) the threshold voltage is shifted to higher values, if the distance to the seed is increased and 2) the scattering is increased considerably with increasing distance. This is due to the rotation of the crystallographic axis along the scan [1] and the higher probability of grain boundaries in the channel area, respectively. The result is, that in 3D circuits each SOI MOSFET should have an own seeding window to obtain reproducible electrical characteristics.

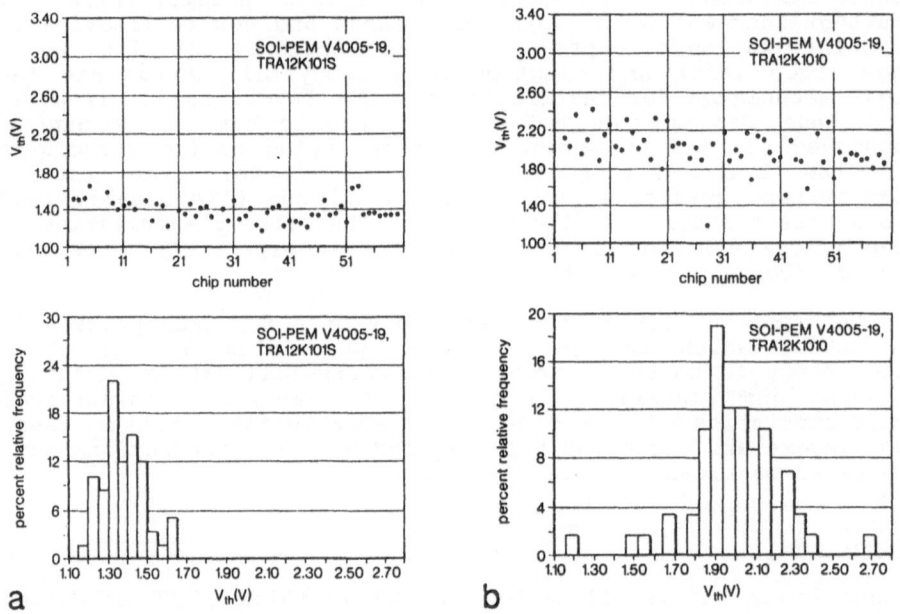

Fig.1. Distribution of the threshold voltage of n-channel SOI MOSFETs (W/L=21/um/3/um, SOI thickness=350 nm) with different distances to the seeding windows (a:15/um; b:415/um).

The thickness of the SOI film was varied between 100 nm and 350 nm. As expected, in thick SOI films a pronounced kink effect could be observed, which was suppressed by biasing the body of the n-channel MOSFET. In thin SOI films ,a strong influence of the back gate bias on the electrical behavior of the transistor was observed. In Fig. 2 the subthreshold characteristics of a n-channel MOSFET (W/L = 21/um/3/um) is shown, which is fabricated in a 100 nm thick film. The application of a negative back gate bias leads to a significant shift of the threshold voltage. Simulations using the newly developed device simulator GALENE II [3-5] showed that the electric field of the back gate penetrates the fully depleted body of the very thin film SOI MOSFET and changes the threshold voltage at the front channel.

In Fig.3 the simulated subthreshold behavior of a n-channel SOI transistor is compared with the experimental data. In this example the channel length was 2.3/um and the thickness of the SOI layer 300 nm. The doping profiles of the source/drain regions and the SOI layer are taken from the two-dimensional process simulator LADIS [6,7].The excellent agreement is obtained without any fit parameter.

By device simulation not only single SOI transistors can be investigated, but also multi-layered CMOS structures. As an example, the electrical coupling between different stacked layers of a CMOS inverter has been studied to find out suitable measures to prevent cross talk between active device areas. It was shown, that an additional polysilicon layer with a suitable doping concentration can be used. Contacts must be applied to keep the potential of this polysilicon shield on ground potential.

Conclusions

The electrical behavior of MOSFETs fabricated in argon recrystallized SOI films demonstrated that a precise control of the crystal quality is mandatory for reproducible device characteristics. If laser recrystallized layers are thinned to about 100 nm for realization of very thin SOI devices, planarization techniques must be applied to obtain sufficiently homogeneous film thickness. Device simulations showed that cross talk in 3D structures can be prevented when polysilicon shields between the active regions are applied.

Acknowledgements

The authors wish to thank all their collegues from the 3D projekt, namely R.Lemme for the SEM investigations and M.Steger and M.Kreitmaier for the electrical characterization. The work was supported by the German Federal Ministry for Research and Technology. For the contents the authors alone are responsible.

876

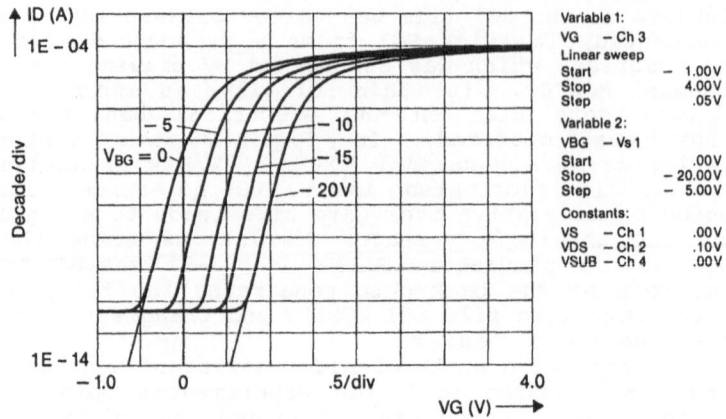

Fig.2. Subthreshold characteristics of a 100nm thick n-channel SOI MOSFET (W/L=21/um/3/um) at different backgate bias.

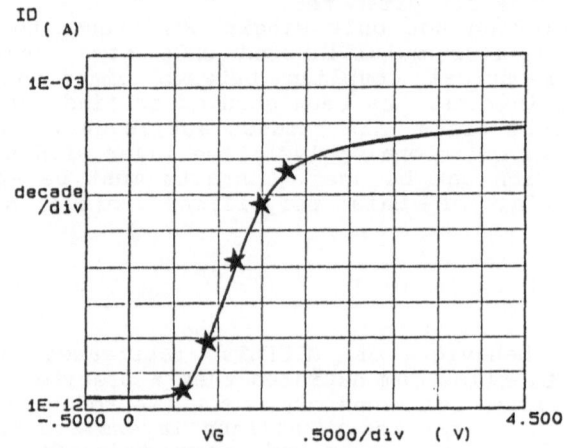

Fig.3. Subthreshold characteristics of a 350nm thick n-channel SOI MOSFET (W/L=21/um/2.3/um);
(- experimental, ★ simulation)

References:
[1] Sugahara,K.; Kusunoki, S.; Inoue, Y.; Nishimura, T. and Akasaka,Y. : J.Appl.Phy.62,4178, (1987)
[2] Maekawa, M. and Koba, M.: Ext.Abstr.5th Internat.Workshop on Future Electr.Dev.,Miyagi-Zao,171, (1988)
[3] Engl, W.L.; Kircher, R.; Bach, K.H.; Götzlich, J.: VLSI Process/Device Modeling Workshop ,Tokyo (1987)
[4] GALENE II User's Guide ,RTWH Aachen, West Germany (1988)
[5] Kircher, R. and Götzlich, J.: European SOI Workshop, Meylan,Digest of Technical Papers,p.B-02, (1988)
[6] Tielert, R.: ,IEEE ED-27,No.8,p1479,(1980)
[7] Lau, F.and Gösele,U.: Appl.Phys.A40,101,(1986)

Stacked CMOS Technology by Local Overgrowth (LOG)

R. P. Zingg, B. Höfflinger and G. W. Neudeck*

Institute for Microelectronics, D-7000 Stuttgart 80, FRG
*Purdue University, W. Lafayette, IN 47907, USA

Abstract

A new process is presented for building stacked CMOS transistors with high device quality. Device deterioration on bulk devices was minimized by reduced temperature processing for the SOI device, and the use of epitaxial lateral overgrowth to produce the silicon film for top devices improved film quality. The application of chemo-mechanical polishing allowed realization of $0.7\mu m$ silicon films with better than 10% uniformity. NMOS transistors built in the bulk material exhibit $700cm^2/Vs$ mobility and 105mV/dec subthreshold slope, while PMOS devices in the SOI film have surface mobilities of $165cm^2/Vs$ and a subthreshold slope of 109mV/dec. Threshold voltages are 0.5V and -1.5V, respectively.

Introduction

Increasing circuit complexity has been satisfied to date by reducing device dimensions. This scaling of dimensions is coming close to physical limitations, imposing an ever increasing process complexity to gain further miniaturization. Another approach to packing density is the vertical stacking of several devices, which up to now has only been possible with zone-melt recrystallization (ZMR, [1-3]). We present a new technology with improved device performance and the potential for further scaling: Local Overgrowth (LOG), a process related to the Epitaxial Lateral Overgrowth (ELO, [4]).

To realize useful silicon-on-insulator (SOI) devices, a good quality insulator is necessary to prevent leakage and provide dielectric isolation. For suitable device performance a good quality semiconductor is needed to obtain good carrier mobility and thus high transconductance. Finally, the silicon to insulator interface is crucial to prevent junction leakage and pinning of the Fermi-potential. Our process uses thermally grown oxides as insulators. It also uses selective epitaxy and lateral overgrowth in a hydrogen and chlorine atmosphere to yield very low interface states between oxide and overgrowth [5]. The high crystalline quality of the epitaxial overgrowth is well known [4,6].

In order to build 3D-CMOS circuits which give up to five-fold increase in circuit density [7], the process has to be optimized for planarity [1], and connectability to the lower devices has to be insured. LOG is especially suited for these requirements, since the seeded growth yields connections to the substrate devices and the local nature of the overgrowth allows for further contacts to the bottom layer. Additionally, the planarization process necessary for the thinning of the overgrowth yields a surface more planar than in a bulk CMOS process with a standard LOCOS field oxide.

Process

The processing for 3D LOG circuits starts with a standard NMOS in a (100) oriented substrate. Epitaxy is grown selectively with local overgrowth (LOG) [8,9] on a device scale to allow connections to the substrate transistors. The epitaxial growth is seeded by rectangular windows in the masking oxide, which were aligned with <100> directions. Growth progresses at about the same rate along the <100> directions vertically as laterally along the oxide interface, producing a silicon layer on this dielectric insulator. The lateral overgrowth step is performed at reduced temperatures of 850-900°C to minimize the thermal budget on the substrate device and eliminate potential damage to oxide layers [5]. Due to the limited aspect ratio between lateral and vertical growth of about one, the overgrowth has to be planarized by chemo-mechanical polishing from 13μm to about 1μm thickness [10]. This polishing step is made self-limiting by an uniformly applied CVD oxide. The final SOI thickness is controlled by the amount of oxide deposited and therefore better than 10% uniform across the wafer. After planarization, a standard PMOS process is implemented on the SOI film. With the self-aligned process for source and drain definition it is possible to produce a 3μm channel CMOS process with only 7 masks and very short processing time. Compared to a bulk process, masks for well definition and source/drain masking of the opposite channel type is eliminated. As CVD epitaxy and polishing can be considered standard processing technology (although both steps are often performed at the wafer manufacturer), no new processes have to be introduced in device fabrication. Contrary to ZMR, the LOG process becomes simpler as device dimensions are scaled, as the amount of overgrowth is reduced and therefore also the planarization process is simplified. In contrast, grain boundaries limit the scaling of ZMR devices. Contact definition for subsequent metalization is slightly complicated by the varying thickness of the isolating oxide, but the improved surface planarity simplifies lithography for metalization.

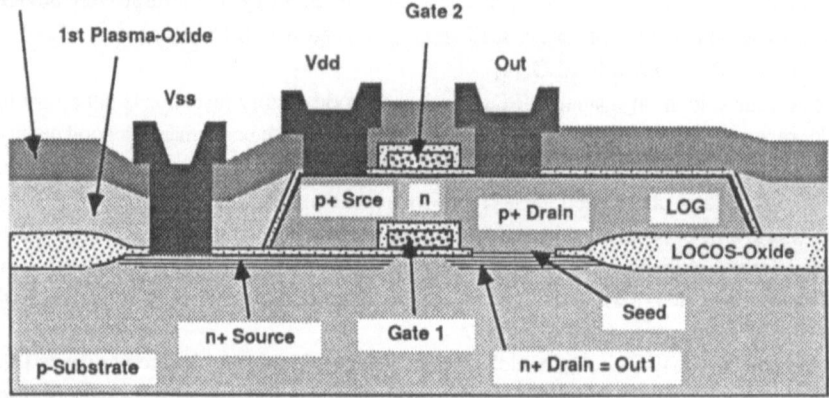

Figure 1: Cross-section of stacked CMOS by the LOG process.

Device Characterization

The performance of this new technology has been verified with devices built in the substrate and the SOI film. NMOS transistors in the substrate demonstrate minimal effects on the impurity profiles by the processing of the top devices. Remaining parameter shifts were compensated by tailoring the implantation parameters. Optimized transistors showed characteristics given in Figure 1, mobilities were 700cm^2/Vs, subthreshold slope was measured to be 105mV/dec, and junction leakage of the source and drain areas were less than 10nA/cm^2. Table 1 summarizes the device parameters. PMOS devices with surface channels were realized in the SOI film. Contrary to frequently used buried channel devices, channel mobility is reduced by the surface scattering, but subthreshold slope is improved by the reduction of fixed charge. Junction leakage current of the SOI devices were about 0.5μA/cm^2, but the junction area is reduced to about 1/20 by driving the source and drain regions throughout the SOI film. Mobilities for holes was measured at 165cm^2/Vs, which agrees well with mobilities observed in reference devices built on bulk material with similar doping levels. Subthreshold slope was measured at 109mV/dec, in good agreement with the doping level of 8.8·10^{15}/cm^3 and 40nm gate-oxide thickness. Threshold voltages are not quite symmetric yet, but can be corrected by corresponding implants. They are 0.5V and -1.5V, respectively. These properties indicate low defect density in the SOI material and excellent silicon-to-oxide interface quality. Mobilities exceed values reported in recrystallized material by a factor of 4-8 except for [11], and show less variation across the wafer as reported in [1], which we associate with the absence of sub-grain boundaries in our material.

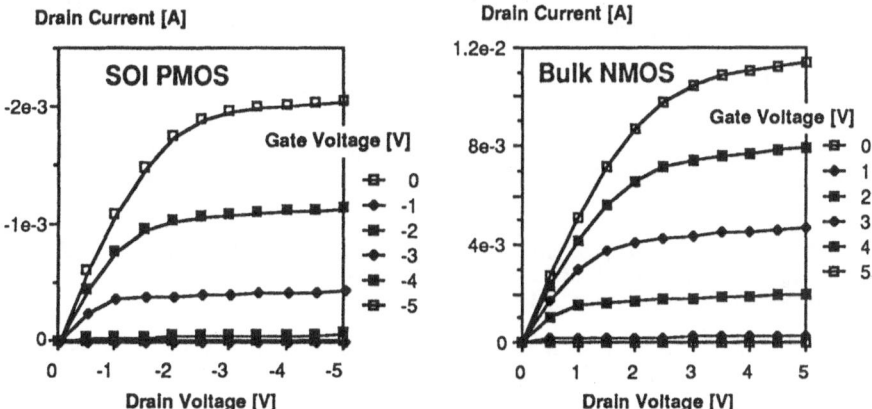

Figure 2: Output characteristics of SOI PMOS and bulk NMOS transistors built by the LOG process.

Table 1: Device parameters of the stacked transistors.

		SOI PMOS	Bulk NMOS
Channel Width	W	90μm	80μm
Channel Length	L	3μm	3μm
Under-diffusion	L_d	0.5μm	0.5μm
Gate Oxide Thickn.	t_{ox}	40nm	40nm
Threshold Voltage	V_t	-1.5V	0.5V
Channel Mobility	μ_o	165cm^2/Vs	700cm^2/Vs
Doping level	N_b	8.8·10^{15}/cm^3	5·10^{15}/cm^3
Srce/Drain Resistiv.	$R_{s,d}$	50Ω	25Ω
Subthreshold Slope	S	109mV/dec	105mV/dec
Junction Leakage	J_j	0.5μA/cm^2	20nA/cm^2

Conclusion

A novel process has been described to produce stacked complementary MOS devices. The second active layer was grown crystalline, circumventing defects often observed with recrystallized material. Film thickness for this SOI material is controlled by a self-limiting polishing step, avoiding variations observed in ZMR material due to occurrence of "teardrops". The planarity of the process, which in fact is better than in a bulk CMOS process, eases metalization lithography and enables the realization of additional semiconducting films for device fabrication. Initial device characteristics are very encouraging for applying this process to VLSI and ULSI integration. The increase in circuit density, which was reported to be up to five-fold, is especially beneficial for complex circuits.

References

1. Y. Akasaka, "Three-Dimensional IC Trends", Proc. of the IEEE 74 (12), pp1703-1714, December 1986

2. J. F. Gibbons and K. F. Lee, "One-Gate-Wide CMOS Inverter on Laser-Recrystallized Polysilicon", IEEE EDL-3 (8), pp117-118, June 1980

3. J. P. Colinge, E. Demoulin and M. Lobet, "Stacked Transistor CMOS (ST-CMOS), an NMOS Technology Modified to CMOS", IEEE SC-17 (2), pp215-219, April 1982

4. L. Jastrzebski, "SOI by CVD: Epitaxial Lateral Overgrowth (ELO) Process - Review", Journal of Crystal Growth, Vol. 63, pp493-526, 1983

5. J. A. Friedrich and G. W. Neudeck, "Interface Characterization of Silicon Epitaxial Lateral Growth over Existing SiO2 for Three-Dimensional CMOS Structures", IEEE EDL-10 (4), pp144-146, April 1989

6. S. T. Liu, K. Newstrom, M. Hibbs-Brenner, R. J. Stokes, B. Hoefflinger, G. Neudeck, R. Zingg, L. Bousse and J. Meindl, "Morphology of Silicon Islands Grown by Selective Epitaxy over Silicon Dioxide", MRS Symp. Proc, pp169-174, 1986

7. B. Hoefflinger, S. T. Liu and B. Vajdic, "A Three-Dimensional CMOS Design Methodology", IEEE SC-19 (1), pp37-39, February 1984

8. R. P. Zingg, G. W. Neudeck, B. Hoefflinger and S. T. Liu, "Epitaxial Lateral Overgrowth of Silicon over Steps of Thick SiO2" J. of the El-chem. Soc, Vol 133 (6), pp1274-1275, June 1986

9. R. P. Zingg, G. W. Neudeck, B. Hoefflinger and S. T. Liu, "Epitaxial Lateral Overgrowth of Silicon over Oxidized Polysilicon for Silicon-On-Insulator FET's" ESSDERC Abstr. pp226-227, September 1986

10. R. P. Zingg, H. G. Graf, W. Appel, P. Voehringer and B. Hoefflinger, "Thinning Techniques for 1µm ELO-SOI", IEEE SOS/SOI Workshop Proc. p. 52, October 1988

11. S. D. S. Malhi, H. W. Lam and R. F. Pinizzotto, "SOI CMOS Circuit Performance on Graphite Strip Heater Recrystallized Material", IEDM'82, pp441-443, 1982

Buried and Surface n-channel MOS Transistors in SOI

M. HAOND
Centre National d'Etude des Télécommunications
BP 98, 38243 Meylan Cedex, France

ABSTRACT

We compare buried- and surface-channel NMOS transistors made in SOI films. We show that buried channels can be used in SOI for improving the performances of the transistors. The kink is avoided. The mobility is increased by 20% as is the drain current. This can be obtained without drastically thinning the SOI film. This is important if we consider the technological problems related with very thin films, such as thickness uniformity and homogeneity, interface quality... However, improvements must be found in the short channel behaviour of the buried channels if we want to use them in the deep submicron range.

INTRODUCTION

In the conventional CMOS technology, a N+polysilicon gate is used, which provides a surface channel (SC) in the NMOS transistors and a buried channel (BC) in the PMOS after the threshold voltages have been adjusted. In bulk Si, BC transistors have shown to be more prone to punchthrough than SC devices. Therefore a lot of work has recently been devoted to the fabrication of SC PMOS by a P+ doping of the gate polysilicon. On the other hand, one of the major advantages of a BC relies on the important increase of the mobility, as a result of reduced vertical field /1/. Consequently, a compromise must be found to optimise the devices for high speed IC's.

The purpose of our work is to carry out similar analysis in SOI. Both SC and BC NMOS transistors are compared in terms of subthreshold slope, mobility, and short channel effects. Emphasis is put on the drive capability of these devices since this is the key parameter for speed.

EXPERIMENT

The SOI material was obtained by lamp-ZMR /2/. The SOI film thickness is 200 nm. The lateral isolation was performed by Reactive Ion Etching (RIE) the silicon film on oxide resulting in the formation of mesa structures. The well was obtained by a full-sheet boron implantation. A 25nm thick gate oxide was then grown at 950°C. The BC and SC were defined by an arsenic and a boron implant respectively. The 420 nm deposited polysilicon layer was then implanted with boron at a dose of 1E16 cm-2 at 30 KeV for the P+gate of the BC devices,

and doped in a POCl3 drive-in for the N+gate of the SC transistors respectively.
A 4E15 cm-2 phosphorous self-aligned implant at 50 KeV is provided for the drain/source
formation. The only high temperature step was a RTA reflow of the deposited PSG
at 1130°C for 20 seconds. This allows the BC to be in-depth adjusted, close to
the SiO2/Si interface, by the ion implantation energy. TITAN 5 /3/ was used to

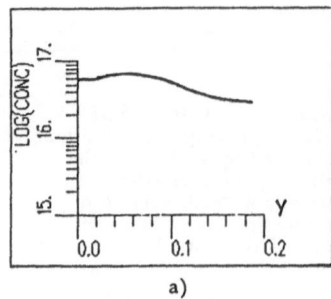
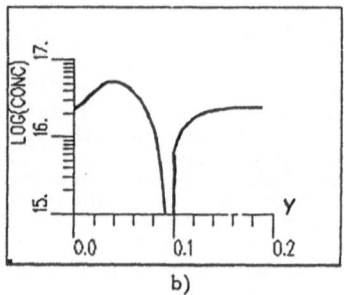

a)   b)

Fig.1: TITAN 5 simulations of doping concentrations in the channel region for
SC- (a) and BC- (b) NMOS transistors.

simulate the buried N-P junction in the BC devices (Fig.1a), to derive the
surface doping level in the SC devices (Fig.1b) and to adjust the threshold
voltage of each device. In this case, 5E11 cm-2 As at 140 KeV and 6E11 cm-2 B
at 25 KeV were implanted for the BC and SC devices respectively.

RESULTS AND DISCUSSION

By comparing the threshold voltages measured on natural N+gate and P+gate
transistors respectively, we derive a work function difference of 1.15 to 1.2V
between the N+- and P+polysilicon gates, which is in agreement with reported
values /4/ . The threshold voltage of both SC and BC transistors is 0.8V as
measured at VD = 0.1V. Fig.2 presents subthreshold characteristics of each type
of device for W/L = 20/20 um and VD = 0.1V and 4V. First notice the low leakage
current in both types (below .1 pA/um). This confirms the good quality of both
the SOI material and its Si/SiO2 interfaces. Second, the subthreshold slopes of
the BC device remain almost similar at low and high VD (S = 75 mV/decade), whereas
a difference is visible for the SC transistor: S = 80 mV/decade at VD = .1V and
S = 54 mV/decade at VD = 4V. This can be explained by the "kink effect" observed
in Id-Vd characteristics of SC transistors /2/. Fig.3 shows output characteristics
of both SC and BC transistors. The kink is visible for the SC device and is
explained as follows: the holes created by impact ionization at the drain end
bias the neutral floating zone present below the depletion region under the
channel. This bias of the substrate lowers the threshold voltage. This induces

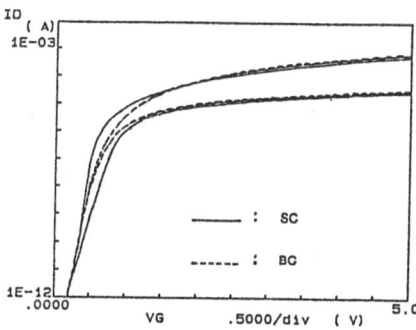

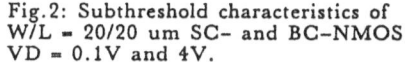

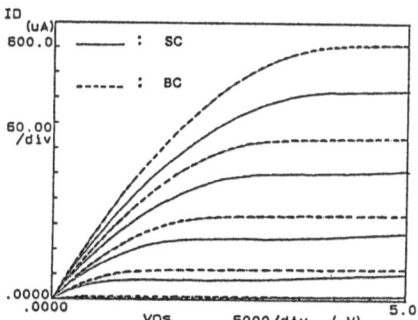

Fig.2: Subthreshold characteristics of
W/L = 20/20 um SC- and BC-NMOS .
VD = 0.1V and 4V.

Fig.3: Output characteristics of
W/L = 20/20 um SC- and BC-NMOS.
VG ranges from 0 to 5V by 1V steps.

an abrupt drain current increase at high Vd /2/ which is visible on the
subthreshold slope as indicated above. In the BC device, the depletion region on
the P-side of the buried junction extends down to the buried oxide. This avoids
any floating region below the channel and depletion regions. This is similar to
deep depletion in SC transistors made in very thin SOI films.

   Moreover, in Fig.3, it is worth noting the increase of the saturation
current of more than 15% in the BC transistor. This is attributed to the
increase of the mobility which reaches 20% in the low field values for BC
devices as compared to SC transistors: 500 cm2/V.s and 600 cm2/V.s for the SC
and BC devices respectively for a doping level of 6E16 cm-3 and 2.3E16 cm-3
respectively. This is the expected result of the reduction of the vertical electric
field along the N-channel. We wish to point out that this is obtained in a 200 nm
thick film, whereas similar results are expected for transistors made in films
thinned well below 100 nm /5/. Thinning below 100 nm presents lots of
technological problems in the film preparation. This leads to yield problems
related to the uniformity and the reproducibility of the film thickness. The
interface quality also has greater influence on the device behaviour when the film
is thinned far below 100 nm. We therefore believe that for yield and reliability
reasons, the SOI film thickness should be kept somewhere between 100 and 200 nm.
The device structure must then be adapted to that thickness and designed in
connection with the application in mind. If speed is to be considered, BC
transistors are good candidates owing their high drivability.

   The short channel behaviour of BC and SC transistors is illustrated in
Fig.4. Fig.4a and Fig.4b depict the subthreshold slope variation and threshold
voltage shift vs the effective channel length (Leff) at Vd = 3V. We see that,
as in bulk devices, buried channels are more prone to punchthrough than surface

884

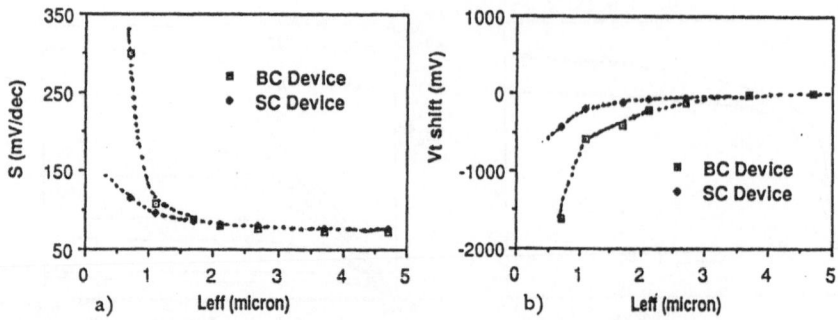

Fig.4: a) Subthreshold slope and b) threshold voltage shift of SC- and BC-NMOS as a function of effective channel length, measured at VD - 3V.

channels. This is a consequence of the presence in the BC devices, of a surface N-type region which has a low barrier height to electrons. The drain induced barrier lowering (DIBL) /6/ has therefore a major impact and leads to surface punchthrough. SC devices are more susceptible to bulk punchthrough which occurs at lower Leff. In SOI, it is expected that this effect can be limited or even avoided in very thin films /7/ since the buried oxide will limit or even avoid the extension of the space-charge regions at the drain/source ends. Obviously, surface punchthrough is hard to reduce by thinning the SOI film but, as in bulk, the devices might be optimized by adjusting the doping levels and the buried channel depth or also by an adapted drain-engineered structure.

CONCLUSIONS

We have shown that SOI films as thick as 200 nm can be used for VLSI circuit applications if a proper device structure was chosen. For instance, if speed is the issue, BC devices are good candidates, since they have a high drivability related to an increased carrier mobility. We have noticed, however, that some improvements should be provided in the short channel behaviour of these devices, in order to use them in deep submicron circuits.

REFERENCES

/1/ D.T.Amm, H.Mingam, P.Delpech, and T.Ternisien d'Ouville, IEEE Trans. Electron. Dev.,ED-36, 963 (1989).
/2/ M.Haond, D.Dutartre, D.Bensahel, A.Monroy-Aguirre, S.Thouret and D.Chapuis, Microelectr. Engineering, 8, 201 (1988).
/3/ TITAN, a process and device simulator. To be published in Proceedings of the Nasecode 89 short course.
/4/ G.J.Hu, and R.H.Bruce, IEEE Trans. Electron. Dev., ED-32, 584 (1985).
/5/ M.Yoshimi, H.Hazama, M.Takahashi, S.Kambayashi, H.Tango, Electron. Lett., 24, 1078 (1988).
/6/ R.Troutman, IEEE Trans. Electron Dev., ED-26, 461 (1979).
/7/ T.Sekigawa and Y.Hayashi, Solid-State Electron., 27, 827 (1984).

Explanation for the Negative Differential Resistance in SOI-MOSFETs

[*]L.J. MCDAID, [*]S. HALL, [*]P.H. MELLOR and [*]W. ECCLESTON, [+]J.C. ALDERMAN

[*]The University of Liverpool, Department of Electrical Engineering and Electronics, Brownlow Hill, P.O. Box 147, Liverpool, L69 3BX.
[+]Allen Clark Research Centre, Plessey Research (Caswell), Towcester, Northants, NN12 8EQ.

Abstract

The negative resistance region seen in the output characteristics ($I_{DS}(V_{GS})$ vs V_{DS}) of silicon-on-insulator MOSFETs is shown to be a result of the relatively poor thermal properties of the buried insulator compared with the Si-substrate. At sufficiently high power levels, the device island heats up causing degradation of the carrier mobility and a consequent limiting of the output current. This thermal self-limiting is also responsible for the departure from the ideal MOSFET "square" law which should be evident for these transistors having minimal "body" effect.

Introduction

MOS transistors fabricated on silicon-on-insulator (SOI) substrates offer attractive features for future VLSI circuits, particularly if this technology can be realised by a process without an epitaxial growth stage. Thin-film (<100nm) SOI-transistors already show superior performance over their thick-film counterparts. For instance, in thin film devices the kink effect is eliminated, short channel effects are suppressed and a higher transconductance is evident [1]. An effect which is common to both thick and thin film transistors is the negative differential resistance (NDR). The effect is seen for both p- and n-channel, short and long channel devices. Similar behaviour has been observed in power devices and was explained by a mobility reduction caused by local heating in the channel region [2]. This explanation will be shown from both theoretical analysis and experiment, to be applicable to SOI MOSFETs. In addition the deviation of experimental characteristics from the expected ideal MOSFET behaviour is explained by the model.

Theoretical

Figure 1 shows the temperature distribution in the buried insulator (oxide) and substrate obtained from a finite-element simulation using the commercially available package PE2D. The plot is for an annular transistor assuming a uniform temperature in the body (50°C) and an ideal heat sink at the back of the wafer (0°C). Because of axial symmetry and the previous assumption, it is only necessary to model the temperature profile of a radial plane from the centre of the film to the drain. The contours in fig.1. represent lines of constant temperature and the spacing between each contour is constant in temperature difference. The figure demonstrates that the majority of the temperature is dropped linearly across the buried oxide owing to its poor thermal conductivity. The heat diffusion equation can therefore be solved in one-dimension to relate power dissipated P_t (= $I_{DS} \cdot V_{DS}$) to the temperature increase in the channel relative to ambient, ΔT_c, giving

$$\Delta T_c = \frac{P_t t_{ob}}{K_{OX} A} \qquad (1)$$

Here t_{ob} is the buried oxide thickness, K_{OX} is its thermal conductivity and A is the area over which the heat is dissipated. The temperature dependence of mobility is taken from a model for power MOSFET devices[2] and incorporated into the ideal MOSFET equation to give

$$I_{DS} = \mu_o C_{of} \frac{W}{L} \frac{1}{(1+\Delta T_c/T)^m} [(V_{GS}-V_T)V_{DS} - \frac{V_{DS}^2}{2}] \qquad (2)$$

Here m is a constant in the range $1.5 < m < 1.8$, μ_o is the mobility at absolute temperature T and other symbols have their usual meaning. Figure 2 shows output characteristics predicted by equation 2 plotted as a function of gate length, L for a gate voltage of 6V. The current saturation or pinch-off point is seen to shift to lower drain voltages with reducing channel length: an effect observed experimentally. This shift is more pronounced in short channel devices due to the higher current drive and the smaller volume of silicon film (reflected in the area A in equation 1). The heating therefore induces a self-limiting effect on the output current.

Experimental

Transistors were fabricated on SIMOX substrates produced by implanting the equivalent of 1.8×10^{18} O+ ions cm^{-2} at 200keV and a subsequent anneal at $1405^{o}C$ for two hours. This was followed by standard SOI-CMOS processing to produce thin-film transistors with a range of gate lengths. Evidence for the thermal affect was provided by the following experiments.

Figure 3 shows the change in slope of the NDR as the external temperature was increased using a heated chuck. The slope dI_{DS}/dV_{DS} is seen to decrease as the absolute magnitude of drain current and therefore the power dissipated and associated temperature, decrease. Also the rate of change of mobility with temperature decreases with increasing temperature.

Figure 4 shows the output characteristic of an n- channel SOI-MOSFET with W = $40\mu m$, L = $4\mu m$ and a substrate voltage of -20V. The substrate bias causes the positive kink to reappear in the output characteristics. As the top gate voltage increases, the kink disappears to be replaced by the NDR. This is explained by the negative temperature coefficient of the multiplication process which gives rise to the kink. The device heats up and the kink is suppressed so that the NDR becomes dominant.

The strongest evidence for the temperature model is provided by thermal imaging and table 1 shows results for a $1.5\mu m \times 100\mu m$, annular transistor produced by porous Si (FIPOS) technology with a $1\mu m$ buried insulator. ΔT(theory) is obtained from equation (1) for an effective heat generating area A of 750 μm^2; K_{ox} is 1.4 W/m ^{o}C. These data represent a direct measurement of the device operating temperature. It can be seen that the temperatures are in excess of $70^{o}C$, assuming ambient room temperature operation, at nominal levels of power dissipation. The drain voltage dependence of the current is evident from the last entry in the table. Using the mobility model in equation (2) with a worst case of m=1.8, shows the mobility reduced by about one third (for Pt = 23 mW) which reflects the self-limiting drain current phenomenon. The NDR is apparent from the continuing reduction in mobility for higher drain voltages.

I_{DS}	V_{DS}	P_t	ΔT(expt)	ΔT(theory)	μ/μ_o
(mA)	(V)	(mW)	(^{o}C)	(^{o}C)	
9.2	2.5	23	17	22	0.34
9.8	3	29.4	55	28	0.24
9.9	3.5	34.7	32	33	0.18
10	4	40	37	38	0.16
9.9	4.5	44.6	49	43	0.11
9.9	5	49.5	55	47	0.10

Table 1: Summary of thermal imaging experiment showing measured (ΔT(expt)) and theoretical (ΔT(theory) = ΔT_c from equation 1) temperature rise for different power levels, P_t

Conclusions

Strong evidence has been presented for mobility degradation with temperature due to Joule heating in the channel region of SOI MOSFETs. The model explains the NDR region seen in the output characteristics and also the deviation from ideal 'square-law' MOSFET behaviour predicted by theory.

Acknowledgements

Thanks are owing to Peter Hemment and Karen Reeson of Surrey University for SIMOX substrates and to George Celler of AT & T Bell Labs. for the 1405°C anneals. This work has been partly supported by the procurement executive, Ministry of Defence (RSRE) under contract RP009/335.

References

[1] J.C. Sturm, "Performance advantages of submicron silicon-on-insulator devices for ULSI", Fall MRS meeting, Boston, December 1987.

[2] D.K. Sharma, J. Gautier and G. Merckel, "A high voltage circuit for driving liquid crystal displays", IEEE Journal of Solid-State Circuits, Vol.SC-13, No.3, pp.373-378, June 1987.

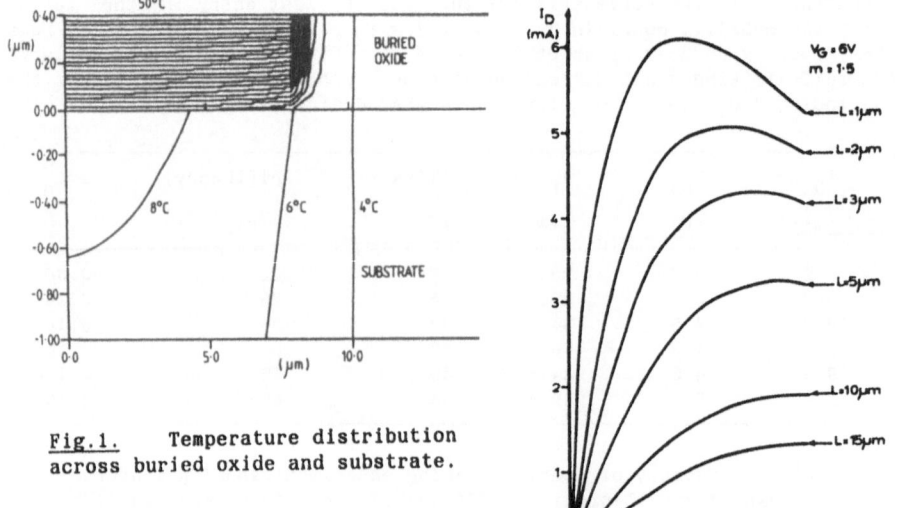

Fig.1.   Temperature distribution across buried oxide and substrate.

Fig.2.   Theoretical MOSFET output characteristics with channel length as a parameter, showing the shift in pinch-off point.

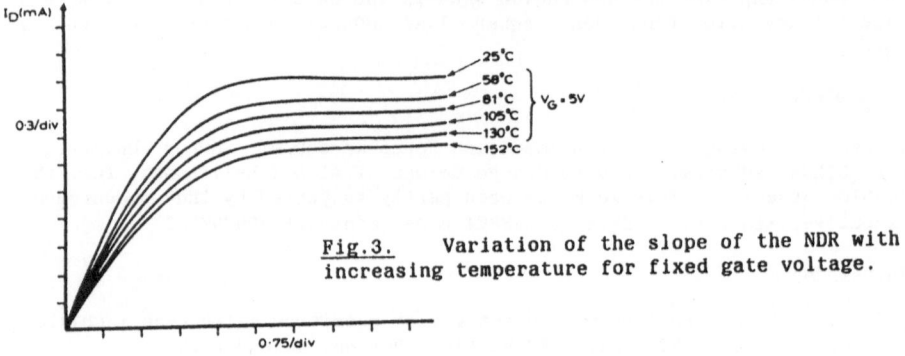

Fig.3.   Variation of the slope of the NDR with increasing temperature for fixed gate voltage.

Fig.4.   Suppression of positive kink by increasing power dissipation

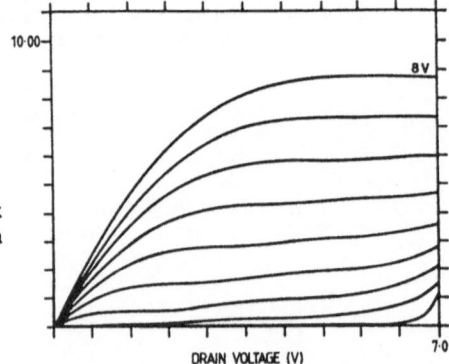

Modeling of Single and Double Gate Thin Film SOI MOSFETs

F. BALESTRA, G. GHIBAUDO, M. BENACHIR, J. BRINI

Laboratoire de Physique des Composants à semiconducteurs (UA-CNRS)
ENSERG/INPG, 23 av. des Martyrs, 38016 Grenoble (France)

Introduction

The understanding and the optimization of the electrical properties for Silicon-On-Insulator MOSFETs necessitate the development of reliable analytical models. Some models have been previously established in strong or weak inversion for a single gate control [1,2,3] but do not take into account the whole parameters of the SOI structure (for instance the third interface buried oxide/silicon substrate) or cannot represent analytically the subthreshold behaviour. In this paper, we propose analytical models for thin film SOI transistors controled by one (conventional operation) or two gates (Volume-Inversion type, which presents a number of interesting properties [4]). They give the electrical characteristics in ohmic operation, in particular the weak inversion slope and the threshold voltage. These models are based on the following approximations: for the single gate operation (classical MOSFET) the potential is supposed to vary linearly in the silicon film, and for the double gate operation (Volume-Inversion MOSFET, for which both front and back gates are biased simultaneously) the potential is supposed nearly constant. The numerical simulation (ISIS 1 program for SOI devices [5]) is compared with these analytical modellings, and justifies these approximations for transistors fabricated on sufficiently thin film.

1. Analytical modelling of single gate SOI MOSFET

a) Two interfaces case:

Figure 1 shows the electrostatic potential inside the silicon film for a transistor controled by a single gate. We can observe that for a thin silicon layer ($\leq$ 200 nm in this case) the potential presents a quasi-linear variation in the major part of the film. By using this approximation, analytical models for the subthreshold swing and the threshold voltage are obtained. The model of the swing takes into account two cases:

i) (rigorous analysis) : $S = Ln(10).dV_{g1}/d(LnI_d) = Ln(10).dV_{g1}/d\phi_{s1}.d\phi_{s1}/dQ_i.Q_i$, where I_d is the drain current, V_{g1} the applied gate voltage, Q_i the channel inversion charge, and ϕ_{s1} the front surface potential;

ii) (approximated formula) : we suppose that $d\phi_{s1}/dQ_i.Q_i \simeq kT/q$, as it is the case in bulk MOSFETs, thus giving : $S_o = Ln(10).kT/q.dV_{g1}/d\phi_{s1}$, kT/q ($=1/\beta$) being the thermal potential, which only represents the ratio of the various capacitances of the SOI structure given by the resolution of the charge conservation equation (for the ISI (Insulator-Semiconductor-Insulator) case):

$$\frac{d\phi_{s1}}{dV_{g1}} = \frac{C_{ox1}(C_{Si} + C_{ss2} + C_{ox2})}{C_{Si}(C_{ss1} + C_{ox1}) + (C_{ss2} + C_{ox2})(C_{ox1} + C_{ss1} + C_{Si})} \qquad (1)$$

where C_{ox1} and C_{ox2} are the front and back gate oxide capacitances, C_{ss1} and C_{ss2} the front and back interface state capacitances, and C_{Si} ($= \epsilon_s/t_{Si}$, ϵ_s is the silicon permittivity and t_{Si} the Si film thickness) the silicon film capacitance (the depletion capacitance is equal to zero for a fully depleted Si film).

After calculation, the rigorous subthreshold swing S can be evaluated as:

$$\frac{1}{S} = \frac{1}{Ln(10)}\frac{kTn_o}{Q_i}\left[\frac{AC_{ox1}}{\epsilon_s} + \frac{d\phi_{s1}}{dV_{g1}}\left[-\frac{A}{\epsilon_s}(C_{ox1} + C_{ss1}) + B\right]\right] \qquad (2)$$

where $n_o = n_i^2/N_a$ (n_i being the intrinsic Si charge density and N_a the Si film doping), Q_i, A and B are found as:

$$Q_i = -\frac{kTn_o}{E_s} e^{\beta\phi_{s1}} \left[\exp(-\beta E_s t_{Si}) - 1 \right] \tag{3}$$

$$A = \frac{\exp(\beta\phi_{s1})}{E_s^2} \left[\exp(-\beta E_s t_{si}) - 1 \right] + \frac{\beta\, t_{Si}\exp(\beta\phi_{s1})}{E_s} \exp(-\beta E_s t_{Si}) \tag{4}$$

$$B = \frac{\beta\,\exp(\beta\phi_{s1})}{E_s} (-\exp(-\beta E_s t_{Si}) + 1) \tag{5}$$

The electric field is given by:

$$E_s = \frac{qN_a t_{Si} - C_{ox2}(V_{g2} - \phi_{MS2} - \phi_{s1}) - Q_{f2} + C_{ss2}\phi_{s1}}{\varepsilon_s + C_{ox2}t_{Si} + C_{ss2}t_{Si}} \tag{6}$$

The average value of the front surface potential in the weak inversion regime can be taken as $\phi_{s1} = 1.5\,\phi_F$, ϕ_F being the Fermi potential. The threshold voltage can be calculated as (for $\phi_{s1} = 2\,\phi_F$):

$$V_t = V_{FB1} + \frac{2\phi_F}{C_{Si} + C_{ox2} + C_{ss2}} \left[C_{Si} + C_{ox2} + \frac{C_{ox2}}{C_{ox1}}(C_{Si} + C_{ss1}) + \frac{C_{ss1}C_{Si}}{C_{ox1}} + \frac{C_{ss2}}{C_{ox1}}(C_{ox1} + C_{ss1} + C_{Si}) \right]$$

$$\dots + \frac{q\,N_a t_{Si}C_{Si}}{C_{Si} + C_{ox2} + C_{ss2}} - \frac{C_{ox2}C_{Si}}{C_{ox1}(C_{Si} + C_{ox2} + C_{ss2})}(V_{g2} - V_{FB2}) \tag{7}$$

where V_{FB1} and V_{FB2} are the front and back gate flat band voltages, V_{g2} is the back gate voltage.

In figure 2 are plotted the results obtained by using the analytical models and the numerical simulation for a thick back insulator. The subthreshold swing decreases when the Si film thickness is lowered (for a low back interface state density) both for the numerical simulation and the rigorous analytical model (given by S), which are in good agreement. Nevertheless, by using the approximate formula (given by S_o), the swing increases slightly for thinner film, which points out the difference between the two analytical models. Fig. 3 represents the variation of the subthreshold swing in the case of a thin back insulator. In this situation, an increase of the swing is obtained by decreasing the Si film thickness both for the numerical simulation and analytical models (S agrees very well with the numerical simulation, and S_o within 10%). Therefore, these simulations put forward the importance of the back oxide thickness, and emphasize that enhanced performances can be obtained for Ultra Thin Film transistors but only with a sufficiently thick back insulator. The importance of the substrate capacitance will be emphasized after. It is worth noting that, for a fully depleted film, S_o is independent of the Si doping and S depends slightly on the doping, which points out the validity of such models even for high doping. On the other hand, the threshold voltage is constant in the case of a thick back insulator, and, in the case of a thin back insulator both the numerical simulations and the analytical model show an increase of the threshold voltage when the Si film thickness is lowered.

b) Three interfaces case:

For the ISIS (Insulator-Semiconductor-Insulator-Semiconductor) structure the variation of the surface potential as a function of the gate voltage becomes:

$$\frac{d\phi_{s1}}{dV_{g1}} = \frac{C_{ox1}[(C_{Si} + C_{ss2})\,C + C_{ox2}(C_{sub} + C_{ss3})]}{(C_{ox1} + C_{ss1})[(C_{Si} + C_{ss2})C + C_{ox2}(C_{sub} + C_{ss3})] + C_{Si}[C_{ss2}C + C_{ox2}(C_{sub} + C_{ss3})]} \tag{8}$$

where $C = C_{ox2} + C_{ss3} + C_{sub}$, C_{ss3} is the interface state capacitance at the buried oxide/ Si substrate interface and C_{sub} is the substrate capacitance. This quantity depends on the ratio of the various capacitances of the SOI structure, which can be represented by the equivalent circuit of Fig. 4. The subthreshold swing S and S_o can then be obtained with the previous eqs. (2-5) but with the following Si film electric field:

$$E_s = 1/(\epsilon_s + C_{ss2}t_{Si}) \cdot (-Q_{f2} - Q_{f3} + C_{ss2}\phi_{s1} + C_{ss3}\phi_{sub} + qN_a t_{Si} - Q_{sub}) \qquad (9)$$

where Q_{f2} and Q_{f3} are the fixed charge at the Si film/buried oxide and buried oxide/Si substrate interfaces respectively, Q_{sub} is the substrate charge and ϕ_{sub} is the surface potential at the buried oxide/Si substrate interface. On the other hand, the threshold voltage in the ISIS case becomes:

$$V_t = V_{FB1} + 2\phi_F \ D \left[1 + \frac{C_{ss1}}{C_{ox1}} + \left[\frac{C_{ss2}}{C_{ox1}} + \frac{C_{ss3}}{C_{ox1}}(1 + \frac{C_{ss2}}{C_{ox2}}) \right] \right] \left[1 + \frac{C_{ss1}}{C_{Si}} + \frac{C_{ox1}}{C_{Si}} \right] + \frac{C_{ss3}}{C_{ox2}}(1 + \frac{C_{ss1}}{C_{ox1}}) \right]$$

$$\dots + D \left[\frac{qN_a t_{Si}}{C_{ox1}} - \frac{Q_{sub}}{C_{ox1}} - \frac{Q_{f2}}{C_{ox1}}(1 + \frac{C_{ss3}}{C_{ox2}}) - \frac{Q_{f3}}{C_{ox1}} \right] \qquad (10)$$

With D being given by: $D = C_{Si}C_{ox2}/(C_{ox2}(C_{Si} + C_{ss2} + C_{ss3}) + C_{ss3}(C_{ss2} + C_{Si})) \qquad (11)$

ϕ_{sub}, and therefore Q_{sub} and C_{sub}, can be calculated by considering that the Si substrate is depleted, which is often the case. Therefore, S_o, S and V_t can be obtained analytically for the ISIS case by using the previous equations. Nevertheless, when ϕ_{sub} approaches zero the substrate tends to the accumulation regime and this treatment is no more valid. However, it can be shown that in this case both V_t and S for the three interfaces model tend to the previous equations developped for the two interfaces model. The equivalent circuit of the ISIS case (Fig. 4) becomes that of the ISI one by taking a very high C_{sub}.

For a thick buried oxide, similar variations for the ISI and ISIS cases are obtained, whatever the substrate regime. For a thin buried oxide, when the substrate is depleted ($N_{sub} = 10^{15}$ cm^{-3}), a strongly different behaviour is obtained compared with the ISI case. Indeed, we obtain a decrease of the swing by decreasing the Si film thickness like in Fig. 2. Nevertheless, when the substrate is accumulated (for $N_{sub} = 10^{17}$ cm^{-3}), a similar variation is obtained for the subthreshold swing than in the two interfaces case.

On the other hand, when the buried oxide thickness decreases, the swing is degraded whatever the substrate regime. However, it is worth noting that for a depleted substrate, the swing is always better than for an accumulated one. The same behaviours are obtained for the threshold voltage.

2. Analytical modelling of double gate SOI MOSFET

For a double gate control, two limiting situations are considered : the Ultra Thin Film (UTF) and the Double MOS (DM) cases. Figure 5 is a plot of the variation of the electrostatic potential inside the silicon layer. For a sufficiently thin Si film, a quasi-constant potential is obtained (UTF approximation). On the other hand, for a thick Si film the device can be trivially represented by the sum of two conventional MOSFETs (DM case). By using these considerations, analytical models are obtained for the MOSFET ohmic operation:

$$S = q/kT \cdot 2C_{ox}/(2C_{ox} + 2C_{ss}), \quad \text{and} \quad V_t = 2\phi_F + V_{FB} + qt_{Si}N_a/2C_{ox} + Q_{ss}/C_{ox} \qquad (12)$$

for a symmetric case with $C_{ox1} = C_{ox2} = C_{ox}$, $C_{ss1} = C_{ss2} = C_{ss}$ and $Q_{ss1} = Q_{ss2} = Q_{ss}$.

For instance, in Fig. 6 we show the transfer characteristics $I_d(V_g)$ obtained for the UTF and DM models compared with the numerical simulation. This figure emphasizes particularly the validity of the UTF approximation for Si film lower than 100 nm. The effect of a dissymmetry between front and buried oxides can also be taken into account.

In conclusion, satisfactory analytical models have been developped for single and double gates SOI MOS-FETs operated in ohmic region, which enables the electrical properties and the performances of such devices to be obtained.

References
1 - H.K. Lim and J.G. Fossum, IEEE Trans. Electron Dev., ED-30, p. 1244 (1983)
2. K.K. Young, IEEE Trans. Electron Dev., ED-36, p. 504 (1989)
3. E.R. Worley, Solid-St. Electron., 23, p. 1107 (1980)
4. F. Balestra, S. Cristoloveanu, M. Benachir, J. Brini, T. Elewa, IEEE Electr. Dev. Let. 8, p.410 (1987)
5. F. Balestra, J. Brini, P. Gentil, Solid-St. Electron., 28, p. 1031 (1985)

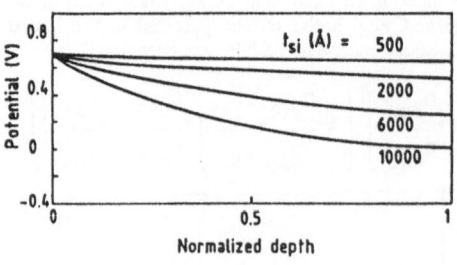

Fig. 1 : Electrostatic potential variations with normalized silicon film thickness (single gate: $V_{g1}=1V$, $V_{g2}=0$, $t_{ox1}=27nm$, $t_{ox2}=350nm$, $N_a=10^{15}cm^{-3}$)

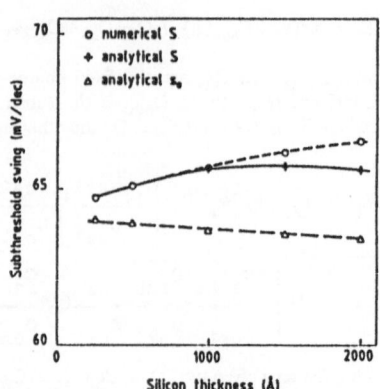

Fig. 2 : Variations of the subthreshold swing as a function of silicon film thickness; $t_{ox1}=27$ nm, $t_{ox2}=350$ nm, $N_a=10^{15}$ cm^{-3} (single gate)

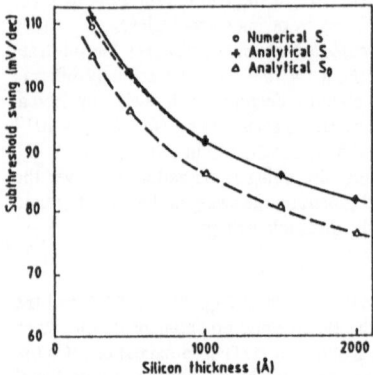

Fig. 3 : Subthreshold swing vs silicon film thickness; $t_{ox1}=t_{ox2}=27$ nm, $N_a=10^{15}$ cm^{-3} (single gate)

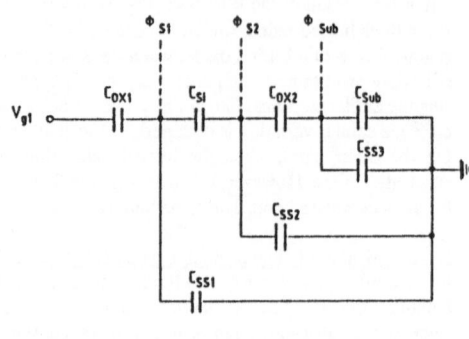

Fig. 4 : Equivalent circuit of the SOI MOSFET in the three interfaces case

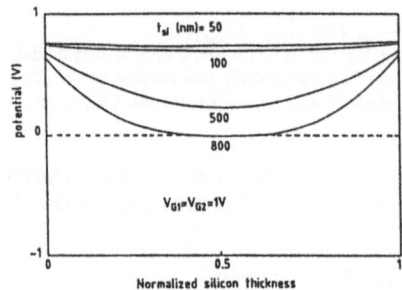

Fig. 5 : Electrostatic potential variations with normalized silicon film thickness; $V_{g1}=V_{g2}=1V$, $t_{ox1}=t_{ox2}=27$ nm, $N_a=10^{16}$ cm^{-3} (double gate)

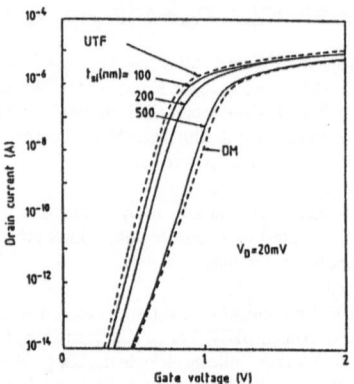

Fig. 6 : Drain current versus gate voltage; dashed lines : analytical models, full lines : numerical simulations (double gate)

Gate Oxide Breakdown in a SOI CMOS Process Using MESA Isolation

M. HAOND, O. LE NEEL(*), G. MASCARIN and J.P. GONCHOND
CNET, BP98 38243 Meylan Cedex, France

ABSTRACT

We have studied the gate oxide breakdown behaviour in a CMOS process on SOI substrates, where the lateral isolation is obtained by the definition of individual islands (mesa) by using Reactive Ion Etching. Multiedge and edgeless transistors are compared to study the corner effects, such as early breakdown of the gate oxides. The breakdown voltage of the gate oxide at the corners has been improved by using an adequate local oxidation procedure of the edges of the mesa, resulting in a softening of the upper corners of the mesa.

INTRODUCTION

Thanks to its intrinsic isolation property, SOI material has many advantages over bulk silicon for the fabrication of CMOS devices in a VLSI/ULSI technology. It allows an efficient solution to the problems of lateral isolation and latch-up encountered in bulk-Si devices when the dimensions are shrinked. Within a CMOS process on Silicon On Insulator (SOI) substrates, the fabrication of insulated islands (mesa) by Reactive Ion Etching (RIE) the SOI film should allow a higher density of integration and less narrow channel effect than a LOCOS process. This leads however to early breakdown of the gate oxide at the edges of the mesas /1/. It has been shown /2/ that oxidation performed below or close to the SiO2 viscous flow temperature (965°C) results in a stress-induced thinning of the oxide at the upper corners of etched silicon steps. This is namely the case for gate oxides which are generally grown in the 900–980°C range. We present the results of our investigations on the gate oxide breakdown voltage when a mesa isolation technique is used, and a procedure to improve it by a specific local oxidation of the edges of the mesa.

EXPERIMENTAL

The SOI substrates were prepared by the lamp-Zone Melting Recrystallisation technique /3/: a deposited polysilicon film on oxidised silicon substrates was recrystallised by scanning a molten zone across. After recrystallisation, the silicon film was thinned by oxidation down to a thickness ranging from 250 nm

to 300 nm. The N- and P-wells were then implanted separately. The lateral
isolation (mesa formation) was performed by etching off the field silicon regions
in a RIE reactor, using a Si3N4/SiO2 mask. The run was then split in two parts:
on some wafers, a local sidewall 100nm oxide was grown in a wet atmosphere,
before mask removal. The mask was then removed from all wafers, and a 25 nm
dry gate oxide was grown at 950°C. The following steps correspond to a classical
2 um N+-Polysilicon-gate single-metal (aluminum) process. Fig.1 presents SEM

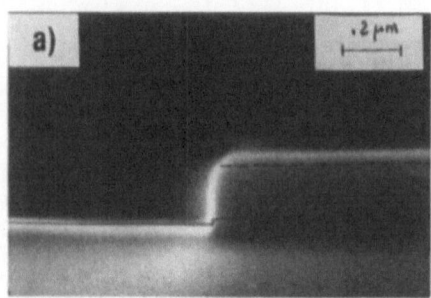 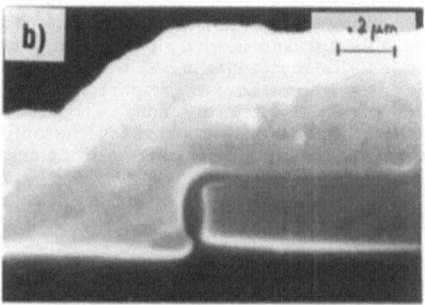

Fig.1: a) Mesa structure after silicon RIE. The Si3N4/SiO2 was not removed.
b) Same as a) but after an oxidation step at 1000°C. A poly-Si layer was added
to better reveal the shape of the structure.

micrographs of the mesa structure before (Fig.1a) and after (Fig.1b) the local
oxidation at 1000°C. In Fig.1, the Si3N4/SiO2 mask has not been removed. In
Fig.1a, one can notice that some overetch occured in the underlying SiO2, which
is also visible in Fig.1b. In Fig.1b, a polysilicon layer has been added over
the structure to reveal the shape of the oxidised mesa structure. It is clearly
visible in Fig.1b, that the bird's beak resulting from the local oxidation has
cut the upper corner angle: there is no reentrant angle as discussed in Ref.2.
We have tried different oxidation temperatures: 920°C, 950°C and 1000°C, but we
have noticed only slight changes: a better rounding of the upper corner occurs
and the extension of the bird's beak both at the upper and lower corners is
shorter at higher temperatures. Notice, however that the bird's beak at the upper
corner disapppears during the Si3N4/SiO2 mask removal, but not the sidewall oxide.

RESULTS AND DISCUSSION

    In order to study the electrical effects of the corner, we have used two
types of transistors: an edgetype transistor made of an array of 5 parallel
transistors (W/L=20/5) and a circular edgeless transistor of the same size

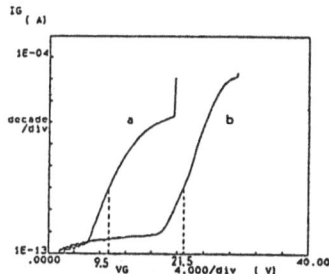

Fig.2: Injection current characteristics in edgetype (a) and edgeless (b)
transistors of the same size: W/L=100/5.

(W/L=100/5). The gate current is recorded vs the gate voltage, drain and source
shorted to 0V. The breakdown voltage is defined,here, as the inititiation of high
injection (Fig.2).Fig.2a shows the measurements performed on an edgetype
transistor from a wafer where the mesa has not been oxidised. The breakdown
voltage is 9.5V, whereas it is 21.5V for an edgeless transistor of the same
wafer (Fig.2b). This gives a strong evidence of the dramatic effect of the corners
of the mesa. One can also notice that the deleterious oxide breakdown of the
edgeless transistor occurs at 29V, i.e. for a field of more than 11 MV/cm which
is definitely as good as that measured in bulk devices for a 25 nm gate oxide.
Fig.3 presents 2 histograms of the breakdown voltage measured on edgetype
transistors made in non oxidised and reoxidised mesa structures. Notice the

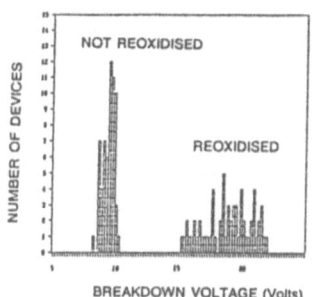

Fig.3: Histograms of edgetype transistors (W/L=100/5) made in non oxidised and
reoxidised mesa.

drastic improvement in the breakdown voltage behaviour: it has increased by 8 Volts.
The breakdown field (high injection field) has shifted from 4 MV/cm to 8 MV/cm.
It shows that no (or a reduced) thinning of the gate oxide occured at the upper
corners of the mesa after the sidewall oxidation. This oxidation has cut and
increased the angle of the upper corners of the mesa. This has reduced or avoided

any possible compressive stress that is supposed to arise at step corners during an oxidation made at a temperature lower than the oxide viscous flow temperature of 965°C /2/. This is namely the case of the gate oxide which is grown at 950°C. We believe that this happens in the non reoxidised wafers and leads to a thinning of the gate oxide at the edges with the associated lower breakdown voltage.

By comparing Fig.2b and Fig.3, on can notice that the breakdown field of the edgetype transistors of reoxidised islands (Fig.3b) is still lower than that of edgeless devices (either on oxidised or non oxidised structures). We have attributed this to the formation of thin oxinitride filaments (Kooi effect/4/) during the local oxidation step with the presence of the Si3N4/SiO2 mask. This leads to a local thinning of the gate oxide as in bulk devices /4/. Further improvements in the above described technique should be provided by the suppression of the oxinitride filaments.

CONCLUSION

We have demonstrated that the use of an adapted sidewall oxidation in a mesa isolation SOI CMOS process would result in a drastic improvement of the gate oxide breakdown voltage. This is obtained by a softening of the upper corners of the mesa structure which avoids the thinning of the gate oxide. This allows the use of mesa isolation in future VLSI CMOS processes on SOI.

ACKNOWLEDGEMENTS

We wish to thank D.Dutartre (CNET) and J-C.Rival(••)for the provision of the SOI wafers.

REFERENCES

/1/ R.K.Smeltzer and C.W.Benyon, Jr., RCA Review, 45, 196 (1984).
/2/ R.B.Marcus and T.T.Sheng, J. of Electrochem. Soc., 129, 1278 (1982).
/3/ M.Haond, D.Dutartre, D.Bensahel, A.Monroy-Aguirre, S.Thouret and D.Chapuis, Microelectr. Engineering, 8, 201 (1988).
/4/ E.Kooi, J.G. Van Lierop and J.A. Appels, J. Electrochem. Soc., ul1;123, 1117 (1976).

(•) from Matra-MHS Semiconducteurs, France.
(••) Micropolish, Z.I. des Pradeaux , 13850 Gréasque, France.

Analytical Modelling of the Kink Effect in MOS Transistors

I.M. Hafez, G. Ghibaudo and F. Balestra
Laboratoire de Physique des Composants a Semiconducteurs,
ENSERG, 23 rue des martyrs, B.P. 257, 38016 Grenoble, France.

Abstract
An analytical model of the kink effect in MOS transistors is proposed. This model procures a comprehensive view of the kink effect in bulk silicon MOSFETs and, subsequently, in SOI devices. It is shown both experimentally and theoretically that the excess drain current induced by the kink effect is proportional to the body transconductance of the device operated at room as well as liquid helium temperatures.

1. INTRODUCTION
The analysis of the kink effect in MOS devices has been the subject of much research [1–3]. Despite of some attempts for SOS (Silicon On Sapphire) devices [4,5], no analytical approach of the kink effect has already been developed. Only two dimensional numerical simulations have been recently proposed [1–2].

In this work, a simple analytical model for the kink effect in a MOS transistor is presented. This model which provides a quantitative account for the kink effect, is not only suitable for MOS devices made on Silicon On Insulator (SOI) operating at room temperature but also to bulk MOSFETs operated at very low temperature.

2. MODEL AND ANALYSIS
The appearance of an excess of drain current in the saturation region of MOSFET/SOI (the so called kink effect) arises from the threshold voltage shift due to the forward biasing of the source–substrate diode caused by the substrate impact ionization current flowing from drain to source [4,5]. A similar situation occurs in bulk silicon MOSFETs operating at very low temperature [6] or when disconnecting the substrate electrode at room temperature [7].

The excess drain current due to the parasitic substrate biasing is, at first order, proportional to the shift in the internal substrate bias V_b near the source and to the body transconductance, $g_b = dI_d/dV_b$, such as :

$$\Delta I_d = g_b \, V_b \qquad (1).$$

It is easy to show from the gate charge conservation equation that the body transconductance is linked to the device transconductance, $g_m = dI_d/dV_g$, as :

$$g_b = (C_d/C_{ox}) \, g_m \qquad (2),$$

where C_d and C_{ox} are the depletion and gate oxide capacitances respectively.

The shift in the internal substrate bias V_b is related to the substrate current I_{sub} passing into the forward biased source–substrate junction. At room temperature, the internal substrate voltage is given by :

$$V_b = (kT/q) \, Log(I_{sub}/I_r + 1) \qquad (3),$$

where I_r is the reverse junction saturation current, kT/q being the thermal voltage. The subtrate current I_{sub} generated by the avalanche multiplication near the drain can be empirically related to the drain voltage V_d as [8] :

$$I_{sub} = A\ I_d\ exp[-\ B/(V_d - V_{dsat})]$$ (4),

where V_{dsat} is the drain saturation voltage and A and B are empirical constants.

At liquid helium temperature, no simple relation is available for the p—n junction current [9]. However, the I(V) chracteristics present, in this case, a very sharp transition at a turn—on voltage nearly equal to the built in potential V_{bi} [9] and will be empirically represented hereafter by an expression of the form :

$$I_{sub} = I_0\ Arg\ tanh\ (V_b/V_{bi})$$ (5),

where I_0 is a critical current. At T=4.2K, the built in potential of the junction is nearly equal to the band gap i.e. $V_{bi} \simeq 1.17V$.

3. RESULTS AND DISCUSSION

The previous model has been tested on n channel MOS transistors operated at room and liquid helium temperatures. We have used different MOS devices fabricated at the LETI (Grenoble) with the parameters : channel width to length ratio W/L=20/2, oxide thickness t_{ox}=26nm, subatrate doping $N_a = 10^{16}/cm^3$ for room temperature, and, W/L=100/100 or W/L=50/10, t_{ox}=120nm, $N_a = 10^{15}/cm^3$ for liquid helium temperature.

Fig. 1 (a) shows how a typical kink effect can be obtained on standard bulk MOSFETs operating at room temperature when the substrate electrode is disgrounded. Fig. 1 (b) displays typical $I_d(V_d)$ characteristics obtained on bulk MOSFETs operating at liquid helium temperature. In this case, the bulk is so resistive because of the impurity freeze—out that the substrate is practically at a floating potential [6].

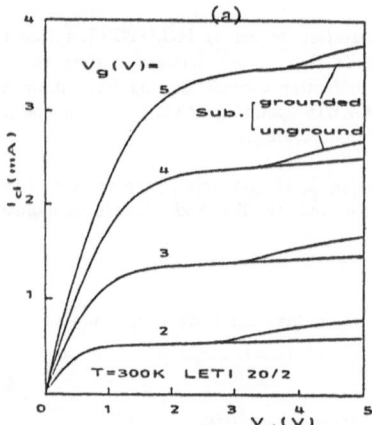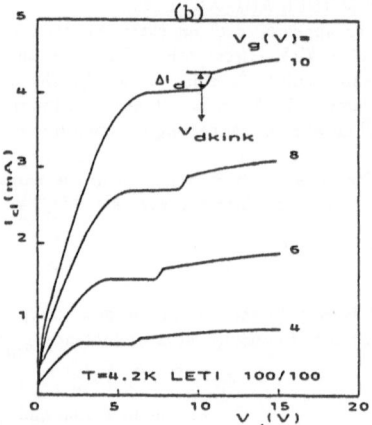

<u>Fig.1</u> : Typical $I_d(V_d)$ characteristics for a bulk MOSFET at T=300K (a) and T=4.2K (b).

Figs. 2 (a) and (b) give the corresponding variations of the excess drain current ΔI_d (defined in

Fig.1) with gate voltage V_g. On the same figures are also reported the body transconductance characteristics. As predicted from Rel. 1, an excellent correlation between the excess drain current ΔI_d (due to the kink effect) and g_b is observed both for room and liquid helium temperatures. Moreover, note that the coefficient of proportionality between ΔI_d and g_b which is about 0.8 at room temperature and 1.1 at liquid helium temperature compares favorably to the corresponding diode built in potential.

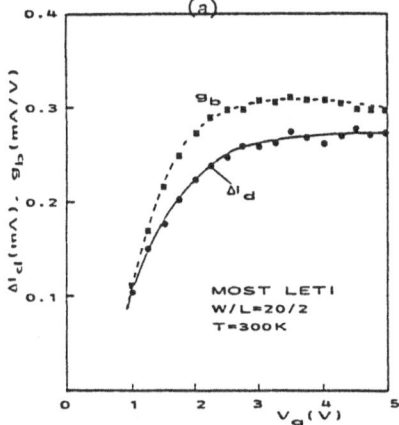 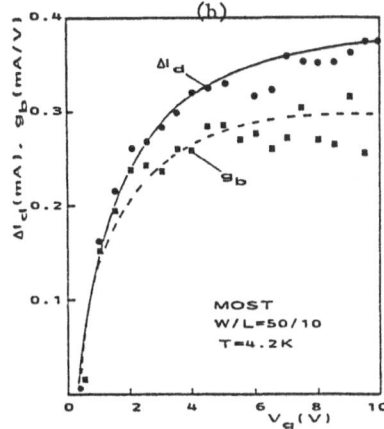

<u>Fig.2</u> : Excess drain current ΔI_d and body transconductance g_b vs gate voltage at T=300K (a) and T=4.2K (b).

Fig. 3 shows the variations of the drain voltage position V_{dkink} (defined in Fig.1) at which the kink effect starts to appear, as a function of gate voltage for room and liquid helium temperatures. In both cases, V_{dkink} has practically a linear dependence with gate voltage, which has been checked to be well correlated to the turn—on drain voltage of the substrate current.

<u>Fig.3</u> : Variation of V_{dkink} with gate voltage V_g for room and liquid helium temperatures.

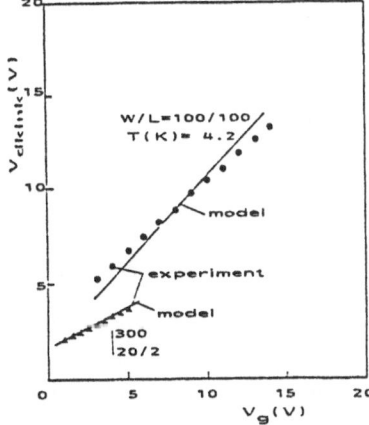

Fig. 4 displays a global simulation of the output characteristics of Fig. 1 obtained using the previous analysis and specific models for non ohmic operation at room [10] and liquid helium [11]

temperatures. Such simulations were obtained by solving proper strong inversion drain current equations self—consistently with the threshold voltage shift induced by the internal substrate bias :
$$\Delta V_t = -C_d / C_{ox} \cdot V_b.$$

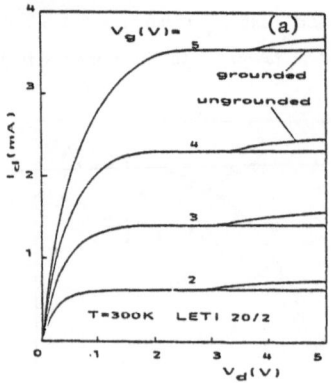
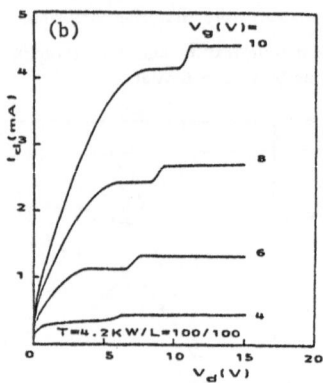

<u>Fig.</u>4 : Simulated $I_d(V_d)$ characteristics corresponding to the device of Fig.1 (Fitting parameters :

maximum mobility $\mu_0 = 800 \text{cm}^2/\text{Vs}$, $V_t = 0.6\text{V}$, attenuation mobility factor $\theta = 0.05/\text{V}$, A=200,

B=39V, $I_r = 10^{-12}\text{A}$ for room temperature and maximum mobility $\mu_m = 14000 \text{cm}^2/\text{Vs}$, $V_t = 0.3\text{V}$,

attenuation mobility factor $\theta_{lt} = 0.12/\text{V}$, A=2x10^{-4}, B=5.25V, $I_0 = 10^{-8}\text{A}$, $V_{bi} = 1.1\text{V}$ for liquid

helium temperature).

4. CONCLUSION

An analytical model for the kink effect in MOS transistors, which procures a comprehensive view of the kink effect in bulk silicon MOSFETs and, subsequently, in SOI devices has been presented. This model enables a quantitative description of the excess drain current both for room and liquid helium temperatures. In particular, we have shown that the kink effect in a MOS transistor can be primarily attributed to a body effect and not solely to the generation and recombination phenomena [1].

REFERENCES
1. K. Kato, T. Wada, K. Tanigushi, IEEE Trans Electron Devices, vol ED—32, p. 458, 1985.
2. J. P. Colinge, IEEE Electron Device Letters, vol EDL—9, p. 97, 1988.
3. B. Dierickx, L. Warderman, E. Simoen, J. Vermeiren, C. Claeys, IEEE Trans Elec. Dev., vol ED—35, p. 1120, 1988.
4. J. Tihanyi, H. Schlotterer, Sol State Electron, vol 18, p. 305, 1975.
5. C. Ipri, in Silicon Integrated Circuits, Edited by D. Kahng (Academic Press, New York, 1981) p. 253.
6. F. Balestra, L. Audaire, C. Lucas, Sol State Electron, vol 30, p. 321, 1987.
7. M. Nakahara, H. Iwasawa, K. Yasutake, Proc. IEEE, p. 2088, nov 1968.
8. T.Y. Chan, P.K. Ko, IEEE Electron Device Letters, vol EDL—5, p. 505, 1984.
9. A. Jonscher, Brit J. Appl. Phys., vol 12, p. 363, 1961.
10. G. Ghibaudo, Sol State Electron, vol 32, p. 87, 1989.
11. I.M. Hafez, F. Balestra, G. Ghibaudo, Sol State Electron, in press (1989).
12. J. Brini, M. Benachir, G. Ghibaudo, F. Balestra, NASECODE Conference, Dublin, Ireland (july 1989).

Memories

Electrical Characterization of a Submicron Titanium Silicide Local Interconnect Technology

M. G. Pitt, A. G. M. Jonkers, H. G. Pomp and R. de Werdt

Philips Research Laboratories,
P.O. Box 80000, 5600JA Eindhoven,
The Netherlands

Summary

In advanced CMOS technologies the use of a local interconnect technology under the intermediate oxide provides a means of increasing circuit packing density at the cost of only one additional mask layer. A scheme involving the creation of titanium silicide interconnect has been incorporated in a submicron CMOS technology used for the production of 1 Mbit static RAMs. This paper describes the full electrical characterisation of this local interconnect technology as it is used in this process. It is shown that the titanium/amorphous silicon thickness ratio must be less than 0.38 if silicon suckout from the active area regions is to be avoided. Suckout of silicon may result in increased junction leakage dependent on the silicide strap layout. An increase in the p-channel transistor series resistance may also occur.

1 INTRODUCTION

The use of local interconnection layers at the transistor level in submicron integrated circuits is advantageous as a means of increasing packing density. Such a scheme involves the definition of a conducting layer after source/drain implantation and before deposition of an intermediate oxide at the cost of only one additional mask layer. It provides a means of creating buried contacts between active area and polysilicon and of allowing conventional contacts to overlap the edges of field oxide regions. Use of a $TiSi_2$ local interconnect technology provides an attractive means of realising such a scheme [1] as it can be readily integrated into a process requiring the use of a self-aligned silicide for polysilicon and active area regions. The implementation of such a process scheme in a CMOS technology is shown schematically in figure 1. After source/drain implantation thin layers of titanium and α-silicon are sequentially sputtered in one deposition cycle. The α-silicon is patterned through the application of a lithographic mask and etched using a fluorine plasma. The wafers are then given a 675°C, 30s rapid thermal anneal in nitrogen in order to react the titanium with silicon to form $TiSi_2$ in the C49 phase [2]. This will occur in active areas, on polysilicon and also over oxide where there was α-silicon present. Unreacted titanium is then removed using NH_3/H_2O_2 solution and the wafers given a second rapid thermal anneal at 875°C for 30s in order to convert the silicide to the lower resistivity C54 phase. Figure 2 shows a transmission electron micrograph of a cross-section through a strap connecting active area to polysilicon.

This local interconnect (strap) technology has been implemented in a 0.7μm CMOS process for the production of 256k and 1Mbit static RAMs [3]. Figure 3 shows the implementation of local interconnect in a memory cell.

2 RESULTS

Electrical characterisation was primarily carried out through parametric testing of purpose built test structures using a Keithley 350 test system. Using the silicidation process described above a sheet resistance of 2.3 $\Omega/\square$ was obtained for the titanium silicide strap on field oxide, as compared to sheet resistances of active area (without strap) of 3.6 $\Omega/\square$ for n^+ regions and 2.4$\Omega/\square$ for p^+. The polysilicon sheet resistance without strap was 3.9$\Omega/\square$. The sheet resistance of the strap is as expected from complete reaction of the titanium and is similar to that obtained on p^+ active area regions. The sheet resistances obtained on the n^+ doped polysilicon and n^+ active area regions were higher due to a retardation of the silicidation reaction by phosphorus and arsenic [4].

2.1 Strap continuity/isolation

Figure 4 shows the yield of strap/active area strings as a function of overlap and active area width. For $1.2\mu m$ wide active areas close to 100% yield was obtained for a minmum overlap of $0.6\mu m$. As the active area width was reduced to $0.6\mu m$ the minimum overlap was increased to $0.8\mu m$ due to shortening of the active areas by lithographic proximity effects and from an increase in the extended birds beak which occurs at the ends of narrow active area regions. This dependence of minimum overlap on active area width is incorporated in the design rules. One effect of this is that the most effective means of making contact to narrow active area regions is to provide a 'dogbone' shape at the end of the active area. Figure 5 shows the isolation of the silicide strap to unrelated polysilicon. Here the yield is determined primarily by linewidth and layer to layer misregistration control and acceptable yields for separations of greater than $0.5\mu m$ were obtained.

2.2 Silicon suckout

The Ti/α-Si thickness ratio was found to have a strong influence on silicide formation [6] and hence on junction leakage and transistor series resistance. If there is too thick a titanium layer then not all of the titanium is converted to silicide during the first $675°C$ anneal. This titanium is not nitrided, however, as the α-silicon prevents diffusion into the titanium. Thus where a large strap makes contact with a small active area region silicon can be 'sucked out' of the source-drain region into the titanium; there being no nitrogen incorporated in the titanium layer to suppress diffusion. This effect is illustrated in figure 6 which shows an SEM picture of a region where pitting occurred. Table 1 shows results obtained for different Ti/α-Si ratios. It shows that the thickness ratio $t_{Ti}/t_{\alpha-Si}$ should be less than 0.38 to prevent suckout from occuring. This corresponds to a Ti/Si atomic ratio of 0.43 indicating that there should be an excess of silicon over that required for stoichiometric formation of $TiSi_2$.

Electrical results were obtained for devices processed with titanium/α-silicon thicknesses of (a) $35nm/80nm = 0.44$ and (b) $35nm/100nm = 0.35$. In situation (a) some increase in junction leakage was observed, especially where straps overlapped narrow active area regions as is illustrated in figure 7. With the increased α-Si thickness no such leakage was observed. In addition on special test structures designed to reflect a 'worst case' situation, the series resistance of p-channel drain regions where the strap connected to polysilicon increased from typically $900\Omega.\mu m$ to $2300\Omega.\mu m$, possibly due to some suckout of the boron source/drain dopant (figure 8). With the lower Ti/α-Si ratio (b) no increase in series resistance was observed.

3 CONCLUSIONS

Silicidation of a titanium/α-silicon structure provides a very effective means of achieving local interconnection at the cost of only one mask layer. Good control of sheet resistances was obtained and high yields of strings were observed provided there was sufficient overlap of the strap over active area to accomodate lithographic effects and the residual enlarged birds beak at the ends of narrow active area regions. The thickness ratio of titanium to α-silicon is critical in determining junction leakage and the resistance of the silicide/silicon contact.

References

1. S. S. Wong, D. C. Chen, P. Merchant, T. R. Cass, J. Amano and K. Y. Chiu: HPSAC—A silicided amorphous-silicon contact and interconnect technology for VLSI. IEEE Trans. Electron Devices ED-34(1987)587–591.

2. A. G. M. Jonkers, H. J. W. van Houtum and A. Moet: Self aligned $TiSi_2$ for submicron CMOS. Le Vide, Les Couches Minces **42(236)**(1987)103–105.

3. R de Werdt, P. van Attekum, H. den Blanken, L. de Bruin, F. op den Buijsch, A. Burgmans, T. Doan, H. Godon, M. Grief, W. Jansen, A. Jonkers, F. Klaassen, M. G. Pitt, P. A. van der Plas,

A. Stolmeijer, R. Verhaar and J. Weaver: A 1M SRAM with full CMOS cells fabricated in a 0.7µm technology. 1987 IEDM Technical Digest, 532–555.

4. H. K. Park, J. Sachitano, M. McPherson, T. Yamaguchi and G. Lehman: Effects of ion implantation doping on the formation of TiSi$_2$. J. Vac. Sci. Technol. **A2(2)** (1984)264–268

5. P. A. van der Plas, N. A. H. Wils and R. de Werdt: Geometry dependent bird's beak formation for submicron LOCOS isolation. To be presented at ESSDERC 1989.

6. H. J. W. van Houtum, A. A. Bos, A. G. M. Jonkers and I. J. M. M. Raaijmakers: TiSi$_2$ strap formation by Ti-amorphous Si reaction. J. Vac. Sci. Technol. **B6(6)** (1988)1734–39.

Table 1. *Effect of various Ti and α-Si thicknesses on silicon suckout. Thicknesses were measured using RBS and pitting evaluated using SEM examination of an active area/strap string.*

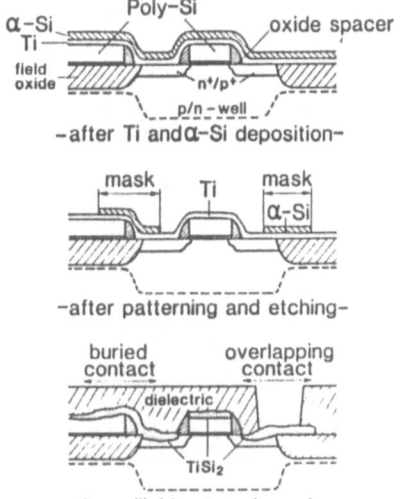

Figure 1. *Schematic cross-sections showing Ti/α-Si strap process.*

t_{Ti} (nm)	$t_{\alpha-Si}$ (nm)	$t_{Ti}/t_{\alpha-Si}$	atomic ratio (Ti/Si)	suck-out
37.0	66.0	0.56	0.64	Yes
37.0	76.5	0.48	0.55	Yes
37.0	87.5	0.42	0.48	Yes
38.0	98.0	0.39	0.45	Yes
33.0	87.5	0.38	0.43	Some
35.0	100.0	0.35	0.40	No
38.0	121.0	0.31	0.35	No

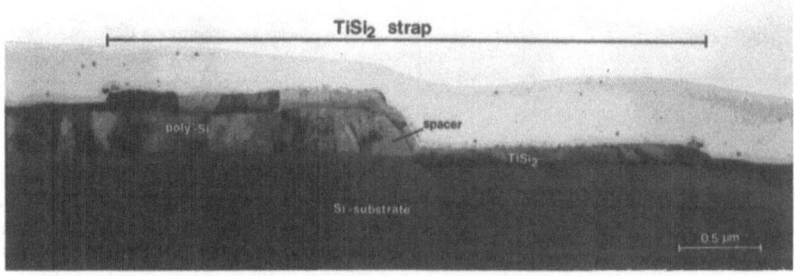

Figure 2. *Cross-sectional TEM showing strap connecting active area to polysilicon.*

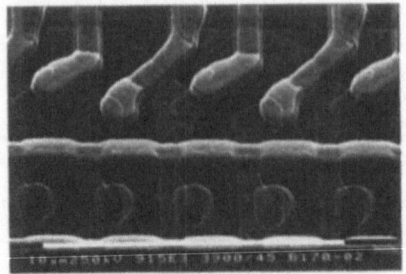

Figure 3. *SEM showing implementation of strap in 1M SRAM cell.*

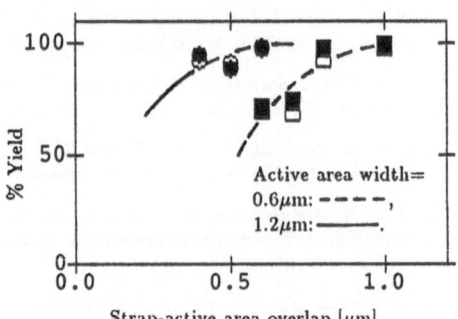

Figure 4. *Yield of* TiSi$_2$ *straps versus overlap over* n^+ *(*$\bigcirc$, $\square$*) and* p^+ *(*$\bullet$, $\blacksquare$*) active area regions.*

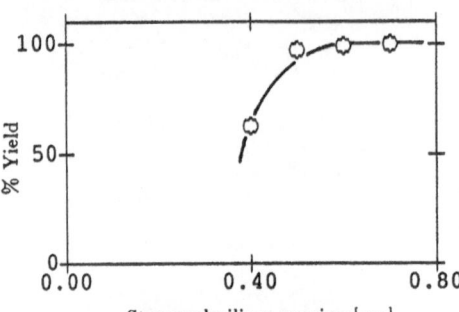

Figure 5. *Yield of* TiSi$_2$ *on field oxide to polysilicon versus* TiSi$_2$/*polysilicon separation.*

Figure 6. *SEM showing pitting in active area* $(t_{Ti}/t_{\alpha-Si} = 35nm/80nm = 0.44).$

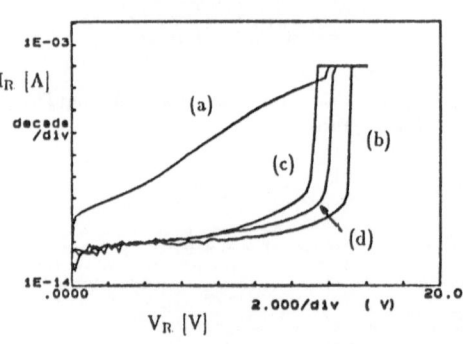

Figure 7. *Junction leakage for 0.6μm wide* n^+ *active area regions where (a)* t_{Ti}/t_{n-Si} *= 35nm/80nm = 0.44; (b)* $t_{Ti}/t_{\alpha-Si}$ *= 35nm/100nm = 0.35; (c)* n^+ *active area region without strap from same wafer as (a); (d) as (a) but active area width = 3μm.*

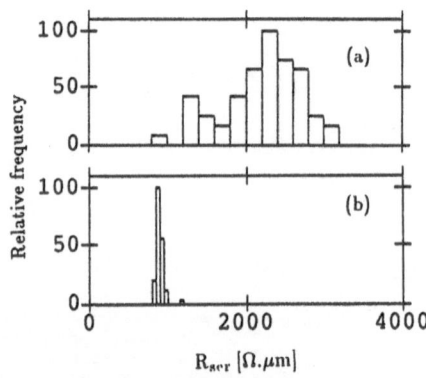

Figure 8. *Histograms of the series resistance of p-channel transistor drains. (a) With large overlapping strap connected to polysilicon, (b) conventional drain structure. Note that* $t_{Ti}/t_{\alpha-Si}$ *= 35nm/80nm = 0.44.*

Stacked Capacitor in Trench Cell for 16M-DRAM

K.H.Küsters, L.DoThanh, F.X.Stelz, W.-U.Kellner, H.M.Mühlhoff, W.Müller

Siemens AG, HL T 1
Otto-Hahn-Ring 6, 8000 München 83, West-Germany

Abstract:
A 16 Mbit DRAM cell based on a Stacked Capacitor (STC) in an isolated trench has been investigated. The storage node of the capacitor consists of an As doped poly Si spacer in Trench. The poly Si spacer is connected to source of the transfer gate by a buried trench contact. The "STC in trench" cell allows trench to trench/active area distances of 0.7µm. The ONO dielectric (d_{eff} = 9 nm) on the As doped poly Si spacer exhibits similar properties as ONO on Si substrate. Using the reported "STC in trench" process a 5.12 µm² DRAM cell (design rules: 0.7µm) has been fabricated.

1.Introduction:

High density DRAM cells are based on three dimensional cell structures. Trench type capacitor cells have been intensively investigated [1-3] and used for 4Mbit DRAM production.

However, when this cell type is shrunk for 16 Mbit DRAM, the isolation of diffusion areas around trenches can be a serious problem: the distances trench to adjacent trench/active area are in the range of 0.7 to 1.0µm. The isolation of trench diffusion areas can be improved by a highly doped p well, however, at the cost of side effects on transistor performance. Furthermore, trench depth is limited, because the p well concentration at the trench bottom has to be sufficient to avoid punch through.

To overcome these isolation problems, a stacked capacitor (STC) in an isolated trench has been proposed and applied to 4 Mbit VSRAM [4]. The formation of a 16 Mbit DRAM cell based on a "STC in trench" requires new process technologies, which are reported in this paper.

2.Process flow:

The process flow for the capacitor of the "STC in trench" cell is shown in fig. 1:

a) After definition of LOCOS isolation areas trenches (5µm depth) are etched in the Si substrate. The trenches are partly overlapping isolation areas. An isolation oxide is grown on the trench sidewalls.

b) The isolation oxide is removed at the upper trench edge, where the trench is adjoining diffusion area.

c) As doped poly Si spacers are formed in trenches by anisotropic poly Si back etch.

As doping of poly Si is performed by diffusion from an As doped glass.

The poly Si spacers are the lower plate of the capacitor, they are connected by buried trench contacts (defined in b)) to the source of the transfer transistor.

d) After deposition of the capacitor dielectric on the poly Si spacer the upper poly Si plate of the capacitor (P-doped poly Si cell plate) is patterned

Fig. 2 shows a cross section of a 5.12µm² "STC in trench" cell.

3. Electrical characterization

3.1 Leakage current between buried trench contacts:

No remarkable leakage current is observed between adjacent trenches on the same bitline even for a trench-trench separation as small as 0.6µm. Trench to trench leakage can only occur between different buried contacts. Fig. 3 shows the dependence of leakage current on the distance between neighbouring buried contacts, which are opposite to each other in wordline direction. Punch through (V_{pt} = 1pA per trench) is observed for a voltage of 7V between different buried contacts at a distance of 0.7µm. To prevent punchthrough under the field oxide a channel stopper doping concentration of $8E16cm^{-3}$ below field oxide has been used. The results shown in fig.3 have been taken on trenches in a p well (surface concentration $4E16cm^{-3}$), however, comparison to results without p well exhibit no remarkable difference. Compared to a conventional trench cell trench to trench distance can be reduced. A further reduction of trench to trench distance is possible for a cell layout, in which buried trench contacts are not arranged directly opposite to each other.

3.2 Leakage current buried trench contact to active area:

Fig.4 shows results of teststructures with a constant distance trench to active area of 0.6µm. The position of the buried contact has been shifted along the trench side. The results show that the leakage current depends only weakly on the displacement of the buried trench contact. In contrast, the trench shape strongly affects the isolation between trench and active area: much larger leakage current was found for rectangular trenches whereas no punch through occurs for T-shaped trench.

3.3 Diode leakage current

Enhanced leakage current due to gated diode as reported in [5] has not been observed, most probably because of different formation of trench contact.

3.4 Contact resistance of buried trench contact:

less than 1.5 K Ω, even for a contact depth of 100 nm (minimum contact depth)

3.5 Capacitor dielectric on As doped poly Si Spacer:

For the capacitor dielectric a d_{eff} = 9 nm oxide/nitride/oxide (ONO) is used.The storage capacitance is larger than 50 fF for 5µm trench depth (T shaped trenches, perimeter of capacitor = 2,8µm.

Fig. 5 shows the ONO dielectric on top of the poly-Si spacer. The process is optimized to avoid poly-Si roughness. The quality of the ONO dielectric on the poly-Si spacer is similar to an ONO dielectric an Si substrate. Fig. 6 shows the I (V) characteristics. The leakage current is negligible at 16M operation voltage (V $_{poly\ Si}$ plate = ± 1.65 V) Also the mean time to breakdown is comparable to 9 nm ONO films on Si substrate (see [6])

References:

[1] H.Sunami et al, IEDM Tech. Dig., 1982, p. 806
[2] S.Yoshikawa et al., VLSI Symposium 1989
[3] S.Röhl et al., ESSDERC 1989
[4] S.Yoshioka, ISSCC Dig. Tech. Papers, 1987, p. 20
[5] F.Horiguchi et al., IEDM Tech. Dig., p. 324, (1987)
[6] J.Hirschler et al., ESSDERC 1989

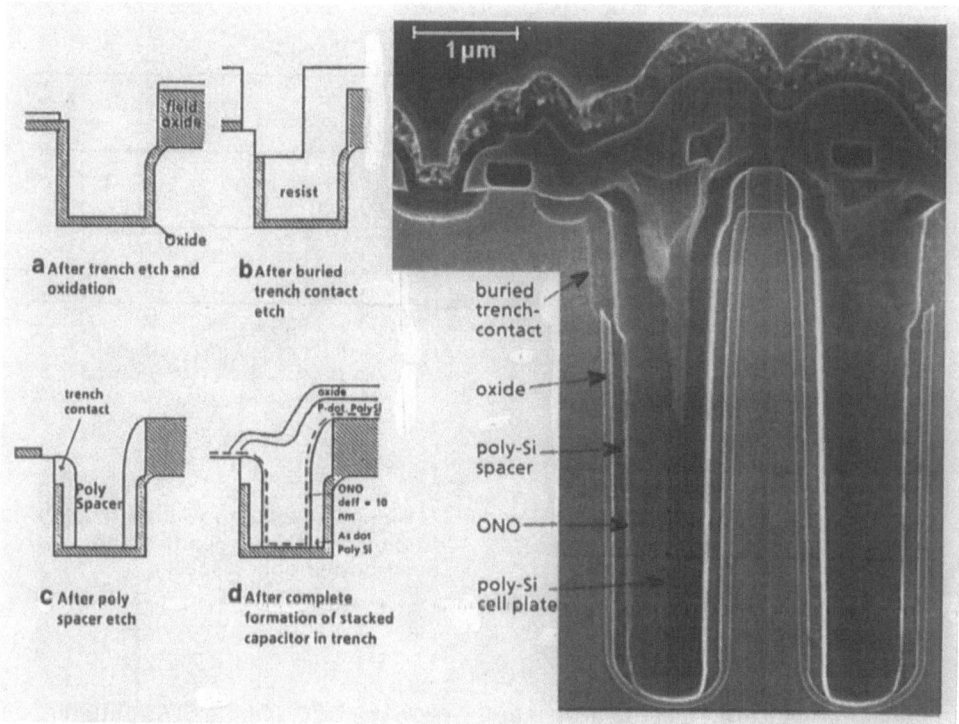

Fig 1: Process flow of "STC in Trench" capacitor

Fig. 2: SEM cross section along bitline direction Stacked Capacitor in Trench

910

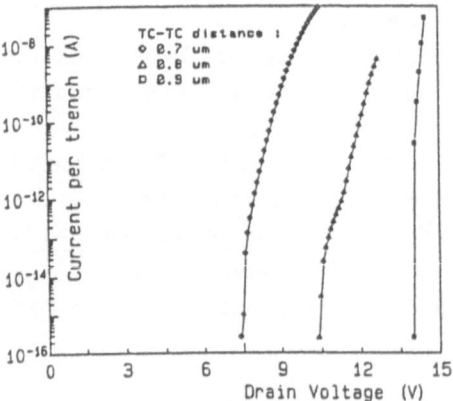

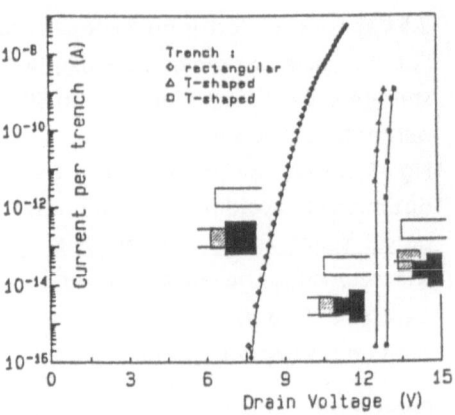

Fig. 3: Leakage current between buried trench contacts vs distance of buried contacts (as determined by SEM) Trench contacts are opposite to each other in wordline direction
($V_{cell\ plate}$ = 1.65 V, V_{sub} = - 1.5 V)

Fig. 4: Leakage current buried contact to active area of adjacent cells.for T-shapedand rectangular trenches. Minimum distance trench to active area: 0.6µm.
($V_{cell\ plate}$ = 1.65 V, V_{sub} = - 1.5 V)

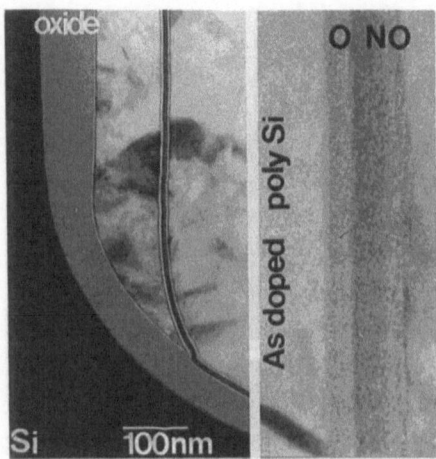

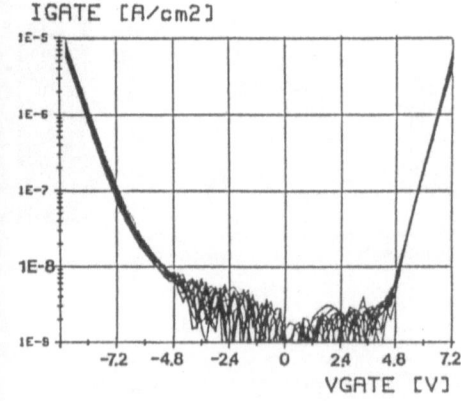

Fig. 5: TEM cross sections of an ONO dielectric on a poly Si spacer

Fig. 6: Leakage current of an ONO dielectric (d_{eff} = 9 nm) on an As doped poly-Si spacer.

Acknowledgements:
The authors are grateful to J.Barth, H.Bolze, A.Gutmann, D.Sarlette, A.Strohbach, B.Wild for process development, to I.Bunge, I.Neumann for SEM analysis, to H.Cerva, H.Oppolzer for TEM analysis to J.Hirschler, R.Vollertsen for ONO evaluation.and to M.Elahy for rewarding discussions.

0.5 μm CMOS Devices and Circuit Characterization

G.Guegan, M.Lerme, S.Deleonibus
F.Martin, S.Tedesco, J.Gautier

D.LETI CENG

85 X
38041 GRENOBLE Cedex FRANCE

INTRODUCTION

CMOS circuits with feature sizes scaled down to $0.5\mu m$ [1],[2] and even less [3],[4] have already been reported. In this paper, we report the fabrication of half micron CMOS using mixed e-beam/optical lithography and Rapid Thermal Annealing (RTA) for both BPSG reflow and junction activation. The aim of this work is to characterize devices and to compare speed performances of demonstrator circuits such as ring oscillator and 16kx1 SRAM fabricated using either $0.5\mu m$ CMOS advanced technology at a 3.3V supply voltage or a $0.8\mu m$ CMOS at 5V.

FABRICATION PROCESS AND DEVICE STRUCTURE

The CMOS circuits were fabricated using a $0.8\mu m$ CMOS technology P^-/P^+ epitaxial substrate and twin tub as base process. The transistors were designed with N^+ polysilicon gate material, a gate length of $0.5\mu m$, a gate oxide thickness of 12 nm and a power supply voltage of 3.3V. The design of the buried P-channel was performed with BF_2 counter doping and a deep arsenic implant. A LDD structure with $0.15\mu m$ sidewall oxide spacers was incorporated for both N and P channel devices. The $TaSi_2$ gate was e-beam patterned while the other layers were printed using an optical stepper. A BPSG layer, reflowed using RTA(1050°C 10s) forms the dielectric underneath the aluminum layer. The intermetal dielectric sequence employs a tri-level glass with SOG etch back. This cold end process allows us to obtain shallow N^+ and P^+ junctions with a sheet resistance of about $80\Omega/sq$.

DEVICE AND CIRCUIT PERFORMANCE

Figures 1 and 2 show the $I_{DS}-V_{DS}$ characteristics of $0.5\mu m$

gate length transistors. Subthreshold characteristics for different drain voltages (fig.3-4) show two important points: first, short channel effects and DIBL are weak; second, off currents are in agreement with design criteria. Important characteristics of both N and P channel FET's are given in table 1. Threshold voltage variations versus effective channel length for $V_D=0.1$ and 3.3V are shown on Fig.5. Short channel effects are well controlled for channel length larger than 0.35μm and 0.45μm for respectively NMOS and PMOS.

In order to point out the dynamic performances of such a technology, ring oscillators and a 16K x 1 SRAM have been employed, using 0.8μm CMOS design rules except for the gate which was undersized to 0.5μm. Figures 6 and 7 show dynamic performance comparisons between this work and a previously optimized 0.8μm CMOS technology, with L_{eff} of 0.7μm and a gate oxide of 18nm. Two characteristics can be pointed out: first, for a given power supply, gate delay and access times are nearly two times shorter with this 0.5μm technology; second, at nominal supply voltage, this 0.5μm CMOS technology gives a 30% speed improvement: 6ns at 3.3V compared to 8.5ns at 5V. This SRAM operates from 2V to 4.5V with an access time of respectively 18.5 and 5ns. A minimum speed power product of 1.22 fJ was measured at 1.5V supply voltage.

CONCLUSION

A high performance CMOS process using half micron gate has been demonstrated. Good device behaviour with high drivability and negligible short channel effects were observed. Gate delays and SRAM access times were compared to those of a previous 0.8μm CMOS technology. The process was well controlled and a 16K x 1 SRAM was fabricated and access times as low as 6ns were obtained at 3.3V.

ACKNOWLEDGEMENTS

The authors would like to thank the different teams of D.LETI SMSC for device processing and testing.

REFERENCES

[1] R.A. Chapman, et al.
    IEDM 1988 Tech. digest
[2] P.H. Woerlee, et al.
    ESSDERC 1988
[3] B. Davari, et al.
    IEDM 1988 Tech. digest
[4] M. Kinugara, et al.
    Symposium on VLSI Technology 1988

Parameter	N-MOS	P-MOS
L_{eff} (μm)	0.43	0.50
Threshold voltage at low V_D (V)	0.76	-0.94
Threshold voltage at $|V_D| = 3.3V$	0.70	-0.82
Mobility at low field ($cm^2/V/S$)	360	150
Subthresold slope (mV/dec)	88	93
gm sat (mS/mm)	115	86
Rseries (both side) (Ω.mm)	1.25	2.75
Junction depth (μm)	0.15	0.22

Table 1: Key device parameters for 0.5μm MOS

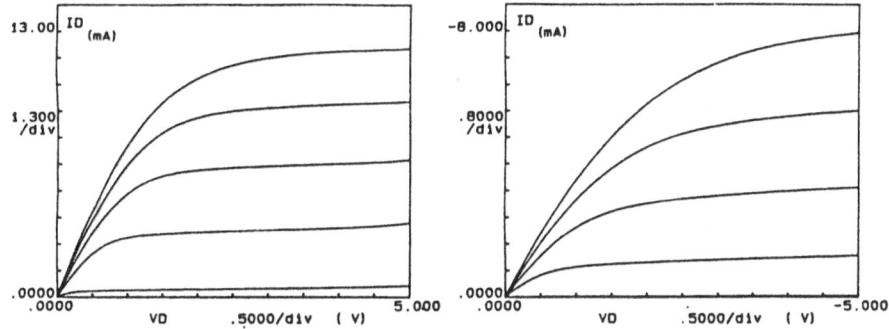

Fig.1. $I_D(V_D)$ characterisics
NMOS W/Leff = 25/0.43μm
VG = 1,2,3,4,5 V.

Fig.2. $I_D(V_D)$ characteristics
PMOS W/Leff = 25/0.50μm
VG = -1,-2,-3,-4,-5 V.

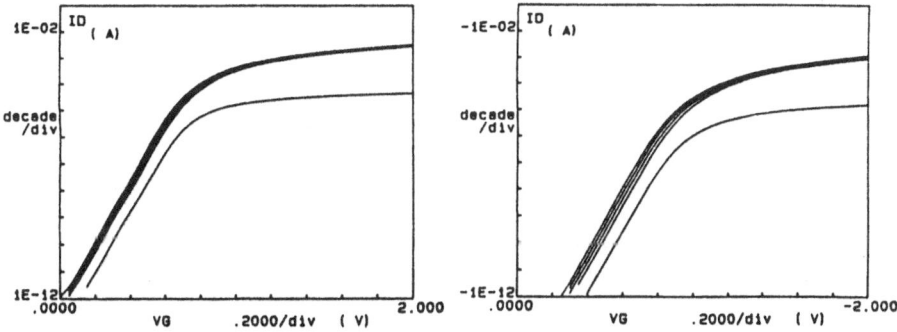

Fig.3. $I_D(V_G)$ characterisics
NMOS W/Leff=25/0.43μm
VD = 0.1,1,2,3,4 V

Fig.4. $I_D(V_G)$ characterisics
PMOS W/Leff=25/0.50μm
VD= -0.1,-1,-2,-3,-4V

914

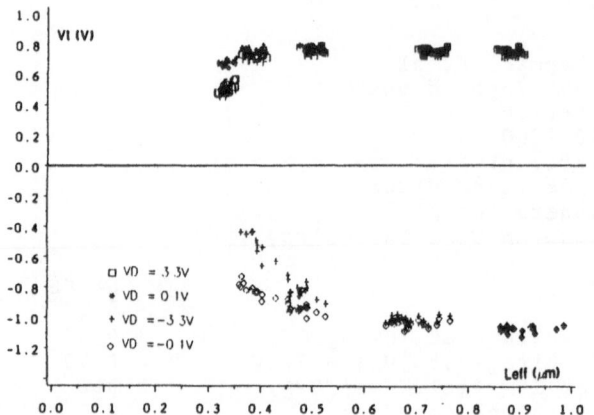

Fig.5   Linear and saturation threshold voltage versus ef-
        fective channel length for N and P-channel

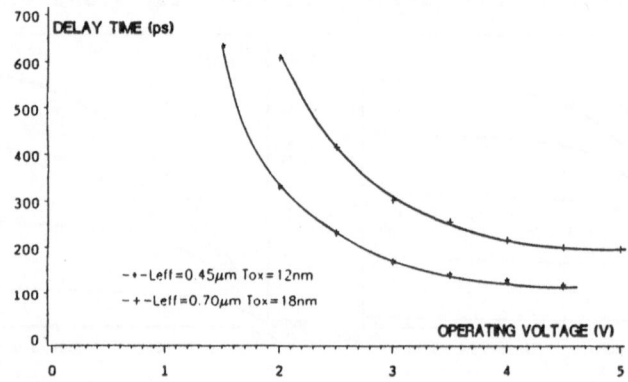

Fig. 6. CMOS ring oscillator performance comparisons.

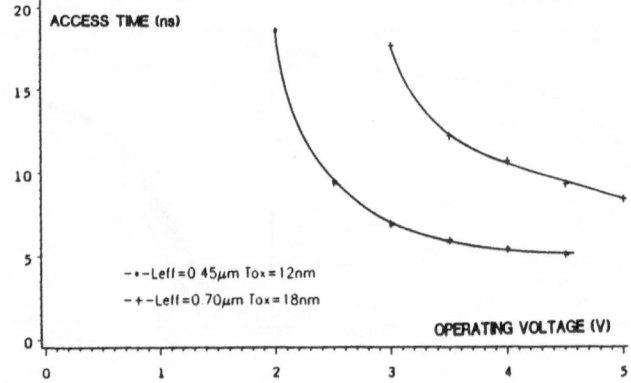

Fig. 7. 16kx1 SRAM performance comparisons

VIPMOS, a Buried Local Injector for EPROMs

R.C.M. Wijburg, G.J. Hemink, L. Praamsma, J. Middelhoek

University of Twente, IC-Technology and Electronics department
P.O. box 217, 7500 AE Enschede (NL)

Abstract

A highly effective substrate hot electron injection EPROM device can be made using a buried injector, which operates in punch-through mode. The buried injector is formed by a local overlap of the N-well and P-well of a retrograde twin-well CMOS-process. As the VIPMOS-EPROM is compatible with VLSI-processing and the danger of latch-up doesn't exist, the VIPMOS-structure may be used in VLSI-applications. Due to an efficient electron supply mechanism as well as a high injection probability, programming rates of 1V/μs can be obtained.

Introduction

Substrate Hot Electron (SHE) injection is an attractive alternative to the conventional Channel Hot Electron (CHE) injection for programming EPROM's. SHE injection offers a higher injection efficiency and less stress on the gate oxide because of the distributed charge transport through the gate oxide (figure 1 & 2). However, SHE injection has been unsatisfactory until now, because an efficient mechanism to supply electrons underneath the gate was not available [1], [2].

Buried Injector

A Vertical Injection Punch-through based MOS-structure (VIPMOS) can be used for a highly effective SHE injection EPROM (figure 3). The EPROM has an injector, which is formed by a local overlap of the N-well and P-well of a retrograde twin-well CMOS-process. The VIPMOS-EPROM is compatible with this process. During programming, the injector is grounded and a high voltage (e.g. 15V) is applied

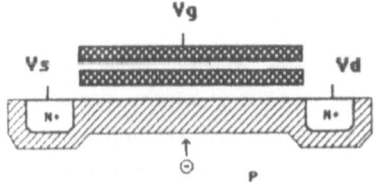

fig. 1: SHE injection

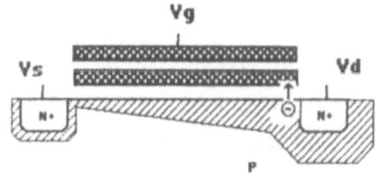

fig. 2: CHE injection

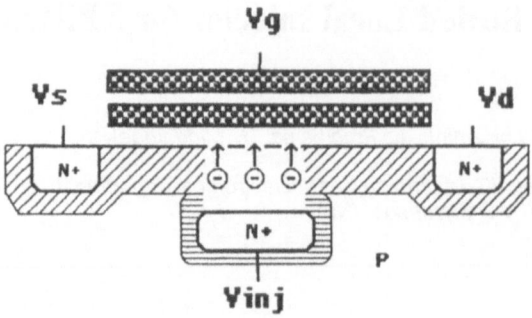

fig. 3: VIPMOS-EPROM with the buried local n-injector

to the control gate. Increasing the voltages of both source and drain will extend the depletion layer under the gate into the direction of the injector. At a certain voltage Vpt, punch-through occurs and electrons are injected into the depletion layer under the gate. Some electrons become hot and will surmount the Si/SiO2 potential barrier. Simulation results [3] show (figure 4), that the VIPMOS-structure very well supplies the electrons, which should be accelerated.

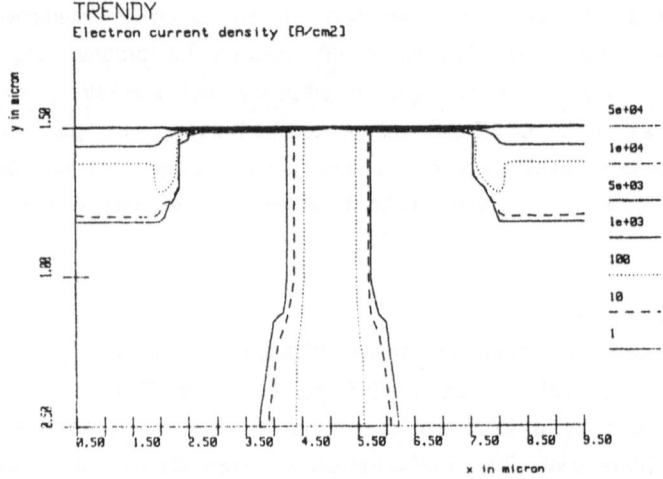

fig. 4: Simulated electron current density in programming mode
(Vd=Vs=5V, Vinj=0, Vg=10V, Vb=0V)

The electron current density from injector towards the channel easily exceeds 100A/cm². The injection only occurs in vertical direction, which makes this device suitable for VLSI-application without the danger of latch-up. As the potential of the floating gate decreases sufficiently, the punch-through condition cannot be satisfied any more and the injecting process automatically stops. During the read mode, the injector is connected to Vdd. Therefore, the potential barrier between the channel and the injector is raised and electrons are precluded from flowing into the direction of the gate.

Measurements

In figure 5a, the measured drain/source-current (Ids) and the "floating" gate current (Ig) as a function of the drain/source to bulk voltage (Vds) are shown. The voltage applied to the "floating" gate (Vg) is 12.5V. The source- and drain potential are equal. In figure 5b, Ids and Ig as a function of Vg, with Vds is 5.5V, are shown. The injection area of this device is $2.5*5.0\mu m^2$. The voltage, at which punch-through occurs, is about 4.8V. However, Vds has to be increased with a small amount to attract the electrons, which are not able to surmount the Si/SiO2 barrier, from the channel.

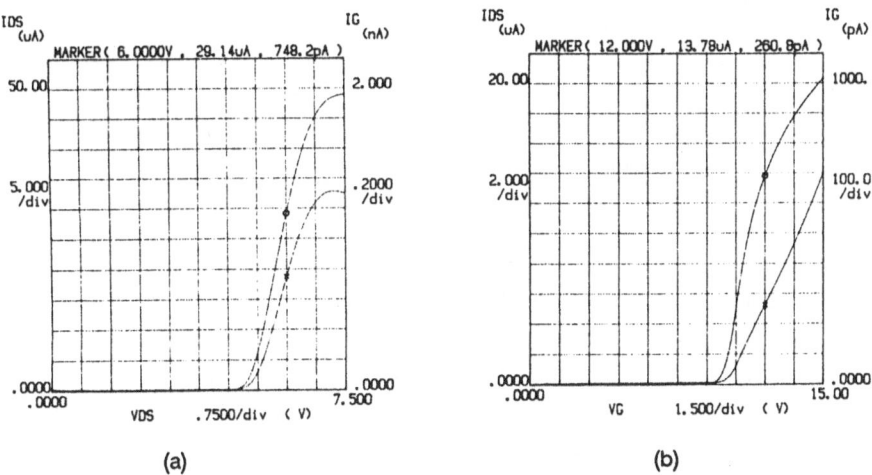

(a)   (b)

fig. 5: Drain/source and gate current as a function of Vds (a) and Vg (b)

By varying the doping-profile of the P-well, the punch-through voltage and thus the injection probability (Pinj=Ig/Ids) can be varied. Figure 6 shows the measured injection probability at two different electrical-field strength's (Eox) in the oxide. The solid lines in the figure present the theoretical curves according to the lucky electron model [4]. The parameters are the same as used in [4], while the doping concentration has been fitted by a parameter extraction program [5].

Due to an efficient electron supply mechanism as well as a high injection probability, a high electron current density through the gate oxide can be obtained. We have measured current densities up to $1A/cm^2$ for the devices with a punch-through voltage of 8V. This implies, that depending on the ratio of the injection area versus the floating gate area, programming rates of 1V/μs can be obtained for the devices with a punch-through voltage of 8V. On the other hand, Vpt can be adjusted to a value of 4V. As a result, a 5V-only VIPMOS-EPROM should be possible.

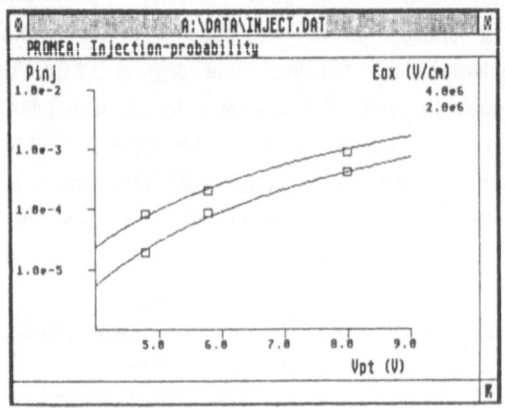

fig. 6: The measured and theoretical (solid lines) injection probability as function of the punch-through voltage

Conclusion

A local buried injector offers new possibilities for programming EPROM's using SHE injection. By means of punch-through, the local buried injector can effectively supply the electrons, which should be accelerated. Therefore, the injection of electrons doesn't occur into the lateral direction. This makes this device suitable for VLSI-applications without the danger of latch-up. The injector is formed by a local overlap of the N-well and P-well of a retrograde twin-well CMOS process and is compatible with VLSI-processing. Oxide current densities of $1A/cm^2$ have been measured as a result of the effective injection mechanism and the high injection probability.

Acknowledgements

The authors would like to thank the Dutch Foundation for Research on Matter (FOM) for the financial support, G. Boom for his valuable discussions and support, L.F.P. Warmerdam and F.W. Ragay for their encouragements.

References

1. J.F. Verwey, J. Appl. Phys. 44, 2681 (1973)
2. B. Eitan et al., IEEE Trans. Elect. Dev. ED-31, 934 (1984)
3. E. van Schie et al., Nasecode VI conf. (1989)
4. T.H. Ning et al., J. Appl. Phys. 48, 286 (1977)
5. P.B.M. Wolbert et al., Nasecode VI conf. (1989)

Leakage Current Limitations of Trench Cell for 16M DRAMs

S.Roehl, L.Kusztelan, P. Kuepper, B. Hasler
Siemens AG, Components Group
Otto-Hahn-Ring 6, 8000 Muenchen 83, West Germany.

I. Summary

One of the primary factors controlling 16M DRAM cell concept choice is that of sufficiently small leakage. Trench cell leakage to the substrate, to the adjacent trench and to neighboring transistor and bitline contact have been determined here for reduced geometry Trench Cells using a manufacturable 4M process. It is shown that the "Super Shrink Trench Cell" is a viable 16M DRAM candidate insofar as leakage limitations are concerned.

II. Trench to Trench Punch Through

A depletion cell concept is used, with n+ doped trenches of size .9 x 1 x 4.5 µm formed within a Pwell of 5E16 cm^{-3} surface concentration and 6 µm deep. A .5 µm thick LOCOS isolation technique with 4E13 cm^{-2} field implant is used, and the bitline contact is fabricated with a selfaligned FOBIC technique[1].

The primary geometry dependant intercell leakage path is from trench to trench and is punchthrough limited, due to the influence of the trenches. Punch through Voltage (Vpt) as a function of Trench-trench separation was measured using a Comb type Test Structure (Fig 1) where refresh and Cell capacitance considerations set an upper criterion of about 1pA per trench. Fig 2 indicates a Vpt below 5V, under 4M operating conditions, only at .8 µm actual Trench-trench separation (0.9 µm as drawn). Furthermore, Vpt was not substantially influenced by reduced Plate or Substrate Voltages, which are typical for 16M DRAM's. Process spreads, viz. during Trench Etching, could, however, have major influence on Vpt due to the slope of the Vpt versus Trench separation curve. This parameter was evaluated over a number of batches (Fig 3), which data confirmed Vpt's of approx 3V, with spreads of about 1V for .9 µm (as drawn) separation devices.

3D-simulation of bulk punch through with experimental doping profiles [2] for trench n+ and p-well show a Vpt onset at 0.25 μm larger distances and predict punch through at the trench bottom. Enhancing trench bottom doping via higher doped substrate (2.5 to 6.5E15 cm^{-3}) did, however, not improve punch voltage indicating that surface leakage is still the dominating mechanism.

III Inter bitline Leakage

An interbitline leakage path exists between trench and neighboring planar diffusion regions. This was evaluated by Vpt and Threshold Voltage measurements down to FOX length of .6 μm (0.8 μm as drawn). Individual device data (Fig 4) indicated Vt's above 8V and avalanche breakdown before punch, again confirmed across several batches.

A further critical leak path is from trench to the neighboring bitline contact since lateral field oxide etch occurs on both sides of the LOCOS. Worst case test structures with selfaligned contacts opposite to trenches showed excessive leakage only at 0.5 μm LOCOS width (0.8 μm as drawn) for 5 Volts on bitline node (Fig 5). Substrate bias is useful here as punch currents strongly decrease with increasing bias. Reduction of bitline and capacitor plate voltage also suppress leakage so that isolation at 16M conditions is better than for 4M (Fig 5). Allowing for a trench to LOCOS misalignment and lateral field oxide etch of 0.2 μm each, a trench to transistor isolation distance of 0.9 μm appears feasible allowing 16M bitline pitches of 1.5 μm.

IV Trench to Substrate Leakage

In a viable 16M Cell concept, the trench will overlap Field Oxide. The Trench Etch process can then expose the p+ field implant/n+ varactor junction, and capacitor plate controlled diode effects are possible. Substrate leakage at various LOCOS-Trench separations, with and without pn junction exposure (Fig 6), always remains below the 1pA/trench limit. Enhanced leakage occurs only for negative plate voltages, and is identified as Zener Tunneling by its temperature dependence.

V Conclusion

We therefore conclude that present cell separation design rules of 1.5-1.8 μm trench separation and 2.0-2.3 μm bitline pitch for conventional 4M DRAM trench technology may be reduced to approx. .8 µm and 1.5 µm ,respectively, allowing a 4M supershrink cell concept for the 16M generation. Optimization of Pwell and n+ trench doping profiles, for instance, should allow further improvement.

References
1. K. H. Kuesters et al., Symposium on VLSI Technology, Digest of Technical Papers, 1987, p. 93
2. W. Bergner, R. Kircher, SISDEP 88, Bologna, 1988, p.165

This work has been supported by the Federal Department of Research and Technology of the Federal Republic of Germany (sign NT 2696).

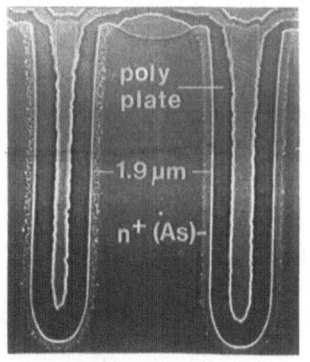 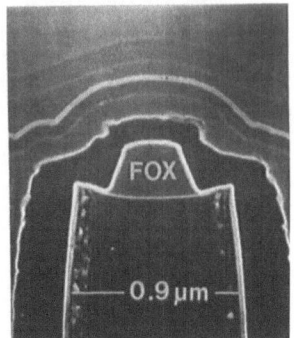

Fig. 1: SEM cross sections of trench to trench leakage test structure. Field oxide width is reduced due to trench processing for small trench to trench separations.

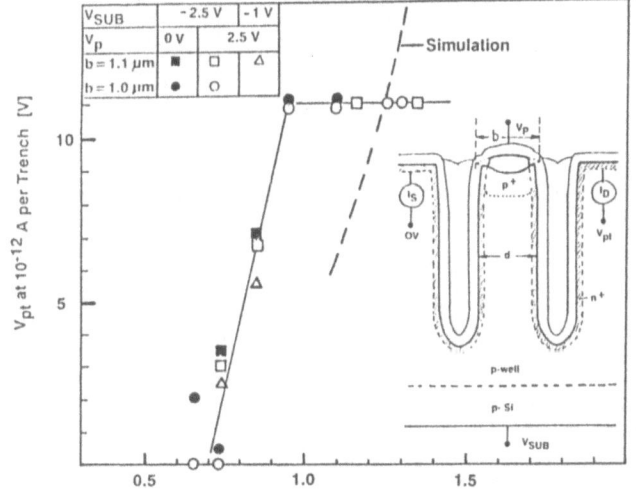

Fig. 2:

Punch through voltage V_{pt} versus trench to trench distance for different field oxide lengths and different substrate and poly plate voltages.

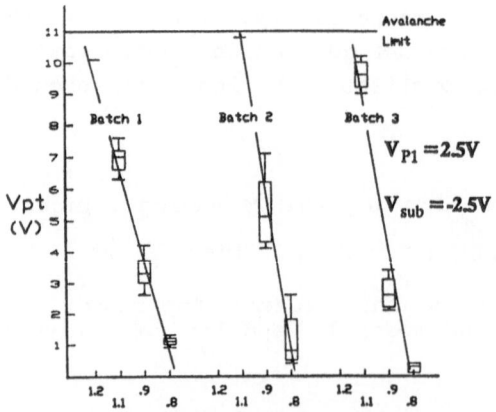

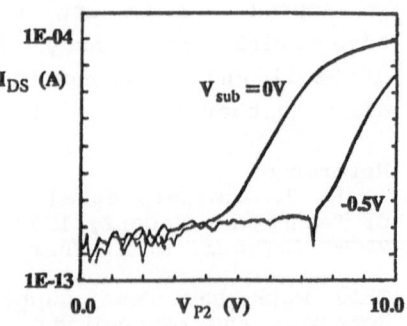

Fig. 3:
Batch to batch comparison of Punch-through spread. The figures refer to as drawn trench to trench separations.

Fig. 4:
Subthreshold Characteristic for Trench to Active Area Field Region under Wordline. FOX length=.8 μm (as drawn) Vds=4V, Vsub=0 & –.5V

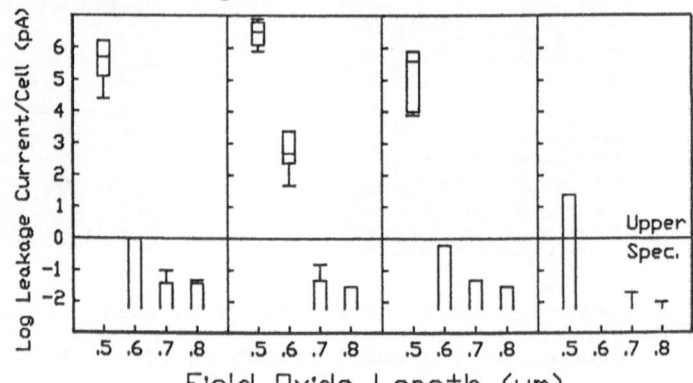

Vpl (V)	2.5	2.5	1.5	1.65
Vsub (V)	-2.5	-1	-1	-2.5
Vbl (V)	5	5	5	3.3

Fig. 5: Trench to bitline contact leakage under 4M (left) and 16M (right) operating conditions. Leakage dependence on worst case substrate bias and reduced Poly plate voltage. Bitline voltage Vbl is 5 or 3.3 Volts.

Fig. 6:

Box plots of leakage current of the trench n+ to substrate diode at 7.5 V and at a poly plate voltage of 2.5 V. Three different trench to field oxide separations leading to exposure (right) or non exposure (left) of pn-junction (see insert and Fig 1) are shown.

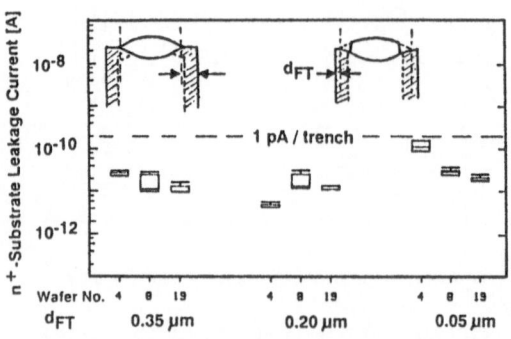

A Highly Reliable Gate/n⁻ Overlapped Transistor for Mega-bit DRAMs

M.Nagatomo,Y.Okumura,K.Mitsui,I.Ogoh,H.Genjoh,M.Inuishi
and T.Matsukawa

LSI R&D Laboratory, Mitsubishi Electric Corp.
4-1 Mizuhara,Itami,664,JAPAN

SUMMARY

A noble gate/N⁻ overlapped Tr. fabricated using oblique rotating ion implantation technique was developed. It is confirmed that this Tr. meets high performance M bit DRAMs' requirements, that is, high drain current, enough punchthrough voltage and low substrate current. The mechanism of this Tr.'s action is analyzed by simulation and is concluded that peak position of electric field is located far from the drain current pass in this Tr.'s structure. The maximum electric field is also relaxed by formation of N⁻ layer using oblique ion implantation.

INTRODUCTION

Submicron gate MOS FETs used for M bit DRAM require (1) high drain current (I_d) for high speed operation, (2) punchthrough voltage (BV_d) over 10V for wide operational margin of power supply voltage (V_{cc}), (3) low substrate current (I_{BMAX}) for suppression of the hot carrier degradation and latch up. Since, I_{BMAX} increase proportionally with increase of I_d, it is difficult to pursue the improvement of current driving capability with the reliability maintained at the same level as in conventional submicron MOS FETs.[1] We developed a noble gate/N⁻ overlapped Tr. and confirmed that it meets those above mentioned requirements for M bit DRAMs.

EXPERIMENTS and RESULTS

The gate/N⁻ overlapped structure [2][3] has been fabricated using oblique rotating ion implantation techniques [4][5], as shown in Fig.1. The relationship between I_{BMAX} and oblique angle is shown in Fig.2. It is found that I_{BMAX} decrease with increasing of oblique angle, on the other hand, I_d and B_{vd} do not depend on this parameter up to 60 degrees. The relationship between I_{BMAX} and another implanted parameter, dose rate and incident angle, are also measured. And it is found that I_{BMAX} have minimum point against the dose rate and I_{BMAX} decrease with increasing of incident ion energy. Overlapped parasitic capacitance is estimated using ringoscilator, as shown in Fig.3. This result shows that this parasitic capacitance does not affect device access time, significantly.

ANALYSIS of MECHANISM

To analyze these phenomena, simulation of the two-dimensional impurity distribution and the electric field at the drain edge was performed. At first, N^- impurity distribution was estimated, as shown in Fig.4(a)-(d). W(x) and Wmod(x) are the weight functions describing shadowing effect of gate electrode and penetrating effect through polysilicon at the gate edge of oblique implanted ions , respectively, as a function of the position x along channel direction. $\rho(z)$ and ρmod(z) are the profiles of oblique rotating implanted ions corresponding W(x) and Wmod(x), respectively, as a function of the depth z. Using these profiles, lateral electric field and carrier concentration at the drain edge were calculated, as shown in Fig.5. This simulation result shows that the peak position of the electric field is located far from the drain current pass in this newly developed structure. And the electric field is relaxed as compared with that of conventional LDD Tr. Moreover, the location of the peak position can be controlled by varying the implantation parameters, such as dose rate, oblique angle and energy of incident ions. This is the reason why this Tr. can reduce I_{BMAX} independently of I_d and B_{vd}.

DEVICE TEST

As shown in Fig.2, using this technique, I_{BMAX} of the MOS FETs whose L/W=0.9μ/10μ, can be varied from 50μA to 120μA. In this range, hot carrier degradation was estimated by low temperature(-20.c) test using actual 4MDRAM. As the result, it was confirmed that 4MDRAM does not degrade due to hot carrier effect if only the value of I_{BMAX} is reduced less than 100μA.

CONCLUSIONS

It is concluded that the newly developed MOS FETs have enough performance to realize high speed and high reliability Mega bit DRAMs at $V_{cc}=5v$.

ACKNOWLEDGEMENT

The authors wish to thank Drs.K.Shibayama T.Nakano and T.Katoh for their continuous encouragement. They would like to thank Drs.K.Tsukamoto and M.Yamada for their helpful discussions.

REFERENCES

[1].Toyoshima,Y;Nihira,H;Wada,M; and Kanzaki,K;"Mechanism of Hot Electron Induced Degradation in LDD NMOS FET,"IEDM Tech.Dig.,1984,p.786.

[2].Mayaram,K;Lee,J;Chan,T.Y;and Hu,C;"An Analytical Perspective of LDD MOSFETs,"Proc.of VLSI Symp.1986,p.61.

[3].Izawa,R;Kure,T;Iijima,S;and Takeda,E;"The Impact of Gate-Drain Overlapped LDD(GOLD) for Deep Submicron VLSI's,"IEDM Tech.Dig.,1987,p.38.

[4].Eimori,T;Ozaki,H;Oda,H;Ohsaki,S;Mitsuhashi,J;Satoh,S;and Matsukawa,T;"The Improvement of LDD MOSFET's Characteristics by the Oblique-Rotating Ion Implantation,"SSDM.Extended Abstracts,1987,p.27.

[5].Hori,T;Kurimoto,K;Yabu,T;and Fuse,G;"A New Submicron MOSFET with LATID(Large-Tilt-Angle Implanted Drain) Structure,"Proc.of VLSI Symp.1988,p.15

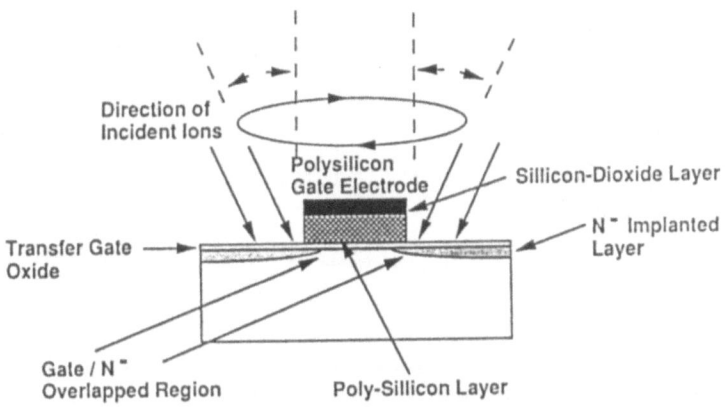

Fig.1 Cross sectional view of the process forming gate / N⁻ overlapped structure in nMOS Tr. by oblique rotating implantation

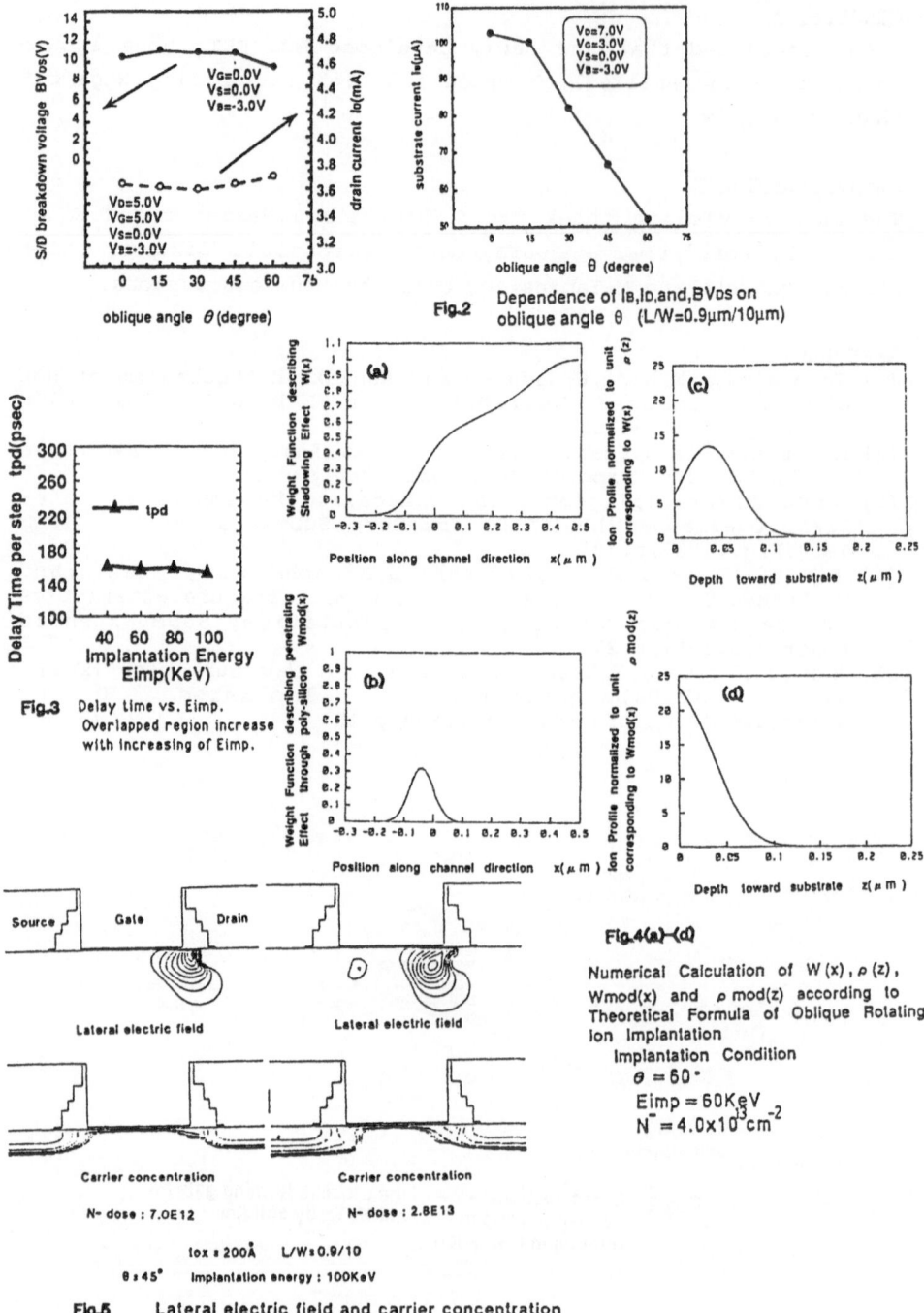

Fig.2 Dependence of I$_B$,I$_D$,and,BV$_{DS}$ on oblique angle θ (L/W=0.9µm/10µm)

Fig.3 Delay time vs. Eimp. Overlapped region increase with increasing of Eimp.

Fig.4(a)-(d)

Numerical Calculation of W(x), ρ(z), Wmod(x) and ρmod(z) according to Theoretical Formula of Oblique Rotating Ion Implantation

Implantation Condition
θ = 60°
Eimp = 60KeV
N$^-$ = 4.0×10^{13}cm^{-2}

Fig.5 Lateral electric field and carrier concentration at maximum of substrate current near the drain
Applied voltage : V$_D$=7.0V, V$_G$=3.0V, V$_S$=0.0V, V$_B$=-3.0V

Characterization

Characterization of Shallow Junction Ion Implantation Profiles: Correlation Between a Noncontact Photodisplacement Thermal Wave Technique and Rutherford Backscattering Analysis

G.M. Crean*, C. Jeynes+, M.G. Somekh', and R.P. Webb+

* National Microelectonics Research Centre, University College, Lee Maltings, Prospect Row, Cork, Ireland.

' University College London, Department of Electrical Engineering, Torrington Place, London WC1E 7JE, England.

+ University of Surrey, Department of Electrical Engineering, Guildford, Surrey GU2 5XH, England.

Abstract

This paper correlates photodisplacement thermal wave characterization of ion implanted silicon wafers with the lattice information provided by Rutherford Backscattering Spectrometry.

Introduction

The fabrication of fine geometry very large scale integration (VLSI) complementary metal oxide semiconductor (CMOS) devices requires the formation of shallow source-drain junctions in order to minimize short channel and punch through effects. The fabrication of shallow p+ junctions is particularly challenging due to the problems of boron channeling during implantation and the high diffusivity of boron during subsequent thermal processing. As this junction depth is scaled, it becomes necessary to accurately control the implanted impurity profile both before and after annealing. Ideally, this information is required in a real-time and nondestructive manner.

Recently it has been demonstrated that photodisplacement (PD) thermal wave analysis may be a response to this process control problem. Photodisplacement microscopy uses a modulated laser to thermally excite the sample, this results in a period surface expansion which is detected with an interferometer focused onto the same spot as the heating source, the resulting signal provides information about local thermal and (for semiconductor materials) electronic properties [1,2]. However while several research groups have presented experimental ion implantation characterization results using this and other thermal wave techniques [2,3,4], the relationship between the measured signal and the actual microstructure of the as-implanted samples has not been discussed in detail. The purpose of this paper is to correlate the measured photodisplacement thermal wave signal with the lattice information provided by Rutherford Backscattering Spectrometry (RBS) [5]. In particular, the study will demonstrate the sensitivity of the PD probe to changes in the implanted impurity profile.

Experiment

Single crystal (100) orientated n type silicon wafers with 3 ohm cm resistivity were implanted with BF_2 ions at energies of 50, 100 and 200 keV. Implantation parameters are presented in Table 1. Implants were performed at nominal room temperature. Ion dose was varied from 5E14 to 3E15 ions/cm2.

Rutherford backscattering spectrometry (RBS) channelling analysis was performed using a 1.5 Mev helium beam to investigate the lattice damage induced by the BF_2 implantation process. The thermal wave measurements were made using an intensity modulated semiconductor laser (maximum power output 20mW) as the excitation source. The modulation frequency was 10kHz and the spot size was two micron.

Discussion

Figure 1 shows normalised RBS channelling spectra from the as-implanted samples. These spectra have two distinct features, both due to implantation damage. The first is seen most clearly in the 50keV series, and is a well defined "amorphised" layer extending down from the surface. Higher doses will deposit sufficient energy for "amorphising" deeper in the material, which explains why the 3E15 dose gives a thicker surface layer than the 5E14 or 1E15 dose implants. We have given the thicknesses of these surface damage peaks in Table 2.

The other damage feature can be seen clearly in the high dose 100keV implant, where there is sub-surface damage peaking at about the range of the implant. There is also a knee in the back edge of the surface peak signal for the low dose 100keV sample, but this is not in the right position for the range. We interpret these effects as self-annealing during implantation, where the annealing cannot be complete for the high dose sample because of the high concentration of impurities (F). Again, we present numerical values in Table 2.

The corresponding measured photodisplacement signals for these samples are shown in figure 2. One immediate conclusion that can be drawn from the above data is that while the photodisplacement signal increases as a function of increasing lattice disorder (and is thus a means of monitoring implantation dose), the relationship between the photodisplacement signal and implantation dose is not linear and this reflects the structure of the samples. The two low dose 50keV implants look very similar on both PD and RBS despite there being a factor of 2 in their doses. On the other hand, the high dose 100keV implant has a structure totally different from either the medium dose 100keV implant or the corresponding high dose 50keV implant. Not only is the PD signal not linear for 100keV series, but the 50keV implant gives a higher signal than the 100keV implant in the high dose (3E15) series. As we explained above, we interpret this as reflecting the thermal artefacts of the implant.

Finally, we have also observed that PD is sensitive to heavy metal contamination in low dose (10^{12}) implants into silicon. This will be reported in detail elsewhere.

Conclusion

We have shown that the PD technique is sensitive to the microstructure of silicon samples implanted with BF_2. An apparently uniform series of doses and energies gave PD readings which were subsequently found to correlate well with very non-uniform damage profiles in the material, as determined by RBS. It has been shown that PD is sensitive not only to dose, but also to the crystalline structure of the sample, and surface contamination. Clearly, PD requires complementary analysis to be valuable as a diagnostic tool, but the examples presented underline the possibility of using this thermal wave technique as a process control instrument for monitoring the ion implantation process, or for incoming wafer inspection.

References

1. S. Ameri, E.A. Ash, V. Neuman and C.R. Petts, Elec. Letts., Vol. 17, 337, 1981.
2. G.M. Crean, S.J. Sheard, C.W. See and M.G. Somekh, Proc. IEEE Ultrason. Symp., 597, 1987.
3. W.L. Smith, A. Rosencwaig and D.L. Willenborg, Appl. Phys. Lett., 47, 584, 1985.
4. G.M. Crean, M.G. Somekh, S.J. Sheard and S.W. See, Mater. Sci. Eng., B2, 207, 1989.
5. W.K. Chu, M.A. Nicolet and J.W. Mayer, Backscattering Spectrometry, Academic, New York, 1977.

	B			F		
BF_2 ion energy (keV)	50.0	100.0	200.0	50.0	100.0	200.0
Atom energy (keV)	11.2	22.4	44.9	19.4	38.8	77.6
Range (nm)	45.8	89.1	172.4	44.3	86.0	−
Straggle (nm)	26.2	43.1	67.3	21.9	37.1	−
Amorphisation Dose ($*10^{15}$ atoms/cm^2)	4.5	5.4	6.9	2.0	2.3	2.7

Table 1: Implantation Parameters

Sample		Damage	
Energy (keV)	Dose ($10^{14} B/cm^2$)	Surface Peak (nm)	Sub-surface Peak (nm)
50	5	49.1	0
	10	52.0	0
	30	113.0	0
100	5	59.5	0
	10	84.0	0
	30	48.2	25.2
200	5	−	47.8
	10	18.6	30.5
	30	73.1	158.0

Table 2: RBS Measurements of Surface and Sub-surface Damage, attributable to the implantation, expressed in nm where $10^{16} Si/cm^2 = 2nm$

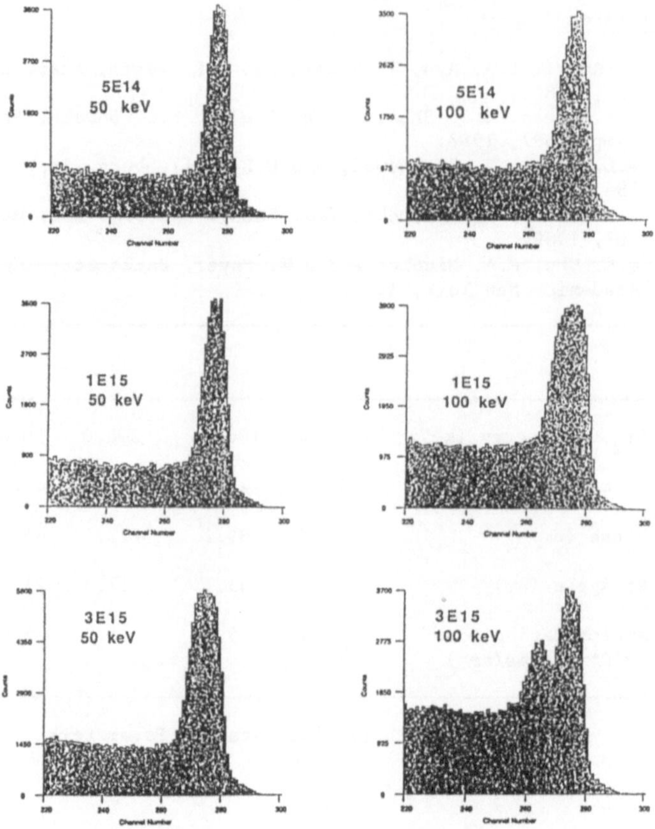

Figure 1

Normalised RBS channelling spectra from the as-implanted samples, depth scale is 6nm/channel.

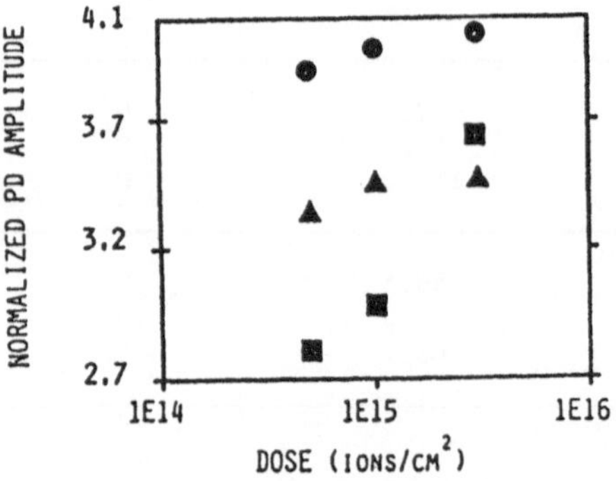

Figure 2

Photodisplacement signal amplitude as a function of BF_2 dose and energy implant parameters: 50 keV (■), 100 keV (▲) and 200 keV (●).

Non Destructive Electrical Characterization of Semiconducting Layers by a Novel Microwave Method

P.TABOURIER, C. DRUON, N. BOURZGUI and J.M. WACRENIER

Université des Sciences de Lille Flandres-Artois
Centre Hyperfréquences et Semiconducteurs, U.A. CNRS 287
Bâtiment P3 - 59655 - VILLENEUVE D'ASCQ CEDEX (France)

ABSTRACT

We propose a microwave device using a novel cell which allows one to measure the sheet resistance ($R_\square$), the carrier density (n) and the mobility (μ) of epitaxial layers. The electrical contacts between the sample and the cell are capacitive. The method is non-destructive and doesn't require any technological process. Measurements can be performed within the following ranges :

$$5 \ \Omega < R_\square < 2500 \ \Omega \ ; \quad 100 \ cm^2/V.s. < \mu \ ; \quad 5 \ 10^{15} cm^{-3} < n < 5 \ 10^{18} cm^{-3}$$

INTRODUCTION

It is very useful to be able to measure some main electrical characteristics of semiconducting layers quickly i.e. the sheet resistance ($R_\square$) ; the majority carrier concentration (n) and mobility (μ). In addition, mappings for n and μ can be desirable. For these properties, absolute measurements and great accuracy are often not needed, but good reproducibility is necessary to allow the comparison of different samples. Standard methods, such as the polaron profiling or the van der Pauw method, usually degrade the sample or require some technological processes before operating [1-3]. Moreover when non-destructive, they seldom give $R_\square$, n and μ at the same time [4-8]. To perform this without special sample preparation, we propose a novel cell using non-destructive contacts between the tested layer and the measuring apparatus.

DEVICE REALIZATION

To study magnetoelectrical effects, electrodes are generally placed on several points of the sample under a magnetic field B. These electrodes are used to measure the voltages linked to a set current flow. In our device, the electrodes are made up of four microstrip lines laid crosswise to each otheras shown in figure 1. Each of the four line ends are covered, once for all, with a thin dielectric layer. In the operating disposition the sample is placed at the centre of the cross and lightly pressed to it.

So, the dielectric layer is the insulator of a MIS contact and the

934

contacts between the four electrodes and the sample are capacitive. The sample size and the working frequency are chosen in order :

-to get small values for the capacitive impedances with respect to $R_\square$.

-to analyse the sample as a lumped element circuit.

Therefore the cell has the following characteristics :

- 600 μm wide, golden, 50 Ω microstrip lines on a 1"x 1" polished alumina substrate.

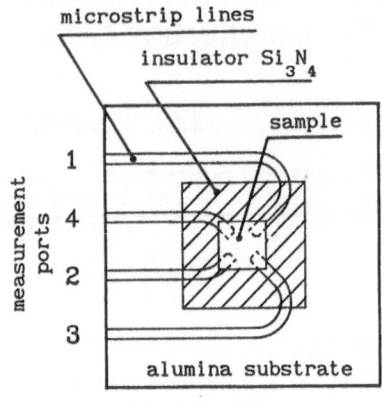

Fig.1- Microwave measurement cell

- a 1800 μm gap between two opposite lines ends.

- a Si3N4 insulator thickness of about 2000 Å.

Square shaped samples of 3x3mm are used and the working frequency ranges between 1 and 2 GHz.

MEASUREMENTS AND EXPERIMENTAL DATA PROCESSING

Measurements and computer flow chart leading to the determination of $R_\square$, μ and n are shown in figure 2.

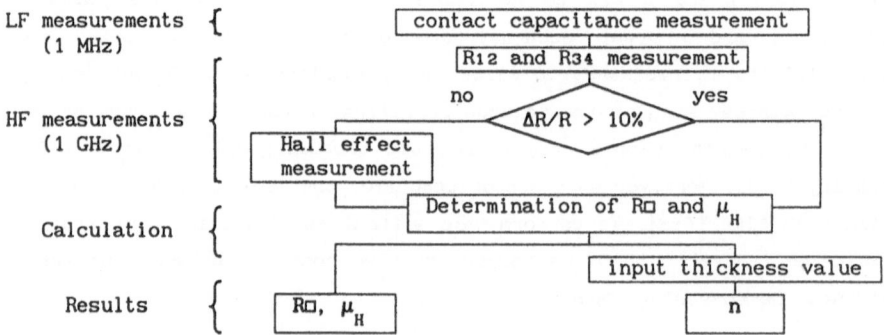

Figure 2 : Flow chart for determination of $R_\square$, μ and n

This chart can be divided into three parts :

- In the first part, the contact capacitances are measured. These values (typically 20 pF) are used as corrections in the third part.

- In the second part, the resistance R_{12} (R_{34}) between opposite ports (see figure 1) is measured without B. Then the measurements of magnetoresistance and of Hall effect are performed. In each case it is only necessary to measure the modulus of the transmission parameters between the different ports. So getting rid of the scattering parameters phases leads to a notable simplification of the experimental set up lowering its cost. Moreover, the choice of a symmetrical structure allows to perform redundant measurements and to check the sample homogeneity.

.The magnetoresistance results are used if the relative deviation of R (R12,R34) noted ΔR/R is greater than 10 % when B is raised from 0 to 1 T.

.The Hall measurement consists in measuring, for example, the difference between the transmission parameters S13 and S14 when the incident power is sent on the port 1; all the other ports being loaded by 50 Ω.

- In the third part, R□ and μ_H (Hall mobility) are calculated. The computer program we have created takes into account (by means of geometrical factors) the shape of the sample, the shape of the electrodes and the existence of a 50 Ω impedance between the ports used for the Hall effect measurement. These factors depend on the Hall angle α (defined by: $tg\alpha = \mu_H . B$) and their value has been determined experimentally and confirmed by analytic calculation.

MEASUREMENT CONDITIONS AND RESULTS

We have checked, by a microscopic investigation, that the surface is not damaged when the electrodes are applied on the semiconductor. Repeated measurements on the same sample show that the reproductibility is better than 5 %. Setting the sample is an easy operation and it takes about two minutes to perform the characterization. The measurement ranges, with GaAs samples at ambient temperature are as follows :

$$5 \ \Omega < R□ < 2500 \ \Omega \ ; \quad 5 \ 10^{15} cm^{-3} < n < 5.10^{18} cm^{-3} \ ; \quad 100 \ cm^2/V.s < \mu$$

These ranges correspond with practical situations usually encountered in laboratories and in industry. Figure 3 shows the results obtained for n and μ with GaAs samples. It allows a straight comparison with the van der Pauw method considered usually as a reference method.

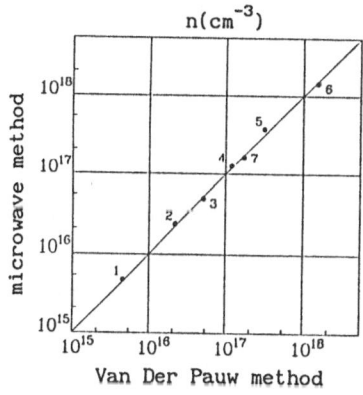
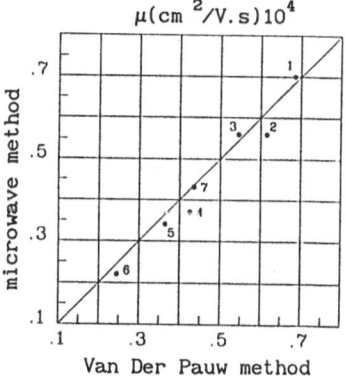

Fig.3- Results for n and μ.

Interconnect Heating by Pulsed DC

G.Röska[1] and C.Mazuré[2]

Siemens AG,Semiconductor Division[1] and
Corporate Research and Development[2], Microelectronics
Otto-Hahn-Ring 6, D-8000 München 83, F. R. G.

Summary
The line heating and cooling in a pulsed DC mode of AlSiTi
conductors on thermal and CVD oxides is investigated. A novel
measurement technique is described for direct monitoring of
transient temperature changes. The rise and fall times for Joule
heating are experimentally measured. The self-heating and
relaxation values for different isolation layers is discussed.

Introduction

The down scaling of device geometry increases the MOSFET output
current and reduces the line cross section. The consequence is
an overproportional increase of current density and of Joule
heating effects. On the other hand, the metallization complex
requires thicker isolation oxides in order to minimize parasitic
capacitances. This leads to a higher average line temperature.
In general Joule heating accelerates line degradation. The
failure of Al interconnects due to electromigration is strongly
temperature dependent. In the case of pulsed DC stress Joule
heating will start at the onset of the current pulse and in the
off-state damage relaxation might lead to interconnect self-
healing [1-3]. In order to better understand and estimate the
line degradation behavior it is necessary to separately characterize the Joule heating effects from the current density
dependent component.

Experimental

The test samples were packaged unpassivated AlSi(1%)Ti(0.2%)
lines patterned on oxide/Si substrates. Different isolation
underlayers were considered: boron-phosphorus-silicon glass
(BPSG), BPSG on thermal wet oxide (FOX) and a plasma oxide
(POX)-BPSG-FOX system. The test conductors were single straight
lines (500μm long, 1.5μm wide). The AlSiTi thickness was 800nm.

The time resolved signal analysis was performed with a Boxcar Averager and a signal processing unit (EG&G Model 4400). The current change was inductively measured with a current transformer probe (Tektronix CT-2/P6041-Probe) in order to avoid any modification of the stress pulse by the signal detection probes. The sample resistance at time t was obtained from the ratio voltage to current amplitudes of the in-going and out-going monitor pulses, respectively.

The DC current pulses for interconnect heating (2.5×10^6 A/cm^2) were generated by a high power pulse generator with a frequency of 1kHz. The pulse width was varied between 0.1 and 5 µs. The duty cycle was chosen sufficiently low for the interconnects to cool down totally between the heat pulses. The interconnect temperature change was monitored after the heating pulse by measuring the transient resistance with a low amplitude pulse train to avoid any subsequent heating (0.2×10^6 A/cm^2; 0.2µs). During measurement no noticeable interconnect degradation occurred. The resistance change was significantly lower than 1%.

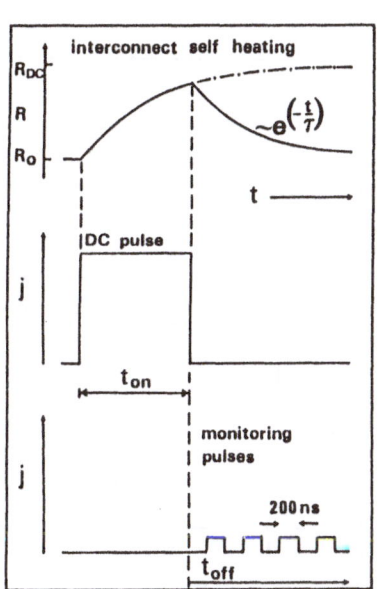
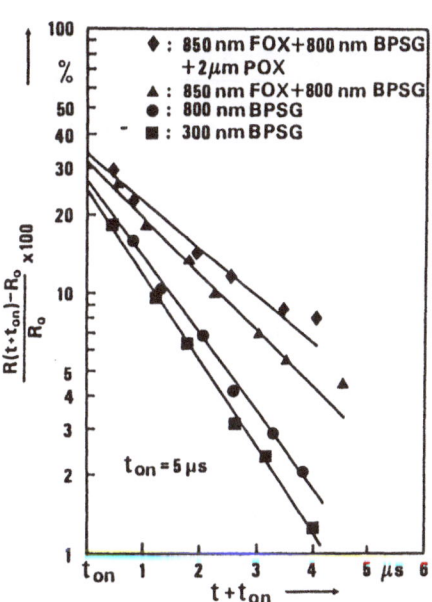

Fig. 1: Schematical drawing of a) line heating and cooling, b) of the DC pulse-on and pulse-off times and c) of the monitoring pulse train.

Fig. 2: Relative interconnect resistance decrease after current pulse vs. $t_{on}+t \leq t_{off}$ for different oxide thicknesses.

Results and Discussion

The temperature increase and relaxation due to Joule heating is normally assumed to follow an exponential time law [4], $(1-\exp(-t/\tau))$ and $\exp(-t/\tau)$ for the temperature increase and decrease, respectively , where τ is a function of sample geometry and of the thermal conductances of the substrate, dielectric and conductor (fig. 1). The temperature change is linearly proportional to the line resistance variation. The interconnect self-heating occurs during the time interval $0 \leq t \leq t_{on}$. $t_{on} \leq t \leq t_{on}+t_{off}$ is the pulse-off time interval. t_{on} and t_{off} define the pulse-on and pulse-off times, respectively.

The relative resistance decrease $[R(t_{on}+t)-R_0]/R_0$ after the current pulse as a function of $t_{on}+t \leq t_{off}$ is shown in fig. 2. The measurements were done for $t_{on}= 5\mu s$ for different oxide underlayers. R_0 is the interconnect resistance at ambient temperature. $[R(t_{on}+t)-R_0]/R_0$ follows an exponential law. The extrapolation to $t= t_{on}$ gives the line resistance at the end of the DC pulse. To determine the temperature increase during a

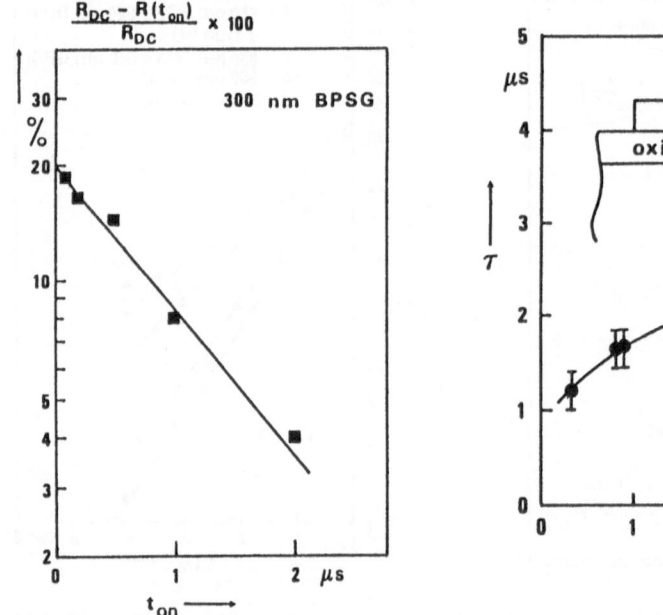

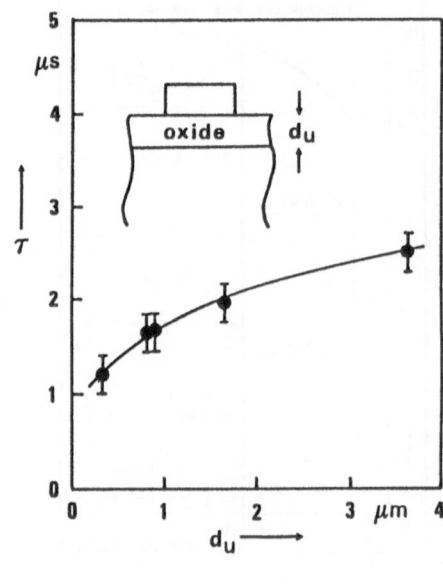

Fig. 3: Relative interconnect resistance increase vs. t_{on}.
Fig. 4: Influence of the oxide underlayer thickness on relaxation time τ.

current pulse t_{on} was varied and the transient resistance change was measured after the heating DC pulse. As shown in fig. 3 the relative resistance increase $[R_{DC}-R(t)]/R_{DC}$ vs. t_{on} is described by an exponential time law. The line resistance saturates at R_{DC} which is the value reached under constant current conditions (Fig. 1).

For 300 nm BPSG the temperature rise and fall times are found to be equal $\tau_{rise}= \tau_{fall}= (1.2 \pm 0.1)$ µs as expected. Furthermore for samples with different oxide types (FOX and BPSG) but with the same oxide thickness (850nm) the same τ is found. This suggests that τ is mainly given by the total oxide thickness. The oxide thickness vs. τ is shown in fig. 4 neglecting the specific differences of the oxide underlayers. Notice that τ increases very weakly with the oxide thickness. As a consequence almost the same self-heating and relaxation behavior is to be expected for the first metal level (oxide thickness range: 0.5 - 1.0µm) as for the second metal (oxide thickness range: 2 - 4µm).

Conclusion

The self-heating and relaxation times for different dielectric underlayers were investigated. All the results suggest that an exponential law for the temperature calculation of metal IC interconnects is valid. The rise and fall times for Joule heating were experimentally measured. The heating and cooling time constants were found to be equal and to be mainly a function of the total isolation thickness. Furthermore the measuring technique permits the characterization of the transient temperature dependence during and after the current pulse. This is necessary to better understand the different mechanisms leading to line degradation during dynamic stress.

References

[1] J.M. Schoon, J. Appl. Phys. 51, p. 508-512 (1980).
[2] L. Brooke, IEEE Proc. 25th IRPS, pp. 136-139 (1987).
[3] B.K. Liew, N.W. Chueng, C. Hu, IEEE Proc. 27th IRPS, pp. 215-219 (1989).
[4] J.A. Maiz, IEEE Proc. 27th IRPS, pp. 220-228 (1989).

Split-conductance, -capacitance and -current Measurements in MOSFETs

M. EL-SAYED[1] and H. HADDARA[2]

1) Electrical Eng. Dept. Alexandria Univ., Alexandria, Egypt.
2) Electronics and Computer Eng. Dept., Ain-Shams Univ., Cairo, Egypt.

Summary

In this paper, we propose a unified analytical model which describes small signal split-admittance as well as large signal split-current in relatively short channel MOSFETs. In the case of large signals, the average value of the split-current is what we classically know as the charge pumping current. Unlike what is known untill now, our model indicates that the charge pumping current is not simply the net electron or hole recombination current through the interface traps. Instead, this current is greatly influenced by the capacitive coupling between the interface traps and the device terminals. This model is validated using small signal split-admittance and large signal split-current measurements.

Introduction

Since the introduction of the idea of split-capacitance measurements in MOSFETs [1], it has been mostly carried out under quasistatic or low frequency conditions and the influence of the interface traps was not adequately considered. In a recent work, we presented a new approach to admittance measurements on relatively short channel MOSFETs [2,3]. It was shown that the admittance measurements in the gate circuit at various frequencies, can be used to study interface trap properties in most of the Si bandgap by direct measurements on a single MOSFET. In the present work, we introduce a new and unified approach to split- current and admittance measurements in MOSFETs taking into account the interface trap dynamics when an arbitrary time varying gate signal is applied. The interest of this approach resides in three main aspects: 1) Its universality in explaining generation-recombination processes under small and large gate signals. 2) It allows an accurate evaluation of small signal split-admittance which is very useful in the determination of interface trap properties [3], effective channel mobility [4] and other device parameters. 3) It provides a precise representation of time dependent split-current under large signal conditions. This allows a better understanding of the origin of the charge pumping current [5,6].

Model of Split-Current in MOSFETs

We consider in our analysis the case of an n-channel MOSFET in which an arbitrary time varying signal is applied to the gate while the source, drain and substrate are grounded. At any instant t the gate voltage $V_g(t)$ is expressed as

$$V_g(t) = \Phi_{ms} + V_{ox}(t) + \Psi_s(t) \tag{1}$$

Also, the charge neutrality requirement gives

$$Q_g(t) + Q_B(t) + Q_{inv}(t) + Q_{it}(t) + Q_{ox} = 0 \tag{2}$$

By differentiating (1) and (2) with respect to t and expressing the term dQ_{it}/dt as the sum of the electron and hole recombination currents, $I_n(t)$ and $I_p(t)$, through the interface traps, we can show that the gate current is given by

$$I_g(t) = C_g \frac{dV_g}{dt} + \frac{C_{ox}}{C_{ox}+C_B+C_{inv}} (I_p + I_n) \tag{3}$$

where, $C_g = C_{ox}(C_B+C_{inv})/(C_{ox}+C_B+C_{inv})$, is the total gate capacitance.

The split-current measured in the substrate and source/drain circuits, $I_{sub}(t)$ and $I_{s,d}(t)$, are given by

$$I_{sub}(t) = C_{gb} \frac{dV_g}{dt} + (1- \frac{C_{gb}}{C_{ox}}) I_p - \frac{C_{gb}}{C_{ox}} I_n \tag{4}$$

$$I_{s,d}(t) = C_{gc} \frac{dV_g}{dt} + (1-\frac{C_{gc}}{C_{ox}}) I_n - \frac{C_{gc}}{C_{ox}} I_p \tag{5}$$

where $C_{gb}=C_{ox}C_B/(C_{ox}+C_B+C_{inv})$, is the gate-substrate capacitance and $C_{gc}=C_{ox}C_{inv}/(C_{ox}+C_B+C_{inv})$, is the gate-channel capacitance. In (3), (4) and (5), the first term on the right hand side is the displacement current while the other terms are the contributions of I_n and I_p measured in the gate, substrate and source/drain circuits. It is clear that the latter terms are not the pure recombination currents; instead, I_n and I_p are multiplied by weighting factors which include capacitive coupling combinations between the device terminals. This point will be discussed in detail when the model is applied to the charge pumping technique.

The above model can be extended to the case of an ac small signal applied to the gate while the device is biased either in depletion or in inversion regime. Hence, expressions for the gate-substrate admittance and gate--channel admittance can be obtained. The same results can be obtained from the MOSFET small signal equivalent circuit developped in [3]. For example, the gate-substrate conductance G_{gb} and gate-channel conductance G_{gc} in inversion are given by

$$G_{gc} = \frac{C_{ox}(C_{ox}+C_B)\ G_{pn}}{(C_{ox}+C_B+C_{inv}+C_{pn})^2 + G_{pn}^2/\omega^2} \tag{6}$$

$$G_{gb} = - \frac{C_B}{C_{ox}+C_B}\ G_{gc} \tag{7}$$

where G_{pn} and C_{pn} are the conductance and capacitance of the interface traps interacting with the conduction band.

It is interesting to note that G_{gb} in inversion is not zero but it is a negative quantity. This is due to the fact that G_{gc} is reflected in the substrate through the capacitive coupling factor $C_B/(C_{ox}+C_B)$ in (7).

It seems that the capacitive coupling between the device terminals is of great importance when MOSFET split-admittance and large signal split-current are used to extract device parameters. The results of the above model are verfied experimentally by performing split-admittance and large signal split-current measurements on relatively short channel MOSFETs.

Experimental Details

Measurements have been performed on poly-silicon gate n-channel MOSFETs having substrate doping of 2×10^{16} cm^{-3}, gate oxide thickness of 25 nm, channel length of 3 μm and channel width of 6000 μm. The devices were irradiated to increase interface trap density in order to magnify their influence on the electrical performance of the device. Admittance measurements are carried out using the HP4192A impedance analyzer at frequencies between 1 kHz and 10 MHz. Large signal split current measurements are performed in an identical way as those in the charge pumping technique [5,6]. Yet, the instantaneous substrate and source/drain currents, whose average are equal to the well known charge pumping current, are visualized on an oscilloscope and recorded as a function of time using the setup shown in Fig. (1).

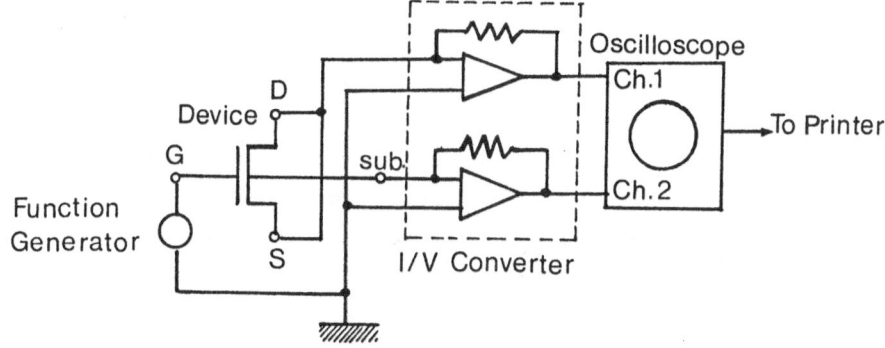

Fig.(1). Schematic diagram of the large signal split-current measuring circuit.

Results and Discussion

Figure (2) shows the obtained experimental results for the split-conductances as a function of V_g at 1 MHz. It is seen that G_{gb} exhibits a peak in depletion attributed to the interaction of the interface traps with the valence band, but it is not zero in inversion. Instead, G_{gb} goes through a negative peak at the same V_g value as that of the positive peak of G_{gc}. This is in accordance with (6) and (7).

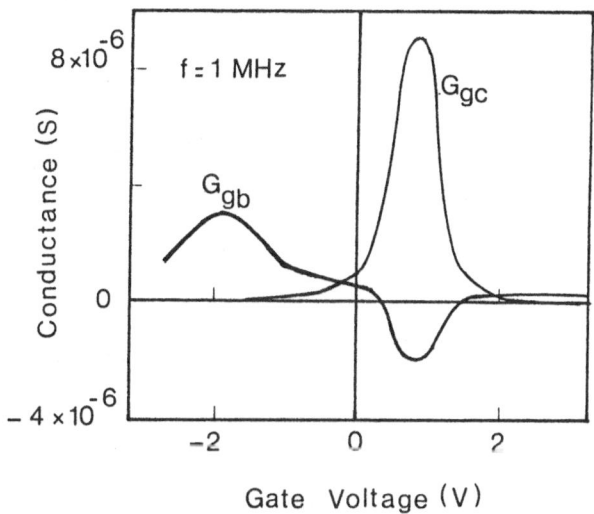

Fig.(2). Measured split-conductance vs gate voltage at f=1MHz.

Figure (3) shows the obtained results of I_{sub} and $I_{s,d}$ for the two sweep directions for a large triangular signal of 1 kHz frequency. The time axis is translated to a gate voltage axis for a better interpretation of the data. It can be seen that during the positive sweep, $I_{s,d}$ is simply a superposition of two components: a displacement current which is directly proportional to C_{gc} and a contribution of I_n resulting from the capture of electrons by the interface traps from the rapidly formed channel. This contribution is significant in weak inversion and is negligibly small in strong inversion. This can be explained in the light of (5) since C_{gc} approaches C_{ox} when the device enters the strong inversion regime. On the other hand, I_{sub} in depletion, is simply the displacement current which is directly proportional to C_{gb} while in weak inversion a negative contribution of I_n is added to the displacement current, in accordance with (4).

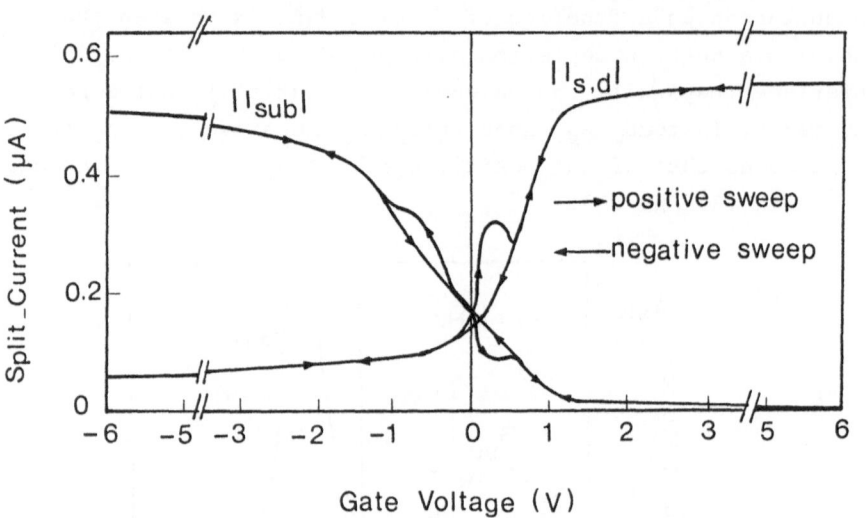

Gate Voltage (V)

Fig.(3). Measured split-current vs gate voltage for a large triangular signal (f=1 kHz) for the two sweep directions.

During the negative sweep, $I_{s,d}$ is nearly a pure displacement current since most of the interface traps interact with the substrate. This is clear in I_{sub} curve where a contribution of I_p is added to the displacement current in depletion. This contribution becomes negligibly small in strong accumulation since C_{gb} approaches C_{ox}, in accordance also with (4).

In the light of our model and the above discussion it is clear that the contributions of the electron and hole recombination currents in I_{sub} and $I_{s,d}$ are greatly influenced by the capacitive coupling between the device terminals. Knowning that the average value of either I_{sub} or $I_{s,d}$ over one complete cycle is what we know as the charge pumping current I_{cp}, we can see that I_{cp} is not simply the average value of either I_p or I_n as classically known. Instead, I_{cp} is a cobination of both of them and includes capacitive terms which are in general, time dependent functions especially when triangular or trapozoidal signals are used. This could lead to significant errors in evaluating interface trap properties. Therefore, it is necessary to make corrections for I_{cp} to obtain the average values of the pure I_n and I_p. The kind of measurement proposed in this work could be a possible solution for this problem.

References

1. Koomen, J; Solid St. Electron. 16 (1973) 801.
2. Haddara, H.and El-Sayed, M; Solid St. Electron, 31 (1988) 1289.
3. Haddara, H.and El-Sayed, M; Proc. Essderc87, Pologna (1987) 695.
4. Liang, M.S.; Choi, J.Y.; Ko, P.K.; Hu, C., IEEE Trans. Electron Devices, ED 33 (1986) 409.
5. Groeseneken, G.; Maes, H.E.; De Keersmaeker, R.F., IEEE Trans. Electron Devices, ED 31 (1984) 42.
6. Haddara, H. and Cristoloveanu, S., Solid St. Electron. 29 (1986) 767.

Determination of the Emitter Lateral Doping Profile for Self-aligned Bipolar Transistors

M.Ohnemus and M.Miura-Mattausch

Siemens AG, Corporate Research and Development, Otto-Hahn-Ring 6, 8000 Munich 83

The lateral arsenic doping profile of the emitter outdiffused from the polysilicon layer has been estimated. It has been done by a combination of measured E/B depletion junction capacitance C_{EB} and that calculated by 2D simulation. Since the peripheral C_{EB} is very sensitive to the lateral doping profile, it is varied until the simulated C_{EB} reproduces the measurement. The result shows that the lateral diffusion length is very much dependent on the drive-in condition for the diffusion. The drive-in temperature of 1050 °C / 10 sec gives the lateral diffusion of about 75 %. That of 1075 °C / 10 sec gives nearly 100 % of the vertical one.

By decreasing the transistor size the link part, where the base contact p^+ and the emitter n^+ lateral diffusions are merged underneath the oxide spacer, becomes increasingly important for transistor performance. Therefore an optimization of this part is a key issue. Though vertical doping profiles are available from SIMS measurements, the dependence of the lateral diffusion profile on the process technology is still lacking. We will show here for the first time results of the emitter lateral doping profile derived by a combination of measured capacitances at E/B junction with 2D simulation results.

Device Preparation :

In a polysilicon self-aligned scheme, the extrinsic base and emitter are outdiffused of and contacted with highly doped silicon layers separated by a residual oxide spacer (see Fig. 1) [1]. The drive-in is performed by rapid thermal annealing at 1050 °C and 1075 °C both for 10 sec. The intrinsic base is implanted prior to the formation of the spacer for this study to avoid an incomplete overlap of active and inactive base regions. The spacer width is kept large enough so that no overlap between n^+ and p^+ occurs. It has been observed that the encroachment of the extrinsic base with the active base region leads to a reduction in current gain [2]. This is not the case for devices studied here. A measured reverse I-V characteristics at the E/B junction is shown in Fig. 2. The moderate contribution of tunneling current is due to the relatively high doping concentration of the active base region.

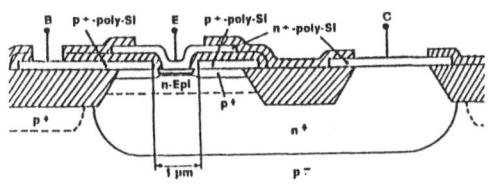

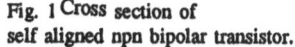

Fig. 1 Cross section of
self aligned npn bipolar transistor.

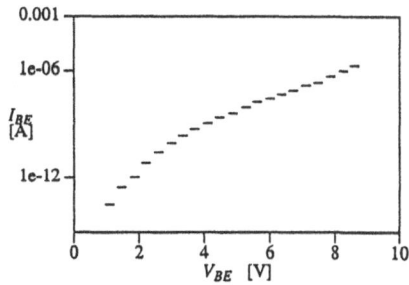

Fig. 2
Measured reverse I-V characteristics.

Measured and Calculated Capacitance :

Fig. 3 shows measured capacitance at the E/B junction C_{EB} as a function of applied voltage V_{EB}. A steep increase of C_{EB} for $V_{EB} > 0.4$ V refers to the domination of the neutral capacitance [3]. However, for low V_{EB} the total capacitance is equal to that of the depletion capacitance. Thus, capacitances at $V_{EB} = 0$ V are taken as measured values later on. Measured is C_{EB} for 8 different emitter sizes ranging from 20×20 μm^2 to 1.4×20 μm^2. From the size dependence of C_{EB}, a partitioning of C_{EB} into an area part $C_{EB,a}$ and a peripheral part $C_{EB,p}$ is possible (see Fig. 4), under the approximation $C_{EB} = A \times C_{EB,a} + P \times C_{EB,p}$. In the approximation A and P denote the area size and the periphery length, respectively. If the transistor becomes too small, an additional non-linear effect occurs . The value of $C_{EB,p}$ is not constant along the junction but is large where the junction curvature is large as shown in Fig. 5. Such non homogeneous distribution in $C_{EB,p}$ gives a steep increase for small transistors as seen in Fig. 4. For $C_{EB,a}$ and $C_{EB,p}$ partitioning, thus, the approximation above has to be modified [4]. Although $C_{EB,a}$ is determined by the vertical doping profile, $C_{EB,p}$ is expected to be influenced by the lateral profile which may be sensitive to the drive-in conditions. The partioned result is shown in Table 1.

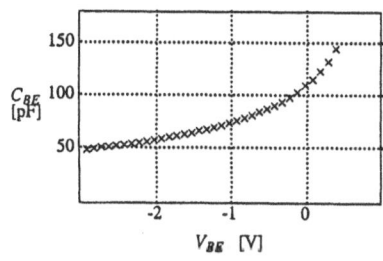

Fig. 3
Measured capacitance at E/B junction.

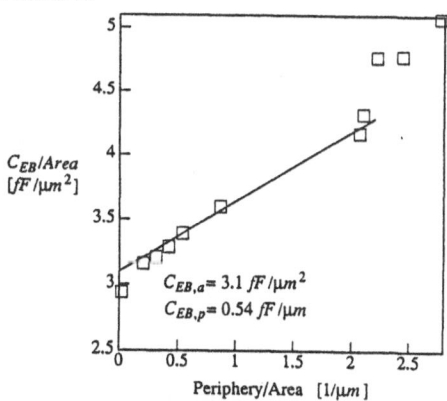

Fig. 4 Partioning of measured C_{EB}
into $C_{EB,a}$ and $C_{EB,p}$.

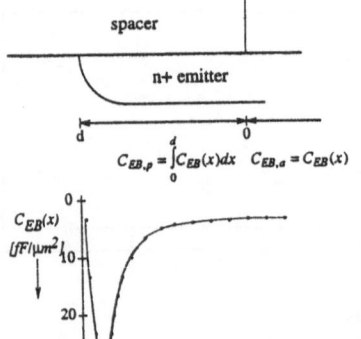

Table 1	Measured C_{EB}	
Temperature [°C]	$C_{EB,a}$ [fF/μm²]	$C_{EB,p}$ [fF/μm]
1050	3.4	0.42
1075	3.1	0.54

$$C_{EB,p} = \int_0^d C_{EB}(x)\,dx \qquad C_{EB,a} = C_{EB}(x)$$

Table 2	C_{EB} dependence on d	
d [μm]	$C_{EB,a}$ [fF/μm²]	$C_{EB,p}$ [fF/μm]
(a) 46	3.4	0.42
(b) 15	3.4	0.27

Fig. 5 Calculated C_{EB} distribution under the spacer.

A 2D simulation is performed by GALENE II [5] to calculate C_{EB}. The calculated forward I-V characteristics are compared with the measured ones. The activated n^+ emitter concentration is fitted to reproduce measured current gain. Since the measurement is not available for $V_{EB} < 0.2$ V and the accuracy of simulation is no more precise enough, simulation is performed at $V_{EB} = 0.2$ V. As the calculated $C_{EB,p}$ shows no dependence on V_{EB} below $V_{EB} < 0.3$ V, this condition gives no influence on determing the lateral doping profile. The area capacitance $C_{EB,a}$ is a function of the vertical doping profile, which is fitted to the SIMS measurements (see Fig. 6). The lateral doping profile is varied until the measured $C_{EB,p}$ is reproduced. We have varied both the lateral diffusion length d and the gradient of the doping profile. Two examples of the $C_{EB,p}$ dependence on d value and their p/n-junction profiles are shown in Table 2 and Fig. 7. About 10 nm change causes 0.05 fF/μm change. The gradient is modified according to the change of the lateral diffusion length d by keeping the condition that the ratio of the vertical diffusion to d is equal to that of the vertical doping gradient to the lateral doping gradient.

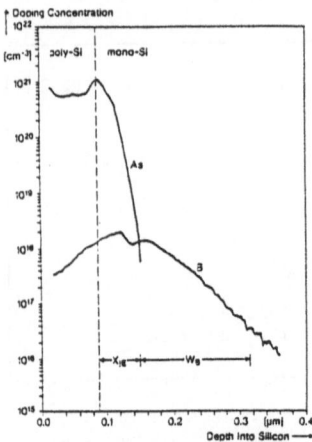

Fig. 6 Vertical E/B doping profile measurement by SIMS

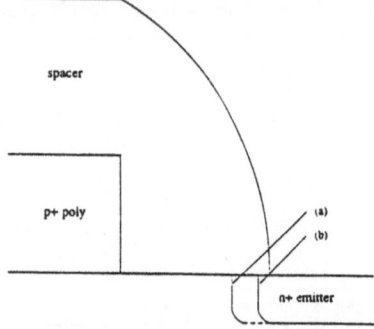

Fig. 7 Two examples of lateral diffusion: calculated C_{EB} values are seen at Table 2.

Resulting lateral diffusions :

Resulting doping profiles for different drive-in cycles are shown in Fig. 8. The estimated lateral diffusion d is listed in Table 3 and compared with the vertical one. The accuracy of this method is dependent on that of the measured $C_{EB,p}$, which is the gradient of the straight line in Fig. 4 and expected to be below 5 %. If the measured $C_{EB,p}$ has 5 % error, the inaccuracy of the estimated lateral diffusion becomes about 5 nm. Thus this method is promising in the sense of accuracy.

Usually about 70 % of the vertical diffusion is assumed for the lateral diffusion in simulations. However, it is shown here that the lateral doping profile is very much dependent on the drive-in conditions. By increasing the temperature of the drive-in cycle the lateral diffusion is more enhanced, and becomes nearly equal to the vertical diffusion. In order to reduce the relative lateral diffusion to keep low peripheral capacitance, a drive-in temperature of 1075 $^{\circ}C$ is already too high.

Table 3

Temperature	Diffusion length		
	vertical	lateral	lateral/vertical
1050 $^{\circ}C$	60 nm	45 nm	75 %
1075 $^{\circ}C$	80 nm	80 nm	100 %

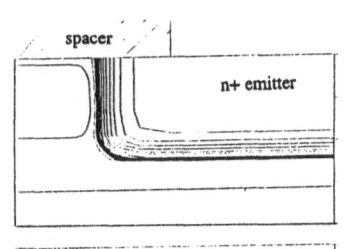

(a)

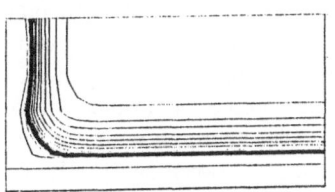

(b)

Fig. 8 Emitter diffusion profile
(a) 1050 $^{\circ}C$
(b) 1075 $^{\circ}C$

Acknowledgment :

The authors would like to express their thanks to R. Kircher and E. Pfoser for their help and valuable discussions.

References:

[1] H.Schaber, H.Bieger, T.Meister, K.Ehinger, and R.Kakoschke,
    IEDM Techn. Digest, p.170 (1987).

[2] G. P. Li, C. T. Chuang, T. C. Chen, and T. H. Ning,
    IEDM Techn. Digest, p.174 (1987).

[3] R. S. Sirsi and A. R. Boothroyd,
    IEEE Trans. Electron Devices, p. 348 (1976).

[4] J.Fertsch, J.Weng, and M.Miura-Mattausch,
    ICMTS in Edinburgh, p. 333 (1989).

[5] GALENE II User's Guide (1989), RWTH Aachen.

A Novel Approach for an Electrical Vernier to Measure Mask Misalignment

A.J. Walton, D. Ward, J.M Robertson, R.J. Holwill

Edinburgh Microfabrication Facility,
Department of Electrical Engineering,
Kings Buildings, University of Edinburgh,
Edinburgh, EH9 3JL, UK.

Summary

A novel interconnect scheme is presented which reduces the number of pads required by electrical verniers to measure mask misalignment. It makes the use of a shift register no longer necessary to keep the pad count to a reasonable number and the process is only required to support the fabrication of diodes. The vernier can be measured using any test equipment which can test for continuity.

Introduction

The misalignment that occurs between layers during the IC fabrication process must be allowed for in the design rules and this ultimately reduces the packing density. As a consequence it is important that the misalignments can be monitored in a routine manner. The use of optical verniers has long been commonplace [1] but their measurement is time consuming, tedious and error prone. Another approach is to use electrical structures that are based on the measurement of voltages between taps [2,3]. These devices all assume a uniform sheet resistivity and with misalignment tolerance being reduced as circuits become smaller so the accuracy of the voltmeter become more of a limiting factor.

The Electrical Vernier

One method of overcoming the above problem is to use an electrical vernier [4-7] which operates on the principle illustrated in figure 1. Connection between the teeth is electrically tested and the degree of misalignment extracted. This has the advantage that its resolution is only limited by the lithography and etch of the process. One other advantage of this structure is that information on the combined over-etch between two layers can also be extracted from the contact pattern. This pattern also provides a check on data integrity since the number of teeth which are not connected should always be an odd number and always adjacent to one another. An even number of open circuits or the presence of outliers indicate that there are potential problems with the integrety of the data. To individually test the connection between each tooth of the vernier obviously requires a large number pads. The number can be reduced by using a parallel load shift register to clock out the measurements (see figure 2). However, this has the disadvantage that a fully functional process is required and the test structure designer may not have the necessary skills required for the layout. The test structure proposed in this paper overcomes the limitations of all the above designs.

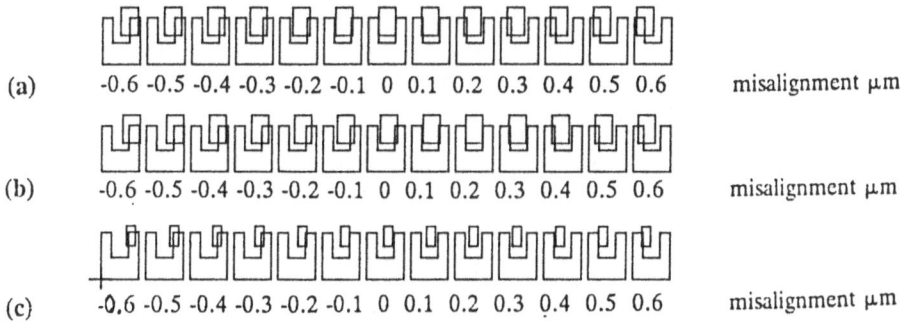

(a) -0.6 -0.5 -0.4 -0.3 -0.2 -0.1 0 0.1 0.2 0.3 0.4 0.5 0.6   misalignment μm

(b) -0.6 -0.5 -0.4 -0.3 -0.2 -0.1 0 0.1 0.2 0.3 0.4 0.5 0.6   misalignment μm

(c) -0.6 -0.5 -0.4 -0.3 -0.2 -0.1 0 0.1 0.2 0.3 0.4 0.5 0.6   misalignment μm

Figure 1. An electrical vernier which compensates for over-etch. (a) No over-etch with zero misalignment. (b) No over-etch with +0.2μm misalignment. (c) 0.3 over-etch with +0.2μm misalignment.

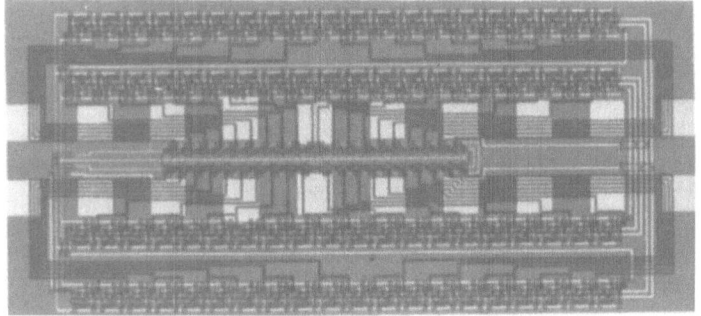

Figure 2. Electrical vernier that uses a shift register for the measurement.

The Diode Vernier

In this new diode vernier design the pads are connected to a number of teeth on the vernier in such a way to ensure that the information can be decoded in a very simple manner. Figure 3 shows a schematic illustrating the interconnect matrix and a typical contact pattern between the vernier teeth. Contact between the teeth is tested by sequentially forcing a voltage between every combination of group and access pads. The diodes which are in series with the bottom set of teeth are required to prevent current flowing via other parallel paths which may exist if the teeth are connected in a certain pattern. Figure 4 shows the layout of a diode vernier for measuring metal to polysilicon misalignment. This design performs the same function as the shift register vernier shown in figure 2 and the reduced area required can be observed since the pad size in both cases is the same. Table 1 gives the contact matrix between the pads for the vernier of figure 3 and from this it is possible to easily decode the position of the teeth that do not contact each other. Table 2 summarises the capabilities of the test structure and its superiority in terms of pad count to that where every tooth is connected to an individual pad is self evident. It can be observed that the pad to teeth ratio improves as the number of teeth increases. It is consequently advantageous to combine the x and y verniers on one interconnect matrix.

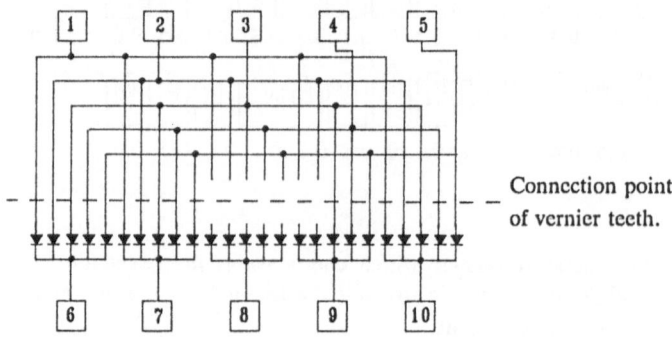

Figure 3. Schematic diagram of the diode vernier showing the pad connection.

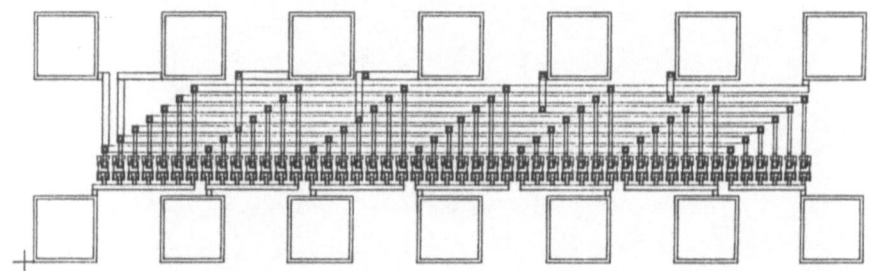

Figure 4. Layout of a diode vernier in nMOS technology.

Access	Group Pads				
Pads	6	7	8	9	10
1	1	1	0	0	1
2	1	1	0	0	1
3	1	1	0	1	1
4	1	1	0	1	1
5	1	1	0	1	1

Table 1. The contact matrix for the vernier schematically represented in figure 3.

No. of teeth per layer		16	25	36	49
No. of pads for diode vernier		8	10	12	14
No. of pads for standard vernier		17	26	37	50
Maximum range of misalignment and overetch	0.1μm resolution	1.6μm	2.5μm	3.6μm	4.9μm
	0.05μm resolution	0.8μm	1.25μm	1.8μm	2.45μm

Table 2. Comparison of the standard and diode vernier specifications.

Conclusions

In conclusion a test structure has been proposed which can measure misalignment with a resolution of that commensurate with that of the lithography and etch capabilities of the process. It has all the advantages of the vernier compared with the analogue structures but since it does not require a shift register its operation and measurement are much simpler. As a result it has more in common with other test structures and can be measured on an unmodified parametric test system. Furthermore, since the measurement is digital in nature, the accuracy will not be compromised if pads are commoned with those of other test structures to reduce the pad count still further.

References

1. W.C. Schnieder, "Testing the Mann Type 4800DSW Wafer Stepper", SPIE Developments in Semiconductor Microlithography IV, vol 174, 1979, pp 6-14.

2. D.S. Perloff, "A Four-Point Electrical Measurement Technique for Characterizing Mask Superposition Errors on Semiconductor Wafers", IEEE J Solid-State Circuits, vol sc-13, no 4, Aug 1978, pp 436-444.

3. C.M. Cork, "Off-line Photolithographic Parameter Extraction using Electrical Test structures", Proc ICMTS, Edinburgh, 13-14th March 1989, pp 7-14.

4. B.M.M Henderson, A.M. Gundlach, A.J. Walton, "Integrated Test Structure Which uses a Vernier to Electrically Measure Mask Misalignment", Electronics Letters, vol 19, no 21, 13th Oct 1983, pp 868-869.

5. A.J. Walton, W.R. Gammie, R.J. Holwill, B.M.M. Henderson, "Digital Measurement of Polysilicon to Diffusion Misalignment for a Silicon gate MOS process", Electronics Letters, vol 20, no 23, 8th Nov 1984, pp 951-952.

6. B.M.M Henderson, A.J. Walton, "A Complete Vernier Tool for the Measurement of Mask Misalignment", IEEE VLSI Workshop on Test Structures, Long Beach, USA, 17-18th Feb 1986, pp 34-50.

7. R. Yamaguchi, K. Komatsu, S. Moriya, K. Harada, "Integrated Electrical Vernier to Measure Registration Accuracy", IEEE Electron Device Letters, vol EDL-7, pp 463-464.

Nuclear Reaction Analysis of Electronic Materials – Part I: Metal-oxide-semiconductor Structures

M. A. BRIERE, F. WULF, D. BRÄUNIG

Hahn-Meitner-Institut Berlin GmbH
Glienicker Straße 100, 1000 Berlin 39, Germany

Summary

Use of the resonant nuclear reaction of ^1H (^{15}N, α,γ) ^{12}C at 6.4 MeV is applied to the analysis of hydrogen in materials used in the fabrication of microelectronic devices. Specifically, thin thermally grown silicon dioxide and glow discharge deposited amorphous silicon, have been chosen due to the critical role of hydrogen in determining the electrical behavior of devices incorporating these films. The characterization technique is capable of absolute hydrogen concentration determination to ~ 10 ppma as well as depth profiles with better than 10 nm resolution. A brief review of the measurement system is presented followed by some results for the aluminum-oxide-silicon system, including the first measurements of the accumulation of hydrogen at the silicon-silicon dioxide interface.

Analysis Technique

As is often the case, Nuclear Reaction Analysis (NRA) here refers to the ion beam technique which uses a resonant nuclear reaction between hydrogen and heavy nuclei (^{15}N), producing characteristic charged particles and/or gamma radiation, measurement of which yields information about the number and spatial distribution of the reaction site. Energy loss by charged particles while exiting the sample, provides direct depth information, the attenuation of gamma radiation is a quantum process, therefore only intensity information is gathered, yielding concentrations within the reaction volume. The reaction depth is determined by the energy of the resonance and the incident energy of the ion beam. As the incident ion penetrates the solid, it looses energy in a characteristic way and, for incident energies greater than that of the reaction, the ion's energy passes through the peak of the reaction cross section, yielding the characteristic signal. Thus a measure of the gamma radiation intensity as a function of incident ion energy provides a direct depth profile of the reaction site distribution. In the work discussed here, the resonant nuclear reaction involves hydrogen and the ^{15}N isotope, producing an alpha particle and a 4.43 MeV gamma ray. For the measurement of hydrogen, a beam of ^{15}N is accelerated to the resonant energy of 6.4 MeV, which would occur at the surface of a sample. The natural width of the resonance is very narrow, 1.8 keV, and even though beam energy spread and doppler broadening (due to the vibrational motion of the hydrogen), significantly increase this width (to 10-13 keV), a resolution of about 3 nm is possible in the near surface region. It not only looses energy, but its energy distribution broadens due to energy loss fluctuations (straggling). Straggling increases as the square root of the depth and quickly dominates the effective resonant width, causing poorer resolution. Even so, the depth resolution is better than 10 nm at a depth of 100 nm and 25 nm at 1μm in Si. The gamma-ray intensity is monitored using a (6" x 6") NaI (Tl) scintillator detector, integrating the signal under the photo peak and the first escape peak (3.7-4.6 MeV). Detectability is limited by natural background (.18 counts/s), which is dominated by cosmic ray induced showers of muons [1]. Anti-coincidence as well as heavy metal shielding can reduce this level by more than an order of magnitude [1,2], though no shielding is used in the arrangement used here. The ultimate detectability of the technique is estimated at 5 ppma [1], and very low level systems are under construction [2]. However, without shielding, the detectability of the system used here is better than 150 ppma. Errors are dominated by counting statistics. The nitrogen beam is produced by the 6 MeV single ended Van deGraff injector at the VICKSI fascility of the HMI in Berlin [3]. Energy changes are currently performed in discrete steps, adjusting the analyzing field in accordance with a NMR probe. All beam optics are automatically adjusted by computer control.

Hysterisis and NMR resolution limit the energy reprocibility to ± 1-2 keV. A hysterisis free system, utilizing electrostatic plates, as demonstrated elsewhere [4,5], is planned for installation within the next year. A schematic of the measurement chamber is shown in Figure 1. The sample is enclosed in a LN_2 cold trap to reduce hydrocarbon and water vapor residues as well as to provide sample cooling. The sample is mounted on the back copper plate with silver paste to insure good thermal contact. The distance from the sample to the detector is minimized so as to reduce signal losses. Beam currents of 80-100 nA on areas of 1 cm^2 are used to maximize sensitivity but minimize radiation damage to the sample.

Measurement of the concentration as well as the dynamic behavior of hydrogen in the MOS system is necessary for a more complete understanding of the instability phenomena which dominate the stress related response of such structures (ie. interface states and oxide trapped charge generation due to radiation or bias-stress). The technique of NRA combines inherent accuracy and resolution with a fundamentally non-invasive nature. Thus, this method presents the possibility of dynamic measurements of the hydrogen behavior in MOS devices (capacitors). In principle, in-situ measurements of radiation or other stress related effects on the hydrogen distribution as well as the initial distribution in as-processed films are possible.

Previous attempts to investigate the MOS structure using NRA have been only moderately successfull [6,7]. Issues concerning the stability of the hydrogen during the measurement as well as system sensitivities have hampered these efforts. The first complete characterization of such structures using NRA is presented here.

Experimental Results

Figure 2 is a depth profile of hydrogen concentration in a thick (2 μm)SiO_2 film that was implanted with 1E16 cm^{-2} H at 40 keV. The results are in excellent agreement with theoretical predictions. The distribution is found to be stable at room temperature over a period of months. Figure 3 is a profile across the aluminum-oxide interface of an MOS structure, showing the accumulation of hydrogen at the interface. The hydrogen is actually distributed within some 50 nm of the interface, as the measured profile is dominated more by beam straggling than hydrogen diffusion in the oxide [7]. It was found that most of the hydrogen has diffused away from the interface after a period of two months. Figure 4 shows some of the first NRA measurements of the accumulation of hydrogen at the SiO_2-Si interface, repeated twice for different parts of the same thermal oxide. The results are in general agreement with the SIMS data of Feigl et. al [8], indicating a measured peak concentration of ~4E19 cm^{-3} and an areal density on the order of 1E14 cm^{-2}. The oxide thickness was measured by ellipsometry to be ~ 48 nm. The baseline level is thought to be due to off-resonance contributions from the surface layer, though oxide thickness variations appear to be significant in at least one of the measurements. Later measurements have confirmed these findings with better statistical resolution.

Conclusion

The application of NRA to the investigation of electronic devices incorperating thin thermal oxides holds promise for a better understanding of the role of hydrogen in the long term stabilitiy of these systems.

References

[1] H.Damjantschitsch,et. al.:Nuclear Instruments and Methods 218 (1983),p.129-140
[2] F. Rausch: Private Communications
[3] K. Ziegler: IEEE Trans. on Nucl. Science, p- 1872, Vol. NS-76, # 2, April 1979
[4] G. Amsel, et. al.: Nuclear Instruments and Methods 205 (1983), p. 5-26
[5] M. Zinke-Allmang,et. al.:Nuclear Instruments and Methods 315 (1986),p. 563-568
[6] A. D. Marwick and D. R. Young: Journal Applied Physics 63 (7), (1988), p. 2291
[7] R. E. Benenson, et. al.: Nuclear Instruments and Methods 168 (1980), p. 5-47
[8] F. J. Feigl et. al.: Nuclear Instruments and Methods 131 (1984), p. 348-354

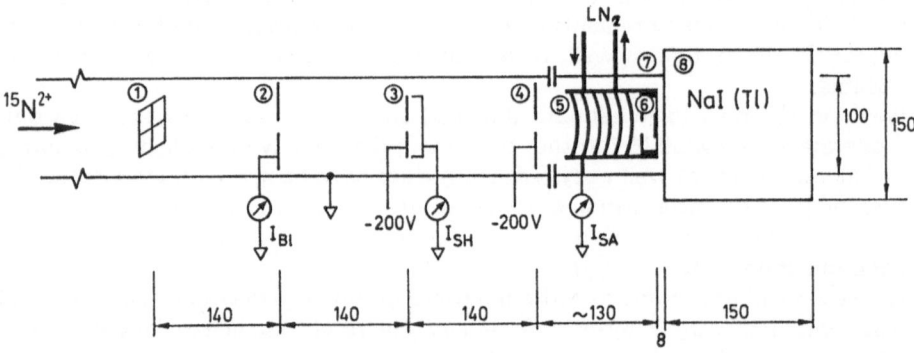

Fig. 1   Measurement Chamber: 1. Beam Monitor Quartz; 2. 7 mm Ø Aperture;
3. Standard Faraday Cup; 4. Supressonplate (14 mm Ø); 5. Coldtrap-Cu-coils,
6. Sample Location; 7. Sample Faraday; 8. Scintillation Detector

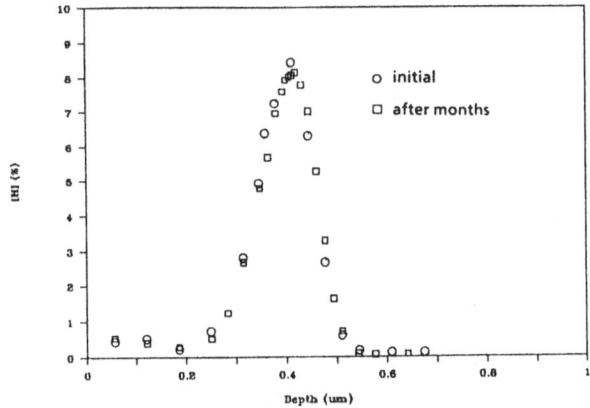

Fig. 2   Depth Profile of Hydrogen Concentration in a Thick (2 μm) SiO$_2$ Film
Dose 3E16 cm^{-3}, Energy 40 keV

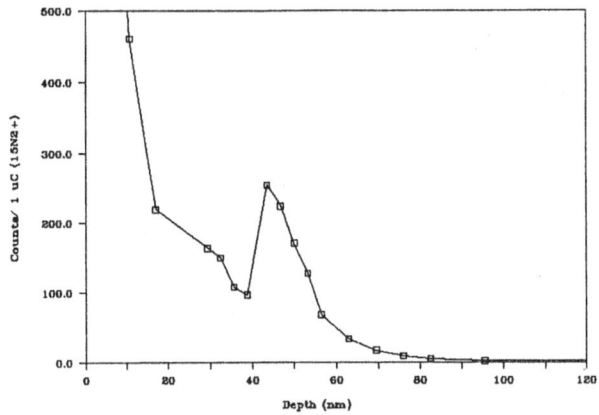

Fig. 3   4.43 MeV Gamma-ray Yield vs Depth across Al-SiO$_2$ Interface, Demonstrating the Pile Up of Hydrogen at this Boundary

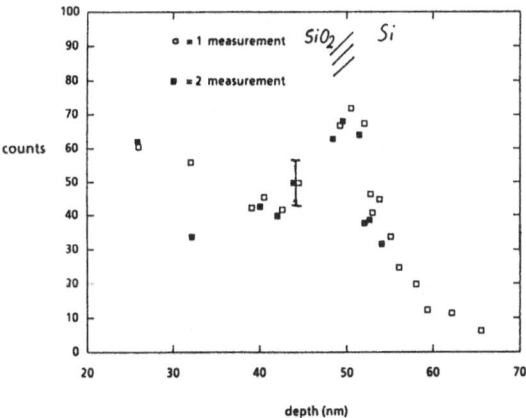

Fig. 4   Accumulated Hydrogen at the SiO$_2$-Si Interface

Author Index

R. Unbehauen, A. Cichocki

MOS Switched-Capacitor and Continuous-Time Integrated Circuits and Systems

Analysis and Design

1989. XIII, 631 pp. 284 figs. (Communications and Control Engineering Series) ISBN 3-540-50599-7

Contents: Fundamentals of Sampled-Data Systems. – MOS Devices for Linear Analog Integrated Circuits. – Basic Properties and Systematic Analysis of Switched-Capacitor Networks. – Basic Building Blocks of Linear SC Networks. – Synthesis and Design of SC Filters. – Design of Adaptive and Nonlinear Analog CMOS Circuits: Building Block Approach. – CMOS Analog to Digital and Digital to Analog Conversion Systems. – Subject Index.

C. M. Snowden (Ed.)

Semiconductor Device Modelling

1989. VIII, 259 pp. 111 figs.
ISBN 3-540-19545-9

This book describes a variety of approaches to modelling modern semiconductor devices. The text contains a review of essential device and semiconductor physics followed by a summary of scattering processes which influence the operation of devices. Classical and semiclassical physical device models are summarized and illustrated. The application of physical device models to VLSI (silicon) devices is discussed. State–of–the–art developments in compound semiconductor devices and modelling are reviewed. The Monte Carlo modelling technique is presented, together with essential data and examples. Equivalent circuit modelling techniques are described in separate chapters for silicon (bipolar and MOS) and high frequency devices. Modelling of noise processes in semiconductor devices is discussed, together with several examples. The modelling of optoelectronic devices, – semiconductor lasers and photo-diodes, is discussed. Quantum modelling techniques and their application are presented, extending the text into the regime of ultra-small-scale devices. The book includes a summary of modern CAD software available to model devices and comments on important features in developing simulations. Lastly there is a chapter which discusses the practical aspects and applications of semiconductor device modelling.

Springer-Verlag Berlin
Heidelberg New York London
Paris Tokyo Hong Kong

H. Heinrich, G. Bauer, F. Kuchar (Eds.)

Physics and Technology of Submicron Structures

Proceedings of the Fifth International Winter School, Mauterndorf, Austria, February 22–26, 1988

1988. X, 287 pp. 190 figs. (Springer Series in Solid-State Sciences, Vol. 83) ISBN 3-540-19109-7

M. L. Cohen, J. R. Chelikowsky

Electronic Structure and Optical Properties of Semiconductors

1988. XII, 264 pp. 161 figs. (Springer Series in Solid-State Sciences, Vol. 75) ISBN 3-540-18818-5

K. Seeger

Semiconductor Physics
An Introduction

4th ed. 1989. XIV, 480 pp. 301 figs. (Springer Series in Solid-State Sciences, Vol. 40) ISBN 3-540-19410-X

L. Treitinger, M. Miura-Mattausch (Eds.)

Ultra-Fast Silicon Bipolar Technology

1988. IX, 167 pp. 125 figs. (Springer Series in Electronics and Photonics, Vol. 27) ISBN 3-540-50638-1

H. Ryssel, H. Glawischnig (Eds.)

Ion Implantation Techniques

Lectures given at the Ion Implantation School in Connection with the Fourth International Conference on Ion Implantation: Equipment and Techniques

Berchtesgaden, Fed. Rep. of Germany, September 13–15, 1982

1982. XII, 372 pp. 245 figs. (Springer Series in Electrophysics, Vol. 10) ISBN 3-540-11878-0

Y. Tarui (Ed.)

VLSI Technology
Fundamentals and Applications

Springer-Verlag Berlin
Heidelberg New York London
Paris Tokyo Hong Kong

1986. XIV, 450 pp. 377 figs. (Springer Series in Electrophysics, Vol. 12) ISBN 3-540-12558-2